2017 China Semiconductor Technology International Conference (CSTIC 2017)

Shanghai, China
12 – 13 March 2017

IEEE Catalog Number: CFP1760Y-POD
ISBN: 978-1-5090-6695-7

**Copyright © 2017 by the Institute of Electrical and Electronics Engineers, Inc
All Rights Reserved**

Copyright and Reprint Permissions: Abstracting is permitted with credit to the source. Libraries are permitted to photocopy beyond the limit of U.S. copyright law for private use of patrons those articles in this volume that carry a code at the bottom of the first page, provided the per-copy fee indicated in the code is paid through Copyright Clearance Center, 222 Rosewood Drive, Danvers, MA 01923.

For other copying, reprint or republication permission, write to IEEE Copyrights Manager, IEEE Service Center, 445 Hoes Lane, Piscataway, NJ 08854. All rights reserved.

*** This is a print representation of what appears in the IEEE Digital Library. Some format issues inherent in the e-media version may also appear in this print version.

IEEE Catalog Number:	CFP1760Y-POD
ISBN (Print-On-Demand):	978-1-5090-6695-7
ISBN (Online):	978-1-5090-6694-0

Additional Copies of This Publication Are Available From:

Curran Associates, Inc
57 Morehouse Lane
Red Hook, NY 12571 USA
Phone: (845) 758-0400
Fax: (845) 758-2633
E-mail: curran@proceedings.com
Web: www.proceedings.com

Table of Contents

Preface

Chapter I - Device Engineering and Memory Technology

III-N Heterostructure Devices for Low-Power Logic**.....1   I-31
P. Fay[1*], W. Li[1], D. Digiovanni[1], L. Cao[1], H. Ilatikhameneh[2], F. Chen[2], T. Ameen[2], R. Rahman[2], G. Klimeck[2], C. Lund[3], S. Keller[3], S. M. Islam[4], A. Chaney[4], Y. Cho[4] and D. Jena[4]
[1] *Dept. of Electrical Engineering, University of Notre Dame, Notre Dame, IN, USA*
[2] *Dept. of Electrical and Computer Engineering, Purdue University, West Lafayette, IN, USA*
[3] *Dept. of Electrical and Computer Engineering, Univ. of California, Santa Barbara, CA USA*
[4] *Dept. of Electrical and Computer Engineering and Dept. of Materials Science and Engineering, Cornell University, Ithaca, NY, USA*

Monolithic 3D (M3D) Reconfigurable Logic Applications Using Extremely-Low-Power   I-7
Electron Devices*.....4
Woo Young Choi
Department of Electronic Engineering, Sogang University, Seoul, Republic of Korea

A Study of 28 nm LDMOS HCI Improvement by Layout Optimization.....8   I-2
Ruoyuan Li, Yongsheng Yang, Fang Chen, Ling Tang, Zhengyong Lv, Byunghak Lee, Ling Sun, Weizhong Xu and TzuChiang Yu
Semiconductor Manufacturing International Corporation, Shanghai, China

Reliability Investigations on The Programming Currents of 28 nm Metal E-Fuse.....12   I-24
Guangyan Zhao, Yong Zhao and Wei-Ting Kary Chien
Semiconductor Manufacturing International Corporation, Shanghai, China

Defects and Lifetime Prediction for Ge PMOSFETs Under AC NBTI Stresses*.....15   I-20
J. F. Zhang, J. Ma, W. Zhang and Z. Ji
Department of Electronics and Electrical Engineering, Liverpool John Moores University, Liverpool, United Kingdom

Recent Progress in RRAM Technology: From Compact Models to Applications.....20   I-15
Yue Zha[1], Zhiqiang Wei[2], Jing Li[3]
[1,3] *Department of Electrical and Computer Engineering, University of Wisconsin-Madison, Madison, WI, USA*
[2] *Advanced Devices Development Center, Panasonic Corporation, Osaka, Japan*

Investigation of Trap Profile in Nitride Charge Trap Layer in 3D NAND Flash
Memory Cells*.....24   I-14
Jong-Ho Lee and Ho-Jung Kang
Seoul National University, Republic of Korea

Advanced Logic and Specialty Technologies for VLSI Manufacturing in Fast Expansion at China*.....26   I-4
Min-hwa Chi[1] and Hanming Wu[2]
[1]*Semiconductor Manufacturing International Corporation, Shanghai, China*
[2] *BriteIP Corporation, China*

Analytical Modeling for Substrate Effect of Lateral Power Devices.....30   I-21
Jing Chen[1,2], Yufeng Guo[1,2], Man Li[1,2], Ling Du[1,2], Xinchun Ji[1,2], Changchun Zhang[1,2] and Jiafei Yao[1,2]
[1]*National and Local Joint Engineering Laboratory of RF integration & Micro-Assembly Technology, Nanjing, China*
[2]*College of Electronic Science and Engineering, Nanjing University of Posts and Telecommunications, Nanjing, China*

MOSFET RF Performance Improvement Through Spacer Profile Optimization for 28 nm Poly/SioN SoC Technology.....33   I-29
Hai Liu, River He and Byunghak Lee
Semiconductor Manufacturing International Corporation, Shanghai, P.R. China

A Method to Solve Reverse Tunneling Disturb Issue for SuperFlash® Memory.....36   I-1
Tao Xu, Zigui Cao, Guoqing Han, Hong Chen and Hui Wang
Shanghai Huahong Grace Semiconductor Manufacturing, Shanghai, China

Compact Electrodynamics MEMS-Speaker.....39   I-6
Burhanuddin Yeop Majlis[1], Gandi Sugandi[1,2] and Mimiwaty Mohd Noor[1]
[1]*Institute of Microengineering and Nanoelectronics (IMEN), Universiti Kebangsaan Malaysia (UKM), Malaysia*
[2]*Research Center for Electronics and Telecommunication Indonesian Institute of Sciences (PPET-LIPI),Bandung-Indonesia*

Identification and Solutions for a Novel Particulate Pollution Matter in Wafer Surface Caused by Concentrated Sulfuric Acid.....42   I-22
Lu Sun
Semiconductor Manufacturing International Corporation, Shanghai, China

Optimal Experiment Design in Poly Etch Process for Performance Improvement on Different Type Tool.....45   I-23
Ying Emily Lu, Wei William Guo and Chingluan Jenny Wu
Semiconductor Manufacturing International Corporation, Shenzhen, China

A Reliability Study of a New Embedded Flash to Reduce Charge-Loss Issue.....49   I-26
Lingling Shao, Yong Atman Zhao, Wei Han and Wei-Ting Kary Chien
Semiconductor Manufacturing International Corporation, Shanghai, China

Idle Recipe Automatic Control.....52   I-28
Weiwei Yan, Hongtao Qian, Zhiqian Liu and Jiuzhou Zhao
Semiconductor Manufacturing International Corporation, Shanghai, P.R. China

Investigation of CMOS Image Sensor Dark Current Reduction by Optimizing Interface Defect.....55   I-30

Wuzhi Zhang[1], Zhengying Wei[1], Yansheng Wang[1], Wei Zhou[1], Chang Sun[1], Jun Qian[1] and Yuhang Zhao[2]
[1]*Haili Microelectronics Corporation, Shanghai, China*
[2]*Huahong Group, Shanghai, China*

A Novel 25 V DDD NMOS Design in 500-700 V Ultra High Voltage BCD Process.....58   I-33
Donghua Liu[1,2], Zhaozhao Xu[2,3], Wenting Duan[2], Feng Jin[2], Wenqing Yang[2], Huihui Wang[2], Jun Hu[2], Jiye Yang[2], Wensheng Qian[2] and David Wei Zhang[1]
[1]*State Key Laboratory of ASIC and System, School of Microelectronics, Fudan University, Shanghai, China*
[2]*HuaHong Grace Semiconductor Manufacturing Corporation, Shanghai, China*
[3]*State Key Laboratory of Functional Materials for Informatics, Shanghai Institute of Microsystem and Information Technology, Chinese Academy of Sciences, Shanghai, China*

Metal-Electrode-Dependent Negative Photoconductance Response of the Nanoscale Conducting Filament in the SiO_2-Metal Stack.....63   I-34
T. Kawashima[1,2], Y. Zhou[1], K. S. Yew[1], H. Z. Zhang[1] and D. S. Ang
[1]*School of Electrical and Electronic Engineering, Nanyang Technological University, Singapore, Singapore*
[2]*Corporate Manufacturing Engineering Center, Toshiba Corp., Isogo-ku, Yokohama, Japan*

A Novel 25 V PLDMOS Design in 700 V BCD Process.....66   I-35
Donghua Liu[1,2], Wenting Duan[2] and Wensheng Qian[2]
[1]*State Key Laboratory of ASIC and System, School of Microelectronics, Fudan University, Shanghai, China*
[2]*Huahong Grace Semiconductor Manufacturing Corporation, Shanghai, China*

Application of Resist Profile Model and Resist-etch Model in Solving 28 nm Metal Resist Toploss.....69   I-38
Yiqun Tan, Weiwei Wu, Quan Chen and Shirui Yu
Shanghai Huali Microelectronics Corporation, Shanghai, China

Depletion-Mode MOS Capacitor Modeling Investigation.....72   I-40
Chien-Lung Tseng and Yuan Sheng Wang
Semiconductor Manufacturing International Corporation, Shanghai, China

Simulation on the Performance Comparison for Nanoscale SOI and Bulk Junctionless FinFETs.....75   I-41
Cheng-Kuei Lee, Jin-Yu Zhang and Yan Wang
Department of Microelectronics and Nanoelectronics Tsinghua University, Beijing, China

Analysis and Modeling of Self-Heating Effect in Bulk FinFET.....78   I-42
Shawn Lin, SenSheng Li, Li Shen, Jianhua Ju and Shaofeng Yu
Semiconductor Manufacturing International Corporation, Shanghai, China

Chapter II – Lithography and Patterning

Key Points in 14 nm Photolithographic Process Development, Challenges and Process   II-20

Window Capability.....80
Qiang Wu
Semiconductor Manufacturing International Corporation, Shanghai, China

An Offline Roughness Evaluation Software and its Application in Quantitative   II-1
Calculation of Wiggling Based on Low Frequency Power Spectral Density Method.....86
Libin Zhang, Lisong Dong, Xiaojing Su, Yansong Liu, Lijun Zhao and Yayi Wei
Institute of Microelectronics of Chinese Academy of Sciences (IMECAS), Beijing, China

Level-Set Based ILT With a Vector Imaging Model.....90   II-22
Yijiang Shen
School of Automation, Guangdong University of Technology, Guangzhou, China

Data Analytics and Machine Learning for Design-Process-Yield Optimization in   II-45
Electronic Design Automation and IC Semiconductor Manufacturing*.....93
Luigi Capodieci
Motivo Data Analytics, Sunnyvale, CA, USA

Development of 250 W EUV Light Source for HVM Lithography*.....96   II-18
Taku Yamazaki, Hiroaki Nakarai, Tamotsu Abe, Krzysztof M Nowak, Yasufumi Kawasuji,
Takeshi Okamoto, Hiroshi Tanaka, Yukio Watanabe, Tsukasa Hori, Takeshi Kodama,
Tatsuya Yamagida, Yutaka Shiraishi, Takashi Saitou and Hakaru Mizoguchi
Gigaphoton Inc., Shinomiya Hiratsuka Kanagawa, Japan

The Solutions for 3D-NAND Processes with Canon's Latest KrF Scanner.....100   II-32
Masanori Yamada, Hajime Takeuchi, Kazuhiko Mishima, Keiji Yoshimura and Kazuhiro
Takahashi
Canon Inc., Utsunomiya, Tochigi, , Japan

Application of OPE Master for Critical Layer OPE Matching.....103   II-13
Yuan Tao[1], Yifei Liu[1], Yuanzhao Ma[1], Zhenyu Yang[1], Chun Shao[1], Xuedong Fan[2], Junji
Ikeda[3] and Koichi Fujii[3]
[1]*Nikon Precision Shanghai, China*
[2]*Shanghai Huali Microelectronics Corporation, Shanghai, China*
[3]*Nikon Corporation, Saitama-shi, Japan*

DDR Process and Materials for NTD Photo Resist toward 1X nm Patterning and   II-17
beyond*.....105
Shuhei Shigaki, Wataru Shibayama, Makoto Nakajima and Rikimaru Sakamoto
Nissan Chemical Industries, Fuchu-machi, Toyama, Japan

Novel EUV Resist Development for Sub-14 nm Half Pitch.....109   II-35
Koichi Fujiwara
JSR Shanghai Co., Ltd., Shanghai, China

Advancement in Resist Materials for Sub-7 nm Patterning and Beyond.....111   II-14
Li Li[1], Xuan Liu and Shyam Pa[1]
[1]*Advanced Technology Development, GlobalFoundries, Malta, NY, USA*

Design and Synthesis of Novel Directed Self-Assembly Block Copolymers for Sub-10   II-8
nm Lithography Application.....114

Jie Li, Xuemiao Li and Hai Deng
Department of Macromolecular Science, State Key Laboratory of Molecular Engineering of Polymers, Fudan University, Shanghai, China

DOF Enhancement of 3bar PO Pattern in 28 nm Technology Node.....117
II-2
Bin-Jie Jiang, Shi-Rui Yu and Zhi-Biao Mao
Shanghai Huali Microelectronics Corporation, Shanghai, China

The Incorporation of the Pattern Matching Approach Into a Post-OPC Repair Flow.....120
II-4
Yaojun Du
Semiconductor Manufacturing International Corporation, Shanghai, China

Synthesis and Directed Self-Assembly of Modified PS-B-PMMA for Sub-10 nm Nanolithography.....123
II-9
Xuemiao Li, Jie Li and Hai Deng
Department of Macromolecular Science, State Key Laboratory of Molecular Engineering of Polymers, Fudan University, Shanghai, China

Research of SMO Process to Improve the Imaging Capability of Lithography System for 28 nm Node and Beyond.....126
II-10
Haibin Yu, Yueyu Zhang, Binjie Jiang, Shirui Yu and Zhibiao Mao
Shanghai Huali Microelectronics Corporation Shanghai, China

The Phenomena of Footing's Variation Caused by Oxidation of Nitrogen-containing Substrates.....129
II-11
Song Bai, Hanmo Gong, Chang Liu and Qiang Wu
Semiconductor Manufacturing International Corporation, Shanghai, China

A Study of N-Induced Residue Defect on Gate Oxide after Lithography Rework.....132
II-19
Zhou Fang, Chang Liu and Zhoujun Pan
Semiconductor Manufacturing International Corporation, Shanghai, China

Wafer Edge Treatment in Lithographic Process for Peeling Defect Reduction.....135
II-23
Xiaofeng Yuan, Qiang Zhang and Jing'an Hao
Semiconductor Manufacturing International Corporation, Shanghai, China

Ultrapure Chemical Components for Next Generation Materials.....139
II-24
Hyun Yong Cho[1], Ram Sharma[2] and Jeffrey D. Fogle[2]
[1]Heraeus Korea Corp., Suwon, Republic of Korea
[2]Heraeus Precious Metal North America Daychem LLC, Vandalia, OH, USA

Extension of Photoresist Lifetime at Extreme Conditions.....142
II-25
Qiaoqiao Li, Weiming He and Huayong Hu
Semiconductor Manufacturing International Corporation, Shanghai, China

Illumination Optimization for Lithography Tools OPE Matching at 28 nm Nodes.....145
II-26
Wuping Wang, Long Qin, Zhengkai Yang, Yulong Li, Zhibiao Mao and Yu Zhang
Shanghai Huali Microelectronics Corporation, Shanghai, China

Investigation of Critical Dimension (CD) Variation Control on Unstable Substrate.....149
II-27

Xiaoyan Sun, Chang Liu and Lei Wu
Semiconductor Manufacturing International Corporation, Shanghai, China

Diffraction-Based and Image-Based Overlay Evaluation for Advanced Technology Node.....152   II-29
Jian Xu, Long Qin, Qiaoli Chen, Hui Zhi, Yanyun Wang, Zhengkai Yang and Zhibiao Mao
Shanghai Huali Microelectronics Corporation, Shanghai, China

BLOB Defect Solution for 28 nm Hole Pattern in 193 nm Topcoat-Free Immersion Lithography.....156   II-30
Dan Li, Biqiu Liu, Yulong Li, Zhengkai Yang, Zhibiao Mao and Yu Zhang
Huali Microelectronics Corporation, Shanghai, China

Chapter III – Dry & Wet Etch and Cleaning

The Study of Poly Gate Etching Profile, Micro Loading and Wiggling for NAND Flash Memory.....159   III-10
Zhuo-Fan Chen, Hai-yang Zhang and Yi-Ying Zhang
Semiconductor Manufacturing International Corporation, Shanghai, China

Study of Poly Etch for Performance Improvement with Alternative Spin-On Materials in FinFET Technology Node.....162   III-30
Yan Wang, Qiuhua Han and Haiyang Zhang
Semiconductor Manufacturing International Corporation, Shanghai, China

Challenges and Solutions of 28 nm Poly Etching.....165   III-29
Xiangguo Meng, Lian Lu, Fuhong Chen, Sifei Chan, Quanbo Li, Zhibiao Mao, Yu Zhang and Albert Pang
Shanghai Huali Microelectronics Corporation, Shanghai, China

Dry Plasma Cleaning by Advanced HDRF Technology (High Density Radical Flux).....168   III-40
Yannick Pilloux
Plasma-Therm LCC, St Petersburg, FL, USA

Study and Solution of 28 nm AIO Etch Seal Ring Residue Issue.....170   III-33
Haibo Shen, Lei Sun and Quanbo Li
Shanghai Huali Microelectronics Corporation Shanghai, China

A Study of Silicon Etch Process in Memory Process.....173   III-11
Rong-Yao Chang, Yi-Ying Zhang and Hai-Yang Zhang
Semiconductor Manufacturing International Corporation, Shanghai, China

SiN Removal Process for Poly Damage Control in Memory Flash.....176   III-32
Jia Ren, Haiyang Zhang and Yiying Zhang
Semiconductor Manufacturing International Corporation, Shanghai, China

The Loading Effect Study in Metal Hard-Mask All-In-One Etch With Double   III-12

Patterning Scheme.....179
Kefang Yuan, Junqing Zhou, Zhidong Wang, Minda_Hu, Qiyang He and Haiyang_Zhang
Semiconductor Manufacturing International Corporation, Shanghai, China

SADP Core Etching Performance Comparison for Different CCP Etchers.....182   III-13
Yibin Song, Haiyang Zhang and Yiying Zhang
Semiconductor Manufacturing International Corporation, Shanghai, China

BARC Open In FinFET Technology Node.....184   III-14
Long-Juan Tang, Qiu-Hua Han and Hai-Yang Zhang
Semiconductor Manufacturing International Corporation, Shanghai, China

Cavity Profile Control in Drie Process.....187   III-15
Wang Jing, Nie Miao, Jiang Zhongwei and Huang Yahui
North Microelectronics CO., Beijing, China

Dummy Poly Removal in FinFET Technology Node.....189   III-21
Ruixuan Huang, Shiliang Ji and Qiuhua Han
Semiconductor Manufacturing International Corporation, Shanghai, China

The Line Edge Roughness Improvement with Plasma Coating for 193 nm Lithography.....192   III-24
Erhu Zheng, Haiyang Zhang and Yiying Zhang
Semiconductor Manufacturing International Corporation, Shanghai, China

The Study of Deep Trench Etch Process for PCRAM.....195   III-25
Yiying Zhang, Zhuofan Chen and Haiyang Zhang
Semiconductor Manufacturing International Corporation, Shanghai, China

Improvement of the Pattern Wiggling Profile by PR treatment Methods.....198   III-31
Pan-pan Liu, Cheng-long Zhang and Hai-yang Zhang
Semiconductor Manufacturing International Corporation, Shanghai, China

A Study of AA CD Uniformity Loading Optimization at 28 nm Node.....201   III-34
YiZheng Zhu, Lian Lu, Fuhong Chen, Sifei Chan, Quanbo Li, Yu Zhang and Albert Pang
Shanghai Huali Microelectronics Corporation, Shanghai, China

Producer BARC Open Challenges and the Baseline Set Up for FinFET.....203   III-36
Song Huang, Ying Huang, Judy Wang and Qiang Ge
Applied Materials China, Shanghai, China

Optimization of Wet Strip after Metal Hard Mask All-in-One Etch for Metal Void Reduction and Yield Improvement.....206   III-43
Yong Huang, Jialei Liu, Zhiyong Yang, Jing Zhao and Huanxin Liu
Semiconductor Manufacturing International Corporation, Shanghai, China

Chapter IV – Thin Film, Plating and Process Integration

The Technology Trend of IC Manufacture During Post Moore's Era.....209**   IV-42

Hanming Wu[1], Jin Kang[2,3], Minhwa Chi[2], Xing Zhang[3], Tong Feng[2] and Poren Tang[2]
[1]*BriteIP Corporation, China*
[2]*Semiconductor Manufacturing International Corporation, Shanghai & Beijing, China*
[3]*Peking University, Beijing, China*

Fin Critical Dimension Loading Control by Different Fin Formation Approaches for FinFET Process.....216 IV-17

Qingpeng Wang, Gang Mao, Hai Zhao, Cheng Li, Fangyuan Xiao, Rex Yang and Shaofeng Yu
Semiconductor Manufacturing International Corporation, Shanghai, China

Fin Bending Mechanism Investigation for 14 nm FinFET Technology.....219 IV-18

Cheng Li, Hai Zhao and Gang Mao
Semiconductor Manufacturing International Corporation, Shanghai, China

14nm Metal Gate Film Stack Development and Challenges*.....222 IV-10

Jianhua Xu, Anni Wang, Jun He, Xuezhen Jing, Ziying Zhang and Beichao Zhang
Semiconductor Manufacturing International Corporation, Shanghai, China

Forming a More Robust Sidewall Spacer with Lower k (Dielectric Constant) Value*...225 IV-14

Tao Han, Man Gu, Stephan Grunow, Huang Liu, Sujatha Sankaran and Jinping Liu
Advanced Technology Development, GLOBALFOUNDRIES Inc., Malta, NY, USA

Review of Thin Film Porosity Characterization Approaches*.....228 IV-39

Konstantin P. Mogilnikov[1], Dongchen Che[1], Mikhail R. Baklanov[1,2], Kangning Xu[1] and Kaidong Xu[1]
[1]*Leuven Instruments Co. Ltd (Jiangsu), Pizhou, Jiangsu, China*
[2]*North China University of Technology, Beijing, China*

The Application of the Smoluchowski Effect to Explain the Current-Voltage Characteristics of High-k MIM Capacitors.....232 IV-22

W.S. Lau
Zhejiang University, Hangzhou, China

The Study of the Impurity in HK Film.....235 IV-23

Yingming Liu, Qiuming Huang and Haifeng Zhou
Shanghai Huali Microelectronics Corporation, Shanghai, China

The Study of 28 nm BEOL Cu Gap-Fill Process.....238 IV-28

Yu Bao, Gang Shi, Lin Gao, Yanyan Zhang, Yingming Liu, Peng Tian, Fuchun Xi, Wei Hu, Ying Gao, Zhenhua Cai, Baojun Zhao, Zhigang Yang, Jianghua Leng, Haifeng Zhou and Jingxun Fang
Shanghai Huali Microelectronics Corporation, Shanghai, China

Nickel Silicide Anneal Process Research for 28 nm CMOS Node.....241 IV-25

Zhenping Wen, Xinhua Cheng and Jingxun Fang
Shanghai Huali Microelectronics Corporation, Shanghai, China

Tungsten Voids Improvement by Optimizing MOCVD-Tin Barrier Layer Plasma Treatment at 28 nm Technology Node.....245 IV-29

Lin Gao, Yanyan Zhang, Yu Bao, Bin Zhong, Yingming Liu, Kang Ye, Gang Shi, Haifeng

Zhou and Jingxun Fang
Shanghai Huali Microelectronics Corporation, Shanghai, China

Rotary Spatial Plasma Enhanced Atomic Layer Deposition – An Enabling Manufacturing Technology for μm-Thick ALD Films.....248   IV-36
Sami Sneck, Mikko Söderlund, Markus Bosund, Pekka Soininen
Beneq Oy, Espoo, Finland

Passivation Quality and Electrical Characteristics for Boron Doped Hydrogenated Amorphous Silicon Film.....250   IV-15
Ching-Lin Tseng[1], Yu-Lin Hsieh[1], Chien-Chieh Lee[3], Hsiang-Chih Yu [2] and Tomi T. Li[1]
[1]Department of Mechanical Engineering, National Central University, Taoyuan, Taiwan
[2]Department of Energy Engineering, National Central University, Taoyuan, Taiwan
[3]Optical Science Center, National Central University, Taoyuan, Taiwan

Laser Spike Annealing and SiGe Dummy Pattern Layout Study to Improve Contact Misalignment Overlay Issue.....253   IV-1
Guiying Ma and Tzuchiang Yu
Semiconductor Manufacturing International Corporation, Shanghai, China

Investigation of Intrinsic Hydrogenated Amorphous Silicon (A-Si:H) Thin Films on Textured Silicon Substrate With High Quality Passivation.....256   IV-19
Min-Lun Yu[2], Yu-Lin Hsieh[1], Sheng-Kai Jou[1], Tomi T. Li[1] and Chien-Chieh Lee[3]
[1]Department of Mechanical Engineering, National Central University, Taoyuan, Taiwan
[2]Department of Energy Engineering, National Central University, Taoyuan, Taiwan
[3]Optical Science Center, National Central University, Taoyuan, Taiwan

SiN/SiC$_x$N Stack Film as Cu Capping Layer in Cu/ULK Interconnect for 28LP.....259   IV-26
Yi Hailan, Lei Tong, Kang Ye, Yongyue Chen, Wei Hu , Zhigang Yang, Haifeng Zhou and Jingxun Fang
Shanghai Huali Microelectronics Corporation, Shanghai, China

The Methodology to Reduce Poly Bump Defect.....262   IV-27
Junlong Kang, Xiaogong Fang, Xinhua Cheng, Huaming Luo, Jingxun Fang and Yu Zhang
Shanghai Huali Microelectronics Corporation, Shanghai, China

Residue Defect Study on Amorphous Silicon Film.....265   IV-43
Zhiyong Yang, Jialei Liu, HuanxinLiu, Yonggen He and Yong Huang
Semiconductor Manufacturing International Corporation, Shanghai, China

Fabrication of Single Phase Transparent Conductive Cuprous Oxide Thin Films By Direct Current Reactive Magnetron Sputtering.....268   IV-44
Ruijin Hong, Jinxia Wang, Chunxian Tao and Dawei Zhang
Engineering Research Center for Optical Instruments and System, University of Shanghai for Science and Technology, Shanghai, China

Chapter V – CMP and Post-Polish Cleaning

Study on the Defect of Post Cleaning Step after W CMP and its Improving Solution.....274   V-5

Zigui Cao, Jun Kang, Hao Huang and Hui Wang
Huahong Grace Semiconductor Manufacturing Corporation, Shanghai, China

Abrasive Particle Trajectories and Material Removal Non-Uniformity During Chemical Mechanical Polishing.....277   V-9
Vahid Rastegar and S.V. Babu
Center for advanced Material Processing, Clarkson University, Potsdam, NY, USA

CMP Slurry Metrology to Meet the Industry Demand*.....280   V-23
Rashid Mavliev
Ipgrip Inc, Campbell, CA, USA

Challenges in Chemical Mechanical Planarization Defects of 7 nm Device and its Improvement Opportunities*.....283   V-12
Ji Chul Yang , Dinesh Penigalapati, Tai Fong Chao, Wen Yin Lu and Dinesh Koli
GLOBALFOUNDRIES, Malta, NY, USA

SiOC CMP Developed and Implemented in 7 nm and Beyond.....286   V-7
Haigou Huang, Taifong Chao, Ja-Hyung Han, Dinesh Koli and Qiang Fang
GLOBALFOUNDRIES, Malta, NY, USA

Impact of Wafer Transfer Process on STI CMP Scratches*.....290   V-16
Fan Bai, Zhijie Zhang, Jia Wang and Hongdi Wang
Semiconductor Manufacturing International Corporation, Beijing, China

Research and Solution of STI CMP Dishing and Uniformity Improve for 28LP.....294   V-19
Lei Zhang, Junhua Yan, Kun Chen, Wenbin Fan, Yefang Zhu and Jingxun Fang
Shanghai Huali Microelectronics Corporation, Shanghai, China

A New Acidic ILD Slurry Formulation for Advanced CMP.....296   V-22
Yi Guo, Arun Reddy, David Mosley and Robert Auger
Dow Electronic Materials, CMP Technologies, NJ, USA

Settling of Colloidal Silica Particles in CMP Slurry: Monitoring, Effect, and Handling...299   V-4
Jie Lin and W. Scott Rader
Fujimi Corporation, Tualatin, Oregon, USA

Outsourced CMP for Rapid Development and Efficient Manufacturing*.....302   V-13
Robert L. Rhoades
Entrepix, Inc., Phoenix, AZ, USA

Process Stability and Tool Capacity Improvement with 150 mm Profiler and 200 mm Contour Heads.....305   V-25
Yanghua He[1], Michael Lube[1] and Jamie Leighton[2]
[1]Process Engineering, Qorvo, Inc., Richardson, Texas, USA
[2]Applied Global Services, Applied Materials, Santa Clara, CA, USA

Optimization of Slurry and Process Parameter on Chemical Mechanical Polishing of Cr-Doped Sb_2Te_3 Thin Film.....309   V-20
Ruifang Huo, Fang Wang, Yulin Feng, Yemei Han, Yujie Yuan and Kailiang Zhang
School of Electronics Information Engineering, Tianjin University of Technology, Tianjin,

China

Study of Weakly Alkaline Slurry for Copper Barrier CMP on Manufacture Platform...313   V-26
Jin Kang[1,2], Hanming Wu[2], Xing Zhang[1], Qiang Li[2], Jun Ge[2], Tong Feng[2], Ziqing Yin[2] and Liu Yuling[3]
[1]*Peking University, Beijing, China*
[2]*Semiconductor Manufacturing International Corporation, Shanghai and Beijing, China*
[3]*Institute of Microelectronics, Hebei University of Technology, Tianjin, China*

Chapter VI – Metrology, Reliability and Testing

A System Qualification Platform for FPGA Board Level Testing and Automated   VI-32
Regression.....316
Yibin Sun and Ping Chen
Lattice Semiconductor (Shanghai) Co., LTD. Shanghai, China

Two Indirect Methodologies for Testing FPGA Intrinsic Programmable Logic Cell   VI-41
Timing Performance.....320
Hongpeng Han
Lattice Semiconductor Corporation, Shanghai, China

A Statistically Robust Methodology for Optimized Sample Size Determination for   VI-6
FPGA Post-Silicon Validation.....324
Weijun Qin
Lattice Semiconductor, Shanghai, China

Using VerilogA for Modeling of Single Event Current Pulse: Implementation and   VI-12
Application.....327
Jia Liu[1], Yusen Qin[2], Tiehu Li[3], Yuxin Wang[1], Weidong Yang[1], Jun Liu[1] and Ruzhang Li[1]
[1]*Science and Technology on Analog Integrated Circuit Laboratory, Chongqing, China*
[2]*Chongqing Nan'an Power Supply Bureau, Nan'an District, Chongqing, China*
[3]*No.24 Institute, China Electronics Technology Group Corp., Chongqing, China*

Finger Print Sensor Molding Thickness None Destructive Measurement With   VI-25
Terahertz Technology.....330
Longhai Liu[1], Haitao Jiang[2], Ying Wang[2], Qinghua Shou[2], Jianhua Xie[1] and Yaqi Lu[1]
[1]*Advantest (China) Co., Shanghai, China*
[2]*Amkor Assembly & Test (Shanghai) Co., Shanghai, China*

High Efficiency Test System for Envelope Tracking Power Amplifier.....333   VI-38
Feifan Du[1] and Hui Yu[2]
[1]*Department of Electronic Engineering, Shanghai Jiao Tong University, National Instruments China*
[2]*Department of Electronic Engineering, Shanghai Jiao Tong University, Shanghai, China*

Effective Method to Automatically Measure the Profile Parameters of Integrated   VI-2
Circuit From SEM/TEM/STEM Images.....336

Xiaolin zhang[1], Zubiao Fu[1], Yi Huang[2], Alien Lin[2], Yaoming Shi[1] and Yiping Xu[1]
[1]*Raintree Scientific Instrument (Shanghai) Corporation, Shanghai 201203, China*
[2] *Semiconductor Manufacturing International Corporation, Shanghai, China*

Measurement of Nanoscale Grating Structure by Mueller Matrix Ellipsometry.....339 VI-7
Shiqiu Cheng, Fengjiao Zhong, Huiping Chen, Yutao Jia, Yaoming Shi and Yiping Xu
Raintree Scientific Instruments (Shanghai) Corporation, Shanghai, China

The Study and Investigation of Inline E-Beam Inspection for 28 nm Process VI-39
Development.....342
Yin Long[1,2], Rongwei Fan[2], Hunglin Chen[2] and Haihua Li[1]
[1]*School of Microelectronics, Shanghai Jiao Tong University, China*
[2]*Shanghai Huali Microelectronics Corporation, Shanghai, China*

Stress Control Metrology in Epitaxy.....345 VI-18
Yang Song, Pengwei Fan and Yi-Shi Lin
Semiconductor Manufacturing International Corporation, Shanghai, China

Low Frequency Noise Characterization of 22 nm PMOS Featuring with Filling W Gate VI-9
Using Different Precursors.....348
Liang He[1,2,4], Eddy Simoen[1,3], Cor Claeys[1,2], Guilei Wang[5], Jun Luo[5], Chao Zhao[5], Junfeng
Li[5], Hua Chen[4],Yin Hu[4] and Xiaoting Qin[4]
[1]*Imec, Leuven, Belgium*
[2]*Dept. Electrical Engineering, KU Leuven, Leuven, Belgium*
[3]*Dept. Solid State Sciences, Ghent University, Gent, Belgium*
[4]*School of Advanced |Materials and Nanotechnology, Xidian University, Xi'an, China*
[5]*Key laboratory of Microelectronic Devices & Integrated Technology, Institute of
Microelectronics of Chinese Academy of Sciences, Beijing, China*

Exploration of Poly Irms Based on 40nm Technology Node.....351 VI-3
Xiang Fu Zhao, Wei Ting Kary Chien and Kelly Yang
Semiconductor Manufacturing International Corporation, Shanghai, China

Effect of High Temperature Storage on Fan-Out Wafer Level Package Strength.....353 VI-31
Cheng Xu[1,2], Z.W. Zhong[1] and W.K. Choi[2]
[1]*School of Mechanical & Aerospace Engineering, Nanyang Technological University,
Singapore*
[2]*JCET STATS ChipPAC Pte Ltd, Singapore*

Effects of Copper Line-Edge Roughness on TDDB at Advanced Technology Nodes of 28 VI-14
nm And Beyond.....356
Dongyan Tao, Jinling Xu, Yanhui Sun, Wei-Ting Kary Chien, JS Chen and Guan Zhang
Semiconductor Manufacturing International Corporation, Beijing, China

Study of Safe Operating Area and Improvement for Power Management Integrated VI-15
Circuit.....359
Sarah Zhou, Yongliang Song, Kary Chien and Canny Chen
Semiconductor Manufacturing International Corporation, Shanghai, China

Using Static Voltage Propagation Approach to Assist Full Chip LUP and TDDB VI-23

Physical Verification.....362
Yi-Ting Lee and Frank Feng
Mentor Graphics Corp, Wilsonville, OR, USA

Deep Level Investigation of InGaAs on InP Layer.....365   VI-24
Chong Wang[1,2,4], Eddy Simoen[1,3], Alian AliReza[1], Sonja Sioncke[1], Nadine Collaert[1], Cor
Claeys[1,2] and Wei Li[4]
[1]Imec, Leuven, Belgium
[2]KU Leuven, EE Dept., Leuven, Belgium
[3]Dept. Solid State Sciences, Ghent University, Gent, Belgium
*[4]University of Electronic Science and Technology of China, School of Optoelectronic
Information, Chengdu, China*

Practical Wafer Level Threshold Voltage Stability Measurement Methodology for the   VI-4
Fast Evaluation of Flash Technology.....369
Gang Niu, Wei-Ting Kary Chien, Jack_Chen, Dennis Zhang, Susie Yu, Daniel Zhao，Silvia
Duan, Ming_Li and Alicia_Ding
Semiconductor Manufacturing International （Tianjin）Co, Xi Qing Econ. Dev. Area, China

GDI Failure Mechanism Investigation and Improvement in HK Process.....372   VI-10
Lingxiao Cheng, Lijuan Yang and Kai Wang
Semiconductor Manufacturing International Corporation, Shanghai, China

Highly Effective Low-k Dielectric Test Structures and Reliability Assessment for 28 nm   VI-11
Technology Node and Beyond.....375
Zhijuan Wang, Yueqin Zhu, Kai Wang, Yuzhu Gao and Wei-Ting Kary Chien
Semiconductor Manufacturing International Corporation, Shanghai, China

Fail Mechanism of Program Disturbance for Erase Cells Vt Positive Shift in NAND   VI-17
Flash Technology.....378
Chunmei Zou[1], Yong Zhao, Wei-Ting Kary Chien and Junyao Tang
Semiconductor Manufacturing International Corporation, Shanghai, China

The Research of Intelligent Feedback Mechanism Between Document Control and   VI-1
Production System.....381
Zhou Zhenlin and Guo Yingying
Semiconductor Manufactory International Corporation, Tianjin, China

A Study on Problem Solving Strategy Using Experiment of Design.....386   VI-13
Xinyuan Ji and Sheng Kang
Semiconductor Manufactory International Corporation, Shanghai,, China

Improvement on the Stress Migration in Tungsten-Plug Via.....392   VI-16
Juan Wen, Wei-Ting Kary Chien, Guan Zhang and Yanhui Sun
Semiconductor Manufactory International Corporation, Shanghai,, China

High-k Metal Gate In-line Measurement Technique Using XPS.....395   VI-19
Yang Song, Yi Huang and Yi-Shi Lin
Semiconductor Manufactory International Corporation, Shanghai,, China

Application Study of Qmerit Function on Overlay Accuracy Verification.....398   VI-36
Huayong Hu, Lei Ye, Qiaoqiao Li, Weiming He and Zhicheng Liu
Semiconductor Manufactory International Corporation, Shanghai,, China

Optimization for the Measurement Parameters of CD SEM for Several Specific Situations.....401   VI-43
Hanmo Gonga, Song Bai, Qian Ren and Qiang Wu
Semiconductor Manufactory International Corporation, Shanghai,, China

Systematic maintenance and applications of failure modes and effects analysis (FMEA) in semiconductor manufacturing.....404   VI -5
Hongtao HT Qian, Ziqian Javaer Liu, Yuhong Betsy Xu
Corporate Quality & Reliability Center, Semiconductor Manufacturing International Corporation, Shanghai, China

Investigation of multiple soft breakdown during time-dependent dielectric beakdown...408   VI-21
Qiwei Wu, Binfeng Yin, Ke Zhou, Jiong Wang, Jinde. Gao
Huali Microelectronics Corporation, Shanghai, China

Chapter VII – Packaging and Assembly

Technology Advancement of Laminate Substrates for Mobile, IoT and Automotive Applications*.....411   VII-32
Ken Lee, Min Sung Kim, Peter Shim, Ica Han, Jack Lee, Jeffery Chun and Samuel Cha
Simmtech Co., Ltd., Cheongju, Republic of Korea

High-Bandwidth IC Interconnects with Silicon Interposers and Bridges for 3D Multi-Chip Integration and Packaging.....415   VII-7
Boping Wu
Intel Research, Hillsboro, OR, USA

Development of Wafer Level Hybrid Bonding Process Using Photosensitive Adhesive and Cu Pillar Bump.....418   VII-23
Mingjun Yao[1], Daquan Yu[2], Ning Zhao[1], Jun Fan[2], Zhiyi Xiao[2] and Haitao Ma[1]
[1]Dalian University of Technology, Dalian, China
[2]Huatian Technology Electronics, Kunshan, China

Novel Leveling Materials for Copper Deposition in Advanced Packaging.....421   VII-10
Tao Ma , Jiang Wang, Zifang Zhu and Peipei Dong
Shinhao Materials LLC, Suzhou, Jiangsu, China

Production-scale Flux-free Bump Reflow Using Electron Attachment.....424   VII-15
C. Christine Dong[1], Richard E. Patrick[1], Gregory K. Arslanian[1], Tim Bao[1], Kail Wathne[2], and Phillip Skeen[2]
[1]*Air Products and Chemicals, Allentown, PA, USA*
[2]*Sikama International, Inc., Santa Barbara, CA, USA*

Fine Feature Solder Paste Printing for SIP Applications*.....427   VII-38

Sze-Pei Lim, Kenneth Thum and Andy Mackie
Indium Corporation, Malaysia/USA

Investigation of Thermal Interface Materials Reinforced with Micro- and Nanoparticles...432 VII-2
Kamil Janeczek[1], Aneta Araźna[1], Yan Zhang[2], Shiwei Ma[2], Janusz Sitek[1] and Jingyu Fan[3], Johan Liu[2,4] and Krzysztof Lipiec[1]
[1]*Tele and Radio Research Institute, Warsaw, Poland*
[2]*SMIT Center, Shanghai University, Shanghai, China*
[3]*SIAMM, Shanghai University, Shanghai, China*
[4]*MC2, Chalmers University of Technology, Gothenburg, Sweden*

A Fast and Low-Cost TSV/TGV Filling Method.....435 VII-16
Jiebin Gu, Bingjie Liu, Heng Yang and Xinxin Li
Shanghai Institute of Microsystem and Information Technology, Shanghai, China

Development of Plating Resist for FO-WLP.....438 VII-12
Kenji Okamoto
Advanced Electronic Materials Laboratory, JSR Corporation, Yokkaichi, Japan

In-Situ TEM Observation of IMC Evolution at Atomic Scale*.....440 VII-37
Chaolun Wang and Xing Wu
Dept. of Electrical Engineering., East China Normal University, Shanghai, China

The Study on the Moldability and Reliability of Epoxy Molding Compound.....443 VII-4
Wei Tan[1], Hongjie Liu[1], Yangyang Duan[1], Lanxia Li[1], Xingming Cheng[1], Dong-en zhang[2] and Junyan Gong[2]
[1]*Jiangsu HHCK Advanced Materials Co., Ltd., Lianyungang, Jiangsu, China*
[2]*Dept. of Chemical Engineering, Huaihai Institute of Technology, Lianyungang, China*

High Productivity PVD Solution for an Ever-Evolving Advanced Packaging Industry...446 VII-5
Frantisek Balon, Patrick Carazzetti, Juergen Weichart, Mohamed Elghazzali and Mike Hoffmann
Evatec Ltd., Trübbach, Switzerland

Electrostatic Discharge Failure Control of IC Package by Epoxy Molding Compound Modification.....449 VII-8
Byung-Seon Kong, Sang-Sun Lee, Da Eun Lee, Hyung Ouk Choi and Hyun Woo Kim
Central Research Institute, KCC Corporation, Yongin, Republic of Korea

New Enabling Laser Applications in Advanced Chip Packaging *.....452 VII-11
Dirk Müller[1], John Kennedy[1], Dietrich Tönnies[2] and Rainer Pätzel[1]
[1]*Coherent Inc., Santa Clara, CA, USA*
[2]*Coherent-Rofin GmbH, Bergkirchen, Germany*
[3]*Coherent Laser Systems GmbH & Co. KG, Göttingen, Germany*

Multi Beam Full Cut Dicing of Thin Si IC Wafers.....456 VII-20
Jeroen van Borkulo, Richard van der Stam, Won Chul Jung and Paul Verburg
ASMPT Laser Separation International B.V., Beuningen. The Netherlands

Morphology Control of Copper Nanomaterials for IC Bonding.....460 VII-17
Jiayue Wen, Yanhong Tian and Zhi Jiang

State Key Lab of Advanced Welding and Joining, Harbin Institute of Tech., Harbin, China

Latest Material Technologies for Fan-Out Wafer Level Package.....463   VII-14
Itaru Watanabe, Masaya Kouda, Koji Makihara and Hiroki Shinozaki
Sumitomo Bakelite Co., Ltd. Nogata-City, Fukuoka-Pref, Japan

Using DOE to Improve COB Bonbability.....466   VII-3
Wei Xin*, Sherry Chen and Wei-Ting Kary Chien
Semiconductor Manufactory International Corporation, Shanghai,, China

Study of White Epoxy Molding Compound for LED Bracket.....471   VII-6
Hongjie Liu[1], Wei Tan[1], Lanxia Li[1], Xiaojuan Jiang[1], Liang Cui[1], Yangyang Duan[1], Xingming Cheng[1], Dongen Zhang[2] and Yuhui Luo[2]
[1]Jiangsu HuaHaiChengKe Advance materials Co. Ltd., Lianyungang, China
[2]Department of Chemical Engineering, Huaihai Institute of Technology, Lianyungang，China

Chapter VIII – MEMS, Sensors and Emerging Semiconductor Technologies

Novel Electron Devices Based on Laser Scribed Graphene*.....474   VIII-16
Lu-Qi Tao[1], Dan-Yang Wang[1], He Tian[2], Ning-Qin Deng[1], Yi Yang[1] and Tian-Ling Ren[1]
[1]Institute of Microelectronics & Tsinghua National Laboratory for Information Science and Technology (TNList), Tsinghua University, Beijing, China
[2]Ming Hsieh Department of Electrical Engineering, University of Southern California, Los Angeles, CA, USA

Materials Screening Workflow Methodologies for Metal Oxides and Chalcogenides for Use in Novel Devices*.....478   VIII-23
Tony Chiang, Karl Littau, Stephen L. Weeks, Ashish Pal, Vijay Narasimhan, Greg Nowling, Michael Bowes, Sergey V. Barabash and Dipankar Pramanik
Intermolecular, Inc., CA, USA

Neutral Beam Technology for Future Nano-device*.....482   VIII-1
Seiji Samukawa
Advanced Institute for Materials Research, Tohoku University Innovative Energy Research Center, Institute of Fluid Science, Tohoku University, Sendai, Miyagi, Japan

8 Inches Monolithic CMOS-MEMS Manufacturing Platforms for Consumer Products...485   VIII-6
Wai Soon and Liew Shanghai
Huahong Grace Semiconductor Manufacturing Corporation, Shanghai, China

Thin-film Processing of "Exotic" Phase-Change and Ferroelectric Materials for IoT Applications*.....489   VIII-17
K. Suu , I. Kimura, H. Kobayashi, Y. Miyaguchi, T. Masuda, Y. Kokaze and T. Jimbo
ULVAC, Inc., Hagisono, Chigasaki, Kanagawa, Japan

Wafer Size MoS$_2$ with Few Monolayer Synthesized By H$_2$S Sulfurization.....493   VIII-25

Yen-Teng Ho[1], Yung-Ching Chu[1], Lin-Lung Wei[1], Tien-Tung Luong[1], Chih-Chien Lin[2], Chun-Hung Cheng[2], Hung-Ru Hsu[3], Yung-Yi Tu[1] and Edward Yi Chang[1]
[1]*Department of Materials Science and Engineering, National Chiao Tung University, Hsinchu, Taiwan,*
[2]*Center for Nano Science technology, National Cheng Kung University Tainan, Taiwan*
[3]*Industrial Technology Research Institute, Hsinchu, Taiwan*

Buffer-Optimized Improvement in RF Loss of AlGaNGaN HEMTs on 4-Inch Silicon (111).....496 VIII-26
Tien Tung Luong[1], Franky Lumbantoruan[1], Yen-Yu Chen[1], Yen-Teng Ho[1], Yueh-Chin Lin[1], Shane Chang[1] and Edward-Yi Chang[1,2]
[1]*Department Materials Science and Engineering, National Chiao Tung University, Hsinchu, Taiwan*
[2]*Department of Electronics Engineering, National Chiao Tung University, Hsinchu, Taiwan*

Sputter Deposition Technology for $Al_{(1-x)}Sc_xN$ Films with High Sc Concentration.....499 VIII-2
Bernd Heinz[1], Stefan Mertin[2], Oliver Rattunde[1], Marc Alexandre Dubois[3], Sylvain Nicolay[3], Gabriel Christmann[3], Maurus Tschirky[1] and Paul Muralt[2]
[1]*Evatec AG, Trübbach, Switzerland*
[2]*Electroceramic Thin Films Group, Ecole Polytechnique Fédérale de Lausanne, Switzerland*
[3]*CSEM SA, Neuchâtel, Switzerland*

A 10 Bit Analog Counter in SPAD Pixel.....502 VIII-3
Bin Li[1], Yue Xu[1,2] and RuiMing Luo[1]
[1]*College of Electronic Science & Engineering, Nanjing University of Posts and Telecommunications, Nanjing, China*
[2]*National and Local Joint Engineering Laboratory of RF Integration and Micro-assembly Technology Nanjing, China*

Controllable Shrinking of Silicon Oxide Nanopores by High Temperature Annealing...505 VIII-19
Jian Chen[1], Tao Deng[2], Zewen Liu[3] and Haizhi Song[1]
[1]*Department of Laser Photoelectric Technology, Southwest Institute of Technical Physics, Chengdu, China*
[2]*School of Electronic and Information Engineering, Beijing Jiaotong University, Beijing, China*
[3]*Institute of Microelectronics, Tsinghua University, Beijing, China*

A Novel Dual-Frequency Terahertz Antenna in Standard CMOS Technology.....508 VIII-24
Jingyu Peng, Xiaoli Ji, Xingxing Zhang, Yiming Liao and Feng Yan
Institute of the electronic Science and Engineering, Nanjing University, Nanjing, China

Multi-Sensory Combined Integrated Low Power Tunable-Gain Interface Circuit.....511 VIII-11
Chun-Te Tung and Kuei-Ann Wen
National Chiao Tung University, Hsinchu, Taiwan

Chapter IX – Circuit Design, Systems and Applications

TSV Inductor Optimization and Its Design Implication*.....515

Cheng Zhuo[1] and Baixin Chen[2]

[1]*College of Information Science & Electronic Engineering, Zhejiang University, Hangzhou, China*

[2]*School of Microelectronics, Xidian University, Xian 710126, Xian, China*

IX-27

Energy Efficient SoC Power Delivery Using Fully-Integrated Voltage Regulators with High-Frequency Switch Control*.....519

Boping Wu

Intel Research, Hillsboro, OR, USA

IX-8

Geometry Effect with Respect to ESD and Radiative Charged Particles in SoC.....522

C.-Z. Chen[1] and David Y. Hu[2]

[1]*Qualchip Technologies, Inc., Wuxi, Jiangsu, China*

[2]*MetroSilicon Microsystems, Kunshan, Jiangsu, China*

IX-16

A Low Noise SPAD Pixel Array with Analog Readout Method.....525

Ruiming Luo[1], Yue Xu[1,2] and Bin Li[1]

[1]*College of Electronic Science & Engineering, Nanjing University of Posts and Telecommunications, Nanjing, China*

[2]*National and Local Joint Engineering Laboratory of RF Integration and Micro-assembly Technology, Nanjing, China*

IX-10

A Novel FMEA Tool Application in Semiconductor Manufacture.....528

Lijuan Sun, Liping Peng, Guihong Deng and Kary Chien

Semiconductor Manufacturing International Corporation, Shanghai, China

IX-9

Ka-Band Low Noise Amplifier Using 70 nm MHEMT Process for Wideband Communication.....532

Xu Cheng, Liang Zhang and Xianjin Deng

Microelectronic and Terahertz Research Center, CAEP, Chengdu, China

IX-11

Design of K/Ka-Band Passive HEMT SPDT Switches with High Isolation.....534

Liang Zhang[1,2], Xu Cheng[1,2], Xianjin Deng[1,2] and Xinxin Li[13]

[1]*Institude of Electronic Engineering, China Academy of Engineering Physics, Mianyang Sichuan, China*

[2]*Microsystem and Terahertz Research Center, China Academy of Engineering Physics, Chengdu Sichuan, China*

[3]*Patent Examination Cooperation Center of The Patent Office, SIPO, Sichuan, Chin*

IX-12

The Multi-Segment Adaptive Control of the High Temperature Heat Source in a MOCVD Vacuum Reactor.....537

Jung- Ching Chiu[1], Chih-Kai Hu[2], Pi-Cheng Tung[2] and Tomi T. Li[2]

[1]*Graduate institute of Opto- Mechatronics Engineering, National Central University, Taoyuan, Taiwan*

[2]*Department of Mechanical Engineering, National Central University, Taoyuan, Taiwan*

IX-5

Hybrid thermal aware reconfigurable 3D IC with dynamic power gating architecture.....541

Chun-chen Liu[1], Yilei Li[2], Yuan Du2, Li Du2 and Tianchen Wang3*

[1]*University of California, Los Angeles, USA*

IX-6

[2]Kneron, Inc., San Diego, California, USA
[3]University of Notre Dame, Notre Dame, Indiana, USA

The Application of a Heating Baffle in a High Temperature Vacuum Reactor.....544

Kuei-Fang Chen[1], Jun-Ching Chiu[1], Chih-Kai Hu[2], Tomi T. Li[2] and Pi-Chen Tung[2]

[1]Department Opto-Mechanical Engineering, National Central University, Taoyuan, Taiwan
[2]Department of Mechanical Engineering, National Central University, Taoyuan, Taiwan

IX-7

Evaluation of Ultra-Low Power Tunneling Field Effect Transistor Power Management Unit.....547

Haifang Lu, Xin'an Wang, Jipan Huang, Zhiqiang Yang, Yuqian Huang and Jijia Guo

The Key Laboratory of Integrated Micro-system Science and Engineering Applications, Peking University Shenzhen Graduate School, Shenzhen, China

IX-1

Design and Implementation of a Digital HBC Coordinator for Body Area Network.....550

Ying Zhang[1], Hao Chen[1], Zhongmin Lin[1], Xin-an Wang[1] and Xing Zhang[2]

[1]The Key Lab of Integrated Microsystems Peking University Shenzhen Graduate School, Shenzhen, China
[2]Institute of Microelectronics, Peking University, Beijing, China

IX-2

The Design and Implementation of a Reconfigurable Convolution Operator Based on APU.....553

Yuqian Huang, Jipan Huang, Xin'an Wang, Zhiqiang Yang, Haifang Lu and Miren Tian

The Key Laboratory of Integrated Micro-system Science and Engineering Applications, Peking University Shenzhen Graduate School, Shenzhen, China

IX-3

A Novel OLED-On-Silicon Microdisplay Drive Circuit with The Digital Analog Hybrid Scan Strategy.....556

Yongnan Chu, Tingzhou Mu, Yuan Ji, Yunsen Yu, Feng Ran and Jiao Li

Microelectronic Research and Development Center, Shanghai University, Shanghai, China

IX-13

A Concise and Precise Model of the Gate Delay for EDA Simulation.....559

Zhipeng Yue, Zhuoquan Huang, Dihu Chen and Tao Su

School Electronics and Information Technology, Sun Yat-sen University, Guangzhou, China

IX-14

Design of a Novel AC LED Driver with No Current Glitch Based on Soft Switching Operation.....562

Yang Boxin, Liang Zhiming, Wu Zhaohui and Li Guoyuan

School of Electronic and Information Engineering, South China University of Technology Guangzhou, China

IX-15

High-Performance Single-Phase Full-Bridge Inverter Using Gallium Nitride Field Effect Transistors.....567

Chih-Chiang Wu[1] and Shyr-Long Jeng[2]

[1]Department of Mechanical Engineering, National Chiao Tung University, Hsinchu, Taiwan
[2]Department of Electrical and Electronic Engineering, Ta Hua University of Science and Technology Hsinchu, Taiwan

IX-17

The Air Quality Evaluation Based on Gas Sensor Array.....570   IX-18
Chang-Yong Chiu and Zhun Zhang
College of Optoelectronic Engineering, Shenzhen University, Shenzhen, China

Design and Implementation of a High Quality R-Peak Detection Algorithm.....575   IX-19
Zhongmin Lin, Bo Wang, Hao Chen, Ying Zhang and Xin-An Wang
The Key Lab of Integrated Microsystems, Peking University Shenzhen Graduate School, Shenzhen, China

China Semiconductor Technology International Conference 2017 (CSTIC 2017)

Editors:

Cor Claeys
Imec
Leuven, Belgium

David Huang
Pall Inc.
Port Washington, NY, USA

Hanming Wu
Semiconductor Manufacturing International Corporation - SMIC
Shanghai, China

Qinghuang Lin
IBM Thomas J. Watson Research Center
Yorktown Heights, NY, USA

Ying Shi
The Integrated circuit Materials Industry Technology Innovation Alliance-ICMTIA
Beijing, China

Steve Liang
Jiangsu Changjiang Electronics Technology Co. Ltd
Wuxi, China

Ru Huang
Peking University
Beijing, China

Kafai Lai
IBM Research Division
Hopewell Junction, NY, USA

Ying Zhang
Applied Materials
Santa Clara, CA, USA

Zhen Guo
Intel
Santa Clara, USA

Yuchun Wang
Anji Microelectronics
Shanghai, China

Peilin Song
IBM Thomas J. Watson Research Center
Yorktown Heights, NY, USA

Viyu Shi
Missouri University of Science and Technology
Columbia, MO, USA

PREFACE

This issue contains a selection of the accepted papers presented at *China Semiconductor Technology International Conference 2017 (CSTIC 2017)*, March 12-13, 2017 in Shanghai, China. The memory stick version of these papers serves as the CSTIC 2017 conference proceedings and is distributed to all registered conference attendees. After reviewing a selection of the presentations will be considered for publication in IEEE Xplore.

CSTIC is the largest and the most comprehensive annual industrial semiconductor technology conference in China. It aims to provide a platform for executives, managers, engineers and researchers from around the world to exchange the latest developments in semiconductor technology and manufacturing and related fields. It also offers an opportunity for those who are interested in investing and collaboration opportunities in the semiconductor industry in Asia, particularly in China.

CSTIC covers all the aspects of semiconductor technology and manufacturing, including circuit design, system integration, devices, materials, patterning (lithography and etching), processes, integration, testing, reliability, device physics and manufacturing as well as emerging semiconductor technologies, including clean energy such as light emitting diodes (LEDs), III-V semiconductors, sensors and micro-electromechanical systems (MEMS).

CSTIC 2017, organized by Semiconductor Equipment and Material International (SEMI), imec, and The Integrated circuit Materials Industry Technology Innovation Alliance (ICMTIA) and technically sponsored by the IEEE Electron Devices Society relies on a long time tradition, which started in 2001. The original International Semiconductor Technology Conference (ISTC) merged in 2009 to become CSTIC, aiming for a broad international representation and increased paper submissions from around the world. For CSTIC 2017 the papers came from all major semiconductor manufacturing regions in the world, including Belgium, Brazil, China, Finland, France, Germany, Indonesia, Japan, Korea, Malaysia, Poland, Singapore, Sweden, Switzerland, Taiwan, The Netherlands, United Kingdom and the United States of America. About 249 papers have been selected for oral presentations and approximate 89 papers for poster presentations after careful reviews by the conference organizing committee.

Almost *180* papers are included in these Proceedings after peer reviews. They represent a snapshot of the recent developments in semiconductor technology and manufacturing in the world. In particular, they offer a glimpse into the state-of-the-art of semiconductor technology and manufacturing in China. These papers are divided into nine (9) chapters according to the nine symposia of CSTIC 2017:

• Device Engineering and MemoryTechnology
• Lithography and Patterning
• Dry & Wet Etch and Cleaning
• Thin Film, Plating and Process Integration
• Chemical-Mechanical Polishing (CMP) and Post-Polish Cleaning
• Metrology, Reliability and Testing
• Packaging and Assembly
• MEMS, Sensors and Emerging Semiconductor Technologies
• Circuit Design, Systems and Applications

These Proceedings are very valuable to engineers and researchers in the fast-moving and growing semiconductor industry. It will give readers a clear understanding of the status of semiconductor technology and manufacturing in China. Furthermore it will also serve as a useful reference for those who are interested in nanofabrication, micro- and nano-fluidics, micro- and nano-photonics, organic electronics, bio-chips, light emitting diodes (LEDs) and other clean energy technologies.

We thank the invited speakers and the authors, particularly the conference plenary speakers, Hiroshi Amano, Nobel Laureate in Physics 2014 and Professor at Nagoya University, Japan, Dr. Tzu-Yin Chiu,

CEO and Executive Director Semiconductor Manufacturing International Corp (SMIC), China, Dr. Jong Shik Yoon, Executive Vice-President Foundry Business Systems-LSI, Samsung Electronics, Korea and Rao Tummala, Professor Georgia Institute of technology, Atlanta, USA, for their valuable contributions to CSTIC 2017. We also thank the more than 120 organizing committee members, particularly the symposium chairs, for their dedication and hard work to help improve the quality and to broaden the reach of CSTIC. These committee members are experts in their respective fields of semiconductor technology and are from well-known companies or prestigious institutions. They all have demanding day jobs, yet they have volunteered to help organizing this conference and to critically review papers presented in these Proceedings. Their contributions were crucial for the success of the conference. We are also indebted to the financial support from the sponsors of CSTIC 2017. Finally, we extend our sincere thanks to SEMI for their tireless efforts and their meticulous organizational skills to help organize CSTIC 2017 and to assemble and publish these CSTIC 2017 proceedings.

David Huang, General Chair CSTIC 2017 2017
Pall Inc, Port Washington, NY, USA

Cor Claeys, Co-Chair, CSTIC
Imec, Leuven, Belgium

CSTIC 2017 Organizing Committee

12 March 2017, Shanghai, China

III-N HETEROSTRUCTURE DEVICES FOR LOW-POWER LOGIC

P. Fay[1]*, W. Li[1], D. Digiovanni[1], L. Cao[1], H. Ilatikhameneh[2], F. Chen[2], T. Ameen[2], R. Rahman[2], G. Klimeck[2], C. Lund[3], S. Keller[3], S. M. Islam[4], A. Chaney[4], Y. Cho[4], and D. Jena[4]

[1] Dept. of Electrical Engineering, University of Notre Dame,
Notre Dame, IN 46556, USA
[2] Dept. of Electrical and Computer Engineering, Purdue University,
West Lafayette, IN 47907 USA
[3] Dept. of Electrical and Computer Engineering, Univ. of California
Santa Barbara, Santa Barbara, CA 93106 USA
[4] Dept. of Electrical and Computer Engineering and Dept. of Materials
Science and Engineering, Cornell University, Ithaca, NY 14853 USA
*Corresponding Author's Email: pfay@nd.edu

ABSTRACT

Future generations of ultra-scaled logic may require alternative device technologies to transcend the limitations of Si CMOS; in particular, power dissipation constraints in aggressively-scaled, highly-integrated systems make device concepts capable of achieving switching slopes (SS) steeper than 60 mV/decade especially attractive. Tunneling field effect transistors (TFETs) are one such device technology alternative. While a great deal of research into TFETs based on Si, Ge, and narrow band gap III-Vs has been reported, these approaches each face significant challenges. An alternative approach based on the use of III-N wide band gap semiconductors in conjunction with polarization engineering offers potential advantages in terms of drain current density and switching slope. In this talk, the prospects for III-N based TFETs for logic will be discussed, including both simulation projections as well as experimental progress.

INTRODUCTION

Continued advancements in transistor scaling, performance, density, and functional integration are becoming increasingly difficult to achieve as CMOS approaches fundamental limits. To continue scaling in power-constrained applications, devices offering steeper switching slopes (SS) than the 60 mV/decade limit of conventional FETs is highly desirable; tunneling FETs (TFETs) are one potential option that promises to allow the SS to transcend the thermionic limit. Extensive work on Si, Ge, and narrow band gap III-V TFETs has been reported (see e.g. [1-3] and references therein). In many cases, however, these devices face fundamental challenges. With the group IV materials, the on-current is typically small, due to inefficient interband tunneling in

indirect semiconductors. On the other hand, for narrow band gap III-Vs the small band gaps lead to ambipolar conduction that degrades the off current, and high oxide-semiconductor interface state densities that limit SS.

The use of III-N heterostructures using polarization engineering to facilitate interband tunneling is a potentially attractive option that has not been widely explored. The candidate materials—GaN, InN, and associated alloys—are direct-gap materials (for efficient tunneling), with wider band gaps than conventional materials used for TFETs (to suppress ambipolar conduction), and also have been demonstrated with low-interface-state-density oxide/semiconductor interfaces. By combining these features with the additional design freedom afforded by polarization in these wurtzite materials, TFET designs based on III-Ns appear as a promising candidate for future low-power systems [4-6].

In this talk, the prospects for III-N based TFETs for logic will be discussed, including both simulation projections as well as experimental progress. Simulations suggest that TFETs capable of delivering average SS of below 30 mV/decade over 4 decades of drain current with on-current densities exceeding 100 µA/µm, and in some cases approaching 1 mA/µm, should be possible. Both simulation projections as well as recent experimental developments in demonstrating nanowire-based devices will be presented.

DEVICE STRUCTURE

The fundamental principle of operation for a III-N based TFET centers on field-effect modulation of interband tunneling. In order to achieve appreciable

978-1-5090-6695-7/17 $31.00 © 2017 IEEE

tunneling in III-N heterostructures, a quasi-broken gap heterostructure is required, since direct interband tunneling currents in III-N homojunctions are typically quite small due to the large band gaps [4]. One possible implementation that combines the use of polarization engineering (to develop favorable band alignments at the tunnel junction) and the favorable electrostatics of a gate-all-around structure is illustrated in Fig. 1 [5]. In this design, a p-type GaN source and n-type GaN drain are used, with a (nominally undoped) InN interlayer to adjust the band alignments between the source and drain layers. In addition, to improve the source carrier injection, a step-graded InGaN layer is included to augment the p-type chemical doping in the source (implemented with Mg doping) with "polarization doping" [7].

In this device concept, the gate bias controls the tunneling window between the valence band edge of the p-GaN and the conduction band edge, as illustrated by the band diagrams in Fig. 1(b) and (c). As illustrated in Fig. 1(b), in the off state the valence band edge in the source lies lower in energy than the conduction band edge in the drain, leading to no direct source-drain tunneling and low device current. However, by applying a more positive gate bias (Fig. 1(c)) the conduction band edge in the drain is brought below the valence band edge in the source, opening a window for interband tunneling and leading to appreciable source-drain tunneling current. Although a gate-all-around nanowire configuration is shown in Fig. 1, similar concepts can also be applied to in-line and sidewall geometry devices [4].

Simulations of these devices suggest that on-currents of over 1 mA/μm, switching slopes below 30 mV/decade (averaged over 4 decades of drain current), and on/off current ratios in excess of 10^7 should be attainable. However, while these simulations do include carrier scattering and thermalization effects, they do not include the impact of defects or other imperfections in the materials or interfaces that can degrade device performance. Recent experimental results have, however, indicated that interband tunneling can be achieved in GaN/InGaN/GaN heterostrutures of the types proposed here, with tunneling current densities consistent with simulation [8].

In this talk, the latest experimental and simulation-based results for these devices will be discussed. The performance of device structures based on both MBE and MOCVD-grown heterostructures will be presented, and theoretical projections of ultimate device performance will be shown, suggesting the potential of these devices for low-power logic applications.

ACKNOWLEDGEMENTS

This work is sponsored in part by LEAST, a funded center of STARnet, a Semiconductor Research Corporation (SRC) program sponsored by MARCO and DARPA.

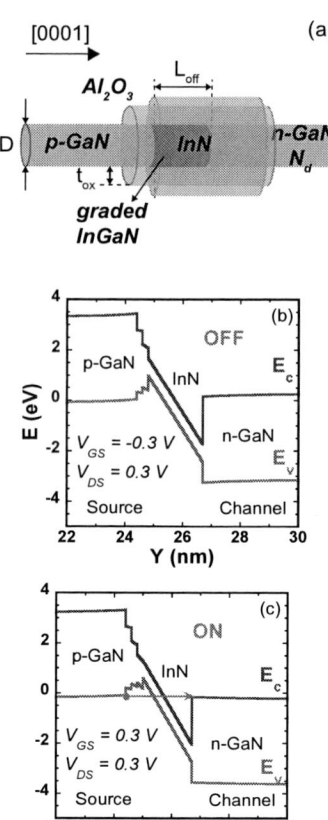

Figure 1: Prototype example of a III-nitride nanowire-based gate-all-around TFET: (a) schematic device structure (b) off-state band diagram and (c) on-state band diagram [5].

REFERENCES

[1] A. C. Seabaugh and Q. Zhang, "Low-voltage tunnel transistors for beyond CMOS logic," *Proc. IEEE*, vol. 98, no. 12, pp. 2095-2110, 2010.

[2] A. M. Ionescu and H. Riel, "Tunnel field-effect transistors as energy-efficient electronic switches," *Nature*, vol. 479, no. 7373, pp. 329-337, 2011.

[3] G. Dewey et al., "Fabrication, characterization, and physics of III-V heterojunction tunneling Field Effect Transistors (H-TFET) for steep sub-threshold swing," *IEEE Intl. Electron Devices Meeting*, pp. 33.1.1-33.1.4, 2011.

[4] W. Li et al., "Polarization-engineered III-Nitride Heterojunction Tunnel Field-Effect Transistors," *IEEE J. Exploratory Solid-State Computational Dev. And Circuits*, vol. 1, pp. 28-34, 2015.

[5] W. Li et al., "Performance projection of III-nitride heterounction nanowire tunneling field-effect

transistors," *Physica Status Solidi A*, DOI 10.1002/pssa.201532564, 2015.

[6] W. Li et al., "GaN Nanowire MISFETs for Low-Power Applications," *Proc. Compound Semiconductor Manufacturing Technology Conf. (CS-MANTECH)*, 2016.

[7] D. Jena et al., "Polarization-engineering in group III-nitride heterostructures: New opportunities for device design," *Physica Status Solidi A*, vol. 208, pp. 1511-1516, 2011.

[8] X. Yan et al., "Polarization-induced Zener tunnel diodes in GaN/InGaN/GaN heterojunctions," *Appl. Phys. Lett.*, vol. 107, pp. 163504-1-5, 2015.

MONOLITHIC 3D (M3D) RECONFIGURABLE LOGIC APPLICATIONS USING EXTREMELY-LOW-POWER ELECTRON DEVICES

Woo Young Choi

Department of Electronic Engineering, Sogang University, Seoul 04107, Republic of Korea
*Corresponding Author's Email: wchoi@sogang.ac.kr

ABSTRACT

CMOS and nano-electromechanical (NEM) hybrid reconfigurable logic (RL) circuits are implemented by using monolithic three-dimensional (M3D) integration process. Their operation and feasibility are discussed based on simulation and experimental results.

INTRODUCTION

Reconfigurable logic (RL) has attracted researchers' attention thanks to its design flexibility and development time reduction. Its most well-known example is a field-programmable gate array (FPGA) as shown in Fig. 1a. The connection blocks (CBs) connect adjacent logic blocks (LBs) with each other and consist of many one-to-n or n-to-one multiplexers (MUXes). The switch boxes (SBs) connect the CBs which surround them and consist of many n-to-n MUXes. Using the CBs and SBs, the flexible data signal paths between each LB is implemented. Conventional CMOS-only RL circuits have some limitations such as low speed, high energy consumption and large chip area, because they use MOSFETs in CBs and SBs [1], [2]. In order to address these problems, CMOS-NEM hybrid RL circuits are proposed which replace MOSFETs in the CBs and SBs with NEM memory switches as shown in Fig. 1b. Nonvolatile property is obtained by using adhesion force [3], [4]. The proposed NEM-memory-based RL circuits show higher speed, lower power consumption and smaller chip area than conventional MOSFET-based RL circuits. Their advantages are confirmed based on simulation and measurement results.

SIMULATION RESULTS

Fig. 2a shows the structure of a NEM memory switch for RL applications. Its operation principle referring to [3] and [4] has already been confirmed experimentally in the case of silicon beam material [5], [6]. It consists of two metal line electrodes which are called Selection Line 1 (L_1) and Selection Line 2 (L_2), respectively, and a movable cantilever beam attached to the bit line (BL) which switches between L_1 and L_2. Although any conductive materials can be used for the electrodes, titanium nitride (TiN) has been assumed in this work. The surfaces between movable beam and electrodes are assumed to be smooth. It has been reported that the smooth surfaces of the beam and electrodes can be implemented by using the

focused ion beam (FIB) process with low beam current [7]. Fig. 2b shows the initial state when the BL is not connected to either L_1 or L_2. If either V_{BL-L1} or V_{BL-L2} becomes nonzero, F_{elec} is induced between L_1 and BL or between L_2 and BL. If high V_{BL-L1} is applied, the beam is pulled down to L_1 (State 1) as shown in Fig. 2c. Likewise, if V_{BL-L2} is high, the beam is pulled down to L_2 (State 2). It means that data signal paths are determined by V_{BL-L1} and V_{BL-L2}. Once State 1 or 2 are determined, even if both V_{BL-L1} and V_{BL-L2} become zero, the state is maintained thanks to F_{ad} [8]. If the beams and selection lines have smooth contact surfaces, F_{ad} can be larger than F_r [9]. In this case, the connection between the BL and selection line is maintained without supply voltage and the R_{co} between them decreases [10].

CMOS-NEM hybrid RL circuits are compared with conventional CMOS-only RL circuits in terms of performance, energy consumption and chip area by using H-SPICE and Verilog-A. The beam motion is determined by multi-physics solvers implemented by Verilog-A [11]. Then, the calculated beam motion parameter x is transferred to the equivalent circuit model of a NEM memory switch for CMOS-NEM hybrid circuit simulation. Each FPGA is compared in the three operation modes: configuration, standby, dynamic mode. The configuration mode means switching process of each switch component of CBs and SBs. The data signal path of an FPGA is determined in this mode. In the standby mode, neither data signal path change nor data signal transfer is performed while supply voltage is applied. After data signal paths are determined in the configuration mode, data signals are transferred in the dynamic mode. For an example, the CBs are discussed in this paper. As shown in Fig. 3a, four one-to-two MUX cases have been considered for the CBs: pass gates, transmission gates, tri-state buffers and NEM memory switches. The first three cases correspond to conventional CMOS-only RL circuits. The fourth case proposed in this paper consists of one NEM memory switch and two selection MOSFETs. The operation voltage of a NEM memory switch is applied to either L_1 or L_2 through a selection MOSFET. In the case of conventional CMOS-only CBs, MOSFETs directly connect or disconnect data signal paths following control signals. However, in the case of the proposed NEM-memory-based CBs, data signal paths are maintained regardless of data signal. Thus, no control

978-1-5090-6695-7/17 $31.00 © 2017 IEEE

signal is necessary to maintain data signal paths in the dynamic mode and less energy is consumed. Also, because data signals are transferred through low-resistive metal lines and large contact areas rather than semiconductor channels of MOSFETs, NEM-memory-based CBs implement smaller time delay than conventional CMOS-only CBs. Figs. 3b to e show the circuit simulation results of each CB.

EXPERIMENTAL RESULTS

In order to confirm the proposed CMOS-NEM hybrid RL circuits, the first M3D CMOS-NEM hybrid reconfigurable circuits are fabricated by using standard CMOS process. Fig. 4a shows the schematic of the fabricated 3D CMOS-NEM hybrid reconfigurable circuit. Logic and routing parts consist of CMOS and NEM memory switches, respectively. The data signals of CMOS logic gates are transferred through NEM memory switches which determine data signal paths. NEM memory switches are located in the metal layers over CMOS devices by using standard CMOS BEOL process. A NEM memory switch acts as a one-to-two multiplexer connecting two stages of CMOS inverters. Data signal paths and logic functions are determined by the beam position which varies as a function of V_{L1} and V_{L2}. Key fabrication steps are shown in Fig. 4b. First, CMOS logic circuits are fabricated on a silicon substrate by using standard CMOS FEOL process. Second, metal interconnection lines and NEM memory switches are formed by using standard CMOS BEOL process. FIB patterning has been used only to define small air gap patterns and can be easily replaced by conventional optical photolithography for mass production. Finally, the intermetal dielectric (IMD) layers surrounding NEM memory switches are selectively removed to release the beam by using hydrofluoric acid (HF) vapor etch.

Fig. 5 shows the SEM images of the fabricated 3D CMOS-NEM hybrid reconfigurable circuit. The output of the first CMOS inverter is transferred to the beam of a NEM memory switch through vias. In this work, the L_1 of the NEM memory switch is connected to Out 1 node while L_2 is connected to the input of the second CMOS inverter through vias. It means that the output of the first inverter is transferred to either Out 1 node or the input of the second inverter depending on the beam position of the NEM memory switch.

Fig. 6a shows the measurement results of State 1. V_{in} is transferred to V_{out1} through the first inverter and the NEM memory switch, which leads to "$V_{in}=/V_{out1}$" logic function. Subsequently, the state of the NEM memory switch is converted from State 1 into State 2 and then the same procedure is repeated except that V_{L2} ramps up this time rather than V_{L1}. In this case, the logic function becomes "$V_{in}=V_{out2}$". Fig. 6b shows the measurement results of State 2 when V_{in} is transferred to V_{out2} through two stages inverters and one NEM memory switch.

SUMMARY

M3D CMOS-NEM hybrid RL circuits have been proposed, simulated and implemented. NEM memory routing switches are vertically integrated over CMOS logic circuits by using standard CMOS process. The proposed NEM-memory-based RL circuits feature higher speed, lower energy consumption and smaller chip area than conventional CMOS-only RL circuits.

ACKNOWLEDGEMENTS

This work was supported in part by the NRF of Korea funded by the MSIP under Grant NRF-2015R1A2A2A01003565 (Mid-Career Researcher Program), NRF-2015M3A7B7046617 (Fundamental Technology Program), NRF-2016M3A7B4909668 (Nano·Material Technology Development Program) and in part by the MOTIE/KSRC under Grant 10044842 (Future Semiconductor Device Technology Development Program).

REFERENCES

[1] I. Kuon, R. Tessier, and J. Rose, "FPGA architecture: survey and challenges," Foundations and Trends in Electronic Design Automation, vol. 2, no. 2, pp. 135-253, Feb. 2008.

[2] K. Compton, and S. Hauck, "Reconfigurable computing: a survey of systems and software," ACM Computing Surveys, vol. 34, no. 2, pp. 171-210, Jun. 2002.

[3] C. G. Smith, R. Kampen, J. Popp, D. Lacy, D. Pinchetti, M. Renault, V. Joshi, and M. A. Beunder, "Nano-mechanical cantilever arrays for low-power and low-voltage embedded nonvolatile memory applications," in Proc. SPIE, 2007, p. 646406.

[4] W. Y. Choi, "Laterally-actuated electromechanical memory cell and its fabrication," Korean Patent 10-2012-0080281, July 23, 2012.

[5] B. Soon, E. J. Ng, N. Singh, J. M. Tsai, Y. Qian, and C. Lee, "A Bistable Electrostatic Silicon Nanofin Relay for Nonvolatile Memory Application," J. Microelectromech. Syst., vol. 22, no. 5, pp. 1004–1006, Oct. 2013.

[6] B. W. Soon, E. J. Ng, Y. Qian, N. Singh, M. J. Tsai, and C. Lee, "A bi-stable nanoelectromechanical non-volatile memory based on van der Waals force," Appl. Phys. Lett., vol. 103, no. 5, pp. 053122 - 053122-5, Aug. 2013.

[7] Y. Park, S. Park, U. Lee, and S. H. Choi, "Nano-scale lateral milling with focused ion beam for ultra-smooth optical device surface," Recent Patents on Space Technology, vol. 2, pp. 51–58, July. 2010.

[8] F. W. DelRio, M. P. De Boer, J. A. Knapp, E. D. Reedy, Jr., P. J. Clews, and M. L. Dunn, "The role of van der

Waals forces in adhesion of micromachined surfaces," Nat. Mater., vol. 4, no. 8, pp. 629–634, Aug. 2005.

[9] T. J. W. Wagner, and D. Vella, "Switch on, switch off stiction in nanoelectromechanical switches," Nanotechnology, vol. 24, no. 27, pp. 275501–275511, July. 2013.

[10] T.-J. King Liu, L. Hutin, I.-R. Chen, R. Nathanael, Y. Chen, M. Spencer, and E. Alon, "Recent progress and challenges for relay logic switch technology," in Symp.VLSI Tech. Dig., 2012, pp. 43-44.

[11] H. Toshiyoshi, T. Konishi, K. Machida, and K. Masu, "A multi-physics simulation technique for integrated MEMS," in Proc. IEEE IEDM, Dec. 2012, pp. 123–126.

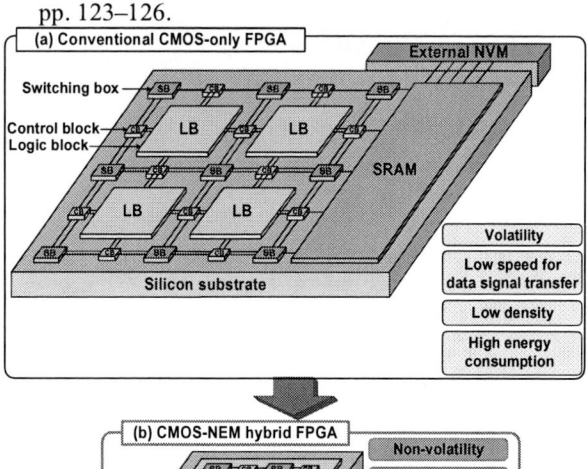

Fig. 1. Block diagrams of (a) conventional CMOS-only FPGA and (b) the proposed CMOS-NEM hybrid FPGA.

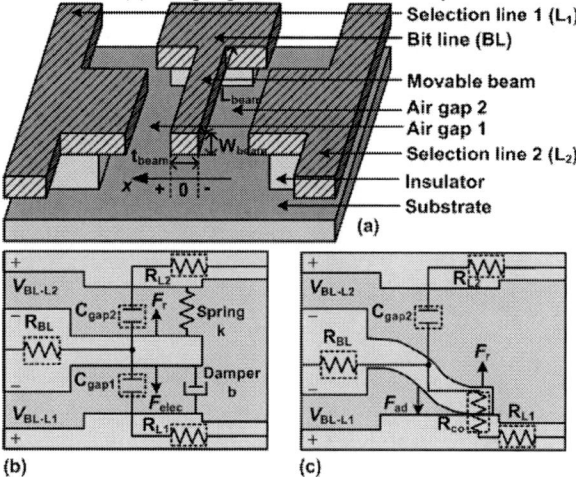

Fig. 2. (a) Proposed NEM memory switch. (b) Its plan view in the initial state and (c) in State 1.

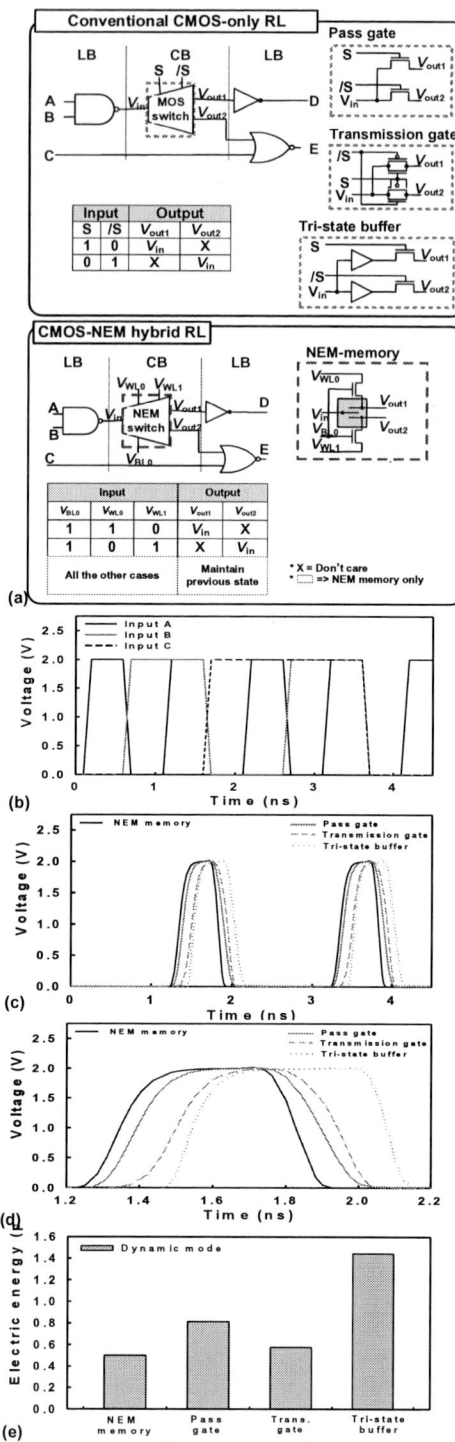

Fig. 3. (a) CBs of pass gates, transmission gates, tri-state buffers and NEM memory switches. (b) Input data signals of the CBs. (c) Output data signal D of the CBs. (d) Enlarged output data signal D shown in (c). (e) Energy consumption of various CBs per cycle.

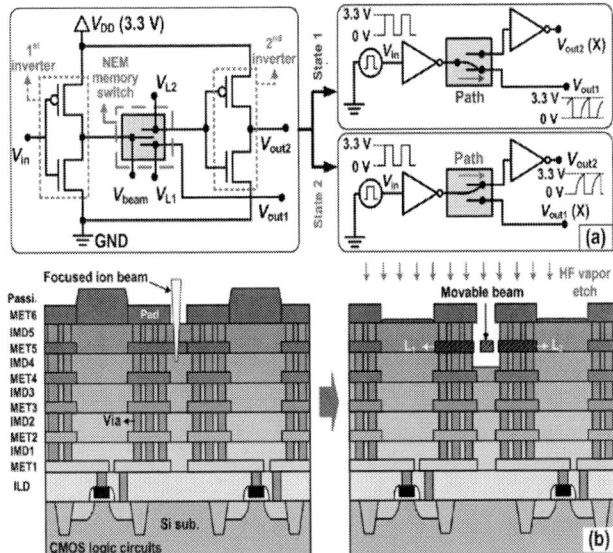

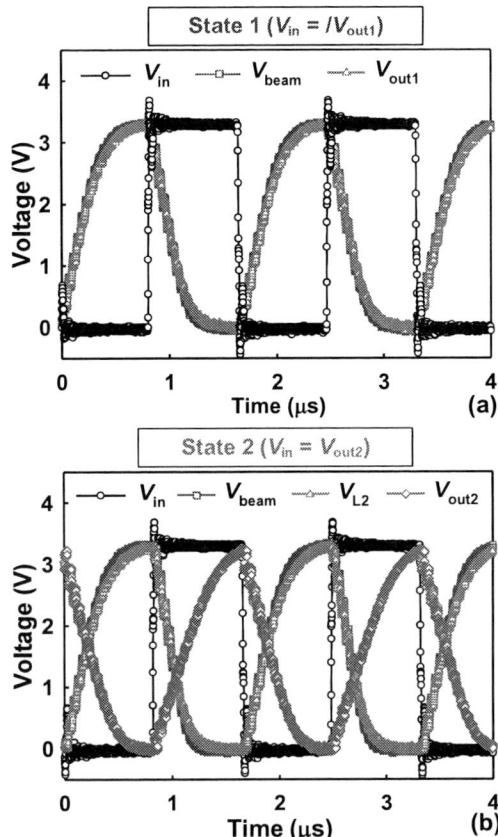

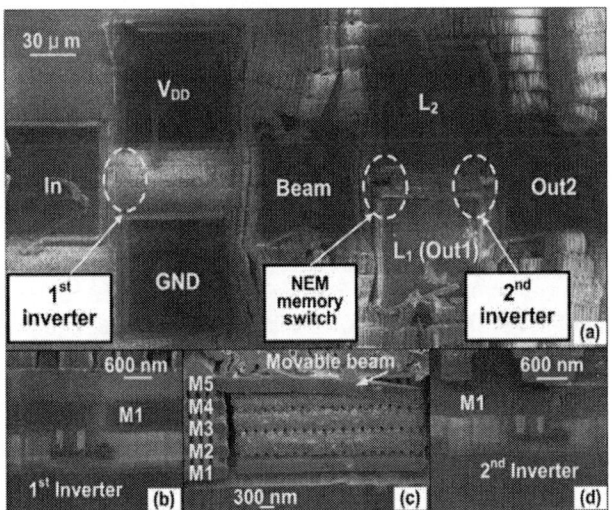

Fig. 4. (a) Schematic and (b) key fabrication process of the M3D CMOS-NEM hybrid RL circuit.

Fig. 5. (a) Plan view of the fabricated M3D CMOS-NEM hybrid reconfigurable circuit. The circuit schematic is shown in Fig. 4a. The cross-sectional view of (b) the first inverter, (c) the NEM memory switch and (d) the second inverter.

Fig. 6. Data signal transfer of the fabricated M3D CMOS-NEM hybrid reconfigurable circuit (a) in State 1 and (b) in State 2.

978-1-5090-6695-7/17 $31.00 © 2017 IEEE

A STUDY OF 28NM LDMOS HCI IMPROVEMENT BY LAYOUT OPTIMIZATION

Ruoyuan Li*, Yongsheng Yang, Fang Chen, Ling Tang, Zhengyong Lv, Byunghak Lee, Ling Sun, Weizhong Xu, TzuChiang Yu

Technology R&D Center, SMIC
18 Zhangjiang Road, Pudong New Area, Shanghai, P.R. China 201203
E-mail: Nikki_Li@smics.com

ABSTRACT

LDMOS (lateral diffused MOS) is an important class of device finding applications in high voltage and smart power management due to their compatibility with the standard CMOS process. However, high operational drain voltage makes LDMOS devices highly vulnerable to the damage caused by hot-carrier injection (HCI). In this paper, the various layout parameters of NLDMOS with shallow trench isolation (STI) are systematically studied to check HCI performance using 28nm Poly/SiON logic process, including effective channel length (Lc), a drift region and poly gate overlap (Lp), a drift region and Pwell overlap/space (Lw) and STI width (Ls). Extensive TCAD simulations and experiments reveal that small Lp and large Lw overlap can greatly improve NLDMOS substrate current and HCI performance without any additional process step or process modification. The physical mechanism behinds the results should be that the impact ionization has been driven further away from the Si/SiO2 interface with a reduction in magnitude, which can improve substrate current and HCI performance.

INTRODUCTION

LDMOS (lateral diffused MOS) transistors are widely implemented in high voltage, smart power management and display drivers, which are easily integrated within existing technology to handle a wide range of operating voltages without significant process changes to decrease not only a manufacturing cost but also responding time to market. Since these processes are originally oriented for low voltage operation, it is a challenging work to overcome hot-carrier injection (HCI) issue using thin gate oxide without a process modification [1]-[6]. As technology is scaled to 28nm, LDMOS optimization becomes more challenging to get the comparable HCI performance with the previous nodes because well implantation has a higher concentration and is closer to Si/SiO2 interface, which results in the electric field stronger and HCI worse. Among the LDMOS HCI improvement methods, layout optimization is the favorite due to without any process change and cost issue.

It is well known that the hot carriers, the high kinetic energy carriers due to high electric field in the channel region, can generate electron-hole pairs through impact ionization. The holes are collected by substrate, which is defined as substrate current (Isub). Consequently, Isub is usually used to judge the severity of hot-carrier damage and a good indicator of HCI degradation. The key factors to minimize hot carrier generation and injection are to reduce the electric field magnitude and to push impact ionization region deep into silicon, which can result in the lower possibility of hot carriers reaching the Si/SiO2 interface [7]-[9].

For 28nm Poly/SiON logic technology, LDMOS shares all process with standard CMOS. This work focuses on the analysis of NLDMOS layout optimization, not process change, to improve HCI performance. Accordingly, the various layout parameters of NLDMOS with shallow trench isolation (STI) to check Isub are systematically studied, including effective channel length (Lc), a drift region and poly gate overlap (Lp), a drift region and Pwell overlap/space size (Lw) and STI width (Ls). Based on TCAD simulations and experiments, it can be seen that small Lp and large Lw overlap size can greatly improve NLDMOS Isub and HCI performance.

EXPERIMENTS

The MOSFET samples in this paper were fabricated by 28nm Poly/SiON CMOS process on P-type (100) silicon substrates. After STI process, the poly silicon gate with gate oxide and offset spacer were formed, followed by the selective deposition of in situ boron-doped SiGe on PMOS source/drain. The nickel silicide and Cu backend is applied to complete the process.

The schematic cross section of NLDMOS transistor with layout variables, Lc, Ls, Lp and Lw, is shown in Fig.1. In order to investigate HCI degradation of NLDMOS very well, substrate current (Isub) and saturation drive current (Idsat) are measured. And for all the plots of Isub vs. layout parameters, there is the exactly same Isub scale to analyze the experimental results.

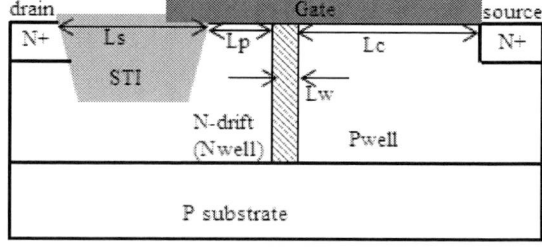

Fig.1: The schematic cross section of NLDMOS with layout parameters Lc, Lp, Lw and Ls

RESULTS AND DISCUSSIONS

Lp dependence of HCI performance

As shown in Fig.1, Lp is defined as N-drift region and poly gate overlap size. In this part, Lp dependence of HCI performance is studied with fixed other layout parameters.

978-1-5090-6695-7/17 $31.00 © 2017 IEEE

N-drift and poly gate overlap, that is, Lp, is a critical layout parameter for LDMOS key electrical characteristics, such as HCI, the device resistance and breakdown voltage. As indicated in Fig.2, with the reduction of Lp, NLDMOS Isub is obviously improved. But when Lp is continually reduced to very small size, Isub does not decrease any more. On the contrary, it does slightly increase. HCI test results show that the lifetime can be improved about one order and meet the criteria.

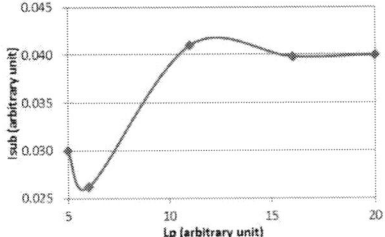

Fig.2: NLDMOS Isub vs. Lp.

The results can be clearly explained by TCAD simulations. Fig.3 is NLDMOS impact ionization simulated by TCAD, which exhibits that the impact ionization peak locates at N-drift and Lp region. When Lp is reduced, the impact ionization has been driving further away from the Si/SiO2 interface with a reduction in magnitude as observed in Fig.3 (b), which has implied a lower rate of hot carrier generation. It is likely because N-drift region with small Lp can be fully exhausted by depletion area, which can make N-drift region higher resistance and take larger drain voltage. As a result, Isub and Idsat degradation can be improved. If Lp is reduced too much, though, current crowding effect becomes worse, which will increase electric field magnitude and deteriorate Isub as observed in Fig.3 (c).

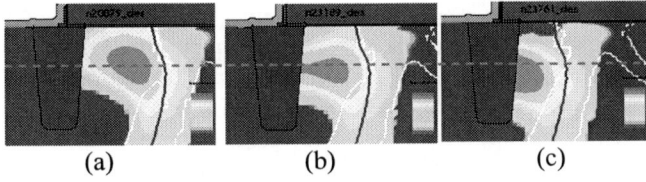

(a)　　　　　(b)　　　　　(c)

Fig.3: Impact ionization of different NLDMOS Lp simulated by TCAD. (a) large Lp (b) middle Lp (c) small Lp

Lw dependence of HCI performance

As shown in Fig.1, Lw is defined as N-drift region and Pwell overlap or space. In this part, Lw dependence of HCI performance is studied with fixed other layout parameters.

Fig.4 is NLDMOS Isub vs. Lw correlation. Negative Lw value means N-drift region and Pwell space that is substrate region without any implantation. Positive Lw value means overlap that has both Nwell and Pwell implantation. Fig.5 exhibits that when Lw is positive value, Isub can be significantly improved with the increase of Lw. The result is within the expectation. Nwell and Pwell have the contrary dopant type. The larger overlap of Nwell and Pwell is, the more the contrary dopant compensates. It can reduce dopant concentration and increase the resistance of overlap region. Accordingly, the overlap area can take much more electrical field and make effective channel less impact ionization. As a

result, NLDMOS Isub can be improved.

In addition, Fig.4 and Fig.5 display that when Nwell and Pwell is separated by substrate region, Isub has no strong correlation with Lw. The root cause may be that the substrate region with very low dopant concentration will prevent Nwell and Pwell with contrary dopant type to diffuse each other. The larger the space is, the more difficult the diffusion is. The conclusion is different from the reference paper [1], but is consistent with our TCAD simulation as indicated in Fig.5, which is possible due to the different implantation scheme from the reference paper.

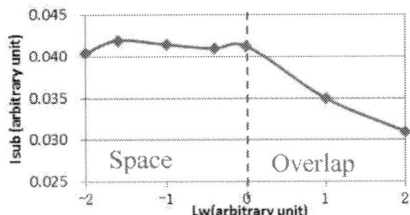

Fig.4: NLDMOS Isub vs. Lw (Negative Lw means N-drift and Pwell space and positive Lw means overlap).

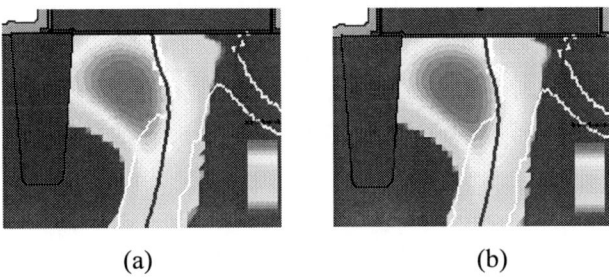

(a)　　　　　　　　(b)

Fig.5: Impact ionization of different NLDMOS Lw space simulated by TCAD. (a) Small Lw space (b) Large Lw space

Lc dependence of HCI performance

As shown in Fig.1, Lc is defined as effective channel length. In this part, Lc dependence of HCI performance is studied with fixed other layout parameters.

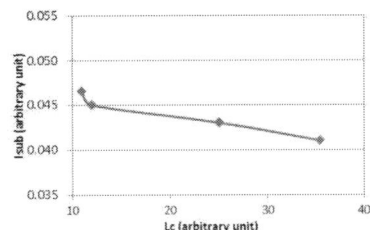

Fig.6: NLDMOS Isub vs. Lc.

As we know, the substrate current is related to the impact ionization generation rate that is well connected to drain edge electric field (Em). Em contributes to the generation of hot carrier [4].

$$Isub = Idsat \times \alpha \times \Delta Lc \tag{1}$$

$$Em = (Vds-Vdsat)/Lc \approx [Vds-(Vg-Vt)]/Lc \tag{2}$$

In the equations (1) and (2), α is impact ionization

generation rate that is directly proportional to Em. Vds is drain voltage and Vg is gate voltage. Vt is threshold voltage. Lc is LDMOS effective channel length as mentioned in this paper. When Lc increases, Em will decrease, which leads to impact ionization and Isub improvement as indicated by Fig.6, The results are perfectly matching TCAD simulations of Fig.7. With the increase of Lc, electric field becomes weaker and deeper into the substrate.

But Fig.6 exhibits that Lc has a fewer impacts on Isub compared to Lw and Lp. It may be because in our implantation scheme, well dopant concentration is so high that dominates electric field and impact ionization. The change of effective channel length within process window cannot obviously improve Em and Isub.

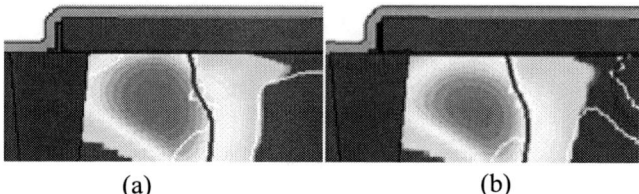

(a)   (b)

Fig.7: Impact ionization of different NLDMOS Lc simulated by TCAD. (a) Small Lc (b) Large Lc

Ls dependence of HCI performance

As shown in Fig.1, Ls is defined as STI width. In this part, Ls dependence of HCI performance is studied with fixed other layout parameters.

When integrating high voltage LDMOS devices into an existing process, the STI is often incorporated into the extrinsic drain as an integral part of the drift region designed to increase on-state and off-state breakdown voltage and HCI performance. The current flow along the STI profile imposes new requirements regarding depletion region shape that is well associated with Isub [3], [6]. And electric field locates in N-drift region and close to STI exhibited by TCAD simulations. Thus, STI is another key factor for Isub. In this work, the impact of STI width to Isub is studied.

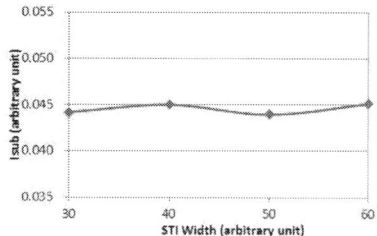

Fig.8: NLDMOS Isub vs. STI width

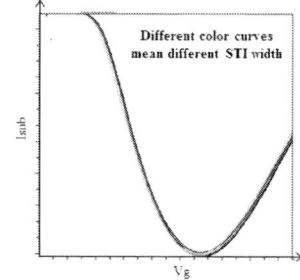

Fig.9: The impact of STI width on Isub simulated by TCAD

In general, the wider STI is, the better HCI performance is, which is because the wider STI can take much more drain voltage and make effective channel less impact ionization. But Isub vs. STI width plot of Fig.8 and TCAD simulations of Fig.9 show that Isub has no obvious correlation with STI width. The reason should be that the well dopant concentration dominates the electric filed and Isub. And the impact of STI width change within process window can almost be neglected.

Actually, as reported in [3], increasing STI trench sidewall angel by process optimization can improve NLDMOS HCI lifetime as well, which is perfectly supported by our TCAD simulations of Fig.10. But in this work, layout optimization is the focus and any process change will not be discussed.

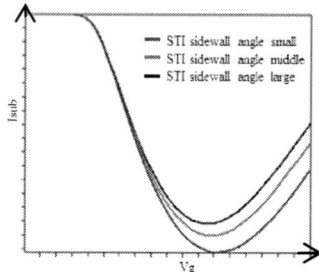

Fig.10: The impact of STI sidewall angle on Isub simulated by TCAD

It has to be mentioned that another important electrical parameter of LDMOS, Ron, that is defined as the resistance of the devices in on-state is a trade-off with HCI performance. The reduction of electric field to improve HCI generally increases Ron, which is not preferred.

SUMMARY

In this work, a systematical study of NLDMOS layout optimization to improve HCI performance is presented. The layout parameters include effective channel length (Lc), a drift region and poly gate overlap (Lp), a drift region and Pwell overlap/space (Lw) and STI width (Ls). It can be found that small Lp and large Lw overlap can greatly improve NLDMOS Isub and HCI performance without any additional process step or process modification. And HCI lifetime can be improve about one order since the impact ionization has been driven further away from the Si/SiO2 interface and the magnitude of electric field is reduced.

REFERENCES

[1] Huixiong Zheng, Huaqiang Wu, Bin Wang, *Electron Devices and Solid-State Circuits (EDSSC), 2014 IEEE International Conference, pp.1-2, 2014.*

[2] Susanna Reggiani, Gaetano Barone, Stefano Poli, *et al. IEEE Transactions on Electron Devices, vol. 60, pp.691-698, 2013.*

[3] S. Haynie, A. Gabrys, T. Kwon, et al. *2010 22nd International Symposium on Power Semiconductor Devices & IC's (ISPSD), pp.241-244, 2013.*

[4] Kyuheon Cho, Seonghoon Ko, Fumie Machida, et al. *2015 IEEE 27th International Symposium on Power Semiconductor Devices & IC's (ISPSD), pp.69-72, 2015.*

[5] Douglas Brisbin, Philipp Lindorfer, Prasad Chaparala, et al. *2006 IEEE International Integrated Reliability Workshop Final Report, pp.44-48, 2006.*

[6] S. Poli, S. Reggiani, G. Baccarani, et al. *2011 IEEE 23rd International Symposium on Power Semiconductor Devices and ICs,* pp.152-155, 2011.

[7] E. Takeda, IEE Proceedings I - Solid-State and Electron Devices, *vol. 131,* pp.153-162, 1984.

[8] Chen Min-Liang, Leung Chung-Wai, W. T. Cochran, *IEEE Transactions on Electron Devices, vol. 35,* pp.2210-2220, 1988.

[9] Jong Mun Park, Martin Knaipp, Hubert Enichlmair, et al. *2012 24th International Symposium on Power Semiconductor Devices and ICs,* pp.189-192, 2012.

RELIABILITY INVESTIGATIONS ON THE PROGRAMMING CURRENTS OF 28NM METAL E-FUSE

Guangyan Zhao, Yong Zhao, Wei-Ting Kary Chien

Product Reliability Engineering, Semiconductor Manufacturing International Corporation

18 Zhangjiang Road, PuDong New Area, Shanghai 201203, China

(Email: Guangyan_Zhao@smics.com)

ABSTRACT

The reliability performance of 28nm metal e-Fuse programmed with different current is investigated. High temperature stress (HTS) or temperature cycling (TC) may cause the shift of metal-e-fuse element resistance and shape. In this paper, we find that 28nm metal e-Fuse programming with low current reliability performance is more stable than metal e-Fuse programming with high current; Resistance shift was only observed on fuses programmed in the over-programmed mode. In addition, the SEM profile of metal e-Fuse programming with low current is obviously better than high current SEM profile.

Keywords - 28nm metal e-Fuse, programming current, reliability, resistance stability.

INTRODUCTION

Electrical fuses (e-FUSEs) are used for redundancy implementation, chip self-repair, and reconfiguration in the factory as well as in the field to improve chip yield because of its ability to integrate with sophisticated built-in self-test and built-in self-repair approaches. There have been several publications [1]-[5] covering WSi2 and CoSi2 e-FUSE reliability, which are applied to 90-nm technology and above. Publications [6] covering NiPtSi reliability, which are applied to 28-65nm technology.

For the 28-nm technology node, both NiPtSi e-FUSE and metal e-FUSE are used. The metal e-FUSE, as in Fig.1, is embedded in the HK (High-k) process, which aims for high speed devices.

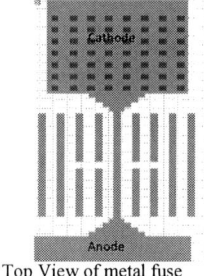 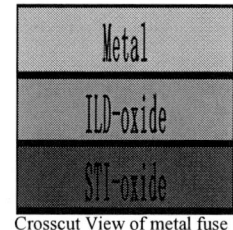

| Top View of metal fuse | Crosscut View of metal fuse |

Fig 1: The structure of 28nm metal e-FUSE.

The component of the e-fuse is anode, cathode, and fuse. The fuse consists of oxide and topper metal. When a e-Fuse is programmed, the topper metal copper will be burned. We have conducted extensive reliability evaluations of Metal e-FUSEs in 28nm technology, including both high temperature stress (HTS) and temperature cycling (TC). The reliability evaluation of programmed fuses covers fuses programmed with different programming currents. 28nm Metal e-Fuse material is copper which ruptured the fuses by local self-heating induced by very high current. High current which make copper melt profile out of charge and stress ambient dummy area to destroy reliability performance. For the fuses programmed with a high current, both resistance shift and functional sensing fails were observed. However, a low-programming current fuse of 28nm has a good reliability performance both resistance shift and functional sensing fails were not observed.

EXPERIMENT

The goal of this experiment is to evaluate and compare the 28nm Metal e-FUSE reliability of different programming currents. Results of three different currents are presented under TC and HTS quality stress in this paper. The reliability evaluations were conducted to fuse resistance stability.

Based on the voltage window of the applications, we design three testing voltages: NV (target current), LV (90% target current) and HV (110% target current). The reliability tests include HTS, which evaluates the influence of drawn-out high temperature, and TC, which assesses the influence of high speed temperature variety. We judge pass/ fail from both resistance shifts and SEM profiles.

As shown in Fig. 2, the programming current is determined by VPRG and VGS. We change programming current by varying these 2 voltages. The reliability test details are summarized in Table1.

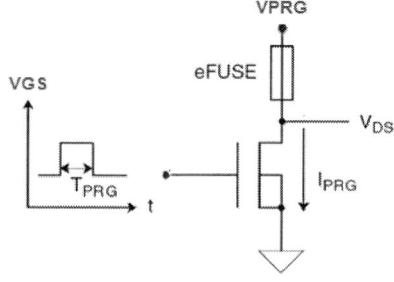

Fig 2: Schematic of the discrete fuse structure.

978-1-5090-6695-7/17 $31.00 © 2017 IEEE

TABLE I: Reliability Stress Item and Read-point

Stress Item	Stress Condition	Read Point
T/C	-65℃ ~ 150℃	0/25/250/500cycles
HTS	150℃	0/25/250/500hrs

RESULTS

Reliability Performance of E-fuse about TC with Different Programming Current

As shown in Fig. 3, the resistance degradation of low current programming e-fuse is at 10kohm after 500 hours temperature stress. But the resistance degradation of high current is above 30kohm. This means e-fuse of low programming current of 28nm have a better reliability performance abort high speed temperature variety stress than E fuse with high-programming current.

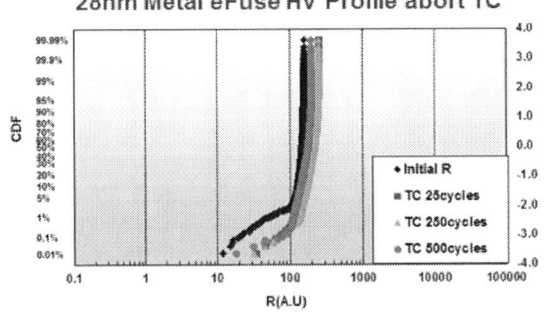

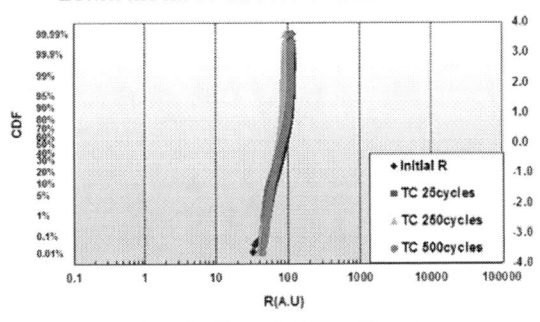

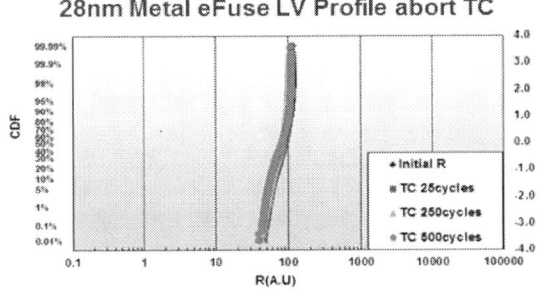

Fig 3: the resistance profile by programming current after TC 500-cycling stress.

Reliability Performance of E-fuse about HTS with Different Programming Current

As shown in Fig. 4, the resistance degradation for the e-fuse of low programming current is at 1kohm after temperature cycling 500 cycling stress. But the resistance degradation of high current is above 40kohm. This means e-fuse of low programming current of 28nm have a better reliability performance abort drawn-out high temperature stress than E fuse with high-programming current.

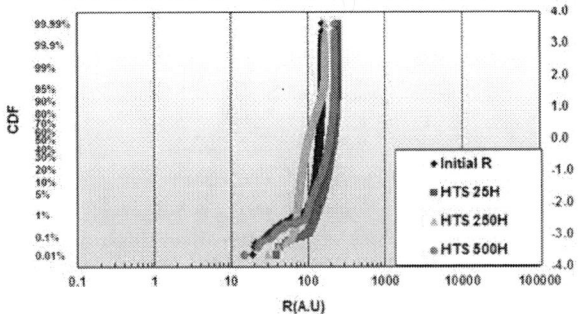

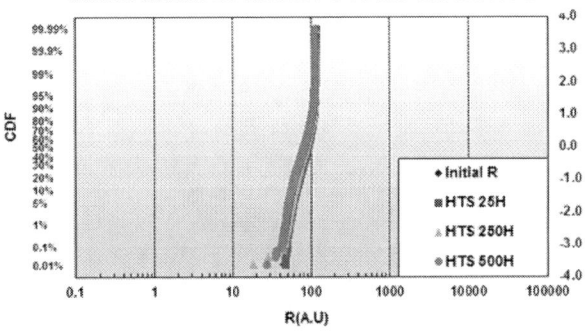

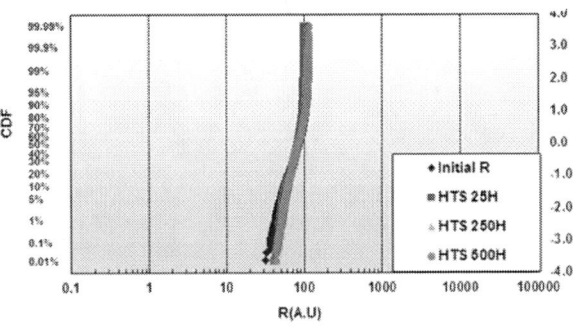

Fig 4: the resistance profile by programming current after 500-hours HTS stress.

DISCUSSIONS

The e-fuse SEM picture after low current programming is shown as Fig. 5. The SEM profile is very clean and in charge after high temperature and temperature cycling stress.

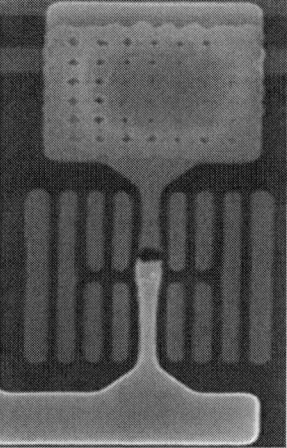

Fig 5: The low current programming e-fuse SEM

The e-fuse SEM picture after high current programming is shown as Fig. 6. The e-fuse line is open, but the high current induced high thermo shock to ambient dummy area. The SEM profile is not clean and will out of charge after high temperature and temperature cycling stress.

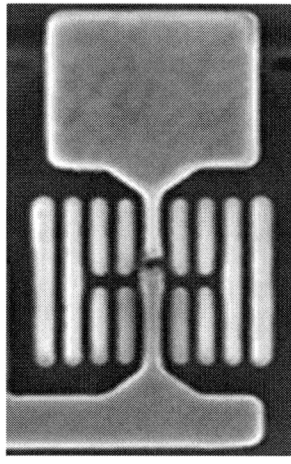

Fig 6: The high current programming e-fuse SEM

For copper, the melting point is 1083°C and thermal expansion parameter is 0.175*10-4 M/°C. The oxide melting point is 1670°C. Copper fuse of 28nm will melt when the programming current is between 30~50mA. In this current range, lower current programming fuse will have better SEM profile and qualification performance. High thermal stress brought by high current will uncontrollably melt the copper fuse, producing irregular ambient dummy area. Future work will include a more comprehensive investigation on how the copper melting profile changes with programming current.

CONCLUSIONS

The reliability performance of low-current programming fuse is much better than that of high current programming fuse, after temperature cycling and long-time high temperature stress. The reason is that the fuse burning profile at low current will still keep its shape, while it will deforms severely at high current. Higher thermal induced by high current make copper melt profile out of charge and stress ambient dummy area to destroy reliability performance. For e-Fuse programming current range invested in this paper, lower current programming fuse has the better SEM profile and reliability performance. Future extensions include the improvements on the design for a smaller programming current at the same VPRG and VGS or the process changes to adapt higher currents.

ACKNOWLEDGMENTS

The authors would like to express sincere thanks to Tony Li, Tyler Tang, Xianfeng Chen, and Yulin Ding from SMIC Q&R for the valuable technical discussions and strong test supports.

REFERENCES

[1] C. Kothandaraman, et al., "Electrically Programmable Fuse (eFUSE) Using Electromigration in Silicides," IEEE Electron Device Letter, Vol. 23, no. 9, pp. 523–525, Sep. 2002.

[2] A. Kalnitsky, et al., "CoSi2 Integrated Fuses on Poly Silicon for Low Voltage 0.18 um CMOS Applications," IEDM Tech. Dig., pp. 765–768, 1999.

[3] W. R. Tonti, et al., "Reliability and Design Qualification of A Sub-micron Tungsten Silicide e-fuse," IRPS 2004, pp. 161–165.

[4] J. Fellner, et al., "Lifetime Study for A Poly Fuse in a 0.35 um Polycide CMOS Process," IRPS 2005, pp. 446–449.

[5] C. Tian, et al., "Reliability Qualification of CoSi2 Electrical Fuse for 90nm Technology," IRPS 2006, pp. 392–397.

[6] S. Iyer, et al., "Reliability Investigation of NiPtSi Electrical Fuse," IRPS 2008, pp. 446–449.

[7] T. Sasaki, et al., "Melt-segregate-quench Programming of Electrical Fuse," IRPS 2005, pp. 347–351.

[8] T. S. Doorn, et al., "Ultra-fast Programming of Silicided Polysilicon Fuses Based on New Insights in The Programming Physics," in IEDM Tech. Dig., 2005, pp. 683–686.

[9] Y. Li, et al., "Blowing Polysilicon Fuses: What Conditions Are Best," IRPS Workshop, 2006, pp. 194–197.

DEFECTS AND LIFETIME PREDICTION FOR GE PMOSFETS UNDER AC NBTI STRESSES

J. F. Zhang, J. Ma, W. Zhang, and Z. Ji

Department of Electronics and Electrical Engineering, Liverpool John Moores University,
Byrom Street, Liverpool L3 3AF, UK
E-mail: j.f.zhang@ljmu.ac.uk

ABSTRACT

Germanium has higher hole mobility and is a candidate for replacing silicon for pMOSFETs. This work reviews the recent progresses in understanding the negative bias temperature instability (NBTI) of Ge pMOSFETs and compares it with SiON/Si devices. Both Ge and SiON/Si devices have two groups of defects: as-grown hole traps (AHT) and generated defects (GDs). The generation process, however, is different: GDs are interface-controlled for SiON/Si and dielectric-controlled for Ge devices. This leads to substantially higher GDs under DC stress than under AC stress for Ge, although they are similar for SiON/Si devices. Moreover, GDs alter their energy levels with charge status and can be reset to original precursor states after neutralization for Ge, but these processes are insignificant for SiON/Si. The impact of these differences on lifetime prediction will be presented and the defects and physical mechanism will be explored.

INTRODUCTION

Hole mobility in Ge is ~4 times of that in Si, making Ge pMOSFETs faster than Si. The Negative Bias Temperature Instability (NBTI) of Ge devices varies substantially: the Si-capped Ge device can have longer lifetime than Si devices, while GeO_2/Ge has much shorter lifetime. This has attracted a lot of attentions [1-6] and a review will be given for the recent progresses in understanding the NBTI of Ge devices, based on the authors' works [2-6]. We will use SiON/Si devices as the benchmark and explore the similarity and differences between Ge and SiON/Si devices in terms of defects, generation mechanism, and lifetime prediction.

Table 1: Gate stack
a) 2.3nm or 2 nm plasma-N SiON/Si
b) 4nm Al_2O_3/1.2nm GeO_2/Ge
c) 2nmHfO$_2$/~0.4nmSiO$_2$/ Si-cap/Ge

DEVICES AND EXPERIMENTS

The gate stack used is given in Table 1. Tests follows the 'stress-and-sense' procedure [7,8]. After a preset stress time, a gate pulse with an edge time of 5 µs was applied and the threshold voltage shift, ΔVth, was measured at a constant source current of $100 \times W/L$ nA at Vd= -100 mV [7,8]. Unless otherwise specified, the tests were carried out at 125 °C.

NBTI DYNAMICS

To investigate the NBTI dynamics under both AC and DC stresses, the gate bias, Vg, waveform in Fig. 1 was used. Initially, AC stress was applied at a frequency of 10 kHz and a duty factor of 50%. This was followed by a DC stress, where Vg has the same value as the AC amplitude. Finally, the AC stress was reapplied.

1st AC stress -- DC stress -- 2nd AC stress

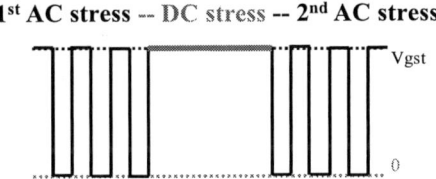

Fig. 1 The waveform of stress gate bias.

A typical result for SiON/Si devices is given in Fig. 2. When the AC stress was replaced by DC stress, Fig. 2a shows that NBTI become substantially higher [9]. As the AC stress was reapplied, however, the DC-enhanced degradation quickly recovers and Fig. 2b shows that NBTI kinetics returns to the same power law line, when plotted against "effective stress time", i.e. the AC stress time multiplied by its duty factor.

For Ge devices, Fig. 3a shows that the NBTI also increases substantially when switched to the DC stress and there is a recovery after AC stress was reapplied. Unlike the 'full recovery' in Fig. 2a, the recovery in Fig. 3a, however, is 'partial' and there is a substantial DC-induced 'additional generation' that did not recover, leading to an up-shift of the power-law in Fig. 3b [5]. To explain this difference, the generation process and defect properties will be explored.

978-1-5090-6695-7/17 $31.00 © 2017 IEEE

GENERATION PROCESS AND DEFECTS

We have proposed that the NBTI in Si devices follows the As-grown-Generation (AG) model [10,11] that divides defects into two groups: As-grown hole traps (AHTs) and Generated defects (GD),

$$\Delta Vth = AHT + GD. \qquad (1)$$

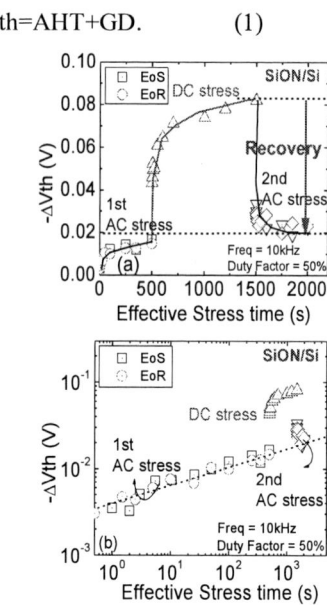

Fig. 2 In SiON/Si device, (a) The DC-enhanced charging recovers during 2nd AC stress. (b) The AC-DC-AC stress follows the same generation kinetics [5].

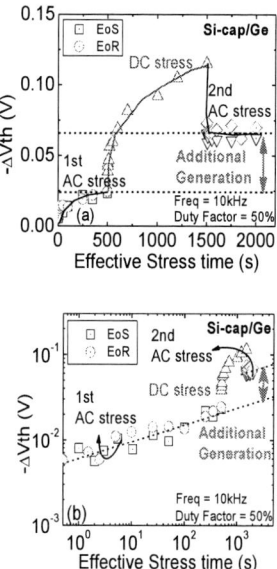

Fig. 3 In Ge devices, (a) The 2nd AC stress cannot fully recover the DC-enhanced NBTI (b) AC-DC-AC stress does not follow the same generation kinetics [5].

AHTs are located below the top edge of Si valence band, Ev, as shown in Fig. 4a. They are charged up under DC stresses, but neutralized under Vg=0, and dominates the recovery when switched to AC stress in Fig. 2.

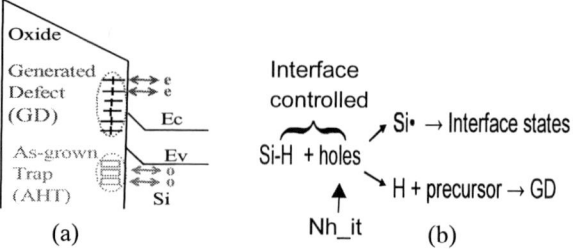

Fig. 4 SiON/Si: (a) AG model and (b) Interface controlled generation: Interface states and GD are two products of the same controlling reaction.

On the other hand, GDs have higher energy level, are more difficult to neutralize, and dominate the NBTI under AC stress. Fig. 2 indicates that the generation process is controlled by the accumulative time under a stress bias, Vgst. The interruptions of Vgst during AC stress have little effects on GDs. It has been reported that for every generated defect in gate dielectric, there is a generated interface states [9]. We speculate that the GDs in dielectrics and the created interface states are the two products originating from the same controlling electrochemical reaction at the SiON/Si interface, as illustrated in Fig. 4b. It is possible that the reaction starts from breaking a Si-H bond at the interface and the breaking rate depends on the oxide field and hole density at the interface, Nh_it. For both DC and AC stresses, the same Vgst gives the same Nh_it, and in turn the same GD in Fig. 2. In this way, one may call the generation in Si device as 'interface-controlled'.

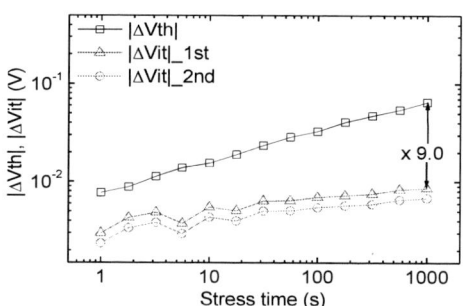

Fig. 5 Ge devices: A comparison of the shift induced by generated interface states, ΔVit, with the total ΔVth [3].

For Ge devices, there are also AHTs and they dominate the recovery in Fig. 3, similarly to the AHTs in SiON/Si devices. The differences in Ge and SiON/Si devices are mainly in the GDs, which are responsible for the 'additional generation' marked out in Fig. 3. Unlike

SiON/Si devices, there is no one-to-one correlation between GD in dielectric and created interface states and Fig. 5 shows that the GDs in dielectric can be substantially higher than the generated interface states.

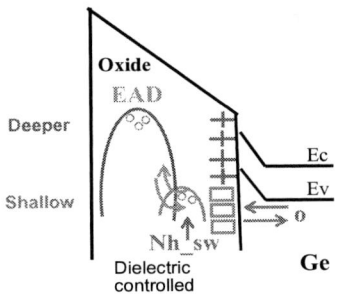

Fig. 6 Ge devices: Dielectric-controlled two-step generation: holes are captured by shallow well and then move to deep well through relaxation. The energy level alters with charge status.

We propose that the generation in Ge is a 'dielectric-controlled' process. It has two steps. In the first step, a defect in the dielectric captures a hole into a shallow well from substrate. This initiates a structure relaxation process, which ends when the captured hole overcomes a barrier and trapped stably in a deeper well, as illustrated in Fig. 6. The generation rate here is controlled by the number of holes in the shallow well of the dielectric, Nh_sw, rather than at the interface. Under AC stress, Nh_sw can be lower than that under DC stress, since the holes in the shallow well is not stable and can tunnel back to the substrate during the $Vg=0$ phase of the AC stress. As a result, the DC stress induced 'additional generation' in Fig. 3 can originate from the higher Nh_sw under DC stress. The generation in Ge devices is a 'dielectric-controlled' relaxation process [5].

The physical process described above for Ge devices involves the defect energy alternation during the generation process: the neutral precursor has a shallow well, but the charged GD settles down in a deeper well [3]. To support such energy alternating defects (EADs) indeed existing in Ge devices, we compare the discharge-then-recharge of GDs in Ge and SiON/Si devices.

In Fig. 7a, we first charged up the defects by stress. They were then progressively discharged by sweeping Vg in steps in the positive direction. This is followed by a recharge, where Vg was swept back towards negative [2,10,11]. For the SiON/Si devices, it can be seen that the difference between the recharge and discharge is small, indicating that the energy level of the GD changes little after discharge, so that the defect can be recharged as it moves above Fermi level again. For fresh SiON/Si, there is little defects above Ev(Si) [10,11]. The presence of defects above Si(Ev) after stress indicates that a neutralized GD does not return to its precursor state.

Fig. 7b shows that the Ge device behaves differently: when Vg and (Ef-Ef_FB) was swept toward negative direction, there is little recharge until Ev(Ge) was reached. This is because neutralizing a defect alters its energy level back to the shallow well at ~Ev(Ge), so that they cannot be recharged until reaching Ev(Ge). In another word, the energy level of EADs in Ge alters with their charge status: shallow when neutral and deeper when charged. A neutralized GD can be reset to its original precursor statue, therefore.

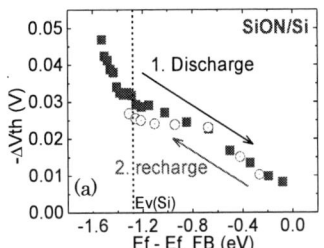

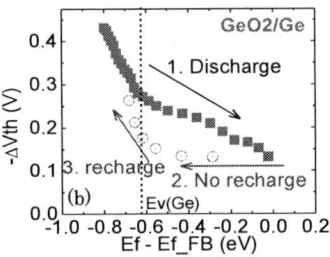

Fig. 7 Differences in defects: (a) Recharge starts as soon as energy sweeping negatively, well above ~Ev(Si) for SiON/Si. (b) Recharge is negligible when biased above ~Ev(Ge) for GeO$_2$/Ge [5].

IMPACT ON LIFETIME PREDICTION

According to the As-grown-Generation (AG) model, the AHTs typically saturate in seconds and only the GD component follows the power law [10,11]. For silicon devices, AHTs contribution is insignificant under AC stress [10,11], as they are efficiently neutralized during the $Vg=0$ phase. Under DC stress, AHTs can also be effectively neutralized if there is a measurement delay of ~10 ms [9,12]. Fig. 8a shows that the DC NBTI with a delay can be used as an approximation of AC NBTI. As a result, the lifetime of AC NBTI can be estimated from the DC NBTI with a measurement delay, as shown in Fig. 8b.

Although AG model is also applicable to Ge devices and the AHTs are efficiently neutralized for DC stress with a measurement delay, the DC NBTI with a delay should not be used for estimating the lifetime of AC NBTI, since the dielectric-controlled generation of energy alternating defects (EADs) introduces an additional generation under DC stress, as shown in Figs. 3 and 9a.

978-1-5090-6695-7/17 $31.00 © 2017 IEEE

Fig. 9b shows that the use of DC NBTI even after a delay will underestimate the AC device lifetime of Ge.

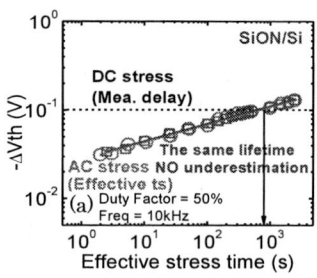

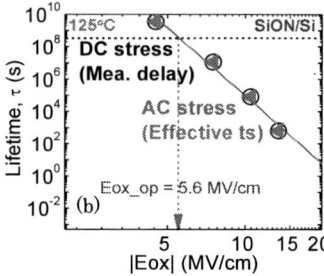

Fig. 8 SiON/Si: A comparison of AC and DC stress with a measurement delay (a) kinetics and (b) lifetime [5].

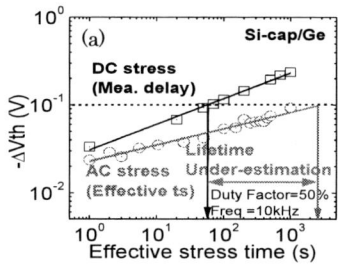

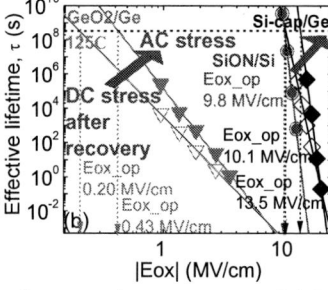

Fig. 9 Ge devices: A comparison of AC and DC stress with a measurement delay (a) kinetics and (b) lifetime [5].

CONCLUSIONS

In this work, we reviewed the recent progresses in understanding the NBTI defects in Ge devices and compared them with those in Si devices. Both Si and Ge devices follow the As-grown-Generation (AG) model. The GDs in SiON/Si devices are interface-controlled, similar under DC and AC stresses, their energy level

change little with their charge status, and they do not return to their original precursor state after neutralization. In contrast, GDs in Ge devices are dielectric-controlled, with additional generation under DC, their energy level alternates with their charge status, and they can be reset to their precursor state following neutralization. As a result, DC stress will substantially underestimate the AC lifetime of Ge devices even after a measurement delay and must not be used.

ACKNOWLEDGEMENTS

The authors thank J. Mitard, J. Franco. Kaczer and G. Groeseneken of IMEC, Belgium, for supply of test samples used in this work. This work is supported by the EPSRC of UK under the grant no. EP/L010607/1.

REFERENCES

[1] M. Caymax, G. Eneman, F. Bellenger, C. Merckling, A. Delabie, G. Wang, R. Loo, E. Simoen, J. Mitard, B. De Jaeger, G. Hellings, K. De Meyer, M. Meuris, and M. Heyns, *Proc. IEDM*, 2009, pp. 1-4.

[2] J. Ma, J. F. Zhang, Z. Ji, B. Benbakhti, W. Zhang, J. Mitard, B. Kaczer, G. Groeseneken, S. Hall, J. Robertson, and P. Chalker, *IEEE Elec. Dev. Lett.*, vol. 35, 2014, pp.162-164.

[3] J. Ma, J. F. Zhang, Z. Ji, B. Benbakhti, W. D. Zhang, X. F. Zheng, J. Mitard, B. Kaczer, G. Groeseneken, S. Hall, J. Robertson, and P. R. Chalker, *IEEE Trans. Electron Dev.*, vol. 61, 2014, pp. 1307-1315.

[4] J. Ma, W. Zhang, J. F. Zhang, B. Benbakhti, Z. Ji, J. Mitard, J. Franco, B. Kaczer, and G. Groeseneken, *Proc. IEDM*, 2014, pp. 820-823.

[5] J. Ma, W. Zhang, J. F. Zhang, Z. Ji, B. Benbakhti, J. Franco, J. Mitard, L. Witters, N. Collaert, G. Groeseneken, *Proc of IEEE VLSI Tech. Symp.*, 2015, pp.34-35.

[6] J. Ma, W. Zhang, J. F. Zhang, B. Benbakhti, Z. Ji, J. Mitard, H. Arimura, *IEEE Trans. Electron Dev.*, vol. 63, 2016, pp. 3830-3836.

[7] Z. Ji, J. F. Zhang, M. H. Chang, B. Kaczer, and G. Groeseneken, *IEEE Trans. Electron Dev.*, vol. 56, 2009, pp. 1086-1093.

[8] Z. Ji, L. Lin, J. F. Zhang, B. Kaczer, and G. Groeseneken, *IEEE Trans. Electron Dev.*, vol. 57, 2010, pp. 228-237.

[9] M. H. Chang and J. F. Zhang, *J. Appl. Phys.*, vol.101, 2007, art.no.024516.

[10] Z. Ji, S. F. W. M. Hatta, J. F. Zhang, J. G. Ma, W. Zhang, N. Soin, B. Kaczer, S. De Gendt, and G. Groeseneken, *Proc. IEDM*, 2013, pp.413-416.

[11] Z. Ji, J. F. Zhang, L. Lin, M. Duan, W. Zhang, X. Zhang, R. Gao, B. Kaczer, J. Franco, T. Schram, N. Horiguchi, S. De Gendt, and G. Groeseneken, *Proc of IEEE VLSI Tech. Symp.*, 2015, pp.36-37.

[12] J. F. Zhang, *Microelectron. Eng.*, vol. 86, 2009, pp.1883-1887.

RECENT PROGRESS IN RRAM TECHNOLOGY: FROM COMPACT MODELS TO APPLICATIONS

YUE ZHA[1], ZHIQIANG WEI[2], JING LI[3]

[1,3]Department of Electrical and Computer Engineering
University of Wisconsin-Madison, Madison, WI 53706, USA
{[1]yzha3, [3]jli587}@wisc.edu
[2]Advanced Devices Development Center, Panasonic Corporation, Osaka 570-8501, Japan
[2]wei.zhiqiang@jp.panasonic.com

ABSTRACT

We survey the recent progress in the research and development of RRAM technology in the past decade, ranging from compact models to involving applications. In particular, we first present an overview of the representative compact models that have been developed to capture essential electrical/chemical/thermal properties of RRAM such as IV characteristics, switching dynamics, variability and reliability, etc. We then present the product development and commercialization progress of RRAM in traditional applications as a drop-in replacement to Flash including both embedded and standalone memory, and in emerging applications as storage class memory (SCM). We finally survey work in the nascent field of developing alternative non-Von Neumann architectures using RRAM, opening up broad opportunities beyond memory and storage.

INTRODUCTION

The last decade has seen significant progress in emerging nonvolatile memory technologies including Spin Torque Transfer RAM (STT RAM), phase change memory (PCM) and resistive RAM (RRAM). Among them, RRAM becomes the front-runner in the race of development and commercialization, primarily driven by the perceived advantages of RRAM over PCM and STT RAM, such as fab-friendly materials, simple cell designs, good scalability (<10nm [1][2]), low switching current (~nA [3]), and ease of manufacturing due to the decoupling between switching volume and lithographic patterning.

Due to diverse material options, RRAMs are generally classified into a number of types based on various categorization methods. One widely used categorization is to divide RRAMs into two types, depending on their intrinsic switching mechanisms: 1) filamentary RRAM, based on the formation and disruption of either oxygen vacancy-based filament in oxides such as TaO_x and HfO_x (also known as OxRAM [4]), or metallic conductive filament comprising of metal ions such as Cu or Ag (also known as conductive-bridge RAM or CBRAM), and 2) non-filamentary RRAM, within which ions move across the entire aperture of the active region of the device. In this review, we specifically focus on the first type - filamentary RRAM.

Despite of tremendous research activities in developing more insightful understanding of the underlying physics of electrical transport and switching dynamics in RRAM, compact modeling is widely considered as a key to bridge the gap between technologist and IC designers. It provides a good balance between simulation speed and accuracy, and thus, enables efficient design space exploration and performance/power evaluation of RRAM-based circuits. A good model will not only provide an effective guidance to optimal adoption of RRAM in traditional memory and storage applications but also will drive new directions of exploring non-conventional architectures, opening up new opportunities in emerging applications.

In this paper, we will first present an overview of the recent progress on RRAM compact models, which are developed to capture the essential properties and to facilitate efficient evaluation of RRAM-based circuits and architectures. Then we will focus on discussing the status of RRAM-based applications, including both traditional and emerging applications.

COMPACT MODEL

In recent years, lots of compact models are proposed to facilitate the evaluation of RRAM-based circuits and architectures. Moreover, these models can provide physical insights on RRAM characteristics, and thus, can be used as guidelines to improve RRAM performance, including reliability (retention) [5], endurance [6], etc. In this section, we overview the recent progress in the development of compact models for RRAM, and categorize these models based on two criteria, physical models and simulated features, as summarized in Table 1.

Physical Models

In prior art, two main physical models are widely applied. One of them explains the switching behavior of RRAM as the formation and rupture of the conductive filament (CF) [7][8]. The current in low resistance

978-1-5090-6695-7/17 $31.00 © 2017 IEEE

Table 1. A summary and categorization of the recent developed RRAM compact models.

Criteria	Category	Work
Physical Model	Conductive filament	[10][11][12][15][16][17][19][20][21]
	Oxygen vacancy concentration	[5][6][13][14]
Simulated Feature	Resistance states	[5][13][14][15][16][17]
	Switch dynamic	[6][10][11][12][18][19][20][21]
	Random telegraph noise	[23][24][25][26][27]

state (LRS) is attributed to the electron drift through the CF, while the current in high resistance state (HRS) is attributed to the trap-assisted tunneling (TAT). Additionally, this model can also be extended to explain the random telegraph noise (RTN) [9]. RRAM compact models [10][11][12] using this physical model can fully (or partially) explain the experimental observation, but one limitation is the difficulty to capture the inherent stochastic feature in this physical model.

Another physical model attributes the switching dynamic between LRS and HRS to the change of oxygen vacancy concentration [13], as the measurement on the temperature dependence of resistance indicates that conductions in both LRS and HRS conform to electron hopping model. More specifically, it is variable range hopping at low temperature, while it is fixed range hopping at high temperature. This model successfully explains the LRS retention characteristics [13] and the related endurance issue [6]. Moreover, it can be combined with stochastic differential equation to describe the intrinsic RRAM variations [14].

Simulated Features
Resistance States
Multiple compact models are proposed to simulate the different resistance states of RRAM cells. One widely accepted simple one is that the RRAM resistance is equivalent to that of two resistors connected in series. One resistor is used to describe the CF, while the other resistor is for the dielectric barrier caused by the rupture of CF. Many compact models [15][16][17] use this simple assumption to simulate RRAM resistance states. But it is difficult to consider the intrinsic variation of RRAM in these models.

Panasonic performed measurement on the temperature dependence of resistance, and reported that the conductions in both LRS and HRS are caused by electron hopping in their early work [13][5]. In the following work [14], they propose the SDE-PH model that combines the stochastic differential equation with percolation hopping to simulate resistance distribution of

LRS and HRS. This model successfully explains the resistance variation, and provides useful guidelines to improve RRAM retention.

Switching Dynamic
By applying voltages on the two terminals of an RRAM cell, it can be switched between HRS and LRS. Mathematically, we can describe the switching dynamic by a voltage controlled switching rate (dR/dt) [18]. This switching rate describes the change of RRAM resistance, and is only related to the applied voltage and current resistance value. This compact model is simple and compute-efficient, but it cannot provide microscopic information of RRAM cells.

As we mentioned, the switching dynamic can be explained as the formation and rupture of CF. Therefore, we can describe this switch dynamic through simulating the growth of CF, and it has been used in many compact models [19][10][20][12][11][21]. A set of differential equations are used to describe the growth of CF, therefore, the computation complexity of these models is high. Additionally, these compact models are difficult to explain the variation of the switching dynamic. Although stochastic terms are included in some models [19][10], it is still not enough to describe the intrinsic variation.

The switching dynamic can also be modeled as a dynamic flow of oxygen vacancy through the filament [6]. In contrast to previous model, in this model, the CF never breaks, but only become narrower. In addition, HRS is caused by the current-blocking potential barrier, which is raised because of quantum mechanical confinement [22]. This model provides a new perspective to analyze the switching dynamic, and may provide unique insights.

Random Telegraph Noise
As the random telegraph noise (RTN) hinders the device scaling and the multi-bit storage implementation, a growing research effort is devoted to developing compact models to get better understanding on RTN.

Although the physical mechanism of RTN is not completely understood, one typical explanation is that it is due to the charge carrier capture and emission processes caused by active traps [23]. The factorial hidden markov model can be used to analyze this RTN effect [23][24][25]. Other models are also proposed to simulate this RTN process [26][27]. These models are insightful for understanding the RTN effect, and are helpful for designing multi-bit storage units.

APPLICATIONS

In this section, we survey the potential applications of RRAM, from traditional data storage as a drop-in replacement to NAND flash, new memory/storage as storage class memory (SCM), to emerging non-Von Neumann architectures. We will provide insights into the associated challenges and highlight some future directions.

NAND Flash Replacement

In worldwide memory community, RRAM is typically viewed as a compelling replacement for NAND flash in both embedded and standalone applications.

Embedded Non Volatile Memory (eNVM) is a key component of many modern systems, such as micro controller, wearable devices, smart card and automobile. Compared with the conventional eNVMs (e.g. EEPROM, NOR and NAND), RRAM-based eNVM has low operating voltage and thus low power, and simple fabrication process. Panasonic announced the first mass production of RRAM mounted micro controller in 2013 [28], which achieves about 50% reduction in energy consumption compared with flash-based micro controller. They also demonstrated over 10 years data retention at 85°C [29], which is an important property in the automotive application [30]. In 2012, Adesto also announced their low-power embedded memory product [31], which is based on conductive-bridge RAM (CBRAM) technology.

In addition to low-density embedded storage, RRAM technology can also be used as high-density stand-alone memory. With the compact cell size of $4F^2$, the multi-layer structure and multi-bit storage of RRAM further improve its storage density and reduce cost-per-bit. In 2013, a 32 Gb RRAM chip with two-layer structure is developed at 24 nm technology node by SanDisk [32]. In the next year, a 16 Gb RRAM device has been demonstrated at 27 nm technology node by Micron [33]. It is also worth noting that, numerous selector devices [34][35][36] have been developed in recent years to suppress the leakage current on the sneak paths, thereby improving the read/write performance and energy efficiency of the standalone large-scale memory devices [37].

Storage Class Memory

Besides a drop-in replacement to NAND flash, RRAM has also been considered as a new "storage class memory" (SCM), which promises an even more fundamental change in state-of-the-art Von Neumann computer as a new layer of memory or storage. But the promise cannot be fulfilled unless it can tackle the challenges of scale - both practical and economical, beyond just the technology development.

One big issue for the wide adoption of SCM is developing the systems that can truly take advantage of it. Current systems, everything from the applications to the operating systems to memory and storage subsystems and interface protocols, are designed for the traditional division between memory operated with loads and stores, and persistent storage programmed in blocks. Nonetheless, all that will have to be redesigned for SCM to become mainstream. Thus, standardization is a key and will require collaborative efforts from memory manufactures and computer industry to build a holistic ecosystem to support it.

Non-Von Neumann Architecture

More aggressive adoption of RRAM has also been explored in recent years, in the context of building next-generation non-Von Neumann architectures. The motivation behind is to address the "memory wall" challenges in existing computer systems. One implementation is to utilize the inherent dot-product capability of its crossbar structure to accelerate matrix multiplication [38], which is a key computational kernel in a wide array of applications including deep learning, optimization, etc. Another interesting direction is to implement a neuromorphic system which can mimic the fuzzy, fault-tolerant and stochastic computation of the human brain, without sacrificing its space or energy efficiency [39][40][41][42]. Key challenges may include the lack of fully understanding of brain function, high design complexity and cost due to extensive use of mixed-signal circuits in these architectures.

Alternative promising directions include 1) non-volatile ternary content addressable memory (nv-TCAM [43]), and 2) nonvolatile field programmable gate array (nv-FPGA [44][45]). Compared to the CMOS-based implementation, both RRAM-based TCAM and FPGA show great promise in improving integration density and thus reducing cost. More importantly, they do not need expensive mixed-signal circuits and thus their adoption barrier is lower than the matrix multiplication accelerator and the neuromorphic architecture. However, it poses other challenges in operational robustness due to the limited ON/OFF resistance ratio of RRAM, which can be mitigated by advanced material engineering [46], cell design [47][48][49], and coding technique [50]. Most recently, an interesting reconfigurable architecture is developed to combine the best advantages of TCAM and FPGA [51]. By programming the peripheral CMOS circuits, one RRAM crossbar can be configured to implement computation, search, routing or data storage. Such superior programmability blurs the boundary between computation and storage and may open up a new path towards intelligent machines.

CONCLUSION

In this paper, we present an overview of the recent development of RRAM, from compact models to applications. The potentials and limitations of RRAM technology and RRAM-enabled applications have been discussed to facilitate the future research and development of the technology in the era of big data.

REFERENCES

[1] C. Ho et al., in *IEDM*, 2010, p. 19.1.1-19.1.4.
[2] K.-S. Li et al., in *2014 Symposium on VLSI Technology: Digest of Technical Papers*, 2014, pp. 1–2.
[3] Q. Luo et al., in *IEDM*, 2015, p. 10.2.1-10.2.4.
[4] H. S. P. Wong et al., *Proc. IEEE*, vol. 100, no. 6, pp. 1951–1970, Jun. 2012.
[5] S. Muraoka, et al., in *VLSIT*, 2013, pp. T62–T63.
[6] R. Degraeve et al., in *2012 IEEE International Integrated Reliability Workshop Final Report*, 2012, pp. 3–7.
[7] L. Vandelli et al., in *IEDM*, 2011, p. 17.5.1-17.5.4.
[8] G. Bersuker et al., in *IEDM*, 2010, p. 19.6.1-19.6.4.
[9] D. Veksler et al., in *IEDM*, 2012, p. 9.6.1-9.6.4.
[10] H. Li et al., in *DATE*, 2015, pp. 1425–1430.
[11] J. F. Kang et al., in *IEDM*, 2015, p. 5.4.1-5.4.4.
[12] P. Huang et al., *IEEE Trans. Electron Devices*, vol. 60, no. 12, pp. 4090–4097, Dec. 2013.
[13] Z. Wei et al., in *IEDM*, 2011, p. 31.4.1-31.4.4.
[14] Z. Wei et al., in *IEDM*, 2015, p. 7.7.1-7.7.4.
[15] F. M. Puglisi et al., *IEEE Electron Device Lett.*, vol. 34, no. 3, pp. 387–389, 2013.
[16] J. Noh et al., *IEEE Electron Device Lett.*, vol. 34, no. 9, pp. 1133–1135, 2013.
[17] L. Larcher et al., *IEEE Trans. Electron Devices*, vol. 61, no. 8, pp. 2668–2673, 2014.
[18] I. Messaris et al., *Authors Orig.*, pp. 1–8, 2016.
[19] Z. Jiang et al., in *2014 International Conference on Simulation of Semiconductor Processes and Devices*, 2014, pp. 41–44.
[20] P. Y. Chen and S. Yu, *IEEE Trans. Electron Devices*, vol. 62, no. 12, pp. 4022–4028, Dec. 2015.
[21] A. Padovani et al., *IEEE Trans. Electron Devices*, vol. 62, no. 6, pp. 1998–2006, Jun. 2015.
[22] R. Degraeve et al., in *VLSIT*, 2012, pp. 75–76.

[23] F. M. Puglisi and P. Pavan, in *2013 IEEE International Conference of Electron Devices and Solid-state Circuits*, 2013, pp. 1–2.
[24] F. M. Puglisi et al., in *IRPS*, 2014, p. MY.5.1-MY.5.5.
[25] F. M. Puglisi and P. Pavan, *ECTI Trans. Electr. Eng. Electron. Commun.*, vol. 12, no. 1, pp. 24–29, Dec. 2014.
[26] B. Guan et al., in *IRPS*, 2016, p. MY-5-1-MY-5-4.
[27] S. Balatti et al., in *IRPS*, 2014, p. MY.4.1-MY.4.6.
[28] "Panasonic Starts World's First Mass Production of ReRAM Mounted Microcomputers," *http://news.panasonic.com/global/press/data/2013/07/en130730-2/en130730-2.html*.
[29] A. Kawahara et al., *JSSC*, vol. 48, no. 1, pp. 178–185, Jan. 2013.
[30] R. Strenz, in *IEDM*, 2011.
[31] "Adesto CBRAM Memory Technology Demonstrates Ultra Low-Power Operation," *http://www.adestotech.com/news-detail/adesto-cbram-memory-technology-demonstrates-ultra-low-power-operation/*.
[32] T. Liu et al., in *ISSCC*, pp. 210-211, 2013.
[33] R. Fackenthal et al., in *ISSCC*, pp. 338–339, 2014.
[34] S. H. Jo et al., in *IEDM*, 2014, p. 6.7.1-6.7.4.
[35] K. Gopalakrishnan et al., in *VLSIT*, 2010, pp. 205–206.
[36] S. G. Kim et al., in *IEDM*, 2015, p. 10.3.1-10.3.4.
[37] A. Chen, *IEEE Trans. Electron Devices*, vol. 60, no. 4, pp. 1318–1326, Apr. 2013.
[38] M. Hu et al., in *DAC*, New York, NY, USA, 2016, p. 19:1–19:6.
[39] A. Shafiee et al., in *ISCA*, 2016.
[40] P. Chi et al., in *ISCA*, 2016.
[41] C. Liu et al., in *DAC*, New York, NY, USA, 2015, p. 14:1–14:6.
[42] Y. Wang et al., in *GLSVLSI*, New York, NY, USA, 2015, pp. 189–194.
[43] L. Zheng et al., in *ISCAS*, 2016, pp. 1382–1385.
[44] Y. Tsuji et al., in *VLSI-Circuits*, 2016, pp. 1–2.
[45] J. Cong and B. Xiao, *IEEE Trans. Very Large Scale Integr. VLSI Syst.*, vol. 22, no. 4, pp. 864–877, Apr. 2014.
[46] H. Y. Lee et al., in *IEDM*, 2008, pp. 1–4.
[47] S. H. Jo et al., *IEEE Trans. Electron Devices*, vol. 62, no. 11, pp. 3477–3481, Nov. 2015.
[48] J. Zhou et al., *IEEE Trans. Electron Devices*, vol. 61, no. 5, pp. 1369–1376, May 2014.
[49] S. H. Jo et al., in *VLSIT*, 2015, pp. T128–T129.
[50] J. Li et al., *JSSC*, vol. 49, no. 4, pp. 896–907, Apr. 2014.
[51] Y. Zha and J. Li, in *ICCAD*, 2016.

Investigation of Trap Profile in Nitride Charge Trap Layer in 3-D NAND Flash Memory Cells

Jong-Ho Lee and Ho-Jung Kang

Department of ECE and ISRC
Seoul National University, Seoul 151-742, Korea
Tel.: +82-2-880-1727; E-mail: jhl@snu.ac.kr

Abstract

We extract the trap density (N_t) profile of the nitride storage layer in 3-D NAND flash memory cells. The adjacent cells which are programmed suppress significantly the lateral diffusion during retention measurement so that we can extract accurate N_t profile. The AC-g_m method makes the N_t profiling in an E_C-E_T range of 1~1.2 eV possible, and provides a Gaussian N_t profile together with the retention model. The threshold voltage shift with trapped electron profiles is firstly modeled as a parameter of channel radius and its model is verified.

Keywords—3-D NAND flash memory, nitride trap, retention

Introduction

Recently, 3-D NAND flash memory has been attractive for low cost and high density memory. Word-line (WL) stacked structure in 3-D NAND flash have already been applied to mass production [1], [2]. In WL stacked structures, the gate dielectric stack including nitride storage layer surrounds the poly-Si body standing vertically. The extraction of accurate trap density (N_t) profile of the nitride layer in 3-D NAND flash memory is very important, because the profile affects P/E speed, retention characteristics and threshold voltage (V_{th}) window. In this work, we characterize the N_t profile in the nitride layer in 3-D NAND flash memory by applying derived retention model and AC-transconductance (g_m) method. And the ΔV_{th} with trapped electron density (N_e) is modeled by changing tube radius

Results and discussion

Fig. 1 explains schematically electron loss mechanisms of the programmed target cells. To understand the influence of the lateral diffusion, retention measurements with three modes of adjacent cell states are performed as shown in Fig. 2 (a). Here, initial V_{th} of the target cells is 4 V and the V_{th}s of adjacent cells at erase (E), neutral (N) and program (P) states are 0 V, -1.5 V and 3 V, respectively. Fig. 2 (b) shows ΔV_{th}s over retention time are measured with three different modes at 200 °C. At P-P-P mode, the ΔV_{th} reduces remarkably as compared with those of other modes, because the diffusion of electrons stored in the target cells is suppressed by the electrons stored in adjacent cells. N_t profile with respect to E_C-E_T can be extracted by using retention characteristics at high temperature (T) [3]. To extract the N_t in 3-D NAND flash memory cells, we modified the equation explaining charge loss during retention at a high T in Cartesian coordinate to that in cylindrical coordinate. Fig. 2 (c) shows extracted N_t profiles from using derived equation and measured retention data of Fig. 2 (b). The extracted N_t profile at P-P-P mode shows the peak N_t decreased by ~50 % compared to those of other modes and the peak at an E_C-E_T of 1 eV. From these results, we can observe that the extracted peak N_ts at E-P-E and

N-P-N modes are overestimated due to significant lateral diffusion, because the lateral diffusion in the nitride layer is not considered in derived equations [4]. Deep N_t profile in the nitride layer is also extracted by using AC-g_m dispersion with frequency (f). During measuring AC-g_m of cell, channel carriers interact with traps in gate stack by a small signal applied to the control gate (CG). The number change of channel carriers interacting with the small signal makes the f dispersion of AC-g_m [5]. Fig. 3 shows normalized AC-g_m dispersion versus f at V_{CG} of 0.2 V. In the gate stack, the tunneling oxide layer of fresh cells has too little N_t to change AC-g_m, but the nitride layer has a large N_t. As a result, the f of the small signal increases, the carriers interacting with traps decrease and the channel carriers interacting with the small signal increase as shown in the inset of Fig. 3. Therefore, we can say that the traps in the nitride interact with the signal up to ~200 Hz because the normalized AC-g_m saturates above ~200 Hz [6]. The N_t versus E_C-E_T extracted by using retention data of P-P-P mode and AC-g_m measurement is shown in Fig. 4. The N_ts at energy lower than an E_C-E_T of 1 eV were obtained by using retention model and the N_ts at a higher energy than 1 eV were extracted by applying AC-g_m method. The extracted N_t profile with respect to energy looks like Gaussian function. The peak N_t observed at the E_C-E_T of 1 eV is ~1.2×10^{19} cm^{-3}eV^{-1}. The inset in Fig. 4 depicts the trap energy level scanned by the retention model and AC-g_m method. The N_e profile in the nitride layer seems to be uniform, but the profile can be non-uniform depending on P/E time. To see ΔV_{th} with the amount of N_e and its profile in the nitride layer, the ΔV_{th} was modeled as a parameter of tube radius as shown in Figs. 5 and 6. The ΔV_{th} model showed excellent agreement with the simulation data as shown in Figs. 5 (b) and 6 (b). The ΔV_{th} difference with different N_e profiles decreases as the tube radius decreases. Using the model, we can calculate N_e from measured V_{th}.

Conclusion

We have extracted accurately N_t profile in 3-D NAND cells by considering lateral diffusion of charges in nitride layer, utilizing derived retention model and applying AC-g_m method. Extracted N_t profile shows continuity near E_C-E_T of 1 eV when plotted with the N_ts extracted using retention model and AC-g_m method. ΔV_{th} model with N_e profiles in the nitride layer showed a good agreement with simulation data as a parameter of tube radius.

Acknowledgements

This work was partially supported by SK hynix Inc. 2016, the Brain Korea 21 plus in 2016, and the National Research Foundation of Korea (NRF-2016R1A2B3009361).

References

[1] J. Jang et al., *VLSI Tech. Dig.*, p. 192, 2009.
[2] E. S. Choi and S.-K. Park, *IEDM Tech. Dig.*, p. 211, 2012.
[3] Y. Yang and M. H. White, *SSE.*, vol. 44, p. 949, 2000.

[4] H-J. Kang et al., *VLSI Tech. Dig.*, p. 182, 2015.
[5] X. Sun et al., *IEDM Tech. Dig.*, p. 462, 2012.
[6] M-K. Jeong et al., *EDL.*, vol. 36, no. 6, p. 561, 2015.

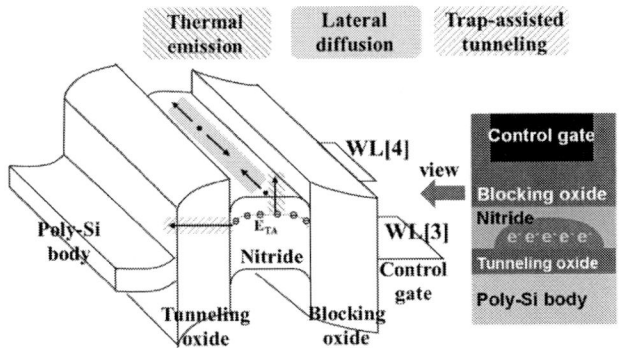

Fig. 1. Schematic energy band diagram of a programmed cell of 3-D NAND flash memory, explaining trapped charge loss mechanisms during retention.

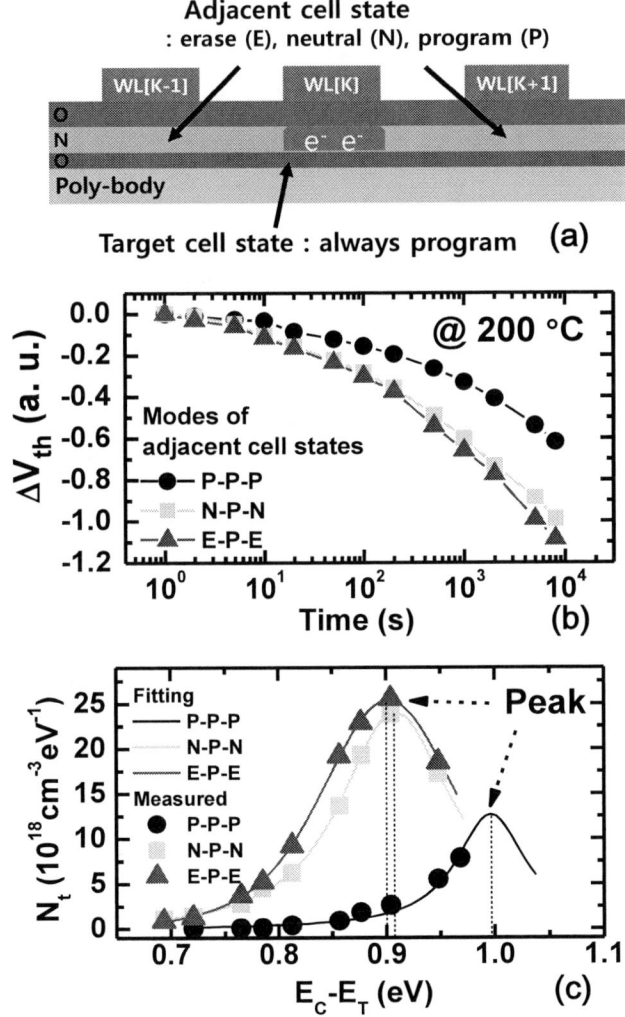

Fig. 2. (a) Schematic of NAND flash cell string. The target cell in the center is always programmed and the state of two adjacent cells has one of three modes which are E, N, and P. (b) V_{th} shifts versus retention time with various adjacent cell states at 200 °C [4]. (c) Extracted N_t profiles from retention characteristics at three modes.

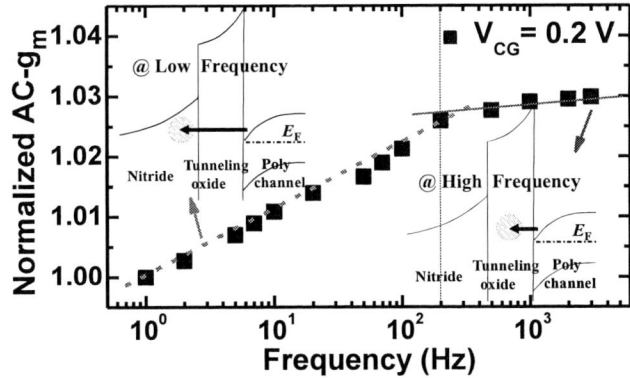

Fig. 3. Normalized AC-g_m dispersion versus frequency at V_{CG} of 0.2 V. The insets show the probing depth depending on the frequency of AC signal.

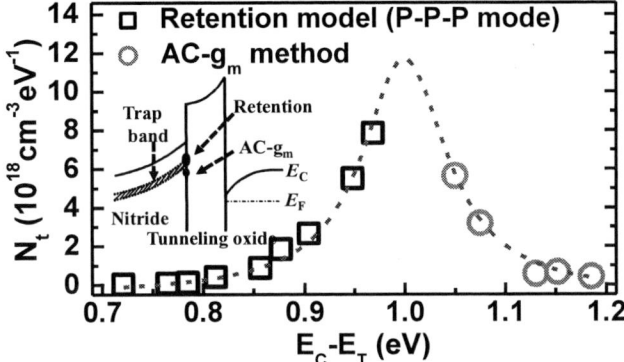

Fig. 4. Nitride trap density profile extracted by retention model and AC-g_m method, respectively. The inset shows the energy range of N_t with two different methods.

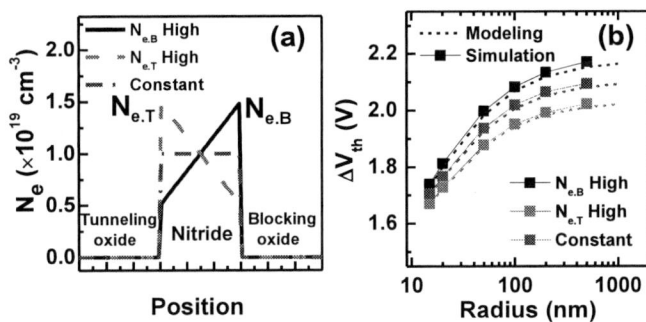

Fig. 5. (a) The trapped electron density (N_e) profiles with three different cases used in simulation. (b) ΔV_{th} versus tube radius as a parameter of the electron profiles as shown in (a).

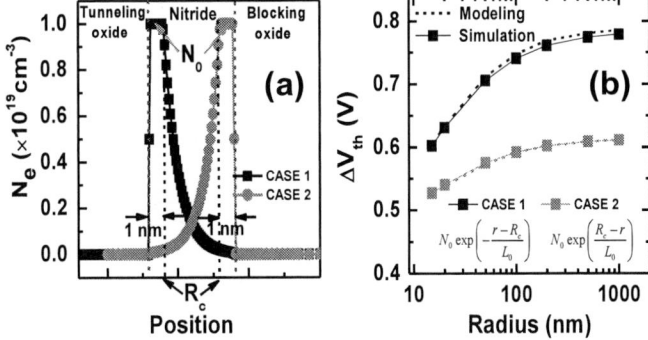

Fig. 6. (a) Non-uniform N_e profiles in nitride storage layer. The L_0 in the inset of (b) is 1 nm. (b) ΔV_{th} versus tube radius for two profiles as shown in (a).

978-1-5090-6695-7/17 $31.00 © 2017 IEEE

Advanced Logic and Specialty Technologies for VLSI Manufacturing in fast expansion at China

Min-hwa Chi[1] and Hanming Wu[2]

[1]Technology Development, SMIC, Shanghai, 201203, China

[2]BriteIP Corporation, China

*Email: min-hwa_chi@smics.com

ABSTRACT

As the electronics production at China continuously growing, the semiconductor chip consumption already exceeds 50% of global total since 2012. This attracts investment of wafer Fab's as well as advanced technologies development at China from both domestic and international partners. In turn, all semiconductor manufacturing related development and services (i.e. tools, materials, packaging, testing, bumping, assembly, etc.) as well as design services (i.e. EDA, fabless, IP design, etc.) are also in fast expansion mode into an efficient eco-system with competitive advantages. This further fueled the growth of electronic production at China.

INTRODUCTION

As the electronics production at China continuously growing, the semiconductor chip consumption already exceeds 50% of global total since 2012 (*Fig. 1*). The local semiconductor chip manufacturing only can meet <30% of the total consumption, with growing need of importing semiconductor chip. This attracts investment of new wafer Fab's (*Fig. 2*) as well as advanced technologies development at China from both domestic and international IDM's. In turn, all semiconductor manufacturing related services (i.e. tools, materials, packaging, testing, bumping, assembly, etc.) as well as design services (i.e. EDA, fabless, IP design, etc.) are also in fast expansion mode into an efficient service chain with competitive advantages (*Fig. 3*). This is further accelerating the growth of electronic production at China.

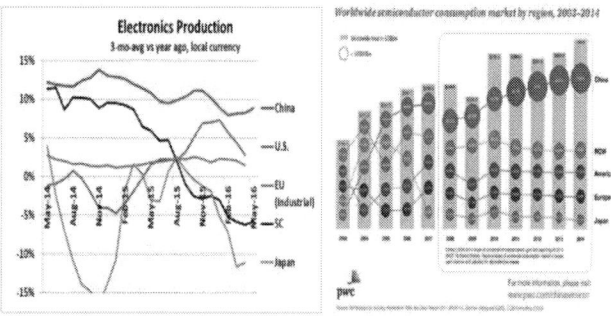

Figure 1: The Electronics production at China is the largest globally and the consumption of semiconductor chip at China exceeds 50% of global total since 2012 [1].

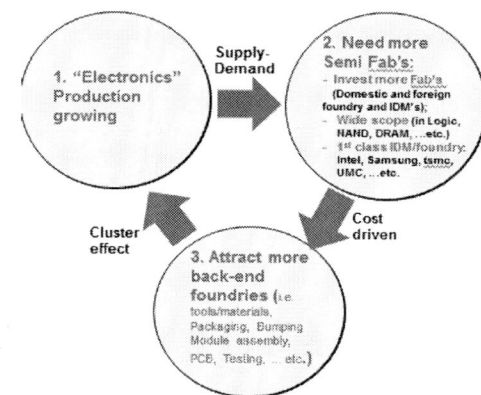

Source: World Fab Forecast report (November 2016, SEMI)

Figure 2: There are 26 new Fab's to be operating in China (2017-20); ~42% of the world total (source from SEMI).

Figure 3: The electronics production triggers semiconductor chip consumption and investment of new fab's, and also Design service/IP houses, and backend foundries (for packaging, bumping, 3DIC, assembly, etc.) as a positive loop.

FinFET PLATFORM TECHNOLOGY

The manufacturing of state-of-art FinFET technology at 16/14nm node at China is planned by the leading foundries, e.g. TSMC (2017 at NanJing), SMIC (~2018 at Shanghai), etc. The FinFET technology has enhanced performance and yet good compatibility with conventional planar CMOS. The process starts with fin formation similarly as the formation of active area for planar CMOS, followed by shallow-trench-isolation (STI) gap-fill and planarization by chemical-mechanical polishing (CMP); then recessing oxide to reveal the fins. The rest of flow proceeds as similar steps (e.g. well, gate,

978-1-5090-6695-7/17 $31.00 © 2017 IEEE

epi- grown on source/drain (S/D), etc.) as the gate-last high-k and metal-gate (HKMG) for planar CMOS. There are a few new challenges in manufacturing 14/16nm FinFET: *Firstly*, the Si surface on fins may have different characteristics than in bulk. The profile of Si fins is well controlled by optimizing wet clean (*Fig. 4a*), oxidation, and etching [2] (*Fig. 4b*). *Secondly*, the tapered shape of fin leads to wider Vt spread, but possibly tightened by vertical implant of N2 (for altering local WF) (*Fig. 5*). The multi-Vt scheme for system-on-chip (SOC), as usually implemented by multiple work-functions (WF) for n/pFET selectively; may also be implemented [3] by using multiple tapered fins with vertical implant dopants (N, Al, C, etc.) into WF layers (for altering WF). *Thirdly*，the usual stress engineering at S/D to enhance the mobility of electrons and holes is typically applied for good device performance.

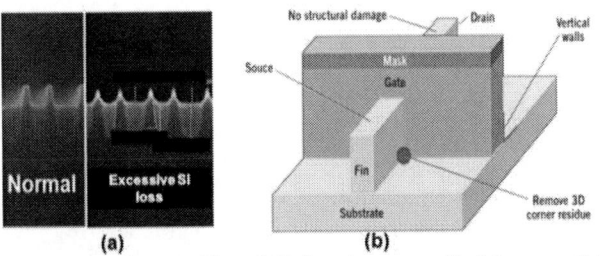

Figure 4: The profile of Si fins is controlled by careful wet clean (a), oxidation, and etching [2] (b).

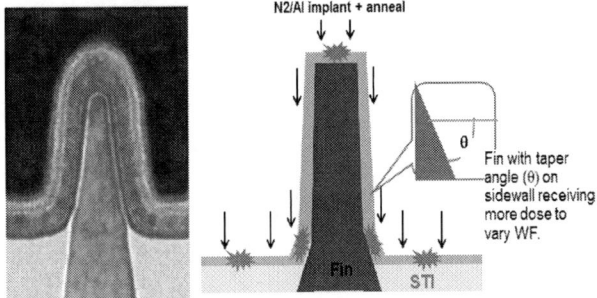

Figure 5: The tapered shape of fin leads to wider Vt spread, thus possibly tightened by vertical implant of species [3]. A multi-Vt scheme can also be implemented by forming multiple level of tapered fins.

3D-NAND AND DRAM TECHNOLOGY

The state-of-art 3D-NAND [4] flash memory (*Fig. 6*), already revolutionized the world for solid-state data storage, is in mass production by Samsung at Xi'an Fab since 2014. Intel at Dalian fab is also in ramping up of 3D-NAND in 2016. SK Hynix at Wuxi is in mass production of state-of-art 25/20nm 4/8Gb DRAM and stackable by TSV-based 3D integration technology into high bandwidth memory (HBM) package.

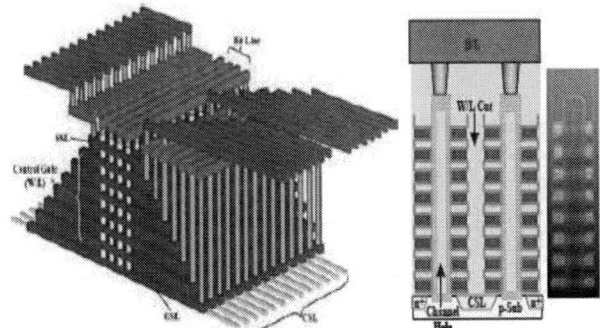

Figure 6: 3D-NAND in TCAT (Terabit Cell Array Transistor) architecture with 3D and cross-section view [4].

EMERGING NVM TECHNOLOGY

The non-volatile memory (NVM) technologies are progressed toward full compatibility at front-end-of-line (FEOL) or entirely formed in back-end-of-line (BEOL) as in *Fig. 7*, e.g. Phase-Change-RAM (PCRAM), Resistive-RAM (RRAM), Magnetic-RAM (MRAM), and Nanotube-RAM (NRAM). The NVM at BEOL is particularly interesting, as they can serve not only for the embedded applications with simpler process, but also capable of stacking-up into 3D array for high density and high performance system applications. More details are described below. *Firstly*, the FinFET transistor with high-k metal-gate stack itself can be used as one-time-only-programmable (OTP) memory [5] and reprogrammable RRAM [6] by the breakdown or the resistive switching of the high-k (Hf-oxide) gate-stack. *Secondly*, advanced RRAM [7] and PCRAM [8] cells are formed within via, contact, or MIM layers with simpler process (e.g. 2-3 extra masks) in BEOL. By applying suitable electrical pulses the resistance can be switched between high and low resistance states (with orders of magnitude difference) by forming or rupturing conductive paths in RRAM or phase change in PCRAM. *Thirdly*, the magnetic-tunnel-junction (MTJ) based MRAM [9-10] and magnetic logic [11] is a promising non-volatile technology by its almost infinite endurance, high switching/sensing speed. There is also magnetic spin-based device possibly formed at BEOL for nonvolatile and reconfigurable logic circuits. The combination of spin-logic (at BEOL) and CMOS logic circuits (at FEOL) can enhance logic function by simpler circuits and smaller foot-print. *Fourthly*, the **NRAM [12]** is based on electromechanical switch using spin-on Carbon nanotube (CNT) material with excellent characteristics demonstrated (e.g. >10^{11} endurance, low power and fast array program <5ns). *Fifthly*, various embedded NVM at BEOL can be stacked-up into high density 3D array as enabled by recent advances in access devices (or selectors) (*Fig. 8*), e.g. Ovonic Threshold Switch (OTS) in Intel/Micron's 3XPoint technology [13] and C-nanotube FET [14].

978-1-5090-6695-7/17 $31.00 © 2017 IEEE

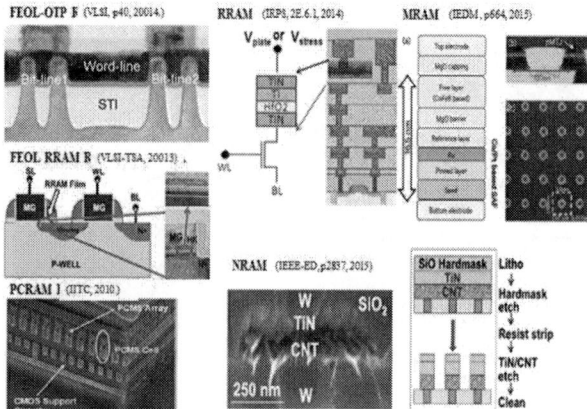

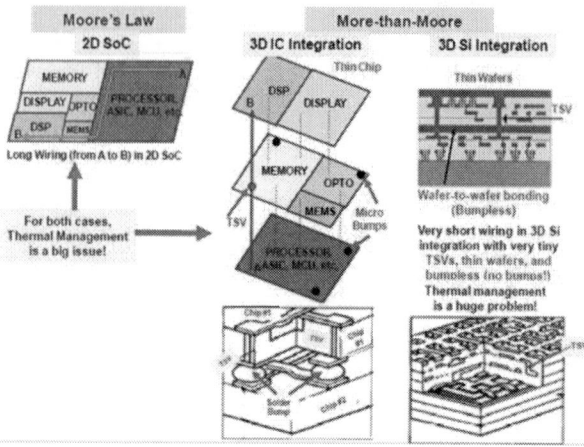

Figure 7: Various NVM technologies [7-12] e.g. RRAM, PCRAM, MRAM and NRAM formed at BEOL for embedded applications. FinFET transistors can also be used as OTP and 2T RRAM [5-6].

Figure 9: The conventional 3DIC integration includes TSV-based chip-to-chip or wafer-to-wafer stacking-up as well as the 3D system-in-package (SIP) technology [15].

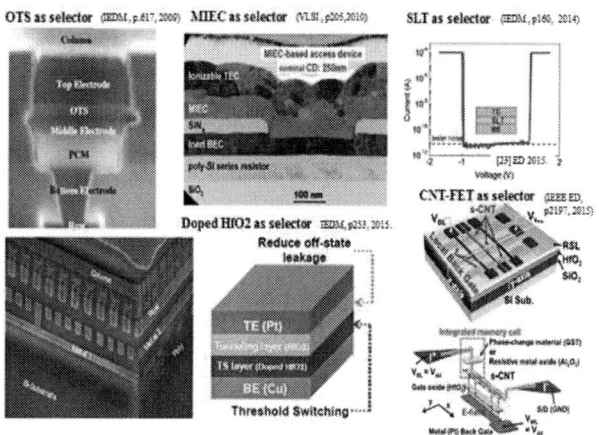

FOUNDRY R&D AND BUSINESS MODELS

The foundry industry may evolve rapidly into business models of "super-foundry" (or virtual IDM) and "Foundry chain-store" (*Fig. 10*) with high quality operation in shared R&D and manufacturing on advanced CMOS platform with various specialty technologies for customers applications. The new types of foundries have strong capability for effective collaboration with partners in R&D (*Fig. 11*) and manufacturing facilities among international partners (e.g. IMEC, design and system houses, tool/material vendors, ...etc.), Licensing technologies from advanced industry leader, as well as merging companies for multi-site Fab facilities. There are foundries for specialized manufacturing in areas of FEOL, BEOL, or NVM at BEOL, 3DIC integration, and system-in-package. These new foundry business models already started appearing in China and result in accelerated pace of advanced manufacturing and product development with lower cost as well as technology development with high quality and efficiency.

Figure 8: Various selector technologies for stacking-up NVM into cross-point 3D array, e.g. OTS in Intel/Micron's 3XPoint [13], C-nanotube FET [14, and others.

3DIC TECHNOLOGY

The conventional 3DIC integration technology [15] as based on chip-to-chip or wafer-to-wafer stacking by using through-Si-via (TSV) technique (*Fig. 9*) or based on system-in-package (SIP) can offer significant advantages in low power, high speed, data security, etc. Further, the monolithic 3DIC integration [16] including emerging NVM (e.g. RRAM, PCRAM, MRAM, C-nanotube RAM, ...etc.) as well as logic circuits (e.g. magnetic spin-logic, C-nanotube FET, etc.) at BEOL together with Si-based CMOS at FEOL promises revolutionary digital systems with massive connectivity and data communication with high security, high speed, and low-power. 3DIC integration technology is considered as an effective way to extend the Moore's law (i.e. scaling of transistors in 2D manner) into "More-than-Moore" era.

Foundry: as "Chain-store"

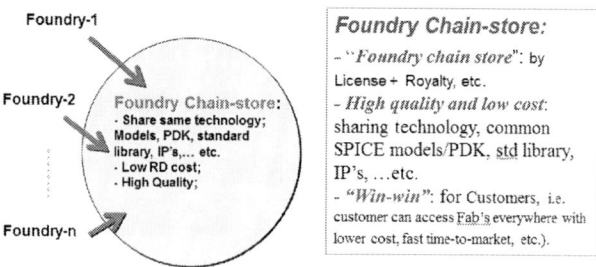

978-1-5090-6695-7/17 $31.00 © 2017 IEEE

Foundry: "Super-foundry" or "Virtual-IDM"

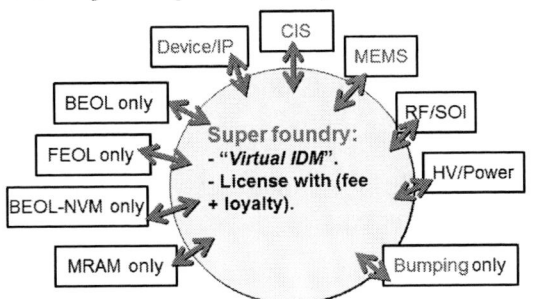

Figure 10: The foundry industry may evolve rapidly into models of "super-foundry"(or virtual IDM) and "Foundry chain-store".

Foundry: Serving as R&D center

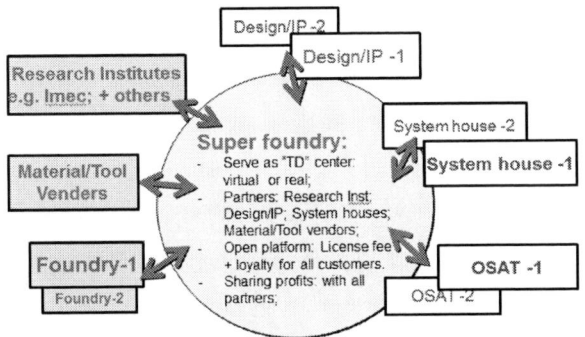

Figure 11: The foundry may serve as collaborated R&D center for establishing advanced CMOS platform and specialty technologies.

CONCLUSIONS

As China's semiconductor chip consumption exceeding 50% of global total since 2012, the investment of wafer Fab's at China is significantly accelerated. In turn, the development of semiconductor manufacturing related technologies as well as design services are also in fast expansion mode into an efficient service chain with competitive advantages. *Firstly*, the state-of-art 14/16nm FinFET platform technology (as planned for manufacturing at China) has new features including the profile control of Si fins by careful wet clean, oxidation, and etching, low-doping of fin and vertical profile minimizing Vt variations, multi-Vt scheme for SOC using multiple WF layers. *Secondly*, the state-of-art 3DNAND and DRAM are manufactured at China by leading companies of Samsung and Hynix. *Thirdly*, the emerging specialty technologies are in active R&D at China, including: non-volatile memory (NVM) technologies progressed toward full compatibility at FEOL or in BEOL, e.g. Phase-Change-RAM (PCRAM), Resistive-RAM (RRAM), Magnetic-RAM (MRAM), magnetic spin-logic, and Nanotube-RAM (NRAM). The NVM and non-Si logic at BEOL is particularly interesting for monolithic 3D

integration of high density memory and Logic for high performance systems. *Finally*, foundry business is rapidly evolving into models of "super-foundry" (or virtual IDM) and "Foundry chain-store" possibly appearing in China firstly. This not only accelerates advanced manufacturing and product development, but also technology development with high quality.

REFERENCES

[1] PwC, "China's impact on the Semiconductor industry: 2015 update", 2016. (Source: www.pwc.com/chinasemicon).

[2] K. J. Kanarik, G. Kamarthy, and R.A. Gottscho "Plasma etch challenges for FinFET transistors", Solid State Technology, v.55, No.3, April 2012.

[3] Y. Shen, M. Chi, *et. al.* US Patents 9,419,899, 2016.

[4] J. Jang, et.al. "Vertical Cell Array using TCAT (Terabit Cell Array Transistor) Technology for Ultra High Density NAND Flash Memory", VLSI-T, p.192, 2009.

[5] Y. Liu, et.al. "Anti-Fuse Memory Array Embedded in 14nm FinFET CMOS with Novel Selector-Less Bit-Cell Featuring Self-Rectifying Characteristics", VLSI-T, p.40, 2014.

[6] C. Y. Mei, et.al.,"28nm high-k metal-gate RRAM with fully compatible CMOS logic processes", IEEE，VLSI-TSA, 2013.

[7] H.-S. Philip Wong, et.al., "Metal–Oxide RRAM", IEEE Proceedings, V.100, No. 6, p.1951, 2012.

[8] H. Y. Cheng, et.al., "Novel Fast-switching and High-data Retention Phase-change Memory Based on New Ga-Sb-Ge Material", IEDM, p.56 ,2015.

[9] W. S. Zhao, et.al., "Embedded MRAM for High-speed Computing", 19th International Conference on VLSI and System-on-Chip, p.37, 2011.

[10] C. Park, et.al., "Systematic Optimization of 1 Gbit Perpendicular Magnetic Tunnel Junction Arrays for 28nm Embedded STT-MRAM and Beyond", IEDM, p.664, 2015.

[11] J. Kim, et.al., "Spin-Based Computing: Device Concepts, Current Status, and a Case Study on a High-Performance Microprocessor", IEEE Proc., v.103, No.1, p.106 , 2015.

[12] S. Ning, et.al., "Investigation and Improvement of Verify-Program in Carbon Nanotube-Based Nonvolatile Memory", IEEE ED, p.2837, 2015.

[13] D. C. Kau, et.al., "A stackable cross point phase change memory", IEEE，IEDM, p.617, 2009.

[14] C. Ahn, et.al., "1D Selection Device Using Carbon Nanotube FETs for High-Density Cross-Point Memory Arrays", IEEE ED, v.62, No.7, p2197, 2015.

[15 J. lau, "Evolution, Challenge, and Outlook of TSV (Through-Silicon Via) and 3D IC/Si Integration", Keynote at IEEE Japan ICEP, 2011.

[16] M. M. Shulaker, et.al. "Monolithic 3D integration: a path from concept to reality"; DATE, p.1197, 2015.

ANALYTICAL MODELING FOR SUBSTRATE EFFECT OF LATERAL POWER DEVICES

Jing Chen[1,2], Yufeng Guo[1,2]*, Man Li[1,2], Ling Du[1,2], Xinchun Ji[1,2], Changchun Zhang[1,2], Jiafei Yao[1,2]

[1]National and Local Joint Engineering Laboratory of RF integration &
Micro-Assembly Technology, Nanjing 210023, China

[2]College of Electronic Science and Engineering, Nanjing University
of Posts and Telecommunications, Nanjing 210023, China

*Corresponding Author's Email: yfguo@njupt.edu.cn

ABSTRACT

A two-dimensional(2-D) analytical model for substrate effect of lateral power devices is developed. By solving the 2-D Poisson's equation, an analytical model is proposed and verified by the agreements between the analytical results and numerical simulation results using MEDICI. It suggests that the optimized substrate voltage modules the distributions of the surface potential and electrical field, whose effect is equivalent to changing the concentration of the drift region. As a result, the breakdown voltage is improved.

Keywords—lateral power devices; substrate effect; model; breakdown voltage

INTRODUCTION

Lateral power devices are widely applied due to its high breakdown voltage, easy integration[1]. Providing the analytical model to analyze the performance is the popular method[2-3]. The substrate impacts greatly on the performance of the devices[4-5], but there is no report on model of this aspect. In this paper, an analytical model for substrate effect of lateral power devices is presented. The model indicates that substrate voltage modulates the distributions of the surface potential and electrical field. The 2-D device simulator Medici[6] is employed to perform the numerical results. A good agreement between the analytical and numerical results is obtained.

MODEL

Fig.1a) and Fig.1b) show the cross section of the lateral power diodes with an applied voltage on the silicon-on-insulator(SOI) or bulk silicon substrate, respectively. By solving the 2-D Poisson equation in the full-depleted drift region, the surface potential and electric field of the devices can be expressed as

$$\varphi_s = \varphi_{s0} + V_{sub} \left\{ 1 - \frac{\sinh(x/t) + \sinh\left[(L-x)/t\right]}{\sinh(L/t)} \right\} \quad (1)$$

$$E_s = E_{s0} + \frac{V_{sub}}{t} \left\{ \frac{\cosh\left[(L-x)/t\right] - \cosh(x/t)}{\sinh(L/t)} \right\} \quad (2)$$

with

$$t = \begin{cases} \sqrt{t_s^2/2 + \varepsilon_s t_s t_{ox}/\varepsilon_{ox}} & \text{for SOI} \\ t_s\left\{1/4 + 1/2\left[\varepsilon_s V_{app}/\left(qP_{sub}t_s^2\right) + N_d/P_{sub}\right]\right\}^{\frac{1}{4}} & \text{for Bulk Silicon} \end{cases} \quad (3)$$

where φ_{s0} and E_{s0} are the surface potential and electric field when the substrate is grounded[7-9]. t is the characteristic thickness, L is the length of the drift region with a uniform doping concentration N_d, the permittivity ε_s and the thickness t_s. t_{ox} is the buried oxide thickness of SOI with the permittivity ε_{ox}. P_{sub} is the substrate concentration of bulk silicon with the substrate depletion region thickness t_b. V_{app} and V_{sub} are the voltages applied to the cathode and substrate, respectively. x and y represent the horizontal and vertical position relative to the surface edge of the p region, respectively.

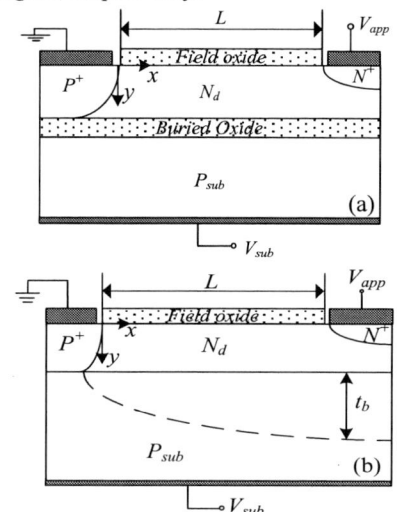

Fig.1: Cross section of the lateral power devices
(a)SOI and(b) Bulk Silicon

978-1-5090-6695-7/17 $31.00 © 2017 IEEE

DISSCUSSIONS

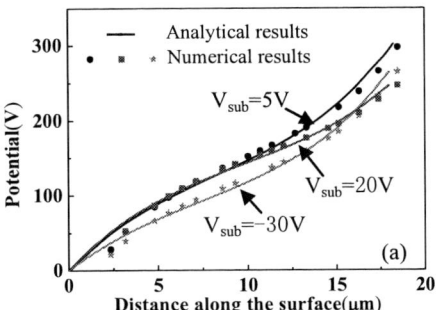

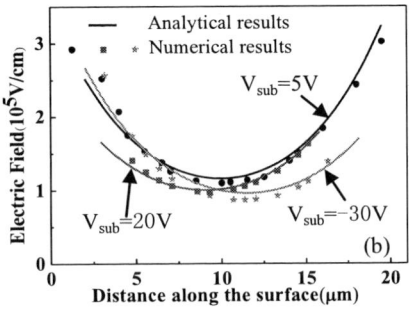

Fig.2:Analytical and numerical results of surface (a) potential and(b) electric field distributions for SOI(t_s=3μm,L=20μm,t_{ox}=3μm,N_d=2.5×10^{15}cm^{-3})

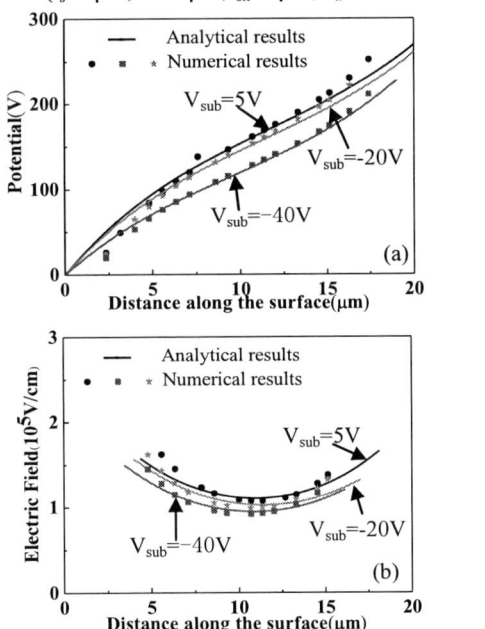

Fig.3: Analytical and numerical results of surface (a) potential and(b) electric field distributions for Bulk Silicon (P_{sub}=1×10^{15}cm^{-3},t_s=5μm,N_d=2×10^{15}cm^{-3},L=20μm)

Fig.2 and Fig.3 show the analytical and numerical results of the SOI/Bulk lateral diodes under different substrate voltage. The distributions of potential and electric field are different under different substrate voltages, the optimized substrate voltage can modulate the distributions of the surface potential and electrical field. Also, as the figures show, the analytical results are in good agreement with the simulation results in general. There is a certain deviation around the P$^+$N$^-$ junction and N$^+$N$^-$ junction, because the effects of curvature radius of P$^+$N$^-$ and N$^+$N$^-$ junctions are neglected in the derivation of the model.

Fig.4 shows the relationship between the breakdown voltage and the substrate voltage at different concentration of the drift region. As shown in the figure, only under the conditions of the optimum substrate voltage, the maximum breakdown voltage can be obtained. For a low V_{sub}, the breakdown point is located at the N$^+$N$^-$ junction, leading to an increase of the breakdown voltage with the increase of V_{sub} by reducing the electric field of N$^+$N$^-$ junction. On the contrary, the breakdown occurs at the P$^+$N$^-$ junction and the breakdown voltage reduces by increasing larger V_{sub}, which deteriorates the distribution of the surface electric field. Hence, substrate bias can be equivalent to the change of the drift doping concentration, which satisfies

$$N_d{}' = V_{sub}\varepsilon_s / (qt^2) \tag{4}$$

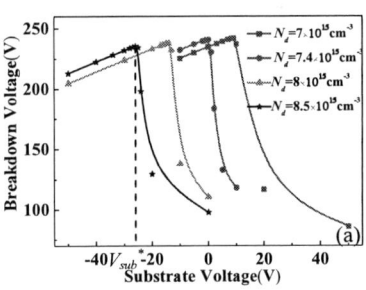

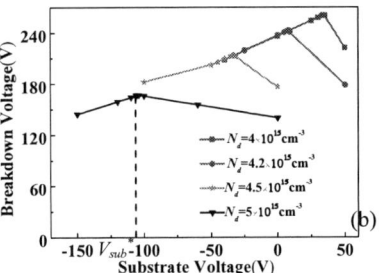

Fig.4: The relationship between breakdown voltage and substrate voltage (a) SOI(t_s=2μm,t_{ox}=2μm,L=20μm) (b) Bulk Silicon (t_s=5μm,P_{sub}=1×10^{15}cm^{-3},L=20μm)

Usually, there is no vertical breakdown, because a large amount of interfacial charge is generated on the surface of the substrate for SOI and the ionization charge in the depletion region of the substrate increase for bulk silicon due to the substrate voltage, which shielding the high electric field beneath the N^+N^- junction. Also, with the increasing of N_d, the optimum substrate voltage decreases. A condition for optimizing the substrate voltage for a given drift doping concentration can be derived as

$$V_{sub}{}^* = \tanh\left[L/(2t)\right]E_c t - qN_d t^2/\varepsilon_s \qquad (5)$$

Where E_c is the critical electric field[8], $V_{sub}{}^*$ is the optimized substrate voltage.

Hence, the breakdown voltage can be improved by a negative/positive substrate voltage if the structure of the drift region is heavily/lightly doped.

CONCLUSIONS

An analytical model for substrate effect of lateral power devices is proposed in this paper. The comparison between analytical model and numerical simulation is performed. The results show the model provides a simple, clear and accurate physical insight to study the breakdown characteristics of lateral power devices when the substrate bias is considered. In the lateral direction, the substrate voltage increases the electric field at the P^+N^- junction while reduces the electric field at the N^+N^- junction. The effect is equivalent to changing the drift region doping concentration. An optimal substrate voltage can compensate the heavy or light doping concentration in the drift region, leading to the highest breakdown voltage.

ACKNOWLEDGEMENTS

This research was supported by Natural Science Foundation of Jiangsu Province of China(No. BK20141431), Key Technology Support Program of Jiangsu Province of China (No. BE2013130), and National Natural Science Foundation of China (No. 61574081).

REFERENCES

[1] H Akiyama, N Yasuda, J Moritani, K Takanashi, G Majumdar. Proceedings of The 16th International Symposium on Power Semiconductor Devices & ICs, Japan, May 24-27, 2004, pp. 375-378.

[2] J Zhang, Y F Guo, Y Xu, H Lin, H Yang, Y Hong. Chinese Physics B, vol. 24, 2015, pp. 028502.

[3] S Yuan, B X Duan, Z Cao, H J Guo, Y T Yang. Solid-State Electronics, vol. 123, 2016, pp. 6-14.

[4] B Zhang, W T Zhang, Z H Li, M Qiao, Z J Li. IEEE Transactions on Electron Devices, vol. 61, 2014, pp. 525-532.

[5] P Buccella, C Stefanucci, J M Sallese, M Kayal. IEEE Transactions on Power Electronics, vol. 31, 2016, pp. 6586-6595.

[6] User's Manual of Two-Dimensional Device Simulation Program MEDICI, Version 2001.2, Synopsys MEDICI User's Manual, Synopsys Inc., Mountain View, CA, USA, 2010.

[7] J F Yao, Y F Guo, Li M, X F Huang, H Lin, X C Ji. Japanese Journal of Applied Physics, vol. 54, 2015, pp. 024301.

[8] J He, X Zhang. Microelectronics journal, vol. 32, 2001, pp. 655-663.

[9] Y F Guo, Z J Li, B Zhang. 2007. ICCCAS 2007, Japan, July 11-13, 2007, pp. 1278-1282.

MOSFET RF PERFORMANCE IMPROVEMENT THROUGH SPACER PROFILE OPTIMIZATION FOR 28NM POLY/SION SOC TECHNOLOGY

Hai Liu*, River He, Byunghak Lee

Logic and Device Engineering Division, Technology R&D Center, SMIC
18 Zhangjiang Road, Pudong New Area, Shanghai, P.R. China 201203
E-mail: Hai_Liu@smics.com

ABSTRACT

As the MOSFET scales down, with keeping increasing of speed (ft), minimum noise figure (NF) is difficult to scale down where increasing gate resistance (Rg) is one of crucial impact factors. In this paper, we present a novel approach to boost MOSFET RF performance by spacer profile optimization. It can help reduce Rg about 12% and increase Fmax about 7% for critical size device (L=27nm) without degrading device DC performance on 28nm Poly/SiON technology platform. This approach does NOT introduce any extra steps into process flow and can be easily combined with those traditional Rg reduction methods.

INTRODUCTION

In the recent years, the fast growth of IOT (internet of things) market has put RF circuits in high demand. CMOS technology has already been proved to have great advantages of rapid technology evolution and cost reduction, resulting in significant improvement of RF performance [1-2], and it may also provide the possibility of a single-chip solution. Various approaches have been carried out for MOSFET RF performance improvement [3-5]. However, as MOSFET continuously scales down, increasing gate resistance will be one crucial factor dragging device RF performance improvement.

In this paper, we present a novel method to boost MOSFET RF performance by spacer profile optimization.

EXPERIMENTS

The MOSFET samples in this paper were fabricated with SMIC 28nm Poly/SiON process platform. After STI process, the poly silicon/SiON gate stack are formed and patterned. Selective deposition of in situ boron-doped SiGe source/drain is used to boost PMOS performance, while tensile contact etching stop layer (CESL) layer and stress memory technique (SMT) are used to boost NMOS performance. The nickel silicide and Cu backend is applied to complete the process.

Besides control split A with baseline process flow, another 3 experimental splits are fabricated with different process conditions to evaluate their impact on device RF performance. Split B is with increasing silicide rapid thermal annealing (RTA) thermal temperature compared to sample A; Split C is with the same silicide RTA process change as split B and, furthermore, it also has optimized

spacer profile tuning process; split D has different combination of RTA process change and spacer profile optimization process compared to split C.

RESULTS AND DISCUSSION

Multiple finger (NF=16) NMOS devices with width/length=2.7um/27nm are tested under Vg=1.05V, Vd=1.05V, Vs=Vb=0 bias condition. Y parameters are extracted at 5GHz with standard SOLT characterization and de-embedding procedure. For each split wafer, nine tests at different locations are tested to rule out with in wafer variation.

From Fig.1, it shows that Ft performances are comparable among 4 splits, and split C shows about 9.5%/7% fmax improvement to split A/B, respectively.

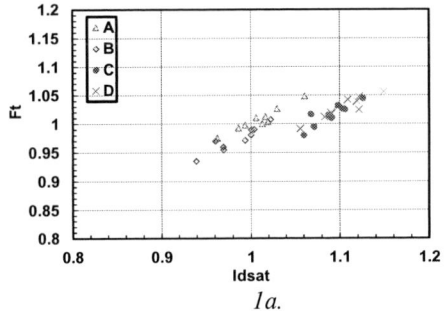

1a.

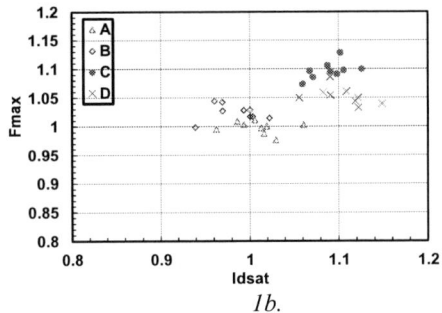

1b.

Figure 1: NMOS device RF performance
1a. ft-Idsat correlation, 1b. fmax-Idsat correlation
(all data are normalized by control split A

Gate resistance Rg characteristics, which can also be presented as Re(1/Y11) from Y parameter extraction[6], is shown Fig. 2. Split B/C show about -9%/-20% Rg reduction compared to split A.

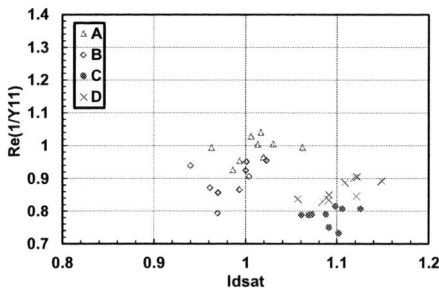

Figure 2: Device RF performance: Re(1/Y11)-Idsat correlation (all data are normalized by control split A)

And from Fig. 3, Im(Y11/w) (Cgg) and ImY12/w (Cgd) performance are comparable among all splits.

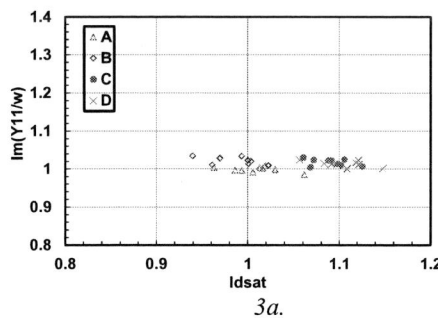

3a.

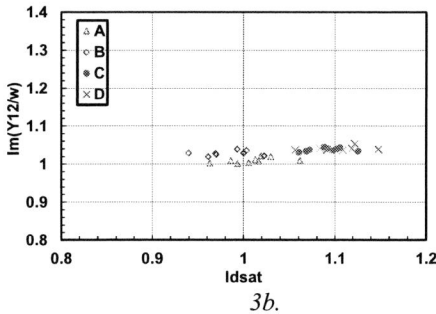

3b.

Figure 3: NMOS device RF performance
3a. Im(Y11/w)-Idsat correlation,
3b. Im(Y12/W)-Idsat correlation
(all data are normalized by control split A

The above test results are summarized in Table I. With spacer profile optimization (C vs B), MOSFET RF performance is obviously improved ~7% in fmax and ~12% in Rg. And it is worth to mention that this spacer profile tuning method does not bring any additional steps and cost to the process.

Compared split B to control split A, increasing silicide RTA can also help to reduce gate resistance about 9% and improve Fmax about 3%. However, increasing silicide RTA thermal budget will increase nickel piping

risk and degrade yield performance. While, compare split B with split C, spacer profile optimization approach will not increase any silicide RTA thermal budget.

Table1. Device RF performance Summary

	ft	fmax	Rg	Im(Y11/w)	Im(Y12/w)
A	-	-	-	-	-
B	-3.2%	2.7%	-8.8%	2.6%	1.9%
C	1.3%	9.6%	-20.6%	2.5%	2.8%
D	2.1%	5.0%	-14.6%	1.7%	2.9%

This clear device RF performance boost by spacer profile optimization can be understood from the below schematic in Fig. 4.. By spacer profile optimization, the silicide profile and volume can be maximized, which gives out smaller Relec and results in smaller Rg from equation (1).

$$R_g = R_{elec} + R_{channel} \qquad (1)$$

where Relec is distributed gate electrode resistance and Rchannel is channel coupling resistance from non-quasi-static (NQS) effects.

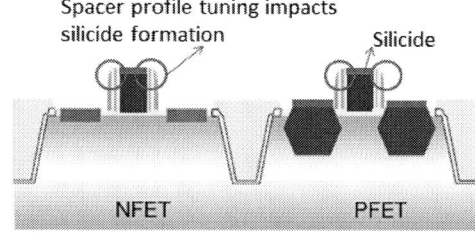

Figure 4: Schematic of spacer profile impact on poly silicide formation

This model can also be evidenced by measured Rg is well correlated Rs measured from N+ poly silicide resistor structure with same polyline width to NMOS presented above, as shown in Fig. 5.

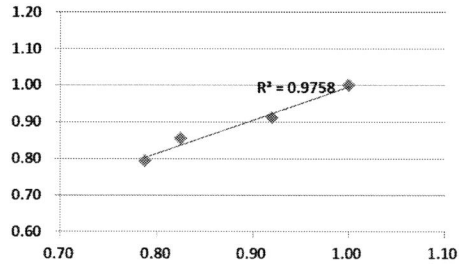

Figure 5: Rg-Rs Correlation (all data are normalized by control split A)

Thus, for Fmax improvement, it can also be well understood from equation (2), where Rg plays a significant role on Fmax.

$$f_{max} = f_t \, / \, 2 \sqrt{\frac{R_g + R_i + R_s}{R_{ds}} + 2\pi f_t R_g C_{dg}} \qquad (2)$$

And from equation (3), it is expected that minimum noise figure performance will also be improved.

$$NF_{min} = 1 + K \frac{f}{f_t} \sqrt{g_m \left(R_g + R_s\right)} \qquad (3)$$

Finally, it is worth to mention that the above device RF performance improvement does not sacrifice device DC performance, as shown in figure 6. It is clear that the split C with optimized spacer profile has comparable or slightly improved Idsat-Ioff and Gm (transductance) – Rout (output resistance) performance compared to control split.

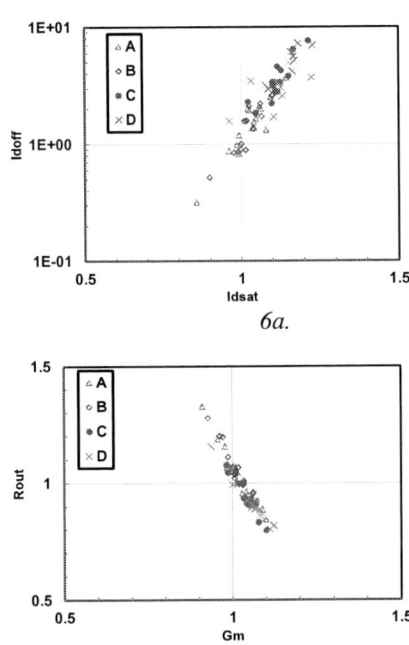

6a.

6b.

Figure 6: NMOS device DC performance
 a)   *DC Idsat-Ioff Correlation, b) Gm-Rout Correlation*
Gm, Rout are measured under (Vg=Vtgm+0.2V,
Vd=Vdd/2), (all data are normalized by control split A)

SUMMARY

In this work, a novel method to boost MOSFET RF performance by spacer profile optimization is presented. This method is fully compatible with current 28nm Poly-SiON CMOS process without induce any additional process steps. With this method, critical dimension device with gate length L=27nm will have Fmax improved about 7% and gate resistance reduced about 12%.

REFERENCES

[1] C.-H. Jan, M. Agostinelli, H. Deshpande, M. A. El-Tanani, W. Hafez, U. Jalan, et al. , *IEDM 2010*, pp 604-607.

[2] [2] H. L. Kao, Albert Chin, C. C. Liao Sean P. McAlister, J. Kwo and M. Hong, *DRC 2006*, pp 64-65.

[3] [3] M.C. King, M.T. Yang, C.W. Kuo, Yun Chang, and Albert Chin, *Microwave Symposium Digest IEEE 2004*, Vol 1, pp. 9–12.

[4] [4] Han-Su Kim, Jedon Kim, Chulho Chung, Jinsung Lim, Joohyun Jeong, Jin Hyoun Joe, et al. , *IEEE Transactions on Electron Device 2008*, PP 2712 -2718.

[5] [5] Jagar Singh, Ciavatti Jerome, Andy Wei, Roderick Miller, Bousquet Arnaud, Cheng Lili, Hui Zang, et al. , *VLSI-Technology 2014*, PP1-2.

[6] [6] Y Ickjin Kwon, Minkyu Je, Kwyro Lee, and Hyungcheol Shin, *IEEE Transactions on Microwave Theory and Techniques, 2002*, vol 50, No6, PP 1503-1509

A Method to Solve Reverse Tunneling Disturb Issue for SuperFlash® Memory

Tao Xu, Zigui Cao, Guoqing Han, Hong Chen, Hui Wang,
Department of Process Integration, Shanghai Huahong Grace Semiconductor Manufacturing
Corporation, Shanghai 201203, China
Corresponding Author's Email: Tao.Xu@hhgrace.com

ABSTRACT

Oxygen gas is conventionally used for photoresist ashing process during the fabrication of Integrated Circuits. However, Reverse tunneling disturb fail of flash memory happened frequently if we only use Oxygen gas during photoresist ashing process after floating gate nitride etch. Through physical failure analysis, we found it is reverse word line tip that induce RTD fail. By analysis and comparison, an additional forming gas (N_2H_2) that can enhance polymer removal during Photo resist ashing process was introduced to solve RTD fail for flash products.

INTRODUCTION

FLASH memory with nonvolatile and reprogrammable property has become widely used as memory storage, among these, Split gate cells with higher program efficiency and over-erase immunity [1-4] than other designs of flash memory has been widely used in various consumer electronic products. Furthermore, SuperFlash® technology provided by SST Company has become the most successful flash memory, which not only bases on a proprietary split-gate flash memory cell but also provides a cost effective and high performance programmable SOC solution.

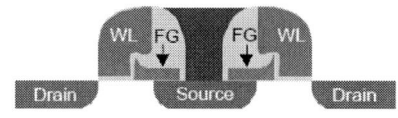

Figure1. Self-aligned SuperFlash® cell structure

The SuperFlash® (Figure 1) is a split-gate flash memory cell utilizing source-side channel hot electron injection for program, and poly-to-poly Fowler-Nordheim (F-N) tunneling for erase [5, 6]. It has three operation mode: Program (PGM), Erase and Read. In this paper, we only focus on how to solve Reverse Tunneling Disturb (RTD) issue during Program, so we will only give a brief description for Program and RTD condition.

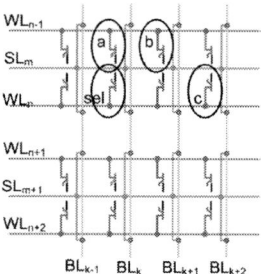

Figure2. SuperFlash® Cell array [7]

Figure2 shows SuperFlash® Cell array [7]. When a cell (sel. Cell in Fig.2) is selected to PGM, not only WLn (the Nth Word Line) will be forced with a voltage to open the WL channel and BLk (the Kth Bit line) will be forced with a fixed current, but also SLm (the Mth Source line) will be forced with a high voltage; At this time, the cell (b cell in Fig.2) that in different row and column with sel. cell share the same high voltage SL and low voltage WL, so cell b is prone to be disturbed: electrons tunneling from WL to FG (floating gate), Which is opposite to erase operation, we named this phenomenon as Reverse Tunneling Disturb (RTD) fail.

During fabrication, there are many processes that can affect flash memory operation, among these, there are some process that may induce RTD fail. In this work, we have done electrical failure analysis (PFA) and physical failure analysis (PFA) for the RTD fail samples and found it is reverse WL tip that induce RTD fail. Through a series of trial and error experiments, the fail model was clarified and the forming gas (N_2H_2) that can enhance the polymer remove was introduced into FGSN ET PR RM process to solve flash memory RTD fail.

FAILURE ANALYSIS AND FAIL MODEL

Figure3 shows RTD fail map and EFA data for the fail samples; Base on Figure3 EFA data, we can see that Cell current after reverse tunneling stress test (Icell after RTST) is significantly reduced compared with cell current after chip erase (Icell after chip erase), so we can confirm it is really RTD fail.

Figure4 shows SEM images (PFA) of RTD fail and normal bit, from the images, we can see that there is a reverse WL tip for RTD fail bit compared with normal bit.

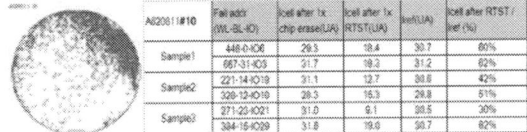

Figure3. RTD fail map (left picture) and EFA (right)

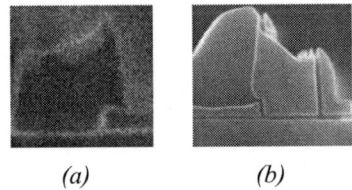

(a)   (b)

Figure4. SEM pictures (PFA) of RTD fail bit (a) and Normal bit (b)

Basing on our work experience, we provide a fail model for this issue; Figure5 shows the fail model for reverse WL tip [Left side shows the abnormal profile and right side shows the normal profile after FGSN ET and FGSL ET (Floating gate sloped etch) process]; From figure5 (a), we suspect that there is some polymer residue after FGSN ET process for abnormal cell while there is no polymer residue for normal cell, which may affect isotropic FGSL ET process and finally affect FG tip profile. Figure5 (b) shows the effect of polymer residue on FG tip, on one hand, FG tip height (H) will be higher than normal cell, on the other hand, reverse WL tip will be formed after polymer was removed by tunnel oxide deposition pre-clean process; Both of these two factors will induce flash cell RTD fail.

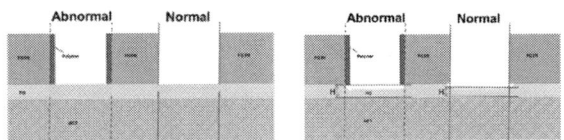

(a)After FGSN ET process   (b)After FGSL ET process

Figure5. Fail model for reverse WL tip: (a) after FGSN ET process, (b) after FGSL ET process

EXPERIMENT

The samples were full loop silicon wafers in our fabrication line. Figure 6 shows the samples during FGSN ET process.

For clarification the root cause of reverse WL tip and find a way to solve this problem, two conditions of FGSN ET photoresist (PR) ashing process that with or without forming gas (N_2H_2) were tried.

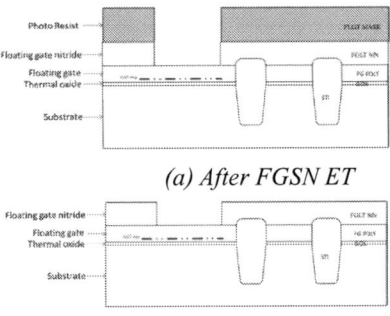

(a) After FGSN ET

(b) After FGSN ET PR ashing and remove

Figure6. Diagrams of the samples: (a) after FGSN ET, (b) after FGSN ET PR ashing and remove

RESULTS AND DISCUSSION

In order to clarify our provided fail model, we divide the samples into two groups, for one group, we use the baseline photo resist (PR) ashing process after FGSN ET, for the other group, we add forming gas (N_2H_2) into PR ashing process besides Oxygen gas which can enhance the polymer remove.

Figure7 shows the RTD fail ratio of these two groups, we can found that RTD fail ratio is significantly decreased by adding forming gas in FGSN ET PR ashing step, which has verified our fail model.

In the following, we will give a detailed description about this case combination with process

978-1-5090-6695-7/17 $31.00 © 2017 IEEE

flow, Figure 8 shows the related process flow about how FGSN ET polymer affect RTD performance. From figure8, we can see that the polymer that formed during FGSN ET process will be removed by tunnel oxide deposition (dep) process for baseline condition, which will induce reverse word line tip after word line poly dep and etch. However, after we add forming gas in FGSN ET PR ashing process, there is no polymer residue after PR remove, so there is no reverse WL tip after word line poly dep and etch.

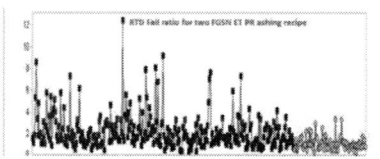

Figure7. RTD fail ratio split by FGSN ET PR ashing recipe: blue spot represent baseline Oxygen gas ashing recipe, green spot represent ashing recipe added with forming gas (N_2H_2)

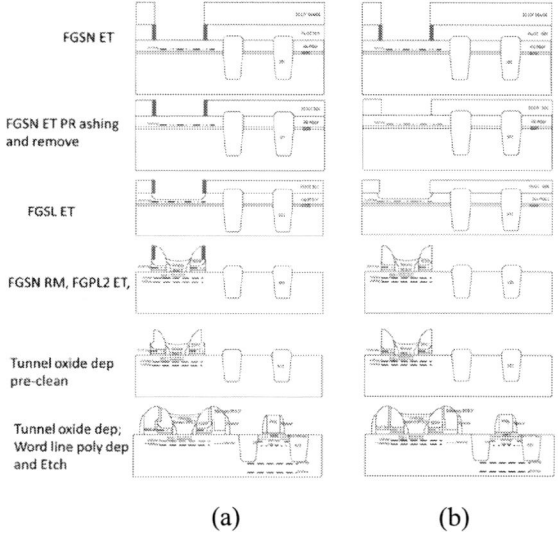

(a)　　　　　　　　(b)

Figure8. (a) Process with baseline condition (Only Oxygen gas for FGSN ET PR ashing); (b) Process with improved condition (add forming gas for FGSN ET PR ashing)

CONCLUSIONS

In summary, we have investigated and clarified the root cause of Reverse Tunneling Disturb (RTD) fail for flash memory. Through physical failure analysis, we found it is reverse Word Line tip that induce RTD fail. By analysis and comparison, an additional forming gas (N_2H_2) that can enhance polymer remove during photoresist ashing process was introduced to solve RTD fail for flash products.

ACKNOWLEDGEMENTS

Author would like to thank our team member Guoqing Han, Hong Chen and Hui Wang for the experiment support and our manager Zigui Cao for the technical discussion.

REFERENCES

[1] J. V. Houdt, G. Groeseneken, and H. E. Maes, "An analytical model for the optimization of source-side injection flash EEPROM devices," *IEEE Trans. Electron Devices*, vol. 42, pp. 1314–1320, 1995.

[2] Y. Ma, C. S. Pang, J. Pathak, S. C. Tsao, and C. F. Chang, "A novel high density contactless flash memory array using split-gate source-side injection cell for 5 V-only application," in 1994 Symp. *VLSI Technology Dig. Tech. Papers*, vol. 5A, pp. 49–50.

[3] K. Chang, W. Chen, C. Swift, J. M. Higman, W. M. Paulson, and K. Chang, "A new SONOS memory using source-side injection for programming," *IEEE Electron Device Lett.*, vol. 19, pp. 253–255, 1998.

[4] A. T. Wu, T. Y. Chan, P. K. Ko, and C. Hu, "A novel high-speed, 5-volt programming EPROM structure with source-side injection," in *IEDM Tech. Dig.*, 1986, pp. 584–587.

[5] S. Kianian, A. Levi, D. Lee, and Y. W. Hu, Proc. Symp. *VLSI Technol.*, p.71 (1994).

[6] R. Mih, J. Harrington, K. Houlihan, H.K. Lee, K. Chan, J. Johnson, B. Chen, J. Yan, A. Schmidt, C. Gruensfelder, K. Kim, D. Shum, C. Lo, D. Lee, A. Levi, C. Lam, Symp. *VLSI*, p.120 (2000).

[7] H.-C. Sung, T. F. Lei, T.-H. Hsu, et al. Novel program versus disturb window characterization for split-gate flash cell. *Electron Device Letters, IEEE*, 2005(3):194–196.

COMPACT ELECTRODYNAMICS MEMS-SPEAKER

Burhanuddin Yeop Majlis[1], Gandi Sugandi[2], and Mimiwaty Mohd Noor[1]*

[1]Institute of Microengineering and Nanoelectronics – The National University of Malaysia
(IMEN-UKM), Bangi-Selangor, Malaysia
[2] Research Center for Electronics and Telecommunication- Indonesia Institute of Sciences
(PPET_LIPI), Bandung-Indonesia
*Corresponding Author's Email: burhan@vlsi.eng.ukm.my

ABSTRACT

This paper describes the design, fabrication and characterization of an electrodynamic MEMS speaker. The miniaturized speaker consists of a deposited single turn microcoil on a 25 μm thick suspended polyimide diaphragm with a 2.5 mm diameter and a small volume permanent magnet Neodymium-Iron-Boron (Nd-Fe-B). A significant improvement of frequency response to sound pressure level of MEMS-speaker in sealed condition was achieved. The sealed measurement performed in a 1500 mm^3 silicon rubber tube resulted in a peak amplitude of 90 dB-SPL at a frequency range of 1, 5 and 10 kHz and experienced magnitude boosting with a value of 25 to 30 dB at frequency below 1 kHz compared to unsealed condition. Peak magnitude level of 92.5 dB-SPL and 110 dB-SPL were achieved at the frequency of 200-500 Hz and 20-60 Hz respectively. The results shows that the fabricated MEMS speaker can be applied in ear canal applications such as hearing aids and music player earphones.

INTRODUCTION

Micro electro-mechanical systems (MEMS) technology has been proved to be successful in the fabrication of micromechanical structures such as micro sensors and actuators which are small in size, lightweight and low power consumption with low cost production devices (1). Driven by the development of multimedia devices and mobile consumer electronic products such as the smart phone, tablet, music player and hearing aids to be small and slim products, research efforts to fulfill the demand of reducing microspeaker size that can reproduce sound source using MEMS technology has become an exciting investigation. The feasibility of MEMS-speakers have been studied by several works with different actuation mechanism, such as an electrostatic (2), piezoelectric (3,4), electro-magnetic (5,6) and electrodynamics (7-11) actuator. The generated force for the diaphragm displacement in an electrodynamic speaker requires higher current and thus higher power consumption than in piezoelectric speaker and electrostatic speaker but can be generated with low voltage bias. Therefore, the electrodynamic speaker can

potentially be used in mobile electronics consumer product devices.

Our goal in this work is to develop an electro-dynamic MEMS-speaker using silicon micro fabrication batch processing. The MEMS-speaker structure consists of a metal planar micro coil driver deposited on a flat flexible polyimide diaphragm with a disc permanent magnet placed along the symmetric axis below the metal planar microcoil.

DESIGN OF MEMS SPEAKER

Design Structure and Priciples Work

A schematic of the MEMS speaker is shown in figure 1. The MEMS-speaker design consists of two silicon wafer; which is the top wafer functions as a micro machined where a flat flexible polyimide diaphragm with a metal planar micro coil deposited on it is attached to the wafer whereas the bottom wafer functions as a permanent magnet platform.

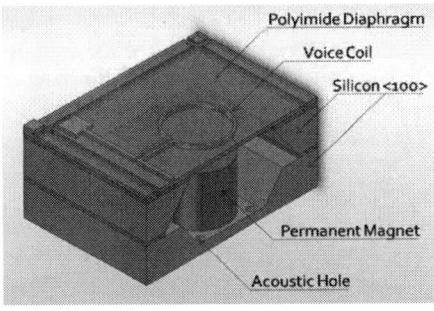

Figure 1: The schematic structure of thr 3D MEMS Speaker(11)

The MEMS-speaker design consists of two silicon wafer; which is the top wafer functions as a flat flexible polyimide diaphragm with a metal planar micro coil deposited on it is attached to the wafer whereas the bottom wafer functions as a permanent magnet platform. The total dimension of device is 5mm x5mm x 1.5 mm. The Permanent magnet Neodymium-Iron-Boron, (K&J Inc.) has diameter of 1.6 mm, 0.8 mm in thickness, and remnant flux of 1.4 T. The gap between the magnet and the micro coil is about 100 μm.

978-1-5090-6695-7/17 $31.00 © 2017 IEEE

Electrodynamic Actuator Analysis

The permanent magnet function is to generate a uniform magnetic field surrounding the planar coil. At the location of the coil, the magnetic field makes an angle θ with the vertical, the total Lorentz force acting on the current carrying planar coil is generally given by (12):

$$\vec{F} = \sum_{i=1}^{N} 2\pi R_i\, I \times \overrightarrow{B_r}(R_i) \times \sin\theta \quad (1)$$

where I is the coil current, R_i the radius of each turn coil, B_r the radial component of magnetic field in the coil plane, and θ angle direction of magnetic field to vertical axis.

Acoustic Analysis

According to (9) the dimension of the MEMS speaker and an ear canal is small compared to the sound wave length. This is to ensure the pressure is distributed uniformly in the volume. The generated sound pressure level (SPL) of MEMS speaker in ear canal application is expressed by

$$SPL = 20 \log\left[-\frac{1.4\, p_0}{p_{ref}} \frac{\Delta V}{V_0}\right] (dB) \quad (2)$$

where P_{ref} is 20 μPa.

FABRICATION

The MEMS-based microspeaker device was constructed from two component : the first component, silicon wafer frame with planar coil suspended on polyimide membrane, and the second component, the acoustic hole which is the permanent magnet NdFeB bonded. Each component was fabricated by MEMS technology, and they assembled together by adhesive bonding. The main techniques process for fabrication of components, include; photolithography, anisotropic wet etching, deposition metal using sputtering and electroplating method. The fabrication process for the components are shown in Figure 2.

For the first silicon wafer, wafer (100) with the thickness of 650 μm coated with 200 nm thick silicon nitride (Si_3N_4) was used as starting material. The substrate was etched using an anisotropic wet etching to form a 50 mm thick silicon diaphragm. The silicon diaphragm was then coated with 25 μm thick polyimide, as shown in figure 2(a). The pattern for the coil was transferred by photolithography process, sputtered with 1 μm of Ti/Au followed by the lift-off process as shown in figure 2(b). Next, the silicon diaphragm was released from the reverse surface by continuing the anisotropic wet etching process with low temperature as shown in figure 2(c). The remaining etch-stop silicon nitride was removed by HF or BOE solution, as shown in figure 2(d). The second component was fabricated to form a platform for the permanent magnet. Firstly, the KOH etch window at the both side of the substrate was aligned and patterned. The opening area was then etched by the anisotropic wet etching from both sides of the wafer as shown in figure 2 (e and d). Next, the permanent magnet was bonded at the center of the wafer using glue as shown in figure 2(f). Finally, Figure 3 shows photograph of the MEMS-speaker packaged in the earphone shell.

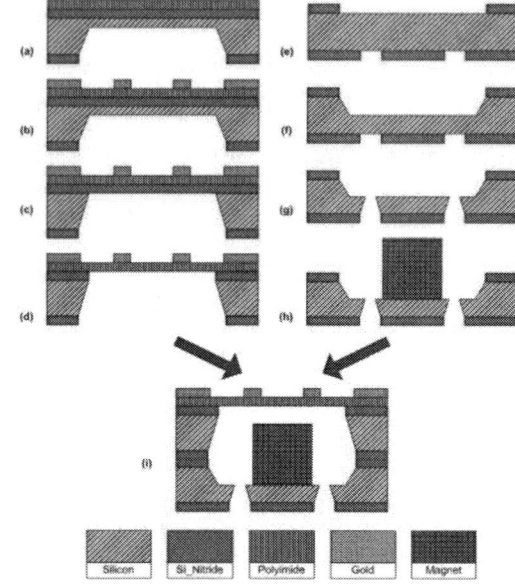

Figure 2 : Schematic fabrication processes of electrodynamics MEMS-Speaker.

Figure 3 : The photograph of MEMS-speaker in earphone package

MEASUREMENT RESULTS

The measurement set up for measuring the performance of the MEMS speaker prototype is shown in figure 4. The measured result of frequency response versus sound pressure level of MEMS speaker in sealed and unsealed tube condition, is shown in figure 5.

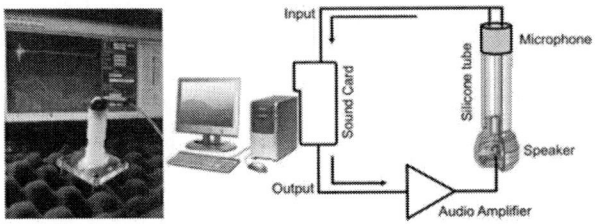

Figure 4 : The measurement setup for the characterization MEMS Speaker

Compared to the unsealed condition, the sound pressure level of speaker in the tight condition experienced excellent sound pressure boosting around 25dB and 30dB at the frequency of 200Hz and 50Hz respectively. The peak level of 92.5 dB-SPL and 110 dB-SPL was achieved at the frequency of 200-500 Hz and 20-60 Hz respectively.

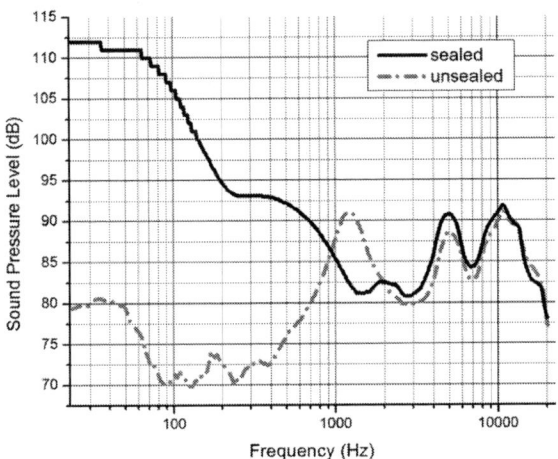

Figure 5 : Measured results of the frequency response versus SPL of the MEMS Speaker sealed and unsealed condition

CONCLUSION

The performance of the fabricated MEMS speaker was measured in a 1500mm3 silicone rubber tube. The results showed a significant and excellent performance of the frequency response to the sound pressure level for the MEMS speaker in an ear canal condition. The peak amplitude is around 90 dB-SPL at the frequency of 5 and 10 kHz and the MEMS speaker experienced boosting level around 25 to 30 dB SPL at the frequency below 1 kHz. The peak amplitude of the sound pressure level are 92.5 dB-SPL and 110 dB-SPL at 200-500 Hz and 20-60 Hz respectively. The characteristic of this MEMS speaker has best performance for hearing of music player and speech.

ACKNOWLEDGEMENTS

The authors would like to thank Universiti Kebangsaan Malaysia (UKM) under Institute of Microengineering and Nanoelectronics (IMEN-UKM) for sponsoring this work under project UKM-GUP-NBT-08-25-084, UKM-AP-NBT-10-2009 and DIP-2012 - 16.

REFERENCES

[1] Madou, M.J., *Fundamentals of Microfabrication: The Science of Miniaturization, Second Edition.* 2002: Taylor & Francis

[2] Neumann Jr, J.J. and K.J. Gabriel, *CMOS-MEMS membrane for audio-frequency acoustic actuation.* Sensors and Actuators A: Physical, 2002. **95**(2–3): p. 175-182.

[3] Ren, T.-L., et al., *Micro Acoustic Devices Using Piezoelectric Films.* Integrated Ferroelectrics, 2006. **80**(1): p. 331-340.

[4] Seo, K., et al., *Micromachined Piezoelectric Microspeaker Fabricated With High Quality AlN Thin Film.* Integrated Ferroelectrics, 2007. **95**(1): p. 74-82.

[5] Harradine, M.A., et al. *A micro-machined loudspeaker for the hearing impaired.* in *Solid State Sensors and Actuators, 1997. Transducers '97 Chicago., 1997 International Conference on.* 1997.

[6] Rehder, J., R. Pirmin, and H. Ole, *Balanced membrane micromachined loudspeaker for hearing instrument application.* Journal of Micromechanics and Microengineering, 2001. **11**(4): p. 334.

[7] Ayatollahi, F.L. and B.Y. Majlis. *Parametric simulations of a micromachined mesoscopic acoustic speaker.* in *Semiconductor Electronics, 2008. ICSE 2008. IEEE International Conference on.* 2008.

[8] Chen, Y.C. and Y.T. Cheng. *A low-power milliwatt electromagnetic microspeaker using a PDMS membrane for hearing aids application.* in *Micro Electro Mechanical Systems (MEMS), 2011 IEEE 24th International Conference on.* 2011.

[9] Ming-Cheng, C., H. Wen-Sheh, and H. Star Ruey-Shing, *A silicon microspeaker for hearing instruments.* Journal of Micromechanics and Microengineering, 2004. **14**(7): p. 859.

[10] Shahosseini, I., et al., *Microstructured silicon membrane with soft suspension beams for a high performance MEMS microspeaker.* Microsystem Technologies, 2012. **18**(11): p. 1791-1799.

[11] Sugandi, G. and B.Y. Majlis, *Fabrication of MEMS Based Microspeaker Using Bulk Micromachining Technique.* Advanced Materials Research, 2011. **254**: p. 171-174.

[12] Shahosseini, I., et al. *Towards high fidelity high efficiency MEMS microspeakers.* in *Sensors, 2010 IEEE.* 2010.

IDENTIFICATION AND SOLUTIONS FOR A NOVEL PARTICULATE POLLUTION MATTER IN WAFER SURFACE CAUSED BY CONCENTRATED SULFURIC ACID

Lu Sun

Department of Material Quality Engineering, Semiconductor Manufacturing International Corporation (SMIC)
Shanghai, China
86-21-20812097; Summer_Sun@smics.com

Biography

This paper analyzed a new particulate pollution on the surface of wafer, and confirmed that the particulate pollution was generated from carbonization of siloxane, which was formed from the concentrated sulfuric acid of the SPM solution in wet process for wafer treatments.

Abstract

The concentrated sulfuric acid is a necessary chemical for IC-Industry, especially in the wet process. The electronic grade concentrated sulfuric acid offered by major suppliers satisfies the requirements of the new technologies in metal ion impurities and insoluble particulate matters, but during the use of that offered by a core supplier in wet process, a new kind of particle contamination appeared. In this paper, the particulate pollution matter has been confirmed as carbide formed from Siloxane, which was produced at high temperature and strong acidic conditions. In order to control the generation of this new particulate pollutant, TOC (Total Organic Carbon) and Silicon content (Sc) were measured in electronic grade concentrated sulfuric acid, and used to establish the relevant mathematical model for calculating the Sr (Siloxane Relative Value) in electronic grade concentrated sulfuric acid, which was first proposed as a new quality demand for the concentrated sulfuric acid.

Keywords—concentrated sulfuric acid; IC-industry; siloxane; particulate pollution; Sr Value

Introduction

The worldwide IC-Industry has generally entered into the technology below 90nm; some advanced technology has reached the 10nm or less. This requires the wafer surface cleanliness to achieve a higher level, so that the IC-Industry supporting chemical products were required for higher purity [1]. In order to appease the requirements of the new IC process for high-purity clean chemicals, the world leading chemical suppliers have improved purification technique to respond to this demand [2].

SPM method is an important wet process in Front End of Line for wafer treatment. This method is designed to remove organic pollutants and the photo resist on the wafer surface [3]. The concentrated sulfuric acid is one of the main chemicals in SPM method. When using BV-V grade concentrated sulfuric acid produced by one leading chemical supplier to clean wafer applied in advanced-technology, a new particulate pollution was detected. After technical analysis and identification, it was confirmed that the particulate pollution was derived from carbonization of siloxane. Furthermore, the introduced reason of siloxane was analyzed and determined [4]. At last, the relative content of siloxane (Sr value) was calculated as a new standard for evaluating the electronic ultra-clean concentrated sulfuric acid, so as to solve the contaminative particles in the process of carbonization. The Sr value was obtained, through an establishment of relevant mathematical model which based on the measurement of TOC and Silicon content (Sc) in the concentrated sulfuric acid.

Produces and carbonlization of siloxane

When the particulate pollution appeared in the wet process, all of unit operations in the wet process were analyzed; it was found that, after the SPM cleaning, on the surface of the wafer, the carbide particle pollution appeared. The possibility of bringing pollution from a variety of raw materials and equipment contamination were ruled out. At last the focus was locked on concentrated sulfuric acid used in SPM method. The electronic grade concentrated sulfuric acid generally doesn't do TOC measurement, so it is most likely to cause carbonation pollution. However SPM process itself is to clean organics on the wafer surface through oxidization of organic pollutants into CO_2 and H_2O.

$$H_2SO_4 + H_2O_2 \rightarrow HO\text{-}(SO_2)\text{-}O\text{-}OH + H_2O$$

$$HO\text{-}(SO_2)\text{-}O\text{-}OH + (CH_2)\,n \rightarrow CO_2 + H_2O$$

Equation 1 The reaction of oxidization of SPM

What caused this oxidation process only at the stage carbon but CO_2, and adhere to the surface of the wafer? In this regard, the following derivation was taken.

The oxidation reaction stopped on the stage of carbon without reaching the higher CO_2 stage, it can be explained if some organic materials difficult to oxidize exist in the process, these organic materials have high reduction potential, and the SPM condition was insufficient to be oxidized to CO_2. And then the carbonized particles from organic materials easily attached to the wafer surface indicated that the carbide particles

978-1-5090-6695-7/17 $31.00 © 2017 IEEE

have a strong affinity for the surface of the wafer.

Based on the determination of these two facts, it was hypothesized that the formation of the carbide particles were silicon-containing organics, since the reduction potential of the silicon-containing organics were generally higher than ordinary carbohydrates. And the carbide particles formed from silicon-containing organics have strong affinity for the surface of the wafer.

In order to verify this hypothesis, the concentrated sulfuric acid samples in each line system of SPM method were analyzed with GCMS, the results really validated our vision. The siloxane present of concentrated sulfuric acid in SPM process was far higher than that in blank concentrated sulfuric acid sample. The test results also indicated that the siloxane was generated after concentrated sulfuric acid entering into the SPM process. The possible mechanism how siloxane generated was speculated as follows: On the very strong chemical conditions of SPM method, some particular organics and silicon-containing impurities interacted to form siloxane in concentrated sulfuric acid

The production of siloxane was in concentrated sulfuric acid rather than the presence of organic matter in the surface of the wafer, since the amount of siloxane already increased significantly before SPM liquid contact wafer, according to the results of GCMS.

Quality control of concentrated sulfuric acid and carbide particle pollution prevention

In order to solve the wafer surface particulate pollution problem, it was proposed for the detection of total organic carbon (TOC) and Silicon content (Sc) in concentrated sulfuric acid, and then calculates the siloxane relative value (Sr) to control quality of concentrated sulfuric acid, so as to solve particulate pollution.

Determination of toc and silicon content (S_C)

According to the hydrogen and carbon source of siloxane, TOC value is a visual representation of this factor. TOC analyzer with detection limit of ppm was used. For determination of silicon content (Sc) in concentrated sulfuric acid, taking into account that the silicon elements are in the presence of electronic grade concentrated sulfuric acid with trace amounts. The Inductively coupled plasma mass spectrometry (ICP-MS) was used to detect trace silicon element with the detection limit of the silicon element ppb [5].

Sr values made and calculated

The siloxane generated carbide particle pollution was caused by some uncertain silicon-containing impurities and organics. However, if the silicon content (Sc) and TOC were as the quality index of concentrated sulfuric acid at the same time, the cost will be very high and unnecessary. Forming of carbonized particulate contaminants was relevant to the content of siloxane in concentrated sulfuric acid. If the TOC and Sc value can be used to characterization and calculation of the relative content of siloxane (Sr), to bring about a double detection but single indicator, it can greatly facilitate the production and saving cost.

It was believed that only when the value of TOC and Sc within a certain range, the siloxane could form. This relationship can be expressed as the formula:

$$Sr = A * TOC/Sc$$

In the formula, A is a relative coefficient to generate siloxane, the coefficient associated with the particulate concentration Pc inspected on the surface of wafer that is a function as following:

$$A = f \ (Pc)$$

Depending on the resulting data from different production processes, the function between relative coefficients (A) and the particulate concentration (Pc) can be got, so that the concentrated sulfuric acid products can provide Sr value as a quality indication.

Conclusions

This paper analyzed a new particulate pollution on the surface of wafer, and confirmed that the particulate pollution was generated from carbonization of siloxane, which was formed from the concentrated sulfuric acid of the SPM solution in wet process for wafer treatments. As the generation of siloxane mechanism was relative to TOC and Silica content (Sc), this paper proposed that the siloxane relative value (Sr) in concentrated sulfuric acid could be calculated by detecting the TOC and Sc, so as to achieve quality control of electronic grade concentrated sulfuric acid with double detection but one indicator.

This indicator is also extended in electronic-grade reagents with strong chemical reactivity, such as the concentrated nitric acid, concentrated phosphoric acid, and so on. Using this indicator can greatly reduce the possibility of particulate contamination to the wafer surface.

Illustrations

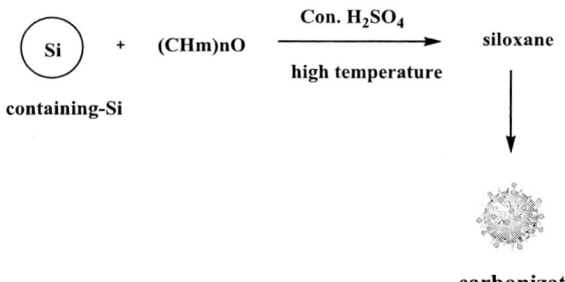

Fig.1. Formation of carbonization particle.

References

[1] Yole developpment. Market Forecast for SiC Devices [EB/OL].http://www.yole.fr/2014-galery-CS.aspx, 2014-09-23

[2] Ho~t alek Mr e' t i n. Buett ner Wemer, Haf ner Rol f, et al. Procedure for the producion of highly pure sulfuric acid: US, 20021 921 44A1[P] 20021 119

[3] PARK J G, HAN J H. The behavior of ozone in wet cleaning chemical: Cleaning technology in semiconductor device manufacturing[J]. J Electrochem Soc Proc, 1998, 197(35):237-234

[4] J Wang T Wang G Pan, et al.Mechanism of GaN CMP Based on H2O2 Slurry Combined with UV Light[J]. ECS Journal of Solid State Science and Technology 2015,4(3):112-117.

[5] CAROLINE F, BERNARD B. Silicon isotope evidence against an enstatite chondrite eart [J]Science,2012,335(23);1477-1480

OPTIMAL EXPERIMENT DESIGN IN POLY ETCH PROCESS FOR PERFORMANCE IMPROVEMENT ON DIFFERENT TYPE TOOL

Ying Emily Lu[1], Wei William Guo[2], and Chingluan Jenny Wu[3]
Corporate Q&R Center, Semiconductor Manufacturing International Corporation,
Shenzhen 518118, China
Corresponding Author's Email: Emily_LY@smics.com

ABSTRACT

In this paper, the solution to optimize poly etch condition on different type tool was studied. In this course, we investigated the potentiality of recipe optimization on different type tool by applying an optimal experiment design (OED) to the poly etch step. In our case, the experiments were carried out with four factors/three levels response surface method (RSM). Base on experiences, we select a, b, c, d parameter as four factors to do the design of experiment. After DOE (design of experiment) data analysis, we finally take the process condition of 64/56/130/6.6 (a / b / c / d) as the optimal poly etch condition for the different type tool.

INTRODUCTION

With the development of semiconductor manufacturing in the world, the competition of low price and high technology became white-hot. In order to control the cost, the most effective method is to increase the capacity at the most. However, the capacity extension always have bottle neck on different type tool qualify, must go experimental design to solve the problem as soon. The big challenge is different tool having different hardware and software setting, and process tuning experience & window before is not able to meet current requirement.

The baseline performance on different type tool can't meet the requirement for production, as show in TABLE I. The bad performance influence the tool's released schedule, and the different tool's process recipe cannot be tuned up immediately with the experience before.

TABLE I BASELINE PROCESS PERFORMANCE BEFORE OPTIMIZATION

Baseline Process Performance before Optimization		
Response Name	Criteria/spec	Measure Data
A	85-89	86
B	85-89	85
C	182+/-3	168
D	30+/-2	27
E	3% (minimize)	5%

In view of the control of cost and efficiency, the optimal experiment design (OED) is an appropriate method when many variables/levels are required to be investigated [1]. OED is an appropriate method for the screening experiments of both linear and non-linear parameters with the minimal experiment times.

EXPERIMENT SETUP

Before implementing DOE, the factors and levels, response, and DOE experiment conditions should be considered firstly.

Factors and levels

Four factors are screened from poly etch step's parameters, as show in TABLE II.

TABLE II DOE FACTORS DESIGN

DOE Factors Design			
PARAMATER	Low	Middle	High
a	60	65	70
b	55	60	65
c	70	100	130
d	5	8	11

Response

Five responses are selected to estimate the performance of poly etch process, as show in TABLE III.

TABLE III DOE RESPONSE

Response	
Response Name	Criteria/spec Measure Data
A	85-89
B	85-89
C	179-185
D	<3% (minimize)
E	28-32

Experiment table

Considering the experiment time, we design 16 times of experiments with 2 center point, and the optimal design can minimize the experiment times [2], as show in Figure

1. The designed experimental points are symmetrical on each plane from Figure 2.

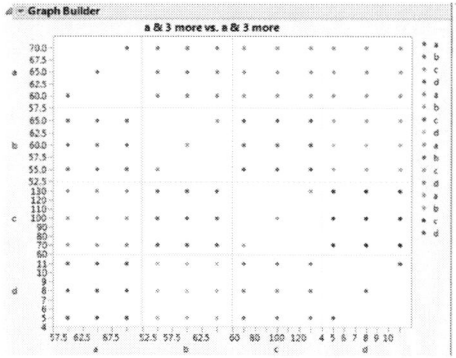

Figure 1: DOE table by JMP

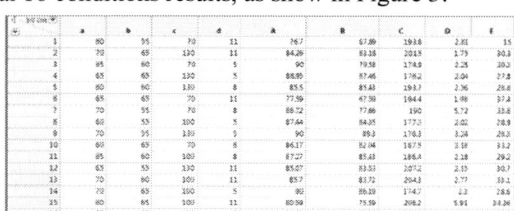

Figure 2: DOE Experiment Chart

RESULTS

Following the methodology described in experiment table, we process the wafers on the settled tool and use a single metrology for all wafers' measurement, and collect total 16 conditions results, as show in Figure 3.

Figure 3: Experiment Result

As shown in Figure 4, the RSquare is more than 0.75, the p-value of ANOVA is less than 0.05, which means that the linearity model of fit is good and successful.

For the lack of fit analysis, the p-value is larger than 0.05, which means no lack of fit, and it's enough to setup model with RSM (response surface method).

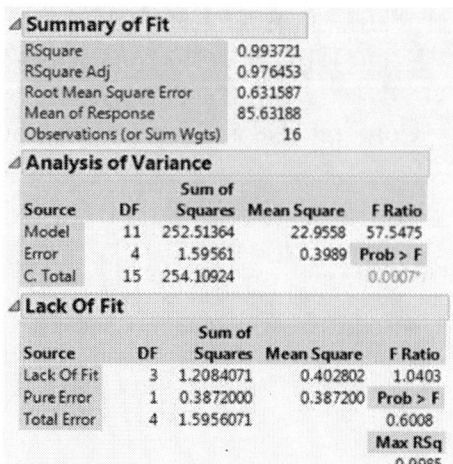

Figure 4: DOE Simulation Status Check

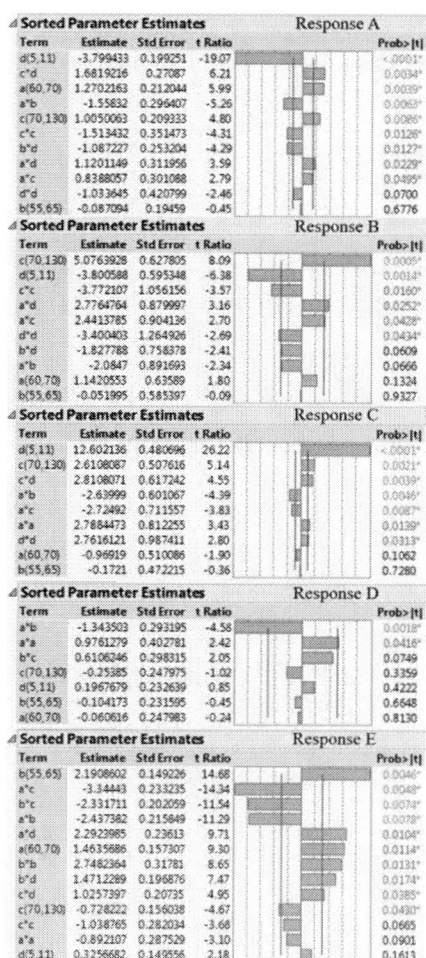

Figure 5: Key Factors and Interaction Checking for Response A, B, C, D, E)

978-1-5090-6695-7/17 $31.00 © 2017 IEEE   46

The sorted parameter estimates shows the parameters which have significant effect from big to small influence, and not all the factors or interactions are significant in the model. From the results above, the parameters which p-value less than 0.01 have significant effect on the relevant response, and c vs. d, a vs. b, b vs. d, a vs. c, b vs. c and a vs. d have strong interaction effects, as show in Figure 5.

According to p-value which less than 0.01, the effective parameters for response A are d, a, c vs. d, a vs. b and c; c and d are the effective parameters for response B; d, c, a vs. b, c vs. d, a vs. c are effective for response C; b, a vs. c, b vs. c, a vs. b are effective for response E; and a vs. b is effective parameter for response D.

According to the model and set desirability for the responses, the best process condition 64/57/130/6.7 (a / b / c / d) can be determined with prediction profiler by maximize desirability, as shown in Figure 6.

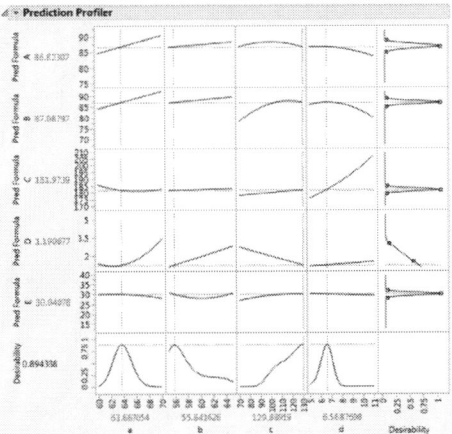

Figure 6: DOE Best Condition Analysis

From the contour profiler charts in Figure 7, we can see that, when selecting the best condition of 64/56/130/6.6 (a / b / c / d), we can also get the wanted response value, and the white area in the picture show that the process window is enough.

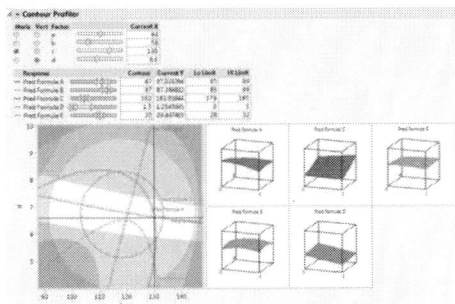

Figure 7: Process Window Analysis

Checking with residual of predicted plot in Figure 8, the residual seems to be distributed randomly near zero,

not show particular structure, and there is no more information to extract. This means that the fitting regression model is adequate.

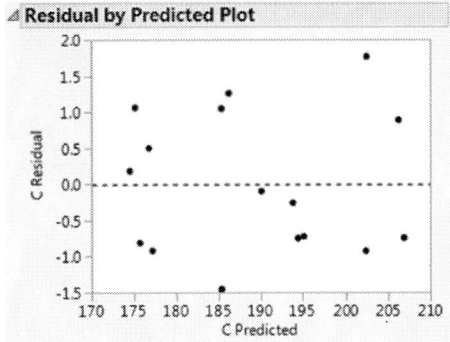

Figure 8: Residual by Predicted Plot

DISCUSSION

According to the DOE analysis results in Section 3, a vs. b interaction is the significant factor to influence response D, c and d main factors are the significant factor to influence response C. Data were collected to verify response C under the best process condition 64/56/130/6.6 (a / b / c / d). The charts in Figure 9 show that the data on new tool meet the requirement and is comparable with baseline tool, which matches the previous DOE results. And the results are proofed by the good WAT (Wafer Acceptance Test) & CP (Circuit Probe) data collected in mass production too. Since the parameter a and parameter b are hard to tune in the actual inline operation, suggest keeping it as fixed parameter. In addition, high level of c or low level of d will induce AA damage for high plasma bombardment and less polymer protection. In order to achieve good yield and avoid side effect, all experimental results must be validated in practice.

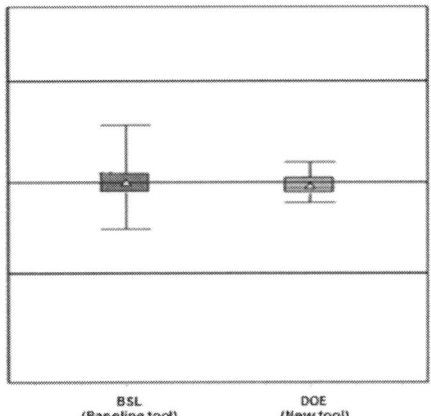

Figure 9: Response verification under DOE condition (New tool vs. baseline tool: result comparable)

CONCLUSION

With the DOE analysis results, the effect of the main factors and their mutual interaction are proofed. The significant factors and window check data will be used for reference in production and point out the direction for process fine tuning. From the contour profiler charts, when fix the parameter of a and b, the window of c and d are enough too, which is 90-140 and 6-8. The effective parameter for Response A/ Response B/ Response C / Response D and Response E will provide the guidance for recipe tuning in following jobs. The optimal design is helpful to find the optimal process conditions [3] so that the different type tool can be qualified quickly and released successfully in advance. We have used DOE to determine the finally process condition of 64/56/130/6.6 (a / b / c / d) as the optimal poly etch condition for the different type tool in mass production. In this case, using the optimal design with the minimal runs save the quantity of wafers and shorten the time of tool release, having important practical significance and sound economic benefits. OED has played a pivotal role in engineering experimental design and the method will be popular for DOE optimization in semiconductor industry in the future.

ACKNOWLEDGEMENTS

Here, I want to appreciate etch engineers' support to implement the experiment and collect the data. And I would like to acknowledge the helpful statistical guidance given by Siyuan Frank Yang, Sheng Randy Kang, Xinyuan Serena Ji.

REFERENCES

[1] Erhu Zheng, Yi Huang, and Haiyang Zhang. *China Semiconductor Technology International Conference (CSTIC)*, Shanghai, March 2012，pp.1-3.

[2] A.C. Atkinson, and V.V. Fedorov. *Biometrika*, vol. 62(2), 1975, pp.289-303.

[3] F.Z. Zeng, *Method and Practice*, Beijing, Weapon Industry Press, 2004.

A Reliability Study of A New Embedded Flash to Reduce Charge-Loss Issue

Lingling Shao, Yong Atman Zhao, Wei Han, Wei-Ting Kary Chien
Semiconductor Manufacturing International Corporation
18 Zhangjiang Road, PuDong New Area, Shanghai 201203, China
(Email: Shani Shao@smics.com)

Abstract

We investigated the mechanism of read stress and standby with power-on after more than 20 program/ erase cycles, which cause conventional embedded flash memory read "0" fail. To solve this, a new e-flash with reversed drain-source cell device was introduced. In this paper, we studied the reliability performance of conventional and the new e-flash. Experimental results proved that the newly designed e-flash exhibits superior performance in terms of data retention, endurance, and the potential at multilevel operations.

INTRODUCTION

In recent years, the market of digital consumer electronics such as mobile smart phones, tablet computers, and digital cameras continue to explode. The excellent properties of flash memories in terms of small sizes, light weight, and strong shock immunity excel the traditional hard disks, and the market demand for flash memories greatly drives the technology development. Compared with conventional NOR flash memories, the conventional embedded flash (e-flash) memory demonstrates higher reliability performance due to the better bit line disturbance, bit line leakages, and over-erase. Because its cell structure contains a split gate [1-2], which serves as an essential building block in microcontroller unit (MCU), it provides a secure non-volatile storage for program, code, and system parameters during periods when the chip is not powered [1].

The conventional e-flash memory uses the floating gate (FG), which is surrounded by dielectrics, to store charges when power is off. The data stored in a flash cell can be determined by measuring the threshold voltage (Vt) of its FG MOS transistor. The way to evaluate data-storage performance is to read the source-drain current driven by the cell at a fixed gate (control gate) bias. As shown in Fig. 1, the current-voltage (Ids-Vcg) plane shows two cells, logic "1" and "0", exhibiting the same trans-conductance curve with a shift by a quantity-the threshold voltage shift (ΔVt) which is proportional to the stored electron charges, Q. Hence, once a proper change and a corresponding ΔVT is defined, it is possible to fix a reading voltage at which the reading current of logic "1" is very high, while the reading current of logic "0" is zero. In this way, it is possible to define the logical state "1" as no electron charge stored in FG and as large reading current, and vice versa for the logical state "0".

The Q that determines ΔVt match is key to erase or program the state parameters. The leakage of the electric field influences the outcome of Q that is used to judge the performance of flash cells. The reason is that the typical state of flash structure at read or standby to the control gate (CG) has always been stressed in an electric field. If this electrical field is applied to FG for a long time, FG will lose the electrical charges, particularly after the cycling stress.

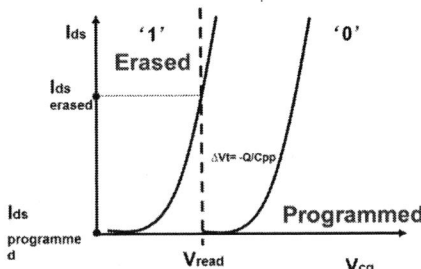

Fig. 1. Read Operation of a Flash Cell

To overcome the charge-loss weakness, researchers developed a new e-flash design with reverse source-drain cells. Experimental data show the conventional e-flash has a strong e-field stress on the tunnel oxide when power is on; whereas the new e-flash has no e-field on the tunnel oxide at standby or at the weak e-field mode under read operations.

In this paper, we studied the reliability performance of the new e-flash.

METHODOLOGY

The reliability test items and conditions were based on the JEDEC standard. After the precondition of 10K Program/ Erase (P/E) cycles at high temperature operation life (HTOL) and low temperature operation life (LTOL), a continuous read stress was applied to the entire cells of the new flash. A continuous read stress post 100 P/E cycles at room temperature (LTOL) was applied to a conventional flash as a comparison.

A summary of test conditions are in Table I.

TABLE I TEST CONDITIONS OF EXPERIMENTAL TRAILS

Item	# P/E cycles	Stress pattern	Operation voltage/V	Time /hrs	Temp. /°C	Reference Standard JESD
Cycling	10Kc for new IP	All0/All1	1.1*Vcc	NA	25	A117
	100c for old IP					
HTOL	NA	CKB	1.2*Vcc	500	125	A108
LTOL	NA	CKB	1.2*Vcc	500	25	A108

978-1-5090-6695-7/17 $31.00 © 2017 IEEE

RESULTS

A. Failure Mechanism of Conventional Flash

As shown in Fig. 2, the read cell FBC's at different read points and different Vt voltages indicate the V_T degradation of the conventional e-flash is at 0V after LTOL 500 hours (LTOL-500) stress post 100 P/E cycles. This implies there is nearly no margin when reading data '0' from bits with V_T close to 0V.

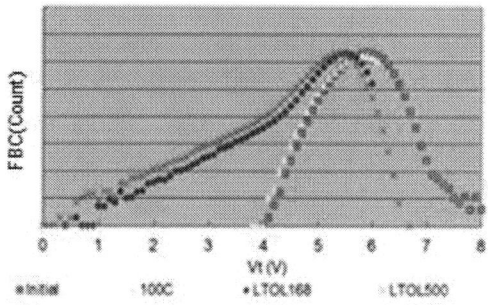

Fig. 2. The V_T distribution of a conventional e-flash with LTOL stresses post 100 P/E cycles

The charge loss mechanism of a conventional e-flash is shown in Fig. 3. Tunnel oxide traps are formed after P/E cycling. When the traps occasionally "line up", they form a channel between FG and the silicon substrate in order to promote tunneling. In addition, an e-field stress on the tunnel oxide results in electron run away from FG to silicon through the leakage channel. This mechanism is also called stress induce leakage current (SILC) [4]. Electrons or holes tunnel from one electrode to the other through traps generated under stress. The P/E cycling generates traps, and the resulted SILC then causes VT drifts. Programmed flash cells may lose electrons (charge loss) and erased Flash cells may gain electrons (charge gain).

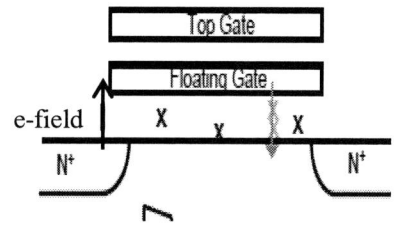

Fig. 3. The SILC mechanism of an e-flash with conventional e-flash cells

B. Reliability Performance of New Flash

As shown in Fig. 4, read cell fail bit counts (FBC's) at different read points and different Vt's show the V_T distribution of LTOL-500 post 10K cycling for new e-flash cells. The V_T degradation of LTOL-500 is at 2.2V. The main degradation is caused by the SILC induced tail bits. The new e-flash cells have a sufficient margin after LTOL-500.

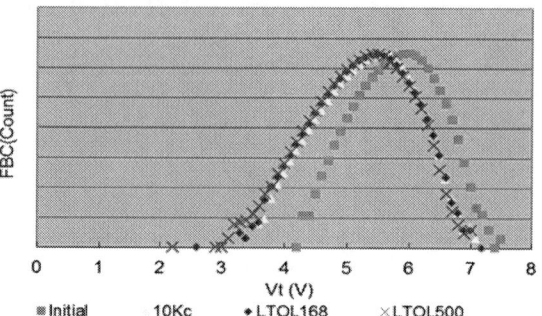

Fig. 4. The V_T distribution of new e-Flash with LTOL post 10K P/E cycles

Fig. 5 illustrates the V_T distribution of HTOL 500 hours (HTOL-500) post 10K cycling for the new e-flash cells. V_T degrades at 3.3V after HTOL-500. Compared with LTOL-500, a main peak shift rises to the V_T degradation of HTOL-500. As traps generated in the tunnel oxide post P/E cycling can be de-trapped at high temperatures [5], no degradation of the tail bits appears after HTOL stress. The main peak shift of HTOL after 10Kc cycling is caused by some intrinsic degradation mechanisms rather than SLIC: electrical charges stored in the FG is activated at high temperatures and are able to run away from FG [6].

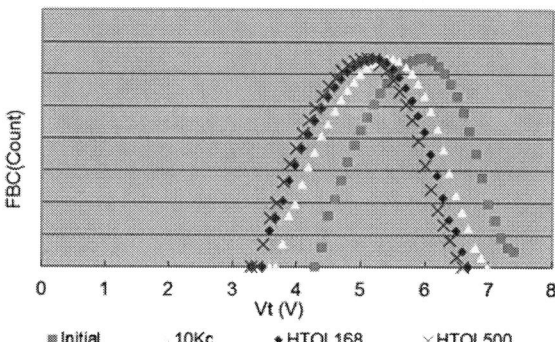

Fig. 5. The V_T distribution of new e-flash with HTOL post 10K P/E cycles

DISCUSSIONS

Fig. 6 shows the read operation and standby bias condition of conventional embedded flash cells. The e-field stress stays strong at the tunnel oxide when power is on. So the e-flash with conventional cells inevitably suffers charge loss.

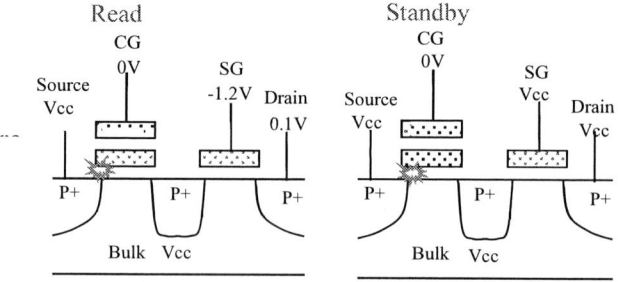

978-1-5090-6695-7/17 $31.00 © 2017 IEEE   50

Fig. 6. The schematic of read operation and standby bias condition of conventional embedded flash cell

A schematic plot of Vcc stress for new embedded flash cells with the reversed drain-source [7] is shown in Fig. 7. There is no electrical field on the tunnel oxide at standby, and the strength of the e-field is also greatly reduced at the read operation due to the distance between the source and the FG. As a result, the new e-flash IP is immune to the charge loss problem at read or standby mode after P/E cycling. Compared with conventional e-flash cells, the V_T degradation caused by SILC under LTOL is greatly improved for the newly designed e-flash.

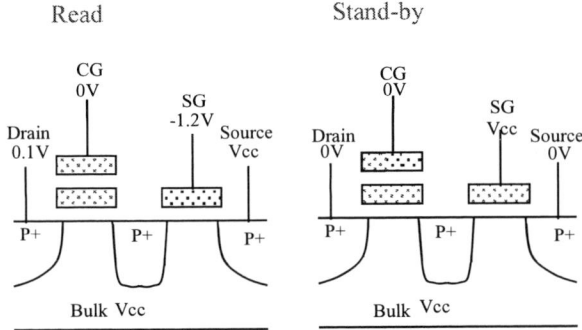

Fig. 7. The schematic of read operation and standby bias condition of new embedded flash cell

CONCLUSION

In this paper, we conducted reliability tests to compare the reliability of a conventional and of a new e-flash cell. The experiments showed that the V_T degradation is mainly induced by tail bit under LTOL post 10K cycling. However, the main peak shift gives rise to the V_T degradation under HTOL post 10K cycling. These results serve as the evidences that new e-flash cells can overcome the charge-loss weakness due to the reverse source-drain cell. The reliability of the proposed new e-flash is much better than that of a conventional e-flash. This is because the decrease of Vcc stress on the tunnel oxide in the new e-flash cell can prevent charge loss at read or standby post cycling. The new e-flash with the reverse source-drain cell device exhibits superior reliability in terms of longer data retention, better endurance, and has great potentials in multilevel applications.

ACKNOWLEDGMENTS

The authors would like to express sincere thanks to Zhen Yang, Lei Zhao from SMIC TD and Xianfeng Chen, Yulin Ding from SMIC Product Reliability Department for helpful technical discussions and the strong supports on tests.

REFERENCES

[1] R. Strenz, "Embedded Flash Technologies and Their Applications: Status & Outlook", *Proc. IEEE Int. Electron Devices Meeting (IEDM)*, pp. 211–214, Dec. 2011.

[2] S. T. Kang, B. Winstead, J. Yater, M. Suhail, G. Zhang, C. H. Hong et al, "High Performance Nanocrystal Based Embedded Flash Microcontrollers with Exceptional Endurance and Nanocrystal Scaling Capability", *Proc. 4th IEEE Int. Memory Workshop (IMW)*, pp. 1–4, May 2012.

[3] R. Bez, E. Camerlenghi, A. Modelli, A. Visconti, "Introduction to Flash Memory", *Proceedings of the IEEE*, Vol. 91, Issue 4, pp. 489-502, 2003.

[4] N. K. Zous, T. Wang, C. C. Yeh, C. W. Tsai, C. Huang, "A Comparative Study of SILC Transient Characteristics and Mechanisms in FN Stressed and Hot Hole Stressed Tunnel Oxides", *IEEE 37th Annual International Reliability Physics Symposium*, pp. 405-409, 1999

[5] N. Mielke, H. P. Belgal, A. Fazio, Q. Meng, N. Righos, "Recovery Effects in the Distributed Cycling of Flash Memories", *IEEE International Reliability Physics Symposium Proceedings*, pp. 29-35, 2006.

[6] R. Fastow, K. Ahmed, S. Haddad, M. Randolph, C. Huster, P. Hom, "Bake Induced Charge Gain in NOR Flash Cells", *IEEE Electron Device Letters*, pp. 184-186, 2000.

[7] Y. Chan, K. Kim, G. Zhao, X. Ye, Z. Yang, "Method for Setting A Flash Memory for HTOL Testing", China Patent 201410195902.5 [P]. 14th May, 2014.

978-1-5090-6695-7/17 $31.00 © 2017 IEEE

IDLE RECIPE AUTOMATIC CONTROL

Weiwei Yan[1], Hongtao Qian[1], Zhiqian Liu[1] and Jiuzhou Zhao[1]*

[1]Department of Process Quality Engineering, Semiconductor Manufacturing International
Corporation, Shanghai, P.R. China
*Corresponding Author's Email: Vivian Yan@SMIC.com

ABSTRACT

For semiconductor manufacturing, EQ (equipment) performance maybe shift to different process level after long idle time. Wafer quality will be ventured into worse condition processing idle recipe. System is thus set up to reduce the potential risk. Long time idle recipe, which idle more than specific idle spec, is under control by setting up Idle Recipe Control (IRC) system. All the process tools will be involved in this function and recipe status is real time refreshed and updated according to EAP (Equipment Automation Program) successfully processed lot information. Idle recipe will be hold and alert to engineer, and lot couldn't be processed unless risk assessment had been done. Wafer quality thus can be effectively controlled and improved.

KEYWORDS

Idle recipe control, quality, methodology

INTRODUCTION

For semiconductor manufacturing, EQ and recipe are indispensable and most important factors. And what we focus on is well controlling tool and recipe performance to keep and improve production quality.

Most effort had been done on EQ management [1-2], such as parts quality control, SPC (Statistics Process Control) application to real time monitor tool performance, life time definition [3], periodic procedure maintenance [4], and so on. However, EQ performance maybe shift to different process level after long idle time due to parts aging, function decay, or other unquenchable reason along with time changing. It's high risk for mass production proceeding idle recipe, and lots can be scrapped or big yield loss can occur.

Also, recipe management is developed along with manufacturing techniques becoming more and more complex and advanced. However, Most of the existing recipe management methods [5-8] restrict on the contents of recipes. Thus, idle recipe need under control and wake up mechanism need be set up to early detect or prevent the abnormal situation.

In this paper, Idle Recipe Control (IRC) system with control methodology is introduced.

FORMATTING THE PROCEDURE OF IDLE RECIPE CONTROL
Idle Recipe Related Items Definition

(1) Idle time:

Recipe idle time can be calculated according to production lot track in time, which is linked with tool or chamber. That's the interval time of production recipe last track in time and now time. Detail is as below:

Recipe1 idle time in tool/chamber A= Current time-last track in time of recipe 1 of production lot in tool A.

Recipe2 idle time in tool/chamber B= Current time-last track in time of recipe 2 of production lot in tool B.

To verify recipe safe period of idle time, long term performance of fab is analyzed including inline SPC, defect, WAT (Wafer Acceptance Test), yield, reliability, considering process, tool type and other parameters impacting wafer quality.

Also idle period should be considered to be refined based on fab real time performance.

(2) Handling Procedure for Idle Recipe

For idle recipe release procedure control, same recipe body is defined base on key parameters valuation. All parameters of recipe and tool hardware configure related were listed for risk assessment. Recipe group were thus classified according to key parameters setting, and among which recipes with same recipe body will be all included. In other words, recipe group is set up for recipes with same recipe body, and idle recipes can be released base on each other in the same group. Take ETCH for example, all the key parameters like helium pressure, temperature, gas flow, need be exactly same exclude etch time, which is defined as ETCH same recipe body.

For risk assessment of idle recipe, ECN (Engineer Chang Notice) needs be issued if there is no recipe with the same recipe body running in a specific period of time. ECN evaluation items need be defined for process release standard. STR (Special Test Request) and MSTR (Mass STR) lots will be approved accordingly to run the idle recipe, and more data will be collected for verification. For example, if WAT is comparable using statistic method of Harmonization Confidence. And idle recipe can only be released after all data pass the criteria.

The Figure 1 shows the brief handling procedure of idle recipe.

System Introduction and Main Functions

To realize automatic control, system had been set up to control idle recipe. The Figure 2 describes the design of IRC system with brief architecture, and detailed design will be introduced later.

All the process tools will be included in the system for

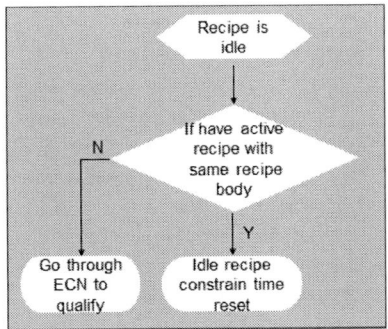

Figure 1 Flowchart of Idle Recipe Handle

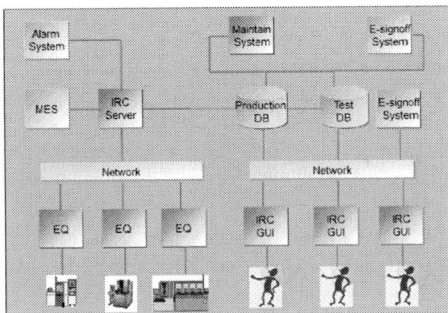

Figure 2 System Architecture

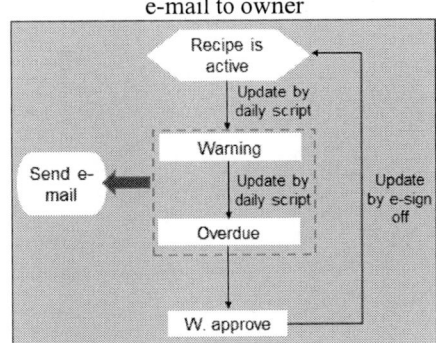

Figure 3 Time Chart of Idle Recipe

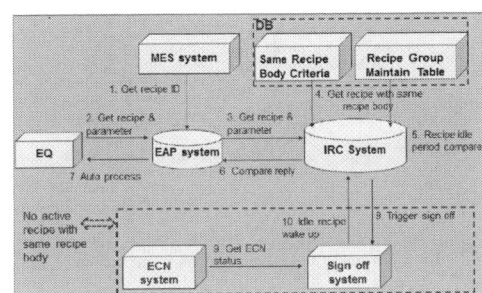

Figure 4 IRC System Design Flow

production recipe control, and system's main functions are as below:

(1) All production lot is in the control loop, by defining production lot type and setting mapping table to check the flag. IRC need be checked before lot production if the flag is true.

(2) Real time refreshes and updates recipe process time through IRC server.

(3) Calculation will then automatically done to set constrain time adding the safe period.

(4) Automatically refresh constrain time of the recipe group synchronously.

(5) DB (Database) update needs be through e-sign off flow following same recipe body definition criteria.

(6) Control STR/MSTR lot count and run period following ECN procedure.

(7) Reset the recipe idle time and wake up the idle recipe according to e-sign off flow when related ECN released.

(8) Send warning e-mail to related sponsor before constrain formally in advance.

The Figure 3 shows the time line of recipe changing process, which will be idle if no lot processed in a safe period.

To realize functions being implemented automatically, system integration is needed. Information needs be collected from other production systems, to verify tool, chamber and recipe status. Also dictating actions will be sent for idle recipe control: send warning

based on screen EQ/recipe information comparison; constrain recipe and stop lot track in for recipes out of idle spec; reset recipe idle time following wake up criteria through sign off system and so on, as expressed in figure 4.

(1) EAP system: As we all know, production flow is controlled by MES (Manufacturing Execution System system) to auto select recipe, which is combined with EQ to form EAP system. In EAP system, recipe and parameters are automatically collected. IRC system will record every single production recipe according to EAP successfully processed lot information, setting constrain time at the same time. Also lot can't be tracked in and will be hold by EAP system when recipes were constrained.

(2) DB: Database is maintained for recipe group, to automatically refresh recipes with the same recipe body. That's other recipes status in the same group, namely recipes with the same recipe body, can be automatically refreshed at the same time. There are two ways for DB maintenance, and defining the same recipe body criteria is one method. Take IMP for example, key parameters are defined in recipe name following standard naming rule. Thus IRC system can automatically compare recipe body and search for the recipes through recipe name. However, the method can only be applied to recipes, whose key parameters can be directly identified by the system for recipe body comparison. We could maintain recipe names into the group and update through sign off system when new recipes arrive.

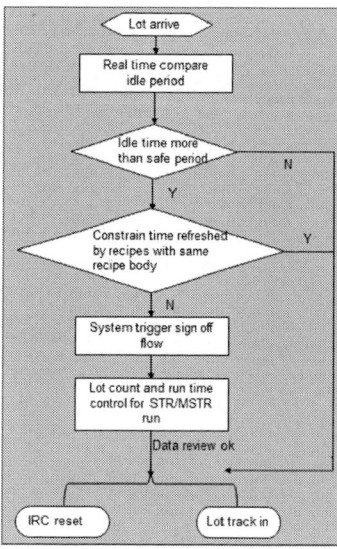

Figure 5 Work flow

(3) Alarm system: Weekly compare screen EQ/recipe information of which the idle time is in warning stage (will be out of idle spec in one month), and send reminding E-mail to related sponsor before formal constraint in advance.

(4) Sign off system: IRC will trigger e-sign off to release constrain recipe, following standard wake up rule and implementing in ECN system.

Work flow Introduction

In Figure 5, the flow chart of the control method is illustrated with the main processes, combing procedure and system function introduced before.

When lot arrives at tracked in, IRC system will compare the idle time with the safe period maintained in DB,

(1) If the idle time is less than safe period, lot will successively track in in the EAP system and the new track in time will be recorded accordingly. Also constraint time will be refreshed accordingly for the whole group.

(2) If the idle time is more than safe period, and recipe is constrained as there is no active recipe in the group, lot will be hold before recipe qualification. For recipe constrained by IRC system, waking up rules need be followed to release the recipe. Customer's ECN rules are followed to define risk level and implement related qualification plan through ECN system. And sign off system are auto linked to partially release idle recipe running STR/MSTR. After idle recipe qualification finished, constraint time will be refreshed according to the recipe release time.

Also reminding function is included in IRC system to send alert mail to owner ahead of constraint time. And lot won't be hold if the idle recipe can be released following waking up rules, preventing unnecessary move loss

CONCLUSIONS

This paper designs and optimizes production recipe control method. The proposed method improves the recipe management and the optimized system is useful in semiconductor manufacture and other applications where parts aging or function decay occur along with mass production. In the future work, system enhancement will be continually done to better and more smartly control production recipe, and wafer quality is guaranteed.

Using the Idle Recipe Control system, idle recipe is automatically controlled and production quality is improved to a high level.

REFERENCES

[1] Z. Binghai, X. Lifeng and C. Jianguo, "Knowledge-based decision support system for tool management in flexible manufacturing system," Journal of Systems Engineering and Electronics ,Year: 2004, Volume: 15, Issue: 4 , Pages: 537 - 541

[2] G. Wang, Y. Yan, H. Nakajima, X. Zhang and H. Li, "A visualized cutting tool management pattern for flexible manufacturing systems," Industrial Engineering and Engineering Management, 2009. IEEM 2009. IEEE International Conference on, Year: 2009 , Pages: 1925 – 1929.

[3] G. Wang , H. Nakajima, Y. Yan and X. Zhang, "A methodology of tool lifecycle management and control based on RFID," Industrial Engineering and Engineering Management, 2009. IEEM 2009. IEEE International Conference on Year: 2009 , Pages: 1920 - 1924,

[4] J. R. Morrison, Hungil Kim and A. A. Kalir "Mean cycle time optimization in semiconductor tool sets via pm planning with different cycles: A G/G/m queueing and nonlinear programming approach," Simulation Conference (WSC), 2014 Winter Year: 2014, Pages: 2466 – 2477.

[5] Qian Wang and Zhijie Wang, "Design and Implementation of a Novel Recipe Management Component," 2010 International Conference on Information, Networking and Automation (ICINA).

[6] Hossam A. Gabbar, Atsushi Aoyama, Yuji Naka, "Automated solution for control recipe generation of chemical batch plants," Computers & Chemical Engineering, vol. 29, pp. 949-964, April 2005.

[7] D. de Kerf and M. Jhaveri, "The benefits of intelligent recipe management," Pulp and Paper Industry Technical Conference, 2002. Conference Record of the 2002 Annual.

[8] Aoyama A., Yamadai Isao, "Multi-dimensional object oriented approach for automatic generation of control recipes," Computer and Chemical Engineering, vol. 24, pp. 519-524, July 2000.

978-1-5090-6695-7/17 $31.00 © 2017 IEEE   54

Investigation of CMOS Image Sensor Dark Current Reduction by Optimizing Interface Defect

Wuzhi Zhang[1], Zhengying Wei[1], Yansheng Wang[1], Wei Zhou[1], Chang Sun[1], Jun Qian[1]*
Yuhang Zhao[2]

[1]Haili Microelectronics Corporation, Shanghai 201203, China
[2]Huahong Group, Shanghai 201203, China
*Corresponding Author's Email: zhangwuzhi@hlmc.cn

ABSTRACT

Dark current (DC) was one of the most critical parameters of CMOS image sensors (CIS), and interface defects during semiconductor fabrication process dominate the DC performance. The research investigated Tx Negative-Bias / P-Well and P+ IMP in this paper, and achieved extreme low DC at high temperature of 60 °C. Firstly, Tx negative bias was used to restrict the Poly/Gate OX/Si substrate interface defects. The DC reduced 83.9% while -0.7 V Negative-Bias implemented on Tx. Secondly, P-Well IMP conditions were studied for reducing the Interface defects of shallow trench isolation (STI). The DC could decrease 39.8 mV/s by increasing Boron dosage of P-Well. Thirdly, photodiode surface IMP (P+) was researched. The suppression of DC induced by PD surface interface defects would decrease 20 mV/s with experimental condition.

Index Terms: CMOS Image sensor (CIS), dark current, negative bias, P-Well, P+ IMP, four–transistor pixel, pinned photodiode (PPD).

INTRODUCTION

COMS Image Sensor, as lower power consumption, higher signal processing speed and better technology compatibility with CMOS device, had been widely applied for many field, such as cell phone, automotive equipment, traffic monitoring, digital cameras, and unmanned equipment [1]. However, dark current levels in CIS's were still more than an order of magnitude larger than those of the CCD sensors, as the result of that, dark shot noises caused by dark current deteriorated the image quality obviously, especially, the noise got worse with the pixel size shrink-down [2][3][4].

Based on different recent studies, several theories were provided to explain the contributor for dark current. First, trap of Si/SiO$_2$ interface defects was the main source of dark currents in the normal operation mode, thermal generation or recombination of the free minority carriers were existed as the trap of interface defects create an energetic step in the band gap of semiconductor [5][6], while the interface in traditional CIS included: Poly / Gate OX / Si substrate and shallow trench isolation (STI) showed as region A and region C in Fig.1. , They were the main origins of dark current [7][8][9][10][11]. Second, Defects of pinned photodiode (PD) surface as region B in Fig.1. was another significant resource for dark current. There were some factors could induce such defects, one of them was metallic contamination, metals introduced deep levels in photodiodes silicon bandgap that generates dark current levels. It [12] reported that tungsten contamination degrades the minority-carrier recombination lifetime and there was a correlation between the minority-recombination lifetime degradation and the leakage current in a photodiode. The other one for surface defects was Plasma Induced Damage, research [13][14] explained that photo generation phenomenon in the dielectric nitride layer, induced by the plasma UV, and assisted by the wafer surface charge, and this mechanism leaded to a positive fix charge creation on the pixel surface and induces dark current.

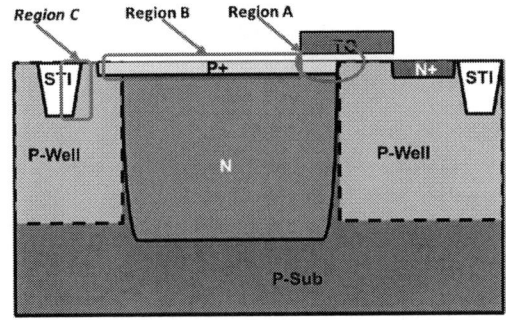

Fig. 1. Cross section of CIS pixel area, Region A, B and C was the dominate resource for DC.

EXPERIMENTAL

A 4-T pinned N-photodiode (PD) active pixel sensor was prepared by 55 nm technology, Front-Side

Illumination (FSI) sensor, the device pixel count was 728(H) × 1288(V) with 4.2 um x 4.2 um in pitch size.

The dark signal of pixels was tested under totally no luminous environment with 60 ^{O}C, and the cumulative populations of the dark signals for pixels were plotted with time expand, then slope of dark signal/time curve was defined as dark current. The detection and analysis tool was Teradyne's IP750Ex.

RESULTS AND DISCUSSION

A. Negative-Bias of Tx

Fig. 2 showed the slope of Vout of PD by exposure time under nonluminous environment, which was defined as dark current, dropping as the increasing of Tx negative bias. The initial dark current value was 632.8 mV/s without Tx negative bias, and that value dropped drastically as the negative bias implementing, the values was 535.1, 275.6,178.7 and 101.5 mV/s corresponding to -0.25, -0.5, -0.6 and -0.7V of Tx negative bias respectively. The reduction of the dark current was attributed to the accumulated holes near the P+ implant surface under the Tx transistor gate area (region A), The result indicated that the region A was the main reason for dark current generation.

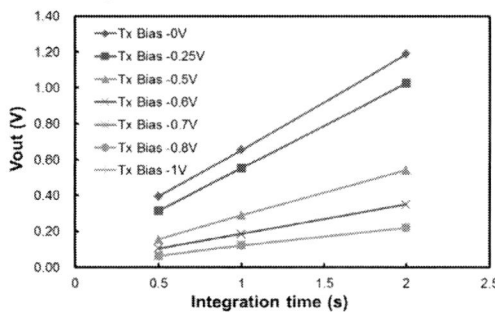

Fig. 2. Vout of PD by exposure time with changing Tx negative bias under nonluminous environment.

But meanwhile, the dark current did not decrease any more as the negative bias got higher than -0.7V, it was caused by the limited electrons generated by defects at the Si/SiO$_2$ surface, the count of hole formed by Tx negative bias become more and more as the voltage increased, and the recombination got saturation eventually, no more active dark electrons would produce the dark current.

B. P-Well IMP

The dark current was mainly generated from the interface defects, and some of them located in the sidewall of the STI. Fig. 3. showed the dark current value with different P-Well condition under -0.7 V of Tx

negative bias. Dark current value was 101.5 mV/s when the P-Well condition was B1E12 (Boron, Dose 1.0E12), while the value changed to 80.6 and 61.7 mV/s when the P-Well condition was B3E12 and B6E12. Dark current reduced obviously by the P-Well dose increasing, it caused by the defects located in the sidewall of STI was recombined by hold of P-Well, and this passivating force was enhanced as the P type dose increasing.

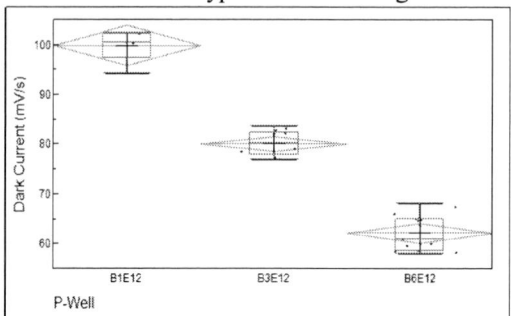

Fig. 3. Dark current value with different P-Well condition under -0.7 V of Tx negative bias.

C. PD surface P+ IMP

It was reported [14]that other significant dark current's resource was located at PD surface, which had relation with heavy metal contamination, interface trap, lattice damage and plasma-induced damage.

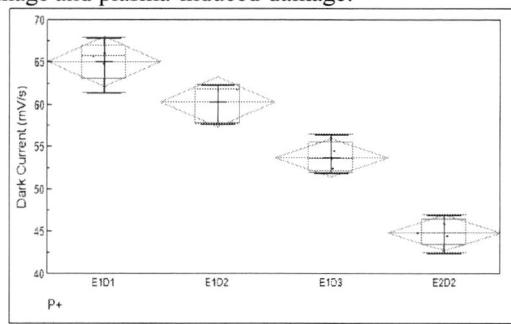

Fig. 4. Dark current performance under different P+ condition with -0.7 V of Tx negative bias.

Fig. 4. was the dark current performance under different P+ condition with -0.7 V of Tx negative bias. The dark current mean value was 65.0 mV/s with P+E1D1, and the value reduced to 60.4 and 53.8 mV/s when only raised dose of P+ to D2 and D3 respectively. The experimental result also showed that the energy (Depth of P+) could impact the dark current performance obviously, the higher energy of P+ (E2) obtained the lower dark current (44.9 mV/s). The result matched the previous statement, whatever higher energy or more doses would strengthen the PD surface internal electric field to overcome the impact by heavy metal contamination, plasma-induced damage and et al.

SUMMARY AND CONCLUSION

Based on the theory of dark current generated by Si/SiO$_2$ interface defects, three actors were considered and implemented on 4.2 um x 4.2 um pixel 4-T CIS development. Firstly, Poly / Gate OX / Si substrate interface was the dominate resource, the dark current deceased seriously when the Tx negative bias raised, it total reduced 531.3 mV/s as the implemented of -0.7 V Tx negative bias. Secondly, P-Well IMP could suppress the dark current producing from sidewall of AA/STI, 39.8 mV/s was reduced when the P-Well Boron dose changed from 1.0E12 to 6.0E12, more effort would be taken under this conclusion to improve the dark current further. Thirdly, P+ IMP for PD surface dark current suppressing also showed remarkably effective, not only concentration of dosage, but also the doping depth would impact the DC, more detail attempt of this achievement would be experimented under next stage.

ACKNOWLEDGMENT

The authors were grateful to W. Zhou, X. Q. Gu, C. H. Fan and P. C. Xi from Shang Hai ICR&D for their help with packaging, testing and experimental issues. The authors also thank D. M. Sun, C. Li and J. X. Wen from ICR&D for their help with simulation and circuit design.

REFERENCE

[1] E. Gamal and H. Eltoukhy, "CMOS image sensors," IEEE Circuit. Devic., vol. 21, pp. 6-20, May-Jun. 2005.

[2] H. Kwon, I. Kang, B. Park, J. Lee, and S. Park, "The analysis of dark signals in the CMOS APS," IEEE Trans. Electron Devices, vol. 51, no. 2, pp. 178–184, Feb. 2004.

[3] B. Pain, T. Cunningham, B. Hancock, C. Wrigley, and C. Sun, "Excess noise and dark current mechanisms in CMOS imagers," in Proc. IEEE Workshop Charge-Coupled Devices Adv. Image Sensors, pp. 145–148, 2005.

[4] N. V. Loukianova, H. O. Folkers, J. P. V. Maas, D. W. E. Verbugt, A. J. Mierop, W. Hoekstra, E. Roks, and A. J. Theuwissen, "Leakage current modeling of test structures for characterization of dark current in CMOS image sensors," IEEE Trans. Electron Devices, vol. 50, no. 1, pp. 77–83, Jan. 2003.

[5] V. Goiffon, C. Virmontois, P. Magnan, Member, S. Girard, and P. Paillet, Senior, "Analysis of Total Dose-Induced Dark Current in CMOS Image Sensors From Interface defects and Trapped Charge Density Measurements," IEEE Trans. Nuclear Science, vol. 57, no. 6, Dec. 2010.

[6] S. S. Park, and J. D. Lee, "The Effects of Deuterium Annealing on the Reduction of Dark Currents in the CMOS APS Hyuck In Kwon, O. Jun Kwon, Hyungcheol Shin, Byung-Gook Park," IEEE Trans. Electron Devices, vol. 51, no. 8, Aug. 2004.

[7] Y. Kunimi and B. Pain, "Consideration of dark current generation at the transfer channel region in the solid state image sensor," in Proc. Int. Image Sensor Workshop, pp. 66–69, 2007.

[8] S. H. Park, J. D. Bok, H. M. Kwon, W. I. Choi, M. L. Ha, J. I. Lee, and H. D. Lee, "Decrease of Dark Current by Reducing Transfer Transistor Induced Partition Noise With Localized Channel Implantation," IEEE Trans. Electron Device Letters, VOL. 31, no. 11, pp. 1278–1280, Nov. 2010.

[9] I. Inoue, N. Tanaka, H. Yamashita, T. Yamaguchi, H. Ishiwata, and H. Ihara, "Low-leakage-current and low-operating-voltage buried photodiode for a CMOS imager," IEEE Trans. Electron Devices, vol. 50, no. 1, pp. 43–47, Jan. 2003.

[10] M. W. Seo, S. J. Kawahito, K. Yasutomi, K. Kagawa, and N. Teranishi, "A Low Dark Leakage Current High-Sensitivity CMOS Image Sensor With STI-Less Shared Pixel Design," IEEE Trans. Electron Devices, vol. 61, no. 6, Jun. 2014.

[11] C. R. Moon, J. W. Jung, D. W. Kwon, J. Yoo, D. H. Lee, and K. Kim, "Application of Plasma-Doping (PLAD) Technique to Reduce Dark Current of CMOS Image Sensors," IEEE Trans. Electron Device Letters, vol. 28, no. 2, Feb. 2007.

[12] S. H. Song, I. H. Kim, G. S. Lee, and J. G. Park, "Impact of tungsten contamination on the sensing margin of a CMOS image sensor cell," Jpn. J. Appl. Phys. 54, pp.016501-1–016501-6, 2015.

[13] O. Mitsuru, et al, "Ultraviolet-induced damage in fluorocarbon plasma and its reduction by pulse-time-modulated plasma in charge couple device image sensor wafer processes," Journal of Vacuum Science&Technology B, vol.22, no 6, p.2818–2822, 2004.

[14] K. Tokashiki , K. H. Bai, K. H. Baek, Y. J. Kim, G.J. Min, C. J. Kang, H. K. Cho, J. T. Moon "Study of plasma charging-induced white pixel defect increase in CMOS active pixel sensor," Thin Solid Films 515, pp.4864–4868, 2007.

978-1-5090-6695-7/17 $31.00 © 2017 IEEE

A NOVEL 25V DDD NMOS DESIGN IN 500-700V ULTRA HIGH VOLTAGE BCD PROCESS

Donghua Liu[1,2], Zhaozhao Xu[2,3], Wenting Duan[2], Feng Jin[2], Wenqing Yang[2], Huihui Wang[2], Jun Hu[2], Jiye Yang[2], Wensheng Qian[2], David Wei Zhang[1]*

[1] State Key Laboratory of ASIC and System, School of Microelectronics, Fudan University, Shanghai 200433, China

[2] HuaHong Grace Semiconductor Manufacturing Corporation, Shanghai 201206, China

[3] State Key Laboratory of Functional Materials for Informatics, Shanghai Institute of Microsystem and Information Technology, Chinese Academy of Sciences, Shanghai 200050, China

*Corresponding Author's Email: donghua.liu@hhgrace.com

ABSTRACT

This article reports a high voltage DDD NMOS with 25V operating voltage and 40V breakdown voltage. Usually the DDD MOS in the logic process has a typical breakdown voltage lower than 18V. The further application requirement in higher breakdown voltage adopts LDMOS structures. In order to achieve high enough on-state breakdown voltage, the DDD NMOS presented in this paper has relatively high (about 1e13/cm2) doping level in the diffused drain region. In the case of the diffusion drain region with high level doping, the off-state breakdown voltage is affected, but the DDD NMOS in this paper adopts a novel design. Thus, the on/off-state breakdown voltage can reach 40V. The optimized device has excellent input/output characteristics and excellent on-resistance characteristics. The Ioff of the device is slightly larger and will be further optimized.

INTRODUCTION

As the widely use of panel display and portal electric devices, the high voltage IC is getting more and more attention due to their application in LCD driver and power management. In the latest applications, the DDDMOS (double diffused drain MOSMET, DDDMOS) is adapted to control the high voltage signal of the IC chips. The channel of DDDMOS uses metal-oxide-semiconductor structure and which is same as normal MOSFET. In order to sustain high voltage, the drain adopts lightly doped structure, which likes the drift region of power MOSFET. It uses two times or above ion implantation to form this drain structure. That's why it's called double doped drain MOSFET. For different application request, the channel length, Vt and other parameters can be tuned separately[1] [2] [3].

The breakdown voltage of common DDDMOS is around 18V. Some researchers had reported the DDDMOS with breakdown voltage higher than 30V. But it focuses on the study of device performance degradation. The detailed data of electric performance was not found in that paper [1]. This paper studies a DDDMOS, its off-state breakdown voltage (off-BV) can be higher than 40V and

on-state breakdown voltage (on-BV) can be higher than 30V. Input-output performance is excellent and it can satisfy the application above 30V.

DDDMOS DEVICE STRUCTURE

When a bias is applied to the drain, the resistance of drift region will increase due to depletion region widening and kirk effect while the channel voltage is unchanged. Thus, the peak electric field is reduced and eventually we can increase the breakdown voltage[4].

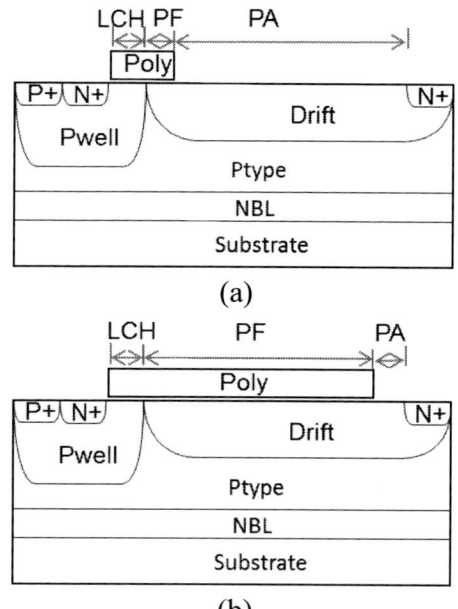

Figure 1: DDDMOS device cross section (a) conventional; (b) this work

Figure 1 shows the cross section of the DDDMOS in this paper. Figure 1(a) is a conventional DDDMOS structure which is shown here as comparison. Figure 1(b) is the novel structure in this work. This device is fabricated with ultra-high voltage 700V BCD process. As with the

700V BCD device, the gate oxide thickness of our DDDMOS is 850A. The thicker gate oxide is one of the key factors for achieving high voltage DDDMOS. For example, a device with thinner gate oxide cannot operate at or above 30V gate voltage. Meanwhile, thicker gate oxide is also helpful to improve breakdown voltage by reducing the electric field at poly gate edge in drain side.

This device also is a drain isolated DDDMOS. Drift region is formed by n-type epitaxial layer and n-implantation region. Under drift region, there are Ptype layer and NBL layer. The Ptype layer isolates the drain (including drift) from the bulk. The reference length of channel and drift region is 1.5um and 2um, respectively.

RESULT AND DISCUSSION

In this process platform, firstly we made a 18V DDDMOS which has excellent performance. But when we try to design a higher breakdown voltage DDDMOS with the same structure, it's hard to get an ideal performance. Especially we hardly can get higher off-state and on-state breakdown voltages simultaneously. When the off-state breakdown voltage meet requirement, the on-state breakdown voltage is smaller than operating voltage. As shown in Figure 2, 20V DDDMOS have excellent electronic characteristics. Figure 3 is characteristics curves of the 30V DDDMOS. The on-state breakdown voltage is not high enough.

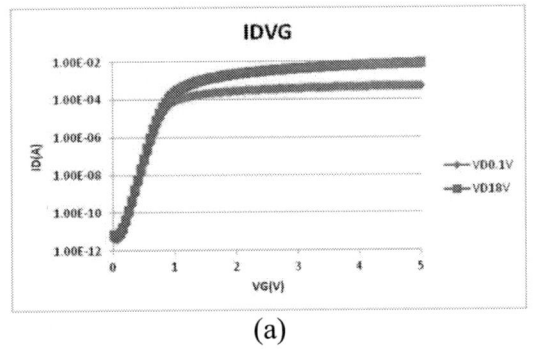

(a)

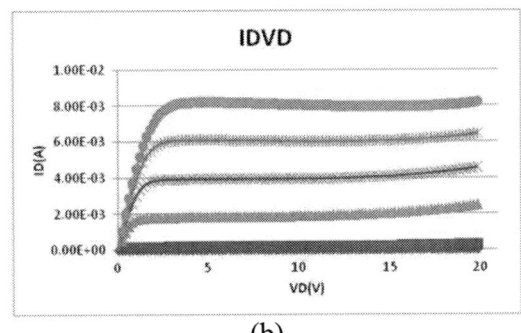

(b)

Figure 2: 18V DDDMOS characteristics curves (a) IDVG, (b) IDVD

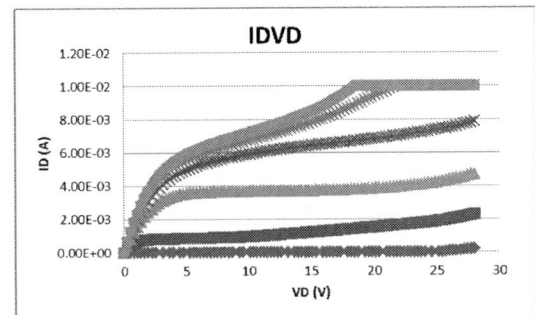

Figure 3: 25V DDDMOS IDVD curve

Dose experiment

In order to improve the off-BV of transistor, large enough device size is required and then we can optimize the doping of drift region for higher off-BV. The common approach of drift region optimization is RESURF technology [5] [6]. Table 1 shows the simulation result with different drift doping split. Device channel length is 1.5um. The drift length equals PF+PA which PF is 0.3um and PA is 1.5um. Split1 is the baseline condition, the off-BV is 49V and on-BV is lower than operating voltage (not listed in the table). As dose increasing, the off-BV decrease dramatically. For example, the breakdown voltage drops from 49V to 31.3V when the implant dose increases from 1.5e12cm^{-2} to 5e12cm^{-2}. For different implant energy with same dose, we found that the lower the energy, the lower the breakdown voltage. The impurity which distributes more close to the surface has higher effect on breakdown voltage.

Table 1: Result of drift implantation experiment

Split	1	2	3	4	5	6	7	8
Drift imp 1	*500keV /1.5e12*	*500keV /1.5e12*	*350keV /1.5e12*	*350keV /1.5e12*	*350keV /1.5e12*	*350keV /1.5e12*	*350keV /6e12*	*350keV /3e12*
Drift imp 2	*50keV /1.5e12*	*50keV /1e13*	*50keV /5e12*	*50keV /3e12*	*50keV /1e13*	*100keV /1e13*	*100keV /5e12*	*100keV /8e12*
Off-BV	49	20.6	31.3	33.2	20.2	18.9	18.9	19.1

978-1-5090-6695-7/17 $31.00 © 2017 IEEE   59

Figure 4(a) illustrates the impact ionization location of split 1, 2 and 3 listed in table 1. In split 1 and 2, the breakdown both happens in drift region surface and near the edge of poly gate. The breakdown of split 3 happens in PN junction between P-well and N-type drift. The figure 4(b) shows the lateral electric field distribution of the three splits. Their location of electric field peak in three conditions is same as that ion impaction shows in figure 4(a).

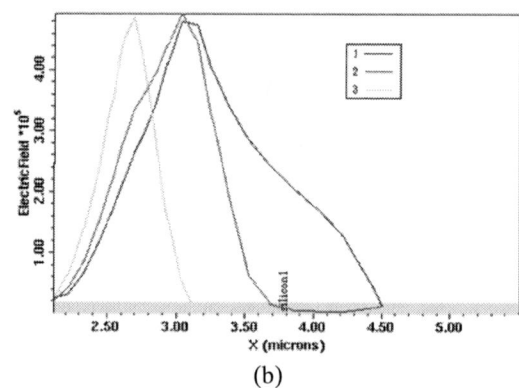

(b)

Figure 4: Experiment of drift implantation dose (a) impact ionization; (b) electric field distribution

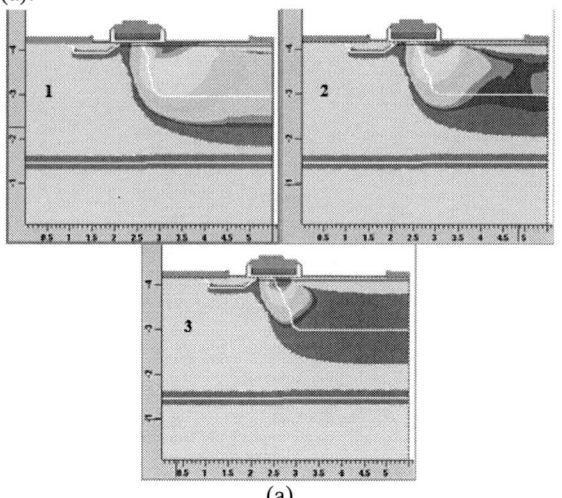

(a)

Optimization of device size and drift region doping: with big PF size

Table 2 lists the simulation result of device size and drift doping. The PF size has increased a lot and total drift length increases to 2.3um. The PF size in this table is 1.8um and 1.3um. The relevant PA is 05um and 1um, respectively.

Table 2: Result of drift implant with big PF size

Split		1	2	3	4	5	6	7	8
Drift imp 1		500kev /1.5e12	500kev /1.5e12	500kev /1.5e12	500kev /1.5e12	500kev /1.5e12	500kev /1.5e12	500kev /1.5e12	500kev /1e12
Drift imp 2		50kev /1e13	50kev /5e12	50kev /3e12	50kev /1.5e12	100kev /3e12	100kev /1.5e12	300kev /1e12	300kev /1e12
Drift imp 3								P/100k ev/1e12	P/50ke v /1e12
PF=1.8um	PA=0.5um	32.2 (*1)	28.7	29.9	29.4 (*2)	28.3	29.6	29.2	
PF=1.3um	PA=1um	22.5	25	29	38.7 (*3)	29	38.2	36.8	38.8 (*4)

* Results marked 1~4 are same as that in figure 5.

When PA=0.5um and drift dose decreases from $1e13cm^{-2}$ to $1.5e12cm^{-2}$, impact ionization location shift from inner side to outer side of poly edge and BV become smaller.

Due to breakdown location is close to drain side, we reduced the poly length (reducing the PF size and increasing the PA size) while the total drift length is unchanged. As show in figure 5(a) and (b), BV increases from 29.2V to 38V. This off-BV satisfies the requirement of 30V application, but its on-BV is not high enough.

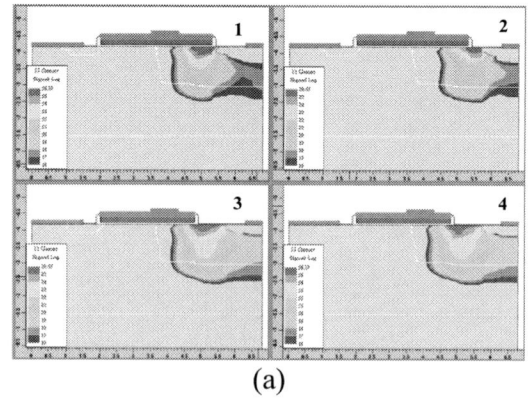

(a)

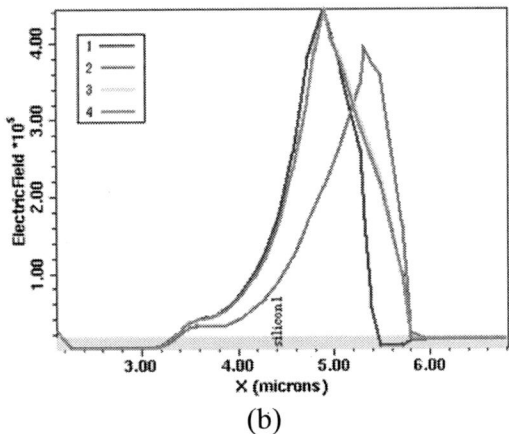

(b)

Figure 5: Experiment of drift implantation dose and PF size experiment (a) impact ionization; (b) electric field distribution. The mark 1~4 is marked

When drift region has higher doping, the bigger PF size leads higher breakdown voltage. The reason is that longer PF can help to deplete the drift region which is beneath it. This extends the width of depletion region even though the drift region has high doping. If there is no this large PF, off-BV will be lower than application requirement due to drift region is not depleted enough wide.

Big PF size and space between drift and Pwell experiment

Higher implant dose of 8e12cm^{-2} is adopted in Table 3 and different space between drift and P-well are used in our experiments. Upon the result, we found that the shorter the distance of N-drift implant, the higher the off-state breakdown voltage. But on-BV has a peak value at space=0.8um.

Table 3: Result of space between drift and Pwell

Split		1	2	3	4	5	6
Drift imp 1		500kev /1.5e12					
Drift imp 2		50kev /8e12					
Drift to Pwell space (um)		1.4	0	0.5	0.8	1	1.2
PF = 1.8um	Off-BV	37.5	47	46.5	45.7	44.7	42
	On-BV	35.3	30	36.3	38.7	38.4	36

Experiment of gate oxide thickness

Increasing the gate oxide thickness, the device breakdown voltage can be further improved but the improvement is not too much, as table 4 shows. The on-BV will saturate as gate oxide thickness rises.

Table 4: Result of gate oxide thickness experiment

Split	1	2	3
PF (um)	1.9		
Drift to Pwell Space	0.5		
Gate Oxide (A)	800	900	1000

Off-BV	45	48.7	49.3
On-BV	36.9	37.8	38.8

Silicon experiment result

Figure 6 is the demonstrated silicon result. Its off-BV is higher than 35V and on-BV higher than 30V. The input and output performance is good. Thus, this device meet 25V ~ 30V application.

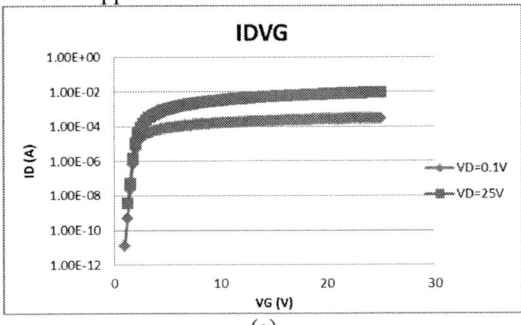

(a)

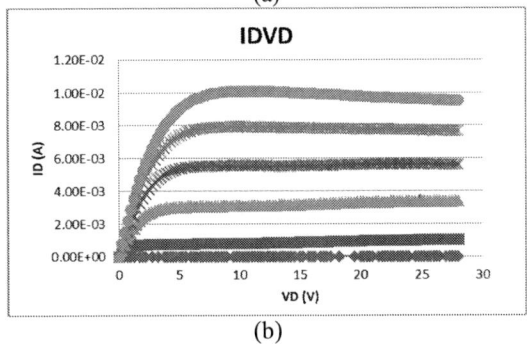

(b)

Figure 6: Demonstrated 25V DDDMOS performance (a) IDVG; (b) IDVD

CONCLUSIONS

This paper reports a DDDMOS, higher on-state BV can be achieved with highly doped drift region; meanwhile the novel design of poly overlap drift region enables the high off-state breakdown voltage. The mechanism is that the highly doped drift region capped by poly (PF) can be depleted with the aid of poly. This extends the width of depletion region even the drift region has high doping. If there is no this large PF, off-BV will be lower than application requirement due to depleted drift region is not wide enough. In this demonstrated DDDMOS, Its off-BV is higher than 35V and on-BV higher than 30V. The input and output performance is good. This device can meet 25V ~ 30V application.

ACKNOWLEDGEMENTS

The authors would like to thank all members of device group and high voltage group of HHgrace for their great support on this work.

978-1-5090-6695-7/17 $31.00 © 2017 IEEE

REFERENCES

[1] J.Wang, L.Wang, Y.M.Dong, X.Zou, *Mechanism and impact of the double-hump substrate current in high-voltage double diffused drain MOS transistor*, ACTA PHYSICA SINICA, vol57 no.7, 2008, pp.4492-4496

[2] C. E. Chen, T. C. Chang, H. M. Chen, B. You, K. H. Yang, S. H. Ho, J.Y. Tsai, K. J. Liu, Y. H. Lu, Y. J. Hung, Y. H. Tai, and T. Y. Tseng, *On the Origin of Anomalous Off-Current Under Hot Carrier Stress inp-Channel DDDMOS Transistors With STI Structure*, IEEE Electron Device Lett., vol. 35, no. 6, 2014, pp. 651–653

[3] B. J. Baliga, *Trends in Power Semiconductor Devices*, IEEE Trans. Electron Devices, vol. 43, no. 10, 1996, pp. 1717–1731

[4] M.D.Ker, C.Y.Wu, *Modeling the positive-feedback regenerative process of CMOS latch up by a positive transient pole method—Part I:Theoretical derivation, IEEE Trans.*, Electron Devices, vol. 42, no. 6, 1995, pp. 1141–1148

[5] J.A.Appels, H.M.J. Vaes, *HV thin layer devices (Resurf devices)*, IEEE IEDM, 1979, pp. 238-241

[6] D.H.Liu, X.M. Xu, F.Jin, W.T.Duan, H.H.Wang, J.Shi, Y.Yao, J.Hu, W.S.Qian, P.F.Wang, D.W.Zhang, *The Investigation of Field Plate Design in 500V High Voltage NLDMOS*, Advances in Condensed Matter Physics, 2015, pp. 1-6

METAL-ELECTRODE-DEPENDENT NEGATIVE PHOTOCONDUCTANCE RESPONSE OF THE NANOSCALE CONDUCTING FILAMENT IN THE SIO₂-METAL STACK

T. Kawashima[1,2]*, Y. Zhou[1], K. S. Yew[1], H. Z. Zhang[1], and D. S. Ang[1]*

[1] School of Electrical and Electronic Engineering, Nanyang Technological University, Singapore 639798, Singapore

[2] Corporate Manufacturing Engineering Center, Toshiba Corporation, 33, Shin-Isogo-cho, Isogo-ku, Yokohama 235-0017, Japan

*Corresponding Author's Email: tomohito.kawashima@toshiba.co.jp; edsang@ntu.edu.sg

ABSTRACT

Nanoscale resistance reset of the SiO₂/M stack (where M=Cu, Ni, Ti, Al, p-type Si) was investigated via a conductive atomic force microscope (C-AFM). Visible-light illumination triggers a resistance reset for Ti, Al and p-type Si electrodes, however such a behavior is not always observed for the Cu and Ni electrodes. Conversely, electrical reset is possible for Cu and Ni, but not for the others. The observed variations in optical and electrical induced resistive switching behaviors may be caused by a metal-electrode-dependent conducting filament.

INTRODUCTION

The filamentary resistive switching memory device has drawn significant interests on account of its promising application for big-data-storage and neuromorphic computing technology [1][2]. Apart from the traditional electrical induced resistance switching operation, we have recently demonstrated visible-light-induced negative photoconductance response or resistance reset (LIR: Light-induced reset) in the HfO₂, ZrO₂ and SiO₂ resistive memory [3][4]. This finding may pave the way for an optical control of these devices based on wide-bandgap oxides – an aspect that has not received much attention to-date. In this study, we investigate the metal-electrode dependence of LIR to obtain further insights into the underlying mechanism.

EXPERIMENTAL

A SiO₂/M/Ti/p-type Si (where M=Cu, Ni, Ti and Al) stack was prepared via DC magnetron sputtering or thermal deposition on a HF-cleaned p-type Si substrate, followed by the formation of a 5-nm-thick SiO₂ via plasma-enhanced chemical vapor deposition. The thickness of M is more than 30 nm. For comparison, a no-metal sample, i.e. SiO₂ (5 nm)/p-Si, was also prepared. Electrical measurement was performed using an ultra-high vacuum C-AFM system with the diamond-coated Si probe connected to a Keithley SCS4200 parameter analyzer (Figure 1). The bias voltage was applied to the probe and the substrate was grounded. A commercially available white LED light was used to investigate LIR. The measurement set-up has already been reported in detail elsewhere [3][4].

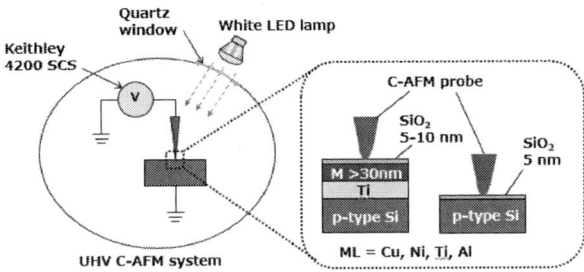

Figure 1: Schematic illustration of the C-AFM setup and the test sample

RESULTS AND DISCUSSION

Typical current-voltage curves for the various SiO₂/M samples are shown in Figure 2. In all samples, a current jump is observed under the first negative voltage ramp (Black Line). In a subsequent positive voltage ramp (Red Line), the current is decreased in the Ni and Cu samples, indicating the occurrence of a reset, while no clear reset is observed in the Ti, Al and no-metal samples. The reset in the Ni and Cu samples is also confirmed by a current hysteresis in a following negative voltage ramp (Blue Line). To evaluate electrical reset, the degree of voltage-induced reset (R) is calculated, which is:

$$R = \frac{V_2 - V_r}{V_1 - V_r} \qquad (1)$$

where V_1 and V_r denote the voltage at 1 nA in the first negative forward-sweep and reverse-sweep, respectively, and V_2 denotes the corresponding voltage in the third negative forward-sweep after the second positive-sweep. R calculated from more than 20 random locations on each sample is summarized in TABLE I. In the Ni and Cu sample, the resistance can be electrically changed between set and reset. On the other hand, an electrical reset is mostly impossible (low R) in the Ti, Al and p-Si samples. It is believed that various intrinsic defects (e.g., oxygen vacancies and dangling bonds) created in the SiO₂ by

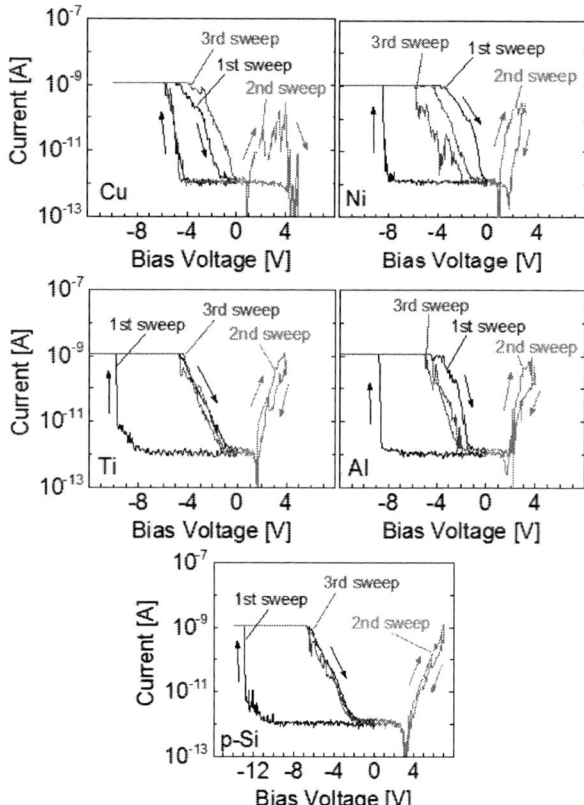

Figure 2: Current-voltage (I-V) curves of the SiO₂/M (M=Cu, Ni, Ti, Al) and SiO₂/p-Si (no-metal) samples.

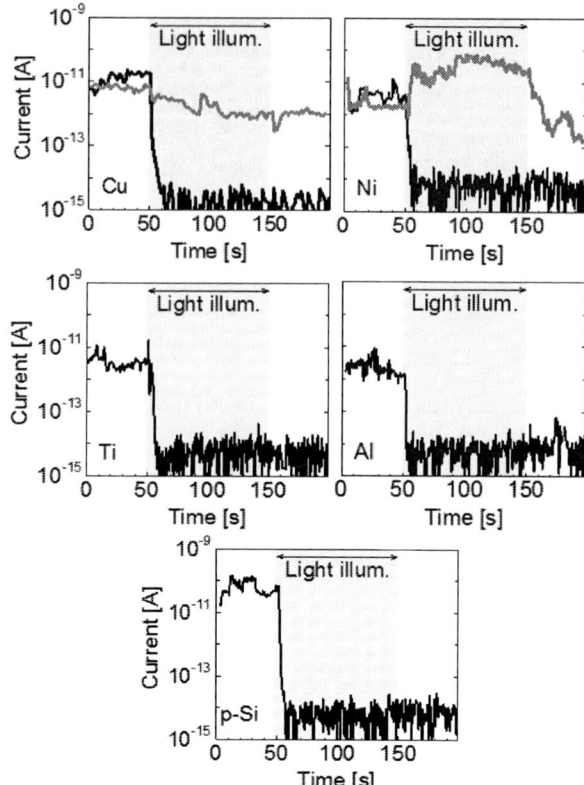

Figure 3: Current-time (I-t) plots of the SiO₂/M (M=Cu, Ni, Ti, Al) and SiO₂/p-Si (no-metal) samples.

TABLE I.   ELECTRODE DEPENDENCE OF THE ELECTRICAL PROPERTY AND LIR

Electrode	Cu	Ni	Ti	Al	p-Si
Forming Voltage (V)	−5.7*	−7.8	−9.5	−9.9	−13
Degree of Electrical Reset, R	89%	40%	12%	25%	3%
LIR probability @ −1.5V	77%**	37%	87%	100%	95%

* The thickness of SiO₂ is 10 nm for the Cu sample
** The monitoring voltage is −1 V for the Cu sample

voltage ramp cause resistance change in the Ti, Al and no-metal samples [4]. This change may be irreversible because oxygen ions emitted from the SiO₂ lattice during filament formation may move towards the anode to form stable Al-O, Ti-O and Si-O bonds. Therefore, the vacancies in the filament cannot be removed by a reverse polarity voltage sweep [5]. On the contrary, Cu and Ni are

considered to be relatively diffusive in SiO₂ [6][7], implying that ionized Cu or Ni may migrate into the SiO₂ bulk during forming/set and move back to the electrode under an opposite-polarity voltage sweep, leading to the reversible resistive switching.

After an electrical set to the low resistance state, the SiO₂/M and SiO₂/p-Si samples are subjected to light illumination with the current monitored at a low voltage. Figure 3 depicts the typical time dependence of the current. The current shown by the Black Line steeply decreases upon light exposure at 50 s and remains low after the light was turned off at 150 s. The percentage of successful LIR for 20 random locations on each sample is shown in TABLE I. In the Ti, Al and p-Si samples, nearly 90% or more of the locations show LIR. In the Ni sample, however, only 37% show LIR. In the Cu sample, the percentage is 77%, also lower than those of the Ti, Al and p-Si samples. The current depicted by the Red Line in Fig. 3 is an example of no-LIR response observed in the Cu and Ni samples.

From these observations, we propose a model in Figure 4, which may explain the metal-dependent current behavior under light. During forming for Ti, Al, p-Si samples, oxygen ions emitted from the filament move towards the anode by electric field, but some of them may remain in the interstitial site in the vicinity of the filament due to Joule heating. It has been suggested that LIR occurs by the photo-assisted migration of these interstitial oxygen ions back to the vacancies in the filament [3][4]. However, the filament in the Cu and Ni samples is considered to partly consist of the corresponding metal ions [8][9]. Owing to the higher diffusivity of Cu and Ni ions [6][7], there would be a greater tendency for these metal ions to migrate into the SiO_2 to form metal filaments during forming. This is evident from the considerably lower forming voltages of the Cu and Ni samples for the same SiO_2 growth condition. Disruption of a metal filament may require higher energy than that for the LED, because both the cleavage of metal-metal bonds and ionization would be involved. Figure 5 depicts an example of the current-time plot in the Cu sample under the light of higher intensity. The current does not drop until the intensity reaches $27mW/cm^2$. Therefore, under the $1mW/cm^2$ LED, the Cu and Ni samples show a lower percentage of LIR than the Ti, Al and p-Si samples. Nevertheless, in the Cu and Ni samples, oxygen-vacancy type of filaments may be still created during forming, and therefore some locations show LIR.

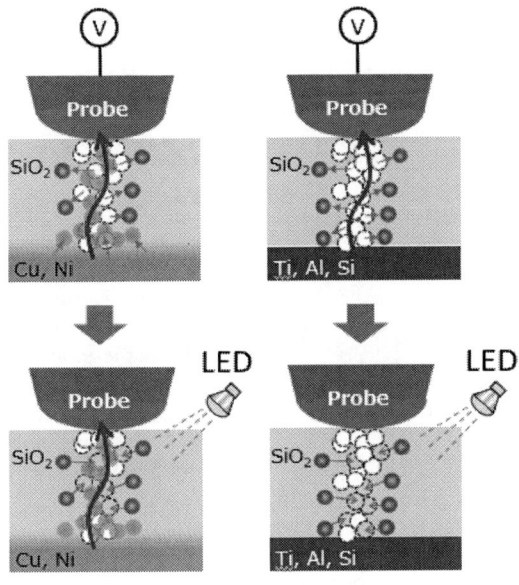

○ Oxygen vacancy

◉ Oxygen ion

◉ Cu or Ni atom/ion

Figure 4: Proposed mechanism of the reactivity to light in the SiO_2/M (M=Cu, Ni, Ti, Al) and SiO_2/p-Si (no-metal) samples

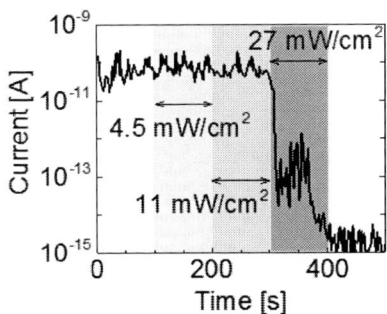

Figure 5: Current-time (I-t) plot of the SiO_2/Cu sample under stronger light

CONCLUSION

Optical and electrical induced resistance reset in SiO_2 is shown to depend on the type of metal electrode used. Samples with Cu and Ni electrodes are found to exhibit relative ease of electrical induced resistive switching but a low percentage of LIR. On the other hand, the Al, Ti and Si counterparts do not exhibit electrical switching but can readily show LIR. The difference may be ascribed to the relative ease for Cu and Ni ions to migrate into the SiO_2 to form metal filament during forming. As a result, the likelihood of LIR is reduced due to higher energy required for metal-metal bond dissociation and ionization. This work suggests the possibility of controlling the negative photoconductance response of the filament in a resistive memory device via the selection of metal electrodes.

REFERENCES

[1] J. S. Meena, S. M. Sze, U. Chand, and T.-Y. Tseng, Nanoscale Res Lett. 9:526 (2014).

[2] Y. F. Wang, Y. C. Lin, I. T. Wang, T. P. Lin, and T. H. Hou, Sci. Rep. 5, 10150 (2015).

[3] Y. Zhou, K. S. Yew, D. S. Ang, T. Kawashima, M. K. Bera, H. Z. Zhang, and G. Bersuker, Appl. Phys. Lett. 107, 072107 (2015).

[4] T. Kawashima, K. S. Yew, Y. Zhou, D. S. Ang, M. K. Bera, and H. Z. Zhang, IEEE Electron Device Lett. 36, 748 (2015).

[5] N. Raghavan, K. L. Pey, X. Wu, W. Liu, X. Li, M. Bosman, and T. Kauerauf, IEEE Electron Device Lett. 32, 252-254 (2011).

[6] J. D. McBrayer, R. M. Swanson, and T. W. Sigmon, J. Electrochem. Soc., 133(6), 1242-1245 (1986).

[7] R. N. Ghoshtagore, J. App. Phys., 40(11), 4374-4376 (1969).

[8] W. A. Hubbard, A. Kerelsky, G. Jasmin, E. R. White, J. Lodico, M. Mecklenburg, and B. C. Regan, Nano Lett. 2015, 15, 3983−3987

[9] J. Sun, Q. Liu, H. Xie, X. Wu, F. Xu, T. Xu, S. Long, H. Lv, Y. Li, L. Sun, and M. Liu, Appl. Phys. Lett. 102, 053502 (2013)

A NOVEL 25V PLDMOS DESIGN IN 700V BCD PROCESS

Donghua Liu[1,2], Wenting Duan[2], Wensheng Qian[2]

[1] State Key Laboratory of ASIC and System, School of Microelectronics, Fudan University, Shanghai 200433, China

[2] HuaHong Grace Semiconductor Manufacturing Corporation, Shanghai 201206, China

*Corresponding Author's Email: Wenting.Duan@hhgrace.com

ABSTRACT

This article reports a high voltage PLDMOS with 25V operating voltage and 40V breakdown voltage. Usually the PLDMOS adopts N-well as channel, and the N-well is as the drift region of NLDMOS. The PLDMOS presented in this paper adopts N-type EPI as channel, which is not affected by NLDMOS drift region. The low doping concentration of N-type EPI causes the punch through (punch through for short) in PLDMOS between the source and drain when the channel length is short than 2.5um. The PLDMOS in this paper adopts blanket phosphorus implant to suppress punch through between source and drain, and blanket boron implant to adjust threshold voltage. Without adding mask, the resulting device still has excellent characteristics under short channel length.

INTRODUCTION

Recently, LDMOS has widely been used in Power Management Circuits for high operating voltage, high current driving capability, ultra-low power and capability with CMOS process[1][2].

Generally, PLDMOS and NLDMOS are fabricated in an IC chip simultaneously. N-well is adopted to formalize the drift region of NLDMOS. N-well is formed in n-type epitaxial layer (NEPI). The breakdown voltage (BV) and the on resistance of NLDMOS can be adjusted by changing the doping of N-well. PLDMOS and NLDMOS are integrated together and PLDMOS channel is n-type. Theoretically, we can use the n-type drift region of NLDMOS as the channel of PLDMOS. However, PLDMOS performance will be affected by the doped n-type drift region (for adjusting the BV and on resistance of the NLDMOS)[3][4]. Thus, in this paper, we adopted the n-type epitaxial layer as the channel of the PLDMOS and PLDMOS performance is not affected by doped n-type drift region (N-well). Additionally, NLDMOS performance can be adjusted independently. The PLDMOS in this paper can suppress the punch through between the source and drain (punch through) and has an unaffected threshold voltage (VT). Moreover, scaling PLDMOS can easily be achieved with this design.

DEVICE STRUCTURE

Conventional PLDMOS is shown in Fig. 1. Channel region is formed by NEPI and drift region by P-well. Punch through will be induced when the channel length is less than 2.5um for lightly doped NEPI.

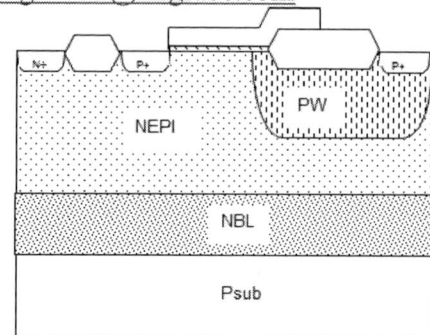

Figure 1: PLDMOS device structure

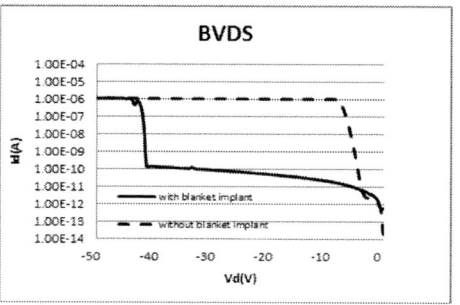

Figure 2: BV curves of before and after blanket implant

Punch through of PLDMOS, in this work, is suppressed by blanket implant of phosphorus (energy of 100 KeV to 500 KeV and dose of 1e11 to 1e13 cm-2) after the growth of field oxide. Fig. 2 shows the silicon data before and after blanket phosphorus implant. We can observe that the punch through is disappeared and that BV curve is normal. The VT of PLDMOS will increase with the implant of phosphorus. Therefore, blanket implant of boron (energy of 10 KeV to 100 KeV and dose of 1e11 to 1e13 cm-2) is required to adjust the VT.

RESULTS AND DISCUSSION

The Correlation between the Device Performance and Blanket Phosphorus Conditions (Simulation Data)

Threshold Voltage

The blanket phosphorus conditions and simulation data are listed in the above table. It indicated that, lower energy and higher dose of the phosphorus will higher the VT of PLDMOS. High energy (300KeV) implant will penetrate the LOCOS and then the performance of other devices will be influenced. Thus, 200KeV implant energy lower implant dose are adopted.

978-1-5090-6695-7/17 $31.00 © 2017 IEEE

Conductive Current

After the blanket implant of phosphorus, the conductive current (Ion) is lower than the NA condition which has not the blanket implant of phosphorus.

Table1: Device Characteristic after the phosphorous blanket implant

Blanket implant	NA	P/100k/ 2e12	P/100k/ 1e12	P/100k/ 5e11	P/200k/ 2e12	P/200k/ 1e12	P/200k/ 5e11
Vt	-1.34	-5.24	-3.4	-2.67	-4.3	-2.38	-2
Ion	368	228	289	328	230	288	323
BV	14.2	33.4	36.1	25.8	35.7	37.8	38.4

Breakdown Voltage (BV)

Before the blanket phosphorus, the punch through BV of 14.2 V can be observed in our device. The BV are increased to above 30 V with the implant and we observed that the higher the dose of phosphorus, the higher the BV. The current distributions of before and after blanket implant are shown in the below figures. It indicated that punch through is disappeared with the blanket implant.

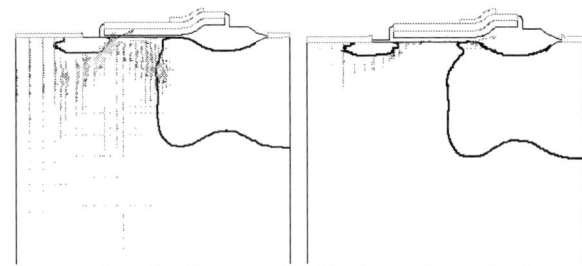

Figure 3: (a). Current distribution of punch through before blanket implant of phosphorus. (b). Current distribution of punch through after blanket implant of phosphorus

The Correlation between the Device Performance and Blanket Boron Conditions (Simulation Data)

Table2: Device Characteristic after the phosphorous and boron blanket implant

blanket implant : phos	P/200k/5e11			
blanket implant : boron	NA	B/30k/2e11	B/30k/3e11	B/30k/4e11
Vt(V)	-2	-1.82	-1.6	-1.32
Ion(uA/um)	323	333	339	344
BV(V)	38.4	38.7	35	30

Phosphorus implant will increase the VT of PLDMOS. Thus, boron implant is adopted. Simulatively, VT is decreased as the increase of boron dose. High boron dose will lower the BV of our PLDMOS. Therefore, trade-off between the VT and BV is required. From the above table, we found that Ion is nearly independent of the boron implant dose.

The Comparison of Surface Doping Concentration before and after Blanket Implant.

Buried-channel device is formed after the phosphorus implant. Structure simulation and surface doping concentration are illustrated in the below figures.

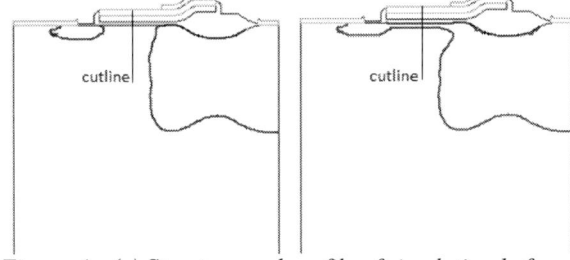

Figure 4: (a) Structure and profile of simulation before blanket implant. (b) Structure and profile of simulation after blanket implant

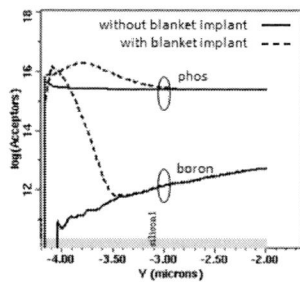

Figure 5: Doping concentration of boron and phosphorus along the cutline

After the blanket implant of phosphorus, the surface concentration of boron and phosphorus is higher than the NA condition. The peak value of the boron is located the surface of the channel and is higher than the concentration of phosphorus. However, the peak value of phosphorus is lightly away from the channel surface, which can account for the bulk rather than the surface punch through.

CONCLUSION

25 V PLDMOS based on NEPI has been investigated in this paper. Lightly doped NEPI will induce the punchthrough between the source and drain. Thus, we presented a solution of blanket implants which can eliminate the punchthrogh without additional masking layer. Blanket implant of phosphorus eliminate the punchthrough issue and VT is unaffected by blanket implant of boron with moderate energy and low dose. Of course, the performance of other devices is unaffected with this solution.

ACKNOWLEDGEMENTS

The authors would like to thank all members of device group and high voltage group of HHgrace for their great support on this work.

REFERENCES

[1] M.J.Zhou, A.D.Bruycker, A.V. Calster, J. Witters, G. Schols, *Breakdown walkout and its reduction in high-voltage pLDMOS transistors on thin epitaxial layer*, IEEE Electron Device Letters, vol. 28, no. 16, 1992, pp. 1537-1538.

[2] X.Luo, Qiao Tan, J.Wei, K.Zhou, G.Q.Deng, Z.J.Li, B.Zhang, *Ultralow ON-Resistance High-Voltage p-Channel LDMOS With an Accumulation-Effect Extended Gate*, IEEE Transactions on Electron Devices, vol. 63, no. 6, 2016, pp. 2614-2019.

[3] M.Abouelatta-Ebrahim, C.Gontrand, A.Zekry, *Complementary LDMOSFET in 0.35μm BiCMOS technology-characterization and modeling*, IEEE International Symposium on Industrial Electronics, 2010, pp. 736-741.

[4] M.Abouelatta-Ebrahim, C.Gontrand, A. Zekry, *Physical understanding and modelling of new hot-carrier degradation effect on PLDMOS transistor*, IEEE International Reliability Physics Symposium (IRPS), 2012, pp. 111-116.

Application of resist profile model and resist-etch model in solving 28nm Metal resist toploss

Yiqun Tan, Weiwei Wu, Quan Chen, Shirui Yu

Shanghai Huali Microelectronics Corporation, No.497 Gaosi Rd. Zhang Jiang Hi-Tech. Park,

Shanghai 201210, China,

* Corresponding Author's Email: tanyiqun@hlmc.cn

Abstract

As critical dimensions decrease to 28 nm node and beyond, more etching failures are induced by the resist loss increases. Only two-dimensional (XY) contours are considered by traditional optical proximity correction (OPC) models, while vertical direction diffusion is neglected, resulting in inaccuracy in valuation of the resist loss. Rigorous resist simulators can simulate a three-dimensional (3-D) resist profile, but they are not fast enough for correction or verification on a full chip, which restrict their usage in technology development below 28nm node. However, for one hand, resist loss for positive-tone resists is mainly driven by optical intensity variations, which are accurately modeled by the optical portion of an OPC model. For the other, resist loss can be reflected by the quickly shrinking in process window condition. In this paper we show that a compact resist model can be used to determine resist loss by properly selecting the optical image plane for calibration or by introduction of CDs data after etching. Both these two models can be used to identify toploss hotspots on a full chip.

Keywords—resist toploss; etch model; profile model; OPC

1. Introduction

Usually, the reduction of resist thickness is not considered by traditional optical proximity correction (OPC) models.[1,2] It just generate simulated contour from optical intensity in the plane selected. The development process introduces small resist losses since 180nm tech-node, but, in general, less attention has been put on height variations of resist. Low k1 imaging systems, however, may have degraded images with higher intensity in dark mask regions, resulting in resist loss at the top of the final resist structure. Large resist losses may weaken the resist's ability to withstand the etching and result in hotspots. For a narrow line, the intensity leaks into the dark areas, resulting in the center intensity rising. This leakage of light can exceed the threshold of resist development and cause resist loss. OPC models consist of an optical model based on the sum of coherent systems approximation and a semiempirical resist model. Most of the optical parameters can be set according to the values in the actual scanner. However, two important parameters are to be calibrated although they are known.[3] The first one is the beamfocus, standing for the focus location of the optical system in the absence of a film stack. The second is defocus, which is a cross-section plane in the resist on which the optical intensity is calculated. With these two parameters, the optical model outputs intensity. The intensity is used by resist model to calculate the resist contour, commonly known as CD for critical dimension, after resist development. Resist processes can be rigorously simulated based on physics and chemistry, but two things limited its usage. Firstly, such a model takes too long time to calibrate. Secondly, a model calibrated by rigorous physics and chemistry process will cost so much resource during OPC simulation in verification that people use it for analyze defect only after they find it in the real chip, suggesting that high cost of mask making will be present. Instead, the rigorous model is approximated by a more compact, faster model with the simplest type being a threshold model as shown below:[4]

$$T = \sum C_i M_i(I)$$

$$M = \left((\nabla^k I_{z,b})^n \otimes G_{z,p} \right)^{\frac{1}{n}}$$

$$= c_0 I + c_1 \nabla I + c_2 \nabla^2 I + c_3 I_{-b} \otimes G_{z3} + c_4 \sqrt{I_{-b}^2 \otimes G_{z4}} + c_5 I_{-b} \otimes G_{z5} + c_6 \sqrt{I_{-b}^2 \otimes G_{z6}} + c_7 I \otimes G_{z7}$$

In this model, Gaussian functions are convolved with optical intensity to simulate diffusion effect in the resist development process, and a threshold is applied to the intensity to calculate the resist CD. The threshold is calculated by finding the one that minimizes the difference between simulated CDs with a set of measured CDs from a wafer exposed in a scanner. More sophisticated compact resist models still use the threshold operation but with a modified intensity function that simulates the resist process. The additional parameters for k, b, s, p and n of this compact model are also optimized to match measured. Such a model is less resource consumed and fast enough for OPC development at the price of 3D information loss. So, the problem is just how to apply this 3D information-less model to resist toploss prediction.

In nodes > 40 nm, the DOF (depth of focus) were large enough to meet the manufacturing need and there is no need to calibrate the beamfocus parameter very well. However, when tech-node develop to 28nm nodes and beyond, correctly setting these parameters becomes important in producing accurate models that predict resist loss and resist profiles for the latest processes. Section 2 shows how the right fitted beamfocus and defocus can predict resist toploss defect. Since resist toploss is actually CD shrinkage in dense patterns, we found another way to calibrate a model consist of data after etch, and predict a threshold CD for recognize defect, which is described in

Section 3. Section 4 will pay more attention on the usage of these two models.

2. Resist Profile modeling

A resist profile for a chemically amplified resist is formed after several processing steps, including exposure, post-exposure bake (PEB), and development. Although the process is complex, the location of the resist edge can be tracked to the first order by applying a threshold to the intensity as discussed in Introduction. The model is based on the dose to clear experiment, where a large clear area is removed completely when the dose exceeds the dose to clear. When mask features are present, the dose is modulated by the intensity. And when the intensity exceeds a threshold, ideally given by the dose to clear divided by the dose, the resist is completely removed. In locations where the intensity is less than a threshold, the resist remains.

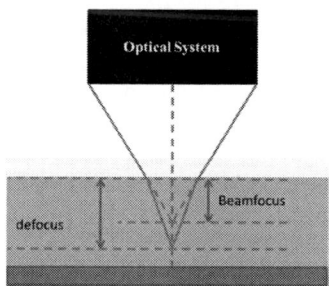

Fig.1 Beamfocus and defocus parameters in optical system.

To model this process, two sequential models are to be calibrated. The first one is optical model which describe the distribution of optical intensity in the resist, and the second one is compact resist model which simulate transformation of resist from aerial image to resist contour.

The optical model has two important parameters that need to be calibrated: beamfocus and defocus These are shown in Fig. 1. Generally, the beamfocus range can include regions outside the resist, while the defocus will typically lie within the resist. The intensity is calculated at the defocus plane in the resist. In general, the optical intensity within the resist can be written as a function of the lateral position and these two optical parameters. For example, the intensity with a good antireflection coating (bottom antireflective coating) can be approximately written as shown in equation below: [5]

$$\Phi_{kr} = \frac{f}{\lambda}(n_r \cos\theta_{kr} - n_i \cos\theta_{ki})$$

$\theta_{ki/r}$ stands for angles of incidence / refraction light with k_{th} diffraction order and $n_{r/i}$ stands for index of refraction for resist and media. The intensity is modulated by a term that depends on the beamfocus and defocus. For a given beamfocus, there is a corresponding defocus at which the modulating term vanishes and the intensity contrast is maximized.

To properly set the simulation beamfocus and defocus parameters, wafer line CD measurements at multiple focus and dose settings are needed to select a combination of beamfocus and defocus with which the model error will be symmetric. Although the beamfocus and defocus parameters defined by this method may not result in smallest RMS, this method will calibrate a model closer to the real process.

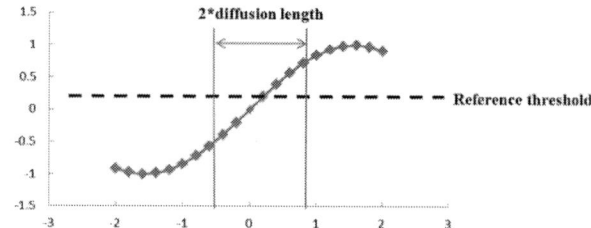

Fig 2. When calibrating resist model, Gaussian function with diffusion length s are convolved with optical intensity distribution function to generate another resist distribution, and a threshold is determined to get resist contour

The compact resist model transforms the optical intensity into resist intensity by linearizing and approximating the system of equations. Like the simple threshold model, a threshold is also applied to the resist intensity to obtain the final CD as shown in Fig. 2. It should be noted that beamfocus and defocus parameters in optical model are also tuned during resist model calibration slightly. They all have to be calibrated by using measured CDs and determining which values minimize the root mean square (RMS) CD error of simulated CDs and measured CDs as shown in Fig.3. A useful characteristic of the compact model is that when calibrated with enough CDs and with the center focus set correctly, for a fixed resist process, the model can then predict CDs at different center focus values and dose values. With the defocus for calibration properly set, the compact model can now also predict CDs at other heights and the resist loss through different best focus and dose values.

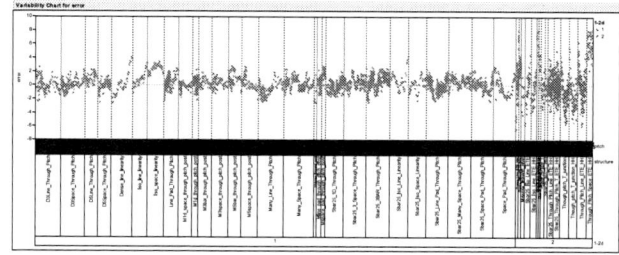

Fig.3 Resist Profile model fitting results.

3. Resist-etch modeling

Resist profile modeling described above has a severe dependence on the accuracy of best focus and defocus setting in optical model. However, in most cases, the settings of these two parameters are determined by RMS calculated with resist model which is semiempirical since the values of the coefficients, threshold, truncation values, and diffusion lengths cannot generally be traced back to the physical quantities. As a result, it takes a lot of time for calibration and verification of the resist profile model. On the other hands, all the toploss defects are found in dense pattern, which means that the accuracy of semi and iso patterns will not influence the prediction of toploss occurrence and more Gaussian term can be involved to simulate the etch effects. At this time, CDs of patterns after etch process are collected and used in the same way as resist profile modeling. The difference is that the weight of semi and iso data are decreased to 10% of that in resist profile modeling. The results are shown in the Figure below.

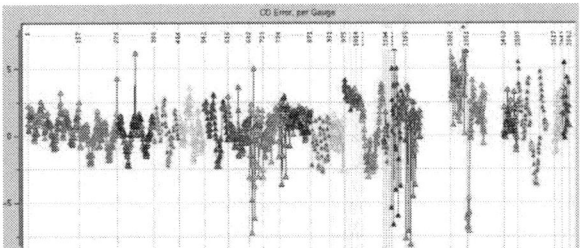

Fig.4 Resist-etch model fitting results

4. Comparison of resist profile model and resist-etch model

To compare and verify the two models above, a rigorous simulator was used to produce the CDs and of some dense patterns commonly used in real chip. A C-quadrapole source with a split x-y polarization was used along with a mask with binary transmission. The settings were chosen to resemble the settings of a 28-nm metal layer. Fig. 5a and 5d show the contours generated through resist profile model resulted from heights at nominal process conditions (dose 1, defocus 0) and process window conditions (dose 1.04, defocus 40 nm). When used for OPC and OPC verification, the resist profile model will be used to produce contours at a height close to the top of the resist. It is not too computationally expensive to compute full resist profiles for an entire layout. In verification, the resist profile model can indicate etching hotspots by showing where the CD contour becomes too thin or breaks away completely.

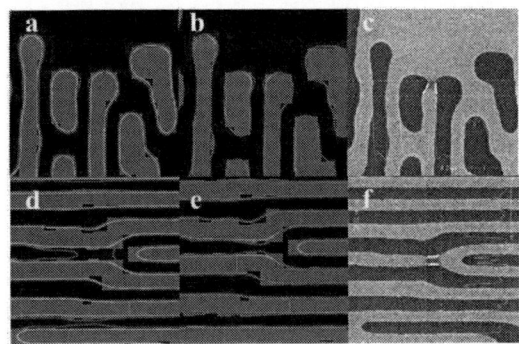

Fig.5 a, b: Image contour of position 1 simulated by resist profile model and resist-etch model respectively. d, e: Image contour of position 2 simulated by resist profile model and resist-etch model respectively. c, f: SEM photo of Position 1 and 2 at PW condition on real wafer after etch.

Fig. 5b and 5e exhibit the simulation contour of a potential defects predicted by resist-etch model. The regular OPC contour shows no breaks and typically reflects the nominal edge of the resist. This contour does not show a problem. However, the contour simulated in the PW condition shows a quick shrinkage, indicating a large resist loss in that area. The SEM results in Fig. 5c and 5f showed that both resist profile and resist-etch model can predict potential defects and be used for full chip search.

5. Conclusion

This paper has shown that the prediction of potential toploss

defect can be realized by two different models. Resist profile model is more accurate for semi and iso patterns, when the sampling height and center focus is set properly. Resist-etch model calibrated with CDs after etch is accurate for dense patterns by sacrifice semi and iso pattern accuracy. The resulting optical and compact resist model can then be used to identify possible etching hotspots in full chip run.

Reference

[1] Cobb N., Fast Optical and Proximity Correction Algorithms for Integrated Circuit Manufacturing, PhD Dissertation University of California at Berkeley, 1998
[2] Granik Y. Medvedev D. and Cobb N. "Towards standard process models for OPC", Proc. of SPIE, Vol. 6520, pgs. 652043-1 652043-6, 2007
[3] Mack C Fundamental Principles of Optical Lithography, Wiley Jan 2008
[4] Klostermann, et al "Calibration of physical resist models: methods, usability and predictive power" J. Micro/Nanolith Jul-Sep 2009 Vol 8(3)
[5] Chen A., et. al "Evaluation of compact models for negative tone development layers at 20/14 nm nodes" Proc. SPIE 9426 (2015)

DEPLETION-MODE MOS CAPACITOR MODELING INVESTIGATION

Chien-Lung Tseng [*], *Yuan Sheng Wang*

Technology R&D, Semiconductor Manufacturing International Corporation,
18 Zhang Jiang Rd., Shanghai 201203, China
*Email: bruce_tseng@smics.com

ABSTRACT

A depletion-mode MOS (DMOS) capacitor modeling methodology with high accuracy and feasibility is proposed. Currently, it is lack of compact model relevant DMOS capacitor modeling but it is significant importance in A/D converters (ADCs) for CMOS image sensor circuit. This modeling methodology not only could provide good accuracy on geometry scaling but also with voltage and temperature sensitivity.

INTRODUCTION

In CMOS image sensor back-end process, 14 bits ADCs resolution or more is needed [1]. DMOS capacitor plays a crucial role in ADCs circuit because of its high density capacitance compared with Metal Oxide Metal (MOM) and Metal Insulator Metal (MIM) capacitors. With regard to traditional CMOS process, DMOS capacitor is compatible except an additional channel implant. The cross section of traditional NMOS capacitor [2] with N-type DMOS capacitor is compared as shown in Fig. 1.

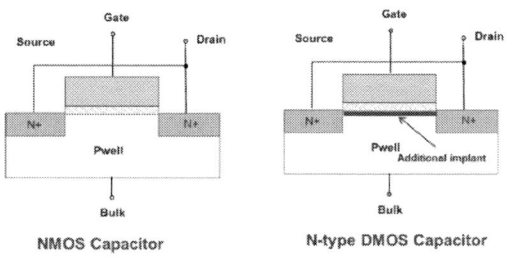

Fig. 1 Cross-section of NMOS capacitor and N-type DMOS capacitor

Due to this additional channel implant for DMOS capacitor, it causes operation mode from normal off to normal on. Consequently, DMOS capacitor has better linearity in capacitance vs. voltage characterization

rather than traditional MOS capacitor as shown in Fig. 2.

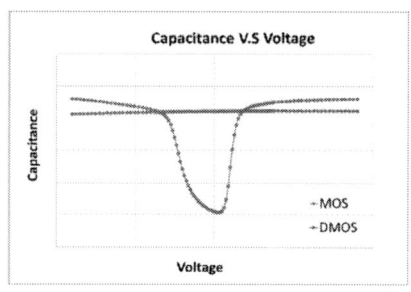

Fig. 2 Capacitance vs. voltage for N-type DMOS and traditional MOS capacitor

MODELING METHODOLOGY

In general, the DMOS capacitance can be expressed as following:

$$C(V, T) = C_0 \times V_c \times T_c \qquad (1)$$

Where C_0 denotes capacitance biasing at $V_{gs} = 0$ and V_c, T_c is the voltage and temperature coefficient respectively. Furthermore, C_0 can be expressed associated with geometry dimensions of channel width (w), channel length (l) and multiplier (m).

$$C_0 = Cgg_0 \times [(w-dw) \times (1 - dl) + Cgg_1 \times pwr ((w-dw) \times (1 - dl), Cgg_2)] \times m \qquad (2)$$

Where Cgg_0 is the unit area capacitance and is extracted from large dimension device and Cgg_1, Cgg_2 account for modified parameters of geometry for small dimension device. The parameters extraction flow is shown in Fig.3.

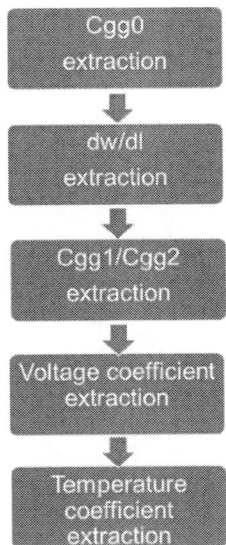

Fig. 3 Parameters extraction flow

First, obtain Cgg_0 form slope of capacitance vs. area for large dimension as shown in Fig.4.

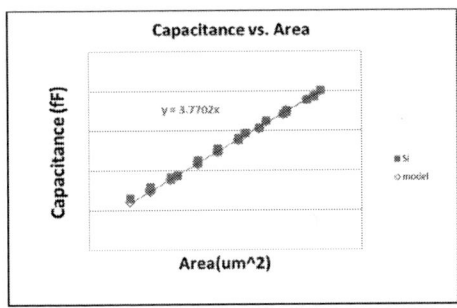

Fig. 4 Capacitance vs. Area

From variable substitution, assume $L = l - dl$ and obtain the dl parameter which is the intercept of x-axis in capacitance/width vs. L plot as shown in Fig.5.

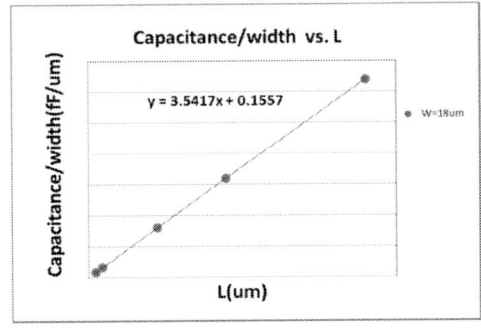

Fig. 5 Capacitance/width vs. L

Similarly, dw could be got with the same methodology in capacitance/length vs. W plot as shown in Fig.6.

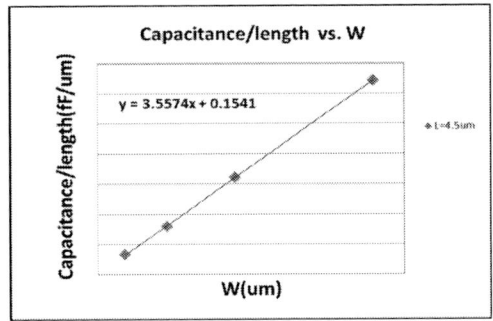

Fig. 6 Capacitance/length vs. W

Subsequently, obtain Cgg_1, Cgg_2 which are modified parameters for small dimension as shown in Fig.4.

For capacitance of voltage dependent modeling, implement empirical formula as below:

$$V_c = \left(1 + V_{c1} \times V_{gs} + V_{c2} \times V_{gs}^2 + V_{c3} \times V_{gs}^3\right) \quad (3)$$

Normalizes it to $Vgs = 0V$ and depict the fitting plot in Fig.7.

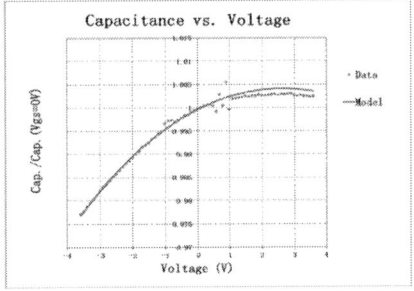

Fig. 7 Capacitance vs. Voltage

Furthermore, consider the temperature coefficient as below:

$$T_c = \left(1 + T_{c1} \times (T - 25) + T_{c2} \times (T - 25)^2\right) \quad (4)$$

Normalizes it to $Vgs = 0V$, $T = 25°C$ and illustrate the fitting result in Fig.8.

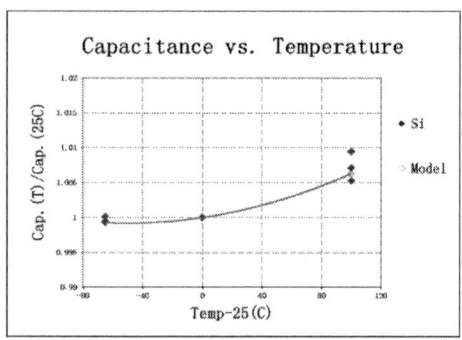

Fig. 8 Capacitance vs. Temperature

CONCLUSION

A novel model for DMOS capacitor has been proposed. This modeling methodology could easy implement to DMOS capacitor and provides good accuracy on geometry scaling, and better voltage and temperature dependency.

ACKNOWLEDGMENT

The authors would like to appreciate the SMIC modeling team for their assistance in this paper.

REFERENCES

[1] Toshio Yasue, Kazuya Kitamura,Toshihisa Watabe, Hiroshi Shimamoto, Tomohiko Kosugi, Takashi Watanabe,Satoshi Aoyama,Makoto Monoi, Zhiheng Wei, Shoji Kawahito, "A 1.7-in, 33-Mpixel, 120-frames/s CMOS Image Sensor With Depletion-Mode MOS Capacitor-Based 14-b Two-Stage Cyclic A/D Converters ", IEEE *Trans.* Electron Devices, vol. 63, no. 1, JANUARY 2016, pp. 153-161

[2] Yuan Taur, and Tak H. Ning, "Fundamentals of Modern VLSI Devices",pp.72-88

SIMULATION ON THE PERFORMANCE COMPARISON FOR NANOSCALE SOI AND BULK JUNCTIONLESS FINFETS

Cheng-Kuei Lee[1]*, Jin-Yu Zhang[1], and Yan Wang[1]

[1]Department of Microelectronics and Nanoelectronics Tsinghua University, Beijing 100062, China
*Corresponding Author's Email: lzk15@mails.tsinghua.edu.cn

ABSTRACT

This paper adapts 3-D quantum transmission model to evaluate the performances of the junctionless (JL) FinFET devices for silicon-on-insulator (SOI) and bulk FinFET structures using high –k/materials as gate oxide dielectrics materials. Silicon dioxides (SiO_2), silicon nitride (Si_3N_4), hafnium dioxide (HfO_2) are used as the dielectric materials of the simulated JL FinFETs. Various device characteristics, such a subthreshold swing, drain-induced barrier lowering and On/Off current, are also explored under different gate lengths and fin widths. The simulated results reveal the device characteristic degrades with the gate length decreases from 20 to 10 nm dramatically under any configurations. The JL bulk transistor have greater performance than SOI structure under the short channel by adjusting the device's threshold voltage with the doping concentration of the substrate. Considering the transistor performance, HfO_2 is the only acceptable gate oxide dielectric material as the gate length is less than 14 nm.

INTRODUCTION

The fin field-effect transistor (FinFET) has become a mainstream trend, of which the process has reached more than ten nanometers. However, the short channel effects (SCEs) such as drain-induced barrier lowering (DIBL) and subthreshold swing (SS) are big challenges for this nano-scale transistor. In recent study, the junctionless transistor is proposed [1] to mitigate above issues. Junctionless transistor has many potential advantages: 1) The source and drain doping are the same as the channel to avoid fabricating abrupt PN junctions. 2) It is able to reduce thermal budget issues during the process. 3) A lower process cost is required [2]. Recently highly-discussed High-k material and stress-strain are not only focus on the device structure improving, but also on the process advance to obtain higher device performance [3]. In this paper, we use the technology computer aided design (TCAD) software, Sentaurus, to simulate various the electric characteristics of junctionless fin field-effect transistor with Bulk and SOI structure (JL SOI & JL BULK).The fin width is changed under 10 nm - 20 nm channel length. High-k materials are adopted to enhance the overall performance of the device. To evaluate the device performance, DIBL, SS, ON-state current (I_{ON}) and OFF-state current (I_{OFF}) are calculated. For the junctionless device, $I_{OFF}$is is a very important index to determine the device power.

DEVICE STRUCTURE AND SIMULATION APPROACH

The study simulates two different structures and two different fin cross sections. Figure 1 shows JL SOI and JL BULK structures. Figure 2 shows two different fin cross sections $10 \times 10 nm^2$ & $10 \times 5 nm^2$ (fin height × fin widths). Besides, different work functions are used in different structures for current comparison consistency. Work functions of JL BULK are *n*-type=4.7eV, *p*-type= 4.45eV; and work function of JL SOI are *n*-type = 4.92eV, *p*-type = 4.3eV. All devices adopt three different dielectric materials, with the oxide thickness of 10 Å. Both and *n*-type of JL BULK and JL SOI structures doping concentrations are $10^{19} cm^{-3}$ and the substrate concentration of JL BULK structures are $10^{18} cm^{-3}$ with *p*-type and *n*-type. We take three channel lengths of 20 nm, 14 nm and 10 nm and three dielectric materials, of silicon dioxide (SiO_2), silicon nitride (Si_3N_4), and hafnium dioxide (HfO_2). Two different physical models are considered, i.e., drift-diffusion (DD) model and hydrodynamic (HD) model. However, DD model in short channel will overrate the device characteristics [6].

With the constant decrease in the device size, the quantum effect cannot be neglected [5]-[7], and thus it is necessary to add density gradient (DG) model to approximate this quantum effect, with the expression shown as equation (1) (2).

$$n = N_C F_{1/2}\left(\frac{E_{F,n} - E_c - \Lambda_n}{kT_n}\right) \quad (1)$$

Where *n* is the electron concentration; $F_{1/2}$ is the Fermi-Dirac integral; E_c is the conductive band; N_C is the conductive band state density; T_n is the electron temperature; Λ_n is defined as follows.

$$\Lambda_n = \frac{\gamma \hbar^2}{12 m_n}\left[\nabla^2 \ln n + \frac{1}{2}(\nabla \ln n)^2\right] \quad (2)$$

Where $m_n$is the electron effective mass; γ is a fit factor. More details of the above expression can be found in [10]. In this work, we adopt HD model combined with DG model.

978-1-5090-6695-7/17 $31.00 © 2017 IEEE

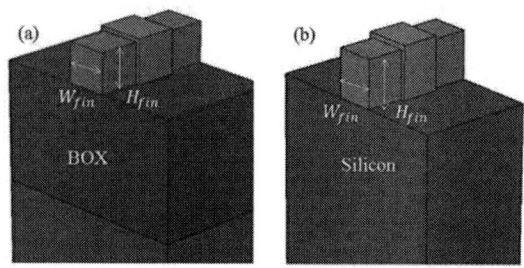

Figure 1: Different structures for the Junctionless FinFETs The structures include: (a) JL SOI (b) JL BULK structures

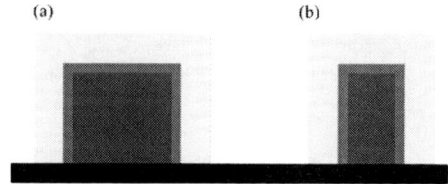

Figure 2: Two fin cross sections: (a) 10×10nm2, (b) 10×5nm2 for the simulate JL BULK & JL SOI FinFETs.

RESULTS AND DISCUSSION

The supply voltage used during the simulation process is |0.8|V. DIBL is defined as the difference between threshold voltage when the drain voltages are 0.05V and 0.8V respectively. Figure 3 shows $I_D - V_G$ curves of JL SOI and JL BULK structures using HfO_2 as dielectric materials when the channel length is 10 nm and the fin width is 5 nm/10 nm. For 5 nm fin width, the DIBL and SS for JL SOI and JL BULK are almost same as shown in Figure 3 (b). However, for 10nm fin width, JL BULK structure is superior to JL SOI structure, as shown in Figure 3 (a). Although JL BULK substrate produce built-in potential barrier with fin in junction, the JL BULK structure has a better SCE inhibition than JL SOI structure if the fin width is 10 nm. When the fin width is smaller, such as 5 nm, carrier injection in channel of source drain may be fully controlled by gates [10] resulting in better SS and DIBL for both SOI and BULK structures. However, for 5 nm fin, larger threshold voltage reduces I_{ON}, which is not desired under the low-voltage system. Increase of the fin height can promote I_{ON}, while it also leads to increase I_{OFF}, as shown in Figure 4. Figure 5-6 present the simulation results of DIBL and SS values under three dielectric materials with the channel lengths of 10 nm to 20 nm and different fin cross sections. When the fin cross section is small, the threshold voltage is not sensitive [11]. The structures have satisfactory DIBL and SS values. As shown in Figure 5, when using larger the dielectric constant material, the parasitic capacitance between gate and channel increases, which reduces capacitance effect between gate and drain arising from bias voltage, and lowers down the leak current to implement better characteristics. In this work, using HfO_2 we can achieve

the best performance in both 10×10 nm^2 & 10×5 nm^2 structures, as shown in Table 1.

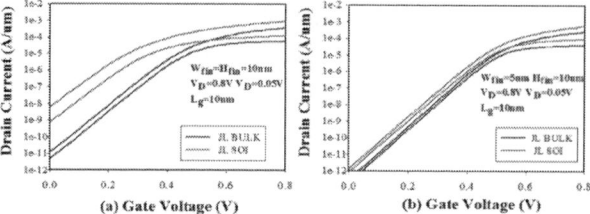

Figure 3: The $I_D - V_G$ curves of the JL BULK & JL SOI with (a) 10 nm or (b) 5 nm fin width and 10 nm gate length under same dielectric material (HfO_2).

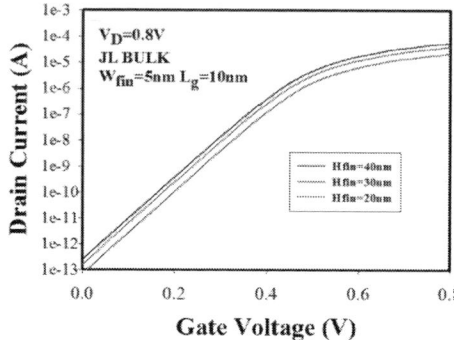

Figure 4: The $I_D - V_G$ curves of the JL BULK 5 nm fin width and 10 nm gate lengths under different fin high.

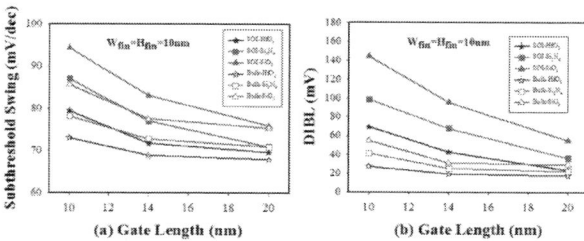

Figure 5: (a) DIBL and (b) SS of the of the JL BULK & JL SOI with the fin cross section 10×10 nm^2 as the gate length decreases from 20 nm to 10 nm under different dielectric materials.

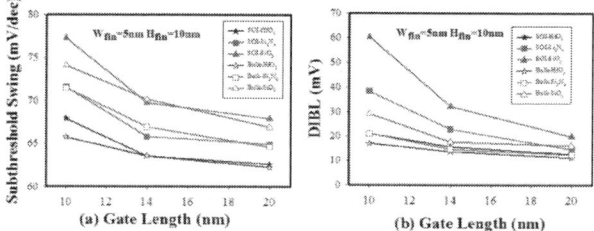

Figure 6: (a) DIBL and (b) SS of the of the JL BULK & JL SOI with the fin cross section 10×5nm^2 as the gate length decreases from 20 nm to 10 nm under different dielectric materials.

978-1-5090-6695-7/17 $31.00 © 2017 IEEE

TABLE I. THE SIMULATION RESULTS

Dielectric Materials		HfO$_2$	Si$_3$N$_4$	SiO$_2$
The JL SOI structure of the fin cross section 10×10 nm^2				
V$_{TH}$ (V)	n: 0.239	n: 0.159	n: 0.061	
	p: -0.101	p: -0.024	p: 0.089	
DIBL (mV)	n: 69.325	n: 98.732	n: 144.476	
	p: 67.027	p: 86.235	p: 121.008	
SS (mV/dec)	n: 79.478	n: 87.053	n: 94.379	
	p: 80.327	p: 87.962	p: 96.576	
I$_{ON}$ (A)	n: 1.54e-5	n: 1.36e-5	n: 1.25e-5	
	p: 5.14e-6	p: 4.75e-6	p: 4.01e-6	
I$_{OFF}$ (A)	n: 6.61e-10	n: 2.10e-9	n: 8.10e-9	
	p: 4.63e-10	p: 1.46e-9	p: 9.79e-9	
The JL SOI structure of the fin cross section 10×5 nm^2				
V$_{TH}$ (V)	n: 0.385	n: 0.345	n: 0.289	
	p: -0.261	p: -0.197	p: -0.145	
DIBL (mV)	n: 21.148	n: 38.367	n: 60.287	
	p: 27.468	p: 37.794	p: 43.724	
SS (mV/dec)	n: 67.988	n: 71.604	n: 77.285	
	p: 66.103	p: 70.922	p: 76.016	
I$_{ON}$ (A)	n: 3.11e-6	n: 2.97e-6	n: 2.83e-6	
	p: 2.28e-6	p: 1.58e-6	p: 1.39e-6	
I$_{OFF}$ (A)	n: 7.79e-13	n: 2.81e-12	n: 1.00e-11	
	p: 7.19e-12	p: 8.16e-12	p: 2.44e-11	
The JL BULK structure of the fin cross section 10×10 nm^2				
V$_{TH}$ (V)	n: 0.378	n: 0.349	n: 0.309	
	p: -0.226	p: -0.203	p: -0.186	
DIBL (mV)	n: 27.431	n: 41.019	n: 54.731	
	p: 36.882	p: 65.482	p: 78.364	
SS (mV/dec)	n: 73.098	n: 78.161	n: 85.653	
	p: 69.476	p: 71.621	p: 78.824	
I$_{ON}$ (A)	n: 6.52e-6	n: 5.55e-6	n: 4.70e-6	
	p: 5.77e-6	p: 3.27e-6	p: 1.00e-6	
I$_{OFF}$ (A)	n: 4.97e-12	n: 1.30e-11	n: 3.49e-11	
	p: 6.11e-11	p: 1.10e-10	p: 1.18e-10	
The JL BULK structure of the fin cross section 10×10 nm^2				
V$_{TH}$ (V)	n: 0.427	n: 0.407	n: 0.384	
	p: -0.321	p: -0.285	p: -0.251	
DIBL (mV)	n: 17.099	n: 21.079	n: 29.197	
	p: 23.179	p: 29.581	p: 37.515	
SS (mV/dec)	n: 65.793	n: 71.512	n: 74.115	
	p: 66.635	p: 69.753	p: 74.898	
I$_{ON}$ (A)	n: 2.46e-6	n: 1.90e-6	n: 1.68e-6	
	p: 1.96e-6	p: 1.46e-6	p: 1.14e-6	
I$_{OFF}$ (A)	n: 8.87e-13	n: 1.49e-12	n: 3.30e-12	
	p: 1.42e-11	p: 1.11e-11	p: 1.16e-11	

CONCLUSIONS

In both JL SOI structure and JL BULK structures, SS and DIBL are influenced by SCE, and as a result, the device characteristics are too poor, especially, JL SOI structure with the fin width of 10 nm. The structure with the fin width of 5 nm has satisfactory SS and DIBL values; however, the large threshold voltage reduces I$_{ON}$, and this is a bad phenomenon under the current low-voltage system. The increase in the fin height can promote the I$_{ON}$, while it also leads to increase in I$_{OFF}$. Therefore, it is necessary to consider fin width, fin height, dielectric material and doping concentration of substrate under the precondition of fin width smaller than 10 nm for outstanding device performance in the future.

REFERENCES

[1] C. W. Lee, A. Afzalian, N. D. Akhavan, R. Yan, I. Ferain and J. P. Colingea. "Junctionless multigate field-effect transistor," *Applphys Lett.*, vol. 94, 2009, p. 053511.

[2] J. P. Duarte, S. J. Choi, D. I. Moon and Y. K. Choi. "Simple analytical bulk current model for long-channel double-gate junctionless transistors," *IEEE Electr Device L.*, vol. 32, 2011, pp. 704-706.

[3] M. H. Han, C. Y. Chang, H. B. Chen, J. J. Wu, Y. C. Cheng and Y. C. Wu. "Performance comparison between Bulk and SOI Junctionless Transistors," *IEEE Electr Device L.*, vol. 34, 2013, pp. 169-171.

[4] K. M. Liu and Y. Y. Hsieh. "Investigation of the dimension effects of 30-nm below multiple-gate SOI MOSFETs by TCAD simulation," *IEEE 4th International Nanoelectronics Conference2011*, Taiwan, June 21-24, 2011, pp. 1-2.

[5] M. P. Anantram, M. S. Lundstrom and D. E. Nikonov. "Modeling of nanoscale devices," *Proceedings of the IEEE*, vol. 96, 2008, pp. 1511-1550.

[6] Z. Ren, R. Venugopal, S. Goasguen, S. Datta and M. S. Lundstrom. "nanoMOS 2.5: A two-dimensional simulator for quantum transport in double-gate MOSFETs," IEEE T Electron Dev., vol. 50, 2003, pp. 1914-1925.

[7] M. Aldegunde, A. Martinez and J. R. Barker. "Study of discrete doping-Induced variability in junctionless nanowire MOSFETs using dissipative quantum transport simulations," *IEEE Electr Device L.*, vol. 33, 2012, pp. 194-196.

[8] Sentaurus Device User Guide, Synopsys Inc., Version D-2010.03.

[9] S. Migita, Y. Morita, M. Masahara and H. Ota. "Electrical performances of junctionless-FETs at the scaling limit (L CH=3 nm)," *IEEE International Electron Devices Meeting2012*, San Francisco, December 10-13, 2012, pp. 8-6.

[10] J. T. Park, J. P. Colinge and C. H. Diaz. "Pi-gate soi mosfet," *IEEE Electr Device L.*, vol. 22, 2011, pp.405-406.

[11] M. H. Han, C. Y. Chang, H. B. Chen Y., C. Cheng and Y. C. Wu. "Device and circuit performance estimation of junctionless bulk FinFETs," *IEEE T Electron Dev.*, vol. 60, 2013, pp. 1807-1813.

Analysis and Modeling of Self-Heating Effect in Bulk FinFET

Shawn Lin, SenSheng Li, Li Shen, Jianhua Ju, Shaofeng Yu
Technology Research and Development Center, SMIC, Shanghai, China
Shawn_lin@smics.com

Abstract

Self-heating effect in bulk FinFET is briefly studied in this paper. The layout dependence of thermal resistance and thermal capacitance are assumed and the correction method is proposed to improve the accuracy of thermal parameter. Measuring methods and their corresponding test structures are also summarized. Besides, two realization methods of self-heating effect for circuit simulation are also discussed.

Keywords—Self-heating, Model, SPICE

Introduction

Compared with planar MOSFET technology, FinFET technology has poor heat dissipation conditions due to its fin structure. Thus self-heating effect (SHE) becomes apparent in FinFET than in planar technology [1]. The temperature increment is reported to be several dozens of degrees during DC measuring which shifts the actual device temperature and shorten device lifetime [2, 3]. The delay time of the logic gates are also changed due to the current variation when devices are at higher temperature. Thus, it is crucial to characterize SHE of devices.

Analysis of the SHE Model

The industrial standard FinFET model BSIM-CMG uses thermal resistance (Rth) and thermal capacitance (Cth) to capture SHE when SHMOD is set to "1" [4]. Cth represents the thermal inertia of a device that the device temperature will not respond to signal when the frequency is high (See Fig 1). These two parameter can characterize SHE as follows,

$$\Delta T(f) = \frac{\Delta P(f) \cdot R_{th}}{\sqrt{1 + (2\pi f \cdot R_{th} C_{th})^2}} \quad (1)$$

However, these two parameters are affected by fin numbers, gate length, finger number and back-end contact numbers. Based on the assumption that the thermal heating length is smaller than gate length, the internal R_{th} of BSIM-CMG are independent of gate length [5], which is not valid when gate length is scaling down. Gate length dependence of R_{th} and C_{th} should be added.

BSIM-CMG uses two exponential parameters, BSHEXP and ASHEXP to cover the R_{th} and C_{th} dependence of finger number (NF) and fin number (NFIN). The dependence originates from two parts. One is the thermal coupling of channels locates at different fin and finger within a device. The other one is the thermal conduction via contacts. The additional contacts that a large device has will help to reduce the R_{th}.

Besides thermal coupling within a single device, it also takes place between devices. Every device will receive power

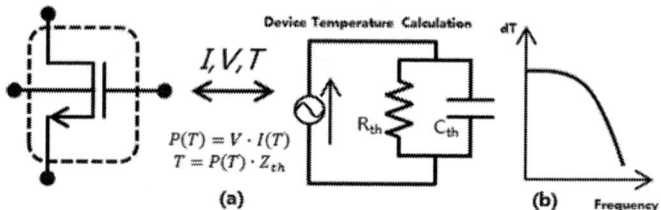

Fig. 1. Equivalent RC Network for Temperature Calculation (a) and the frequency response of the temperature increment (b).

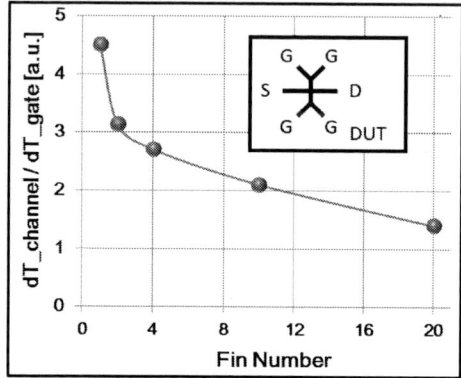

Fig. 2. Fin number dependence of the channel dT relative to gate dT. The inset is the schematic of the gate-detector dut.

spreading from its surrounding devices. Simulation tool should calculate the total power and the relative temperature increment. However, the heat spreading is related to layout design and is difficult to be evaluated. Sometimes, an empirical formula could be used,

$$T = R_{th} * P_0 * (1 + Ratio) = P_0 * [R_{th} * (1 + Ratio)] \quad (2)$$

where *Ratio* is the lumped coefficient to account the additional temperature increment.

SHE Measurement Technique

Several measuring methods have been proposed to characterize SHE. Some researchers use gate thermal detector to monitor the gate temperature (see Fig. 2) [6]. The gate resistance is measured by kelvin structure and the temperature is calculated from the resistance. However, the channel temperature is not the same with the gate temperature. The simulated ratio is shown in Fig.2. The method could be used only after the ratio is verified in a specific technology platform. Meanwhile, the method is unable to extract C_{th}.

The second method is the RF extraction method [7]. The method is based on equation (1). R_{th} and C_{th} could be extracted from the frequency response of gds.

$$R_{th} = \frac{\Delta G_{ds}}{(I_{ds} + V_{ds} g_{ds}) \cdot \partial I_{ds} / \partial T} \quad (3)$$

$$C_{th} = \frac{1}{2\pi f_{th} \cdot R_{th}} \quad (4)$$

978-1-5090-6695-7/17 $31.00 © 2017 IEEE

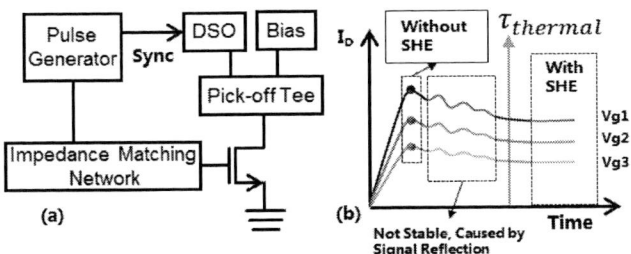

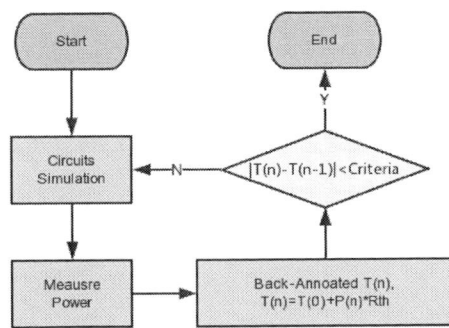

Fig. 3. Schematic of fast-IV system (a). The measured ID reaches its peak value without SHE, and reduces to the stable value when SHE occurs (b).

It should be mentioned that some authors neglect to eliminate the SHE from $\partial I_{ds}/\partial T$ which will lead to about 5~10% error for R_{th} extraction.

Another method is the ultra-fast IV measuring method [8]. The ultra-fast IV measuring could perform an IV spot measurement within several nano-seconds, which is much shorter than the thermal constant time. Thus, a SHE free IV data table containing different device temperature could be made. This technique could also be used to extract the SHE free $\partial I_{ds}/\partial T$ for RF extraction method. To capture SHE, the IV data with SHE is measured after biasing the device for a period of time until the temperature reaches its stable value. Comparing the data with the SHE free IV data table, the device temperature could be extracted. In modern technology, the fast-trap density of silicon surface is low, so trap charging/discharging will not disturb the gate transient voltage during measurement.

Model Realization

To simulate SHE, model engineers should set proper R_{th} and C_{th} values in the model card, and turn on the thermal model switch in BSIM-CMG. SPICE will iterate until the convergence criteria of the current, voltage and temperature are reached [4]. It could simulate the transient behavior of SHE. However, compared with the traditional simulation, the cost of simulation with SHE is large since the additional temperature criteria is present.

Usually, the transient and ac heating behavior is not important. Circuit designers only want to know the final temperature increment caused by SHE. Thus, a quasi-static method could be adopted and it saves plenty of simulation time. The flow is described in Fig. 4. The average power of every device during simulation is fetched. The temperature increment is calculated and back-annotated to the netlist for the next iteration. The procedure is stopped until the temperature difference between two adjacent iterations reaches the criteria. As shown in Fig. 5, the error of dT after second iterations is only -1%, and the third iteration leads to only 0.01% deviation. Thus, the iteration times could be set to a limited value to reduce the simulation time. Three runs are suggested for the flow (2+1, two iterations to calculate temperature increment, and one run containing the last calculated temperature).

Summary

In this paper, we investigate the SHE, summarize the measuring techniques and propose the realization methods briefly. The R_{th} and C_{th} should be modified to add the dependence of fin numbers, gate length, finger numbers and back-end contact numbers. To extract SHE parameters, three

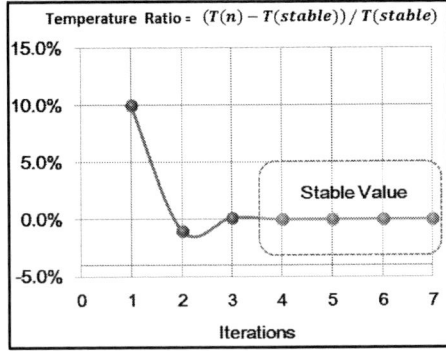

Fig. 4. The flow of quasi-static method. The dT is back- annotated to circuit netlist after each iteration.

Fig. 5. The temperature ratio versus simulation iterations. The temperature reaches stable value after three iterations.

methods are summarized and examined. Using gate thermal detector is most convenient, but the measured temperature is actually the gate temperature, and only R_{th} could be extracted. Ultra-fast IV measuring method is intuitive, but is most difficult in terms of technique. RF extraction technique is suitable to extract both R_{th} and C_{th}. To improve accuracy, ultra-fast measuring technique is suggested to measure $\partial I_{ds}/\partial T$. To integrate SHE during model simulation, users could turn on the BSIM-CMG internal support of thermal features. Or user could use the quasi-static flow to increase simulation speed by giving up the transient and ac characteristics.

References

[1] D. Jang et al, "Self-heating on bulk FinFET from 14nm down to 7nm node",IEDM,2015

[2] S. E. Liu et al, "Self-heating effect in FinFET and its impact on devices reliability characterization", IRPS, 4A.4.1-4A.4.4, 2014.

[3] C. Prasad et al, "Self-heat reliability considerations on Intel's 22nm Tri-Gate technology," IRPS, 5D.1.1-5D.1.5, 2013

[4] S. Khandelwal at el, BSIM-CMG 110.0.0 Muliti-Gate MOSFET Compact Model Technical Manual, Department of Electrical Engineering and Computer Sciences, University of California, Berkeley, CA 94270,2015

[5] L. T. Su, J. E. Chung, D. A. Antoniadis, K. E. Goodson, and M. I. Flik, "Measurement and modeling of self-heating in SOI nMOSFET's," IEEE Transactions on Electron Devices, vol. 41, no. 1, pp. 69–75, 1994.

[6] E. Bury et al., "Characterization of self-heating in high-mobility Ge FinFET pMOS devices," VLSI Technology Symposium, T60-T61, 2015..

[7] S. Makovejev, S. Olsen, and J.-P. Raskin, "RF Extraction of Self-Heating Effects in FinFETs,", IEEE Trans. on Electron Devices, vol. 58, no. 10, pp. 3335–3341, 2011.

[8] J. Lu et al., "PBTI-Induced Random Timing Jitter in Circuit-Speed Random Logic," IEEE Trans. Electron Devices, vol. 61, no. 11, pp. 3613–3618, 2014.

KEY POINTS IN 14 NM PHOTOLITHOGRAPHIC PROCESS DEVELOPMENT, CHALLENGES AND PROCESS WINDOW CAPABILITY

Qiang Wu

SMIC Advanced Technology Research & Development (Shanghai) Corporation
18 Zhangjiang Road
Pudong New Area, Shanghai, P. R. China 201203
+8621-20816818, Ken_wu@smics.com

ABSTRACT

After 20 nm, the 14 nm process has become a major technology node that may stay in market for some extended period of time. In the 14 nm, 2 major technologies have been introduced to the 193 nm immersion photolithography, i.e., the source-mask co-optimization (SMO) and the negative tone developing (NTD). The former will maximize the process window for a given set of design rules, while the latter will further improve the process window through the utilization of bright field imaging together with the use of attenuated phase shifting mask (Att-PSM). Though SMO has a distinct advantage in balancing the process window for various patterns when compared to the parameterized illumination condition, e.g., cross poles or annular, it may cause severe troubles if the input pattern group does not have good representation of all critical designs. Therefore, some trade-offs must be present in order to produce a more accommodating process. The use of negative tone developing, on one hand, has significant advantage in printing vias and trenches through converting the low contrast dark field imaging to the higher contrast bright field imaging. On the other hand, the chemistry of NTD is different compared to the traditional positive tone developing in the fact that it needs more chemical reactions or the acid diffusion-reaction process has to take place on a greater scale. More diffusion-reaction process will mean longer photo acid effective diffusion length, which will damage imaging resolution. This paper will discuss various aspects of 14 nm photolithography process.

INTRODUCTION

When the semiconductor integrated circuit technology enters the 28 nm modes and beyond, the photolithography has reached the single exposure limit. This diffraction limit has set the minimum pitch of 90 nm for bi-directional patterns and 80 nm pitch for uni-directional patterns. Which means should the 193 nm immersion lithographic technology ever be used for more advanced technology nodes, there will be no change for the limitation of single exposure resolution. All smaller dimensions will be relied on multiple patterning, including multiple etch process.

Due to the existence of optical proximity effect (OPE), even the smallest pitch can be printed, there is no guarantee that the all the pitches between the minimum and the more isolated can all be printed with enough process window. Therefore, theoretically it is entirely up to the designer to choose the pitch range that are really needed out of an entire set from the minimally printable to the wider or isolated ones. As a result, the choice of lithographic condition becomes highly dependent on the design rule set up. And because the use of regular parametric illumination condition has limitations in the shape of poles and geometric constraint, the use of flexible illumination becomes inevitable. Early studies are applied to a single shape [1], but the principles are the same.

Another important aspect of the lithography for the more advanced technology nodes is overlay. Since the linewidth can be further shrunk through pitch doubling or quadrupling, etc., The overlay is, to some extent entirely rely on the mechanical performance, such as mechanical positioning stability and precision of wafer stages, lens distortion and distortion stability, etc. does not shrink. The feature to feature separation is getting closer and closer even when the linewidth is shrunk in proportion to the design rule, or pitch. Therefore, if nothing is to be done, the separation related electrical parameter, such as the breakdown voltage (V_{bd}) or time dependent dielectric breakdown (TDDB) can become worse and worse. The introduction of negative tone developing (NTD) is found to able to reduce the after developing inspection (ADI) linewidth, so that it can reduce the amount of linewidth shrink in the etch process, which can improve the risk of shorting in otherwise condition when the ADI linewidth is reaching a limit dictated by optical and resist sensitivity. The negative tone or positive tone study dates back to 19th century [2]. In the 1990's, Mack et al has studied the difference in aerial image [3] and found that there is fundamental difference when partially coherent illumination is used. At the beginning of this century, Brunner et al. has explored optimum tone for various features [4]. Recently, this study has been revisited in negative tone developing by Yang et al. [5]. There are other works associated with the NTD, such as the study by Bae et al. which studied developer effects with different formulation.

In this paper, we will provide our study on SMO concept as well as NTD. In the SMO part, we will summarize all different types of patterns that are SMO sensitive, which means that it can affect the result of the SMO significantly. In the NTD part, this paper will focus on the physics in the aerial image as well as the striking difference in the aerial image and its impact to the choice of mask type and design rules.

SOURCE-MASK CO-OPTIMIZATION (SMO)

Before the introduction of off-axis illumination, the k_1 factor had never become smaller than 0.5. For example, in 0.25 μm technology node, the gate pitch is 600 nm and the critical dimension (CD) is 250 nm. At that time, the Krypton Fluoride（KrF）wavelength (248 nm) lithography has first been used with a numerical aperture (NA) of 0.6. One of the first tools is a SVG Micrascan. The K_1 factor of 0.6 NA can be calculated as,

$$K_1 = \frac{CD}{\lambda} NA = \frac{0.25}{0.248} 0.6 = 0.6 \qquad (1)$$

When KrF tool is used for 0.18 μm technology node at 0.68NA, the representative tool is a Nikon S203/204, the K_1 started to become less than 0.5, as follows,

$$K_1 = \frac{CD}{\lambda} NA = \frac{0.18}{0.248} 0.68 = 0.493 \qquad (2)$$

It is first demonstrated that the use of off-axis illumination, such as, 0.85/0.42 Annular, can significantly improve dense pattern resolution, as well as depth of focus (DoF), such as the one used in 0.18 μm polysilicon gate pitch of 430 nm. However, there is disadvantage associated with the use of off-axis illumination in contrast to the traditional so-called "conventional" illumination.

This effect is understood in the optical community much long time ago as "dark field imaging", or off-axis illumination imaging, which can possess 2X the resolution capability compared to the par-axial illumination. Shown in Figure 1 is a schematic plot of image contrast as a function of spatial frequency for three typical illumination conditions: the small sigma (par-axial, or coherent), medium sigma (partially coherent), and big sigma (non-coherent).

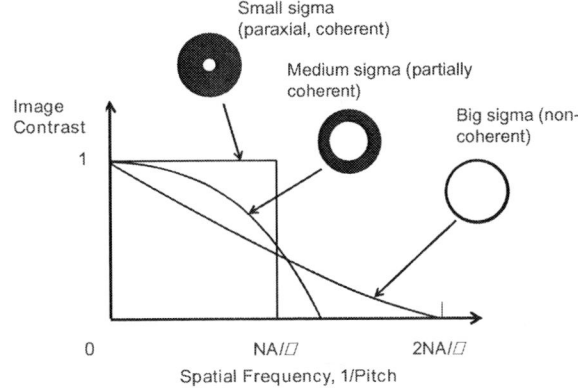

Figure 1: Different Modulation Transfer Function (MTF) for small sigma, medium sigma, and big sigma illumination conditions.

As indicated by Figure 1, the big sigma, or more off-axis illumination can significantly improve the image contrast of spatial frequencies around $2NA/\lambda$, where λ represents the wavelength. The spatial frequency of NA/λ corresponds to a K_1 factor of 0.5. Figure 1 also shows that when the image contrast at pitches below λ/NA is improved, the image contrast at wider pitches will be reduced. Such fact is the first clue for the need to balance amount of off-axis illumination component in the illumination pupil with that of par-axial.

When the technology node enters the 0.13 μm, the K_1 factor becomes deeper into the sub-0.5 regime, as,

$$K_1 = \frac{CD}{\lambda} NA = \frac{0.13}{0.248} 0.68 = 0.356 \qquad (3)$$

The loss of imaging contrast at semi-dense, or intermediate pitches become evident so that a mask bias correction is needed to maintain linewidth uniformity through pitch.

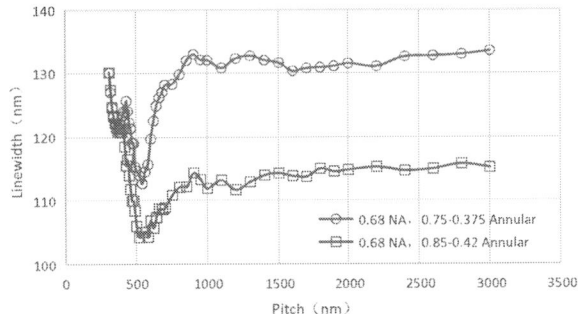

Figure 2: Simulated linewidth through pitch for a 0.13 μm line from a minimum pitch of 310 nm to the 3000 nm. The illumination condition is: 0.68NA with 20 nm effective photo-acid diffusion length and two different partial

coherent settings: 0.75-0.375 annular and 0.85-0.42 annular.

A simulated linewidth through pitch is demonstrated in Figure 2. It is shown that linewidth at the pitches near 500 nm can be significantly smaller than other pitches. In Figure 3, it is shown that the exposure latitude (EL) or the image contrast at pitches near 500 nm also possesses a local minimum. This indicates that the off-axis illumination can improve the contrast at pitches smaller than λ/NA while it may reduce process image contrast at around some intermediate pitches.

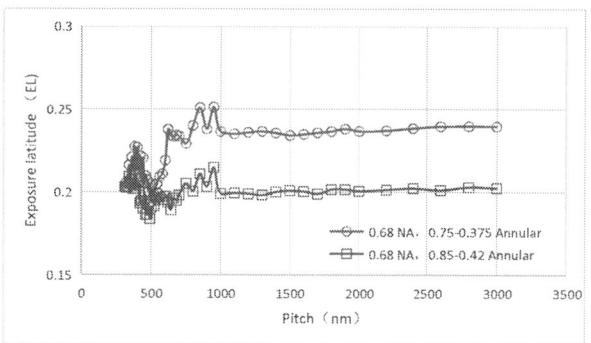

Figure 3: Simulated exposure latitude (EL) through pitch for a 0.13 μm line from a minimum pitch of 310 nm to the 3000 nm. The illumination condition is: 0.68NA with 20 nm effective photo-acid diffusion length and two different partial coherent settings: 0.75-0.375 annular and 0.85-0.42 annular.

The above simple theory is to remind us some history in the illumination development to satisfy the process window needs for the ever shrinking of lithographic linewidth.

TABLE I.   SUMMARY OF DIFFERENT PATTERNS AND THE OPTIMUM ILLUMINATION CONDITIONS FOR THEM

Lithography Process Window Parameters	1D Patterns			2D Patterns
	Big Pitches (>wavelength/NA)	Middle Pitches (around wavelength/NA)	Small Pitches (<wavelength/NA)	
	Big, or isolated	Near forbidden pitches	Critical dimension, dense	Line-end to line-end to trench end to trench end
Achieve linewidth target	Small sigma (paraxial, coherent)	medium sigma	Off-axis	Off-axis (to reduce line/trench end shortening)
Contrast/Exposure latitude/Image intensity gradient (NILS)	Small sigma (paraxial, coherent)	medium sigma	Off-axis	Small sigma (paraxial, coherent)
Depth of focus (DoF)	Small off-axis	medium off-axis	Off-axis	Off-axis (to achieve linewidth target)
Mask error factor, MEF	Big sigma (non-coherent)	medium sigma	Off-axis	Big sigma (non-coherent)

Table I has summarized the illumination condition types for a variety of typical patterns. As we can see, the off-axis illumination condition is favored by the majority

of boxes (shaded) formed by pattern types and a list of 4 different process window parameters: a total of 12 out of 16. This is understandable in the sense that the mission of lithography is to print small features, the capability of being able to imaging at spatial frequencies around $2NA/\lambda$ is definitely needed. **Therefore, the overall illumination is an off-axis one no matter how complicated the source-mask co-optimizing (SMO) result is.** This is the first conclusion of this paper.

TABLE II.   SUMMARY OF DIFFERENT PATTERNS AND THE WAYS TO MAINTAIN THEIR PROCESS WINDOW SHOULD OFF-AXIS ILLUMINATION IS USED

Ways to maintain process window should off-axis illumination is used				
Lithography Process Window Parameters	1D Patterns			2D Patterns
	Big Pitches (>wavelength/NA)	Middle Pitches (around wavelength/NA)	Small Pitches (<wavelength/NA)	
	Big, or isolated	Near forbidden pitches	Critical dimension, dense	Line-end to line-end, Trench end to trench end
Achieve linewidth target	increase linewidth, adding SRAF	increase linewidth, adding SRAF	No need	No need
Contrast/Exposure latitude/Image intensity gradient (NILS)	increase linewidth	increase linewidth	No need	Adding Serif, hammerheads, and increase gap width
Depth of focus (DoF)	increase linewidth, adding SRAF	increase linewidth, adding SRAF	No need	No need
Mask error factor, MEF	increase linewidth	increase linewidth	No need	No need

While the off-axis illumination is generally contributing to process window improvements in most cases, there are still some exceptions as you find in table I.   Hence, we need to find ways to restore lost process window. Shown in Table II is a collection of recipe for such purpose. The result is simple. Most of the recipe is to increase linewidth, or adding sub-resolution assist features (SRAF), serifs, or hammerheads.   The adding of SRAF, serif, and hammerheads are typical optical proximity correction methods for the relatively speaking isolated features and 2-dimensional (2D) features.

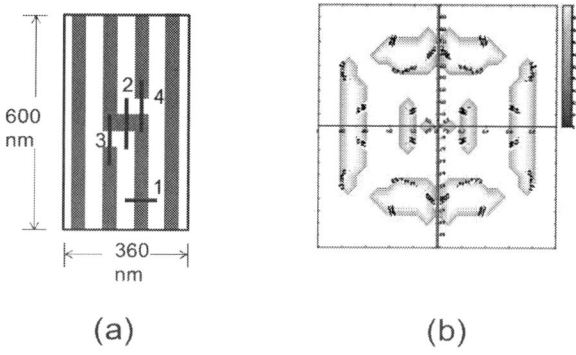

Figure 4: (a) test features to be printed with line cuts 1,2,3, and 4 requesting maximum process window, (b) a calculated optimized illumination source (partial coherence scale within 0-1) that has balanced process

978-1-5090-6695-7/17 $31.00 © 2017 IEEE

windows for all above 4 line cuts. The NA is 1.35 and the wavelength is 193 nm.

Shown in Figures 4(a) and (b) are a test pattern and an optimized illumination source map, respectively. The source calculation is done by an in-house tool. The source map is optimized for all 4 line cuts shown in Figure 4(a) (The linecuts 3 and 4 are redundant if we assume pattern symmetry). This is just a simple example as we can see that the source is just a cross pole with dipoles in both X and Y axes. There are also some small components inside the heavily weighted 4 poles. This is because there are two relatively speaking isolated line ends (cuts 3 and 4). If we check in Table I, they liked small sigma for image contrast. But they need off-axis component for depth of focus, mask error factor, and to achieve less line end shortening.

In real application, such as 14 nm lithography layers, the SMO is a compromise of all patterns, or called clips, that are put into the computation. And the source is a compromise between all different input features.

NEGATIVE TONE DEVELOPING (NTD)

Here we discuss about the advantages and disadvantages of the use of negative tone developing (NTD). From the theoretical perspective, the positive tone developing (PTD) is like "dismantling" a house since the exposed photoresist material is removed by a developing process. To the contrary, the negative tone developing is like "building" a house, in which the exposed photoresist material stays through the developing process which removes the un-exposed photoresist material. As we all know that the dismantling is easier than the building since it only takes to destroy the key structures before the house will fall apart, while the building is far more difficult since it needs to brick-by-brick and frame-by-frame to make to the entirety of the house. Therefore the positive tone developing is naturally easier to do than the negative tone developing. But why the industry wants to move to the seemly disadvantageous negative tone developing?

Let us first consider the aerial image. Shown in Figures 5 (a) and 5 (b) are two simulations for a 45 nm (on mask) line and a space, respectively at 90 nm pitch. The illumination is a 1.35 NA, 0.9-0.7 annular with no polarization and no mask scattering considered. Everything is just pure aerial image. The data indicate that for both the line and the space, they see the identical aerial image distribution. If Figure 5(a) represents the negative tone developing, the photoresist sees the underlined area as the active energy above the photo-chemical reaction threshold. If Figure 5(b) represents the positive tone developing, the photoresist sees the underlined area as the

active energy above the photo-chemical reaction threshold. In this case, both resist platform see the same aerial image intensity. If we use the previous reasoning that the NTD is like "building" a house while the PTD is like "dismantling" a house, then for the dense pitch described here, NTD is in a disadvantageous position. Therefore, for NTD to work at this tight pitch, either it needs much more quantum mechanically efficient photo absorber, or it needs more efficient chemical amplification. From the photoresist evaluation we perform in our company, we found that the NTD photoresists possess longer photo-acid diffusion length when compared to their PTD counterpart.

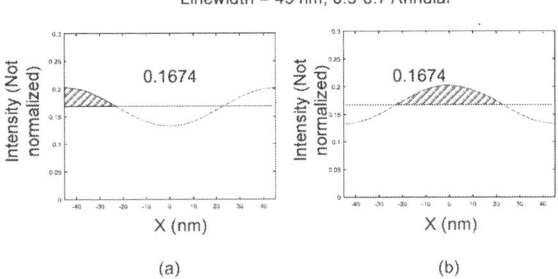

Figure 5: aerial image simulated with 90 nm thick photoresist with zero acid diffusion and 1.35NA with no polarization for a 45 nm (a) line (b) space at 90 nm pitch. The partial coherence is 0.9-0.7 annular with 9 x 9 segmentation in the 1st quadrant of the illumination pupil. The mask is a binary with no Mask 3D scattering considered. The threshold is 0.1674 for both (a) and (b)..

We have also performed a simulation for the 200 nm pitch with all the conditions unchanged. The result is shown in Figures 6(a) and 6(b).

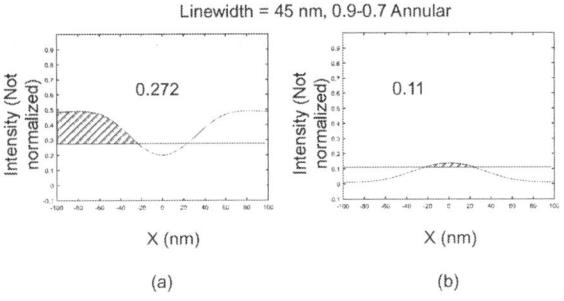

Figure 6: aerial image simulated with 90 nm thick photoresist with zero acid diffusion and 1.35NA with no polarization for a 45 nm (a) line (b) space at 200 nm pitch. The partial coherence is 0.9-0.7 annular with 9 x 9 segmentation in the 1st quadrant of the illumination pupil. The mask is a binary with no Mask 3D scattering considered. The threshold is 0.272 for (a) and 0.11 for (b).

Figure 6(a) shows that for the NTD, the active energy above the photo-chemical reaction threshold is much

978-1-5090-6695-7/17 $31.00 © 2017 IEEE

higher compared with that of PTD case (Figure 6(b)). The average intensity above the threshold for the NTD is about **10.2X** that of PTD! This is very interesting saying that the NTD is favored at semi-dense pitches, or the commonly referred to as the "forbidden" pitch.

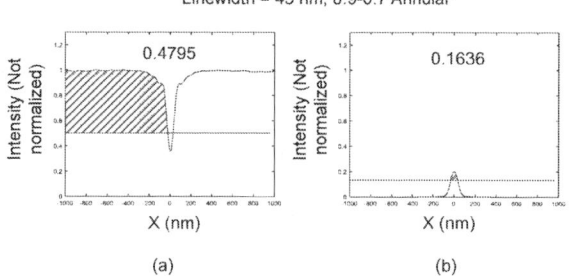

Figure 7: aerial image simulated with 90 nm thick photoresist with zero acid diffusion and 1.35NA with no polarization for a 45 nm (a) line (b) space at 2000 nm pitch. The partial coherence is 0.9-0.7 annular with 9 x 9 segmentation in the 1st quadrant of the illumination pupil. The mask is a binary with no Mask 3D scattering considered. The threshold is 0.4795 for (a) and 0.1636 for (b).

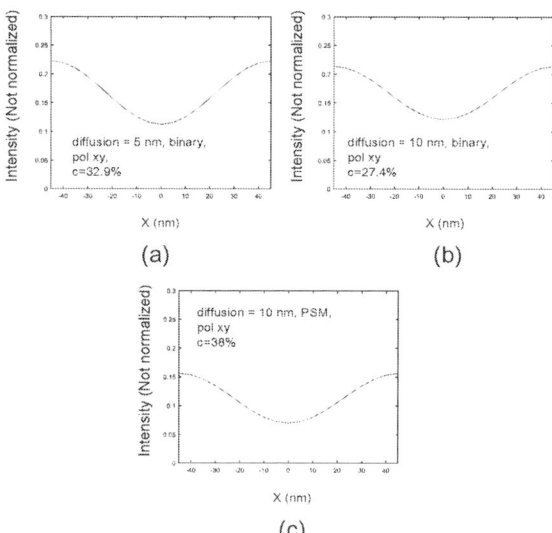

Figure 8: aerial image simulated with 90 nm thick photoresist at 1.35NA with XY polarization for a 45 nm (a) line (b) space at 90 nm pitch. The partial coherence is 0.9-0.7 annular with 9 x 9 segmentation in the 1st quadrant of the illumination pupil. (a) has a photo-acid diffusion length of 5 nm and a binary mask, (b) has a photo-acid diffusion length of 10 nm and a binary mask, (c) has a photo-acid diffusion length of 10 nm and a 6% PSM mask. The mask scattering is not considered.

For the more isolated lines and spaces, the situation remains pretty much the same. Shown in Figures 7(a) and 7(b) are simulations for the pitch of 2000 nm. With the same conditions, the average intensity above the threshold for the NTD is about **20X** that of PTD! This simulation indicates that the NTD is favored by the more isolated patterns, which can provide an order of magnitude more illumination for photo-chemical reaction at the same exposure dose. But all this is gained with 1cost, which is the aerial image contrast at the dense pitches, such as the 90 nm pitch, the NTD needs more photo-chemical reaction where it can only obtain through material innovation or through more photo-acid diffusion.

Unfortunately, most NTD photoresists used in the production today uses the latter approach: through lengthen the effective photo-acid diffusion length. Demonstrated in Figures 8 (a), 8 (b), and 8 (c) are 3 simulations for 3 typical conditions. Figure 8(a) represents a PTD situation, which has a typical 5 nm acid diffusion length with a binary mask and XY polarization. The image contrast c, which defined as,

$$c = \frac{I_{MAX} - I_{MIN}}{I_{MAX} + I_{MIN}}, \qquad (4)$$

for the situation in Figure 8(a), is 32.9%, which approximately corresponds to a 0.314x32.9%=10.3% exposure latitude at +/-10% linewidth variation [7]. If we switch to NTD with all other condition unchanged through making the diffusion length longer to 10 nm, the image contrast, shown in Figure 8(b) becomes reduced to 27.4%, which is equivalent to about 8.6% exposure latitude. The image contrast loss about 16.7% will amount to about 2^3 more chemical amplification. Shown in Figure 8(c) is a simulation with a 6% PSM mask and the 10 nm photo-acid diffusion length. The image contrast has restored back to 38%, or an equivalent EL of 11.9%. This improvement, however, is not without a price. The price is the exposure energy will increase to about 48%, as indicated by the threshold decrease from 0.1645 in Figures 8(a) and 8(b) to the 0.1132 in Figure 8(c).

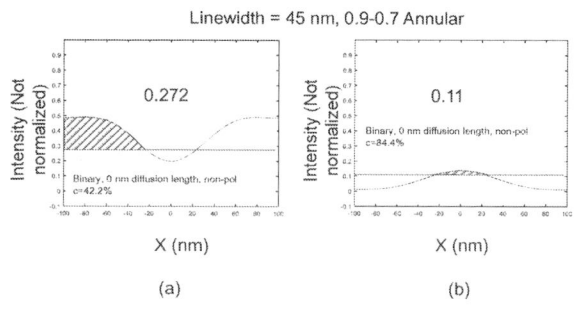

Figure 9: re-plot of Figures 6(a) and 6(b) showing image contrast = (a) 42.2%, (b) 84.4%. The threshold is 0.272

978-1-5090-6695-7/17 $31.00 © 2017 IEEE   84

for (a) and 0.11 for (b).

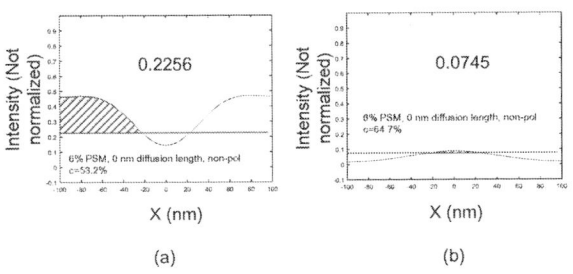

(a)　　　　　　(b)

Figure 10: everything is same to Figures 9(a) and 9(b) but with a 6% PSM mask showing image contrast = (a) 53.2% , (b) 64.4%.

It seems that the 6% PSM mask is a must for the use of NTD at dense pitches, which can gain back some image contrast lost by the need for more chemical amplification. The use of phase shifting mask, however, has consequence of enhancing the line performance, while damaging the performance of spaces. For more isolated features, such as when the pitch = 200 nm, the image contrast with a binary mask, shown in Figures 9(a) and 9(b), is 42.2% for a line and 84.4% for a space, and is equal to 53.2% for a line and 64.4% for a space in case of a 6% PSM mask, shown in Figures 10(a) and 10(b). It looks like that is has improved the unbalanced original image contrast between line and space. However, from the photoresist point of view, remember we have calculated the average intensity above the threshold. At pitch=200 nm, the average intensity above the threshold for the NTD is 10.2X more than that of PTD. In the case of 6% PSM, the situation is getting more favorable toward the line; the number becomes **30.7X, A 3 time increase! This means that with the use of NTD, there is almost no chance in printing small isolated lines, or the lines have to be significantly sized up. This is same to the situation where PTD sees difficulty in printing isolated small contacts or narrow trenches.** This is the second conclusion of this paper.

CONCLUSIONS

We have studied 2 major challenges in 14 nm photolithography process, namely the flexible illumination and negative tone developing. We found that the optimized source must be of off-axis nature, where line end to line end, or trench end to trench end has to be separated to certain distance to enable largest overall process window. We have also found from simulation as well as photoresist evaluation experience, that the 6% PSM is a must for the application of negative tone developing photoresist process in order to have good dense pitch image contrast. The price to pay is that the resist line performance is damaged and the only way to repair is to make lines wider.

ACKNOWLEDGEMENTS

We would like to thank SMIC ATD colleagues and management for the support of this work.

REFERENCES

[1] A. Rosenbluth, S.J. Bukofsky, M. Hibbs, K. Lai, A.F. Molles, R.N. Singh, and A.K.K. Wong, "Optimum mask and source patterns to print a given shape," Proc. SPIE 4346, pp. 486-502, 2001.

[2] C.K. Ober, C. Ouyang, J-K. Lee, M. Krysak, " Solvent development processing of chemically amplified resists: chemistry, physics, and polymer science considerations", Proc. SPIE 7972, 797205-1 (2011).

[3] C. A. Mack, and J. E. Connors, "Fundamental difference between positive and negative tone imaging: chemistry, physics, and polymer science considerations", Proc. SPIE1674, 328-338 (1992).

[4] T. A. Brunner, and C. Fonseca, "Optimum tone for various feature types: positive versus negative" Proc. SPIE 4345, 30-36 (2001).

[5] S-H. Yang, E.S. Kim, S. Moon, S. Lee, S-W. Choi, J. Choi, "Optical Performance Comparison between Negative Tone Development and Positive Tone Development", Proc. SPIE 8325, 832504-1 (2012)

[6] Y. Bae, S-H. Lee, R. Bell, L. Joesten, G. Barclay, "Developer effect on the negative tone development process under low NILS conditions under low NILS conditions", Proc. SPIE 7972, 797207-1 (2011)

[7] R. Chang et al., "Nanoscale Integrated Circuits--the Manufacturing Process", (纳米集成电路制造工艺), pp124, Equation (7.13), Tsinghua university press, July 2014.

AN OFFLINE ROUGHNESS EVALUATION SOFTWARE AND ITS APPLICATION IN QUANTITATIVE CALCULATION OF WIGGLING BASED ON LOW FREQUENCY POWER SPECTRAL DENSITY METHOD

*Libin Zhang, Lisong Dong, Xiaojing Su, Yansong Liu, Lijun Zhao and Yayi Wei**

Institute of Microelectronics of Chinese Academy of Sciences (IMECAS),
University of Chinese Academy of Sciences, Beijing 100029, China
*Corresponding Author's Email: weiyayi@ime.ac.cn

ABSTRACT

It is greatly important to accurately characterize the critical dimension (CD) and the roughness of line patterns in advanced lithography and etch process. Thus we wrote an offline roughness evaluation software and did lots of relevant researches. Two different edge finding algorithms are compared to reduce the CD uniformity. Power Spectral Density (PSD) algorithm is used to analyze the LER and LWR. Palasantzas function is used for fitting the PSD data and giving three important parameters. Such software could evaluate the line wiggling in practical applications. The origin of line wiggling and qualitative methods based on low frequency PSD is studied. With the help of the models and algorithms, we can figure out the existence and strength of wiggling for a SEM image.

INTRODUCTION

With the continuous reduction of the integrated circuit (IC) manufacturing technology node, the critical dimension (CD) has already been reduced to only tens of nanometers. The commonly used measurement method of inline CDSEM has gradually raised measurement uncertainty errors for the state of the art IC node. Therefore, it is significant to find a better accurate analysis method for evaluating the line width and roughness. For the purpose of calculating line roughness quickly and accurately, power spectral density (PSD) method is one mainstream method, attracting lots of research interests [1]-[5].

In IC manufacturing, especially in lithography and etch process, line edge roughness (LER) and line width roughness (LWR) are the two critical parameters, especially in electrical sensitive layers, such as fin layer of finFET and 1X metal layers. There are some main drawbacks of the inline roughness measurement, such as the very limited measurement numbers, large machine time consuming, and structure damage during long time electronic interaction. Hence an offline roughness software is very essential in aforementioned measurement.

CALCULATION FLOW

The interface of our software is shown in Fig.1. There are four main functions, including image analysis, best edge finding, PSD calculation and PSD Palasantzas fitting. The interface displays a practical example and the work flow is plotted in figure 2. First, one CDSEM image could be imported into the software. Then the software conducts some image analysis, such as giving the dimension of each pixel in x and y directions, choosing the property region, and/or doing some denoising for the image. Then an edge finding method is chosen to find the properly edges of all lines. After that, PSD of LER and LWR could be calculated and plotted in left and right figures, respectively. Moreover, the best Palasantzas function will be given for fitting curves of LER and LWR PSD data, meanwhile, the values of three important characterized parameters will be shown in the interface.

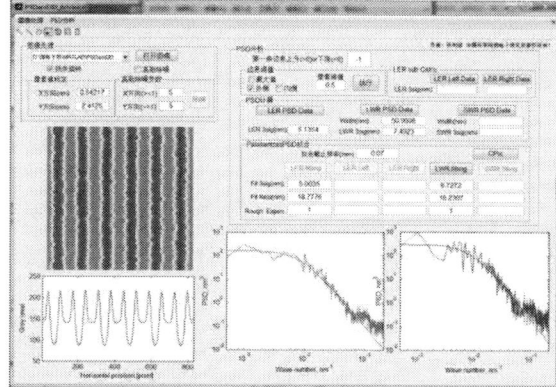

Figure 1: The interface of offline roughness software. One SEM image was used for finding the line edge position and its LER (red) and LWR (blue) PSD data and fitting curves.

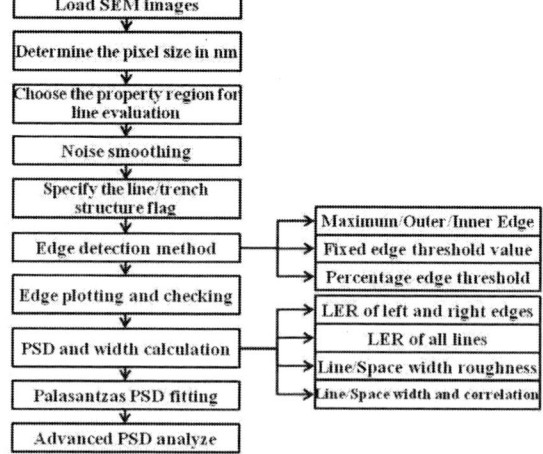

Figure 2: Flow of roughness PSD analysis.

EDGE FINDING METHOD

In real CDSEM metrology, some errors such as scanning direction or electron accumulation may cause different edge grey levels of single line or different lines, as shown in figure 3. Such phenomenon is hardly found from the top-down view of CDSEM image. What's worse, it is even more difficult to read this phenomenon directly from grey curve of every horizontal line. Therefore it is necessary to find the averaging grey level curve along with the line direction. With the help of such averaging curve, two edge finding methods were put forwards in this paper to study the line width uniformity of all lines, using fixed edge threshold value of 120 and percentage edge threshold of 50%. All the performances of the line widths have been plotted in Figure 4. The latter method has a smaller width range than the former one. On the basis of previous analysis, the percentage edge threshold method is adopted to determine the line edges.

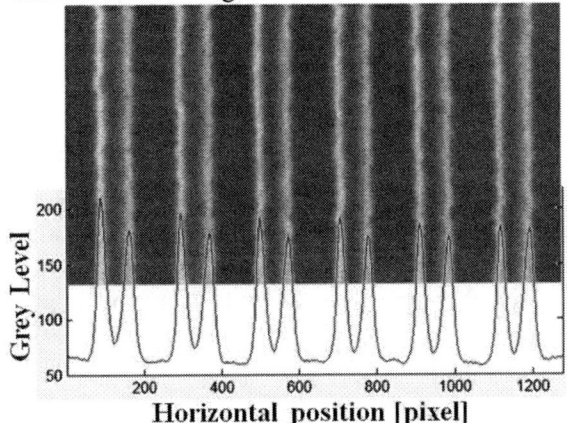

Figure 3: CDSEM image of dense lines and its averaging grey level curves.

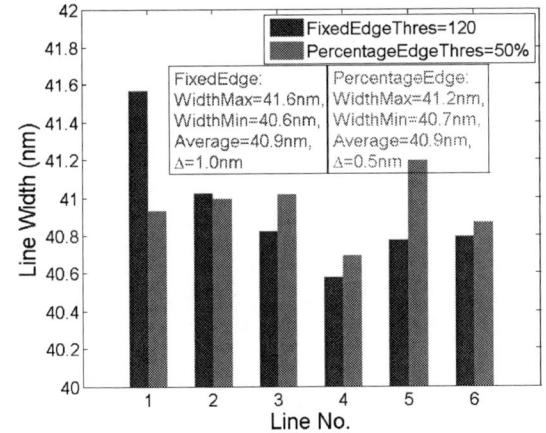

Figure 4: Different edge finding methods and the line width uniformity performance

PSD ANALYSIS AND CURVE FITTING

Power spectrum density method is a well used evaluation method in advanced metrology, especially in 193 immersion and EUV lithography where LWR is critical important. PSD curves indicate the line roughness in frequency domain, correlation length of the edges and the roughness index. So a relative smooth and accurate PSD curve features is important in quantitatively evaluating the line quality.

International Technology Roadmap for Semiconductors (ITRS) has pointed out that the length of line should be larger than 2μm for PSD analysis. It is one of the major reasons that the truncation error of Fourier period analysis method decides the previous result. However, it is seldom to measure such long lines for real industry metrology, especially for the advanced lithography. As a result, a detailed inspection of PSD curves of different line length and numbers is very practical and essential for real metrology.

Figure 5 gives the LER PSD curves of different line length and numbers. For 2μm length single line, the PSD curve show severe error amplitude. Averaging all lines makes the PSD data error range much smaller. Decreasing the length of lines only losses some low frequency information, displaying in the figure. But the trends of PSD curves keep the same with the longer length. To weaken the truncation error of PSD curves, a new method is used, plotting three or more PSD curves of different zone such as Z1, Z2 and Z3 in figure 5, meanwhile, averaging these curves. Thus, we conclude that it is not necessary to keep each line larger than 2μm for line quality comparison and evaluation.

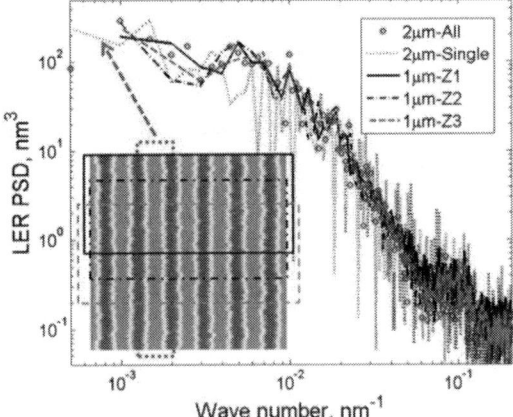

Figure 5: LER PSD curves of different line length and line number. The insert chart is line top view CDSEM image which has 2000nm length and 450nm width.

What is the minimum line length in real industry evaluation? When shortening the line length, the LER is reduced from the PSD curve calculation. For example, the LER of 500nm and 2000nm length in the same CDSEM image give 5.0nm and 5.14nm respectively in figure 5. We recommend that the line length should be larger than the correlation length at least one order of magnitude. The correlation length is mainly associated with the photo-acid diffusion length during photoresist development and PEB processes. Such correlation length can be obtained by fitting PSD data with Palasantzas function as shown in equation (1).

$$PSD(f) = \frac{PSD(0)}{\left[1 + \left(2\pi f \xi\right)^2\right]^{H+1/2}} \quad (1)$$

Where $PSD(0) = 2\sigma^2 \xi \left[\dfrac{\sqrt{\pi} \Gamma\left(H + \dfrac{1}{2}\right)}{\Gamma(H)} \right]$, σ is the line

roughness, H is the roughness exponent, and ξ is the correlation length [6].

The interface shown in Figure 1 gives a Palasantzas function fitting of both LER and LWR PSD curves. The correlation length of LER PSD curve is about 18.8nm. So the minimum line length must be larger than at least 188nm. Besides, the pixel number along the line should be large enough to keep an acceptable frequency step, which the recommended value equals 256 or more.

Palasantzas function is suitable for the lines featuring both edges of line without correlation. For some cases, such as the line wiggling discussed in next section, it is not suitable to fit PSD curves with Palasantzas function.

WIGGLING CALCULATION

Line wiggling is one critical challenge during patterning for large height-width ratio and/or narrow line width structures [7]. Some relevant literatures have pointed out that wiggling is the result of high inner stress [8], such as the film stacks with high different Young modulus [9], or atomic exchange during dry etching process that brings high stress in the structure [10].

The method to calculate the leveling of wiggling is based on LER, from low to high frequency roughness [11][12] . However, from our research, line wiggling is a low frequency (LF) roughness phenomenon. Thus the evaluation method should only focus on the LF LER and LWR.

The first question is to check the existence of wiggling. For an obvious wiggling depicted in Figure 6, two edges of each line have high correlation. As a result, the correlation coefficient between two edges should be close to 1. Besides, LWR of each line should be smaller than LER.

The line images in figure 6 and figure 7 have the same length and pixel numbers. The correlation coefficient is 0.98 and 0.33, respectively. What's more, the ratio of LER to LWR is 3.64 and 0.94, respectively. The above two parameters could describe the strong wiggling phenomenon shown in figure 6.

Sometimes, it is difficult to point out the difference of

similar CDSEM images in Fig. 8, which is difficult to directly determine how strong of the wiggling for P1 and P2 line structures. Using our roughness software, the PSD curves of LER and LWR of P2 image are plotted in Figure 9. In LF, LWR is less than LER, which means two edges of line have same changing trends. However, in middle frequency (MF) and high frequency (HF) areas, LER is less than LWR, which means the roughness of both edges in MF and HF is random. Such PSD comparison confirms that only low frequency roughness should be considered during line wiggling analysis.

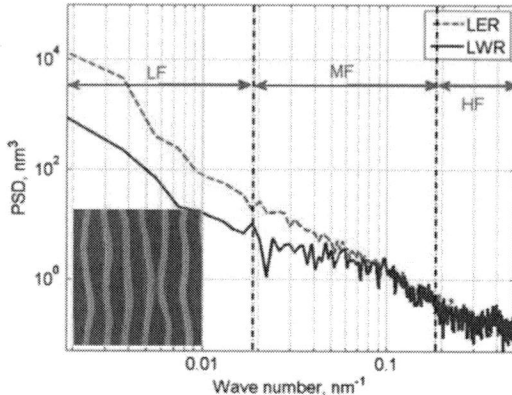

Figure 6: LER and LWR PSD curves with obvious line wiggling.

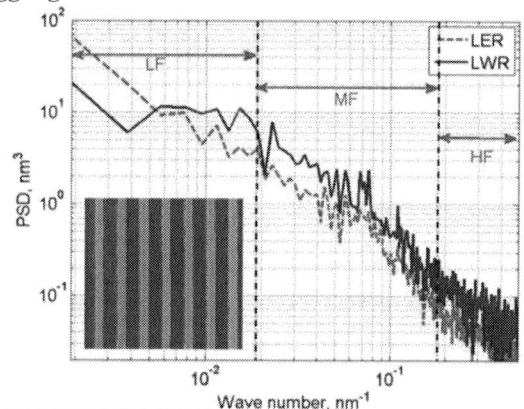

Figure 7: LER and LWR PSD curves without line wiggling.

Therefore, four evaluate parameters are used and summarized in Table 1. Parameter ① (LER LF) means low frequency of LER. Parameter ② gives the ratio of the LER LF to the LER MF and HF. For an obvious wiggling, such parameter value should be larger than 2. Parameter ③ is a quickly evaluation parameter that large value means serious wiggling. Parameter ④ strengthens the wiggling evaluation value and can be used to compare different images. In this table, two different figures were calculated using these parameters. In reality, P1 has stronger wiggling than P2, and all the four parameter values gave the same results.

Besides the quantitative evaluation of line wiggling, this offline roughness software could provide other applications, such as the comparison of different edge threshold, different noise smoothing methods and so on. Furthermore, lithography or etch process performance can be quantitatively evaluated with the help of PSD curves as well.

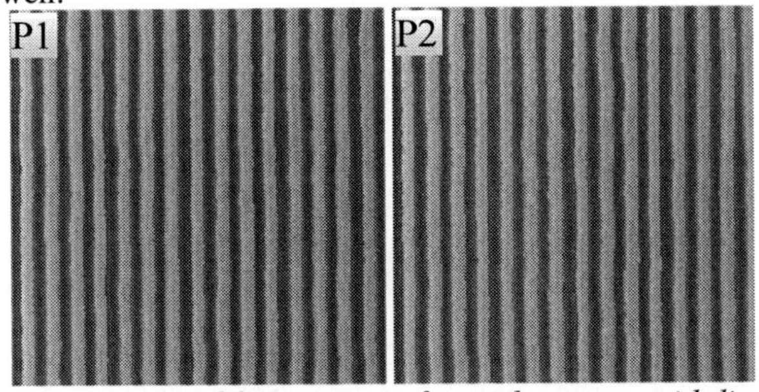

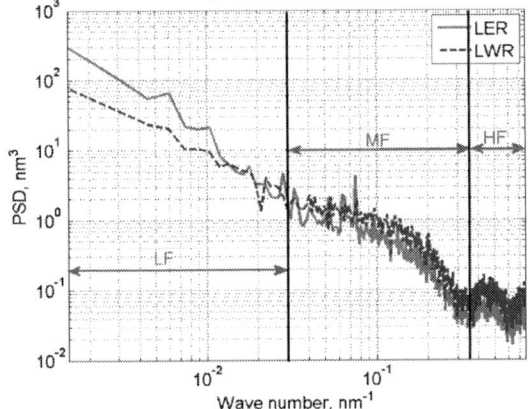

Figure 8: A real line picture after etch process with line wiggling. All line edges are also plotted together.

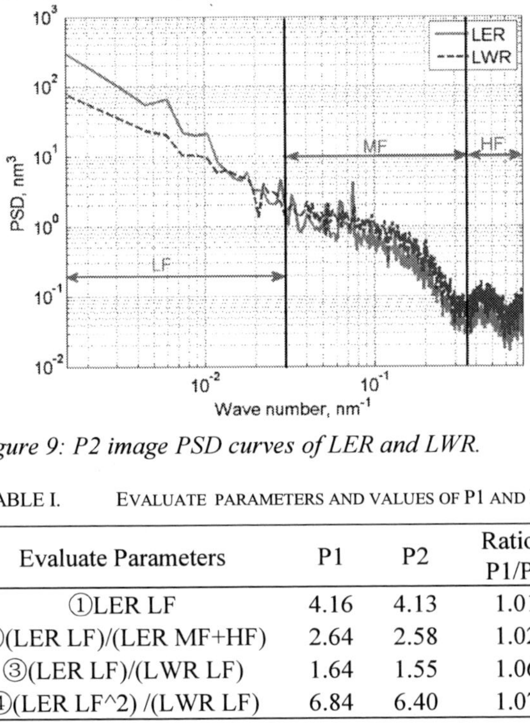

Figure 9: P2 image PSD curves of LER and LWR.

TABLE I.   EVALUATE PARAMETERS AND VALUES OF P1 AND P2

Evaluate Parameters	P1	P2	Ratio= P1/P2
①LER LF	4.16	4.13	1.01
②(LER LF)/(LER MF+HF)	2.64	2.58	1.02
③(LER LF)/(LWR LF)	1.64	1.55	1.06
④(LER LF^2) /(LWR LF)	6.84	6.40	1.07

CONCLUSION

Line width calculation and edge roughness evaluation are very important for advanced lithography and patterning. Based on a research flow, an offline roughness evaluation software is built. Some important comparisons were done to find a better quantitative flow theoretically. Line length and image pixel number are the two important parameters for PSD curve calculation. However, it is not necessary to make each line length be larger than 2μm. A property length which is larger than the correlation length at least one order of magnitude is recommended, together with large enough pixel numbers. Evaluating the wiggling quantitatively is an important application of PSD analysis method. With the low frequency LER and LWR PSD data, four parameters and methods were proposed. These methods can be used for real industry application, both for lithography and etching departments. Lots of other applications can also be analyzed based on the software.

ACKNOWLEDGEMENTS

This work is supported by the National Natural Science Foundation of China (No. 61604172 and 61504161) and Chinese National Science and Technology Major Project (No. 2016ZX02301001). The authors would like to thank the Integrated Circuits Advanced Process Center and Key Laboratory of Microelectronics Devices & Integrated Technology of IMECAS.

REFERENCES

[1]. C. A. Mack, *Applied Optics*, 52(7), 1472-1480 (2013).

[2]. C. A. Mack, *Proc. SPIE* 9424, 942403 (2015).

[3]. L. Sun, N. Saulnier, G. Beique et al., *Proc. SPIE* 9778, 977822 (2016).

[4]. S. Chauhan, M. Somervell, M. Carcasi et al., *J. Micro/Nanolith. MEMS MOEMS*, 13(1), 013012 (2014).

[5]. T. Verduin, P. Kruit, and C. W. Hagen, *J. Micro/Nanolith. MEMS MOEMS* 13(3), 033009 (2014)

[6]. C. A. Mack, *J. Micro/Nanolith. MEMS MOEMS*, 12(3) 033016 (2013).

[7]. M. Weigand, V. Krishnamurthy, Y. Wang et al., *Proc. SPIE*, 8685, 86850R (2013)

[8]. N. Kofuji, N. Negishi, H. Ishimura et al., *Japanese Journal of Applied Physics*, 53, 03DE01 (2014).

[9]. H. Tanaka, T. Hidaka, S. Izumi et al., *Journal of Applied Mechanics-Transactions of the Asme*, 81, 091009 (2014).

[10]. N. Kofuji, H. Ishimura, H. Kobayashi et al., *Japanese Journal of Applied Physics*, 54, 06FH04 (2015).

[11]. Y.-Y. Lin, C.-C. Chen, C.-Y. Li et al., *Proc. SPIE*, 8682, 86821C (2013)

[12]. L. Sun, X. Zhang, S. Levi et al., *Proc. SPIE*, 9780, 97801S (2016).

LEVEL-SET BASED ILT WITH A VECTOR IMAGING MODEL

Yijiang Shen[1]*

[1]School of Automation, Guangdong University of Technology, Guangzhou 510006, China

*Corresponding Author's Email: yjshen@gdut.edu.cn

ABSTRACT

Entering the 22nm realm and beyond, immersion lithography systems with hyper-NA cannot ignore the vector nature of electromagnetic fields, making optical proximity correction (OPC) approaches developed under scalar imaging models inadequate. In this paper, we present a level-set based inverse lithography technology with a vector imaging model; we first establish the forward vector imaging model of aerial images and wafer images, and then we treat photomask design as an inverse imaging problem interpreted as a time-dependent model solved with finite-difference schemes.

INTRODUCTION

The shrinkage of integrated circuit (IC) devices to 22nm technology node and beyond is pushing the potential of resolution enhancement techniques (RETs). Application of off-axis illumination [1] and immersion technology [2] in the state-of-the-art RETs enables highest possible spatial frequency contents in the optical signal which is projected by the lithography optics onto the wafer.

Photomask optical proximity correction (OPC) and inverse lithography technology (ILT), a strong candidate for 22nm and below and low k_1 regime, involves forward modeling known as lithography simulation and an inverse procedure inverting the imaging model and directly synthesizing the optimized mask pattern without the constraint of the topology of the original design [3]. Early works include branch and bound algorithm [4] and "bacteria" algorithm [5] with very limited applicability due to their high computation complexity. Poonawala and Milanfar designed a model-based OPC system and developed an efficient optimization algorithm using steepest descent for coherent imaging systems [6, 7] which is further generalized by Ma and Arce to multiphase phase-shifting mask (PSM) optimization under partially coherent illumination using conjugate gradient methods [8-10]. Applications of ILT are also reported in manufacturability enhancement [11] and photomask designs with variation robustness [12]. Level-set method offers a feasible alternative to inverse lithography whose application to ILT has been explored in [3] and in [13,14], Shen and his coworkers presented systematic level-set optimization formulations and developed finite difference schemes to solve them with high fidelity and robustness. While enriching the weaponry for solving ILT problems, these approaches are based on the scalar imaging models, although proven sufficiently accurate with numerical apertures (NAs) less than approximately 0.4, and are inadequate in describing the vector nature of optical propagation in lithography systems with NAs larger than 0.6. Replacing the air gap between the final lens and the wafer by a liquid medium with a greater than 1 refractive index, the immersion lithography systems are equipped with hyper-NAs, therefore, the imaging process must account for the vector nature of the electromagnetic field propagating the projection system.

Previously, in [15], Yeung developed the Hopkins formulations incorporating thin film effect which simplifies his earlier work [16] proposing a vector imaging and film-stack model that includes high-NA effect. Flagello investigated the high-NA phenomenon in a homogenous film stack and developed a thin-film model in matrix format in coherent systems [17] which was extended to partially coherent imaging systems by Pistor using Abbe method [18]. Peng et al. summarized the above developments in high-NA projection optics into a uniform and consistent formulation with further simplification [19].

This paper focuses on the development of algorithms for photomask synthesis using level-set-based inverse lithography with vector imaging. The advantage of the proposed method is twofold: the vector imaging model is considerably more accurate than the scalar imaging model; the level-set approach treats the mask as a continuum instead of discretizing masks into pixels generating assistant features "naturally" and eliminating unwanted small block objects in the synthesized mask.

VECTOR IMAGING MODEL

A schematic of image formation based on a vector imaging model is illustrated in Figure 1. Consider a monochromatic wave propagating in the direction $\vec{k}$, where $\vec{k} = (\alpha, \beta, \gamma)^T$ is the direction cosine with $\rho = \sqrt{1 - \alpha^2 - \beta^2}$. It is found convenient to use the polarization systems when tracking the projection of light, while a spatial coordinate system is preferred when computing the actual field or intensity. With TE- or s-direction and TM- or p-direction defined as $e_\perp$ and $e_\parallel$, the mapping between the two systems are computed as

$$\begin{bmatrix} E_x \\ E_y \\ E_z \end{bmatrix} = \begin{bmatrix} -\frac{\beta}{\rho} & -\frac{\alpha\gamma}{\rho} \\ \frac{\alpha}{\rho} & -\frac{\beta\gamma}{\rho} \\ 0 & \rho \end{bmatrix} \begin{bmatrix} E_\perp \\ E_\parallel \end{bmatrix} = T \begin{bmatrix} E_\perp \\ E_\parallel \end{bmatrix}. \quad (1)$$

For a point source (α_s, β_s) emanating an electric field $E_0 = (E_\perp, E_\parallel)^T$, according to (1), the electric field in

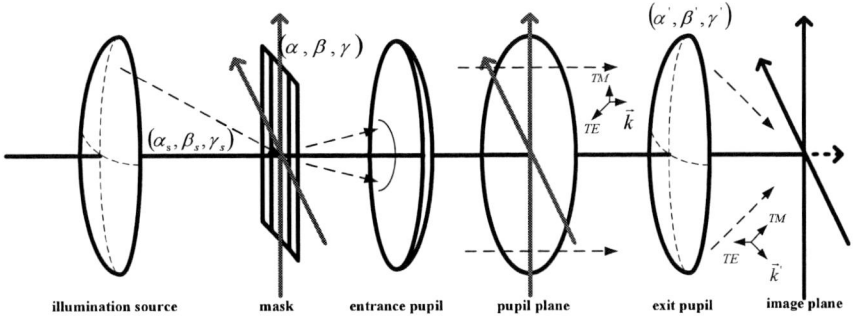

Figure 1: Schematic of vector imaging model

spatial coordinate is computed as $E_0' = TE_0$. Denoting the mask as a scalar matrix $M(r) \in \mathcal{R}^{N \times N}$, the mask near field can be represented as

$$E = E_0' \odot B \odot M, \qquad (2)$$

where $\odot$ is the entry-by-entry multiplication, B is the mask diffraction matrix and we drop r which denotes spatial coordinates where there is no ambiguity. In general, the mask near field has to be computed with a rigorous Maxwell-equation solver, such as FDTD and RCWA methods, which is literally impractical for every source point. In [20], Ma *et al.* adopted a simple embodiment of the CSCA from [21], where the entry in B is defined as

$$B(m,n) = e^{\frac{j \times 2\pi \times \beta_s \times m}{N}} \times e^{\frac{j \times 2\pi \times \alpha_s \times n}{N}}, m,n=0,1,\cdots N-1 \qquad (3).$$

Although B in (3) by no means gives the best mask diffraction matrix, it is indicative of how CSCA can be applied to approximate the mask near field where more accurate scattering coefficients can be introduced into (2).

Following the derivation in [20], the aerial image intensity can be computed as

$$I = \frac{1}{N_s} \sum_{\alpha_s} \sum_{\beta_s} \sum_{p=x,y,z} \left\| H_p^{\alpha_s \beta_s} \otimes (B^{\alpha_s \beta_s} \odot M) \right\|_2^2, \qquad (4)$$

where H_p, B are both functions of (α_s, β_s), $H_p^{\alpha_s \beta_s}$, $p = x, y, z$, are referred to as the equivalent filters of the x, y, z components. The resist effect can be approximated by a constant threshold resist model applying the logarithmic sigmoid function

$$sig(I) = \frac{1}{1 + e^{-a(I - t_r)}}, \qquad (5)$$

with a being the steepness of the sigmoid function and t_r the threshold. Putting (4) and (5) together, we derive $Z = \mathcal{T}(M) = sig(I) =$

$$sig \left(I = \frac{1}{N_s} \sum_{\alpha_s} \sum_{\beta_s} \sum_{p=x,y,z} \left\| H_p^{\alpha_s \beta_s} \otimes (B^{\alpha_s \beta_s} \odot M) \right\|_2^2 \right), \qquad (6)$$

where $\otimes$ stands for convolution operation.

LEVEL-SET-BASED MASK SYNTHESIS

The objective of inverse lithography with a forward model $\mathcal{T}$ in (6) is to find a predistorted input intensity function M^* which minimizes its distance with a desired output I_0, i.e.,

$$M^* = \arg \min_{M \in \mathcal{R}^{N \times N}} d \{I_0, \mathcal{T}(M)\}, \qquad (7)$$

in which $d\{\cdot, \cdot\}$ is an appropriately defined distance metric such as the ℓ_2 norm. In the level-set-based framework of mask optimization, a level set function ϕ relates to the intensity function M as

$$M = \begin{cases} m_{int} & \text{for } \{r: \phi(r) < 0\} \\ m_{ext}, & \text{for } \{r: \phi(r) > 0\} \end{cases}, \qquad (8)$$

where $m_{int} = 1$ and $m_{ext} = 0$ if we are dealing with binary masks. Solving (7) with a least square fit is equivalent to seeking the minimizer of

$$F(M) = \frac{1}{2} |\mathcal{T}(M) - I_0|^2, \qquad (9)$$

where the boundary of M is governed by the zero level set of ϕ, thus the inverse lithography problem is reformulated to handle the level set function ϕ instead of the pixelated mask M. Following the steps in [13] and [14], we arrive at the time-dependent model

$$\frac{\partial \phi}{\partial t} = -|\nabla \phi| \upsilon(r, t), \qquad (10)$$

where t is the artificial time and $\upsilon(r, t)$ is defined as

$$\upsilon(r, t) = -\mathcal{J}(M)^T (\mathcal{T}(M) - I_0) = \frac{1}{2} \frac{\partial}{\partial M} (I - I_0)^2 =$$

$$-\frac{2a}{N_s} \sum_{\alpha_s} \sum_{\beta_s} \sum_{p=x,y,z} \text{Real} \left[(B^{\alpha_s \beta_s})^* \odot \left(\left(H_p^{\alpha_s \beta_s} \right)^{*\circ} \otimes \right. \right.$$

$$\left. \left. \left\{ \left[H_p^{\alpha_s \beta_s} \otimes (B^{\alpha_s \beta_s} \odot M) \right] \odot (I_0 - Z) \odot Z \odot (1 - Z) \right\} \right) \right]. \qquad (11)$$

In (11), $\mathcal{J}(M)$ is the Jacobian of $\mathcal{T}(M)$ at M, $\{\cdot\}^*$ is the conjugate operation, $\{\cdot\}^\circ$ is to flip the matrix in the argument in both up-down and left-right directions. Equation (11) is a partial differential equation (PDE). Once ϕ and υ are defined at every point on the Cartesian grid, (11) can be readily solved using finite-difference methods with first-order temporal and second-order spatial accuracy as suggested in [13], [14].

NUMERICAL RESULTS

We apply the level-set-based inverse lithography technique outlined above to design various circuit patterns with vector imaging emphasis. System parameters used in the simulations are: $\lambda = 193$ nm, $NA = 1.35$, spatial resolution $\Delta x = 4$ nm/pixel, steepness of the sigmoid function $a = 85$, threshold $t_r = 0.3$, and the system is illuminated by a conventional source with partial coherent factor $\sigma = 0.6$. It should be noted that the proposed method is tested on binary masks in this paper, however, the same framework can be conveniently applied to PSMs.

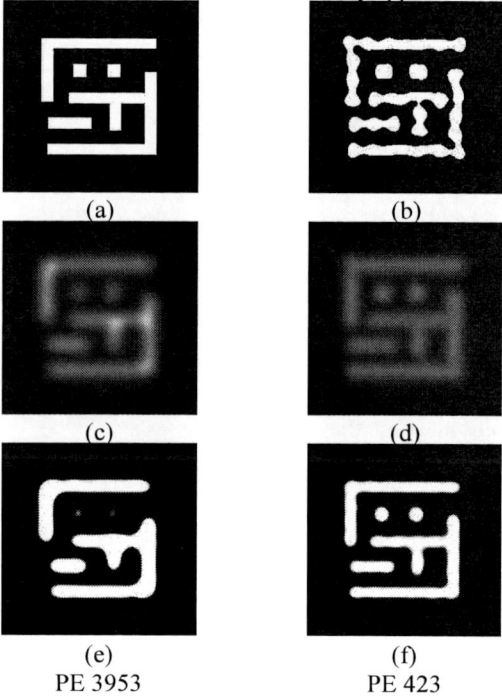

(a)   (b)

(c)   (d)

(e)   (f)
PE 3953   PE 423

Figure 2: Simulation of photomask synthesis using the proposed level-set-based ILT. (a) and (b) use the desired pattern and the synthesized pattern as inputs, with (c) and (d) being respective aerial images, (e) and (f) being respective wafer images.

In Figure 2, the desired pattern is given in (a) with image size 257×257; the synthesized pattern using the proposed method after 200 iterations is shown in (b); (c), (e) and (d), (f) are the aerial images and wafer images of (a) and (f), respectively. It is observed that assistant features are generated as the zero level-set of $\phi(r)$ evolves with a speed of $\upsilon(r,t)$, greatly improving the pattern fidelity from pattern error (PE) 2953 in (e) to PE 423 in (f). It is also worth mentioning, however, that the computation efficiency is severely limited by the computation of $\upsilon(r,t)$ in the spatial domain, resulting hours of simulation time.

CONCLUSIONS

A level-set approach for photomask design in optical microlithography emphasizing the vector nature of image formation in sync with the development of hyper-NA optical systems is formulated. The level-set optimization framework describes the inverse problem as a time-dependent model PDE which is solved by finite-difference temporal and spatial schemes. Experimental results show that the assistant features are generated spontaneously with the evolution of the level-set function while eliminating small unwanted blocks. Future research will lead to incorporating more

accurate mask diffraction matrix and accelerating methods to improve convergence.

ACKNOWLEDGEMENTS

The work in this paper was partially supported by Natural Science Foundation of Guangdong Province, China (2016A030313709, 2015A030310290) and Guangzhou Science and Technology Project, China (201607010180) . The author greatly acknowledges the suggestions from Edmund Lam (Imaging System Lab, the University of Hong Kong).

REFERENCES

[1] A. Wong, *Resolution Enhancement Techniques in Optical Lithography*, SPIE Press, 2001.

[2] B. Streefkerk, J. Baselmans, W. Gehoel-van Ansem, J. Mulkens, C. Hoogendam, M. Hoogendorp, D. Flagello, H. Sewell and P. Graupner, *Proceedings of the SPIE 5377*, May 24, 2004, pp. 285-305.

[3] L. Pang, Y. Liu and A. Dan, *Proceedings of the SPIE 6607*, May 15, 660739,.

[4] Y. Liu and A. Zakhor, *Proceedings of the SPIE 1264*, June 1, 1990, pp. 401-402.

[5] Y. Liu and A. Zakhor, *Proceedings of the SPIE 1264*, June 1, 1991, pp. 382-399.

[6] A. Poonawala and P. Milanfar, *Proceedings of the SPIE 5674*, 2005, pp. 114-227.

[7] A. Poonawala and P. Milanfar, *IEEE Trans Image Process.*, vol. 16, 2007, pp. 774-788.

[8] X. Ma and G. Arce, *J. Opt. Soc Am.*, vol. A 25, 2008, pp. 2960-2970.

[9] X. Ma and G. Arce, *Opt. Express.*, vol. 16, 2008, pp. 20126-20141.

[10] X. Ma and G. Arce, *J. Opt. Soc Am.*, vol. 19, 2011, pp. 2165-2180.

[11] N. Jia, A. Wong and E. Lam, *Proceedings of the SPIE 7520*, 2009, 753032.

[12] N. Jia, A. Wong and E. Lam, *Proceedings of the SPIE 7140*, 2008, 7140w.

[13] Y. Shen, N. Wong and E. Lam, *Opt. Express.*, vol. 17, 2009, pp. 12259-12268.

[14] Y. Shen, N. Wong and E. Lam, *Opt. Express.*, vol. 19, 2011, pp. 5511-5521.

[15] M. Yeung, *Proceedings of the SPIE 922*, 1988, pp. 149-167.

[16] M. Yeung, D. Lee, R. Lee and A. Neureuther, *Proceedings of the SPIE 1927*, 1993.

[17] D. Flagello, *Ph. D. Dissertation*, 1993.

[18] T. Pistor, *Ph. D. Dissertation*, 2001.

[19] D. Peng, P. Hu, V. Tolani and T. Dam, *Proceedings of the SPIE 7640*, 2010, 76402Y.

[20] X. Ma, Y. Li and L. Dong, *J. Opt. Soc Am.*, vol. 29, 2012, pp. 1300-12.

[21] T. Pistor, A. Neureuther, and R. Socha, *Proceedings of the SPIE 4000*, 2000, pp. 228-237.

DATA ANALYTICS AND MACHINE LEARNING FOR DESIGN-PROCESS-YIELD OPTIMIZATION IN ELECTRONIC DESIGN AUTOMATION AND IC SEMICONDUCTOR MANUFACTURING

Luigi Capodieci, Ph.D.[1*]

[1]Motivo Data Analytics , Sunnyvale, California 94086, USA

*Corresponding Author's Email: lmc@motivo.ai

ABSTRACT

In response to the current challenges of end-of-Moore scaling, a systematic analysis of the *data information flows* in the Design-to-Manufacturing pipeline highlights opportunities for the introduction of (big) data analytics and machine learning solutions. In this paper we review the eco-system components and describe the fundamental data-flows in the IC Design-to-Manufacturing chain, highlighting both the well-established and functioning sub-systems, as well as the critical bottlenecks. A quantitative definition of *physical design space coverage* is proposed, as the unifying abstraction available for all components of the Design-to-Manufacturing flow, allowing for the construction of a computational framework where Data Analytics and Machine Learning methodologies and tools can be successfully applied. The juxtaposition of Design-Technology-Co-Optimization (DTCO) with the novel paradigm of DFM-as-Search and their necessary integration in the DFM computational toolkit, clearly exemplify how the all the advanced IC nodes (14, 10, 7 and 5nm) definitely *require* the adoption of a new class of correlation extraction algorithms for heterogeneous data sets.

INTRODUCTION

The accelerated and greatly publicized growth of Data Analytics and Machine Learning methodologies across many application domains and in virtually every industry seems to have missed almost completely the semiconductor integrated circuit (IC) space. With the 14nm process technology node currently in production and both 10nm and 7nm nodes at different stages of development, the IC eco-system is being restructured and consolidated across its four traditional chain components: Fabless Design Companies, Electronic Design Automation (EDA) and IP suppliers, Process and Metrology Tools Suppliers and Silicon Foundries. Intrinsic technology factors, such as the continuing slow-down in geometric scaling and the delayed introduction of key patterning technologies are a primary source of disruption. But there are also critical hidden gaps

and bottlenecks in the Design-to-Manufacturing *data information pipeline*. The deployment of carefully selected data analytics techniques (with and without machine learning algorithms) represents a strategic opportunity to enable a 2-year/node ("more-Moore") cycle at 10nm and below.

Data Analytics for Semiconductor Manufacturing

Since the terms "data analytics" (sometimes also popularly referred to as "Big Data") and "machine learning" comprise a vast constellation [1] of mathematical methodologies, computational platforms and system-level solutions, only few specific examples will be shown in this article with a novel application for Physical-Design-Process-Yield Co-Optimization.

Design-Technology Co-Optimization (DTCO) is a well-established approach [2] which leverages collaboration and cross-domain expertise among physical designers, lithographers, process integration and yield engineering in order to simultaneously define fundamental layout components of cell-architecture, generate geometric design rules and interconnect router rules, while constraining the physical design to be manufacturable, within the expected capabilities of (mostly) patterning and other critical process modules. Even in the presence of "restricted" (and/or highly restricted) Design Rules (DR) the main drawback of traditional DTCO is the limited number of layout configurations, which can be considered, evaluated and optimized against process capabilities (both by simulation and silicon test vehicles). This in turn results in a limited predictability of manufacturing yield for actual IC products, which are not adequately *represented* by the DTCO layout variants.

Physical Design Space Coverage

From a combinatorial perspective the space complexity of layout variants is very large (typically with 10^20 cardinality for single layer layouts to 10^80 for multi-layer). A rigorous quantitative methodology for fully characterizing systematic layout variability has been recently proposed [3], with the definition of *Physical*

Design Space Coverage, (DSC) a unifying abstraction for all components of the Design-to-Manufacturing data-flow, making it the necessary mathematical foundation for a practical data analytics computational framework.

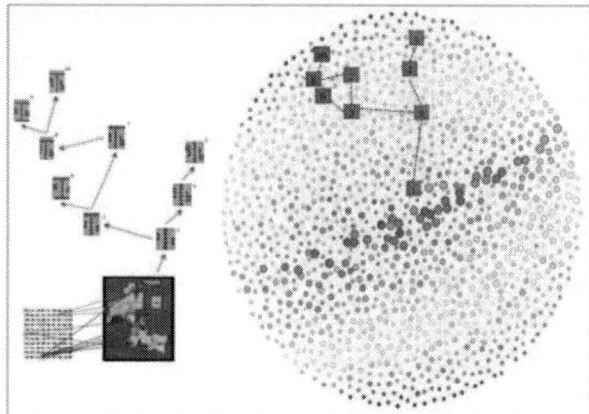

Figure 1: Generation of a topological network map for semiconductor integrated circuit (IC). The physical domain (i.e., a layout) is broken down into abstract element components that are then represented as a set of "connected" multidimensional points in a data-cloud.

In spite of the apparently intractable space complexity, nevertheless, various algebraic combinatorics techniques [4] can be used to show that it is possible to compute all possible layout design variations in less than quadratic-time (and often, for most practical cases, for real IC product layouts, in *almost* linear-time). Coverage can be understood intuitively as a measure of all combinatorial variations in a layout, at all length scales (a direct correlate of its intrinsic geometric entropy). It must be observed that these combinatorial techniques for the quantitative characterization of an abstract space are already commonly adopted in several "big data" domains, from web-search, to genomics, to sentiment text analysis, to cyber-security and anomaly detection.

Physical Design Analytics in Topological Space

For IC R&D and production, the most immediate application of coverage is the quantification of how much a given layout design A is *similar to* (or dissimilar from) another layout B and therefore how *representative* a layout is of any other layout, with respect to process manufacturability and yield. A new (analytics-based) implementation of DTCO is thus made possible, where human engineering expertise about layout components (cell architecture and routed interconnections) is greatly augmented by computational tools for determining "how-much" of a design space has been evaluated (and optimized). This is done by computationally mapping the physical domain space (geometric layout features, edges, polygons, layers, etc.) onto a high-dimensional

topological network map, using well-established (albeit somewhat esoteric) topological data analysis (TDA) techniques [5]. The conceptual construction of a topological network map is shown in Figure 1, where the physical domain (a layout or a set of layouts) is broken down into abstract elemental components, which are then represented as a set of multi-dimensional points (data cloud). This data cloud is then further organized as a topological network (multi-graph) by identifying and storing the intrinsic data relationships among the points, together with extrinsic (domain specific) connections. An entire class of well-established algorithms can then be deployed for the computations of topological network properties and sub-structures, which are used to extract (machine-driven) insights back into the physical domain space.

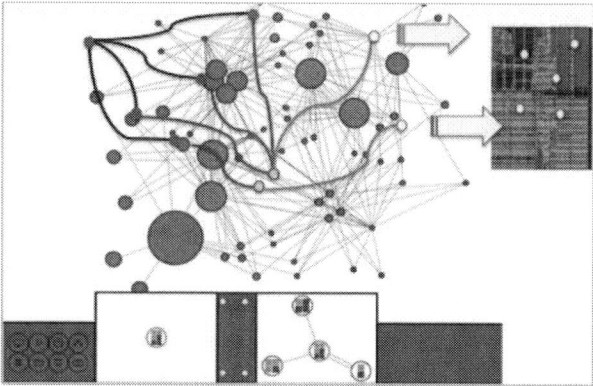

Figure 2: Machine learning, graph traversal, and search to predict the locations of yield detractors. Both supervised and unsupervised machine learning are applied to predict yield-detractor patterns, via graph traversal and node clustering. A known problematic pattern (in blue on the left) is algorithmically broken down into fundamental components (also in blue), which are then clustered into key components (red). The clustered key components can be extrapolated to identify new problematic nodes (yellow), which are finally mapped back onto specific locations in the physical design layout.

Yield Optimization through Graph-Search

To demonstrate how Design-Process-Yield optimization is transformed into a graph-traversal and search problem, let's consider, as a simple example, the prediction of yield-detractors (hotspot layout configurations, under certain process conditions) based on a set of physical measurements (for instance SEM metrology and/or Failure Analysis). In the physical space (geometric layout) domain, empirical observations are typically clustered using various physical attributes, but experimental noise and information loss during dimensional reduction can severely limit the proper identification (and prediction) of actual yield-detractor clusters. In topological network space, physical data points are represented by sub-networks (sub-graphs, with specific sets of nodes and edges) where commonalities and root-causes can be algorithmically identified. Specifically, several machine-learning methodologies [6] (supervised and unsupervised learning) are directly applicable to graph traversal and search for "common causal nodes", as shown in Figure 2. This is just one simple example of the many applications of data analytics and machine learning to the semiconductor design-to-manufacturing space. The novelty of this methodological approach consists in the use of a topological network space where analytics and machine learning is implemented, in contrast to traditional and noise-prone approaches [7], where raw physical data is directly analyzed and clustered (i.e. "machine-learned").

CONCLUSIONS

In summary in this work we have illustrated how advanced data analytics and machine learning methodologies could (and should) be deployed in the joint optimization of Physical Design, Manufacturing Process and IC Product Yield. While the application field is rich in opportunities, there seems to be still only a very small number of semiconductors specific solutions being offered. A substantial and promising innovation has been the recent introduction of quantitative techniques for the characterization of the complete Design Space Coverage, based on rigorous combinatorics theory. This approach, when applied jointly with graph search and machine learning has the potential to open up an entire new class of yield optimization solutions for 10, 7, 5nm and beyond.

ACKNOWLEDGEMENTS

The author would like to thanks and recognize the valuable contribution of the R&D and Engineering teams at Motivo Data Analytics for providing real industrial application examples for the new methodological approaches of data analytics and machine learning for semiconductor manufacturing optimization presented in this work.

REFERENCES

1. P. Warden, Big Data Glossary: A Guide to the New Generation of Data Tools, *O'Reilly Media*, September 2011.
2. L. Liebmann, K. Vaidyanathan, L. Pileggi, Design Technology Co-Optimization in the Era of Sub-Resolution IC Scaling, *SPIE Vol. No.: TT104*, January 2016
3. V. Dai, E. KC Teoh, J. Xu, B. Rangarajan, Optimization of Complex High Dimensional Layout Configurations for IC Physical Designs using Graph Search, Data Analytics and Machine Learning,

Lithography Workshop, Hawaii, November 2016
4. P. Flajolet, R. Sedgewick, Analytic Combinatorics, *Cambridge University Press*, 2009
5. G. Carlsson, Topology and Data, *Bull. Amer. Math. Soc. 46, 255-308*, 2009
6. X. Wu, V. Kumar et al., Top 10 Algorithms in Data Mining, *Journal of Knowledge and Information Systems, Volume 14 Issue 1, pp. 1-37, Springer-Verlag New York*, December 2007
7. N. Rana, Y. Zhang, T. Kagalwala, T. Baley, Leveraging advanced data analytics, machine learning, and metrology models to enable critical dimension metrology solutions for advanced integrated circuit nodes, *Journal of Micro/Nanolithography, MEMS, and MOEMS Volume 13, Issue 4*, December 2014

DEVELOPMENT OF 250W EUV LIGHT SOURCE FOR HVM LITHOGRAPHY

Taku Yamazaki, Hiroaki Nakarai, Tamotsu Abe, Krzysztof M Nowak, Yasufumi Kawasuji, Takeshi Okamoto, Hiroshi Tanaka, Yukio Watanabe, Tsukasa Hori, Takeshi Kodama, Tatsuya Yamagida, Yutaka Shiraishi, Takashi Saitou and Hakaru Mizoguchi

Gigaphoton Inc., Hiratsuka facility: 3-25-1 Shinomiya Hiratsuka Kanagawa, 254-8567, JAPAN

*Corresponding Author's Email: taku_yamazaki@gigaphoton.com

ABSTRACT

Gigaphoton has been developing extreme ultra violet (EUV) light source with the method of laser produced plasma (LPP) which is the most promising solution of the high power light source with 13.5 nm wavelength for high volume manufacturing (HVM) EUV lithography. Unique and original technologies such as combination of pulsed CO_2 laser and tin droplets, dual wavelength laser pulses shooting and debris mitigation with magnetic field have been developed. The theoretical and experimental data have clearly showed the advantage of our proposed method and it is promising to realize 250 W EUV power by 20 kW level pulsed CO_2 laser.

In the proto #2 light source, 130 W with 100 kHz, 50% duty cycle operation during 120 hour was achieved and recently short term operation at 250 W with closed loop operation was demonstrated.

In 2016 the first practical source for HVM; "GL200E" was constructed and high power CO_2 laser power more than 20 kW was already confirmed in collaboration with Mitsubishi electric cooperation. As a test run, five hours operation with 100 W EUV power and conversion efficiency (CE) 5% was confirmed. The next tests are being conducted continuously.

INTRODUCTION

Spread of "internet of things (IoT)" gives us various businesses to improve convenience, efficiency, security and so on. Some of them have already showed their great performance in various fields and the news media reports them every day. IoT has been realized by the technology progress in synactic combination of data, software and hardware. At the moment, hardware devices applied to "things" are relatively simple modules which consist of sensor, display, communication and low-end processor. But it is very easy to think that if "things" have smarter processor and larger memory with low power consumption, "IoT" will break new ground such as a self-driving car and autonomous robot, smart factory, etc.

The semiconductor industry assumes an important role to popularize this advance by achieving innovated device with higher density integrated circuits with lower cost. The EUV lithography is the most promising method to realize not only the ultimate miniaturization in the integrated circuits in terms of continuity from the conventional lithography technology but also the drastic reduction of manufacturing cost.

Gigaphoton has been developing the EUV light source of CO_2-Sn-LPP since 2002.[1]–[6] In the past various methods for EUV light emission including discharge produced plasma (DPP) had tried and tested in light source makers. After several researches and developments, Eventually CO_2-Sn-LPP method, that is generating the light of 13.5 nm wavelength from plasma produced with small tin droplet evaporated by high power carbon dioxide (CO_2) laser, was chosen because of its high efficiency and power scalability which can be applicable to semiconductor road map for some years. Another advantage is spatial freedom around plasma available to use for mitigation of tin debris emitted from tin plasma.

Now several lithography systems with LPP light sources have been set up and operated in advanced factories.[7] At these sites, some trial production with several tens watts are being demonstrated. This status means that the EUV lithography has become reliable as the method of the next generation that supports semiconductor manufacturing. On the other hand, speaking of the output power, it's still much smaller than the power advanced factories' requests, which means that 250 W is necessary to realize reasonable cost for production of logic and memory as 7 nm node generation in coming several years. Furthermore 500 W might be necessary in five years for the high NA lithography to achieve 5 nm node. Therefore power scaling is crucial issue for light source suppliers.

To cope with this situation, Gigaphoton light source has unique and novel concepts as explained later in this paper. The performance of the latest light source are also explained with experimental and measurements data.

LPP EUV LIGHTSOURCE SYSTEMS AND KEY COMPONENTS

System concept

The life cycle of tin which explains EUV emission and debris mitigation is shown in fig.1. In fig.2. the location of components in chamber is presented. Tin droplets with 20 micrometers diameter are continuously ejected in 100 kHz with speed of 300 km/h. Every single

978-1-5090-6695-7/17 $31.00 © 2017 IEEE

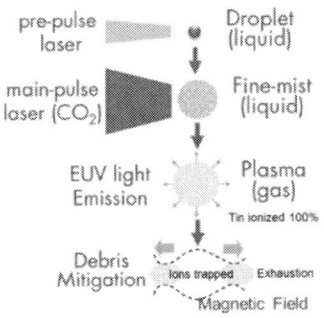

Fig.1 Tin life cycle

droplet is shot firstly by pre-pulse at the plasma point to squeeze and scatter as fine mist. After a certain delay, the mist expands and is heated by main-pulse beam of pulsed CO_2 laser. The tin mist is converted to ions as high temperature plasma and tin ions have several numbers of electrical charges. During recombination process, tin plasma emits EUV light of 13.5 nm wavelength. Most of tin ions are trapped by the magnetic field by Larmor movement and eventually exhausted form the chamber.

EUV light emitted homogenously from the plasma point and the half of them is reflected by collector and then it is sent to the intermediate focus point which is the

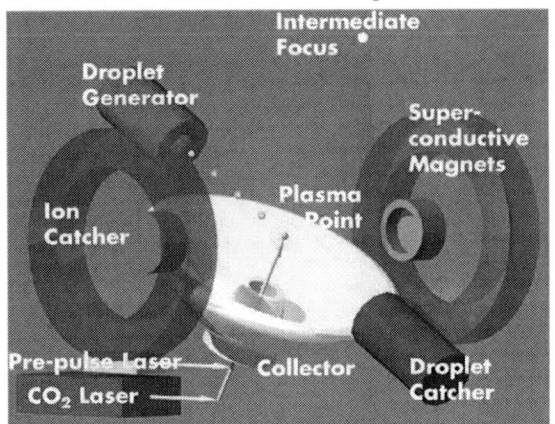

interface to the lithography tool.

Fig.2 Location of components

Pre-pulse technology

For evaluation of the efficiency of the system, conversion efficiency (CE) is generally used as an index, that is, CE = EUV energy / CO_2 laser energy [%].

In order to improve CE, tin configuration before main-pulse shot affects largely. The configuration is totally different in duration time in pre-pulse laser as shown in fig.3. In case of nsec duration, pre-pulse affects tin droplet in low energy with longer time, therefore tin droplet is liquid so that it shifts and deforms into flat disc. Main-pulse cannot penetrate all the tin and only the shallow portion on the irradiated surface contribute to EUV emission.

On the other hand, in psec duration case, pre-pulse attacks tin droplet in higher peak energy with triple-digit short time, this cause shock waves in whole of the droplet and it splashes in every direction and deform into mist

dome. The timing of main-pulse is adjusted for the dome size not to expand larger than beam diameter, therefore, all the tin is going to use for EUV emission.

This difference results drastic improvement in CE as

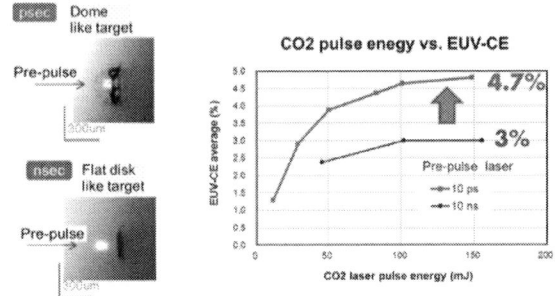

Fig.3 Pre-pulse technology

indicated in the right part of fig.3.

Powerful CO_2 laser

CO_2 laser system is the first key component which leads directly to the EUV output power. In order to shoot every tin target, pulsed laser with uniform quality in high energy and stable beam profile is required. The combination of one oscillator and four amplifiers is applied[8] to achieve this requirement as shown in fig.4.

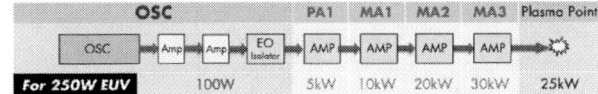

Fig.4 CO_2 laser system configuration

The oscillator module generate 14nsec pulsed laser with averaged power of 100 W. Four semiconductor lasers are applied as seeds to generate multiple wavelengths corresponding to each CO_2 oscillatory level. They are combined and delivered to a regenerative amplifier and it is shaped as a 14nsec pulse laser light by high speed shutter units.

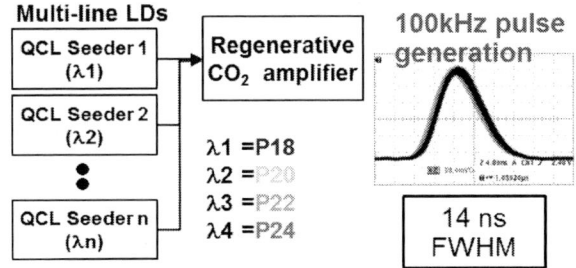

Fig.5 Oscillator configuration

As the amplifiers, fast axial flow (FAF) type was replaced with transvers gas flow (TGF) type to rise up efficiency (fig.6). Volume between electrodes in TGF is spacious so that large amount of excited gas can contribute nearly two times higher gain than FAF. Also TGF is more stable than FAF because light path length is one forth shorter than FAF and no bending mirror is necessary inside path in TGF.[9]

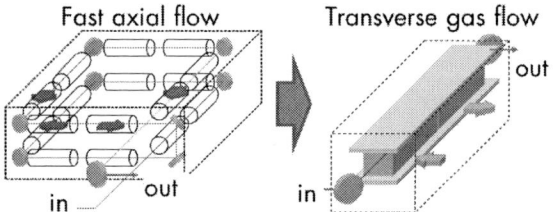

Fig.6 Amplifier configuration

Magnetic mitigation

Tin debris mitigation is crucial technology in terms of collector life time and system availability in cost of operation point of view.

Gigaphoton introduces the magnetic mitigation mentioned in the "System concept" section, but in reality not all the tin particles can be trapped in the magnetic field. Additionally tin disassociation from stannane gas (SnH_4) is another cause to pollute collector. Stannane gas is generated from tin etching by interaction of hydrogen and EUV light inside chamber. In order to reduce tin contamination as much as possible, operation parameters are being adjusted towards the perfect condition. In fig.7 the improvement status is shown as results from the proto EUV system. It improved quickly in these years and it's close to achieve the collector life time from semiconductor fabs.

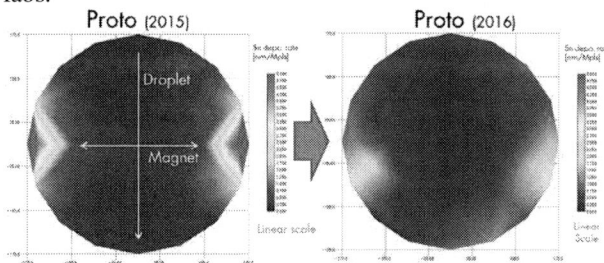

Fig.7 Debris mitigation improvement

SYSTEM TEST AND RESULTS
Overview

Three light source systems are currently working in

Table 1 Proto & Pilot systems

Operational Specification		Proto #1 Proof of concept	Proto #2 Power scaling	Pilot HVM readiness
Target	EUV Power	25 W	> 100 W	**250 W**
	CE	3%	3.5%	**4%**
	Pulse rate	100 kHz	100 kHz	100 kHz
	Output angle	Horizontal	62°upper (matched to NXE)	62°upper (matched to NXE)
	Availability	1 week operation	1 week operation	> 75%
Technology	Droplet generator	20 – 25 μm	20 μm	< 20 μm
	CO₂ laser	5 kW	20 kW	**27 kW**
	Pre-pulse laser	picosecond	picosecond	picosecond
	Debris mitigation	validation of magnetic mitigation	10 days	> 3 month

Gigaphoton. Proto #1 was designed for proof of concept, especially debris mitigation using magnetic and operation parameters for higher CE or automatic controls with high power condition are being brushed up with Proto #2. The most recent system is Pilot #1 which is designed considering commercial application in semiconductor fab of high volume manufacturing (HVM) and it started operating in summer of 2016.

Proto system

In Proto#2, high EUV power more than 250 W and long-time-continuous operations were achieved as shown in fig.8 and 9. The current achievement of Proto #2 is adjustment of debris mitigation parameters as mentioned in "Magnetic mitigation" section.

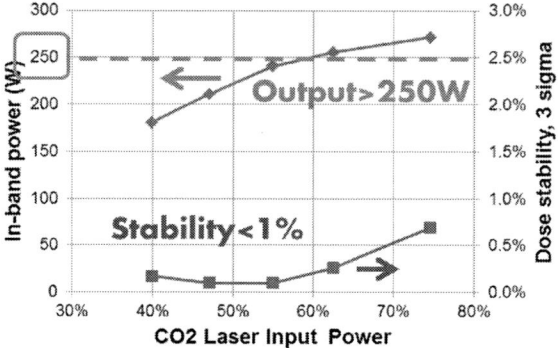

Fig.8 EUV power and stability

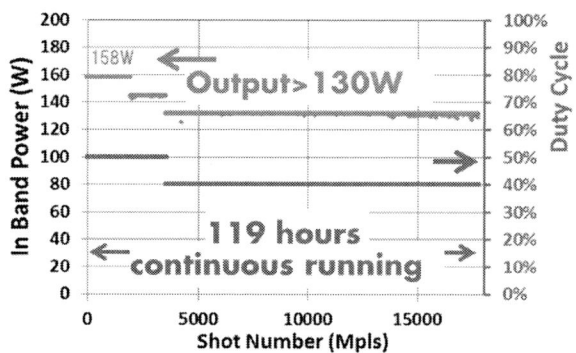

Fig.9 Long-time-continuous operation

Pilot type

In fig.10, the overview is presented. The driver laser system is going to be installed in sub-fab area in factory. This system consists of CO_2 lasers, pre-pulse laser and

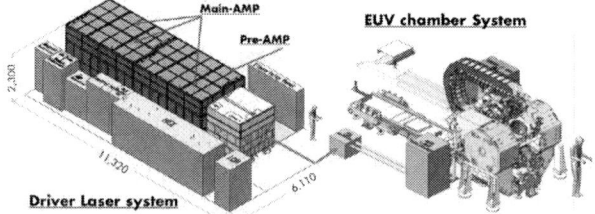

Fig.10 Pilot #1 overview

utility such as water cooling cabinets which are packaged compactly in a few boxes. The EUV chamber system includes beam combine-focusing optics, plasma chamber with super-conductive magnets. It is placed in semiconductor manufacturing fab. as a part of lithography tool. The two main systems are connected by beam transfer unit to convey pre and main pulse lasers from one to another.

The operation tests were started in Pilot #1 and as the first running test, EUV output power 100W with 95% duty, 5% CE over 5 hours continuous operation was resulted. This system achieves higher conversion efficiency as presented in fig.11. The major contribution of improving efficiency is beam profile from CO_2 laser generated by TGF amplifier. This laser delivers solid profile close to Gaussian shape and it interact dome mist tin evenly. On the other hand, Proto's CO_2 beam is caldera like shape not to react whole of tin dome. In FAF type, electrodes are placed in glass tube and gain in the profile rim tends to be higher than the center. This difference might produce different beam profile.

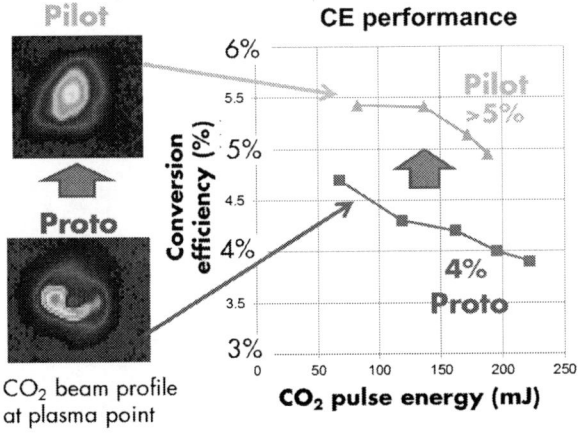

Fig.11 Pilot #1conversion efficiency

Currently the confirmation of debris mitigation performance is becoming the main mission of Pilot as a final preparation for the commercial shipment.

CONCLUSION

Gigaphoton has developed Proto light source and demonstrated more than 250 W EUV power. Also 119 hours continuous operation with more than 130W was resulted.

Pilot system has started in operation which has TGF type CO_2 amplifiers. The conversion efficiency was improved and more than 5% was achieved in the daily test in Pilot. Currently the confirmation of debris mitigation performance is becoming the main mission of Pilot as a final preparation for the commercial shipment.

ACKNOWLEDGEMENTS

This work was partly supported by the New Energy and Industrial Technology Development Organization (NEDO), Japan. We acknowledge their continuous support. We acknowledge to following researchers and organizations; Plasma simulation is supported by Dr. Atsushi Sunahara, Prof. Katsunori Nishihara, Prof. Hiroaki Nishimura, and others in Osaka University. Plasma diagnostics is supported by Dr. Kentaro Tomita, Prof. Kiichiro Uchino and others in Kyushu University. We also acknowledge many companies and engineers; EUV collector mirror collaboration by collector mirror suppliers –especially CO_2 laser amplifier development is supported by Mitsubishi electric CO_2 laser amplifier development team: Dr. Junichi Nishimae, Dr. Shuichi Fujikawa and others.

REFERENCES

[1] Akira Endo, et. al.: "Laser produced EUV light source development for HVM", Proc. SPIE 6517 (2007)

[2] Hakaru Mizoguchi, et. al.: "First generation laser-produced plasma source system for HVM EUV lithography", Proc. SPIE 7636, (2010) [7636-08]

[3] Hakaru Mizoguchi, et. al.: "100W 1st Generation Laser-Produced Plasma light source system for HVM EUV lithography", Proc. SPIE 7969 (2011) [796908]

[4] Hakaru Mizoguchi, et. al.: "Sub-hundred Watt operation demonstration of HVM LPP-EUV Souece", Proc. SPIE 9048 (2013) [90480D]

[5] Hakaru Mizoguchi et al: "One Hundred Watt Operation Demonstration of HVM LPP-EUV Source" (2014 International Workshop on EUV Lithography, Jun.23-27, 2014, Maui, Hawaii)

[6] Hakaru Mizoguchi et al: "Performance of One Hundred Watt HVM LPP-EUV Source" (EUV Symposium 2015, Oct.27-29.2014, Washington D.C.)

[7] Christophe Smeets et al: "EUVL Lithography Industrialization Progress" (2016 International Symposium on Extreme Ultraviolet Lithography, Oct.24-26, 2016, Hiroshima)

[8] Krzysztof M Nowak, Yoichi Tanino et.al.: "EUV driver CO2 laser system using multi-line nano-second pulse high-stability master oscillator for Gigaphoton's EUV LPP system"(EUV Symposium 2013, Oct.6-10.2013, Toyama)

[9] Koji Yasui et al: "Stable and scalable CO_2 laser drivers for high-volume-manufacturing extreme ultraviolet lithography applications" (2016 International Symposium on Extreme Ultraviolet Lithography, Oct.24-26, 2016, Hiroshima)

The solutions for 3D-NAND processes with Canon's latest KrF scanner

Masanori Yamada, Hajime Takeuchi, Kazuhiko Mishima, Keiji Yoshimura, Kazuhiro Takahashi
Canon Inc.
20-2, Kiyohara-Kogyo-Danchi, Utsunomiya, Tochigi, 3213292, JAPAN
Corresponding Autohor's Email: yamada.masanori574@canon.co.jp

ABSTRACT

The NAND type flash-memory is now used not only on smart phones or tablet PCs, but also adopted on infrastructures such as servers, etc.

To enlarge the data capacity, the planer type NAND flash memory will need further miniaturization of the circuit pattern, whereas 3D-NAND allows far more capacity growth per wafer, without stressful miniaturization, since the transistor is formed vertically.

However, there are a lot of technical challenges for 3D-NAND processes.

In this report, we would like to precisely share Canon's approach, on our lithography scanner system, FPA-6300ES6a, towards the above mentioned challenges, which enables low-cost 3D-NAND flash memory mass-production.

Keywords—3D-NAND; KrF Scanner; focus system; productivity improvement; alignment accuracy

INTRODUCTION

With the IoT era awaking, the volume of data to be stored in each and every server will explode, bringing more and more attention to the 3D-NAND technology, which enables enlarging capacity for data storing without chip size change, of the flash memory.

The conventional NAND type flash-memory is called the "planer" type technology, where the transistor is formed horizontally on the planer surface of the wafer. On the other hand, the 3D-NAND technology builds the transistor vertically.

To enlarge the data capacity, the planer type NAND flash memory will need further miniaturization of the circuit pattern, whereas 3D-NAND allows far more capacity growth.

Since miniaturization is not the key factor for 3D-NAND processes, capacity enlargement can be enabled through middle layers to low–end layers exposed with KrF and i-line exposure tools. This kind of low cost operation obviously pushes 3D-NAND up to the major and essential player for the capacity enlargement technology.

However, even with KrF and i-line processes, there are a lot of technical challenges for 3D-NAND processes as follows.

➢ Finer focus control technology is needed, since the physical step is huge.
➢ Extended alignment technology is required to cope with the increase in number of process layers.
➢ Measures against productivity decline due to thicker resist layers, which require higher exposure dose, are demanded.

We would like to report Canon's approaches to these challenges on FPA-6300ES6a for low-cost 3D-NAND flash memory mass-production.

FOCUS CONTROL TECHNOLOGY ENHANCMENTS

The technical challenges for focus control on 3D-NAND processes are, (i) Wafer surface height measurement methods, to deal with huge "um" order device steps, and (ii) compensation methods to deal with wafer in-plane unevenness of resist surface topography.
Figure 1 shows a schematic diagram of FPA-6300ES6a's surface height monitoring system.

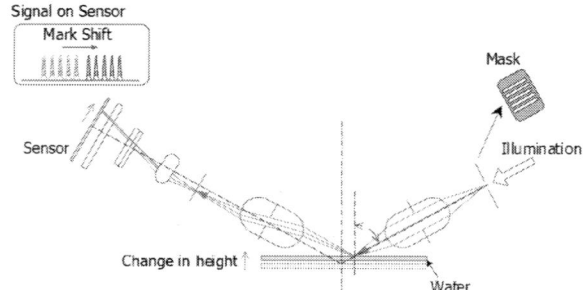

Fig1. Surface Monitor System

The detection mark, which is projected by an oblique-incidence to the wafer, reflects to the photoelectric sensor, and forms the shift amount of the image wave shape at the sensor, then the height (focus amount) is calculated. But when the step height exceeds the detectable range, the adjacent wave shapes overlap, thus interferes correct calculation. (Fig.2(a)). Other incidents are, for example, the steps which are oriented orthogonally against the oblique incidence, often leads to incidence light source loss, and/or light scattering, which both cause deformation of wave shape, what disturbs correct height calculation.

To deal with these symptoms, we optimized the detection mark shape, which more than doubled the height detection range (10um -> >20um), and enhanced the height measurement accuracy at large physical steps. [1] (Fig.2(b))

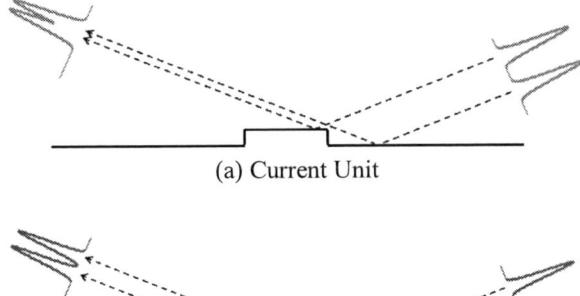

(a) Current Unit

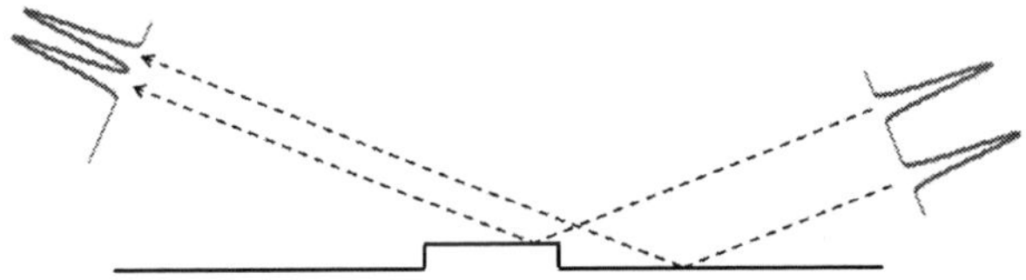

(b) New Unit

Fig2. Height Detection Range Enhancment

OVERLAY ACCURACY IMPROVEMENT

On 3D-NAND processes, as a result of increasing stacks of layers, the wafer warps due to stress difference between the deposited/spattered materials of the wafer. Because of this warp, deformation occurs when the wafer chuck vacuums the wafer. This deformation, not only causes center of the shot offset, but distorts the shape of the shot itself, and even more, the distortion shape differs by the position of the shot, which enlarges overlay error.

To cope with these issues, FPA-6300ES6a equipped so called the "ESHOC" function, which enables high order compensation for each shot, at any position of the wafer, leading to outstanding improvement on overlay accuracy.

The shot shape compensation is achieved through drive control of both wafer and reticle stages, plus controlling

the projection image of the projection optics while scan-exposure is done,

The compensation of projection image is achieved by active lens drive, where the compensation lens configured within the projection optics comes to use, and the projection image is altered.

Figure 3 shows the types of shot shapes which can be corrected through projection optics and stage drive control. As you see, almost every shot shape that may happen on 3D-NAND process is under-cover.

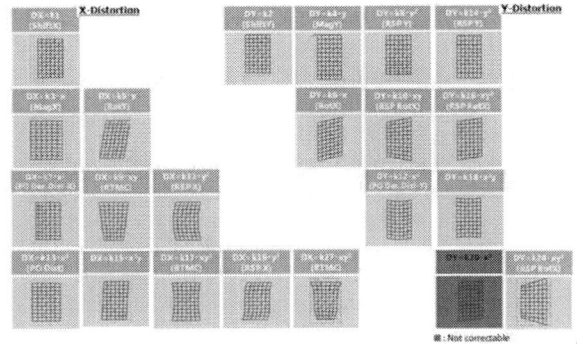

Fig3. ESHOC Correctable Shot Shape Table

Figure 4 is an example of results after adopting ESHOC function. The 1st pattern was exposed by another machine, the 2nd by FPA-6300ES6a, and shows each shot shape of the entire wafer. The overlay error was lower than 5nm @3sigma (Table.1), which shows the improvement in overlay accuracy by using ESHOC function. For your reference, this function uses parameters calculated before-hand, which is inputted to each shot.

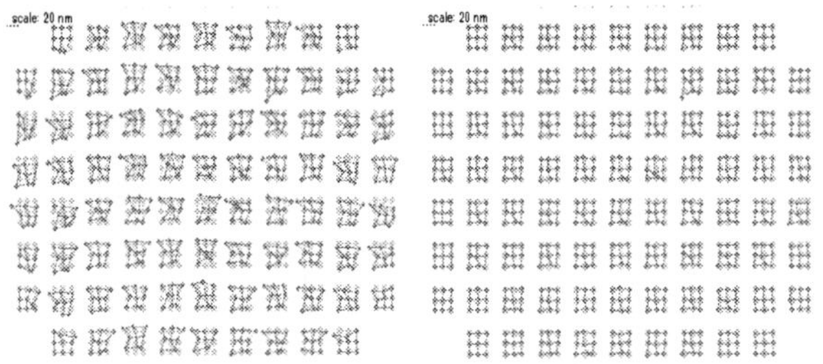

(a) Linear Compensation   (b) Linear+ESHOC Compensation

Fig4. ESHOC Results

Table 1. Overlay Error after ESHOC

Overlay Error	Linear		Linear+ESHOC	
	X	Y	X	Y
3 σ (nm)	12.8	13.7	4.5	4.1

BOOST IN PRODUCTIVITY

Due to the thick film resist used in 3D-NAND technology, productivity is also a challenge for our system, since more exposure load is needed.

To reduce the impact of scan-speed lowering caused by increase in exposure load, the FPA-6300ES6a has enhanced its illumination intensity. We have also focused on continuous development of productivity up-grade options, in order to support the needs for overall productivity improvements from our customers.

These productivity improvements are enabled through software upgrade on the same platform, which allows "easy to do" upgrade in the field, with minimum downtime and minimum impact to production.

The latest upgrade option "Grade6" consists of a learning function for stage control, which prevents degeneration of driving accuracy due to stage speed enhancements, and also allows 255 wafers per hour at 98shots on a 300mm wafer. Table.2 shows the major performance results for Grade 6.

Table.2 Grade6 performance

Item	Results
SMO	x = 4.8 nm y = 4.8 nm
MMO	x = 5.4 nm y = 7.0 nm
Focus Accuracy	32.6 nm
Throughput 98 shot	255wph

SUMMARY

In this paper, we explained Canon's approaches to focus control technology, overlay accuracy improvement, boost in productivity on FPA-6300ES6a for low-cost 3D-NAND flash memory mass-production.

We are committed to support 3D-NAND development through continuous productivity enhancement, and continuous improvement on focus ability, overlay accuracy and CD uniformity, which are foreseen to be impacted by further increase in 3D-NAND layers.

ACKNOWLEDGEMENTS

.We would like to acknowledge T.Miyaharu, M.Imai, A.Takagi and K.Mizumoto for their contribution of this paper. We would like to acknowledge Y.Niijima and T.Ebihara for their useful advices.

REFERENCES

[1] T.Matsumoto. *A wave pattern choice method, a position revision method, an exposure device and a device manufacturing method*, Japan patent JP2006-86450

Application of OPE Master for Critical Layer OPE Matching

Yuan Tao[1], Yifei Liu[1], Yuanzhao Ma[1], Zhenyu Yang[1], Chun Shao[1], Xuedong Fan[2], Junji Ikeda[3], and Koichi Fujii[3]*

[1]Nikon Precision Shanghai, Rm.601 Xin Jin Qiao Tower, No.28 Xin Jin Qiao Road, Pudong New District, Shanghai, 201206, China

[2] Shanghai Huali Microelectronics Corporation, No.497 Gao Si Rd. Zhang Jiang Hi-Tech Park Pudong, Shanghai, 201203, China

[3] Nikon Corporation, 65-2 Naka-cho, 2-chome, Omiya-ku, Saitama-shi 330-0845, Japan

*Corresponding Author's Email: Yuan.Tao@nikon.com

Biography

Yuan Tao is a senior application engineer of semiconductor lithography business unit of Nikon Precision Shanghai. He is primarily responsible for overlay and imaging issues of exposure tools and customer trainings. As an application engineer, he also assists sales engineers to promote Nikon scanners to customers from technical points of view.

Abstract

Tool-to-tool matching of optical proximity effect (OPE) properties is required and the procedure is called OPE matching. Nikon has developed a software called OPE Master for the purpose, which can decrease OPE errors with emphasis placed on critical dimension (CD) errors by optimizing exposure tool's parameters, such as lens numerical aperture (LNA), pupilgram intensity distribution, pupilgram distortion. Thanks to its high affinity to the Nikon NSR series scanners, the software ensures higher accuracies and short turn-around-time (TAT) as it can directly communicate with exposure tools. One secondary benefit of such bilateral communication is that it can realize high data security as we have no need to send data used during OPE matching to the outside of the fab.

In this paper, we are going to introduce OPE Master and report one successful use case. which is a critical layer in 55nm node in which OPE errors has been improved by about 33% which is well within the goal of the process requirements.

Keywords—optical proximity effect; pupilgram; CD errors; Lens NA; Illumination NA; optimization; Zernike polynomial; free form pupils; IIU; 55nm

Introduction

OPE can cause inevitable imaging error, such like CD errors, corner roundings, line end shortenings, etc., in lithography processes, due to limitations posed by physical principles, such as, diffraction, finite LNA, slight variations in pupilgrams, as well as resist process effects. As the result, even if one uses a same exposure condition with an identical mask, the OPE errors may vary, when he or she uses different exposure tools. A projection process can be regarded as a combination of diffractions from mask patterns and interferences on wafer planes. Because different patterns have different diffraction features, changing the illumination pupil and LNA will have different influences on different patterns.

Nikon has developed a software named OPE Master for

tool-to-tool OPE matching [1-4]. In this paper, we describe benefits of OPE Master for OPE matching process and introduce two field cases of OPE matching using OPE Master.

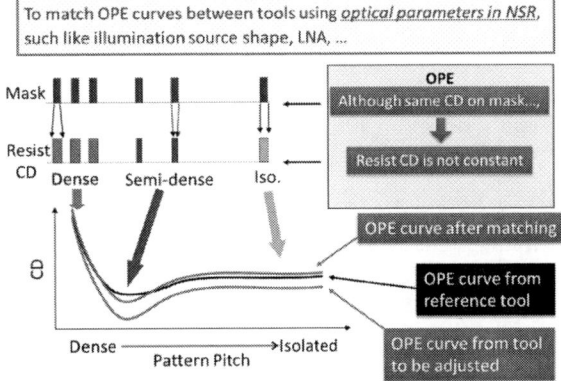

Figure 1: OPE matching concept

Remarkable features of the OPE Master

A. High accuracy

Firstly, tuning variations of freeform pupils could be arbitrary combinations of 36 Zernike intensity modulation (ZIM) functions and 72 Zernike distortion modulation (ZDM) functions which are inspired from the widely known Zernike polynomials used to describe wave front aberrations.

Secondly, the Nikon intelligent illumination unit (IIU) is capable of generating any kind of pupilgrams with outstandingly high fidelities. The combination of IIU and OPEM realizes very flexible matching.

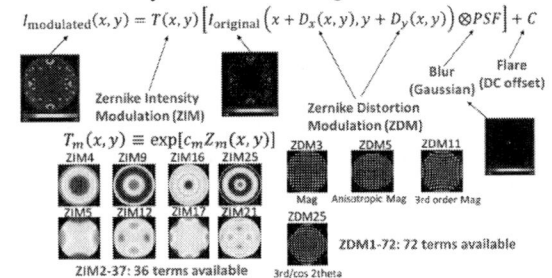

Figure 2: Pupilgram Modulation Model

On the other hand, in case of conventional illumination unit

(IU), OPE Master can precisely predict a modulated pupilgram under consideration of IU optics properties.

B. Short TAT

With the increment of adjustment parameters, traditional grid-searching optimization, that means calculating OPE errors for all combinations of all the parameters is just like a carpet bombing, will be no longer suitable, because if a parameter is added, the computing time will be squared.

To cope with the dimensionality problem, the OPEM provides two new optimization modes called the rigorous mode and the sensitivity mode, within which a prediction and correction algorithm has been implemented, to rid of meaningless combinations of adjustment parameters and to find out the best OPE point in a substantially short time.

In addition, as optimization by OPE Master is based on the technology of computational lithography, actual exposures and CD measurements for OPE sensitivity evaluation are not required.

C. High affinity to NSRs

It is easy to convert NSR's parameters to OPE Master's parameters before matching and vice versa.

The merit of direct import and export of parameters are in three folds. Namely,
1. It can reduce the chance of human error might occur during copy parameter between the OPEM and NSRs.
2. It can further reduce TAT.

It makes NSR property aware optimization possible by considering all finger prints of NSRs.

D. Simplicity of operation and data's security

Once a connection between the OPE Master server and NSRs has been established via intra-net, scanner specific parameters and the result files of pupilgram measurement, both of which are needed during OPE optimization, can be downloaded automatically. The new optical condition can also be sent back to NSRs automatically after an optimization. The security of mask data is guaranteed, as all the optimization cycle can be completed within an intra-net environment and it is not necessary to send mask pattern info to the outside of the fab.

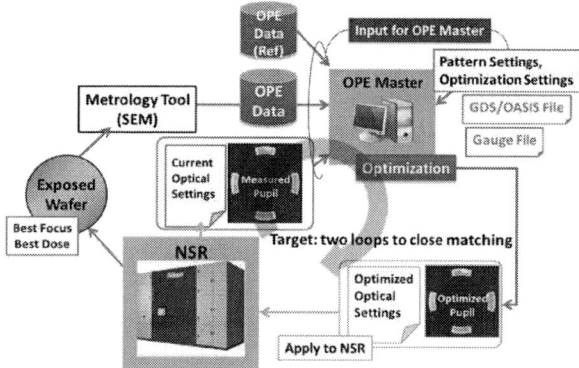

Figure 3: Flow of OPE Matching

Experiments and results

55nm critical layer

Scanner：S620D (An ArF immersion scanner)
Illumination: A parametric pupilgram with conventional illumination unit

231 patterns (116 1D patterns and 115 2D patterns) are select for matching.

Before OPE matching, max CD difference are 6nm and After OPE matching, there is no pattern whose OPE error is over 4nm, 33% reduction of OPE error was achieved.

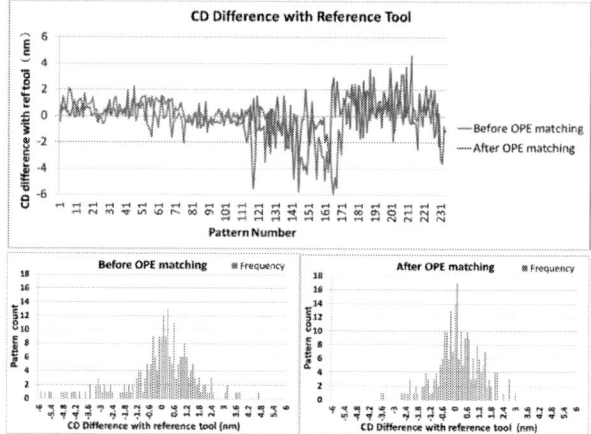

Figure 4: 55nm OPE matching result

Conclusion

The purpose of this paper is to evaluate the Nikon OPE Master which can improve Tool-to-tool CD matching. From the evaluation result, CD errors were improved about 33% from max 6nm to max 4nm. It shows that tool difference can be compensated by using illumination parameter adjustment with the OPE Master system. The result leads to Nikon OPE Master system is a good effect to improve the Tool-to-tool CD errors.

Acknowledgement

The authors wish to thank to OPE Master Team of Nikon for evaluation support. And also thank to our customer for wafer exposure and CD measurement supports.

References

[1] Tomoyuki Matsuyama, Taro Ogata, Yasushi Mizuno, Yasuhiro Ohmura, "Imaging optics setup and optimization on scanner for SMO generation process" Proc. SPIE **8326**-23 (2012).

[2] Shinichi Mori, Hajime Aoyama, Junji Ikeda, Taro Ogata, Ryota Matsui, Tomoyuki Matsuyama, "Imaging Application tools for extremely low-k1 ArF immersion lithography" Proc. SPIE **8683**, 86830A (2013) .

[3] Tomoyuki Matsuyama, Naonori Kita, Yasushi Mizuno, "Pupilgram adjusting scheme using intelligent illuminator for ArF immersion exposure tool" Proc. SPIE **7973**-52 (2011) .

[4] Hajime Aoyama, Toshiharu Nakashima, Taro Ogata, Shintaro Kudo, Naonori Kita, Junji Ikeda, Ryota Matsui, Hajime Yamamoto, Ayako Sukegawa, Katsushi Nakano, Masayuki Maruyama, Kazuo Kasai, Tomoyuki Matsuyama, "Scanner performance predictor and optimizer in further low k1 lithography" Proc. SPIE **9052**-09 (2014).

DDR Process and Materials for NTD Photo Resist
toward 1Xnm Patterning and beyond

Shuhei Shigaki, Wataru Shibayama, Makoto Nakajima, Rikimaru Sakamoto

Semiconductor Materials Research Department,
Materials Research laboratories, Nissan Chemical Industries, Ltd.

635 Sasakura, Fuchu-machi, Toyama 939-2792 Japan
*Corresponding Author's Email: shigakis@nissanchem.co.jp

ABSTRACT

We developed the novel process and material which can prevent the pattern collapse issue perfectly. The process was Dry Development Rinse (DDR) process, and the material used in this process was DDR material. DDR materials were containing siloxane polymer which could be replaced the space area of the photo resist pattern. And finally, the reversed pattern would be created by dry etching process without any pattern collapse issue.

This novel process was useful not only in positive tone development (PTD) process but also in negative tone development (NTD) process. We newly developed DDR material for NTD process. Novel DDR materials were consist of special polymer and it used organic solvent system. Novel DDR materials showed no mixing property with NTD photo resist and it has enough etch selectivity against NTD photo resist.

Image reverse was demonstrated by combination of NTD and DDR process with EUV scanner. The resolution limit was significantly improved and reversed pattern at hp 14nm was obtained without any pattern collapse issue, which couldn't be created in just only normal NTD process.

INTRODUCTION

Extreme Ultra Violet Lithography (EUVL) is one of the most promising candidates for 1X nm patterning and beyond. However, this technology has several issues to be overcome, especially development of high power EUV light source, manufacturing and inspection of multi layer mask and development of photo resist which has good balance for RLS trade off are the most significant issues. Regarding the development of PR for EUVL, improvement of LWR(10% of CD size), dissolution limit below 16nm hp, high sensitivity for high through put (Eop<10mJ/cm2) and pattern collapse improvement are the development target.

To overcome the RLS trade-off, Negative Tone Development (NTD) process is one of the promising processes. NTD process has advantage for trench or contact hole printing due to higher optical contrast compared to PTD process in ArF immersion lithography[3]. In EUVL, some simulation indicated that negative tone imaging has larger photon density than positive tone imaging. It means that the LCDU or LWR can be improved in NTD process with similar dose to size of PTD process because of the reduction of photon shot noise effect[3].

Recently both PTD process and NTD process can be chosen as we like but there is one more limit in EUVL, it is the mask limitation. In EUVL, the preferred mask tonality is dark field because using dark field can reduce defect printing on EUV mask[5]. And dark field mask also prefers due to minimize flare effect[5]. In EUVL, both PTD process and NTD process are limited to specific layout in spite of their optical or material superiority because of the mask issue.

Pattern collapse is also common issue to disturb fine pitch patterning. Especially, in the case of CD size getting smaller and smaller this problem can be more severe. The possible major cause of pattern collapse is the surface tension of the rinsing liquid and the shrinkage of resist pattern's surface in the spin drying process of the rinsing liquid. The influence of surface tension for very small pitch pattern is particularly severe.

Several approaches based on material improvement and process improvement have been tried to solve this issue. One of the most effective solutions for this problem is thinning of the resist film thickness however this method is reaching to its limits in terms of substrate etching process anymore. Recently the tri-layer resist process or hard mask processes have been used with thin thickness resist, but there is a limit to the thinning of resist film and there is no essential solution for this problem. Because, hard mask thickness have to be thinner when PR thickness get thinner because of the etch budget. In that case, hard mask and substrate etch budget can be smaller margin.

On the other hand, the supercritical drying method has been known as an ultimate way to suppress the pattern collapse issue. The supercritical drying method is a dry process advanced to the vapor phase from the liquid phase via supercritical, and the supercritical drying method can dry the rinsing liquid without making the vapor-liquid coexistence state. This method is very useful for

978-1-5090-6695-7/17 $31.00 © 2017 IEEE

improving pattern collapse issue, however, this process is not applied to the mass production process because it requires the introduction of the special equipment. So, there are no primal solutions for avoiding the pattern collapse issue up to now.

In order to solve this issue primly, we newly developed the novel process and material which can prevent the pattern collapse issue perfectly without using any special equipment. The process is Dry Development Rinse process (DDR process), and the material used in the process is Dry Development Rinse material (DDR material). Basically, this DDR process is using conventional procedure for lithography, like PR coating, Prebake, Exposure, Post Exposure bake, development (Alkaline or Organic solvent) and rinse finally. DDR process is the process using DDR material in the rinse step. DDR material is containing the special polymer which can replace the exposed and developed part. And finally, the resist pattern will be developed by Dry etching process without any pattern collapse issue. The key requirement for DDR material are non mixing with resist pattern, high selectivity of dry etching between resist pattern and substrate, planarization property between dense and isolated pattern area, and so on.

We newly developed DDR material for NTD PR and demonstrated the pattern collapse improvement by combination of NTD process and DDR process.

In this paper, we will discuss the approach for preventing the pattern collapse issue and propose DDR process and DDR material as a solution.

EXPERIMENTAL

2.1 Material

We developed and used DDR materials for NTD process. Exposure areas are remaining as pattern and un-exposure area are soluble for organic solvent in NTD process. DDRMs were summarized in Table 1.

Main difference of these materials is polymer system and solvent system. DDRM-A and DDRM-Bare used alcohol as solvent on the other hands, normal butyl acetate is used as solvent in case of DDRM-B. The base polymers of DDR materials are containing Si atom, in order to get the good dry etch selectivity versus photo resist or under layer. Especially, DDRM-B contains special unit to show good solubility and stability for normal butyl acetate. The material design on each process is summarized in table 1.

Table 1 material designs on target resist type

Material	Polymer	Solvent	Remark
DDRM-A	Si containing system A	Org. solv.-A	Ref. DDRM. The property is similar to typical spin on glass hard-mask (SOG).
DDRM-B	Si containing system A	Org. solv.-B	DDRM for ArF NTD PR.
DDRM-C	Si containing system B	Org. solv.-C	Novel DDRM for EUV NTD PR.

2.2 Lithography

DDR process is applicable to most of lithography process such ArF-extended, EUV, EB, DSA and so on. In this paper, we investigated DDR process by ArF and EB lithography process. ArF lithography was carried out on a Nikon NSR-S307E (NA: 0.85) as exposure tool, and TEL CLEAN TRACK ACT8 as coater and developer. EB lithography was carried out on F-125, Elionix, as exposure tool. In both case, BARC (Bottom Anti-Reflective Coatings) or Spin on carbon hard-mask (SOC) were coated on bare silicon wafer and baked at 205degC/60 seconds or 240degC/60seconds to obtain the thin film in 20 to 25nm film thickness. Photo resists were coated, exposed and post-exposure-baked at the recommended conditions. Then photo resists were developed for 30 seconds by normal butyl acetate or DP301 (Fujifilm). Then, DDR material was applied for it. DDR material was dispensed during developer spin drying step.

2.3 Etching

The RIE was used to etch the DDR material and PR materials.

DDR material and PR materials were etched by Reactive Ion Etching (RIE) system. The etching conditions were shown as follows. The etching tool was RIE-10NR (Samco).
[Etching condition: Etching back (recess etching) DDR material]
Gas: CF4=50 sccm, Ar=200 sccm, RF-power=100W, Pressure=15Pa
[Etching condition: PR removal]
Gas: O2=10 sccm N2=20 sccm, RF-power=300W, Pressure=1Pa

RESULTS AND DISCUSSION

3.1 Damage test against NTD PR

At first, we checked damage against NTD PR after DDRM coating. NTD PR pattern made by ArF-dry lithography was used in these tests. Target CD was 65nm L/S pattern. Figure 1 shows the process flow. Three type of DDRMs were used, DDRM-A, DDRM-B and DDRM-C. NBA solvent type. In DDRM removal step, NBA was used as a stripper of DDRMs.

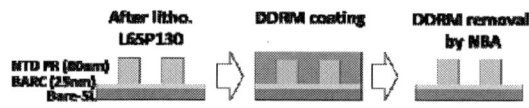

Figure 1 Process flow of PR damage test

The damage against NTD PR was checked on each step. Figure 2 shows X-section image after lithography, wafer view after DDRM coating and X-section image after DDRM removal by NBA. In case of DDRM-A, Pattern missing was observed after DDRM-A removal. It indicated that PR could be resolved by Org. solv.-A in DDRM coating step. On the other hands, PR pattern kept original shape and height after DDRM removal without pattern resolving.

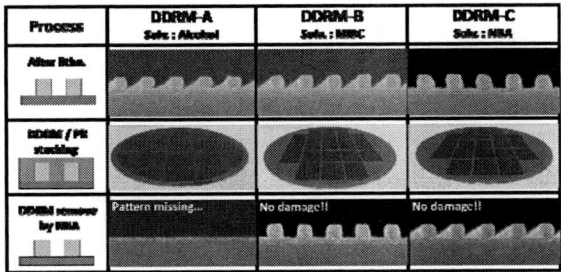

Figure 2 Result of PR damage test

3.2 Analysis of mixing layer between EUV NTD PR and DDRMs

Novel DDRMs which gave lower damage for PR were developed. However, in 1X nm generation, pattern quality after image reversal would be greatly affected by dissolution or mixing at interface of PR and DDRM, even if it was within only few nm orders. Then the dissolution and formation at PR/DDRM surface were analyzed. The procedure was shown in Figure 3.

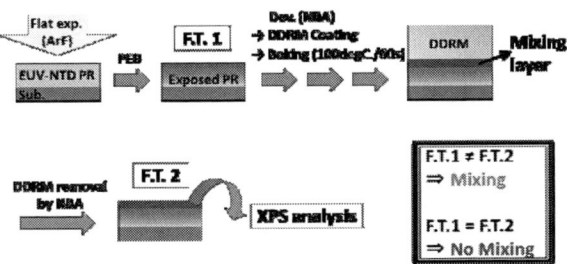

Figure 3 procedures for analysis of dissolution and mixing at PR/DDRM surface

NTD PR for EUV generation was PR1 which was provided Fujifilm. This PR1 was exposed by ArF-dry exposure tool and Post-Exposure-Baked full of coated film. After film thickness measurement, development by NBA and DDRM coating were done by DDR process. Then DDRM was removed by NBA and re-measured film thickness of PR1. Film thickness would be changed if PR was dissolved by DDRM. In addition the contamination of

Si atom was measured on PR1's surface by XPS analysis. The result was summarized in Figure 4.

In DDRM-B case, film thickness was significantly decreased after DDRM-B removal and a lot of Si atom was detected on PR1's surface. It indicated that the mixing layer was formed in wide area due to dissolving of PR1 by DDRM-B's polymer or solvent. On the other hands, there was no significant film thickness change in DDRM-C case and only tenth part of Si atom was detected on PR1's surface compared to DDRM-B case. It suggested that DDRM-C didn't form the serious mixing layer since DDRM-C's polymer and solvent gave no affect for PR1's surface.

Sample	F.T.1 Inital	F.T.2 DDRM removal	XPS (atom%)		
			C	O	Si
NTD PR	68.4nm	-	76.7	23.3	0.0
DDRM-B	-	-	17.9	59.4	22.7
DDRM-C	-	-	30.7	45.6	23.7
DDRM-B/NTD PR	68.4nm	24.4nm	57.9	26.7	15.4
DDRM-C/NTD PR	68.4nm	67.2nm	72.6	26.1	1.3

Figure 4 Demonstration results of preventing pattern collapse in EUV lithography

3.3 Creation of image reversal pattern with NTD-DDR process.

We checked performance of NTD-DDRM with PR1. PR1 was provided by FujiFilm and EB tool was used Elionix F-125. Film thickness of PR1 was 50nm and DP301 (Fujifilm) was used as developer. In development step, space area of PR1's pattern was replaced by DDRM-C. After etch back and PR removal, reversed DDRM's pattern was obtained.

Resolution was compared with original NTD PR pattern and reversed DDR pattern at hp18nm. The comparison data was shown in Figure 5. In normal NTD process, PR pattern couldn't be obtained at hp18nm due to severe pattern collapse but the combination of NTD and DDR process, reversed pattern could be obtained without any pattern collapse. The superiority of DDR process was proved as collapse free process for not only PTD process but also NTD process.

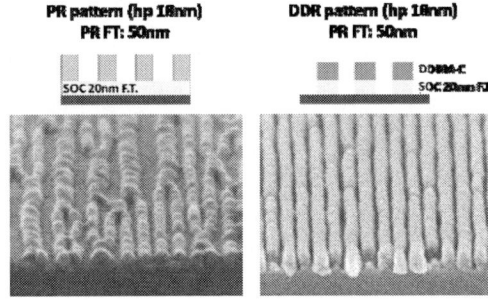

Figure 5 Comparison of pattern quality of PR pattern and image reversal pattern with EB tool

NTD-DDR process was also useful to create fine trench pattern. Figure 6 showed creation of semi-trench pattern by NTD-DDR process. Semi-isolated PR pattern which had 5 times wider space were prepared by EB tool at first and image reversal pattern, semi-trench pattern, was created by NTD-DDR process. Fine semi-trench pattern at hp 20nm could be created perfectly and semi-trench at hp 18nm, hp 16nm could be resolved partially. The CD size of created trench pattern became almost smaller size compared to original PR. It was responsible for the shrinkage of trench size due to deposition.

Dark field is preferred mask tonality because of mask defectivity and flare as I mentioned. In this way, PTD process would be selected to make fine trench pattern in EUVL normally. However, we newly introduced the possibility of NTD-DDR to create fine trench which couldn't be obtained current PTD process.

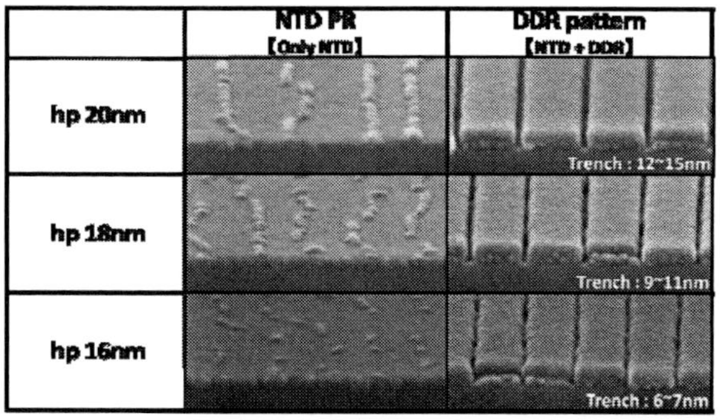

Figure 6 Creation of semi-trench pattern by NTD-DDR process

3.4 Pitch dependence after image reversal.

In DDR process, planarity at several layouts which has various pattern densities is one of the most important properties. We compared pattern height after image reversal at 4 different layouts such a L20P40, L20P120, L50P50 and L50P300. Figure 7 showed comparison data of pattern height after image reversal at 4 different layouts with X-SEM. The Maximum bias was below 4nm at 4 different layouts and it indicated that DDRM-C has higher planarity.

Figure 7 Comparison of pattern height at 4 different layouts.

SUMMARY

DDR material for NTD process was newly developed. New DDRMs (DDRM-B and DDRM-C) showed no damage property for NTD PR pattern made by ArF-dry lithography. The formation of mixing layer was investigated with EUV NTD PR and DDRMs by XPS analysis. The formation of wide mixing layer was indicated in combination of EUV NTD PR and DDRM-B on the other hands, almost no mixing layer was observed in DDRM-C case.

The combination of Negative tone development and DDR process was demonstrated with EB tool. Pattern collapse was perfectly prevented and reversed pattern consist of DDRM was successfully obtained at hp18nm. NTD-DDR process showed possibility to create fine trench pattern. Semi-trench pattern with sub 10nm could be created by NTD-DDR process.

ACKNOWLDGEMENT

The authors would like to thank to Fujifilm corporation for providing photo resist and evaluation of DDR material.

REFERENCE

[1] Koutaro Sho, et al., "Application of reversed pattern transfer process for sub 90nm technology" *Proc. SPIE*, 5039, 1289-1297 (2003).
[2] Yasushi Sakaida, et al., " Development of reverse materials for Double patterning process" *Proc. SPIE*, 7639, (2010).
[3] Sarohan Park, et al., Proc. of SPIE Vol. 9422 94220S-1-9
[4] Hideaki Tsubaki, et al., Proc. of SPIE Vol. 9422 94220N-1-12
[5] Anne-Marie, et al., 2015 EUVL symposium.

Novel EUV Resist Development for Sub-14 nm Half Pitch

Koichi FUJIWARA

JSR Shanghai Co., Ltd.
606 SMEG Plaza, 1386 Hongqiao Road, Shanghai, China
+86-21-6278-7600, Kouichi_fujiwara@jsrworldwide.com

Biography

He is a General Manager of JSR(Shanghai)Co,.Ltd. and has responsibility for semiconductor materials business in China region. He graduated from Kyoto University in 1997 and got doctor degree of organic chemistry at the graduated Kyoto University in 2002. He spends entire 14 years carrier in JSR R&D and business section for semiconductor chemical materials such as photo-resist for frond-end and packaging wafer processes.

Abstract

Extreme ultraviolet (EUV) lithography has emerged as a promising candidate for the manufacturing of semiconductor devices at the sub-14nm half pitch lines and spaces (LS) pattern for 7 nm node and beyond. The success of EUV lithography for the high volume manufacturing of semiconductor devices depends on the availability of suitable resist with high resolution and sensitivity. It is well-known that the key challenge for EUV resist is the simultaneous requirement of ultrahigh resolution (R), low line edge roughness (L) and high sensitivity (S). In this paper, we investigated and developed new chemically amplified resist (CAR) materials to achieve sub-14 nm hp resolution. We found that both resolution and sensitivity were improved simultaneously by controlling acid diffusion length and efficiency of acid generation using novel PAG and sensitizer. EUV lithography evaluation results obtained for new CAR resist are described.

Keywords: EUV Lithography, Resolution, Sensitivity, Novel EUV resist, Chemically Amplified Resist, Sensitizer

Introduction

Extreme ultraviolet (EUV) lithography that extends photolithography to extreme shorter wavelength (13.5nm) is capable to achieve sub 14 nm half pitch resolution.[1] Therefore, EUV lithography is the leading candidate to succeed 193nm immersion lithography. However, EUV lithography presents new challenges in nature of the EUV radiation and requires development of new infrastructure which is different from traditional optical lithography system. Due to the delay of the development of infrastructure including EUV source, multiple patterning technique using 193 nm immersion lithography is widely applied to the mass production of 20 nm and 14 nm node device, and would be also applied to 10 nm node. For 7

nm node, 193 nm multiple patterning technique will require several mask division for some layers which leads to the higher fabrication cost and lower productivity. Therefore application of EUV lithography to 7 nm node device fabrication is strongly anticipated, and EUV resist materials with ultrahigh resolution (sub-14nm) and high sensitivity are required.

To improve EUV resist lithography performance, many research groups were focused on the development of EUV resist with new and novel materials[2-4]. We have developed new materials such as new resist components showing short acid diffusion length, silicon type under layer, and molecular resist and etc[5-7]. EUV resist lithography performance requirements should depend on the EUV source power development. For example, super sensitive novel resist would be required in case of lower EUV source power. On the other hand, ultra-high resolution EUV resist will be required when high EUV source power is achieved. In any case, resolution and sensitivity are key requirements for EUV resists. In this paper, we will discuss the recent progress in EUV resist materials development at JSR for simultaneous resolution and sensitivity improvement.

Results

New CAR resist was developed by combination of higher Tg resin, shorter acid diffusion length PAG and resist formulation optimization. Figure 1 shows the exposure result of new high resolution Resist-C. Resist-C resolved 13 nm half pitch line and space pattern with 35.5 mJ/cm^2. Resist-C also shows resolution modulation at 12 nm half pitch pattern but was not completely resolved. However, these results suggest that 12 nm half pitch would be resolved by further formulation optimization.

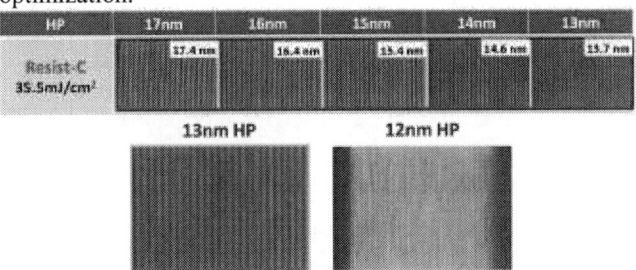

Figure 1. High Resolution Resist-C Imaging Performance on EUV MET Tool (NA0.30).

Detailed lithography performance of Resist-C was evaluated on NXE3300 for 16 nm half pitch line and space and 20nm iso trench pattern. Figure 2 shows the SEM images of 16 nm hp LS and 20 nm hp iso trench pattern obtained at best dose and focus.

978-1-5090-6695-7/17 $31.00 © 2017 IEEE

Resist-C shows good process window for both 16 nm half pitch line and space and 20 nm iso trench. In addition, ultimate resolution performance of Resist-C was evaluated. Figure 3 shows the results of 15 nm half pitch, 14 nm half pitch and 13 nm half pitch line and space patterning performance. Resist-C resolved 15 nm half pitch patterns and has potential to resolve 14 and 13 nm half pitch patterns. However, pattern line collapse was observed at 13nm half pitch pattern and it should be improved with both resist formulation and process optimization.

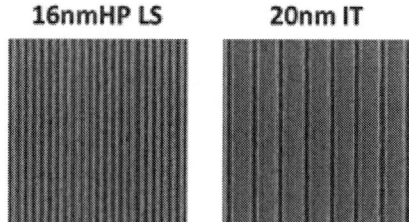

Figure 2. Best dose and focus images of 16 nm half pitch line and space pattern and 20 nm iso trench pattern using Resist-C on NXE3300 (Dipole 90 illumination).

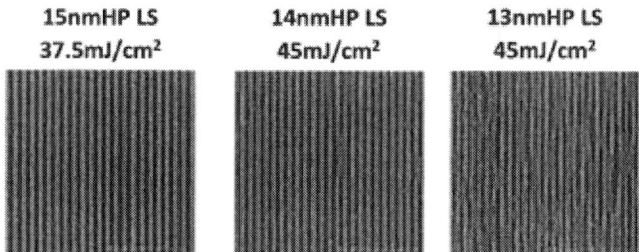

Figure 3. Exposure results of 15 nm, 14 nm and 13 nm half pitch line and space pattern using Resist-C on NXE3300 (Dipole 45 illumination).

It is well known that EUV photo-absorption is important for efficient secondary electrons generation and some atoms such as Fluorine show higher EUV photoabsorption.[8-10] In EUV the PAG is activated by secondary electrons. Therefore addition of sensitizer including some higher EUV photo-absorption atoms to EUV chemical amplified resist would accelerate the secondary electrons generation, which in turn enhance acid generation. Hence, sensitizer development is one of the promising approaches for improving the sensitivity of EUV chemical amplified resist without LWR and resolution degradation. To confirm the concept of sensitizer addition, we applied the new sensitizer including some higher EUV photo-absorption atoms. Figure 4 is the exposure results of two resists evaluated on Berkeley MET. These results demonstrate that the addition of new sensitizer to conventional chemical amplified resist was effective approach for improving sensitivity without degrading other lithography performance.

Pattern HP	Conventional CAR		CAR + New Sensitizer		Sensitivity improve
18nmHP		Resist D 54.2mJ CD=17.5 LWR=2.7		Resist G 45.8mJ CD=18.0 LWR=2.8	16%

Figure 4. EUV lithography result of conventional chemical amplified resist and CAR + new sensitizer.

Summary

In this paper, the development of chemically amplified resist (CAR) containing new materials such as short acid diffusion length PAG and new sensitizer including some higher EUV photo-absorption atoms were described. EUV lithography performances of new high resolution EUV chemical amplified resist containing new PAG or sensitizer were studied. New high resolution chemical amplified resist containing short acid diffusion length PAG resolved 13 nm half pitch line and space pattern on Berkeley MET and has the potential to resolve 14 nm half pitch line and space pattern on NXE3300. Addition of new sensitizer to conventional chemical amplified resist showed the 16 % sensitivity improvement without degradation in resolution and LWR compared to conventional chemical amplified resist. Further sensitivity and resolution improvement would be enabled by the new material development and resist formulation optimization, which would encourage the implementation of EUV lithography process for mass production of 7 nm node and beyond.

References

[1] ITRS website, http://www.itrs.net/

[2] Owendi Ongayi, Matthew Christianson, Matthew Meyer, Suzanne Coley, David Valeri, Kwok Amy, Mike Wagner, Jim Cameron, Jim Thackeray, "High Sensitivity Chemically Amplified EUV Resists through Enhanced EUV Absorption", Proc. SPIE 8322, 83220T-1 (2012).

[3] Kenji Hosoi, Brian Cardineau, William Earley, Seth Kruger, Koichi Miyauchi, Robert Brainard, "Synthesis of Stable Acid Amplifiers that Produce Strong, Highly Fluorinated Polymer Bound Acid", Proc. SPIE 8325, 83251S (2012).

[4] Markos Trikeriotis, Marie Krysak, Yeon Sook Chung, Christine Ouyang, Brian Cardineau, Robert Brainard, Christopher K. Ober, Emmanuel P. Giannelis, Kyoungyong Cho, "A new inorganic EUV resist with high-etch resistance", Proc. SPIE 8322, 83220U-1 (2012).

[5] Ken Maruyama, Makoto Shimizu, Yuuki Hirai, Kouta Nishino, Tooru Kimura, Toshiyuki Kai, Kentaro Goto, and Shalini Sharma, "Development of EUV resist for 22nm half pitch and beyond", Proc. SPIE 7636, 76360T (2010).

[6] Ken Maruyama, Ramakrishnan Ayothi, Yoshi Hishiro, Koji Inukai, Motohiro Shiratani, Tooru Kimura, "Novel EUV Materials and Process for 20 nm Half Pitch and Beyond", Proc. SPIE 8682, 86820N-1 (2013).

[7] Motohiro Shiratani, Takehiko Naruoka, Ken Maruyama, Ramakrishnan Ayothi, Yoshi Hishiro, Kenji Hoshiko, Andreia Santos, Xavier Buch, Tooru Kimura, "Novel EUV resist materials for 16 nm half pitch and EUV resist defects", Proc. SPIE 9048, 90481D-1 (2014).

[8] B. L. Henke, E. M. Gullikson, J. C. Davis, Atomic Data and Nuclear Data Tables, 54, 181-342 (1993)

[9] T. Kozawa, S. Tagawa, Japanese Journal of Applied Physics, 49 (2010) 030001

[10] H. Yamamoto, T. Kozawa, S. Tagawa, H. Yukawa, M. Sato, J. Onodera, Appl. Phys. Express, 1 (2008) 047001.

ADVANCEMENT IN RESIST MATERIALS FOR SUB-7 NM PATTERNING AND BEYOND

Li Li[1]*, Xuan Liu[1], and Shyam Pal[1]

[1]Advanced Technology Development, GlobalFoundries, Malta, New York, 12020, USA

*Corresponding Author's Email: li.li1@globalfoundries.com

ABSTRACT

The rapid development in dense integrated circuits requires significant advancement in small scaling patterning technology. EUV technology is considered as a powerful solution for the sub-7 nm node pattering and beyond. The high performance resist development is required for the practical applications of the EUV patterning for high volume manufacturing. In the current work, the requirements for the development of next generation resist materials is reviewed and summarized to propose the design criterion for high performance photoresist materials.

INTRODUCTION

Advanced lithography techniques are becoming more and more critical for the fabrication of modern microelectronics and energy devices, and high performance resist materials are very significant for transitioning those devices from the research lab to the industrial manufacturing.[1-2] With consideration of the high volume manufacturing (HVM) of wafer-based microelectronics devices, patterning technology with extremely small lithography dimensions and improved cost is required for fast integrated devices manufacturing.[3] Extreme ultraviolet (EUV) lithography is considered as one of the most promising technologies to achieve fabrication of devices smaller than 10 nm nodes.[4] Compared with traditional lithography technologies, EUV lithography has stringent material requirements including high etch resistance, high sensitivity and resolution, low line edge roughness (LER) or line width roughness (LWR), appropriate UV absorption, and small molecular size, among others.

It is reported that the EUV technology has begun to be introduced in trial for the processing in some critical layers in 7 nm technology and replaced the traditional ArF immersion technology combined with multi-patterning. [5] For sub-7 nm technology which is close to the atomic level patterning, the use of EUV can greatly decrease the processing complexity if its throughput and processing availability can meet the manufacturing requirements. Performance requirements for EUV resists will require the development of entirely new resist platforms, which including the development of resist materials, auxiliary materials and the optimization of materials related processing. Herein, we will provide a brief review about the development for the resist materials required for sub-7 patterning and beyond.

RESIST REQUIREMENTS

Absorbance

A challenge in designing new photoresists for EUV wavelengths is the selection of molecular structures that have minimal absorbance and superior characteristics in imaging and etch performance. As we know, at EUV wavelengths, the absorption of all materials is very strong and only dependent on their atomic composition. Figure 1 shows the photo absorption cross-section of different elements under EUV. Note that elements commonly used in photoresists at other wavelengths, such as fluorine, are highly absorbing at ~13 nm, rendering them problematic for EUV applications. Other elements including carbon, silicon, zirconium and hafnium have very high transmission, allowing EUV photons to pass through the entire resist film. Ober and Giannelis in Cornell University are now focused on the development of Hf/Zr based hybrid inorganic/organic resist materials and have made an excellent progress for EUV patterning.[6]

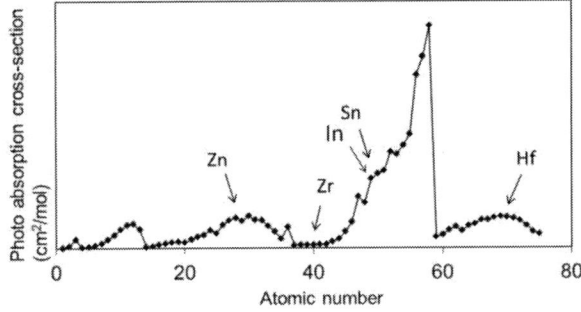

Figure 1. EUV absorption of different elements.

Solubility

For the high performance resist development, solvent selection is also very crucial considering its processing stability and cost used in the industrial fab. A systematic approach is required for the designing and optimizing the resists. Negative tone resists where the solubility characteristics of the exposed and unexposed areas are fairly similar, leading to the challenging selection of an appropriate developer. The Hildebrand solubility parameter is an excellent tool for predicting the solubility of the resist materials in various solvents and can be used to evaluate if the resist material is eligible for use. It is defined as

$$\delta = (E/V_m)1/2 \qquad (1)$$

where E and V_m are the cohesive energy and molar volume, respectively. The Hildebrand solubility parameter can be divided into three components for contribution and they represent the dispersive, polar and hydrogen bonding interactions. These components are the Hansen solubility parameters and are related to the Hildebrand parameter by the equation:

$$\delta_{t2} = \delta_{d2} + \delta_{p2} + \delta_{h2} \qquad (2)$$

Using the Hansen three solubility parameters the solubility behavior of any material can be systematically determined. In this approach the solubility behavior of any material is plotted in a three dimensional graph with each Hansen parameter represented in each of the three axes. The interaction radius R of the spherical volume of solubility, can be used to judge if the solvent is a good developer for the specific resist material or not (shown in Figure 2). The use of Hansen solubility parameters can be extended to investigate how the chemistry of the resists affects solubility and different processing/development conditions (e.g. post apply bake, PAB, and post exposure bake, PEB) as well as various additives on solubility.

exposure,[2, 8-10] and particle size typically affects solubility and dispersion of resist materials. For example, small nanoparticles will be easily dissolved and removed while larger agglomerated ones will not be dissolved. Therefore in parallel to the solubility studies, the influence of processing and development conditions including additives on the particle size should also be studied. Charge stabilization and the corresponding Debye electrostatic double layer involved have been exploited in dispersing and stabilizing nanoparticles. Simply, the electrostatic double layer acts as a shield which prevents the nanoparticles from getting close enough for the van der Waals attractive interaction to set in. When the latter commences the particles are forced to aggregate. The key of the electrostatic stabilization is to develop a repulsive interaction before the attractive interaction begins. This requirement typically translates in having an electrostatic double layer with the required thickness so that at any given point of the interaction vs. distance plot the total interaction is repulsive. For a fixed charge on the particles the electrostatic double layer is inversely proportional to the concentration of ions and to the square of the ions present. In other words as the concentration of the ions increases or higher charge ions are present, the double layer decreases and aggregation can take place. The interaction energy as a function of distance for well dispersed particles and aggregated particles are shown in Figure 3.

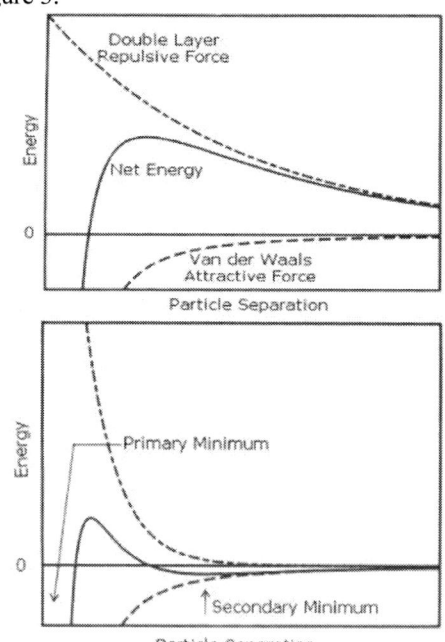

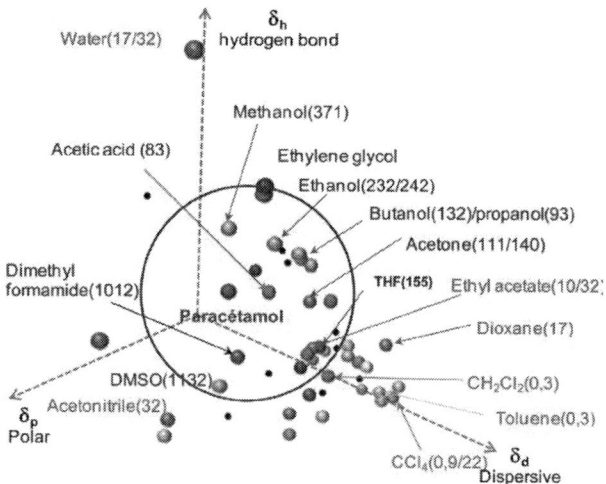

Figure 2. Hansen solubility parameter plot for a material.[7]

Dispersion

The inorganic/organic based nanoparticle EUV resist is reported to show the particle size change after the UV

Figure 3. Interaction energy as a function of distance for well dispersed particles (top) and aggregated particles (bottom). The key is to have a repulsive interaction because of charge commencing earlier than the attractive

interaction due to van der Waals forces. From www.malvern.com

Defect

Defect level is also an important parameter for the resist evaluation and resist outgassing issue also should be carefully analyzed. The resist outgassing can lead to the deformation of patterning and more importantly contaminate the EUV optics. Witness sample testing has been proposed by ASML to qualify the photoresist materials for not excessively contaminating the scanner optics and other parts in the vacuum environment of the tool.[4, 11] The candidate resist is firstly irradiated on a witness substrate and a nearby resist-coated wafer with EUV radiation simultaneously before HVM use. Currently, more work needs to be focused on the detection and control of the defect level and outgassing issue of the resist materials for their use in sub-7 nm patterning.

OUTLOOK

The advancement in novel resist materials for sub-7 nm patterning materials requires the candidate materials should meet the requirements in absorbance, solubility, dispersion and outgassing from the aspect of materials design. Furthermore, the concerns in etch resistance, UV out of band, resist homogeneity, shelf-life and pattern collapse, etc., also need to consider for their use for HVM. Additional efforts should be focused on the resist materials design and optimization to balance and improve the triangle relationship between resolution, LER/LWR, and sensitivity (RLS).

ACKNOWLEDGEMENTS

Thanks for the support from GlobalFoundires for publication.

REFERENCES

1. L. Li, S. Chakrabarty, J. Jiang, B. Zhang, C. Ober, and E. P. Giannelis: *Nanoscale*, vol. 8, 2016, pp. 1338-43.
2. L. Li, S. Chakrabarty, K. Spyrou, C. K. Ober, and E. P. Giannelis: *Chem. Mater.*, vol. 27, 2015, pp. 5027-31.
3. C. Wagner, and N. Harned: *Nat. Photon.*, vol. 4, 2010, pp. 24-26.
4. T. Itani, and T. Kozawa: *Jpn. J. Appl. Phys.*, vol. 52, 2013, p. 010002.
5. *http://www.anandtech.com/show/10704/globalfoundries-updates-roadmap-7-nm-in-2h-2018.*
6. M. Trikeriotis, M. Krysak, Y. S. Chung, C. Ouyang, B. Cardineau, R. Brainard, C. K. Ober, E. P. Giannelis, and K. Cho: *J. Photopolym. Sci. Technol.*, vol. 25, 2012, pp. 583-86.
7. E. Lesellier: *J. Chromatogr. A*, vol. 1389, 2015, pp. 49-64.
8. M. Yu, E. P. Giannelis, and C. K. Ober, In *SPIE Advanced Lithography*, (International Society for Optics and Photonics: 2016), pp 977905-05-5.
9. B. Zhang, L. Li, J. Jiang, M. Neisser, J. S. Chun, C. K. Ober, and E. P. Giannelis, In *SPIE Advanced Lithography*, (International Society for Optics and Photonics: 2015), pp 94251E-51E-6.
10. K. Kasahara, V. Kosma, J. Odent, H. Xu, M. Yu, E. P. Giannelis, and C. K. Ober, In *SPIE Advanced Lithography*, (International Society for Optics and Photonics: 2016), pp. 977604-977604.
11. S. Grantham, C. Tarrio, S. Hill, L. Richter, T. Lucatorto, J. Van Dijk, C. Kaya, N. Harned, R. Hoefnagels, and M. Silova, In SPIE Advanced Lithography, (International Society for Optics and Photonics: 2011), pp 79690K-90K-8.

DESIGN AND SYNTHESIS OF NOVEL DIRECTED SELF-ASSEMBLY BLOCK COPOLYMERS FOR SUB-10 NM LITHOGRAPHY APPLICATION

Jie Li, Xuemiao Li, Hai Deng*

Department of Macromolecular Science, State Key Laboratory of Molecular Engineering of Polymers, Fudan University, Shanghai200433, China
*Corresponding Author's Email: haideng@fudan.edu.cn

ABSTRACT

A series of novel block copolymers, PS-b-PPDMA, were synthesized via anionic polymerization. Small-angle X-ray scattering (SAXS) spectra and Transmission electron microscope (TEM) images indicated lamella or hexagonal structures with a sub-10 nm half-pitch formed under mild thermal annealing condition. The assembly condition is as quick as 5 min or less at 100 ℃ thermal annealing. The smallest lamellar D spacing is 11.8 nm. These block copolymers show the potential as DSA material with high intrinsic resolution for sub-10 nm and beyond nodes.

INTRODUCTION

As a challenge to Moore's Law, patterning technology is approaching resolution limit after 10 nm node. As a competing technology, EUVL is getting more mature and widely considered as the next generation lithography technology for 7 nm and 5 nm nodes. However, the high cost and low throughput of EUV optic system might limit its applications.

While pure optic resolution is facing big physical and economical challenge, people now look for assistant chemical approaches for sub-10 nm litho. Block copolymers (BCPs) are excellent candidates for sub-10 nm lithography due to their capability of self-assembly into periodic structures at the nanometer scale with a range of controllable morphologies [1-2]. Directed Self-Assembly (DSA) of block copolymers has been a promising advanced lithography methodology due to its high potential resolution and extremely low cost compared to EUVL [3-6].

PS-b-PMMA has been the most widely employed system for DSA studies [7-12]. However, PS-b-PMMA has domain period limit to about 20 nm, due to its small and temperature -insensitive χ_{eff} parameter [9]. However, tens of hours high-temperature thermal annealing is usually needed by previous DSA process with known BCPs. For PS-b-PMMA, Very short annealing time of less than 5 minutes is sufficient only when the temperature was high than 200℃. So a rapid low-temperature BCP system is becoming an imperative requirement. Here a series of chemically modified PS-b-PMMA copolymers was synthesized to obtain Sub-10 nm features. The assembly time is as quick as 5 min or less below 100 ℃ annealing,

which is a record fast low-temperature thermal annealing DSA system.

EXPERIMENT SECTION

Solvent and Materials: Styrene (>99% stabilized), sec-butyllithium (sec-BuLi, 1 M in hexane), dibutylmagnesium (MgBu$_2$, 1 M in hexane), and calcium hydride (CaH$_2$) powder were purchased from Energy chemical. All the solvent (THF, methanol) was from Titan.

Reagent Purification. Styrene and THF were distilled over CaH$_2$ and then over MgBu$_2$ just before use. All other reagents and solvents were used as received unless otherwise noted.

RESULT AND DISCUSSION

In this study, a family of PS-b-PPDMAs was synthesized via anionic polymerization or ATRP, and PS-b-PPDMAs were obtained with narrow polydispersity indices (PDI<1.15). The structures of the resulting block copolymers were characterized by NMR. Molecular weight and composition of PS-b-PPDFMA samples were determined by size-exclusion chromatography and ^{1}H-NMR, respectively.

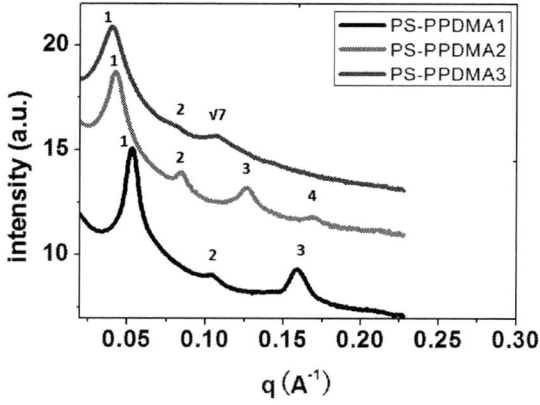

Fig.1. SAXS 1-D patterns for PS-b-PPDMA1~PS-b-PPDMA3 annealed under 160℃ for 5 min.

Micro-phase separation morphologies were investigated using SAXS and TEM. Bulk samples of copolymers were dip-coated on oxidized smooth Si substrates from 5 wt% solution of polymers in toluene.

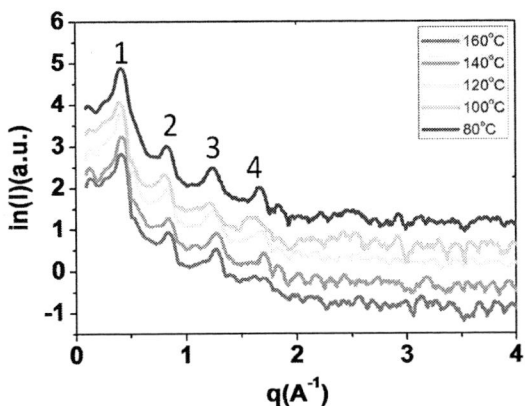

Fig.2. SAXS 1-D patterns for PS-b-PPDMA2 annealed under 160℃ ~ 80℃ for 5 min.

The 2-D SAXS scattering profiles for PS-b-PPDMAs at 25℃ were azimuthally integrated to 1-D patterns shown in Figure 1. Domain spacings (D=2π/q) were calculated from the principal scattering peak position (q).The sample was thermally annealed in air at 160 ℃ for 5 min, followed by rapid cooling. A well-ordered lamellar morphology (scattering peaks of 1, 2, 3), with a D spacing of 11.8 nm, was observed for PS-b-PPDMA1. PS-b-PPDMA2, with a larger molecular weight, also exhibited a well-ordered lamellar morphology, with a much smaller D spacing of 15.2 nm. As the increase of styrene, PS-b-PPDMA3 show hexagonally packed cylindrical structures, which displayed sharp peaks with q/q* of 1, 2, √7. The hexagonal d-spacing calculated from SAXS was 16.5 nm.

Fig. 3. TEM image for bulk sample of PS-b-PPDMA2

Interestingly, further results showed that annealing temperatures lower than 100°C (Figure 2) could be used to achieve the well-ordered lamellar morphology, which is similar to those of solvent annealing [13-15] and lower than those for many other block copolymers that are compatible with thermal annealing [16-21]. Besides, the annealing time (5 min or less) here was also much shorter than that for most other copolymers, which is usually more than 10h. The short thermal annealing process in mild conditions is highly compatible with manufacturing requirements.

TEM image (Figure 3) provided further evidence of the formation of well-defined lamellar morphology. Full-pitches of PS-b-PPDFMA2 calculated from TEM were about 13.5 nm, which is consistent well with SAXS data.

SUMMERY

We obtained a record fast low-temperature thermal annealing DSA system, which is comparable to those solvent annealing systems. PS-b-PPDMA were synthesized via anionic polymerization, with PDI<1.15. The assembly time is as quick as 5 min or less below 100 ℃ annealing. SAXS spectra and TEM images showed that our copolymers exhibited well-ordered lamellar morphology with sub-10 nm half-pitches. The smallest half-pitch is 5.9 nm. The Lamellar and hexagonal morphology was well-controlled by changing the composite of the copolymers

Next step, we will establish collaborations with IC companies to develop sub-10 nm DSA patterning technology for China.

ACKNOWLEDGEMENTS

This work was supported by the China Postdoctoral Science Foundation (2016M600280).

REFERENCES

[1] Ruiz, R.; Kang, H.; Detcheverry, F. A.; Dobisz, E.; Kercher, D. S.; Albrecht, T. R.; Pablo, J. J. de; Nealey, P. F. Density Multiplication and Improved Lithography by Directed Block Copolymer Assembly. *Science* **2008**, *321*, 936.

[2] Park, S.-M.; Liang, X.; Harteneck, B. D.; Pick, T. E.; Hiroshiba, N.; Wu, Y.; Helms, B. A.; Olynick, D. L. Sub-10 nm Nanofabrication via Nanoimprint Directed Self-assembly of Block Copolymers. *ACS Nano* **2011**, *5*, 8523.

[3] Cheng, J. Y.; Ross, C. A.; Smith, H. I.; Thomas, E. L. Templated Self-Assembly of Block Copolymers: Top‐Down Helps Bottom‐Up. *Adv. Mater.* **2006**, *18*, 2505.

[4] Luo, M.; Epps, T. H. III Directed Block Copolymer Thin Film Self-Assembly: Emerging Trends in Nanopattern Fabrication. *Macromolecules* **2013**, *46*, 7567.

[5] Kim, H. C.; Park, S. M.; Hinsberg, W. D. Block Copolymer Based Nanostructures: Materials, Processes, and Applications to Electronics. *Chem. Rev.* **2010**, *110*, 146.

[6] Darling, S. B. Prog. Directing the Self-assembly of Block Copolymers. *Polym. Sci.* **2007**, *32*, 1152.

[7] Stoykovich, M. P.; Müller, M.; Sang, O. K.; Solak, H. H.; Edwards, E. W.; Pablo, J. J.; Nealey, P. F. Directed Assembly of Block Copolymer Blends into Nonregular Device-Oriented Structures. *Science* **2005**, *308*, 1442.

[8] Ruiz, R.; Kang, H.; Detcheverry, F. A.; Dobisz, E.;

Kercher, D. S.; Albrecht, T. R.; Pablo, J. J.;Nealey, P. F. Density Multiplication and Improved Lithography by Directed Block Copolymer Assembly. *Science* **2008**, *321*, 936.

[9] Wan, L.; Ruiz, R.; Gao, H.; Patel, K. C; Albrecht, T. R.; Yin, J.; Kim, J.; Cao, Y.; Lin, G. The Limits of Lamellae-Forming PS-b-PMMA Block Copolymers for Lithography. *ACS Nano* **2015**, *9*, 7506.

[10] Cheng, J. Y.; Rettner, C. T.; Sanders, D. P.; Kim, H.; Hinsberg, W. D. Dense Self - Assembly on Sparse Chemical Patterns: Rectifying and Multiplying Lithographic Patterns Using Block Copolymers. *Adv. Mater.* **2008**, *20*, 3155.

[11] Tsai, H.; Pitera, J. W.; Miyazoe, H.; Bangsaruntip, S.; Engelmann, S. U.; Liu, C. C.; Cheng, J. Y.; Bucchignano, J. J.; Klaus, D. P.; Joseph, E. A.; Sanders, D. P.; Colburn, M. E.; Guillorn, M. A. Two-dimensional Pattern Formation Using Graphoepitaxy of PS-b-PMMA Block Copolymers for Advanced FinFET Device and Circuit Fabrication. *ACS Nano* **2014**, *8*, 5227.

[12] Ruiz, R.; Ruiz, N.; Zhang, Y.; Sandstrom, R.; Black, C. Local Defectivity Control of 2d Self-Assembled Block Copolymer Patterns. *Adv. Mater.* **2007**, *19*, 2157.

[13] Jung, Y. S.; Ross, C. A. Orientation-Controlled Self-Assembled Nanolithography Using a Polystyrene−Polydimethylsiloxane Block Copolymer. *Nano Lett.* **2007**, *7*, 2046.

[14] Jeong, J. W.; Park, W. I.; Kim, M.-J.; Ross, C. A.; Jung, Y. S. Highly Tunable Self-Assembled Nanostructures from a Poly(2-vinylpyridine-b-dimethylsiloxane) Block Copolymer. *Nano Lett.* **2011**, *11*, 4095.

[15] Park, W. I.; Kim, K.; Jang, H.-I.; Jeong, J. W.; Kim, J. M.; Choi, J.;Park, J. H.; Jung, Y. S. Directed Self-Assembly with Sub-100 Degrees Celsius Processing Temperature, Sub-10 Nanometer Resolution, and Sub-1 Minute Assembly Time. *Small* **2012**, *8*, 3762.

[16] Welander, A. M.; Kang, H.; Stuen, K. O.; Solak, H. H.; Müller, M.; de Pablo, J. J. and Nealey, P. F. Rapid Directed Assembly of Block Copolymer Films at Elevated Temperatures. *Macromolecules* **2008**, *41*, 2759.

[17] Kennemur, J.; Yao, L.; Bates, F.; Hillmyer, M. Sub-5nm Domains in Ordered Poly(cyclohexylethylene)-block-poly(methyl methacrylate) Block Polymers for Lithography. *Macromolecules* **2014**, *47*, 1411.

[18] Jung, Y. K.; Ross, C. A. Orientation-Controlled Self-Assembled Nanolithography Using a Polystyrene−Polydimethylsiloxane Block Copolymer. *Nano Lett.* **2007**, *7*, 2046.

[19] Cheng, J. Y.; Ross, C. A.; Thomas, E. L.; Smith, H. I.;.Vancso, G. J. Templated Self-Assembly of Block Copolymers: Effect of Substrate Topography. *Adv. Mater.* **2003**, *15*, 1599.

[20] Chai, J.; Wang, D; Fan, X; Buriak, J. M. Assembly of Aligned Linear Metallic Patterns on Silicon. *Nature Nanotechnology*, **2007**, *2*, 500.

[21] Hirai, T.; Leolukman, M.; Liu, C. C.; Han, E.; Kim, Y. J.; Ishida,Y.; Hayakawa, T.; Kakimoto, M.-a.; Nealey, P. F.; Gopalan, P. One-Step Direct-Patterning Template Utilizing Self- Assembly of POSS-Containing Block Copolymers. *Adv. Mater.* **2009**, *21*, 4334.

DOF ENHANCEMENT OF 3BAR PO PATTERN IN 28NM TECHNOLOGY NODE

Bin-Jie Jiang*, Shi-Rui Yu, Zhi-Biao Mao

Shanghai Huali Microelectronics Corporation, No.497 Gaosi Rd. Zhang Jiang Hi-Tech. Park, Shanghai 201212, China,
* Corresponding Author's Email: jiangbinjie@hlmc.cn

Abstract

In 28nm technology node, developing an enough lithographic process window of PO layer is one of the most basic requirements. Bigger PO DOF means better PO CD uniformity, which is very important to the device performance.

Sizing-up is a very common method for DOF enhancement of non-gate PO. It is convenience and always effective in OPC treatment. But now it is facing a challenge from 3bar PO pattern in 28nm technology node: when the side lines are sized, the DOF of the middle line is sacrificed, while the DOF of the middle line is much more critical in most cases as it is always the real gate PO.

This paper studies the 3bar PO pattern and tries to find a better way to enhance the DOF of PO bars.

Keywords—OPC; PO; lithography; DOF; retargeting; SRAF;

Introduction

As the technology node comes to 28nm, many new applications become very important to make lithography getting higher resolution limitation, such as OPC (Optical Proximity Correction), PSM (Phase Shift Mask), OAI (Off-axis illumination)[1].

OPC is the step which must be passed from design to manufactory for critical layers of 28nm technology node. It mainly includes two steps of mask correction: rule-based OPC and model-based OPC, wherein rule-based OPC contains retargeting and insertion of SRAF[2] (Sub Resolution Assist Feature). For example, 28nm PO(poly-silicon layer) OPC jobs will global size PO layout drawing first to compensate the pattern size bias between ADI (After Development Inspection) and AEI (After Etch Inspection). Then in order to enhance the PO layers' DOF, retargeting step will be applied to the non-gate PO lines, after which SRAF will be inserted beside all PO patterns when the space is enough without SRAF print risk. Finally OPC tool will call for a fine fitted OPC model to correct PO mask based on the target and SRAF.

Ideally, an isolate non-gate PO line's DOF is obviously enhanced after sizing up and SRAF inserting. 3BAR pattern is very common in 28nm PO layout, which has two isolate lines on the two sides and a dense line in the middle. The two isolate ones will be applied big retargeting in conventional OPC rule to make sure enough DOF, but the middle one is not considered, because in old technology nodes, the dense line never has DOF

problem. But now the problem comes from the middle one.

Fig.1 shows the 3bar pattern layout drawn and the target after global bias and conventional retarget. The pitch is 117nm and the widths of the three lines are all 27nm, while the targets of the side lines are 9.5nm bigger than that of middle line. And that the bias between the mask sizes of the lines is further enhanced after model-based OPC. It seems that the middle line is squished by the side lines due to extreme optical proximity effect. This behavior is particularly obvious in 28nm PO OPC process due to the highly polarized lithography source.

The simulated DOF of the side lines is 120nm, while that of the middle line is only 60nm, which cannot meet 28nm lithography process.

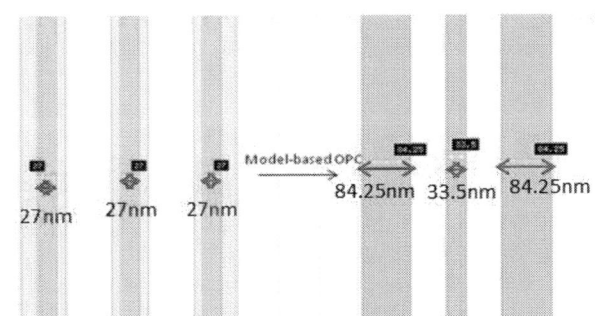

Fig.1 28nm PO 3bar pattern drawn, target and the mask shape after model-based OPC

This paper will study the influence of retargeting and SRAF insertion to the DOF of 3bar PO pattern, and try to find a better way to enhance the DOF.

Results and Discussion

Balance the lines' width

At first, retarget size of the side lines in isolate direction is decreased from base line 9.5nm. Fig.2 shows the DOF results, the DOF of side lines trend down as the size is decreased, while the DOF of the middle line trend up. But the DOF enhancement of the middle line is much smaller than the sacrifice of side lines. When the retarget size is 3.5nm, the DOF of side lines and middle line is 120nm and 80nm respectively, which is chosen for the following tests.

978-1-5090-6695-7/17 $31.00 © 2017 IEEE

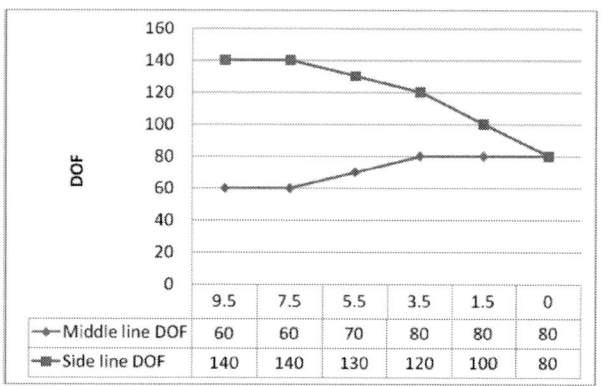

Fig.2 The relationship between retarget size (x-axis/nm) of side lines and pattern DOF (y-axis/nm)

Then, sizing up of middle line is tried, in which step the space between the middle line and the side lines is must be considered, as the sizing up action may cause bridge defect. The relationship between the DOF and the sizing up number is shown in Fig.3. The space DOF is big enough, and when the sizing up number is 6nm, the DOF of 3bar pattern achieves the biggest.

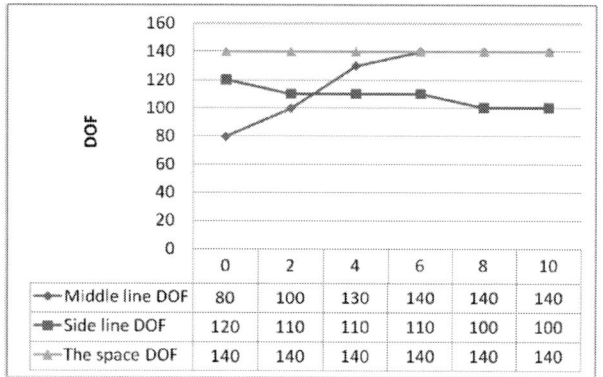

Fig.3 The relationship between middle line sizing up value (x-axis/nm) and pattern DOF (y-axis/nm, DOF 140 in the table means the DOF is >= 140nm)

The 3bar pattern's common DOF is increased from 60nm to 110nm by balancing the three lines' width, which is very important in PO OPC treatment of 28nm technology. The mask size is shown in Fig.4, as you can see, the size of the middle one seems much more robust than the baseline.

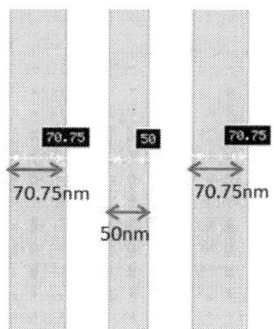

Fig.4 The mask size after 3bar pattern target balancing

Move the side lines

In this part, the side bars will be moved a little further away

from the middle one to study pitch's impact on the DOF of the 3bar pattern.

Fig.5 demonstrates that when the movement number is smaller than 16nm, it only has little impact on the DOF of side lines. But the enhancement of the DOF of middle line is significant.

So moving the side lines is also a useful method, because when the pitch is enlarged along with the movement, the optical proximity effect is decreased, in other words, the middle line achieves less optical impact from the side ones, which results in bigger mask size.

But its application is usually constrained by the real layout design. The movement number should be carefully considered based on the environment of 3bar PO pattern.

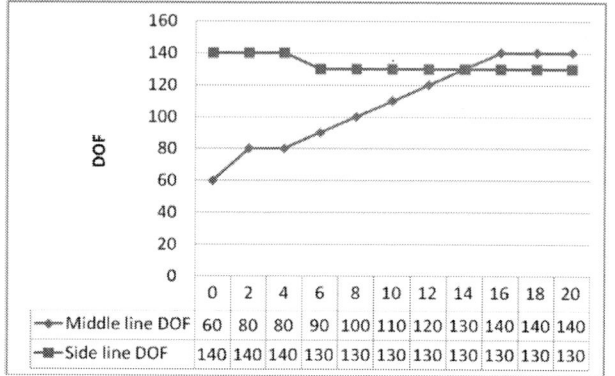

Fig.5 The relationship between movement value (x-axis/nm) of side lines and pattern DOF (y-axis/nm)

SRAF optimization

SRAF insertion is one of the best methods to enhance lithography DOF, and proper SRAF width and position is very important. SRAF rule is always defined after the lithography conditions are fixed, including SRAF width, SRAF space to main pattern and SRAF space to SRAF. SRAF rule is a global setting and a reasonable SRAF rule should satisfy the DOF requirement of all kinds of patterns on the premise of unprintable within the whole layout. Here the baseline SRAF width is 22nm, SRAF space to main pattern is 90nm and the space between SRAF is 70nm. But the SRAF rule may not be the best option for a given pattern, 3bar pattern for example, and special SRAF handle can be induced to enhance its DOF. In this part, the baseline SRAF rule will be adjusted to study the impact on 3bar pattern's DOF.

Fig.6 shows the middle line DOF slightly trends up along with increasing SRAF width. But if 80nm DOF is needed, SRAF width should be sized 23%, which has high risk of printing out.

Fig.7 and Fig.8 show either SRAF space to side line or SRAF space to SRAF has a little impact on the DOF of middle line, as the impact is indirect and the result is within the expected. When the SRAF space to side line is 100nm, the DOF is 70nm as a peak value, and this peak value is also got when the SRAF space to SRAF is 80nm.

978-1-5090-6695-7/17 $31.00 © 2017 IEEE   118

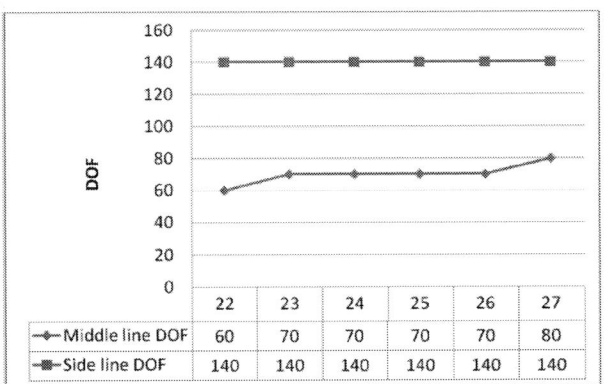

Fig.6 The relationship between SRAF size (x-axis/nm) and pattern DOF (y-axis/nm, DOF 140 in the table means the DOF is >= 140nm)

	22	23	24	25	26	27
Middle line DOF	60	70	70	70	70	80
Side line DOF	140	140	140	140	140	140

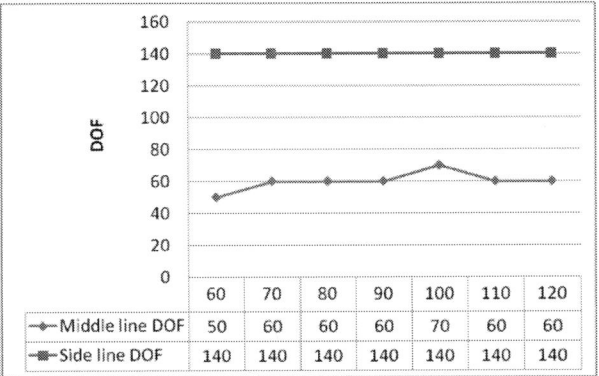

Fig.7 The relationship between SRAF space (x-axis/nm) to side line and pattern DOF (y-axis/nm, DOF 140 in the table means the DOF is >= 140nm)

	60	70	80	90	100	110	120
Middle line DOF	50	60	60	60	70	60	60
Side line DOF	140	140	140	140	140	140	140

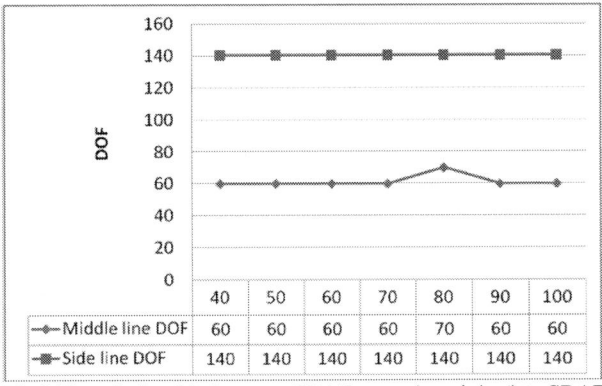

Fig.8 The relationship between SRAF space (x-axis/nm) to SRAF and pattern DOF (y-axis/nm, DOF 140 in the table means the DOF is >= 140nm)

	40	50	60	70	80	90	100
Middle line DOF	60	60	60	60	70	60	60
Side line DOF	140	140	140	140	140	140	140

Conclusion

This paper studies the DOF enhancement of 3bar PO pattern in 28nm technology node. Balance the lines' width between the side lines and the middle one is the best way and has little risk of side effect. Increasing the space between the side lines and the middle line is also an available method for the DOF enhancement, but its application is usually constrained. Sizing up SRAF width can be used to enhance the DOF, but it should be carefully studied to avoid SRAF printing out. Optimizing SRAF space to side line or SRAF space to SRAF can also be considered, though the impact is not as obvious as the other methods above.

Acknowledgements

The authors would like to thank vendors' useful discussions and help from Mentor Graphics.

References

[1] P. Gupta and A. B. Kahng, "Manufacturing-Aware Physical Design", Proc. IEEE/ACM ICCAD, pp. 681-687 (2003).

[2] Petersen JS, Analytical description of anti-scattering and scatteringbar assist features [J]. Proc SPIE, 4000: 77-89 (2000).

The incorporation of the pattern matching approach into a post-OPC repair flow

Yaojun Du

Semiconductor Manufacturing International Corporation
18 Zhangjiang Rd, Shanghai 201203, China
86-21-20816914, yaojun_du@smics.com

Abstract

The model based optical proximity correction (OPC) systematically computes the mask compensation that will be applied to the main features of circuits with sub-wavelength sizes. Even a sophisticated OPC recipe could render thousands of weak points, below the specs. An automatic repair flow may correct most of these post-OPC weak points. The remaining errors will have to demand engineers' visual inspections and subsequent manual fixings; and it might cost a considerable amount of human efforts and hence compromise the turnaround time (TAT). After performing several tape-outs, we have also noticed some weak points that need to be fixed afterward share certain commonalities. This inspires us to incorporate the pattern matching (PM) approach into our post-OPC repair flow. For the previous tape-outs, the remaining weak points will be fixed manually or be fixed by a special OPC recipe. Thus our old knowledge can directly provide proper OPC solutions for these known weak points. For a new tape-out, the design patterns associated with these weak points scan the post-OPC layer and find the match. Then, the proper OPC solutions will be pasted to these matched locations to complete repair process, allowing us to avoid repeatedly performing the manual fixings for the same types of weak points. This approach will also help identify certain OPC weak points that are proven to be fine by the wafer data. This type of weak points can be automatically waived by the OPCV verification. The incorporation of the PM approach into our repair flow can significantly reduce the TAT for a new tape-out.

Keywords—pattern matching, OPC, repair

Introduction

The major semiconductor foundries are striving to employ more advanced tech nodes such as 28 nm and 14 nm tech nodes to gain the competitive edges. The associated sizes of the main feature are significant smaller than the source wave length; the resulting pronounced optical proximity effects require a robust optical proximity correction (OPC) [1] to the mask, in order to transfer the target patterns to the wafer with the maximum fidelity. Tuning the OPC recipe within a short time window to keep up with the fab schedule has become a daunting task. For each new tape-out, an experienced OPC engineer could compose a rather sophisticated recipe which would still produce thousands of weak points. An automatic repair flow[2,3] should come to help fixing most of these post-OPC weak points. The remaining errors that could be up to hundreds require further manual fixings by engineers. Undoubtedly, the manual fixings could cost a huge amount of human efforts and increase the TAT. We have noticed that the post-OPC weak points from different tape-outs share quite a bit similarity or even are the same. Repeatedly performing the manual fixings for the same types of weak points are not desired; so it is essential to recognize these weak points in our repair flow automatically. The pattern matching (PM) approach [4-9] can serve this purpose well.

The pattern matching (PM) approach has been employed for improving productivities of the foundries. Tamer Desouksy and Omnia Saeed [5] integrated the PM into the OPC verification flow. They classify the hotspots into two categories: real hotspots and waivers. The engineers can then focus on the real hotspots and hence reduce the TAT for tape-outs. Daehyun Jung and co-workers [6] applied the PM to tackle the process related weak points. In another case, the PM helped reduce the run time for the lithography compliance check.[7] Moreover, the PM has been used to improve the efficiency of the double patterning for the advanced technology nodes.[9]

In this work, we build up the PM library and integrate the PM approach into our post-OPC repair flow. The PM library consists of various weak points and associated OPC solution from the previous tape-outs. During the repair stage, each weak point within the library will scan the whole chip to find matches. Once a match is found, the associated OPC solution will be pasted to the relevant position so as to complete the repair process. Thus, we can avoid performing repeated manual works for the same type of weak points.

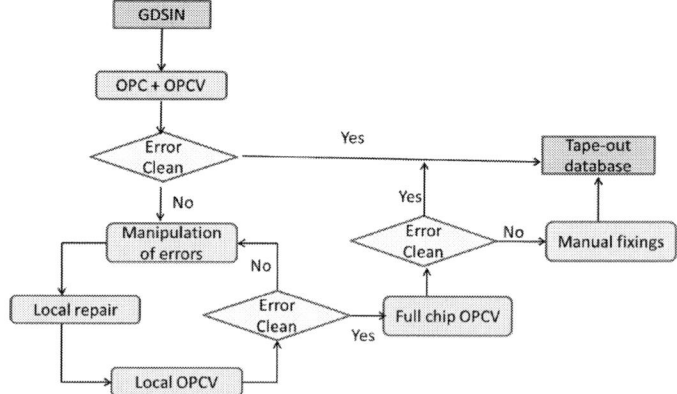

Fig. 1: The flow chart of our repair flow.

978-1-5090-6695-7/17 $31.00 © 2017 IEEE

Method

We implemented the PM manipulations into our repair flow. As shown in Fig. 1, our repair flow adopts the local repair and local OPC verification (OPCV) so that we can achieve a rather fast repair. After the repair, there could exist about a hundred remaining weak points. These weak points usually require engineers' visual inspection and manual fixings. Certain weak points will be fixed manually, while the other will be waived since these OPC weak points are eventually proven to be fine by the wafer data. We perform all simulations that adopt the CM1 resist model on the contact layer of a 28 nm tech-node chip. The 193 nm lithographic process uses an annular source and a high NA. For the PM functionality, the PM tool with Mentor Graphics Calibre[5,8] is employed.

Results

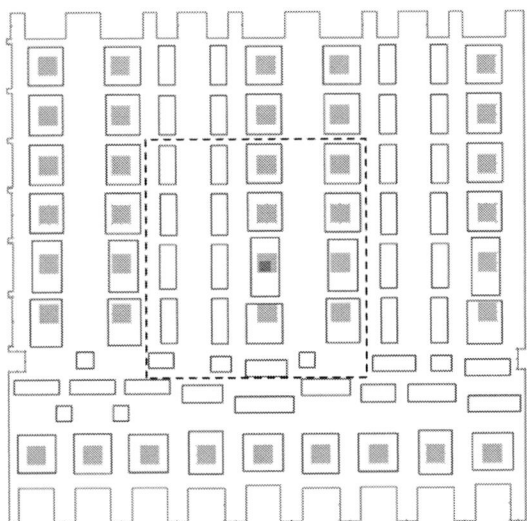

Fig. 2: A schematic diagram of a logic weak point stored in the PM library. The red frame is the entire scope associated with the weak point. The exact location of the weak point is highlighted by the green block. The orange block represents the original design, while the blue frames are the proper OPC patterns. The OPC patterns within the black dash-line frame will be used.

Fig. 2 illustrates the typical structure of the pattern library for the weak points and associated proper OPC solutions. The orange blocks are the design pattern, and the blue polygons represent the pre-calculated OPC patterns which are proper. The red region defines the PM window. The design patterns within the PM window will be used to scan the whole chip. If the exactly matches are found, the pre-calculated OPC polygon within the black frame will be pasted to the relevant position. The black region is significantly smaller than the red PM window. Hence, the patterns that are located between the inner region and outer region can serve as a buffer. The buffer can effectively reduce the rippling effects due to the possible different designs that are farther away from the weak point,

since the patterns outside the PM window could be different between the GDS file from which the PM library is constructed and the current GDS file. From previous tape-outs, we have tackled many weak points. Some of these weak points are carefully manipulated and the proper OPC solutions are found. These weak points and the associated OPC solutions are stored in our PM library. On the other hand, some other weak points are very difficult to find a proper OPC solution, but they are proven to be fine by the wafer data and subsequent processes. These weak points are referred to as waivers. In the past tape-outs as illustrated in Fig. 1, we performed the conventional OPC firstly. The subsequent OPC repair flow would try to fix remaining errors. However, a post repair layout still reports about a hundred weak points. The engineer will review these weak points one by one. Some of these weak points should be fixed manually, while others will be waived. The whole process could be quite lengthy. Our PM library will also include waivers; thus we can automatize the above manual review processes as illustrated in Fig. 3. This PM based repair operation will be incorporated into our OPC repair. It will be placed as the last step of the flow before the final OPC verification, replacing the manual review processes. It is worth noting that in the early stage of a new tech node development, each tape-out will serve as a learning cycle revealing new weak points and waivers. Eventually, our PM library is sufficiently complete, the manual operations can be avoided; hence the TAT of a tape-out can be greatly reduced.

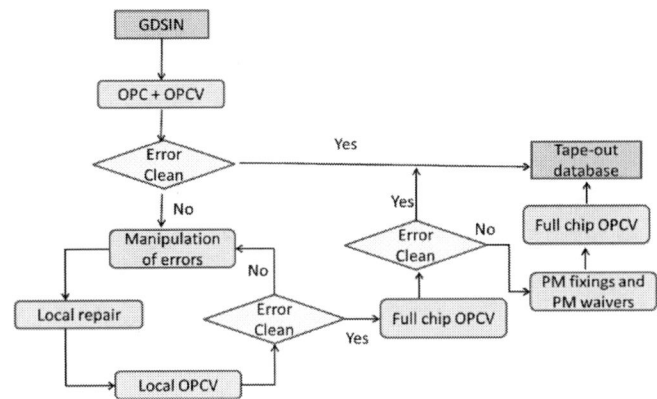

Fig. 3: The flow chart of our repair flow with the PM approach.

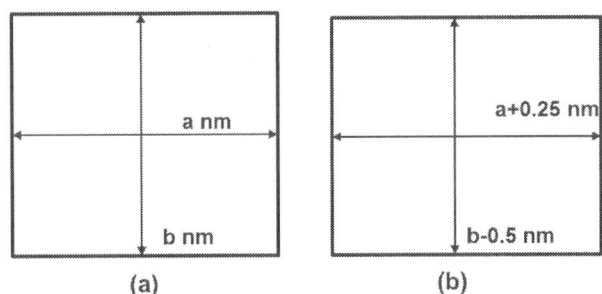

Fig. 4: A schematic diagram of multiple OPC solutions for the same weak point. The dimension of the OPC polygon (b) is 0.25 nm longer in the x direction and 0.5 nm shorter in the y direction than that of the OPC polygon (a). Nevertheless, they are both legitimate OPC solutions for the relevant weak point.

978-1-5090-6695-7/17 $31.00 © 2017 IEEE

Fig. 4 demonstrates that there could be multiple OPC solutions for the same weak point, which could be problematic for the PM-OPC. As shown in Fig. 3, the OPC polygon (b) is 0.25 nm longer in the x direction and 0.5 nm shorter in the y direction than that of the OPC polygon (a). Nevertheless, they are both legitimate OPC solutions for the relevant weak point. If these two OPC solutions are both stored in the PM library, the relevant weak points will be matched twice by these two solutions. The final OPC solution will be a superposition of these 2 solutions with a dimension of a+0.25 nm in the x direction and b nm in the y direction. The resulting OPC solution will be oversized undesirably and yield a new type of weak point. In real practice, we will keep a unique OPC solution for a weak point to avoid the above error.

Summary

In this work, we established the PM library and integrated the PM approach into our OPC flow. The PM library consists of various weak points and associated OPC solution from the previous tape-outs. It also includes the waivers that are obtained from the previous tape-outs. After performing the conventional OPC and OPC repair steps, we will use the PM base repair flow to tackle the remaining errors. Each weak point within the library will scan the whole chip to find matches. Once a match is found and this match is determined to be a weak point, the associated OPC solution will be pasted to the relevant position. Note that the PM feature has an outer region which is used for matching. Once the match is found, the OPC solution within the inner region will be pasted to the relevant location on the clip. This approach can help reduce the ripple effects that are caused by possible varying patterns outside of the outer region. On the other hand, if the match is determined to be a waiver, no error will be reported. In the early stage of the tech node development, each tape-out will help us augment our PM library. Once the PM library is sufficiently complete, we can avoid performing repeated manual works for the same type of weak points, hence reducing the TAT of a tape-out.

References

[1] Nicholas B. Cobb, "Fast optical and process proximity correction algorithms for integrated circuit manufacturing". PhD thesis, University of California, Berkeley, (1998).

[2] Yaojun Du and Qing Yang "The comparison of various strategies of setting up an OPC repair flow with respect to process window constraints", SPIE, 9426, 94261W (2015).

[3] Qing Yang and Yaojun Du, "OPC Solution by Implementing Fast Converging Methodology", SPIE, 9426, 94261V (2015).

[4] Jen-Yi Wuu, Fedor G. Pikus, Malgorzata Marek-Sadowska, "Efficient approach to early detection of lithographic hotspots using machine learning systems and pattern matching", Proc. of SPIE Vol. 7974, 79740U (2011)

[5] Tamer Desouky, Omnia Saeed, "Integration of Pattern Matchinginto Verification Flows", Proc. of SPIE Vol. 8326, 83261X (2012)

[6] Daehyun Jang, Naya Ha, Junsu Jeon, Jae-Hyun Kang, Seung Weon Paek, Hungbok Choi, Kee Sup Kim, Ya-Chieh Lai, Philippe Hurat, Wilbur Luo, "In-Design Process Hotspot Repair Using Pattern Matching", Proc. of SPIE Vol. 8327, 83270S (2012)

[7] Satomi Nakamura, Tetsuaki Matsunawa, Chikaaki Kodama, Takanori Urakami, Nozomu Furuta, Shunsuke Kagaya, Shigeki Nojima and Shinji

Miyamoto, "Clean Pattern Matching for Full Chip Verification", Proc. of SPIE Vol. 8327, 83270T (2012)

[8] ZeXi Deng, ChunShan Du, Lin Hong, LiGuo Zhang , JinYan Wang, "An Efficient lithographic hotspot severity analysis methodology using Calibre PATTERN MATCHING and DRC application", Proc. of SPIE Vol. 9427, 94270Y (2015)

[9] Lynn T.-N. Wang, Vito Dai, Luigi Capodieci, "Pattern matching for identifying and resolving non-decomposition friendly designs for Double Patterning Technology (DPT)", Proc. of SPIE Vol. 8684, 868409 (2013)

SYNTHESIS AND DIRECTED SELF-ASSEMBLY OF MODIFIED PS-B-PMMA FOR SUB-10 NM NANOLITHOGRAPHY

*Xuemiao Li, Jie Li, and Hai Deng *

Department of Macromolecular Science, State Key Laboratory of Molecular Engineering of Polymers，Fudan University，Shanghai 200433, China
* Corresponding Author's Email: haideng@fudan.edu.cn

ABSTRACT

The directed self-assembly (DSA) of block copolymers has attracted a great deal of interest due to its potential applications in sub-10 nm lithography [1-3].

The conventional organic-organic DSA materials such as poly-(styrene-block-methyl methacrylate) (PS-b-PMMA) have been extensively studied [4,5], however, the low etch contrast between two blocks and the difficulty to reduce L_0 limit its application. In this study, we designed and synthesized the novel DSA materials based on PS-b-PMMA. Through the modifying of acrylics part, segment−segment interaction parameter (χ) can be significantly increased, which leads to rapid self-assembly and high etch contrast.

INTRODUCTION

As we all know, photolithography has been the workhorse of the industry, which is crucial in the manufacture of smaller devices with excellent storage capacity. However, the patterning technology is approaching resolution limit after 10 nm node. Directed Self-Assembly (DSA)lithography has attracted more and more attention due to its high potential resolution and extremely low cost when it is compared with EUVL [6-8].

The traditional DSA materials have relatively low etch contrast between two blocks and the resolution limitation, such as PS-b-PMMA, the resolution is up to nearly 20 nm. In this study, we synthesized a series of novel and rapid self-assembly block copolymers, which based on modifying of traditional DSA materials by anionic polymerization [9]. Also we synthesized different brush polymers to modify the surface of silicon wafers as neutral layers for the following DSA process [10-11]. In order to realize sub-10 nm patterning capabilities, we precisely controlled the ratio of each component to reach max resolution.

All resulted materials show extremely high resolution up to 17.7 nm separated lamella morphology and 14.7 nm hexagonal morphology by Small Angle X-ray Scattering (SAXS) measurement and TEM (transmission electron microscopy), indicating potential application as sub-10 nm full pitch DSA materials. (Figure 1)

EXPERIMENT AND DISCUSSION
Synthesis of block copolymers and characterization

We synthesized a series of modified PS-b-PMMA block copolymers following the standard process. [12] And the polymer synthesis details will be disclosed later. After the polymerization the polymer was purified by dissolution and precipitation, following with the nuclear magnetic resonance (NMR) and gel permeation chromatography (GPC) measurement.

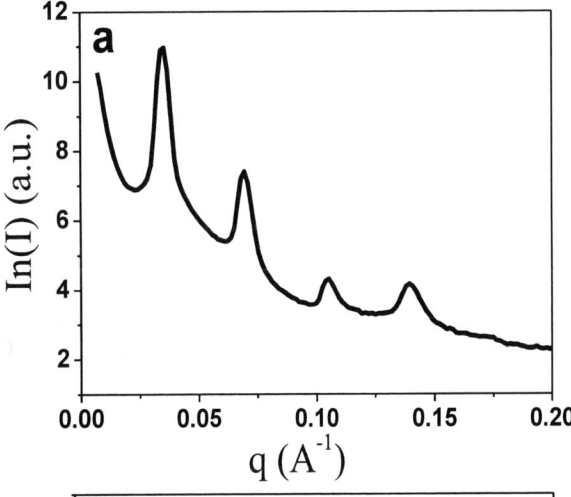

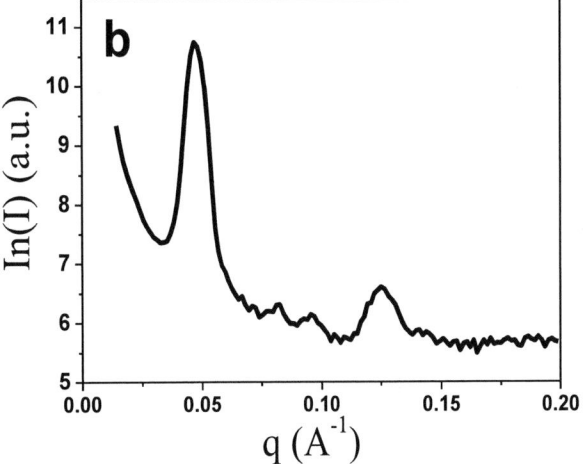

Figure 1. Bulk morphologies at room temperature, after thermal annealing of typical lamella structure and hexagonal structure of sample one (a) and sample two (b) respectively, which were defined by SAXS measurement.

^{1}H-NMR spectra were acquired on a 400 MHz AVANCE III instrument using CDCl$_3$ as solvent and TMS as internal standard. From the ^{1}H-NMR spectra we can identify the characteristic peak of each component.

GPC measurements were used to calculate the molecular weight and PDI. (Polydispersity index) Narrowly distributed polystyrene samples were used as calibration standards. All the block copolymers show a narrow distribution up to 1.13 during this synthesis process.

By controlling the amount of the initiator and the ratio of monomers, we can precisely alter the pitch dimension and morphology. For the hexagonally packed cylindrical morphology (Hex): L$_0$ is the center-to-center spacing and for the lamellar morphology (Lam): L$_0$ is the domain spacing (full-pitch).

Annealing process of material

The novel modified PS-b-PMMA block copolymers were dissolved in toluene, and coating on the subtract, which were annealed at 100 °C for 5 min (or even less) in N$_2$ protection. The modification of this material remarkably reduced the annealing time when compared with traditional PS-b-PMMA. And this demonstrates a potential prospect for this DSA material in the industry application.

TEM image of novel DSA materials

The TEM image can further illustrate the morphology of these novel DSA materials. As can be seen from the picture, the lamella morphology can be obtained after a short time annealing at a relatively low temperature. The clear light and dark lines represent two different components in the block copolymer respectively. When the scale bar is 50 nm, nearly three repeated units in this scale, which is consistent with the SAXS result. The full-pitch of this material is 21.4 nm (SAXS) and 18.6 nm (TEM). Also the long range order of this material is clearly in Figure 2.

In the following work, we can selectively etch one component to get the sub-10 nm straight line, and then transfer the pattern into the silicon wafer or any other substrate for its litho application.

CONCLUSIONS

In this work, we first synthesized a series of novel and rapid self-assembly block copolymers, which based on modifying of traditional DSA materials PS-b-PMMA. The structure of materials and the morphology of the self-assembly sample were measured with ^{1}H-NMR, GPC, SAXS and TEM. All these shows the reasonable structure and excellent performance.

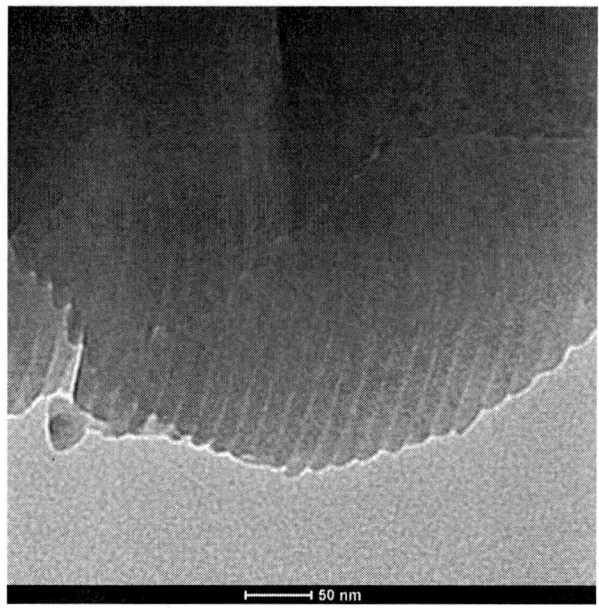

Figure 2. TEM image of this rapid assembly material after the thermal annealing at 100 ℃ for 5 min. The regular stripes show that the lamella morphology forming in long range order.

By precisely controlling the ratio of each component we got different domain spacing and morphology. Also these novel materials easy to self-assembly in relatively low temperature and extremely short time. Finally, we managed to obtain the sub-10 nm patterning by using this high resolution DSA material.

ACKNOWLEDGEMENTS

Funding for this research was provided by the "1000 Talent plan" in China. The authors also acknowledge experimental support from State Key Laboratory of Molecular Engineering of Polymers in Fudan University.

REFERENCES

[1] Kim, H.-C.; Park, S.-M.; Hinsberg, W. D. *Chem. Rev.* 2010, 110, 146.

[2] Herr, D. J. C. *J. Mater. Res.* 2011, 26, 122

[3] Christophe S, Frank S. B, Marc A. Hillmyer *ACS Macro Lett.* 2015, 4, 1044−1050

[4] Benoit, H.; Wu, W.; Benmouna, M.; Moser, B.; Bauer, B.; Lapp, A. *Macromolecules* 1985,18, 986

[5] Bang, J.; Kim, S. H.; Drockenmuller, E.; Misner, M. J.; Russell, T. P.; Hawker, C. J. *J. Am. Chem. Soc.* 2006, 128, 7622−7629

[6] Jeong, S.-J.; Kim, J. E.; Moon, H.-S.; Kim, B. H.; Kim, S. O. *Nano Lett.* 2009, 9, 2300.

[7] Kennemur, J. G.; Yao, L.; Bates, F. S.; Hillmyer, M. A. *Macromolecules* 2014, 47, 1411.

[8] Bates, C. M.; Maher, M. J.; Janes, D. W.; Ellison, C. J.; Willson, C. G. *Macromolecules* 2014, 47, 2.

[9] Saam, J. C.; Gordon, D. J.; Lindsey, S. *Macromolecules* 1970, 3,

[10] Bang, J.; Bae, J.; Lowenhielm, P.; Spiessberger, C.; Given-Beck, S. A.; Russell, T. P.; Hawker, C. J. *Adv. Mater.* 2007, 19, 4552

[11] Ryu, D. Y.; Shin, K.; Drockenmuller, E.; Hawker, C. J.; Russell, T. P. *Science* 2005, 308, 236

[12] T.Thumbrecht, J.Schotter, G.A.Ka, sile, N.Emley, T.Shibanchi, L.Krusin-Elbaum, K.Guarini, C.T.Blaek, M.T.Tuomlnen, T.P.Russelll, *Science*, 2000,290, 2126

Research of SMO process to improve the imaging capability of lithography system for 28nm node and beyond

Haibin Yu, Yueyu Zhang, Binjie Jiang, Shirui Yu, Zhibiao Mao
Shanghai Huali Microelectronics Corporation
Shanghai, China
yuhaibin@hlmc.cn

Biography

Haibin Yu received his Ph.D. in Department of Physics and Astronomy from Shanghai Jiao Tong University in 2015. He is now an OPC engineer of Technology Development Division, in Shanghai Huali Microelectronics Corporation.

Abstract

The source-mask optimization (SMO) solution has become one of the most important branches of Resolution enhancement techniques (RET) to extend the imaging process window with next generation computation lithography, which improve the imaging capability of lithographic systems in the integrated circuit foundry manufacturing. Based on the SMO software RET Selection provided by Mentor Graphics Corporation, we have researched the SMO process to improve the imaging capability of lithographic systems for 28nm node and beyond: choosing the key patterns, confirming the process window conditions and so on. In this paper, the parameters PV band, MEEF, NILS and DOF have been used to evaluate the free form illumination sources, and the final illumination source have been verified, which generated by ASML scanner.

Keywords— Integrated Circuit Foundry Manufacturing, Source Mask Optimization (SMO), Resolution Enhancement Technology (RET), Optical Proximity Correction (OPC)

Introduction

The lithography technology, which is the key technology in the integrated circuit foundry manufacturing, has been rapidly developed under the Moore's law in the past decades. In the case of special illumination wavelength, the resolution enhancement techniques (RET) are the most important approaches to extend the imaging process window and reduce the process factors, including optical proximity correction (OPC) [1], sub-resolution assist feature (SRAF) [2], inverse lithography technology (ILT) [3], double patterning (DP) [4], and so on. However as critical dimensions decrease to 28 nm node and beyond, the conventional source could not satisfy the future requirements in the lithography technology. And the special lighting condition also limits the OPC ability to improve the lithography resolution. Meanwhile the lighting systems of lithography machine have made breakthrough progress in technology--the Flex Ray of ASML and Intelligent Illumination Unit (IUU) of Nikon could both realize the illumination source in the form of freedom. Combined with

OPC, the freeform illumination source could provide a new technology--source mask optimization (SMO) [5]: optimize the illumination source and pattern mask at the same time. The SMO technology, a new branch of RET, could improve the lithography printing quality and the lithography resolution. SMO has not only been widely applied for 20nm node and beyond, but also been used to solve some weak points in 28nm node in the integrated circuit foundry [6].

In SMO technology, there is no single definition of best results, but more trade-offs than just accuracy versus run time, the optimization process could be divided into the following steps: (1) selecting some key patterns from full chip, including general 1D pattern, 2D pattern and some weak points; (2) optimizing the illumination source and the mask of key patterns simultaneously by using ILT; (3) evaluating the optimization source by checking performance including edge placement error (EPE), process variation band (PV band), mask error enhancement factor (MEEF), normalized image log slope (NILS), depth of focus (DOF) and so on.

Experiment

In the experiment, we used the SMO software RET selection provided by Mentor Graphics Corporation to optimize the source and mask simultaneously.

In SMO technology, selecting key patterns from full chip is the first and important step [7, 8]. The selected key patterns must be on behalf most of the patterns on full chip and include some weak points. Table 1 is an example list of selected patterns.

Table.1. SMO selected key pattern list

Label	Dimension	Pattern	Target CD	Pitch
1	1D	Line	50	100
2	1D	Space	60	100
3	2D	Line	50	100
4	2D	Space	60	100
5	WP	Line	50	100
6	WP	Space	60	100
…	…	…	…	…

Before using RET selection to optimize the freeform source and pattern mask based on the key patterns, we have to do following work: (1) preparing an initial litho model, or at least an optical model, which is a mathematical representation of the lithographic exposure system and models constituents of the system such as numerical aperture, lasing wavelength, lens

978-1-5090-6695-7/17 $31.00 © 2017 IEEE

aberration and illumination source; (2) setting the process window conditions (PW conditions), whose goal is to describe the conditions that cause the most variation in the equipment's wafer printing, there are parameters weight, defocus, dose and mask bias in every condition, as shown in table 2.

Table.2. SMO PW conditions

PW Name	Weight	Defocus (nm)	Dose	Mask Size(nm)
nominal	1.0	0	1.0	0
SMO_1	1.0	50	1.05	-0.5
SMO_2	1.0	50	0.95	0.5
PxOPC_1	1.0	45	1.04	-0.5
PxOPC_2	1.0	45	0.96	0.5
OPCV_1	1.0	40	1.03	-0.5
OPCV_2	1.0	40	0.97	0.5
...	...	...	...	...

The optimization sources are strongly related with the input key patterns. For example, the source images in Fig.1 are the optimization results based on key patterns with different target CD and pitch. The sources A to F are respectively according to vertical patterns with different structure parameter, the sources G and H sources according to horizontal patterns. Therefore, the step of selecting key patterns is very important and the selected patterns should be on behalf most of patterns on full chip.

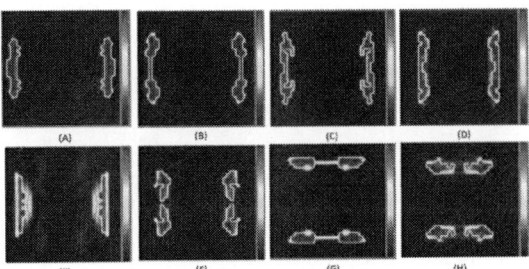

Fig.1. Optimization source based on different input pattern

In the experiment, we prepared dozens of patterns including 1D and 2D patterns. We adjusted the weight of input patterns in order to obtain the appropriate result. There is an optimization source as shown in Fig.2. In the result, the freeform source has both horizontal and vertical symmetry, the maximum sigma is 0.84, the minimum sigma is 0.66, and the illumination angle is 110 degree. The RET selection also output the corresponding litho model (or optical model) which is matching the output freeform source.

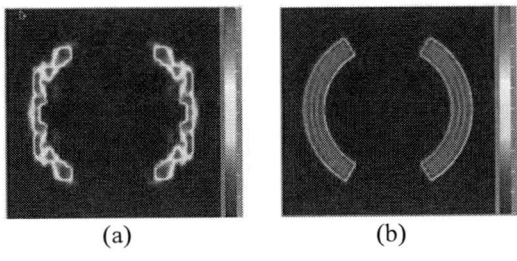

Fig.2. Two illumination sources (a) an optimization freeform source; (b) a conventional dipole source

There are several pre-defined criteria and a means to calculate how close the optimization freeform source comes to meet the layout specification. The EPE is the difference between the ideal edge locations versus the simulated edge locations, and the results are shown in Fig.3, in which the red one is the result by using the conventional dipole source, and the blue one is the result by freeform source. In Fig.3, the EPE by using freeform source are less than the results of conventional source for most selected patterns.

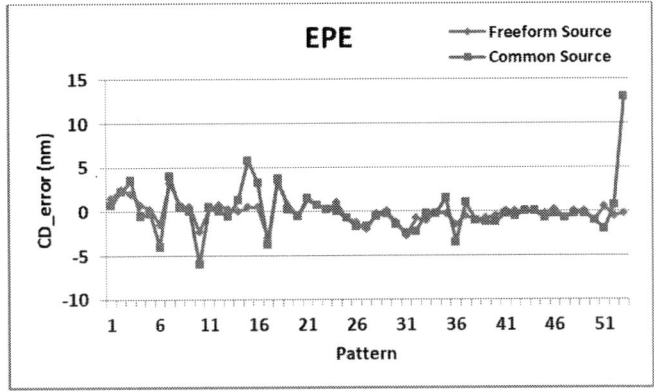

Fig.3. EPE comparison between freeform and conventional dipole source

The PV Band is the difference between the smallest and largest contour created by different process window condition, which is a measure of how consistently a mask will print given normal variation in lithography equipment, and the results are shown in Fig.4, in which the red one is the result by using the conventional dipole source, and the blue one is the result by freeform source. In Fig.4, the PV bands with freeform source are less than the conventional source.

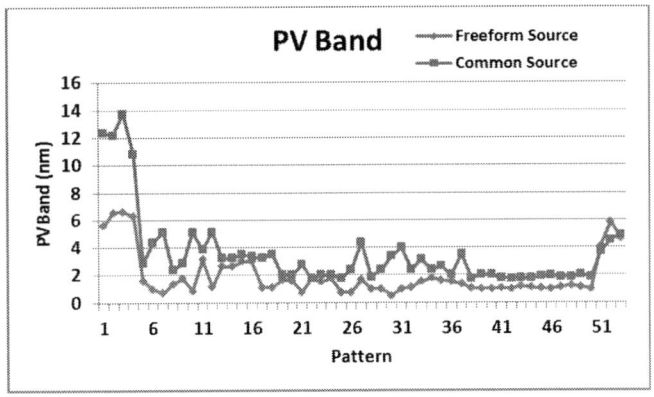

Fig.4. PV Band comparison between freeform and conventional dipole source

As discussion above, the EPE and PV band with freeform source have a better performance than the conventional source, and the other criteria MEEF, NILS, and DOF also have the same performance.

The information of freeform source exported by RET selection have to transmit into lithography machine in order to generate the illumination source, which is used to exposure the wafer. The nonzero differences between freeform source exported by SMO software and generated by ASML scanner are shown in Fig.5.The influence could be accepted because the different is unavoidable, and meanwhile the effect is small.

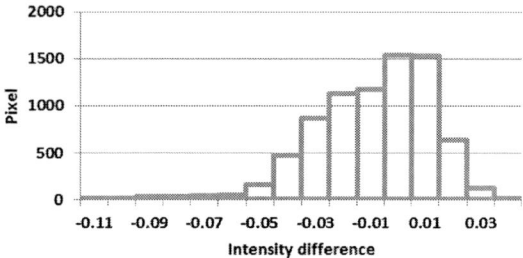

Fig.5. the nonzero differences between SMO source and source generated by ASML scanner

Conclusion

The SMO technology, which optimizes the source and mask simultaneously, has become one of the most important branches of RET to extend the imaging process window, and it has been widely applied for 20nm node and beyond. In this paper, we have researched the SMO process to improve the imaging capability of lithographic systems for 28nm node and beyond: it is very important to select key patterns because of the strong relationship between optimization sources and input patterns; there are differences between SMO source and illumination source generated by ASML scanner, therefore, it is necessary to evaluate the source in order to meet the expectation; the performance of optimization results have been evaluated by several pre-defined criteria EPE, PV Band, MEEF, NILS, DOF, and freeform source have a better performance than the conventional source.

References

[1] Randall J N, Aton T J, Palmer S R. Optical proximity correction: US, US 6634018 B2[P]. 2003.

[2] Wallace C H, Nyhus P A, Sivakumar S. Sub-resolution assist features[J]. 2010.

[3] Granik Y. Fast pixel-based mask optimization for inverse lithography[J]. Journal of Microlithography Microfabrication & Microsystems, 2006, 5(4):851-855.

[4] Wallace C H, Tingey M, Sivakumar S. Double patterning techniques and structures: US, US 20090267175 A1[P]. 2009.

[5] Jia N, Lam E Y. Pixelated source mask optimization for process robustness in optical lithography.[J]. Optics Express, 2011, 19(20):19384-98.

[6] Nagahara S, Yoshimochi K, Uchiyama T, et al. SMO for 28-nm logic device and beyond: impact of source and mask complexity on lithography performance[J]. Proceedings of SPIE - The International Society for Optical Engineering, 2010, 7640(1):511-519.

[7] Yuyang Sun, Yee Mei Foong, etal. Optimizing OPC data sampling based on "orthogonal vector space". Proc. SPIE 7973, Optical Microlithography XXIV, 2011, 79732K.

[8] Dmitry Vengertsev, Kihyun Kim, Seung-Hune Yang, et al. The new test pattern selection method for OPC model calibration, based on the process of clustering in a hybrid space. Proc. SPIE 8522, Photomask Technology 2012, 85221A.

The phenomena of footing's variation caused by oxidation of nitrogen-containing substrates

Song Bai [a], Hanmo Gong [a], Chang liu [b], Qiang Wu [a]

[a]Technology R&D, SMIC Advanced Technology R&D (Shanghai) Corporation
[b]Technology R&D, Semiconductor Manufacturing International Corporation

Pudong New Area, Shanghai, P. R. China 201203
+8621 38610000x15978, Cypress_Bai@smics.com

ABSTRACT

With the substrate suffering from the contamination of nitrogen, the variation of after-developing-inspection (ADI) linewidth with time could be monitored by CDSEM. This phenomenon has been attributed to the footing produced by acid loss. Moreover, the oxidation of nitrogen-containing substrates can cause the variation of capability of trapping acid, so the profile of photoresist (PR) changes with time, which has been verified by top-down CDSEM picture and SEM cut. It is reported that plasma and chemical treatment for the substrate can passivate the trapping sites for acid to remove this instability. In this paper, simulation and contrast experiment will be applied to illustrate the mechanism of footing and to verify this explanation. First, reflectivity and standing wave have been calculated based on the mask pattern and substrate. Second, light intensity distribution in photoresist has been calculated through the simulation of image in resist. At last, we add photoacid diffusion to the image intensity distribution. Such depiction by diffusion equation also goes across the boundary between the photoresist and the substrate, where several effective diffusion constants are used. The variation caused by the oxidation of surface is explained and analyzed based on the model above. The method of monitoring the variation of footing only by top-down CDSEM picture is proposed.

INTRODUCTION

For the lithography of implant layer, the critical dimension (CD) is still relatively large for 28nm and beyond technology node so that the process window such as exposure latitude (EL) or depth of focus (DOF) is not the main problem. But the complicated film stack of front end of the line (FEOL) leads to the challenge of profile controlling over footing, undercut and standing-wave shape. Weak profile control of PR will impair the capability of defining the zones to be implanted and further reduce process window. The traditional simulation of PR profile calculates the energy distribution formed by the beam incident and reflected at first [1]. Secondly the linear diffusion or higher-order model is inserted to give the final shape of latent image. However, the real situation is that similar patterns located in different region within one shot, which don't have enough difference of film stack to give distinct optical response, will also have different PR profile. It was researched that the acid in resist layer can be lost by BARC layer [2] or other inorganic substrates during post exposure bake (PEB) treatment. In this case, only natural oxidation of

film stack could totally change the final PR profile. Considering that only 1-2nm variation of oxidation can be brought by natural oxidation, the phenomena could not be attributed to the optical reason. In our model, the effective length of diffusion in substrates is introduced to describe the degree that the photoacid diffuses on boundary from photoresist to substrates. This disappearance could be normally attributed to the neutralization with chemical from substrates or the real diffusion into substrates. In our representation, they are calculated by the effective length of diffusion and absorption constants in substrate. Hence in this work, we will discuss the competence of algorithm for the calculation of aerial image and the influence of the constants from diffusion equation. Moreover, for process control, we will provide the method to eliminate the confusion between the real variations of CD and the degree of footing, from the view of CDSEM principle.

SIMULATION APPROACHES

Optical latent profile

As we illustrate above, the optical latent image should be calculated as the starting profile of following diffusion equation through the combination of aerial image and standing wave. The combination of aerial image and standing wave could be divided into two methods. The first method calculates the energy distribution in the certain height directly by considering the interference between different diffraction orders of incidence and reflection one by one, which travel the corresponding optical path and own their defocus based on their height of PR. The reflection should be separated in to S and P vector at first due to the calculation of amplitude and phase. This method owns higher accuracy but also higher amount of calculation. The second method takes the two dimension of PR profile separately, which is much easier to calculate and is also capable of being the starting profile for linear diffusion equation with relatively large CD and lower accuracy. The aerial image with defocus of different heights of PR is calculated for horizontal dimension, while the total standing wave is calculated by taking all incident directions into account for vertical dimension. But for the second method, two points could still be implemented to enhance accuracy for the calculation of starting profile for diffusion.

The swing curve provides the guidance for the choice of thickness for PR and bottom anti-reflection coating (BARC). It's for all the patterns with different pitch and CD of this film stack. But it is unreasonable to use the reflectivity for this film

978-1-5090-6695-7/17 $31.00 © 2017 IEEE

stack for any specific optical setup. We should give different incident angles weights based on the mask pattern and illumination condition. P and S vector should also be separately treated to some extent due to their bias for the interference among different orders, as equation (1) shows.

$$I_S(X)=|E_y(x)|^2=a_0^2+2a_1^2+4a_0a_1\cos(2\pi x/p)+2a_1^2\cos(4\pi x/p)$$
$$I_P(X)=|E_x(x)|^2+|E_z(x)|^2=a_0^2+2a_1^2+4a_0a_1\cos(2x/p)+2a_1^2\cos(4\pi x/p)\cos(2\theta) \qquad (1)$$

This equation calculates the three-beam image at best focus with amplitude a_0 for zero order, amplitude a_1 for two first orders and diffraction angle θ decided by pitch p. The calculation of P vector should take the weakening by angle θ into account. With the similar principle, P and S vector also owns bias in the calculation of standing wave.

Furthermore our film stack, between resist and substrate, exists very thin oxide which also has the approximate refractive index with resist, which could be neglected. The phase change caused by the reflection of this film stack comes mainly from the reflection between oxide and substrate, as figure 1 shows. When incidence angle is zero, the phase shift between P vector and S vector is 180 degrees and when incidence angle increases, the phase shift increases slowly at small incidence angle. It explains why when there is a crest of standing wave in P vector, there is almost a trough in S vector, for our film stack and illumination condition.

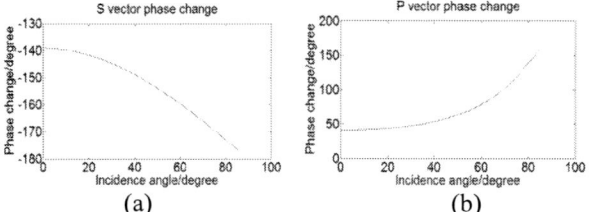

(a)　　　　　　　　　(b)

Figure 1: (a) the phase change of S vector between oxide and silicon; (b) the phase change of P vector between oxide and silicon.

Because of the different profiles for our film stack and the bias of interference between incidence and reflection, P and S vector should be calculated separately.

The profile of diffusion

We could only take two dimensions of diffusion into account due to the uniformity of the third dimension. The diffusion is expressed by equation (2), while H is the concentration of photoacid. D and K are the constants of acid diffusion and absorption respectively, depended on the property of material and temperature.

$$\frac{\partial(\frac{\partial H}{\partial r}\times D)}{\partial r}=\frac{\partial H}{\partial t}+K\times H \qquad (2)$$

The simulation has the periodic boundary condition on left side and right side, while isolated boundary is used on top and total absorption condition is taken at the bottom of substrates. The continuum of acid flow between two materials is implemented, PR and substrates. The diffusivity in PR is D_1 and the diffusivity in or into substrates is D_2. Initial distribution of acid concentration comes from the energy distribution we discuss before. The diffusion length d_1 or d_2 in PR or substrates and the corresponding time is set at first to give the actual value of D_1 and D_2, and then we could observe the acid flow with time.

SIMULATION RESULTS

By the algorithm for optical latent image, we could obtain the starting profile for diffusion equation as figure 2. The dimension is thickness 240nm and pitch 290nm.

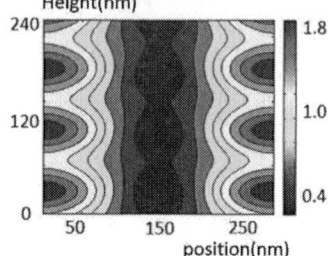

Figure 2: the energy distribution of PR for one pitch, with thickness 240nm and period 290nm. The absolute value of latent image here is nominal and only the relative value is meaningful.

For chemical amplification PR, acid exists as the catalyzer, so de-protection reaction doesn't consume them but have direct proportion with them at every moment. The final shape after PEB should be related to the integral of distribution for acid with time. So the two graphs are the integral of acid distribution for every time step. The unit of difference for diffusion equation used below is the square with 5nm widths.

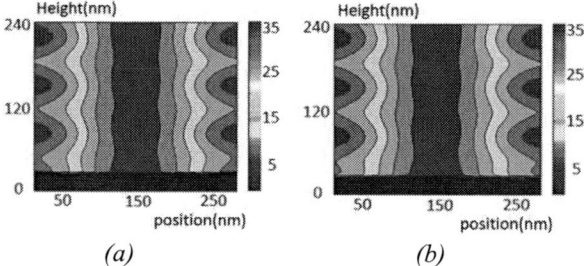

(a)　　　　　　　　(b)

Figure 3: (a) the time integral of photoacid distribution for 90s with PR diffusion length d_1 20nm and substrates diffusion length d_2 2nm, which appears no footing. (b) the time integral of photoacid distribution for 90s with PR diffusion length d_1 20nm and substrates diffusion length d_2 5nm, which appears a little footing.

Moreover, if you keep the value of D_2 and reaction time but improve the diffusion length of PR from 20nm to 35nm, you could observe the shape of standing wave is weakened. The footing for different energy splits shows more consistent results. Setting the diffusion length of acid in substrates could not only give the continuous description for the loss of acid on boundary but also provide the opportunity to discuss the effect of the shape of boundary.

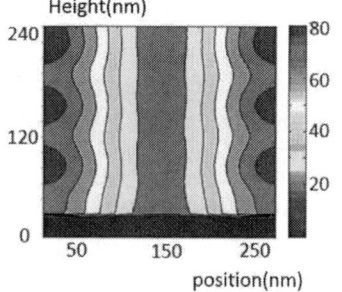

Figure 4: the time integral of photoacid distribution for 90s with PR diffusion length d_1 35nm and substrates diffusion length d_2 5nm.

In our case, we measure the thickness of oxidation on our surface with different time of suspension before coating PR. Without suspension before coating, we measure the average thickness about 13.26A with 3sigma 2.24A. The stable surface with saturation of oxidation owns the value about 21.51A with 3sigma 0.31A. It's known that the principle of measuring thickness is through the detection of optical reflection, which represents the average thickness of a region. So we could deduce that in the progress to the saturation of oxidation, the roughness of surface is also reduced much. At the beginning of oxidation, the oxide is so non-uniform and loose to provide more contact with photo resist and larger space to diffuse, so both the diffusion and neutralization should be ongoing at higher speed. This effect could be simulated by the boundary shape like 'grating'. You could conclude that the diffusion with 'grating' surface have more serious footing even with the same D_2.

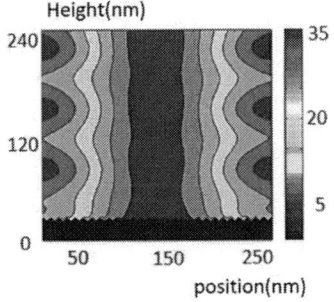

Figure 5: the time integral of photoacid distribution for 90S with photoresist diffusion length d_1 20nm and substrates diffusion length d_2 5nm. The boundary between photoresist and substrates is the shape like 'grating', with the hollow of square with width and depth 5nm.

After the starting of diffusion 80S, the typical value in substrates is about 0.1-0.15, while the concentration in photo resist is about 0.6-1.0 for diffusion length of substrates 2nm. Hence the acid absorption item in diffusion equation could not influence the diffusion in photo resist much, because the difference of concentration from photo resist to substrates doesn't alter much with the variation of acid absorption constant. If the diffusion length into substrate increases observably and then the diffusivity increases with proportional to the square of diffusion length, the acid absorption will become much more vital in the equation.

PROCESS CONTROL

Under the monitor of CDSEM, the footing variation will be interpreted into the variation of critical dimension wrongly, which provides the problem of monitor and control of process. If you measure the trend of CD value with different threshold value of CDSEM settings, you could get the feature as follow, using both the line mode and space mode to measure the space. It illustrates that the real variation of critical dimension due to exposure energy variation will have the shift of the whole curve such as the shift between curve B and C with oxidation saturation. But the curve A shares the same starting value with the curve B and the same ending value with the curve C. From the understanding of CDSEM principle, it represents that the natural oxidation changes the photo resist profile but not the real critical dimension, which is also verified by SEM cut but not shown here. Hence you could monitor the process by adding the use of line mode to measure space, which don't add any time of measurement. This value will be interpreted as the real variation of exposure energy and its difference with inline value could be illustrated as the extent of footing. For our case, we still try the method of wet and plasma to deal with the surface to realize the saturation of oxidation in advance.

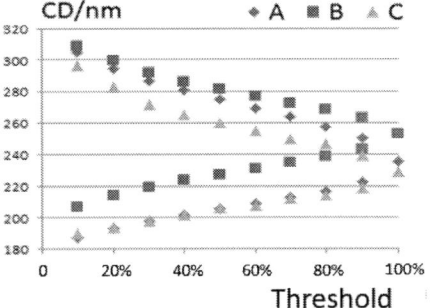

Figure 6: the trend of measurement value with different threshold value to slice the CD. The half curve with CD larger than 240nm is measured by the line mode for measuring space. Curve A is the one with exposure energy 30.5mJ and not suspension before coating. Curve B is the one with 30.5mJ and oxidation saturation before coating. Curve C is the one with 25.5mJ and oxidation saturation before coating.

SUMMARY

We build a model with several effective constants to depict the continuum change of photoresist profile with durative natural oxidation. For the starting profile for diffusion equation, we calculate the optical latent image by separating S and P vector and then adding them together. It could be attributed to the bias of P and S vector for calculating the interference among the different diffraction orders of incidence and reflectance. For diffusion part, the influence of those effective constant has been evaluated through the comparison of the shape for footing. Higher diffusivity into substrates will enhance the footing remarkably, but the higher acid absorption in substrates will not influence evidently. Finally, we add the CDSEM measurement for space with line mode to monitor the real CD variation and footing variation separately. The algorithm above could also be used for the profile analysis with BARC substrates. It could be achieved after the evaluation of diffusion length for certain BARC and PR under certain PEB temperature.

REFERENCES

[1] Zhimin Zhu, Emil Piscani, Kevin Edwards, Brian Smith, *Proc. of SPIE* Vol. 6924 69244A-7
[2] Struyf, H. Hendrickx, D.Van Olmen, J. Iacopi, F. et al, *Proc. of SPIE* Vol. 7273 727331-8

A STUDY OF N-INDUCED RESIDUE DEFECT ON GATE OXIDE AFTER LITHOGRAPHY REWORK

Zhou Fang [1], Chang Liu [1], and Zhoujun Pan [1]*
[1] Technology R&D, Semiconductor Manufacturing International Corp.
Pudong New Area, Shanghai 201203, P. R. China
*Corresponding Author's Email: Zhou_Fang@smics.com

ABSTRACT

In order to improve the semiconductor device performance, decoupled plasma nitridation (DPN) process was used to form the ultra-thin gate oxide film. But we recently found serious residue defect on gate oxide film if we did lithography rework with chemical method. This defect was like a circular-pattern about several-micron in diameter and hard to be removed. The results also showed that the thickness decrease and photoresist (PR) footing phenomenon would become worse after rework. After performing some experiments, we found that the N element doped in the gate oxide film could be one possible origin for this defect. And a model was proposed to explain the generation mechanism of this residue defect based on above analysis. Finally, an optimized lithography rework method was used to avoid the generation of the defect successfully.

Keywords—Defect; DPN; Rework; Gate oxide

INTRODUCTION

As semiconductor industry is being driven into more advanced node, the device critical dimension is getting increasingly smaller. However, the gate current leakage will also increase significantly at the same time and it will reduce the device performance seriously. Silicon oxide incorporated with nitrogen element is considered to be one effective way to reduce leakage current due to its relatively high dielectric constant, efficient diffusion barrier, and relative ease of integration into conventional CMOS flow [1]. And DPN process is one of the most effective gate oxide nitridation methods [1-3].

On the other hand, the defect becomes one fatal factor to wafer yield with continues development of semiconductor industry. Defects may come from the tools, process, material and environment. And even a very small defect will cause the failure of device. In this paper, one kind of circular-pattern defect was found on the surface of gate oxide with DPN. The generation mechanism of such defect was discussed and one effective method was adopted to avoid the generation of the defect successfully.

EXPERIMENT AND RESULTS

The typical gate dielectric process with DPN process consists of three steps, including thin oxide formation, plasma nitridation and post nitridation anneal [3]. First, oxide with DPN is deposited on wafer surface, lithography process is then used to define the useful pattern area. The rework process will be carried out if the lithography can't achieve the target. Afterwards, the second lithography is added into the process after rework. But we recently found serious defect on oxide film surface after the second lithography process.

As we can see in Figure 1, a large amount of defect is found on oxide/DPN surface of rework wafer, whereas there's almost few defect if we don't rework. This defect is like a circular-pattern about several-micron in diameter, as shown in Figure 2. It's also found from the defect image that there's material residue around the defect. After careful analysis of the defect location, we found that basically all the residue defects were located on the area where was covered by PR after lithography develop. Furthermore, the chemical composition of this defect was also characterized by EDS micro-analysis technique, and only normal Si, O, and C element were found in the defect area (the result not shown here). We can conclude from the characteristic of the defect that the defect may come from the residue of PR.

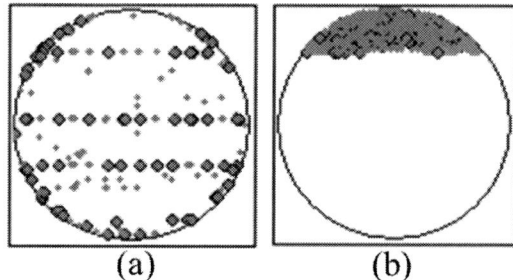

Figure 1: The defect map of wafer with Oxide/DPN substrate, (a) not rework, (b) rework

Figure 2: A typical circular-pattern residue defect image

Wet chemical rework treatment, here H_2SO_4, is used to remove PR. The optimized wet chemical rework method and surface clean method were also tried, but residue defect still existed on oxide/DPN surface. In order to investigate the effect of lithography rework, the thickness of oxide film and critical dimension (CD) of PR were measured. As we can see in Figure 3, the oxide film thickness decreases about 11% after lithography rework. It means that the surface oxide film will be also removed, accompanied with PR remove after rework process. Figure 4 shows the CD SEM image of PR. The top and bottom location of PR are marked with blue dotted line and red real line respectively in order to analyze the CD variation. It's found that the CD difference between top and bottom location of PR increases after lithography rework. Combining with the SEM cut image result (not shown here), it's considered that the PR footing phenomenon become worse after rework.

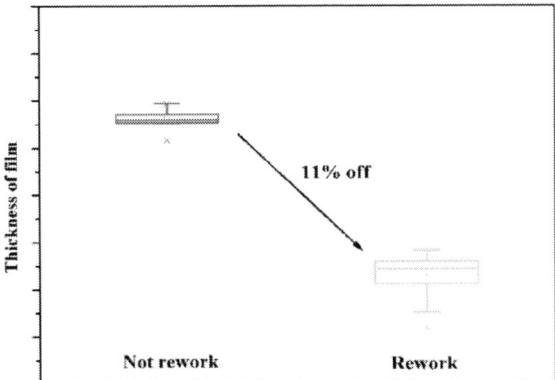

Figure 3: The thickness of Oxide/DPN film

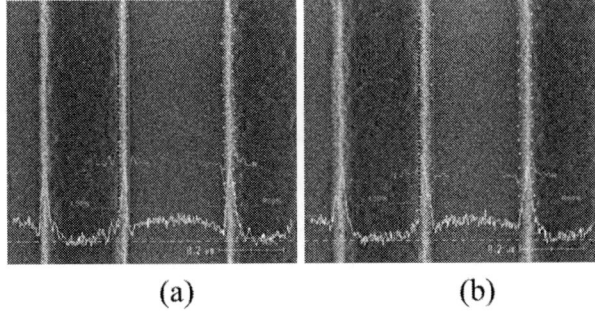

Figure 4: The SEM image of PR with Oxide/DPN substrate, (a) not rework, (b) rework

From the above results, the residue defect on oxide/DPN surface was definitely come from lithography rework process. And rework will result in the decrease of oxide film thickness and the worse PR footing. It's concluded that the N from DPN process may be the origin of the residue defect after comparing different process flow. Therefore, the oxide without DPN was adopted to clarify whether this defect generate from N element. Figure 3 shows the defect map of wafer on oxide surface

(without DPN). The main difference between these two wafers was lithography rework, and the same wet chemical rework method was used. It's found that the rework don't result in the generation of residue defect if we remove DPN process from oxide formation.

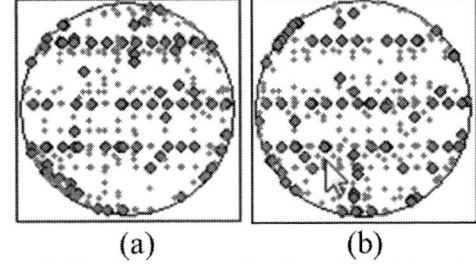

Figure 5: The defect map of wafer with Oxide substrate, (a) not rework, (b) rework

Based on the above results, we can deduce that the residue defect should be originated from N of DPN process. According to the shape and size of the defect, the change of oxide thickness and the PR CD, a simple model is built to clarity the generation mechanism of this defect, as described in Figure 6. After DPN process, N element is doped into oxide film. But the distribution of N is not uniform along the oxide film, and the concentration of N is relatively less at oxide surface [Figure 6 (a)]. After wet chemical rework of PR, the surface of oxide film will be also removed. So the thickness of oxide decreases, and the surface N concentration increases significantly [Figure 6 (b)]. During the lithography process, N will react with PR and therefore generate residue material at the interface of PR and oxide film surface. Meanwhile, N will also trap the photo-acid that diffuses down to the substrate, resulting in worse condition of PR footing [Figure 6 (c)]. Finally, the residue material is hard to be removed and will stay on oxide film surface [Figure 6 (d)].

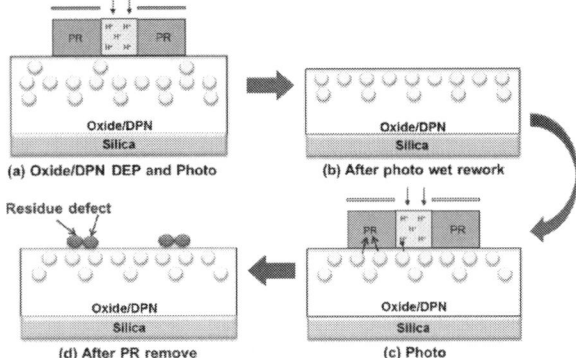

Figure 6: Schematic of the defect formation mechanism

Based on the residue defect generation mechanism, we have tried another targeted lithography rework method. Figure 7 shows the defect result. The defect count is significantly decreased and no residue defect is found if we carry out the new optimized rework method, as shown in Figure 7 (b). The defect result also in turn confirms the defect generation mechanism that we proposed.

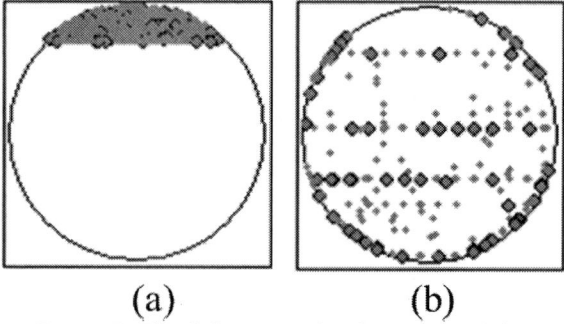

(a)   (b)

Figure 7: The defect map of wafer with Oxide/DPN substrate, (a) with baseline wet rework method, (b) with new optimized rework method

CONCLUSION

In this paper, we have investigated a circular-pattern residue defect on oxide/DPN surface after lithography rework. The results showed that the thickness of oxide film decrease and the PR footing phenomenon became worse after rework. After performing some experiments, a model was proposed to explain the generation mechanism of this residue defect, and N from DPN process may be the origin of the residue defect. Based on above analysis, an optimized rework method was carried out to avoid the generation of this defect successfully.

ACKNOWLEDGEMENTS

The authors would like to express their gratitude to SMIC TD and FAB colleagues who have contributed to this work through discussion, experiment and tool support.

REFERENCES

[1] H.-H. Tseng, Y. Jeon, P. Abramowitz, T.-Y. Luo, L. Hebert, J. J. Lee, J. Jiang, P. J. Tobin, G. C. F. Yeap, M. Moosa, J. Alvis, S. G. H. Anderson, N. Cave, T. C. Chua, A. Hegedus, G. Miner, J. Jeon, and A. Sultan. *Electron Device Letters, IEEE.* vol. 23, 2002, pp. 704-706.

[2] A. Veloso, F. N. Cubaynes, A. Rothschild, S. Mertensl, R. Degraeve, R. O'Connor, C. Olsen, L. Date, M. Schaekers, C. Dachs, M. Jurczak. *European Solid-State Device Research, 2003, Conference On. Essderc,* 2003, pp. 239-242.

[3] K.Saki, T.Shimizu, S.Mori and A.Yamamoto. *IEEE International Conference on Advanced Thermal Processing of Semiconductors,* 2006, pp. 15-19

WAFER EDGE TREATMENT IN LITHOGRAPHIC PROCESS FOR PEELING DEFECT REDUCTION

Xiaofeng Yuan[1], Qiang Zhang [1], Jing'an Hao[1],*

Technology R&D, Semiconductor Manufacturing International Corp.
Pudong New Area, Shanghai 201203, P. R. China

*Corresponding Author's Email: Roger_Yuan@smics.com

ABSTRACT

As advanced technology nodes such as 28nm and below ramp up to volume manufacturing, the treatment of wafer edge becomes more and more important to enhance yield performance. Peeling defect in wafer edge is a key yield killer, which is caused by wafer edge complex film stacks especially in bevel area. We observed the peeling defect after inter metal dielectric (IMD) films deposition. After lithographic process the BARC material will accumulate at the bevel area and is about 10 times thicker than that at the center area, the thick BARC forms a circle at the wafer bevel area. The much thicker BARC at the bevel area cannot be etched clear at the etch process and becomes residue. During the high temperature treatment in the following film deposition process, the BARC residue becomes to be the peeling source because it will be easy to peel off under high temperature. We found that the BARC accumulation at bevel area is caused by backside rinse. But the backside rinse is a necessary step of process at lithographic process to avoid scanner contamination, and it cannot be removed directly. We come up with two solutions to solve the problem. One is bevel rinse and another is BARC Edge Bead Removal (EBR). With the two solutions, though the circle of much thicker BARC shrinks inward by a less than 1 mm from bevel area, the defect review data shows that the peeling defect level drops down obviously. It shows that the wafer edge treatment at the bevel area is very marginal; any small variation of the wafer edge treatment will cause high level of peeling defect. The precise control of wafer edge treatment in bevel area is extremely important to guarantee yield.

Keywords— Peeling Defect; Wafer edge; Bevel rinse; EBR

INTRODUCTION

It's well known that the wafer edge situation in CVD, Lithography, Etch, and CMP process has a great impact on wafer edge correlated yield performance. The peeling defect is one key source of the wafer edge correlated defect. The poor adhesion between the films and the abnormal residua in Lithography and CMP process both tend to generate the peeling defect [1,2,3]. For lithography, to improve the resist situation at the wafer edge, Edge Bead Removal (EBR) and Bevel Rinse is the usual treatment solutions [4,5,6]. In this paper, we observed that the backside rinse can cause the BARC accumulating at the bevel area 10 times thicker than at the wafer center area in metal layers. As shown in Fig1(a), after BARC coating and spinning, some BARC will flow down to the wafer backside, so backside rinse is needed to clean the wafer backside in case of contaminating the equipment of track and scanner. In Fig1(b), the solvent spurt from the backside rinse nozzle to the wafer backside. Some solvent stream up along the wafer bevel and mix with BARC. Because of the stream force, the mixed liquid mainly exists on the up-bevel area. After spin dry, the much thicker BARC is left at the bevel area as shown in Fig1(c).

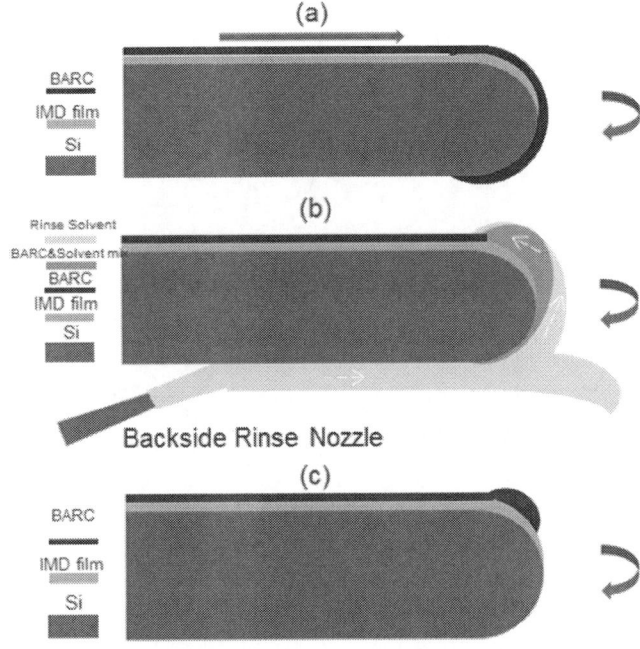

Figure 1: (a) BARC flow down to the wafer backside, (b) Backside rinse stream push the mix liquid to the bevel area, (c) Thicker BARC remain at the bevel area after spin dry.

As shown in Fig2, the thicker BARC at bevel area cannot be removed clearly by etch process and become residua for the following film deposition process. For the process temperature of the following film deposition is about 300℃ and is higher than the safe temperature of BARC, there is outgassing from the BARC and the outgassing lead to the film rupture and the ruptured film become the peeling source. The peeling film may scatter to the chip area. After photo and etch process of next layer, the pattern cannot be printed and etched in the area where peeling material exist as shown in Fig3, so the peeling material cause the pattern fail and it is called peeling defect.

The much thicker BARC is the root cause of this peeling case. The main solution is to remove the thick BARC at the bevel area. For lithography, both the EBR and the Bevel Rinse are tried in the BARC coating process and the results will be discussed later.

978-1-5090-6695-7/17 $31.00 © 2017 IEEE

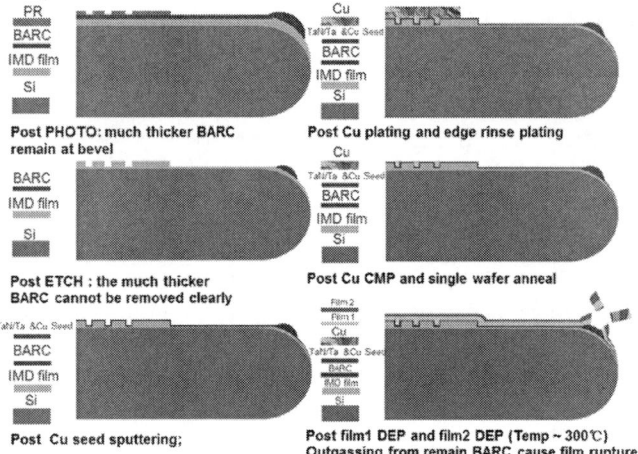

Figure2: The scheme of how much thicker BARC after PHOTO cause the peeling defect after film deposition.

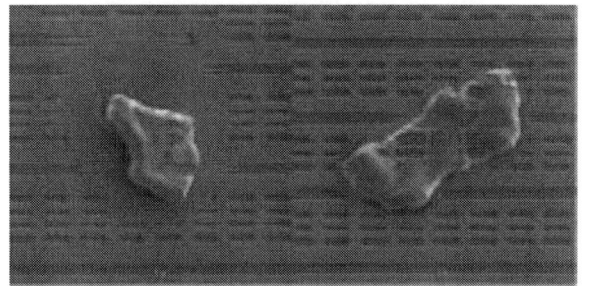

Figure3: The peeling defect in the chip area after ETCH of next layer

EXPERIMENT

The experiment wafers are based on Dual-damascene Cu processing with TiN hard mask and trench first integration scheme, 300mm wafers (the bevel size is 0.3mm) with porous (ULK) material as the dielectric film were prepared. UV cure processing post-ULK deposition is done to densify the ULK film followed by trench/via patterning with a tri-layer scheme of photoresist layer, BARC coating and a flattening material. Etched patterns were then deposited conformably with PVD TaN/Ta and Cu seed for the barrier layer, Cu electroplating for the bulk Cu fill that were polished via CMP and finally capped with several low-k dielectric barrier films. Visual bevel edge inspection tool was used to capture bevel images. KLA defect inspection tool and SEM review tool were used to determine the defect count.

RESULTS AND DISCUSSION

In order to remove the thick BARC at the wafer bevel area in lithography process, both EBR and Bevel Rinse methods were tried in BARC coating process. As shown in Fig4, we compared the BARC residue situation at bevel area among treatments of backside rinse only, backside rinse combined with bevel rinse and backside rinse combined with EBR (size 0.6mm). Fig4(a1) showed the wafer bevel image after BARC coating with backside rinse only, and the SEM cut images of the bevel cross-section are also showed in Fig4(a2). From the bevel image, the backside rinse can clean the area up to 0.45mm from the wafer edge, but the rinse side is rough, the

rinse size of the most area is about 0.3mm. From the SEM cut images, at the bevel area 0.25mm the BARC accumulate to the thickness about ten times of the thickness at the center area. Fig4(b1) showed the wafer bevel image after BARC coating with bevel rinse and backside rinse , and the SEM cut images of the bevel cross-section are also showed in Fig4(b2). From the bevel image, the bevel rinse can clean the area down to 0.35mm and up to 0.75mm from the wafer edge, but the rinse side is also rough, the rinse size of the most area is about 0.5mm. From the SEM cut images, at the wafer edge area of 0.5mm the BARC also accumulate to the thickness about ten times of the thickness at the center area. Fig4(c1) showed the wafer bevel image after BARC coating with EBR (size 0.6mm) and backside rinse , and the SEM cut images of the bevel cross-section are also showed in Fig4(b2). From the bevel image, the EBR (size 0.6mm) can clean the area up to 0.6mm from the wafer edge; the rinse side is serrated and a little neater than backside rinse and bevel rinse. From the SEM cut images, at the EBR rinse side size of 0.6mm the BARC also accumulate to the thickness about ten times of the thickness at the center area.

Figure4.a1 The wafer bevel image after BARC coating with backside rinse only

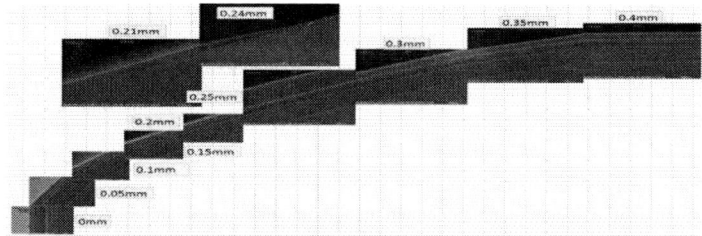

Figure4.a2 The cross section SEM cut images of Fig4.a1

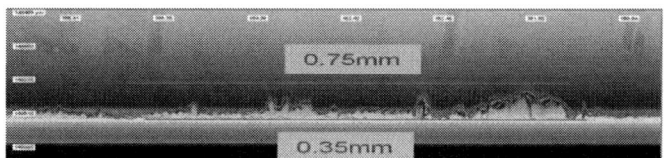

Figure4.b1 The wafer bevel image after BARC coating with bevel rinse and backside rinse

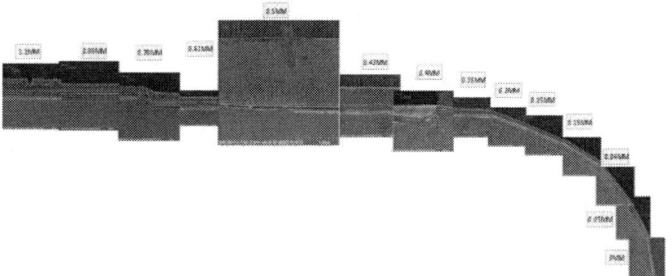

Figure4.b2 The cross section SEM cut images of Fig4.b1

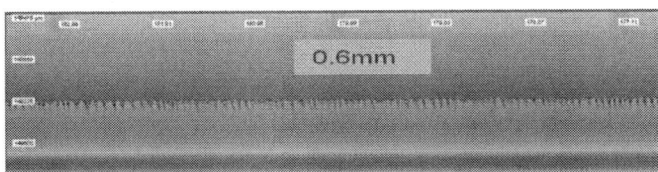

Figure4.c1 The wafer bevel image after BARC coating with EBR (size 0.6mm) and backside rinse

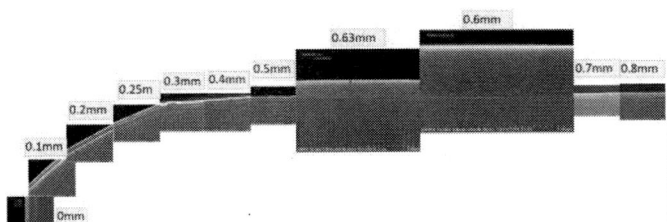

Figure4.c2 The cross section SEM cut images of Fig4.c1

No matter is any treatment of backside rinse or backside rinse combined with EBR or bevel rinse, the much thicker BARC all exists and forms a circle at the wafer edge. The difference among them is the position where thick BARC accumulate. All the positions where thick BARC accumulate are the rinse side area. For backside rinse only, the thick BARC accumulate at the backside rinse side, about 0.3mm; for bevel rinse, the thick BARC accumulate at the bevel rinse side, about 0.5mm; for EBR (size 0.6mm) , the thick BARC accumulate at the EBR size of 0.6mm. We suspect that the thick BARC accumulating effect is correlated with the solubility of BARC in solvent. In order to prove this assumption, we test two other types of BARC in the condition of the same solvent E. As shown in table1, type A is the BL BARC, the solubility in the same solvent is type-A<type-B<type-C, and the Type-C is close to 100%. We showed the thickness ratio of rinse side area to wafer center area among the three types of BARC rinsed by the same solvent. From the table, the thickness ratio of rinse side area to wafer center area is type-A>type-B>type-C and the thickness ratio of type-C is close to 1. It means that the better solubility between BARC and solvent is the smaller thickness ratio of rinse side area to wafer center area is. And if the BARC can totally be dissolved in solvent, there will be no thicker BARC accumulation at the rinse side. So the thick BARC accumulating effect is caused by the poor solubility of BARC in solvent.

BARC type	Solvent	The thickness ratio
Type-A	E	~10
Type-B	E	~4
Type-C	E	~1

Table1 The thickness ratio of rinse side area to wafer center area among the three types of BARC rinsed by the same solvent; the solubility in the same solvent is type-A<type-B<type-C

After backside rinse and bevel rinse and EBR splits action in lithography, we release the wafers to the following process. After trench etching, Cu seed sputtering, Cu plating and CMP, the wafers went into the CVD process. Inter metal dielectric (IMD) films were deposited and the process temperature of one film deposition is as high as up to 300°C. After this film deposition, we use the KLA defect inspection tool and the SEM review tool to calculate the peeling defect count. As shown in table 2, we compared the peeling defect count of the three splits and list the rinse clear area difference among them. From the result, it seems that the peeling defect count will be close to zero if the clear area by rinse up to 0.35mm.As mentioned at the beginning the bevel size of the wafer is 0.3mm as the red line shown in table2 and the wafer is flat from bevel size 0.3mm to the wafer center. If the much thicker BARC exist at the bevel area, the peeling defect count raise up and if the much thicker BARC exist at the flat area of the wafer, the peeling defect almost disappear.

Splits	Peeling defect count	Rinse clear area from wafer edge	
Backside Rinse	42	0~0.45mm	
Bevel Rinse & Backside Rinse	1	0.35mm~0.75mm	
EBR (size 0.8mm) & Backside Rinse	0	<0.6mm	

Table2 The peeling defect count and rinse area difference of three methods. The red line is bevel size 0.3mm from the edge of the wafer.

From the result above, the generation of the peeling defect is not only corrected with the accumulating of the much thicker BARC at wafer edge but also corrected with the position where the muck thicker BARC accumulates. Maybe the film stress at the flat area of the wafer is lower than at the bevel area of the wafer.

CONCLUSIONS

In this work we observed that the much thicker BARC will accumulate at the rinse side area no matter backside rinse or bevel rinse or EBR due to the poor solubility of BARC in solvent. The much thicker BARC cannot be clear after etch process and become to be the peeling source after film deposition process of high temperature. Otherwise, the position where the much thicker BARC accumulate is a key factor to the generation of peeling defect maybe because of the different film stress between the flat area and bevel area. So the precise control of wafer edge treatment in bevel area is extremely important to guarantee yield.

ACKNOWLEDGMENTS

The authors would like to thank Mr. Junjie Xu, Mr. Weihua Sang of SMIC and Mr. Weifeng Tang of TEL China for the technical discussion.

REFERENCES

[1] Jami, Kalyan, et al. , "Optimization of edge die yield through defectivity reduction", Solid State Technology, October 2009.
[2] Jaime D. Morillo, Thomas Houghton, Joseph M. Bauer, Randy Smith, Randy Shay, IEEE/SEMI Advanced

Semiconductor Manufacturing Conference, 2005, pp. 1-4.

[3] Christine Bunke, Thomas Houghton, Kenneth Bandy, George Stojakovic, Grace Fang, ASMC, 2012, PP. 63-66.

[4] L.Tedeschi, et al., "Immersion Lithography Bevel Solutions", Proceedings of SPIE Vol. 6922, 2008.

[5] M. Fujita, et al. , "Defectivity Improvement by Modified Wafer Edge Treatment in Immersion Lithography", Proceedings of SPIE Vol. 7274, 2009.

[6] M. Silvestre, et al. , "Bevel Rinse Optimization for Reduced Edge Defectivity and Improved Edge Yield", ASMC, 2016, pp. 124-128.

ULTRAPURE CHEMICAL COMPONENTS FOR NEXT GENERATION MATERIALS

Hyun Yong Cho[1]*, Ram Sharma[2], and Jeffrey D. Fogle[2]

[1]Heraeus Korea Corp., Suwon, Republic of Korea

[2] Heraeus Precious Metal North America Daychem LLC, Vandalia, OH, United State

Corresponding Author's Email: hyuny.cho@heraeus.com

ABSTRACT

Ultrapure chemical components for next generation materials for semiconductor manufacture are required due to chip yield enhancement. Some components for photoresist, cross linker, monomer, and photo acid generator (PAG) can be provided as representative ultrapure chemical components which have below 10 ppb level ionic metal impurities by particular purification methods.

INTRODUCTION

The materials for lithography process on semiconductor manufacture, which are photoresist, anti-reflective coating, and hard masks, etc., require ultrapure components due to chip yield enhancement. Therefore, many process material providers of lithographic are managing to minimize metal impurities on process materials themselves by specialized impurities removing processes. [1-3]

Normally, removing dissolved impurities, particularly ionic metal, on process chemicals is more difficult than insoluble impurities due to the fact that they are not easily removed by physical filtration process. The product may be contaminated easily by many sources such as chemical reactor, process, raw materials, etc. [4]

In order to minimize ionic metal impurities in photoresist formulations which are one of the main lithographic process materials, the producers implement minimizing metal impurities from raw materials. Chip manufactures are heading towards more tighten metal impurities on specification. Photoresist producers have to consider not only photoresist purification but also each raw material. Photoresist providers request ultrapure organic chemicals as raw materials from chemical suppliers. In here, we introduce one of the solutions to minimize metal impurities from raw materials by various purification methods with good results.

PURIFICATION

Classic purification methods for organic chemicals are considering, in order to reduce ionic metal impurities, liquid-liquid extraction, solid-liquid extraction, filtration, distillation, recrystallization, adsorption (e.g. chromatography, active carbon, etc.), ion exchange, sublimation, modification chemical reaction pathway, etc. The effective purification methods are selected with considerations based on many factors of chemical properties which are polarity, solubility, melting/boiling point, chemical reactivity/stability with specified chemicals, thermal stability, etc. Even though there is the most effective method in terms of chemistry, purification yield, process cost, and environment factors should be considered as well. Sometimes, a particular method should be avoided by specialized customer request. On actual purification methods, several methods can be combined and (or) modify due to above mentioned factors.

RESULTS

Cross linkers: PL-1174 and XL-1170

Cross linker is one of the important components of negative tone photoresist. Glycoluril derivatives are widely used as cross linkers in photoresist and other applications in lithography processes. Product name PL-1174 and XL-1170 with structures are shown in Fig. 1.

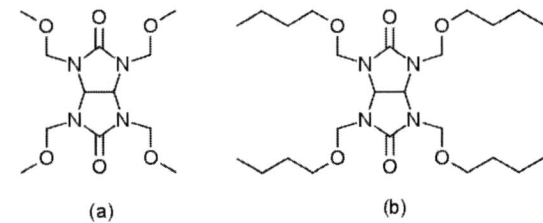

Fig. 1. Structures of (a) PL-1174 and (b) XL-1170

These products were purified with liquid-liquid extraction and adsorption combination. Especially, XL-1170 is a liquid state at room temperature. Therefore, it is not easy to adapt a recrystallization method.

Metal impurities results of PL-1174 are shown in TABLE I from crude, purified (PL-1174), and ultra-purified (UP PL-1174) respectively. Regarding XL-1170, it is shown in first column of TABLE II.

Monomer: MNLMA

Aliphatic acrylate polymers are widely used as a base resin for ArF photoresist. MNLMA is one of the typical monomers for ArF resins. The structure is shown in Fig. 2.(a).

Liquid-liquid extraction, filtration, and adsorption were combined and adapted to purify, furthermore, reaction scheme was designed to minimize impurities.

Metal impurities results of MNLMA are shown in second column of TABLE II.

TABLE I.
METAL IMPURTIES RESULTS OF PL-1174 BY PURIFICATION

			Unit: ppb
Ion	Crude*	PL-1174**	UP PL-1174**
Na	150000	45.1	0.5
K	343	22.8	0.9
Al	1061	19.4	3.0
Ca	1472	1.0	1.0
Fe	1337	68.2	0.2
Cu	167	1.0	1.0
Cr	16	3.9	1.0
Mg	708	0.5	0.5
Mn	91	0.2	0.2
Ni	606	8.9	0.3
Pb	5	10.9	0.7
Sn	33	28.2	0.2
Zn	393	0.5	0.5
Li	n/a	0.2	0.2
Au	n/a	0.2	0.4
Co	n/a	0.2	0.2
Cd	n/a	2.1	0.2
W	n/a	4.6	1.0
Ba	n/a	0.3	0.8
As	n/a	0.5	0.5
V	n/a	0.3	0.3
Ti	n/a	1.9	3.0
Ag	n/a	0.2	0.2

* Measured by GFAAS
**Measured by ICP-MS

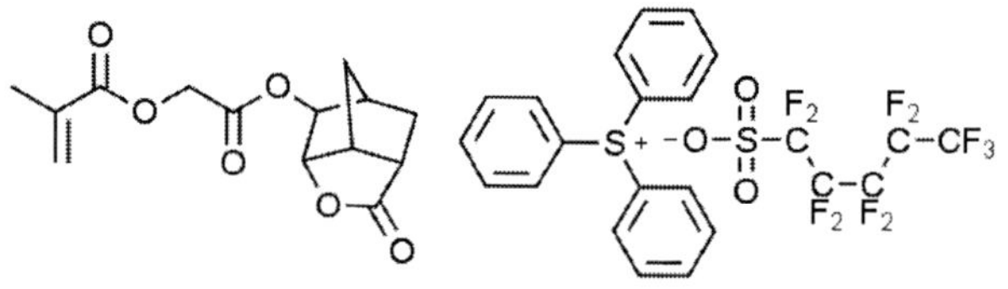

and adapted.

Metal impurities results of TPS-Nf are shown in TABLE III from crude (before purification) and purified (after purification) as well.

TABLE II.
METAL IMPURTIES RESULTS OF XL-1170 AND MNLMA BY PURIFICATION

		Unit: ppb
Ion	XL-1170*	MNLMA*
Na	0.8	2.1
K	8.0	3.0
Al	3.0	9.2
Ca	7.3	2.2
Fe	0.2	2.0
Cu	3.9	2.1
Cr	3.1	5.0
Mg	5.7	2.1
Mn	0.6	2.3
Ni	2.3	2.1
Pb	1.7	5.4
Sn	0.8	6.1
Zn	3.4	5.2
Li	0.4	0.2
Au	0.9	0.5
Co	0.2	0.2
Cd	0.4	0.3
W	0.5	0.5
Ba	1.1	0.4
As	0.5	0.8
V	0.3	0.3
Ti	0.2	0.2
Ag	0.2	0.7

*Measured by ICP-MS

Other analytical results

Purities were measured by HPLC. Cross linkers, PL-1174 and XL-1170, monomer, MNLMA, and PAG had 99% and higher purity.

Residual acid and base values were measured by acid/base titration method. Each value had below 0.001 mol/kg or not detected.

Fig. 2. Structures of (a) MNLMA and (b) TPS-Nf

PAG: TPS-Nf

Triphenysulfonium (TPS) salt and its derivatives are widely used as a photo acid generator (PAG). TPS-Nf is one of the PAGs for KrF photoresist application. The structure is shown in Fig 2. (b). TPS-Nf is a typical ionic product, so removing metal ions from the product requires more complex techniques relatively. Liquid-liquid extraction, ion exchange, and adsorption were combined

CONCLUSION

Typical chemical components for photo resist, which are PAG, monomer as precursor for resin, and cross linker are required lower metal impurities and higher purity more and more.

Cross linkers, PL-1174 and XL-1170, monomer, MNLMA, and PAG, TPS-Nf were yielded below 10 ppb level metal impurities which were measured by ICP-MS and over 99% purity which was measured by chromatography. Additionally, these components had very low level (under 0.001 mol/kg) or no detection level of residual acid and base which are important factors due to the fact that they can directly effect to photo resist performance.

All purifications were not only one method but also combinations of several methods to increase purification efficiency.

TABLE III.
METAL IMPURTIES RESULTS OF TPS-Nf BY PURIFICATION

Unit: ppb

Ion	Before purification*	After purification**
Na	100	9.1
K	200	0.5
Al	235	8.8
Ca	1103	6.1
Fe	34	4.1
Cu	1.9	1.0
Cr	14.3	1.0
Mg	45	0.5
Mn	2	0.2
Ni	5	0.3
Pb	41	0.2
Sn	2	1.8
Zn	38	3.5
Li	2	0.2
Au	n/a	0.2
Co	n/a	0.2
Cd	n/a	0.2
W	n/a	7.6
Ba	n/a	0.5
As	n/a	1.0
V	n/a	9.1
Ti	n/a	6.3
Ag	n/a	0.2

*Measured by GFAAS
**Measured by ICP-MS

REFERENCES

[1] W. A. Burke, *US Patent 5525315*, 1996.
[2] K. Honda, E. A. Fitzgerald, and L. Ferreira, *US Patent 5446125*, 1995.
[3] K. Honda, *US Patent 5446126*, 1995.
[4] C. R. Szmanda, R. J, Carey, *US Patent 5443736*, 1995.

EXTNETION OF PHOTORESIST LIFETIME AT EXTREME CONDITIONS

Qiaoqiao Li[1], Weiming He[1], and Huayong Hu[1]*

[1] Technology R&D, Semiconductor Manufacturing International Corp.

Pudong New Area, Shanghai 200433, China

*Sarah_Li@smics.com

ABSTRACT

Normally, with fixed recipe body, the profile, and thickness of photoresist film, is mainly dependent on the main spin speed to get target thickness with good uniformity. However, in some cases, we have to get some special thickness target with main speeds close either to the lower limit (1000rpm) or to the higher limit (2000rpm). And in such extreme conditions, only adjusting main speed cannot simultaneously achieve both resist thickness target and uniformity. To obtain extreme PR thickness in productions, we usually have to find another viscosity to replace the current one, which adds one more pipeline and costs more. Here, we use a design of experiment (DOE) by JMP to extend the lifetime of photoresist at the extreme thickness conditions, without the introduction of new materials. In this experiment, resist homogenizing speed, as another key parameter, is added into the baseline coating recipe to maximally achieve both thickness target and uniformity. Based on JMP's designs and data analysis, we have achieved our predictions with silicon wafer data. The above results will be presented in this paper.

INTRODUCTION

In lithography process, chemicals, such as photoresist (PR), bottom anti-reflective coating (BARC) and under-layer (UL), are dispensed onto wafer surfaces and spun out to form a thin photoresist film. And usually, the thickness performance of film coating is mainly dependent on the main spin speed [1][2][3][4][5]. Of course, the observation mentioned above is based on the fact that the dispense conditions are fixed and main speeds are kept at middle levels close to 1500 rpm.

Figure 1 shows a typical spin coating process, which contains pre-wet, dispense, casting, and Edge Bead Removal (EBR). In this coating process, PR thickness and uniformity depend on the dispense and the spin coating processes. In general, the relations of PR thickness and main speed is given by the formula as follows,

$$Thickness^2 * RPM = Constants \qquad (1)$$

which was applied in many studies. [1][2][3]

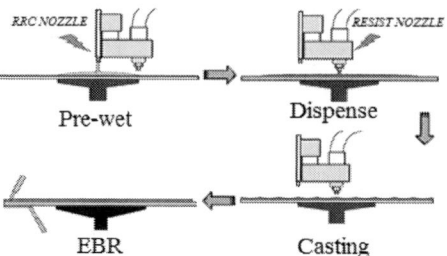

Figure 1: Schematic diagram of typical spin coating process [6].

To guarantee thickness uniformity, normally it is suggested that the main spin speed could not exceed 2000 RPM, or below 1000 RPM, preferably be close to 1500RPM in 300mm wafers coating process.

However, when main speeds go down below 1000 RPM, or rise up above 2000 RPM, the relation of thickness mean and main speeds will not follow Equation (1) any more. Even if the extreme thickness mean can be achieved to its target number, its uniformity may be below our targeted value. So if the viscosity of one chemical is fixed, its thickness mean will be limited to a small range. This is to make sure that enough thickness uniformity exists to avoid coating defects. Therefore, one viscosity version of chemicals can be only applied to limited thickness targets, and new version of chemicals have to be introduced to meet new requirements of thickness, which leads to extra spending. In this paper, we studied the key factors on PR thickness in extreme spin conditions using the design of experiment (DOE) method. In our experiments, we found it is beneficial to add one more step before main spin speed, to maximally achieve both the PR thickness mean and thickness uniformity. We also analyzed the key effects and its theory for thickness mean and uniformity. Finally, we will demonstrate that we have obtained optimal conditions to meet our thickness target with good uniformity and enough process window. The results will be presented in this paper.

MATERIALS AND METHODS

We have chosen a photoresist for the experiment. its normal thickness range can cover >=5000A. But in production, we are requested to get the PR thickness target down below 5000A without introducing new version of the photoresist, for cost concern. However, with current PR version, only tuning main speed in coating recipe can not get our targets with acceptable uniformity, thus we have to add one step of distribution steps in high rotation speed before the casting steps with main speeds, and then fine tune distribution speed and main speed together to get targeted thickness and acceptable thickness range. In this coating recipe, thickness mean vs main speed will not follow normal formulation as shown in Equation (1), we have to find one appropriate combinations of distribution speed and main speed to meet our thickness target with good uniformity.

TABLE I.   SPECIFY SPECS LIMIT FOR RESPONSE, FACTORS AND LEVELS

Parameter	Level	Lo	Me	Hi
Distribution Speed	3	2800	3000	3200
Main Speed	3	1000	1500	2000

We designed our experiment with RSM methods, based on previous trials. In our designs, distributions speed and main speed will be regarded as dominating factors in reasonable ranges, as shown in table I. Plus, thickness mean with +/-20A window and thickness uniformity below 40A will be treated as the responses, as showed as in table II. Figure 2 showed the scatterplot matrix maps of RSM design points.

In above DOE tests, we first coated PR on bare wafers by the TEL Lithius Pro. Track. Then, we measured the full map PR thickness by KLA-Tencor tools: KT Aleris 8350. Finally, we got all thickness mean and 3-sigma from every wafer, and put into JMP to do data analysis.

TABLE II.   OPTIMIZATION OF RESPONSE VARIABLE, T1 REPRESENTS TARGET 1

Response name	Purpose	LSL	Target	USL
thickness Mean	on Target	T1-20	T1	T2+20
thickness Uniformity	Minimize	0	25	40

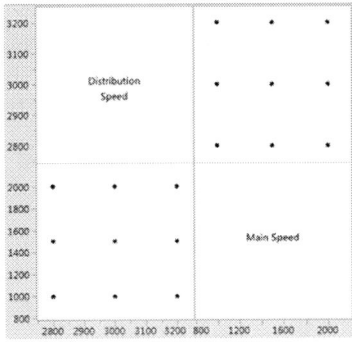

Figure 2: RSM 2 center points of main speed and distribution speed

RESULTS AND DISCUSSION

Based on JMP analysis results, thickness mean and uniformity are respectively given by following Equations (2) and (3), in which A1 to A5 and B1 to B5 are all constants. Compared with Equation (1), these two equations are complex, in the way that they have not only the distribution speed and main speed in linear terms, but also have a cross product of these two factors as well as a single term that is proportional to the square of the each two above mentioned factors. It means although the thickness mean is tuned to target first, thickness uniformity may not be in spec. Therefore, how to balance distribution speed and main speed by DOE become the key to get to the targeted thickness and acceptable thickness range.

Thickness mean=A1+A2*distribution speed+A3*main speed+A4* distribution speed*main speed+A5*main speed^2 (2)

Thickness uniformity=B1+B2*distribution speed+B3*main speed+B4* distribution speed*main speed+B5*distribution speed^2 (3)

From figure 3, the suggested conditions from DOE indicates a distribution speed and a main speed at 2985 and 1980, respectively. When the distribution speed is 2985, main speed is 1980, the model has predicted the thickness Mean to reach to T1, or the first targeted thickness, the thickness uniformity (3sigma) can reach 6A. This theoretical result meets our expectation. However, considering wind mark issue when main speed is close to 2000 RPM, distribution speed>=3050 is preferred. With respect to the thickness Mean spec of T1+/-10A and thickness uniformity<=20A (Log(Uniformity)<=1.3), process window of distribution speed is 3082~3122 and main speed is 1868~1904, as shown by figure 4.

978-1-5090-6695-7/17 $31.00 © 2017 IEEE

Concerning wind mark defects for PR coating with high main speed (close to 2000 RPM), when thickness target to T1 with 10A thickness uniformity, one of optimal recipe has been obtained: the distribution speed is 3100 RPM and the main speed is 1890 RPM. This optimal recipe with distribution speed 3100 RPM and main speed 1897 RPM was confirmed by 4 wafers for thickness mean on target with 3-sigma equal to 12A, matching our model predictions.

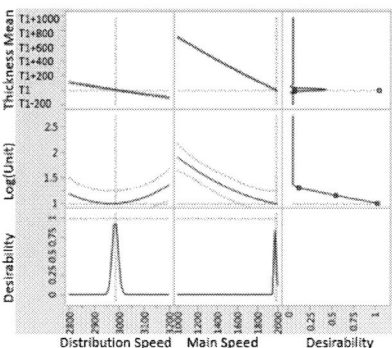

Figure 3: Suggested condition from DOE for T1 photoresist thickness.

This result says that when the distribution speed is added in the coating process, the main speed and PR thickness will not follow Equation (1). Though the PR thickness depends on the main speed, the distribution speed also needs to be considered. Higher distribution speed can be used to get thinner PR thickness by spinning out residual PR. To guarantee PR thickness uniformity, the distribution speed need to be kept in a suitable range. At the same time, this spin coating process can extend PR to wider range of thickness application, and thereby reduce manufacturing costs.

CONCLUSION

It is beneficial to add one more step before main spin speed, the distribution speed, to maximally achieve both the targeted PR thickness mean and thickness uniformity requirement. We have found that both the distribution speed and main speed can affect coating result (thickness mean and thickness uniformity) with strong mutual interactions. We have demonstrated that we have successfully made a photoresist thickness mean (normal thickness mean covered >=5000A) to reach previously believed impossible new target with a good thickness uniformity as well as with wind mark defects minimized by DOE study .

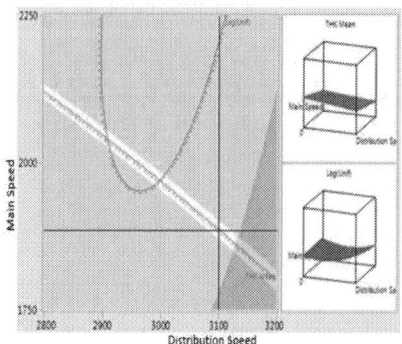

Figure 4: Process window of distribution speed and main speed to get response into target spec performance.

ACKNOWLEDGEMENTS

The author would like to show our appreciation to TD Litho colleagues as well as SMIC SH300-P1 colleagues for the wafer support.

REFERENCES

[1] A. G. Emslie, F. T. Bonner, and L. G. Peck, *J. Appl. Phys.* 29, 858, 1958.

[2] A. Acrivos, M. G. Shah, and E. E. Petersen, *J. Appl. Phys.* 31, 963, 1960.

[3] G. F. Damon, *Proceedings of the Second Kodak Seminar on Microminiaturization* (Eastman Kodak Co., Rochester, New York), 1967, pp.36.

[4] D. Meyerhofer. *J. Appl. Phys. 49*, vol. 7, 1978, pp. 3993-3997.

[5] S. Wolf and R. N. Tauber, *Silicon Processing for the VLSI Era*, vol. 1, Sunset Beach, CA: Lattice, 1986.

[6] TOKYO ELECTRON SHANGHAI LTD. CLEAN TRACK B.U. , *Training material – The Track Process*

ILLUMINATION OPTIMIZATION FOR LITHOGRAPHY TOOLS OPE MATCHING AT 28 NM NODES

Wuping Wang*, Long Qin, Zhengkai Yang, Yulong Li, Zhibiao Mao, Yu zhang

Shanghai Huali Microelectronics Corporation, No.497 Gaosi Rd. Zhang Jiang Hi-Tech. Park, Shanghai 201210, China

*Corresponding Author's Email: wangwuping@hlmc.cn

ABSTRACT

The CD (critical dimension) of large scale integrated circuit was dominated by lithography process. The 193 nm immersion lithography nowadays has been widely used in chip manufacturing at 28 nm nodes. With the application of Nikon 193 nm immersion lithography tools, it is significant to match the Nikon immersion with ASML through OPE (Optical Proximity Effect). Good scanner matching will be beneficial for extending Nikon 193 nm immersion lithography tools and effectively improving production efficiency. In this paper, based on the OPE research between Nikon immersion tool and ASML immersion tool, we have developed a set of matching method for both immersion tools at 28 nm node and realized the 28 nm lithography process transfer from ASML immersion tool to Nikon immersion tool.

Keywords— OPE; OPC; Nikon ArF Immersion; ASML ArF Immersion

INTRODUCTION

With the IC (integrated circuit) CD reach to submicron and deep-submicron, it becomes a great challenge for pattern design with the dedicated structures, which demand more advanced lithography tools with adaptable capability. For the higher resolution, the wavelength of illumination source was used from g-line (436 nm), i-line (365 nm), 248 nm, to 193 nm and nowadays EUV, X-ray and even non-optical exposure method e.g. E-beam (electron beam lithography) and I-beam (ion beam lithography). Lithography tools are always the bottleneck for the mass production. Wide application of 193 nm immersion lithography tools at 28 nm nodes made the corresponding production available to HVM. In order to meet the requirement of CD accuracy and CD uniformity, the OPE matching between different lithography tools need to be considered. OPE matching refers to performing the lithography process on different tools with same quality of process window, comparable OPC (Optical Proximity Correction) model to realize the demand of process design and good electrical parameters of device. However, the demand would not be easily reached only using the same PR (Photo Resist) materials, exposure and develop parameters through adjusting dose and focus etc. on the two lithography tools. Consequently this method for the single site pattern matching is difficult to cover all patterns across the exposure field, which owing to the different lithography tools possessing differential optical module, including: distribution of laser wavelength, wave differ in lens, synchronous precision error during exposure etc. In this paper, based on the OPE research between Nikon ArF immersion and ASML ArF immersion, we have developed a set of matching method for both immersion tools at 28 nm node and realized the 28 nm lithography process transfer from ASML to Nikon immersion tools.

A. Optical proximity effect (OPE) matching model

Optical proximity effect (OPE) matching model is based on illumination pupil modulation distortion model, which is expressed by linear combinations of Zernike intensity modulation functions and Zernike distortion modulation functions. These functions are orthogonal and can be expressed with combinations of Zernike polynomials, as shown in following equation. The illumination pupil intensity and distortion in Zernike polynomial functions is illustrated in Fig. 1.

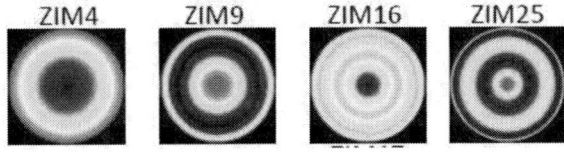

a. Illumination pupil intensity distribution modulation function.

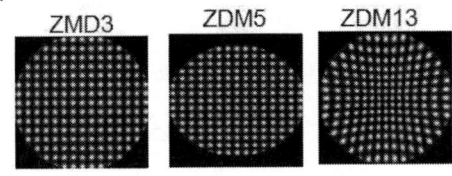

b. Illumination pupil distortion modulation function.

Fig.1. Zernike intensity modulation function and Zernike distortion modulation function.

Fig. 1a shows symmetric illumination pupil intensity function in X and Y direction. Fig/ 1b shows 3, 5 and 13 order illumination pupil distortion function. These Zernike polynomials are suitable for Zernike linear combination analysis method to predict the OPE response to the change in a pupil. Nikon OPE master software is based Zernike polynomial function through setting the initial illumination condition to simulate target pattern profile and CD value. The optimized Zernike intensity modulation function and Zernike distortion modulation function can be obtained through comparing simulated result with target pattern profile and CD value. The final combination of best illumination pupil intensity distribution and shape distortion can be settled while we

978-1-5090-6695-7/17 $31.00 © 2017 IEEE   145

minimized the difference of all the target patterns.

$$I_{Optimized}(x,y) = T(x,y)[I_{Initial}\left(x + D_x(x,y), Y + D_y(x,y)\right) \times PSF] + C$$

Where (x, y) is pupil coordinates,
$T(x, y) = \exp[\Sigma_i C_i Z_i (x, y)]$ is intensity modulation term,
$D_x(x, y)$ is the distribution function in x,
$D_y(x, y)$ is the distribution function in y,
PSF is a Gaussian point-spread function in pupil which generates blur,
C is a constant to express a background intensity offset,
$Z_i (x, y)$: Zernike intensity modulation function (ZIM), XY symmetric intensity modulation functions were selected as follows:
$ZIM4 = 2\rho^2 - 1,$
$ZIM9 = 6\rho^4 - 6\rho^2 + 1,$
$ZIM16 = 20\rho^6 - 30\rho^4 + 12\rho^2 + 1,$
$ZIM25 = 70\rho^8 - 140\rho^6 + 90\rho^4 - 20\rho^2 + 1.$

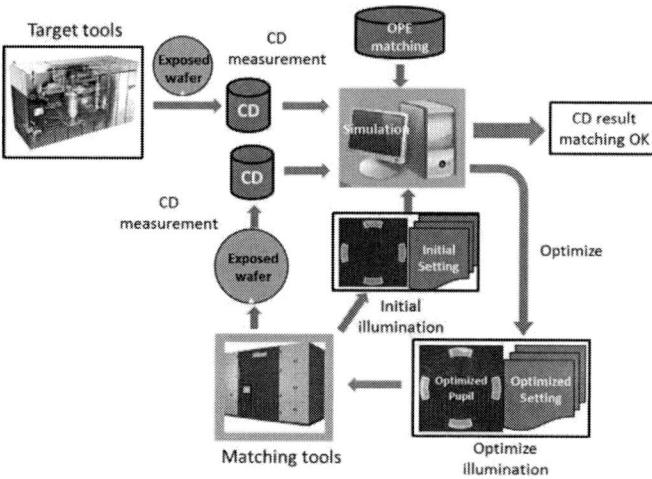

Fig.2. *Process flow of the OPE matching.*

B. Optical proximity effect (OPE) matching method

Optical proximity effect (OPE) matching is performed under the same PR species, film thickness, PEB temperature and development conditions. As shown in Fig.2, a cyclic process is used to realize the OPE matching. The main matching process contains selection of the matching patterns, CD measurement of the matching patterns, OPE matching calculations, exposure and CD check after correcting on the illumination source.

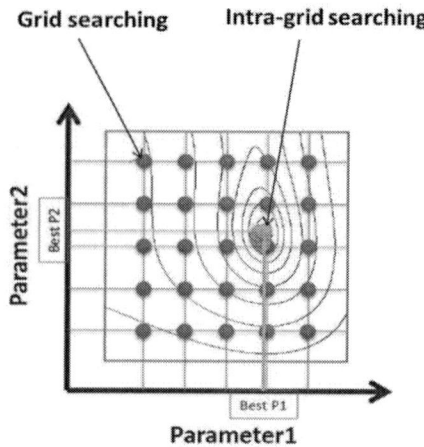

Fig.3. *OPE matching calculation method.*

Firstly, it is the most important procedure to pick out the representative optimal patterns for the OPE matching, since the matching patterns need to be able to represent the other patterns and simultaneously reflect the tiny optical difference of the exposure tool clearly. Nikon OPE Master software was used to simulate the matching patterns. Based on the capacity of calculation and measurement, about 300 matching patterns are totally selected. About 70% of selected matching patterns are from the OPC Modeling patterns which can reflect the optical difference. The electrical performance test patterns and the customer-designed patterns in SRAM account for about 20% and the weak points are about 10%. The sensitivity of about 15,000 OPC modeling pattern to Zernike function was calculated by Nikon OPE Master optical sensitivity analysis module simulating the 23 Zernike intensity modulation functions and distortion modulation functions. We finally got about 345000 sensitivity sites and divided those data into 210 types. Each sensitivity value corresponding to the representative OPC modeling pattern was selected from those 210 groups and used as reference pattern for OPE matching. On the other hand, the CD measurement is carried out by the HITACHI CD-SEM. The CD result for each pattern is the average value of 20 measurement results on the same tool in order to minimize the test error.

Nikon OPE Master software is also used for the OPE matching calculations and the matching flow includes the following steps. 1. Collect target CD of matching pattern on golden tools. 2. Perform pattern simulation by Nikon OPE Master under the defined original illumination pupil and setting, pattern structures and coordinate information. 3. Get the simulated CDs using the corresponding Zernike intensity and distortion functions. 4. Find CD difference between golden tool and matching tool. Fig.3 shows the two searching methods for getting the optimal condition, grid searching and intra-grid searching, on Nikon OPE Master. The grid distribution is performed via selecting calculation condition. The calculation is based on these conditions on grid to find an approximately optimal condition after optical simulation. And then, the optimizing calculation is carried out around the approximately optimal condition to finally obtain optimum condition.

C. OPE matching for BEOL of 28 nm nodes

978-1-5090-6695-7/17 $31.00 © 2017 IEEE   146

The OPE matching for L/S (line/space) and hole-connection layers of 28 nm is performed through the above method. Fig.4 shows the optimizing process of illumination pupil for L/S-connection layer. The optimizing process mainly concerned intensity function ZIM4 and distortion functions ZDM25, ZDM43, ZDM45 and ZDM65, since the illumination pupil for L/S layer just varies along XY directions. So, the calculation is based on the symmetrical Zernike parameters along XY direction. Total 294 matching patterns for L/S layers of 28 nm are selected to optimize. The specification of CD difference is ≤ 1 nm and ≤ 2 nm for one dimensional pattern and two dimensional patterns, respectively. Total 266 matching patterns were finally found in the specification, as shown in Fig.5. The similarity degree of matching patterns reaches 90.5%, and the standard deviation of CD is improved from 1.37 to 0.67. Optimization rate is 51.1%.

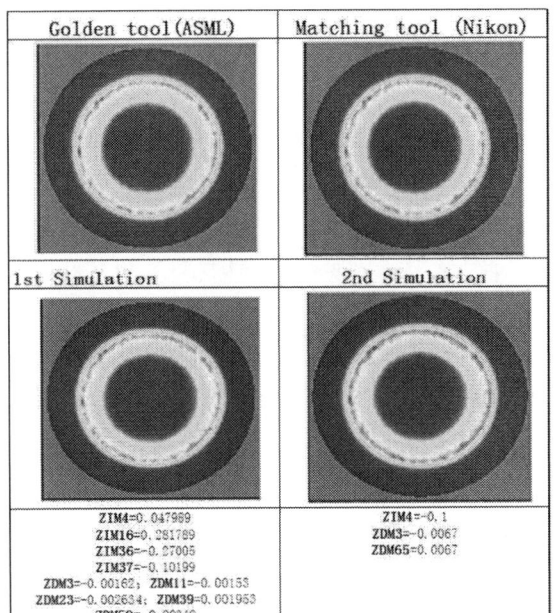

Fig.6. Illumination pupil optimizing of 28 nm hole layer.

D. Conclusion

The lithography tools' matching is a quite complex problem. In this paper, we focused on the OPE matching between Nikon and ASML ArF immersion and found a set of matching method applied on the actual matching program. The OPE matching result for BEOL of 28 nm nodes shows that the matching rate reaches more than 90% and the final yield has been well verified. With the continuous development of lithography, OPE matching between different lithography tools will become necessary. A variety of hybrid matching and more complex pattern matching will become a new research topic.

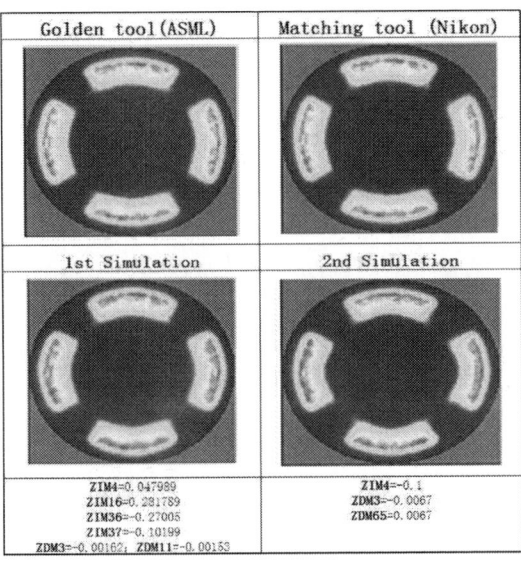

Fig.4. Illumination pupil optimizing of 28 nm L/S layer.

Based on the same method, the optimization of illumination pupil for hole layer is also performed, as shown in Fig.6 and Fig. 7. Total 318 matching patterns were selected and total 266 matching patterns were found in the specification. The similarity degree of matching patterns reaches 93.7%, and the standard deviation of CD is improved from 1.2 to 0.9. Optimization rate is 25%.

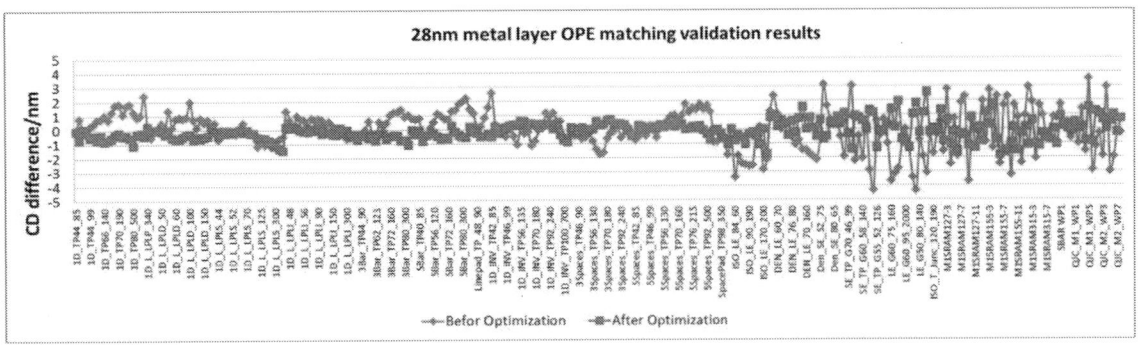

Fig. 5. Optimized OPE matching validation result of 28 nm L/S layer. Blue line is CD difference before optimization and red line is CD difference after optimization.

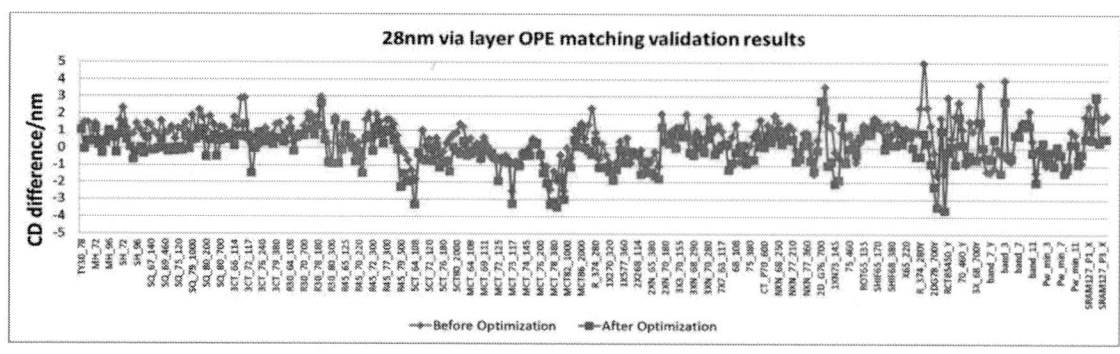

Fig. 7. Optimized OPE matching validation result of 28 nm hole layer. Blue line is CD difference before optimization and red line is CD difference after optimization.

REFERENCES

[1] T. Matsuyama, N. Kita, R. Matsui, and J. Ikeda, "Appliation of illumination pupilgram control method with freeform illumination". SPIE Vol.8326, (2013)

[2] A. Rosenbluth, S. Bukofsky, M. Hibbs, K. Lai, A. Molless, R. Singh, and A. Wong,"Optimum Mask and Source Patterns to

[3]

[4] Print a Given Shape," Proc. SPIE Vol. 4346, (2001)

[5] T. Matsuyama, N. Kita, T. Nakashima, and S. Owa, "Tolerancing Analysis of Customized Illumination for Practical Applications of Source &Mask Optimization," Proc. SPIE Vol. 7640, (2010)

INVESTIGATION OF CRITICAL DIMENSION (CD) VARIATION CONTROL ON UNSTABLE SUBSTRATE

Xiaoyan Sun[1], Chang Liu[1], Lei Wu[1]*

[1] Technology R&D, Semiconductor Manufacturing International Corp.

Pudong New Area, Shanghai, P. R. China 201609

*Corresponding Author's Email: Jenny_Sun@smics.com

ABSTRACT

With the development of the semiconductor devices, advanced technology node requires both tight critical dimension (CD) control and enough photo process window. In many circumstances, the unstable substrate can have great influence on photo CD, causing large variation in the performance of device. Although there have been intensive studies on this topic, we attempt to discuss the physical mechanism underneath within this paper, and to illustrate the phenomenon observed in the function layers.

This paper is to investigate the contributions of several different factors on photo CD variation and discuss the possible controlling methods. Different substrate deposition methods and various surface treatment methods are applied in our experiment and the corresponding influence on CD variations are investigated. Results indicate that the surface treatments play important roles in controlling CD variation. The underlying mechanism is discussed.

INTRODUCTION

Advanced semiconductor industry requires smaller critical dimensions (CD). Generally speaking, the process window requirement of the typical implant layers in FEOL can be met with current process condition. But the main challenge is the controlling of the large CD variation with Queue-time (the time from step 1 track out to step 2 track in) on complicated film stack.

In this paper, we focus on the oxidation process of nitrogen-containing substrate. The variations in oxide thickness and content of nitrogen have great influence on the CD of after-developing-inspection (ADI) [1] [2]. This model was introduced in this paper to illustrate the ADI CD variation with time. Different substrate deposition methods are also utilized to investigate the corresponding influences on CD variation. Results indicate that the film density does not have significant influence on the CD variation. The CD variation effect is mainly attributed to the nitrogen diffusion. At last, the influence of the plasma-related process is discussed in detail as the possible controlling method.

EXPERIMENT

Figure 1 shows the larger CD variation with the Queue-time from previous step to photo step. The vertical scale has been normalized the capability of our process as 100%.

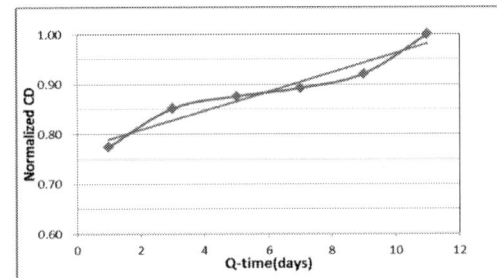

Figure 1: Photo CD distribution with Q-time

The large photo CD variation is generally attributed to the under-layer film thickness variation or the elements content variation on the interface. In our model, both of these two factors contributed to the surface property changes, leading to the large CD variation in photo process, as shown in Figure 2, 3. The detail of the model is as following:

- The larger oxide thickness

Considering that variation of the oxide thickness introduced in Queue-time, larger Queue-time results in a thicker oxide thickness, which blocks the diffusion of the nitrogen from the sub-film., leading to a smaller nitrogen-content on the interface. Hence less photo acid was neutralized on the interface, and the footing of photo resist, which is the main contributor to the CD variation, is eliminated.

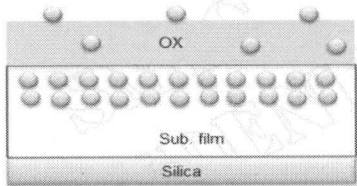

Figure 2: the nitrogen-content distribution of the larger oxide thickness

- The smaller oxide thickness

Similarly, the thinner oxide thickness leads to the interface more nitrogen-contents, so that more photo acid

978-1-5090-6695-7/17 $31.00 © 2017 IEEE

diffuses on boundary from photoresist to substrates. Hence in this model, the photo resist profile with time is the root cause for CD variation, which has been verified by top-down CDSEM picture.

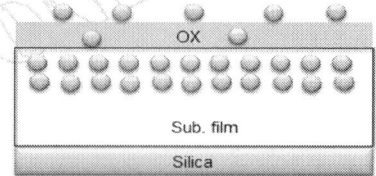

Figure 3: the nitrogen-content distribution of the smaller oxide thickness

SOLUTION

1. The different deposition methods of sub-film

As we illustrated above, the diffusion of nitrogen from the sub-films caused the CD variation. Changing the film density may help to reduce the diffusivity of nitrogen and give a better control on CD variation. Three different film deposition methods A, B and C are used to deposit the nitrogen-containing film and the CD variation performances are checked.

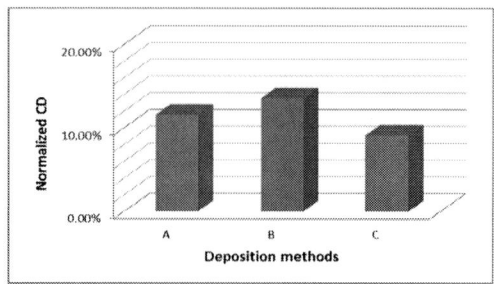

Figure 4: the CD variation with three different nitrogen-containing substrate deposition methods

As shown in Figure. 4, after exposure with the same photo condition, the percent of CD variations with a rather long Q-time in all three conditions is close 10%. Little improvement could be achieved by changing the film deposition method and hence changing the film density.

2. Plasma-related treatment

Another possible solution is to do surface treatment before photo process to block the nitrogen diffused from under layers. In order to avoid great change in device performance, the plasma-related treatment, which is a rather soft method, was chosen to be added before photo process. In this case, only oxide film thickness will be slight changed. The results are shown in Figure. 5. After plasma-related treatment was added, the CD variation of lot to lot is improved to 3% in comparison with the baseline (BSL) performance, which has a CD variation of more than 10%.

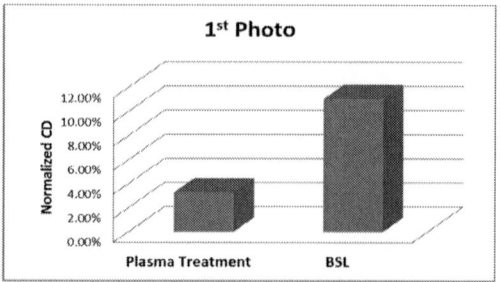

Figure 5: the CD variation with plasma-related treatment and BSL performance

From the data shown above, we could clearly see that the CD variation in one layer could be effectively controlled in a smaller range. However, whether one plasma-related treatment could solve the CD variation in the subsequent implant layers is still another important concern that should be addressed. Two different experiments were carried out. The first experiment only applied one plasma treatment before the first photo process, while in the second experiment, two plasma treatments are applied before the first photo process and the second photo process, respectively. The results are shown in Figure. 6, together with the BSL performance as a reference, the wafer having only one plasma treatment before the first photo process shows a similar performance with the Baseline wafer, which has no plasma treatment at all. Only by adding a second plasma treatment directly before the second photo process, the CD variation could be effectively controlled in a much smaller range of around 1.5%. Hence only one step plasma-related treatment at the first photo stage cannot solve the CD variation issues of all implant layers. Additional plasma-related treatment is still needed before every photo process.

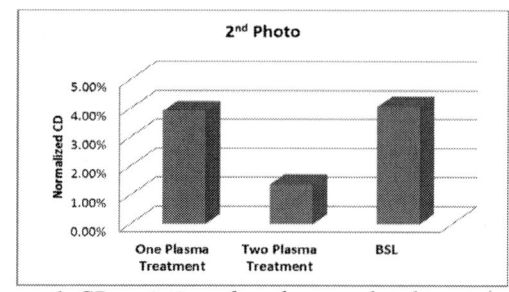

Figure 6: CD variation after plasma-related treatment

CONCLUSION

In this paper, a nitrogen-diffusion based model is built to illustrate the CD variation with Q-time in implant layers of the advanced technology. Experiments are carried out based on this model to check the influences of several factors. The different density of nitrogen-content film has little influence on CD variation. Plasma-related treatment before photo exposure could effectively control the CD variation by building up a block layer. But only one step treatment cannot solve CD variation issue in all implant layers, other additional plasma-related treatments are still needed before other following photo processes.

ACKNOWLEDGEMENTS

The authors would like to show our appreciation to SMIC TD PIE and FAB 8 PIE colleagues for the preparation of the test wafers.

REFERENCES

[1] Zhimin Zhu, Emil Piscani, Kevin Edwards, Brian Smith, *Proc. of SPIE* Vol. 6924 69244A-7

[2] Guerrero,D.J.,Smith,T.,Kato,M.,Kimura,S., and Enomoto, Proceeding of SPIE,vol.6153,242-249(2006)

DIFFRACTION-BASED AND IMAGE-BASED OVERLAY EVALUATION FOR ADVANCED TECHNOLOGY NODE

Jian Xu*, Long Qin, Qiaoli Chen, Hui Zhi, Yanyun Wang, Zhengkai Yang, Zhibiao Mao

Shanghai Huali Microelectronics Corporation, No.497 Gaosi Rd. Zhang Jiang Hi-Tech. Park,
Shanghai 201210, China,
* Corresponding Author's Email: xujian@hlmc.cn

ABSTRACT

The overlay control is one of the main challenges for advanced lithography in sub-28 nm technology node. There are two kind of overlay metrology inuse in semiconductor industry: most conventional image-based overlay (IBO) metrology and advanced diffraction-based overlay(DBO) metrology. In this paper we will compare these two methods through 3 critical production layers, focusing on the accuracy and the total measurement uncertainty (TMU) for the standard overlay targets of both techniques. The results show that both the accuracy and TMU of DBO method are superior to the traditional IBO method, which makes DBO method applicable at the 28nm and below technology node.

INTRODUCTION

Lithography performance has been a limited factor for advanced semiconductor manufacturing, and the decreasing of the critical dimension (CD) puts forward very tight requirement of overlay control. If the overlay between photo layers fails to meet the design rule, the device function and connecting function would also fail, leading to the loss of the production yield. At the 28nm node, the overlay of photo critical layers such as poly and contact should be controlled under 6nm, which is almost the limitation of the equipment and process. On the other hand, for advanced technology like 28nm and below, where stack, thickness or illumination settings are not frozen, adaptability of the measurement tool to this process variation is mandatory, to provide correct measurement to better monitor the process[1, 2].

Precision[3] and accuracy[4] are the two parameters to evaluate on product overlay performance. Total measurement uncertainty is used to present the precision, while the linear relationship between overlay measurement value and real value is used to present the accuracy, as shown in *Figure 1*. The aim is smaller TMU and better linear relationship. Two overlay techniques are commonly used in semiconductor industry: the Diffraction based Overlay (DBO) and the Image Based Overlay (IBO). This paper compares IBO and DBO methods through 3 critical production layers: poly, contact and metal on the performance of precision and accuracy in

detail, providing important information for overlay measurement below 28nm node.

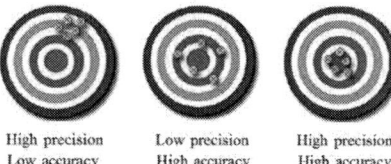

High precision | Low precision | High precision
Low accuracy | High accuracy | High accuracy

Figure 1: Schema of Precision and Accuracy

OVL TECHNIQUES: IBO AND DBO

The IBO measurements presented in this paper are performed using Archer 300 metrology tool from KLA-Tencor, while DBO measurements presented in this paper are performed using YieldStar S250 metrology tools from ASML.

IBO method

Bright field microscopy modes are used for measuring the targets. Measurement is performed using white light while auto-focus is performed via the interferometer. IBO method can be applied to measure the overlay of various overlay mark, among which BIB (Box In Box or Bar In Bar) mark is the most commonly used one, as shown in *Figure 2 (a)*. BIB mark consists of inner (current layer) and outer (previous layer) structures printed on two subsequent layers. The metrology tool obtains the mark image, locates the edges of each target on the previous layer and the current layer, determines the center of each target and finally determines the vector displacement (Dx, Dy), i.e., the X and Y direction overlay.

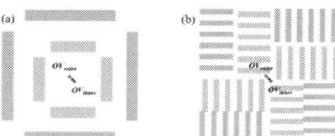

Figure 2: IBO overlay mark: (a) BIB mark; (b) AIM mark

Figure 2(b) shows the advanced overlay mark AIM (Advanced Imaging Metrology), which also consists of inner and outer structures printed on two subsequent layers. The

AIM mark is characterized by periodic series of lines and spaces. As the same with BIB mark, AIM mark's overlay is the vector between its two centers. Nevertheless, AIM mark owns refined repeated image structures, which leads to smaller measurement error as compared with BIB mark.

DBO method

The concept of the YieldStar overlay metrology technique is based on angle resolved scatterometry[5], as shown in *Figure 3(a)*. In angle resolved scatterometry, the illumination branch irradiates the overlay target over a large band of incidence space. The diffracted light from the grating is then captured (see *Figure 3(a)*). As we can see in *Figure 3 (a)*, zero order intensity varies symmetrically as a function of overlay whereas the +/-1 order intensities vary asymmetrically as a function of overlay. To define the overlay we will use the difference of intensity of the first orders. This asymmetry, named "As", is function of the first order intensities as shown in Equation (1). For small overlay the measured asymmetry is linearly proportional to overlay as shown in Equation (2), where K is the slope of the graph as shown in *Figure 3(b)*.

$$As = I_{+1} - I_{-1} \qquad (1)$$

$$As = K \times OV \qquad (2)$$

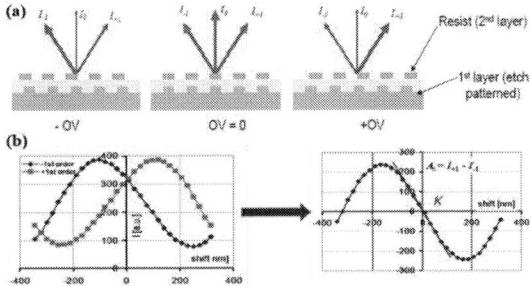

Figure 3: DBO Overlay Measurement Principle

The problem is that K in Equation (2) is not known. The solution is to eliminate K by measuring As with two overlay Overlay+d and Overlay-d where d is a known bias between the two levels measured. In this condition $As+d$ and $As-d$ could be obtained as shown in Equations (3) and (4), thus being able to obtain the overlay value as shown in Equation (5).

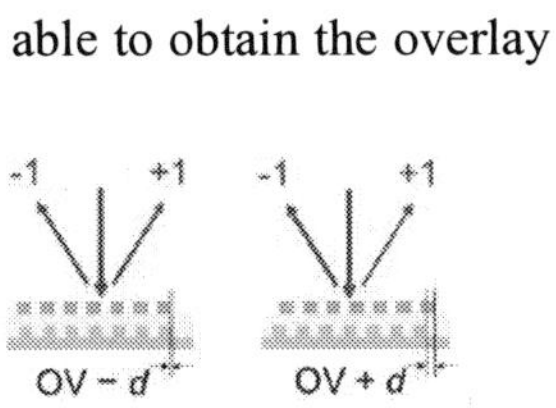

$$As^{+d} = K \times (OV + d) \qquad (3)$$

$$As^{-d} = K \times (OV - d) \qquad (4)$$

$$OV = d \times \left(\frac{As^{+d} + As^{-d}}{As^{+d} - As^{-d}} \right) \qquad (5)$$

Figure 4: Yield Star DBO Overlay Measurement Principle

DESIGN FOR OVL TARGETS

This paper investigate the performance of different overlay targets designed for critical layers (Poly, Poly line end cut, contact and metal layer) in 28nm technology node. All overlay targets are separated in two types: IBO and DBO targets. In which, we utilize a worldwide used Bar in Bar(BIB) target and two advanced imaging metrology(AIM) targets as IBO candidates. AIM target contains outer structure with prelayer pattern and inner structure with present-layer pattern, while AIMR possess opposite location between prelayer and present-layer.Table I gives all overlay target applied in this paper and their feature size.

TABLE I. OVERLAY TARGETS OF CRITICAL LAYERS

OVL	BIB	AIM	AIMR	uDBO
P1-AA				
	2.0/2.0/3.0#	1.1/1.1/2.4*	1.5/1.7/2.4	0.49/0.53/0.70
P2-AA				
	2.0/2.0/3.0	1.1/1.1/2.4	1.5/1.7/2.0	0.54/0.08/0.65
P2-P1				
	2.0/2.0/3.0	1.1/1.1/2.4	0.8/1.2/2.0	0.23/0.08/0.60
CT-P1				
	2.0/2.0/3.0	1.1/1.1/2.4	1.0/1.7/2.4	0.43/0.12/0.75
M1-CT				
	2.0/2.0/3.0	1.1/1.1/2.4	0.8/1.2/2.0	0.37/0.40/0.80

#L1/L2/P: L1 refers Bar width of prelayer; L2 refers width of present layer; P refers the gap width between prelayer and present layer.
*L1/L2/P: L1 refers CD of prelayer overlay target, L2 refers CD of present layer

RESULTS AND DISCUSSION

In this part, we will evaluate the two measurement methods in detail. To be fair, intra-field and intra-wafer measurement mapping are the same and also the same wafers were measured. All overlay targets are placed on adjacent location, and 20 repeating targets dispersing in one exposure field, see in *Figure 5*.

978-1-5090-6695-7/17 $31.00 © 2017 IEEE

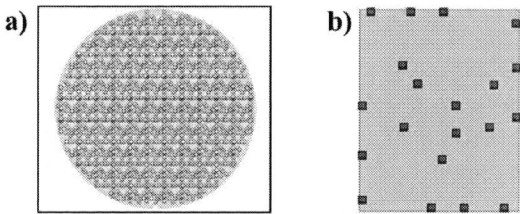

Figure 5: Intra-wafer (a) and intra-field (b) measurement mapping

For measurement, precision and accuracy are two of the most important assessment criteria. So we will evaluate all targets (P1-AA, P2-AA, P2-P1, CT-P1 and M1-CT) by these two criteria in the following paper.

Overlay Precision comparison of IBO and DBO metrology

Overlay precision of IBO and DBO metrology was compared through total measurement uncertainty (TMU). The TMU(1) is a function of the standard deviation of the dynamic reproducibility, the mean TIS and the dispersion of TIS[1]. *Figure 6* shows the TMU comparison of IBO and DBO overlay measurement for 28nm critical layers studied in this paper.

$$TMU = \sqrt{(3\sigma_{dynamic\ reproducibility})^2 + (Mean\ TIS)^2 + (3\sigma_{TIS})^2} \quad (6)$$

For P1-AA layer, TMU in X direction of BIB target is 1.04, which is far beyond the requirement for 28nm node (< 0.6nm). The other two IBO target (AIM and AIMR) also cannot well meet the requirement. On the other hand, uDBO measurement performs very good TMU (0.14nm). TMU in Y direction also shows the same conclusion that uDBO with much less TMU than IBO metrology, which indicate the uDBO metrology with higher precision than its counterpart IBO.

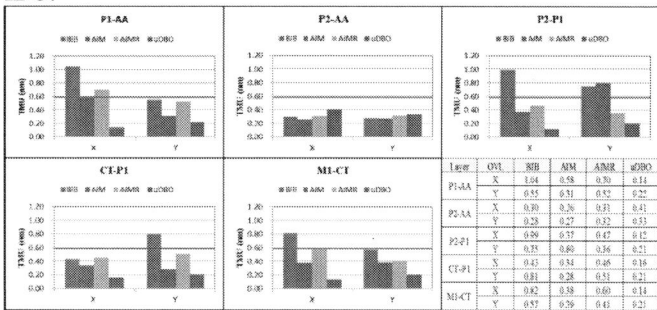

Figure 6: Overlay precision comparison of IBO and uDBO for 28nm critical layers.

For Poly line end cut (LEC) layer, all four overlay targets of P2-AA perform good TMU and meet the requirement. While as for P2-P1, there is a great advantages for uDBO in both X and Y direction.

For contact layer (CT-P1), considering both X and Y direction performance, uDBO metrology has more competitiveness in precision compared with IBO.

Finally, we compared the TMU of uDBO and IBO metrology for metal layer (M1-CT). The TMU of uDBO metrology show very good result (X/Y = 0.14nm/0.21nm), however the TMU of other three targets based on IBO metrology is much worse than uDBO.

Considering the previous result, we can conclude the TMU of all the four overlay targets as below: BIB > AIM ≈ AIMR > uDBO. Therefore, uDBO performs very good measurement precision compared with IBO in all key layers of 28 nm technology process.

Overlay accuracy comparison of IBO and DBO metrology

To evaluate overlay accuracy of different metrology, we also performed overlay measurement using CD-SEM[4, 6]. *Figure 7*. Shows the overlay measurement principle by using CD-SEM. We design present-layer CD target (grey line) located on the center of prelayer CD target (dark line). Take X direction as an example, we calculate the gap between present-layer and prelayer target W_{XL} and W_{XR}. The overlay value could be calculated with equation (7).

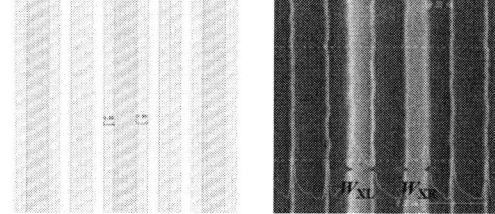

Figure 7: Overlay measurement using CD-SEM graph

$$OV_{CD-SEM} = \frac{W_{XL} - W_{XR}}{2} \quad (7)$$

In order to obtain a wide range of overlay for easily analysis the accuracy, all measured wafers were exposed with additional expansion forming radial type overlayer fingerprint as *Figure 8(a)*. We measured overlay with IBO and uDBO metrology after wafer exposure. Then we measured overlay with CD-SEM after wafer etching. The correlation between optical measured overlay and CD-SEM measured overlay were processed by linearly fitting (as

978-1-5090-6695-7/17 $31.00 © 2017 IEEE   154

Figure 8(b)) to acquire slope and correlation coefficient R^2.

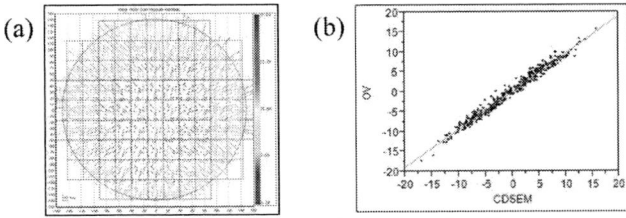

Figure 8: (a) Overlay fingerprint of wafer for accuracy evaluation; (b) The correlations between optical measured overlay and CD-SEM measured overlay by linearly fitting.

TABLE II shows the correlation results between optical measured overlay of four targets and CD-SEM measured overlay. We define high accuracy (green area) with correlation slope 0.9~1.1 and correlation coefficient $R^2 >$ 0.90. Next, we will analysis in detail the accuracy comparison between uDBO and IBO.

TABLE II. OVERLAY ACCURACY COMPARISON OF IBO AND UDBO FOR CRITICAL LAYERS

Layer	OVL Mark	BIB		AIM		AIMR		uDBO	
	direction	X	Y	X	Y	X	Y	X	Y
P1-AA	Slope (0.9-1.1)	1.32	1.38	1.22	1.26	0.89	0.90	0.93	0.97
	R^2 (>0.90)	0.93	0.89	0.95	0.91	0.94	0.96	0.98	0.98
P2-AA	Slope (0.9-1.1)	1.98	2.02	1.56	1.59	0.99	1.04	0.98	1.00
	R^2 (>0.90)	0.88	0.77	0.90	0.79	0.89	0.90	0.93	0.92
P2-P1	Slope (0.9-1.1)	1.11	1.19	1.08	1.12	1.03	1.04	1.05	1.05
	R^2 (>0.90)	0.95	0.93	0.96	0.95	0.98	0.98	0.98	0.98
CT-P1	Slope (0.9-1.1)	1.08	1.08	1.07	1.08	1.08	1.10	1.06	1.10
	R^2 (>0.90)	0.93	0.92	0.94	0.96	0.95	0.97	0.95	0.98
M1-CT	Slope (0.9-1.1)	1.25	1.25	0.97	0.95	0.97	0.95	1.00	1.00
	R^2 (>0.90)	0.81	0.77	0.92	0.90	0.93	0.89	0.93	0.90

For P1-AA layer, the three IBO targets could not meet the high accuracy criteria. While the correlation between uDBO measured overlay and CD-SEM measured overlay show good linearly fitting with slope X/Y = 0.93/0.97 and R^2 X/Y = 0.98/0.98.

Likewise, the correlation slope and R^2 of P2 layer for uDBO measured overlay all lie within the good accuracy range. However, the BIB and AIM targets have very bad results either for correlation slope or for correlation coefficient R^2. On the other hand, AIMR possess better accuracy than AIM target, which could be accounted that the reverse overlay pattern morphology in AIMR target kept better compared with AIM target.

For CT-P1 layer, all targets' overlay measurement meets the accuracy criteria which may contribute to the well tolerance of this layer. But for M1-CT layer, overlay measurement of BIB target is really bad in accuracy. Other three targets show good result of accuracy, especially for

uDBO metrology with correlation slope (X/Y = 1.00/1.00).

In general, overlay measurement with uDBO metrology performs very good overlay accuracy for all critical layers compared to measurement with IBO metrology. Specifically, the accuracy of the four targets can be concluded as below: BIB < AIM < AIMR < uDBO.

CONCLUSION

This paper systematically analyze the two main overlay measurement method (uDBO and IBO), especially comparing their measurement precision and accuracy. Total measurement uncertainty (TMU) is used to present the precision and the linear relationship between optical measurement overlay and CD-SEM measured overlay is used to present the accuracy. The result show that uDBO measurement performs very small TMU compared to other IBO metrology, make it be a more precise overlay metrology candidate. In addition, the correlation result reveal that overlay measured with uDBO metrology perform very good overlay accuracy for all critical layers, while IBO metrology could not simultaneously meet the accuracy acquirement for all critical layers. In conclusion, overlay measurement with uDBO metrology has a huge advantages than IBO metrology not only in precision level but also in accuracy level, which will provide a promising technic for overlay measurement of 28nm and below technology node lithography.

ACKNOWLEDGEMENTS

The authors would like to thank vendors' technical support and useful discussions from ASML. We also thank HLMC TD team for providing the valuable support.

REFERENCES

[1] Y. Blancquaert and C. Dezauzier, *Proc. of SPIE.*, vol. 8681 (2013)

[2] P. Leray, D. Laidler, et al., *Proc. of SPIE.*, vol. 7638 (2010).

[3] C.W.Yeh, C-T. H. Huang, et al., *Proc. of SPIE.*, vol. 8324 (2012)

[4] C-M. Ke, G-T. Huang, et al., *Proc. of SPIE.*, vol. 7971 (2011)

[5] C. S. Saravanan, Y. Liu, P. Dasari, *Proc. of SPIE.*, vol. 6922 (2008).

[6] O. Inoue, Y. Okagawa, et al., *Proc. of SPIE.*, vol. 9778 (2016).

BLOB DEFECT SOLUTION FOR 28 NM HOLE PATTERN IN 193 NM TOPCOAT-FREE IMMERSION LITHOGRAPHY

Dan Li[*], Biqiu Liu, Yulong Li, Zhengkai Yang, Zhibiao Mao, Yu Zhang

Shanghai Huali Microelctronics Corporation, No. 568 Gao Si Rd., Pudong New Area, Shanghai, 201203, P.R. China

*Corresponding Author's Email: lidan@hlmc.cn

ABSTRACT

In this paper, we studied the blob defect arising in the development of 193 nm topcoat-free immersion lithography for 28 nm node hole-pattern and we discussed the root cause of blob defect fundamentally. Finally, we found changing the content of hydrophobic addictive in topcoat-free resist could reduce the dynamic receding contact angle and then could eliminate the blob defect effectively.

Keywords—Embedded Barrier Layer; blob defect; Dynamic Receding Contact Angle (RCA); hydrophobicity

INTRODUCTION

193nm immersion lithography has been adopted in key lithography layers when the semiconductor manufacturing technology steps into 40 nm node or below. In theory (as equation (1) suggested), the resolution (R) of 193nm immersion resist could be greatly improved by increasing numerical aperture (NA). However, apart from the improvement of scanner and the lithography process, the performance of 193 nm immersion lithography resist also need to be improved to realize the high resolution of 193nm immersion process[1].

$$R = k1\lambda/NA \quad (1)$$

Now the research tends to use self-assembly addictive into resist to adjust the contact condition between water and resist [1]. It is during the spin coating step that the phase segregation takes place with self-assembly addictive forming the embed barrier layer (EBL) material distributed mainly on the surface of the resist film to form a self-assembled hydrophobic layer. This embedded barrier layer (EBL) material provides a slightly higher receding angle and lower PAG/quencher leaching rate on the surface of resist film [2]. Compared with topcoat technology, EBL reduce the manufactory cost and increase the throughput of the device production. And all these advantages make topcoat-free resist widely used.

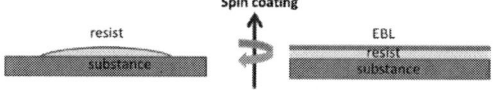

Fig.1 The formation mechanism of EBL material

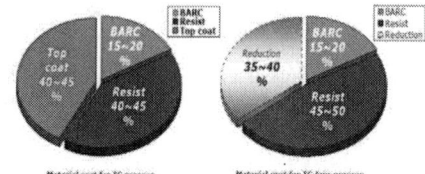

Fig.2 Material cost for ArF immersion lithography

HLMC applies the advanced topcoat-free resist on active area, poly, and poly line end cut (P1LEC) layers. However, a lot of blob defects are detected in the P1LEC layer, in which mostly patterns are hole-like. The blob defect is distributed randomly on the wafer, as suggested in Fig.4. At the same time there are no blob defect detected at other photo layer (such as active area and poly layer). The only difference is that the pattern of P1LEC is mostly hole like. The previous reports also found the similar issue that hole-pattern layers tend to suffer blob defect due to the high surface contact angle after alkali development [3, 4].

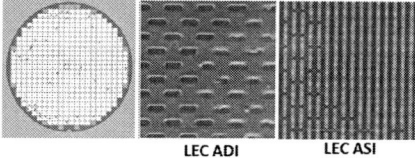

Fig.3 the distribution and the image of blob at ADI and ASI stage

In order to resolve this defect, a partial etch defect trace method was performed to find the defect source and an effective solution was adopted to reduce the defect density to improve the production yield. Thereby the new topcoat-free resist is successfully applied to HLMC 28 nm node logic device products.

RESULTS AND DISCUSSION

The bottom filling material and the bottom anti-reflective coating (BARC) material

Tri-layer structure is applied on P1LEC layer. As Fig. 5(a) suggested, the SOC (spin on carbon) is located in the bottom of film stack and then the BARC (BARC A). The resist is in the top of the film stack. To make sure where the defect source come from, we etched the film partially in which make the SOC as the end point. From the TEM image, the blob area appears in the BARC film, as Fig 5(b) show. So we could conclude that the blob defect comes from BARC or the resist film.

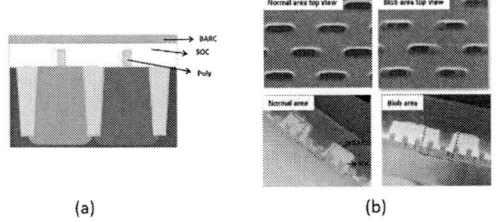

(a)   (b)

Fig.4 (a) P1LEC Tri-layer structure; (b) Partial ETCH area of both normal image and blob defect image

We carried out an orthogonal experiment in which the BRAC and resist type were changed. As table 1 result shows, the blob defect source come from resist surface.

Table 1 the orthogonal experiment result of resist of BARC

Split condition	Normalized blob defect count
BARC A + Resist A	36
BARC B + Resist A	34
BARC B + Resist B	0
BARC A + Resist B	0

The impact of post development rinse on blob defect

Since the blob defect comes from the resist surface, the surface condition is become an important factor which affect the defect formation. Different from line/space photo layer, the bulk exposed area is enclosed shallow well for hole type photo layer. The depth of the well is closed to the thickness of photo resist after development. The shallow well area includes the unprotected resist sidewalls and BARC surface which enriched hydroxy group. Whereas the unexposed areas that abut the wells are all hydrophobic. During the post rinse spin, residual water is easily trapped in the hydrophilic wells since the energy is less favorable for water to advance from a hydrophilic region into a hydrophobic. Thus, once the residual water dries in the wells, the solutes in it will deposit on the areas where water residues used to reside, and form blob defects. The distribution of the defects in the wells depends on the amount of water trapped in the wells as illustrated on the Fig 5.

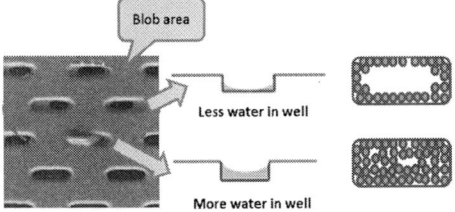

Fig 5 the relationship between water residual and blob defect distribution of hole pattern at exposed area

Since the contact angle of BARC surface is different from resist surface, it is necessary to reduce the resist surface hydrophobicity maximum during the develop rinse and spin dry step. Currently HLMC adopt ADR (Advanced Defect Reduction) process after develop, as Fig.6 suggested. This rinse technology could reducing the water remain sufficiently by applying N_2 assisting and controlling spin speed.

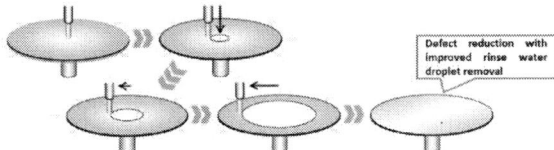

Fig.6 ADR rinse process flow

Based on ADR rinse, we added pre wet and idle spin off stage to reduce the resist surface water residual further. The results showed in table 2 suggest that the blob defect could not optimized by adding pre wet or spin off develop process.

Table 2 the impact of pre wet & spin off on blob defect

Split	Pre Wet	Spin Off	ADR rinse	Blob defect count/wafer
1	NA	NA	42s	37
2	5s	NA	42s	35
3	NA	2s	42s	37
4	5s	2s	42s	36

Developer Switchable EBL Materials

Based on the previous experiment result, the blob defect could not be reduced effectively by optimizing the lithography process. So we focused on resist itself. EBL materials that can perform the change from hydrophobic to hydrophilic in a time frame of seconds during aqueous alkaline development has been demonstrated to be an effective means for reducing the blob defect count and overall defectivity. The developer switchable EBL materials are able to provide a resist surface with a high receding angle as needed during the exposure scanning, and to reach a desired hydrophilicity in the wet processing to ensure an excellent dynamic wetting of the resist surface.

There are mainly three types addictive in topcoat-free resist. As Fig.7 shows, unit A ensures moderate dissolution rate of resist in developer; unit D is alkaline active and could make the surface contact angle low. Unit E plays a role as hydrophobic agent. We could get different resist performance through adjusting the proportion of three addictive respectively. To meet the requirement of low contact angle after development, the D unit proportion is increased and at the same time the E unit is decreased, as Fig.7 shows. Table 3 shows the contact angle results of two resist types after adjusting the two unit proportion. In resist type B the EBL material realized the transformation from hydrophobicity to hydrophilicity perfectly.

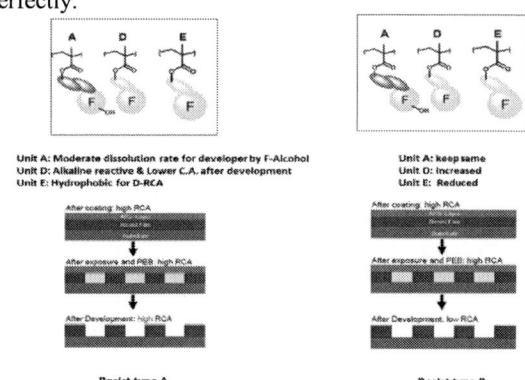

Fig. 7 Different addictive content in resist type A & type B

Table 3 the comparison between resist type A & type B after coating and develop

	After coat		After development	
	ACA	**RCA**	**ACA**	**RCA**
Resist A	95	75	90	55
Resist B	91	80	75	30

We carried out split experiment based on the new resist type. As the Fig.8 suggested, the blob defect were totally disappeared after changing the proportion of addictive in EBL material.

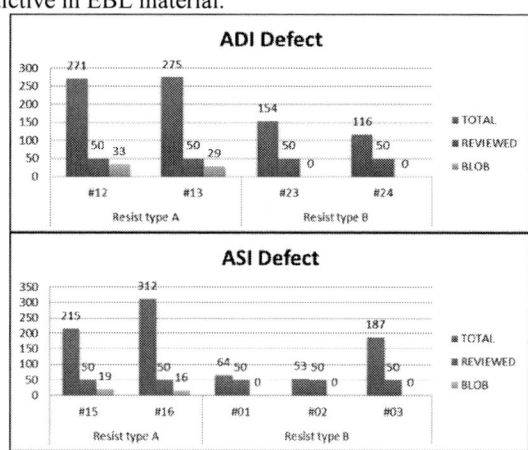

Fig.8 Blob defect count of resist type A & B @ADI

Before resist B applying on production, we evaluate the process window, through pitch and CD uniformity of the two resist. Figure 9 shows the OPE (optical proximity effect) matching results. The through-pitch OPE matching is 0.8 nm or less, which is a very good result. The receding

contact angle is less than 50° suggest the wettability of resist surface is good.

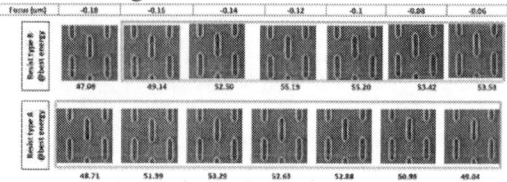

Fig 9 Resist type A & resist type B process window comparision

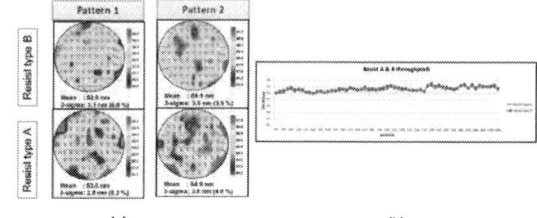

Fig 10. resist type A & resist type B CDU and through pitch comparision

CONCLUSION

The above experiment results show that the blob defect source is from resist surface. The blob defect tend to form in the hole-pattern due to the hydrophilicity difference between holes and the abut areas. The dynamic contact angle of resist surface is significantly reduced and controlled within 50° by adjusting EBL addictive proportion in the resist. Through this way the blob defect is effectively eliminated and the new topcoat-free resist is successfully applied on HLMC 28 nm node logic device products.

ACKNOWLEDGEMENTS

The author gratefully thanks lithography model, etch model, integration model of TD2 and Topco Scientific Co,, LTD for technology support.

REFERENCES

[1] Masafumi Fujita, Takayuki Uchiyama, Tetsunari Furusho, Takahisa Otsuka, Katsuhiro Tsuchiya, Proceedings of SPIE, vol.763923-1, 2010.

[2] M. Kocsis, D. Van Den Heuvel, R. Gronheid, M. Maenhoudt, D.Vangoidsenhoven, G. Wells, N. Stepanenko, M. Benndorf, H.-W. Kim, S.Kishimura, M. Ercken, F. Van Roey, S. O'Brien, W. Fyen, P. Foubert, R.Moerman, and B. Streefkerk, Proc. SPIE 6154, 615409, 2006.

[3] U. Okoroanyanwu and H. Levinson, Brugge, Belgium, Sept. 14, 2005.

[4] N. Shirota, et al., Proceedings of SPIE, vol. 6519-4, 2007.

The Study of Poly Gate Etching Profile, Micro Loading and Wiggling for NAND Flash Memory

Zhuo-Fan Chen, Hai-yang Zhang, Yi-Ying Zhang*

Technology R&D, Semiconductor Manufacturing International Corp.

No.18 Zhang Jiang RD., Pudong New District, Shanghai, 201203, P.R. China

*Email: Elaine_Zhang@smics.com

ABSTRACT

Advanced NAND flash memory requires higher cell density [1-2]. With the gate pitch critical dimension (CD) of NAND flash memory drops to sub-50nm or even lower, numerous process problems occur [3]. Poly gate etching, evolving CG and FG formation, as the dominator for the poly gate profile, confronts critical challenges as the line fluctuation known as wiggling, side wall bowing, depth micro-loading between dense-pattern and iso-pattern area and tapered profile, especially when the aspect-ratio (AR) goes up to 10:1. In this work, several special schemes are applied to avoid CG poly side wall bowing profile and to reduce over 50% depth micro -loading of FG etching between two different areas. Based on the improved depth loading, the etching profile can be optimized by the adjustments of etching parameters. Additionally, the pattern wiggling, which can be judged by line edge roughness (LER) measurement, is reduced by the fine hard-mask (HM) profile resulting from the improved corresponding etching step condition to enhance the cell line robustness.

INTRODUCTION

NAND flash memory poly gate, including control gate (CG) and floating gate (FG), is quite crucial for NAND flash performance. For CG etching, it defines the HM and CG poly profile, while FG decides the final word-line (WL) profile. Since the pattern wiggling, correlated to the HM performance, often occurs after the CG etching, the tuning is concentrated on CG HM etching step. Besides, the CG poly bowing profile needs discussion as well. On the other hand, the micro-depth loading and the problematic profile of FG are two main challenges. The conventional CF-based process on inductive coupled plasma (ICP) etcher is hard to fulfill the total demands. Thus, new type of etching scheme is necessary for our process development.

EXPERIMENTAL

Followed by self-aligned double patterning (SaDP) scheme, CG etching starts to define the HM as well as CG poly profile and then comes FG etching. Fig.1 (a) shows a portion of process flow, while Fig.1 (b) indicates the CG and FG etching with the spacer as the initial mask. Both CG and FG process are performed on a commercial ICP poly etcher. The corresponding LER data are collected on industrial CD-SEM equipment. After etching and corresponding strip and wet clean process, the cross sectional samples are imaged by a conventional scanning electron microscopy (SEM) tool.

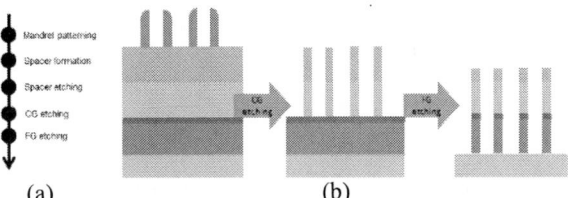

Fig. 1(a) SaDP flow; (b) CG and FG etching notification;

RESULTS AND DISCUSSIONS

Wiggling enhancement

Pattern wiggling, or line fluctuation, according to some previous analysis [4-6], can be detected by LER characterization. Several works has been reported to improve LER or LWR and most of them are based on mandrel or photo resist (PR) treatment as HBr plasma treatment [7-9]. In our work, however, the spacer on both sides of the mandrel is the real mask rather than the PR or the mandrel itself. As a result, we assume that the robustness of the CG line is the knob. Since the cell line CD drops considerably, its mechanical property cannot afford the continuous ion bombardment during conventional C-F based plasma etching, especially when HM is consumed too much during the procedure. Tapered HM profile, shown in Fig 2 (a), cannot provide appropriate protection for the etching afterwards because more etching amount is needed to achieve the same bottom CD. According to our previous work [10], decreasing the ratio can achieve vertical profile. Here we change the ratio of source/bias of ICP etching from initial 3:1 to 1:4 in this step, together with some additional CHxFx polymer gas to enhance the etching selectivity. As displayed in Fig. 2 (b), the HM profile shows the SWA from 86deg to 88deg. In addition, the new process has a better landing effect on CG poly due to the polymer gas mentioned above. As a result, HM consumption is reduced (Total height of HM increases about 30%, not shown here.) and thus increasing the line robustness. Finally, the LER data are collected after the whole WL formation (post final FG etching and clean), which shows that the optimized condition has

978-1-5090-6695-7/17 $31.00 © 2017 IEEE

about 22% LER reduction with the similar line CD and the pattern wiggling is diminished noticeably.

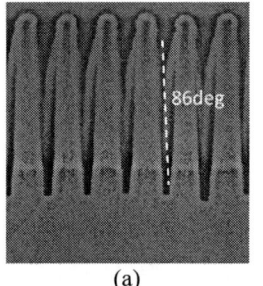 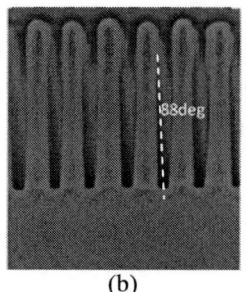

(a) (b)

Fig. 2 HM SWA of cross sectional SEM: (a) Initial condition. (b) Optimized condition.

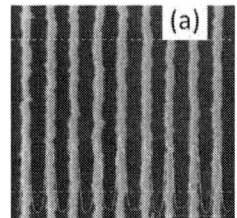 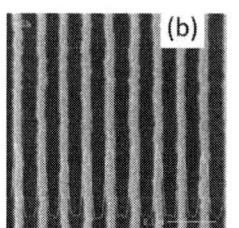

Fig. 3 LER images: (a) Initial condition. (b) Optimized condition.

CG poly profile optimization

As shown in Fig. 4 (a), after the conventional CFx-based main etching (ME), the poly side wall remains quite vertical, yet after soft landing and the final over etching process, the side wall shows bowing deformation as noticed by the arrow pointed in Fig. 4(b). It is conspicuous that the etching after ME causes such issue and therefore we apply additional passivation plasma with neutral gas to protect the etched poly side wall, notified in Fig.4 (c). Such protection step inserted between ME and the steps afterwards reveals a ideal result, shown in Fig.4 (d). The CG profile after the full etching keeps vertical as indicated by the arrow.

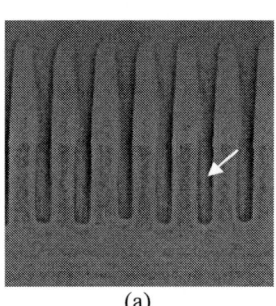 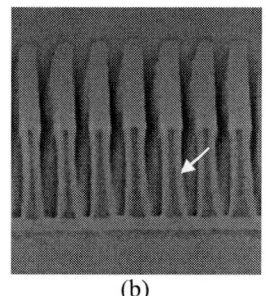

(a) (b)

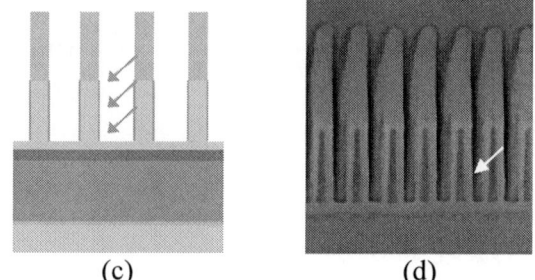

(c) (d)

Fig. 4 The etching profile: (a) After ME step; (b) After initial full etching including SL and OE steps; (c) Passivation step notification; (d) Optimized full etching condition.

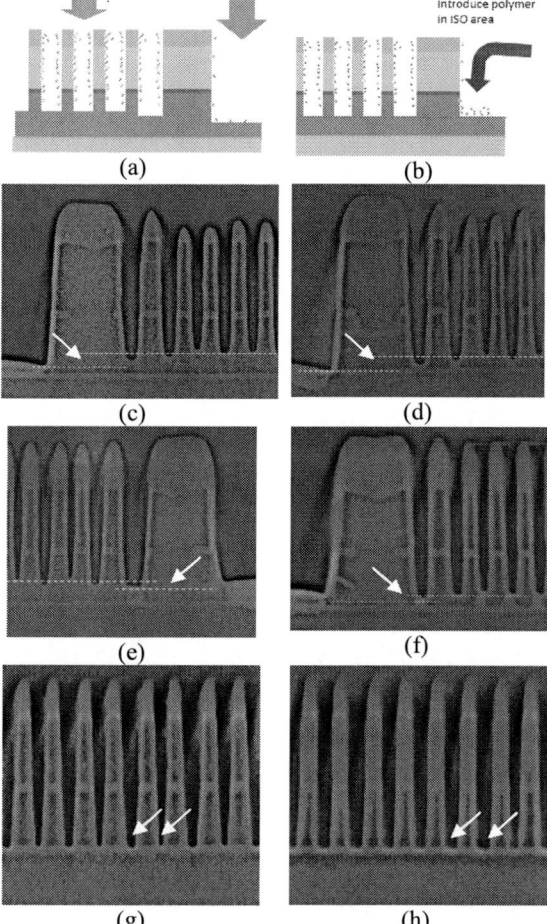

Fig. 5: (a) Depth micro-loading (RIE lag) notification; (b) Optimized approach notification; (c)ME partial etching of initial condition; (d) Condition A;(e) Condition B; (f) ME partial etching of 5x initial pressure; (g) FG full etching of initial pressure(BL); (h) FG full etching of 5x initial pressure condition.

FG etching depth micro-loading reduction

Since the pitch CD drops, the etching by-products or polymer cannot be removed rapidly by the gas flow and thus reduce the etching rate (ER) among the dense area, while the iso-pattern area remains the original ER, which leads to the etching depth micro-loading or RIE-lag performance (Fig. 5(a)). For NAND FG etching, the most difficulty comes from the ER balance of dense cell and the iso-pattern area, which are next to each other separated by the selective gate (SG) as the boundary. One possible solution is to reduce polymer generated during the process. Accordingly, one type of low-polymer etching gas without carbon is introduced considering that carbon is the main source of polymer. Together with some neutral dilute gases, the condition A reduces about 20% etching depth loading between the two areas (Fig. 5(c)&(d)). Nevertheless, it is still far away from the ideal performance. Our next approach is to introduce fluoro-methane (CHxFy) based gas to deposit additional layer among the iso-pattern area before the original ME step to reduce the ER among iso-pattern intentionally. The idea assumes that the polymer is more likely to accumulate among the iso-pattern while it cannot reach inside the dense cell space because of the very limited space CD, thus to balance the ER of two areas on both sides of SG since the dense cell would not be influenced by the new-introduced step heavily (Fig. 5(b)). After such special treatment come the original ME step and an in-situ strip for cleaning purpose. The condition B, verified our assumption above, demonstrates an impressive result: the depth difference of different patterns reduced over 50%, as the arrow indicated in Fig. 5(e).

FG etching profile enhancement

Given that micro-loading of FG etching is diminished as mentioned above, it is possible to further enhance the FG etching profile. Based on our reported work [10], both low ratio of source power/bias voltage and high pressure are crucial to achieve vertical profile, the optimization is concentrated on the pressure test for the source/bias ratio of initial ME is already quite low. Firstly, it is necessary to verify the potential influence of pressure addition for the micro-loading. As shown in Fig. 5(f), the depth difference almost remains the same even with the pressure reaches as high as 5 times of the initial value. Then the comparison of the full etching SEM of initial condition and 5x pressure condition depicts the effect of pressure adjustment. As predicted, the higher pressure can achieve very improved profile, supported by the enlarged bottom space CD pointed by the arrows in the cross sectional SEM images of FG full etching (Fig. 5(g)&(h)).

CONCLUSIONS

We demonstrate a group of scheme for the optimization of NAND poly gate etching to alleviate several problematic phenomena, including cell line wiggling, CG sidewall bowing, FG etching micro-loading and the tapered profile. For CG etching, wiggling can be reduced by the HM profile improvement while side wall passivation can rule out bowing profile. FG etching, on the other hand, can be optimized by the specialized polymer deposition treatment for micro-loading problem and increased ME pressure which is effective to achieve vertical profile.

ACKNOWLEDGEMENTS

The authors would like to thank the technical discussion with NHD NAND group of SMIC T-R&D, the dry etch group of Lam Research, China, and the failure analysis center of SMIC SH for SEM sampling.

REFERENCES

[1] K. Prall, et al., IEDM 2010, pp.102-103.
[2] C. Lee, et al., IEDM 2010, pp. 98-101.
[3] K. Prall, et al. IEDM 2010, pp. 521-524.
[4] D. Reid, et al. IEDM 2010, pp. 2801-2807.
[5] D. Reid, et al. IEDM 2010, pp. 2808-2813.
[6] L. Sun et al. SPIE 2015
[7] X. Meng et al. Ecs Trans. 2012 pp 325-330
[8] X. Meng et al. CSTIC 2014
[9] L. Zhang et al. CSTIC 2015
[10] Z.Chen et al. CSTIC 2016

STUDY OF POLY ETCH FOR PERFORMANCE IMPROVEMENT WITH ALTERNATIVE SPIN-ON MATERIALS IN FINFET TECHNOLOGY NODE

Yan Wang[1], Qiuhua Han[1], and Haiyang Zhang[1]*
[1]Semiconductor Manufacturing International Corporation
No.18 Zhang Jiang Road, Pudong New Area, Shanghai 201203, China
*Corresponding Author's Email: Steven_Z@smics.com

ABSTRACT

In this paper, we systematically investigate the poly etch performance with three different kinds of spin-on materials (BL organic coating material, and two other spin-on materials A and B) in P2 cut process from the point view of defect and process control. Our results show that with one kind of new spin-on material B, the bubble defect performance can be significantly improved. With different gas ratio control, we can achieve comparable etch rate selectivity of B to the hard mask. Thus, the P2 cut process can be well controlled. Finally, we delivered one process with B coating material.

INTRODUCTION

The FinFET structure is becoming more and more attractive for IC shrinking demands in 14nm technology node and beyond[1-2]. As a key layer, poly etch is very critical for yield enhancement roadmap. Since 28nm technology node, the gate etch has been separated into two masks for line end shortness (LES) prevention, namely P1 (line definition) and P2 (space cut process).

In P2 process, the coating material choice is very critical for P2 photo process. Current hard mask is easy to adsorb the water vapor during coating process, and then reveals bubble defects after poly etch. Special pre-treatment can prevent such bubble defect formation in planar technology node. However, it does not work in current FinFET technology node. The alternative coating materials are therefore utilized for bubble defect prevention.

EXPERIMENTAL DETAILS

All etch experiments in this course were completed on one commercial ICP etcher. 300-nm wafers on p-type silicon substrates were prepared for etch process optimization. Three kinds of spin-on materials including BL material were utilized for comparison. The inline defect scan and thickness characterization were performed with traditional measurement tools and transmission electron microscopy (TEM) was employed for poly gate physical profile characterization, respectively.

RESULTS AND DISCUSSION

The etch rate on the selected spin-on materials were

compared with BL material as shown in Table I. The etch rate of spin-on A is comparable with BL material, while the spin-on B is much faster. For etch rate match concern,

TABLE I   NORMALIZED ETCH RATE COMPARISON OF TWO KINDS OF SPIN-ON MATERIALS

Process Step	BL	Spin-on A	Spin-on B
Coating Recess	1	1.04	1.52
Hard mask Cut	1	1.07	1.34

the spin-on A should be a good choice to replace BL material. Figure 1 shows a defect comparison of BL and spin-on A and B conditions The spin-on B shows the best defect performance. Bubble defects decreased 50%. With special treatment on spin-on B, the bubble defect can be free. Overall concern, spin-on B is chosen for a candidate replacing BL organic coating material.

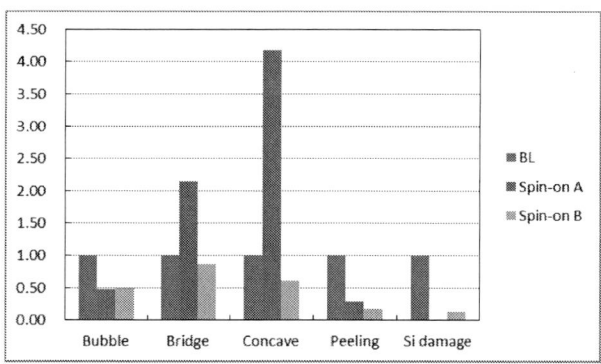

Figure 1: Normalized defect density with different spin-on materials

In P2 cut process, the spin-on material plays as an important role not only for lithography process, but also for etch process. Figure 2 shows a traditional cut process. In Fig2(d), during the hard mask remove step, the spin-on material plays as a mask role during the dummy poly removes. To control the selectivity of hard mask to spin-on material is important for process control. In traditional cut process, the design rule usually restricts the cut area on STI. However, there may be some test key or CD bar areas locating on AA area, such as CD bar-iso. The low selectivity may cause substrate loss as marked in the figure. Once the selectivity becomes much lower, the hard mask on poly may be damaged.

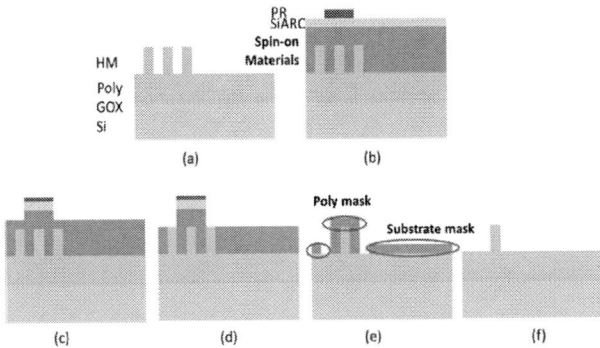

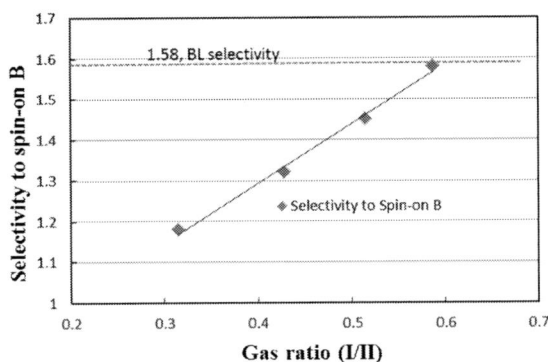

Figure 2: Schematic process for P2 cut process. (spin-on materials as masks)

The spin-on B shows much lower selectivity to hard mask, thus it is possible to suffer substrate damage during the useless hard mask removal. More seriously, the hard mask after poly-etch might be damaged. Our results show by changing gas ratio of the hard mask cut process, the selectivity can become comparable with BL organic coating material. Figure 3 shows the fluorine based gas ratio (I/II) dependent selectivity with the same total flow. The selectivity to spin-on B materials shows a linear dependent on the gas ratio (I/II). The BL selectivity marked in Fig. 3 is 1.58, when gas ratio (I/II) reaches 0.6, the selectivity can meet BL performance.

Figure 3: Dependence of selectivity of hard mask to spin-on B on the fluorine based gas ratio (I/II)

The gas I and gas II all are all fluorine based etchant gas with different polymer containing, in which, gas I contains higher polymer. The introduction of more polymer gas might induce the tapered poly head-to-head profile, and affect the following gap filling performance in metal gate loop[3]. Figure 4 shows the TEM images of poly gate profile with spin-on B material. The gate line shows vertical profile as in Figure 4(a). Besides, despite introducing more polymer gas during the hard mask cut process, the poly line end head-to-head profile also shows vertical, as shown in Figure 4(b), which means that it will not be an obstacle to metal gate gap filling.

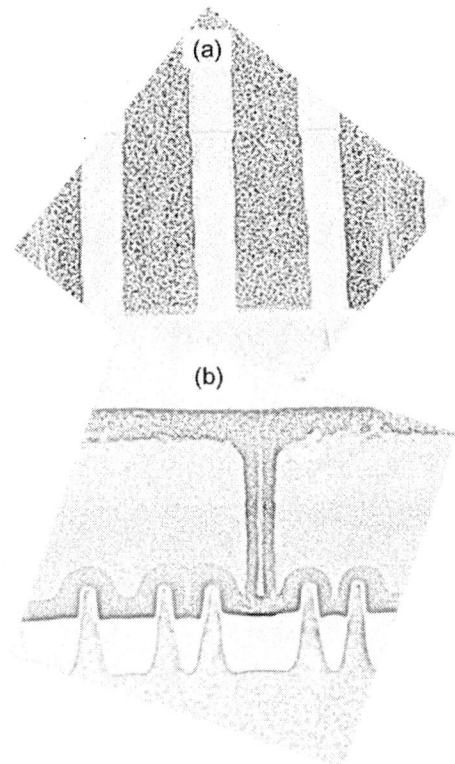

Figure 4: (a) Poly line profile (b) Head-to-head space profile

CONCLUSION

We successfully delivered one run-path for poly etch process with alternative spin-on material B. Considering the etch rate and selectivity difference, we tuned process to achieve a comparable physical performance with BL organic coating material. The defect performance was significantly improved with optimized processes.

ACKNOWLEDGEMENT

This work was partially sponsored by Program of Shanghai Technology Chief Scientist (B type). The authors would like to gratefully thank Mr. Jiangang Liu, and Dr. Yushan Chi for the constructive discussion during the experiments. Integration team members Zhengling

Chen and Huijun Zhang are also appreciated for wafer preparation.

REFERENCES:

[1] D. Hisamoto, W. C. Lee, J. Kedzierski, H. Takeuchi, K. Asano, C. Kuo, E. Anderson, T. J. King, J. Bokor, and C. M. Hu, *IEEE TRANSACTIONS ON ELECTRON DEVICES,* vol. 47, 2000, pp. 2320-2325.

[2] J. G. Fossum, M. M. Chowdhury, V. P. Trivedi, T. J. King, Y. K. Choi, J. An, and B. Yu, *IEDM '03 Technical Digest*, Washington DC, December 8-10, pp. 29.1.1-29.1.4.

[3] Jianhua Xu, Xuezhen Jing, Xiaoniu Fu, Xiaona Wang, Jingjing Tan, Ziying Zhang, and Beichao Zhang, *China Semiconductor Technology International Conference*, Shanghai, March 15-16, 2015.

Challenges and Solutions of 28nm Poly Etching

Xiangguo Meng*, Lian Lu, Fuhong Chen, Sifei Chan, Quanbo Li, Zhibiao Mao, Yu zhang, Albert Pang

Shanghai Huali Microelectronics Corporation, Shanghai City, China

*Corresponding Author's Email: mengxiangguo@hlmc.cn

ABSTRACT

Gate formation for 28nm node is LELE (2 times Litho, 2 times etch process) approach, which is different from traditional poly LE (Litho-Etch) process. Poly line and poly LEC (line end cut) formed during the second Litho etch process. It is great challenge to get appropriate LEC CD (Critical Dimension), meanwhile balance LEC position to achieve residue & pitting defect free. This paper introduces challenges and solutions of 28nm poly LEC etching, especially studies on LEC CD &defect control.

INTRODUCTION

The formation of 28nm poly includes 2 Litho and 2 Etch processes. Figure 1 is a schematic diagram. The 1st process is to form poly hard mask line; the second Litho-Etch process is to form poly line & poly line end cut. We need to make sure poly line profile vertical, additionally line end CD meet target and defect free at LEC process. Especially at LEC position, defect free requires residue and pitting free, which need keep a good balance between residue direction and pitting direction. In the face of challenges, this process requires etching step time, selectivity, gas reasonable.

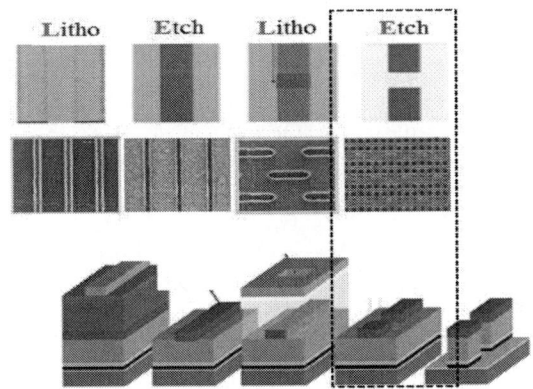

Fig.1. LELE process

Poly LEC etching process is strict with LEC AEI CD, mainly because of the size of the LEC CD affect the electrical performance (large LEC CD may lead to Poly leakage, small CD will cause merge issue and high residue risk, adverse to the subsequent spacer etching). Figure 2 is pictures include post 2nd Litho and etching. Especially the marked area, which is important &weak

Some etch steps are very important for LEC CD and performance. Such as SOC etch, Hard mask over etch (HM OE) etch step &Poly main etch step. Influence factors have been pointed out and studied in this paper.

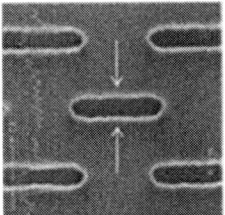 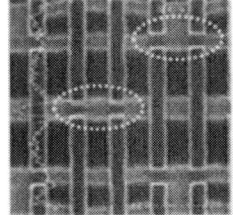

2.1> post 2nd Litho   2.2> post 2nd etch

Fig.2 2nd Litho-Etch process

EXPERIMENT

This work was performed using TCP plasma (Transformer coupled plasma) .Etch chamber name is from LAM Corporation.This type chamber is famous for uniformity for Transformer Coupled Capacitive Tuning and tunable ESC (Electrostatic Chuck).Both productivity and capability are improved.

Test wafer condition is as blow: Wafer diameter size is 300mm, film stack is patterned PR/Si-arc/SOC/OX/SIN/Poly as Figure 3.

A TEM JEM-2100F from JEOL Corporation is used for TEM image.

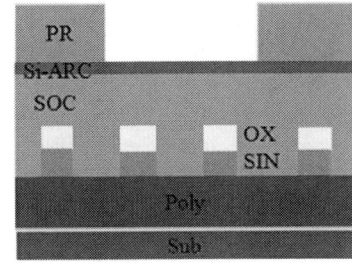

Fig.3. Poly LEC film stack

RESULT AND DISCUSSION

A. Key etch step: SOC etch

As this layer film stack is PR/Si-arc/SOC/OX/SIN/Poly, the 1st step is etching Si-arc, and then etching SOC. When etching SOC, we need consider SOC coating on oxide HM& poly film are different. SOC on poly is protector for LEC HM etching process. Considering this aspect, in fact SOC etch is partial etch. So SOC etch amount is a very important factor besides SOC etch profile. We studied SOC remain amount with LEC CD &defect correlation. Figure 4 is SOC remain thickness & defect correlation. SOC remain thick will cause fence issue, which will be residue issue. This case pointed in Fig.4.1, Fig.4.2, Fig4.7.Fig.4.1 is thick SOC, Fig.4.1 is possible residue post poly etch, Fig.4.7 is residue appears at LEC position. But SOC remain thin means SOC as mask not enough, which will

978-1-5090-6695-7/17 $31.00 © 2017 IEEE   165

be AA pitting issue.Fig.4.3 is thin SOC remain and Fig. 4.4 is possible pitting post poly etching. Fig. 4.8 is AA pitting appears at LEC position. We need control SOC remain thickness appropriate as Fig.4.5, Poly LEC position will be residue & pitting defect free as pointed in Fig.4.6.Fig.4.9 is final LEC position image. Our study show SOC remain thickness is about half HM total thickness is reasonable.

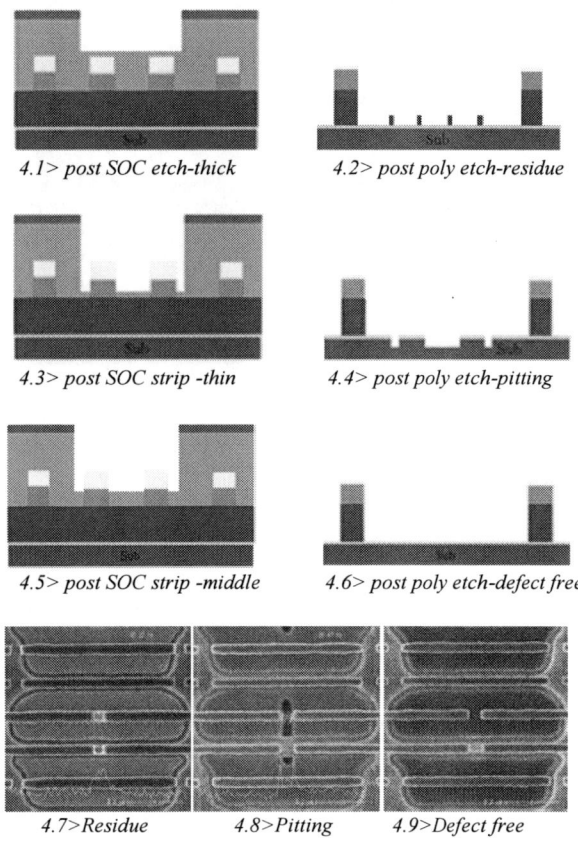

4.1> post SOC etch-thick   4.2> post poly etch-residue

4.3> post SOC strip -thin   4.4> post poly etch-pitting

4.5> post SOC strip -middle   4.6> post poly etch-defect free

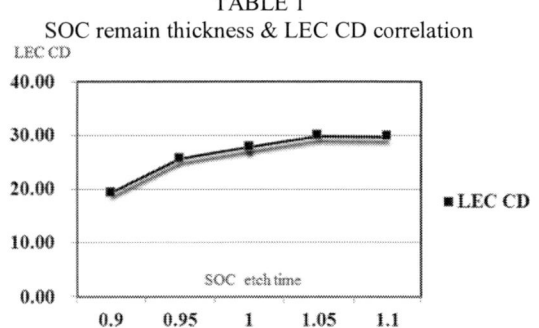

4.7>Residue   4.8>Pitting   4.9>Defect free

Fig.4. SOC Remain thickness & defect correlation

Beside SOC remain thickness effect LEC position defect, SOC remain effect LEC CD directly. We set SOC remain thickness is 50% HM total thickness as center point (etch time set 1), when we add SOC etch time to down SOC thickness, or we down SOC etch time to increase SOC thickness, We can see LEC CD is adjustable. We can fine tune little SOC time if HM etch step acceptable. Table 1 is SOC remain thickness& LEC CD correlation, It can be a reference for LEC CD fine tune and it can be a guide line for process window check

TABLE 1
SOC remain thickness & LEC CD correlation

SOC profile is an important factor for LEC position defect. Fig.5 show different profile, Fig.5.1 SOC profile is middle thick and edge thin at local position, middle and center bias is large. This profile may cause LEC on AA position pitting for not enough mask, Fig. 5.2 is post HM etch profile, it shows Poly loss on SOC thinner area, this pattern transfer to sub-film will lead to pitting issue.Fig.5.3 is improved profile, SOC thickness bias is lower than Fig.5.1, Poly film is more smooth than Fig.5.2, this profile will down pitting risk, which is helpful for enlarge process window.

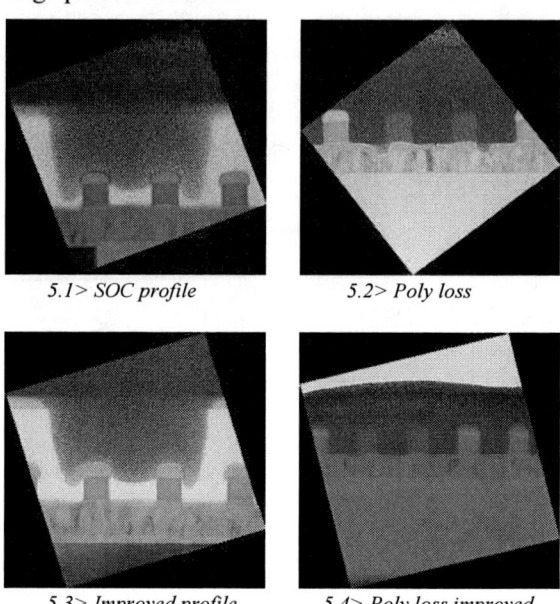

5.1> SOC profile   5.2> Poly loss

5.3> Improved profile   5.4> Poly loss improved

Fig.5 SOC profile & poly loss

B. Key etch step: HM etch

HM etch is a key step for LEC CD control &defect control. Because HM etch especial HM OE (hard mask over etch) affect Poly recess. As we known, HM film is oxide and nitride, when etching this kind film we use CF_4, CH_2F_2, CH_3F etc. F content gas. Selectivity is a must considered factor. We need consider both HM to SOC selectivity and HM to sub poly selectivity. They are opposite directions for gas choosing. When using CH_3F and O_2 as etching gas, poly recess thin but SOC consumes fast. Using CF_4 as etching gas, HM to poly lower selectivity cause high pitting risk.Table2 is HM to SOC and HM to Poly selectivity and typical gas comparison. As a result we use CF_4 mixed with CH_2F_2 plasma gas, this approach can balance pitting and residue defect.

TABLE 2
HM ME etch gas and selectivity

HM OE step	Selectivity	Range	Residue risk	Pitting risk	Typical etch gas
HM/SOC	Low (SOC fast)	<0.5	Low	High	CH3F/O2
	Middle	0.5~2	Middle	Middle	CF4/CH2F2
	High (SOC slow)	>3	High	Low	-
HM/Poly	Low (poly fast)	<1	-	High	CF4
	Middle	1~3	-	Middle	CF4/CH2F2
	High (poly slow)	>10	-	Low	CH3F/O2

HM etch etching use CF_4/CH_2F_2 still has some pitting risk because HM to poly selectivity is lower than 3, HM to poly high selectivity is needed for HM OE step. CH_3F and O_2 are perfect gases as OE step.Fig.6 is post HM etch, Fig. 6.1 is use CF_4 and CH_2F_2 as etch gas, this approach cause some poly recess. Changing gas to CH_3F and O_2 as OE step can improve poly loss greatly as Fig.6.2 pointed. In fact this approach can improve pitting issue.

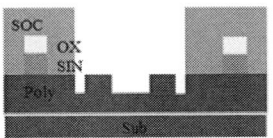 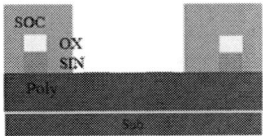

6.1> HMOE: CF_4+CH_2F_2   *6.2> HMOE: CH_3F+O_2*

Fig6. Post HM etch

C. Key etch step: Poly ME etch

Poly ME (main etch) amount is always a traditional topic. Because Poly remains thickness can directly cause pitting issue or profile issue. Poly remains thicker need more OE amount which will cause Poly profile bowing or top poly necking issue. Poly remains thinner will cause AA pitting directly. Overall, we need combine step height & poly total height, poly remain ~20~30% is suitable.

CONCLUSIONS

In our paper we studied 28nm poly LEC etching challenges and solutions, special for LEC position CD control and defect control. SOC etch amount, SOC etch profile, HM etch gas and Poly ME amount are discussed. To get a perfect poly LEC etching process, step need well matched each other.

ACKNOWLEDGMENTS

Thanks for all the authors for great work and collaboration. Thanks for FA (failure analysis) for TEM image support.

REFERENCES

[1] Quanbo Li; Xiangguo Meng; Jun Tian; Jun Huang; etc. Methods of line end cutting of small CD for 28nm technology ,2015 China Semiconductor Technology International Conference, Pages: 1 - 3, DOI: 10.1109/CSTIC.2015.7153373

[2] Zhonghua Li; Runling Li; Tianpeng Guan;etc. The study of 28nm node poly double patterning integrated process,2015 China Semiconductor Technology International Conference Pages:1-4, DOI:10.1109CSTIC.2015.7153427

[3] Liang, C. W., Chen, M. T., Jenq, J. S., Lien, W. Y. etc. A 28nm poly/SiON CMOS technology for low-power SoC applications, 2011 Symposium on VLSI Technology - Digest of Technical Papers > 38 - 39 Proceedings Article March 29, 2013

[4] Hubert Hody; Vasile Paraschiv; etc. Double patterning with dual hard mask for 28nm node devices and below.DOI: 10.1117/12.2010951,Special Section on Advanced Plasma-Etch Technology November 08, 2013

[5] Makoto Satake; etc. Effect of oxygen addition to an argon plasma on etching selectivity of poly (methyl methacrylate) to polystyrene.Micro/Nanolith.MEMS,MOEMS.2013; 12(4):041309. DOI: 10.1117/1.JMM.12.4.041309

[6] Jianliang Li; etc.; Multi-layer model vs. single-layer model for N and P doped poly layers in etch bias modelingLawrence S. Melvin III Proc. SPIE. 7823, Photo mask Technology 2010, 78233V. (September 30, 2010) DOI: 10.1117/12.866040

DRY PLASMA CLEANING BY ADVANCED HDRF TECHNOLOGY (HIGH DENSITY RADICAL FLUX)

Yannick Pilloux

Plasma-Therm LCC

St Petersburg Florida, USA

email: yannick.pilloux@plasmatherm.com

ABSTRACT

Semiconductor industry has been focused over decades on Dry etch and deposition technologies, while cleaning was mainly wet technology.

Die size are shrinking in main applications, as well as reducing packaging size. Thus, limitation of wet chemistry is leaving space to dry plasma cleaning.

This paper will focus on how dry Plasma technology named HDRF, for High Density Radical Flux, can provide efficient cleaning solution without damaging sensitive layer like GaN, TiN, and keep low temperature processing.

In addition, moving from wet to dry cleaning technology, allow to eliminate all dirty processes and keep green environment.

DRY PLASMA CLEANING BY ADVANCED HDRF TECHNOLOGY

Dry plasma cleaning/stripping was introduced in the market many years ago with barrel Asher technology, for which wafers are located in the middle of plasma chamber and are in contact with ions energy & photons. This effect can create damage on sensitive layers like GaN, TiN, as well as damaging sensitive devices by having e-charging effect.

HDRF technology was introduced in the market to strip and clean any organics layers or residues. This unique technology, ICP (Inductive Couple Plasma) source is downstream, and is ideal solution for low damage, low temperature photo-resist stripping without electrical charging effect.

HDRF is high density plasma source higher than of 1E17 cm^{-3} ("Fig. 1") density, while conventional ICP source is in range of 1E13 cm^{-3}. HDRF is able to clean any organic layer or residues in high aspect ratio structure up to 30:1

HDRF plasma characteristic is based on O* radicals free of ions ("Fig. 2") while conventional Asher have ions and photons. Such high plasma density of O* radicals allow to clean organics layers and/or to remove photo-resist for multiples applications in LED, MEMS, Photonics & wireless markets without damaging devices. Damage is identified as electrical charge effect which can generate collapse of Mems membrane, for example.

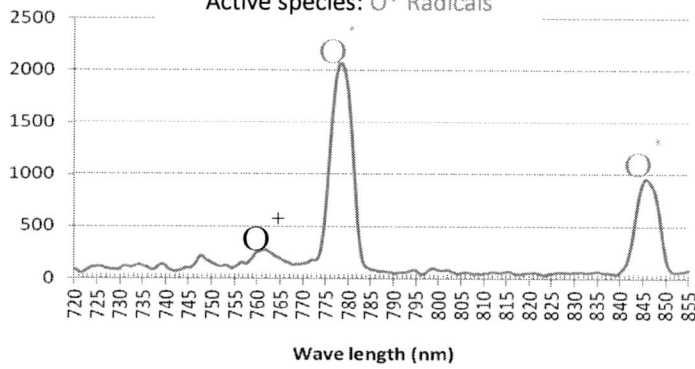

Fig.2. HDRF characteristic

While high temperature on wafer surface is critical, HDRF can remove organics residues below 80°C without having film delamination.

Photo-resist (PR) stripping at low temperature is also possible below 80°C by using HDRF, to avoid damage on GaN film, even with high strip rate performances; this is done by adding F* containing gas in current process gas flow ("Fig. 3")

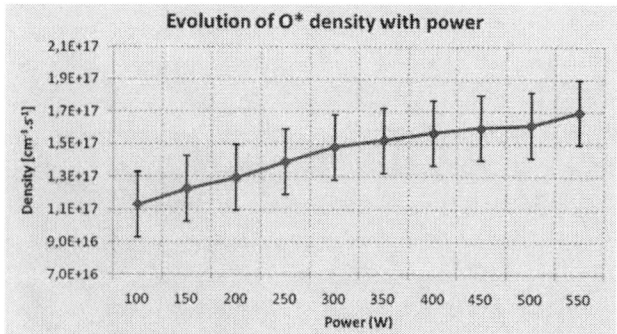

Fig.1. High plasma density versus RF power

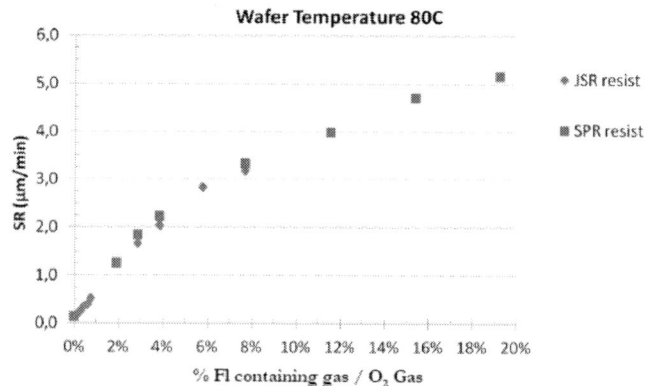

Fig.3. Strip rate vs F % containing gas*

978-1-5090-6695-7/17 $31.00 © 2017 IEEE

Lastly, shaping MEMS devices is commonly done by using Deep Reactive Ion Etching (DRIE) technology and Bosch process, which is alternate process of isotropic etch and sidewall passivation steps (called loop); Alternated Bosch process generate scalloping, which are peaks and valleys on the sidewall structure.

Bosch process, by definition is polymerized process, adding polymers on sidewall which are C* and F* (Carbon and Fluor) to be removed. While traditional wet technology can remove polymers, there is still limitation in high aspect ratio structure, and/or when it is thick polymers.

Due to HDRF high plasma density, polymers are removed in any structure with O* radicals plasma, efficiency measured by Energy Dispersive X-ray Spectroscopy (EDX) ("Fig. 4") for a strip rate 10 times higher than conventional wet clean.

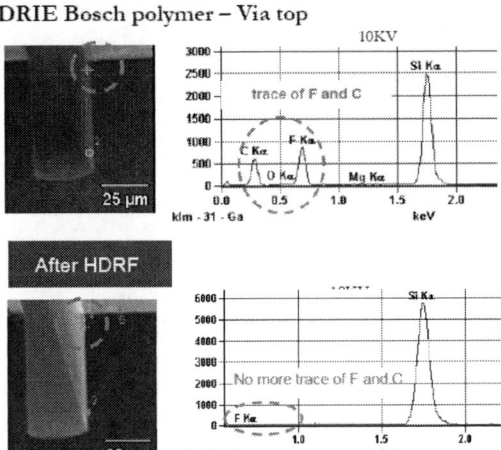

Fig.4. Bosch polymers removal

Then, a 2nd process step is required to smooth scalloping as it can create electrical arcing, in case of isolation layer requirement.

By using fluorine based chemistry, HDRF is able to smooth scalloping in high aspect ratio structure 30:1. It is called micro-isotropic etch step, able to etch mainly peak without impacting structure dimension width. Scalloping from 50 to 800nm (peak to valley) can be reduced to few nanometer by HDRF plasma, within less than 4 minutes. ("Fig. 5")

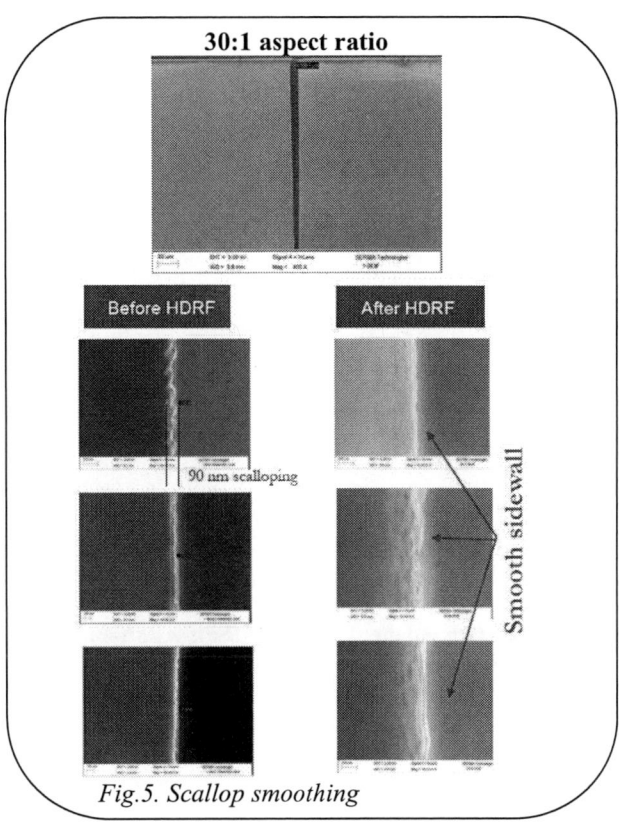

Fig.5. Scallop smoothing

CONCLUSION

Semiconductor industry started to reduce die size, to add more dies per wafer, for which cleaning quality needs to be improved. While wet technology has limitations in high aspect ratio structure, high plasma density of HDRF is able to clean polymers, even in complex 30:1 aspect ratio structure.

In addition, while low temperature processing is required for cleaning/stripping, HDRF is a perfect fit to avoid devices damage, due to low charging effect.

Finally, Bosch polymers removal after DRIE can be cleaned by HDRF, follow by micro-isotropic etch step to smooth scalloping.

REFERENCES

[1] Pr Laifa Boufendi. University of Orleans France, "Projet Gastineau/Lacroix – characterization by actinometry"

Study and solution of 28nm AIO etch seal ring residue issue

Haibo Shen, Lei Sun, Quanbo Li

Shanghai Huali Microelectronics Corporation
Shanghai City, China
Phone: +86 21 61871212 E-mail Address: shenhaibo@hlmc.cn

Biography

Haibo Shen is an etch PE. He works in Shanghai Huali Microelectronics Corporation, TD2 department. He got bachelor's degree of physics from Jilin University. He already worked in etch more than 5years. He focuses on All-In-One etch.

Abstract

BEOL pitch is less than 90nm in 28nm and below technology node. As a result of the reduction of size, the etching of AIO is a huge challenge. This paper introduces the issue of seal ring residue in the development of 28nm AIO etch process and the balance of SAV (Self-Align-Via) profile control, Via CD control. We put forward theoretical analysis and give the final solution to etch issue on this basis.

Keywords—AIO; seal ring; plasma; u-loading.

Introduction

At the 28nm and below technology node, in order to further reduce the post-section of the resistance and capacitance delay effect, dual-damascene process and ultra-low-k dielectric material (BD2) has been widely used. Metal hard mask all in one etch process has shown great advantages in reducing the effect of the plasma etch process on the dielectric constant of ultra-low dielectric materials and overcoming the challenge of lithography with critical size reduction. This paper mainly deals with the problems of seal ring residue in AIO etching process development and the research on a series of problems arising from the development of 28nm low power chip in Huali Microelectronics Co.Ltd.

AIO(All-In-One) etching process shown in fig.1, the black part of the process is the AIO etching—Via etch, PR Strip and Trench etch three steps in the same process step.

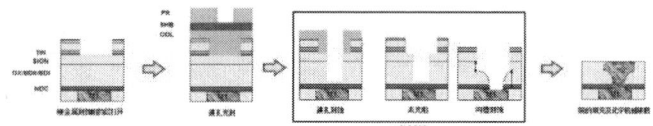

Fig.1. AIO etching process description

1. Seal ring residue issue and solutions

A. AIO seal ring residue defect

Seal ring is widely used to release the stress reduction on the functional areas within the chip. Seal ring residue can cause stress release issue. As can be seen from the fig.2, the seal ring region open only near the side wall, and most of the region has a dielectric film residue.

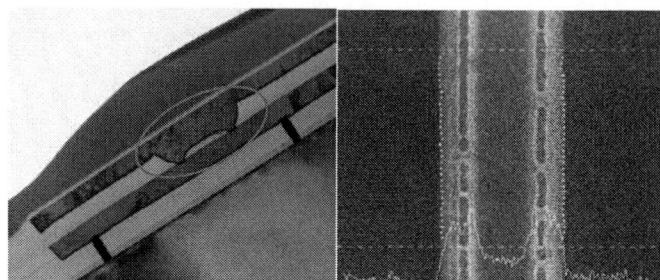

Fig.2. seal ring residue section and top view

B. AIO seal ring residue source

Relationship between etching/polymerization effect and fluorine/carbon ratio in the etch process is shown in fig.3. When the fluorine/carbon ratio is less than 3, the process is dominated by polymerization which may block the fluorine-based radical with dielectric film. The region near side wall gets more bombardment due to ion reflection, but blanket area suffers etch stop because of weaker bombardment. The thickness of the surface polymer is proportional to the gas ratio, pressure, temperature and so on.

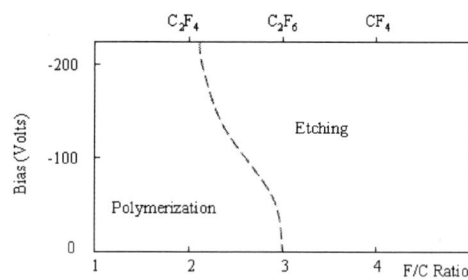

Fig.3. Relationship between etching/polymerization effect and F/C ratio

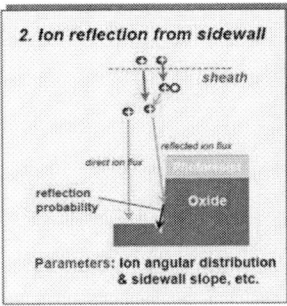

Fig.4. sidewall ion reflection

C. solution of seal ring residue

According to the issue mode, take the corresponding measures: in the plasma control, the plasma energy increasing can maintain good ion bombardment direction. In gas control, the fluorine/carbon ratio increasing can reduce polymer deposition and prevent large line width etching residue. In addition, the appropriate temperature can also be adjusted to accelerate the evaporation of polymers to prevent the etching stop. Finally, we chose to increase the ion energy optimization (pressure down and LF up), the section and top view shown in Fig.5. The optimization of the seal ring residue also need to balance other structures performance include the SAV profile and the via hole CD.

Fig.5. Optimized seal ring section and top view

2. Balance of the optimization

A. SAV profile control

SAV (Self-Align-Via) process, SAV and non-SAV etching method is shown in Fig.6. It can improve the bright issue, take into account the different structures for via key dimensions and reduce the risk of VIA open.

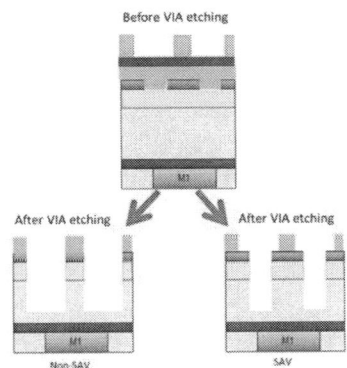

Fig.6. SAV and non-SAV etching method schematic

As shown in Fig.7a, the TDDB structure space does not meet the process requirement after the optimization of seal ring residual. It mainly led by more damage of titanium nitride which is from stronger physical bombardment and more ion energy. We can decrease ion energy or increase polymer deposition to protect TiN from damage. As above we need more ion energy to solve the seal ring residual. Finally, we chose to increase DC to increase surface polymer deposition optimization. As shown in Fig.7b, TDDB structure profile meet process requirement.

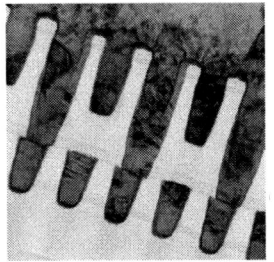 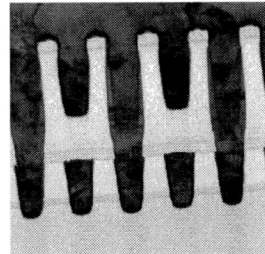

Fig.7a Initial condition TDDB   Fig.7b Optimized TDDB

B. VIA CD control

VIA CD control is also very important. There are mainly 2 methods for etching VIA CD control: ①SHB/ODL CD control, ②PV etching control. CO_2/CO is usually selected for ODL step. Because of no fluorine radical in plasma, the selectivity to silicon or dielectric is very high. Etching time of ODL step can be adjusted to control no confine CD. PV etching influence copper exposure time on trench step which enlarge CD. The polymer deposits more when plasma touches copper than low-k. It resists etching in sidewall. As shown in Fig.8, the more PV etching time is, the less VIA CD is.

Fig.8. relationship between the PV time and the VIA CD

As mentioned above, more surface polymer deposition of optimization led the VIA profile straight. We need more chemical etching to enlarge bottom CD. SAV CD changed less than no confine VIA because hard mask is difficult to etch. It is necessary to shrink SHB/ODL CD. We decrease F ratio in BT step and then make sure ODL over etch enough.

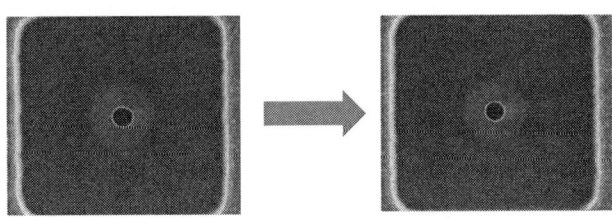

Fig.9. BT optimization VIA top view

Conclusion

This paper mainly analyze the resource and solution of seal ring residue in the development of 28nm AIO etch process and the balance of SAV (Self-Align-Via) profile control, Via CD control. It is significant for AIO etch development of 28nm and below technology node.

Acknowledgements

The authors would like to thank vendors' technical support and useful discussions from TEL. We also thank HLMC TD team for providing the valuable support.

References

[1] R. Brain, et al., "Low-k interconnect stack with a novel self-aligned via patterning process for 32nm high volume manufacturing," Proc. of IEEE International Interconnect Technology Conference (IITC), pp 249- 251, 2009.

[2] C. B. Labelle, et al., "Plasma etch challenges for porous low-k materials for 32nm and beyond," Proc. of SPIE Adv. Litho., Vol. 8328, 2012.

[3] Y. Loquet, et al., "56 nm pitch Cu dual-damascene interconnects with self-aligned via using negative-tone development Lithography-Etch-Lithography-Etch patterning scheme," Microelectronic Engineering, Vol. 107, October, 2013, pp. 138-144.

[4] E. Hudson, et al., "Vacuum ultraviolet plasma emission in a dielectric etch reactor," 31st International Symposium on Dry Process (DPS), 2009.

[5] B. Zhao, IC interconnect technology-challenges and opportunities. In: International Conference on Solid-State and Integrated Circuit Technology Proceedings (ICSICT2001), 2001:337

A STUDY OF SILICON ETCH PROCESS IN MEMORY PROCESS

Rong-Yao Chang, Yi-Ying Zhang, Hai-Yang Zhang*

Technology R&D, Semiconductor Manufacturing International Corp.

Pudong New Area, Shanghai, P. R. China 201203

*Steven_Z@smics.com

ABSTRACT

With the shrinkage of pattern CD (Critical Dimension), pattern collapse, micro-loading effect and silicon to silicon dioxide selectivity become more challenging in STI (Shallow Trench Isolation) patterning. Pattern collapse is closely related to micro-loading effect. To enhance Si to SiO_2 selectivity and suppress micro-loading effect, bias RF pulsing and cycle etch are used [1]. In this paper, the influence of space CD difference on micro-loading effect and bias RF pulsing function in silicon etch process is discussed.

INTRODUCTION

The accelerating scaling down of the flash memory leads to the rapid shrinkage of pattern CD [1]. As for STI, SADP (Self-Align-Double-Patterning) and SAQP (Self-Align-Quadruple-Patterning) techniques are utilized for 20nm and sub 20nm node. The AR (Aspect Ratio) of STI structure is much higher in such process, which may bring several problems.

Micro-loading effect is the depth difference of STI within the area of micron scale. This phenomenon may come from the CD difference or local non-uniformity in patterning. According to T. Matsushita, the deposition removal with specific etching chemical in cycle etch process could suppress the micro-loading effect [1]. It would be interesting to quantify the influence of CD difference on the micro-loading effect. Pattern collapse, as shown in Figure 1, refers to the adjacent line collapse to each other, especially after wet treatment, which is closely related to the micro-loading effect. The thin line pattern also requests higher Si to SiO_2 selectivity. In the experiments, the effects of selected parameters are discussed.

EXPERIMENTAL

The schematic view of the sample preparation process is shown in Figure 2. The line CD was determined by the thickness of spacer film. The adjacent spaces' CDs were determined by the core CD and the thickness of spacer film. The commercial Lam chamber was applied in this experiment. The chemicals used were CH_xF_y, HBr, O_2, Ar, He, N_2 and Cl_2 based gas. The depth was checked through cross-section analysis with SEM and TEM.

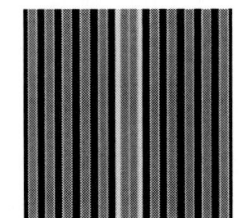

Fig. 1. Line collapse sketch

RESULTS AND DISCUSSION

1. Micro-loading effect study

In the cycle etch process, the STI depth seems to have a correlation with the width of incoming space CD. During pattern transfer with SADP process, the adjacent spaces' CDs are determined by core CD and core space CD respectively. The spaces' CDs difference post spacer etching shown in Figure 1(b) is inevitable and even enlarged during pattern transfer, which could lead to micro-loading effect during silicon etch process. From T. Matsushita's point of view, the deposition in etch process may contribute to this effect [1]. According to E. Pargon, less neutral flux and ion flux could reach the deep trench as the space CD shrinks, which may lead to the decrease of etch depth [2]. Therefore, to achieve acceptable STI structure, the adjacent spaces' CDs difference should be controlled within certain range.

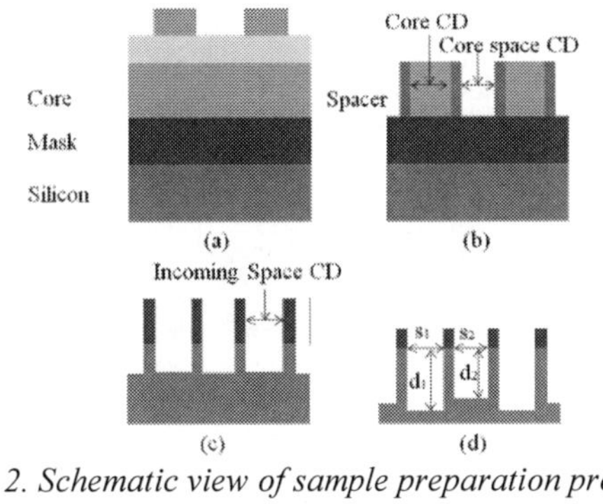

Fig. 2. Schematic view of sample preparation process

As shown in Figure 2, $|s_1-s_2|$ is defined as the adjacent spaces' CDs difference. $|s_1-s_2|$ is not the incoming space difference, but $|s_1-s_2|$ could reflect it to a certain extent. Micro-loading effect is evaluated by $|d_1-d_2|$.

978-1-5090-6695-7/17 $31.00 © 2017 IEEE   173

Through the control of core CD and spacer deposition process, a series of $|s_1\text{-}s_2|$ and corresponding $|d_1\text{-}d_2|$ could be obtained after etch. In Figure 3, the correlation between adjacent STI spaces' CDs difference $|s_1\text{-}s_2|$ and the corresponding Si depth loading $|d_1\text{-}d_2|$ is shown. In this figure, it can be seen that the depth loading increases by approximate 43A with 1nm increase in adjacent spaces' CDs difference. Based on this trend, the micro-loading effect could be fully suppressed if there is no difference between the adjacent spaces' CDs. The optimization of SADP process has the potential to alleviate the micro loading effect in STI.

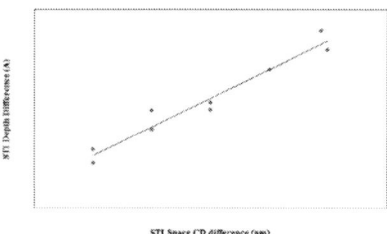

Fig. 3. STI depth loading measured by SEM as a function of adjacent STI space CD

2. Bias power pulsing study

Power pulsing refers to the successive turning ON and OFF alternation of the power, whereas all the other parameters are kept constant [3]. Through power pulsing, the modification of neutral species concentration and charged particles concentration could be achieved. According to literature, the power pulsing has the potential to achieve better uniformity, higher selectivity and less plasma induced damages [2,3]. Power pulsing can be generally divided into three groups, bias power pulsing, source power pulsing and synchronized pulsing. In this section, only bias power pulsing is studied.

In bias power pulsing plasma, the neutral species and charged particles concentration is relatively stable compared to the other two groups. In RF ON and OFF period, ions energy could have a significant difference. High-energy ions are the majority when bias RF is ON (t_1) and silicon etching mainly proceeds in this period as shown in Figure 4. During RF OFF period (t_2), there may be passivation deposition onto the SiO_2 mask surface [2]. This deposition mechanism is one explanation for the improvement of selectivity between SiO_2 and Si.

Pulsing duty cycle refers to the ratio of RF ON time to total cycle time. The reduce in duty cycle means less RF ON time and less time in which ions are at high energy level as well. Therefore, it is expected that the etch rate would decrease as duty cycle decreases. As shown in Figure 5, the blanket silicon etch rates are measured with

zero bias power, 10%, 20%, 40%, 80% duty cycle and continuous plasma. The Si etch rates are normalized by the etch rate measured in continuous plasma. The relative Si etch rate is negligible under zero bias power compared to other conditions.

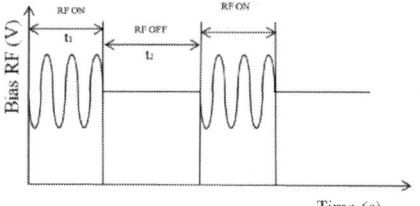

Fig. 4. Schematic view of bias power pulsing

According to M. Darnon [3], the halogen species flux during RF OFF period enhances the Si etch during the RF ON period. The time compensated etch rate [3] which equals to $(\frac{etch\ rate}{duty\ cycle})$ is introduced to reflect the etch rate during RF ON period. In Figure 5, the time compensated etch rate increases linearly with the decrease of duty cycle. There is a sharp increase when duty cycle decreases from 20% to 10%.

The Si etch rates for STI pattern in 15%, 25% and 40% duty cycle pulsing plasma are shown in Figure 6. The etch rate is normalized by the etch rate measured in 40% duty cycle. The results tell that for 10% numerical value decrease in duty cycle, there will be about 10% decrease in Si etch rate within the range of 15%~40%. The compensation etch rate increases with the drop of duty cycle. The trend seems to be similar to blanket silicon.

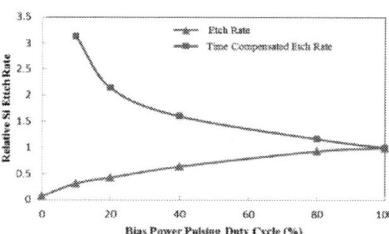

Fig. 5. Etch rate and time compensated etch rate as a function of bias power pulsing duty cycle in blanket silicon wafer

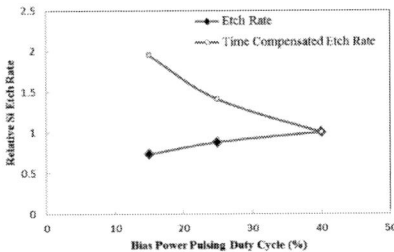

Fig. 6. Etch rate and time compensated etch rate as a function of bias power pulsing duty cycle in STI pattern

CONCLUSION

In STI etch process, the control of SADP process has the potential to reduce the pattern collapse and the micro-loading effect. The relation of micro-loading effect and adjacent spaces CD is quantized. The control of SADP process has the potential to eliminate the micro-loading effect. During STI patterning, the etch rate decreases with the increase of duty cycle. Meanwhile, the time compensated etch rate increases. The optimized duty cycle could be utilized for process improvement.

ACKNOWLEDGEMENTS

The authors would like to greatly appreciate the team members in the Etch department of Technology Development and Research Center in SMIC for their continuous help and useful discussion.

REFERENCES

[1] T. Matsushita, T.Matsumoto, H. Mukai, S. Kyoh and K. Hashimoto. "Dry etch challenges for CD shrinkage in memory process": Proc. SPIE Vol. 9428 (2015).

[2] E. Pargon, M. Dranon, O. Joubert, T. Chevolleau, L. Vallier, L. Mollard and T. Lill. "Towards a controlled patterning of 10nm silicon gates in high density plasmas": J. Vac. Sci. Technol. B, Vol. 23, No. 5 (2005).

[3] M. Darnon, G. Cunge, C. Petit-Etienne, M. Haass, P. Bodart, M. Brihoum, R. Blanc, E. Despiau-Pujo, S. Banna and O. Joubert. "Pulsed plasmas for etching in microelectronics": PESM (2014)

SIN REMOVAL PROCESS FOR POLY DAMAGE CONTROL IN MEMORY FLASH

*Jia Ren[1], Haiyang Zhang[1], Yiying Zhang[1] ***

[1] Technology R&D, Semiconductor Manufacturing International Corp.
Pudong New Area, Shanghai, P. R. China 201203
*Corresponding Author's Email: *Elaine_Zhang@smics.com

ABSTRACT

As the NAND Flash dimension scales down, many scaling problems were caused, such as the poly damage issue during cell STI recess process and SiN remove process. The problems include the cell poly loss uniformity due to the smaller CD and Peri area poly damage issue after cell SiN remove process. Further scaling down beyond 2x generation, it is also facing new scaling limitations, more severe CG (control gate) poly-Si filling problem between FGs (floating gate). We utilized an alternative etching process to the current plasma dry etch. The process is based on the film modification by light ions implantation followed by a selective removal of the modified layer. In this work, we demonstrated the process can remove silicon nitride without poly damage. Meanwhile, it can achieve the FG slimming profile, which benefits for the void-free filling of CG poly-Si.

INTRODUCTION

With the constant scaling down in dimensions, plasma etching requirements become more and more stringent. The problem is using conventional plasma etching, high energy ions are key in controlling etch anisotropy, but they also jeopardize the etch precision required when layers are involved in the stack to pattern by transferring the reactive layer formed on the top of the etched film into the under-layer, damaging the film [1]. Meanwhile, while the CD scaling down to beyond 2x generation, the uniformity would be more sensitive to the etch environment, as showed in Figure 1. Figure 2 shows the traditional Si_3N_4 film removal flow in flash memory, and cross-section (Figure 2d) of the cell in the X-axis direction shows void for a non-optimized CG poly filling process, which can lead to the reduction in program (PGM) speed due to CG poly depletion between FG and FG. It is reported that the floating gate slimming process can achieve the void-free filling of CG poly-Si and wider active area CD which can obtain large cell current. An electrical depletion in CG poly-Si is greatly suppressed by this void-free process. As a result, BL interference is successfully improved 20% compared with conventional process [2].

To overcome the issues discussed above, we utilized light ions implantation to modify remaining Si_3N_4 after cell STI oxide recess process. The modified Si_3N_4 film can be selectively removed by HF solution, which matched with the process flow of NAND flash memory well, as shown in Figure 3. Using this method, traditional Si_3N_4 remove process by hot H_3PO_4 can be replaced and no other process was introduced. Meanwhile, we can control the poly uniformity well and achieve the FG slimming profile, which is beneficial for device performance. This method has been utilized in nitride spacer etch stopping on silicon germanium with silicon germanium film consumption less than 6A [3].

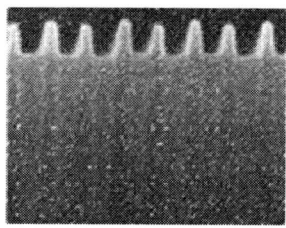

Figure 1: The poly damage issue: (a) X-SEM image of cell FG poly damage after STI recess

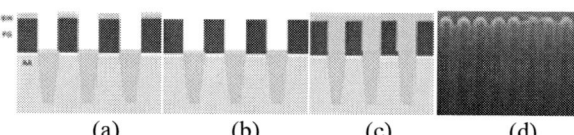

(a)   (b)   (c)   (d)

Figure 2: Cross-section of traditional Si_3N_4 removal flow: (a) after cell STI oxide recess; (b) SiN removal by H_3PO_4; (C) ONO deposition and cap CG poly; (d) X-SEM image after CG cap poly along the WL direction.

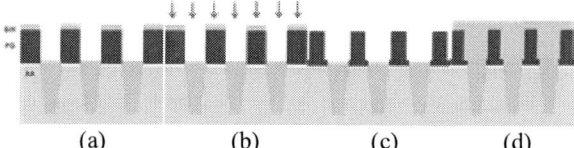

(a)   (b)   (c)   (d)

Figure 3: Cross-section of proposed Si_3N_4 removal process: (a) after cell STI oxide recess; (b) light ions implantation; (c) selective removal of modified Si_3N_4 layer; (d) ONO deposition and cap CG poly.

EXPERIMENTS

In this context, about 10nm remaining silicon nitride film after cell STI oxide recess process needs to be modified and removed. We evaluated the Si_3N_4 film (with and without special gas plasma exposure) behavior in capacitive coupled plasma (CCP). And special gas treatment experiments on control wafer and pattern wafers were carried out. The impacts of power and treatment time on the thickness of modified Si_3N_4 were studied.

The silicon nitride film etching traditionally used hot phosphoric acid (H_3PO_4) with high operational temperature, which could lead to peri area poly damage during cell remaining Si_3N_4 remove process (Figure 4). Aqueous HF solution was chosen to remove modified Si_3N_4 film, as the following ONO film deposition pre-clean also used the diluted HF solution. 0.5% HF concentration impact has been investigated on Si_3N_4 film with and without treatment (with pressure setting at 50mTorr), and the impact of DHF dip time was also investigated on the performance of modified Si_3N_4 film remove. The difference between pre- and post-measurements, measured at 49 points using a spectroscopic ellipsometer, gives the film consumption by the HF solution.

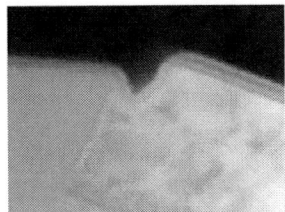

Figure 4: X-SEM image of Peri area FG poly damage after cell SiN removal process.

Before and after the modified Si_3N_4 film removal with patterned wafer, the Transmission Electron Microscopy (TEM) was used to check whether modified Si_3N_4 film can be removed by HF solution.

RESULTS AND DISCUSSIONS

The Si_3N_4 film behavior with special gas treatment in CCP chamber was studied on control wafer firstly to check the etch amount performance of modified Si_3N_4. When exposed to HF dip, the pristine Si_3N_4 film consumption is linear with the exposure time (Figure 5). In our experimental conditions, the etch rate of Si_3N_4 is estimated to ~6Å/min. After special gas treatment, two phases of modified Si_3N_4 film consumption are observed: a non-linear increase between 0 and 30 s HF dip followed by a linear Si_3N_4 film consumption with a speed similar to the pristine Si_3N_4 film exposed to HF dip. The result indicates that the modified silicon nitride by special gas treatment can be etched by an HF dip with selectivity to the pristine Si_3N_4. In this case, the modified layer is etched during the first phase, while the second phase corresponds to the remaining non modified Si_3N_4 film consumption. In the current operating condition, increasing special gas treatment time from 60s to 300s, the thickness of modified

Si_3N_4 film only increased about 20A, as shown in Figure 6. The ion penetration depth should be related with the settled power.

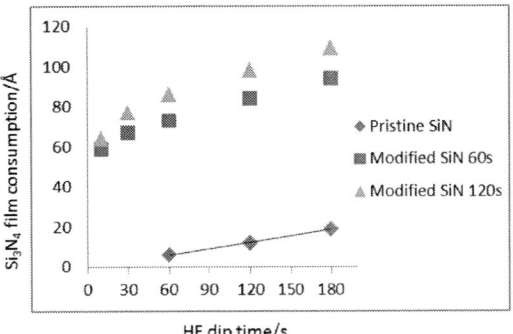

Figure 5: Evolution of the Si_3N_4 film (with and without treatment) consumption as a function of HF 0.5% dip time

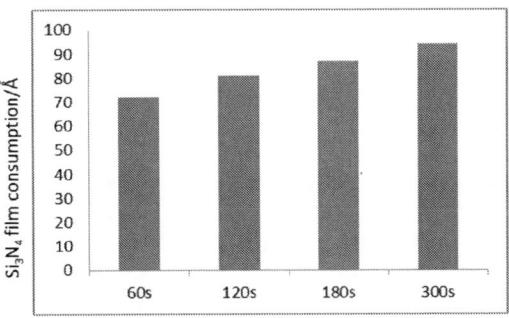

Figure 6: Evolution of the Si_3N_4 film consumption with different special gas treatment time (HF 0.5% dip 60s)

The impact of the source power and bias power on the thickness of modified Si_3N_4 film has been investigated using capacitive coupled plasma, with special gas treatment time 60s, 120s, 180s, and 300s under each condition respectively. As shown in Figure 7, bias power has obvious impact on the thickness of modified film compared with increasing source power. The thickness of modified film increased by ~30% by increasing bias power, however, the thickness only increased only several Å with increasing source power.

This method has been evaluated on patterned wafers where the remaining silicon nitride after cell STI oxide recess process that needs to be removed was about 100A (Figure 8a). The special gas treatment condition has been adjusted to target 120 Å silicon nitride film removal, corresponding to 20% over-etch. Figure 8b showed that remaining SiN can be removed by HF solution after special gas treatment, and poly has no loss, therefore poly damage not existed. Meanwhile, we got the slimming FG profile, FG top CD shrink by ~20%. This is beneficial for

978-1-5090-6695-7/17 $31.00 © 2017 IEEE

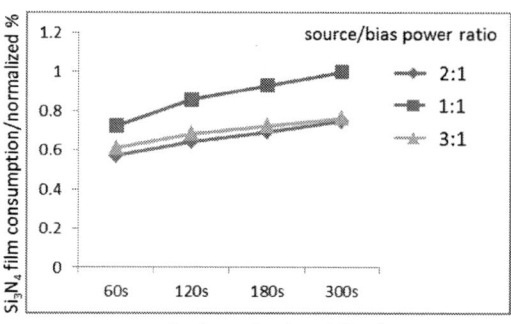

Figure 7: Evolution of the Si_3N_4 film consumption under different power conditions with special gas treatment time 60s, 120s, 180s, and 30s (HF 0.5% dip 60s)

CG poly void-free filling. Silicon nitride film modified with special gas treatment can be understood in the following two aspects. One is that the ion bombardment probably breaks the Si-N bonds enhancing the reaction with the HF dip. Besides this, exposed silicon nitride film could be oxidized, as the intensity of N-H peak of Attenuated Total Reflection infrared spectroscopy (ATR) analyses increased after special gas treatment [3].

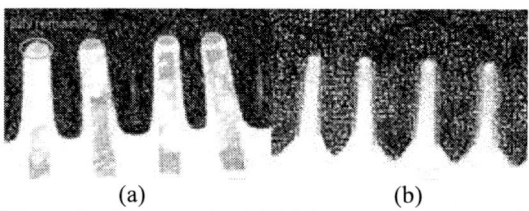

Figure 8: Cross-section TEM image with proposed SiN removal process of: (a) cell before SiN removal; (b) cell after SiN removal along the WL direction.

CONCLUSIONS

In this paper, we have demonstrated that silicon nitride film can be modified by special gas treatment, and can be selectively removed by HF solution without poly damage. Meanwhile, FG slimming profile could be achieved, thus it's beneficial for CG poly void-free filling process. Bias power had obvious impact to the modified thickness of silicon nitride film in this case. And this method could also be utilized in other thin layer etching process.

ACKNOWLEDGEMENTS

The authors would like to thank the technical discussion with the dry etch group of Tokyo Electron Limited, China, and team members of Etch department of SMIC T-R&D.

REFERENCES

[1] R. Blanc, F. Leverd, T. David, and O. Joubert, *J. Vac. Sci. Technol. B* 31(5), 051801 2013.

[2] J. Hwang, J. Seo, Y. Lee, S. Park and H. Kkang etc. *IEDM*, 2011, pp. 199-202.

[3] N. Posseme, O. Pollet, and S. Barnola, *Appl. Phys. Lett.*, vol. 105 (051605), 2014, pp. 1-4.

THE LOADING EFFECT STUDY IN METAL HARD-MASK ALL-IN-ONE ETCH WITH DOUBLE PATTERNING SCHEME

Kefang Yuan[a], Junqing Zhou[a], Zhidong Wang[a], Minda_Hu[a], Qiyang He[a], Haiyang_Zhang[a]*

*Steven_Z@smics.com

Technology R&D, SMIC Advanced Technology R&D (Shanghai) Corporation

Pudong New Area, Shanghai, P. R. China 201203
+8621 38610000x10968, Amy_Yuan@smics.com

ABSTRACT

In advanced CMOS technology node with Cu/low-K interconnection, double patterning scheme with Trench First Metal Hard-Mask (TFMHM) approach All-In-One (AIO) etch is used to define smaller scale via and trench. The loading effect between different patterns, such as via chain and via slot, will cause different final over-etch (OE) amount under etch stop layer. Chip performance will be affected if it suffers severe loading between various patterns. In this paper, we will study the effect of different film stack, etching gas, plasma condition and etch stop layer (ESL) in loading control between different patterns which will both contribute to final loading. Different methods' loading results and the basic theory for explaining the loading phenomenon will be compared and discussed in following part.

INTRODUCTION

Loading performance will affect final performance in following ways. For example severe loading will lead to tiger-tooth effect when landing metal top critical dimension (CD) is smaller than via bottom CD in region with large over-etch amount. Meanwhile, the longer exposure time of under layer Cu to plasma in via slot area during etch stop layer opening step will have bad influence on trench depth and residual defect. The smaller loading between different patterns are necessary to ensure beneficial VBD and TDDB performance.

For sub20nm technology, post lithography CD can't continue to be scaled down because of lithography limit. Therefore, appropriate MHM selectivity which can ensure via totally confined by MHM is imperative in pattern definition. Trench and via profile should also be taken into consideration during process tuning for its big influence on electrical and reliability performance too.

Except for traditional ways of changing etching main gas, pressure or power, different combination of inter-layer film stack, choice of etch stop layer (ESL) and pulsing scheme instead of CW can also be used in different via pattern loading tuning. Different methods are implemented together in order to get a better loading performance with least sacrifice of a selectivity, profile, RE performance and etc. [1]

In this paper, we will evaluate different patterns' loading

performance by three representative patterns. Via chain pattern is dense via array with minimum CD and pitch in narrow trench. The via is fully confined by MHM in directions perpendicular with trench and in line end direction- it is semi-confined which is partial decided by post lithography CD. Device Test-key represents isolation-via with in wide trench. Only one-side is confined by MHM for device TK pattern. Seal ring stands for via slot which is trench-like pattern. It is used to prevent chip from moisture and mechanical force impact when die sawing. The CDSEM top-view of these three structures is shown in Figure 1 and corresponding cross section sketch image is shown in Figure 2.

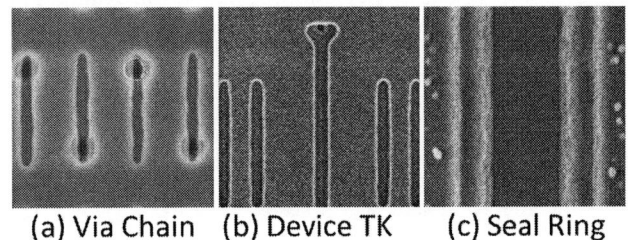

Figure 1: Top-view of (a) Via Chain (b) Device Test-key and (c) Seal Ring

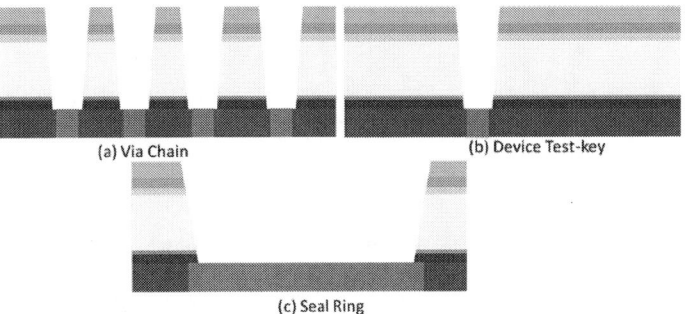

Figure 2: Cross section sketch map of (a) Via Chain (b) Device Test-key and (c) Seal Ring

Usually there exists depth and profile loading between isolated and dense pattern in AIO etch. When etch dense pattern, it will generate more polymer byproduct which will make local etch rate (ER) slower. Loading effect between different via CD is also obvious in this case. For via with large CD, plasma is rich around etching front so that the ER is faster in CD larger region. In this paper, we will try to lighten the performance mentioned above to reduce pattern loading.

EXPERIMENTS

The experiments in this paper are using TFMHM approach. The MHM etch step is performed on one commercial inductively coupled plasma (ICP) etcher. AIO etch is performed on commercial capacitively coupled plasma (CCP) etcher followed by wet clean. After barrier/seed deposition and partial Electrical Cu Plating (ECP), the wafer will be sent to cross section analysis by Transmission Electron Microscope (TEM). Three different via patterns will be observed in TEM and we will focus on comparing the maximum depth loading between these three patterns. [2]

RESULTS AND DISCUSSIONS

Etching Gas and Plasma Condition

Modifications of power, pressure and gas ratio are traditional ways to adjust different patterns' loading. By decreasing HF power and increasing LF power, adjusting the ratio of two main CxFy etch gases, increasing two types of diluted gas in via etch recipe, we can reverse the loading trend of via chain and seal ring pattern. During the etch recipe tuning history, we find that decrease of plasma density, more polymer and more bombardment are both beneficial methods to balance via chain and seal ring loading. By adjusting plasma gas and condition, we can control pattern to pattern loading to a very small extent. As shown in Figure 3, loading performance can be improved from 67.9% to 15.0% by changing etch recipe. In etch recipe 2, loading between via chain and seal ring are decreased to 8.4%.

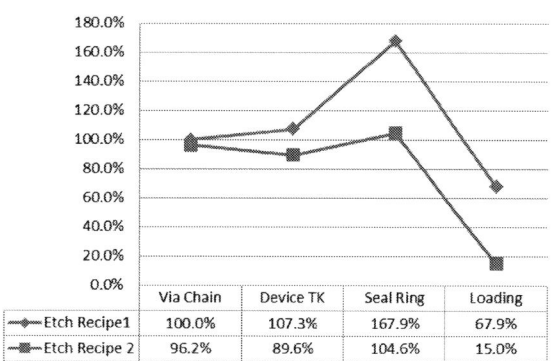

	Via Chain	Device TK	Seal Ring	Loading
Etch Recipe 1	100.0%	107.3%	167.9%	67.9%
Etch Recipe 2	96.2%	89.6%	104.6%	15.0%

Figure 3: Loading performance after traditional modification

Different Inter-layer Film Material

In this experiment, we use two different inter-layer material between MHM and Ultra-Low K (ULK). The layer we change is marked in film stack by taper black line shown as Figure 4. We use same etch recipe to etch through two wafers with different material. We can see the loading trend between via chain and device test-key is even changed after the alternation of the inter-layer material. Material 1 shows better loading performance between via chain and seal ring as shown in Figure 5.

In AIO etch process, we first break through inter-layers under HM and then plasma will etch lower ULK. The ER of ULK is faster than inter-layer. A strong inter-layer can delay

plasma a little before ULK starts to be etched which will show benefit in final loading performance

Figure 4: The alternative layer in experiment

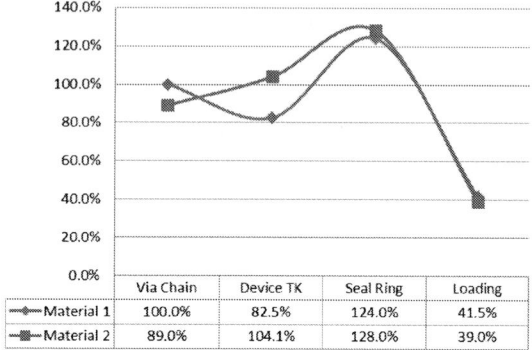

	Via Chain	Device TK	Seal Ring	Loading
Material 1	100.0%	82.5%	124.0%	41.5%
Material 2	89.0%	104.1%	128.0%	39.0%

Figure 5: The alternative layer's influence in loading tuning

Different Etch Stop Layer (ESL)

ESL can provide high selectivity to oxide in trench etch steps so that different kind of via can be landed on it and then we use a Liner Removal (LRM) recipe with low selectivity to open ESL. This is a traditional method to reduce loading to minimum. The powerful ESL can cover loading problem induced by previous etch step. In this experiment, we use two different ESLs. For ESL2, it has around 10 times higher selectivity than ESL1 using our LRM etch recipe body. High selectivity ESL means thinner layer to achieve stop layer capability which will be helpful for via bottom profile continuity.

From Figure 6, we can see the loading performance is 10% improved by different ESL using the same etch body. This ESL can have even better performance after we adjust the etch recipe body more appropriate for its property.

	Via Chain	Device TK	Seal Ring	Loading
ESL 1	100.0%	107.3%	167.9%	67.9%
ESL 2	77.0%	85.7%	134.5%	57.5%

Figure 6: Loading performance in different ESL

Pulsing Plasma Scheme

Pulsed plasma has been used in industry for several years. In this method, we can change two extra parameters rather than

978-1-5090-6695-7/17 $31.00 © 2017 IEEE

continuous plasma which are pulse frequency and pulse duty cycle. The extra tuning knobs give us more chance to optimize the plasma properties such as ion/electron densities, electron temperature, ion/neutral flux ratio, plasma dissociation, plasma potential and so on. [3]

We use a synchronous pulsing approach in order to decrease seal ring depth as well as balance pattern loading. By using pulsing scheme, we can even make seal ring depth less than via chain as shown in Figure 7. After optimization, pulsing plasma can surely be a quite powerful scheme to control loading of different patterns for its flexibility in two extra parameters mentioned before.

Meanwhile, Pulsing also shows more vertical profile which will be good for lowering via contact resistance. Via etching to MHM selectivity can be higher by using pulsing plasma which will contribute to better final reliability result.

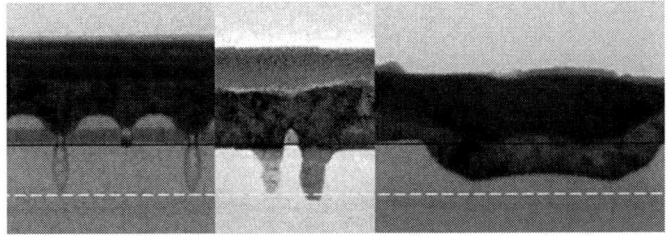

Figure 7: Post Synchronous Pulsing TEM Result

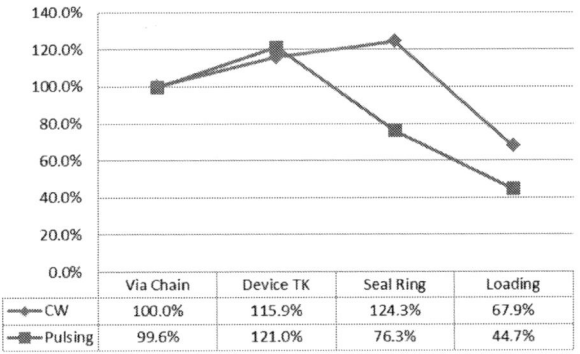

	Via Chain	Device TK	Seal Ring	Loading
CW	100.0%	115.9%	124.3%	67.9%
Pulsing	99.6%	121.0%	76.3%	44.7%

Figure 8: Comparison between pulsing and CW scheme

CONCLUSIONS

In sub20nm technology node and beyond, the partial via etch depth loading between different patterns is a critical problem in BEOL. In this paper, we demonstrate four kinds of methods we can do for better depth loading control between three different patterns. Besides the traditional ways to balance loading, we have more choice of inter-dielectric layer, ESL and even scheme of etch recipe. All sessions are important for final loading performance. Meanwhile, it is also important to keep high MIIM selectivity and acceptable via bottom CD in non-SAV direction to make sure final TDDB performance on target during the process tuning of depth loading.

All of the methods showed in this paper show significant in loading tuning. After more optimization and combination of the mentioned method, we can achieve ideal loading between different patterns within wafer.

ACKNOWLEDGEMENTS

I would like to thank SMIC TD etch group and PIE group for providing me resource and some guidance ideas. Also thank for technical support from TEL staff.

REFERENCES

[1] Jun-qing Zhou, Min-Da Hu, Qi-Yang He, Hai-Yang Zhang, et al, "A Study of Self-Aligned-Via based All-in-one Etch", *CSTIC*, 2015, pp1-3

[2] Junqing Zhou, Minda_Hu, Haiyang_Zhang "The Impact of Metal Hard-mask AIO Etch on BEOL Electrical Performance", *CSTIC*, 2016

[3] Samer Banna and Ankur Agarwal, et al, "Pulsed high-density plasmas for advanced dry etching processes", 《*Journal of Vacuum Science & Technology A*》, 2012, 30(4):040801 - 040801-29

SADP CORE ETCHING PERFORMANCE COMPARISON FOR DIFFERENT CCP ETCHERS

Yibin Song[1], Haiyang Zhang[1] and Yiying Zhang[1]*

[1]Semiconductor Manufacturing International Corporation, No.18 Zhang Jiang Rd., Pudong New Area, Shanghai 201203, China

*Corresponding Author's Email: Elaine_Zhang@smics.com

ABSTRACT

As technology developed, semiconductor moves to 10nm and the beyond. Since pitch becomes tighten and tighten, double patterning and triple patterning attracted researchers' attention [1]. Self-aligned double patterning (SADP) is one of the most popular double patterning methods. The primary idea of SADP is the core and spacer fabrication [2,3]. Herein, three commercial CCP etchers are employed for core etching tuning. Results show that both temperature and pressure are propitious to CD and beneficial to profile control.

INTRODUCTION

In the era of 21 century, the art of semiconductor manufacture come into multiple pattering era. As the most widely employed double patterning technics, SADP becomes main stream in the technology node of 40nm and the beyond. Both core and spacer fabrications are significant during SADP. Besides, core definition is more critical since they can effect spacer definition. Therefore both CD and profile of core become critical. After core pattern lithography, for optimal result acquired, several parameters need to develop during core etching process, including chemistry, pressure and temperature.

EXPERIMENT

Ordinary SADP fabrication process was implemented in this experiment. Core patterns are fabricated by using try-layer lithography, followed by spacer deposition and spacer etching process. For core etching tuning, three commercial capacitively coupled plasma (CCP) etchers are investigated. Several parameters are surveyed, such as pressure, power, temperature etc.

RESULTS AND DISCUSSIONS

1. The Effect of Time for Barc Open

As one of the pattern transfer layers, barc open is extremely significant for CD definition and profile control. In this study, results show that etching time is a crucial parameter. As shown in Fig. 1(a) and (b), vertical profile is attained by shortly etching time, which can get vertical core profile finally (Fig 1. (c)). On the contrary, tapered profile is attained by longer etching time, which caused tapered core profile (Fig 1. (d)). During shorter etching time, profile can be transferred from PR to barc perfectly. As time goes on, barc etched too much, which is bad for

pattern transfer. Therefore etching time control is one of the key factors during barc open process.

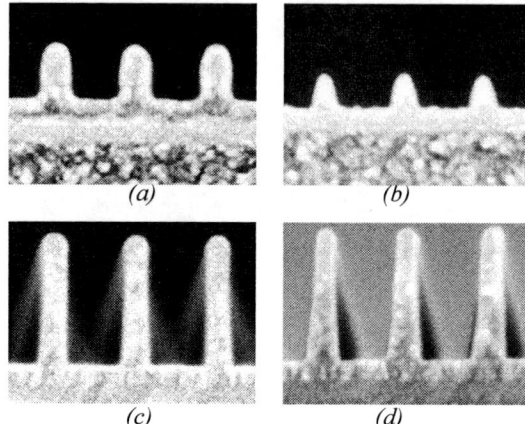

Figure 1 The effect of etching time (a) shorter time barc open (b) longer time barc open (c) full etch by using shorter time barc open; (d) full etch by using longer time barc open

2. The Effect of Temperature for Barc Open

Nearly all the etching processes are polymer dynamic procedures. Temperature is a critical parameter during polymer reaction. Fig.2 (a) illustrates the profile control by temperature. Bowing profile is attained during lower temperature. However higher temperature is apt to get vertical profile. Higher temperature is helpful for byproduct removal.

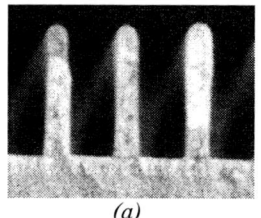 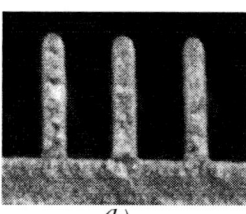

Figure 2 The effect of temperature for barc open (a) lower temperature; (b) higher temperature

3. The Effect of Chemistry for Core Etch

Since core is organic material, so the etching chemistry selection is significant. In this study, 4 kinds of chemistry mixture are investigated, CO_2, CO_2/CO,

CO_2/N_2 and N_2/H_2 respectively. As shown in Fig. 4, whether CO_2 only or CO_2 based mixture, bowing profile is acquired. Nevertheless vertical profile is attained when N_2/H_2 is employed. For this case, we believe CO_2 based mixture is polymer rich chemistry, which is not good for organic core etching process. However, N_2/H_2 mixture is poor polymer chemistry, which is helpful for organic core etching.

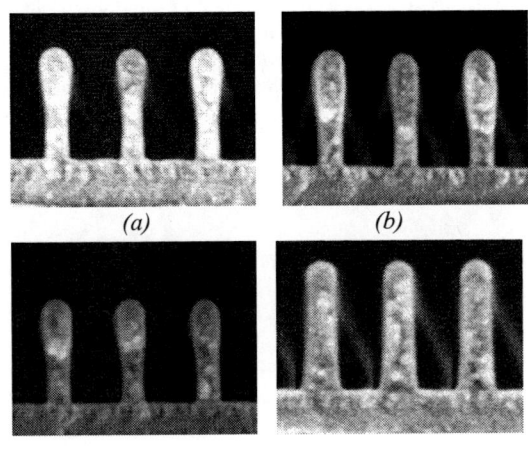

(a)   *(b)*

(c)   *(d)*

Figure 3: The effect of chemistry for core etch (a)CO_2; (b) CO_2+CO; (c)CO_2+N_2; (d)N_2+H_2

4. The Effect of Pressure for Core Etch

The pressure effect of core etch result is shown in Fig 4. As shown in Fig. 4, lower pressure is apt to vertical profile. There are many kinds of byproduct during etching process. Lower pressure is beneficial to byproduct removal. Furthermore, lower pressure is also inclined to get smaller CD since the byproduct remains are carried off within lower pressure environment.

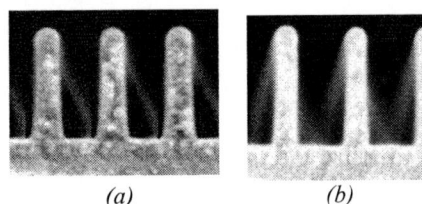

(a)   *(b)*

Figure 4: The effect of pressure for NFC open (a) higher pressure (b) lower pressure

CONCLUSIONS

The SADP core etching process is investigated by three commercial CCP etchers. The critical 2 etch steps, including barc and core etch, are studied. Several key parameters, such as etching time, temperatures, chemistry mixture and pressure, are investigated. Proper etching time during barc open is helpful for final core profile

control. Higher temperature during barc open process is apt to get vertical profile and smaller CD. For core etching process, CO_2 only will cause bowing profile and remarkable mushroom head. CO or N_2 added can improve bowing profile and mushroom head. However the improvement is limited. N_2 and H_2 mixture is inclined to vertical profile without mushroom head.

ACKNOWLEDGEMENTS

The authors would like to thank Mr. Zhou Mo from TEL and Mr. Jiagao Wang from Lam Research for the technical discussion. They also wish to acknowledge the support of Failure Analysis Lab of SMIC in providing TEM images.

REFERENCES

[1] A. Hara et. al., Advanced self-aligned DP process development for 22nm node and beyond, Proc. Of SPIE 7639-79 (2010)

[2] H. Zhang, Y. Du, M. D. F. Wong, and R. O. Topaloglu, Self-aligned double patterning decomposition for overlay minimization and hot spot detection, Proc. DAC, pp. 71-76, 2011.

[3] S. Y. Chang, Y. C. Chen, A. C. Wei et al., Advanced floating gate CD uniformity control in the 75nm node NOR flash memory, IEEE/SEMI ASMC, 2011

BARC OPEN IN FINFET TECHNOLOGY NODE

Long-Juan Tang, Qiu-Hua Han, Hai-Yang Zhang*

Technology R&D, Semiconductor Manufacturing International Corp.

Pudong New Area, Shanghai, P. R. China

*Steven_Z@smics.com

ABSTRACT

Single PR (photoresist) photo works well in implantation process for planar transistors, however, it suffers severe PR residue issue in FinFET technology node due to the three dimensional fin/gate structure and large wafer surface topography. New integration scheme with BARC (Bottom Anti-Reflective Coating) coating/etching was developed to solve this problem. In this work, BARC etching process was compared on two different types of commercial etcher. Key factors which influenced the BARC etching uniformity, remaining PR thickness and BARC profile were discussed. Finally, we delivered a BARC etching process with desired physical profile and improved device performance.

INTRODUCTION

PR serves as the pattern-defining layer and implantation mask in planar transistor process. However, when transistors continuously scale down and the structures transform to three dimension, lithography process encounters new challenges. Because of the higher gate and smaller pitch in FinFET technology which affect the exposure of PR, severe PR residue is found at the bottom of the space. It will block ion implantation and lead to device degradation. PR /BARC scheme with BARC coating/etching is adopted to solve this problem. BARC fills the gap and provides a relatively flat surface, which makes light much easier to enter PR film. In this paper, BARC etching process is studied to obtain an expected PR/BARC profile and good etching uniformity.

EXPERIMENTS

All the dry etching experiments were performed in two different types of 300mm commercial etcher. One is CCP (Capacitively Coupled Plasma) etcher equipped with adjustable cathode gap and switchable RF power configuration, called Etcher A. The other is ICP (Inductively Coupled Plasma) etcher quipped with tunable ESC (Electrostatic Chuck) and TCCT (Transformer Coupled Capacitive Tuning) designs, called Etcher B. PR and BARC were coated on top of HM (Hard Mask) and poly-silicon gate. Top view images were collected via Hitachi CD-SEM (Critical Dimension-Scanning Electron Microscopy). Cross section pictures of PR/BARC were taken using SEM (Scanning Electron Microscope).

RESULTS AND DISCUSSIONS

Figure 1 shows the schematic flow of BARC etching process. In order to get good device performance, we need to consider all kinds of cross-linked aspects of the etching process, such as etch rate uniformity, CD bias, remaining PR thickness, effective height (the total thickness of BARC and PR shoulder thickness), BARC profile and fin damage. Process parameters must be carefully specified in order to satisfy these requirements simultaneously.

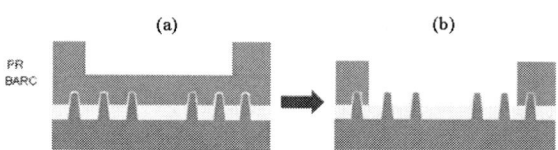

Figure 1: Schematic flow of BARC etching process (a) before etching; (b) after etching.

Etch Rate Uniformity

BARC etch rate uniformity is a very important check item in this process. It will result in BARC residue or severe undercutting unless uniformity is good enough. DOE (Design of Experiment) was carried out for BARC etch rate uniformity improvement. Gap, pressure and gas ratio are proved to be the significant parameters to affect etch rate uniformity, and their impacts are symmetrical. On both tools, an uniformity as low as 2% can be achieved by an optimized recipe. The normalized BARC etch rate maps are shown in Figure 2.

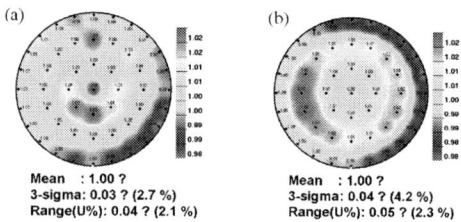

Mean : 1.00 ?
3-sigma: 0.03 ? (2.7 %)
Range(U%): 0.04 ? (2.1 %)

Mean : 1.00 ?
3-sigma: 0.04 ? (4.2 %)
Range(U%): 0.05 ? (2.3 %)

Figure 2: BARC etch rate (a) Etcher A; (b) Etcher B.

Effective Height

Effective height, measured from the top of PR shoulder to the bottom of BARC, is considered to be the actual PR/BARC thickness that obstructs the ion implantation. It cannot be too high, otherwise shadowing effect will occur under off-axial implantation. It cannot be

978-1-5090-6695-7/17 $31.00 © 2017 IEEE

too low either, in which condition PR will be insufficient to protect the covered fin. Since BARC thickness is defined by BARC coating process, efforts are taken to gain more PR remains.

In Etcher A, high pressure, low gap or additional polymer gas will be helpful to gain PR. High pressure is unexpected for BARC profile control, thus, low gap has been taken into consideration. In this process, a H-contained gas and a N-contained gas are used for BARC etching. Etching gas will be ionized after glow discharge, H and N radicals are generated simultaneously. H radical works as an etchant to etch PR/BARC while N radical forms a passivation layer to protect the PR sidewall. This passivation layer could not be etched without ion bombardment. As illustrated in Figure 3, in low gap condition, higher density of N radical is produced, while ion assisted etching is weaker compared to high gap condition for its lower voltage drop across the sheath. More PR remains is obtained by lowering the gap. PR/BARC profile of Etcher A is shown in Figure 4(a).

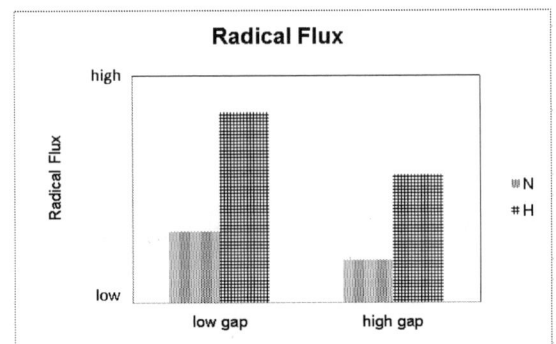

Figure 3: Radical flux in low gap and high gap conditions of Etcher A.

Due to the intrinsic difference of tool configuration and the way of plasma generation [1], it's much easier for Etcher B to get sufficient remaining PR thickness and effective height. The PR shoulder thickness of Etcher B is almost 2 times thicker than Etcher A, as shown in Figure 4(b), which is beyond our expectation. Due to the shadowing effect concern, thinner PR can be used for Etcher B, which can widen lithography process window at the same time. Our recent study shows with the addition of some special gas, the remaining PR thickness of Etcher A can be increased tremendously and become comparable to Etcher B.

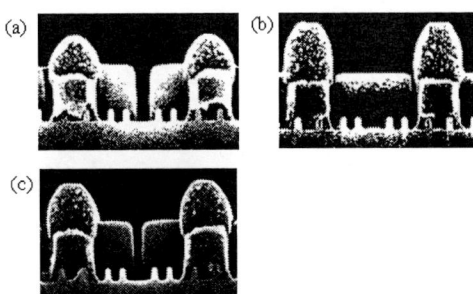

Figure 4: PR/BARC profile after etching (a) Etcher A; (b) Etcher B; (c) Etcher A with the addition of special gas.

BARC Profile

Vertical BARC profile is required for better overall process control, but there seems to be unavoidable undercutting in this process. BARC/PR selectivity may be the root cause of it. For both Etchers, BARC etch rate is much faster than PR, which causes the undercutting profile. We attempted to use heavier passivation to protect the sidewall, however, this passivation layer also deposited on the sidewall of PR, resulting in an expanded PR profile and leading to CD shift. High power and low pressure are effective to alleviate the undercutting profile in Etch A, while they will lower the effective height at the same time. Tradeoff must be taken into consideration. In Etcher B, medium pressure and high total flow are employed to reduce the undercutting effect. Finally, we can get comparable BARC profile from both tools.

Fin Damage

As shown in Figure 1, a thin layer of offset spacer is capped on the fin. In order to avoid fin damage during plasma etching, we need to minimize spacer loss. Generally speaking, the bombardment of CCP tool is higher than ICP tool. In our study, high frequency bias power was applied on Etcher A to minimize the plasma damage. No spacer loss is observed from the TEM images of both etchers.

Device Performance

PR/BARC scheme delivered much better device performance than single PR scheme, as demonstrated in Figure 5. Ion-Ioff curve shows around 10% Idsat improvement.

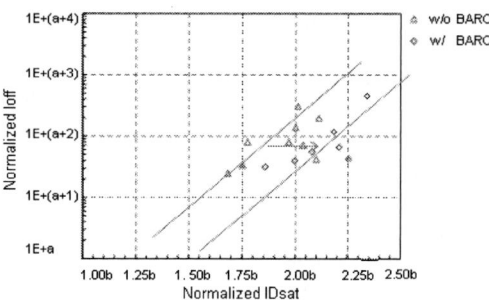

Figure 5: Ion-Ioff of Etcher A.

CONCLUSIONS

In this paper, BARC etching process along with its key process requirements is studied on commercial CCP and ICP ethers. Good etch rate uniformity, sufficient effective height and comparable BARC profile are obtained on both tools. We delivered a BARC etching process with expected physical profile and improved device performance.

ACKNOWLEDGEMENTS

The authors would like to gratefully thank Mr. Song Huang, Ms. Judy Wang and Mr. Jiagao Wang for the constructive discussion and information during the experiments, Mr. Xiaopeng Yu from the integration team of SMIC Technology R&D center for wafer preparation and result analysis.

REFERENCES

[1] V. M. Donnelly, and A. Kornblit, *J. Vac. Sci. Technol. A* 31(5), Sep/Oct 2013"

CAVITY PROFILE CONTROL IN DRIE PROCESS

Wang Jing, Nie Miao, Jiang Zhongwei, Huang Yahui

North Microelectronics CO., Ltd No.8 Wenchang Road, Beijing, 100176, P.R. China

WangJ@bj-nmc.cn

ABSTRACT

Deep Reactive Ion Etching (DRIE) has revolutionized a wide variety of advanced package applications. Cavity etch process is an important step for fan-out wafer level package (WLP), which general fabrication by DRIE. In this paper, we investigated the influence of process parameter on the profile and etch rate in square-hole cavity etch. Sidewall angle was controlled by fluorine isotropic etch. So the sidewall angle was increased with the etch rate, which can be increased by raise source and bias power. It was shown that bias power drastically impact on sidewall angle in our study. High etch rate with optimized profile were obtained by controlling the plasma density and ions bombardment energy independently in two steps. Vertical profile was obtained when auxiliary gas was used in the Si main etching step. Based on the above learning, a cavity etch process be optimized. Both good profile and high etch rate were obtained.

INTRODUCTION

In the fan-out wafer level packaging process, cavity etching is an important step. The cavity is the chip carrier, so need to get to the cavity side wall etching vertical using of BOSCH technology, and the etching rate as high as possible. Generally, for BOSCH process, it has been shown that, fluorocarbon-based passivation has some benefits to control the profile of side wall. High etch rate provide a greater sidewall angle, only by reducing the etching rate to get better sidewall morphology. By properly adjusting the etching and deposition steps and controlling the pressure and power, the morphology of the BOSCH process can be changed to obtain a suitable cavity etching process.

EXPERIMENTAL

The experiments were performed in a 300 mm inductively coupled plasma reactor, NMC APE300 Series.

First, blanket tests were performed on Si wafers. The etch rate and etch uniformity of the substrate material were obtained using thickness meter. To minimize the etch variance, the process steps were optimized by tuning etching time ratio, plasma gradient, and gas composition. Best recipes were chosen to perform on the patterned wafer.

The study was performed to measure rectangular hole width 2350um/2500um. The patterning stack was composed of the following layers: Resist/Si. During the etching process of the patterned wafer test, side wall angle was studied through SEM analysis. Etch rate was calculate by recipe time and depth.

RESULTS AND DISCUSSION

It is often observed that the side wall angle of the BOSCH process decreases with the etch depth [1]. This phenomenon was observed after Si etching over several decades of microns. But in the cavity etching process, with the increase of etching depth, sidewall angle will increase with depth, as shown in Fig. 1(a). This makes it very difficult to control the sidewall angle while obtaining high etch rates. For the 100um target for the cavity etching, if the etch rate is controlled around 5um/min, the sidewall angle will reach 93 degrees, for the angle of 200um will continue to increase to the angle of 95 degrees.

In order to study the control of sidewall morphology, the effect of source power and bias power was investigated separately. It was shown that if the lower source electrode power is used, the sidewall angle will not increase with the increase of the etching depth as shown in Fig. 1(b), but the etching rate will be relatively slow. By increasing the source electrode power, the etching rate increases, but the sidewall angle will gradually increase, thus limiting the etching time step, thereby limiting the higher etching rate. The reason why the sidewall angle is difficult to control is that with the increase of cavity etching depth, the sidewall polymer is more likely to be stripped by ions, resulting in an angle increase.

Figure 1: SEMs of cavity sample with different recipe: angle increase with the depth(a). An example of low source power (b). An example of two step bombardment process (c), and an effect of adding argon (d).

In order to effectively guarantee the sidewall profile under the condition of etching rate on the optimization, the polymer on the side wall should be control. By selecting two different etching steps, increase the use of different

---

978-1-5090-6695-7/17 $31.00 © 2017 IEEE

bombardment process to control the process to remove the side wall of polymer. The side wall angle can be controlled effectively. Using this approach, keep the higher power of source power, the bias power of polymer etching is increased and the bias power of Si etching is reduced. It was shown that the bias power is very sensitive to the removal effect of polymer, the controllability of sidewall angle is greatly improved. In particular, when one bias power is used at 0, the etch rate can be increased to 7um/min, and side wall angle can be controlled within 92 degrees, as shown in Fig.1(c).

In addition, an auxiliary process gas is added to the etching step to adjust the chemical etching process of the Si. This process may have a similar effect of the control of bias, making it possible to control the side wall polymer. It is found that the etching rate of sidewall polymer is significantly reduced when the amount of argon is mixed into the etching gas, which has a great influence on the sidewall angle, as shown in Fig.1(d). Argon is chemically inactive and not only decrease the etch rate of polymer at the base of the trench, but also promotes the anisotropic removal of the polymer by promoting sputtering of the polymer molecules from the surface exposed to the accelerated ions [2]. So the introduction of argon can make the sidewall morphology protect and the etching rate increase at the same time. When the etching depth is 100um, the angle is controlled at 91.5 degrees, and the etching rate can reach 8.5um/min.

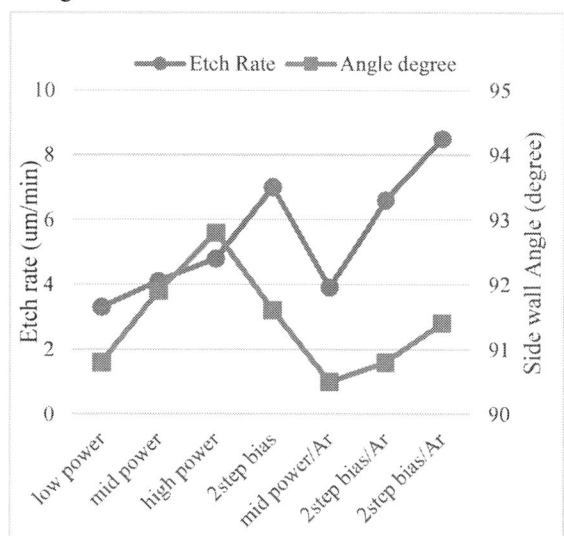

Figure 2: Normalized side wall angle and etch rate for different power and gas.

The normalized side wall angle and etch rate was calculated, in order to eliminate the effect of etch depth difference during comparison (Fig. 2). As is shown below, the side wall angle was reduced as the etching bias power split changed. The etching of argon added was shown that, compared with argon free, sidewall angle has higher

availability for source power.

CONCLUSIONS

The cavity etch side wall profile was controlled by both bias power and auxiliary process gas. High etch rate with optimized profile were obtained by controlling the plasma density and ions bombardment energy independently in two steps. Vertical profile was obtained when argon was used in the Si main etching step. Based on the above learning, a cavity etch process be optimized. Both good profile and high etch rate were obtained.

ACKNOWLEDGEMENTS

The authors wish to express their gratitudes to Wu Yunfei for SEM pictures. Special thanks to Prof. Yu Daquan for his expertise, feedback and support during the whole work. The authors gratefully acknowledge the support by HuaTian Technology.

REFERENCES

[1] M. A. Blauw, T. Zijlstra, and E. Van der Drift. J. Vac. Sci. Technol. B 19(6), 2001, pp.2930-2934

[2] R. Abdolvand, F. Ayazi, Sencors and Actuators A, 144, 2008, pp. 109-116

DUMMY POLY REMOVAL IN FINFET TECHNOLOGY NODE

Ruixuan Huang[1], Shiliang Ji[1], Qiuhua Han[1]*

[1]Technology R&D, Semiconductor Manufacturing International Corp. Shanghai, China

* Jack_Huang2@smics.com

ABSTRACT

When CMOS technology reaches 14nm and beyond, FinFET is implemented to further improve the device performance. Dummy poly removal process works as a key process to control the work function of metal gate, threshold voltage, and gate leakage.

In this paper, we compared the dry etch process on 3 different commercial tools with 2 different approaches which shows that the gate leakage could have more than an order's improvement with proper process and approach.

INTRODUCTION

The technology of HKMG has led to the continuity of device scaling and enabled the extension of Moore's Law towards 45 nm and beyond nodes. The gate-last approach has been widely adopted since it can better control the transistor threshold voltage (Vt) and yield better electrical performance.[1-5]

As known, both dry etching and wet etching can effectively remove polysilicon. However, both schemes has weak points in some aspects. For example, traditional dry etching scheme might degrade electrical performance by damaging ILD and gate oxide layer, ending up with worse NBTI, PBTI and gate leakage performance. On the other hand, wet clean process is hard to control profile and cannot effectively remove the native oxide of polysilicon and the top part of which is doped by implant process. To balance these factors, partial dry etching followed by wet clean process was wildly used for dummy poly removal process. However, this scheme could NOT fully avoid the disadvantage of both wet etching and dry etching process, such as PID to degrade the device performance, ILD/SiN spacer loss, and the discontinuous sidewall profile.

We introduced one full dry etching scheme to remove dummy gate polysilicon at FinFET. Both partial etching followed by wet etching and full dry etching are demonstrated on a commercial CDE (chemical downstream) etcher and two normal ICP etchers. The CDE etcher (Etcher B) for full dry etching is more effective for this process, because it can not only effectively filter high energy ions and electrons, reduce the electron temperature to zero, but also show the most magical and remarkable difference with other remote plasma etcher by renewing the fresh radical after filtering out and right before etching process. By controlling the type of radical, ratio, flow rate, energy distribution, even angle distribution, and the selectivity could be significantly improved. At 3D structure dummy poly gate

removal process, this approach can realize almost zero material loss for both gate oxide and spacer SiN. The ILD loss could be neglected, mainly coming from the break-through step. As compared with Etcher B, the ICP etchers (etcher A & C) was also tested for the purpose of dummy poly gate removal, both the selectivity and the overall physical results could meet process specifications, however, the final electrical performance especially gate leakage is worse than etcher B.

In this paper, we focused on the full dry etching to remove dummy poly gate by commercial CDE etcher B, the corresponding advantages. The results show that such scheme is significantly outperformance the traditional single dry etching, single wet etching or the hybrid scheme from the point of view of less gate leakage, well-controlled material loss, queue time control and throughput.

RESULTS AND DISCUSSIONS

Partition etching method

For FinFET dummy poly gate removal process, the standard scheme is a hybrid dry-wet approach, consisting of the removal of native oxide and the bulk of the poly-si by plasma process, followed by the removal of the remaining poly-si by wet process, was selected as the final approach because it offers the best combination of the effective removal with minimal substrate impact.

Because of the doping impact, the top part of dummy poly-si must be removed by dry etch process. Besides, for the sake of plasma damage concern, the remaining poly-si is removed by wet process.

The trench profile must keep smooth to reduce the gap-fill difficulty and deliver better device performance. For partition etch process, the best profile is flat front surface in order to deliver a smooth profile. This is very difficult for dry etch process as it's a bulk etch without etching stop layer.

In dry etching, the generated polymer/byproduct will deposit on sidewall which will result in slower etch rate at corner/sidewall area. It will induce corner residue and bowl-shaped front surface profile. As shown in Fig1, perfect flat front surface profile was achieved (Fig 1c and d) by Etcher B. However, either generic ICP etcher (Etcher C) or generic ICP etcher with synchronous pulsing plasma (Etcher A) could only delivers bowl shaped front surface profile, namely, some poly-si remains at sidewall. In Fig 1a and b, whatever we tuned the process, the poly-si thickness at corner area is thicker than middle area of the poly-si trench. This means flat surface cannot be achieved.[8]

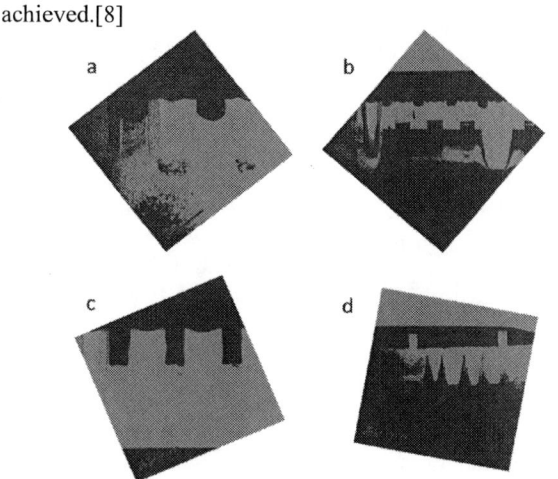

Figure 1: Partition-etch profile (a) ICP tool using synchronous pulsing plasma; (b) generic ICP tool. (c) and (d) Etcher B

Full etching method comparison

In dummy poly-si removal process, break-through is an integrant step to remove native oxide. In this approach, the process must keep enough OE (over etch) to avoid oxide residue for block etch. But the step is mainly related to physical etching, needs stronger bombardment to define the profile. Too much OE will induce more ILD loss and/or damage. These kinds of defect will degrade electric performance and lead to the overheating problem. In addition, ILD loss has n/p MOS bridge risk because of metal residue after CMP (chemical mechanical planarization). In our previous work, we can accurately control etch process to complete break-through and not induce too much OE amount. As shown in Fig2, the insufficient OE could induce poly residue issue (Fig 2a). As we need to control the oxide loss, the poly main etch step will have high selectivity to oxide. Thus, when there's oxide residue after break-thorough, most time of the main etch process is etching the oxide residue, which will end up with poly-Si residue.

If too much OE is adopted to ensure oxide open fully (Fig 2c), the ILD loss will increase dramatically, therefore leading to lower gate height. As it required more CMP polish amount to make sure that there's no metal residue. This is because that break-through step can't achieve high selectivity between native oxide and ILD as they are both oxide film. Nevertheless, if we appropriately control the break-through amount and use synchronous pulsing plasma on etcher A, we could also achieve lower ILD loss. However, as shown in Fig 2c, the profile is not smooth enough which is caused by different etch rate polymer deposition amount of break-through and poly main etch steps.

We relied on one kind of remote plasma etcher

(Etcher D) to do the break-through process. Fig 2b shows a good balance performance, only 5% loss of original gate height, no poly-Si residue in trench and trench profile is smoother. Therefore, the combination of Etcher D and Etcher B does NOT induce gate oxide loss. We can see gate oxide is NOT consumed at both sidewall and fin top. We got prefect trench profile with zero gate oxide loss after full dry etch poly removal process, and ILD loss is almost same as native oxide thickness, it is a surprising result. Another advantage of full dry approach is smooth profile like original poly gate. Fig 2 c shows inflexible boundary for the sake of the performance difference between dry etching and wet etching. Besides, for the hybrid approach (dry etching followed by wet etching), we must control Q-time from dry to wet process, otherwise wet etching will fail to remove the oxide (form native oxide again in atmosphere) on appealed poly-si surface.[8]

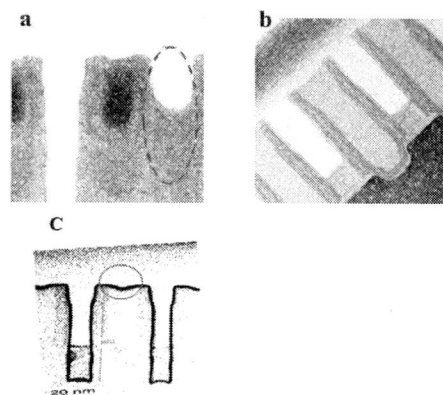

Figure 2: Full etch profile (a) Insufficient OE; (b) full dry etch (Etcher B). (c) Dry-wet hybrid approach on etcher A

Electrical performance comparison

Plasma induced damage (PID) has become a particular important reliability issue in advanced high-k metal gate technology.The complicated inner matter interaction in HK gate stacks makes physical origins of PID remain puzzle and capable of deteriorating gate oxide integrity. For instance, it has been reported that plasma damage induces interface traps and HK bulk traps in SiO2 and HfO2 devices, respectively [9]. The transient charge trapped in Hf-silicate films was also shown to generate latent damage after plasma exposure [10]. Thus, avoiding PID damage is critical for process solutions. Below 14nm node, Fin will be exposed during processing, the PID issue will become difficult to solve.

In dry etching process, even for the partition etch method, there's still PID risk as plasma may penetrate the remaining poly to damage the gate oxide. Thus, etcher A with synchronous pulsing plasma could reduce such PID issue as the ion emerge is very low which could limit the penetrate ability of plasma. We could see that the gate leakage of etcher A will be an order less than normal ICP tool (Etcher C).

TABLE I.   GATE LEAKAGE

Tool	Etcher A	Etcher B	Etcher C
Gate leakage	E	E	10E

Etcher B could also deliver soft plasma as etcher A, as it could effectively filter out high energy ions and electrons, reduce the electron temperature to zero.

In the full dry etching method, we found that longer dry etch process time will result in higher gate leakage. Obviously, it implies that full-dry process would degrade electrical performance of device.

CONCLUSIONS

In this paper, we analyzed the challenges of dry dummy poly removal process at sub20nm node on traditional ICP and CDE ethers, pointed out lots of critical technical indices have to be considered such as post-removal trench profile, ILD loss, trench poly residue and PID . Finally, we investigated the potential feasibility of full dry plasma scheme. Etcher B has been successfully demonstrated on all the above indices, on both partition etching and full dry etching method.

Current result shows that etcher B with partition etching method is the best choice for sub 20nm node dummy poly removal process.

ACKNOWLEDGEMENTS

The authors would like to greatly appreciate the team members in the Etch department of Technology Development and Research Center in SMIC for their continuous help and useful discussion. The help discussion with Dr. Huang Song from Applied Materials is always appreciated.

REFERENCES

[1] Naoki Tega, Hiroshi Miki, Zhibin Ren, Christoper P. D'Emic, Yu Zhu, David J. Frank, Michael A. Guillorn, Dae-Gyu Park, Wilfried Haensch, and Kazuyoshi Torii, IEEE, IRPS11(2011).

[2] Hyohyun Nam, Changhwan Shin. IEEE ELECTRON DEVICE LETTERS, vol. 34, NO. 4, APRIL (2013).

[3] Samer Banna, Ankur Agarwal, J. Vac. Sci. Technol. A 30(4), Jul/Aug (2012).

[4] Demetre J Economou, J. Phys. D: Appl. Phys. 47, (2014).

[5] Ruixuan Huang, Xiao-Ying Meng, Qiu-Hua Han, Hai-Yang Zhang, SPIE, (2015).

[6] Shih Chang Tsai; San Lein Wu; Bo Chin Wang; Shoou Jinn Chang; Che Hua Hsu; Chih Wei Yang; Chien Ming Lai; Chia Wei Hsu; Cheng, O.; Po Chin Huang; Chen, J.F.. Electron Device Letters, IEEE, Volume: 34, Issue: 7 Pages: 834 - 836, (2013)

[7] Demetre J Economou, *J. Phys. D: Appl. Phys.* 47, (2014).

[8] Shi-Liang Ji, Qiu-Hua Han, Hai-Yang Zhang, ECS, (2016).

[9] C.D. Young, G. Bersuker, F. Zhu, K. Matthews, R. Choi, S.C. Song, H.K. Park, J.C. Lee, and B.H. Lee International Reliability Physics Symp. Proc., (2007).

[10] S. Cimino, L. Pantisano, M. Aoulaiche, R. Degraeve, D.H. Kwak, F. Crupi, G. Groeseneken, and A. Paccagnella, International Reliability Physics Symp.Proc, (2005).

[11] Guang-Yaw Hwang, J. H. Liao, S. F. Tzou, Mark Lin, Autumn Yeh, David Lou, Eason Chen, Weien Huang, Gowri Kamarthy, Kaidong Xu and Amulya Athayde, A Hybrid Dry-Wet Approach for Removal of a Dummy Polysilicon Gate in a Replacement Metal Gate Scheme, Solid State Phenomena Vol 187,57-60, (2012).

THE LINE EDGE ROUGHNESS IMPROVEMENT WITH PLASMA COATING FOR 193NM LITHOGRAPHY

*Erhu Zheng, Haiyang Zhang, Yiying Zhang 1**

[1]Technology R&D, Semiconductor Manufacturing International Corp.,
Pudong New Area, Shanghai, P.R.China 201203
*Corresponding Author's Email: Elaine_Zhang@smics.com

ABSTRACT

Incorporation of self-aligned multiple patterning (SaMP) techniques have had limited uses in the industry due to a number of issues including: pitching walking, initial line width roughness (LWR) of photoresist, line edge roughness (LER) degradation of subsequent layer patterning. Utilizing plasma coating for PR hardening is attractive for 193nm lithography application. This paper presents the design of experiments (DOE) to optimize the parameters of pressure, RF power and chemistry ratio to achieve the optimal condition on the LER improvement. As a result, the LER of 1st layer is improved 32% at dense pattern region comparing to initial condition.

INTRODUCTION

Though the extreme ultraviolet (EUV) scanners were introduced to the semiconductor industry [1], the high-volume manufacturing and cost efficiency is still a subject of concern. Current state-of-the-art photo ArF 193 nm immersion lithography has reached limited of half pitch 40 nm, one of most promising extension technique must be self-aligned multiple patterning (SaMP) at the present. The SaMP can easily form fine and repetitive line/space patterns but also introduces pitch walking (systematic variations such as the even/odd effect between the space of core area and non-core area) [2] and challenging line roughness issues.

Previous researches have demonstrated that line width roughness (LWR) performance can be improved through some lithography related factors optimization, such as mask roughness transferred to the photo-resist (PR), the size of polymer aggregates, and the interaction of the exposed resist with the developer [3]. And the plasma curing method can achieve further decrease through specific plasma generated vacuum ultraviolet (VUV) light (110-210nm) to reflow resist [4]. Other solution, like plasma induced polymer formation (CH4 coating) or ion implantation into photoresist for graphite-like structure formation, is also effective but suffers serious PR and critical dimensions (CD) loss [3].

Most of the studies are assuming the line edge roughness (LER) would be improved along with LWR decreasing. But the recent research have revealed that the LER becoming worse while LWR become better in the self-aligned double patterning (SaDP) flow for 7nm node (shown in Fig. 1) [5]. As illustrated in Fig. 2, it can be contributed to the aggressive attack of F radical or physical bombardments from the subsequent fluorocarbon contained etch process. It also indicated that the initial LER need to be minimized.

In this work, we investigated a novel plasma coating on 193nm PTD resist (popular platform of SaMP) to the reduction of post core etch LER. The method has been reported on NTD resist of Litho-Freeze-Litho-Etch process [7] and EUV resist [8]. To understand the mechanism and obtain minimized LER, the design of experiments (DOE) statistic method is chosen.

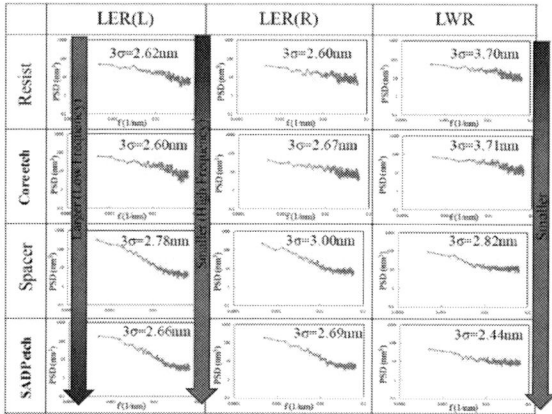

Figure 1. As the SaSP process proceeds, LER becomes larger while LWR becomes smaller. Data and figures cited from [5].

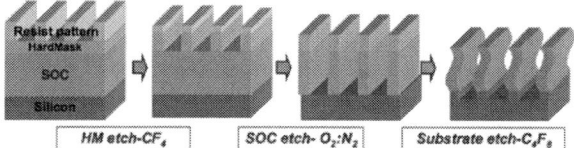

Figure 2. LER degradation illustration in a typical SaDP flow. The figure cited from [6].

EXPERIMENT

All plasma coating experiments were performed in a 300-mm TEL capacitively coupled plasma (CCP) etcher. The resist under investigation was 193nm PTD resist. The samples were <1000Å photoresist patterns on the following film stack: Barc/Darc (dielectric anti-reflective coating)/a-C (amorphous carbon)/ silicon oxide etch stop

978-1-5090-6695-7/17 $31.00 © 2017 IEEE

layer (ESL). The a-C core etch were performed in a Lam inductively coupled plasma (ICP) etcher. The LER data were collected post a-C core etch using Hitachi CD-SEM (critical dimension-scanning electron microscopy) on 200K magnification (vertical direction) images with 32 frames to reduce random noise. The measurement length was selected to be 2μm, based on the standard in ITRS. Cross section profiles of the photo resist patterns were imaged via transmission electron microscopy (TEM).

The DOE was carried out with a six factors/two levels response surface method (RSM) by JMP software (version 12). Three center points are considered in the design table. Six factors are screened from plasma coating recipe's parameters. The specific factor range was tailored based on previous learning. The optimal plasma coating condition was finally identified among the designed 19 DOE splits. A baseline plasma coating split and a no plasma coating split are used for control group.

RESULTS AND DISCUSSIONS
Basic Mechanism

The plasma coating is conducted in a CCP reactor where a negative voltage is imposed upon the top electrode with Ar- and H-contained plasma chemistry. As shown in Fig. 3(b), it will generate a constant flux of VUV light onto the wafer to reflow PR. PR loss is obvious but the surface also become much more smooth. Moreover, the negative voltage (illustrated in Fig. 3(a)) accelerates high energy ion from plasma towards the top electrode, the ion bombardment can causes physical sputtering of Si from the top electrode to resist surface. The Si coating on the PR gets instantaneously oxidized upon exposure to air after transfer out from the vacuum environment of the CCP reactor and results high etching durability during subsequent under layer etch[7][9].

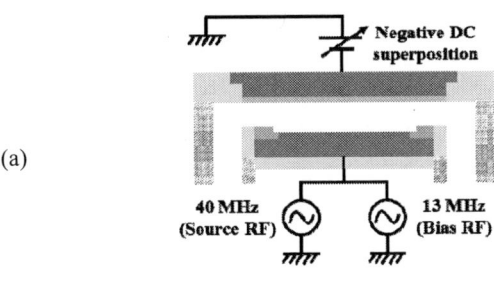

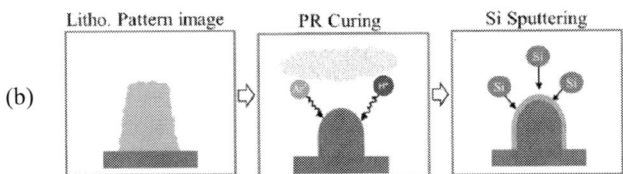

Figure 3. Illustration of the plasma coating etcher (a) and procedure (b). The figure referenced from [9].

DOE Experiment

To investigate etch parameter influence on final core LER and further improve the plasma coating performance, DOE technology is applied.

Six process parameters in plasma coating recipe were selected as the experimental variables including pressure, source power, negative voltage, Ar flow, H-contained gas flow and time. The values of all these variables fell in a reasonable range to ensure a healthy process, as shown in Table I. The factor level was set to 2 to control the overall experiment number, three center points were added for model setup, and a DOE model of 19 runs was thus formulated. The CD, LWR and LER were selected as the responses. Fig. 4 shows such a table of experimental model.

H-contained gas flow was screened out after parameter screening in JMP, negative voltage, pressure, source power, Ar flow ratio and interaction between source power and time were identified as the most statistical significant parameters (P value<<0.05) for the LER improvement. At last, the validation test was performed for the above predicted condition to check the LER performance. Such a recipe showed the best LER performance (32% improvement compared with the initial baseline condition) while keeping comparable CD/LWR. The TEM results (shown in Fig. 5) confirmed that the optimized plasma coating condition's profile was also acceptable.

The experiment validated the plasma coating method can significantly improve the line roughness, the optimal condition can even achieve 60% LER improvement than no plasma coating condition (data is not shown). The different trend for LWR and LER changing is also confirmed. We can improve the LER through plasma coating without LWR variation.

From the model analysis, the H-contained gas flow is irrelevant, so the PR curing (shown in Fig. 3(b)) should be mainly contributed to Ar* plasma. It is a different conclusion from previous study [8]. The interaction between source power and time is important but time itself is rather irrelevant, so for a specific source power, the LER value is improved during the initial short time. In other words, plasma coating time can be keep reasonably short to obtain high throughput.

CONCLUSIONS

In this study, we investigated plasma coating parameters correlation with LER through DOE method for the core etch of a SaDP flow. The optimal condition was finally identified by means of statistically scientific analysis and delivered 32% LER improvement than the initial baseline condition. The DOE model indicated that negative voltage and pressure were most significant parameters. H-contained chemistry and the process time are not dominating factors.

978-1-5090-6695-7/17 $31.00 © 2017 IEEE

TABLE I. DOE factors & experimental specification

No.	Parameters	Level	Low	Baseline	High
1	Pressure	2	A1	A2	A3
2	Source power	2	B1	B2	B3
3	Negative voltage	2	C1	C2	C3
4	Ar	2	D1	D2	D3
5	H contained gas	2	E1	E2	E3
6	Time	2	F1	F2	F3

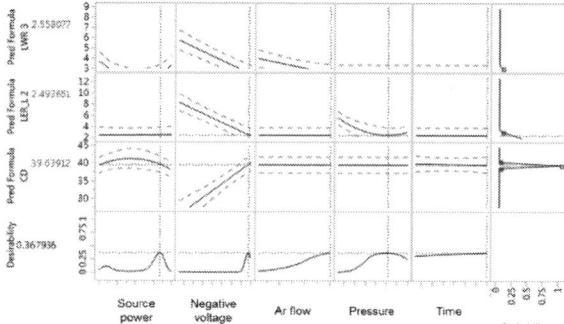

Figure 4. Plasma coating DOE Prediction Profiler.

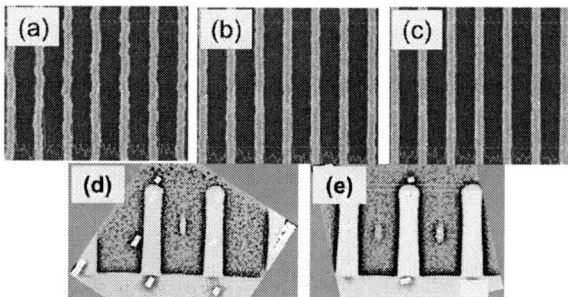

Figure 5. Inline topview result of dense line pattern w/o plasma coating (a), baseline condition of plasma coating (b), optimal condition of plasma coating (c); TEM result of baseline plasma coating condition (d) and optimal condition (e).

ACKNOWLEDGEMENTS

The authors would like to thank their colleagues of etch department, Ms. Yi Huang, Ms. Jialei Feng and Dr. Siyuan Yang of MMD department from SMIC technology research and development center for technical assistance and useful discussions during the course.

REFERENCES

[1] N. Felix, et al., "EUV Patterning Successes and Frontiers", Proc. SPIE, Vol. 9776, 977610, 2015.

[2] Y. Zhang, et al., "SaDP Process Challenges and Solutions on NAND Flash Cell", ECS Trans., Vol. 60, pp 395-400, 2014.

[3] X. Meng, et al., "The Improvement of Poly-Si Gate Line Width Roughness", ECS Trans., Vol. 60, pp 367-372, 2014.

[4] E. Pargon, et al., "Mechanisms involved in HBr and Ar cure plasma treatments applied to 193 nm photoresists", J. Appl. Phys. 105, 094902, 2009.

[5] E. Sanchez, et al., "Self-aligned-quadruple-patterning for N7/N5 silicon fins", SPIE, 2016

[6] M. Weigand, et al., "Evaluating spin-on carbon materials at low temperatures for high wiggling resistance", Proc. SPIE, Vol. 8685, 8685R, 2013.

[7] J. Smith, et al., "Incorporation of Direct Current Superposition as a Means for High Quality Contact and Slotted Contact Structures utilizing Litho-Freeze-Litho-Etch", ASMC, pp 377-383, 2014.

[8] M. Honda, et al., "Dielectric Etch Challenges and Evolutions", AVS, 2014.

[9] H. Yaegashi, et al., "Photoresist performance modification through plasma treatment", SPIE, Vol. 9428, 94280H, 2015.

THE STUDY OF DEEP TRENCH ETCH PROCESS FOR PCRAM

Yiying Zhang, Zhuofan Chen, and Haiyang Zhang[*]

Technology R&D, Semiconductor Manufacturing International Corp.
No.18 Zhang Jiang R.D., Pudong New District, Shanghai, 201203, P.R. China
*Corresponding Author's Email: Steven_Z@smics.com

ABSTRACT

The deep trench (DT) is the key process to form the diode array in the diode-selected Phase Change Random Access Memory (PCRAM). In this work, the DT has been successfully developed with common etch chamber. We investigated the influence of different etch schemes on the DT profile. It is demonstrated Si etch with hard mask scheme can deliver better DT profile than that with soft mask. The gas impact on the DT profile is also studied. Bowing free profile is got after F-based gas decrease and necking profile is much improved with N2 gas decrease. At last the correlation between Si damage and DT profile is also showed.

Keywords— PCRAM, diode, deep trench; profile;

INTRODUCTION

Phase Change Random Access Memory (PCRAM) is attractive as it has kinds of advantages over traditional flash memory [1-4]. PCRAM has high scalability as it is compatible with standard Complementary Metal-Oxide-Semiconductor (CMOS) technology, which makes it to be one of the best candidates for next generation nonvolatile memory (NVM) at 40nm and beyond. However, the size of the PCRAM cell is difficult to shrink when NMOS is used as the selector due to the reset current requirement. To meet both the requirements of cell size and reset current, a diode used as the selector (diode-selected) PCRAM has been designed [5]. For diode-selected PRCAM, deep trench (DT) etch is the key process to form the diode array. In this work, the influence of different etch schemes and etch parameters on the DT profile are studied.

EXPERIMENTS

The DT etch scheme comparison is schematically presented in Fig. 1. In both schemes, the film stack including resist, bottom anti-reflect coating (BARC), tri-layer mask layer such as DARC and a-C in this work, oxide-nitride hard mask (ON HM) layer and substrate Si. In scheme 1 (Fig. 1(a)) which is called in-situ HM and Si etch, HM and Si are etched with a-C as the mask layer. The remaining a-C is stripped after Si etch. This scheme can be also called as soft mask scheme. In scheme 2 (Fig. 1(b)) named as ex-situ HM and Si etch, ON HM is etched with a-C as the mask layer while Si is etched using oxide as the mask layer. The remaining a-C is removed right after HM open. So scheme 2 is HM scheme as Si etch use

oxide instead of a-C as the mask layer.

Both schemes etch process are implemented in a Lam inductively coupled plasma (ICP) poly etcher, which is not designed for deep trench etch process. After the dry etch and post clean process, the cross sectional DT etching profile are measured using commercial scanning electron microscopy (SEM) tool.

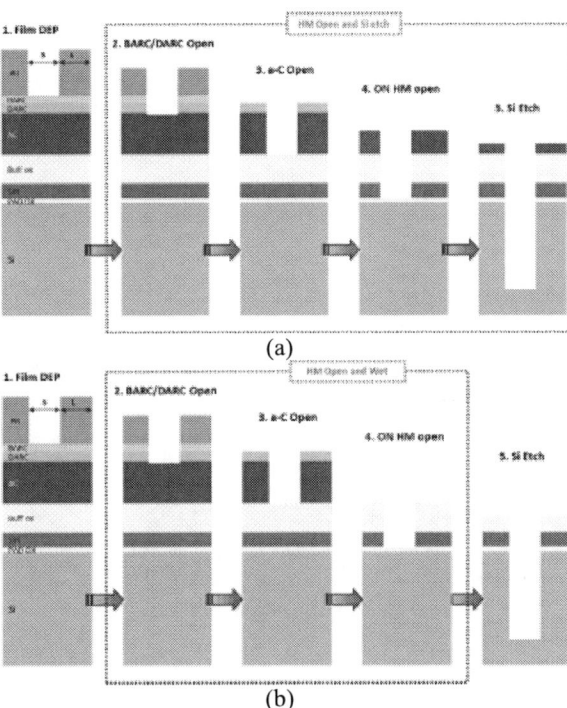

Fig. 1 DT process scheme comparison: (a) Scheme 1: in-situ HM and Si etch. (b) Scheme 2: ex-situ HM and Si etch.

RESULTS AND DISCUSSIONS

Comparison between two schemes

Fig. 2 shows the partial and full etch profile comparison of two schemes discussed above. In scheme 1, there is no HM loss during Si etch using a-C as the mask layer. Thus, oxide HM thickness can be much decreased. However tri-layer a-C mask thickness need increase to cover ON HM and Si etch. Fig. 2(a) shows the partial profile of scheme 1. The HM profile is a little tapered while partial Si profile is vertical. And a-C remaining is enough for the following Si full etch. However in Fig. 2(b) full etch image, top Si profile is bowing and HM top is almost merged. DT Si etch with long etch time will generate heavy polymer on the HM top and sidewall. Separating Si etch into several steps and inserting additional post etch treatment (PET) process can reduce the polymer deposition. However PET process will consume the soft mask a-C, which will induce ON HM loss with insufficient a-C remaining. Comparing with HM scheme shown in Fig. 1(b), tri-layer soft mask a-C is also the polymer resource. Additionally, the remaining etching gas in the deep trench will introduce lateral etch then induce the Si bowing profile.

To reduce the polymer generated from a-C mask, we propose HM scheme as shown in Fig. 1(b). A-C soft mask is in-situ removed right after ON HM open then Si etch is implemented with oxide as the mask layer instead of a-C soft mask. Therefore, a-C mask thickness can be decreased but OX HM thickness need increase to ensure no SIN HM loss during following Si etch. And no more oxide loss will occur even adding additional PET process to clear heavy polymer during long time Si etch. As shown in Fig. 2(c), scheme 2 Si partial profile is much tapered than scheme 1. But full etch shows bowing-free profile in Fig. 2(d).

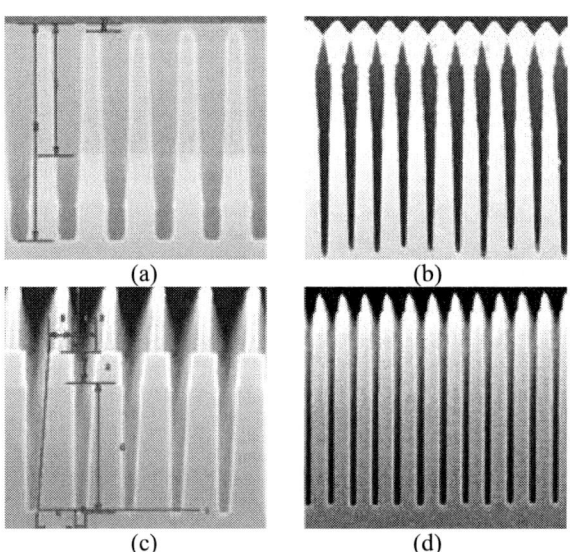

Fig. 2 DT profile comparison: (a) Scheme 1 partial etch profile. (b) Scheme 1 full etch profile. (c) Scheme 2 partial etch profile. (d) Scheme 2 full etch profile.

DT profile enhancement

DT Si etch with HM scheme can deliver better profile than soft mask scheme. However, Si bowing or necking profile also can be observed if the etch condition is not optimized. As illustrated in Fig. 3(a), the left image shows obvious bowing profile. From left to right the Si profile get more and more vertical if continue decreasing F-based

etch gas. The Si lateral attack decreases with the etching gas reduction. Fig. 3(b) shows the impact of the polymer gas N2 on the DT profile. Left image shows necking profile at the top-middle Si as more N2 gas protect the sidewall. When N2 gas continue to decrease necking-free profile can be got as demonstrated in Fig. 3(b) from left to right. The optimal DT Si profile will be got only when etching gas and polymer gas reach the balance.

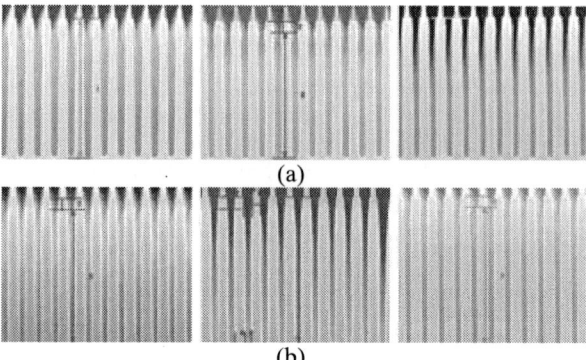

Fig. 3 The gas impact on the DT profile: (a) F-based gas decrease. (b) N2 gas decrease.

DT Si damage reduction

During DT profile improvement, another issue is raised. It is observed random sidewall damage at the top-middle of deep trench. As shown in Fig. 4, it seems the Si damage is strongly related with the DT profile. Fig. 4(a) shows the SEM image of Si sidewall cut along DT direction. Fig. 4(b) is the corresponding DT profile image. The Si damage is easily occurs at the sidewall slope area and serious when the double slope get obvious. It is concluded that the plasma bombardment is the root cause of this damage issue. The bombardment is strong at the large-angle slope area. So the damage is only decreased (Fig. 4(a) left to right) when the DT profile get vertical as illustrated in Fig. 4(b) from left to right. Increasing the

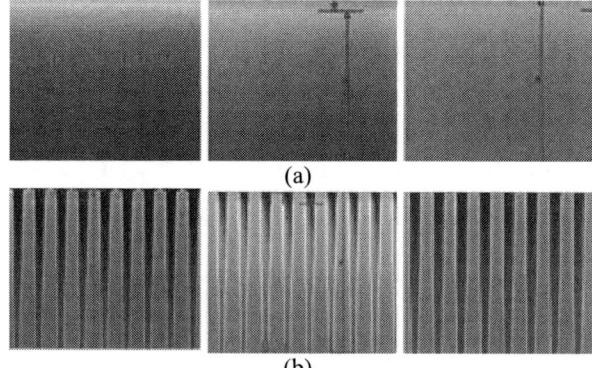

Fig. 4 The correlation between Si damage and DT profile: (a) Si damage get better. (b) DT profile get more vertical.

e-chuck temperature can deliver more vertical DT profile with damage free.

From above discussion, DT profile is influenced by kinds of parameter. The optimal profile without Si damage will be got after etching gas, polymer gas, e-chuck temp, etc. get balance.

CONCLUSIONS

In this paper, we investigated two etch schemes for the DT etch process in the diode-selected PCRAM. Compared with soft mask scheme, the proposed HM scheme is proven to be superior for more smooth DT profile delivery with less polymer generated. Besides, we also studied the influence of kinds of etch parameters on the DT profile of the proposed HM scheme. Bowing free profile can be got after F-based etch gas decrease and necking profile is much improved with less polymer gas N2. At last the sidewall Si damage issue is obvious improved with less plasma bombardment on the sidewall under vertical DT profile.

ACKNOWLEDGEMENTS

The authors would like to thank the etch members in the Etch department of Technology R&D center in SMIC and the dry etch group of Lam Research, China, for their continuous help and useful discussion.

REFERENCES

[1] C. Zhang, et al. *IEEE Electron Device Letters,* vol. 32, No. 8, 2011, pp. 1014-1016.

[2] L. Li and M. Chan, *2008 IEEE International Conference on Electron Devices and Solid-State Circuits*, 2008, pp. 1-4.

[3] S. Ahn, et al. *2005 Symposium on VLSI Technology,* 2005, pp. 98-99.

[4] Y. Li, et al. *2010 10th IEEE International Conference on Solid-State and Integrated Circuit Technology,* 2010, pp. 1127-1129.

[5] D. Cai, et al, *IEEE 11th Annual Non-Volatile Memory Technology Symposium Proceeding,* 2011, pp. 1-4.

Improvement of the Pattern Wiggling Profile by PR treatment Methods

Pan-pan Liu, Cheng-long Zhang, Hai-yang Zhang

Semiconductor Manufacturing International Corporation,
No.18 Zhang Jiang Rd., Pudong New Area, Shanghai, 201203, P.R.China
021-38610000-15972, Penny_Liu@smics.com

Biography

Pan-pan Liu, has been with Semiconductor Manufacturing International (ShangHai) Corp, P.R.China since August, 2014. Pan-pan Liu received her Ph.D. degree in Plasma Physics from Zhejiang University, Hangzhou, P. R. China.

Abstract

For advanced node such as 14nm technology and beyond, tradition immersion photolithography transforms to double patterning, EUV or negative tone development (NTD) technology to improve pattern transfer techniques. Even through, a pattern wiggling issue has arisen with the reduction of pattern dimension during substrate dry etch, which prevents the successful pattern transfer. In this paper, two kinds of photoresist (PR) treatment methods, instead of the lithographic resolutions, are adopted during dry etch process in commercial etchers. It is found that pattern wiggling profile has been drastically improved. What's more, the relationship between parameters in PR treatment and dense/isolated pattern loading have been studied.

Keywords—NTD, wiggling, PR treatment, dry etch

Introduction

Efficient pattern transformation is key requirement in semiconductor manufacturing technology. RIE (reactive ion etch) is an important part in which the pattern was transferred from the photolithography process into substrate. Multilayer processing [1] which consists of PR, Si-containing layer (Si-ARC or Si-Oxide) and Carbon-containing under-layer (UL) films, has been developed and implemented in dry etch to achieve smaller node avoiding pattern collapses issue during development of PR. However, pattern wiggling issue comes out as the aspect ratios of the UL material increase and device size decrease, which ultimately prevents pattern transfer from UL to substrate [2].

It is reported that pattern wiggling occurred due to the substitution of hydrogen by fluorine atoms that introduced by dry etch on the surface of the films (Fig.1). The etch pattern distortions occurs by material stress of F as the volume of F is larger than that of H [3]. Many lithographic resolutions have been proposed to improve pattern wiggling according to the mechanism, such as increasing UL baking temperature to give lower hydrogen content film [4-6], higher hardness UL to resist

wiggling performance.

In this paper, we focus on improve pattern wiggling in dry etch process, in which special gases are used to PR treatment before Si-containing is open. Two kinds of gases are introduced. Then the relationship between parameters in PR treatment and dense/isolated pattern loading have been studied.

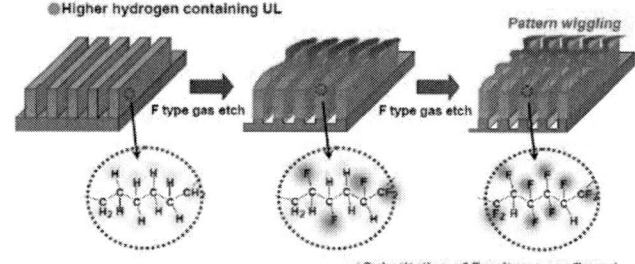

Fig.1. Hypothetical cause of wiggling [3].

Experimental

In Fin-FET patterning definition etch process, negative tone development(NTD) photolithography technology is applied in which special solvents is used for pattern reversal to form high-resolution pattern. Besides, multilayer including PR, Si-containing layer, and SOC(spin-on-carbon-hardmask) is used. The pattern (trench) suffers wiggling issue which can be seen in Fig.2.

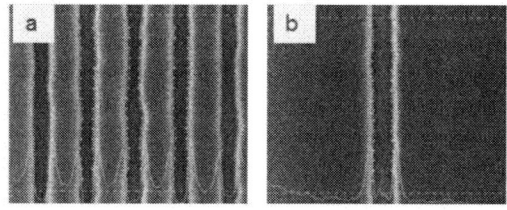

Fig.2. AEI results show wiggling and dense/isolative loading issue.

A. PR treatment method 1

Gas1/gas2 is used for PR treatment before Si-containing film is open and wiggling issue is improved.

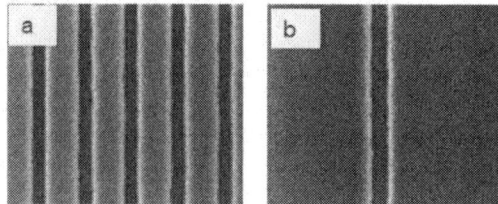

Fig.3. AEI results show wiggling improved..

In PR treatment method 1, the impact of several parameters on the etch performance has been studied.

1. PR treatment time.

Keeping the other operating parameters (including pressure, power, gases and temperature) constant, we vary the PR treatment time to elucidate the effect of this parameter on the process. The results are illustrated in Fig.4. It is found that the etch performance gets better(LER gets smaller) when treatment time time is shorter.

Time	Dense	Isolation	LER
Split 1			2.64
Split 2			2.58
Split 3			2.44
Split 4			2.30

Fig.4. Results comparison of different treatment time.

2. Gas ratio of gas A and gas B.

The analogical analysis in which the gas ratio is the only variable and the other parameters are content. Different gas ratio (gas A and gas B) is studied and the result can be seen in Fig.5, where lower gas ratio results in the better performance. It can be calculated that gas B plays the key role in the PR treatment process.

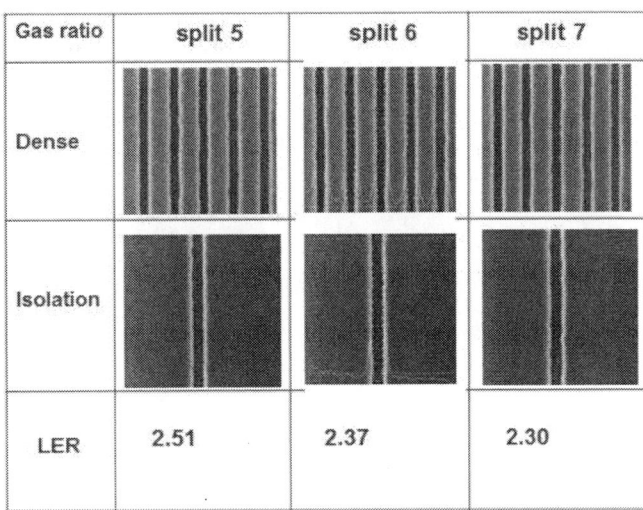

Gas ratio	split 5	split 6	split 7
Dense			
Isolation			
LER	2.51	2.37	2.30

Fig.5. Results comparison of different gas ratio..

3. The action of gas D.

There are dense and isolated pattern in the etch process and through pitch (dense/isolated loading) issue exists. The loading is improved but still exists with PR treatment. We do the test in which gas D is adding in the PR treatment. The result shows dense and isolated loading free. Fig.6 illustrates the loading comparison between PR treatment with and without gas D. It seems that gas D has more effect on isolated pattern .

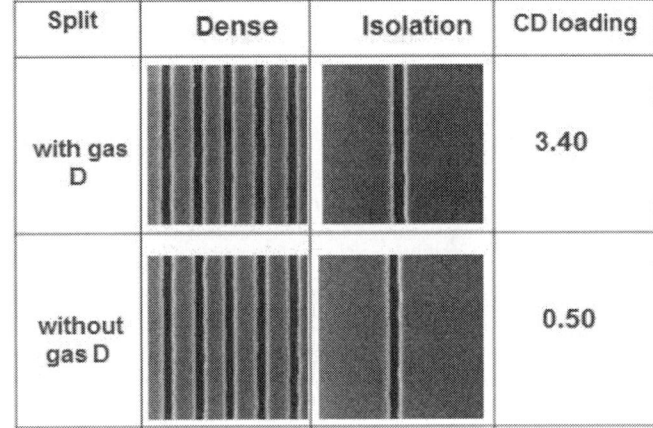

Split	Dense	Isolation	CD loading
with gas D			3.40
without gas D			0.50

Fig.5. Results comparison of with/without gas D..

B. PR treatment method 2

In PR treatment method 2, a kind of gas, C is adopted. The result shows that pattern wiggling also improved. Furthermore, parameter—PR treatment time has impact on the result.

Time	Dense	Isolation	LER
Split 8			2.36
Split 9			2.51

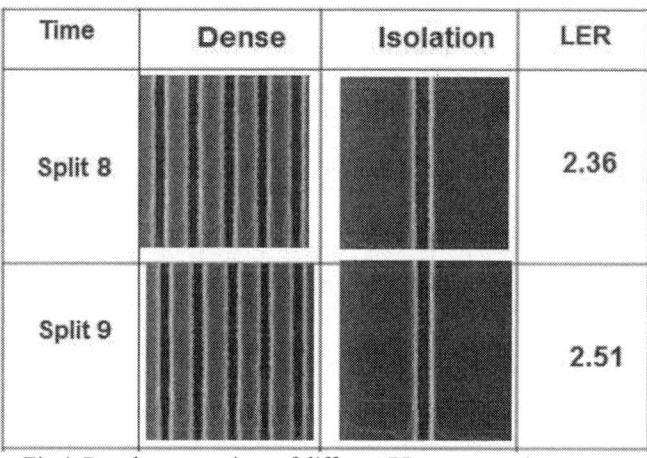

Fig.4. Results comparison of different PR treatment time.

Illustrations

Two kinds of photoresist (PR) treatment methods, instead of the lithographic resolutions, are adopted during dry etch process in commercial etchers. It is found that pattern wiggling profile has been drastically improved. What's more, the impact of parameters such as treatment time, gas ratio of treatment gases on the etch performance is studied, and specially, it is found that in the PR treatment method 1, adding gas D can solve the dense/isolated pattern loading issue.

References

[1] Burns, S., Pfeiffer, D., Mahorowala, A., Petrillo, K., Clancy, A., Babich, K., Medeiros, D., Allen, S., Holmes, S.,Crouse, M., Brodsky, C., Pham, V., Lin, Y., Patel, K., Lustig, N., Gabor, A., Sheraw, C., Brock, P. and Larson, C., "Silicon containing polymer in applications for 193 nm high NA lithography processes", Proc. SPIE 6153, 61530K-1 –61530K-12 (2006).

[2] Peters, L., "Photoresists Meet the 193 nm Milestone", Semiconductor International 28(2), 38-64 (2005).

[3] Goji Wakamatsu, Kentaro Goto, Yoshi Hishiro, Taiichi Furukawa al., "Investigation of pattern wiggling for spin-on organic hardmask materials," Advances in Resist Materials and Processing Technology XXIX,, Proc. of SPIE Vol. 8325 83250T-1,2012

[4] [2] Makoto, M., Mitsuaki, I., Takashi, K., Hisashi, H, and Seiji, F., "Etch durable Spin-on hard mask," Proc. SPIE 7972, 797226 (2011).

[5] Tadokoro M., Yonekura K., Yoshikawa K., Ono Y., Ishibashi T., Hanawa T., Fukiwara N., Matsunobe T., MatsudaK., "Improvement of the wiggling profile of spin-on carbon hard mask by H2 plasma treatment," J. Vac. Sci.Technol. B 26, 67-71 (2008).

[6] Someya, Y., Shinjo, T., Hashimoto, K., Nishimaki, H., Karasawa, R., Sakamoto, R., Matsumoto,T., "Spin-on-carbon-hardmask with high wiggling resistance," Proceedings of the SPIE, vol.8325, 83250U (2012).

A STUDY OF AA CD UNIFORMITY LOADING OPTIMIZATION AT 28NM NODE

YiZheng Zhu[1]*, Lian Lu[1], Fuhong Chen[1], Sifei Chan[1], Quanbo Li[1], Yu Zhang[1], Albert Pang[1]

[1]Shanghai Huali Microelectronics Corporation, Shanghai City, China

*Corresponding Author's Email: zhuyizheng@hlmc.cn

ABSTRACT

At 28nm technological node, shallow trench top CD becomes narrower for tighten transistors density. Smaller space CD makes process control more difficult, especially uniformity control. On the early stage of our STI etching process developing, dense/ISO structure has opposition within wafer CD distribution map, which leads to device compromise. From our study, dense/ISO uniformity loading comes from accumulation of the whole pattern transfer process, mainly DARC/APF etching step. The etch scheme is studied in this paper, focused on PR remain, DARC profile integrity optimization and soft mask(APF) profile control, which can affect loading greatly and need precisely control.

INTRODUCTION

With continuously scaling down of the critical dimension, process complexity increasing at a great pace. STI etching as one of the most important process, which defines the active area, is critical to device performance. For the concern of device matching and yield enhancement, the CDU should be as better as possible on both dense/ISO areas. As illustrated in Fig.1, in order to keep device performance the AA CD and uniformity should be precisely controlled.

In this paper, we focused on dense/ISO within wafer CD uniformity loading issue we encountered during the process developing, as showed in Fig. 2. There is a tradeoff between dense and ISO uniformity, which is unacceptable for device performance. In order to solve this problem, we demonstrate several measures to identify dense/ISO CD uniformity loading root causes, which may come from the accumulation of the whole pattern transfer process. The challenge here is define the contribution of each step. In this case PR remaining, DARC profile integrity and soft mask (APF) profile were studied thoroughly.

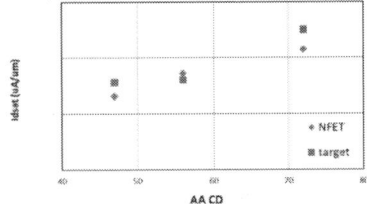

Fig.1 AA CD and Idsat correlation

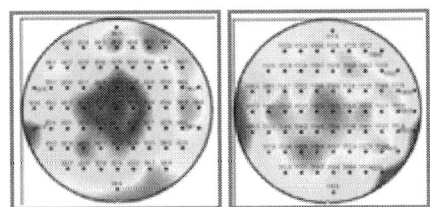

a) Dense structure b) ISO structure
Fig.2 Dense/ISO CDU loading

EXPERIMENT PROTOCOL

In this paper, all dry etching experiments were performed in a 300-mm commercial Transformed coupled plasma (TCP) reactor from Lam Research Corporation. The resist under investigation was 193nm resist. The samples patterns printed by ArF lithography on the following film stack: bottom anti-reflection coating, DARC, amorphous carbon, silicon nitride and pad oxide as showed on Fig.3. Fig.3 also shows the whole dry etching flow. The first step is designed to open BARC. Then DARC is etched. Subsequently APF and SIN HM are etched. Process stops after the whole trench is formed.

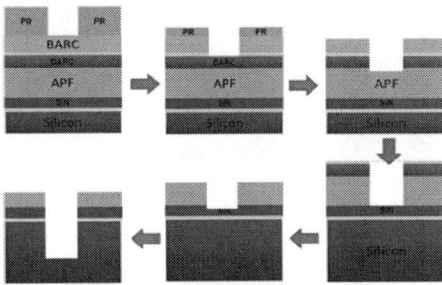

Fig.3 Steps of STI etching process

RESULT AND DISCUSSION

A. Partial profile check

Based on the dense/ISO map distribution difference, the profile must be changed in different way during the pattern transfer process. In order to decouple the contribution of each step, the partial check was performed. As Fig.4a) depicts after the APF etch step, the DARC profile is not robust enough to hold the pattern transfer. The damage on the top of APF is different between dense and ISO structure. Then the difference was magnified at SIN step (Fig.4b). Dense area transfer pattern with the whole integrity soft mask, but ISO area transfer with the top damage part. Mask profile difference may be the biggest contribution to this loading effect. That means the DARC/APF profile should be carefully controlled.

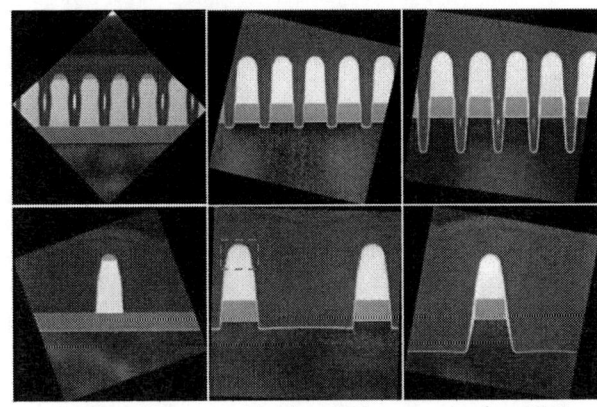

a) After APF   b) after SIN   c) after STI1
Fig.4 TEM of partial etching profile

B. Effect of DARC profile integrity

During the etching process, DARC profile integrity issue was firstly observed. There are two key process factors were studied in the DARC open steps: PR remaining and bias power. Both were selected as the experimental variables in this paper. In our recipe design, PR remaining is used to protect the DARC integrity especially the corner part. More PR remaining is beneficial for DARC corner protection, so resister hardening was adopted to increase etching selectivity. Since bombardment is also crucial for DARC corner damage control, reducing bias power is helpful to maintain DARC profile.

Fig.5 shows the corresponding experimental results, the DARC profile integrity was preserved after APF etching by adopting PR hardening and reducing bias power.

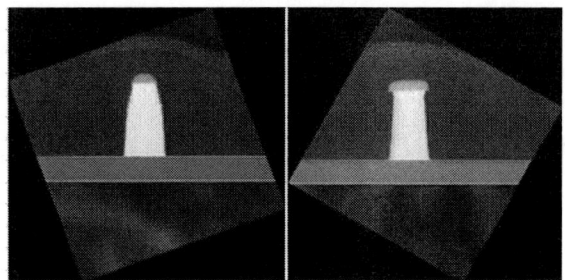

a) Before optimization   b) after optimization
Fig.5 TEM of wo/wi DARC profile integrity optimization

C. Effect of APF profile improvement

With the concept of dense/ISO HM profile difference induced CDU loading issue during pattern transfer, we design a second round of experiment to minimize the mask difference on both areas. In our study, SO2/O2 gas was used at APF etching step. The gas ratio is critical to profile control, with less passive gas SO2 will produce bowing profile during plasma etching; but if there is too much passivation on sidewall will produce taper profile. It's easier to eliminate dense/ISO profile difference if we get vertical profile. As Fig.6 showed, by adjusting the gas ratio, vertical APF profile was obtained on both areas.

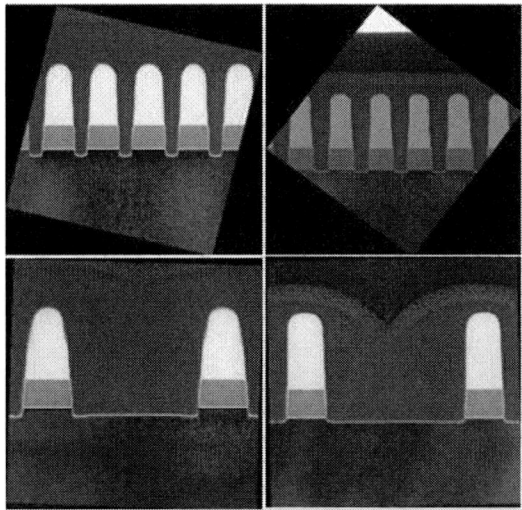

a) Before optimization   b) after optimization
Fig.6 TEM of wo/wi APF profile optimization

Combined with both optimizations, we minimized the differences between dense and ISO areas. The corresponding experimental results as showed in Fig.7. Dense and ISO CD uniformity was improved and distribution is more uniform.

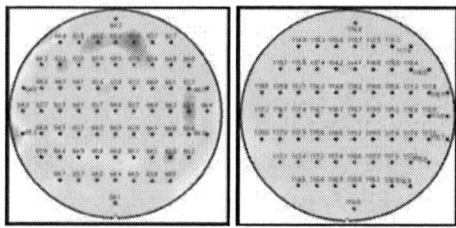

a) Dense structure   b) ISO structure
Fig.7 after optimization Dense/ISO CDU

CONCLUSIONS

In this paper, we described CD uniformity loading issue encountered during STI etch process development. Firstly, the scheme and causes was investigated, which is proved to be dense/ISO mask profile difference induced during pattern transfer. By optimizing the mask profile, mainly DARC/APF etching step, the CD uniformity was greatly improved and distribution opposite free map was obtained.

ACKNOWLEDGMENTS

The authors would like to greatly appreciate the etch/lithology team members and PIE in the Technology Development II department for their continuous help and useful discussion..

REFERENCES

[1] K. Fujita, et.al., "Advanced Channel Engineering Achieving Aggressive Reduction of VT Variation for Ultra-Low-Power Applications", IEDM, p.749, 2011.

[2] Michael Quirk, Julian Serda, "Semiconductor Manufacturing Technology".

[3] E.Altamirano-Sanchez, et.al. "Dry etching fin process for SOI finFET manufacturing: Transition from 32 to 22nm node on a 6T-SRAM cell" Microelectronic Engineering (2011)

[4] M. Ercken, E. Altamirano-Sanchez, C. Baerts. Challenges building a 22nm Node 6T-SRAM Cell Using Immersion Lithography. 6th Int. Symp. on Immersion Lithography Extensions. 2009.

PRODUCER BARC OPEN CHALLENGES AND THE BASELINE SET UP FOR FINFET

Song Huang, Ying Huang, Judy Wang, Qiang Ge
Applied Materials China, Shanghai 201204, China
Corresponding Author's Email: song_huang@amat.com

ABSTRACT

Multiple Vt tuning is necessary for SOC (system-on-a-chip) FinFET devices at 16nm technology node and beyond. To achieve this, BARC open, as a mask, for NMOS/PMOS implantation process is needed. Vertical profile with enough PR effective height is the key challenge of this process. Besides, WPH for run cost reduction is another major consideration from customer perspective. So gas N2/NH3 chemistry is adopted for its high etch rate, and thus high throughput. For this recipe to get a good etch uniformity on blanket wafer first to ensure good performance on the whole pattern wafer, each parameter was studied, that is, bias power, pressure, NH3 flow, N2 low, FRC, gap, and backside helium pressure. Finally, etch rate >1000A/min with good uniformity <4% on blanket BARC wafer was got using optimized bias power and gap. Results on pattern wafer show performance is almost acceptable, it is near to vertical and PR effective height is enough.

INTRODUCTION

Today, as the device moves into 16nm and beyond, a lot of new processes are introduced to meet new and tight requirements. Among them, BARC etch back is one of the key processes for multiple Vt tuning with the following implant. Ion implantation commonly modifies the characteristics of the surface layer by introducing either p-type or n-type dopant ions into the host material. The implant can be precisely customized to reach specific areas within the die, using patterned photoresist mask to protect other regions of the die from the implant, thereby enabling local electrical modification to form various circuit device elements. Thus, vertical profile with enough PR effective height is the key challenge of this process. It is the very important parameter and should be appropriate. If too low, then some area will not be sufficiently covered. And if too high, PR will block the some of the beam when implantation occurs at an angle. In this paper, detail studies were carried out to get an appropriate results as above mentioned.

EXPERIMENT

The plasma is operated at 50-110mT, gap 1.1 -3.8 inch, with a N2/H2 or NH3/H2 chemistry gas flow, and a bias power 600-1200W. Those plasma conditions are generally used in CCP chamber to have a stable process window for mass production run.

The etching of BARC blanket wafer is investigated for various conditions. 300 mm silicon wafer coated with 240nm BARC are used for BARC etch rate measurement in KLA tool. Then best condition was selected for pattern wafer etching.

The experiments are performed in a 300mm producer CCP etch tool from Applied Materials. The plasma is excited by 60 MHz rf in this application, actually it can also be excited by 13.56 MHz if needed. To get a repeatable data, there is an in situ chamber clean after each wafer processed.

RESULTS AND DISCUSSIONS

Figure 1 shows the BARC open process film stack. Open area BARC need to be removed completely without any residue, while lateral etch is not allowed, as shown in the figure 2. PR effective height is the very important parameter and should be appropriate. If too low, then some area will not be sufficiently covered. And if too height, PR will block the some of the beam when implantation occurs at an angle.

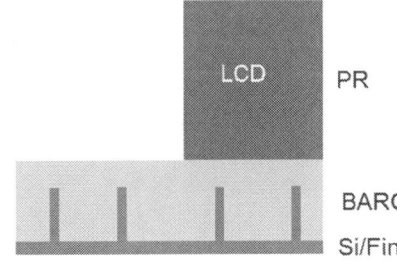

Figure 1: BARC open film stack pre etch

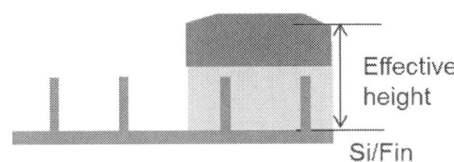

Figure 2: BARC open film stack post etch

In the beginning, N2/H2 chemistry process was tried on pattern wafers, however, through put is merely 25pcs/hour per twin regardless of physical profile. It cannot show the superiority of producer's high throughput and surpass competitor chamber. So N2/NH3 chemistry was adopted for its high etch rate, and thus high through put.

As usual, etch rate study was carried out before processing pattern wafer. It is essential to get a good etch uniformity first to ensure good performance on the whole pattern wafer. For starting up recipe with NH3/N2 based chemistry, each parameter was studied, that is, bias power, pressure, NH3 flow, N2 low, FRC, gap, and backside helium pressure. Among them, low gap process are most important and preferred since gap trend study show it helps gain PR effective height a lot as shown in figure 3 and figure 4. XSEM images show PR profile was improved at low gap. Effective height gets improved from 62nm to 117nm with gap changed from 3.5 to 1.15.

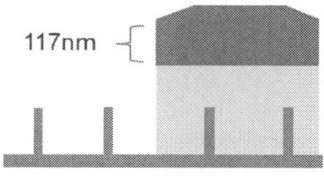

Figure 3: PR effective height at gap 1.15

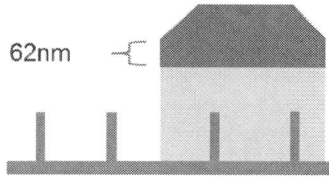

Figure 4: PR effective height at gap 3.5

Nagai et al. have studied etching of organic film FLARETM in high density NH3/N2 plasma [1, 2]. H radicals isotropically etch organic films. N radicals form a passivation layer on the sidewall against the etching by H radicals. This passivation layer could not be etched without ion bombardments. Here, in our case, simulation data shows higher voltage drop across the sheath in large gap, which means higher ion bombarding energy in large gap. In other words, higher total ion density produced in smaller gap due to same power delivery as shown in figure

5. Simulation data also tells that H, NH2 (amidogen) and N are dominant radical flux. Radical flux reduces to 60% - 70% at larger gap as shown in figure 6. Etch rate tests show PR to BARC selectivity is almost the same with various parameters tuning. PR shoulder profile depends on lateral etch and ion bombardment. At low gap, there's a thicker passivation layer (richer N radicals) against etching by H radicals. And ion assisted etch is weaker compared to high gap, for later has the higher voltage drop across the sheath. Thicker passivation layer and less ion bombardment help to maintain shoulder profile at low gap conditions, that is, less PR loss compared to high gap condition.

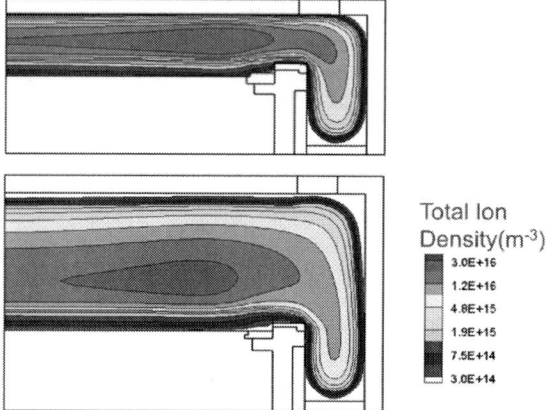

Figure 5: Total Ion Density (m-3)

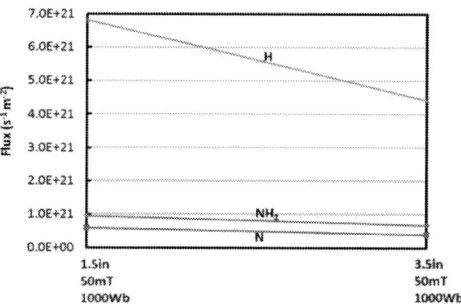

Figure 6: Radical Flux (s-1 m-2) vs Gap

Finally, etch rate 1008A/min with good uniformity 3.15% on blanket BARC wafer was got using bias power 1000W and gap 1.5 condition. This is a desired result because etch rate is fairly fast, which helps to obtain the high throughput. Results on pattern wafer show performance is almost acceptable, it is near to vertical and PR effective height is enough.

CONCLUSIONS

In summary, key challenges BARC vertical profile, PR effective height, through put are discussed and fixed and meet process requirements using optimized bias power and gap. This Producer Barc open process base line was ready for FinFET. At the same time, etch mechanism are addressed with the combination of simulation data and experimental results. PR shoulder profile depends on lateral etch and ion bombardment. It is gained at low gap for low ion bombardment and thicker passivation layer formed by N radicals.

REFERENCES

[1] Nagai et al., J. Appl. Phys. vol 91, 2002, pp. 2615 – 2621.

[2] Nagai et al., J. Appl. Phys. vol. 94, 2003, pp. 1362 -1367.

Optimization of Wet Strip after Metal Hard Mask All-in-One Etch for Metal Void Reduction and Yield Improvement

Yong Huang, Jialei Liu, Zhiyong Yang, Jing Zhao, Huanxin Liu*

Technology R&D, Semiconductor Manufacturing International Corp.

PuDong New Area, Shanghai, P. R. China 201203

*Corresponding Author's Email: HX_Liu@smics.com

Abstract

As semiconductor technology node continuously shrinks, Wet strip process works as a more important role beyond 45m. For RC delay concern, Ultra Low-K material is introduced to BEOL ILD (Interlayer Dielectric). After Trench First Metal Hard Mask All-in-One Etch, ULK film sidewalls are exposed during Wet strip. Wet strip needs to take care of not only no ULK K value shift, but also HM TiN pull back / remove for better gap fill capability. We systematically study Wet strip process parameters, clean efficiency is evaluated by metal void defect by defect scan after Cu CMP step. Device electrical test like Kevin-Via resistance and Via-Chain resistance are compared between old and optimized conditions. By optimization of Wet strip, remarkable metal void defect density is reduced and 20% yield improvement is achieved when device reliability is qualified.

Keywords – MHM AIO Etch, Clean efficiency, HM TiN, ULK K shift, Copper recess, Interconnect resistance, Device yield

Introduction

When semiconductor technology node continuously shrinks, Metal interconnect critical dimension of trench/via at BEOL (Back End of Line) becomes smaller. Increasing trend/via aspect ratio makes Wet strip more difficult to remove polymer residue. Meanwhile Lithograph pattern weak points, Trench hard mask etch CD variation around wafer and Double Patterning odd/even effect will increase Wet process difficulty.

For lower RC delay, Ultra Low-K is introduced into CMOS manufacturing. ULK material property is porous and low-density, Dry etching process use TFMHM approach to achieve acceptable low-K damage [1]. As shown in Fig.1, after Dry etching process, trench/via ULK sidewall and via bottom copper is exposed in Wet strip process, when ULK structure easily absorbs moisture, Wet strip process needs to avoid ULK K value shift. On the one hand polymer residue must be removed clear, on the other hand hard mask TiN needs to be pulled back or removed for better barrier layer gap filling. At ESL liner remove step in AIO etch, process must has some OE (Over-etch) for via open concern which will leads to underneath Cu expose in fluorocarbon plasma. The F content will catalyzes Cu-oxidation [2], for lower Via interconnect resistance, this damage layer also needs to be removed during Wet strip. But this Copper loss must be well controlled, or it will causes device reliability issue when this "smiling curve" is too big.

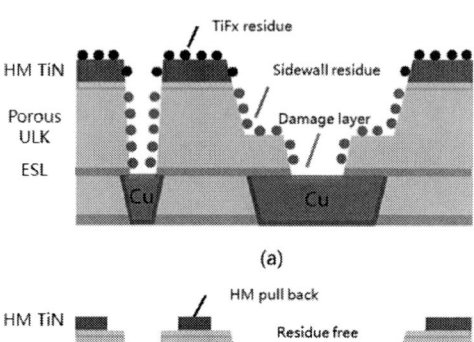

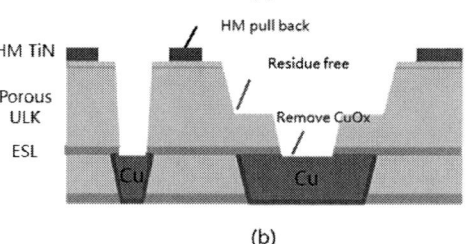

Fig.1 MHM Wet strip process sketch map: (a) after All-in-One etch, (b) after Wet strip process.

In this paper, we investigate Wet strip parameters to improve cleaning capability. By the introduction of Wet strip optimized condition, we make much lower metal void defect density and get better device yield.

Experiments

After All-in-One Etch including partial via etch, trench etch and etch stop liner etch, Wet strip process is performed on a commercial single wafer clean tool. Then barrier layer and Cu seed layer is deposited by PVD tool. After copper electroplating and CMP planarization, metal void defect is detected by a bright field scan tool and the electrical character is measured on a commercial wafer acceptance test tool.

For HM TiN pull back and fully removal approaches, we study two types of TiN etch rate by the adjustment of Wet strip parameters. Wet strip's effect on ULK film K value is evaluated with measurement of ULK blanket wafers K-value before and after Wet strip. Not only N2 dry mode, but also IPA dry mode is tested, IPA dry will benefits pattern collapse probability as lower surface tension [3]. Copper loss of Via bottom is judged by cross section TEM (Transmission Electron Microscope).

Results and Discussions

Polymer Residue Removal

All-in-One Etch is performed on a commercial CCP

(capacitive coupled plasma) etcher with CxFy based gas. Etch loading effect makes polymer residue distribution difference from wafer center to wafer edge. Non-optimized Wet strip condition can't remove wafer center polymer that will induce to metal void defect. As shown in table 1, Wet strip process usually adopts 2 chemicals solution.

Wet Conditions	Step1		Step2		Dry mode	Defect
	Chemical 1	Parameters	Chemical 2	Parameters		
1	Type A	Setting I	Type B	Setting II	N2	Worse
2	Type B	Setting II	Type A	Setting I	N2	Worse
3	Type A	Setting III	/	/	N2	Better
4	Type A	Setting III	Type B	Setting IV	N2	Best
5	Type A	Setting III	Type B	Setting IV	IPA	Best

Table 1 Wet conditions tuning table, two chemicals (type A & B) are adopted in Wet strip, tuning parameters setting leads to very low level of metal void defect

By tuning sequence of cleaning chemical and process parameters (like Temperature, chuck speed, liquid flow, nozzle scan, mix ratio, etc.), Wet condition 4 & 5 show much better cleaning efficiency that leads to remarkable metal void defect reduction. Fig.2 shows metal void defect center map and typical defects, metal void defect center map is obviously alleviated by the introduction of Wet condition 4. To Verify Wet condition #04 cleaning capability, Dry etch increases PV (partial via) etch time to make much richer polymer residue [4], optimized Wet condition still is capable to remove it completely.

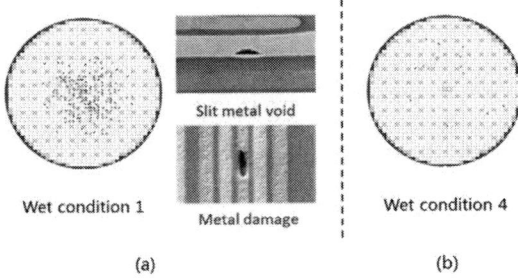

Slit metal void

Metal damage

Wet condition 1   Wet condition 4

(a)   (b)

Fig. 2 After Cu CMP defect scan map
(a) Non-optimized Wet condition has wafer center defect (b) Optimized Wet condition eliminates center defect

Wet strip process final step has two choices: High-spin + N2 and High-spin + IPA/N2. For high AR pattern collapse concern, IPA (Isopropyl Alcohol) Dry mode reduces the collapse force by ~3x compared to N2 Dry, as the surface tension of IPA is around one third of DIW. Moreover IPA Dry has watermark free benefit.

HM TiN Pull Back / Removal

After Wet strip process, the barrier layer is deposited by PVD. At advanced technology nodes narrow trench Gap-fill process window is very marginal, HM pull back or total removal is proposed to reduce Via hole's aspect ratio. So Wet strip process must be tunable for etch rate of TiN, that value is around 20~150A/min. Two types of TiN wafer A and B (TiN B is tougher than TiN A) are tested, TiN ER meets target after tuning Wet strip conditions as shown in Fig. 3.

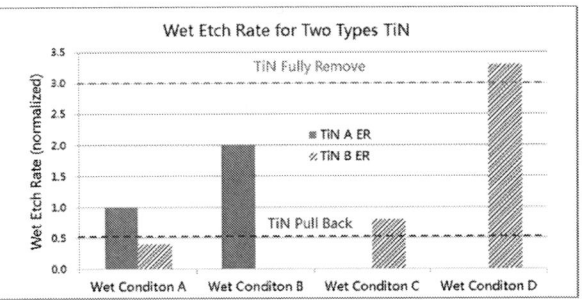

Fig. 3 Tuning Wet conditions for TiN pull back and fully remove approach (A/B/C/D stands for different Wet conditions)

ULK K-value Shift

Ultra Low K physical structure is porous and low density, and it easily absorbs moisture that causes K-value increase. As ULK sidewalls expose during Wet strip, it is very critical to control ULK material K-shift from Wet strip process. As shown in Fig.4, Fined Wet strip process barely has influence on ULK K-value, meanwhile single wafer anneal process is generally adopted to recover ULK film property.

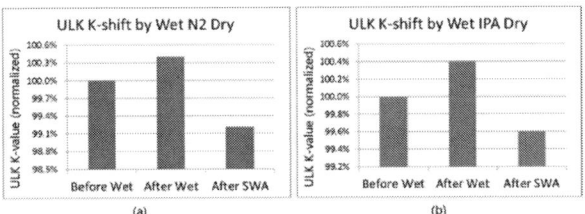

Fig.4 ULK K-value shift, after Wet process increase 0.3%, after Single wafer anneal K-value recover
(a) Wet N2 Dry, (b) Wet IPA Dry

Cu Damage Layer Removal

AIO etch must has OE (over-etch) amount at final liner open step. This OE makes underlying Cu expose in fluorocarbon plasma that leads to Cu damage layer. It's important to control dry etch OE amount for the reduction of plasma damage, but dry etch is a trade-off between Cu damage layer and Via open risk. This loose Cu is significantly oxidized as F element in polymer residue catalyzes this oxidation. Wet strip process need to remove this copper oxide, otherwise Cu interconnect resistance will incidentally increases.

As shown in Fig. 5, Via bottom copper is slightly recessed that like smile curve. This copper recess also needs to be well controlled, for too much recess will causes undercut on via bottom that will be a disadvantage for device reliability.

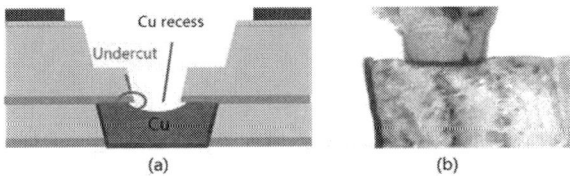

Fig. 5 Via bottom copper recess. (a) Too much Cu recess sketch map, (b) Controlled Cu recess TEM

Copper Interconnect Resistance

It is a common sense to reduce Cu interconnect resistance for RC delay reduction. Usually Dry etch profile, barrier layer thickness dominates this resistance, but insufficient cleaning also has negative effect. We use single via and via chain structures to evaluate Wet strip process under same Etch / PVD / ECP/ CMP conditions. When single via open in via chain structure that generally blocked by residue at via bottom, it will leads to high resistance. If ULK sidewalls are rough, all via structures will widely suffer increasing resistance. As shown in Fig. 6, Single via resistance can be slight improved by tuning Wet condition 4, no outliers in via chain structure indicates all via bottom residue is sufficiently removed.

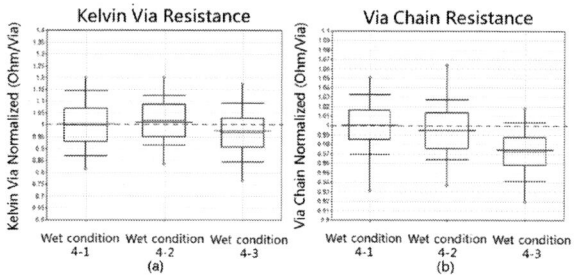

Fig. 6 Copper interconnect resistance comparison of tuning Wet condition 4 (a) Single Via structure, (b) Via Chain structure

Device Yield and Reliability

Wet optimized condition sufficiently eliminates polymer induced killer defect—metal damage, wafer center no longer suffers severe yield loss. As shown in Fig. 7, optimization of Wet strip achieves increasing 20 percent of device yield, no matter the different Front End of Line conditions.

Device reliability by Wet optimized condition is qualified, downstream EM is a good method to evaluate process integration of via bottom interface, downstream EM MTTF shows Wet optimized condition gets a little improvement. Upstream EM is slightly meliorated and TDDB performance is comparable that means ULK sidewalls roughness also is well recovered.

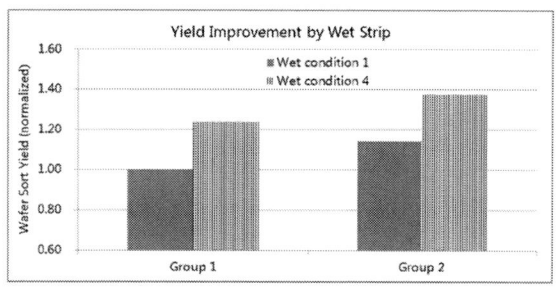

Fig. 7 Wet optimized condition achieves 20% yield improvement. (Group 1 & 2 is Front End of Line split)

Conclusions

Cleaning efficiency is enlarged and killer defect - metal damage is almost eliminated by systematically study of Wet strip process parameters. HM TiN etch rate is tunable for different Wet conditions, so HM pull back and fully removal are choices for process integration. ULK material K-value is well controlled during Wet process, K-value added~0.02 will be recovered after single wafer anneal. Copper damage layer is slightly removed, because this layer that easily be oxidized will results in copper interconnect resistance increase. Copper interconnect resistance can be slightly improved by Wet optimized condition, and no Via open fail happens in Via chain structure. Device yield adds 20% when metal void is obviously reduced; Device RE character is also qualified of Wet optimized condition.

Acknowledgements

The authors would like to thank SMIC BEOL PIE and Technology R&D Modules for the technical discussions.

References

[1] Struyf, H. et al, Low-damage damascene patterning of SiOC(H) low-k dielectrics, 2005

[2] Takao Kamoshima, et al, Ambient gas control in slot-to-slot space inside FOUP to suppress Cu-loss after dual damascene patterning, 2007

[3] Vos, I. Hellin, D. et al, Silicon Nano-Pillar Test Structures for Quantitative Evaluation of Wafer Drying Induced Pattern Collapse, 2011

[4] Junqing Zhou, et al, The Cu Exposure Effect in AIO Etch at Advanced CMOS Technologies, 2016

The Technology Trend of IC Manufacture During Post Moore's Era

Hanming Wu[1], Jin Kang[2,3], Minhwa Chi[2], Xing Zhang[3], Tong Feng[2] and Poren
[1] BriteIP Corporation, China
[2] Semiconductor Manufacturing International Corporation, Shanghai and Beijing, China
[3] Peking University, Beijing 100871, China
*Corresponding Author's Email: Hanming_Wu@smics.com

ABSTRACT

The Post-Moore era implies that the feature size is no longer a single parameter to indicate an advanced technology node, instead, variety of functions and low power consumption become top priority index. Therefore, more innovation must come out in near future to promote the manufacture technology development. In the present paper, the most promising new technologies have been reviewed, i.e. More Moore, More than Moore and Beyond Moore.

INTRODUCTION

The Moore's Law era has indicated by a long run of lithographically-enabled pitch shrinking for more than 50 years. It is directly reduced the cost per (von Neumann) function, as well as system power and performance improvements [1, 2].

In each technology node, there were a couple of critical process technologies to support manufacturing technology moving forward as Moore's law rhythm. Main issues are discussed in part II. Several studies of future technologies are briefly reviewed in part III.

MAIN CHALLENGES IN MOORE'S LAW ERA

There are five main challenges for chip process technology of manufacturing during the Moore's law era.
(1) Patterning Transfer
How to achieve 20nm CD with 193nm wave length become extremely difficult. Conventional 193 nm immersion lithography has a difficulty to resolve features below 80 nm pitch in a single exposure [3]. The advanced optical proximity correction (OPC) and multi-patterning with immersion provides the approaches to the success of process technology. Three metal level 56nm-pitch Cu dual damascene interconnects in k2.7 low-k ILD have been demonstrated by using sidewall-image-transfer (SIT) patterning scheme to investigate the feasibility of the SIT process for sub50nm-pitch technology node. Figure 1 shows schematics and cross-sectional STEM images of different pitch and constant pitch lines Cu line pitch investigation [4].

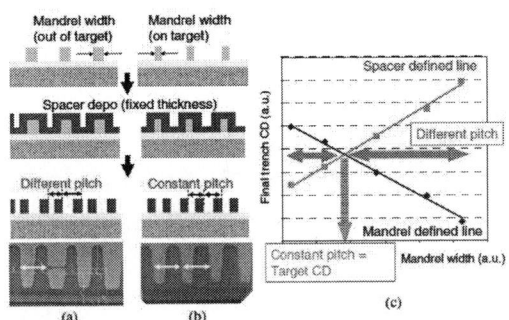

Figure 1: Cu line pitch investigation. Schematics and cross-sectional STEM images for (a) different pitch and (b) constant pitch lines, and (c) Cu line pitch optimization by mandrel resist width control.

(2) New Materials And Processes
A growing numbers of new elements are applied by chip maker since this century. There are 64 elements have been used in either device or process tuning. The fact is these new materials that supported the 70% performance improvement. All new material application needs numerous process and integration experiments. A study to grow graphene directly on patterned Cu wires below 400 °C, within the thermal budget of back-end-of-line processes (BEOL), which exhibits 2× lower resistivity, 1.4× higher breakdown current density and 40× longer electromigration (EM) lifetime than as-deposited Cu. Figure 2 shows the process steps for fabricating the test structure [5].

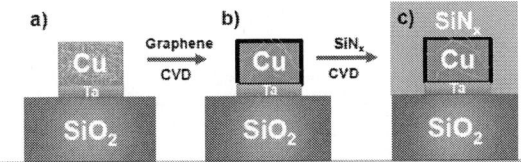

Figure 2: Cross-sectional view of the process steps for fabricating the test structure. a) Ta/Cu is sputtered onto the SiO2/Si substrate followed by metal lift-off. b) Graphene is grown directly on the patterned Cu wires by CVD below 400 °C. c) The whole structure is then capped with SiNx as a protection layer against oxidation under high temperature stress testing
(3) Process Fluctuations

978-1-5090-6695-7/17 $31.00 © 2017 IEEE

Following the CD shrinking, the process fluctuation becomes very critical. For example, FinFET Vt dopant fluctuation should be less than 100 atoms. This fluctuation should be controlled accurately and precisely. The small aspect ratio device has greater immunity of random dopant fluctuation (RDF), while suffers from process variation effect (PVE) and work function fluctuation (WKF). Figure 3 shows the comparison of fluctuation sources. Figure 4 (a)-(c) shows the fluctuated ID-VG curves by RDs in the channel and penetration from S/D extension and the extracted DC parameters. The cumulative probabilities of Ion, Ioff and Vth for different RDF mechanisms are shown in Figs. 4(d)-(f) [6].

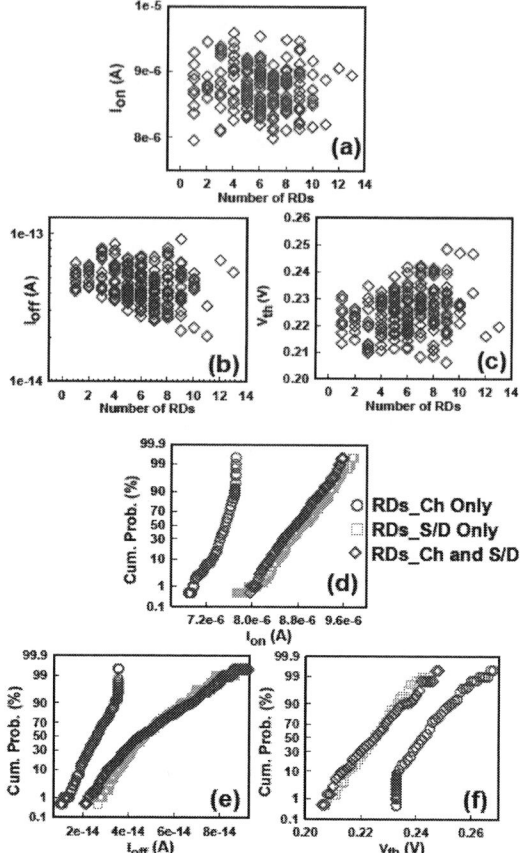

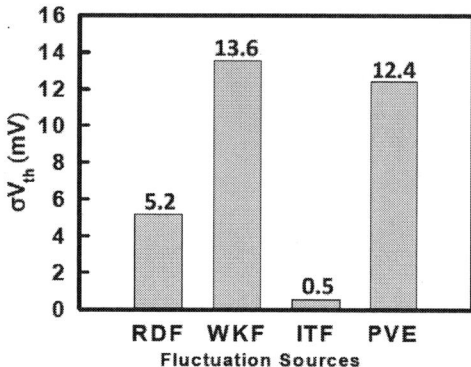

Figure 3: The fluctuation sources comparison; where the variation induced by interface trap fluctuation (ITF) (D_{it} < 10^{12} $eV^{-1}cm^{-2}$) is much less than the variation induced by RDF, WKF, and PVE, respectively. Hence, we neglect the ITF effect in the following discussion.

(4) New Architecture

Beginning in early 21 century, the traditional transistor's channel length shrink into 20/14nm technology node, the step of the farther breakthrough becomes slowed to a crawl. Fortunately, we still created several great CMOS inventions to continue reduce the size and improve its performance simultaneously.

a) 3D transistor FinFET has been developed to suppress the short channel effect. It is proved that 3D transistor can effectively reduce Vdd and leakage while increase the Ion significantly. FinFET turns the channel to vertical from horizontal, and allows the gate to wrap around three sides of this block of silicon, which the ratio of its height and width makes it looks exactly like a fish fin, we called this type device FinFET transistor. It were widely using for the advanced technology, two key features of FinFET transistor fueled its spectacular growth of marketing occupation on semiconductor manufacturing field. The special structure improves its electrostatics

Figure 4: The extracted DC parameters such as (a) Ion, (b) Ioff and (c) Vth based on the simulated ID-VG curves induced by combined random dopant in the channel and from source/drain, simultaneously. The cumulative probability of (d) Ion, (e) Ioff and (f) Vth with respect to different RDF mechanisms, where RDs_Ch, RDs_S/D, and RDs_Ch_and_S/D represent dopants in the channel, dopants from S/D, and the combination of them, respectively. The trend of their fluctuation is governed by penetration dopants from S/D extension

performance, a fully depleted channel, which leads good SCE, Lower DIBL and SS. In addition, it also only needs very low operation voltage: Vdd < or = 0.8V. Another feature is the benefit of its dynamical performance, caused by the strong controlling ability of the gate. Comparing with other devices, the FinFET process speed improve more than 35% (at the same power), or power reduce more than 55% (at the same speed).

b) HKMG, We can simply replace the gate material and gate dielectric by Metal and Hf-based High-K material for achieve a HKMG transistor. Comparing with general CMOS, High-K can reduce by orders of magnitude of the gate leakage. Surface phonon scattering in high-K is primary source of mobility degradation. Metal gate is effective for screening phonon scattering and improves channel mobility.

c) Strained Si, To keep the downscaling trend of Moore's law, staining engineering is an effective approach to improve the mobility of electrons and holes. We knew that electron mobility increases with tensile strain and decreases with compressive strain. On the other hand, the hole mobility increases with both signs of strain and the effect is more significant for compressive strain because the hole effective mass decreases with compressive strain but increases with tensile strain [7].

Besides, backend 3D package is an option to cope with the increasing RC delay power consumption. A WOW process and its prospects for realizing Tera-scale memory capacity with high energy efficiency, counter to conventional scaling, and combination with three-dimensional stacking to overcome the problems associated with scaling. Figure 5 shows a comparison of conventional wafer process and a use of WOW 3DI assuming 10-stack dies from 300-wafer having 1000-chips.

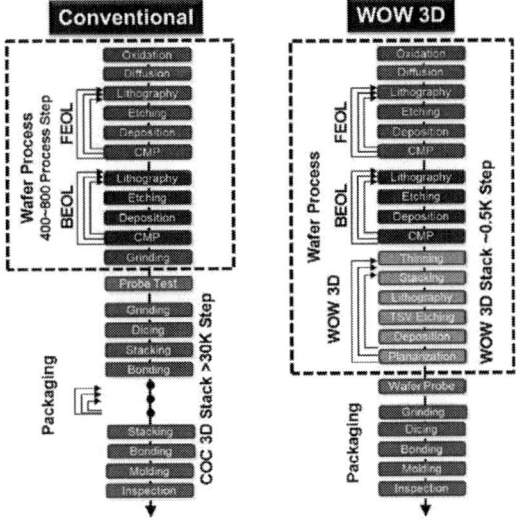

Figure 5: A comparison of conventional wafer process and a use of WOW 3DI assuming 10-stack dies from 300-wafer having 1000-chips

(5) Process Integration and Yield Enhancement

As we can see that the process steps increased tremendously as the technology moving forward. For example a typical 65nm technology consists of about 600 process steps while 14nm requires at least 2000 process steps. More and more in-line metrology and yield enhancement tools should be developed to cope with the issue. In the past more than 40 years, transistor and interconnect pitch scaling has been used successfully to drive significant density and performance benefits in integrated circuits. Transistor performance has continued to improve due to pitch scaling combined with other process enhancements. Interconnects represent a much larger portion of the overall delay and cost of integrated circuits today than in the past [8].

All these five process challenges become a barrier for future post Moore's technology. Figure 6 shows the trends of density, speed, power and cost per transistors, and we can see that the speed and power improvement slow down, and cost goes to flat and even reversal. We can see that the Moore's law is going to the end. From the technology side, G. Moore predicts that the end of Moore's law will come from 2015 to 2025(see Fig. 7). It is expected that physical gate length 10nm would be the end of shrinking (see Fig. 8).

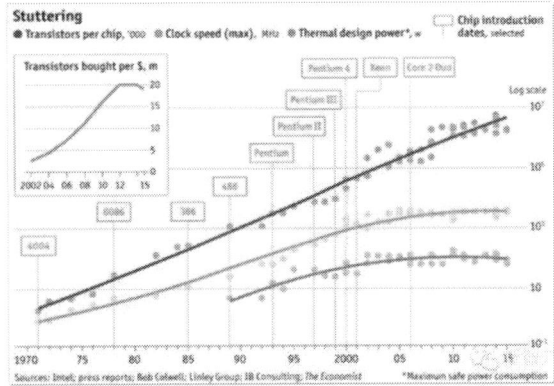

Figure 6: Feature of Post-Moore's Law.

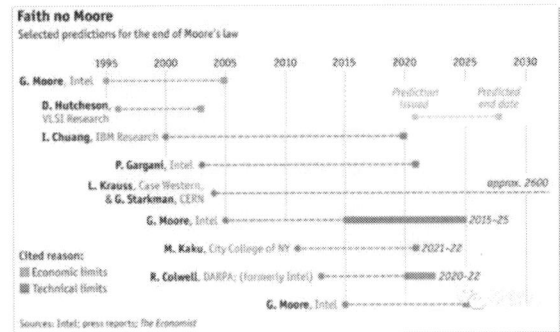

Figure 7: Prediction of Moore's Law.

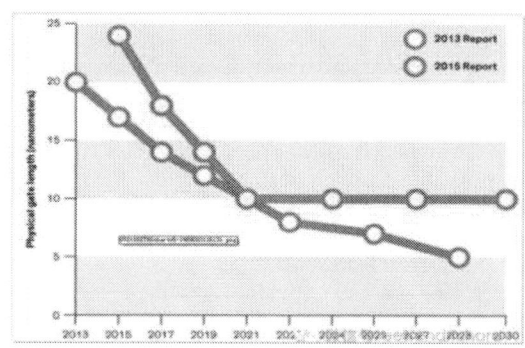

Figure 8: Gate physical length would stop at 10nm as ITRS2015 predicted.

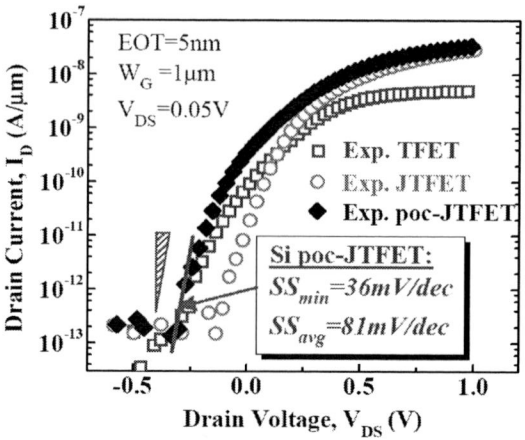

Figure 9: Measured transfer characteristics of the fabricated Si poc-JTFET, Si JTFET and Si TFET in the same die. Poc-JTFET shows the steepest SS_{min} of 36mV/dec within one decade of drain current.

INNOVATIONS IN POST-MOORE'S ERA

It is realized that some innovations become necessary conditions to continue CMOS technology development. Recently numbers of leading companies and universities have made a great progress. The following part roughly summarizes these promising technologies.

(1) New Architecture

a) TFET is believed one of the best for the ultra-low power devices. It is because of the steep S factor that can be well below 60mv/dec while Vdd can be as low as 0.1V. The innovative 2D heterojunction TFET is capable of scaling up to Lg=3nm, and can be proposed to high-density 3D logic integration as a promising pathway for beyond Lg = 3 nm, in terms of alleviating the challenging issue of thermal dissipation in 3D ICs [9].Some experiments shown that Based on traditional CMOS-compatible technology on bulk silicon substrate, the fabricated Si JTFET with an optimized pocket shows SSmin of 36mV/dec. (see Fig. 9) [10]. Another experiment shown that a highly manufacturable Complementary TFET foundry technology on Si CMOS baseline platform is presented, showing the superior characteristics with high manufacturability. By new TFET device design, improved performance and variability simultaneously are experimentally achieved, and circuit-level implementation shows significant operation speed enhancement (up to 93%) and energy reduction (by 66%) at VDD of 0.4V, as well as remarkably suppressed variation, indicating its great potential for ultralow-power applications [11].

However, there are three major issues need to be solved, i.e. JNT fabrication, huge process fluctuation and switch speed slow.

b) All Gate Around is another candidate for its better device current control. The ratio of Ion/Ioff can be as high as 2.8×10^8. Some analog circuit has been fabricated. It is shown that the application potential is quite optimistic. Recently, some experiments have done to demonstrate on the vertically stacked gate-all-around (GAA) silicon nanowire MOSFETs with matched threshold voltages (VT,SAT ~ 0.35V) for N- and P-type devices. The VT setting is enabled by nanowire-compatible dual-work-function metal integration in a high-k last replacement metal gate process [12].

(2) New Materials

a) SiGe and III-V seem very promising technology. As we mentioned, there are more than 70% performance improvements base on new material applications. SiGe promotes ion mobility while III-V helps electron moving. SiGe for channel material has been explored as a major technology element after the introduction of FINFET into CMOS technology. For instance, some experiments shown that the optimized integration of SiGe FIN for pFET into a standard 10nm platform process flow and capable to extend to 7nm (see fig. 10), which the CMOS FINFET with Si-FIN nFET and SiGe-FIN pFET is demonstrated as a viable technology solution forboth server and mobile applications at 10nm node and beyond. Some critical barriers for manufactured applications still need to study, such as defect related atomic level dynamic study, and charge transportation mechanism in nano-scale, etc.

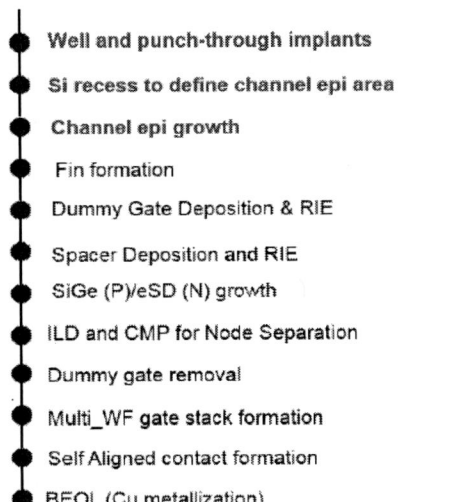

- Well and punch-through implants
- Si recess to define channel epi area
- Channel epi growth
- Fin formation
- Dummy Gate Deposition & RIE
- Spacer Deposition and RIE
- SiGe (P)/eSD (N) growth
- ILD and CMP for Node Separation
- Dummy gate removal
- Multi_WF gate stack formation
- Self Aligned contact formation
- BEOL (Cu metallization)

Figure 10: 10nm integration flow with optimized process steps for SiGe FIN.

b) CNT is promising candidate for future computer chip because of its high frequency and low power property [13]. It is shown that based on 5nm node, CNTFET has more advantage than FinFET (see Fig.11). Recently, 3D computer system was proposed (see Fig.12) [14]. It is founded that there is still long way to go for manufacture due to serious fluctuation during fabrication (see Fig.19). Some experiments demonstrate the—Fully wrap-gated carbon nanotube (CNT) transistors with vertically suspended (VS) semiconducting single-walled CNTs, purified up to 99.9%, which approach will increase the applicability of CNTs toward high-performance emerging materials [15].

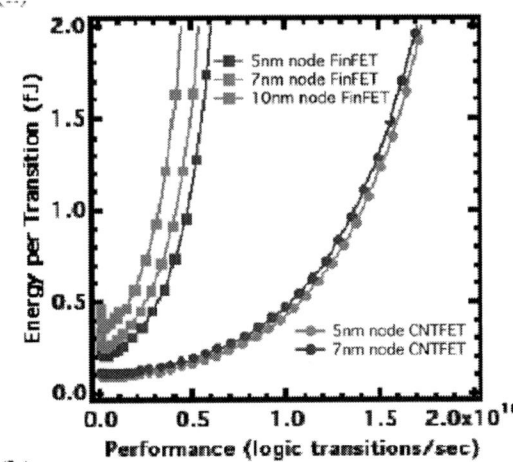

Figure 11: Comparison of FinFET and CNFET.

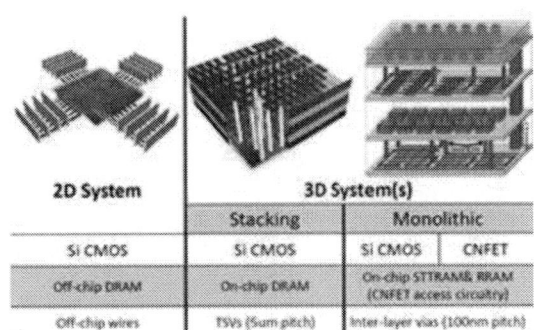

Figure 12: N3XT: Nano-Engineered Computing Systems Technology.

c) 2D materials are very attractive to future application. The current study mainly focus on the material property, such as 2D-crystal materials, e.g., MoS2/WSe2/BN/graphene. It is shown that high work function MoOx contact can play a role of efficient hole inject layer to MoS2 WSe2. With this contact they demonstrated P-type transistor [16]. Some experiment shown that A U-shape MoS2 pMOSFET with 10nm channel and poly-Si source/drain is demonstrated. Fig. 13 shows process flow of MoS2 pMOSFET fabrication [17].

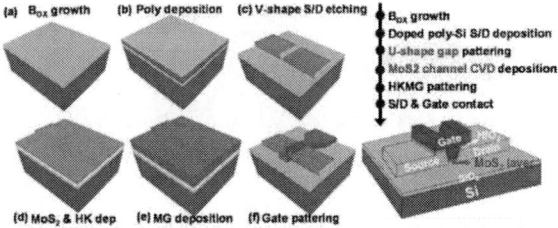

Figure 13: Process flow of MoS₂ pMOSFET fabrication. 10nm channel length was achieved.

d) New Process Materials

It is well known that the process for nano-scale manufacture is very important. Non-equilibrium plasma physics and chemistry need more fundamental research work. New electrode material study for MRAM and RRAM will keep hot for a long time. Ultra-low K material is still to be developed for better mechanical property. These process consumable materials should be key technologies to push the technology progress. Recently, 4Gbit density STT-MRAM using perpendicular MTJ in compact cell was successfully demonstrated through the tight distributions for resistance and magnetic properties, which will brighten the prospect of high-density STT-MRAM [18].

978-1-5090-6695-7/17 $31.00 © 2017 IEEE

(3) New Design Architecture

At the backend interconnection, some new structure come up for application such as airgap technology application for metal connection. As the ultra low-K limitation, air-gap technology was proposed to replace low-K technology. The air-gap can be produced by PECVD process which forms overhead geometry as the deposition rate is high enough (see Fig. 14) [19]. It is reported that Intel used for 14nm node [20] (see Fig.15). Some experiment shown that an air gap integration scheme to achieve 17% RC benefit [21] (see Fig. 16, 17). Recently 3D chip attract industry attention because it can combine slices of memory with slices of processing logic together. Therefore it can reduce the signal transfer delay and achieve high performance. The key issue for 3D package is the thermal diffusion. It is very difficult to make the heat source less than heat transfer and keep the chip cool. Besides, more than 80% metal pins are used for power transfer while only less than 20% for signal transfer.

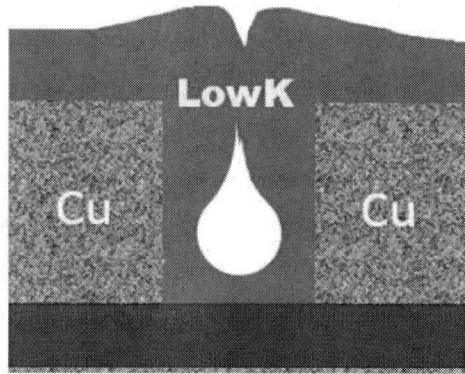

Figure 14: Plasma technology: PECVD feature generate uniform water drop-like void.

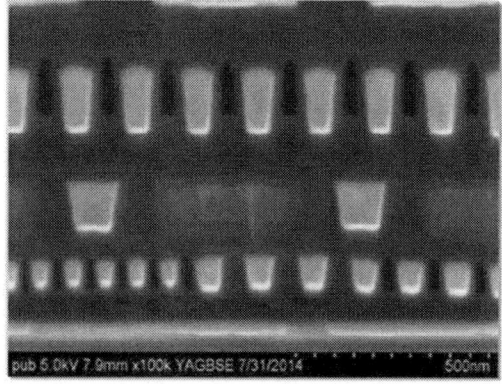

Figure 15: Intel air-gap application. 80nm and 160 minimum pitch layers and provide a 17% improvement in capacitance.

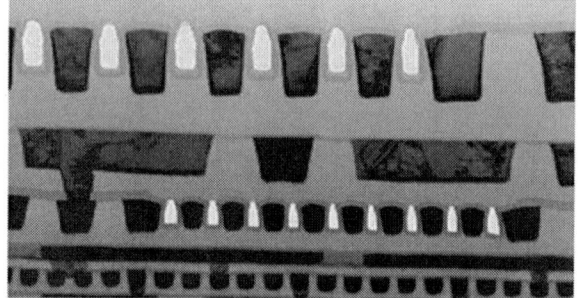

Figure 16: TEM showing air gap at Metal 4 and Metal 6.

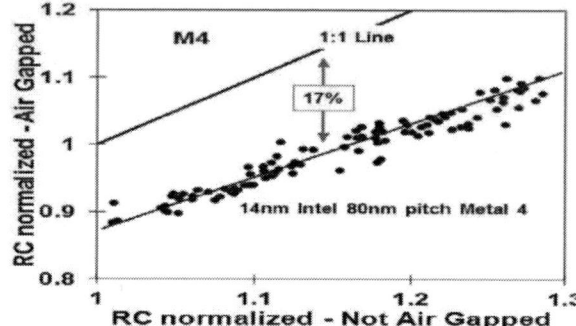

Figure 17: Air gap vs none air gap interconnects demonstrates from 14% to 17% RC benefit.

(4) IP and IP integration

As the feature structure shrinking to nano scale and market coming to clastic, the chip design cost would become really significant to most design houses. Therefore, it is necessary to enhance the construction of design IP platform that would support innovative chip design. As we can see the fact that more than 90% design houses have to purchase 3rd party IP for their products. There is very few design houses owners self-developed IP. From the view of manufacture, it is found that more than 75% wafers includes 3rd party IP. To reduce the cost of design, the public IP platform should play a critical key role during post-Moore's era, especially to those young innovative design houses.

CONCLUSION

As Moore's law rhythm get slow down after more than half century rapid growing, some new technologies come up. Some of them are even destructive innovation which may lead CMOS to a revolution way. Industry expect the new ear coming. More attention to fundamental research is necessary to cope with the uncertainty of the future IC world.

REFERENCES

[1] R Huang, Science in China Series F: information Sciences;

[2] Greg Yeric, IEDM15-4

[3] James Hsueh-Chung Chen et al., pp. 1 - 3, DOI: 10.1109/IITC.2012.6251637

[4] M. Tagami et al., pp. 1 - 3, DOI: 10.1109/IITC.2012.6251664

[5] Ling Li et al., IEDM16-241

[6] Yiming Li et al., pp. 34.4.1 - 34.4.4, DOI: 10.1109/IEDM.2015.7409827

[7] Decai Yu, et al., DOI: 10.1103/PhysRevB.78.245204

[8] Ruth Brain, IEDM16-232

[9] Kaustav Banerjee，Nature 2015, pp.12.3.1 - 12.3.4, DOI: 10.1109/IEDM.2015.7409682

[10] R Huang et al., pp. 8.5.1 - 8.5.4, DOI: 10.1109/IEDM.2012.6479005

[11] R Huang et al., pp. 22.2.1 - 22.2.4, DOI: 10.1109/IEDM.2015.7409756

[12] H. Mertens et al., IEDM16-524

[13] IBM ACS Nano 8(2015) 8730-8745

[14] S Mitra, Nano-Engineered Computing Systems Technology, Computer 24(2015)

[15] Dongil Lee et al., IEDM16-115

[16] Steven Chuang, Nano Lett, 2014, 14

[17] Kai-Shin Li et al., pp. 1 - 2, DOI: 10.1109/VLSIT.2016.7573375

[18] S.-W. Chung et al., IEDM16-659

[19] Yong Liu and Hanming Wu, 2003，patent number ZL 02118928.5；

[20] S. Natarajan et al., pp. 3.7.1 - 3.7.3, DOI: 10.1109/IEDM.2014.7046976

[21] K. Fischer et al., pp. 5 - 8, DOI: 10.1109/IITC-MAM.2015.7325600

[22] K. Cheng et al., IEDM16-444

FIN CRITICAL DIMENSION LOADING CONTROL BY DIFFERENT FIN FORMATION APPROACHES FOR FINFETS PROCESS

Qingpeng Wang[1]*, Gang Mao[1], Hai Zhao[1], Cheng Li[1], Fangyuan Xiao[2], Rex Yang[1], Shaofeng Yu[1]
[1]Advanced Technology Research and Development, SMIC ATD, Shanghai, 201203, China
[2]Technology Research and Development, SMIC, Shanghai, 201203, China
*Corresponding Author's Email: Qingpeng_Wang@smics.com

ABSTRACT

FinFETs become dominant transistor architecture to replace planar metal-oxide-semiconductor field-effect transistors (MOSFETs) for larger effective channel width and better short channel controllability. Fin critical dimension (CD) of cross-fin direction is one of the most important parameters which impact MOSFET subthreshold swing and threshold voltage. However, CD loading between isolated and dense fins are naturally unavoidable in fin etch and STI filling process for different environments in these two areas. This paper mainly focuses on the fin formation process to give a demonstration of device performance and fin CD uniformity between different fin formation approaches.

INTRODUCTION

Continuous scaling down to gain both transistor density and electrical performance for bulk planar CMOS devices beyond 20 nm node become increasingly difficult owing to the poor short channel controllability [1]. FinFETs become dominant transistor architecture to replace traditional bulk planar MOSFET [2]. Its superior performance mainly gains from two parts: 1) The 3D tri-gate structure make the effective channel width much larger and not be strictly limited by the certain gate pitch, thus both the device density and on-state drain current (I_{Don}) could continually increase; 2) The fully depleted fin structure make the short channel controllability much better, thus both the sub-threshold swing (*SS*) and off-stage drain current (I_{Doff}) become lower. For both parts, fin profile, including fin critical dimension (CD), sidewall angle (SWA), height (H) etc., are key parameters which decide the final electrostatic performance of the device.

However, the fabrication process of the FinFETs, especially for the fin CD loading control, face more challenges because of the complicated 3D structure and process variation. CD loading between isolated and dense fins are naturally unavoidable for different environments in these two area. On one hand, etch loading is unavoidable in fin etch process since the ratio of enchant to etched Si are quite different between dense and isolated area, thus the isolated fin always shows larger CD after fin etch. On the other hand, even with uniform fin CD, the isolated fin will loss more in the following STI filling process due to higher oxygen environment in the isolated area.

This paper mainly focuses on the fin formation process to give a demonstration of fin CD uniformity and device performance between different fin formation approaches.

EXPERIMENTS

A schematic process flow for two different fin formation approaches is shown in Figure 1. The self-aligned double patterning process (SADP), including fin core lithography/etching, fin-core spacer deposition/etching, is used for fin formation to achieve more tighten fin pitch [3].

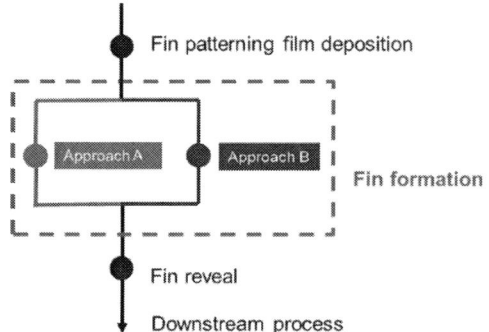

Figure 1: Simple process flow for two different fin formation approaches.

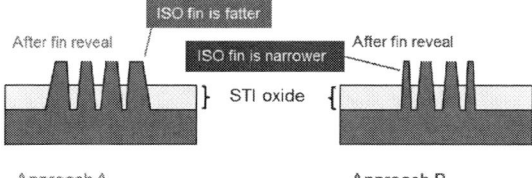

Figure 2: Illustration for two fin formation approaches.

In the following fin formation process, two different approaches are done. Then STI recess process is done to reveal the active fin parts. The fin CD loading is quite different between two fin formation approaches, 1) approach A: Isolated fins are fatter after fin reveal process; 2) approach B: Isolated fins are narrower after fin revealing process. The downstream process are standard FinFET fabrication process including dummy poly gate formation, source and drain strained epitaxy growth,

978-1-5090-6695-7/17 $31.00 © 2017 IEEE   216

replacement metal gate, source/drain contact formation, etc. Figure 2 shows illustration of two fin formation approaches, which lead different fin CD loading.

RESULTS AND DISSCUSSION

Firstly, we checked fin profile after fin revealing by both transmission electron microscope (TEM) and scanning electron microscope (SEM). Two nominal structures with 4 fins and 1 fin were used for demonstration. In the 4-fin structure, the center fins are in relatively dense environment while the edge fins are in semi-isolated environment. The 1-fin structure is in a fully isolated environment. With both 4-fin and 1-fin structures, CD loading among dense, semi-isolated and fully-isolated area could be observed at the same time.

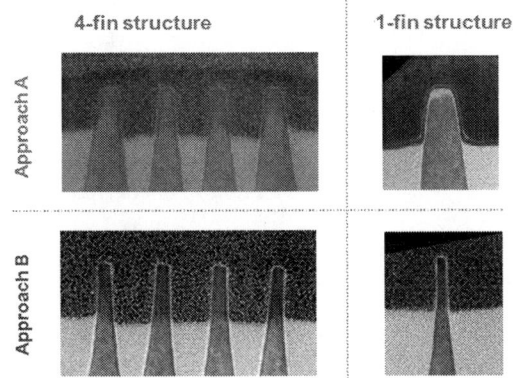

Figure 3: TEM cross-sectional image of 4-fin and 1-fin structures

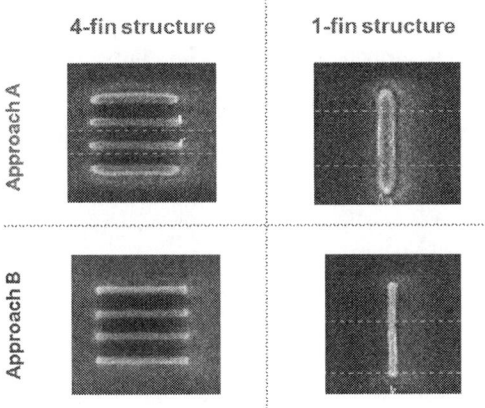

Figure 4: SEM top view of 4-fin and 1-fin structures

Figure 3 shows the TEM profile for 4-fin and 1-fin structure fabricated by both fin formation approaches. Figure 4 shows the SEM top view of these structures after fin revealing. Due to the etch environment difference, the isolated fin shows much larger CD with more taper profile

than the dense fin in approach A. On the contrary, the isolated fin is narrower than the dense fin in approach B.

Electrical test was done to compare performance with different fin CD loadings. Figure 5 shows the universal curve (I_{Doff}-I_{Dsat}) of PMOS and NMOS 4-fin devices with arbitrary unit. Obvious performance gain could be observed in device with approach B comparing with that in approach A for both NMOS and PMOS. The main improvement comes from the reduction of I_{Doff} for devices with approach B.

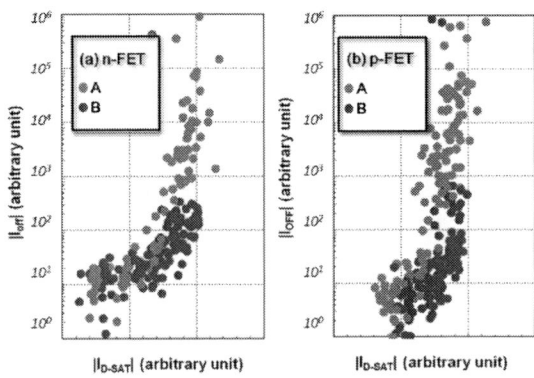

Figure 5: universal curve of 4-fin devices.

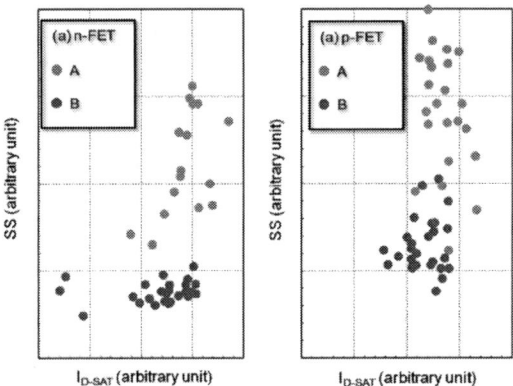

Figure 6: Subthreshold swing of 4-fin devices.

The reduction of I_{Doff} shows the device fabricated by approach B has much better short channel controllability which could be indicated by the key parameter of subthreshold swing (*SS*). The calculated value of *SS* for both NMOS and PMOS of 4-fin device are shown in figure 6. An obvious performance gap between device with approach A and approach B could be observed from *SS* for both PMOS and NMOS.

The higher off-stage drain current may mainly come from the fat fin on the edge side of the 4-fin structure which could not be fully depleted at subthreshold region. Figure 7 shows a simple illustration to describe the cause of the higher I_{off}. On one hand, the un-depleted parts of the fat fin would be influenced more by drain to source electrical filed, thus lead short channel effect and higher I_{off}; On the other hand, the depletion process itself will introduce extra depletion capacitance C_D, lead worse SS and finally increase I_{off}.

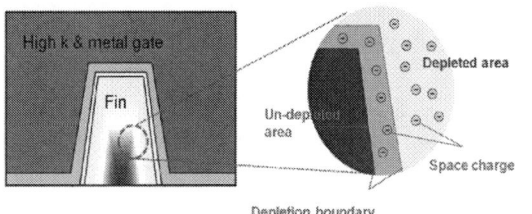

Figure 7: Depletion status in fat fin of n-FET.

Equation (1) shows the relationship between SS, the gate oxide capacitance C_{ox} and depletion capacitance C_D [4], where k the Boltzmann constant, T the room temperature, q the electron charge quantity, C_{it} the interface trap introduced capacitance, ΔV_G the variation of gate voltage, ΔQ_D the variation of depletion charge quantity due to depletion boundary change with ΔV_G. Here, we assume the interface trap introduced capacitance C_{it} is the same for both fat fin and normal fin. Due to depletion width change with gate voltage, an extra C_D would be introduced in fat fin, the final SS would be larger. In a thinner fin, the fin is fully depleted, the quantity of depleted space charge is stable, no much extra C_D be introduced, thus SS would be much smaller and I_{off} could be further reduced.

$$SS = \ln(10)\left(\frac{kT}{q}\right)\left(\frac{C_{ox} + C_D + C_{it}}{C_{ox}}\right) \quad (1)$$

Here,

$$C_D \approx \frac{\Delta Q_D}{\Delta V_G}$$

Beside 4-fin device, we also checked the electrical performance of 1-fin device. Figure 8 shows the $V_{T\text{-}SAT}$-$I_{D\text{-}SAT}$ results of 1-fin device for both n-FET and p-FET with arbitrary unit. An obvious $I_{D\text{-}SAT}$ drop could be observed for 1-fin device fabricated with approach B. Unlike the 4-fin device which shows obvious performance gain due to the reduction of I_{off} by approach B, the 1-fin device is with worse performance due to the obvious $I_{D\text{-}SAT}$ drop. The $I_{D\text{-}SAT}$ drop is mainly caused by the extremely narrow fin CD of the 1-fin structure fabricated by approach B. In 4-fin device, the edge semi-isolated fin is also a little but not obviously narrower than the dense fin (see figure 3 and 4), so the $I_{D\text{-}SAT}$ drop for approach B is also not so obvious (see figure 5).

The experiments results above demonstrate that fin CD loading obviously decides the final electrical performance of the device. The fin CD should be neither too large nor too narrow and must be in certain range, especially for 1-fin device. To obtain better device performance for both dense and isolated device, much effort should be done to control fin CD loading in fin formation process.

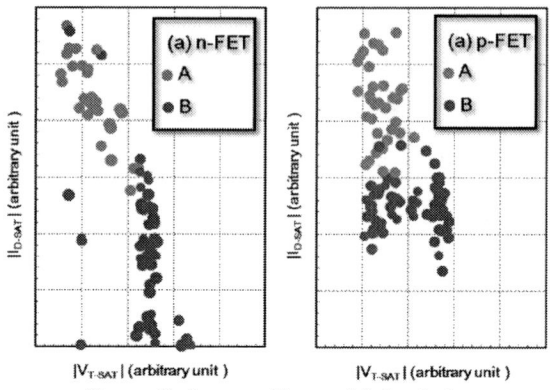

Figure 8: $I_{D\text{-}SAT}$ vs $V_{T\text{-}SAT}$ of 1-fin device

CONCLUSION

Fin CD uniformity and electrical performance by different fin formation approaches are compared. For 4-fin device fabricated by approach B, obvious performance gain could be observed comparing with that fabricated by approach A for both NMOS and PMOS from the universal curve. But for 1-fin device fabricated by approach B, obvious performance drop was observed. The higher off-stage drain current in 4-fin device fabricated by approach A may mainly come from the fatter fin on the edge side which could not be fully depleted at subthreshold region. The lower on-stage drain current in 1-fin device fabricated by approach B is mainly because of the extremely narrow fin CD comparing that of approach A. The experiment results demonstrate that much effort should be done to control fin CD loading in fin formation process to obtain better device performance for both dense and isolated device.

REFERENCES

[1] A. Rastogi, W. Chen, and S. Kundu. *IEEE Design Automation Conference IEEE*, 2007:712-715.

[2] Y. K. Choi, N. Lindert, P. Xuan, et al. *Electron Devices Meeting. iedm.technical Digest.international*, 2001:19.1.1-19.1.4.

[3] E. M. Chong, Y. Z. Zhu, C. Y. Yi, et al. *Semiconductor Technology International Conference. IEEE*, 2015:1-3.

[4] S. M. Sze, K. K. Ng. *Physics of semiconductor devices*, Wiley, 1981:38.

FIN BENDING MECHANISM INVESTIGATION FOR 14NM FINFET TECHNOLOGY

Cheng Li[1]*, Hai Zhao[1], Gang Mao[1]

[1]Advanced Technology Research and Development, SMIC ATD, Shanghai, 201203, China

*Corresponding Author's Email: Cheng_Li@smics.com

ABSTRACT

Beyond 14nm logical technologies, Fin structure becomes one of the common features and the most balance solution to improve MOSFET performance. Fin structure benefits electrical characteristics due to its narrow and tall properties, but it also easily suffers unwanted mechanical failure due to high aspect ratio. One of these typical mechanical failures, named fin bending, usually occurs during STI formation because of stress problem. This paper will discuss the possible factors and demonstrate the mechanism.

INTRODUCTION

In recent years, FinFETs have become popular and mature for state of art advanced CMOS technologies[1], with the advantages of the gain of transistor density and electrical performance. According to the generation node shrinkage as small as 16nm or less, the critical dimension and depth of Fin structures are getting more and more challenged, resulting in the aspect ratio of Fin structures ~6:1. In this case, the Fin structure becomes less robust than planar structures mechanically and any structural imbalance will lead to some unexpected failures.

Fin bending is one of these failures during our technology development. A typical fin bending may lead to electrical failure or physical failure. Its leaning angle causes asymmetric depletion area and makes leakage current larger than expected. Meanwhile, fin bending compresses metal gate deposition process margin and higher possible for void formation.

Some people consider the residual stress of asymmetric film deposition loading upon fin sidewall[2-4] as the main root cause. However, another possible mechanism has been demonstrated after a series of experiments and analyses.

This paper mainly focuses on the fin bending mechanism exploration and the discussion of some possible solutions.

EXPERIMENTS

Before the design of experiments, we checked the fin status by TEM failure analysis and found that no fin bending during fin patterning process. In the other hand, we found the fin bending after STI filling & annealing loop. Therefore, we consider the STI filling & annealing loop highly suspected.

The schematic process flow of STI filling & annealing loop is shows in *Figure 1* to clarify related factors which may contribute to fin bending. As the sequence of process, the suspected factors list as *Table 1* below:

Factor	Material	Usage
STI liner thickness	SiO2	Buffer layer
STI deposition rate	FCVD SiO2	STI gap filling
STI annealing temp.	FCVD SiO2	SiO2 densification

Table 1: The suspected factors list for fin bending

Furthermore, a reasonable criteria to evaluate fin bending status is need for the experiment. After several attempts, fin pitch, or period in other words, is selected. Comparing to STI width, it can eliminate the affection of fin CD variation, even though the former one is more direct and visible. Additively, we focus on SRAM area due to fin bending often performs severely at SRAM. The SRAM asymmetric design is highly suspected.

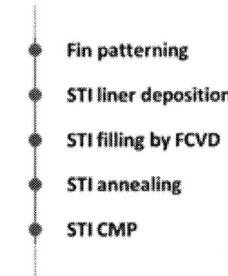

Figure 1: Process flow of STI loop

RESULTS AND DISSCUSSION

Experiment series 1 is STI liner thickness split shown in *Table 2*, which acts as the buffer layer between fin and STI SiO2. The liner is usually deposited by means of furnace or PECVD, which is more densified than FCVD and therefore has stronger mechanical properties than fin, silicon, itself.

Series 1	Condition					
Liner split	+100% THK	BL THK	-60% THK	-50% THK	-40% THK	-30% THK

Table 2: Experiment split table of series 1- STI liner thickness

A series of STI liner split from +100% to -60% has

been carried out. But out of our expectation, the results do not show any obvious trend in *Figure 2*. And the fin bending performance doesn't get any improvement.

For this phenomenon, we consider that STI liner is not a dominated role for fin bending. Although the Young's module of dense SiO2 is about several times higher than silicon, it can hardly sustain the lateral stress and keep fin stand straight.

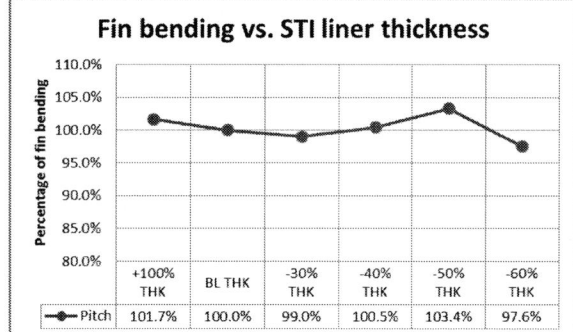

Figure 2: Series 1 experiment result - STI liner thickness

Experiment series 2 is the FCVD deposition rate split shown in *Table 3*. FCVD is a common approach to fill STI gap for 14nm or more advanced generation, due to its excellent gap filling capability. However, FCVD has also the disadvantage that less densified than HARP or HDP process. Here we discuss the deposition rate as one factor to control the densification of FCVD SiO2. That's the slower deposition rate, the more densified SiO2 we got.

Series 2	Condition	
Deposition rate split	BL deposition rate	+50% deposition rate

Table 3: Experiment split table of series 2- FCVD deposition rate

The result in *Figure 3* shows little impact for FCVD deposition rate, which means the densification impacted by deposition is not as much as we has once considered before experiment.

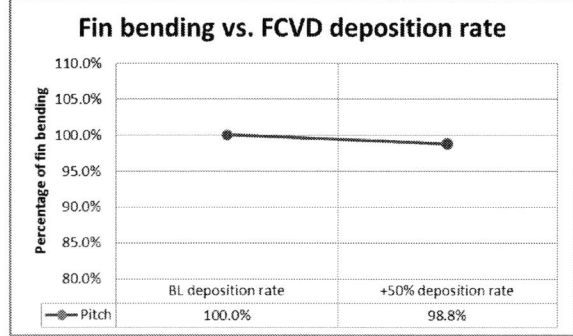

Figure 3: Series 2 experiment result - FCVD deposition rate

Experiment series 3 is the STI annealing temperature

split shown in *Table 4*. The annealing temperature is highly suspected in our hypothesis. The mode is the fin bending takes place by lateral stress generated during annealing process, which is supported by FCVD process characteristic. Because some precursor is used in FCVD, the total SiO2 film will shrink when precursor degas out when annealing. This give fin a lateral, tensile stress and make it bending.

Series 3	Condition				
Anneal split	BL temperature	-10% temperature	-20% temperature	-30% temperature	-50% temperature

Table 4: Experiment split table of series 3- Annealing temperature

The result in *Figure 4* confirmed our hypothesis. As the annealing temperature decreases to BL -30%, the fin bending is as less as about 85% of BL. And furthermore, the bending trend exist a certain limit. That is all precursor in film degas out and the SiO2 film is fully densified. In our experiment result, when annealing temperature is higher than BL -10%, the fin bending performance changes little.

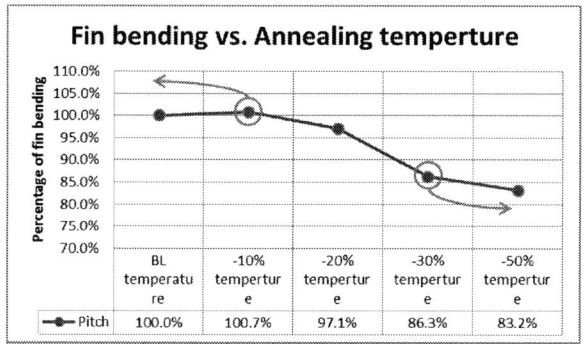

Figure 4: Series 3 experiment result - Annealing temperature

From the discussion above, the dominated factor has been highlighted to be STI annealing temperature, while STI liner thickness and FCVD deposition rate contribute little to fin bending. Besides these factors, there are some more possibilities which may impact fin bending performance, such as fin CD, fin profile etc. However, according to the experiment results so far, we can draw the conclusion that the lateral stress during STI annealing plays the most important role. For FCVD SiO2, shrinkage during annealing is its native characteristic and hard to be eliminated. The possible solution is decreasing annealing temperature, which will also bring the side effect of less film densification and impact following process, e.g. STI CMP. In summary, fin bending is a systemic and complicated issue for 14nm generation or advanced, which is worthy of enough attention.

978-1-5090-6695-7/17 $31.00 © 2017 IEEE

CONCLUSION

A series of experiment and systemic discussion for fin bending has been developed. All of the suspected factors have been analyzed, and the annealing temperature is highlighted as the most dominated one for the fin bending improvement. Furthermore, the hypothesis of lateral, tensile stress has been demonstrated to well explain the experiment phenomenon. As one of the significant feature of future advanced technology, FinFETs will surely be applied to more types of devices, and fin bending will become more and more critical issue. This paper can give a clear guideline to avoid fin bending and homological problems.

REFERENCES

[1] K. J. Kuhn, *International Symposium on VLSI Technology, Systems and Application (VLSI-TSA)*, 2011, pp. 1

[2] Alp H. Gencer, D. Tsamados, V. Moroz, *2013 International Conference on Simulation of Semiconductor Processes and Devices (SISPAD)*, 2013, 109-112

[3] Chun-Hway Hsueha, *Journal of Applied Physics*, 2002, Vol. 91, Num. 12, 9652-9656

[4] K. Prall, K. Parat, *International Electron Devices Meeting*, 2010, 5.2.1-5.2.4

14NM METAL GATE FILM STACK DEVELOPMENT AND CHALLENGES

Jianhua Xu[1]*, Anni Wang[2], Jun He[2], Xuezhen Jing[1], Ziying Zhang[2], Beichao Zhang[2]

2016-09-26

1. Semiconductor Manufacturing International Corporation (SMIC)

2. SMIC Advanced Technology R&D (Shanghai) Corporation

18, Zhangjiang Road, Pudong New Area, 201203 Shanghai, P. R. China

*Correspondent author E-mail:Jianhua_Xu@smics.com

ABSTRACT

As IC technology advances to 16/14 nm and beyond, FinFET architecture with advantage of excellent leakage performance becomes main stream in IC industry. However, it also brings big challenges for integration and processes due to its very aggressive structure and profile, CD shrinkage, shadow effect and gap-fill difficulty.

In this work, atomic layer deposition (ALD) metal films, including TaN, TiN (TiSiN), TiAl and CVD W, were studied for replacement metal gate application. Challenges of step coverage & gap-fill, loading effect and tunable range of work function will be discussed and addressed. Thickness of high K capping layer (TiN or TaN), work function metal (TiN & TiAl), W barrier layer (TiN) all show strong effect on N/P MOS device Vt, and more than 300 mv tunable range of work function can be achieved. Besides, higher Al : Ti ratio process, interfacial special treatment between TiAl & W barrier TiN and different W process can lower down NMOS Vt. At the last, ALD and CVD process ensure good gap-fill performance when CD opening is larger than 5nm (aspect ratio is about 20:1).

Keywords: *14nm; FinFET; Replacement metal gate; Work function; High K capping layer; Barrier layer; ALD; TiN; TiAl; TaN; W*

INTRODUCTION

As IC technology advances to 16/14 nm and beyond, FinFET architecture with advantage of excellent leakage performance becomes mainstream in IC industry. However, it also brings big challenges for integration and processes due to its very aggressive structure and profile, CD shrinkage, shadow effect and gap-fill difficulty. Replacement high-K & metal gate process has been inherited from 28nm generation [1-3]. Comparing with traditional poly silicon gate device, high-K metal gate device can get higher performance and lower leakage. High-K gate dielectric can help to reduce gate leakage and continue scaling down Tox. Meanwhile, metal gate is needed to replace poly silicon gate, to improve electron mobility which is degraded by soft optical phonons scattering, and to solve electron depletion Vt shift issue which is induced by Fermi level pinning [4-5].

In this work, atomic layer deposition (ALD) metal films, including TaN, TiN (TiSiN), TiAl and CVD W, were studied for replacement metal gate application. Challenges of step coverage & gap-fill, loading effect and tunable range of work function will be discussed and addressed. Thickness of high K capping layer (TiN or TaN), work function metal (TiN &

TiAl), W barrier layer (TiN) all show strong effect on N/P MOS device Vt, and about 300 mv tunable range of work function can be achieved. Besides, higher Al : Ti ratio process, interfacial special treatment (IST) between TiAl & W barrier TiN and different W process can lower down NMOS Vt. At the last, ALD and CVD process ensure good gap-fill performance when CD opening is larger than 5nm (aspect ratio is about 20:1).

EXPERIMENT AND DISCUSSION

1. ALD TiN, TaN & Multi-Vt

Beyond 14nm, FinFET architecture and HK last gate last approach becomes main stream in industry. It brings big challenges for integration and processes due to its very aggressive structure. For example, shadow effect (Fig.1) will seriously restrict tilt angle of SRAM implantation, thus it is very difficult for Vt tuning and multi-Vt device fabrication.

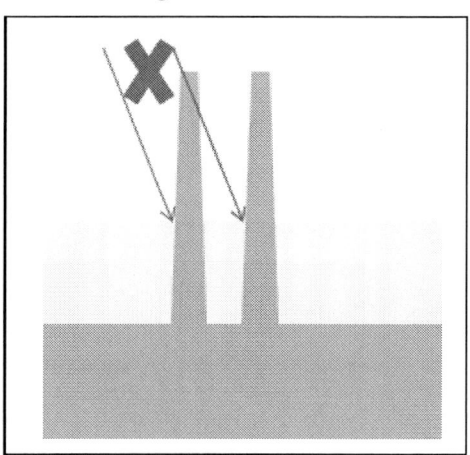

Fig. 1 diagram of shadow effect

ALD TiN is widely used as a robust HK capping layer, due to its good barrier characteristic, process controllability and excellent step coverage. Besides, work function of ALD TiN is close to 5 eV, so different thickness of ALD TiN capping layer can be designed to adjust multi-Vt, according to device requirement. As an optional choice, ALD TaN with middle gap work function, also can be adopted together with ALD TiN, to act as etch stop layer and adjust Vt.

Herein, the effect of TiN and TaN thickness on work function tuning was studied by MOSCAP C-V test, and the results have shown in Fig 2. The detailed MOSCAP structure and process have been reported in our previous work [6].As shown in Fig 2(a), work function value is increasing when the TiN thickness varies from YA to (Y+15)A. Meanwhile, the

978-1-5090-6695-7/17 $31.00 © 2017 IEEE   222

same trend has been observed in Fig. 2(b), i.e. work function shift toward a higher value with the TaN thickness increasing. Compared with Fig. 2(a) and Fig. 2(b), TaN thickness is more sensitive for work function. TiN thickness enhanced 1A will increase work function 4mV, while TaN thickness enhanced 1A will increase work function 15mV.

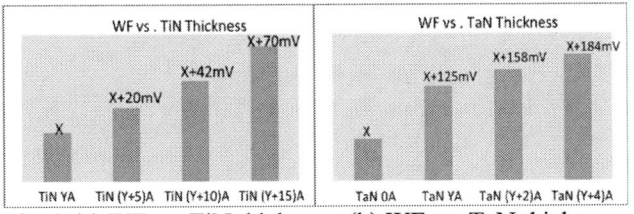

Fig. 2 (a) WF vs. TiN thickness, (b) WF vs. TaN thickness

As mentioned previously, work function of ALD TiN is close to 5 eV, and ALD TaN has a middle gap work function. Thus, both increasing the TiN/TaN thickness can enhance the work function shift toward valence band, i.e. the value of work function enhanced. These results indicate that the work function of metal gate can be tuned by changing HK capping layer TiN and TaN thickness. For P metal gate, the work function of metal layer is required to be tuned toward 5 eV, which can be achieved by increasing the TiN/TaN thickness. For N metal gate, the work function of metal layer is required to be tuned toward 4.1 eV, a thinner TiN/TaN HK capping layer can be realized based on above data. Hence, it is a potential method that multi Vt can be adjusted by tuning HK capping layer TiN/TaN thickness.

2. TiAl thickness and Al/Ti ratio

N metal work function tuning is one of the greatest challenges in 14nm N-type FinFET device, which will affect device Vt. TiAl film as a typical n-type work function layer has been used in HKMG device. The TiAl thickness and Al/Ti ratio are sensitive for N metal work function. The relationship between TiAl thickness and N metal work function is illustrated in Fig. 3(a). The work function is decreasing regularly when TiAl film becomes thicker. Furthermore, Fig. 3(b) also illustrates the influence on Al/Ti ratio for N metal work function. It can be obvious that the work function shift toward lower direction as the Al concentration become richer. As discussed above, TiAl as n-type work function layer, its thickness and element ratio will affect the N metal work function.

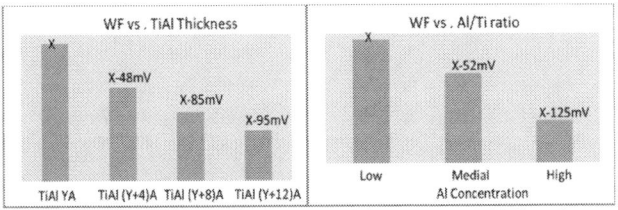

Fig. 3(a) WF vs. TiAl thickness, (b) WF vs. Al/Ti ratio

3. W barrier TiN

As we known that a barrier layer in HKMG device is very important, this barrier layer will help to prevent impurities (for example: F or B) diffusion. F and B element root in W

deposition. In our previously work, we have reported the comparison of barrier capability for TiN and TiSiN [6]. Herein, we will further study barrier capability of TiN film by TiN thickness tuning on n-type FinFET device.

Fig. 4(a) shows the Vt data on different barrier layer TiN thickness. Obviously, thicker TiN film will result in NMOS Vt degraded more than 100mV. One suspected model for thicker TiN can reduces NMOS Vt is that thicker TiN film has a better barrier capability (Fig. 5). As we mentioned above, TiN as a barrier layer aim to prevent impurities (for example: F or B) diffusion. F and B as mobilizable ions pass through barrier layer into work function layer which resulting in metal work function shift. Thus, the thicker TiN barrier layer can improve impurities diffusion.

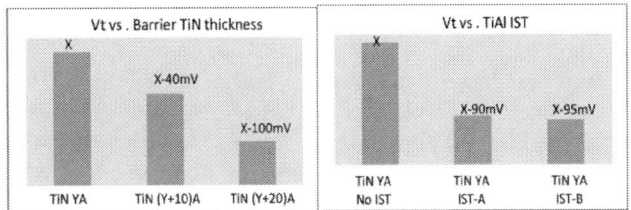

Fig. 4(a) Vt vs. Barrier TiN thickness, (b) Vt vs. TiAl IST

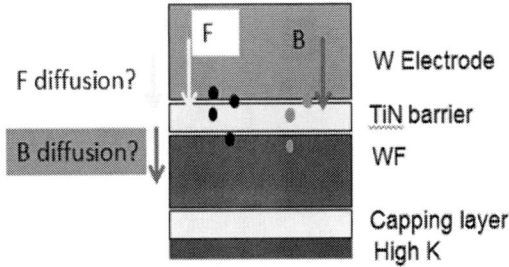

Fig. 5 One Suspected model for thicker TiN can reduces NMOS Vt

To verify this model, two different film stack are prepared, A is HK/cap layer/TiAl/TiN(thin)/W, B is HK/ cap layer /TiAl/TiN(thick)/W. The analysis for element in depth profile is carried out by backside SIMS. From data presented in Fig. 6, it can be obvious seen that F diffusion is reduced in thicker TiN film stack while the distribution of B is the same in two film stack. It means that F diffusion is sensitive to induce Vt shift and thicker TiN barrier layer can block F diffusion.

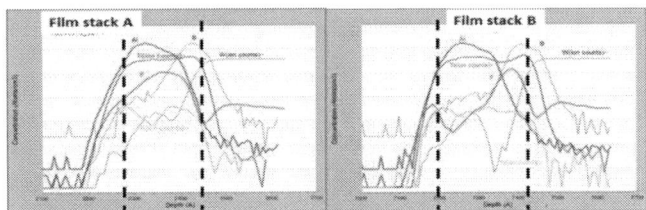

Fig. 6 SIMS depth profile of different TiN thickness

Enhanced barrier TiN thickness is a feasible method to reduce NMOS Vt. However, TiN become thicker will affect W gap fill due to the shrinking of gap fill trench. Another potential method to improve F diffusion is that adding an IST process after TiAl deposition. An IST process can form an special interfacial layer on TiAl. This interfacial layer also can act as a barrier layer to block F diffusion. Fig. 4(b) shows the influence of TiAl IST on NMOS Vt. The results show that both IST-A and IST-B can degrade Vt more than 80 mV. One of the major concerns for adding IST is barrier layer TiN step coverage and metal electrode W gap fill. A special interfacial layer may affect the TiN adhesion on TiAl film, and further influence the capability of W gap fill performance. Fig. 7 shows the TEM cross section image of N metal gate with TiAl IST. A good W gap fill performance can be obvious when CD opening is larger than 5nm (aspect ratio is about 20:1), and each film has a good step coverage and uniformity. Further, the different W deposition process will also result in Vt shift in our study. Fig. 8 shows the Vt data for two different W deposition process which exist about 100 mV gap. The detailed effect mechanism for these two type process will be studied in our further work. It is worth noting that previous data shown in this paper are in type A condition for W deposition.

Based on above discussion, enhanced barrier TiN thickness and adding an IST process after TiAl deposition are potential method for N metal gate integration to achieve Vt target. Different W deposition condition also can make Vt down. Furthermore, a good W gap fill performance is ensured due to the fact that ALD and CVD process are used in the whole metal gate process tuning.

ensure good metal gate gap-fill performance and step coverage.

ACKNOWLEDGEMENTS

The author would like to acknowledge all team members of SMIC TD TF for the instructions and help.

REFERENCES

[1] K. Mistry et al., *IEDM Tech. Dig.,* 2007, pp.247-25.
[2] A. Veloso et al., *VLSIT,* T194 (2013).
[3] C. Auth et al., *Symp. on VLSI Tech.,* 2008, pp. 128-129.
[4] E. Josse et al., *IEDM Tech. Dig.,* 1999, pp. 661–664.
[5] G. D.Wilk et al., *J. Appl. Phys.,* vol. 89, 2001, pp. 5243–5275.
[6] J. H. Xu et al., 2016 China Semiconductor Technology International Conference.

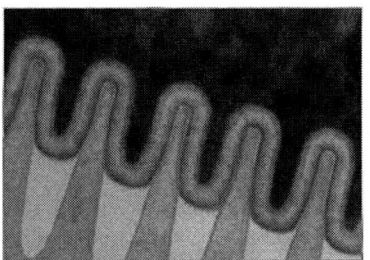 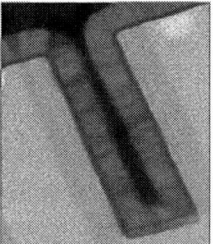

7-(a) multi Fin metal gate   7-(b) 20:1 aspect ratio gap fill
Fig. 7 TEM cross section image

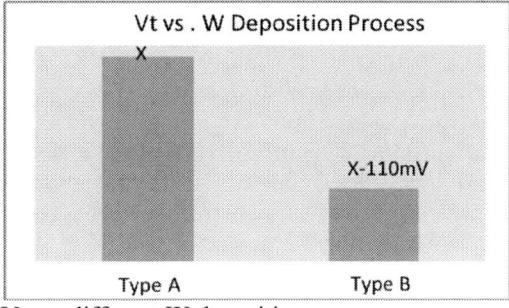

Fig. 8 Vt vs. different W deposition process

CONCLUSION

In this work, Challenges of step coverage & gap-fill, loading effect and tunable range of work function are discussed and addressed. Multi Vt can be adjusted by tuning HK capping layer TiN/TaN thickness. TiAl as n-type work function layer, its thickness and element ratio are sensitive on N metal work function. Enhanced the thickness of W barrier TiN film can degrade NMOS Vt due to the fact that thicker barrier layer can block F diffusion. Another method to decrease NMOS vt is adding an IST process between TiAl and W barrier TiN. Furthermore, we find that different W deposition process will also affect Vt shifting. At the last, ALD and CVD process

978-1-5090-6695-7/17 $31.00 © 2017 IEEE

Forming a More Robust Sidewall Spacer with Lower k (Dielectric Constant) Value

Tao Han, Man Gu, Stephan Grunow, Huang Liu, Sujatha Sankaran, Jinping Liu

Advanced Technology Development, GLOBALFOUNDRIES Inc. US
Malta, New York State, USA
+1-518-305-7405, Email: tao.han@globalfoundries.com

ABSTRACT

Device scaling leads to tough challenges not only in patterning, but also in device performance due to scaled contact area, smaller stressors, and increased parasites capacitance. There is immediate need to implement low k spacers. Low-k materials, however turn out to be weak, especially after going through subsequent integration, such as cleans and etching. Here we report issues with integrating low-k spacers materials in the state-of-the-art CMOS technologies and propose one method to solve these issues.

Keywords— low-k spacer, robust film, forming, various etches or cleans

INTRODUCTION

Spacers may be formed adjacent to the sidewalls of the gate electrode with properties intended to boost device performance. Improved structures for spacers in a device structure and methods for forming spacers in such device structure like a field-effect transistor are needed. Before 20nm technology, the typical spacer material is ALD (Atomic Layer Deposition) SiN (Nitride, process temperature at 550°C to 650°C). Normally, ALD nitride is a robust film with good uniformity (both wafer-to-wafer and within wafer) and step coverage. But its k value (dielectric constant) is high at around 7 (compare to thermal oxide or ALD oxide SiO_2 around 3.9 only). Thus a new film SiOXY (Silicon OxyXY) is introduced into the industry to replace ALD nitride as spacer material from 20nm technology and beyond. This new film can be as robust as ALD nitride or even stronger in various subsequent integration steps (cleans or etching), and most important thing is that this new film can provide much lower k value (dielectric constant value is around 5.5 or below) which helps to improve device performance. There are multiple options on X and Y elements in the new film, such as B, C, N, and so on. It is defined by different process and device requirements accordingly. However, from current existing mature processes, the lower k value of the new SiOXY film, the weaker the film will be. In certain conditions, directly using new lower k

value film will cause more film erosion in downstream various cleans or etching. Sometimes, the new film will also affect EPI growth due to higher carbon concentration that causes more carbon diffusion or out-gassing during EPI growth.

Many methods are researched or implemented to solve the issues when introducing the lower k value film as sidewall spacer. One of the best and well-accepted methods is implementing a thin crust layer to protect the lower k but weaker film. Typical options of the crust layer will be SiN, SiXYN, SiXN, etc. This kind of sacrificial layer is required to be strong (sustain in various etching or cleans) at very thin level (a few nanometers only). The sidewall spacer thickness is getting thinner and thinner in 20nm technologies and beyond. So the crust layer thickness has to be controlled well at very thin level. If it is too thick, the inner lower k value film will be too thin. Then it will be no meaning to introduce lower k film to improve the device performance. Those typical crust layer films have many challlenges during implementation. A new invented film is introduced as crust layer to protect inner lower k value spacer film. It can reach even thinner thickness and less erosion compare to typical crust layer films.

Issues with Integrating Low-k Spacer Materials and Proposal for Solutions

As we all know, from 20nm technology and beyond, more and more lower k value materials are introduced to form sidewall spacer to boost device performance. One of the most popular lower k value film is SiOCN (Silicon Oxycarbonnitride). The C (Carbon) in the film helps to strengthen the film to slower etch rates. The O (Oxygen) in the film helps to reduce k value. One advantage of this process is that the k value can be adjusted or changed by changing the O element composition through recipe conditions changing. Obviously, increasing the O% of the film is the way to get lower k value of the film. But there is conflict between the two elements C and O in the film. Once O% is increased, the C% will be decreased at meantime. Once O% is increased too

---

978-1-5090-6695-7/17 $31.00 © 2017 IEEE   225

much, the film will be pure oxide without C composition and this kind of oxide film will be too weak to sustain in various cleans or etching. When defining spacer film condition, the trade-off between C and O is the key element to consider. From device performance point of view, the k value is required to be lower, which means O composition is getting higher (C composition is lower and lower), then so does the etch rates of the film, Thus directly implementation of the lower k film is difficult and can generate more problems, such as more film loss, worse structure profile.

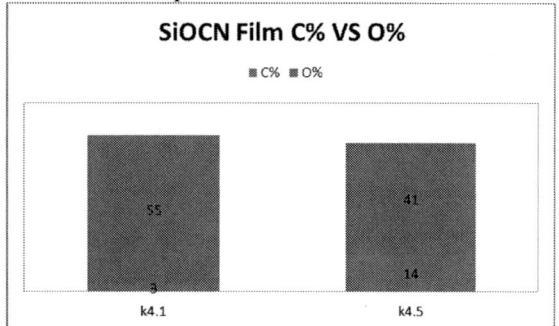

FIG. 1: Elements composition in low-k(SiOCN) film, C and O conflict with each other, to lower the k value of the film, O% has to be increased while C% will be decreased, XPS measurement, atomic percentage

As discussed above, when implementing the lower k value film as sidewall spacer, a good method is using a thin crust layer to protect lower k but weaker film. The typical crust layer films are SiN, SiOCN, SiOC, SiCN, etc. As known, they are robust films and have very low etch rates in various cleans or etches (refer to below Fig. 2)

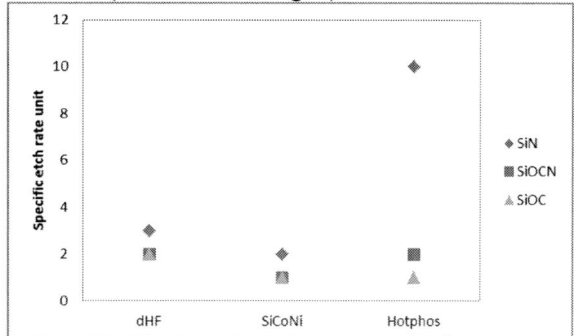

FIG. 2: Etch rates comparison among ALD SiN, SiOCN, SiOC (a). 1:100 dHF (b). hotphos (c). SiCoNi

But above (Fig.2) etch rate data is based on thicker film thickness (~10nm). There is a threshold thickness (>3nm) of these films to be a robust film. If the film thickness is thinner than 3nm, the film will be too weak to protect the inner lower k value film.

Another serious concern of sidewall spacer film erosion is the film oxidization resistivity. A lot of

processes will oxidize and weaken these very thin spacer films, such as dry etch, ashing processes, spike or RTA (Rapid Thermal Annealing) anneal, oxide deposition, etc. If the surface of the spacer film is oxidized, it will be easily removed in cleans (dilute HF) or SiCoNi processes. As shown in Fig.3 below, O2FG (Oxygen and Nitrogen Forming Gas) ashing process can oxidize the film twice or more higher compare with the ratio of H2N2 ashing. Different lower k value films have different O2 resistivity. But in common, they are all easily to be oxidized. Sometimes even the entire lower k film will be oxidized after O2FG ashing. The oxidized film is much weaker to sustain in various cleans or etches. For example, the dilute HF etch rate will be 10 times higher than the film prior ashing process.

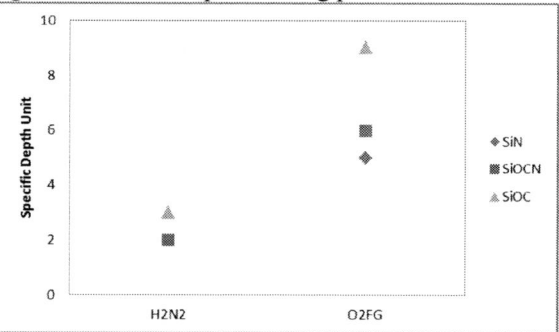

FIG. 3: Comparison of films (SiN, SiOCN, SiOC) oxidation depth by ashing (H2N2 & O2FG). (Specific Depth Unit of thickness, nanometer level)

According to discussion above, the typical crust films (SiN, SiOCN, SiOC, SiCN,etc) are not good enough to be implemented to protect lower k value films. They are not robust and easily to be oxidized at thinner thickness (<3nm) level. As we all know, in 20nm technologies or beyond, the sidewall spacer thickness is getting thinner and thinner due to the device size shrinkage and the canyon space is getting narrower and narrower as well. Thus the sidewall spacer film is required to be strong at very thin level with as less erosion as possible. To solve this issue, a newly invented film is introduced to replace typical crust films to protect lower k film. It can give similar or better etch resistance compare with those typical films (SiN, SiOCN, SiOC, SiCN, etc) even at very thin thickness (~2-3nm).

978-1-5090-6695-7/17 $31.00 © 2017 IEEE

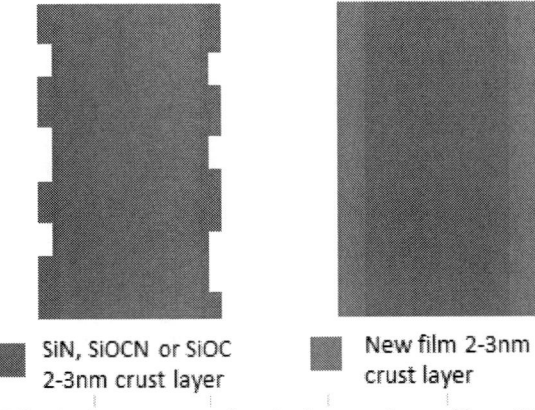

SiN, SiOCN or SiOC
2-3nm crust layer

New film 2-3nm
crust layer

FIG. 4: Comparison of typical crust layer films (SiN, SiOCN, SiCN) profile after various cleans or etches with new film profile (@ very thin thickness level <3nm)

SUMMARY

More and more low k value films or materials are introduced and implemented into 20nm or beyond technologies as sidewall spacer. There is a trade-off between the film k value and film robustness (related to film O% and C% conflict). To keep lower k value film and reduce film erosion in subsequent integration steps (various cleans or etching), a good option is to use a stronger film as crust layer to protect the lower k value but weaker film inside. But at very thin thickness level (<3nm), traditional films (SiN, SiOCN, SiOC or SiCN) are not as strong as when they are at thicker thickness leval (>10nm). O2 resistivity is also a serious concern for sidewall spacer films. Many processes can oxidize the lower k films very much and make the films much weaker and cause more film erosion and affect spacer shoulder profile. An invented film is introduced and shows robustness even at very thin thickness level (<3nm).

REFERENCES

[1] "Semiconductor device with low-k spacer", K Cheng, A Khakifirooz, A Reznicek, US Patent US9349835

[2] "Finfet Device Containing a Composite Spacer Structure", J R.Holt, S Mehta, A Reznicek, GLOBALFOUNDRIES INC. US Patent US9236397

[3] "Chemical-reaction engineering in the semiconductor industry", F. C. Eversteyn, Philips Res. Rep. **29**, 45 (1974).

[4] "Chemical processes in vapor deposition of silicon", V. S. Ban, J. Electrochem. Soc. 122 (10), 1389 (1975).

REVIEW OF THIN FILM POROSITY CHARACTERIZATION APPROACHES

Konstantin P. Mogilnikov[1*], Dongchen Che[1], Mikhail R. Baklanov[1, 2], Kangning Xu[1] and Kaidong Xu[1]

[1] Leuven Instruments Co. Ltd (Jiangsu), Pizhou, Jiangsu, China
[2] North China University of Technology, Beijing, China
*Corresponding Author's Email: kpmog@leuven-instruments.com

ABSTRACT

The most important properties of porous thin films depend on the pore structure. The evaluation of porosity is of great importance for analyzing their pore structure. Some known methods were adapted and proposed for the study of thin films porosity, such as microscope techniques, radiation scattering, wave propagation, gas adsorption. Besides, there are some new approaches developed for thin film porosity, such as X-ray porosimetry, positron annihilation lifetime spectroscopy, quartz crystal microbalance, and ellipsometric porosimetry. In this paper, the possibilities of various methods of studying thin films porosity will be discussed, including the latest developments in this area.

INTRODUCTION

In the last decades, the study of porous thin films is primarily associated with the needs of microelectronics [1, 2]. The most important properties of porous thin films depend on the pore structure, such as density, stiffness/strength, thermal conductivity, and chemical reactivity [3]. And the evaluation of porosity is of great importance for their successful integration. As for three-dimensional porous bulk materials, many different methods have been applied to measure the pore size and porosity [4, 5]. However, these methods were not simply applicable to thin films, mainly due to the very small quantity of the analyte in these films. Thus, existing methods need to be revised/adjusted and new approaches to be designed for thin film porosity characterization.

Some known methods were adapted and proposed for the study of thin films porosity: stereology analysis such as microscope techniques; intrusive methods such as gas adsorption; nonintrusive methods such as radiation scattering [6-8]. Besides, during the last decade, there are some new approaches developed for thin film porosity, such as X-ray porosimetry, positron annihilation lifetime spectroscopy (PALS), quartz crystal microbalance (QCM), and ellipsometric porosimetry (EP) [9-12].

All these methods are based on different physicochemical ideas and each technique comes with its own specific strengths and weaknesses. In this paper, the possibilities of various methods of studying thin films porosity will be discussed, including the latest developments in this area.

MICROSCOPE TECHNOLOGY

Microscope techniques are based on the stereology analysis. One can observe the pores directly by electron microscope, such as scanning electron microscope (SEM) and transmission electron microscope (TEM) [13, 14]. This method gives information in structure and view of the pores. However, it takes long time and some special preparation of samples. In addition, only a small area can be observed under the microscope. If one wants to get the pore size and porosity, additional image analysis techniques are needed. Mahmood et al. used SEM to view the porous GaN samples [14]. After that, they built an algorithm using MATLAB software to obtain the distributions of statistics of the pores such as the maximum, minimum and average radii. The porosity of the structures was obtained by calculating the areas occupied by the pores.

RADIATION SCATTERING

Radiation scattering is a nonintrusive method, using scattering of some beams such as neutron beam or x-ray with small angles [15]. The scattering is related to the number of boundaries crossed by the beam: the more boundaries, the more scattering. As a result, the researchers can get one value: number of boundaries. To find the porosity and pore size, they still need additional information such as matrix density and atomic composition of the sample. Radiation scattering includes small-angle neutron scattering (SANS), small-angle X-ray scattering (SAXS) and so on [16, 17].

The SANS has been applied in transmission to thin porous dielectric films. The measurement result provides information related to the pore size because its intensity is sensitive to density fluctuation correlations. Vogt et al. measured the pore structure and distribution of ordered mesoporous silica films by SANS [18]. They found there were two kinds of pores in the sample, spherical mesopores and micropores. The porosity is about 39%, similar to the X-ray reflectivity results. However, the few available number of neutron sources limits this method's wide use.

Similar to SANS, the SAXS is well established for determining the size and shape of nanostructures inside the materials. But X-ray absorption causes the scattering intensity due to density fluctuations in a thin film deposited on a thick substrate to be extremely weak [19]. Recently, grazing incident small angle X-ray scattering (GISAXS) seem to be an alternative method [20]. In a GISAXS measurement, the scattering volume can be increased by two orders of magnitude over the normal incident geometry. It allows to observe the two dimensional picture of scattering and one can see more clearly the regular structure of the porous film. Ito et al. determined the pore size distributions and porosity of porous low-*k* films using the GISAXS [21]. The results agreed well with that of the N_2 adsorption technique. However, the method cannot give real pore size distribution, but only the mean pore size.

GAS ADSORPTION

Gas adsorption, especially N_2 porosimetry, is the most common method for porous media evaluation. The main feature of this method is the extraction of porosity and pore size from the measured adsorption isotherms. They are followed by calculating the monolayer capacity based on the BET (Brunauer-Emmett-Teller) adsorption isotherm [22]. The method allows to define the porosity of a material in a wide range of mesopores and lower range of macropores (pore diameters 1.7-300 nm) according to IUPAC classification [23]. Sujka et al. analyzed the pore structure of the extruded rice grits products by nitrogen adsorption [24]. From the measurement, they obtained the cumulative pore volume for native material. Lapham et al. measured the porosity of $NiCo_2O_4$ films and powders using this method [25]. The results showed that the materials are low in mesopore, and mainly in the macropore region. The samples had different pore size when prepared with different methods. However, this method needs over 0.1 gram porous materials for measurements, making it difficult to be used on thin films.

X-RAY POROSIMETRY

X-ray porosimetry is based on vapor adsorption, one of intrusive methods. This method shows promise as a general laboratory characterization method that provides detailed structural information of nanoporous thin films [12]. It measures the changes of the film density profile and potentially the film thickness when the film is exposed to a partial pressure of condensable vapors. The change in film density can be directly related to the porosity, if the density of the condensed fluid is known or assumed as bulk-like. X-ray porosimetry directly measures the average porosity and film density of the material separating the pores, which is not intercommunicating pore. The average pore size and pore size distribution from the physisorption isotherm can be determined by existing porosimetry models. Further, the results can be used to determine pore structure and pore size distribution (down to a few nanometers) as a function of depth in non-uniform

thin films. Lee et al. used this method to determine the mass uptake of probe molecules in porous thin films deposited on thick silicon wafers [9]. Then they calculated the pore size distribution of the films by the Kelvin equation without any other additional assumptions.

POSITRON ANNIHILATION LIFETIME SPECTROSCOPY (PALS)

PALS is a pore volume characterization technique whereby the shortening of the annihilation lifetime of Ps due to collisions with the pore walls is directly correlated with the pore size [26]. Ps in vacuum will self-annihilate into gamma-rays with a lifetime of 140 ns, but in a pore its lifetime is further shortened by the positron's annihilation with molecular electrons from the pore wall material. It has been used for about 40 years to probe the sub-nanometer voids in bulk polymers [27]. PALS can be used as a good calibration for the deduced pore size in the diameter range 0.3–30 nm, truly a probe of nanoporous materials [28]. Also, the diffusion of Ps within the pores of a sample can be used to determine which pores are interconnected. But this method needs the positron source and a set of detectors, which are available only in several laboratories in the world, limiting its application in pore size analysis. Recently, PALS has been used for investigation of various free spaces, such as mesopores by He et al., micropores by Kullmann et al. and intermolecular free volumes by Zgardzińska et al. [29-31]. Zaleski et al. had compared the results of pore size and porosity between PALS and N_2 adsorption [32]. The results showed the consistent pore sizes obtained by PALS are wide, which may be caused by the positronium mobility in the system of connected pores. Averaging of the positron porosimetry results lead to underestimation of the contribution of the smallest and the largest pores.

QUARTZ CRYSTAL MICROBALANCE (QCM)

QCM is one of the tools with high sensitivity to measure porosity in thin porous film. The mechanism is to monitor the change of the filling of porous film with some absorbents, such as air or water [33]. The main component of the QCM sensor is a Y-cut piezoelectric quartz crystal plate [34]. The frequency of oscillation of the quartz crystal is dependent upon the thickness of the crystal; any mass deposited onto the surface of the crystal will increase its thickness and so decrease the crystal's resonant frequency. A typical QCM instrument can measure the resonance frequency to ±1 Hz, using crystals with a fundamental resonance frequency of 5 MHz [35]. Therefore, QCM can be used to detect the pore size and porosity on the micro-/nanoscale. There have been many models to interpret the experimental results. Zhang et al. presented a theoretical model for QCM analysis [34].

978-1-5090-6695-7/17 $31.00 © 2017 IEEE

After testing, their model can provide more reasonable porosity compared with the existing models. Mills et al. demonstrated a rapid and simple measurement for the porosity of mesoporous TiO_2 films by QCM [36]. The results are accurate and repeatable.

ELLIPSOMETRIC POROSIMETRY (EP)

A new approach to adsorption measurements is EP. It uses the ellipsometric measurements to obtain the refractive index and thickness of the film before and after being filled by condensate. Monitoring the amount of adsorbed condensate as a function of the partial pressure defines an isotherm of vapor adsorption in the pores. During the measurement, the porosity evaluation is non-destructive and based on analysis of dielectric function that makes the results very suitable for microelectronics [37]. Moreover, this method allows the measurement of pore size distribution at room temperature in thin films directly deposited on Si or any other smooth solid substrate. Intermediate layers between the silicon substrate and the porous film do not create any problem for the measurements if their optical characteristics (n and k) are known. Experimental hardware and software for the pore size distribution and porosity calculations from the ellipsometric measurements of the adsorption of the organic vapors in porous films have been developed [38]. Some instruments have been built based on this method and the measurement is conducted in vacuum environment . Rouessac et al. used this tool to measure mesostructured TiO_2 thin films [39]. The EP results of the pore size and porosity were in agreement with N_2 adsorption and TEM.

Some new tools allow the adsorption measurements to be done at atmospheric pressure. Due to the accurate measurement results and fast test time, EP has become a recognized standard method of study of thin films. Bittner et al. characterized the stacked sol–gel films by the EP tool, SEM and UV–vis spectroscopy [40]. From the report, the EP tool can characterize the multilayer assemblies accurately. Perrotta et al. used the dynamic EP to investigate the permeation pathways in moisture barrier layers on polymers [41]. It was shown to be sensitive to the detection of macroscale defects in SiO_2 layers on polyethylene naphthalate substrate. Lépinay et al. obtained the adsorption isotherms of water, methanol, and toluene on pristine and damaged SiOCH porous materials [42]. The pore size distributions obtained by this method combined with the simulated adsorption data were found to be in good agreement with data established by other techniques such as PALS and GISAXS.

The newest tool using the EP method for adsorption measurements is the Gas Adsorption Monitoring system (GAM, Leuven Instruments Co. Ltd). This tool uses a special nozzle to deliver the gas-vapor mixture to the sample surface. The absorbents can use the isopropanol (IPA), toluene and n-Heptane. It allows fast measurements of adsorption and desorption, as well as kinetic of pore filling. Thus, GAM can characterize the porous thin film and barrier layer, without restrictions on the sample sizes. Moreover, the GAM can measure all the other EP functions such as thickness, refractive index, and density, to fully describe the properties of porous films.

CONCLUSION

This review summarizes and compares the methods to measure the pore structure and size for the porous thin films. Microscope techniques can observe the pores and the structure directly in quite small area. Using image analysis techniques, researchers obtain the porosity. Radiation scattering, as nonintrusive method, provides information related to the mean pore size and porosity. Gas adsorption is the most common method to obtain the porosity and pore size from the measured adsorption isotherms. It measures the pores in large range from 1.7 to 300 nm with more than 0.1 gram materials. X-ray porosimetry measures the changes of the film density profile and potentially the film thickness when the film is exposed to a partial pressure of condensable vapors, which directly relates to the porosity and pore size. PALS is based on the effect of the changing of the lifetime of ortopositronium in pores of different dimensions. It can be used as a good calibration for the nanoporous materials in the range 0.3-30 nm. QCM is sensitive to measure porosity in thin porous film, which needs the deposition of porous layer on the contact sensor. It can be used to detect the pore size and porosity on the micro-/nanoscale. EP is a new approach to adsorption measurements for pore structure. The newest tool using the method, GAM, allows fast film characterization at atomspheric pressure and room temperature in the range 0.5-40 nm. Moreover the GAM can realize all the functions of the ellipsometer to fully describe the properties of porous films and no restrictions on the sample dimensions.

REFERENCES

[1] F. Sittner and W. Ensinger. *Thin Solid Films*, vol. 515, 2007, pp. 4559-4564.

[2] S. S. Manickam, J. Gelb and J. R. McCutcheon. *J. Membr. Sci.*, vol. 454, 2014, pp. 549-554.

[3] K. Maex, M. R. Baklanov, D. Shamiryan, F. lacopi, S. H. Brongersma and Z. S. Yanovitskaya. *J. Appl. Phys.*, vol. 93, 2003, pp. 8793-8841.

[4] S. Thränert, D. Enke, G. Dlubek and R. Krause-Rehberg. *Mater. Sci. Forum*, vol. 607, 2009, pp. 169-172.

[5] L. Kong, A. Uedono, S. V. Smith, Y. Yamashita and I. Chironi. *J. Sol-Gel Sci. Technol.*, vol. 64, 2012, pp. 309-314.

[6] A. J. Ricco, G. C. Frye and S. J. Martin. *Langmuir*, vol. 5, 1989, pp. 273-276.

[7] J. Kobler and T. Bein. *ACS Nano*, vol. 2, 2008, pp. 2324-2330.

[8] A. Gibaud, S. Dourdain and G. Vignaud. *Appl. Surf. Sci.*, vol. 253, 2006, pp. 3-11.

[9] H.-J. Lee, C. L. Soles, D.-W. Liu, B. J. Bauer and W.-L. Wu. *J. Polym. Sci., Part B: Polym. Phys.*, vol. 40, 2002, pp. 2170-2177.

[10] C. He, M. Muramatsu, T. Ohdaira, A. Kinomura, R. Suzuki, K. Ito and Y. Kabayashi. *Appl. Surf. Sci.*, vol. 252, 2006, pp. 3221-3227.

[11] N. Kananizadeh, C. Rice, J. Lee, K. B. Rodenhausen, D. Sekora, M. Schubert, E. Schubert, S. Bartelt-Hunt and Y. Li. *J. Hazard. Mater.*, vol. 322, Part A, 2017, pp. 118-128.

[12] M. C. Fuertes, M. P. Barrera and J. Plá. *Thin Solid Films*, vol. 520, 2012, pp. 4853-4862.

[13] T. B. Landkammer, L. T. Solonariu, I. Cimpoesu and C. Baciu. *Metalurgia International*, vol. 18, 2013, pp. 75-79.

[14] A. Mahmood, N. Mahmoud Ahmed, Y. Fong Kwong, C. Lee Siang, M. Bukhari Md Yunus and Z. Hassan. *J. Exp. Nanosci.*, vol. 9, 2014, pp. 87-95.

[15] S. Dourdain, A. Mehdi, J. F. Bardeau and A. Gibaud. *Thin Solid Films*, vol. 495, 2006, pp. 205-209.

[16] T. Li, S. Karwal, B. Aoun, H. Zhao, Y. Ren, C. P. Canlas, J. W. Elam and R. E. Winans. *Chem. Mater.*, vol. 28, 2016, pp. 7082-7087.

[17] H.-J. Liu, U. S. Jeng, N. L. Yamada, A.-C. Su, W.-R. Wu, C.-J. Su, S.-J. Lin, K.-H. Wei and M.-Y. Chiu. *Soft Matter*, vol. 7, 2011, pp. 9276-9282.

[18] B. D. Vogt, R. A. Pai, H.-J. Lee, R. C. Hedden, C. L. Soles, W.-l. Wu, E. K. Lin, B. J. Bauer and J. J. Watkins. *Chem. Mater.*, vol. 17, 2005, pp. 1398-1408.

[19] A. Terawaki, Y. Otsuka, H. Lee, T. Matsumoto, H. Tanaka and T. Kawai. *Appl. Phys. Lett.*, vol. 86, 2005, pp. 113901.

[20] V. Jousseaume, O. Gourhant, A. Zenasni, M. Maret and J.-P. Simon. *Appl. Phys. Lett.*, vol. 95, 2009, pp. 022901.

[21] Y. Ito and K. Omote. *Meas. Sci. Technol.*, vol. 22, 2011, pp. 024008.

[22] S. Brunauer, P. H. Emmett and E. Teller. *J. Am. Chem. Soc.*, vol. 60, 1938, pp. 309-319.

[23] K. S. W. Sing, D. H. Everett, R. A. W. Haul, L. Moscou, R. A. Pierotti, J. Rouquerol and T. Siemieniewska. *Pure Appl. Chem.*, vol. 57, 1985, pp. 603-619.

[24] M. Sujka, Z. Sokolowska, M. Hajnos and M. Wlodarczyk-Stasiak. *J. Food Eng.*, vol. 190, 2016, pp. 147-153.

[25] D. P. Lapham and J. L. Lapham. *Microporous Mesoporous Mater.*, vol. 223, 2016, pp. 35-45.

[26] L. Liszkay, F. Guillemot, C. Corbel, J. P. Boilot, T. Gacoin, E. Barthel, P. Pérez, M. F. Barthe, P. Desgardin, P.

Crivelli, U. Gendotti and A. Rubbia. *New Journal of Physics*, vol. 14, 2012, pp. 065009.

[27] H.-G. Peng, R. S. Vallery, M. Liu, M. Skalsey and D. W. Gidley. *Colloids Surf. Physicochem. Eng. Aspects*, vol. 300, 2007, pp. 154-161.

[28] J. N. Sun, Y. F. Hu, W. E. Frieze and D. W. Gidley. *Radiat. Phys. Chem.*, vol. 68, 2003, pp. 345-349.

[29] C. He, B. Xiong, W. Mao, Y. Kobayashi, T. Oka, N. Oshima and R. Suzuki. *Chem. Phys. Lett.*, vol. 590, 2013, pp. 97-100.

[30] J. Kullmann, D. Enke, S. Thraenert, R. Krause-Rehberg and A. Inayat. *Colloids Surf. Physicochem. Eng. Aspects*, vol. 357, 2010, pp. 17-20.

[31] B. Zgardzińska and T. Goworek. *Chem. Phys.*, vol. 458, 2015, pp. 62-67.

[32] R. Zaleski, M. Gorgol, A. Kierys and J. Goworek. *Adsorption*, vol. 22, 2016, pp. 745-754.

[33] M. D. Levi, L. Daikhin, D. Aurbach and V. Presser. *Electrochem. Commun.*, vol. 67, 2016, pp. 16-21.

[34] S. Zhang, X. Zhang, Y. Tian and Y. Meng. *Biomicrofluidics*, vol. 10, 2016, pp. 024127.

[35] M. Lilja, U. Butt, Z. Shen and D. Bjöörn. *Appl. Surf. Sci.*, vol. 284, 2013, pp. 1-6.

[36] A. Mills, L. Burns, C. O'Rourke and H. Madsen. *Sol. Energy Mater. Sol. Cells*, vol. 144, 2016, pp. 78-83.

[37] M. R. Baklanov and K. P. Mogilnikov. *Microelectron. Eng.*, vol. 64, 2002, pp. 335-349.

[38] A. Perrotta, S. J. García and M. Creatore. *Plasma Processes and Polymers*, vol. 12, 2015, pp. 968-979.

[39] V. Rouessac, R. Coustel, F. Bosc, J. Durand and A. Ayral. *Thin Solid Films*, vol. 495, 2006, pp. 232-236.

[40] A. Bittner, A. Schmitt, R. Jahn and P. Löbmann. *Thin Solid Films*, vol. 520, 2012, pp. 1880-1884.

[41] A. Perrotta, W. M. M. Kessels and M. Creatore. *ACS Applied Materials & Interfaces*, vol. 8, 2016, pp. 25005-25009.

[42] M. Lépinay, L. Broussous, C. Licitra, F. Bertin, V. Rouessac, A. Ayral and B. Coasne. *Microporous Mesoporous Mater.*, vol. 217, 2015, pp. 119-124.

THE APPLICATION OF THE SMOLUCHOWSKI EFFECT TO EXPLAIN THE CURRENT-VOLTAGE CHARACTERISTICS OF HIGH-K MIM CAPACITORS

W.S. Lau

Zhejiang University, Department of Information Science and Electronic Engineering

No. 38 Zheda Road, Hangzhou 310027, Peoples' Republic of China

E-mail: liuweicheng@zju.edu.cn

ABSTRACT

In this paper, the Smoluchoski effect will be explained and is further used to understand the physics of the current-voltage (I-V) characteristics of high-k MIM capacitors in mixed-signal CMOS technology application.

INTRODUCTION

MIM capacitor structures with high-k dielectric have been widely used in mixed-signal CMOS integrated circuits. However its physics underlying the leakage current vs. voltage (I-V) characteristics is not well understood even though intensive research has been continuing for several decades. In this paper, the Smoluchoski effect will be used to explain the physics of the I-V characteristics of high-k MIM capacitor structures.

Metal has free electrons and its electron cloud tends to protrude slightly beyond the metal surface. In 1941, Smoluchowski pointed out that the electron cloud does not follow the morphology of a non-flat metal surface. In fact, the surface of the electron cloud can be smoother than the metal surface [1]. In this way, the work function of the metal can be modified by the surface morphology of the metal. For example, Li and Li [2] suggested that the work function of a metal tends to decrease with the increase of the surface roughness of the metal.

Previously, Lau [3] proposed an extended unified Schottky-Poole-Frenkel theory to explain the leakage current of MIM capacitors with various high-k dielectric materials. In this paper, the author will apply a combination of the Smoluchoski effect and Lau's extended unified Schottky-Poole-Frenkel theory to understand the physics of the I-V characteristics of high-k MIM capacitors.

Theory

For the P-F mechanism, the leakage current through an insulator is given by

$$J_{PF} = BE\exp\{[\phi_B - ((qE)/(\pi\varepsilon oK_{PF}))^{1/2}]/(kT/q)\} \quad (1)$$

where B is a constant while E, ϕ_B, k, T, q, εo and K_{PF} are the electric field, barrier height of defect state, Boltzmann constant, absolute temperature, electronic charge, vacuum permittivity and the dielectric constant for the P-F effect. Beside the P-F effect, leakage current can also be induced by Schottky emission. According to the Schottky emission mechanism, the leakage current through an insulator is given by

$$J_{SK} = A^{**}T^2\exp\{[\phi_B - ((qE)/(4\pi\varepsilon oK_{SK}))^{1/2}]/(kT/q)\} \quad (2)$$

where A^{**} is Richardson constant while E, ϕ_B, k, T, q, εo and K_{SK} are the electric field, barrier height at metal-insulator interface, Boltzmann constant, absolute temperature, electronic charge, vacuum permittivity and the dielectric constant for the Schottky effect.

It is not easy to distinguish between the P-F mechanism and the Schottky emission mechanism because in both cases the logarithm of leakage current plotted against the square root of voltage is a straight line. In addition, there is also a controversy whether the dielectric constant in eq. (1) and eq. (2) is the dielectric constant at low frequency or that at optical frequency. Previously, the author proposed an unified Schottky-Poole-Frenkel model for MIM capacitor, as shown in Fig. 1 [3]. Three assumptions have been taken into consideration in this model as explained below: (A) The high-k dielectric can be considered as a very slightly n-type large bandgap semiconductor because high-k dielectric has oxygen vacancies, which are deep double donors, as the dominant type of donors. (B) The dielectric constant K_{PF} for the P-F mechanism in eq. (1) and the dielectric constant K_{SK} for the Schottky mechanism in eq. (2) are the same, i.e. $K_{PF} = K_{SK}$. (C) The I-V characteristics of RNL in the proposed model is represented by eq. (1) which applies to all bias voltages and is independent of bias voltage polarity. The last assumption is that the I-V characteristics of the Schottky diodes D1 and D2 in the proposed model is not symmetrical. Instead it has a highly conductive "forward" I-V characteristics and a much less conductive "reverse" characteristics. Eq. (2) represents the reverse characteristics of Schottky diodes D1 and D2 for relatively larger reverse bias voltages. Eq. (2) is not applicable to the forward characteristics; furthermore, Eq. (2) may not be applicable to the reverse characteristics near zero bias voltage. There are three cases as follows. Case A: If the reverse biased D1 is less insulating than RNL, the I-V characteristics follows the P-F theory. Let the slope of log I versus the square root of applied voltage be S_{PF}. Case B: If the reverse biased D1 is about as insulating as RNL, the I-V characteristics is intermediate between the Poole-Frenkel theory and the Schottky theory. Then the slope of log I versus the square root of applied voltage would be between $S_{PF}/2$ and S_{PF} because of the assumption that $K_{PF} = K_{SK}$. Case C: If the reverse biased D1 is more insulating than RNL, the I-V characteristics would follow the Schottky theory. In this case the slope of log I versus the square root of applied voltage would be $S_{PF}/2$ because of the assumption that $K_{PF} = K_{SK}$. According to Lau's model, it is much easier to distinguish between the P-F mechanism and the

Schottky emission mechanism by examining the I-V characteristics of both polarities instead of only one polarity as discussed below.

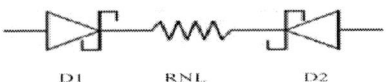

Fig. 1 An MIM capacitor structure involving a high-k dielectric can be thought as two back-to-back Schottky diodes D1 and D2 with a non-linear resistor RNL in between.

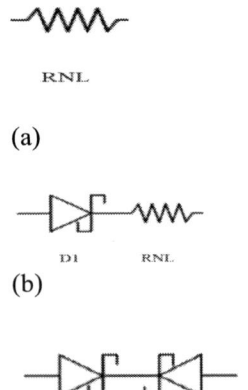

Fig. 2 Three cases of further approximation of an MIM capacitor structure: (a) a non-linear resistor RNL, (b) a Schottky diode D1 in series with RNL and (c) two back-to-back Schottky diodes D1 and D2.

For a MIM capacitor using high work function metal as both top and bottom electrodes, there are not too many electrons injected into the bulk of the high-k dielectric. The MIM capacitor can be modelled as shown in Fig. 1. Fig. 2 shows that there are 3 cases of further approximation of an MIM capacitor structure. For the case as shown in Fig. 2(c), two different work functions will result in two different effective Schottky barrier heights $\phi_{B1 \text{ and }} \phi_{B2}$ such that the leakage current will be given by two different equations as follows:

$$J_{SK1} = A^{**}T^2 \exp\{[\phi_{B1} - ((qE)/(4\pi\varepsilon_0 K_{SK}))^{1/2}]/(kT/q)\} \quad (3)$$

$$J_{SK2} = A^{**}T^2 \exp\{[\phi_{B2} - ((qE)/(4\pi\varepsilon_0 K_{SK}))^{1/2}]/(kT/q)\} \quad (4)$$

The plot of logI (or logJ) against the square root of V (or E) will appear as two parallel lines. Without understanding the Smoluchowski effect, Gaillard et al. [4] tried to explain the higher leakage current for bottom electron injection from a rougher bottom high-k/metal interface by local electric field enhancement effect. However, such a theory cannot be further developed; no equation can be derived. The application of the Smoluchowski effect to high-k MIM capacitors can be used to lead to equations (3) and (4) if the Schottky emission is the dominant leakage mechanism. The Smoluchowski effect is also important for MIM capacitors with thin high-k dielectric when

Fowler-Nordheim tunneling is the dominant leakage mechanism.

Results and Discussion

The Smoluchowski effect can help to understand some historical experimental results which were not adequately explained at the time of publication. Two examples are given below. As shown in Fig. 3, the I-V characteristics of Mo/Ta$_2$O$_5$/Mo capacitors modified from Hashimoto et al. are obviously not the same when the bottom Mo electrode has different thickness [5]. The leakage current is much higher when the bottom Mo electrode is thicker. In addition, the logarithm of the leakage current density versus the square root of the electric field plot for the Mo/Ta$_2$O$_5$/Mo capacitor with the thicker bottom Mo electrode is obviously parallel to that with the thinner bottom Mo electrode. This can be easily explained by the Smoluchowski effect. Another example is shown in Fig. 4. The I-V characteristics of a TiN/Ta$_2$O$_5$/TiN capacitor at various temperatures reported by Deloffre et al. are obviously not symmetrical for gate bias polarity [6]. Deloffre et al. tried to interpret their results by using the Poole-Frenkel mechanism for the situation when positive voltage was applied to the top TiN gate. However they did not manage to suggest a mechanism to explain the leakage current under the other polarity [6]. Using point by point extraction of the experimental data shown in Fig. 3, the author noticed that the experimental data are consistent with the model of two back-to-back Schottky diodes as shown in Fig. 2(c). It has been reported that amorphous Ta$_2$O$_5$ deposited by CVD has a surface smoothing effect [7]. Because of such smoothing effect, the bottom TiN/Ta$_2$O$_5$ interface is rougher than the top TiN/Ta$_2$O$_5$ interface. With the help of the Smoluchowski effect, it can be predicted that the bottom TiN/Ta$_2$O$_5$ interface has a smaller effective Schottky barrier height than the top TiN/Ta$_2$O$_5$ interface due to a smaller effective work function of the bottom TiN as compared to the top TiN. If the Ta$_2$O$_5$ deposited by CVD is relatively thin, the MIM capacitor can be modelled by two back-to-back Schottky diodes as shown in Fig. 2(c).

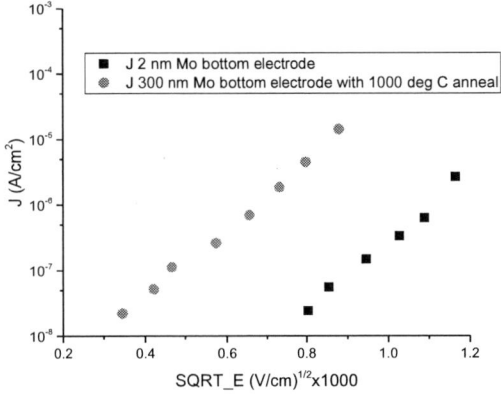

Fig. 3 The J-E characteristics of 2 Mo/Ta$_2$O$_5$/Mo capacitors with 2 different bottom Mo thicknesses modified from Hashimoto et al. [5].

978-1-5090-6695-7/17 $31.00 © 2017 IEEE

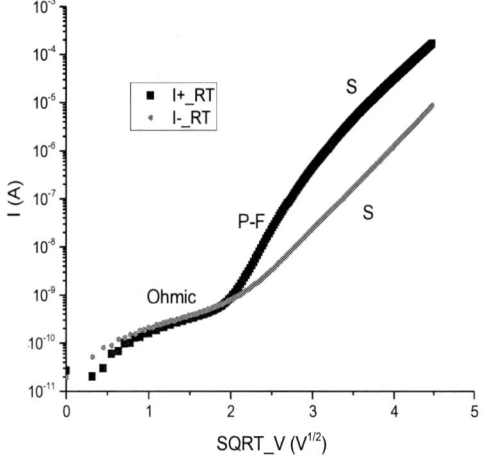

Fig. 4 The J-V characteristics of a TiN/Ta₂O₅/TiN capacitor at various temperatures modified from Deloffre et al. [6]. The bias is the voltage applied to the top TiN gate.

As shown in Fig. 5, the logarithm of current measured from the author's W/Ta₂O₅/W capacitor samples at room temperature is plotted against the square root of applied voltage [7]. The Smoluchowski effect can also be observed at a relatively high bias voltage (> 9 V) such that an approximate model shown in Fig. 2(c) is valid. For lower bias voltage (< 9 V), an approximate model shown in Fig. 2(b) has to be used. For voltage > 4 V, the leakage current can be explained by the Poole-Frenkel and Schottky emission mechanisms; however, the Smoluchowski effect is needed to explain the asymmetry. The temperature dependence of the leakage current has been studied in more details [7]. The effective Schottky barrier height was about 0.67 eV and 0.88 eV for the bottom and top W/Ta₂O₅ interfaces respectively. (Note: For very low voltage, the I-V characteristics can be symmetric and Ohmic.)

Fig. 5 The logarithm of current versus the square root of voltage characteristics of a W/Ta₂O₅/W capacitor [7]. The Ta₂O₅ film (93.3 nm thick) was deposited by CVD on W. The top W was deposited by sputtering. A mild post-metallization anneal was done in nitrogen. The bias is the voltage applied to the top W gate.

An important precaution for sample preparation is that the high-k dielectric film should not be too thick. Otherwise the I-V characteristics will be controlled by the bulk electrical properties of the high-k dielectric instead of the interfacial roughness. Another important precaution for sample preparation is as follows. If there are a lot of defect states in the high-k dielectric, there is some sort of Fermi level pinning such that the effective Schottky barrier height of the bottom interface is equal to that for the top interface even though the bottom metal electrode and the top metal electrode are not the same [8], resulting in a symmetrical I-V characteristics. Lau [8] called this Type B I-V symmetry. The author noticed that Type B I-V symmetry can occur in MIM capacitors with the same top and bottom metal electrodes but an asymmetry of top and bottom interfacial roughness when there are a lot of defect states in the high-k dielectric. The defect states might arise from the top metal deposition process, for example, sputtering due to plasma damage or plasma charging. A mild post-metallization annealing in nitrogen is quite frequently needed to suppress Type B I-V symmetry in order to observe the Smoluchowski effect.

Conclusion

Without the understanding of the Smoluchowski effect, a lot of experimental observations on high-k MIM I-V characteristics cannot be understood. The Smoluchowski effect can strongly influence leakage current due to Schottky emission or Fowler-Nordheim tunneling.

References

[1] R. Smoluchowski, "Anisotropy of the electronic work function of metals," Phys. Rev., vol. 60, pp. 661-674, 1941.

[2] W. Li and D.Y. Li, "On the correlation between surface roughness and work function in copper," J. Chem. Phys., vol. 122, article number 064708, 2005.

[3] W.S. Lau, "An extended unified Schottky-Poole-Frenkel theory to explain the current-voltage characteristics of thin film metal-insulator-metal capacitors with examples for various high-k dielectric materials," ECS J. Solid State Sci. Technol., vol. 1, pp. N139-N148, 2012.

[4] N. Gaillard, L. Pinzelli, M. Gros-Jean and A Bsiesy, "In situ electric field simulation in metal/insulator/metal capacitors," Appl. Phys. Lett., vol. 89, article number 133506, 2006.

[5] C. Hashimoto, H. Oikawa and N. Honma, "Leakage-current reduction in thin Ta2O5 films for high-density VLSI memories," IEEE Trans. Electron Dev., vol. 36, pp. 14-18, 1989.

[6] E. Deloffre, L. Montes, G. Ghibaudo, S. Bruyere, S. Blonkowski, S. Becu, M. Gros-Jean and S. Cremer, "Electrical properties in low temperature range of tantalum oxide dielectric MIM capacitors," Microelectronics Reliability, vol. 45, pp. 925-928, 2005.

[7] D.Q. Yu, W.S. Lau, H. Wong, X. Feng, S. Dong and K.L. Pey, "The variation of the leakage current characteristics of W/Ta₂O₅/W MIM capacitors with the thickness of the bottom W electrode," Microelectronics Reliability, vol. 61, pp. 95-98, 2016.

[8] W.S. Lau, "A new mechanism of symmetry of current-voltage characteristics for high-k dielectric capacitor structures," ECS Trans., vol. 45(3), pp. 151–158, 2012.

THE STUDY OF THE IMPURITY IN HK FILM

Yingming Liu[1]*, Qiuming Huang[1], and Haifeng Zhou[1]

[1] Shanghai Huali Microelectronics Corporation

568 Gaosi Rd., Pudong district, Shanghai, China

*Corresponding Author's Email: liuyingming@hlmc.cn

ABSTRACT

As CMOS size scaling down, HKMG was introduced into CMOS manufacture process to replace Poly-SiON scheme earlier at 45nm node. In many HK materials, HfO2 was finally chosen for its good thermal stability, non-reacting with silicon, no water absorption. Both ALD and MOCVD could be used for HfO2 deposition. From film quality, ALD was the better choice for its film quality, since ALD as self-limited process was not very sensitive to the process parameters like flow, temperature or pressure. Though considered relative stable, HK film properties, especially film impurity should be carefully studies for its dominating influence to device performance. From SIMS analysis, Cl in HK film will reduce when adding water pulse time or raising temperature of the water tank, and H will increase. While the content of Hf and O will keep stable.

Keywords—HfO2, ALD, Impurity, SIMS

INTRODUCTION

As the CMOS device density increasing, traditional process like Poly-SiON scheme met many challenges. In order to match the requirements of CMOS size scaling down, thinner dielectric should be used. But thinner dielectric will lead serious gate leakage, and worse controllability of device. To get rid of the negative effects, new HK dielectric materials were introduced. At the same equivalent oxide thickness (EOT), high K materials have thicker thickness of physical [1]. The relationship of K value and thickness was shown in Equation (1) and Equation (2):

$$\tag{1}$$

$$\tag{2}$$

In the equation, W and L are the gate width and length respectively; μ is carrier mobility; V_G is gate voltage; V_{th} is gate threshold voltage; A is capacitor area; t is dielectric material thickness; is the permittivity of vacuum.

Most of high K materials can be classified into metal oxide. Considering process integration, K value was not the higher the better. First, because many thermal processes were used like S/D active anneal, silicide RTA anneal, most of the high K materials cannot meet the thermal stability requirement. Second, the chosen material cannot react with silicon. Third, the band gap E_g should be larger than 5ev to suppress gate leakage [2].

So, the K value of Al_2O_3 is too low; La_2O_3 is bibulous; ZrO_2 suffers low temperature crystallization, and reacts with silicon; Band gap of TiO_2 is too small. So the best chosen material is HfO_2.

Because between HfO_2 film and poly has serious Fermi level pinning, metal gate is used to instead of poly gate. By choosing suitable metals to match NMOS and PMOS work function, could achieve the purpose of adjusting V_t [3]. Also metal gate has no depletion layer which can reduce T_{inv} and improve I_{dsat}. Based on the metal gate thin film deposition before or after S/D active anneal, HKMG technology can be divided into HK and metal gate both first, HK first gate last, HK and metal gate both last. HK and metal gate both first has the simplest process, but HK and the metal gate will suffer high temperature process, so it is hard to control V_t, also device performance is difficult to achieve high performance requirements. Usually HK and metal gate both first uses HK cap layer to tune flat band voltage (VFB), then meet the work function requirement [4]. Gate last process flow was much more complex, dummy poly gate was first fabricated, and then was removed to fill metal gate. Therefore gate last, also was known as the replacement metal gate. Gate last avoided the high temperature anneal, so it could get the best device performance. Usually to adjust the V_t, different work function metal layers were chosen for NMOS and PMOS [5].

In this paper, we just discuss the film quality of HfO_2 film. Except K value, the film quality and process parameters will impact the reliability like NBTI and PBTI. This is because the impurity of the HK film will diffuse into the interfacial layer and then form a trap center. Finally results in V_t shift and mobility degrade.

ALD PROCESS

Single atomic layer deposition (ALD) was also known as atomic layer deposition and atomic layer epitaxy (atomic layer). This technology allowed gas precursors pulsing injected into the reactor and deposited film on the substrate by chemical adsorption and reaction [6].

The basic steps of ALD were shown in Figure 1. The first precursor was introduced into the reaction chamber and adsorbed until saturated on the substrate surface. Then the rest of gas was pumped away from the chamber. Next, the second precursor was injected into the chamber, and reacted with the first precursor on the wafer surface. The rest of the reactants and reaction by-products were pumped away. By repeating the above processes, ALD process can realize precise thickness control.

ALD process was different from CVD. First, ALD contains a saturated adsorption process. Excessive reactants will be purge clean. Second, ALD is a self-limited reaction, when achieved the saturated adsorption, increasing the reaction gas pulse time, deposition rate will not change [7]. The saturation curve of ALD was shown in Figure 2. In the first tilt range of the saturated curve, the reaction rate increased with pulse time. After achieved saturated adsorption, the curve became flat and reaction rate was not increasing. If continue to increase the gas pulse time, CVD reaction would happen in the chamber. So the precursor pulse time could not set too long to avoid CVD reaction. The critical point is observed deposition rate increasing sharply [8].

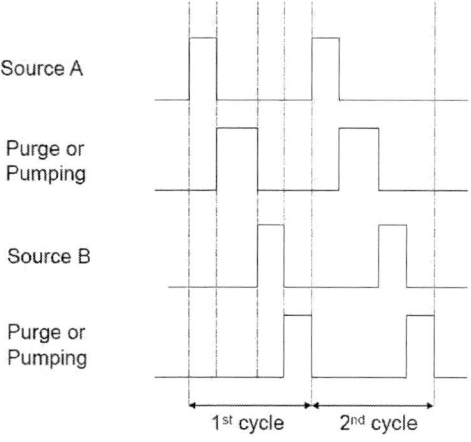

Figure 1: ALD process control

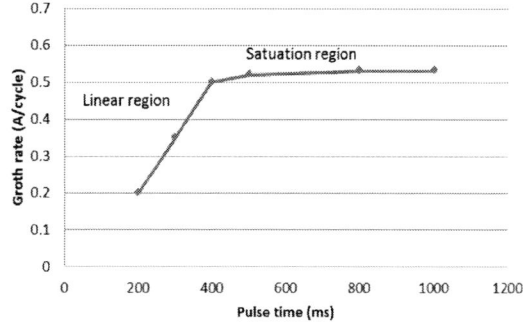

Figure 2: Saturated curve of ALD process

HK FILM DEPOSITION

We used ASM XP4 platform to deposition HK film. Water and $HfCl_4$ were used as HfO_2 film deposition precursors. The reaction temperature was about 300 ℃.

When deposit film, first injected the water vapor into the chamber. The precursors were compound with N_2 which acted as carrier gas. When water was saturated absorbed on the surface, the rest water was pumped away by N_2 flow. Next pulsed the HfCl4 gas flow and purged the unreacted gas and by products. This was a whole growth ALD cycle. By setting the growth cycle count could get desirable film thickness.

The water tank was set at 15 ℃ ~ 50 ℃. $HfCl_4$ source was heated to 150 ℃ ~ 200 ℃. Considering the reaction gas concentration and the using life of the source, usually the temperature was set at around 175 ℃. The film TEM section is shown in Figure 3. In the figure, the white thin layer is the native oxide.

Among the lots of process parameters, gas flow rate, pulse and purge time, chamber temperature, and reaction pressure are the tunable parameters. The gas pulse and purge time, chamber temperature will impact film uniformity. Gas flow impacts particle and deposition rate. Because ALD process is self-limited growth, after saturated adsorption happened, flow and pulse time have little impact on the deposition rate.

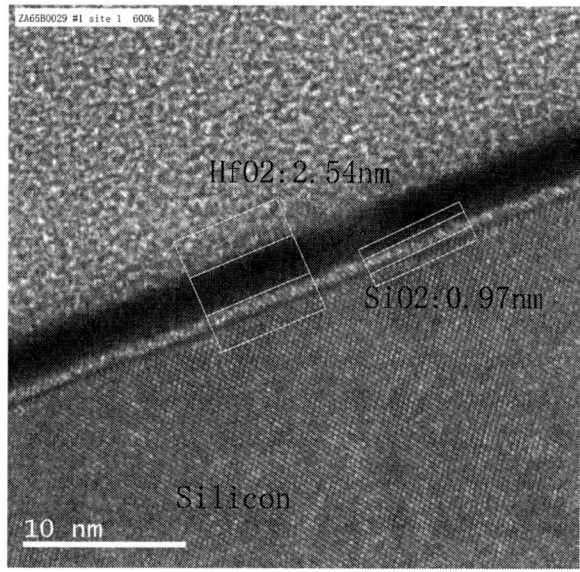

Figure 3: TEM cross section of HfO_2

The film impurity content was related with precursor gas flow rate, pulse and purge time, especially related with water pulse time and flow rate. This is because the substrate has stronger ability to absorb water than $HfCl_4$. So extending water pulse time, or increasing water gas flow can make $HfCl_4$ react more completely. Cl ion content will reduce. But correspondingly H ion concentration will increase. While Hf and O content basically remain unchanged.

RESULTS

Based on the above analysis, we set up several related splits, the water pulse time set as <500ms and <5s two conditions, water source temperature set as <20℃ and <35℃. The split conditions were shown in Table 1:

TABLE 1: The split table of HK film deposition

Split	H2O TEMP	H2O Pulse(ms)
1	<20 ℃,	<500
2	<20 ℃,	>5000
3	< 35 ℃,	<500
4	<35 ℃	>5000

Taking the samples to do SIMS analysis, obtained the following results as shown in Figure 4. In the four conditions, Hf content was almost no changed. O content had little lower only at <20 °C water source temperature combined with pulse time of <500 ms. The Cl and H content had bigger variation. Increasing the pulse of time or the water source temperature could significantly reduce the content of Cl, but H element content would increase extremely. Too much of the Cl and H both had negative effect to device performance. The two elements content should be balanced when at actual usage.

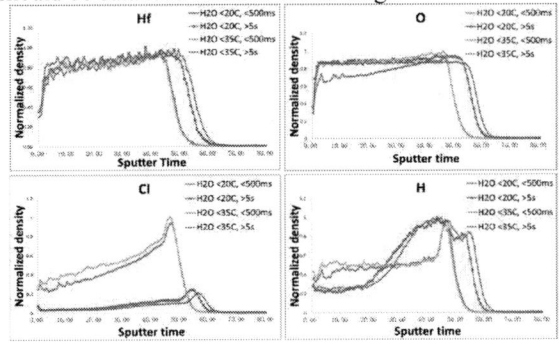

Figure 4 : The elements content of different splits analyzed by SIMS

REFERENCES

[1] M. Leskela, M. Ritala. *Thin Solid Films*, vol. 409, 2002, pp. 138-139.

[2] L. Niinistö, J. Päiväsaari, J. Niinistö, M. Putkonen, and M. Nieminen. *phys. stat. sol.* Vol.201, No. 7, 2004, pp. 1443–1452.

[3] John Robertson, Robert M. Wallace. *Materials Science and Engineering.* vol. 88, 2015, pp. 1–41.

[4] Yukinori Morita, Shinji Migita, Wataru Mizubayashi, Meishoku Masahara, Hiroyuki Ota. *Solid-State Electronics.* vol.84, 2013, pp. 58–64.

[5] A. Kerber, S. A. Krishnan, E. A. Cartier. *IEEE EDL.* 2009, pp. 1347-1349.

[6] E. Cartier, A. Kerber, T. Ando, M. M. Frank, K. Choi, S. Krishnan, B. Linder, K. Zhao, F. Monsieur, J. Stathis, and V. Narayanan. *IEEE IEDM.* Dec. 2011, pp. 18.4.1–18.4.4.

[7] K. Joshi, S. Hung, S. Mukhopadhyay, T. Sato, M. Bevan, B. Rajamohanan, A. Wei, A. Noori, B. McDougall, C. Ni, C. Lazik, G. Saheli, P. Liu, D. Chu, L. Date, S. Datta, A. Brand, J. Swenberg and S. Mahapatra, *Electron Device Letters.* vol. 34, no. 1, 2013, pp. 3 –5.

[8] Satoru Mayuzumi, Shinya Yamakawa, Yasushi Tateshita, Tomoyuki Hirano, Masashi Nakata. IEEE *Transactions On electron Devices.* vol. 56, no. 4, 2009.

THE STUDY OF 28NM BEOL CU GAP-FILL PROCESS

Yu Bao, Gang Shi, Lin Gao, Yanyan Zhang, Yingming Liu, Peng Tian, Fuchun Xi, Wei Hu, Ying Gao,
Zhenhua Cai, Baojun Zhao, Zhigang Yang, Jianghua Leng, Haifeng Zhou, Jingxun Fang
Shanghai Huali Microelectronics Corporation, Shanghai 201203, China
baoyu@hlmc.cn

ABSTRACT

In this paper, the influence of Copper (Cu) barrier and seed process tuning on step coverage was analyzed. TEM images show relatively thinner barrier can improve the opening CD of a metal line structure hence improve the sidewall coverage of Cu seed. Cu Seed adopts the deposition/re-sputter method to improve the step coverage, and a higher ratio of re-sputter/deposition can increase the thickness of Cu seed on the sidewall. According to the post CMP surface defects scan results, the optimization of barrier Cu seed thickness can significantly reduce the copper void defects density.

INTRODUCTION

Copper has been widely used in integrated circuit as the interconnect material because of its low resistivity and good electro-migration (EM) resistance. However, As the CD of integrated circuit scaling down, the Cu gap filling became a big challenge. For Cu interconnect formation, Cu barrier seed process is very important, and a robust condition can provide wider process window for subsequent electro Cu plating process (ECP). [1-3] Combined with the TEM images of post-Cu barrier/seed deposition and the post-Cu-CMP surface defects data, the influence and mechanism of Cu barrier/seed process tuning on Cu-void performance were proposed.

EXPERIMENT

Metal line structures were prepared with ULK/MHM scheme. Ta(N) based barrier and CuMn seed layer were deposited by PVD. To investigate the mechanism of Cu gap filling, the thickness of Cu barrier/seed were measured using TEM images. The gap fill performance was evaluated by surface defects scan at post CMP.

RESULTS AND DISCUSSIONS

Barrier Layer

Bilayer Ta(N) based barrier is deposited by PVD. The whole deposition process is divided into 4 steps, including TaN deposition, Ta deposition, Ar re-sputter, and Ta Flash. TaN is the barrier layer to prevent Cu diffusion. Ta as the wetting layer for Cu seed. Ar re-sputtering can thin down the barrier at the bottom of

metal line structures by ion bombardment. The re-sputtered Ta(N) re-deposited onto the sidewall resulting in better step coverage. Figure 1 showed the TEM images of barrier seed with 3 different barrier conditions.

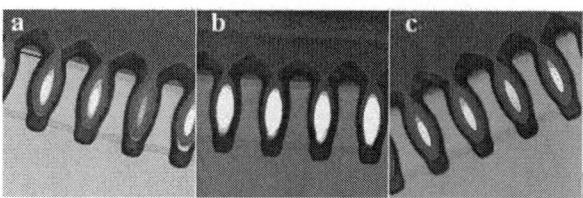

Figure 1. The TEM images of barrier seed with different barrier. (a). Thinner barrier with less Ar re-sputter; (b). Thinner barrier; (c). Control barrier

The step coverage of Cu barrier seed was shown in Table I. It is difficult to distinguish the boundaries between barrier and Cu seed on sidewall, so we measured the total thickness of the barrier and Cu seed. As shown in Figure 2, the thinner barrier improved the opening CD and hence improved the sidewall coverage of Cu seed.

Table I. Normalized step coverage of barrier seed with different barrier

Split	Condition	Barrier+ Seed Overhang	Barrier+ Seed Sidewall
1	Less Ar re-sputter Barrier + Control seed	0.86	1.09
2	Control barrier + Control seed	0.91	1.08
3	Thicker barrier + Control seed	1.00	1.00

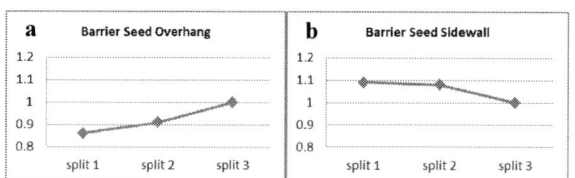

Figure 2. Normalized step coverage with different barrier. (a). Overhang; (b). Sidewall

Cu Seed Layer

Copper seed works as cathode and conductive layer

during Cu ECP process. The quality of Cu seed is critical to ECP. Cu Seed formation can be divided into 2 steps, deposition and Cu re-sputter. The Cu re-sputter step can improve the coverage on the sidewall. Figure 3 showed the TEM images of Cu barrier seed with 3 different seed conditions.

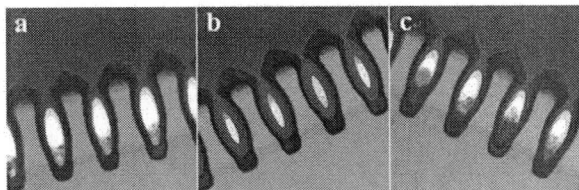

Figure 3. The TEM images of barrier seed with different seed. (a).Less deposition seed; (b).Control seed; (c).More re-sputter seed.

The step coverage of Cu barrier/seed is shown in Table II. The higher the Cu re-sputter/deposition ratio, the higher the Cu seed coverage on sidewall.

Table II. Normalized step coverage of barrier seed with different seed

Split	Condition	Barrier+ Seed Overhang	Barrier+ Seed Sidewall
4	Thicker barrier + Less dep seed	0.89	1.05
3	Thicker barrier + Control seed	1.00	1.00
5	Thicker barrier + More re-sputter seed	1.00	1.13

As shown in figure 4, both less deposition and more re-sputter seed can improve sidewall coverage, and less deposition can reduce overhang. So seed with more re-sputter is preferred to gap filling.

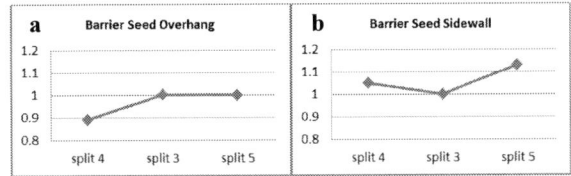

Figure 4. Normalized step coverage with different seed. (a).Overhang; (b).Sidewall

Defect Scan Results

Table III is the design of experiment of Cu barrier/ seed thickness. The gap fill results were evaluated by post Cu CMP void inspection.

Table III. experiments design of barrier seed

Step	Condition	#1	#2	#3	#4
Barrier	Thinner barrier	V	V		
	Control barrier			V	V
Seed	Thinner seed	V		V	
	Control seed		V		V

Post-CMP defect data in Fig.5 showed thinner barrier is beneficial for ECP gap filling due to larger opening CD. However, the thinner seed split can make the Cu void worse significantly.

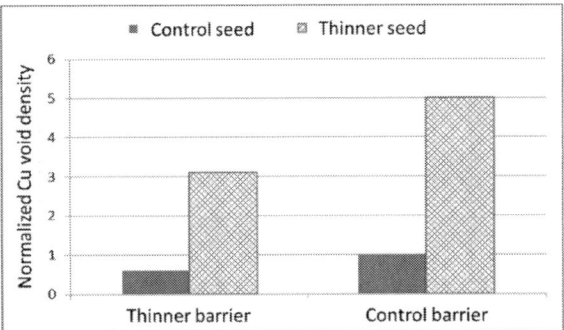

Figure 5. Normalized Cu void density post Cu-CMP with various barrier/seed splits

Typical Cu void post CMP was shown in figure 6, missing Cu can be found on the sidewall of recess structure. It seems that continuous seed on the sidewall is more important for gap filling.

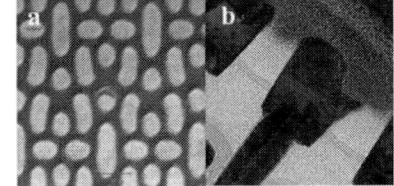

Figure 6. Typical Cu void post CMP. (a).Top view; (b).Cross section

CONCLUSION

In conclusion, Cu-void performance was remarkably improved by the optimization of Cu barrier/seed scheme. Relative thinner barrier combined with thicker continuous seed layer was preferred for subsequent Cu ECP. Improved gap fill performance confirmed by post Cu-CMP defect data. Result is consistent with the good step coverage of Cu barrier/seed shown by TEM images.

ACKNOWLEDGEMENTS

The author would like to acknowledge all the 28nm process integration team members of HLMC for the technical discussions, and the failure analysis lab for supporting TEM work.

REFERENCES

[1] Weiye He, Beichao Zhang, Jian Kang, et. al. The contributions of barrier resputter for BEOL integration. ECS Transactions. 44 (1) 487-492 (2012)

[2] Yu Bao, Xuezhen Jing, Jingjing Tan, et. al. Optimization of Metallization Processes for 28-nm-node Low-k /Cu Multilevel Interconnects. ECS Transactions. 44 (1) 477-480 (2012)

[3] Xuezhen Jing, Jingjing Tan, Jiquan Liu. 32/28NM BEOL CU GAP-FILL CHALLENGES FOR METAL FILM. CSTIC 2015

NICKEL SILICIDE ANNEAL PROCESS RESEARCH FOR 28NM CMOS NODE

Zhenping Wen[1], Xinhua Cheng[1], and Jingxun Fang[2]*

[1] Shanghai Huali Microelectronics Corporation, Shanghai 201210, China

wenzhenping@hlmc.cn

ABSTRACT

A NiPt silicide for CMOS ohmic contact formation process extended from 65nm to 28nm node has been achieved by co-optimization of NiPt alloy deposition thickness, Pt additive amount complementary wet selective etch process. In this study, a lower RTP-1 process has been investigated on nickel rich silicide phase formation and physical defect reduction. A higher millisecond anneals (MSA) RTP-2 has been investigated on its process window for nickel mono-silicide formation without nickel silicide agglomeration and additional nickel piping. Then the optimized RTP program which combines lower RTP-1 and higher RTP-2 by MSA has been demonstrated effective reduction to nickel piping by e-beam inspection count, but at the expense of higher sheet resistance which dominated impact by the thinner silicide thickness.

INTRODUCTION

The use of nickel silicide layer to reduce the contact resistance at the gate or source-drain regions has become standard process in CMOS processing. The conventional nickel silicidation process consists of two separate low temperature rapid thermal processing (RTP) steps (Fig. 1) with a selective etch inserted between the two anneals. The first anneal (RTP-1) is typically a lower temperature (220~300℃) anneal to drive nickel diffuses into the Si to form the nickel rich silicide layer (Ni2Si). The key challenges during this step is the formation of a thin metal-rich silicide layer, meanwhile controlling the Silicide-Si interface quality to avoiding excessive nickel piping diffusion to form short channel nickel pipes or spiking defects, which are a source of leakage current. After a wet selective removal of un-reacted metal, the wafer moves to the second anneal (RTP-2) process, it is typically a higher temperature (400~500℃) process which transforms the nickel silicide from higher resistance Ni2Si phase to the low resistance NiSi phase. The key challenge during this step is achieving a complete and uniform phase transformation without NiSi agglomeration, additional nickel piping diffusion or subsequent NiSi2 formation, which are a source of higher contact resistance and leakage issues.

As logic device scale to the 28nm technology node, introduction of ultra-low temperature anneal (ULT) and millisecond anneal (MSA) into the nickel silicide process is a promising approach for increasing device performance and yield improvement.

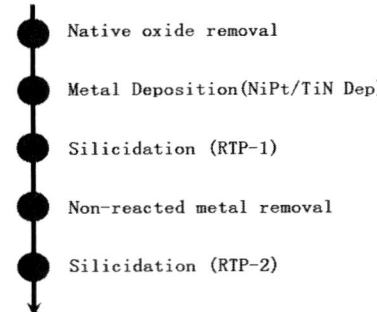

Fig. 1. Conventional Nickel Silicidation Process

In this study, firstly, a lower RTP-1 is characterized its effect to nickel silicide formation and piping defect; then a MSA RTP-2 process window is determined for low resistance NiSi formation; After that, an anneal program which combines a lower temperature RTP-1 and higher temperature RTP-2 by MSA is used to check the inter-metal WAT Rs performance compared to baseline process.

EXPERIMENT

Both blanket and pattern wafers were prepared with 28nm baseline process flow. Firstly, the native silicon dioxide on the wafer was removed before the Ni(Pt)/TiN metal deposition; then, the wafers were moved to RTP-1 soak anneal process splits to form starting nickel rich silicide (Ni2Si) layer and control on the same thickness level. After that, all the wafers combined and proceeded with wet selective etching process in the batch spray processor to remove un-reacted metal; then, the wafers splited into baseline RTP-2 soak anneal and several millisecond anneal (MSA) process conditions to form final nickel mono-silicide (NiSi) and complete the nickel silicide formation process. The major process flow for the pattern wafer study is shown in Fig. 2.

978-1-5090-6695-7/17 $31.00 © 2017 IEEE

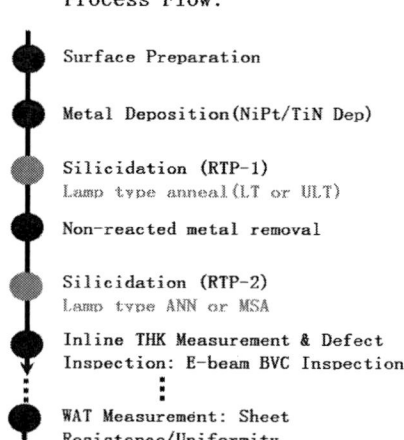

Process Flow:

- Surface Preparation
- Metal Deposition(NiPt/TiN Dep)
- Silicidation (RTP-1)
  Lamp type anneal(LT or ULT)
- Non-reacted metal removal
- Silicidation (RTP-2)
  Lamp type ANN or MSA
- Inline THK Measurement & Defect
  Inspection: E-beam BVC Inspection
  ⋮
- WAT Measurement: Sheet
  Resistance/Uniformity

Fig.2. Process Flow for Experiment

On the blanket wafers, after RTP-2 process, the sheet resistance (Rs) and thickness were measured by four-point probe and KLA-Tencor thickness measurement tool; The Transmission Electron Microscopy (TEM) were used to characterize the nickel silicide interface roughness, agglomeration behavior and microstructure. The RTP-2 millisecond anneal process window is determined by the measured result from blanket wafers.

On 28nm pattern wafers, after RTP-2 process, the wafers continue on the CMOS integration process to form tungsten contact; After W-CMP, the different process split wafers were sampled to proceed e-beam scan for defect inspection of nickel piping by brightness voltage contrast (BVC) count. Also the WAT Rs performance is measured after metal 2 layer formed.

RESULT DISCUSSION

First, Fig. 3 illustrated the NiSi thickness and Rs trends with process time split based on 240℃-RTA1 and 130Å NiPt deposition thickness. It was observed that removing 120sec condition, Rs value went to higher in terms of Rs and thickness value went to stability in terms of thickness which indicated that Ni is fully consumed under 120s condition, so in order to ensure that the excess of NiPt, all the below experiments were based on 180Å NiPt deposition thickness condition which had been approved sufficient enough in our internal studies.

A. RTP Program Impact on Sheet Resistance

On the blanket and pattern wafers, two groups of process split have been explored; It includes a baseline RTP-1 soak anneal with T1 (270°C) temperature and a lower temperature condition (T1 - 30°C) and combine a MSA RTP-2 anneal with the temperature range from 650 to 1000°C, which was summarized as Table1. The results were summarized in figure 7-9. From Fig. 8 and Fig. 9, it

is observed that lower RTP-1 gets higher sheet resistance compared to baseline RTP-1 temperature. Since the silicide thickness is different because of the different soak annealing time, this sheet resistance shift is considered dominated by the silicide thickness and subordinately by the phase composition of formed nickel silicide. From Fig. 7 it is also observed that a stable sheet resistance value over a wide MSA temperature range approaching 980°C on both RTP-1 cases; this indicates that no significant nickel silicide phase change or agglomeration takes place. When MSA RTP-2 temperature approaching 980°C, the Rs begins to increase; this indicates either NiSi agglomeration is taking place or partial nicke silicide phase transformed to higher resistive NiSi2.

The thickness difference between NMOS and PMOS is shown on Fig.6, it is observed that lower RTP-1 gets smaller difference of N/PMOSFET, but it is goes higher with process time increased.

B. NiSi Agglomeration Behavior

Wafers that processed by different RTP programs had been analyzed by TEM micrograph for cross section and the result is shown in Fig. 4. It is observed that the nickel silicide thickness had been controlled to the same level. For the NiSi film roughness, it is noted that process B shows better NiSi film roughness (15%) compared to baseline process A (30%); This suggest higher MSA RTP-2 has attribution to reduce the NiSi agglomeration. By lower RTP-1 temperature, the process C results even better NiSi film roughness (9%); this indicates the lower RTP-1 combines MSA RTP-2 anneal program will effect on reduction of NiSi agglomeration.

C. E-beam Brightness Voltage Contrast for Ni Piping Defect

Another interest of physical defect detection is to check the nickel piping defect, which is believed a source of N_MOS-leakage current. This measurement is done after W-plug CMP process by an e-beam defect scan tool. When an e-beam charged W-plug contacts to a leaky junction or a leaky poly-silicon gate, the electrons from the substrate can neutralize the positive charge, and its SEM image indicates higher brightness. This brightness voltage contrast (BVC) has reported effective to characterize the nickel piping behavior. In this test, the (BVC) measurement result from the wafers that proceeded by different RTPs programs were summarized in Fig. 5.

Fig. 5 shows that the lower RTP-1 combined MSA RTP-2 will have lower nickel piping defects. As the piping diffusion is mainly driven by thermal dynamic during the anneal process, it also indicates the MSA RTP-2 not reduce secondary Ni diffusion.

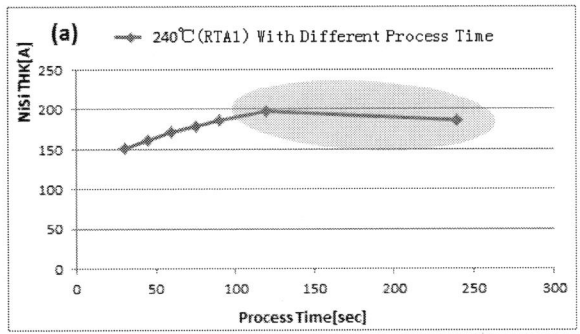

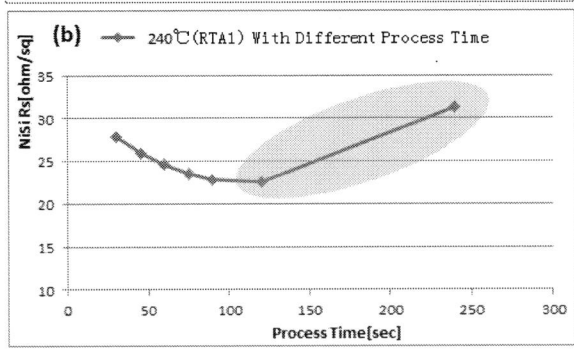

Fig.3. NiSi Thickness and Rs Result with RTA1-240 ℃
Process Time Split Based on 130Å NiPt

TABLE I.   RTP PROCESS CONDITIONS SPLIT

Proces Split	Anneal Temperature	
	RTP-1	RTP-2
A	T1	T2
B	T1	MSA(T2+300℃)
C	T1 - 30℃	MSA(T2+300℃)

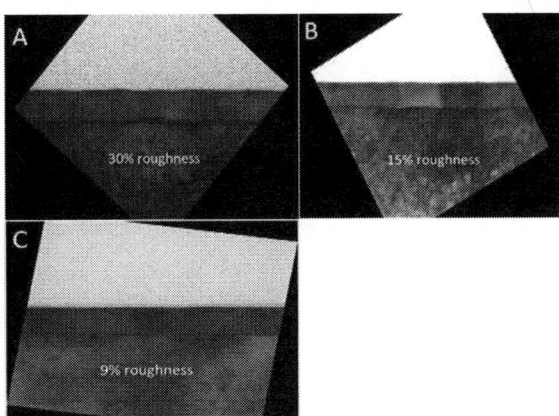

Fig.4. TEM Micrographs of the Films Received RTPs
Process A (B) Process B (C) Process C

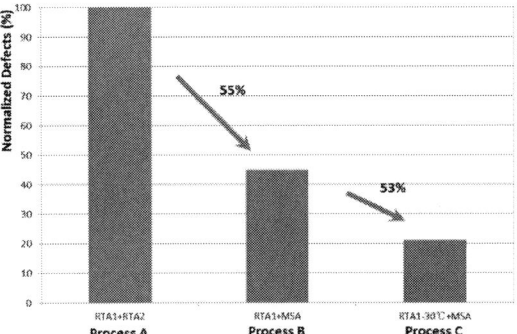

Fig.5. Total BVC counts of E-beam Inspection

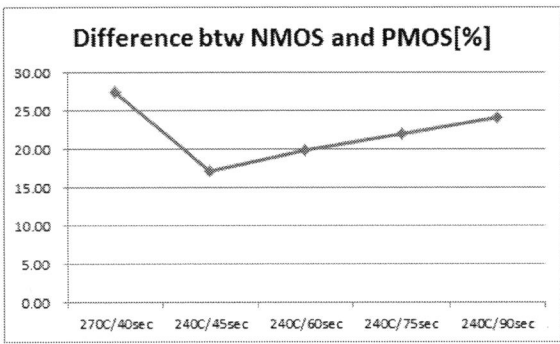

Fig.6 Thickness Difference btw NMOS and PMOS

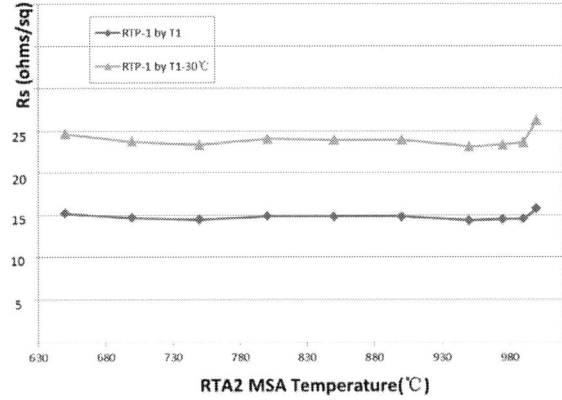

Fig.7. Nickel Silicide Sheet Resistance Trend of MSA
RTA-2 Temperature on Two RTP-1 Programs

978-1-5090-6695-7/17 $31.00 © 2017 IEEE   243

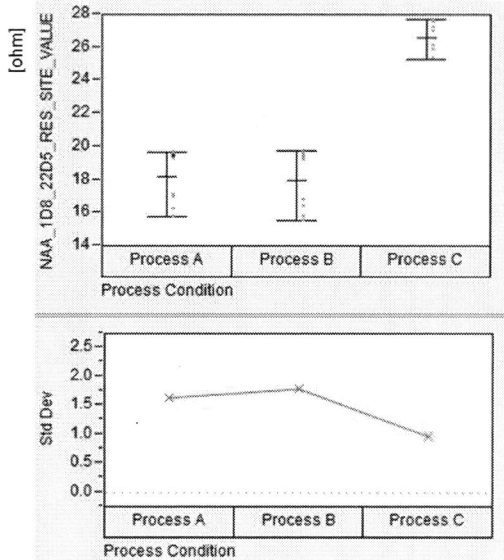

Fig.8. N+ AA NiSi Sheet Resistance of Different RTPs Programs

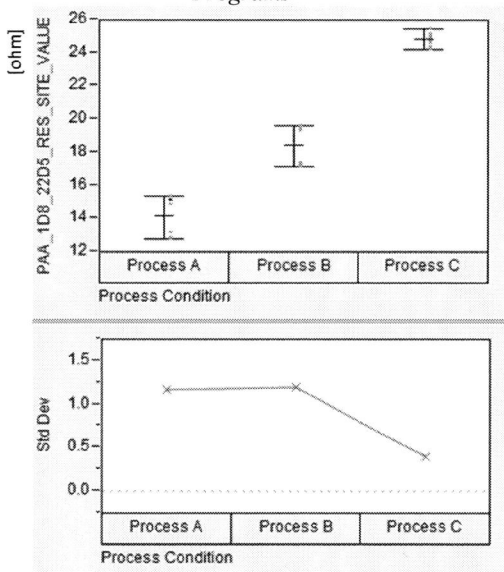

Fig.9. P+ AA NiSi Sheet Resistance of Different RTPs Programs

CONCLUSION

In this work, the potential RTP program optimization is demonstrated for 28nm CMOS NiPt silicide formation. The results show that lower RTP-1 process can lead to physical defect reduction and the smaller thickness difference of N/PMOS, and a higher millisecond anneal (MSA) RTP-2 can improve NiSi roughness, suppress NiSi agglomeration, and NMOS nickel piping. Then the optimized RTP program, which combines a lower RTP-1

and higher RTP-2 by MSA, has demonstrated the effective reduction of nickel piping defect by Brightness Voltage Contrast (BVC) inspection, but at the expense of higher sheet resistance which dominated impact by the thinner silicide thickness.

REFERENCES

[1] Yi Wei Chen, et. al, "The Determination of Annealing Program for NiSi Formation", Applied Materials Front End Products (references), 13th IEEE International Conference on Advanced Thermal Processing of Semiconductors - RTP 2005

[2] Yi-Wei Chen1, Nien-Ting Ho1, Jerander Lai1, T.C. Tsai1, C.C. Huang1, J.Y. Wu1, Ben Ng2, A. J. Mayur2, Alex Tang2, Shankar Muthukrishnan2, Jeremy Zelenko2, and Helen Yang3, "Advances on 32nm NiPt Salicide Process" United Microelectronics Corp., Applied Materials Inc. October 2, 2009.

[3] Bruce Adams, Dean Jennings, Kai Ma, Abhilash J. Mayur, Steve Moffatt,, Stephen G Nagy, Vijay Parihar, "Characterization of Nickel Silicids Produced By Millisecond Anneals", Applied Materials Front End Products, 15th IEEE International Conference on Advanced Thermal Processing of Semiconductors - RTP2007

Tungsten voids improvement by optimizing MOCVD-TiN barrier layer plasma treatment at 28 nm technology node

Lin Gao, Yanyan Zhang, Yu Bao, Bin Zhong, Yingming Liu, Kang Ye, Gang Shi, Haifeng Zhou, Jingxun Fang

Shanghai Huali Microelectronics Corporation
568 Gaosi Rd., Pudong district, Shanghai, P.R. China
021-61871360-5723, gaolin_td2@hlmc.cn,

Biography

Lin Gao, 1982~, received his bachelor degree From Lanzhou University of Technology in 2008, is currently working at Technology Development Division II of Shanghai Huali Microelectronics Corporation as thin film process R&D engineer.

Abstract

As the dimensions of middle-of-line (MOL) contacts shrink, the tungsten (W) gap-filling capability becomes more critical to eliminate function failure in SRAM and logic circuit caused by W-voids. The formation of W-voids is generally related to contact profile, barrier (Ti/TiN) property and W-plug deposition method. The barrier layer may be degraded due to out-gassing from polymer residues underneath which prevent robust CVD W fill and leads to W void issue. In this paper, the impact of post-deposition plasma treatment of the underlying MOCVD-TiN barrier on the subsequent W gapfilling behavior and contact resistance was systematically investigated. Results show that the optimized plasma treatment of the barrier layer can reduce out-gassing during the subsequent W deposition, thus achieve bettergapfilling capability and minize W-voids.

Keywords: Barrier layer, TiN, Plasma treatment, W filling, out-gassing

Introduction

In order to fulfill the device requirements, the critical dimension (CD) of local interconnect has scaled down to 40nm and the aspect ratio (AR) of contact hole is higher than 4:1 for 28nm logical IC. It's a big challenge for W gap-filling at such CD and AR. Therefore W voids can often be observed in MOL contacts at 28nm technology node (shown in Fig.1), althoughCVD W deposition process itself has high step coverage. Moreover, advanced metallization systems need to have low resistivity to guarantee excellent circuit speed. In multi-level metallization systems, contact plugs and via plugs filling are important to the resistance performance. The contact integration scheme generally includes a sequential ion metal plasma

(IMP) Tiglue layer/MOCVD-TiN barrier and CVD-W gap-filling. In the present study, the impact of plasma treatment of the underlying MOCVD-TiN barrier on the subsequent W CVD gap-filling capability and contact resistance was systematically investigated. The DOE was focusing on the response of contact void defect counts and film resistivity depending on the plasma treatment time.

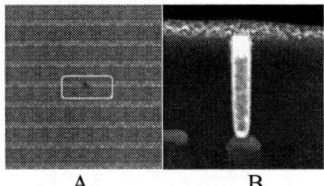

Fig.1. Typical W voids observed by E-beam voltage contrast scan (A) and TEM cross-section (B)

Experiment

A stacked film consisting of a sequential ion metal plasma (IMP) Ti/MOCVD TiN barrier and a W CVD layer was used in the contact metallization with 28nm technology node MOL contact size. TiN films were deposited using terakis-dimethyl-amino-titanium (TDMAT) as precursor in a CVD chamber, followed by an in-situ N2/H2 RF plasma treatment. The TiN deposition and subsequent plasma treatment were conducted in two repeated cycles. Plasma treatment conditions were varied to evaluate their effects on the W-void performance.

W gap-filling capability was examined by E-beam defect-scanning after the W-CMP process inline and further analyzed by cross-sectional transmission electron microscopy (TEM) offline. The sheet resistance was measured using a four-point probe method[1-5].

Results and discussion

In this work, since TiN barrier layer is fabricated by a MOCVD method using TDMAT as precursor, carbon and hydrogen impurities are always present in the as-deposited TiN film leading to the formation of TiCxNyHz. Therefore, a post-deposition plasma treatment is needed to remove carbon and hydrogen from the

978-1-5090-6695-7/17 $31.00 © 2017 IEEE   245

film and lower the film resistivity. The general chemical equations for the deposition process and plasma treatment are shown as below respectively:

$$Ti((NCH3)2)4 \rightarrow TiCxNyHz + NH(CH3)2 + NH2CH3 +$$

$$NH(CH2)2 + other\ hydrocarbons \qquad (1)$$

$$TiCxNyHz + N2 + H2 \rightarrow TiN + CxHy + R2NH + other$$

byproducts \qquad (2)

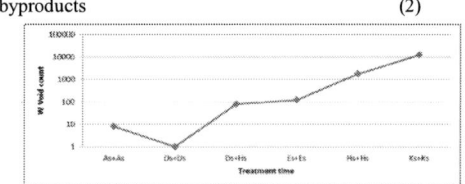

Fig2 Evolution of W void counts with plasma treatment time.

Fig.2 shows that the counts of W void found by e-beam scan drastically increased with plasma treatment time. After 2-cycle D seconds plasma treatment, almost no W void could be found, whereas the W void count dramatically rose to 1200 (~ 3 orders of magnitude increase) after a 2-cycle H seconds treatment.

The Evolution of W void count with plasma treatment time (shown in Fig.2) may be attributed to the crystallinity change of TiN barrier layer during the plasma treatment. With longer plasma treatment time, the TiN film became more crystalline compared to being amorphous as-deposited.

Raghunath Singanamalla studied the effects of in-situ N2/H2 RF plasma treatment on crystallinity change of TDMAT-based Ti(C)N film on Poly-Si/Ti(C)N/SiON film stack[6]. Investigated by Transmission Electron Microscopy (TEM), after in-situ N2/H2 RF plasma treatment, as deposited Ti(C)N film transforms from amorphous structure to polycrystalline structure with 4~7nm grain size.

In our model, it is inferred that crystalline TiN films possess more channels than amorphous films for underlying polymer residues to outgass (Fig.3). The outgassing during the subsequent W deposition is adverse to gapfilling by retarding the precursor's entrance into contact holes and thus causes W voids.

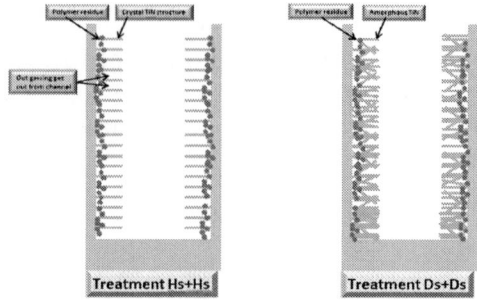

Fig.3 Model of outgassing channel in various TiN films.

The downside of reducing TiN plasma treatment time is that it may cause higher TiN Rs,

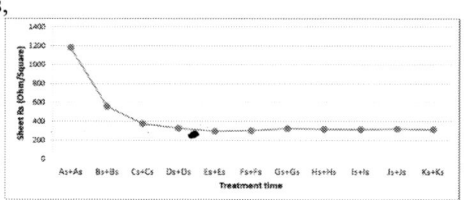

Fig.4 Effect of plasma treatment time on the sheet resistance of MOCVD-TiN films.

Fig.4 shows the effect of N2/H2 plasma-treatment time on the sheet resistance of MOCVD-TiN films. The sheet resistance decreased with increasing plasma treatment time and reached a saturated value of ~ 320 Ω/sqwhen the treatment time was longer than H seconds. From Fig.4, insufficient plasma treatment of TiN films may result in much higher contact resistance, which would degrade the circuit speed. Therefore, the post-deposition plasma treatment time of TiN barrier layer should be carefully optimized to balance contact resistance and W-void performance. In the present study, a good W-gapfilling capability and properly low contact resistance were achieved by a 2-cycle plasma treatment of D seconds based on the response from plasma treatment time Design of Experiment (DOE).

Conclusion

In this work, the impact of N2/H2 plasma treatment of the underlying MOCVD-TiN barrier layer on the subsequent W gap-filling behavior and contact resistance was systematically investigated. It was found that the optimized plasma treatment of the barrier layer can minimizeout-gassing issue during the subsequent W deposition resulting in a better W-fill capability. The post-deposition plasma treatment time of TiN films should be carefully optimized to balance contact resistance for device speed and W-void performance for product yield.

References

[1] D.H. Kim, B.Y. Kim, Jpn. J. Appl. Phys. 38 (1999) L461.

[2] S.H. Whang, J.K. Kim, J.W. Park, D.H. Kim, D.L. Cho, W.J. Lee, Jpn. J. Appl. Phys. 40 (2001) 265.

[3] J.Y.Kim, D.Y.Kim, H.O. Park, H. Jeon, J. Electrochem. Soc. 152 (2005) G29.

[4] V.Melink, D.Wolanski, E. Bugiel, A. Goryachko, S. Chernjavski, D. Kruger, Mater. Sci. Engineer. B102 (2003) 358.

[5] J.K. Huang, C.L. Huang, S. C. Chang, Y.L. Cheng, Y.L Wang, Thin Solid Films 519 (2011) 4948–4951.

[6] R. Singanamalla, et al., Mater. Res. Soc. Symp. Proc. 917 (2006) E12-02.

Rotary Spatial Plasma Enhanced Atomic Layer Deposition – an enabling manufacturing technology for μm-thick ALD films

Sami Sneck*, Mikko Söderlund, Markus Bosund, Pekka Soininen

Beneq Oy
Olarinluoma 9, 02200 Espoo, Finland
* +358 40 764 3413 sami.sneck@beneq.com

Abstract

Atomic Layer Deposition (ALD) is well known for its high film quality and high conformality, but limited by the low deposition rate. Beneq proposes a novel approach using Rotary Spatial Plasma Enhanced ALD process, which can reach deposition rates 10x higher than traditional pulsed ALD. This technology also enables use of PEALD in batch mode with high throughput. This paper describes the technology in more details.

Keywords—ALD; PEALD; rotary; spatial; deposition rate; batch; direct; DC plasma; high volume manufacturing.

Introduction

ALD is currently one of the main process technologies in microelectronics. The technology was first invented in 1970s in Finland to enable manufacturing of Thin Film Electroluminescent (TFEL) displays. This was the first ALD application, and the production started already in mid 1980s. This production in based on micrometer-thick ALD film stacks on large batches. Later semiconductor industry adopted ALD first as a single wafer process and then as a batch process. Current semiconductor-related ALD applications are based on films that are relatively thin, 1-100nm. However, it has been shown in the TFEL display manufacturing, that the high quality ALD films are highly useful also for thicker films [1].

The novel Rotary Spatial PEALD offers a totally new method for wafer-based ALD processing. A process with a batch of 10 wafers and an extremely high deposition rate (>μm/h) makes it possible for the first time, to use PEALD for high volume manufacturing.

The operating principle of Rotary Spatial PEALD is illustrated below in Fig. 1. Wafers are placed on rotating donut-shaped platform and rotated multiple times through Precursor zone and Plasma zone. Between these active zones, there is an inert gas purge zone. One cycle of ALD is deposited during each rotation. The basic principle has been previously described by Dickey [2].

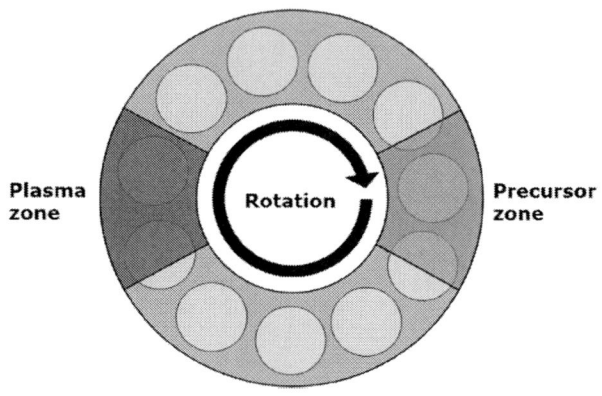

Figure 1. Wafers are rotated through precursor zone and plasma zone multiple times. One rotation equals one cycle of ALD.

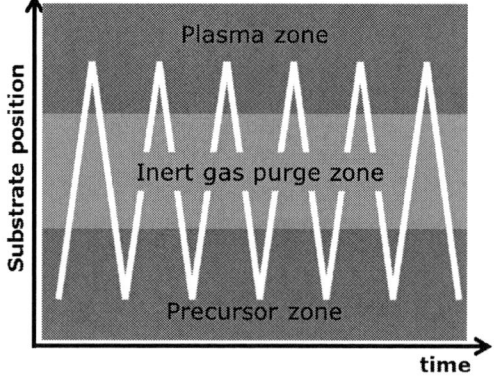

Figure 2. Substrate is moving between the plasma zone and the precursor until the desired film thickness is reached. Typical growth per cycle is ca. 1 A/cycle, so 10 000 cycles are needed for 1μm deposition.

Experimental

All samples were deposited with a prototype rotary spatial PEALD system at process temperature of 100C and pressure range of 1-2 hPa. Precursor materials used for the trials were SAM.24 (tetraethylsilanediamine) for SiO_2, TMA (trimethylaluminum) for Al_2O_3, TTIP (titanium tetraisopropoxide) for TiO_2 and TBTEMT (tert-butylimidotris(ethylmethylamido) tantalum) for Ta_2O_5. Films were deposited on Si-monitor samples and film thicknesses were measured by ellipsometer.

Results and Discussion

Rotating Spatial PEALD can reach very high deposition rates because of the following key advantages. First, spatial ALD enables shorter cycle times than pulsed ALD, where the substrate is static and gases are pulsed successively over the substrate. It is much faster to move the substrate from one gas zone to another, than changing the gas in a chamber. Second, rotation allows faster average linear speeds than a linear spatial movement, which needs to constantly accelerate, stop and change direction. Third, using plasma process instead of thermal process, the purge phase is much faster. The plasma radicals die quickly and cannot enter the precursor zone. Fourth, the plasma process often has significantly higher growth per cycle compared to a thermal process.

Utilizing these advantages, we were able to demonstrate very high deposition rates for common oxide film. The deposition rate results are summarized in Table 1 below.

Table 1. Deposition rates achieved with certain oxide materials using the rotary spatial ALD system.

	rotation speed (rpm)	growth-per-cycle (nm)	dep. rate (nm/min)
SiO_2	200	0.120	24.0
Al_2O_3	200	0.170	34.0
TiO_2	200	0.082	16.4
Ta_2O_5	200	0.054	10.8

Using the Al_2O_3 as an example to compare the novel process with traditional one, we can make the following assumptions. In typical state of the art single wafer PEALD tool, the cycle time is around 2 s and growth-per-cycle is approximately 0.100nm. With these figures, the deposition rate of traditional PEALD is 3nm/min. So, the rotary spatial ALD has more than 10x higher deposition rate. In addition, the rotary system is by nature a batch system, whereas the typical PEALD systems are single wafer tools. With 10 wafer batch size, the throughput of the rotary system is approximately 100x higher than with the single wafer tool. Projected throughput figures for a production system with 10 pcs of 200mm wafers are summarized in Table 2 below.

Table 2. Throughput (wafers per month) with different film materials and film thicknesses for a rotary spatial PEALD system with batch size of 10pcs of 200mm wafers.

w/month	SiO_2	Al_2O_3	TiO_2	Ta_2O_5
10 nm	17 985	18 133	17 864	18 086
100 nm	15 194	16 320	14 369	15 944
500 nm	8 993	11 298	7 686	10 446
1000 nm	5 955	8 160	4 860	7 300

Conclusions

The ultra-high deposition rates achieved with rotary spatial PEALD enable use of ALD for applications, where thicker films are required. This helps ALD to penetrate further into areas currently dominated by Chemical Vapor Deposition (CVD) and Physical Vapor Deposition (PVD). Examples of such applications include optical coatings, transparent conductive (TCO) coatings and insulators for high voltage applications. These films are typically in the μm-thickness range, where ALD has often not been considered as an alternative.

References
[1] T. Suntola, J. Antson, A. Pakkala and S. Lindfors, Digest of 1980 SID International Symposium (1980) 108.
[2] E. Dickey, W. Barrow, B. Aitchison, "Optical coatings by high speed Rotary Spatial ALD", 58th Annual Technical Conference Proceedings of the Society of Vacuum Coaters, 2015.

978-1-5090-6695-7/17 $31.00 © 2017 IEEE

PASSIVATION QUALITY AND ELECTRICAL CHARACTERISTICS FOR BORON DOPED HYDROGENATED AMORPHOUS SILICON FILM

Ching-Lin Tseng [1], Yu-Lin Hsieh [1], Chien-Chieh Lee [3], Hsiang-Chih Yu [2], and Tomi T. Li [1]*

[1] Department of Mechanical Engineering, National Central University, Taoyuan, Taiwan (R.O.C.)
[2] Department of Energy Engineering, National Central University, Taoyuan, Taiwan (R.O.C.)
[3] Optical Science Center, National Central University, Taoyuan, Taiwan (R.O.C.)
*Corresponding Author's Email: ams0829@gmail.com

ABSTRACT

Borons doped amorphous silicon (a-Si:H) that deposited on a n-type silicon substrate was prepared by plasma enhanced chemical vapor deposition (PECVD). The conductivity increases with increasing B_2H_6 flow when the electrode distance, working pressure and total flow rate are fixed. The Ellipsometer, Four Point Sheet Resistance Meter, Hall measurement, Secondary Ion Mass Spectrometer and Photo-conductance lifetime tester were used to obtain the electrical and physical properties of thin films.

The research shows that while changing process parameters, the effect on the film that has the good conductivity and the carrier lifetime are most critical. When the amounts of the boron atoms increase, the conducting properties of the boron-doped hydrogenated amorphous silicon film increase effectively. However, too much boron atoms increase densities of the defects, thus reduce the carrier lifetime and affect the activation of boron atoms in films. Based on the results of the carrier lifetime ratio on intrinsic layer and stacked dopant layer, it is found that the carrier lifetime of the doping layer stacks over intrinsic layer can effectively improve the field effect on passivation film quality.

INTRODUCTION

The development of high efficiency in heterojunction silicon solar cell needs to have an excellent P or N-type doped amorphous silicon layer. Boron-doped amorphous silicon layer processed by PECVD is used as a P-type passivation (emitter) layer stacked on silicon intrinsic layer as part of a silicon heterojunction solar cell. The high conductivity of the film is strongly dependent on the doping element [1]. Excellent electric field can form a field-effect passivation [2] on cell and limit to generate minority carrier for recombination. In addition, an excellent emitter layer can reduce the surface carrier recombination rate and improve the carrier collection rate at interface, thus enhance the short-circuit current (Jsc) [3].

In this study, we investigated the effect on deposition conditions of boron-doped amorphous silicon thin film using PECVD. In order to ensure the high quality and good electrical properties of the deposited P-type amorphous silicon film, the electrical characteristics and physical properties of the film were examined.

EXPERIMENTAL

The a-Si:H thin films were deposited onto single side of N-type CZ wafers (<100> ; 300 μm ; 1-10 Ω cm) using a standard PECVD (13.56 MHz). Before deposition, all wafers were dipped in H_2O_2 : H_2SO_4=1 : 2 for 10 min, followed by cutting the wafer size to 2 cm x 2 cm, and dip in hydrofluoric acid (HF) 2 % for 1 min to remove the chemical oxide. After rinsing and drying, then wafers were immediately transferred into vacuum chamber in order to avoid the native oxide to grow. In this work, we adjusted B_2H_6 flow from 30 to 90 sccm to observe the electrical characteristics and physical properties of the films. The main parameters were listed in Table 1. Before the experiments, wafers were pre-heated for 7 minutes in chamber for thermal equilibrium at the temperature 150 °C using a (B_2H_6/ SiH_4/ H_2) mixture. The thickness for all deposited samples was 20 nm. The thickness of thin film on wafer was measured by spectroscopic ellipsometry (SE) with Tauc-Lorentz (TL) model [4].

Hall measurement results indicated that electrical properties of the films were strongly dependent on the doped elements [5] and the Four Point Sheet Resistance Meter, Secondary Ion Mass Spectrometer (SIMS) were used to obtain the electrical and physical properties of thin films. Finally, Photo-conductance lifetime tester was used to obtain the minority carriers lifetime and to dertermine the passivation quality.

TABLE I. DEPOSITION CONDITIONS OF THE P-LAYER

| $B2H6$ (flow) | Deposition conditions of the P-layer | | | |
	T_{sub} (°C)	Pressure (mTorr)	P_{rf} (W)	Electrode Distance (mm)
30-90	150	300	15	30

RESULTS AND DISCUSSION

B_2H_6 flow dependence

B_2H_6 is the key flow to provide acceptor to complete a P-type semiconductor. An acceptor, having three valence electrons can form a hole in silicon crystal. Increasing

B_2H_6 flow can form most holes and have the excellent conductivity in films. Thus, the B_2H_6 flow can directly affect the concentration of dopant and defect density.

The resistivity (Ohm-cm) and B / Ar ratio of the films from different B_2H_6 flow were shown in Figure 1. The result showed that when the B_2H_6 flow changed from 30 sccm to 80 sccm, the B/Ar ratio showed an increasing trend in the plasma, which indicated that the dissociation of B increased in the plasma with the increase of the B_2H_6 flow. So the volume resistivity was decreased, but the B_2H_6 flow over 90 sccm had turning point. The reason is that the plasma dissociation energy is fixed and the activation of the doped carrier is limited, not all the doped carrier can be activated into the film [6]. In which the boron atoms that cannot be activated, thus causing excessive defects in the film, causing the film resistivity to rise and the film structure is more and more loose.

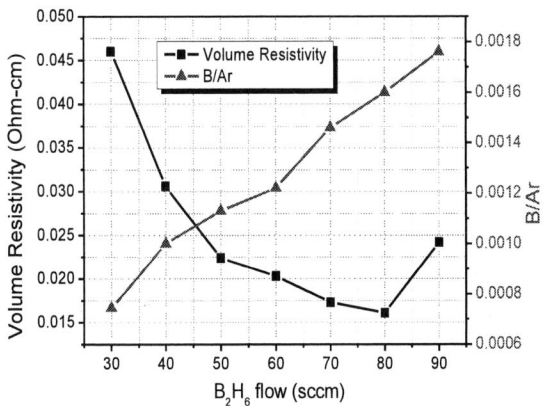

Figure 1: The resistivity (Ohm-cm) and B / Ar ratio of the films from different B_2H_6 flow.

The total content of boron atoms as function of B_2H_6 flow in SIMS measurement was shown in Figure 2. The results showed that the boron content was increased in the films (approximately quadruple) when the B_2H_6 flow increased. In contrast to the doping concentration in Figure 3, it could be seen that the boron content increased in the film could further enhance the doping concentration (approximately triple). However, the dopant concentration was close to saturation at higher B_2H_6 flow (70-80 sccm). It was observed that the carrier concentration increased with the increase of the doping amount in the dopant ion scattering, but too much carrier concentration would form a scattering center, which blocked the carrier movement and led to mobility decreased. If the dopants could not be effectively activated, the neutral atoms or compounds were formed in the lattice resulting in increase of resistivity.

Figure 4 showed the SIMS depth analysis of doping with B_2H_6 flow and Figure 5 showed the activation rate of

boron atoms with B_2H_6 flow. From the above description, the doping concentration increased when the B_2H_6 flow changed from 30 sccm to 70 sccm. Because the B_2H_6 flow increased, the boron atoms excited would also increase in the chamber. It could be effective to bond to boron atoms and silicon atoms. In Figure 5, the activation rate was the highest activation efficiency when the B_2H_6 flow was 50sccm, and the activation efficiency slightly decreased if the B_2H_6 flow increased again. However, the dopant particles were not fully activated into the films, and the carriers cannot be effectively contributed to form defects in the films. The references also pointed out that excessive doping would cause the film $Si\text{-}H_2$ bonding increased, resulting in loose of film structure [7,8].

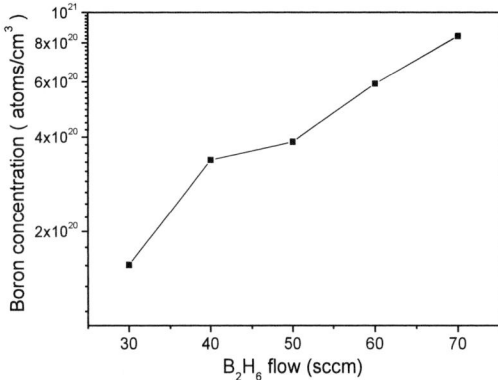

Figure 2: The total content of boron atoms in SIMS measurement.

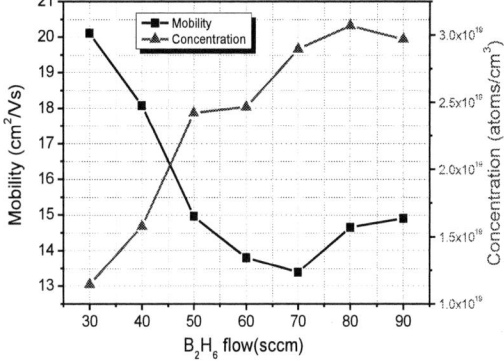

Figure 3: The mobility (cm^2/Vs) and concentration (atoms/cm^3) of films from different B_2H_6 flow.

Finally, the doped layer was stacked over the intrinsic layer (i) with the same carrier lifetime. The carrier lifetime for ratio of intrinsic layer and dopant stacked layer as shown in Figure 6. Based on the above results, because doping more boron atoms caused the film having more defects as B_2H_6 flow changed from 40 sccm to 60 sccm and the carries lifetime would decrease. However, the experimental results showed that, as long as the doped layer stacked on the passivation layer in which all the field effect passivation would increase in carries lifetime.

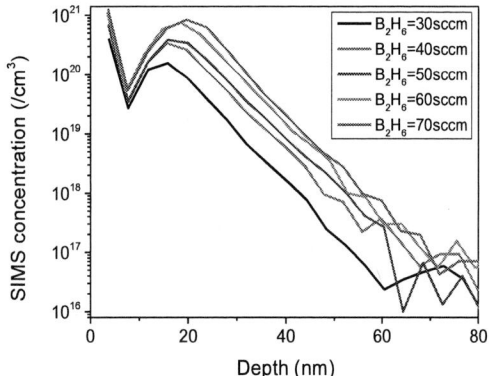

Figure 4: SIMS depth analysis of doping with B_2H_6 flow.

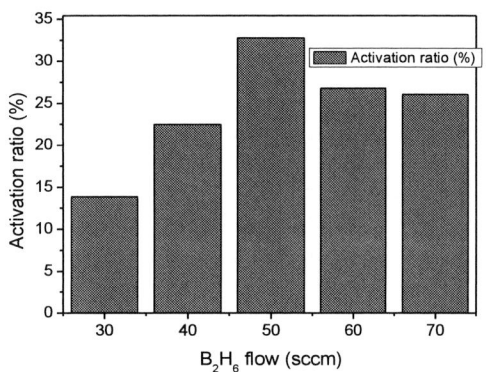

Figure 5: Activation rate of boron atoms with B_2H_6 flow

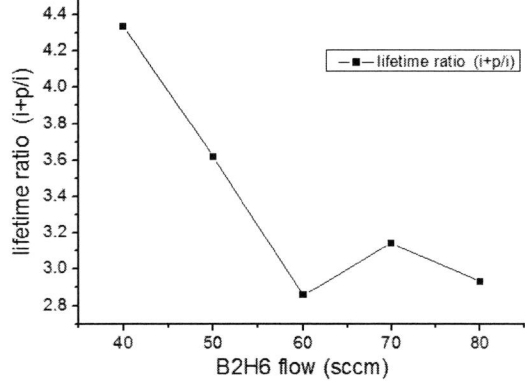

Figure 6: The carrier lifetime ratio of intrinsic layer and dopant stacked layer.

CONCLUSIONS

In this study, when the B_2H_6 flow increased, the boron atoms excited would also increase in the chamber. Therefore boron atoms can be effectively activated into the film to achieve good film conductivity for improving electrical properties. If there is not effectively activated, the conductivity can decrease. Based on the results of the carrier lifetime ratio on intrinsic layer and stacked dopant layer, it is found that the carrier lifetime of the doping layer stacks over intrinsic layer can effectively improve the field effect on passivation film quality.

REFERENCES

[1] M. Taguchi, A. Terakawa, E. Maruyama and M. Tanaka. Prog. Photovoltaics ： Research and Applications, vol. 13, 2005, pp. 481-488.

[2] J. Sritharathikhun, C. Banerjee, M. Otsubo, T. Sugiura, H. Yamamoto, T. Sato, A. Limmanee, A. Yamada and M. Konagai. Jpn. J. Appl. Phys. Lett. , vol. 46, 2007, pp. 3296-3300.

[3] F. Wang, X. Zhang, L. Wang, J. Fang, C. Wei, X. Chen, G. Wang and Y. Zhao. Sol. Energy, vol. 108, 2014, pp. 308-314.

[4] J. G. E. Jellison and F. A. Modine. Appl. Phys. Lett., vol. 69, 1996, pp.371

[5] S. H. Jeong, J. H. Park and B. T. Lee. J. Alloys Compd, vol. 617, 2014, pp.52-57.

[6] K. Oda, M. Miura, H. Shimamoto and K. Washio. Appl. Surf. Sci., vol 254, 2008, pp. 6017-6020.

[7] S. Martín de Nicolás, J. Coignus, W. Favre, J. P. Kleider and D. Muñoz. Sol. Energy Mater. Sol. Cells, vol. 115, 2013, pp. 129-137.

[8] U. K. Das, M. Z. Burrow, M. Lu, S. Bowden and R. W. Birkmire. Appl. Phys. Lett., vol. 92, 2008, pp.063504-1–063504-3.

LASER SPIKE ANNEALING AND SIGE DUMMY PATTERN LAYOUT STUDY TO IMPROVE CONTACT MISALIGNMENT OVERLAY ISSUE

Guiying Ma, Tzuchiang Yu

Date: 2016-12-31

Technology R&D Center, SMIC

18 Zhangjiang Road, Pudong New Area, Shanghai, P.R. China 201203

Tel: 86-21-3861-0000, E-mail: Connie_Ma@smics.com

ABSTRACT

The use of strained SiGe is essential to improve PFET MOS device performance. However, the structure is susceptible to strain relaxation and wafer deformation during thermal annealing. The accumulation of stress in the wafer needs to be controlled to minimize contact misalignment overlay issue. This paper analyzes laser spike annealing impact and SiGe dummy pattern layout impact on contact overlay by experiments on 28nm technology, design guide lines on shape of dummy SiGe patterns for slip free condition have been investigated and clarified. With optimized dummy SiGe patterns and laser spike annealing, random component of contact misalignment has been successfully reduced to normal level.

Keywords—wafer deformation; SiGe dummy pattern; Laser anneal; contact misalignment;

INTRODUCTION

Strain engineering and material innovation have been the critical keys to CMOS device performance improvement on 40nm or 28nm advanced technology. For instance, utilizing selective epitaxially SiGe in source/drain regions, uniaxially compressive stress was introduced into PFETs and the carrier mobility was enhanced significantly, resulting in higher drive current which would be very difficult to achieve by conventional Si technology [1]. Sub-millisecond laser spike annealing (LSA) allows for device fabrication with abrupt, ultra-shallow, and highly-activated low resistivity junctions but LSA Dwell time is necessary to be optimized for stress and photolithography overlay errors [2-4]. At active area level, dummy pattern with SiGe must be placed for both planarity control and controlling SiGe pattern density uniformity in chip for good device performance uniformity [5], it is because SiGe self-epitaxial process is sensitive on SiGe pattern density.

In this study, we have found that slips which are created at the SiGe dummy patterns worsen the contact misalignment. To solve this problem, we have performed a mass of experiment to find design windows on SiGe dummy structure to solve laser spike annealing impact on wafer deformation issue. As a result, wafer deformation was solved by new SiGe dummy pattern and new laser annealing recipe.

EXPERIMENTS AND RESULTS

a. Laser Spike Annealing Experimental

28nm technology TQV2 suffered contact (CT) photo lithography misalignment issue (Fig1) and it was especially worse on wafer edge. We suspect that was induced by wafer deformation since wafer curvature value is -7 times and reverse trend to normal TQV1(Table1) and CT to poly overlay is 6 times of TQV1. Based on process flow sequence that is shown in Fig2, laser spike annealing and SiGe loop were suspected. To reduce wafer deformation curvature value, a series of laser spike annealing split were done on TQV2 product, including LSA temperature, LSA Dwell time, LSA rotation and LSA skipping experiment. Fig3 shows that reducing laser anneal temperature or Dwell time can improve wafer curvature (Fig3-a), but wafer curvature is still different with baseline TQV1 and SAB & CT overlay is still higher than baseline (Fig3-b). Only skipping laser spike annealing experiment got comparable wafer curvature value with TQV1 and SAB & CT overlay are OK, but skipping laser anneal bring some side-effect such as device performance degradation (Fig4-a) and non-salicide Poly resistor high resistance issue (Fig4-b), so it is important to find other approach to reduce wafer deformation.

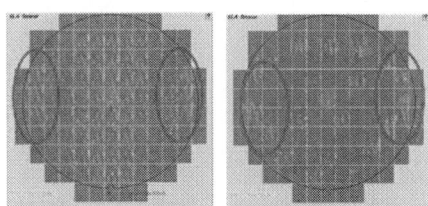

Fig1: TQV2 CT overlay map (old baseline)

Split Condition	Bow-X (a.u.)	Bow-Y (a.u.)	SAB to AA OVL	CT to PO OVL
	Post LSA	Post LSA	SPEC=3 (a.u.)	SPEC= 2.5 (a.u.)
TQV1 Baseline	1.0	1.0	1.0	1.0
TQV2 Baseline	0.6	-7.1	4.8	6.2

Table1: TQV2 wafer curvature value (Bow)

978-1-5090-6695-7/17 $31.00 © 2017 IEEE

- Active area definition
- Gate definition
- eSiGe form
- Spacer loop
- S/D implant
- Spike RTA
- Laser spike annealing
- NiSi
- ILD DEP/CMP
- CT Loop

Fig2: 28nm PolySion process flow

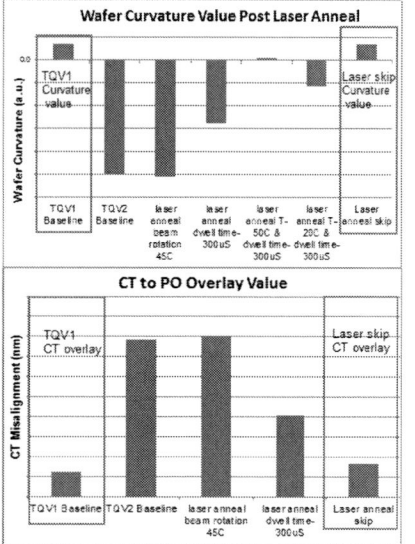

Fig3 (a) Wafer curvature value (bow) and (b) CT to Poly overlay value by laser anneal split

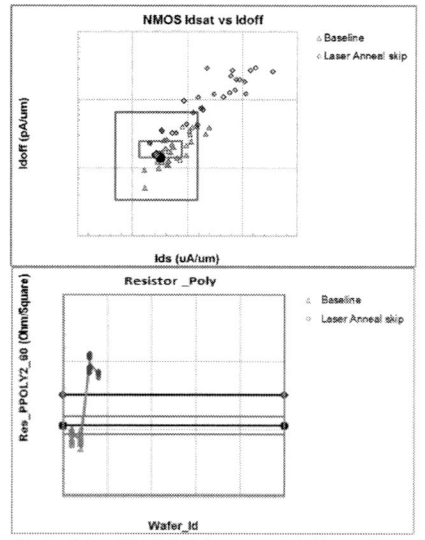

Fig4 (a): NMOS Idsat vs Idoff performance and (b) Poly resistor with laser skip split

b. SiGe Density Experiment

To solve TQV2 wafer warpage issue, we also did SiGe density experiment with 6%, 3%

and 0.5% SiGe density within TQV2 chip. Wafer curvature value show that wafer deformation has strong correlated with SiGe density, and wafer deformation would be better when SiGe density is lower. Without dummy SiGe pattern (0.5% SiGe density), wafer curvature value and CT misalignment value can be close to TQV1 performance (Fig5), but so low SiGe density impact SiGe formation process window and SiGe uniformity, so it is important to study SiGe dummy pattern as below.

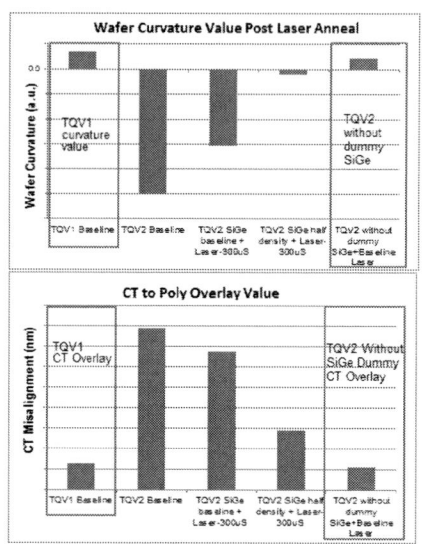

Fig5 (a): Wafer curvature value and (b) CT to Poly overlay value by SiGe density split

c. SiGe Dummy pattern experiment

Although skipping laser anneal or lower SiGe density can solve wafer deformation issue, skipping laser anneal would impact NMOS device performance and low SiGe density would impact SiGe formation process window and PMOS device performance due to worse SiGe uniformity, so another method to solve wafer deformation issue is needed. SiGe dummy pattern study is necessary to solve wafer deformation and CT litho misalignment overlay issue. First we changed part of SiGe dummy pattern from square shape (Fig6-a) to transistor like shape (Fig7-a) and keep 4-7% SiGe density, wafer deformation issue can be solved combined lower laser anneal temperature. Second we changed all SiGe dummy pattern from square shape to transistor like shape and keep 4-7% SiGe density also, wafer deformation issue are solved even with high laser anneal temperature (Fig8), CT misalignment overlay is back to TQV1 level and within wafer uniformity is good also(Fig9).

Full SiGe dummy pattern changed to new layout, device performance was comparable with old baseline (Fig10) and device IDU was

978-1-5090-6695-7/17 $31.00 © 2017 IEEE

obviously better than old baseline especially for local IDU and total IDU (Fig11).The mechanism is annular SiGe shape around poly has strong stress and easily enhance wafer deformation with high temperature laser anneal, especially on TQV product with a mass of auto dummy pattern.

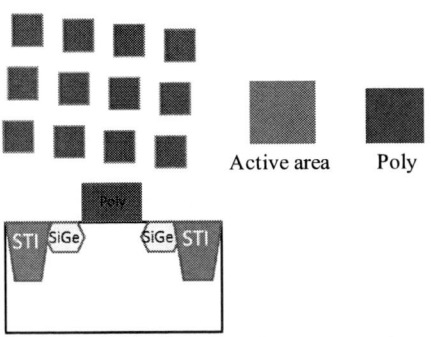

Active area   Poly

Fig6 (a): SiGe dummy pattern square layout and (b) Cross-section of square dummy

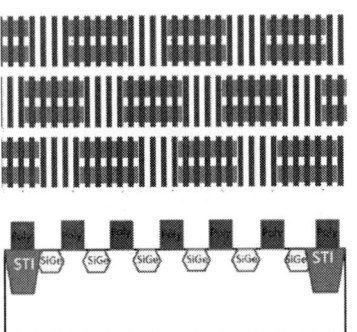

Fig7 (a): SiGe dummy pattern transistor like layout and (b) Cross-section of transistor-like dummy
Note: Active area (red color) without covered by poly (blue color) can form SiGe.

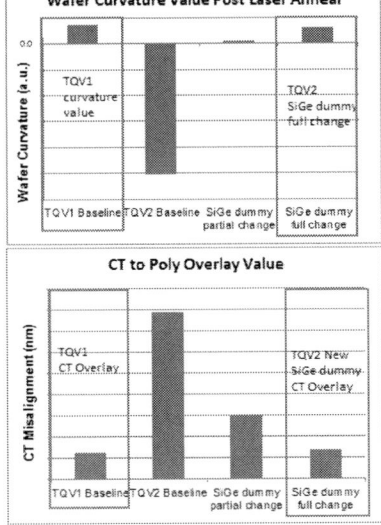

Fig8 (a): Wafer curvature value and (b) CT to Poly overlay value by SiGe dummy Pattern changing

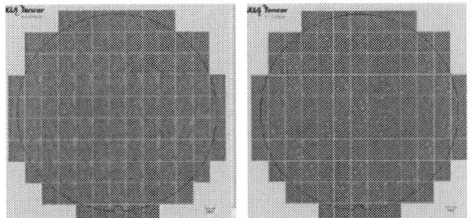

Fig9: CT Overlay Map post SiGe dummy pattern full change

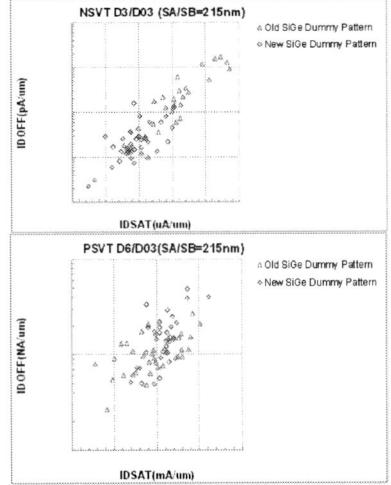

Fig10: Device performance with SiGe dummy split

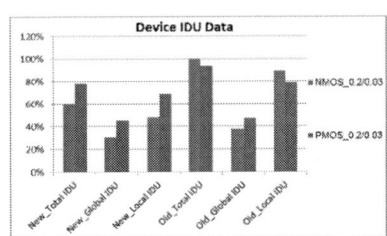

Fig11: Device IDU with SiGe dummy split

CONCLUSION

Higher laser anneal temperature and longer dwell time enhance wafer deformation and causes worse CT to Poly or CT to active area misalignment overlay. To solve wafer deformation issue, SiGe dummy pattern layout are critical, transistor like SiGe dummy pattern can solve CT to Poly or CT to active area misalignment issue and at the same time improve device IDU.

REFERENCES

[1] Scott E.Thompson.et al.,IEEE ED, pl，2004
[2] A.Shima et al., Symp. VLSI Tech. Dig., p. 144, 2005
[3] S.K.H.Fung et al., Symp. VLSI Tech. Dig., p.92, 2004
[4] S.Shetty et al., IEEE Tech. Junction p.119, 2009
[5] O.Fujii et al., symp. VLSI Tech. Dig., p.156, 2009

978-1-5090-6695-7/17 $31.00 © 2017 IEEE

INVESTIGATION OF INTRINSIC HYDROGENATED AMORPHOUS SILICON (a-Si:H) THIN FILMS ON TEXTURED SILICON SUBSTRATE WITH HIGH QUALITY PASSIVATION

Min-Lun Yu [2], Yu-Lin Hsieh [1], Sheng-Kai Jou [1], Tomi T. Li [1]*, and Chien-Chieh Lee [3]

[1] Department of Mechanical Engineering, National Central University, Taoyuan, Taiwan (R.O.C.)
[2] Department of Energy Engineering, National Central University, Taoyuan, Taiwan (R.O.C.)
[3] Optical Science Center, National Central University, Taoyuan, Taiwan (R.O.C.)
*Corresponding Author's Email: poppin1020@gmail.com

ABSTRACT

In this study, the intrinsic hydrogenated amorphous silicon (a-Si:H) thin films deposited by Plasma Enhanced Chemical Vapor Deposition (PECVD) was investigated for the application of the heterojunction silicon solar cell on the textured silicon substrate. During the process, we used the optical emission spectrometer (OES) and quadrupole mass spectrometry (QMS) to analyze the concentrations of free radicals in plasma.

The results showed that the better surface recombination velocity (SRV) and passivation quality of a-Si:H thin films on the textured silicon substrate were obtained when the electrode distance at PECVD was 35mm. Furthermore, while the electrode distance was 35mm, the lowest electron temperature and the same spectrum ratio trend in OES (Si*/SiH*) and QMS (SiH_2/SiH_3) respectively were received.

INTRODUCTION

Among the solar cells, the heterojunction silicon solar cell (HIT) solar cell has received the most attention because it's low process temperature, low production cost and without rare metals [1]. The Plasma Enhanced Chemical Vapor Deposition (PECVD) has already been used for the purpose of growing solar a-Si:H films [2,3], and there are some advantages for plasma studies such as low process pressure, low deposition rate and growing better quality of thin films.

In this study, in order to obtain a high quality of passivation layer, we used PECVD to deposit the a-Si:H film on the textured silicon substrate that could improve the open-circuit voltage (Voc) value. During the processing, we utilized the OES and QMS to analyze the characteristics of species in plasma. The correlation of thin-film property and plasma characteristics was examined. The results of spectrum ratio trends on OES (Si*/SiH*) and QMS (SiH_2/SiH_3) with respect to electrode distances from PECVD were investigated.

EXPERIMENT

In this study, the double polished 1~10 ohm-cm Fz (N-type) silicon wafers (<111> in 260~300 μm thickness) were used. The size of 2x2 cm^2 textured silicon wafer was used for the solar cell substrate study.

Cleaning the wafer is very important before the amorphous silicon layers is deposited on the substrate, the passivation quality is very sensitive to the cleaning wafer surface. In order to remove the organic pollutants, the wafer dipped into the liquid which were mixed with the H_2SO_4 and H_2O_2 (2:1) for 10 min and rinsed with the DI water for 3 min. Before deposition, the wafer dipped in the HF (2%) for 1 min to remove the oxide layer. After rinsing and drying, the wafer was loaded into the chamber immediately. The a-Si:H thin films were deposited on both sides of textured silicon wafer by PECVD.

The electrode distance will affect the passivation quality of the solar cell [4], therefore, we varied the electrode distance from 25 mm to 45 mm, and the RF-power is 30W under the 300mTorr. The Ar, SiH_4 and H_2 for the depositing gases were introduced. The ratio of H_2/SiH_4 is 7.3 and the process temperature was set to be 160℃. All the parameters are listed in TABLE I.

TABLE I. EXPERIMENTAL PARAMETERS OF THE PECVD

Electrode Distance (mm)	Deposition condition of the i-layer			
	Tsub (ºC)	*Pressure (mTorr)*	*Rf power (W)*	*Ratio (H_2/SiH_4)*
25-45	160	300	30	7.3

After the processing, the passivation double layers of the substrates were used for SRV measurement and thickness of the films was measured by Ellipsometer. In OES analysis, the Actinometry technique is utilized to obtain the intensity ratios of species that are correlated to the concentration of Ar (750nm) in plasma [5]. In order to count the content of free radicals, thermal ionization mass spectrometer (TIMS) technique was used for QMS.

RESULTS AND DISCUSSION

Figure 1 showed that the spectrum ratios which were collected from the OES and were the same ratio trends with respective to electrode distance under the pressure at 300mTorr. The ratio of SiH*/Ar* and Ha*/Ar* were higher but the ratio of Si*/Ar* was lower when the electrode distance was 35 mm. The ratio of Si*/SiH* can be used as an indicator of electron temperature [6], and the lower electron temperature can indicate the better quality of a-Si:H [7]. Therefore, the results showed that the lower value of electron temperature (Si*/SiH*) was found at 35 mm under the pressure 300mTorr (Figure 2). In addition, there was a better SRV when the electrode distance was 35mm (Figure 3).

According to the results of SRV in Figure 3, and Si*/SiH* by OES in Figure 2, the better passivation quality could be found when the electrode distance was 35mm. In the meantime, for the plasma species measured by QMS for analyzing the reaction species, the results showed that the ratio densities of SiH_2 and SiH_3 were the lowest when the electrode distance was 35 mm (Figure 4).

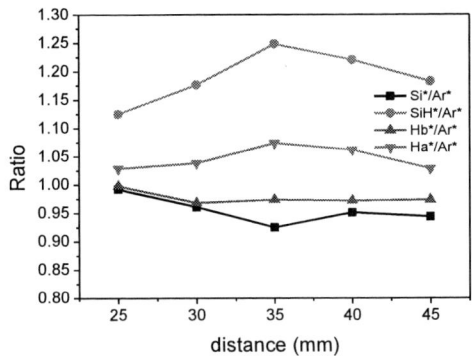

Figure 1: The OES spectrum radio with different electrode distances under 300mTorr

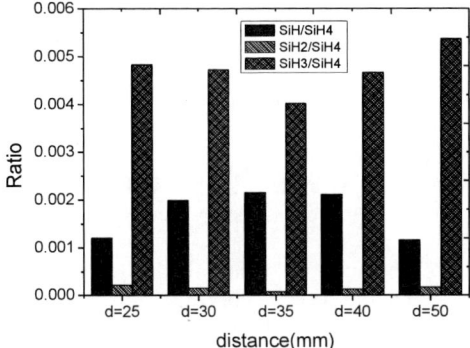

Figure 4: The free radicals of ratio with different electrode distances under 300mTorr measured by QMS

The microstructure factor R*, as in (1), is correlated to the ratio of SiH_2/SiH_3 since the SiH_2 will cause more defects in the films [8]. Thus, the SiH_2/SiH_3 could be used to determine the quality of the thin film.

The results showed that there was a lowest ratio value when the electrode distance was 35mm, and the ratio of SiH_2/SiH_3 increased at shorter or longer electrode distance than 35mm. The change amount of SiH_2 increased and caused changing defects in the film from changing electrical field due to effect of electrode distance (Figure 5) [9].

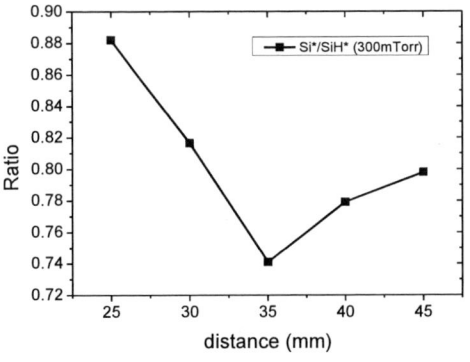

Figure 2: The OES spectrum ratio (Si*/SiH*) with different electrode distances under 300mTorr

$$R^* = \frac{I_H(2090)}{I_H(2090) + I_H(2000)} \qquad (1)$$

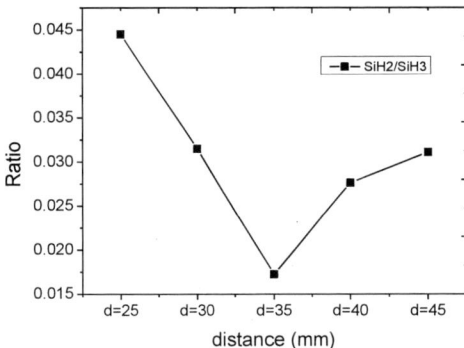

Figure 5: The ratio of (SiH₂/SiH₃) with different electrode distances under 300mTorr measured by QMS

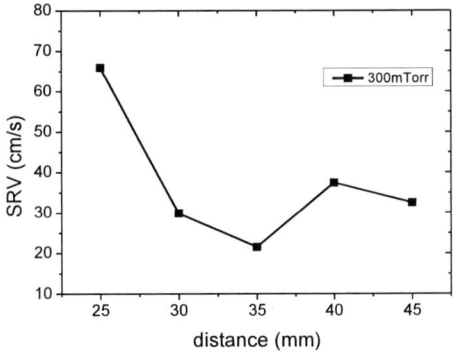

Figure 3: The SRV with different electrode distances under 300mTorr

CONCLUSION

The results showed that there are same ratio trends in OES (Si*/SiH*) and QMS (SiH_2/SiH_3) when the electrode distance was 35mm. The SRV decreased at electrode distances before 35mm, and it raised after 35mm.

The ionization of the species varies with the different distances between electrodes because the electrode distance will affect the electric field. In this study, the better quality of passivation could be found when the electrode distance was 35mm on the textured substrate.

REFERENCES

[1] J. Liu, S. H. Huang and L. He, J. Semicond., vol. 36, 2015, 044010-1.

[2] S. J. Lee, S. H. Kim, D. W. Kim, K. H. Kim, B. K. Kim and J. Jang, Sol. Energy Mater. Sol. Cells, vol. 95, 2011, pp.81-83.

[3] J. Ge, Z. P. Ling, J. Wong, T. Mueller and A. G. Aberle, Energy Procedia, vol. 15, 2012, pp107-117.

[4] P. K. Chang, P. T. Hsieh, F. J. Tsai, C. H. Lu, C.H. Yeh, N. F. Wang and M. P. Houng, Thin Solid Films, vol. 520, 2012, pp. 5042–5045.

[5] R. D. Robertson, H. Catham and A. Gallagher, Appl. Phys. Lett., vol. 43, 1983, pp. 54-56.

[6] A. Matsuda, M. Takai, T. Nishimoto and M. Kondo, Solar Energy Materials &Solar Cells, vol. 78, 2003, pp. 3–26.

[7] M. Takai, T. Nishimoto, M. Kondo and A. Matsuda et al, Appl. Phys. Lett., vol. 77, 2000, pp. 18-22.

[8] L. C. Hu, C. J. Wang, Y. W. Lin, T. C. Wei, C. C. Lee, J. Y. Chang, I. C. Chen, Y. Kawai and T. T. Wei, ECS J. Solid State Sci. Technol., vol. 4, 2015, pp.213-219

[9] S. Martín de Nicolás, J. Coignus, W. Favre, J. P. Kleider and D. Muñoz. Sol. Energy Mater. Sol. Cells, vol. 115, 2013, pp. 129-137.

SIN/SIC$_x$N STACK FILM AS CU CAPPING LAYER IN CU/ULK INTERCONNECT FOR 28LP

Yi Hailan, Lei Tong, Ye Kang, Chen Yongyue, Hu Wei, Yang Zhigang, Zhou Haifeng, Fang Jingxu

Shanghai Huali Microelectronics Corporation

568 Gaosi Rd., Pudong district, Shanghai, P.R. China

yihailan@hlmc.cn

ABSTRACT

The application of SiN/SiC$_x$N stack as capping layer of damascene Cu/ULK interconnect was discussed. The stack film of ultra-thin SiN+SiC$_x$N Cu capping films were fabricated by plasma enhanced chemical vapor deposition (PECVD) method, which consisted of ultra-thin SiN as barrier and SiC$_x$N as etch stop layer. The resistance and capacitance of Cu/ULK and the capping film SIN/SiC$_x$N was measured by regular WAT test, and the RC delay was induced. The structure included 3nm UT-SiN and 28nm C-rich SiC$_x$N and its K_{eff} is about 5.3. Electric and reliability test indicated satisfactory RC and V_{BD} for Cu/ULK interconnects by such capping layer.

Keywords—thin film; PECVD; low k; Cu barrier; capping layer

INTRODUCTION

With the progressive miniaturization of advanced integrated circuits fabricated using the Cu damascene process, an increasingly aggressive current density is applied to BEOL interconnects. Therefore resistance-capacitance (RC) delay is a dominant factor in determining the performance of ultra large scale integrated circuits (ULSI), improving RC delay and reliability of Cu/(ultra-low k film) ULK interconnect become critical issues in ULSI. Reduction of the effective K (K_{eff}) in inter metal dielectric layers is desired so as to bring lessening of crosstalk and propagation delay. But reducing k in ULK film is very difficult because the k value of ULK film has already been obtained as low as 2.4 and been damaged after etching or ashing process. Further reduction would weaken mechanical strength of the low-k film (k < 2.3), then resulting both lower reliability and more serious damage from CMP process.

Cu diffusion barrier layer with low dielectric constant is in urgent need for this integration scheme [1-2]. Amorphous silicon nitricarbide (SiC$_x$N or SiCN) have received much attention for applications as Cu dielectric diffusion barrier and capping in Cu damascene process [3] because of its satisfactory properties such as relatively low dielectric constant and good manufacturability [4]. The thin film of SiC$_x$N was synthesis into amorphous, thus it provide Cu diffusion barrier property. In addition, SiC$_x$N show good thermal stability, it can be used as etch stop layer (ESL).

Cu/inter metal dielectric integrated scheme involves the incorporation of Cu, ULK and barrier/capping layer. In order to solve the exigent issue of poor adhesion between Cu and porous ULK film, an ultra-thin SiN was induced as pre-layer, stack of Ultra-thin SiN and C-rich SiC$_x$N is obtained, this kind of film stack meet 28LP BEOL integration properties, provide good adhesion and moisture blocking .

The preparation techniques of SiC$_x$N thin films have been extensively investigated in the recent years, such as reactive radio frequency sputtering [5], molecular beam epitaxy [6], magnetron sputtering [7], and most widely used chemical vapor deposition method [8]. In this work, plasma enhanced chemical vapor deposited (PE-CVD) SiC$_x$N stack films was applied as capping layer in Cu/SiC$_x$N/ULK structure. Then resistance and capacitance of obtained Cu/SiC$_x$N/ULK structure with different SiC$_x$N stack films were discussed, and the reliability performance of these structures was also investigated.

Application of ultra-thin SiN+ SiC$_x$N stack

1.Adhesion layer

The EM reliability is related to the adhesion between Cu and the capping material [9]. Cu adheres poorly with inter metal dielectric, Cu could affect Si and ULK properties, therefore the application of thin barrier layers for copper interconnects technology solves the problems [10]. In order to strengthen adhesion, an NH_3/N_2 based plasma pre-treatment was introduces to remove surface CuO before ultra-thin SiN deposition. This treatment would increase reliability by bringing good adhesion between Cu and ULK.

2.Lower K_{eff}

The k value can be varied between 5.2 and 6.0 depending on the deposition methods. The conventional Cu capping layer used SiCN with K_{eff} about 5.5. By introducing a thin SiN, low K_{eff} of the ES/Barrier can be gained. The new structure included 3nm UT-SiN and 28nm C-rich SiC$_x$N show K_{eff} about 5.3 (Figure 1).

3.Barrier layer for moisture and Cu diffusion block

Cu/ULK interconnect become common integration scheme for 40nm technology and beyond, since copper replace Al due to its lower bulk resistance (1.68 AV cm) and excellent electro-migration resistance, ULK replaced conventional interlayer dielectric SiO_2 brought lower dielectric constant, i.e. reduction of RC delay. But still copper is a faster diffuser in SiO_2 based film ULK, barrier layer plays a crucial role in Cu interconnects. SiCN show good barrier behavior for Cu/black diamond low k material systems because it's amorphous with no fast diffusion paths since there are no grain boundaries. With dense ultra-thin SiN as a pre-layer, which is excellent barrier, this SiCN stack would prevent Cu diffusion into porous ULK, and blocking H_2O, O_2 diffusion into Cu cause erosion, as show in Figure 2. It can also help ULK resist moisture absorption and maintain acceptable k value.

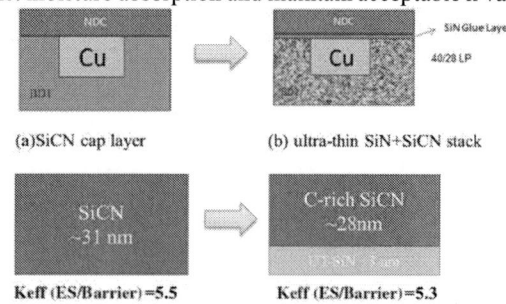

Figure 1: Comparison of ES/Barrier structures (a) SiCN capping layer and (b) ultra-thin SiN+ SiC_xN stack

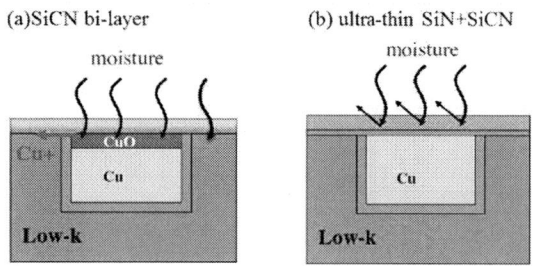

Figure 2: Mechanism of moisture block: (a) bilayer SiCN; (b)Ultra-thin SiN + SiC_xN

4. Etch stop layer

In this ultra-thin SiN+ SiC_xN stack film, SiC_xN plays the role of etch stop layer (ESL), which has high via etch selectivity. SiC_xN film has good resistance to wet chemicals, aggressive etch and chemical mechanical polishing. It helps provide good profile of Cu damascene structure for etching and CMP process.

EXPERIMENTS

The detailed process of ultra-thin SiN+ SiC_xN stack film is shown in Figure 3. Wafer after Cu CMP and post cleaning was prepared, the wafer was transferred into PECVD chamber for Cu capping layer deposition(LAM Vector Express PECVD). In the first step, reducing plasma was applied to remove surface oxidized Cu at around 350 C, this pretreatment can also reduce post-CMP residue and recover k value of ULK while exposure to air, CMP slurry and wet chemical. In the second step, the UT-SiN was deposited in the reactor using SiH_4, NH_3 and N_2 gases and a 13.56 MHz radio frequency power supply. An enhanced nitride treatment with higher plasma power as a post treatment was used in the third step for densification and improving film quality, for instance,

reducing defects due to many hydrogen bonds and dangling bonds. In the final step, C-rich SiC_xN as the etch stop layer was deposited on SiN by plasma using 4MS (tetramethyl silane), NH_3, hydrocarbon gas.

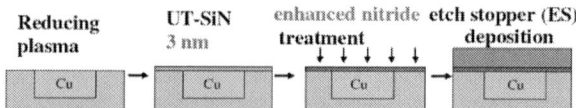

Figure 3: Process flow for ultra-thin SiN+ SiC_xN stack film

RESULTS AND DISCUSSION

In order to compare the RC delay of ultra-thin SiN+SiC_xN stack films with different SiN prelayer thickness, match the device design needs for RC, the PECVD deposition process parameters were set for ultra-thin SiN deposition. The RC requirement of designed BEOL Cu damascene layer is normalized resistance of 0.33, and normalized capacitance of 0.21. These critical parameter included Si source SiH_4 flow, reactive gas NH_3 flow, and deposition time. The condition of each wafer's capping layer split and the capacitance was show in Table 1. Obviously condition NDC-S3 show lowest capacitance according to same normalized resistance value, due to none SiN film existence, but the reliability cannot be accepted. Among all the stack film with ultra-thin SiN, condition NDC-S2d show acceptable capacitance. Nonetheless, the thickness of SiN could not thin down to less than 20Å because there would be poor barrier performance.

Table 1: Process condition of ultra-thin SiN+ SiC_xN stack film

Split table	SiH_4 flow	NH_3 flow	SiN Dep time	SiN	Capacitance (Re=0.33)
NDC-S1	2c	1x	1t	45 A	0.245
NDC-S2	1c	1x	1t	30 A	0.234
NDC-S3	0	0	0	0 A	0.200
NDC-S2a	c	2x	1ts	30 A	0.214
NDC-S2b	0.9c	3x	1t	25 A	0.212
NDC-S2c	0.6c	4x	1t	25 A	0.210
NDC-S2d	0.5c	4x	3t	20 A	0.200

Figure 4 revealed electrical performance, resistance and capacitance of devices of each wafer. The disparity of Cu line width result in linear data distribution of these devices on same wafer, therefore, location of each line indicated the RC delay performance. It's indicated the condition NDC-S2d met requirement, with normalized capacitance of 0.20 while normalized resistance was 0.33. Figure 5 show the voltage of breakdown (VBD) of device of wafer capping different ultra-thin SiN+SiC_xN stack film, the condition NDC-S2d show good reliability.

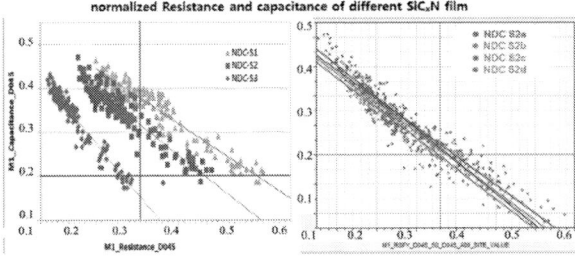

Figure 4: Normalized resistance and capacitance of electrical test of devices on each split wafer sample

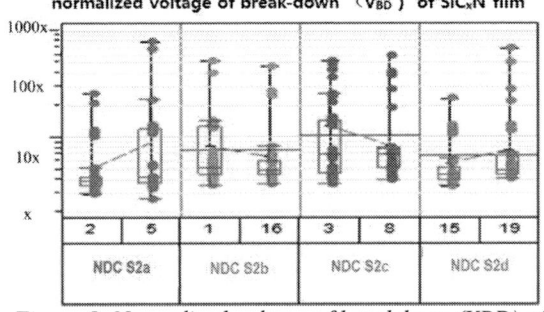

Figure 5: Normalized voltage of breakdown (VBD) of device of wafer capping different ultra-thin SiN+ SiC$_x$N stack film

CONCLUSION

Application of SiN/SiC$_x$N stack as capping layer for electric and reliability performance of damascene Cu/ULK interconnect was discussed. The stack film of ultra-thin SiN+SiC$_x$N capping films were fabricated by chemical vapor deposition method, which consisted of N$_2$ plasma treatment, SiN etch stop layer deposition with post treatment, C-rich SiC$_x$N cap layer deposition. ES/Barrier can be gained. The SiN+ SiC$_x$N stack film with~28 nm C-rich SiN/3 nm UT-SiN show lower K$_{eff}$, it is about 5.3. The critical parameter of ultra-thin SiN+ SiC$_x$N stack film deposition included Si source SiH4 flow, reactive gas NH3 flow, and deposition time. The stack film with ultra-thin SiN thickness of 20 Å and 230 Å SiC$_x$N show acceptable RC performance and good reliability.

REFERENCES

[1] Li Y Z, Zhou J C, Chen H B, et al. Diffusion barrier of Cu metallization in ULSI [J]. Materials Review, 2007, 21(5): 17−18.

[2] Yang L Y, Zhang D H, Li C Y, et al. Comparative investigation of TaN and SiCN barrier layer for Cu/ultra low k integration [J]. Thin Solid Films, 2006, 504: 265−268.

[3] Mitu B, Ddnescu G, Budianu E, et al. Formation of intermediate SiCN interlayer during deposition of CNx on a-Si:H or a-SiC:H thin films [J]. Applied Surface Science, 2001, 184: 96−100.

[4] Vitiello J, Ducote V, Farcy A, et al. New techniques to characterize properties of advanced dielectric barriers for sub-65 nm technology node [J]. Microelectronic Engineering, 2006, 83: 2130−2135.

[5] Peng X, Sond L, Le J, et al. Spectra characterization of silicon carbonitride thin films by reactive radio frequency sputtering [J]. Science & Technology B: Microelectronics and Nanometer Structures, 2002, 20(1): 159−163.

[6] Dashiell M W, Kulik L V, Hits D, et al. Carbon incorporation in Si$_{1-y}$C$_y$ alloys grown by molecular beam epitaxy using a single silicon-graphite source [J]. Applied Physics Letters, 1998, 72: 833−835.

[7] Gao P, Xu J, Piao Y, et al. Deposition of silicon carbon nitride thin films by microwave ECR plasma enhanced unbalance magnetron sputtering [J]. Surface & Coatings Technology, 2007, 201: 5298−5301.

[8] Li Y W, Chen C F. Effect of ammonia plasma treatment on the electrical properties of plasma-enhanced chemical vapor deposition amorphous hydrogenated silicon carbide films [J]. Jpn J Appl Phys, 2002, 41: 5734−5738.

[9] Yi M, Shim C, Lee H C, et al. Effect of capping layer and post-CMP surface treatments on adhesion between damascene Cu and capping layer for ULSI interconnects [J]. Microelectronic Engineering, 2008, 85(3): 621−624.

[10] C.K. Hu, R. Rosenberg, K.Y. Lee, J. Appl. Lett. 74 (1999) 2945

THE METHODOGY TO REDUCE POLY BUMP DEFECT

Junlong Kang*, Xiaogong Fang, Xinhua Cheng, Huaming Luo，Jingxun Fang, Yu Zhang

Shanghai Huali Microelectronics Corporation, Shanghai 201210, China

kangjunlong@hlmc.cn

ABSTRACT

As CMOS technology continues to scale down, one of the most critical factors that affect the device performance is the various defects which are introduced in the process. The poly defect formation was found in un-doped gate poly film. In this paper, the formation mechanism of poly defect was discussed along with methods to reduce the poly defect. The result shows the condition of low temperature and high N_2 gas flow can be used to control the deposition rate and help generate fine grain poly silicon and thus smooth surface. In addition, low temperature and high N_2 gas flow can eliminate the bump defect effectively.

INTRODUCTION

In the advanced Poly/SiON technology, gate poly silicon defect become more critical because it impacts the device yield significantly. Figure 1a and 1b demonstrated two most common defects that impact the device performance. Current gate poly silicon can't meet Poly/SiON technology needs, as shown in Figure 1(a), because larger poly grain size impacts poly line profile after poly silicon dry etching, which will impact the device local mis-match between different transistors[1-4].

(a)

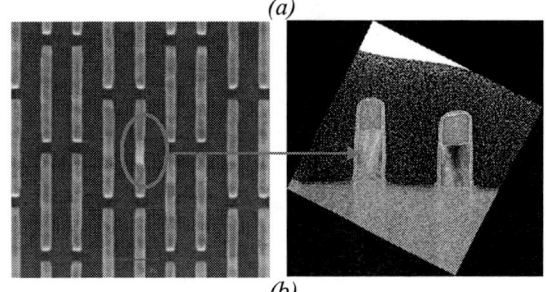

(b)

Figure 1:TEMs of poly line post dry ETCH: (a)Poly line profile, and (b) Poly bump

Figure 1(b) showed that poly bump defects lead to electrical failures of the device. So it is important to reduce or even eliminate the poly bump defects.

In this paper, the mechanism of poly bump defect was discussed along with methods to reduce the poly bump defect.

EXPERIMENTS

The bump defect formation mechanism

The poly silicon film is deposited by low pressure chemical vapor reactor (LPCVD) with SiH_4 gas, with various deposition conditions. The detailed cross-sectional analysis of poly bump was done with TEM technique as shown in Figure 2. No particle was found under the bump. There appears to be some exaggerated poly silicon growth in the bump area with radially oriented grains, in contrast with the normal vertical poly silicon grains. Within the core area of the bump, no discernible foreign elements were detected within the sensitivity of the technique used in conjunction with TEM, we suspect nuclei are formed by the migration of the Si atoms at surface. In general, diffusion of Si along grain boundaries, referred to as surface mobility of un-bound Si atoms during deposition is known to cause poly bump defect [5].

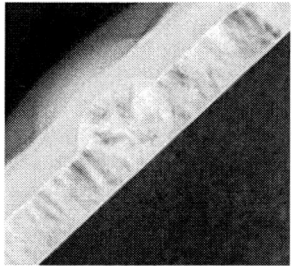

Figure 2: TEM image of the bump defect distribution

Various approaches for controlling poly silicon defect have been developed [5, 6]. Panwar et al [7] proposed that the average grain size of crystallized poly silicon films depend on two parameters-deposition rate and deposition temperature. Therefore, in LPCVD poly silicon deposition, low temperature and low pressure can be used to reduce deposition rate and hence the poly bump of the as-deposited film.

To solve this kind of bump defect, the poly process parameter split tests was performed as shown in Table 1.

TABLE 1. THE POLY PROCESS PARAMETERS

Sample	Temperature (°C)	Pressure (torr)	N2 flow (sccm)
Baseline	620	0.15	228
A	620	0.08	228
B	600	0.15	244

RESULTS AND DISCUSSION
Roughness

X-ray reflectivity (XRR) was used to measure the roughness. Table 2 summarizes the results of the split tests. Sample A showed worse roughness performance, as process pressure decreased, the total heat time of wafer increased, thermal treatment is strongly correlated to poly grain size and the surface roughness. Low temperature poly silicon formed at 600°C, 0.15torr and N_2 flow 244sccm (sample B) with an approximate roughness of 26.54A.

TABLE 2. THE POLY ROUGHNESS RESULT

Sample	Baseline	A	B
Roughness(A)	28.28	35.2	26.54

Grain size

High resolution scanning electron microscopy (HRSEM) and cross-sectional transmission electron microscopy (XTEM) were used to characterize the grain size in poly.

Figure 3 shows SEM (100000X magnification) and TEM (400X magnification) plane view of the poly silicon films formed using the LPCVD process. It seems that low temperature (600°C) and higher N_2 gas flow(244sccm) pre poly silicon deposition resulted in smaller grain size and hence smoother surface, compared with baseline and sample A, which supported the measure results of XRR (Table 2).

Poly bump defect

Inline defect performance was checked on pattern wafer by KLA-Tencor. The bump defect analysis chart result is shown in Figure 4. The bump defect count in sample B condition was reduced slightly by lower pressure in poly silicon deposition, but were effectively eliminated by sample A condition of lower temperature (600°C) in conjunction with higher N_2 gas(244sccm).

CONCLUSION

In this paper, we varied the temperature and N_2 flow to modulate the poly bump defects. The result shows lower temperature and higher N_2 gas flow can be used to slow the deposition rate and help generate fineer grain poly silicon and thus smoother surface. Based on this study we estimate it would be an effective method to eliminate bump defect.

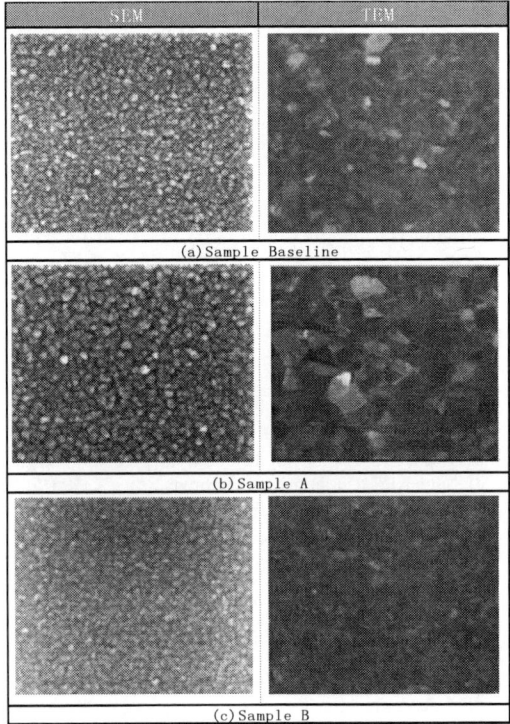

Figure 3: SEM and TEM of images of the poly silicon film surface and poly grain size

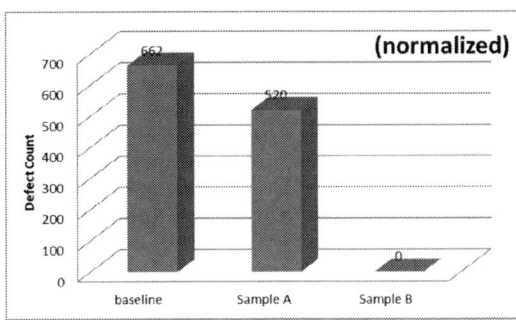

Figure 4: Poly bump defect count of different split condition

ACKNOWLEDGMENTS

This work was completed with help of diffusion process and integration groups in technology development of Huali. The author also appreciated the failure analysis group in Shanghai Huali Microelectronics Corporation.

REFERENCES

[1] SungHyung Park etc, Japanese Journal of Applied Physics, Vol43, p.1709 (2004).

[2] Abu H.M.Kamal etc, IEEE, Vol.15,No.4 (2002).

[3] M. E. McGruer, R.Oikari, "Poly silicon capacitor failure during rapid thermal processing, IEEE Trans. Electron. Devices, vol ED-33, pp 929-932, 1986.

[4] Youfeng He , Haifeng Zhu, Lan Jin, Yonggen He, Jingang Wu," Investigation of Poly Film Application in Advanced MOSFET Technology," ECS Transactions, 60 (1) 507-511 (2014).

[5] Anish Khandekar, Shyam Surthi, Vincent Hou, Niraj Rana, Benjamin Williams,"Mechanism of Surface Bump Defect Formation in Phosphorus Doped Polysilicon-Silicon Nitride Film Stack," IEEE, 978-1-4244-2343, 2008.

[6]Tuung Luoh, Ling-Wu Yang, Tahone Yang, Kuang-Chao Chen, and Chih-Yuan Lu,"Improving the Endurance of Nonvolatile Flash Memory Using Micro-Grain Poly-Silicon Floating Gate" IEEE Transactions on Semiconductor Manufa-Cturing, VOL. 23, NO. 3, AUGUST 2010.

[7] O. S. Panwar, R. A. Moore, S. H. Raza, H. S. Gamble, and B. M.Armstrong, "Comparative study of large grains and high-performance TFTs in low-temperature crystallized LPCVD and APCVD amorphous silicon films," Thin Solid Films, vol. 237, pp. 255–267, Jan. 1994.

Residue defect study on amorphous silicon film

Zhiyong Yang [1*], Jialei Liu [1], Huanxin Liu[1], Yonggen He[1] and Yong Huang[1]

[1] Technology R&D center, Semiconductor Manufacturing International Corporation, No. 18 Zhangjiang Road, Pudong New Area, Shanghai, PRC

*Corresponding Author's Email: Alex_Yang@smics.com

Abstract

In advanced technology nodes, many new materials have been introduced to improve the device capability. This means new particle sources are also introduced through more complicated process. Hence, the defect control will be more important for manufacturing and yield improvement. In this paper we addressed one type of residue defect found on surface of amorphous silicon film. The defect source and forming mechanism were revealed by our experiments. Finally, the cleaning method was shared that can remove this defect.

Keywords- Amorphous silicon; FOUP environment; Residue defect;

I. Introduction

With the need of device scaling, many new materials were introduced into wafer manufacture. This usually brought many defects problems at the beginning of development, such as surface particle, in-film particle, scratch, residue remaining which killed yield and cause production scrap. One new type residue defect was found on amorphous silicon (α-Si) film surface which appeared to be residue remaining on the wafer. Figure1 shows the defect map and image. The defect map was ring type. There are 2 types of defects. One film peeling, the other one was wafer mark defects. Figure2 shows that the EDX [1] data are O and Si element. This defect only occurred after scrubber cleaning post a-Si deposition. And this defect was only found on high k value production, where the new materials of high k value and metal gate are involved.

II. Experiment

An experiment was designed to find out the defect source. The wafer film stack and process flow are shown in figure3. The film stack is Si/SiO$_2$/HK/TiN/α-Si. After α-Si growth, backside clean (BSC) was used to prevent tool cross contamination[2]. Scrubber clean is usually used to remove surface particle after BSC, and then follows by defect scanning. It was found that before scrubber the defect was clean, but after scrubber the defect occurred. Scrubber tool only used deionized water, no defect source could be found. The a-Si film surface condition suspected to be changed after previous process or suffered some contamination issue.

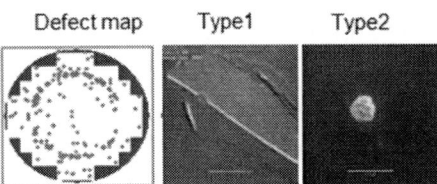

Figure1. Reside defect show ring map, the SEM image have two types: type1 and type2.

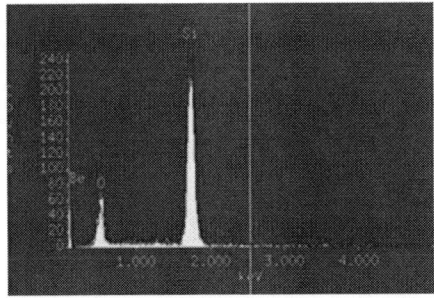

Figure2. Use EDX analysis, the data show O and Si element.

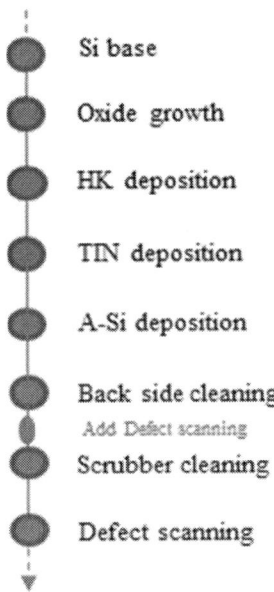

Figure3. Film stack is Si/SiO2/HK/TiN/α-Si. After α-Si growing have back side clean and scrubber clean. Before scrubber, defect is good, but after scrubber clean, the defect occurred.

III. Result and Discussion

978-1-5090-6695-7/17 $31.00 © 2017 IEEE

Simulation was done on control wafers to verify the queue time and the defect correlation. The film stack is the same as structure wafer as Figure 3. The result showed on Figure 4, after long queue time, the defect counter did not increase Figure 4(a), but after use scrubber clean, the defect counter increased, the data showed in Figure 4(b).The SEM image aligned with the inline production residue defect. This concluded that queue time and scrubber clean (only use Deionized Water) was the main factor forming this defect. When use other tool DIW instead of scrubber to clean the long queue time wafers the defect also appeared. Data showed as Figure 4(c). Therefore we have found that the residue defects occurred on α-Si surface after long time queuing in FOUP and touch DIW.

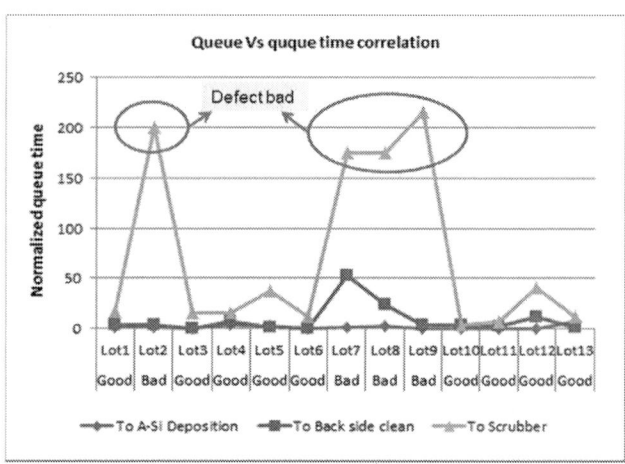

Figure5. Queue time Vs defect performance. After long queue time from BSC to scrubber, residue defect can be seen.

IV. FOUP environment test

Ambient environment, outgassing from wafers, and FOUP contamination were likely reason of the defect formation when wafer stay in FOUP for long time [4]. It was suspected that, the wafer surface reacted with DI water, when queue in FOUP for a long time. FOUP contaminations is undesired for wafers. The ionic (such as NO^{2-}, Br^-, SO_4^{2-}, H^+, Cl^-, F^-, PO_4^{3-} and NH^{4+}) concentration of FOUP surface and wafer surface are much higher than FOUP inner ambient, moreover, the concentration of some ions will increase when queue time increase [5]. The FOUP surface and wafer surface will adsorb the vaporization of these ionic after process and become worse when queue time increase. Afterwards, it will cause defect easily.

From above discuss, FOUP environment status should be an important clue to find out the defect root cause. Adixen POD Analyzer (APA) usually used to monitor the ions contaminants and volatile organic compound (VOC). Use APA to compare the FOUP (POD) inside and outside (Ambient) by the time that wafers stay in it and the result as showed in the Figure 6. After wafers stays in FOUP long time, NH_3 Figure 6(a), Total Acids Figure 6(b), VOC Figure 6(c), increase in the FOUP obviously. After long time, the FOUP environment is stable.

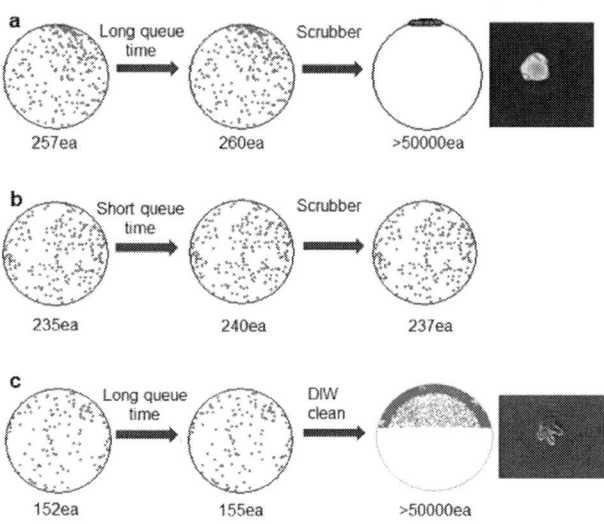

Figure4. Wafers stay in FOUP short and long time (a) and (c), defect result is good. But after scrubber clean, long queue time wafers defect high. And other tool DI clean, defect still high. Defect TEM image is the same as found on production.

The process queue time and the defect result that scanning on production wafers were compared in Figure5. There is a clear correlation of Long queue time between back side clean to scrubber v.s. residue defect on the wafers. We think after wafers stay in the FOUP for long time, the surface will release some byproduct from previous process which adsorb in the FOUP environment and now vaporized as the surface condition changed, when encountered the DIW, the defect occurred.

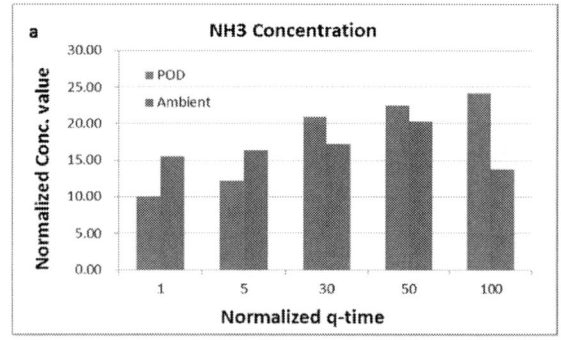

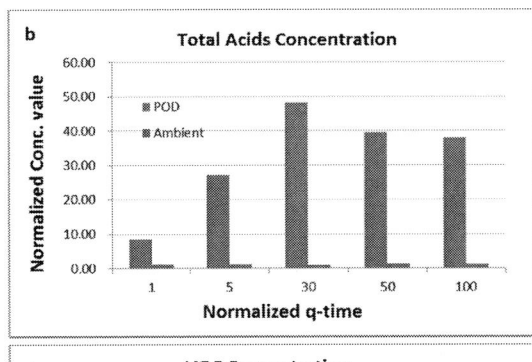

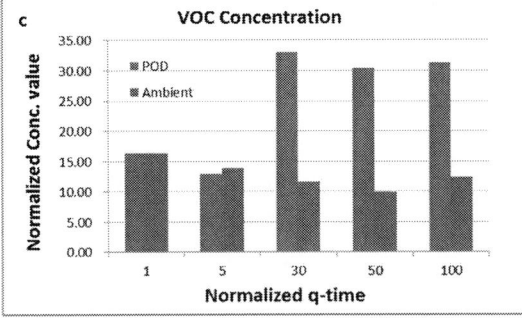

Figure6. FOUP environment test data show after queue in FOUP NH$_3$, Total Acids, and VOC concentration increased.

V. WET Cleaning

Wafer surface will adsorb the vaporization and reacted with the α-Si, and then the surface changed to hydrophobic characteristic, after use DIW clean, the residue defect occurred which looks like water mark. EDX data showed O and Si base. HF and DIO$_3$ (O$_3$ dissolve in deionized water) was used to remove the residue defect. The result showed in Figure7. After HF+DIO$_3$ clean the defect disappeared, only remain pre-layer bump defect.

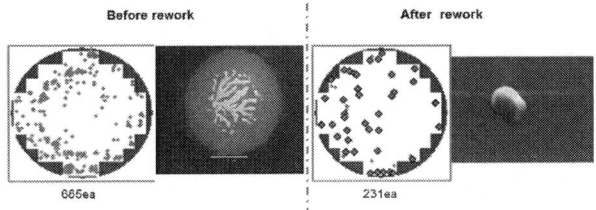

Figure7. After use HF+DIO$_3$ clean residue defect has been removed.

As discussed, there are two key factors caused such defects. So control the queue time, remove the DIW clean is the solution to solve this problem. In fact that there are two other methods are much effective to prevent such defect. One is N$_2$ purge FOUP after process step. The other one is to use HF+DIO$_3$ instead of scrubber clean. The result showed as the Figure 8. HF can remove the native oxide on the A-Si contaminated surface when staying long time inside FOUP, then DIO$_3$ can grow thin oxide film to protect the α-Si surface and make sure the film is stable. SO$_2$ and NOx can be high-effectively removed by DIO$_3$ [6].

Clean Condition	Pre Defect	Post Defect
Scrubber	251 ea	>50000 ea
HF+DIO3	1116 ea	1170 ea

Figure8. HF+O$_3$ clean instead of scrubber defect is good

VI. Conclusion

We found that the residue defects will occur on α-Si surface after long time queuing in FOUP and touch DIW. The reason is FOUP mini-environment getting worse after complicated process. This changed the α-Si surface characteristic and reacted with DIW and leave over the outgrowth. Use N$_2$ purge FOUP after process step or HF+DIO$_3$ clean instead of scrubber are effective ways to remove this defect effectively.

References

[1] Lai Chin Yung, et al, FIB with EDX Analysis use for Thin Film Contamination Layer Inspection, 2013.

[2] Niti Garg, et al, Yield Improvement in 2x Node Technology by Introducing Backside Cleaning, 2015.

[3] FANG Li-jun, et al, Experimental Study on the Performance of Flow & Desulphurization in a Spray Scrubber for Wet Flue Gas Desulfurization,2010

[4] Tuung Luoh, et al, FOUP Mini-Environment Contaminants Analysis in Semiconductor Manufacturing -Yiting Kuo, 2014

[5] Howard Tsao, et al, FOUP Environment Control and Condense Reduction-Yiting Kuo, 2012

[6] Wei Linsheng, et al, Experimental Study on Ozone Generation and Ozone Oxidation to Removal Multi-Pollutant of Flue Gas,2010

Fabrication of single phase transparent conductive cuprous oxide thin films by direct current reactive magnetron sputtering

Ruijin Hong [1], Jinxia Wang [1], Chunxian Tao [1], Dawei Zhang [1]*

[1]Engineering Research Center of Optical Instrument and System, Ministry of Education and Shanghai
Key Lab of Modern Optical System, University of Shanghai for Science and Technology, No. 516
Jungong Road, Shanghai 200093, People's Republic of China
*Presenter; Corresponding author: dwzhang@usst.edu.cn

ABSTRACT

A series of cuprous oxide thin films with various thicknesses were deposited on quartz substrates by direct current reactive magnetron sputtering at room temperature. The crystal structure, element composition, morphology, optical and electrical properties of the samples were characterized using X-ray diffraction (XRD), X-ray photoelectron spectroscopy (XPS), atomic force microscopy (AFM), optical absorption/scattering, and sheet resistance measurement, respectively. XRD patterns indicated that single phase with the preferred orientation in (111) plane could be obtained by controlling the appropriate film thickness and oxygen pressure. The binding energy of Cu 2p and O 1s confirmed that the chemical valence of Cu in the samples is +1. With the increase of film thickness, surface roughness of the samples increased and the optical band gap edge shifted toward longer wavelength.

INTRODUCTION

Cuprous oxide (Cu_2O) is quite a promising semiconductor material [1-3] and has extensively potential applications in photovoltaic devices [4], gas and humidity sensors [5,6], photocatalyst [7], anti-bacterial coatings [8] and resistance swiching memory devices [9] etc. due to its abundant resource, low-cost fabrication, non-toxicity, high absorption coefficient and good photoelectric properties. Particularly, it is reported that Cu_2O thin films are widely used in heterojunction solar cells [10] due to its suitable transmittance and a direct optical band gap value of 2.10eV-2.60eV. In theory, the photoelectric conversion efficiency of Cu_2O is 20% as a p-type semiconductor [11]. In recent years, more research efforts have been focused on the photoelectric characteristic of cuprous oxide. Experimentally, multiphase, mixed phase (the phase of

Cu_2O is mixed with the phase of Cu or CuO), or non-conductive Cu_2O thin films usually cannot meet the demand of industrial application. Therefore, the preparation of single phase transparent conductive cuprous oxide thin films is currently a significant challenge for diverse applications in optoelectronics.

Apart from DC reactive magnetron sputtering [12-14], many other techniques have been used to fabricate Cu_2O thin films, such as RF reactive magnetron sputtering [15], thermal evaporation [16], chemical deposition [17], electrochemical deposition [18], pulsed laser deposition [19], electron beam evaporation [20] and molecular beam epitaxy [21]. For example, Serin et al reported a chemical deposition method for Cu_2O thin films prepared on glass substrates by dipping the microscope glass slide for 20 s each in 1 M NaOH and copper complex solutions. High preparation temperature, high annealing temperature and complex procedure were features of this method.

Hence, comparing to other preparation technologies, DC reactive magnetron sputtering, which can fabricate high-quality thin films with smooth surface, high adhesion and chemical stability because of its high deposition rate, controllability and low deposition temperature, etc, is one of the best preparation technologies. Moreover, as a promising semiconductor material for optoelectronics and solar cells as well as energy converter, the research of single phase transparent conductive Cu_2O thin films is far from enough so far.

In the present work, we proposed a cost-effective technique to prepare single phase transparent conductive Cu_2O thin films using DC reactive magnetron sputtering. The effects of thin film thickness on composition, structural and photoelectric properties were discussed in this paper. Then, the surface morphology, roughness and growth trend of thin films were analyzed as well. finally, the single phase transparent conductive Cu_2O thin film was fabricated on the quartz substrate by controlling the appropriate film thickness.

EXPERIMENTS

Cu_2O films were grown by DC planar magnetron sputtering on silica substrates at room temperature. A disc of copper with 60 mm in diameter with a purity of 99.999% was used as a target. High purity argon and oxygen were used as the sputtering and reactive gas, respectively. The target-to-substrate distance was 70 mm. The cathode was mounted on a water-cooled copper plate. The chamber was pumped to a base pressure of 6×10^{-4} Pa before deposition. Film growth was carried out in the growth ambient with $O_2/Ar+O_2$ ratios ranging from 0.1 to 0.90 at a constant working pressure of 0.8 Pa. Samples with different thicknesses were also prepared, denoted from S1 to S6, with increasing thickness.

The crystal structure of the films was characterized by XRD, with Cu Kα radiation (λ=0.15408 nm). The phase composition of the films was analyzed by XPS (K-Alpha⁺). The optical scattering of the films were measured with an UV-VIS-NIR double beam spectrophotometer (Lambda 1050, Perkins Elmer), scanning range from 200 nm to 1400 nm. The surface morphology and roughness was examined by AFM (XE-100, Park Systems) with scanning area for 3 μm×3 μm. The square resistances of the films were measured with four-point probes resistivity measurement system. All the measurements were conducted at room temperature.

RESULTS AND DISCUSSION

Fig. 1 shows the XRD patterns of the Cu_2O thin films with various thicknesses. It is apparent that the Cu_2O thin films exhibit high crystalline quality with a polycrystalline face centered cubic (fcc) crystal structure of Cu_2O (JCPDS NO. 02-1067). The strongest diffraction peak appeared at around 36.5 ° (2θ), which was corresponding to (111) crystallographic plane, meaning a preferential orientation of the Cu_2O grains along the (111) crystallographic direction. In General, when deposited onto a substrate without the influence of epitaxy, the surface of a film tends to be either (111) or (001) plane, because these planes have minimum surface free energies. This means that the (111) textured film must form in an effective equilibrium state where enough surface mobility is given to impinging atoms under a certain deposition condition. With the thickness increasing, several diffraction peaks at 29.6°, 36.5°, 42.6°, 61.4°,

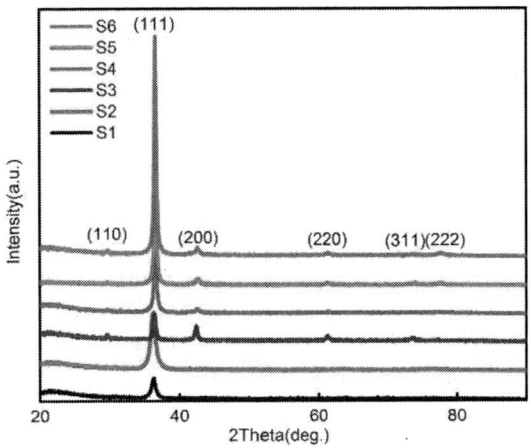

Fig.1. The XRD patterns of the Cu_2O thin films with various thicknesses.

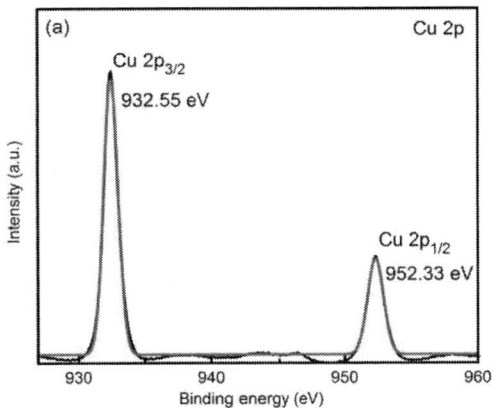

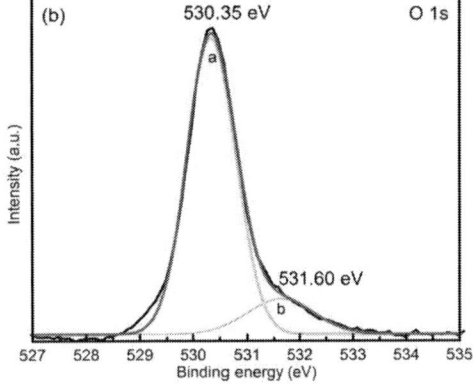

Fig.2. The XPS patterns of the Cu_2O thin films with various thicknesses. (a) Cu 2p; (b) O 1s

73.9 ° and 77.4°, corresponding to (110), (111), (200), (220), (311) and (222) crystallographic planes were observed in the patterns. With the increase of film thickness, the diffraction peak intensity of Cu_2O thin films increased, indicating that grain growth has occurred, and shifting the peaks to higher 2θ angles, as a result of the partial increase of residual stresses within the films.

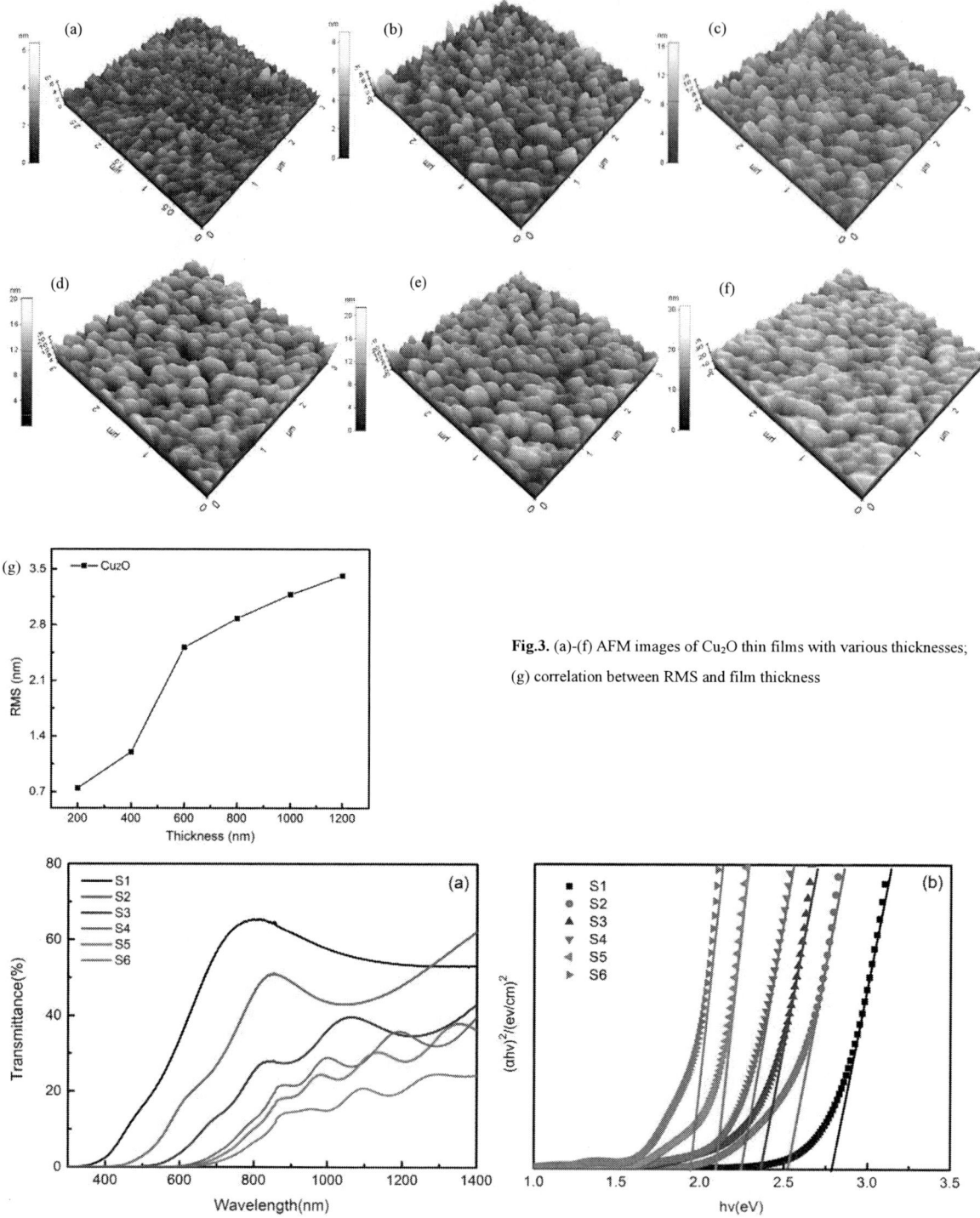

Fig.3. (a)-(f) AFM images of Cu_2O thin films with various thicknesses; (g) correlation between RMS and film thickness

Fig.4. (a) Optical transmittance spectra of Cu_2O thin films with various thicknesses, and (b) Plots of $\alpha\,(h\nu)^2$ versus $h\nu$ for Cu_2O thin films with various thicknesses.

Therefore, to further clarify the presence of the Cu_2O phase, XPS was done, and Fig. 2 represents the X-ray photoelectron spectrum of Cu 2p and O 1s for samples. Two major peaks at 932.55 and 952.33 eV, which were corrected with reference to C 1s (284.6 eV), corresponding to the binding energies of $2p_{1/2}$ and $2p_{3/2}$ states of Cu+1, are in good agreement with data observed for Cu_2O. As shown in Fig. 2 (b), Two O 1s peaks marked as a and b were resolved by using a curve-fitting procedure. Peak (a) at higher energy of 530.35 eV, is in accordance with O^{2-} in Cu_2O. Peak (b) at the lower energy of 531.60 eV, is attributed to O adsorbed on the surface of Cu_2O. Thus, all the above-mentioned structural and chemical compositional analyses indicate that the as-grown samples are single phase Cu_2O with no other impurities.

The AFM micrographs shown in Fig. 3 illustrated the growth process of the Cu_2O thin films with various thicknesses. As shown in the Fig. 3 (g), the root-mean-square surface roughness (Rq) values of the samples are 0.747 nm, 1.201 nm, 2.517 nm, 2.879 nm, 3.179 nm and 3.417 nm, for samples S1-S6. In the Fig. 3 (a)-(b), the nucleation energy may be low, leading to randomly oriented islands and channels but a small columnar structure is forming in the Fig. 3 (b). Further increasing the thin film thickness, grain coarsening, dense spherical structure, coalescence and formation of a continuous and uniform structure are observed according to the Fig. 3 (c)-(f), resulting in the increase of surface roughness. And the grain size is overtly increased and the crystallization is enhanced. The isotropic grain growth was interpreted by a random nucleation mechanism and granule orientation [22, 23]. The increase in crystallite size may be attributed to the change in crystallographic phase from (111) to (110), (200), (220), (311) and (222), as confirmed by XRD data in Fig. 1. It is clear that the Fig. 3 exhibits the microstructural evolution processes during thin film growth [24].

Fig. 4 shows optical transmittance spectra (T) of the Cu_2O thin films with various thicknesses. For all the samples, with the increase of film thickness, the transmittance is successively decreased from about 68% to 20%. The transmittance spectra of S1 and S2 are significantly different from S3-S6, with the higher transmittance and a sole peak in the wavelength of near 800nm. For S3-S6, as is shown in the graph, the transmittance spectra vary in the range of 40%-20% and several peaks appear in different wavelength range. The decrease of transmittance of the samples may be due to the enhancement of absorption and scattering with the increasing film thickness. The defects and interference of the films could contribute to the appearance of more peaks. It has been reported that the Cu_2O thin films are great for heterojunction solar cells for its suitable transmittance and a direct optical band gap.

Therefore，the Tauc plots of the Cu_2O thin films are showed in the Fig. 4 (b). The absorption coefficient α is calculated from the optical transmittance, using the equation (1) below.

$$\alpha = \ln(1/T)/d \qquad (1)$$

Where T is the optical transmittance and d is the thickness of the thin film. The direct optical band gap (Eg) of the films can be calculated by the relationship between the absorption coefficient and the optical band gap, using the following equation (2).

$$(\alpha h\nu)^2 = A(h\nu - Eg) \qquad (2)$$

Where Eg is optical band gap, A is a constant, α is absorption coefficient and hν is incident photon energy. As showed in the Fig. 4 (b), when the curve of $\alpha(h\nu)^2$ versus hν for the Cu_2O thin films is completed, a different color straight line should be drawn corresponding to the S1-S6, respectively. The intersection between the straight line and hν axis represents the direct optical band gap (Eg) value by extrapolation [25]. Then, the optical band gaps of the S1-S6 are 2.757eV, 2.525eV, 2.364eV, 2.235eV, 2.107eV and 1.956eV, respectively. With the increase of film thickness, the optical band gaps of S1-S6 are decreased and manifest the red shift (from the visible light wavelength to the infrared wavelength) in good order. Because of the increasing film thickness will certainly make the density of grain boundaries increases, resulting in the shrinkage of the band gap. Reported that the direct optical band gap value of Cu_2O is about 2.1eV-2.6eV, the band gaps of the samples all fall into the range except the S1 and S6. Well explaining the reasons, then, the optical band gap edge may appear defect states due to too low or too high energy gap. Furthermore, the physical properties, the deposition method and the current deposition conditions are the factors to affect the band gap value.

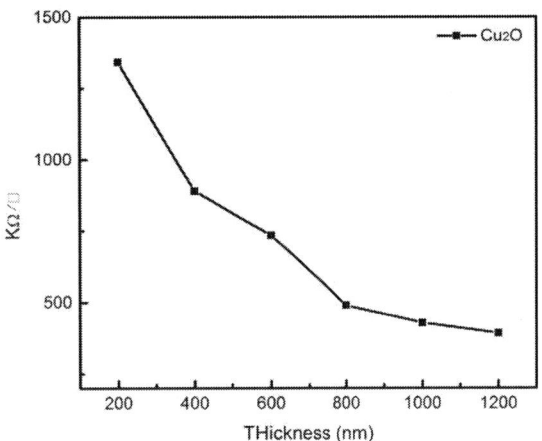

Fig.5. Curve of square resistance value for Cu_2O thin films with various thicknesses.

Fig. 5 showed the curve of square resistance for all the samples. Overall, the thicker the film, the lower the square resistance. This phenomenon is mainly due to the scattering effect of the defects and grain boundary and the

978-1-5090-6695-7/17 $31.00 © 2017 IEEE

concentration of the current carrier. At lower thickness, the thin films are in discontinuous islands and channels with a high hole concentration and hence higher resistance. As the film grows, the increased crystalline size makes grain boundary area relatively decreased and weakens the scattering effect of grain boundary, so the mobility and the concentration of the current carrier are increased. All of the factors contribute to the decrease of the square resistance. As the film continues to grow, the holes and channels are filled with subsequent deposited atoms, leading to the formation of a dense and continuous structure and the crystalline size has little change. As a result, the square resistance change slows down.

CONCLUSION

In summary, we investigated the effects of film thickness on structure, optical and electrical properties of Cu_2O thin films. Single phase transparent conductive cuprous oxide thin films are obtained and analyzed using various analytical techniques. By controlling the appropriate film thickness, the preferred orientation in (111) plane of the Cu_2O thin films are obtained. With the increase of film thickness, a red shift is observed and the electrical conductivity is enhanced. The Cu_2O thin films fabricated in this paper demonstrate excellent optical properties and electrical conductivity.

ACKNOWLEDGMENTS

This work was partially supported by the National key research and development program of China (2016YFB1102303), the National Basic Research Program of China (973Program) (2015CB352001), and National Natural Science Foundation of China (61378060).

REFERENCES

[1] F.L. Weichman, Photoconductivity of cuprous oxide in relation to its other semiconducting properties, Phys. Rev. 117 (1960) 998-1002.

[2] R.N. Briskman, A study of electrodeposited cuprous oxide photovoltaic cells, Sol. Energy Mater. Sol Cells 27 (1992) 361-368.

[3] N.L. Peterson, C.L. Wiley, Diffusion and point defects in Cu_2O, J. Phys. Chem. Solids 45 (1984) 281-294.

[4] C. Xiang, G.M. Kimball, R.L. Grimm, B.S. Brunschwig, H.A. A twater, N.S. Lewis, 820 mV open-circuit voltages from Cu_2O/CH_3CN junctions, Energy Environ. Sci. 4 (2011) 1311-1318.

[5] H. Zhang, Q. Zhu, Y. Zhang, Y. Wang, L. Zhao, B. Yu, inside front cover: one-pot synthesis and hierarchical assembly of hollow Cu_2O microspheres with nanocrystals-composed

porous multishell and their gas-sensing properties, Adv. Funct. Mater. 17 (2007) 2766-2771.

[6] S.T. Shishiyanu, T.S. Shishiyanu, O.I. Lupan, Novel NO_2 gas sensor based on cuprous oxide thin films, Sens. Actuators, B 113 (2006) 468–476.

[7] X. Duan, R. Gao, Y. Zhang, Z. Jian, Synthesis of sea urchin-like cuprous oxide with hollow glass microspheres as cores and its preliminary application as a photocatalyst, Mater. Lett. 65 (2011) 3625-3628.

[8] N.G. Durmus, E.N. Taylor, K.M. Kummer, T.J. Webster, Enhanced efficacy of superparamagnetic iron oxide nanoparticles against antibiotic-resistant biofilms in the presence of metabolites, Adv. Mater. 25 (2013) 5706.

[9] A. Chen, S. Haddad, Y.C. Wu, Z. Lan, T.N. Fang, S. Kaza, Switching characteristics of Cu_2O metal-insulator-metal resistive memory, Appl. Phys. Lett. 92 (2007) 123517.

[10] L. Papadimitriou, N.A. Economou, T. Dan, Heterojunction solar cells on cuprous oxide, Solar Cells. 3 (1981) 73-80.

[11] A.E. Rakhshani, Preparation, characteristics and photovoltaic properties of cuprous oxide-a review, Solid State Electron. 29 (1986) 7-17.

[12] S. Ishizuka, S. Kato, Y. Okamoto, T. Sakurai, K. Akimoto, N. Fujiwara, H. Kobavashi, Passivation of defects in polycrystalline Cu_2O thin films by hydrogen or cyanide treatment, Appl. Surf. Sci. 216 (2003) 94-97.

[13] T. Itoh, K. Maki, Preferentially oriented thin-film growth of CuO(111) and Cu_2O(001) on MgO(001) substrate by reactive dc-magnetron sputtering, Vacuum. 81 (2007) 904-910.

[14] M. Zhu, R. An, C. Wang C, et al. Laser-induced self-propagating reaction in Ti/a-Si multilayer films in proceedings of the 15th International Conference on Electronic Packaging Technology(ICEPT). IEEE, 2013: pp. 1227-1230.

[15] S. Ishizuka, T. Maruyama, K. Akimoto, Nitrogen doping into Cu_2O thin films deposited by Reactive Radio-Frequency Magnetro Sputtering, Jpn. J. Appl. Phys. 40 (2001) 222-226.

[16] M.F. Al-Kuhaili, Characterization of copper oxide thin films deposited by the thermal evaporation of cuprous oxide (Cu_2O), Vacuum. 82 (2008) 623-629.

[17] P.R. Markworth, X. Liu, J.Y. Dai, W. Fan, T.J. Marks, R.P.H. Chang, Coherent island formation of Cu_2O films grown by chemical vapor deposition on MgO(110), J. Mater. Res. 16 (2001) 2408-2414.

[18] M.J. Siegfried, K.S. Choi, Electrochemical crystallization of cuprous oxide with systematic shape evolution, Adv. Mater. 16 (2004) 1743-1746.

[19] S.B. Ogale, P.G. Bilurkar, N. Mate, S.M. Kanetkar, N. Parikh, B. Patnaik, Deposition of copper oxide thin films on different substrates by pulsed excimer laser ablation, J. Appl. Phys. 72

(1992) 3765-3769.

[20] J.H. Ho, R.W. Vook, Misfit dislocations in the epitaxial (111) Cu_2O/(111) Cu system, Phil. Mag. 36 (1977) 1051-1062.

[21] Z.Q. Yu, C.M. Wang, M.H. Engelhard, P. Nachimuthu, D.E. Mccready, I.V. Lyubinetsky, S. Thevuthasan, Epitaxial growth and microstructure of Cu_2O nanoparticle/thin films on $SrTiO_3$(100), Nanotechnology. 18 (2007) 1691-1691.

[22] K. Barmak, E. Eggeling, D. Kinderlehrer, R. Sharp, S. Taasan, A.D. Rollett, K.R. Coffey, Grain growth and the puzzle of its stagnation in thin films: The curious tale of a tail and an ear, Prog. Mater. Sci. 58 (2013) 987-1055.

[23] R.C. Cammarata, Surface and interface stress effects in thin films, Prog. Surf. Sci. 46 (1994) 1-38.

[24] C.E. Murray, R. Rosenberg, C. Witt, M. Treger, I.C. Noyan, Evolution of strain energy during recrystallization of plated Cu films, J. Appl. Phys. 113 (2013) 203515.

[25] T.S. Ahn, A.M. Müller, R.O. Alkaysi, F.C. Spano, J.E. Norton, D. Beljonne, J.L. Bredas, C.J. Bardeen, Experimental and theoretical study of temperature dependent exciton delocalization and relaxation in anthracene thin films, J. Chem. Phys. 128 (2008) 354-368.

STUDY ON THE DEFECT OF POST CLEANING STEP AFTER W CMP AND ITS IMPROVING SOLUTION

Zigui Cao1, Jun Kang1, Hao Huang1, Hui Wang1

1.   Huahong Grace Semiconductor Manufacturing Corporation, 1399 Zuchongzhi Road, Zhangjiang Hi-Tech Park, Shanghai

201203, People's Republic of China

*Email: Steam.cao@hhgrace.com

ABSTRACT

In this paper, water mark like defect was investigated during post clean step after W CMP. Through a series of experiments, we found this water mark defect was caused by condensed WOX residue which was dissolved into De-Ionized (DI) water at acid environment. By an additional dilute HF(DHF) wet clean step after typical W CMP process, we found that most of this defect can be removed with high efficiency to about 99.6%.

Key words: Water mark, CMP, defect, DI water

INTRODUCTION

Tungsten is widely used as the suitable metal plugs in ULSI due to its higher resistance to electro migration in high current density and its better gap filling. During tungsten plug forming, tungsten CMP is introduced to gain the global planarity of tungsten plugs[1]. Simultaneously, surface particles on wafer are produced as the killer to product CP yield. For tungsten CMP induced particles removal, many papers focus on the effectiveness of Post-CMP cleaning such as using chemically enhanced brush scrubbing or using especially selected surfactant[2-4]. However, few investigations are carried on the defect forming of post clean itself.

This work studied the component of defect and put forward one mechanism to demonstrate the procedure for defect forming. Afterwards, an effective integration method for this defect removal has been proposed and applied to optimize process conditions to suppress such product failure.

EXPERIMENTAL

In this work, the testing samples are patterned silicon wafers of 0.13um e-flash short loop. Tungsten CMP and afterward post clean is process on EBARA FREX200 tool. Generally, WCMP post clean is widely used effective method to reduce particles on high-defect-density impacted wafers for 0.18um node and above. The particles generally consist of 80% surface particles and 20% tiny particles. Nevertheless, though the removal efficiency to surface particles is almost 100%, while for the defect of post W CMP clean, water mark like defect is always detected on wafer surface by scanning electron microscope (SEM) occasionally as shown in **Figure 1**.

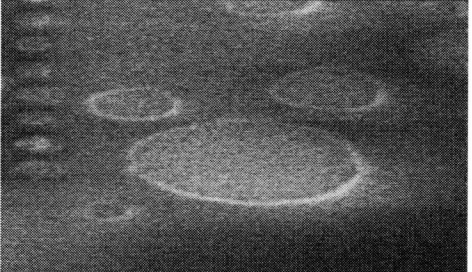

Figure 1 Typical water mark defect reviewed by scanning electron microscope (SEM) .

For clarification the mechanism of defect forming, three samples were prepared with oxide surface and W plug pattern surface respectively, as shown in **table 1**. By the way, our purpose is to clarify the root cause of defect forming during post clean after W CMP and find a way to solve this problem. Therefore, three samples are intended to categorize into 2 groups due to different stack structures: **Group A** is only using silicon substrate and oxide film as surface layer for short loop wafer preparing, plus W CMP post cleaning process. **Group B** is using pattern wafer stack structure with W plug included, with/without additional dilute HF cleaning post normal W CMP post cleaning step. For sample analysis, Transmission electron microscopy (TEM) plus Energy Dispersive X-ray Spectrometer (EDX) methodology were used.

TABLE I. DIFFERENT FILM STACK AND EXPERIMENT PLAN USED FOR WATER MARK CLARIFICATION.

Group	Sample	Film Stack	Post W CMP Clean	Additional Clean
A	1	Si Sub+ Oxide	Yes	No
B	S1	Si Sub+ILD film with Contact WCMP	Yes	No
	S2		Yes	Dilute HF acid

RESULTS AND DISCUSSION

As comparative samples of Group A and Group B S1, **Figure 2**(a) (b) shows defect scan and review result after W CMP post cleaning. From the scan results, we can see that the defect is almost clear in Group A while for Group B S1, serious defect was found and exhibited as circle shape.

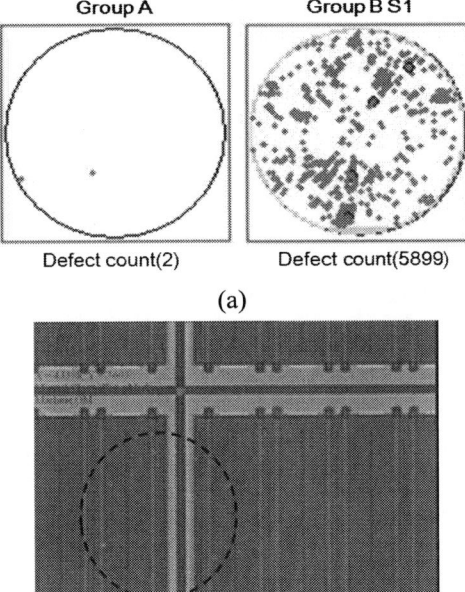

(a)

(b)

Figure 2 Defect scan and review result of Group A and Group B S1: (a) scan map; (b) OM picture.

To understand the defect components, one wafer was prepared base on the process flow and scanned defect by KLA scan tool, then TEM analysis was used for element analysis. **Figure 3**(a) (b) showing the process flow diagrams and the cross-section picture of water mark like defect, from this picture, a white film was found to insert into the normal film stack and EDX shows a tungsten element peak at energy closing to 1.8KeV, implying that water mark defect was generated by W compound.

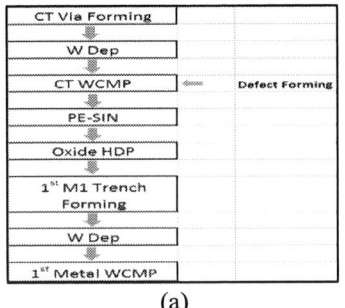

(a)

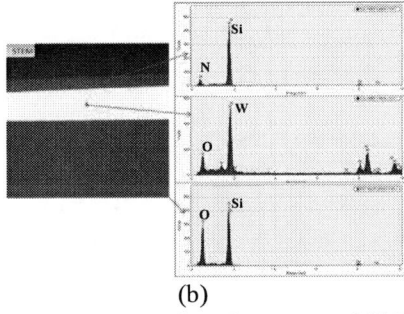

(b)

Figure 3 (a) Process flow diagrams and (b)TEM cross-section and EDX of water mark like defect, W element was found in water mark defect.

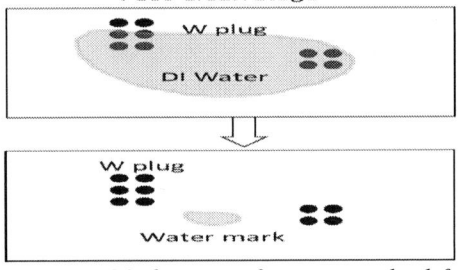

Figure 4 Mechanism of water mark defect forming.

Therefore, defect forming mechanism can be described in **Figure 4**: After W CMP polish, W plug is exposed in atmospheric environment and forms WO_X compound which is remaining at the surface of W plug.

During DI water post clean stage, if there is DI water remaining on the wafer surface, WO_X will be dissolved into water at acidic environment and condensed into water mark like defect during post spin dry process.

To remove this defect, we add one re-work process with dilute HF(DHF) for water mark defect impacted wafer(detail condition shown in Group B S2). The testing wafer is cleaned with the proposed additional DHF clean process and detail result is shown in **Figure 5**,. From the result, we can see that the removal efficiency of additional clean is about 99.6%, which reduced original defect count from 5899ea to 26ea

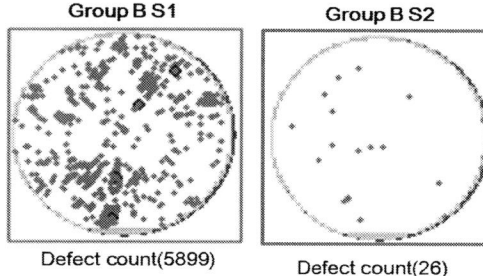

Figure 5 Defect scan result of Group B S1 and S2, an additional DHF cleaning removed water mark defect with 99.6% efficiency.

CONCLUSIONS

In summary, water mark like defect was investigated during post clean step after W CMP. Through a series of experiments, we found this water mark defect was caused by condensed WOX residue which was dissolved into DI water. By additional DHF clean step after typical W CMP process, we found that most of this defect can be removed with high efficiency to about 99.6%.

REFENCES

[1]李薇薇, 周建伟, 尹睿,等. IC 制备中钨插塞 CMP 技术的研究 [J]. 半导体技术, 2006, 31(1):26-28.

[2] Martinez M A. Chemical mechanical polishing : route to global planarization [J] , Solid State Technology ,1995 , (May) :26.

[3]Yzhao E. Post CMP Cleaning Using Chemically Enhanced Brush Scrubbing[J]. Equipment for Electronic Products Manufacturing, 2004.

[4]Wei-Wei L I, Tan B M, Zhou J W, et al. Study of Surface Cleaning Technology in Post CMP[J]. Semiconductor Technology, 2006, 31(3):186-188.

ABRASIVE PARTICLE TRAJECTORIES AND MATERIAL REMOVAL NON-UNIFORMITY DURING CHEMICAL MECHANICAL POLISHING

Vahid Rastegar[1] and S.V. Babu[1,]*

[1]Center for advanced Material Processing,
Clarkson University, Potsdam, NY, USA
*Corresponding Author's Email: babu@clarkson.edu

ABSTRACT

A mathematical model, a continuation of the work by Zhao et al. [1] that describes particle trajectories during chemical mechanical polishing, was extended to account for the effects of larger particles, particle location changes due to slurry dispensing and in-situ conditioning. Material removal rate (MRR) and within wafer non-uniformity (WIWNU) were determined based on the calculated particle trajectory densities in the absence of chemical activity from the slurry. The effect of pad-wafer rotary dynamics and reciprocating motion of the wafer carrier on MRR uniformity were included. It was also shown that in-situ conditioning improves the MRR uniformity of the polished wafers.

Using the model, we also investigated the effect of particle size distribution and large particles (>200 nm in diameter) on WIWNU and scratch growth. It was shown that the presence of even 1 wt.% of larger particles can deteriorate the WIWNU.

INTRODUCTION

Understanding the fundamentals of the material removal mechanisms during CMP can offer guidance to the control and optimization of the polishing processes. Various wafer-level models have been proposed in the literature to attain this goal. Wafer-level models seek to address the non-uniformity of polishing rates across the whole wafer, particularly issues arising as a result of polishing tool and/or process limitations [2].

The earliest and most well-known CMP model for material removal is Preston's equation which is still widely applied as an important theoretical reference in CMP process development in spite of many limitations. It relies on a linear dependence of removal rate on pressure and relative velocity and it does not take into account the parameters of the slurry like chemical composition, pH, particle size, etc. and specific role of the polishing pad, conditioning, etc.

Zhao et al. [3], Kim and Jeong [4], Feng [5], and Kasai [6] proposed the concept of the sliding distance and calculated the removal rate nonuniformity based on the relationships among the sliding distance distribution and the kinematic parameters, also using the Preston's equation. However, Kim and Jeong [4], Feng [5], and Kasai [6] did not consider the oscillatory motion of the wafer carrier in their analysis.

Recently, Zhao et al. [1] simulated the particle sliding trajectories for the CMP process and discussed their effects on polishing rate uniformity after including the wafer carrier oscillatory motion. They calculated the particle trajectory density and used it to evaluate the global MRR uniformity and identified the kinematic parameters that lower nonuniformity. However, the particle size distribution was not considered in their model and consequently the effect of larger particles on rate uniformity was not discussed. In the actual fabrication process, minimizing defects like scratches is of critical importance. It is known that large particles should be removed from the slurry before it is distributed on the pad to avoid such defects.

In this paper, we extend the calculations of Zhao et al. [1] by including the effect of larger particles, particle size distribution, particle location changes due to slurry dispensing and in-situ conditioning. MRR and WIWNU are determined based on the calculated particle trajectory densities. Even though many modifications of the Preston's equation have been proposed, it is still useful to gain many fundamental insight into the polishing process. Hence, we retain Preston's model in our simulation.

PARTICLE TRAJECTORIES

Figure 1 shows a schematic illustration of the parameters used in this analysis using a fixed coordinate system XY and a moving coordinate system xy that are identical to those used by Zhao et al. [1]. The co-ordinate system XY is centered on the pad while the xy system is centered on the wafer. R and r in Figure 1 refer to the radial position of an arbitrary point on the wafer in the XY and xy systems. The parameters e_o, ω_w and ω_p represent the distance between the centers of the pad and the wafer, and angular velocity of wafer and pad, respectively. When a particle at P is anchored on the pad asperity, it rotates with the pad along P_0P, as shown in Figure 1. P_0 here is the intersection of the particle rotary path with the wafer leading edge, and the angle from line OP_0 to the X axis is φ_0. Thus as the particle begins to pass over the wafer surface, its polar coordinates start at $(R, -\varphi_0)$ and change continuously. When the particle leaves the wafer and rotates with the polishing pad, it may come back and enter the wafer-pad gap at a location different from P_0 due to the rotation and vibration of the carrier and can contribute to the polishing again.

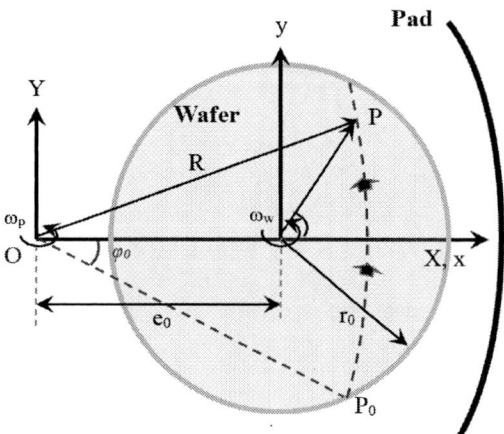

Figure 1: Schematic of motion relationship between the wafer and the pad. XY is the fixed location on the pad and xy is the moving location on the wafer.

Following the approach of Zhao et al. [1], the time dependent trajectory of a particle in the fixed coordinate system can be described as:

$$\begin{bmatrix} X(t) \\ Y(t) \end{bmatrix} = R \begin{bmatrix} cos(-\varphi_0 + \omega_p t) \\ sin(-\varphi_0 + \omega_p t) \end{bmatrix} \qquad (1)$$

The moving coordinate system and the fixed coordinate system have a translational relationship given by [1]:

$$\begin{bmatrix} X \\ Y \end{bmatrix} = \begin{bmatrix} cos(\omega_w t) & -sin(\omega_w t) \\ sin(\omega_w t) & cos(\omega_w t) \end{bmatrix} \begin{bmatrix} x \\ y \end{bmatrix} + \begin{bmatrix} e_0 + \Delta e \\ 0 \end{bmatrix} \quad (2)$$

where Δe corresponds to the reciprocating motion of the wafer carrier. Thus, the particle trajectory with respect to the local coordinate system attached to the wafer can be determined from:

$$\begin{bmatrix} x \\ y \end{bmatrix} = \begin{bmatrix} cos(\alpha \omega_p t) & sin(\alpha \omega_p t) \\ -sin(\alpha \omega_p t) & cos(\alpha \omega_p t) \end{bmatrix} \\ \left(\begin{bmatrix} Rcos(-\varphi_0 + \omega_p t) \\ Rsin(-\varphi_0 + \omega_p t) \end{bmatrix} - \begin{bmatrix} e_0 + \Delta e \\ 0 \end{bmatrix} \right) \qquad (3)$$

where α is ratio of the rotational speed ratio of the wafer and the platen.

The trajectories of all the injected particles during their multiple travels across the wafer will be very dense and cannot be distinguished by the naked eye since billions of particles are involved. Therefore, a numerical statistical simulation of the distributing trajectories is necessary. To quantitatively describe the distribution of the trajectories, the wafer surface is divided into equidistant grids. Following Zhao et al, we divide the wafer into a 300×300 grid for the statistical calculations with each mesh having dimensions of 1×1 mm². The total of the trajectory lengths of all the particles travelling over one mesh area of the wafer divided by the area of the mesh is defined as the local trajectory density and has units of cm⁻¹. A MATLAB program was developed to calculate and discretize the particle trajectories using the kinematic equations. A statistics function was used to measure and add the total trajectory length within each grid area of the wafer surface, and the path density was calculated from it.

According to Preston`s law, the material volume removed is proportional to pressure and relative velocity and since the particles are fixed on the polishing pad and moving with the pad, the trajectory length of particles divided by time is the relative velocity between the wafer and the pad:

$$MRR_v = k_p \, p(x,y) \, v(x,y) = k_p \, p \frac{L(x,y)}{t} \qquad (4)$$

where k_p is Preston constant. Assuming that the pressure and polishing time are constant, the material removal rate is related to the total of the local trajectory lengths. Hence, the material removal with-in-wafer-non-uniformity (WIWNU) can be defined as:

$$WIWNU = \frac{\sigma_{MRR}}{MRR_{mean}} = \frac{\sigma_L}{L_{mean}} \qquad (5)$$

where σ_L and L_{mean} are the standard deviation and the mean trajectory density, respectively. Therefore, the uniformity of the particle sliding trajectories is directly correlated to the uniformity of material removal rates. If the sliding trajectories are not uniformly distributed, the mesh areas covered with denser trajectories would see higher material removal rate and contribute to removal rate nonuniformity.

Figure 2 illustrates the material removal profiles for the wafers polished for different times showing the material removal non-uniformity. As expected, the trajectory density and material removal increase with polishing time. Also it shows that the trajectory density is becoming denser at the edges of the polished wafers with increasing polish time indicating a higher removal rate at the edges, the so-called edge fast removal.

Large Particles Influence on WIWNU

MRR depends on the size of the abrasive particles. According to Qin et al. [7] when film thickness of the modified layer on the surface being polished is larger than the particle penetration depth, the dependence of MRR on particle size can be derived by assuming that the Preston's constant is proportional to square of the particle radius. We this approximation in our calculation of the effects of particle size on MRR nonuniformity.

Figure 3b shows the WIWNU obtained when two different CMP slurries with different particle size distributions shown in Figure 3a were used for polishing. The distribution labelled 2 in the figure could be the result of point of use filtration to lower the large particle tail in the distribution. As can be seen, filtering out the larger particles (>200 nm in diameter in this case) leads to a decrease in the WIWNU of wafers polished for 60 s. The particles deteriorate WIWNU, indicating the importance of the removal of larger particles from the slurry [8]. If several large abrasive particles or debris from conditioner or pad are embedded in the polishing pad and repeatedly slide along the same path, deep scratches can also be generated. This is especially true for the special case of $\alpha = 1$ where each particle travels the same path over and over.

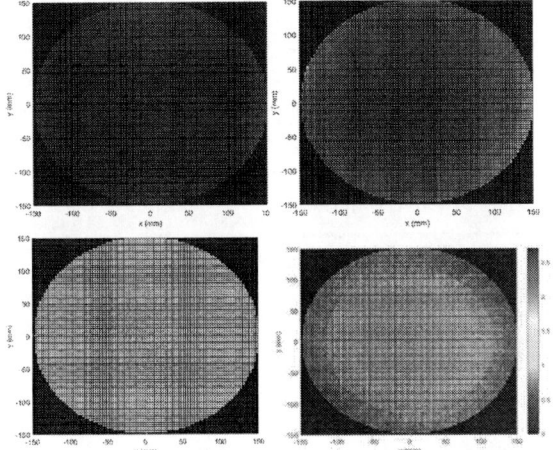

Figure 2: Trajectory density distribution across the polished wafer for 10, 20, 40 and 60 S showing the MRR non-uniformity.

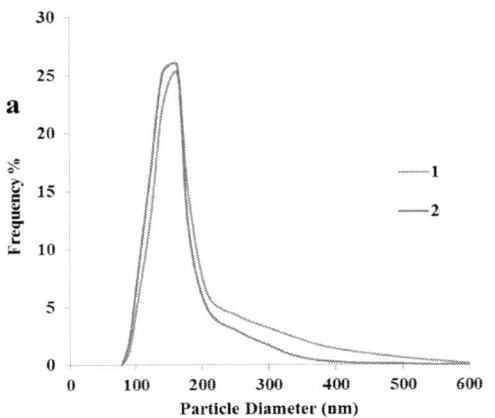

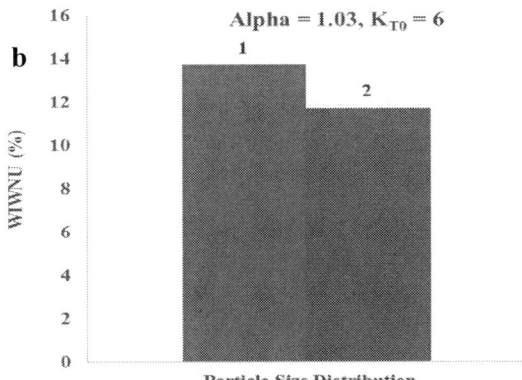

Figure 3: (a) Particle size distribution of CMP slurries (b) WIWNU for CMP slurries with different particle size distributions.

ACKNOWLEDGEMENTS

The authors would like to gratefully thank Prof. Xinchun Lu and Drs. D. Zhao and A. Vahdat for fruitful discussions and useful suggestions.

REFERENCES

[1] D. Zhao, T. Wang, Y. He, and X. Lu, *IEEE Trans. Semicond. Manuf.*, vol. 26, no. 4, pp. 556–563, Nov. 2013.

[2] W. Fan and D. Boning, *Adv. Chem. Mech. Planarization (CMP)*, S.V. Babu (Editor) Woodhead p. 137, 2016.

[3] D. Zhao, T. Wang, Y. He, and X. Lu, *IEEE Trans. Semicond. Manuf.*, vol. 26, no. 4, pp. 556–563, Nov. 2013.

[4] H. Kim and H. Jeong, *J. Electron. Mater.*, vol. 33, no. 1, pp. 53–60, Jan. 2004.

[5] T. Feng, *IEEE Trans. Semicond. Manuf.*, vol. 20, no. 4, pp. 451–463, Nov. 2007.

[6] T. Kasai, *Tribol. Int.*, vol. 41, no. 2, pp. 111–118, Feb. 2008.

[7] K. Qin, B. Moudgil, and C.-W. Park, *Thin Solid Films*, vol. 446, no. 2, pp. 277–286, Jan. 2004.

[8] V. Rastegar, G. Ahmadi and S.V. Babu, *Separation and Purification Technology* (2016), doi: http://dx.doi.org/10.1016/j.seppur.2016.12.017.

CMP SLURRY METROLOGY TO MEET THE INDUSTRY DEMAND

Rashid Mavliev
Ipgrip Inc, Campbell, CA, USA
rmavliev@gmail.com

ABSTRACT

Slurry is one of most critical yield defining components of CMP process and potential source of problems. Timely and proper monitoring of slurry parameters is critical for CMP yield improvement. Slurry is very complex system – the combination of chemistry and nanoparticles with wide range of parameters. Metrology of slurry parameters could be done in 3 levels– supplier manufacturing/delivery site, Slurry Delivery Systems (SDS) in subfab and point-of-use (POU) - CMP tool, with very different time requirements, size and concentration limits on each level.

INTRODUCTION

CMP (Chemical Mechanical Planarization) has become a critical process in modern semiconductor device manufacturing. CMP provides a polishing process by combining mechanical abrasion with a chemical modification providing an ultra-flat, defect free surface. The CMP defects, significantly affecting device yield, can be caused by all process components (slurry, pad, pad conditioner, wafer handling and cleaning) but Large Particle Count (LPC) in CMP slurry has been documented to correlate well with defects[2] making the slurry quality one of the key parameters for yield improvement.

Typical LPC metrology methods use light scattering techniques that are designed to determine the number of particles larger than 0.56 um [3]. These techniques require substantial preliminary dilution of slurry because of high slurry turbidity and particle content. Slurry dilution results in distortion of particle size distribution due to pH shock and limits the detection of particles at low concentration. Controlling LPC in the CMP process is becoming even more important for new materials and applications. Commercially available slurries today may have very low LPC levels at production, but slurry transport, handling and process conditions can create LPC [4]. Thus, data taken from as-delivered slurries are as not reliable predictors of final defect counts; POU measurement and control are required.

The impact of parameters deviation is different on each level of slurry delivery (Figure 1) being highest in POU when slurry touches a wafer. This level also is most difficult from slurry metrology point of view – it will require integration with CMP tool and synchronization with start-stop polishing process. Typical time for slurry to reach a wafer after control point (valve, POU filter, flow control etc) is in range of seconds leaving very limited time to monitor slurry parameters and take an action before it can damage a wafer. Full flow monitoring in real time also desired for this level.

Next impact level of slurry metrology is SDS - typical time for slurry to reach a wafer is in range of minutes or less. Slurry affecting steps such as additional pumping, switch valves, POU filtration could be used inside of CMP tool so monitoring of slurry in SDS level does not guarantee the knowledge of slurry parameters when it touches a wafer. But slurry parameters deviation in SDS may affect many CMP systems in the loop. With slurry flow rate at loop of 10-30 l/min the total flow monitoring is practically impossible so

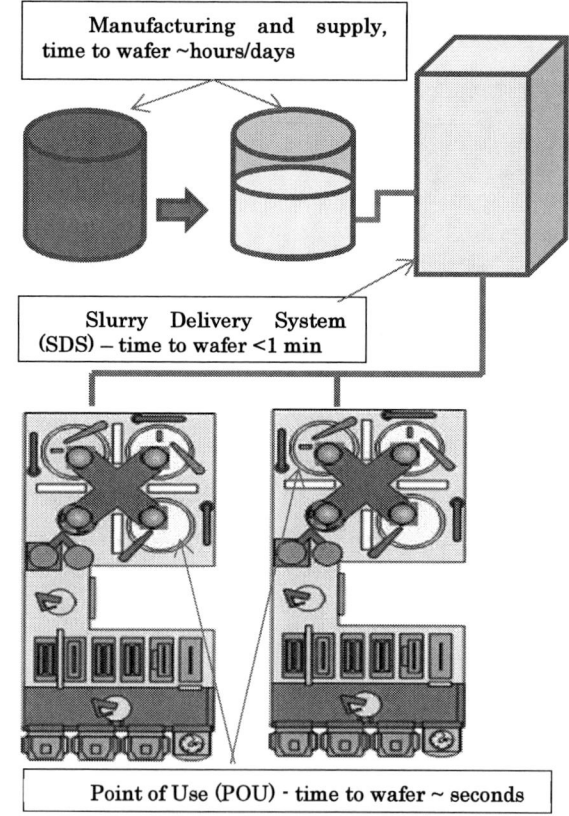

Figure 1. Slurry metrology application points.

slip-flow approach should be used. LPC monitoring is important at this level but more important is monitoring of main mode deviation which is responsible for CMP polishing performance.

Slurry Metrology at Point of Manufacturing

Metrology of slurry parameters at supplier manufacturing / delivery site is less critical timewise. The main issue for metrology at this stage is to obtain the size distribution for main mode and to predict the slurry stability and performance. For example, very important question is how well the size distribution of main mode of particles can be described and extended toward bigger particles. In another words, how LPC data correlates with main mode data of slurry particles.

Most of data on main mode parameters of size distribution are coming from electron microscopy, Dynamic Light Scattering (DLS) and from Scanning Mobility Particle Sizer (SMPS) [5].

All those techniques require sample dilution and different level sample alteration (for example drying, nebulizing etc.) prior to measurements which may result in different level of agglomeration and artificial LPC generation. It is probably one of reasons for limited data on big particles "tail" of main mode size distribution. This information would be very important to understand and reduce the large particle formation which are well reported cause of scratches [1-2]. As result, a methods requiring heavy slurry dilution may be misleading on defect generation performance of slurry.

Slurry Metrology at Point of Delivery

Slurry delivery systems circulate slurry starting from a day tank, through a pump, optional filters, then out to a fab global distribution loop, and then back to the day tank. Most slurry systems in the field will normally turn the slurry over between one hundred and a few hundred times before it is consumed for processing. It is critical to preserve the critical slurry properties during this recirculation period.

Several studies show that large agglomerates form in various slurry types as the number of slurry turnovers increase, good example of such study is [6]. Most of published research was done with Accusizer sensor (Particle Sizing Systems, Santa Barbara, CA) which requires additional dilution to handle high turbidity slurries. Very important results from [6] are presented in Figure 2. There is strong indication that slurry handling is affecting not only LPC responsible for scratch performance reported in [1-2], but also the shape of main mode which is responsible for polishing performance. This data indicates requirement the

measurement of particles preferably in range of 0.2 to 1 um and at concentrations up to 10^9 particles per ml.

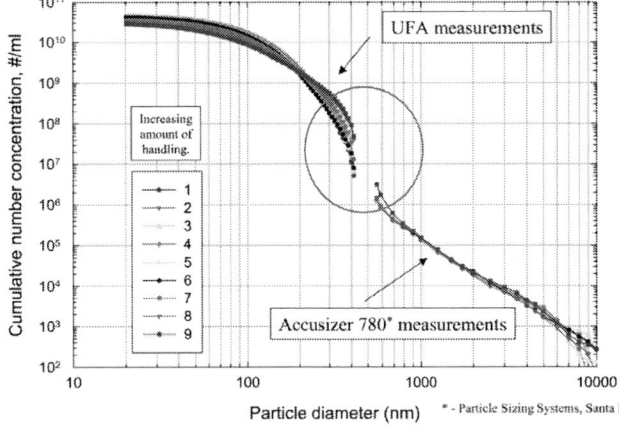

Figure 2. An example of slurry measurements from [6]. Red circle highlights the area of slurry parameters with potential process deviation and limited information.

Single Particle Optical Counter (SPOC) is the only method which satisfies to real-time and non-dilution requirements [7-8]. SPOC method is based detection of individual particles above detection limit by optical scattering means. Number of detected particles directly corresponds to particle concentration and each detected particle is sized based on intensity of scattered light. Described in [7-8] device Slurryscope (from Vanatge Technology Corp. San Jose, CA) only partially satisfies the requirements due to limitations on maximum number concentration of particles detectable without coincidence. Coincidence errors may substantially distort the measured size distribution data [9].

Extending SPOC parameters to high concentration without slurry dilution represent very difficult task. Figure 3 represents examples of slurry parameter measurements in range of 500 – 1000 nm with extended upper concentration range. Several measurements of the same slurry were done at different level of dilution (points). Dilution compensated data points overlapping indicates the absence of coincidence errors. The data indicates that upper concentration limit can be extended to and beyond 10^8 particles per ml. Two curves in figure 3 represent the estimated integral size distribution for ceria slurry with lognormal parameters from [6]. In that work the main mode particle distribution was measured by SMPS, the measured data points are best approximated by lognormal distribution with d_{50} of 92 nm and σ_g of 1.535. Green line represents the same distribution with σ_g of 1.6. This is only an exemplary approach because data do not belong to the same slurry, more detailed research is required.

978-1-5090-6695-7/17 $31.00 © 2017 IEEE

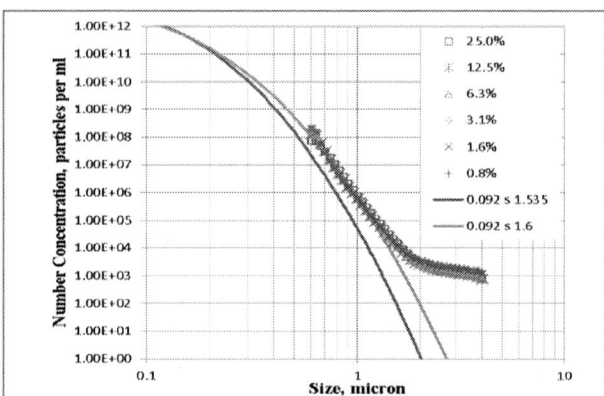

Figure 3. An example of the ceria slurry measurements at different level of dilution (points,). Lines represent estimated lognormal size distribution with shown parameters.

Slurry Metrology at Point of Use

Slurry, delivered to CMP tool, passes through point-of-use filtration and flow control system before reaching the platen and wafer. It is last point slurry parameters can be monitored and controlled. For most CMP systems the distance between available measurement point and slurry delivery to platen do not exceed 1-2 seconds. Thus, only direct measuring method can be used.

To determine the required concentration limits let's assume the CMP tool recipe uses slurry at 200 ml/min. For typical polishing time of 1 min the wafer would be exposed to ~20 ml of slurry assuming 10% of slurry goes under wafer [10]. In this case the detection limit should be better than 0.01 particles per ml to detect a single particle which may cause the "killer" defect. Low concentration ~0.01 particle/ml and requirement for real time monitoring at POU does not allow using the dilution of sample. For example, Accusizer requires dilution of 100 to 1000 times putting low concentration limit to 10^3 particles per ml [11].

Described in [7] device only partially satisfies for real time POU requirement because of slip flow arrangement at 15 ml/min. Very precise time alignment is required to synchronize the particles data between measurement and utilization points if single particle detection is required. Also slip-flow arrangement may result in Cost-of-Ownership (COO) issues if 10-20% of slurry flow is used only for analysis.

CONCLUSIONS

With each new technology node, the slurry parameters deviation impact on CMP becomes higher in terms of tool downtime, process yield and stability. Slurry metrology should be done on each step of slurry life starting from manufacturing to final consumption during wafer polishing.

It is extremely important to timely provide the necessary information on slurry parameters excursions for damage and process deviation preventive actions. Timing requirements for slurry metrology is different for each step of slurry handling being shortest for Point-of-Use. Also POU slurry metrology is practically impossible with slurry dilution. Slurry dilution is not desirable at other points of slurry monitoring because of particles agglomeration and distribution distortion.

Measuring of undiluted slurry in real time with Single Particle Optical Counter (SPOC) is possible on each level of slurry delivery and handling. For slurry monitoring at point of use a SPOC technique should be further advanced to operate at full flow rate of CMP slurry delivery line (up to 200 ml/min). Another critical advancement is required to improve upper detection limit to avoid coincidence effects.

Extending SPOC parameters to high concentration or high flow rate represent significant challenge. At the same time SPOC has advantages of being non-destructive, real-time non-invasive direct measurement technique which makes it best for slurry monitoring application.

REFERENCES

[1] Jae-Gon Choi,Y.N. Prasad, I.-K. Kim, I.-G. Kim, W.-J. Kim, A. Busnaina, Jin-Goo Park, *J. Electrochem. Soc.*, 157, H186-191, (2010)

[2] E. Remsen, S. Anjur, D. Boldridge, M. Camiti, S. Li, T. Johns, C. Dowell, J. Kasthurirangan, P. Feeney, *J. Electrochem. Soc.*, 153, G453-461, (2006)

[3] D. F. Nicoli, P. O'Hagan, G. Pokrajac, K. Hasapidis, *Am. Lab.* 2000, 32, 18–22.

[4] B. Johl, *29th European CMP Users Symposium*, 2013, Zurich, Switzerland

[5] H. Gary Van Schooneveld, M. Litchy, D. C. Grant, *NSTI-Nanotech 2011*, Vol 1. 2011

[6] D. C. Grant *CTA Pub 74 CT Associates, Inc. Presented at the CMP Users Conference*, 2008

[7] R. Mavliev, M. Parkin, and G. Moloney, *Proceedings of International Conference on Planarization/CMP Technology ICPT 2011*, pp 169-172, 2011

[8] Kim, A. Murai, K. ; Parkin, M. ; Mavliev, R. , *Advanced Semiconductor Manufacturing Conference (ASMC)*, Pp: 227 – 230, 2012

[9] B.Tolla, D. Boldridge, *Part. Part. Syst. Charact.* 27 (2010), 21-31

[10] A. Philipossian, E. Mitchell, *J. Appl. Phys.*, 42, 7259, 2003.

[11] M. Bumiller Advances in LPC Monitoring in CMP Slurries, *Semicon West Symposium on CMP Technology and Market Dynamics*, 2016

978-1-5090-6695-7/17 $31.00 © 2017 IEEE

CHALLENGES IN CHEMICAL MECHANICAL PLANARIZATION DEFECTS OF 7NM DEVICE AND ITS IMPROVEMENT OPPORTUNITIES

Ji Chul Yang[1]*, Dinesh Penigalapati[1], Tai Fong Chao[1], Wen Yin Lu[1] and Dinesh Koli[1]

[1]CMP Team, Advanced Technology Development (ATD), GLOBALFOUNDRIES, US 12020
*Corresponding Author's Email: jichul.yang@globalfoundries.com

ABSTRACT

CMP (Chemical Mechanical Planarization) defects are always one of the top yield detractors in IC (Integrated Circuit) devices since CMP processes have been applied in the semiconductor industry. Most of all, new structures and materials in 7nm devices make it challenging for CMP processes to meet device requirements. The CMP process obviously needs to control or contain not only the number of defects but also defect size in accordance with scaling speed. In this paper, the results of fundamental studies to elucidate CMP defects will be introduced and discussed as they pertain to 7nm devices. This paper will cover the phenomena and its research activities about atomic scale scratches, dishing control in uneven surface topography and surface defects with 7 nm logic device.

INTRODUCTION

Every new generation of technology has more aggressive scaling targets than its predecessor; 7nm technology features have highly scaled critical dimensions in all integrated modules. The CMP (Chemical Mechanical Planarization) process must modify its process capability to respond to the shrinking device scale without exception. Most of all, CMP process capabilities are determined by the defect significance of dishing/erosion, micro-scratches and surface particles with respect to both number and size of each defect. When looked into by detailed specifications, defect size must be controlled under 14 nm in FEOL (Front-End of Line) and 36 nm in BEOL (Back-End of Line). Also, dishing is controlled under 4nm, based on ITRS (International Technology Roadmap) for Semiconductors 2.0 2015 Edition.[1]

There have been lots of research and engineering activities to elucidate the root-cause of CMP defects and to improve CMP process defects since this so-called particle-and-scratch generating process was introduced in the fabrication process over 20 years ago. These yielded countless beneficial effects on figuring out issues and problems in a rapidly changing semiconductor industry. Unfortunately however, research results could not cover all sides of current processes in addition to issues occurring in the latest devices. In fact, it is critical point to overcome these barriers with efficient communication between industrials and academia. Therefore, we will introduce three defect modes found in the 7nm CMP process and discuss their research results in this paper: 1) Atomic scale scratches, 2) Dishing phenomena with uneven surface topography, 3) Surface flake defect. The suggested mechanism in this paper will give insight into defect control for sub 7nm device CMP manufacturing in the future.

ATOMIC SCALE SCRATCHES

The CMP process inevitably creates micro scratches since mechanical polishing is done using hard abrasives and physical contact. Fundamentally, the stick-slip friction mechanism and impact of contact surface hardening were investigated to find the root cause of polishing scratches.[2,3] Most papers studied the effect of large slurry particle and proposed mono-size abrasives in the slurry to resolve scratch issues.[4,5]

As the device pitch scale moved under 10nm, nano-roughness on the post-polished wafers became a concern. Fig.1 shows typical images of post-polished wafers in STI (Shallow Trench Isolation) CMP. AFM (Atomic Force Microscopy) clearly showed line-types of scratches which have depth range around 0.5nm to 1nm.

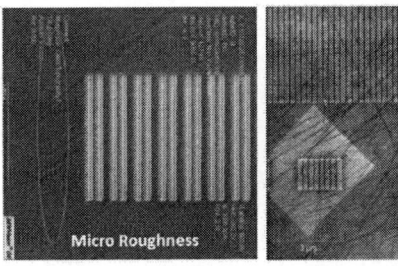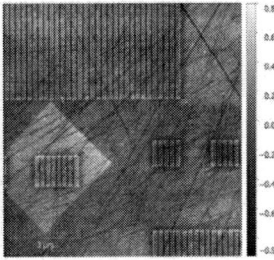

(a) Top View Image   (b) AFM Scanning Image
Figure 1: A post-polished wafer's surface images

Around 1 nm depth of indentation on the post-CMP wafers did not initiate problems in larger scale devices because the material removal of post-process treatments like wet or dry etching could compensate nanoscales of topography. However, as 7 nm structures are limited to specific etching amounts from the top polished surface, CMP-created topography was transferred to the final structure profile and induced device problems.

In this study, we examined how to minimize surface nano-topography for post STI CMP by employing the novel type of slurry particles. Fig.2 shows surface roughness values of wafers processed by three different

978-1-5090-6695-7/17 $31.00 © 2017 IEEE

kinds of slurry abrasives: (a) calcined ceria abrasives (3nm), (b) sol-gel type ceria abrasives (130nm), (c) colloidal silica abrasives which are coated with thin ceria film (120nm), (d) calcined-type ceria (POR slurry, 110nm). Each different slurry had a removal rate of around 200nm/min. Processed wafers were scanned by 2um x 2um area using AFM. All three abrasives resulted in a lower roughness value than the POR value. Above all, the smallest particles showed the lowest numbers. They have the lowest indentation depth during polishing. Interestingly, the other two types of abrasives also improved surface finish under 0.5nm level. There was no limitation to utilize any of these slurries in the device. These new abrasive types will be good candidates for future CMP processes.

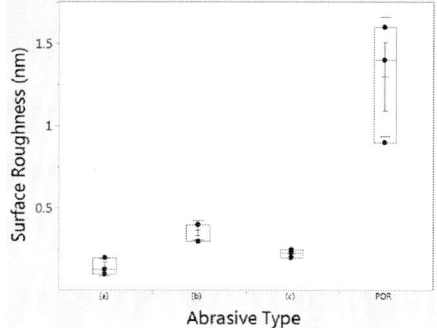

Figure 2: Surface roughness with respect to abrasive type

DISHING PHENOMENA WITH UNEVEN SURFACE TOPOGRAPHY

G. Fu and A. Chandra studied analytical modeling about CMP dishing and step height reduction phenomena and proved this modeling with experimental data.[6] J.C. Yang et al. investigated abrasive size impact for dishing improvement.[7] However, the actual dishing signal of production wafers is quite different from the previous experimental results due to non-uniform patterning density in micro-scale. The density distribution of current logic devices ranges from 10 % to 80 % with 25um pitch scale. Fig.3 is one example of unexpected dishing after polishing a wafer with higher selectivity slurry. As shown in Fig.4, two protrusion points (a,b) before CMP process induced high dishing after polishing. The wafer could not be inspected in a uniform patterned density area but was examined in the transition area from higher density to lower density. This phenomenon is mostly observed in polishing steps using high selectivity slurry and a soft pad. This signal could be elucidated with three dimensional contact problems of material's wedge.[8] As soon as the protruded mask area is exposed to the pad and slurry, the top surface will experience higher contact stress. If it is surrounded by a uniform structure, mechanical stress will

be distributed evenly with the neighboring structure. However, the protruded structures will keep polishing due to higher contact pressure and will generate higher temperature and friction followed by more polishing than in the uniformed area. It is an inevitable situation that is a consequence of the process of removing material by mechanical friction and wear.

To improve dishing, a non-selective buffing process was implemented accompanied by an increase in pre-deposition thickness. The focus was to determine if dishing amount generated by a high selectivity process was diminished with non-selective CMP. Interestingly, the dishing could be reduced to a minimum value but the ideal flat surface could not be created. It is assumed that planarization length is the main contributor for remaining step height.

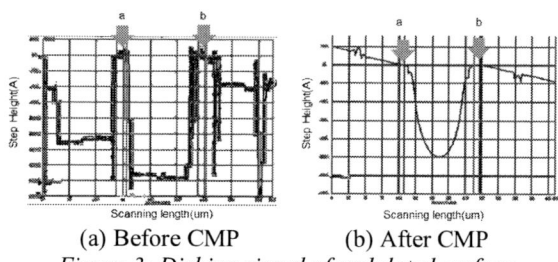

(a) Before CMP   (b) After CMP

Figure 3: Dishing signal of undulated surface

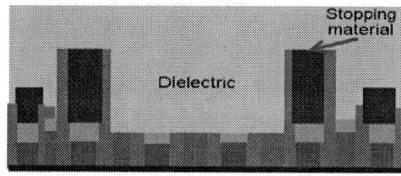

(a) Incoming structure

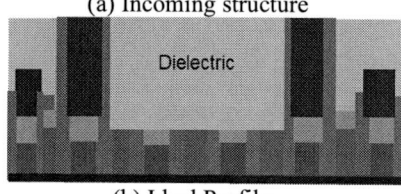

(b) Ideal Profile

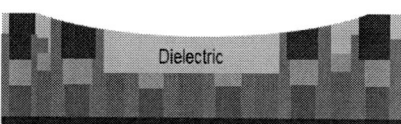

(c) Actual profile

Figure 4: Dishing signal of undulated surface

SURFACE FLAKE DEFECT

Surface defects such as organic residue, ring particles and flake particles are the main yield detractors. Among them, surface flakes in post Cu CMP appeared to be a more serious issue in recent devices.[9-11] Fig.5 shows typical Cu flake surface images and TEM (Transmission electron microscopy) results. In this study, the impact of in-situ cleaner modules was examined. CMP tools' cleaner modules can include a megasonic bath, jet spray, pencil-type brush, horizontal-type brush or vertical type brush, as shown in Fig.6. All polishing experiments were performed on a 300 mm polisher (LKPS, AMAT) with Hitachi Cu Barrier Slurry (HS-915TS). Defect density changes were calculated by two steps: 1) scanned with KLA-Tencor 2915 Bright Field Inspection Tool, 2) reviewed by Applied-Materials SEM-Vision with 30 nm horizontal resolution. In results, vertical type brush had the highest values of flake defects and followed by horizontal brush, pencil brush and megasonic bath. It is assumed that a higher contact-driven cleaning process and lack of cleaning chemical resilient time could be the main contributors to higher particle levels.

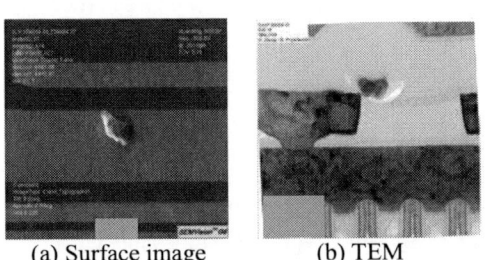

| (a) Surface image | (b) TEM |

Figure 5: Cu surface flake particle

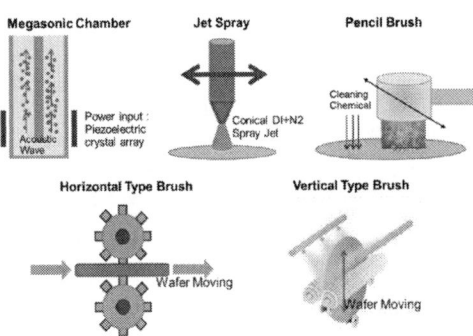

Figure 6: CMP in-situ cleaner modules

When it comes to the efficiency of particle reduction in post-CMP cleaning, many papers and presentations insist that the brush is the most influential cleaning method to obtain high particle removal. On the contrary, based on this study the brush is the worst contributor of surface flakes. It will take time to consider alternative methods to replace brushes.

SUMMARY

New defect modes in new devices have become more complicated than before. This study discussed three different issues in the 7nm CMP process. 1) Atomic scale scratches: Three different abrasive particles were examined to obtain better surface topography. Smaller particle abrasive was the most promising candidate but sol-gel and composite abrasive also showed comparable performance. 2) Dishing phenomena in uneven surface: This signal was explained by the three-dimensional contact problem. Buffing CMP was minimized to reduce the dishing within planarization length. 3) Surface Flake Defect: Cleaner module characteristics were studied to find the mechanism of small surface flake defects in Cu CMP. It was found that the brush module was the dominant factor in creating surface flakes. These research results would like to give more hints toward the right directions for future CMP researchers and developers.

REFERENCES

[1] *http://www.semiconductors.org/main/2015_itrs_itwgs*

[2] H. J. Kim and et. al., *Journal of Nanoscience and Nanotechnology*, Volume 12, Number 7, 2012, pp. 5683-5686(4)

[3] J. C. Yang and et. al., *Journal of Electronic Materials*, Volume 39, Issue 3, pp 338-346

[4] E.E. Remsen and et. al., *Journal of the Electrochemical Society*, Volume 153, Issue 3, pp. G453-G461

[5] J. C. Yang and et. al., *Journal of the Electrochemical Society*, Volume 157, Issue 3, pp. H235-H240

[6] G. Fu and A. Chandra, *IEEE TRANSACTIONS ON SEMICONDUCTOR MANUFACTURING, VOL. 16, NO. 3, 2003*, pp. 477-485.

[7] J. C. Yang and et. al., *Wear*, Volume 268, Issues 3–4, 2010, pp. 505–510.

[8] A.B. Chaudhary and K. Bathe, Computers & Structures, Volume 24, Issue 6, 1986, pp. 855-873.

[9] A. Skumanich and et. al., *SPIE Proceedings*, Volume 3884, 1999, pp. 278-289R.

[10] G. Garteiz and A, Zylberman, *Proceedings of the 32th International Symposium for Testing and Failure Analysis,* 2006, pp. 273-275.

[11] J. C. Yang and et. al., *(Invited Talk) 229th ECS meeting*, 2016, D2-1035.

978-1-5090-6695-7/17 $31.00 © 2017 IEEE

SiOC CMP DEVELOPED AND IMPLEMENTED IN 7NM AND BEYOND

Haigou Huang*, Taifong Chao, Ja-Hyung Han, Dinesh koli, Qiang Fang
Advanced Technology Development, GLOBALFOUNDRIES, 400 Stone Break Drive, Malta, NY,
USA 12020
*Corresponding Author's E-mail: huanghg@globalfoundries.com

ABSTRACT

In this study, new SiOC Chemical Mechanical Planarization (CMP) process is fully developed with the characterization of the blanket wafer selectivity, SiN loss on pattern wafer, within chip SiN uniformity, and topography of CMP house and device areas using Atomic-force microscopy (AFM), Transmission electron microscopy (TEM), high resolution profiler (HRP) and KLA- Aleris. Those results of SiN within-chip uniformity show one step process (only slurry A_bulk + SiN stop) with poor process window, which cannot meet 7nm MOL integration process requirement. And two steps process (Slurry A_bulk + Slurry B_SiN stop) with promising results, good SiN within-chip uniformity (< 2nm) and wide process overpolish margin.

INTRODUCTION

To scale below 10nm node and beyond, new technology elements include FINFETs on bulk or SOI [1], replacement gate process, multi-work function gate stacks, self-aligned contacts and replacement metal contacts [2-3]. Especially in the MOL, with gate pitch scaling, direct TS RIE becomes more challenging due to the etch process selectivity limitations and high aspect ratio requirement.

The replacement metal contact scheme will be benefit to contact resistance reduction by minimizing the spacer thickness requirement and the gauging into Epi to maximum contacting area. The simple MOL flow shows in Figure 1. Wet dHF strips the trench oxide and gate is protected by cap SiN. Then, it fills with poly dummy and CMP planarization to stop on SiN. After the Non-TS location patterning, it will be filled with SiOC and planarized by CMP. In the BEOL, Low-K carbon doped SiO2 (SiOC) as ILD materials have been implemented to reduce RC delay and power consumption in the BEOL because of its low-k value, high mechanical strength, low cost of ownership and compatibility with existing integration schemes [4]. SiOC can be regarded as dielectric material in MOL due to its properties meeting those requirements: good gap fill capability (aspect ratio > 9:1), low temperature deposition or treatment (< 400C), low K value (< 3.9, no worse than oxide), good wet or dry etch resistance during dummy contact pull out (>100:1) and planarization.

Thus, SiOC CMP development is the key element for the replacement metal contact schemes. But, the development of SiOC CMP process has great challenges since this process needs to meet many requirements: good within-chip uniformity (<5nm for all the structures), good dishing control (<100A on CMP house 50X70um), less SiN loss (<5nm), wide process margin and less defectivity performance. In this study, new SiOC Chemical Mechanical Planarization (CMP) process is fully explored with the characterization of the blanket wafer selectivity, SiN loss on pattern wafer, within-chip SiN uniformity, topography evaluation of CMP house and SiOC film shrinkage using Atomic-force microscopy (AFM), Transmission electron microscopy (TEM), and KLA- Aleris. Those results of SiN within-chip uniformity show one step process (only slurry A_bulk + SiN stop) with poor process window, which cannot meet 7nm MOL integration process requirement. And two steps process (Slurry A_bulk + Slurry B_SiN stop) can deliver promising results, good SiN with-in-die uniformity (< 2nm) and wide process overpolish margin.

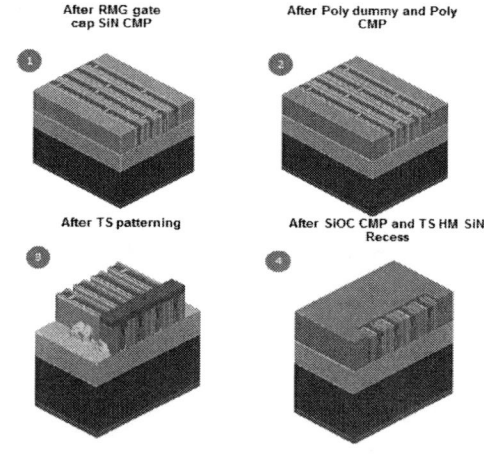

Figure 1 Simple flow of SiOC CMP in MOL.

EXPERIMENTAL

The SiOC films (5000Å) on blanket and pattern wafers are prepared by a commercial flowable CVD system (Applied Materials Inc.). Reflection LK polisher is used for the polishing experiments. Two kinds of slurry (Slurry A and B) were selected for test in this experiments. Slurry A is a novel developed and dedicated for SiOC polish, while slurry B is a SiN stop slurry. The film thickness is characterized by using Aleris (KLA) and transmission electron microscopy (TEM). And the topography data were collected by high resolution profiler (HRP) and AFM. Two kinds of short loop (SL) pattern wafers were used in this experiment, one is from gate module SL with 85nm SiN hardmask and other is from FIN module SL with 40nm SiN.

RESULTS AND DISCUSSION

Figure 2 shows the removal rate of slurry A and B on blanket SiOC, TEOS, SiN and Poly film wafers. Those data show the high selectivity of slurry A on SiOC and SiN / Poly (>40:1), and low selectivity to the normal oxide (TEOS), around 2.5:1. For slurry B, the high selectivity on TEOS to SiN /Poly (>100:1) is observed and SiOC to SiN / Poly is about 25:1. All those data are based on the blanket wafers, which has no pattern density effect. Thus, it is possible different selectivity behavior between blanket and pattern wafers.

Figure 3 shows HRP topography on CMP house macro (including 22 solid pads with 70X50um) with different polish time (0s, 30s, 60s and 90s) using slurry A. Those data are collected just on the same wafer. It is obviously that the topography is significantly plananrized at first 30s polish and almost flat after 90s. Figure 4 shows step height evolution between trench area and active area evolution using slurry A only. The incoming step height is around 215nm without polish and the step height is closed to 0nm after 90s polished. This results show the surface is full planarization around 90s.

Table1 shows the TEM summary of SiN remaining on different macros using slurry A only with 100s, 110s and 120s splits. In this experiment, pattern wafers is gate module short loop with 85nm hardmask SiN and TEM

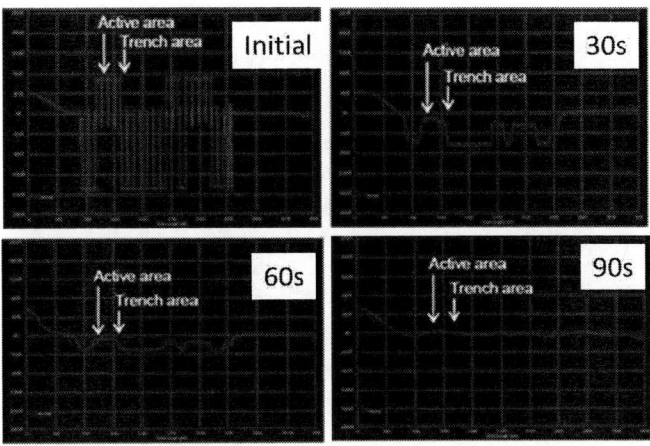

Figure 3. HRP Topography on CMP house macro with different polish time using slurry A.

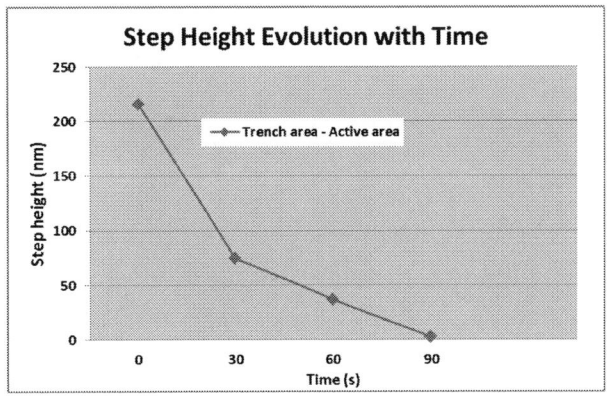

Figure 4. Step Height between trench area and active area evolution using slurry A only.

cut on CMP house , PX (short gate length) and PB (long gate length) location. These wafers are polished with Motor- Toque endpoint control. There is almost no SiN loss on CMP house and slight loss on the other structures in 100s split. However, there is huge SiN loss (50nm) on PB location in 120s split. Figure 5 shows within chip SiN uniformity comparison on time splits using slurry A only. 110s split demonstrates excellent with-in-chip uniformity, only 2.1nm range. But just adding 10s overpolish, PB location SiN thickness is significantly reduced, only 36.5nm remaining, comparing to CMP house (67.4nm). These results indicate that there is much stronger macro-loading dependent, special on PB location. This strong macro-loading is possible related to pattern density. Thus, one step process – slurry A only can be obtained

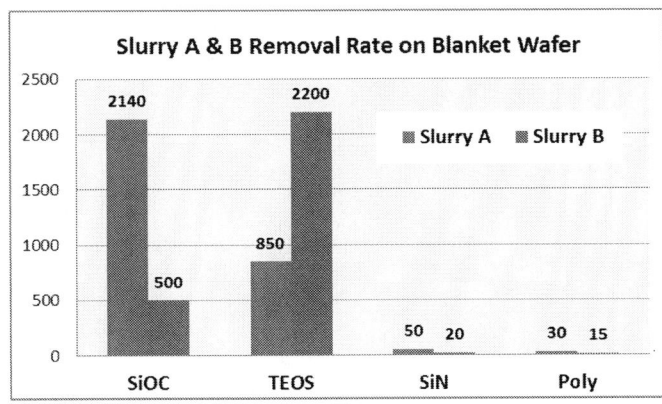

Figure 2 Blanket wafer removal rate (Å/min) of slurry A and B.

good with-in-chip uniformity wafer just at target point, but with narrow process window (even less 10s). It is difficult for process control due to the narrow process window.

Table1: The TEM summary of SiN remaining on different macros using slurry A only with time splits.

Splits	CMP House	PX	PB	WIC Range (nm)
100s	85	74.1	81.1	10.9
110s	71.3	69.3	71.4	2.1
120s	67.4	59.5	36.5	30.9

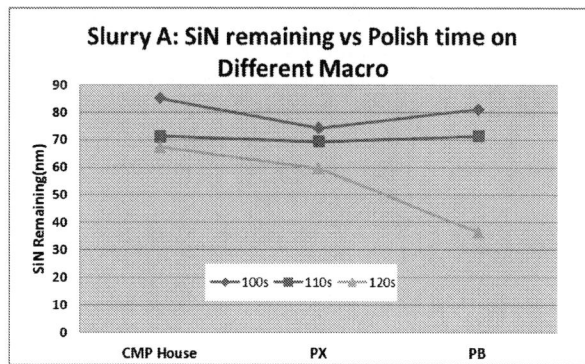

Figure 5.WIC SiN uniformity comparison on time splits using slurry A only.

In order to improve the process window, low down force with slurry A only is tested again; however, the results do not show any improvement. On the other hand, two steps process- slurry A_Bulk + slurry B_stop on SiN was evaluated. Table2 shows the TEM summary of SiN thickness remaining on different macro with two (slurry A_Bulk + slurry B_stop on SiN) and one (slurry A only) steps process. FIN module short loop pattern with 40nm SiN hardmask is used in this experiment. In the two steps process, all splits demonstrate excellent within-chip (SRAM, logic and CMP house) uniformity of SiN thickness, even with 40s overpolish, WIC uniformity doesn't degrade. These results indicate two steps process with super wide process window and excellent WIC uniformity. One step process –slurry A only is also repeated for comparison in Table 2. The conclusion is the same as above, with narrow process window.

The topography is also studied by HRP on CMP house trench pad and AFM on device area (SRAM and Logic). Figure 6 showed dishing performance in CMP house trench pad (50X70um) from both two steps and one step process with on target and 10s overpolish splits. All on-target processes from both one & two step processes show a reasonable dishing performance (less 100Å), which meet the process requirement spec

(<150Å). However, 10s overpolish data show the dishing increasing much for both one and two step processes. Figure7 showed AFM images and profile on both SRAM and logic area with on target pattern wafer.

Table2: The summary of TEM SiN remaining thickness on different macro with two (slurry A_ Bulk + slurry B_stop on SiN) and one (slurry A only) steps process.

Process	Splits	SRAM	Logic	CMP House	WIC SiN Range(nm)
Slurry A_Bulk + Slurry B_stop on SiN	On target	35.5	36	34.5	1.5
	40s OP with slurry B	24	22.5	25	2.5
Slurry A only	On Target	39	36.9	35	4
	10s OP with slurry A	38.4	31.3	31.5	6.9

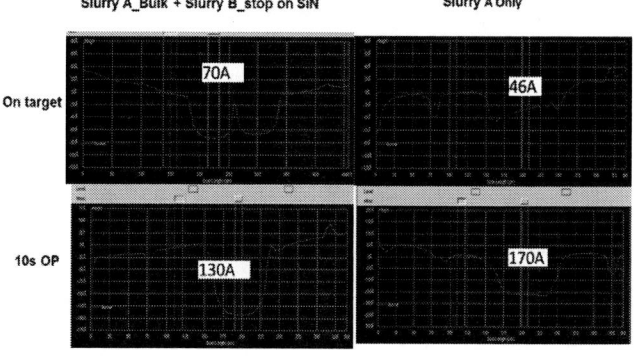

Figure 6.Dishing performance in CMP house trench pad (50X70um) both two steps and one step process with on target and 10s overpolish splits.

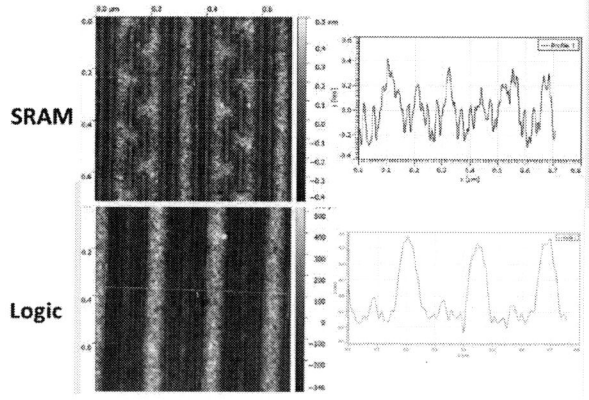

Figure 7 AFM images and profile on both SRAM and logic area with on target pattern wafer.

The perfect topography (less 10Å) is observed both on SRAM and logic area. Thus, both HRP and AFM results indicate the dishing performance of two processes meet the integration requirements.

CONCLUSION

In this study, new SiOC chemical mechanical planarization (CMP) process is fully developed with the characterizations of the blanket wafer selectivity, SiN loss on pattern wafer, within-chip SiN uniformity, topography of CMP house and using AFM, HRP, TEM and KLA- Aleris. The results of SiN within-chip uniformity and dishing performance show one step process (Slurry A only_bulk + SiN stop) with poor process window, which cannot meet 7nm MOL integration process requirement. However, two steps process (Slurry A_bulk + Slurry B_SiN stop) demonstrates the promising results, good SiN within-chip uniformity (< 2nm), wide process overpolish margin and reasonable topography performance. Thus, the SiOC CMP baseline is established and further process improvement will be continued.

ACKNOWLEDGEMENT

Thanks for helpful discussions from many colleagues in GLOBALFOUNDRIES.

REFERENCE

[1] K. Ahmed and K. Schuegraf, "Transistor wars: Rival architectures face off in a bid to keep Moore's Law alive", IEEE Spectrum, Nov. 2011, p.50.
[2] K. Seo, B. Haran, D. Gupta, D. Guo, T. Standaert, R. Xie, H. Shang, E. Alptekin, D. Bae, G. Bae, "A 10 nm platform technology for low power and high performance application featuring FINFET devices with multi workfunction gate stack on bulk and SOI, Dig. Tech. Papers VLSIT (2014) 1–2.
[3] C. Auth, *et.al*. "A 22nm High Performance and Low-Power CMOS Technology Featuring Fully-Depleted Tri-Gate Transistors, SAC and High Density MIM Capacitors",VLSI Technology, 2012, p152.
[4] S. W. King, "Dielectric barrier, etch stop, and metal capping materials for state of the art and beyond metal interconnects", ECS J. Solid State Sci. Technol.4, 2015, p3029.

Corresponding Author:
Haigou Huang
Tel: +1 518-305-6035
Fax: +1 518-305-6178
E-mail: huanghg@globalfoundries.com
Advanced Module Engineering, Globalfoundries
400 Stone Break Rd. Ext. Malta, NY 12020 USA

IMPACT OF WAFER TRANSFER PROCESS ON STI CMP SCRATCHES

Fan Bai, Zhijie Zhang, Jia Wang, Hongdi Wang
Semiconductor Manufacturing International Corporation, Beijing
Corresponding Author's Email: Jason_ZZ@smics.com

ABSTRACT

Shallow trench isolation chemical mechanical polishing (STICMP) technology has been widely applied in the fabrication of ultra large scale integrated (ULSI). In STI-CMP, the defect, topography control, thickness uniformity and so on are all so critical, especially, scratch defect is the major problem. Pad, disk, agglomerated slurry particles and foreign particles are the main sources of the tiny scratch. In this article, impact of transfer process on scratch during STI-CMP, such as pre CMP, bulk polish post treatment, and pre selective polish was studied. Variable down force, DIW rinse time, slurry flow rate, slurry buff treatment were verified respectively. It was found that the pre CMP slurry buff can reduce the scratch by 55%, and bulk polish post step with optimized buff condition also can reduce scratch by 30%. Besides, the backside clean also can reduce the scratch significantly.

INTRODUCTION

With the development of modern integrated circuit (IC) technology, the increasing number of active components has led to a significant decrease in feature dimensions. The nanoscale dimensions of multilevel-interconnection has put forward a higher demand for the planarization[1]. Shallow trench isolation chemical mechanical polishing (STICMP) technology has been widely applied in the planarization of IC chips. This technology possesses an excellent performance in removing unwanted topography and obtaining a flat isolation structure[2]. By using appropriate selectivity abrasive slurry and polish conditions, STICMP can effective stop at a hard mask layer, nitride, for guaranteeing the uniformity of post layers. And it makes STICMP a crucial process to enhance packing density and high degree of planarity[3].

However, in STICMP, besides the topography control and thickness uniformity, the defect, especially, micro-scratch is always a major problem to limit its development[4]. The microscratch defect on the wafer surface can induce the leakage current and short between gates[5]. And how to reduce the microscratch defect is always a hot spot of STICMP research. Actually, Pad, disk, agglomerated slurry particles and foreign particles are the main sources of the tiny scratch.

In this article, transferring process during STI-CMP about pre CMP, bulk polish post treatment, pre selective polish and backside clean was focused on and studied. Variable down force, DIW rinse time, slurry flow rate, slurry buff treatment were respectively verified. It was found that the pre CMP slurry buff can reduce the scratch by 55%, and bulk polish post step with optimized buff condition also can reduce scratch by 30%. Besides, the backside clean also can reduce the scratch significantly.

EXPERIMENT

The pattern wafers were polished at a polishing tool by using a three-step polishing approach (bulk polish, selective polish and DIW buff). Post CMP clean was done on an auto-cleaner of tool. The microscratch defect was measured by a defect scanning tool. And the defect review was performed by using a scanning electron microscopy (SEM).

RESULTS AND DISCUSSION
Pre Treatment Before Main-polish

As we all know, the particles, such as foreign particle and aggregated slurry particle, are the major source to generate scratch defect. In the CMP process, these particles are easily to adhere on the surface of wafers before main polishing process. Once the particle is brought into the polish process, under the action of down force, it can generate the microscale scratch defect. In order to eliminate the effect of these type particles, a pre CMP treatment can be added before main polish, as

978-1-5090-6695-7/17 $31.00 © 2017 IEEE

shown in Figure 1a. On the one hand, in the platen 1 of STICMP, bulk polish, the particles from pre-layer STI-DEP exit on the wafer surface. Then a buff treatment with low pressure and low slurry flow rate can remove the foreign particles to avoid impacting the wafer surface. On the other hand, in the platen 2 of STICMP, selective polish, the particles on the wafer mainly come from the residual slurry particles of platen 1. And the pre buff treatment also can remove the residual particles (as shown in Figure1b and 1c). The defect review results show that the introduction of pre CMP treatment can reduce the microscratch average count about 55% decreasing amplitude (as seen in Figure 2).

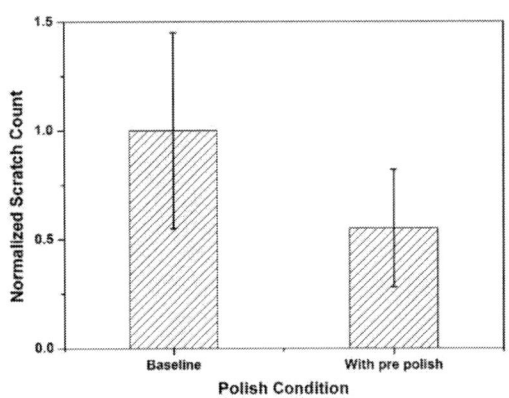

Figure 2. Average scratch count under the condition with pre CMP treatment

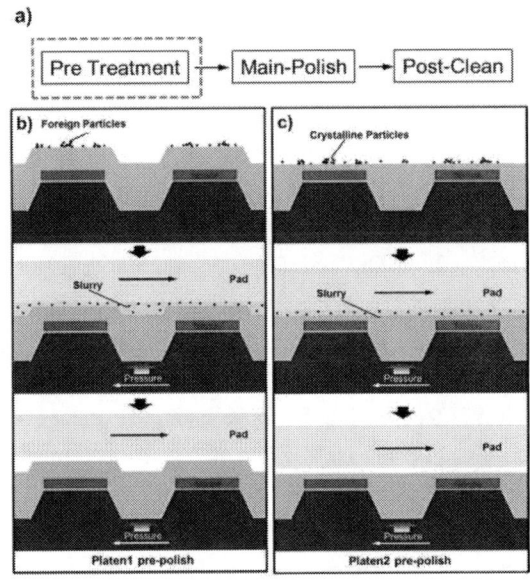

Figure 1. Schematic diagram of the pre CMP treatment before main polish. a) flow chart of pre buff treatment process; b) and c) the pre buff treatment in platen 1 and platen 2 of STICMP.

Post Treatment After Platen 1 Main-polish

In STICMP process, the SiO_2 slurry (pH~12) and CeO_2 slurry (pH~5) are applied into the platen 1 and platen 2, respectively. At the end of P1 process, a deionized water buff is always used to clean the pad, which can induce the sharp down of slurry pH from 12 to 7. And the change of pH can turn the zeta potential of SIN from negative to positive. Then the slurry particles with negative charge can easily adhere to the SIN part (as shown in Figure 3). Under the next polish process, the adhered particles can induce the microscratch on the wafer surface. Thus, in order to eliminate the impact of these adhered particles, a post slurry buff treatment can be added after the platen 1 main polish. This treatment can improve the pH of slurry back to the alkaline condition, which can prevent the adsorption of slurry particles. And the mechanical actions also can remove the surface particles.

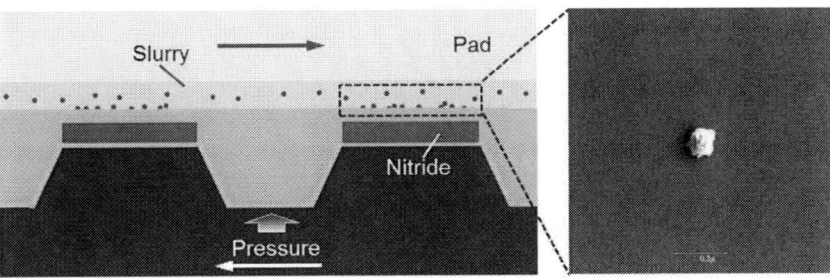

Figure 3. The falling down of slurry pH can make the particles with negative charge adhere to the surface of SIN part. Then, during the post CMP buff treatment, the adhered particles can be removed.

978-1-5090-6695-7/17 $31.00 © 2017 IEEE   291

Post Treatment Conditions Optimizing

In order to verify the optimum conditions of post CMP treatment, different wafer down force, slurry flow rate and time of DIW buff were investigated. First, the down force of wafer is one of main factor to generate the microscratch defect. Under the action of wafer down force, the oxide can be removed and the scratch defect also can be formed on the surface. As shown in Figure 4a, with the increasing of wafer down force, the scratch count under 2.0 psi condition is the least. It can be explained that the low down force is not enough to remove the adhered particles on the wafer surface. And the high down force can improve the risk of generating microscratch defect. The appropriate down force is important for the process of post treatment. Second, the slurry flow rate can also impact the condition. As shown in Figure 4b, with the increasing of slurry flow rate, the normalized scratch is reduced by about 30%. It is because that the slurry with high flow rate can bring out the polished particles easily. Third, the variable DIW buff time is also investigated. The results indicate that the long DIW buff time possessed the least scratch count. The reason is that more water can remove out the polished particles, so that the residual particles can not impact the surface of wafer. Thus, by applying the above optimum conditions into the post treatment, the normalized microscratch can be reduced by about 30% decreasing amplitude. The post treatment can decrease the scratch defect effectively.

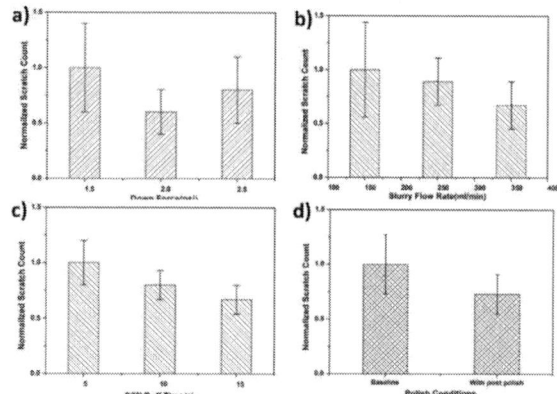

Figure 4. Correlation between different post treatment conditions and normalized scratch

Backside Clean Treatment

Besides the above pre and post CMP treatment, we found that the wafer backside also possesses a lot of foreign particles (as shown in Figure 5). These particles from pre-layers may have a large risk on the formation of microscratch. During the polishing process, the wafer is in high speed rotation. And the backside particles can be spinned out on the polish pad, resulting to generate scratch defect. In order to eliminate the impact of backside particles, the wafer must be cleaned carefully before going into the STICMP process. As shown in Figure 5, after cleaning the wafer again, the average total defect count is reduced by about 60%, and the normalized scratch is reduced by about 40%. The result indicates that the wafer backside clean treatment can reduce the total defect count and scratch defect significantly.

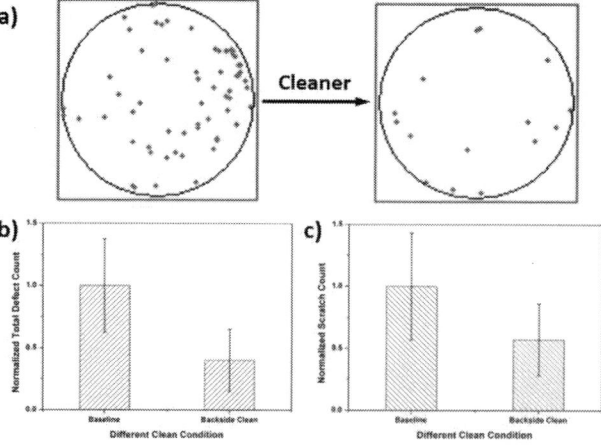

Figure 5. Backside clean of wafer. a) Total defect maps before and after backside clean treatment; b) and c) Total defect count and scratch count before and after backside clean.

CONCLUSION

In this article, some novel treatments, such as pre, post buff treatment and backside clean, have been designed to reduce the microscratch defect. Different down force, slurry flow rate and DIW buff time conditions also have been investigated to optimize the process. The results indicate that the the pre CMP slurry buff can reduce the scratch by 55%, and bulk polish post step with optimized buff condition also can reduce scratch by 30%. Besides, the backside clean also can reduce the scratch by 40%. These all treatments provide a novel idea to improve the STICMP defect performance.

REFERENCES

[1] T. Y. Kwon, M. Ramachandran, J. G. Park. "Scratch Formation and its Mechanism in Chemical Mechanical Planarization (CMP)", Friction, 1(4) 279-305 (2013)

[2] P. Song, D. Yao, J. D. Sun. "STI CMP: Exploration of a Colloidal Silla Based Slurry System", ECS Transactions, 34(1) 113-117 (2011)

[3] B. Reinhold, J. J. Gagliardi, S. Endle, "Defectivity Improvement for Fixed Abrasive Based STI CMP in Advanced Logic Technology", International Conference on Planarization/CMP Technology, October 25-27 2007 Dresden

[4] Y. H. Kim, S. K. Kim, J. G. Park, U. Paik, "Increase in the Adsorption Density of Anionic Molecules on Ceria for Defect-Free STI CMP", Journal of The Electrochemical Society, 157(1) H72-H77 (2010)

[5] L. L. Hwee, S. Balakumar, S. Mahadevan, Z. M. Sheng, A. See, M. Rahman, A. Senthilkumar, "Dishing and Nitride Erosion of STI-CMP for Different Integration Schemes", Journal of Electronic Materials, 30(12) 2001.

RESEARCH AND SOLUTION OF STI CMP DISHING AND UNIFORMITY IMPROVE FOR 28LP

Lei Zhang*, Junhua Yan, Kun Chen, Wenbin Fan, Yefang Zhu, Jingxun Fang

Shanghai Huali Microelectronics Corporation

NO.568, Gaosi Rd. Zhangjiang Hi-Tech Park Pudong Shanghai, 201203, China

*Corresponding Author's Email: zhanglei_td2@hlmc.cn

ABSTRACT

CMP is becoming an enabling technology to meet the demands of precise machining of wafer surface in various applications. In this paper, dishing and uniformity performance of an 28nm STI-CMP process was studied with the influence of machine, slurry, polish pad, polish time, zone pressure and retaining ring force are analyzed, which affects the surface geometric parameter of silicon wafer. The results of experiment indicate that STI dishing, uniformity and wafer loading of silicon wafer are improved, by using new slurry with low SiN removal rate, controlling zone pressure, adjusting retaining ring force.

INTRODUCTION

Currently CMP is commonly used in achieving both local and global planarization in ultra-large-scale integrated circuits. In the CMP process, a wafer is pressed down and rotated against a rotating polyurethane pad that is saturated with abrasive slurry particles and a chemical solution. CMP application is current widely used in interlayer dielectric planarization to metal CMP and silicon oxide trench planarization [1]. During the approaches of the fabrication process in ULSL circuits, the inter-level di electrical layer, most commonly, a SiO_2 layer are used to electrically isolate the different metal layers in the typical multilevel interconnects. As the devices of ULSI scale down, the CMP step requires slurry which produces high silicon oxide remove rates and a planarized surface. Because of the hardness and chemical inter ness of SiO_2 compared with other semiconductor material, it is essential to expedite the remove rates to raise the efficiency of manufacturing process and avoid the surface damage at the same time [2].

Currently, two slurries are widely used in industrial manufacturing for SiO_2 polishing. The ceria based slurry is highly active. However, the cost, broad size distribution and irregular shape of the ceria particles limited their widespread commercial acceptance. Silica sol has also been one choice in the CMP of dielectric surfaces. As one of the famous silica sol CMP products, has been recognized for its low preparation cost and less defects consequence [3]. Owing to high concentration of which is high cost and unstable, we expect a better performance of silica based slurry with high polishing efficiency.

STI-CMP required to polishing the oxide on silicon nitride, at the same time to minimize the groove of dishing.

Early STI CMP use ILD CMP slurry, with silica gel as the abrasive particles (silica -based slurry). The choice of silica gel slurry is low (SiO_2: $SiN_4 \sim 4$), the end of polish control ability is worse, and the process window is very narrow. It had to use a flat before polishing methods, such as reverse mask and other method; however, this has greatly increased the cost of process. So, high selectivity slurry (SiO_2: $SiN_4 > 30$) arises at the historic moment, which using cerium oxide (CeO_2) as abrasive particles (ceria -based slurry). Silicon nitride became a polishing stop layer, process window greatly widened, reverse mask method become history, direct STI CMP come true. STI CMP is greatly took a step forward. Ceria-based slurry process of dishing (200 ~ 600 A), still is its weakness, doesn't meet the requirement of the new technology of defect is increasingly strict [4].

To 28 nm process node, STI-CMP process window narrower than 40 nm, major challenges including dishing requires less than 200 A; Wafer uniformity oxide requires less than 200 A, SIN requires less than 25 A. At the same time, guarantee the defect within the acceptable range. In this paper, in order to solve the problems of dishing and uniformity were studied, by evaluating the machine polishing pad polish parameters such as pressure and polish time, optimizing the 28 nm process.

EXPERIMENT

As shown in figure 1, for shallow trench isolation chemical mechanical polishing process diagram. STI CMP in isolation between the devices on the surface of the wafer. The difficult of the technology is the injury on large area of the active area (AA), silicon nitride residues and target thickness control and so on.

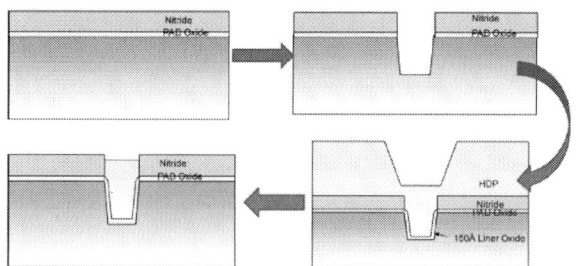

Figure 1. Diagram of shallow trench isolation chemical mechanical polishing process.

978-1-5090-6695-7/17 $31.00 © 2017 IEEE

Process control: polish end point detection, can stability control silicon thickness after polish, as shown in figure 2. The EPD process control method can successfully use in mass production. It greatly helps on STI CMP process capability and production efficiency improvement.

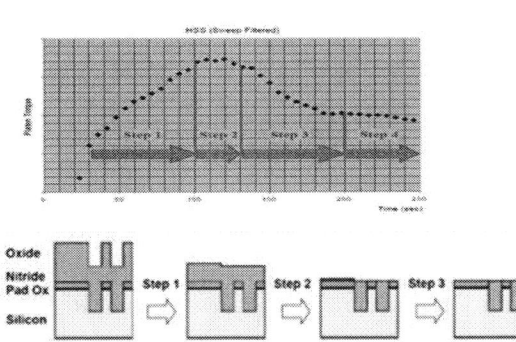

Figure 2. STI-CMP endpoint detection schematic diagram.

In order to solve the problem of dishing, this research choose a more excellent performance of slurry and polish experiment, through the performance optimization, the inline THK test, dishing and uniformity to show the results.

RESULT AND DISCUSSION

Compared with NP 70TS pad, IK 4250 pad has lower dishing both at wafer center and edge; reduce to 124A/63A from 182A/161A, measurement by AFM. Figure 3 shows dishing improve significantly.

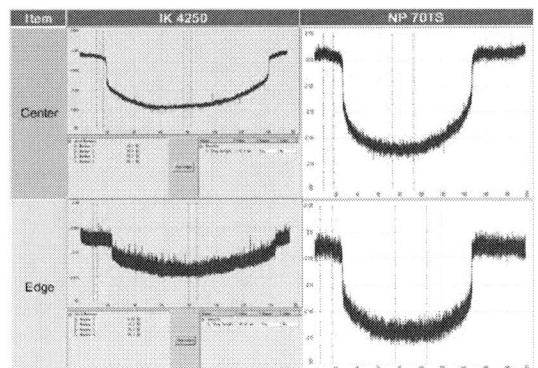

Figure 3. Compared with AFM measurement of different pads.

Compared with 40 nm process, 28 nm process needs by choosing a more suitable for slurry for dishing and improve of the profile. Figure 4 shows dishing contrast situation of different slurry under the same polish condition. Dishing performance decreased to less than 180

A from more than 500 A, which improved significantly; dishing up to ~70%. Experiment shows that the improved condition of new slurry can effectively improve the STI-CMP dishing.

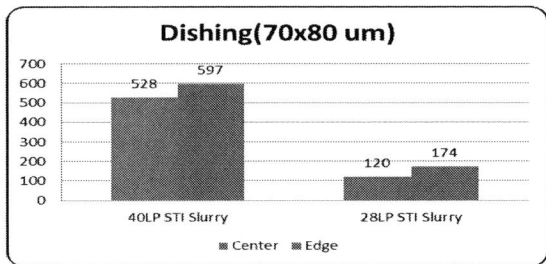

Figure 4. Dishing contrast situation of different slurry under the same polish condition.

CONCLUSION

This paper researched the different machines, polish pad, slurry type and polish parameters on the influence law of STI-CMP. Combined with the inline measurement and TEM analysis and defect characterization methods such as scan, realized with dishing, defect, THK, wafer loading key factors such as the effective control. Experimental results show that the IK pad can realize dishing reduced 70% the defect quantity is decreased obviously. SIN slurry can effectively improve the wafer oxide/SIN uniformity, so as to meet the requirements of industrial production and application of the reliability of the chip. This paper using CMP technology to solve the dishing of STI, uneven, the problem such as defect, so as to improve product yield, reduce the defects, implements the fine flat on the surface of the chip processing, for 28 nm and below nodes process propulsion is substantive.

REFERENCES

[1] R. DeJule. CMP Challenges below a quarter micron. Semiconductor International. 1997.

[2] Michael Quirk, Julian Serda. Semiconductor Manufacturing Technology [M]. Publishing house of electronics industry. 2007.

[3] Zhang Rujing, et al. Nanoscale Integrated Circuits-The manufacturing Process [M]. Publishing house of Tsinghua University. 2014.

[4] C.K. Huang, et al. A Fixed Adrasive STI Process for 200mm Rotary Polishing 11th intern. CMP-MIC Proc., 2006:145-151.

A NEW ACIDIC ILD SLURRY FORMULATION FOR ADVANCED CMP

Yi Guo[*], Arun Reddy, David Mosley and Robert Auger

Dow Electronic Materials, CMP Technologies, Newark, DE, 19711, USA

*Corresponding Author's Email: yiguo@dow.com

ABSTRACT

For advanced interlayer dielectric (ILD) CMP slurry, it is highly desirable to achieve enhanced polishing efficiency and performance with reduced overall cost of ownership (CoO). A colloidal-silica based acidic ILD slurry,[†] was thus designed for accurate manipulation of the particle/wafer interaction by leveraging the use of proprietary additive. NMR studies were conducted to quantitatively understand particle adsorption and zeta potential of abrasives and polished films were characterized to establish the correlation of electrostatic forces to removal rate. The formulation of the slurry is optimized in such a way to reflect the maximized benefits particularly when applied together with Dow leading CMP pads.

INTRODUCTION

Advancement of CMP in Semiconductor Fabrication

The semiconductor industry continues to see growth in both Logic and Memory chips, driven by expanding applications in segments such as mobile, server, data processing, communications, consumer electronics, industrial and automotive. Advanced Logic and Memory device nodes demand significantly greater performance from chemical mechanical planarization/polishing (CMP) processes. The rapid growth of new transistor/device architectures and technologies like 3D FinFETs, 3D NANDs and 3D packaging poses additional opportunities (in terms of CMP process steps) but also presents new challenges for CMP consumables and tool sets.[1] New requirements include substantially enhanced planarization efficiency, near-zero level defectivity and substantial reductions in process cost versus previous device nodes. Highly tunable and dilutable CMP slurries, in conjunction with performance-enabling CMP pads and processes are needed to achieve both technical and economic objectives.

Technical Trends and Challenges in Dielectric CMP

The market-leading ILD slurries have traditionally been alkaline pH-based fumed silica slurries comprising >10 wt% abrasives having >100 nm particle sizes. The oxide removal mechanism is stated to occur through OH⁻ facilitated siloxane bond breakage of TEOS wafers followed by effective particle/wafer collision and abrasion.[2] For advanced-node ILD CMP processes, precise topography control along with both with-in-die (WID) and with-in-wafer (WIW) uniformity are becoming

---

[†]OPTIPLANE™ and VISIONPAD are trademarks of the Dow Chemical Company ("Dow") or an affiliated company of Dow. OPTIPLANE represents a series of slurries designed for selective / non-selective CMP of non-metal layers. OPTIPLANE™ 2118 Slurry is formulated for ILD polishing.

---

critical as overall polishing margins shrink substantially.

Increasingly stringent requirements necessitate new slurry formulations. Herein we present on the design and performance evaluation of a new dielectric CMP slurry, enabled by advanced colloidal silica abrasives and functional additives to offer high removal rates, planarization efficiency and exceptionally low defect levels. The newly developed slurry, provides high performance with low overall cost of ownership (CoO) via point-of-use dilution. .

DESIGN RATIONALE

New ILD Slurry in Acidic Regime

Design of low-abrasive, silica-based CMP slurries in the acidic pH regime requires charge reversal of the silica particles to promote favorable particle/wafer interaction. As demonstrated in the plot of zeta potential vs. pH (Fig.1), colloidal silica abrasives have the same negatively-charged surface as the polished TEOS films (throughout all measured pH ranges from pH 2 to pH 11) and thus exhibit undesirable electrostatic repulsion during polishing. With the introduction of proprietary additives in the formulation, such slurry possesses a significantly shifted isoelectric point (IEP) and creates a positively-charged surface at acidic pH via additive adsorption onto the silica particle surface.

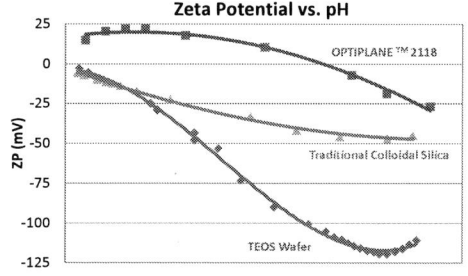

Figure 1: Zeta potential compared against pH for OPTIPLANE™ 2118 CMP slurry, traditional colloidal silica slurry and a TEOS wafer

With the use of proprietary amine-type additives for charge-reversal, the silica particles are intrinsically attracted to the wafer surface such that the point-of-use (POU) abrasives can be significantly reduced without sacrificing removal rate (RR) performance (Fig.2). Chemical additives were screened and down-selected to optimize particle surface charge densities and the corresponding zeta potential of slurries. With appropriate characterization techniques, this formulation reflects precise control of the particle/wafer interface in order to maximize the CMP benefit while maintaining colloidal stability.

DOW CONFIDENTIAL - Do not share without permission

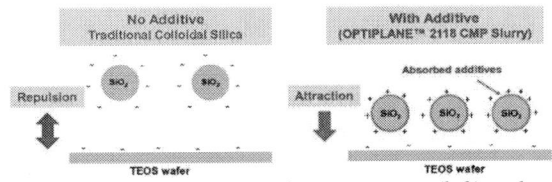

Figure 2: Silica abrasive/wafer interactions (left) without a charge modifying additive and (right) with a charge modifying additive

Quantitative Control of Additive Adsorption onto Particle Surface

CMP is well known as a 'tribological process' with interactions involving pad, wafer and slurry. But once the additives are introduced in the slurry formulation, it is crucial to understand the interfacial interactions and optimize the electrostatic benefit of charge reversal additives, while ensuring favorable particle/wafer attraction. Rather than employing destructive centrifugation approach to separate additives from particles, we established a robust in-situ quantitative ^{1}H NMR method for particle surface characterization as adsorbed molecules at the interface of particle have reduced mobility leading to reduced spin-spin (T2) relaxation times compared with the ones in the bulk fluid (Fig.3). Thus, a nucleus with a long T2 relaxation time gives rise to a very sharp NMR peak which is used for quantitative calculation of free additives in bulk liquid.

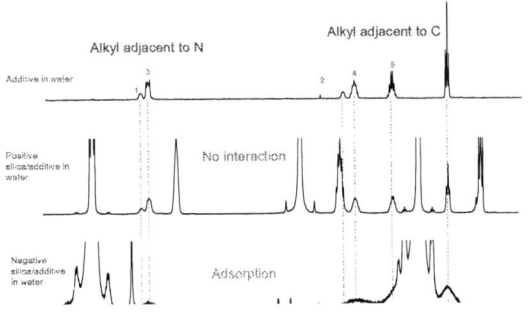

Figure 3: ^{1}H NMR reflecting surface adsorption of additive

As seen in Table 1, the additive adsorption % is nearly 100% until additive level is above 0.15. Although the moles of additives per particle continue to increase, the adsorption percentage decreases pronouncedly. The Langmuir-type adsorption behavior assumes monolayer coverage and molecules adsorbed on the surface interact with each laterally.[3]

In practice, oxide RR exhibited strong correlation with adsorption% but not [adsorbent]. The oxide RR decreases even though additives surrounding abrasives increase, indicating the free additive present in the bulk liquid interfere with particle/wafer interaction (Fig. 4.).

TABLE I. ADDITIVE ADSORPTION BY ^{1}H NMR MEASUREMENT

Sample (slurry contains 6% silica)	Calculation from ^{1}H NMR Spectra			
	Normalized Additive	Slurry NMR integral, %	Additive Adsorption %	Moles of Additive per Particle
Additive	1	100	N/A	N/A
Slurry A	0.02	0	100%	85
Slurry B	0.05	0	100%	213
Slurry C	0.15	1.9	93%	594
Slurry D	0.2	3.1	90%	766
Slurry E	0.5	14.9	68%	1447
Slurry F	1	41.8	58%	1433

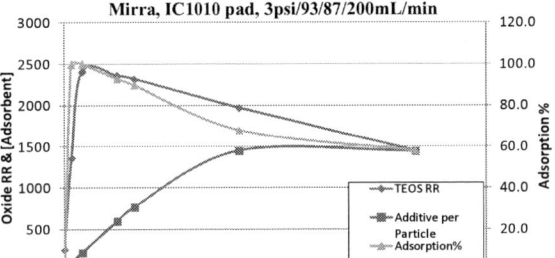

Figure 4: Correlation between TEOS RR and adsorption% as well as [adsorbent] (Slurry: 6% silica)

ADVANCED CMP PERFORMANCE

The new acidic slurry also exhibits reduced dependence of film removal rates on polishing time as shown in Fig. 5. This feature is particularly useful for 20 nm and below advanced nodes where the polishing processes typically involve low "Pressure*Velocity" and/or reduced polishing times. On an IC1010 pad using a fumed silica slurry, TEOS removal rate ramps up slowly and does not reach a steady state even after >60 seconds polishing time. However, under those same pad and process conditions, the new slurry formulation achieves the steady-state removal rate in only about 20 seconds due to the additive chemistry effect. The removal rate ramp time is further reduced to five seconds when used with a VISIONPAD™ 6000 (K7+R32 groove) pad, underscoring the importance of understanding and characterizing slurry/pad/grooving interactions for use with low-solids acidic platform slurries.

A variety of defects are generated during oxide CMP processes, including scratches, particle residues, pad debris and roughness-related non-visible defects. Scratches are widely believed to be the most detrimental to wafer yields. The onset of scratch formation is often a result of an increased number of large particles in the polishing slurry.[4] The use of highly controlled spherical silica particles and advanced filtration technology helps minimize the probability of defect events resulting from these sources. This slurry exhibited ~ 45% scratch reduction when used on undoped silica glass (USG) wafers (Fig.6) and demonstrates > 70% scratch reduction

DOW CONFIDENTIAL - Do not share without permission

978-1-5090-6695-7/17 $31.00 © 2017 IEEE

on TEOS wafers compared to those polished with conventional fumed silica slurry under the same polishing process.

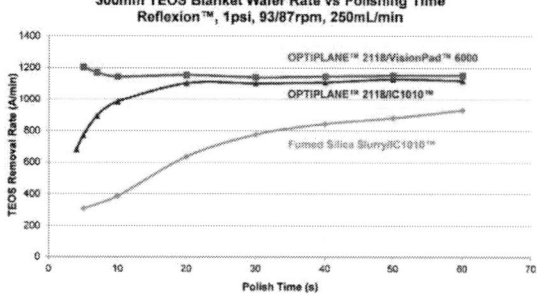

Figure 5: TEOS removal rate with polishing time for OPTIPLANE™ 2118 and fumed silica slurry. Also shown is a comparison of removal rate vs. time for the -2118 slurry with IC1010 and VISIONPAD™ 6000

Such defect benefits, together with remarkably reduced pad wearing and polishing temperature, can be attributed to the use of low POU abrasive content and steric protection from additive adsorption on the polished surface.

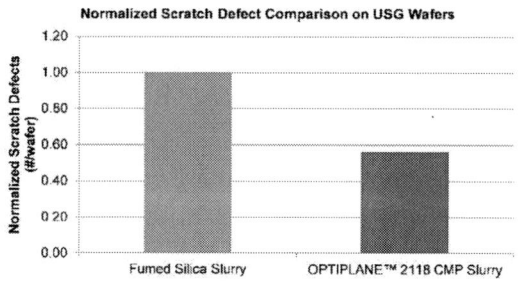

Figure 6: Normalized scratch defect on USG wafers

Feature-scale planarization and die-scale planarity are other CMP performance metrics where understanding of the pad-slurry interactions is critical. All the planarization data shown in this manuscript are generated using commercially available ILD patterned wafers with MIT masks (SKW, Inc.). Fig.7 shows planarization efficiency of three different pad materials with the acidic colloidal silica slurry and an industry standard fumed silica slurry. Planarization efficiency as a function of feature size is also evaluated with three selected pad materials ranging in hardness from 54 to 67 Shore D. With IC1010 (62 Shore D) and VISIONPAD™ 5000 (67 Shore D) pads, the planarization efficiency of both slurries is similar, but the combination of VISIONPAD™ 6000 (54 Shore D) and the colloidal silica slurry clearly outperforms all other pad/slurry combinations for feature-scale planarization. This result contrasts with the typical response, wherein pads with higher hardness tend to deliver higher planarization efficiency. Evidently, the use of VISIONPAD™ 6000 pad results in a contact pressure distribution on the "high" and "trench" features of the

wafer that favors better planarization efficiency. More importantly, for patterned wafers consisting of multiple density features within a die, the die-scale planarity should also be considered for a given CMP consumable set. Further studies on these pad/slurry interactions are currently in progress and will be discussed in the future.

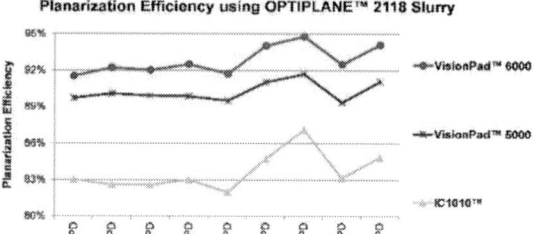

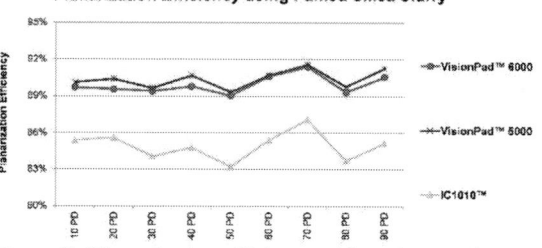

Figure 7: Planarization efficiency with various pads using (top) OPTIPLANE™ 2118 ; (bottom) fumed silica slurry

CONCLUSION

As advanced device manufacturers identify needs for new and additional CMP steps, new slurry solutions can deliver exceptional planarization and defectivity within a stable oxide CMP process. Advanced performance can be achieved while lowering process costs, through low point-of-use abrasive concentration, high removal rates and exceptional process consumable lifetime.

ACKNOWLEDGEMENTS

The authors want to acknowledge Jia-Ni Chu and John Nguyen for providing insights on the CMP slurry market, thank Dr. Kebede Beshah for his contribution to developing HNMR characterization method, and thank Matt VanHanehem and Todd Buley for sharing technical opinions on advanced ILD applications.

REFERENCES

[1] S. Babu *Advances in chemical mechanical planarization (CMP), 1st Edition. Woodhead Publishing.* 2016

[2] F. Zhang, A. Busnainab, G. Ahmadi. *J. Electrochem. Soc.* Vol. 146, 1999, pp. 2665-2669

[3] H.Baeza *Int. J. Electrochem. Sci.,* vol. 8, 2013, pp. 7518

[4] J. Choi, Y.Prasad, I. Kim, W. Kim, A. Busnaina, J. Park, Vol.2, 2010, *J. Electrochem. Soc.,* pp. 157.

DOW CONFIDENTIAL - Do not share without permission

SETTLING OF COLLOIDAL SILICA PARTICLES IN CMP SLURRY: MONITORING, EFFECT, AND HANDLING

Jie Lin[*], and W. Scott Rader*

Fujimi Corporation, Tualatin, Oregon, USA 97062

*Corresponding Author's Email: jlin@fujimico.com

ABSTRACT

The settling of colloidal silica particles in slurries results in changes of turbidity, specific gravity (SG), viscosity, particle size distribution (PSD), and concentrations of particle and chemical as a function of depth. Depending on the characteristics of slurry and mechanism of polishing, these changes may affect the removal rate. A container of slurry should be mixed properly before use in analysis and polishing to mitigate the effects from the settling of particles. Large particle count (LPC) is used as a sensitive method for monitoring the settling and re-dispersion of particles.

INTRODUCTION

A CMP (chemical and mechanical planarization) slurry is usually comprised of particle and chemistry. In a CMP process, chemical and mechanical polishing work together synergistically to produce a desired performance. Change in particle and/or chemical concentration may impact the polishing performance.

In a slurry, particles settle according to their size and the characteristics of the slurry [1, 2]. Presented in Table I are three colloidal silica-based slurries with different physical and chemical properties utilized for CMP of Cu and TEOS (silicon dioxide). In this study, these slurries were tested to determine the effects of particle settling on QC analysis and polishing performance.

Table I. Three colloidal silica-based slurries

Slurry	Particle size	SiO$_2$ (wt%)	Surfactant (wt%)	pH	Viscosity (cP)	CMP
Slurry A	~67 nm	~ 6%	None	6.7	1.6	Cu
Slurry B	~100 nm	~ 7%	~ 1 %	3.6	1.5	Cu
Slurry C	~65 nm	~ 5%	< 0.1 %	11.0	1.2	TEOS

EXPERIMENTAL

Slurry in a 2 L bottle (with sample depth of 15 cm) was taken at 2.5 cm depth intervals by using a 10 mL pipette. Slurry from a 0.5 L bottle was drawn at flow rate of 2 mL/min by a pump (and immediately measured for LPC with a custom laser-based light scattering instrument) [3]. Slurry in a 320 G tote was drawn from specific depths by using a hand pump.

Turbidity of slurry was measured with a Micro PTI turbidimeter (HF Scientific, Inc.). Specific gravity (SG) and viscosity were measured with a DMA 5000 density meter and AMVn viscometer (Anton Paar), respectively. Particle size distribution (PSD) was measured with an LA-950 Laser Diffraction Particle Size Analyzer (Horiba,

Ltd.). SiO$_2$ wt% was determined with an Optima 7000DV ICP-OES Spectrometer (Perkin Elmer, Inc.). Total organic carbon (TOC) was determined with a TOC-L Total Organic Carbon Analyzer (Shimazu Corp.). Surfactant in Slurry B was determined with a Zorbax 300SB C$_{18}$ column and Agilent Infinity 1260 HPLC system.

Wafers were cut into 1.5x1.5" coupons for polishing test on a MultiPrep bench-top polisher (Allied High Tech Products, Inc.) with a Fujibo H7000 pad. The polishing conditions are as follows: pressure = 1.1 psi, head speed = 22 rpm, platen speed = 200 rpm, slurry flow rate = 50 mL/min, and polishing time = 45 sec. Thickness of the Cu film was measured with a ResMap 178 (Creative Design Engineering, Inc.). TEOS film was measured with an F50 Thin-Film Mapper (Filmetrics, Inc.).

RESULTS AND DISCUSSION

Changes as a function of depth

Figure 1 shows depth profiles of physical properties and SiO$_2$ wt% of Slurry A after sitting for 6 months. Settling of particles results in increases in turbidity, SG, and viscosity towards the bottom of the container. The median particle size (d50) and width of distribution (d90-d10) also increase with depth as larger particles settle more quickly. The changes in these physical properties are validated by the depth profile of SiO$_2$ wt%, which increases from 4.4% in the top to 7.6% in the bottom.

The extent of change as a function of depth depends on the characteristics of the slurry. This is exemplified by the depth profiles of SG in slurries A, B, and C shown in Figure 2. Slurry B has the largest extent of particle settling largely because it has the highest solid content (i.e. SiO$_2$ wt%) and the largest particle size (Table I).

Effect on removal rates

Figure 3 shows that the effects of particle settling on removal rates (RR) are different among Slurries A, B, and C. For polishing of Cu, the effect is insignificant on Cu RR with Slurry A (with a small relative standard deviation of 1.3%) but significant with Slurry B. Polishing of TEOS with Slurry C shows a significant effect as a function of depth that is opposite to polishing of Cu with Slurry B.

In Slurry A, despite the significant change in particle concentration as a function of depth, concentrations of the chemicals do not change with depth (Table II). Polishing of Cu with Slurry A is dominated by chemical polishing as illustrated by Figure 4. It shows that Cu RR by chemistry is 6x of that by particle. When chemistry and particle work together, the slurry produces 21x Cu RR. Therefore, Cu RR is not affected significantly by settling of particles (even though particle concentration decreases by ~26% in the top layer and increases by ~29% in the bottom layer because chemical concentrations are not affected).

Changes in chemical concentrations in Slurry B are evidenced by the depth profiles of total organic carbon adsorbed on silica particles and surfactant concentration in liquid phase shown in Figure 5. In Slurry B, the surfactant adsorbs on the surface of silica particles. Slurry in the bottom layer has higher concentration of particle (that increases Cu RR) which also increases concentration of surfactant (that decreases Cu RR). The decrease in Cu RR by higher concentration of surfactant dominates over the increase in Cu RR by higher concentration of particle. Thus, lower Cu RR is found in the bottom layer of Slurry B in the container.

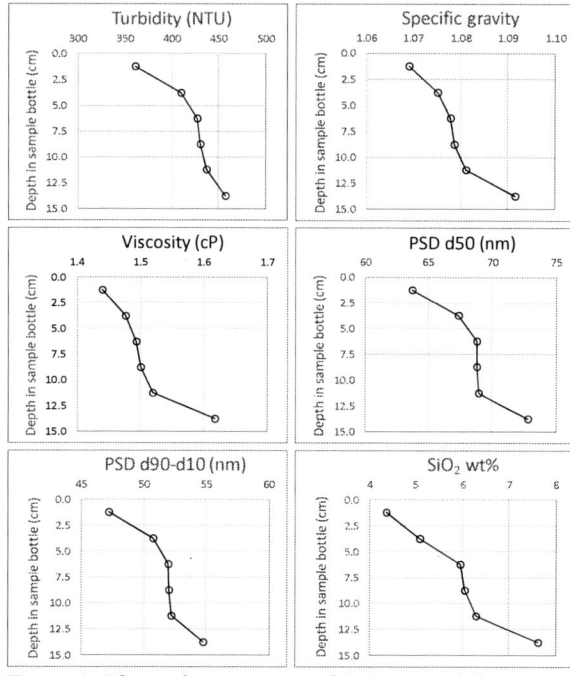

Figure 1: Physical properties and SiO₂ wt% of Slurry A as a function of depth in a 2 L bottle after sitting at room temperature for 6 months. (NTU = Nephelometric Turbidity Units)

Table II. Normalized concentrations of particle and chemicals in a 2 L bottle Slurry A after sitting for 6 months

Depth (cm)	Normalized concentration			
	Particle	Chemical 1	Chemical 2	Chemical 3
0-2.5	0.74	0.98	0.97	1.01
2.5-5	0.86	1.00	1.01	1.02
5.0-7.5	1.01	1.01	1.00	1.01
7.5-10	1.03	1.00	1.00	1.00
10-12.5	1.07	1.00	1.00	0.99
12.5-15	1.29	1.00	1.02	0.97

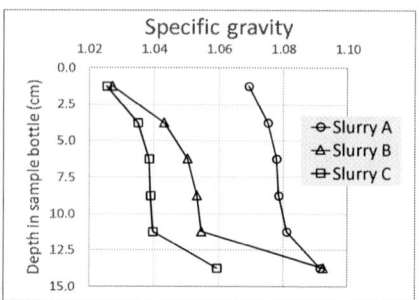

Figure 2. Comparison of specific gravity depth profiles among Slurries A, B, and C in 2 L bottles. Slurries A and C after sitting for 6 months and Slurry B after 5 months.

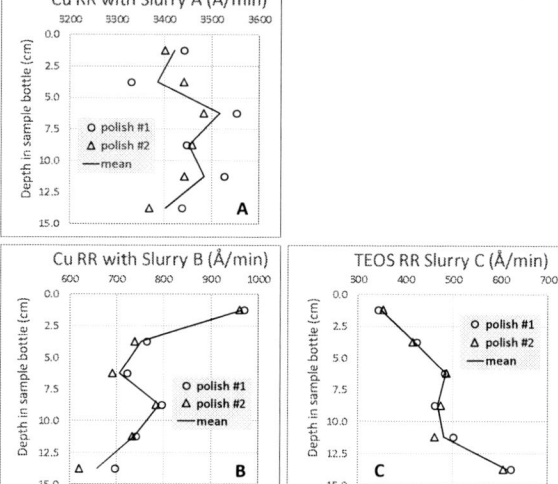

Figure 3: Effect of particle settling on removal rates (RR) as a function of depth in 2 L bottles: (A) Cu RR with Slurry A after sitting for 6 months, (B) Cu RR with Slurry B after 5 months, and (C) TEOS RR with Slurry C after 6 months.

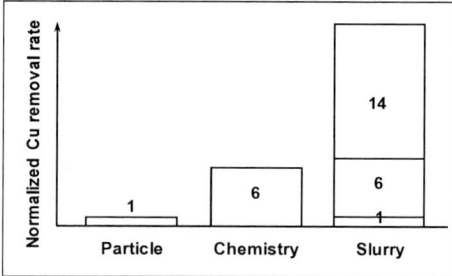

Figure 4: Normalized Cu removal rate with Slurry A: by particle only, chemistry only, and slurry with particle and chemistry working together synergistically.

Polishing of TEOS with Slurry C is mechanically dominated. The increase in TEOS RR from the top to the bottom of the container is largely a result of increased concentration of particle from top to bottom (2.1 to 7.3%).

LPC for monitoring settling/re-dispersion of particles

LPC is commonly used in the CMP industry for the consideration of "health" of slurry and a predictor of mechanical defects such as scratches [3,4]. LPC depth profiles in Figure 6 show that many large particles (>0.56 μm) in Slurry A settle to the bottom after only 22 days. Upon mixing by inverting the bottle, the settled large particles are fully re-dispersed and the original uniform depth profile is recovered. This demonstrates that LPC can be used as a sensitive method for monitoring the settling and re-dispersion of particles in a slurry.

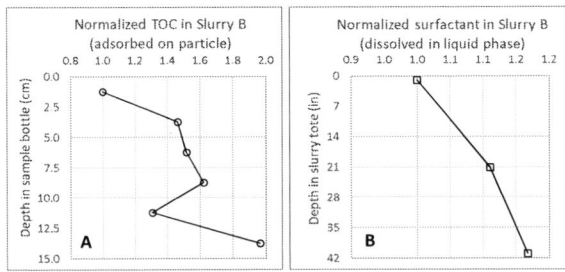

Figure 5: Effect of particle settling on chemical concentrations in Slurry B as a function of depth: (A) Normalized TOC adsorbed on particles in a 2 L bottle after sitting for 5 months, (B) Normalized surfactant concentration in liquid phase in a 320 G tote after 42 days.

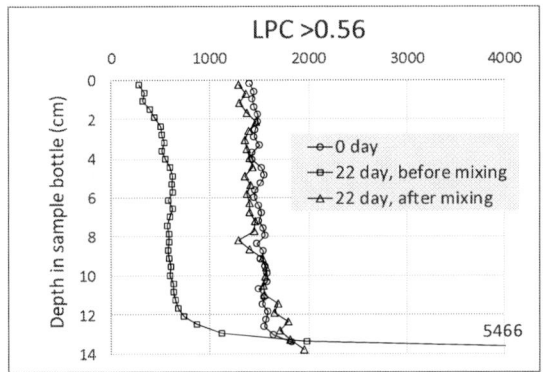

Figure 6: Effects of settling and mixing on LPC depth profile in a 0.5 L bottle of Slurry A based on sampling method described in ref. 3.

Handling of slurry

Bottles of slurry are commonly collected for QC analysis to ensure the quality of the product. The settling of particles will result in changes of physical properties and chemical concentrations as a function of depth in a bottle of slurry. Figure 1 clearly shows that a bottle of slurry, after sitting for a long time, must be mixed properly before QC measurements to avoid erroneous results.

The settling of particles in a large container (e.g. drum and tote) of slurry will also result in changes as a function of depth. Figure 7A shows depth profile of SG a 320 G tote of Slurry B after sitting for 9 months. If this tote of slurry is used without being re-dispersed, it will result in the change of Cu RR as previously discussed. Figure 7B shows the depth profile of SG in a 320 G tote of Slurry C after sitting for 2 months. Similarly, if this tote of slurry is used without proper mixing, TEOS RR will deviate from the expected performance. The uniform depth profiles of SG after mixing show that these two slurries can be re-dispersed by mixing to mitigate or eliminate the effects of particle settling.

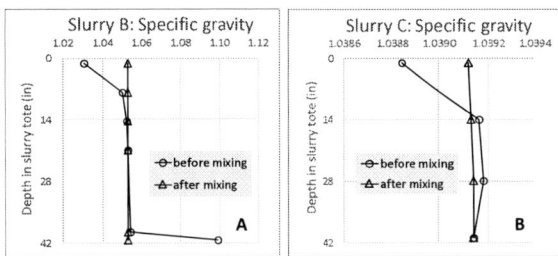

Figure 7: Depth profiles of SG in a 320 G tote of slurry before and after mixing. (A) Slurry B after sitting for 9 months, mixed by pumping from one tote to another, (B) Slurry C after sitting for 2 months, mixed by pumping and circulation in the tote (for one turn-over time).

CONCULSION

This study shows that settling of particles in colloidal silica based CMP slurries will result in changes in physical properties, as well as concentrations of particles and chemicals, as a function of depth. These changes may or may not affect the removal rates depending on the characteristics of slurry and mechanism of polishing. Therefore, a container of slurry (from a small bottle to a large tote) must be properly mixed to re-disperse the particles and chemicals before being used in analysis or polishing to avoid erroneous results.

ACKNOWLEDGEMENTS

The authors thank colleagues at Fujimi Corporation, particularly Karl Ulbricht, Annette Schaper, and Dr. Charles Poutasse, for assistance and discussions.

REFERENCES

[1] W. E. Dietrich, *Water Resources Research*, vol. 19, 1982, pp. 1615-1626.

[2] C. Allain, M. Cloitre and M. Wafra, *Phys. Rev. Lett.* vol. 74, 1995, pp. 1478-1481.

[3] W. S. Rader, J. Lin and T. Holt, "*Large Particle Formation in Silica Based Slurries,*" CAMP's 12th International Symposium on Chemical Mechanical Planarization, August 12-15, 2007.

[4] E. E. Remsenz, S. Anjur, D. Boldridge, M. Kamiti, S. Li, T. Johns, C. Dowell, J. Kasthurirangan and P. Feeney, *J. Electrochem. Soc.* vol. 153, 2006, pp. G453-G461.

OUTSOURCED CMP FOR RAPID DEVELOPMENT AND EFFICIENT MANUFACTURING

Robert L. Rhoades

Entrepix, Inc., 4717 E. Hilton Ave, Phoenix, AZ 85034 USA

Email: rrhoades@entrepix.com

ABSTRACT

Device foundries around the world have clearly demonstrated that they can successfully provide both outsourced technology development and volume production of advanced semiconductor devices. Two of the factors that contribute to successful outsourcing are (i) distributing the enormous cost of technology development across a broad customer base, and (ii) leveraging production economies of scale. Similar advantages can now be realized at the CMP process module level.

DISCUSSION

Integration Development

Companies that have not yet invested in CMP process capability can immediately access both the equipment and process expertise to accelerate implementation timelines for new device integrations. From prototype, development and qualification through volume production manufacturing, all necessary components for the planarized layers can be rapidly defined and tested without any capital investment in polishers or post-CMP cleaning equipment.

One example of integration development involves a medical device manufacturer which had been using traditional sloped-wall aluminum interconnect for many years in their own fabrication facilities. They desired to move to smaller design rules and implement a planarized interconnect scheme. The process and integration development effort employed a segmented approach. First, the pre-metal dielectric planarization and tungsten contacts were developed and in-line tested. Next, the ILD1 layer and first level of metal was characterized and then duplicated to the ILD2 layer. Electrical probe data and SEM cross sections were examined at each of the steps. Data was first collected from a 6-wafer split lot. This initial partial lot was processed through the contact level and then evaluated using contact resistance and SEM data. Based upon the acceptability of these data, the remaining wafers were processed through that step and then held pending the next process step. Prior to committing each partial product lot, the appropriate CMP polishing process was characterized for removal rate, uniformity and defect density on blanket film wafers comprised of the appropriate thin films for that step. By using Entrepix as an outsource provider for integration help, and for oxide CMP and tungsten CMP, the full 3-level interconnect

integration was demonstrated in just over 8 months with zero capital outlay and using fewer than 50 product wafers [1]. It is now ready for new designs in a fraction of the time and expense traditionally associated with developing a planarized interconnect module and introducing CMP into a manufacturing line.

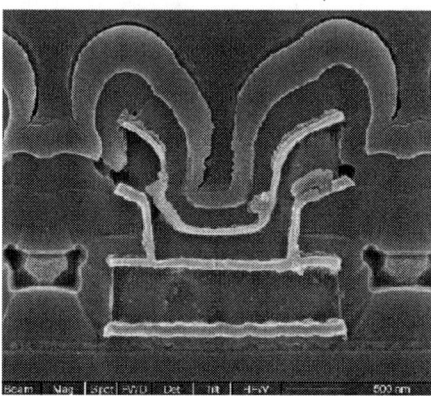

Figure 1: SEM image of single M1-M2 contact fabricated using existing sloped-wall aluminum interconnect module.

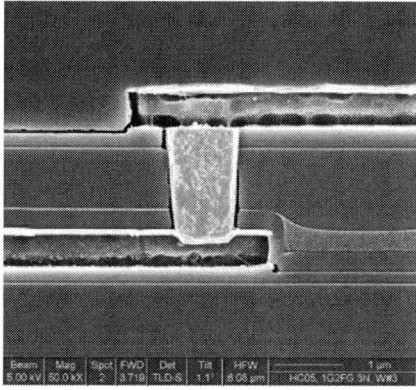

Figure 2: SEM image of the same M1-M2 contact as shown above now fabricated using a planarized interconnect module with both oxide and tungsten CMP developed using outsourced CMP services.

Process Development for New Materials

In addition to helping companies adapt known processes to their device designs, outsourced CMP services can be very effective in developing CMP processes for new materials. An example is polishing platinum (Pt) for high temperature contacts in a MEMS device [2]. A key step in the fabrication of these high temperature vias is Pt CMP. The process requirements are high removal rate of Pt, low selectivity to Ti, high selectivity to oxide, low erosion in plug arrays, no Ti barrier metal corrosion, and low surface defects. An initial round of screening tests were performed on four slurry formulations with a series of unpatterned blanket film wafers of each of the three materials that would be exposed by the end of the polish. All wafers were polished on an IPEC 472 polisher using a process recipe settings of 7 psi downforce and 60 rpm platen speed.

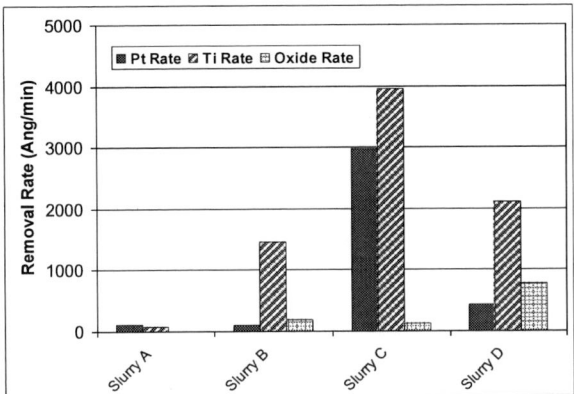

Figure 3: Graph of CMP removal rates for Pt, Ti and oxide for four different CMP slurries.

Slurry C exhibited the best combination of surface finish and desired selectivities, thus it was used for further process optimization then used to polish a first series of patterned wafers. As shown in the figure below, the vias were planarized with very good surface finish and no Ti barrier metal corrosion.

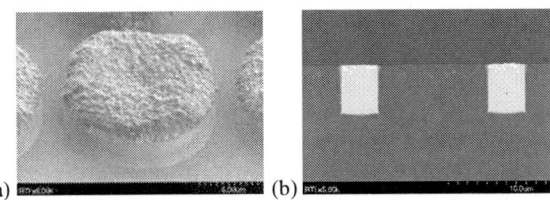

Figure 4: SEM micrographs showing (a) an as-deposited Pt plug with very rough overburden above the oxide, and (b) cross section of two Pt plugs after CMP with slurry C.

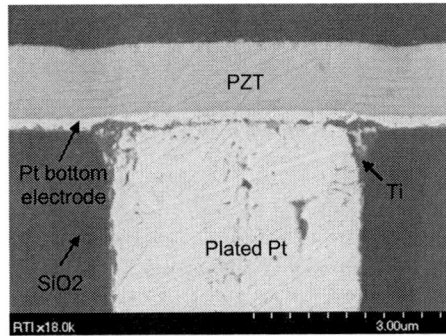

Figure 5: SEM cross section of a Pt via with PZT piezoelectric layer showing no cracking or degradation after 700°C anneal.

Outsource Production

A typical through-silicon via (TSV) module requires a combination of deep via etch, liner deposition, barrier deposition, via fill with Cu (or other conductive material), and damascene CMP to form the vias. A particular version of a TSV module involves electroplating Cu to roughly 4 to 8 micron thickness to fill moderately large via structures. The plating process used for the first several batches of wafers was rather immature and prone to a wide range of defects and inconsistent material properties in the plated film. This created problems during CMP with a random occurrence of one to several regions on some wafers that polished much more slowly than the surrounding Cu film. These slow spots ranged from a few mm to over 2 cm in diameter and required more than double the typical polish time to clear, which was already long due to the thick film. In parallel with efforts to improve plating consistency, Entrepix developed a new CMP process that accomplished two important goals: 1) increased Cu removal rate to about 8500 Ang/min while maintaining good selectivity to the dielectric stop layer, and 2) was much less sensitive to variations in local Cu material properties. The Cu rate qual data shown in the figure below is from multiple runs over a span of greater than six months. These data confirm that the improved CMP process is more robust and repeatable.

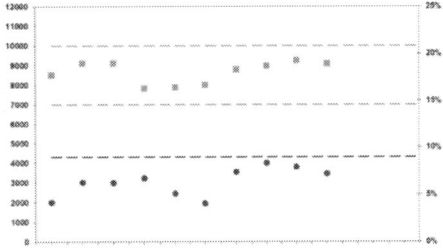

Figure 6: Cu Rate (Ang/min) and WIWNU (% 1-sigma) over 6 months of qualifications for production runs.

In the above example, after the integration was proven and repeat batches were giving consistent results, the customer decided to delay adding internal capacity and continue to outsource the CMP process step for at least the initial portion of the volume production ramp. This decision was driven mostly by financial factors which favor delaying capital expenditures as long as possible, especially when operating costs can be controlled at a reasonable level through continued outsourcing of CMP.

SUMMARY

Outsourced CMP can provide substantial savings in both time and money for companies in a wide range of circumstances. For companies with minimal or no internal CMP capability, outsourcing provides immediate access to the expertise necessary to accelerate integration timelines and reduce cycles of learning. For companies with well-established internal capability, outsourcing is effective at providing an alternative for performing engineering trials, such as new process development or consumables testing, without having to take production equipment off line. In today's competitive business environment, outsourced CMP provides a new opportunity to accelerate timelines and minimize costs for both development and production demands.

ACKNOWLEDGEMENTS

The author gratefully acknowledges the continuing efforts and expertise of the process foundry staff of Entrepix and permission from various customers to share examples and data.

REFERENCES

[1] R.Rhoades and R.Danzl, *Proceedings of the CMP-MIC Conference*, San Jose, CA, Feb 2004.

[2] D.Dausch, C.Gregory and R.Rhoades, Proceedings of International Interconnect Technology Conference (IITC), San Jose, CA, June 2010.

PROCESS STABILITY AND TOOL CAPACITY IMPROVEMENT WITH 150MM PROFILER AND 200MM CONTOUR HEADS

Yanghua He[1], Michael Lube[1], and Jamie Leighton[2]*

[1]Process Engineering, Qorvo, Inc., Richardson, Texas, USA

[2]Applied Global Services, Applied Materials, Santa Clara, California, USA

*Corresponding Author's Email: Yanghua.He@Qorvo.com

ABSTRACT

This paper describes the improvement obtained when using an Applied Materials Mirra CMP system equipped with 150mm Titan Profiler or 200mm Titan Contour wafer-carrier heads, versus a polisher with a basic carrier. The wafer-edge profiles and polish-rate stability were compared between the polishers using different wafer-carrier heads. The polisher usage/capacity and CMP cycle time in production were also compared. Results show that the process stability with 150mm Profiler and 200mm Contour wafer-carrier heads makes daily process qualification unnecessary, and enables the elimination of the time-consuming look-ahead (L/A) step for each product lot. As a result, the polisher usage and CMP capacity was greatly improved to meet the requirement of high volume Bulk Acoustic Wave (BAW) production.

INTRODUCTION

Chemical mechanical planarization/polishing (CMP) processing is affected by various factors, including consumables, process parameters, and polisher hardware design. Among all these factors, the wafer carrier is particularly important because it directly affects the performance of the wafer-polish process (e.g., across-wafer uniformity, removal rate stability). Furthermore, the wafer-carrier head design determines the process capability including extendibility and flexibility in providing wafer polish profiles that meet expectation.

Single Pressure Zone Design (CMP-a)

When Qorvo's 150mm Bulk Acoustic Wave (BAW) product development first started, an oxide CMP process was developed using a conventional polisher with a basic wafer-carrier design, referred to as CMP-a. This carrier has a single pressure zone with a number of vacuum holes in the center area of the carrier plate. These vacuum holes are used not only for wafer handling but also for applying extra pressure to the center area of the wafer backside. The center pressure adjustment is used as a process knob to dial in the wafer center removal.

WAFER-EDGE PROFILES

This type of carrier has worked well in production to address across-wafer uniformity, but the design lacks robust control on wafer-edge removal profiles. A common factor in this, and all CMP carriers, is a wafer-retaining ring to keep the wafer inside the carrier during polishing. This ring has a fixed position in the carrier, determined by the wafer protrusion required for a stable process. In a normal setup the retaining ring is not flush with the wafer surface, and does not contact the CMP pad during polishing. Due to this design feature on such basic carriers, the wafer-protrusion setting manages wafer-edge control. This was proven to have limited control over the wafer-edge removal profiles. Because the retaining ring is designed to be stationary and cannot touch the polish pad, it cannot keep the pad flat around the wafer-edge area during polishing. Therefore, the wafer edge protruding out of the carrier is exposed to the rotating and compressed CMP pad, and is susceptible to the adverse effects of elastic pad rebound. This leads to wafer-edge uniformity and profile fluctuations, as shown in Figure 1. The measurement site 1 in the figure is wafer edge at 3mm edge exclusion, while site 25 is at the wafer center. The profile fluctuations at the wafer edge are a reflection of pad rebound effects.

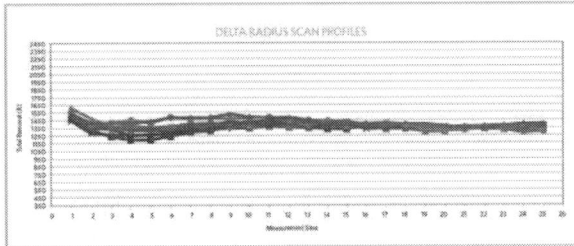

Figure 1: Delta radius scan profiles of oxide pilot wafers polished on CMP-a

Multiple Pressure Zone Design (CMP-b)

Using the same consumables (polish pad, conditioning disk, slurry), removal rate and within-wafer non-uniformity (WIWNU) specifications, oxide pilot type/thickness, and measurement tools/recipes as the data collected on the said CMP-a carrier, a new CMP process has been developed on Applied Materials Mirra polishers, referred to as CMP-b. These polishers are equipped with 150mm Profiler and 200mm Contour wafer-carrier heads, as shown in Figures 2 and 3. They have similar carrier design features except the number of pressure zones differs, depending on the size of wafers they handle. Specifically, there are three pressure zones, labeled P_1, P_2, and P_3 in Figure 2 for the 150mm Profiler wafer-carrier

978-1-5090-6695-7/17 $31.00 © 2017 IEEE

head, and five pressure zones, labeled P_1, P_2, ..., P_5 in Figure 3 for the 200mm Contour. Even though the retaining ring has its own pressure channel, it is not counted as one of the pressure zones. Note that the $P_{carrier}$ represents the software input "retaining ring" pressure. The actual retaining ring surface pressure, usually denoted as $P_{retaining\ ring}$, is the key factor for wafer retaining safety as well as wafer edge profile tuning. The average membrane pressure, denoted by $P_{avg\ membrane}$, is another important factor calculated based on all other pressure settings involved.

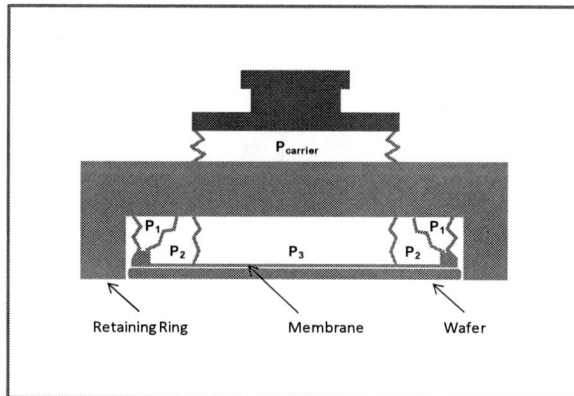

Figure 2: CMP-b 150mm Profiler carrier design

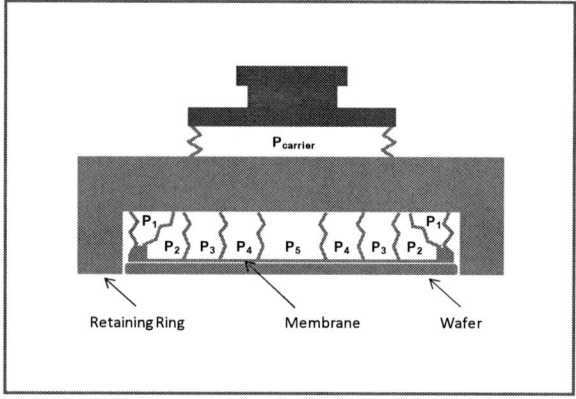

Figure 3: CMP-b 200mm Contour carrier design

Unlike the CMP-a wafer carrier, the CMP-b carriers all have multiple pressure zones, which are designed to provide a better capability for wafer-profile tuning. The retaining rings in these carriers can be controlled by a pressure channel in the ring. This allows the ring to stay flush with the wafer surface and to contact and/or press the polish pad during polishing. When the retaining-ring pressure and the carrier-edge-zone pressure are properly configured, the adverse effects of pad rebound on the wafer edge are greatly reduced or eliminated. As a result, wafer edge profiles are more stable, flat, and controllable,

as illustrated in Figure 4. Some minor fluctuation in the profile traces still exists around the wafer edge relative to the center, but compared to the profiles shown in Figure 1, these edge profile fluctuations are minimized and negligible.

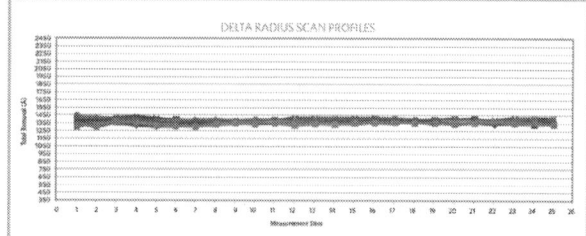

Figure 4: Delta radius scan profiles of oxide pilot wafers polished on CMP-b

POLISH RATE STABILITY

An advantage of the carrier head used in the CMP-a polisher is its basic design and ease of operation; however, it has limitations in wafer-center and -edge polish control. As a result, the CMP-a polish process used in the BAW production line requires daily qualification to ensure stability in both the process removal-rate and WIWNU. This daily process qualification involves small adjustments in polish-pressure settings. For example, the center pressure applied on the wafer backside usually needs to be increased by up to 0.4 psi in order to maintain a wafer-center removal-rate equivalent to that around the wafer center. The primary polish pressure may also need to be slightly increased to keep the overall removal-rate within specifications, especially with increasing wafer count on a polish pad. Both the backside and primary polish-pressure settings have windows for adjustment, and finding an appropriate pressure ratio for the daily process qualification is critical to the stability of the removal-rate and WIWNU. However, this process qualification leads to day-to-day removal-rate changes from the required pressure-setting adjustments. Removal-rates of all daily process qualifications performed over a pad life generally follow a gradually declining trend.

In contrast, the carrier heads used in CMP-b polishers employ a more complex hardware design, but have superior performance in both profile control and process stability. The CMP-b polish process also remains stable with similar removal-rates and WIWNU throughout a pad life without the need for adjustments in process parameters over the pad's life and after consumable changes. Such stability also removes the need for a daily qualification on the CMP-b process. Process qualification on the CMP-b setup has been reduced to a semiweekly interval, mainly to check process performances on all four carrier heads and obtain more accurate removal-rate readings from each of the individual heads. The

978-1-5090-6695-7/17 $31.00 © 2017 IEEE

removal-rate qualification data from the two types of polishers over a one-year period is plotted in Figure 5. Long-term stability in the CMP-b process removal-rate data is demonstrated by the narrower variation as compared to CMP-a. Pad lifetime is a major factor in the wider variability of the removal-rate data in the CMP-a process (i.e., higher rate with a newer pad, decreasing throughout the pad's life), as evidenced by the downward trend in the CMP-a data over time.

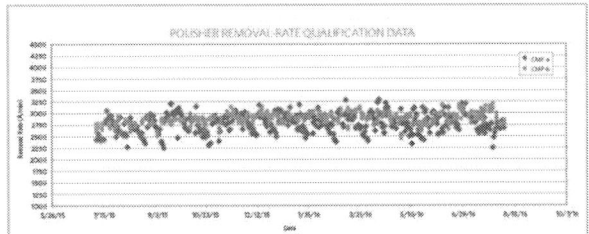

Figure 5: Comparison of CMP-a and CMP-b removal-rate qualification data

STEP HEIGHT CONTOUR PLOTS

To compensate for the larger removal-rate variations in the CMP-a process, a look-ahead (L/A) wafer is polished from each production lot before committing the rest of the wafers in the lot. The polished L/A wafer provides a more accurate on-product polish rate for process-time calculation, ensuring accuracy of the polish process relative to the desired step-height target. Daily process qualification coupled with the L/A methodology works well with the CMP-a process, delivering satisfactory process performance/product quality with the fixed polish-time approach.

Figure 6 shows the step-height contour plot of a typical product wafer polished on CMP-a, which meets the step-height specification on all measurement sites. As previously mentioned, the CMP-b process has superior process stability with more advanced wafer-carrier heads compared to CMP-a. This not only makes daily process qualification unnecessary on CMP-b, it also eliminates the need for an L/A polish step on production material. A production lot therefore is polished with a calculated process time based on the removal-rate from semiweekly process-qualification.

Figure 7 is the step-height contour plot of a typical product wafer polished on CMP-b, which also meets the step-height specification at all measurement sites. The comparison of step-height statistics in Figures 6 and 7 demonstrates that the CMP-b process performance is superior on production wafers, with significantly lower across-wafer step-height variation (e.g., standard deviation and high/low variance).

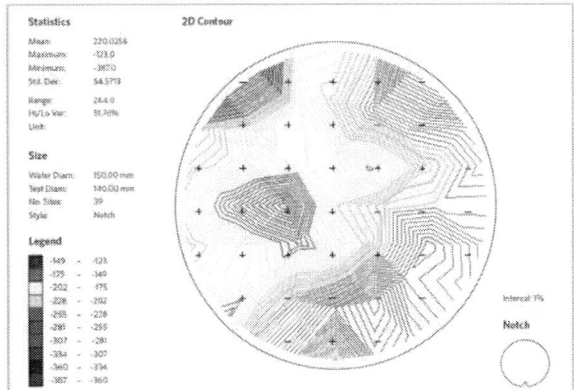

Figure 6: Step-height contour plot of a production wafer polished on CMP-a

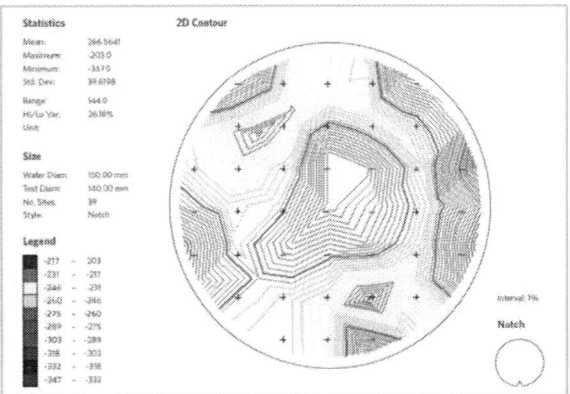

Figure 7: Step-height contour plot of a production wafer polished on CMP-b

EXTENDIBILITY AND FLEXIBILITY

Recall that the CMP-a carrier is basic in hardware design and lacks certain process capability that the CMP-b carrier has. In contrast, the CMP-b carrier has a sophisticated hardware design that enables precise control of pressure channels for retaining ring and multiple pressure zones. It not only makes process development or process transfer from other CMP platforms easy, but also possesses significant process extendibility and flexibility. The CMP-b process actually used the CMP-a process as a starting point for the removal rate match. The recommended sequence to determine pressure parameters for each individual pressure zones is from center to edge. For example, in the case of the 150mm Profiler carrier, P_3 for pressure zone 3 is the first to be determined by the desired polish rate under certain platen/carrier speed combination. P_2 for pressure zone 2 is next, and is fairly close to P_3. P_1 for pressure zone 1 is last to be optimized with the retaining ring pressure $P_{retaining\ ring}$ for a flat wafer edge profile. Note that the $P_{retaining\ ring}/P_{avg\ membrane}$ ratio is important for the CMP-b process. This ratio is not only a wafer handling safety indicator but also a key factor affecting wafer edge profiles. A higher ratio is better for the wafer retaining safety, but it may not be the optimized ratio for edge profile. The general range for the ratio is from 0.50 to 1.20. In reality, the ratio usually falls between 0.51 and 0.91, and 0.71 is near the optimal.

When P_1 is increased or decreased while keeping other pressures unchanged and the $P_{retaining\ ring}/P_{avg\ membrane}$ ratio fixed at 0.71 by adjusting $P_{carrier}$, the wafer edge removal profiles will be changed accordingly, as shown in Figure 8. The slow, flat, and fast edge removal responses can be a practical solution to compensate incoming wafer edge variances of around 10mm from wafer edge.

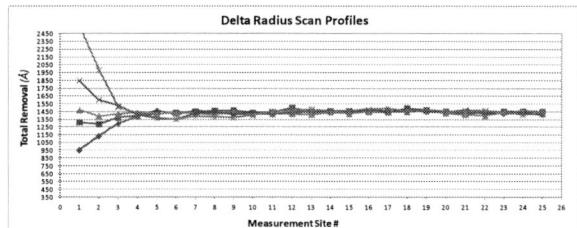

Figure 8: Delta radius scan profiles of oxide pilot wafers polished with P_1 matrix on CMP-b

Similarly, when P_2 is intentionally changed with the $P_{retaining\ ring}/P_{avg\ membrane}$ ratio fixed at 0.71 and other pressures unchanged, the wafer removal profiles in the region corresponding to P_2 will change accordingly, as shown in Figure 9. The slow, flat, and fast removal P_2 settings can be used to deal with incoming wafer issues such as fast or slow "donut" rings located about 25mm from wafer edge. P_3 covers the majority of wafer center of about 110mm in diameter. Like the P_1 and P_2 matrix tests described above, the P_3 matrix test at constant $P_{retaining\ ring}/P_{avg\ membrane}$ ratio of 0.71 is also performed to demonstrate the wafer center removal responses, as shown in Figure 10. The slow, flat, and fast center removal profiles can be very useful to process incoming wafers with center slow/fast signatures.

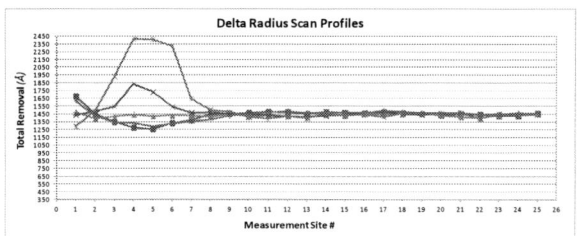

Figure 9: Delta radius scan profiles of oxide pilot wafers polished with P_2 matrix on CMP-b

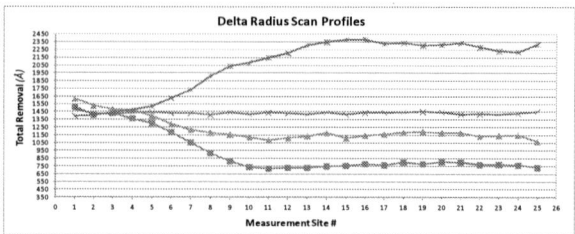

Figure 10: Delta radius scan profiles of oxide pilot wafers polished with P_3 matrix on CMP-b

CONCLUSION

In summary, both CMP-a and CMP-b polish tools and processes have been used in volume BAW production, meeting the same process-specification requirements. Due to the hardware limitations of the carrier heads in CMP-a, the process is required to maintain a daily qualification schedule, which consumes approximately 7 hours per week per polisher. Comparatively, the CMP-b process requires only a semiweekly qualification, which consumes normally less than 2 hours per week per polisher. Thus CMP-b tool usage in production is higher than that of CMP-a by at least 5 hours per week per polisher. CMP-a also requires an L/A wafer-polish step on every production lot, adding about 30 minutes per lot to overall process time. The L/A wafer-polish methodology has been proven adequate for polishing product wafers to the desired target; however, it significantly reduces polisher capacity/usage in production. Because the CMP-b process does not require the L/A wafer polish step, the cycle time is reduced by about **40%** over CMP-a. As for the tool throughput metrics, CMP-b outperforms CMP-a with about **66%** higher mechanical throughput. For the CMP operation in production, CMP-b is nearly **2.5 times faster** than CMP-a in completing a production lot polish.

This improvement in CMP process cycle time and throughput is key to high-volume manufacturing, allowing increased CMP production capacity, faster wafer turns, and cost avoidance in additional tooling to maintain equivalent capacity. CMP-b has many significant advantages over CMP-a to support the continuous ramp of BAW production, given the improvements in process quality and stability as well as polisher usage and capacity. With the demonstrated process extendibility and flexibility, CMP-b is poised as the polisher tool of choice at Qorvo. It undoubtedly can meet any new challenges of processing shrinking device features with more stringent process requirements for future BAW products.

OPTIMIZATION OF SLURRY AND PROCESS PARAMETER ON CHEMICAL MECHANICAL POLISHING OF CR-DOPED SB₂TE₃ THIN FILM

Ruifang Huo, Fang Wang, Yulin Feng, Yemei Han, Yujie Yuan, Kailiang Zhang*

School of Electronics Information Engineering, Tianjin Key Laboratory of Film Electronic & Communication Devices, Tianjin University of Technology, Tianjin, 300384, China

*Corresponding Author's Email: fwang75@163.com, kailiang_zhang@163.com

ABSTRACT

In this paper, we studied the composition of slurry including pH and the oxidizing agent Hydrogen Peroxide (H_2O_2) for Cr-doped Sb_2Te_3 (CST) thin film chemical mechanical polishing (CMP). Also the effects of the process parameters including down force and platen rotation rate were studied in detail. The results demonstrate that Material Removal Rate (MRR）has a relatively large dependence on pH values as well as the concentration of the oxidizing agent. Moreover, the MRR still exists when there is no down force and rotation, indicating that it is a mechanical abrasion assisted by chemical corrosion. Eventually, the root mean square (RMS) roughness was reduced from 4.02nm to 0.425nm and the MRR can be achieved at 100.45nm/min.

INTRODUCTION

Phase change random access memory (PRAM), one of the most promising mainstream of next-generation nonvolatile memories, has lots of advantages such as high speed, low power dissipation, high density, and low cost. CST has been considered as a good candidate chalcogenide material for the application of PRAM by its thermo-chemical stability and lower power dissipation [1, 2]. As the confined cell structure evolved from conventional planar structure is used to reduce the reset current [3], the CMP process becomes an essential process to achieve smooth and defect-free surface of the functional layer in this structure [4].

There is an amassing number of CMP research on the traditional phase-change material $Ge_2Sb_2Te_5$. However, few data about CST CMP has been reported yet. In this work, the influence of CMP process parameters including slurry pH, concentration of oxidizing agent H_2O_2, down force and platen rotation rate on polishing result of CST thin film was investigated in detail.

EXPERIMENTAL

The CST thin film samples were deposited by magnetron sputtering on 4-inch silicon wafers at room temperature with Cr target and Sb_2Te_3 target. The topography, roughness and the thickness of CST thin films were measured by atomic force microscopy (AFM, (5600LS, Agilent Technologies, Inc.) and step profiler (Dektak 150, Veeco Instruments Inc.). The polishing experiments were carried out on nSpire-6EC polisher (Strasbaugh) with double-layer pad (IC 1000/Suba IV). Colloidal silica slurry was used, H_2O_2 was added as the oxidizing agent and the pH values varying from 2 to 10 were adjusted by diluted nitric acid and KOH. All the process parameters were given in Table I.

RESULTS AND DISCUSSION

Figure 1 depicts the MRR and RMS roughness of CST film in relation to slurry pH values. It is evident that the MRR decreases gradually from the maximum (100.45nm/min) to the minimum (35.29nm/min) while the pH ranges from 2 to 10. This result illustrates that the CST CMP process depends on the pH value strongly, which might be interpreted as different dissolution of the CST oxide in acidic and alkaline solution. As a kind of

Table I. Parameter ranges of CMP for CST thin film

Polishing Parameter	value
Slurry pH	2~10
H_2O_2 concentrations (wt%)	0, 0.25, 0.5, 1, 2
Abrasive concentrations (wt%)	10
Down force (psi)	1~5
Platen rotation (rpm)	20~80
Slurry flow rate (mL/min)	120

corresponding oxide, which is apt to dissolve in acidic solution than in alkaline solution.

The RMS roughness also increases with respect to the slurry pH value, which elucidates that mechanical reaction was the main role at alkaline solution rather than the oxide dissolution. Meanwhile, the role of oxide dissolution gradually becomes stronger in the acidic solution. In general, the best result in terms of both MRR and the RMS roughness is obtained with acidic slurry of pH 2.

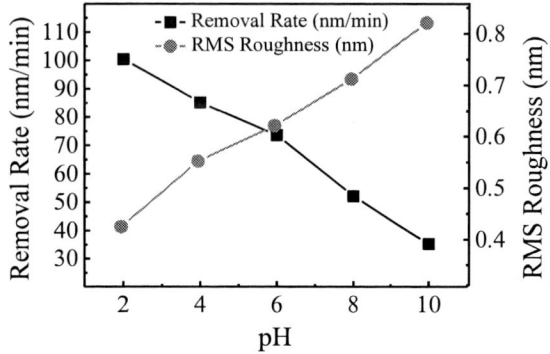

Figure 1: Relationship of removal rate and RMS roughness with respect to pH value

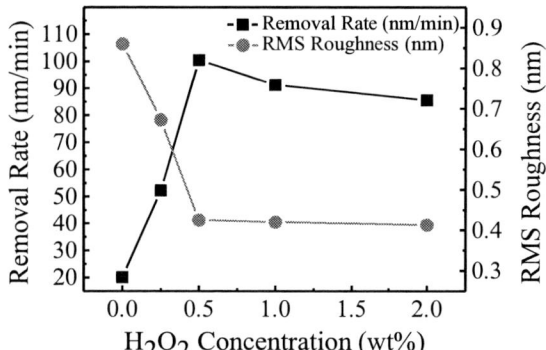

Figure 2: Relationship of removal rate and RMS roughness with respect to H_2O_2 concentration

In figure 2, as the H_2O_2 concentration rising, the MRR increases to a maximum of 100.45nm/min at 0.5wt% H_2O_2. A slight decrease of MRR can be seen at 1wt% H_2O_2 concentration and 2wt%. It is turned out that a small amount of H_2O_2 concentration is more advantageous to the MRR of CST. In the absence of H_2O_2, the MRR is as

low as 20.15nm/min, which means that the mechanical action has little impact on the removal of CST. The possible reason might be that as-deposited CST is harder than the formation of CST oxide layer. While in the presence of H_2O_2, the increase of MRR may be explained by the formation of CST oxide. With the H_2O_2 concentration further increasing, the excess oxide inhibits further increasing of the MRR. Eventually, a low value is received since the insufficient of CST oxide destruction and dissolution.

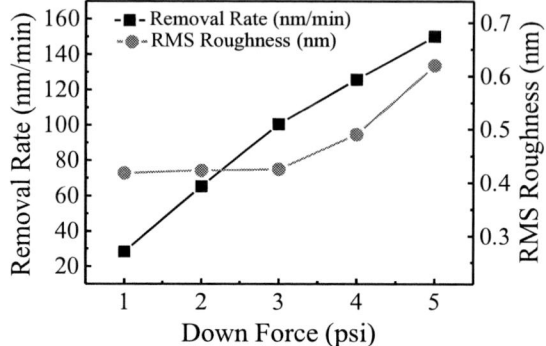

Figure 3: Relationship of removal rate and RMS roughness with respect to down force

Figure 3 shows that the MRR rise perpendicularly with the down force increasing, and the surface roughness gradually ranges from 0.418nm to 0.62nm. The rise of MRR may be attributed to the contact area between CST surface and slurry abrasives, which increases as the down force increasing. This relationship is consistent with Preston equation:

$$RR=KPV \qquad (1)$$

RR is the removal rate, P is the down force, V is the linear velocity with respect to platen rotation, and K is the Preston's constant.

Figure 4 shows that with the rise of platen rotation rate, the MRR increases linearly. It also meets Preston equation. While the platen rotation rate is 80rpm, a relatively high MRR of 100.45nm/min and low RMS roughness of 0.425nm are obtained.

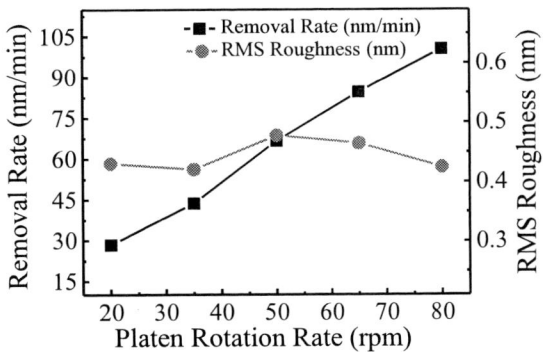

Figure 4: Relationship of removal rate and RMS roughness with respect to platen rotation rate

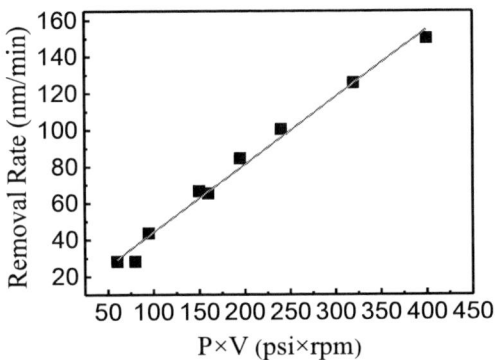

Figure 5: Relationship of removal rate and RMS roughness with respect to platen rotation rate

The MRR with respect to down force multiplied by velocity is plotted to observe whether the polishing behavior of CST is in line with the Preston equation. However, as shown in figure 5, it is discrepant that the MRR shows a positive intercept when both P and V tend to 0. This indicates that there is a small polishing rate under no P or V, which is caused by pure chemical reaction. So the Preston equation is slightly modified as:

$$RR=R_0+KPV \qquad (2)$$

R_0 is a constant removal caused by chemical reaction. According to the linear fitting result, K is equal to 3.27 and R_0 is to 7.10nm/min, which is consistent with the result of the wet etching experiment.

Figure 6 shows the surface topography variation of CST thin film pre- and post-polishing measured by AFM. Obviously, the surface structure has a major change after polishing. However, some corrosion pits could be seen at post-polishing surface. This means that the mechanical reaction is violently, and some corrosion inhibitor might be required to add in the slurry in order to delay the corrosion response in the subsequent experiment.

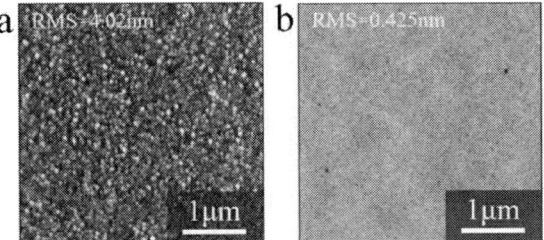

Figure 6: Surface topography and RMS roughness of CST thin films (a) pre- and (b) post-polishing

CONLUSION

The composition of the slurry and the effect of process parameters for CST thin film CMP were investigated in detail. The acidic solution is more suitable for CST-CMP, and the MRR has a relatively large dependence on H_2O_2 concentration. Overall, the RMS roughness is reduced to 0.425nm and the MRR is raised to 100.45nm/min. The process parameter study shows that the MRR still exists when there is no down force and rotation. Hence Preston equation was slightly modified as: $RR= R_0+KPV$, indicating that it is a mechanical abrasion assisted by chemical corrosion.

ACKNOWLEDGEMENTS

This work is supported by the National Natural Science Foundation of China (Grant Nos 61404091, 61274113, 61505144, 51502203，and 51502204），and Tianjin Natural Science Foundation (Grant Nos 14JCZDJC31500, 14JCQNJC00800)

REFERENCE

1. Wang Q, Liu B, Xia Y, et al. *Physica Status Solidi - Rapid Research Letters*, vol. 9(8), 2015, pp. 470-474.
2. Wang Q, Jiang M, Liu B, et al. *ACS Applied Materials*

& Interfaces, vol. 8(32), 2016, pp. 20885-20893.

3. Pirovano A, Lacaita A L, Benvenuti A, et al. *IEEE International Electron Devices Meeting*, 2003, pp. 29.6. 1-29.6. 4.

4. Lee, J.I., Park, H., Ch, S.L. et al., *Symp. on VLSI Tech.*,2007, pp. 102-103.

5. Q. Luo, S. Ramarajan, and S. V. Babu, *Thin Solid Films*, vol. 335, 1998, pp. 160-167.

STUDY OF WEAKLY ALKALINE SLURRY FOR COPPER BARRIER CMP ON MANUFACTURE PLATFORM

Jin Kang[1,2], Hanming Wu[2], Xing Zhang[1], Qiang Li[2], Jun Ge[2], Tong Feng[2],Ziqing Yin[2]Liu Yuling[3]

[1]Peking University, Beijing 100871, China

[2]Semiconductor Manufacturing International Corporation, Shanghai 201203 and Beijing 100176, China

[3]Institute of Microelectronics, Hebei University of Technology, Tianjin 300130, China

Email: Jin_Kang@smics.com

ABSTRACT

This study reports a weakly alkaline slurry (WAS) for copper barrier chemical mechanical planarization (CMP) process in standard 12-inch CMOS manufacture. We do copper barrier CMP process result comparison between current maintream production used slurry (BL) and WAS slurry adopts a unique alkaline macromolecular organic chelating agent with the high activation energy (named FA/O). Based on the same CMP process recipe, WAS shows the better erosion (more than 60% improvement) and dishing (more than 45% improvement) performance than BL. At the same time, WAS keeps the same level performance as BL by inline monitor pad thickness, Rs, and inline defect. All the results show that the WAS has advanced properties and it has potential application for future production line.

INTRODUCTION

In the 1980s, CMP was invented as a planarization technology for supporting scaling of CMOS devices. [1,2,3] And copper was used as interconnect materials from 0.13um technology to14nm production technology. In general, copper CMP process includes two main steps: first is bulk copper CMP which removes almost all copper on the wafer surface and stops on the barrier layer, the second is barrier CMP which to removes all the copper and barrier layer material on the wafer surface (see fig. 1).

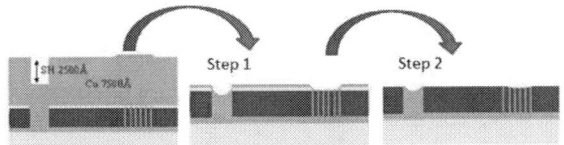

Fig. 1 two steps of Copper CMP process

As to the new WAS slurry proposed by Hebei University, the key technique is the addition of a unique alkaline macromolecular organic chelating agent with the high activation energy to complex metallic ion[4]. FA/O contains multiple functional groups such as tertiary amine, acetic acid and ethanol, and formula of FA/O is shown in Fig.2.

Fig. 2 Formula of a FA/O chelating agent. [5]

In the experiment, we use this WAS as barrier slurry to evaluate the barrier CMP performance under the real manufacture process condition. We also use a blanket copper wafer and a blanket TEOS wafer to study the WAS key components impact on the Cu removal rate (RR)and OX removal rate . To do more exactly process result comparison on the patterned wafers, we use the same step1 process condition (see fig. 1) to keep barrier CMP as the same pre-condition. To minimizethe process difference, we keep the same barrier CMP process recipe body, and only adjust the process time to meet final inline requirement between BL and WAS. We compare the dishing performance, erosion performance, inineRs performance and inline defect performance.

EXPERIMENT

We use the following tools to carry on our study:

12inch polisher: AMAT Reflexion LK.

Oxide Thickness measurement: NOVA SCAN 3090NEXT.

Copper Thickness measurement: RuDOLPH Metal pulse-II 300.

AFM: Veeco Dimension AFM; Atomic Force profiler TM.

Inline Defect Scan: KLA-Tencor KLA-2825I/S.

Inline Defect Review: APPLIED MATERIALS SEMvision TM G4.

We are aware of the fact that the chelating agent and the surfactant in WAS show very key rules during the process, so we do the study based on the table 1 below to check the Cu RR and OX RR impactation.

978-1-5090-6695-7/17 $31.00 © 2017 IEEE

Name	Condition
Process A	Baseline
Process B	Surfactant +60 units
Process C	Surfactant +30 units
Process D	chelating agent +0.2 unit
Process E	chelating agent -0.2 unit

Table 1: chelating agent and surfactant split condition

We use 90nm technology Metal-4 CMP patterned wafers to do the inline result comparison vehicle. Based on the RR test result, we select the one WAS condition to do the patterned wafer test to compare the dishing performance, erosion performance, inline Rs performance and inline defect performance. To minimize the process difference, we keep the same barrier CMP process recipe body, and only adjust the process time to meet final inline requirement between BL and WAS. So that the comparison can be between apple to apple.

RESULTS AND DISCUSSION

Table 2 shows the blanket Cu RR & OX RR obtained from table 1 conditions. We can see that both Cu RR and OX RR of BL is higher than WAS under thesame process recipe.

Process B and process C show that more surfactant does not impact about Cu RR, but will impact OX RR a little slower (5.3%), and more surfactant reduce OX RR, and thesurfactant can reduce the surface tension of solution, enhancing the fluidity and then improve the mass transfer of the slurry.More surfactant means more material transportation. For Cu RR, the key player is the oxygen dissolved in the slurry which has less relationship with surfactant. For OX RR, the more surfactant means lower surface tension, and result in lower mechanical effect, which is a key player in OX CMP process.

Process D and process E show that more chelating agent does not impact about OX RR (lower 0.9%), but will impact Cu OX RR faster (9.3%). FA/O chelating agent plays an important role in the Cu reaction during the Cu CMP process, and it is easy understanding that the more FA/O chelating agent will increasing the Cu RR before this kind of chelating reaction saturation.

Name	Cu RR (A/min)	OX RR (A/min)
Process A	438	817
Process B	333	735
Process C	329	774
Process D	333	689
Process E	364	695

Table 2: Cu RR & OX RR result by different chelating agent and surfactant

The table 3 shows the inline monitor padand erosion pad performance.The monitor pad thickness of WAS is

thicker 148A (5.1%). Fig. 2 shows the metal 4 Rs test result, and we can see that WAS's result is 3.9% lower than BL. As the test pad design we know that Rs test result directly responses the inline monitor pad result. Our experiment result of monitor pad and Rs exactly match.

Condition	Monitor Pad(A)	Erosion Pad(A)
BL	2899	1243
WAS	3047	2038

Table 3: Monitor Pad and Erosion Pad performance comparison

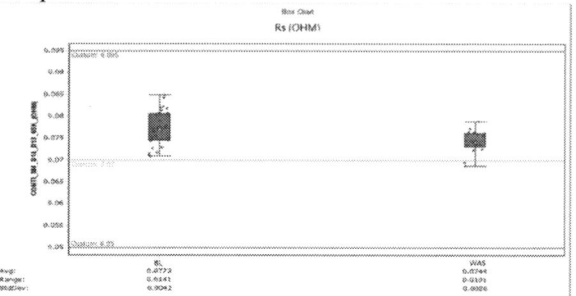

Fig. 2 Metal 4 Rs comparison

The erosion pad thickness of WAS is thicker 795A (64%). We keep the monitor pad thickness match by fine tuning polishing time, and at the same time, the erosion pad result shows that WAS is much more thicker than BL which means WAS has better planarization efficiency than BL. As we know that the patterned wafer surface condition just after step1 (shown in Fig.1) has exist the step high not only in monitor pad or dishing pad or erosion pad area, but also in the whole wafer. Barrier CMP need to remove more in peak area and remove less in valley area, and the copper is the main material in valley area, and WAS's result shows it has lower Cu RR than BL. In general, mechanical effect will be less in the concave region than in the convex region. For WAS, surfactant molecules are adsorbed preferentially on the copper surface, which generate an isolated layer between copper and FA/O alkaline agent, which is a unique alkaline macromolecular organic chelating agent with the high activation energy to complex metallic ion. As we know that chemical reaction will stop if the activation energy is not high enough to overcome the threshold reaction energy. Usually, a high molecular weight possesses the higher activation energy. So in the concave region, WAS will be more difficult to remove copper than in the convex region, which means that WAS should have better planarization ability.

We use AFM to measure the dedicate dishing and erosion monitor pad on the patterned wafers. Fig. 3 shows the measurement result by AFM, and we can see that from the wafer center to the wafer edge, and both in the dishing area and in the erosion area, WAS shows the better erosion (more than 60% improvement) and dishing (more than 45%

improvement) performance than BL.

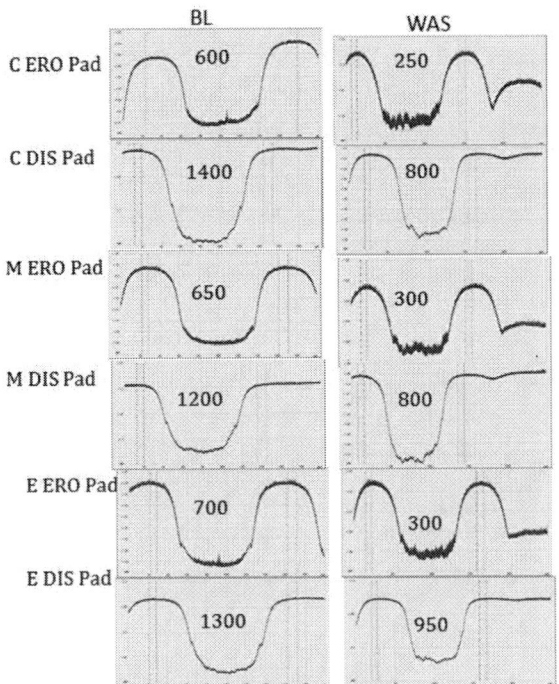

Fig. 3 AFM results of dishing and erosion

As we know that in a production line, defect performance is always a very key index, so we also monitor the inline defect performance of WAS. Fig. 4 shows the inline defect result between BL and WAS, and we can see that both BL and WAS show the very low defect density.

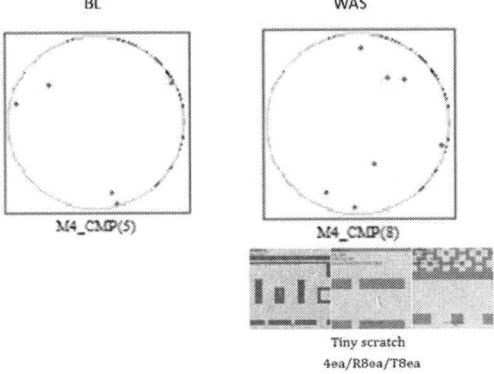

Fig. 4inline defect result between BL and WAS

CONCLUSION

This study reports the performance of Cu CMP process with our unique weakly alkaline slurry. Compared WAS's behavior with manufacture baseline slurry's, WAS seems the lower removal rate on Cu and

oxide blanket wafer, but obviously better erosion and dishing performance on patterned wafer than baseline slurry. And the patterned wafers with WAS process also get the comparable defect level, Rs result to what withbaseline process.

In this paper, the ratio of surfactantand chelating agent in WAS varies to compare the process performance. And the result shows little difference.

Although weakly alkaline slurry is currently not the mainstream slurry in application, as IC node evolution, more and more interest will be focused on this type of slurry, especially in Cu CMP process, since erosion and dishing would become more critical in future. We complete a simple test on it, and more systematic work will go on in future.

ACKNOWLEDGEMENTS

This work was supported by SMIC(BJ) B1 CMP team, SMIC United Lab, and Institute of Microelectronics, Hebei University of Technology.

REFERENCES

[1] S. Wolf. Silicon processing for the VLSI Era (Lattice Press, Sunset Beach, CA, 2002) Vol. 4, p.313.

[2] Y. Li, Microelectronic Applications of Chemical Mechanical Planarization (Wiley, Hoboken, NJ, 2007) p.16.

[3] S.M. Sze, Semiconductor Devices: Physics and Technology (Wiley, New York, 2012) p.205.

[4] H. Y. Wang, Y. L. Liu and D. C. Zhang, Chinese J. of semiconductors, 23(2), 217 (2002)

[5] Xiaodong Luan, Yuling Liu, etc. al. / Microelectronic Engineering 160 (2016) 5–11

A SYSTEM QUALIFICATION PLATFORM FOR FPGA BOARD LEVEL TESTING AND AUTOMATED REGRESSION

Yibin Sun[1][*] and Ping Chen[1]

[1] Lattice Semiconductor (Shanghai) Co., LTD. Shanghai 200233, China

*Corresponding Author's Email: Sun.Yibin@latticesemi.com

ABSTRACT

This paper introduces a practical platform called the System Qualification Platform (SQP). The SQP is designed to execute FPGA Software (SW) and Silicon (HW) co-verification at the board level and becomes an effective complementary solution to current SW validation methods which are based on simulation. With the SQP the user can develop cases, manage test suites/boards, launch testing, and analyze on-board running results on a PC. Moreover, the SQP's flexible architecture makes it a best practice for integrating new SW products and correspondingly-supported HW products, while offering more test cases to improve coverage. Automated regression can be easily fulfilled when the development phase ends.

Keywords—FPGA; platform; wishbone, FTDI, Software; Hardware; Co-verification; Validation; UVM; Automation

INTRODUCTION

Classic FPGA software qualification includes GUI testing, standalone engine testing, integration testing and software based co-verification for HW support. Like most pure software testing, regression testing is one of the most frequently-used methods for SW release. This methodology is very suitable for most cases without real-world hardware and can be used to achieve a high level of automation relatively easily. However, it's not perfect for new silicon (FPGA), even if each implementation stage has been qualified from synthesis to bitgen, for the following reasons:

(1) Software modeling is limited for some silicon features such as mixed-signal function blocks;

(2) Full chip simulation can be time consuming for some features such as OSC and I2C;.

(3) Feature-level testing restricts system-level complexity and diversity in some test cases.

To resolve the testing gap, we propose a complementary validation solution that extends beyond the existing methodology with the following key features:

(1) It must have the capability to qualify software at the system level and speed up the testing cycle (HW acceleration) with the help of hardware;

(2) It must migrate between different software and silicon product lines easily;

(3) It must plug into the new components for specific software / silicon features or be reused / wrapped from existing designs or open source designs in a standard and

simple way;

(4) It must launch easily for regression testing.

Under such requirements the SQP has been evaluated, defined and developed to enhance FPGA Software (SW) and Silicon (HW) co-verification at the FPGA board level with automated regression capability.

A PLATFORM APPROACH

The block diagram of the structure of the SQP as shown in Figure 1 can be divided into a HW platform (SQP_HW), a SW platform (SQP_SW) and an Operation Platform (SQP_OP). All of these SQP sub-platforms follow Lattice's Register Mapping Interface (RMI) [1] standard.

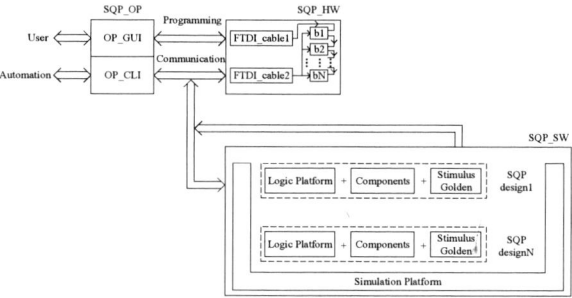

Figure 1: SQP Structure

The typical workflow for using the SQP is to develop a SQP design (RTL design for FPGA) with the help of a SQP_SW, then use a SQP_OP to call specific FPGA software to download the designs to the SQP_HW and qualify the results.

REGISTER MAPPING INTERFACE

The Register Mapping Interface (RMI) is a light standard defined and used by Lattice [2] to help communicate between the outside world (as master) and the FPGA chip (as slave).

As shown in Figure 2, the RMI features three layers and two interfaces. The outside application processor (AP), SQP_OP or simulation platform (SQP_SW) works at the physical layer followed by a Physical Transport (PT) interface which uses a SPI, I2C or UART communication protocol. The logic platform (SQP_SW) works at the Transport layer and a wishbone [3] is used as the Register Transport (RT) interface. The register layer is the core of RMI. It is a register set composed of 256 pages with 256 registers on each page. Page 0 is used for global registers in the logic platform. The additional pages are used for the components.

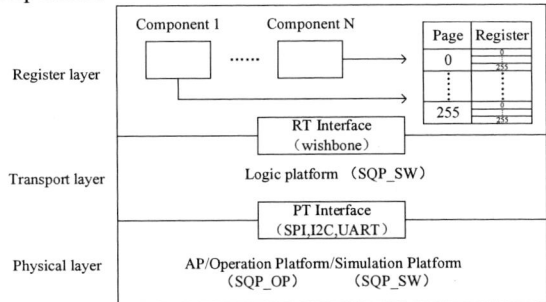

Figure 2: RMI Layer Structure

A simple packet-based protocol comprised of a 3-byte header and data bytes of unlimited length is defined to fit the PT interface (see Figure 3). The first two bytes are used to locate the specific SQP design and the components inside it. The third byte defines the operation type. The rest of the bytes are reserved for read-write data based on that operation. For write mode, the data starts from the fourth byte and is written to a global/component's register(s). For read mode, the data means golden data and is compared with the real data read from SQP design.

Bytes	0-1	2	3 - N
Contents	Addr	Control Set	Data

Figure 3: Simple Packet

HW PLATFORM

The hardware platform is comprised of multiple Lattice Evaluation Kits connected in a daisy chain configuration. Each kit features an on-board FPGA chip (silicon). One FTDI cable is used to download the SQP designs to the FPGAs one by one (order honored). Another FTDI cable with a SPI protocol as a PT interface broadcasts the packets (stimulus) to all the boards and then reads back the real data from the chosen board, if needed. These two FTDI cables are controlled by the operation platform on a PC. Figure 4 shows what the actual HW platform looks like.

Figure 4: HW platform with two boards

Usually a new Lattice Evaluation Kit [2] is added to HW platform when a SQP design for any new silicon (FPGA) has been developed.

SW PLATFORM

The SW platform consists of a logic platform, components and a simulation platform. The first two are used to build SQP designs. The third is used to verify the designs.

Logic Platform

The logic platform is a requirement for all SQP designs. As mentioned before, a specific board address must be assigned inside it. The logic platform works as a bridge and a central controller at the same time as shown in Figure 5.

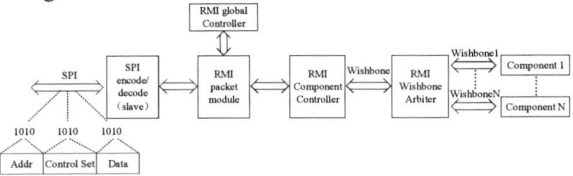

Figure 5: Logic Platform Structure

A typical write operation is used to illustrate the workflow. When the SPI module detects a broadcast, it encodes the SPI protocol (as a slave) based bits into a defined packet and the packet module analyzes the content. If there is a device hit (when the pre-defined board address matches the given address in the packet), then it passes the command to the Global Controller (GC) or the Component Controller (CC). The GC handles operations like system reset, connection test and so on. The CC translates the command to downstream components with reference to the wishbone standard [3]. Meanwhile, a wishbone arbiter decides which component should be connected in real time. Finally, the write data is written to the specific register(s) of a chosen component.

The implementation of the logic platform is pure RTL modeling. It only takes about 0.5 K LUTs. In other words, it can fit into any Lattice silicon product line including the ultra-low density devices (1K LUTs minimum) without any changes to Lattice software support.

Components

Components are the independent function blocks with a self-check feature and wrapped with the wishbone interface. In test standpoint, each component is verified on SQP just like other general test cases.

For component development, the system must consider what kind of features are critical and not easily verified at the software level regardless of whether they are software or silicon related. Meanwhile, valuable existing designs and open source designs must also be taken into consideration in parallel. They are the core logic

978-1-5090-6695-7/17 $31.00 © 2017 IEEE   317

components and will be wrapped with limited efforts by adding some glue logic, such as a simple state machine (FSM), and performing some necessary changes based on the template for self-checking. Each component has one unified component address (monopolizes 1 page of register layer) and a set of predefined registers that can be accessed by the operation platform/simulation platform, as shown in Figure 6.

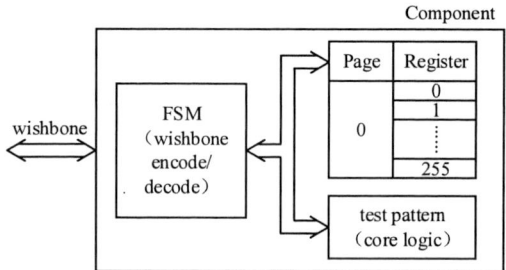

Figure 6: Component Structure

These new components combined with previously released components as well as the logic platform become a new SQP design which can be treated as a set of "read-write registers".

Simulation Platform

The simulation platform was developed by SystemVerilog and is used to verify each SQP design with text files (input.txt) that contain the stimulus (for write operation) and the golden data (for read operation). It emulates SQP_OP's behavior and delivers the detailed results (output.txt) by launching simulation using Questasim.

Input.txt is organized with the following RMI data packet definition, as shown in Figure 7:

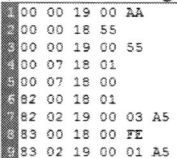

Figure 7: one example of input.txt

OPERATION PLATFORM

Once a SQP design has been fully verified at the software level (usually the RTL level) by using the simulation platform and the corresponding hardware environment has also been setup in SQP_HW, it's time to use the operation platform to perform board-level testing and automated regression.

The operation platform has two versions: OP_GUI and OP_CLI. OP_GUI supports four key features by manual operation in a GUI:
(1) Create a SQP project based on the available SQP designs and SQP_HW;

(2) Configure the run environment: change the design order according to SQP_HW, update input.txt (optional), assign software used;
(3) Run and check: run and download designs using daisy chain mode, broadcast stimulus by reusing the same input.txt used by simulation platform, read back results and compare them with the golden data, display the detailed results;
(4) Save all configurations to a configuration file which can be reused by both OP_GUI or OP_CLI later.

Figure 8 shows what the OP_GUI looks like. It is usually employed to help users setup or update the run environment when a new evaluation board with a SQP design is added, to check the results for the first time, and to prepare regression for the future SW release, as shown in Figure 8. Qualification can also be performed locally using a command line mode with the help of the OP_CLI which needs a configuration file and SQP designs as inputs.

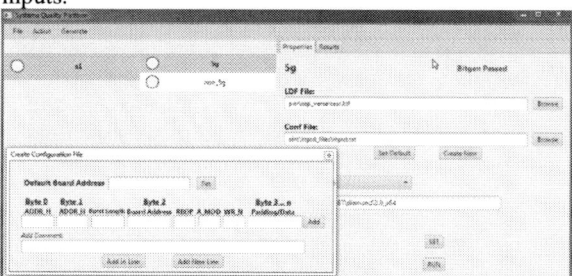

Figure 8: GUI version of SQP_OP

In summary, the operation platform offers maximum flexibility to the regression manager: automated regression can be conducted by either "push button" with the OP_GUI or a "single command" (launched by scripts with a standard API) with OP_CLI.

CURRENT STATUS AND FUTURE DEVELOPMENT

The SQP has been released in two versions. After putting SQP into practice for the release of two major Lattice FPGA software products, the SQP has proven to be an effective complement to the current SW validation method for the following reasons:
(1) It supports regression with a high level of automation;
(2) It allows engineers to finish the complete testing of some specific features in minutes rather than the hours/days required for a full chip simulation during the test cycle;
(3) It also catches some critical issues.

Besides adding more components to improve test coverage, we are also planning to enhance both the simulation platform and the operation platform. As mentioned before, both stimulus and golden must be created manually before a SQP design is under

978-1-5090-6695-7/17 $31.00 © 2017 IEEE   318

development. We cannot achieve high functional coverage this way, so a UVM (Universal Verification Methodology) [4] -based simulation platform that supports both constrained random test vectors and direct test will be implemented with the next SQP release. The new workflow will work in this way: the SQP_OP launches the simulation platform to dynamically generate both stimulus and golden, then it uses these "fresh" stimuli to get real data from the SQP_HW and compare it with the golden data. As such, it will offer real time co-verification between FPGA Software (SW) and Silicon (HW) at the board level.

CONCLUSION

SQP offers engineers the ability to achieve an independent output assessment of FPGA software by comparing the behavior of the physical outputs of the silicon (ending) with their corresponding RTL model simulation results (beginning). In other words, the SQP works at a higher level of abstraction and views the whole system, including the FPGA software and silicon, as a single entity or black box.

Automated regression support is also an important key feature for the SQP considering that it's usually much harder to perform automation at the board level.

REFERENCES

[1] Lattice, "ULD Reference Design Register Mapping Interface", unpublished

[2] www.latticesemi.com

[3] WISHBONE System-on-Chip(SoC) Interconnection Architecture for Portable IP Cores Specification B.3

[4] Accellera.org/community/UVM

TWO INDIRECT METHODOLOGIES FOR TESTING FPGA INTRINSIC PROGRAMMABLE LOGIC CELL TIMING PERFORMANCE

Hongpeng Han

Lattice Semiconductor Corporation, Shanghai, China, 200233

Hongpeng.Han@latticesemi.com

ABSTRACT

Programmable Logic Cell (PLC) timing test is one of the most critical items for post-silicon validation of a new Field Programmable Gate Array (FPGA) product because it determines the fundamental performance of the FPGA chip. However, it has been very difficult to accurately measure all aspects of PLC segment timing for several practical reasons. First, some segments exhibit merely 100ps delay which places severe requirements on the resolution of the measurement system. Second, some segments are intrinsic elements in a FPGA and cannot be accessed directly from an external measurement port.

Clock to data output delay (Tco) in one flip-flop and routing delay within a single Programmable Function Unit) (PFU) are two typical PLC characteristics. Engineers measuring both items must address the above-mentioned challenges. Thus, these measurements can't be covered in traditional post-silicon validation, which results in gaps in silicon test coverage.

This paper introduces innovative indirect test methodologies that overcome those challenges and allow engineers to cover those kinds of intrinsic PLC timing parameters in characterization. The methodologies were implemented and verified on real FPGA products. Results show that those key intrinsic PLC timing parameters can now be measured precisely and reliably with the proposed methodologies.

INTRODUCTION

In integrated circuit development flow, post-silicon validation is required to validate the fabricated chip's actual performance taking into consideration foundry process, voltage, temperature variation and non-ideal environmental effects [1]. FPGAs offer a high degree of flexibility and can be used in many ways. So post -silicon work must guarantee maximum usage, making the post-silicon work even more important for a FPGA.

As a fundamental part of the FPGA, PLC affects the performance of the FPGA blocks using it as a resource. PLC work includes software design pre-silicon and hardware test post-silicon. If the PLC in silicon doesn't behave as designed or the PLC software design cannot reflect the silicon's PLC performance, then the applications based on PLC design will be inaccurate and possibly totally wrong. Therefore, testing the silicon performance of PLC correctly and ensuring the silicon and software design match each other are of great importance.

However, many PLC items cannot be tested directly. Some are short and confined within a small unit. Others always appear with other items and

978-1-5090-6695-7/17 $31.00 © 2017 IEEE

cannot be separated from them which increases the difficulty of the PLC test. Among these items, tco in one flip-flop and short routing within a single PFU while driven by different routing are very typical.

INDIRECT TEST METHODOLOGIES

In a typical FPGA PLC is divided into two parts: PFUs and programmable routing units. A PFU is comprised of a Look Up Table (LUT) and two registers. It can be configured as normal logic mode, fast carry in mode, and RAM mode [2]. Tco in one PFU represents the delay of the clock input port to the data out port of the Flip-flop.

Since tco in one PFU is small and always independent of other PFUs, it cannot be directly fetched and tested. Instead, we use a test methodology in which we accumulate M numbers of tco units and insert a known routing delay between two adjacent tco units. This known routing connects the data out port of the preceding stage tco unit and the clock port of next stage tco unit (see Figure1).

The theory behind this methodology is that the preceding stage Flip-flop drives the next stage Flip-flop via a known routing unit. The known routing unit is carefully chosen to meet driving rules. Logic high 1 is set to the data input port for each Flip-flop used in the design. The initial states of the data input and output are both Logic low 0. Take the first stage Flip-flop, for instance. When power is up and the clock comes, the ready high 1 in data input will be read into the Flip-flop, and the output will become logic high 1. Thus, the output will change from 0 to 1 and will drive the clock port of the next stage Flip-flop via the known routing, and tco will transmit to next stage. The rest may be deduced by this analogy.

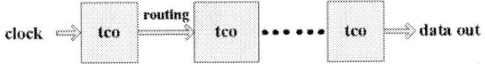

Figure 1. Tco test methodology.

Then use equation (1) below to calculate the tco delay.

$$Tco = [A - (M-1)*B]/M \qquad (1)$$

Where,

A represents the whole path delay.

B represents the known routing delay.

M represents the number of tco units.

Programmable routing units interconnect all fabric units within the FPGA. They can be divided into global routing, long routing, short routing and other dedicated routing types. Different routings are connected by a corresponding switch box. There is one short routing, which lies within a PFU and is driven by another different long routing segment. It is one of the shortest routings.

To test the small routing, N units of the routing and its driver are accumulated to form a whole path. Two Flip-flops are placed at the start point and the end point (see Figure 2). The skew of the two clocks going into the Flip-flops are set within 20ps to ensure test accuracy. By testing the setup time difference of the two Flip-flops, the delay of the whole path can be calculated.

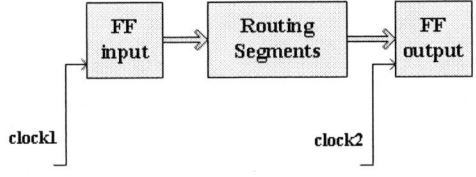

Figure 2. Small routing test methodology.

To eliminate the error caused by the two Flip-flops, we developed two designs. The first, which has more routing segments, is called the long design. The second, which has fewer routing segments, is called the short design. The difference in the routing segments of the long and short designs is carefully chosen to balance the fabric scale and the affection of the Flip-flop.

To eliminate the error caused by the two clock trees, we also swap the two clocks going into the two Flip-flops for both the long and short designs. This means that in one design clock1 goes into the input Flip-flop and clock2 goes into the output Flip-flop while in the other design clock1 goes into the output Flip-flop and clock2 goes into the input Flip-flop. The mean value of the two designs is used to calculate the last small routing.

Next, we use equation (2) to calculate the short routing delay. This equation is based on the difference of the switch box delay of the routing and its driver routing to within 20 ps.

$$Routing = [(L1+L2)/2-(S1+S2)/2]/N-D \qquad (2)$$

Where,

L1 represents the Tsu of clock 1 to clock 2 in long design1,

L2 represents the Tsu of clock 2 to clock 1 in long design 2,

S1 represents the Tsu of clock 1 to clock 2 in short design1,

S2 represents the Tsu of clock 2 to clock 1 in short design 2,

N represents the number of the routing segments difference of the long and short design,

D represents the known driven routing delay.

The designs above were completed using an in-house developed FPGA underlying hardware language, and are fine tuned to match the software model. Finally, we implemented the design into a bit stream and downloaded it to a FPGA to test silicon performance.

RESULTS

Test results of the two items via the indirect methodologies are depicted in Figure 3 and 4. From the graphs we can see that the silicon data using the indirect methodologies matches the software design data both in the trend and value across process corner, core voltage and silicon junction temperature (PVT) conditions, and that the tests are repeatable.

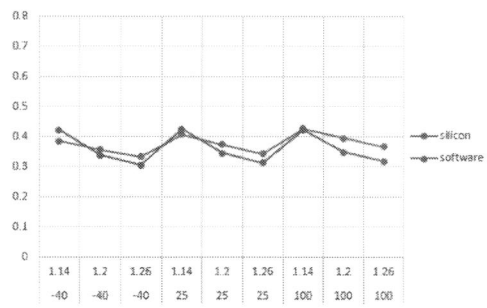

Figure 3. Tco test results vs software value.

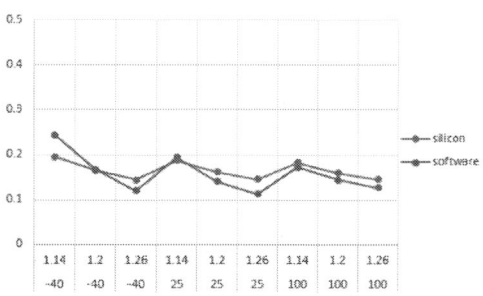

Figure 4. Small routing test results vs software value.

(The blue line shows silicon data by the indirect methodologies. The red line shows software design data. The vertical axis represents the delay with unit ns. The horizontal axis represents the core voltage with unit voltage and the temperature in degrees centigrade.

ACKNOWLEDGEMENTS

The author would like to thank the team members at Lattice for their generous support.

REFERENCE

[1] Gray, R. Post- Silicon Validation Experience: History, Trends, and Challenges. GSRC Workshop on Post- Si Validation, Anaheim, June 9, 2008.

[2] Zhicheng Liang, Lattice FPGA and CPLD Design (Basic), p.4, Posts and Telecom Press (2011).

A STATISTICALLY ROBUST METHODOLOGY FOR OPTIMIZED SAMPLE SIZE DETERMINATION FOR FPGA POST-SILICON VALIDATION

Weijun Qin

Lattice Semiconductor, Shanghai, China

Email: wqin@latticesemi.com

ABSTRACT

Due to sheer design complexity, it is nearly impossible today to detect and fix all bugs before manufacture. Post-silicon validation is becoming extremely important [1]. Since Field Programmable Gate Arrays (FPGAs) use a novel structure to implement programmability, FPGA silicon validation can be very time consuming. It is not unusual to see some measurements take several hours for each unit. It is hard to meet tight time-to-market requirements for a new product release if a redundant sample size is selected. However, an inadequate sample size may give misleading results which can cause serious quality risk. So determining the optimal sample size for FPGA silicon validation has now become a very critical process. This paper introduces a statistical method to determine the best sample size in FPGA post-silicon validation to achieve an optimum balance between the contradicting requirements for product time-to-market and quality coverage.

INTRODUCTION

Sample size determination is often an important step in planning a statistical study—and it is usually a difficult one. An undersized study can waste resources if it doesn't deliver the capability needed to produce useful results, while an oversized one can consume more resources than are necessary. Sample size selection is usually more important for functions that require a lot of time to collect the data [2]. FPGA post-silicon validation is such an area. As product designers rely on FPGAs to implement a rich array of new features, post validation takes longer and has a larger impact on the lead time needed to release a product to mass production. Therefore, determining the right sample size for validation becomes critical because it directly impacts validation cycle time and quality.

Many studies describe methodologies to calculate sample size for point estimator or hypothesis test. But those methods leave some gaps when applied to post-silicon validation of FPGAs. Take, for instance, a website planning to initiate a poll to evaluate the percentage of voters who favor a presidential candidate. How do we determine the optimal sample size if we want a 95% confidence level and +/-5% precision? For each vote, there are only two results: yes or no. It is a binomial distribution.

For populations that are large, Cochran [3] developed an equation to estimate sample size. Calculated sample size is at least 385 for this type of polling. Of course, this method and sample size can't be used for post-silicon validation of FPGA, since most FPGA parameters don't follow a binomial distribution. Besides, 385 samples are too many for FPGA. In one typical FPGA product, there are hundreds of parameters. Some measurements are very time consuming. The data collection for 300+ samples can last around 3 months, even with multiple benches setup in parallel to support data collection.

Clearly a large sample size is not feasible for most of FPGA products. How about selecting just 3 to 5 samples? Validation can meet the requirements of a shorter time-to-market, with even less bench equipment. However, customers may question if the data from a small sample size can adequately represent the true population. Moreover, we observed that the actual failure rate was higher than expected if silicon timing validation was done on limited data only.

How does one determine the proper sample size for post-silicon validation of FPGA? This is a question that must take into consideration not only statistics, but business and engineering considerations as well. In his paper [4] Glenn D recommended considering several factors in sample size selection, including the purpose of the study, population size, the risk of selecting a "bad" sample, and the allowable sampling error. Those guidelines are too general to apply to actual FPGA validation. We still need a methodology to quantify those factors.

This paper proposes a new methodology to fill the gaps described above and determine the optimum sample size for FPGA post-silicon validation. Using this proposed methodology, the impact of sample size on post-silicon validation quality coverage can be quantitatively evaluated.

FACTORS IMPACTING SAMPLE SIZE

If a population has a normal distribution, the sampling mean $\bar{x}$ will be exactly normal distributed, regardless of the size of n. Sampling mean falls into an interval and that interval has relationship to sample size. In his book Will Mendenhall introduces an equation that states if a random sample of n measurements is selected from a population with mean μ and standard deviation σ, the sampling distribution of the sample mean $\bar{x}$ will have mean equal to

978-1-5090-6695-7/17 $31.00 © 2017 IEEE

μ and standard deviation [5]:

$$\bar{s} = \frac{\sigma}{\sqrt{n}} \qquad (1)$$

When talking about sample size we should discuss allowable error first. Error of sampling mean can be estimated with the population's standard deviation using equation (1). Since standard deviation isn't an intuitive indicator, we can introduce a new friendly indicator - DPM (Defect Per-Million) into the discussion. If a population has a normal distribution, DPM can be calculated by using the equation below and Excel functions:

$$DPM = NORM.S.DIST(z, TRUE) \times 10^6 \qquad (2)$$

Where *NORM.S.DIST(z, TRUE)* is an Excel function that returns the cumulative distribution by default the 2nd parameter is *TRUE*, z is the abscissa of the normal curve that cuts off an area at the tails. For instance, when z = -1.64, the area is around 5% at left tail of normal curve and DPM is around 50502.

What DPM level is acceptable? Each application area may differ slightly. For example, Huawei shipped out around 10 million mobile phones each month in first two quarters of 2016. According to summary data from Blanca Technology, the overall defect rate of mobile phones is around 0.2% or 2000 DPM in 4th quarter of 2015. If one assumes 10% of all defects are caused by chips (around 200 DPM), Huawei's chip provider must provide a DPM of less than 50 to meet Huawei's reliability target.

The ultimate DPM target is often driven by business goals and it is a key factor in the determination of sample size. Though DPM is a good tool to describe the business target, Cpk is friendlier for engineering purposes. In mathematical terms, Cpk can be calculated by [6]:

$$Cpk = min\left(\frac{USL - \mu}{3\sigma}, \frac{\mu - LSL}{3\sigma}\right) \qquad (3)$$

where USL and LSL are upper and lower specification limits respectively and the μ and σ are the process mean and standard deviation respectively for individual measurement of characteristic interest.

Parametric (performance) measurement is one important task in post-silicon validation. Parameters are represented in the datasheet as target specs. Generally, parameters in silicon validation are continuous numeric and follow a normal distribution. Parameter margins can be calculated by Cpk. It is generally set to 1.33 (8σ interval). The target Cpk has a relationship to a company's capability or application areas as well. For example, some companies have better process control than others. The Cpk target can go up to 1.67, while some companies reach only 1.0.

DETERMINE SAMPLE SIZE

In this section, we will demonstrate mathematically how to evaluate the impact of sample size with DPM and Cpk. Ideally, if μ and σ of a population are known, DPM is easily derived from Cpk. In this situation, the area falling outside the specification limit is exactly the tail of normal curve that cuts off by abscissa, $3\sigma \times Cpk$ away from μ. To simplify the following discussion, we can assume the parameter has a specification limit on one side only, e.g. *LSL*. Variable z in equation (2) is:

$$z = -(3\sigma \times Cpk - \mu) \qquad (4)$$

After normalized by the standard normal curve, with μ=0 and σ=1, z on left side is simplified to:

$$z = -Cpk \times 3 \qquad (5)$$

In practice, sampling mean $\bar{x}$ from n samples isn't exactly equal to population mean μ. According to equation (1), if the experiment is repeated with n random samples each time, sampling mean $\bar{x}$ also follows normal distribution and it has a much narrower distribution than population. More important, this distribution is dominated by sample size n. When the DPM target is lower, the allowable error is smaller and sample size should be larger. If considering a confidence bound of 95%, the maximal sampling error (abbreviated as *Er*) that can be caused by sample size n can be quantified by the equation:

$$Er = 1.64 \times \frac{\sigma}{\sqrt{n}} \qquad (6)$$

Where σ and n are the same as equation (1), 1.64 is a constant that corresponds to 95% confidence bound on one side. In practice, this equation can also be normalized using standard normal distribution with σ = 1.

The plots below graphically show how DPM is derived from sampling error which is caused by sample size. Consider an example where μ=0 and σ=10, and sample size n=36. If we assume a 95% confidence bound, sampling mean $\bar{x}$ falls into an interval and can be, *1.64 × 10/√36 = 2.73* at worst case. This is the distance between true population-mean (Fig.1 green dotted line) and sampling-mean (Fig.1 red dotted line). Assuming the target spec is -20, if engineers get a parameter sampling mean 2.73, and believe it represents population mean (μ=0), Cpk is 0.76. The defect rate can be calculated around 1.15% (11500 DPM), (refer to Fig.1).

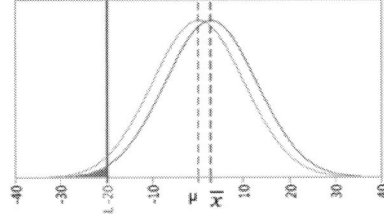

Fig.1. Defect rate without compensation for statistical errors that are caused by sample size

However, the above defect rate calculation doesn't take into consideration the count error of mean. At worst

978-1-5090-6695-7/17 $31.00 © 2017 IEEE

case of a 95% confidence bound, mean from 36 samples can have population mean shift up to 2.73 as seen in the red dotted line in Fig.2. To guarantee silicon performance in a datasheet, it is necessary to compensate for the error of mean. The green dotted line below represents true population mean after compensation for sampling error. The defect rate can be calculated around 2.28% (22800 DMP). It is represented by the grey areas under the green line in Fig.2. Cpk is around 0.67 after compensation.

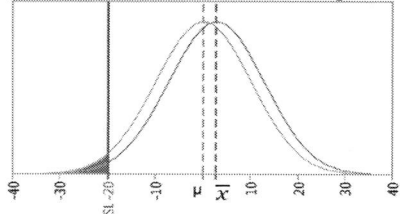

Fig.2. Defect rate after compensation for statistical errors

According to the above analysis of sampling mean and DPM shift, equation (2), (5) and (6) can be combined into the equation below:

$$DPM = NORM.S.DIST\ (\text{-}(Cpk \times 3 + \frac{1.64}{\sqrt{n}})) \times 10^6 \quad (7)$$

Where *NORM.S.DIST* is an Excel function that comes from equation (2) and its 2nd parameter is set to *TURE* by default, *Cpk* comes from equation (3), and n is sample size. In the above equation distribution is normalized with σ = 1. To better understand the relationship among sample size, DPM and Cpk, we can also plot them in Fig.3. To reach the 50 DPM target, sample size can be greater than 300 with Cpk=1.33 or less than 20 with Cpk=1.50.

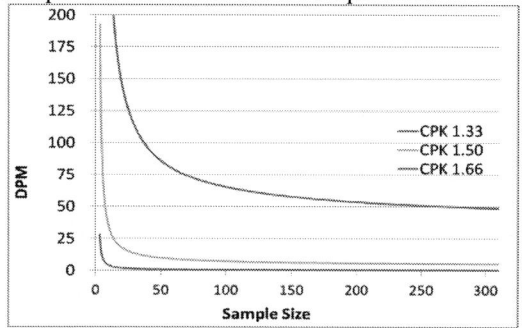

Fig.3. This figure plots DPM vs. sample size up to 300. The three curves correspond to three typical Cpk values.

According to the equation (7), the optimum sample size can be determined clearly for specific Cpk targets and DPM goals. For example, a sample size of 16 to 20 is recommended for post-silicon validation assuming the Cpk target is 1.50 and the DPM goal is 25.

EXPERIMENT AND RESULTS

To test these concepts, we studied a real case in silicon validation in which sample size was 16 and DPM goal was 25. The study had volume test on 1650 chips to measure output frequency of a hard IP. The mean and standard deviation from 1650 chips were 47.597 and 0.2345. Since the sample size of volume test was large, the population mean and standard deviation was assumed to be the static results from 1650 chips, μ = 47.597 and σ = 0.2345.

In this experiment, we randomly selected 16 samples from the above chip pool. Samples were returned to the chip pool after measurement and could be selected again in a subsequent experiment. For each time, we calculated the sampling mean of 16 samples and set the target spec to Cpk=1.5, using sampling mean and population standard deviation. Note that we used the standard deviation from population instead of samples. This is implicit in equation (1). We repeated the random sampling 1000 times, and calculated population DPM according to the target spec, which was set by statistics from 16 samples. In the results, there are 968 counts of DPM less than 25. The results are equivalent to a 96.8% probability that DPM is less than 25. This probability is better than a 95% confidence bound. The histogram of DMP from 1000 times' sampling is shown in figure 4.

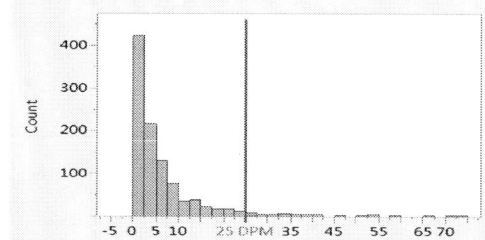

Fig.4. A histogram of DPM shows 968 counts less than 25.

Based on the experiment, a recommended sample size of 16 to 20 can give a 95% confidence level for silicon validation with conditions of Cpk>=1.5 and DPM <=25. This result is consistent with the outcome from the theoretical analysis shown in the above session. With this proposed method, engineers can determine the optimal sample size for post-silicon validation of a FPGA and use that data to achieve the best balance between product quality coverage and product time-to-market competitiveness.

REFERENCES

[1] Subhasish Mitra, Sanjit A. Seshia, "Post-Silicon validation opportunities, challenges and recent advances", 2010.

[2] Russell V. Lenth, "Some practical guidelines for effective sample size determination", 2001.

[3] Cochran, W. G., "Sampling techniques 2nd Ed", 1963.

[4] Glenn D, "Determining sample size", Israel, 2009.

[5] Will Mendenhall, Robert J. Beaver, "Introduction to probability and statistics", 2004.

[6] Stefan Steiner, Bovas Abraham, "Understanding process capability indices", 1998.

USING VERILOGA FOR MODELING OF SINGLE EVENT CURRENT PULSE: IMPLEMENTATION AND APPLICATION

Jia Liu[1], Yusen Qin[2], Tiehu Li[3], Yuxin Wang[1], Weidong Yang[1], Jun Liu[1] and Ruzhang Li[1]*

[1]Science and Technology on Analog Integrated Circuit Laboratory, Chongqing 401332, P.R.China

[2]Chongqing Nan'an Power Supply Bureau, Nan'an District, Chongqing 400060, P.R.China

[3]No.24 Institute, China Electronics Technology Group Corp., Chongqing 401332, P.R.China

*Corresponding Author's Email: liujia_sisc@163.com

ABSTRACT

In the sub-100nm bulk CMOS process technologies, the Single Event Effect (SEE) becomes one of the most critical reliability issues in the semiconductor devices and ICs that are used for the space applications. The modeling of Single Event (SE) current pulse is very important and challenging. In this paper we give the 3D TCAD simulation results of the SE current pulse (SECP) in the devices, and develop a compact model using VerilogA (VA) behavioral language. This model could be used for the validation of the Radiation-Hardened (RH) approaches in the circuit-level simulations.

INTRODUCTION

Today more and more integrated circuits (ICs) have been used in the scientific and technological areas, such as space applications, and working in these areas would expose the ICs in a harsh radiation environment. The ionizing and non-ionizing radiations would result in serious reliability issues, such as Displacement Damage, Total Ionizing Dose (TID) and Single Event Effect (SEE) [1]. The SEEs include the Single Event Latchup (SEL), Single Event Gate Rupture (SEGR), Single Event Upset (SEU), Single Event Transient (SET) and so on. For the sub-100nm bulk CMOS process technologies, due to the fine quality and very thin gate oxide, the MOSFETs are more resistant to the TID effect. Since the MOSFETs are the surface devices, Displacement Damage is not a serious problem. Thus the SEE becomes the most serious reliability issue for the sub-100nm bulk CMOS based ICs that work in the radiation environment.

When the high energy particles (such as heavy ions, protons and so on) strike the ICs, it could deliver adequate energy and generate the electron-hole pairs. If the particles passed near the reverse-biased p-n junctions of devices, the charge would be collected and a transient current pulse would be generated. If this happened in the storage element of a circuit, it would result in the state-upset, which is called SEU. However for the combinational part of the circuit, this would induce a temporary voltage glitch, which is called SET. The SET could propagate and induce an error in the storage cell [1]. Many studies have focused on the modeling of this Single Event Current Pulse (SECP), especially the building of the compact model, which links the device level SEE to the circuit level

simulations. The compact model could be very useful for the Radiation-Hardened (RH) circuit designs [2]. In this paper we will first discuss the 3D TCAD simulation results of the SECP in the NMOSFETs, then give some introductions about the compact model of SECP realized with VerilogA behavioral language [3], and finally discuss the applications of the VA model.

TCAD SIMULATION RESULTS OF SECP

Usually, the current pulse caused by the Single Event in devices can be modeled by a double-exponential-function (DEF) [4-5]:

$$I = Q \cdot (\exp(-t/\tau_f) - \exp(-t/\tau_r)) / (\tau_f - \tau_r) \quad (1)$$

in the formula, τ_f is the collection time constant, τ_r is the track establishment time constant, and Q is the total charge that is generated by the heavy ions in the track path. However, some researchers have indicated that the SECP in devices may be different from the DEF model [6-8], and we should get more information about the SECP.

Since the SECP in the devices (such as in NMOSFETs) cannot be measured easily, the three-dimensional device simulations could be utilized to study the SECP and get the useful information. As we know, Cogenda Ltd. provides integrated solutions for EDA/TCAD/RadHard, which could support the device-level SEE simulations [9]. We could use Cogenda tools (such as Gds2Mesh, Genius, Visual Particle, and so on) to study the SEE in the sub-100nm bulk CMOS process technologies, and in this study we focus on the SECP in NMOSFETs of 65nm technology. First the technology parameters for Cogenda TCAD tool (Genius) were calibrated with typical devices (SMIC 65nm bulk CMOS). Then we used Geant4-based Cogenda-VisualParticle (Gseat) to simulate the passages of particles with different Linear-Energy-Transfer (LET) through the devices, and the heavy ions that we chose for simulations were $^{35}Cl^{11+}$(138MeV,LET~13.9MeV·cm^2/mg), $^{48}Ti^{12+}$(149MeV,LET~22.6MeV·cm^2/mg), $^{63}Cu^{13+}$(161MeV,LET~33.4 MeV·cm^2/mg), $^{79}Br^{14+}$(172MeV,LET~42.0MeV·cm^2/mg), $^{127}I^{15+}$(195MeV,LET~59.0MeV·cm^2/mg). The Gseat/VisualParticle simulation results of these heavy ions were used as the input of the Cogenda-Genius (used as the radiation source), and the physical 3-D device models included drift-diffusion model, band structure model, generation model, carrier recombination (Auger, Shockley

-Read-Hall, and direct recombination), doping-dependent carrier lifetimes, mobility model, and so on. The 3D device SEE simulation flow is shown in Figure 1.

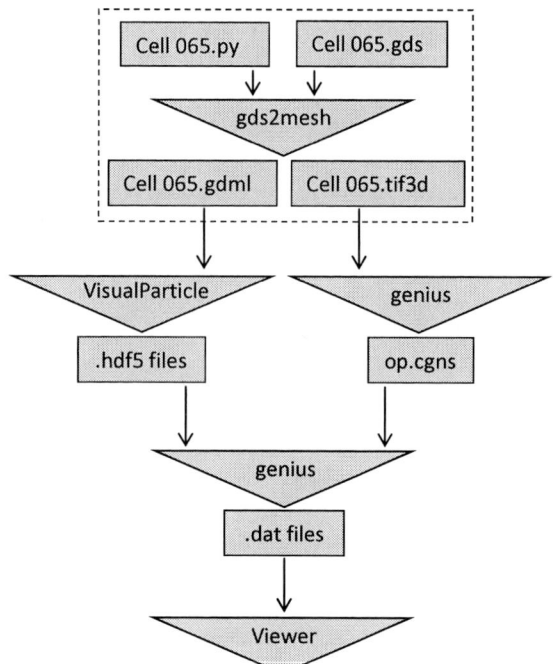

Figure 1: 3D device SEE simulation flow with Cogenda -Tools (based on SMIC 65nm bulk CMOS technology)

The NMOSFETs were in the inverter chain, the sizes were W/L~120nm/60nm and 600nm/60nm, and the heavy ions struck the drain the NMOSFET perpendicularly. The simulation results of the SECP (NMOSFET drain currents) are shown in Figure 2.

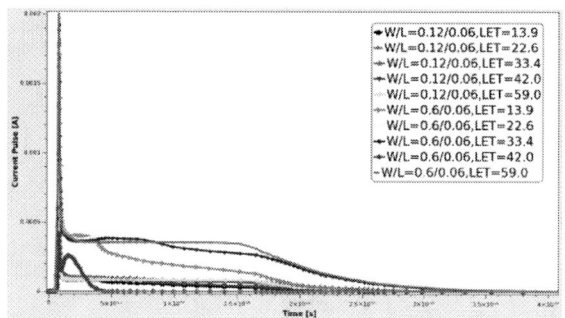

Figure 2: 3D device simulation results of SECP in NMOSFETs with different LETs and different sizes (with Cogenda-Tools, based on SMIC 65nm bulk CMOS technology)

From the simulation results, we can find that the SECP is composed of a short burst of high current

followed by a plateau of low current, which looks like a burst-plateau (BP) current [6-8]. This may because the nodes voltages of the device in the circuit cannot be fixed, the surrounding circuitry can source or sink the Single-Event current [1, 6]. Both the LETs of the heavy ions (particles) and the sizes of the devices have effects on the shapes of the SECP. The 3D device simulation results confirm other researchers' conclusions: for the devices in circuits, the BP model is better than the DEF model, which is more suitable for modeling the SECP in a single-device with fixed terminal-voltages [1].

SECP MODEL DEVELOPMENT WITH VERILOGA

When modeling the SECP, both the DEF model and the BP model should be realized. For the DEF model (1), the generated charge (Q) by the heavy ions can be expressed as the following equation:

$$Q = q \cdot \rho \cdot t \cdot LET / E_{eh} = 0.01036 \cdot LET \cdot t$$

in which q is the value of electron charge (1.6×10^{-19} C), ρ is the silicon density ($2.33 g/cm^3$), t is the particle's track length in silicon (in μm), LET is the Linear-Energy-Transfer of the particle (in $MeV \cdot cm^2/mg$), E_{eh} is the energy for the electron-hole generation (3.6 eV), and Q is the generated charge (in pC) [4-5]. When realizing the BP model, the effects of both the sizes of the devices and the LETs of the particles should be considered.

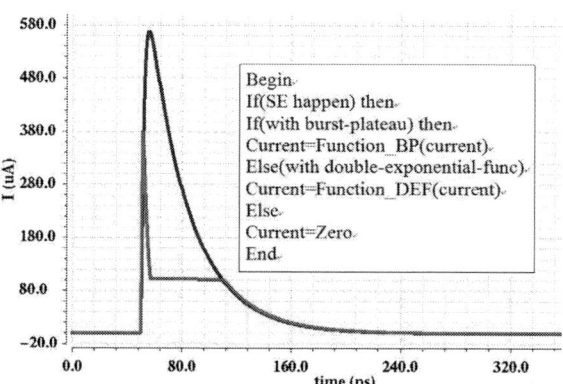

Figure 3: VerilogA modeling algorithm and the realized SE current pulses: DEF model (blue line) and BP model (red line)

VerilogA (VA) behavioral language is an excellent tool for the compact model, and could be used for the modeling of SECP, since both the DEF model and the BP model maybe cannot be realized easily with other modeling methods [3, 7]. The parameters could be extracted using the data from the TCAD simulations or the references [10]. The VA model should support heavy ions with different LETs, devices (NMOSFETs) with different sizes, and so on. A simple modeling flow/algorithm and

978-1-5090-6695-7/17 $31.00 © 2017 IEEE   328

the generated current pulses are shown in Figure 3 (NMOSFET drain current, W/L~120nm/60nm, LET=30MeV·cm^2/mg, SMIC 65nm bulk CMOS technology). The more detailed descriptions of the VA modeling algorithm will be given in another paper.

IMPLEMENTATION AND APPLICATION OF SECP COMPACT MODEL

In the circuit level simulations, the VA model could be used as the SE radiation source and be used for the validation of the RH approaches in the RH circuit designs. The VA model (current pulse) could be injected into any node of the circuit, and the responses of the hardened and unhardened circuits could be compared. The simulation results could tell us whether the RH methods are effective or not.

In order to use the VA model for the circuit level validation, we could use Cadence-SKILL to develop a Pcell, which is shown in Figure 4. The users could specify the parameters of the Pcell for the SECP and the VA model would be generated, which could be used for the circuit level SEE simulations.

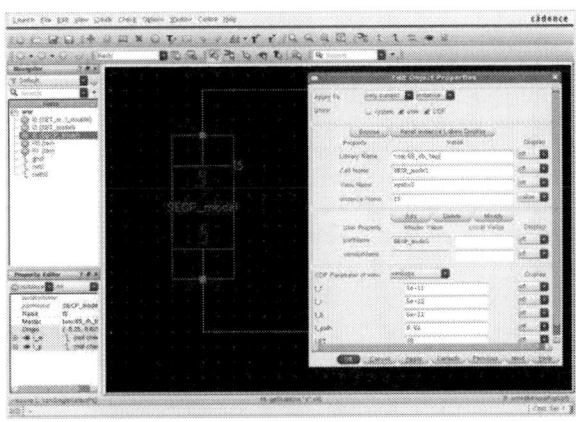

Figure 4: Pcell (based on Cadence-SKILL) of the SECP VA model

CONCLUSION

In this paper we have studied the shapes of SECP in the NMOSFETs (based on SMIC 65nm bulk CMOS technology) using 3D device TCAD simulations, and have developed a compact model of SECP with VerilogA behavioral language, both the double-exponential-function model and the burst-plateau model have been realized. The model could be very useful for the RH circuit designs with the sub-100nm process technology.

ACKNOWLEDGEMENTS

The authors would like to thank the colleagues at No.24 Institute, CETC., for helpful and insightful discussions. This research is sponsored by the Foundation of Science and Technology on Analog Integrated Circuit Laboratory under contracts: 9140C090407150C09045 and 0C09YJTJ1601.

REFERENCES

[1] V.Ferlet-Cavrois, L.W.Massengill, and P.Gouker, Single Event Transients in Digital CMOS - A Review, *IEEE Trans. Nucl. Sci.*, Vol.60, No.3, pp.1767-1790, June 2013.

[2] D.G.Mavis and P.H.Eaton, SEU and SET modeling and mitigation in deep submicron technologies, *Proc.IEEE Int.Rel.Phys.Symp*, Phoenix, AZ, April 2007, pp.293-305.

[3] G.J.Coram, How to (and How not to) Write a Compact model in Verilog-A, *Behavioral Modeling and Simulation Conference 2004*, San Jose, CA, October 2004, pp.97-106.

[4] W. Massengill, M. Alles, and S. Kerns, SEU Error Rates in Advanced Digital CMOS, *Proc. 2nd European Conference on Radiation and its Effects on Components and Systems*, St. Malo, France, June 1993, pp.546-553.

[5] G. Messenger, Collection of Charge on Junction Nodes from Ion Tracks, *IEEE Trans. Nucl. Sci.*, Vol. 29, No. 6, pp.2024-2031, December 1982.

[6] S.DasGupta, A.F.Witulski, B.L.Bhuva, M.L.Alles, R.A.Reed, O.A.Amusan, J.R.Ahlbin, R.D.Schrimpf, andL.W.Massengill, Effect of Well and Substrate Potential Modulation on Single Event Pulse Shape in Deep Submicron CMOS, *IEEE Trans. Nucl. Sci.*, Vol.54, No.6, pp.2407-2412, December 2007.

[7] J.S.Kauppila, A.L.Sternberg, M.L.Alles, A.M.Francis, J.Holmes, O.A.Amusan, and L.W.Massengill, A Bias- Dependent Single-Event Compact Model Implemented Into BSIM4 and a 90nm CMOS Process Design Kit, *IEEE Trans. Nucl. Sci.*, Vol.56, No.6, pp.3152-3157, December 2009.

[8] D.A.Black, W.H.Robinson, I.Z.Wilcox, D.B.L imbrick, and J.D.Black, Modeling of Single Event Transients with Dual Double-Exponential Current Sources: Implications for Logic Cell Characterization, *IEEE Trans. Nucl. Sci.*, Vol.62, No.4, pp.1540-1549, August 2015.

[9] http://www.cogenda.com , http://cn.cogenda.com .

[10] F.Wrobel, L.Dilillo, A.D.Touboul, V.Pouget, and F.Saigne, Determining Realistic Parameters for the Double Exponential Law that Models Transient Current Pulses, *IEEE Trans. Nucl. Sci.*, Vol.61, No.4, pp.1813-1818, August 2014.

FINGER PRINT SENSOR MOLDING THICKNESS NONE DESTRUCTIVE MEASUREMENT WITH TERAHERTZ TECHNOLOGY

Longhai Liu[1]*, Haitao Jiang[2], Ying Wang[2], Qinghua Shou[2], Jianhua Xie[1], and Yaqi Lu[1]

[1]Advantest (China) Co., Ltd, Shanghai 201203, Shanghai, China

[2] Amkor Assembly & Test (Shanghai) Co., Ltd 200131, Shanghai, China

*Corresponding Author's Email: longhai.liu@advantest.com

ABSTRACT

Fingerprint sensor (FPS) becomes rapidly popular due to small size and high safety. The molding thickness of capacitive FPS will affect its performance and needs to be accurately measured and controlled.

Different from cutting and laser drilling method, one none destructive method with Terahertz electromagnetic wave is introduced. Terahertz wave can penetrate into the molding package materials. The molding thickness can be measured through time delay of two pulse Terahertz waves. The measured thickness is correlated with microscope view and the result is within ±4um gap.

INTRODUCTION

Due to small size and high safety, fingerprint chips become increasingly popular in smart phone, security card and other IC. Nowadays there are mainly three kinds of Fingerprint sensor (FPS), i.e., capacitive, optical and ultrasonic, whereas capacitive FPS takes most of the market, especially in smartphone. The capacitive FPS is very sensitive to its package molding thickness. To capture fingerprint pattern, the molding layer thickness must be distributed in an estimated value. It needs high accurate measurement system in production for better quality control.

Conventional method is through cutting or drilling hole on the FPS molding surface, then thickness is measured with microscope. This method will destruct the device and bring device loss, which loss the money for the production line. Meanwhile, due to human observation error, the microscope view method is of low repeatability. High accuracy and high repeatability none destructive molding thickness measurement method is strongly requested in the FPS production line. Hereby, we introduce one none destructive method with terahertz electromagnetic wave to measure the FPS thickness in high accuracy.

When looking into the electromagnetic spectrum, the Electromagnetic wave is divided into electronic wave and photonic wave, where electronic wave is of high transparency, and optical wave is of high resolution. Terahertz wave is within the infrared and millimeter wave gap, whose frequency is 0.1-10 THz. The Terahertz wave holds the advantage of infrared and millimeter wave, i.e, high transparency and high resolution. Terahertz wave can penetrate into non-conducting materials [1], such as plastics, ceramics, paper, etc, which makes it affective to measure the molding thickness none-destructively. Meanwhile, the Terahertz wave is of high resolution and can measure the molding thickness in high accuracy.

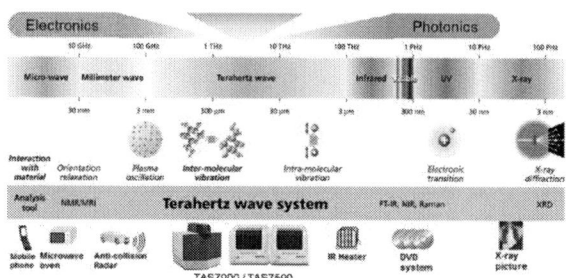

Fig.1. Electromagnetic spectrum

PRINCIPLE

The typical structure of FPS is as in Fig.2. On the substrate, there are die arrays, where the fingerprint sensor is located with specific pattern. The plastic materials are molded on the die array and substrate to protect the chip. The thickness from the die surface to the molding surface is known as molding clearance thickness. The molding clearance thickness is crucial to its performance and need to be measured in high accuracy.

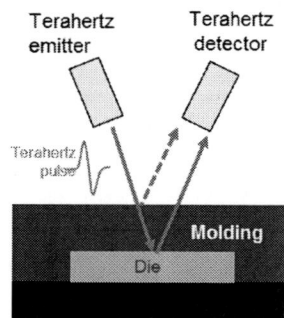

Fig.2. Time of Flight measurement principle

The basic principle of this method is time of flight (TOF) analysis [2, 3]. Terahertz emitter generates very short pulse Terahertz wave. Part of the Terahertz wave will be reflected on the molding surface, while another part Terahertz wave will penetrates into the FPS molding material. Then the Terahertz wave will be reflected on the die surface and transmit the molding materials again and

978-1-5090-6695-7/17 $31.00 © 2017 IEEE

be detected by Terahertz detector. On the Terahertz detector side, two pulse Terahertz waves from molding surface and die surface. The time delay of the two pulsed Terahertz wave can be used to calculate the molding thickness. The two pulsed Terahertz waveforms are shown in Fig. 3.

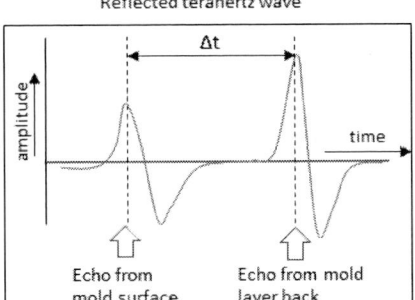

Fig.3. Time difference of two pulse terahertz waves is used to measure molding thickness.

In the molding materials, the Terahertz wave propagates in ray speed. By calculating the time delay of two pulse waves of the molding surface and die surface, the molding thickness can be measured as below,

$$Thickness = \Delta t \ c \ \cos\theta \ /2n$$

Where Δt is time delay of two pulse wave between molding surface and die surface, c is optical ray speed, θ is the incident angle of terahertz wave, n is refractive index of molding material. Δt can be measured by system software. The optical ray speed c is a constant and incident angle θ is determined by system hardware. The refractive index is crucial and needs to be accurately gotten. The above parameter needs to be defined during recipe setup.

The most effective method is using a known thickness flat plate with the same molding materials. The calculated method is slightly different from the above equitation.

$$n = \Delta t \ c \ \cos\theta \ /2thickness$$

Where thickness is known in advance and Δt can be measured by system software.

EXPERIMENT

Advantest TS9000MTA system employs the above Terahertz wave none destructive method to measure IC, especially FPS package molding thickness. The main specification of TS9000MTA system is listed in Table 1.

In real situation, finger print sensor stripe is measured by TS9000MTA. The thickness of each DUT is measured in the left, center and right position. Then the FPS stripe is cut and measured by microscope view. The correlation result of TS9000MTA and microscope view is given in Fig. 4. The measured thickness is around 130um. The

correlation gap of TS9000MTA and microscope view is around ±4um.

TABLE 1. Specification of TS9000MTA system

Items		Specification
Measurement sample		IC strip, Packaged IC unit
Molding thickness measurement specification	Measured points	1～20 points
	thickness range	30 μm～600 μm
	spot size	300μm
	Throughput	> 250 units/hour, (4points by 250 chips)
	Accuracy	±3um

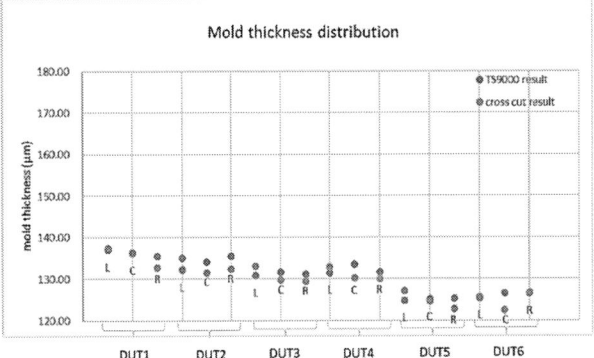

Fig.4. Molding thickness correlation of TS9000 and cross cut result

CONCLUSION

The finger print sensor molding thickness needs to be measured and controlled. A method to measure finger print sensor molding thickness in none destructive is introduced. Terahertz electromagnetic wave can penetrate through the molding materials. And the molding thickness can be measured through time delay of two pulse Terahertz waves.

This method is none destructive and of high accuracy and high throughput. The measured thickness is correlated with microscope view and the result is within ±4um gap.

ACKNOWLEDGEMENTS

The authors would like to acknowledge the help and discuss from the teams working in Amkor Technology Corporation and Advantest Corporation.

REFERENCES

[1] Yun-Shik Lee, Principles of Terahertz Science and Technology, Springer, 2009.

[2] C. Jasen, S. Wietzke, H. Wang, M. Koch, G. Zhao, "Terahertz spectroscopy on adhesive bonds", Polymer Testing, vol. 30, pp. 150-154 (2011).

[3] I. Puperza, R. Wilk, M. Koch, "Highly accurate optical material parameter determination with THz time-domain spectroscopy", Optics Express, vol. 15, pp. 4335-4350 (2007).

HIGH EFFICIENCY TEST SYSTEM FOR ENVELOPE TRACKING POWER AMPLIFIER

Feifan Du [1], Hui Yu [2]

[1] Department of Electronic Engineering, Shanghai Jiao Tong University; National Instruments China
[2] Department of Electronic Engineering, Shanghai Jiao Tong University, Shanghai, China
*Corresponding Author's Email: feifan.du@ni.com

ABSTRACT

Design houses are facing business challenges as improving chip performance and integrating latest technologies now. Particularly in PA design area, envelop tracking (ET) and digital distortion (DPD) are main approaches used in improving the performance of PA. The system is discussed in this paper, which is not only including the regular PA test, but also the additional details for ET, such as the synchronization between Radio Frequency (RF) signal and envelope reference signal, and the extraction of shaping table. This paper discusses a method to improve the performance of Power Modulator, which is used to amplify the reference signal as the power amplifier's DC signal. Normally, ATE is used in production line test, but in lab test, traditional test instruments are used. So this approach consumes a lot of time from engineers in data correlation. This paper promotes a high efficiency test system for ETPA, which is based on an open, modular-based platform, compatible for both lab test and production line test.

INTRODUCTION

Non-constant envelope signal is the mainstream for the current communication system because of the rapid increase in data rate. The complex modulation scheme with high PAPR has become a challenge for RF front end (FEM) designers and test engineers. PA power supply tracks the envelope of RF input signal to improve the performance of PA, which is the concept of envelope tracking. So the Device under test (DUT) can be equal to three-port device, including the PA and PM. It's the big difference between ET Power amplifier and power amplifier with fixed power supply[1].

So the performance of PA is not only related to the RF input, bias circuit etc., also related to the PA power supply. And this system needs two signal sources to generate RF input signal and reference envelope signal. In order to achieve power supply tracking instantaneous envelope of RF input signal, the synchronization mechanism is necessary for these two signal generators. But the envelope of RF input signal cannot directly used as PA power supply because of the knee voltage of PA, so this signal should be processed. Envelope shaping function is also necessary for this ETPA test system.

DESIGN OF ETPA TEST SYSTEM

Test System Architecture

Typically, the instruments for lab and production line are different. The traditional box instruments are used as test bench to improve product design in lab, but in production line, test engineers use automatic test equipment (ATE) to verify the issues and optimize the test speed. So there is a problem for test engineers, different instruments lead to different test results. And data correlation takes a lot of time.

The ETPA test system, which is discussed, is suitable for lab and production line test. It's based on the test platform, which data acquired from instruments are transferred to the host for analyze via PCIe bus on the chassis. The software is developed by using test management software, which calls several test items and DUT control etc. to complete test. Depending on specific test requirements, the test items are developed by using existed instrument drivers and protocol toolkits (e.g., LTE).

The scalability and flexibility of this software-defined test system makes it possible to add DPD and ET as option, and interact with handler, prober or manipulator in Final Test (FT), helping test engineers to speed up testing and reduce the cost of software maintenance.

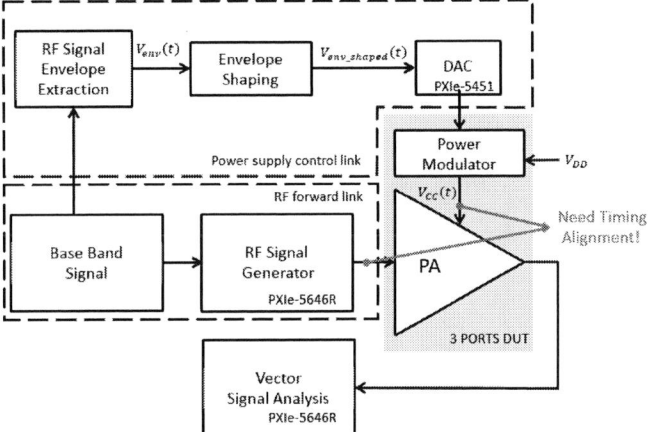

Figure 1: Block diagram of advanced PA test system

The test system structure is shown in figure 1. If RF input signal is 20 MHz LTE signal, after envelope extraction, the shaped RF input envelope signal is sent to DAC, as reference envelope signal. The sampling rate of DAC should be over 100 MHz because this reference envelope signal bandwidth is over 20 MHz. Then the signal is amplified by power modulator as PA power supply. So the ripple of Power Modulator should be calibrated, and the delay between power supply control link and RF forward link is calibrated before using.

The PA output signal is acquired by RF signal analyzer. The PA test items are not only the regular test items such as Gain, Adjacent Channel Power (ACP), but also AMAM and AMPM measurement with/without DPD. To optimize the test speed, PA power servo is implemented using FPGA in this system.

Test Item Details

The typical PA test flow, which shown as Figure 2. Special for envelope tracking PA, DUT is including power modulator, the additional system calibration steps are required, including power modulator calibration, "RF Delay" measurement, and envelope shaping.

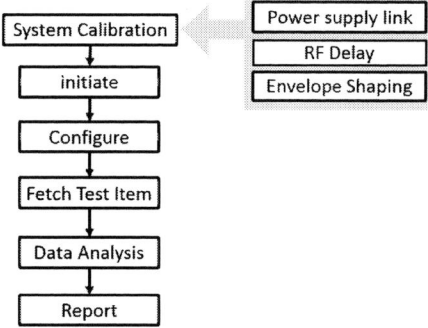

Figure 2: Envelope tracking PA test flow

Power modulator amplify the reference envelope signal from DAC, which input and output transfer function as

$$V_{cc}(t) = V_{env\_shaped}(t) * \text{Gain} + \text{Offset} \qquad (1)$$

Where Gain and Offset are constant. Because the reference envelope signal bandwidth is normal over 10 MHz, which is a big challenge for power modulator. If ripple exists within the band, Gain becomes a variable and PA power supply is distorted.

Pre-equalization is used to solve this issue, as new method to improve the performance, which is realized by using FIR filter. The response of filter is the inverse of power modulator response as

$$H_{filter}(f) = H_{system}^{-1}(f) \qquad (2)$$

Frequency sampling method is used for filter coefficient exaction. As shown in Figure 3, the calibration is only for the required frequency range.

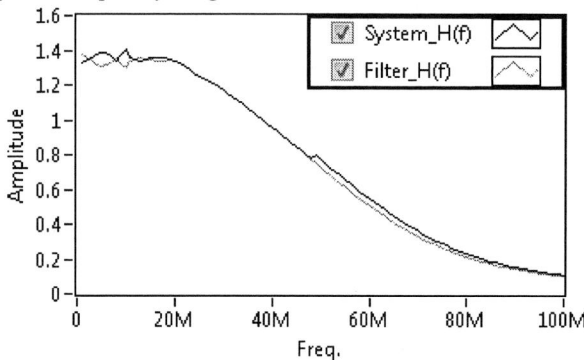

Figure 3: Power modulator spectrum

"RF Delay" measurement is used to calibrate the difference between RF forward link and power supply link. In this system, two signal generator synchronization is based on NI-TCLK technology[2]. Sweeping a certain range of time delay is used to find the optimize link delay in this system. The time step for adjustment is 500ps because 1ns delay means 2dB ACP improvement. As shown in Figure 3, when finding the optimal ACP result, it means that the expected link delay is also found.

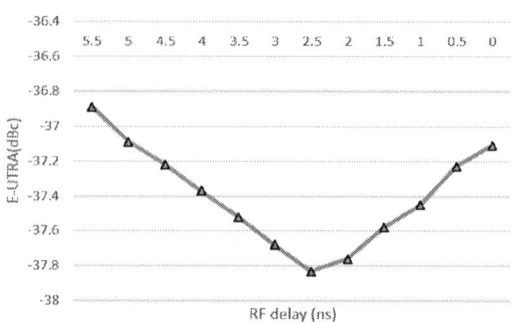

Figure 4: RF delay sweep ACP results

Envelope shaping is the approach to shaping the envelope of RF input signal. In brief, it makes sure that PA power supply is higher than knee voltage to prevent the generation of additional nonlinear components. For this test system, two shaping functions are provided, one is de-trough[3], replace the envelope in low power region to a constant offset voltage level. To prevent the issue of spectrum expansion from the clip of waveform, quadratic function is used to removes the additional frequency components as approximation algorithm, where

$$V_{env\_shaped}(t) = V_{cc\_max}(V_{in}(t) + \alpha e^{-(V_{in}(t)/\alpha)}) \qquad (3)$$

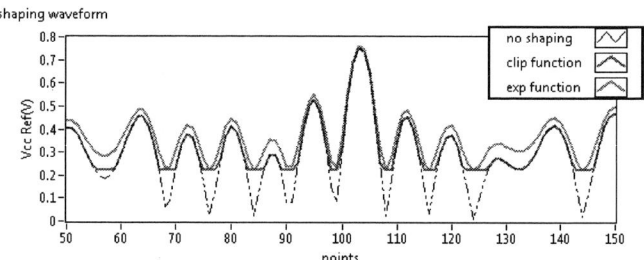

Figure 5 de-trough shaping waveform

This function is easy to use, which only need to configure the output waveform parameter, as shown on Figure 5. But it cannot track the magnitude of PA input power to optimize the efficiency of DUT with the fixed envelope maximum level.

So the shaping function based on measurement is used to optimize the performance of, using shaping table to complete the mapping between PA power supply and RF input signal power level. One of shaping table extraction which is easy to understand is from optimal efficiency curve, based on the instantaneous efficiency measurements, but this method leads to the fact that gain curve is not flat and linearity is worse. Instead, if the table is extracted from the fixed gain curve, called ISO-Gain, linearity of PA would be better and efficiency of DUT only loss 1 ~ 2%, comparing these two extractions[4]. The curve is based on the measurement of AMAM curves, sweeping the PA power supply level within certain range, and the RF input signal is the stepped CW pulse signal. The measurement details are in Figure 6 and the extraction details are in Figure 7.

978-1-5090-6695-7/17 $31.00 © 2017 IEEE   334

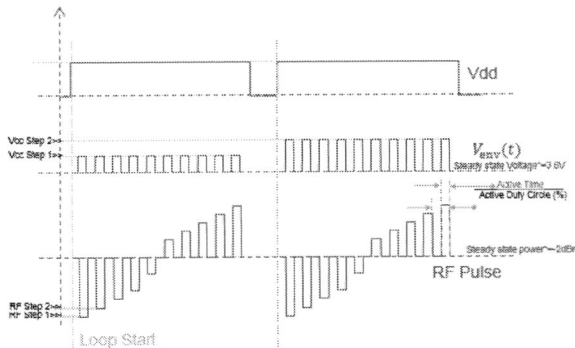

Figure 6: Timing diagram for AMAM curves measurement

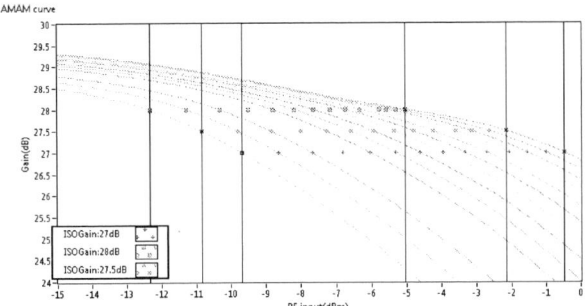

Figure 7: AMAM curves with fixed gain extraction

TEST SYSTEM ANALYSIS

Test Setup

The main part of this ETPA test system is NI PXIe-5646R, vector signal transceiver, and PXIe-5451, arbitrary waveform generator as DAC. Other device is used for DUT control and DC power supply. The software of system is using RFIC Test Software, based on National Instruments RFmx measurement toolkit, to measure ACP, AMAM, AMPM, with/without DPD algorithm. The test signal is 5 MHz LTE signal. The bandwidth and Offset Frequency of ACLR measurement for Evolved Universal Terrestrial Radio Access (E-UTRA) has been defined in the LTE standard, which is 4.5MHz and 5MHz with 1ms capture time. The time for FPGA-based DPD and power servo was benchmarked in production line test environment [5]. The picture of lab characterization system is shown in figure 8.

Figure 8: DUT connection and ET test system capture

Results

A MMMB cellular PA and power modulator are on the evaluation board, which operated at 1950 MHz was used as DUT in lab characterization. The power modulator operated for 5MHz bandwidth signal, so LTE waveform (5 MHz BW, 16QAM) sampled at 245.76 MS/s was used as test signal.

Table 1 shows the envelope tracking with different shaping function lead to the difference of system performance, compare to de-trough function, ISO-Gain function improved 4.1% at the measurement of system efficiency. With DPD function, The ACP results improved 15dB.

Table 1 ACP and PAE results for envelope Tracking PA

	E-UTRA (dBc)	PAE (%)
ETPA with de-trough	-38.16/-36.94	44.1
ETPA with ISO-Gain	-38.29/-37.85	48.29
ETPA with ISO-Gain (w FPGA DPD)	-50.35/-50.24	48.62

In production line test, as illustrated in the figure 9, semiconductor Test System (STS) is used to measure the single site test time. The test item is 20MHz bandwidth LTE signal ACP test, which is including power servo configuration, ACP measurement, ACP and power results fetch. Real-time FPGA-based DPD measurements and PA Power Servo help to reduce serval seconds in the totally project time.

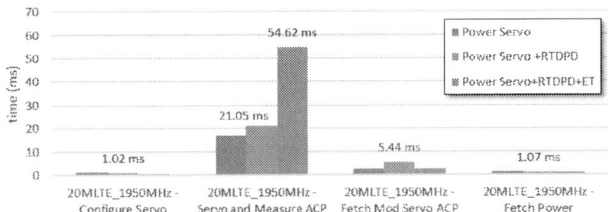

Figure 9: Test time for 20MLTE ACP test items

CONCLUSION

For this ETPA test system, which including the envelope tracking and digital pre-distortion application can be used in both lab and production line test. The software-based and modular test architect concept help engineers to complete the progress fast from lab design to production line when facing the new test challenge.

ACKNOWLEDGEMENTS

The authors would like to thank Peal He for providing help and industry experience for this paper. The authors also wish to thank Sean Ferguson for providing professional technical support all the time.

REFERENCES

[1] Z. Wang, *Envelope Tracking Power Amplifiers for Wireless Communications*: Artech House, 2014.

[2] National Instruments.*National Instruments NI-TClk Technology for Timing and Synchronization of Modular Instruments*.2013. Available: http://www.ni.com/tutorial/3675/en/

[3] B. Kim, J. Kim, *et al.*, "Push the Envelope: Design Concepts for Envelope-Tracking Power Amplifiers," *IEEE Microwave Magazine,* vol. 14, pp. 68-81, 2013.

[4] J. Hendy.*Software defined power amplifiers using envelope tracking*.2014. Available: http://www.radio-electronics.com

[5] S. Ferguson and A. Chopra, "Real-time digital predistortion for radio frequency power amplifier linearization," in *2016 IEEE MTT-S International Wireless Symposium (IWS)*, 2016, pp. 1-4.

EFFECTIVE METHOD TO AUTOMATICALLY MEASURE THE PROFILE PARAMETERS OF INTEGRATED CIRCUIT FROM SEM/TEM/STEM IMAGES

Xiaolin ZHANG[1], Zubiao FU[1], Yi HUANG[2], Alien LIN[2], Yaoming SHI[1], Yiping XU[1]

[1]Raintree Scientific Instrument (Shanghai) Corporation, 68 Huatuo Road, Shanghai 201203, China

[2]Semiconductor Manufacturing International Corporation, 18 Zhangjiang Road, Shanghai 201203, China

E-mail: zhangxiaolin@rsicsh.com

ABSTRACT

An effective image based method to automatically measure the profile parameters (PPs), including the critical dimensions (CDs), the full height and other structural parameters, of the integrated circuit (IC) devices in batch is proposed. In this method, templates are used to indicate the patterns of interest and the regions of the desired PPs; pattern recognition and PPs analysis algorithms are applied to determine the exact PPs from the underdetermined IC device images. In practice, the proposed method was proven of higher efficiency, more accuracy and better repeatability than the traditional manual measurement.

INTRODUCTION

Measuring Critical Dimensions (CDs) and other interested structural parameters from the IC devices under fabrication is one of the critical steps to monitor and control the IC manufacturing process. As the IC device dimension is getting smaller and smaller to below 28 nm, more and more cross-section images are obtained from the scanning electron microscope or transmission electron microscopy or scanning transmission electron microscopy (SEM/TEM/STEM) for reference metrology, although optical CD metrology has been widely used for inline process monitoring. Therefore, there is a strong demand to process and to make measurements from a large batch of digitized SEM/TEM/STEM images. Traditionally, the dimensional measurement of IC device is obtained manually by human operators by pointing cursor lines on the SEM/TEM/STEM image. This manual method is time consuming and of low efficiency and repeatability. In 2013, Dai's group reported to determine the CD at the half of the intensity profile of the TEM image [1]. In 2015, Takamasu and his colleagues proposed a method to measure the CD of FinFET. They firstly extracted the profile of the FinFET structure from its cross-sectional SEM/TEM image and then determined the CD at a proper vertical position [2]. The above two works were limited to measure the CD (width) of a particular kind of IC device image and didn't mention the method to automatically measure the IC device images in large quantities.

This work presents a method to automatically measure the PPs of the IC devices in batch, which involves recognizing the IC patterns of interest and measuring the PPs including the CD, the height and the sidewall angle (SWA) of the components of the underdetermined IC devices. The proposed method is a replacement of the manual measurement and takes the advantages of high accuracy, high efficiency and high reparability. In this paper, the IC device image is limited to the cross-section image obtained by SEM/TEM/STEM; all the following IC device images are real TEM images but colorized in their corresponding graphic format for demonstration.

METHOD

A template is needed in this method, in aid of recognizing the IC pattern of interest and indicating the desired PPs. A template is composed of a template image and a group of template tools, which include interface lines, interface rulers, free rulers, feature points and protractors. See Figure 1. A rectangular frame is used to crop a pattern of interest from an IC device image. The cropped image is the template image, which is preferred to be created in an upright position. Make sure that the template image includes all the features to be measured. Subsequently, template tools are added on the template image to indicate the positions of the underdetermined features (structures) and the methods to measure them. Different template tools corresponds to different measurement algorithms. An interface line (IL) is used to detect the position of an interface; an interface ruler (IR) is used to determine the CD of an interface; a free ruler (FR) is used to determine the CD of any position; a feature point is used to detect the top, bottom, left or right point of a structure; a protractor is used to determine an angle of a structure. In this case (see the right image in Figure 1), the IL is used to determine the position of the top of the IC device; the IR is used to determine the bottom CD of the first layer; the FR is used to determine the CD at the half height of the IC device; the feature point (P) is used to detect the bottom point of the background, which is used to compute the full height of the IC device, together with the IL; the protractors (SWA1 and SWA2) are used to determine the left and the right SWAs.

The main processes to determine the PPs are presented in Figure 2. The template image will be applied to match every underdetermined IC device images in the

978-1-5090-6695-7/17 $31.00 © 2017 IEEE

image database to find the target patterns (the best matches) via pattern recognition. For one image, it may include more than one target patterns, denoted by obj 1, obj 2... obj M. Subsequently, the template tools, which are the guidance of the PPs, will be applied to the target patterns. Using the PPs analysis algorithms, the accurate PPs can be obtained.

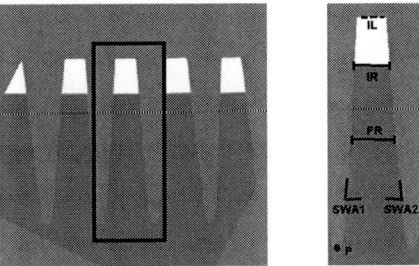

Figure 1: The method to create a template. The left is a cross section IC image, in which the rectangular frame indicates the selected region to create a template. The right is the template.

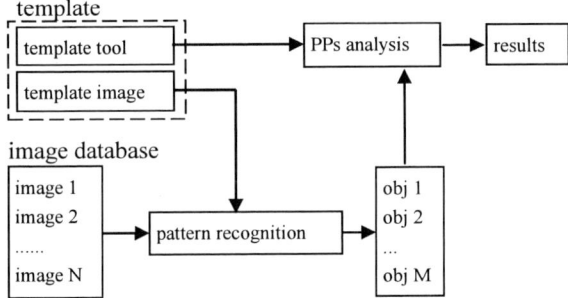

Figure 2: Diagram of auto measurement of IC device images

The spatial correlation algorithm is used for pattern recognition. The correlation of the template $w(s,t)$ with the image $f(x,y)$ is:

$$\gamma(x,y) = \frac{\sum_s \sum_t \left[w(s,t) - \overline{w}\right]\left[f(x+s, y+t) - \bar{f}_{xy}\right]}{\left\{\sum_s \sum_t \left[w(s,t) - \overline{w}\right]^2 \sum_s \sum_t \left[f(x+s, y+t) - \bar{f}_{xy}\right]^2\right\}^{\frac{1}{2}}} \quad (1)$$

where $\overline{w}$ is the average of the template value, $\bar{f}_{xy}$ is the average of the image value in the region coincident with the template. The best match is at the position of the maximum correlation. For an image, it may not include a target pattern or may have more than one target patterns; therefore a reasonable threshold is needed to judge the existence and the number of the targets. If the maximum correlation is smaller than the threshold, it indicates that the underdetermined image doesn't include an IC device

that is the same with the template, which will trigger the image database to provide another image for measurement; otherwise one target is found. Replace the target area with 0 and find the next maximum correlation to compare with the threshold to judge the existence of the other targets, until all the targets are found.

Provided that all the targets in the underdetermined image are obtained, the template tools will be applied to the target patterns. PPs analysis module will choose different algorithms for different template tools. The IL analysis algorithm, corresponding to the IL tool, will compute the gray gradient distribution in the normal direction of the IL in a reasonable region to search the interface, which is supposed at the maximum gray gradient. The FR analysis algorithm, corresponding to the FR tool, will compute the two ends' gray gradient distributions in the line direction of the FR to search the interfaces of the two ends, which are supposed at the maximum gray gradients near the ends. IR is a combination of IL and FR, so the IR algorithm will firstly compute the normal gray gradient distributions to find the interface and then the line-direction gray gradient distributions to find the two ends. To find the feature point, image segmentation algorithm is used to separate the field of interest from the image; then the top, bottom, left or/and right point of the interested field can be obtained. The protractor algorithm uses a line to fit the edge points of the target structure; hence the inclined angle of the line can be obtained.

RESULT

IC device images with the same patterns can be measured using one template. Figure 3 shows the results of four TEM images based on the template shown in Figure 1. Table I tabulates the PPs of the target patterns, in which FH indicates the full height of the objects, which is the distance between the feature point P and the interface line IL. The result accuracy can be roughly judged by eye observation, which is the method of manual measurement; however the presented method detects the interface at the maximum gray gradient; therefore it can avoid the errors caused by eye judgment and avoid the non-repeatability caused by different operators or/and different measurement times from one operator. Table 2 presents the statistic of the PPs, which include the maximum value, the minimum value, the mean value and the mean square error (MSE). The full height of the IC device varies in a relatively bigger range than the other parameters, therefore its MES is greater. Figure 4 displays the running time to detect one image (image (a)) using one thread and the running time to detect four images (image (a) (b) (c) and (d)) using parallel computing (fours threads). The working environment is Intel(R) Pentium(R) CPU G860 @3.00GHz, 4GB RAM, Window7 32 bit. The time to analyze one image is about 20 seconds; the time to simultaneously analyze four images is about 60 seconds, which is roughly 15 seconds per image.

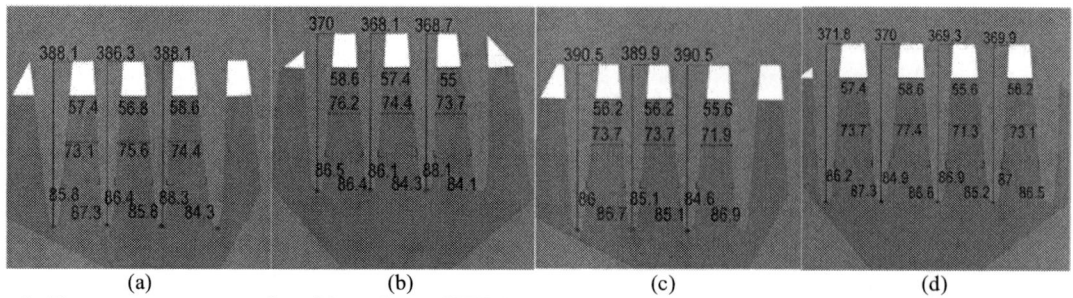

(a)   (b)   (c)   (d)

Figure 3: The measurement results of the colorized TEM images.

TABLE I.   THE PPS OBTAINED FROM THE FOUR TEM IMAGES (UNITS: NM FOR LENGTH, DEGREE FOR ANGLE)

CDs	Image (a)			Image (b)			Image (c)			Image (d)			
	Obj1	*Obj2*	*Obj3*	*Obj1*	*Obj2*	*Obj3*	*Obj1*	*Obj2*	*Obj3*	*Obj1*	*Obj2*	*Obj3*	*Obj4*
IR	57.4	56.8	58.6	58.6	57.4	55.0	56.2	56.2	55.6	57.4	58.6	55.6	56.2
FR	73.1	75.6	74.4	76.2	74.4	73.7	73.7	73.7	71.9	73.7	77.4	71.3	73.1
SWA1	85.8	86.4	88.3	86.5	86.1	88.1	86.0	85.1	84.6	86.2	84.9	86.9	87.0
SWA2	87.3	85.8	84.3	86.4	84.3	84.1	86.7	85.1	86.9	87.3	86.6	85.2	86.5
FH	388.1	386.3	388.1	370	368.1	368.7	390.5	389.9	390.5	371.8	370	369.3	369.9

TABLE II.   PPS STATISTICS

	IR	FR	SWA1	SWA2	FH
Min(nm)	55.0	71.3	84.6	84.0	368.1
Max(nm)	58.6	77.4	88.3	87.3	398.4
Mean(nm)	56.8	73.8	86.2	85.8	377.8
MSE	1.23	1.69	1.12	1.67	9.70

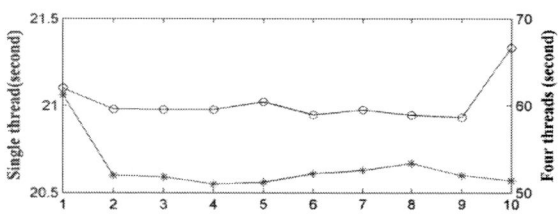

Figure 4: Running time comparison of a single thread and parallel computing. The blue line indicates the time consumptions to compute the PPs of image (a) for 10 times, using one thread; the black line indicates the time consumptions to simultaneously compute the PPs of image (a) (b) (c) (d) for 10 times, using four threads.

CONCLUSION

To summarize, the proposed method possesses five main advantages: (1) it detects the interfaces (or structure boundaries) at the greatest gray level gradient, which will eliminate the measurement errors caused by eye judgment; (2) it uses the same rules to determine the PPs, including the CDs and other interested structural parameters, from a series of digitized SEM/TEM/STEM images, which will avoid the measurement variations caused by different human operators; (3) the measurements based on this method are only determined by the gray distribution of the underdetermined images, which means repeated measurements of an image will deliver the same result; (4) it can be realized in parallel computing and batch processing, which can save labors and time. (5) the present method could automatically compile the statistics of the PPs, which will provide the variation information of the underdetermined IC devices to guide the operators to control the IC manufacturing process. With these advantages, the proposed method has better performances in measurement repeatability, efficiency and accuracy than the traditional manual measurement.

ACKNOWLEDGEMENTS

This work is sponsored by Shanghai Pujiang Program.

REFERENCES

[1] G. Dai, M. Heidelmann, C. Kübel, R. Prang, J. Fluegge, and H. Bosse, "Reference nano-dimensional metrology by scanning transmission electron microscopy," Measurement Science & Technology, vol. 24, pp. 1659-1666, 2013.

[2] K. Takamasu, Y. Iwaki, S. Takahashi, H. Kawada, M. Ikota, A. Yamaguchi, et al., "Line profile measurement of advanced-FinFET features by reference metrology," SPIE Proceedings, pp. 942406-942406-8, 2015.

MEASUREMENT OF NANOSCALE GRATING STRUCTURE BY MUELLER MATRIX ELLIPSOMETRY

Shiqiu Cheng*, Fengjiao Zhong, Huiping Chen, Yutao Jia, Yaoming Shi, Yiping Xu

Raintree Scientific Instruments (Shanghai) Corporation, 68 Huatuo Road, Shanghai, 201203, China

*Corresponding Author's Email: chengshiqiu@rsicsh.com

ABSTRACT

Spectroscopic ellipsometry technology has extensive applications in semiconductor metrology. Compared with spectroscopic ellipsometry, the Mueller matrix ellipsometry can acquire more useful information about the sample. In this paper, we present the construction of Mueller measurement equipment for 300 mm integrated circuit production line application and demonstrate the abilities of the equipment for accurate profile characterization of nanoscale structure. Mueller matrix ellipsometry may become a powerful technology for advanced technology node measurement in semiconductor device manufacturing applications.

INTRODUCTION

There are several technologies to measure the critical dimension of nanoscale structures in semiconductor device manufacturing. CD-SEM technique is a commonly used technique that provides local information about the line width, but without any information about line depth or sidewall angle, etc. 3D-AFM is also very local (only few lines at the time) and scanning over the larger area is very time consuming [1]. Since the year of around 2000, an optical CD measurement technology based on spectroscopic ellipsometry (SE) was introduced to measure the critical dimension (CD) of nanoscale structures in semiconductor manufacturing [2]. Compared with CD-SEM, 3D-AFM, this technology has achieved wide industrial applications due to its attractive advantages, such as low cost, high throughput, non-destruction, and most importantly, providing 3D structure profile information in a single measurement[3]. Among the various types of spectroscopic ellipsometry, Mueller matrix ellipsometry can provide all 16 elements of a 4 by 4 matrix. Consequently, Mueller matrix ellipsometry produce more signals than SE does and can acquire much more useful information about the sample [4].

The construction of Mueller measurement equipment for manufacturing of 300 mm integrated circuit production line is shown in Figure 1.The Polarization State Generator (PSG) consists of a linear polarizer, half-wave plate and quarter-wave plate, each of which can be switched between two different states. As a result, the PSG can generate four different polarization states for the illumination beam. The PSA comprises the same elements in reverse order, and is used to analyze the polarization of the emerging beam over another set of four different polarization states. Finally, the polarimeter subsequently measures a set of 16 raw spectra, each of which is taken with a known state of the PSG and the PSA [5].

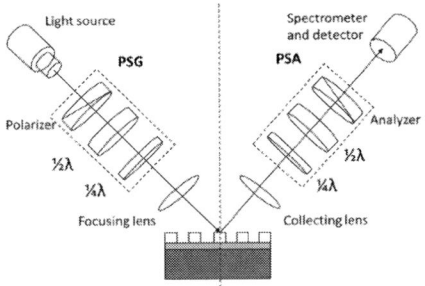

Figure 1: The construction of Mueller measurement equipment

MODELING

In this work, we used our in-house Mueller measurement equipment to measure the structure dimension of nanoscale grating structure sample. Before taking any measurements, a theoretical model was optimized and a spectra database were generated by paralleling computation. We chose trapezoidal profile model with three free parameters directly connected to the important critical characteristics of the profile: the middle CD, the depth and the sidewall angle. All three parameters are illustrated in Figure. 2.

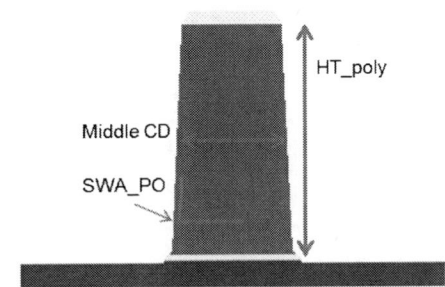

Figure 2. The model of nanoscale grating structure

The grating optical response is modeled using rigorous coupled-wave method (RCWA). RCWA method provides all components of the complex Jones matrix **J**, which can be transformed into the Mueller matrix **M** in a straightforward manner following formula [1]:

$$M=\begin{pmatrix} \left(j_{11}j^*_{11}+j_{12}j^*_{12}+j_{21}j^*_{21}+j_{22}j^*_{22}\right)\Big/2 & \left(j_{11}j^*_{11}+j_{21}j^*_{21}-j_{12}j^*_{12}-j_{22}j^*_{22}\right)\Big/2 & \left(j_{12}j^*_{11}+j_{22}j^*_{21}+j_{11}j^*_{12}+j_{21}j^*_{22}\right)\Big/2 & i\left(j_{12}j^*_{11}+j_{22}j^*_{21}-j_{11}j^*_{12}-j_{21}j^*_{22}\right)\Big/2 \\ \left(j_{11}j^*_{11}+j_{12}j^*_{12}-j_{21}j^*_{21}-j_{22}j^*_{22}\right)\Big/2 & \left(j_{11}j^*_{11}-j_{21}j^*_{21}-j_{12}j^*_{12}+j_{22}j^*_{22}\right)\Big/2 & \left(j_{11}j^*_{12}+j_{12}j^*_{11}-j_{21}j^*_{22}-j_{22}j^*_{21}\right)\Big/2 & i\left(j_{12}j^*_{11}+j_{21}j^*_{22}-j_{22}j^*_{21}-j_{11}j^*_{12}\right)\Big/2 \\ \left(j_{11}j^*_{22}+j_{21}j^*_{11}+j_{12}j^*_{22}+j_{22}j^*_{12}\right)\Big/2 & \left(j_{11}j^*_{21}+j_{21}j^*_{11}-j_{12}j^*_{22}-j_{22}j^*_{12}\right)\Big/2 & \left(j_{11}j^*_{22}+j_{12}j^*_{21}+j_{21}j^*_{12}+j_{22}j^*_{11}\right)\Big/2 & i\left(-j_{11}j^*_{22}+j_{12}j^*_{21}-j_{21}j^*_{12}+j_{22}j^*_{11}\right)\Big/2 \\ i\left(j_{11}j^*_{21}+j_{12}j^*_{22}-j_{21}j^*_{11}-j_{22}j^*_{12}\right)\Big/2 & i\left(j_{11}j^*_{21}-j_{12}j^*_{22}-j_{21}j^*_{11}+j_{22}j^*_{12}\right)\Big/2 & i\left(j_{11}j^*_{22}+j_{12}j^*_{21}-j_{21}j^*_{12}-j_{22}j^*_{11}\right)\Big/2 & \left(j_{11}j^*_{22}-j_{12}j^*_{21}-j_{21}j^*_{12}+j_{22}j^*_{11}\right)\Big/2 \end{pmatrix} \quad (1)$$

Experiment and Result

In the experiment, the incident light's spectral range was varied from 300 to 800 nm with increments of 5 nm. The incidence angle was fixed at 67° and the azimuthal angle was fixed at 45°. Figure.3 illustrates the fitting result of the simulation and the measurement. As observed from Figure.3, we can see the simulation spectra show a good agreement with the measurement spectra. The measurement results of the three parameters are shown in Table.1.

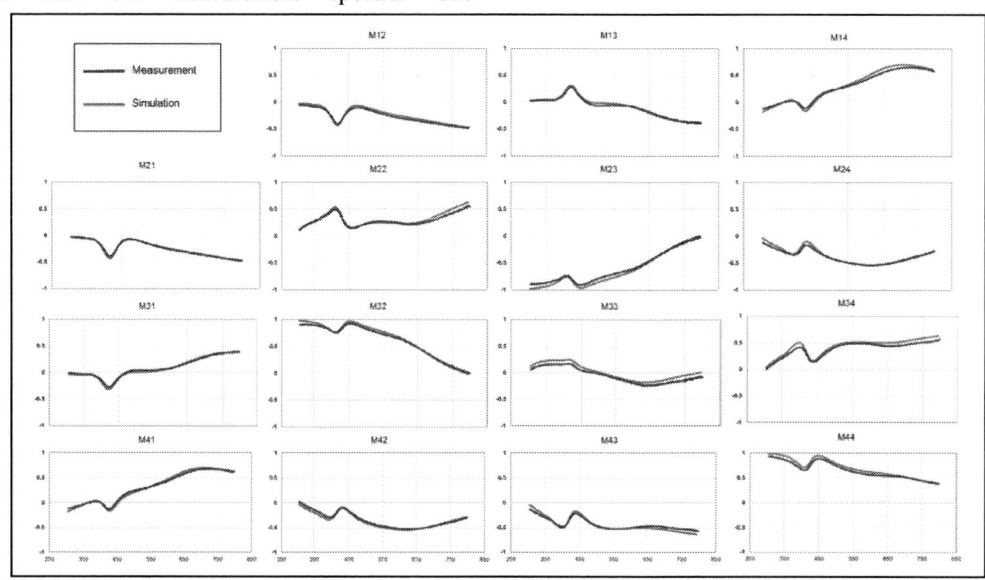

Figure 3. *Experimental spectroscopic Mueller matrix data of a grating structure at azimuth=45⁰*

TABLE I.   THE MEASUREMENT RESULT OF THREE PARAMETERS

Azimuthal angle	Middle CD	HT_poly	SWA_PO
45°	47.2nm	93.1nm	87.1°

CONCLUSIONS

In summaries, we have presented the basic construction of Mueller measurement equipment for integrated circuit production line of 300mm wafers and have demonstrated the abilities of the Muller measurement equipment for accurate nanoscale grating structure profile characterization. The measurement results have demonstrated the capability of Mueller measurement in nanostructure metrology. It is expected that Mueller measurement may be a powerful tool for advanced technological node measurement in the semiconductor device manufacturing applications.

REFERENCES

[1] M. Foldyna, A. De Martinoa, C. Licitrab, and J. Foucherb, Proceedings of SPIE 2010, San Jose, California, United States, February 2010, Vol. 7638, pp.76380-4,

[2] H.T. Huang, W. Kong, and F.L. Terry Jr, Applied Physics Letters, 2001, Vol.78 (25), pp.3983-3985.

[3] M.G. Faruk, S. Zangooie, M. Angyal, D.K. Watts, M. Sendelbach, L. Economikos, P. Herrera, and R.Wilkins, IEEE Transactions on Semiconductor Manufacturing, 2011, Vol.24(4), pp.499-512.

[4] Y.M. Shi, Z.S. Zhang, G.X. Liu, Z.J. Liu, and Y.P. Xu, Ecs Transactions, 2011, Vol. 34(1), pp.955-960.

[5] M. Foldyna1, A. De Martino, E. Garcia-Caurel, R. Ossikovski, C. Licitra, F. Bertin, K. Postava, and B. Drevillon, European Physical Journal Applied Physics, 2008, Vol.42(3) ,pp.351-359.

THE STUDY AND INVESTIGATION OF INLINE E-BEAM INSPECTION FOR 28NM PROCESS DEVELOPMENT

Yin Long[12], Rongwei Fan[2], Hunglin Chen[2], Haihua Li[1]*

[1]School of Microelectronics, Shanghai Jiao Tong University, 200240, China

[2]Shanghai Huali Microelectronics Corporation, 201210, China

*Corresponding Author's Email: longyin@hlmc.cn

ABSTRACT

The research aims at the VC (voltage contrast) of contact-loop defects in 28nm processes. A new type defect, contact W (tungsten) missing, was found during process developments, and the designed inline E-beam inspection was used to detect those defects by their VC features. The mechanism of defects and the method of the enhancement of the detection were described. An extreme high signal E-beam scan recipe is required in corresponding to the unique defect wanted and, in doing so, the partial-opened W missing defects can be found. Since the defect can be detected instantly by E-beam inspection, and therefore an inline monitoring index can be set up. Compared to the end-of-line electrical test, this inline monitor is very much closer to the trouble process and shrinks the response time. The following process experiments and evaluation can be instantly verified and examined. Finally, the W missing defect was fixed by an optimization of CT Etch process and controlling of CT loop minienvironment. Instead to the debug method of failure analysis, E-beam inspection can speed up 28nm development.

W-MISSING DETECTION BY E-BEAM INSPECTION

W with superior step coverage capability is a traditional material used for filling contact holes; however, the contact dimension continually shrinks and W-missing in CT holes were found in 28 nm technology-node wafers.

The defects distributed random on the wafer, showed in Fig.1 (a). The defect detected by electron beam inspection was showed as bright voltage contrast (BVC), showed in Fig.1 (b), and the defect was just only the void at the bottom of the W plug which could not be detected by optical inspection system, showed in Fig.1 (c).

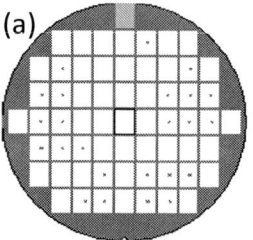

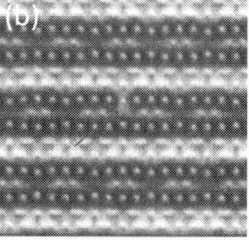

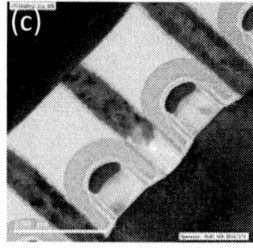

Fig.1. (a) W-missing defect distribution map, (b) W-missing defect image of E-beam review (c) W-missing defect TEM image.

An e-beam scan with VC images

comparison is an effective inspection method and a good alternative to bright and dark field ones as tolerance of defects in the semiconductor process decreases. To achieve the voltage contrast signal of the W-missing defect is a challenge for e-beam inspection since those partial filled contact holes do not really make contact open.

In theory it is necessary to find suitable conditions to enlarge its resistance to high level features can be detected. Only when the charge accumulation rate reaches a certain extent over the charge release rate, W-missing defects can be detected.

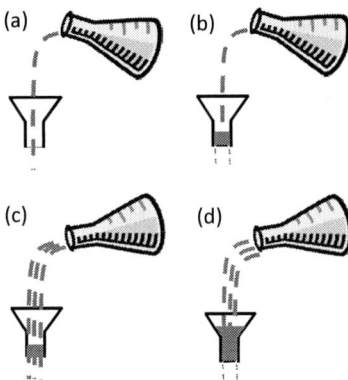

Fig.2 Sketch Map of E-beam scan in different charge accumulation rate （a）Normal CT in normal charge， （b ） W missing in normal charge (c)Normal CT in enhanced charge， （d） W missing defect in enhanced charge

Under the normal inspection mode, shown in Fig.2(b), charge accumulation of W missing is low, then there would be no VC signal difference to the normal CT plug showed in Fig.2(a). However, with the enhanced charge accumulation of EBI tool, shown in Fig. 2(d), there would be VC signal difference between the CT missing and the normal CT plug, since the charge accumulation rate is more than the charge release rate.

Various combinations of electron gun settings, wafer charging modes and scan sequences were tested; a suitable scan method

with "charge enhanced mode" was developed targeted to those partial filled contacts. In this mode, W-missing defect signal is obviously different from background (signal to noise ratio:1.8), shown in Fig.3(c), which meet the e-beam inspection request.

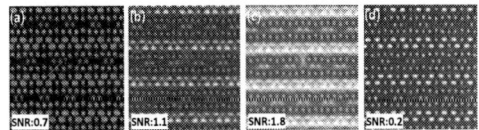

Fig.3 W-missing defect images in different e-beam condition. a) normal positive, b)normal negative, c) charge enhanced mode, d)charge suppress

EXPERIMENTS AND RESULT

Combining other inspection conditions, inline index for W missing defects was built up, which would speed up the development progress of 28nm technology node.

As the contact dimension continually shrinks, the filling of the contact holes were more and more difficulty. There are many factors could impact the W filling process, including CT hole aspect ratio (AR), byproduct polymer performance post CT etch, the minienvironment of the FOUP and the wet clean condition. With the inline index built up by the proposed novel detection method, a series of inline process experiments were then carried out to fix W-missing defects.

From some of the experiment results, high correlation between W-missing count and CT hold aspect ratio (AR) was found, shown in Fig.4. It means that, when AR ratio>4, the higher AR ratio, the more defect count.

However, to get lower AR ratio, we have to enlarge CT CD. That will bring in some side-effect, i.e. the process window of CT to poly overlay will be narrowed which maybe induced CT to poly short.

Combining the AR ratio condition, a

serious of process tuning was carried out in order to optimize the polymer post CT etch, improve the minienvironment of FOUP and enhance the clean efficiency post CT etch.

The split results indicated that, shown in Fig.4, the defect count reduced obviously by the process tuning. The positive results provide the feasible solution to balance the AR ratio and overlay process window.

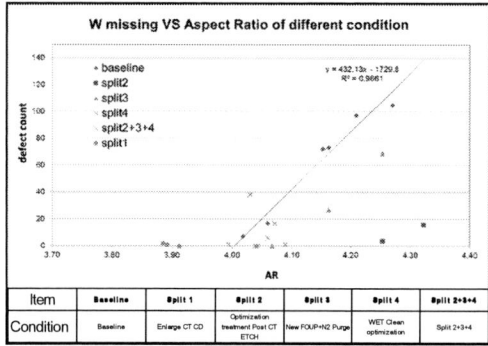

Item	Baseline	Split 1	Split 2	Split 3	Split 4	Split 2+3+4
Condition	Baseline	Enlarge CT CD	Optimization treatment Post CT ETCH	New FOUP+N2 Purge	WET Clean optimization	Split 2+3+4

Fig.4 experiments results of W-missing defect

Finally, with the optimized processes that could reduce the CT hole AR, enhance the treatment post CT ET, combining FOUP inserted gas purge could reduce the W-missing defects nearly to zero. At the same time, the overlay process window of CT photo was acceptable.

CONCLUSIONS

The study builds up an inline index for W-missing defects and speed up the development progress of 28nm technology node. A series of experiments were carried out and showed the defect was obviously reduced by reducing CT AR ratio. The failure model involved the accumulating polymers in the CT hole and the minienvironment of the FOUP. Furthermore, the corresponding improvement actions including reducing CT hole AR, optimizing the processes of CT ETCH and post clean, improving the FOUP minienvironment of adding inert gas purge were executed. The W missing defects were trend low accordingly.

ACKNOWLEDGMENT

The authors would like to thank Fiona Pan of Hermes-Epitek for the support of performing the e-beam inspection. Dr. Fang of Huali for coordinating the split experiments and process direction.

REFERENCE

[1] X.P. Wang, et al, "Dry Etching Solutions to Contact Hole Profile Optimization for Advanced Logic Technologies", CSTIC, 2012

[2] Jing-Yong Huang, "Contact Etch Schemes at Advanced Logic Technology Nodes", CSTIC, 2014

[3] T.R. Cass, D. Hendricks, J. Jau, H.J. Dohse, A.D. Brodie, W.D. Meisburger, Application of the SEMSpec electron-beam inspection system to in-process defect detection on semiconductor wafers, Microelectronic Engineering, Volume 30, Issues 1–4, January 1996, Pages 567-570, ISSN 0167-9317, 10.1016/0167-9317(95)00311-8.

[4] L. Lin, J.Y. Chen, S.D. Luo, W. Y. Wong, S. Oestreich, A. Tsai, I. Yao, and L. Grella, "Residual oxide detection with automated E-beam inspection", IEEE International Symposium on Semiconductor Manufacturing, Sep. 2005, pp. 241-244.

[5] O.D. Patterson, K. Wu, D. Mocuta, and K. Nafisi, "Voltage Contrast Inspection Methodology for Inline Detection of Missing Spacer and Other Nonvisual Defects ", IEEE Transactions on Semiconductor Manufacturing, vol. 21, no. 3, , Aug. 2008 pp. 322-328.

STRESS CONTROL METROLOGY IN EPITAXY

Yang Song, Pengwei Fan, and Yi-Shi Lin*

Technology R&D, Semiconductor Manufacturing International Corp.

Pudong New Area, Shanghai 201203, China

*Corresponding Author's Email: Alien_Lin@smics.com

ABSTRACT

This paper presents a study performed to evaluate the benefits of stress control metrology in Silicon Germanium (SiGe) epitaxy. Various in-line and off-line metrology tools are studied and compared, such as using OCD (Optical Critical Dimension) to monitor SiGe profile, using SE (Spectroscopic Ellipsometry), HRXRD (High Resolution X-ray Diffraction), SIMS (Secondary Ion Mass Spectrometry) to achieve the Ge composition control, using HRXRD to monitor strain status *etc*. The study also demonstrates that using combined metrologies enables better control of epitaxy in research and manufacturing environments.

INTRODUCTION

Stress control metrology of epitaxy is of particular interest as the node goes beyond 28nm. For PMOS transistors, embedded SiGe can significantly enhance the channel carriers' mobility due to germanium induced compressive strain [1]. In addition, sigma shaped SiGe source/drain is proved to enhance the electrical performance. The formation of sigma shape of embedded SiGe, including PMOS silicon dry and wet etch, epitaxial growth, needs to be effectively monitored in order to characterize the structure and strain property. This paper illustrates the implementation of various metrologies for monitoring SiGe.

SiGe profile can be measured with OCD. The sigma shape tip location and other trench parameters are very important in controlling the stress distribution. After epitaxial growth, OCD is also used to monitor the cap silicon thickness and overfill. Meanwhile, SiGe strain and stress can be in-directly monitored by measuring germanium composition, this may be achieved by using SIMS or SE, but it should be noted that ellipsometry is not sensitive to the composition change, and the model will not be appropriate when there exist obvious process changes. HRXRD is a direct method to monitor SiGe strain and relaxation. This paper also illustrates that with the function of RSM (reciprocal space map), it is possible to measure the Si strain in OCD pad, which is more representative to the channel strain status. Strengths and drawbacks of these techniques are compared in this paper. In addition, there exist increasing challenges for metrology techniques as the node goes down, one promising method is to combine the data from different metrology tools in order to obtain more information about the structure. "Hybrid metrology" is often used to describe such combination of data from different metrology tools.

SIGE PROFILE CONTROL

In this section, implementation of OCD is described in order to monitor the SiGe profile, here we take the wet etch profile as an example.

Sample Description

OCD is a widely used technique for in-line mass production. It takes the structure parameters and material properties (n & k optical constant) into account when doing the analysis. OCD is a calibrated technique; the results are related to the TEM qualified libraries. Data is analyzed by comparing the real obtained spectra and model fitting spectra as a function of wavelength.

DoE (Design of Experiment) set of wafers is needed to be prepared to ensure a large measurement range so that the library can cover the process window needed. Usually, OCD spectra are collected on a 50*50μm pad with single periodic structure in each direction, the periodic pattern should be large enough---the line/space ratio has to be larger than 1/10, and the aspect ratio is also suggested to be larger than 1/10. By the way, process should be stable and process behavior within OCD pad need to be the same, as OCD gives an average result of the periodic pattern.

Accuracy

The accuracy is tested by comparing the OCD obtained parameters with a reference characterization technique TEM. Figure 1 shows the trench depth measured data of three DoE wafers, considering 2 points per wafer. The result shows that there is a very good matching between OCD and TEM results, indicating that the OCD technique can obtain accurate measurement with an appropriate model.

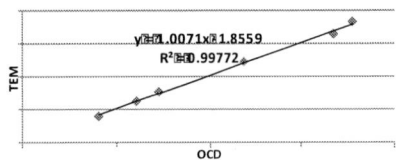

Figure 1: OCD versus TEM of trench depth measurement

Gauge Repeatability and Reproducibility (GRR)

GRR is another key quality criterion for metrology tools. GRR test is performed through a 10 times dynamic repetitive measurements on 13 points of the wafer, as is shown in Figure 2. The result shows very good repeatability measured by OCD (3σ are less than 1%).

Overall, it can be seen that the results from OCD can be qualified as trustable and stable.

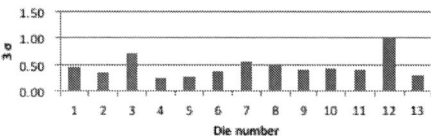

Figure 2: GRR test for trench depth

SIGE GE% CONTROL

For PMOS transistors, SiGe structure can significantly enhance the channel carriers' mobility due to Ge induced compressive strain, as the lattice constant of Ge is larger than Si. Ge% is usually measured to predict the strain status of SiGe structure and thus the channel Si strain status. Higher Ge% usually means a larger strain. The thickness and Ge composition of SiGe film can be analyzed through SE by using a non-linear fit algorithm based on the minimization of the residual χ^2 [2]. Similar to OCD, SE is also a model based technique.

HRXRD and SIMS can also be used to determine the Ge%. Compared with SE, HRXRD is more sensitive to Ge% change, and can monitor the epi-quality. But the challenge of HRXRD is the throughput is lower as R&D mode. HRXRD gives an averaged composition of specific layer, while SIMS is capable to monitor the depth profile of the SiGe layer. But as SIMS is a destructive technique, the cycle time is quite long. Figure 3 shows the correlation of HRXRD and SIMS of Ge% measured on solid pad

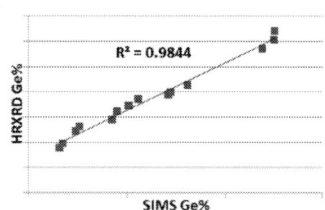

Figure 3: HRXRD versus SIMS of SiGe Ge% on solid pad

SIGE STRAIN CONTROL

HRXRD is a fast, direct, non-destructive, inline metrology method to determine the channel strain, Ge%, relaxation, thickness *etc.* [3]. The measurement principle is Bragg's Law. Ge% is calculated by measuring the lattice parameter change. Thus the Ge% given by HRXRD

is the effective Ge%, that is, the substitutional Ge atoms in SiGe are measured. It should also be noted that once the film is relaxed, the obtained Ge% will not be accurate. According to the strain, the SiGe stress can be calculated by using the Young's modulus and Poisson ratio of the material.

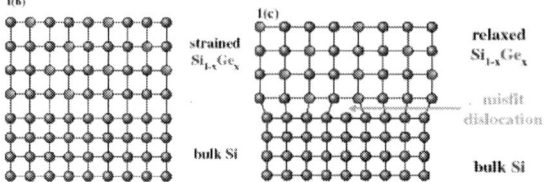

Figure 4: SiGe lattice structure under fully strain and relaxed status [4]

SiGe strain may be changed during subsequent process. In our work, SiGe strain of as deposited condition and after spike anneal are studied and compared, as we can see from Figure 5, both in horizontal (x) and vertical (y) direction, the SiGe stain are lossed obviously with the anealing process. Thus it is suggested to add SiGe measurement steps after subsequent thermal processes.

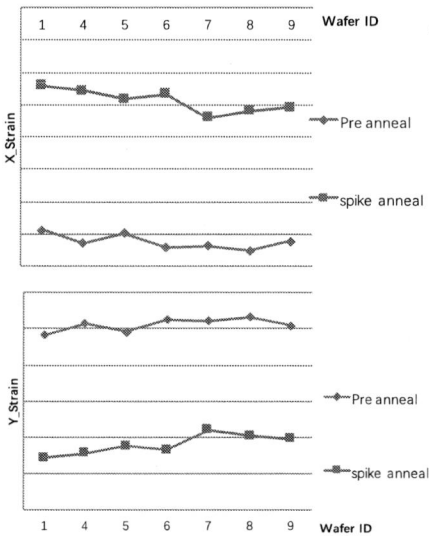

Figure 5: Thermal effect to SiGe solid pad of X (top) and Y (Bottom) direction

SI STRAIN CONTROL

Si strain measurement is of particular interest in this structure as it reflects directly the channel strain status. Previous studies of Ge% and SiGe strain are performed on SiGe solid pad, but OCD pad is the most representative structure of the in-device strain. Implementation of reciprocal space maps (RSM) can be regarded as a diagnostics and debugging tool for epitaxy quality. Si strain is possible to be measured under this condition.

978-1-5090-6695-7/17 $31.00 © 2017 IEEE

SIGE B% MEASUREMENT

PMOS is very sensitive to boron dosage in SiGe process. Below certain boron dosage, increasing boron will help enhance the device performance. In addition, feeding forward the B concentration to HRXRD model can also help to extract a real Ge%, as the boron doping will also influence the strain status of the SiGe structure. So how to effectively measure boron has raised considerable interest among many researchers. Various metrology methods have been proposed and studied. WD-XRF (Wavelength Dispersive X-ray Fluorescence) can be used to measure B%, and has been used for BPSG (Borophosphosilicate Glass) film characterization. However, the detection limit of WD-XRF is relatively high. LEXES (Low Energy X-ray Emission Spectrum) can also be used to do elemental composition analysis with very low detection limit. But similar to WD-XRF, this technique also needs reference sample to calibrate the tool. MBIR (Model Based Infrared Reflectometry) is a model based technique to measure the boron dopants. In 2014, D. Le Cunff *et al.* published one paper about combining MBIR with SE to measure the Boron concentration with good accuracy. [5] SRP (Spreading Resistance Measurement Profiling) is a destructive method and can in-directly measure the doping density, as it is achieved by converting the resistance to resistivity and carrier density. The advantage of SRP is that effective B% can be determined. Similar to SIMS, the challenge is that the sample preparation and measurement time are long, only finite samples can be analyzed outside the fab.

Hybrid metrology is suggested to be used to combine these metrology methods to extract more information, as is mentioned above, feeding forward material properties, such as B% by MBIR, LEXES *etc.* or Ge% by HRXRD to SE/OCD tools, will help setup more accurate SE/OCD models, as SE/OCD obtained spectra are significantly influenced by material n, k properties, which are determined by Ge%, B% etc. In addition, combination of metrology tools will also help improve throughput. For example, the throughput of HRXRD is not high enough, using SE to monitor the process stability and meanwhile use HRXRD to selectively measure the samples will significantly increase throughput.

CONCLUSIONS

Stress control metrology in epitaxy is discussed in this work. The strengths and drawbacks of each technique are presented. SiGe profile can be monitored by OCD with sufficient floating parameters. HRXRD can be used to determine the SiGe effective Ge composition, SiGe strain and Si strain. B% can be measured by WDXRF, LEXES, MBIR, SRP, SIMS *etc*. Hybrid techniques for such applications are recommended to achieve more information and higher throughput.

ACKNOWLEDGEMENTS

The authors wish to express their gratitude to Yi Huang at SMIC metrology department for her feedback and support during the whole work. The authors gratefully acknowledge Dr. Alex Tokar and Dr. Qu Bo at Bruker for the support and their expertise.

REFERENCES

[1] Ghani, T., Armstrong, M., Auth, C. etc. "A 90nm high volume manufacturing logic technology featuring novel 45nm gate length strained silicon CMOS transistors," Electron Devices Meeting, 2003. IEDM '03 Technical Digest. IEEE International, pp.11.6.1,11.6.3, 8-10 Dec. 2003J.

[2] D. Le-Cunff, S. Couvrat and F. Abbate "In-line metrology capability for epitaxial multi-stack SiGe layers", Advance Semiconductor Manufacturing Conference (ASMC), 2012 23rd Annual SEMI

[3] Y. Song, Y. Huang and Y. S. Lin, "Challenges and prospects of X-ray metrology in advanced semiconductor industry," 2016 China Semiconductor Technology International Conference (CSTIC), Shanghai, 2016, pp. 1-3.

[4] D.J. Paul, Physics World 13, pp27-32 (February 2000)

[5] D. Le Cunff et al., "Benefit of combining metrology techniques for thin SiGe:B layers," 25th Annual SEMI Advanced Semiconductor Manufacturing Conference (ASMC 2014), Saratoga Springs, NY, 2014, pp. 31-36.

LOW FREQUENCY NOISE CHARACTERIZATION OF 22NM PMOS FEATURING WITH FILLING W GATE USING DIFFERENT PRECURSORS

Liang He[1,2,4], Eddy Simoen[1,3], Cor Claeys[1,2], Guilei Wang[5], Jun Luo[5], Chao Zhao[5], Junfeng Li[5], Hua Chen[4, *], Yin Hu[4], and Xiaoting Qin[4]*

[1]Imec, Kapeldreef 75, Leuven B-3001, Belgium

[2]Dept. Electrical Engineering, KU Leuven, Leuven B-3001, Belgium

[3]Dept. Solid State Sciences, Ghent University, Gent, Belgium

[4]School of Advanced |Materials and Nanotechnology, Xidian University, Xi'an 710126, China

[5]Key laboratory of Microelectronic Devices & Integrated Technology, Institute of Microelectronics of Chinese Academy of Sciences, Beijing 100029, China

*Corresponding Author's Email: xidianhic@163.com, addal@163.com

ABSTRACT

Low frequency noise behavior of 22 nm Si pMOSFETs featuring with filling W gate using different precursors, *i.e.,* SiH_4 or B_2H_6, have been studied. It is shown that carrier number and correlated mobility fluctuations are the underlying mechanisms for the flicker noise. Moreover, devices with B_2H_6 W processed gate metal were found to have a higher mean trap density and scattering coefficient than SiH_4 W processed gate metal. In addition, both gate-voltage-dependent and gate-voltage-independent Generation-Recombination (GR) noise peaks have been found, which are assigned to traps in the gate oxide or depletion region, respectively.

INTRODUCTION

As chip size scales down continuously, the gate starts to lose control over the channel, moreover, gate leakage has become significant. Both of these have resulted in the development and emergence of a broad range of CMOS technologies including new materials and architectures. Multiple gate devices with high-k dielectric have been considered promising candidates for the future of microelectronics.

The implementation of thicker high-k dielectrics allows leakage reduction without loss in the gate capacitance. In addition to the high-k dielectrics, new gate materials have been studied to satisfy the requirement for a proper threshold voltage setting on both n- and p-channel devices in high-performance CMOS applications. Replacing the poly gate by metal gate electrodes can avoid the degradation caused by high-k/polysilicon stacks on the carrier mobility, suppresses the polysilicon depletion effect and reduces the dopant penetration through the gate dielectric [1, 2]. For gate filling metal, aluminum(Al) was firstly introduced at 32nm node, however it has imposed more and more challenges in filling capability and Chemical Mechanical Polishing(CMP) process. Recently, the application of Atomic Layer Deposition(ALD) Tungsten as gate filling metal has attracted tremendous attentions,

because of its excellent step coverage capability and mature CMP process. Some interesting research on using ALD W with B_2H_6 and SiH_4 precursors as gate filling metal have been carried out[3,4].

Low frequency noise which is sensitive to defects and imperfections in the current path, can be used as a non-destructive diagnostic tool to identify traps in the Si and the gate oxide, thus giving information on the quality of the transistors' fabrication process. It is clear that there are essentially two physical mechanisms behind any fluctuations in the current: fluctuations in the mobility or fluctuations in the number of carriers [5]. Based on that, low frequency noise can reveal details on carrier transport mechanisms in the channel (trapping, scattering), providing a better understanding of the underlying temporal fluctuation phenomena. The aim of this work is to investigate the noise behavior and underlying mechanism for transistors with filling W gate using two different precursors.

EXPERIMENTAL DETAILS

The Si MOSFETs under study are p-type channel with planar architecture. The width is 3μm and the length L=0.1μm. The gate dielectric consists of 0.8 nm SiO_2 oxide (IL), on which 2 nm HfO_2 is deposited by ALD. The gate electrode consists of 2 nm TiN deposited by ALD, 3 nm Ti deposited by Physical Vapor Deposition (PVD), 5 nm TiN deposited by Metal Organic Chemical Vapor Deposition (MOCVD), and 75nm W by ALD, whereby two different W deposition processing approaches were studied, namely B_2H_6 W and SiH_4 W. Low frequency noise investigations were performed with the devices biased in linear operation at a drain bias V_{DS}=-50 mV, with the gate voltage V_{GS} stepped from weak to strong inversion. The substrate contact was grounded.

RESULTS AND DISCUSSION

One can observe significant scatter between two devices with different gate metal processing in fig.1. This can be explained to some extent by the presence of a

978-1-5090-6695-7/17 $31.00 © 2017 IEEE

significant GR noise component, which shows up as a peak around 100 µA.

From fig. 2, one can conclude that the flicker noise of the SiH$_4$ W device is dominated by number fluctuations, since the normalized drain current spectral density shows to be proportional to $(g_m/I_D)^2$ in the studied I_D range. In addition, the corresponding input-referred noise S_{VG} is constant around the threshold voltage V_T [6], as fig. 4 shows. However, obvious deviation can be observed in strong inversion for the B$_2$H$_6$ W device in fig. 2, which could be attributed to the additional correlated mobility fluctuations.

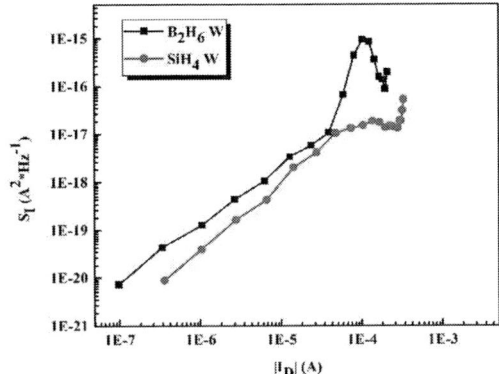

Figure 1: Drain current noise power spectral density (PSD) as a function of absolute drain current at f=10 Hz for two gate metal processing devices, L=0.1µm, W=3µm

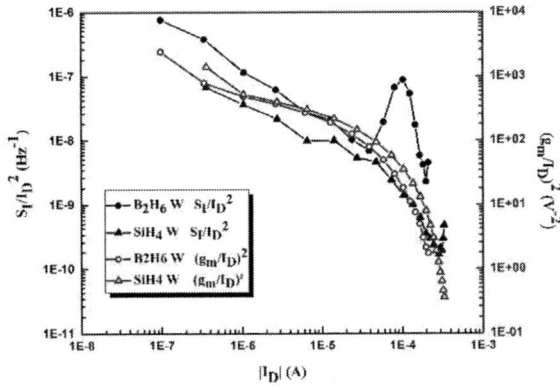

Figure 2: The dependences of S_I/I_D^2 and $(g_m/I_D)^2$ on I_D at f=10 Hz for two gate metal processing devices

Through comparison of fig. 3 with fig. 4, one can notice that in weak inversion, the B$_2$H$_6$ W pMOSFET has a noticeable higher $S^{½}_{VG}$, and in strong inversion, the $S^{½}_{VG}$ presents the same trend as the SiH$_4$ W counterpart.

The PSD shows a higher value, which can be associated with a higher oxide trap density, since the latter is directly proportional to the input-referred voltage noise at the flat-band voltage region [7].

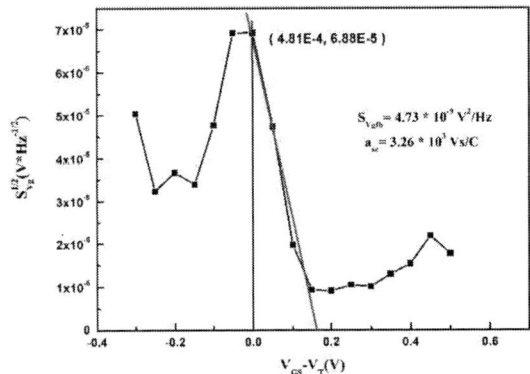

Figure 3: $\sqrt{S_{VG}}$ at 10 Hz versus gate voltage overdrive for a B$_2$H$_6$ W device.

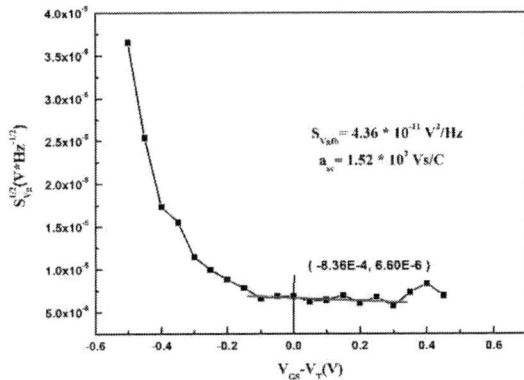

Figure 4: $\sqrt{S_{VG}}$ at 10 Hz versus gate voltage overdrive for a SiH$_4$ W device

Table I is a summary of the flat-band voltage noise (S_{vgfb}) and Coulomb scattering coefficient (α_{sc}) for the B$_2$H$_6$ W and SiH$_4$ W transistors, respectively. It shows that transistors with B$_2$H$_6$ W gate metal have a relatively higher S_{vgfb}, which indicates a higher density of oxide traps, also, the higher α_{sc} in B$_2$H$_6$ W devices reflects a stronger scattering component effect. These can be attributed to the B diffusing through gate stacks into high-k dielectric, which would induce acceptor gap state in the vicinity of conduction band of high-k dielectric.

978-1-5090-6695-7/17 $31.00 © 2017 IEEE   349

TABLE I. *PSD OF FLAT-BAND VOLTAGE NOISE AND COULOMB SCATTERING COEFFICIENT FOR SAMPLES WITH TWO DIFFERENT GATE METAL PROCESSING*

No.	B_2H_6 W		SiH_4 W	
	$S_{Vgfb}(V^2/Hz)$	$\alpha_{sc}(Vs/C)$	$S_{Vgfb}(V^2/Hz)$	$\alpha_{sc}(Vs/C)$
1	4.73e-9	3.26e3	2.37e-10	6.45e2
2	1.54e-9	3.13e3	4.36e-11	1.52e2
3	1.61e-10	6.73e2	6.81e-11	2.61e2
4	6.87e-11	1.15e3	7.71e-11	4.78e2
5	6.13e-11	6.37e2	4.58e-10	6.51e2
mean	1.31e-9	1.77e3	1.77e-10	4.37e2

Except for the flicker noise, GR noises, corresponding with a Lorentzian spectrum, were also present in some devices, as shown in fig. 5 and fig. 6, from which both gate-voltage-dependent and gate-voltage- independent corner frequencies have been found. These are assigned to traps in the gate oxide and the silicon depletion region, respectively[8].

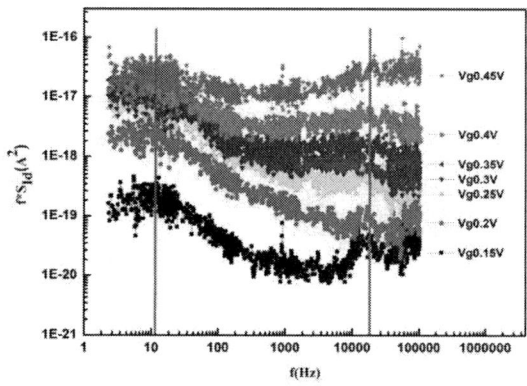

Figure 5: Frequency normalized spectra for a B_2H_6 W device

CONCLUSION

It has been shown that the flicker noise of SiH_4 W device is dominated by number fluctuations mechanism, however, for B_2H_6 W device, the effect of additional correlated mobility fluctuations mechanism should be concerned. Meanwhile, higher values of S_{vgfb} and α_{sc} from B_2H_6 W gate metal devices can be attribute to B diffusing through gate stacks into high-k dielectric, thereby inducing acceptor states in the band gap.

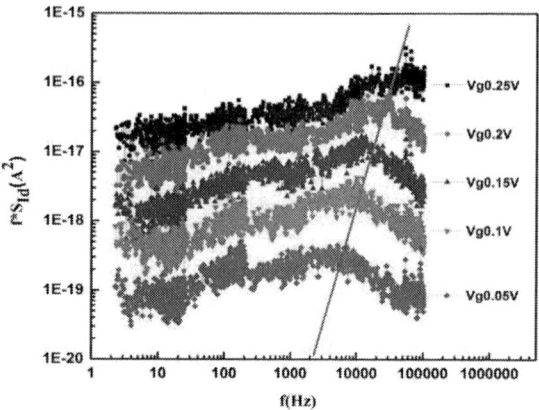

Figure 6: frequency normalized spectra for a SiH_4 W device

ACKNOWLEDGEMENTS

This work was supported by China Scholarship Council, the National Natural Science Foundation of China (Grant No. 61106062) and the Fundamental Research Funds for the Central Universities (Grant No. JB151403).

REFERENCES

[1] M. Rodrigues, M. Galeti, J. A. Martino, N. Collaert, E. Simoen, and C. Claeys, Solid-State Electronics, vol. 62, 2011, pp. 146-151.

[2] M. Rodrigues, J. A. Martino, A. Mercha, N. Collaert, E. Simoen, and C. Claeys, Solid-State Electronics, vol. 54, 2010, pp. 1592-1597.

[3] Q. Xu, J. Luo, G. L. Wang, T. Yang, J. F. Li, T. C. Ye, D. P. Chen, and C. Zhao, Microelectronic Engineering, vol. 137, 2015, pp. 43-46.

[4] G. L. Wang, Q. Xu, T. Yang, J. J. Xiang, J. Xu, J. F. Gao, C. L. Li, J. F. Li, J. Yan, D. P. Chen, T. C. Ye, C. Zhao, and J Luo, ECS Journal of Solid State Science and Technology, vol. 3, 2014, pp. 82-85.

[5] M. Von Haartman and M. Östling, Low-frequency Noise in Advanced MOS Devices, Springer Press, 2007.

[6] G. Ghibaudo, O. Roux, C. Nguyen-Duc, F. Balestra and J. Brini, Phys. Stat. Sol. A, vol. 124, 1991, pp. 571-578.

[7] E. Simoen, M. G. C. Andrade, M. Aoulaiche, N. Collaert, and C. Claeys, IEEE Trans. Electron Devices, vol. 59, 2012, pp. 1272-1278.

[8] E. Simoen, B. Cretu, W. Fang, M. Aoulaiche, J. M. Routoure, R. Carin, S. D. Santos, J. Luo, C. Zhao, J. A. Martino, and C. Claeys, Phys. Status. Solidi C, vol. 12, 2015, pp. 292-298.

Exploration of Poly Irms Based on 40nm Technology Node

Xiang Fu Zhao[*], Wei Ting Kary Chien, and Kelly Yang

Reliability Engineering, Semiconductor Manufacturing International Corporation

Pudong New Area, Shanghai, 201203, China

*Corresponding Author's Email: x-f.zhao@163.com

ABSTRACT

The Root Mean Square current (Irms) has been explored on both N doped un-silicided poly and silicided poly at 40nm technology node. It is found that poly resistance-current (RI) curves show an initially high resistance due to Schottky Rectifying contact. However, poly Irms can be estimated similar to metal Irms. Parameters for poly Irms calculation is provided on the basis of 5℃ Joule heating.

INTRODUCTION

Poly, especially for un-silicided poly, has attracted wide attentions due to its high resistance. Even a small current flow through it, a significant Joule heating can occur, which may affect the performance of the device nearby. Critical thermal issues in nanoscale IC design have been discussed and the related methodology to limit the Joule heating effect has been proposed for better design and reliability evaluations [1]. The failure mechanism was explored on silicided poly-tungsten-copper line structure and it was found that poly can fail under the condition of its high Joule heating, which is associated with current density [2]. The Joule heating of un-silicided poly has been also explored on the similar structure and it is found that the Joule heating can be apparent even at a normal working current density [3].

It has been proved that the thermal coefficient of resistance (TCR) is very stable for undoped polysilicon at the temperature between 300K and 400K [4]. Thus, it is believable that the Joule heating of polysilicon can be calculated using TCR. The Irms rules of BEOL (Back-End Of Line) of 65nm node circuits have been proposed for the Joule heating concern in alternating current (AC) interconnects [5]. However, the Irms of poly is seldom reported. In this paper, we explore the Irms on both N doped silicided and un-silicided poly. We also compare the parameters for poly Irms design rules.

EXPERIMENTAL AND RESULTS

N doped un-silicided poly and silicided poly were selected to estimate Irms; there are five different widths: 0.4/ 1/ 1.5/ 2/ 2.5um and 0.15/ 0.4/ 0.8/ 1.2/ 1.6um for un-silicided and silicided poly, respectively. Poly length is the same for the same kind of poly. Four temperatures (75℃, 100℃, 125℃, and 150℃) were chosen for TCR resistance measurement by Agilent 4072 with TEL P12 prober. The Joule heating was evaluated at 110℃ with a ramped current starting from 0.1mA. The Irms estimation method has been given in Ref. [6] in the final expression below:

$$I_{rms}^2 \approx \Delta T \cdot \left(b \cdot w^2 + c \cdot w \right) \quad (1)$$

Where, ΔT is the rising temperature generated from the Joule heating and is usually set to 5℃. W presents the poly width; b & c are constants to be determined.

Fig. 1 shows the RI curve of N doped un-silicided poly with width 0.4/ 1/ 1.5/ 2/ 2.5um, respectively. It can be seen, from Fig. 1 (b) to 1(e), that there is an initial resistance drop. This is because poly is connected with a copper line via tungsten. Thus, there is a Schottky Rectifying contact. With a rising current, the Schottky Rectifying contact will become a Schottky Tunneling ohm contact, which causes such resistance drop. The initial drop is not apparent in Fig. 1 (a) because the contact resistance is covered within the high resistance of poly with 0.4um width. It can also be clearly seen that the thresh current, which induces apparent resistance raise, becomes larger. This is because of the decreasing poly resistance when the poly width increases. Similar results can be obtained as in Fig. 2, which shows the RI curve of N doped silicided poly with width 0.15/ 0.4/ 0.8/ 1.2/ 1.6um, respectively.

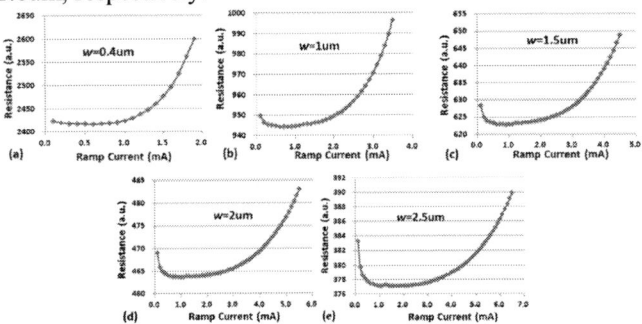

Fig.1. RI curves of N doped un-silicided poly varying with widths (a) w=0.4um, (b) w=1um, (c) w=1.5um, (d) w=2um, (e) w=2.5um

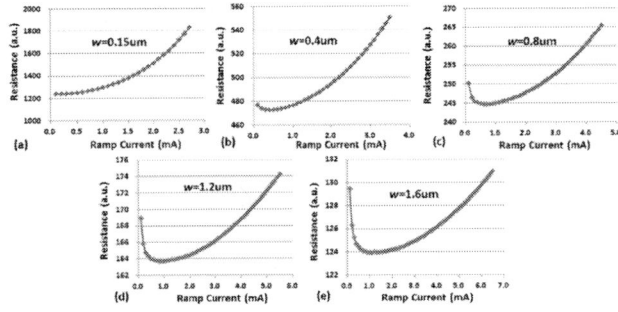

Fig.2. RI curves of N doped silicided poly varying with widths (a) w=0.15um, (b) w=0.4um, (c) w=0.8um, (d) w=1.2um, (e) w=1.6um

Fig. 3(a) and 3(b) display Irms^2 vs width curves of N doped un-silicided and silicided poly, respectively. It can be observed (from Fig. 3) that both simulative curves well coincide with experimental data, especially for the silicided poly. This means the estimation method of metal Irms [6] can also be applied to estimate the Irms of poly. Comparing with Eq. (1), constants b and c can be easily derived from simulated expressions in Fig. 3 as listed in Table 1, where the un-silicided poly has smaller b and c. This is because the former has a higher resistivity, for the same width and 5 ℃ Joule heating temperature, the former bears a lower Irms value.

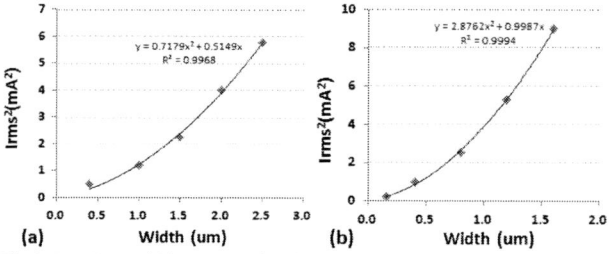

Fig.3. Irms2 vs width curves of N doped poly, (a) un-silicided, (b) silicided

TABLE I
IRMS PARAMETERS FOR N DOPED UN-SILICIDED POLY
AND SILICIDED POLY

Parameter Values	Un-Silicided	Silicided
b	0.14	0.57
c	0.10	0.19

CONCLUSIONS

The Irms of un-silicided and silicided poly were investigated at 40nm technology node. It is found that the metal Irms estimation method can be well applied to estimate the Irms of poly resistance. Schottky Rectifying contact is observed to change into Schottky Tunneling ohm contact when ramping current at 110 ℃ for poly Joule heating measurement. Irms parameters for both N doped un-silicided poly and silicided poly resistance are also provided. Our research indicates that ploy resistance can well serve as a heater or a temperature sensor under AC condition.

REFERENCES

[1] L. Jiang, D. Pantuso, P. Sverdrup, and Wei-kai Shih, 2009 IEEE International Reliability Physics Symposium, 2009, pp. 909 – 912

[2] C. Huang, M. Lin, J. Liang, A. Juan, and K.C. Su, 2011 IEEE International Reliability Physics Symposium, 2011, pp. EM.1.1 - EM.1.3

[3] X. Zhao, M. Zhang, and W. T. Chien, 2016 International Symposium on the Physical and Failure Analysis of Integrated Circuits, 2016, pp. 289-292

[4] K. Z. Lv, J. Yuan, L. Tian,L. Liu, and Z. Li, 2004 International Conference on Solid-State and Integrated Circuits Technology, 2004, pp.2190-2193

[5] D. Ney, X. Federspiel, V. Girault, O. Thomas, P. Gergaud, 2006 IEEE International Reliability Physics Symposium, 2006, pp. 669-670

[6] Harmon et al., IRW Final Report, 1998, pp.1

EFFECT OF HIGH TEMPERATURE STORAGE ON FAN-OUT WAFER LEVEL PACKAGE STRENGTH

Cheng Xu[1, 2,], Z.W. Zhong[1] and W.K. Choi[2]*

[1] School of Mechanical & Aerospace Engineering, Nanyang Technological University
50 Nanyang Avenue, Singapore 639798
[2] JCET STATS ChipPAC Pte Ltd, Singapore
*Email: XUCH0005@e.ntu.edu.sg

ABSTRACT

Fan-out wafer level packaging technology becomes attractive because of its flexibility for integration of diverse devices in a small form factor. In this study, the effect of high temperature storage test on fan-out wafer level package strength was evaluated. There were three different structure fan-out wafer level packages. The high temperature storage reliability test was used to store the specimens up to 1000 hours. The three-point bending test method was conducted to evaluate the specimen flexure strength. The experiment results showed that FOWLP flexure strength increased with the high temperature storage test time increasing.

INTRODUCTION

Fan-out wafer level package (FOWLP) strength is critical to its reliability since the low strength often leads to silicon die crack or package crack issues. There is much research on the evaluation of the silicon die strength [1]. The conclusions state that the silicon die surface condition and edge condition decide the silicon die strength. The wafer grinding process decides the wafer surface condition [2, 3], while the wafer sawing process decides the silicon die edge condition [4].

The high temperature storage (HTS) reliability test is an essential reliability test for FOWLPs. The HTS test is used to store specimens in a high temperature chamber for a long time durian. The aim of HTS test is to evaluate the time and temperature effects on packages. The specimens in the HTS test are not in their working condition. Hence the failure mode after the HTS test is related to the package material failure such as cracks or declamations. It has been proved that the HTS test could change the epoxy molding compound properties [5, 6] such as glass transition temperature.

There is little research about the package strength especially the FOWLP strength. The FOWLP is composed of various materials beyond the silicon die such as epoxy molding compound, passivation layers, redistribution layers and solder balls. The structure effects on FOWLP strength are evaluated [7, 8]. However, the thermal effect on FOWLP strength is uncertain. In this study, the FOWLP package strength was evaluated to understand the high temperature storage effects on the FOWLP strength and reliability.

TABLE I. SUMMARY OF FOWLP SPECIMEN STRUCTURE & THICKNESS

Specimen Name	FOWLP specimen structure (not in scale)	Thickness
Wafer 1		490 um
Wafer 2		200 um
Wafer 3		225 um

METHODOLOGY

The specimens were built by stand FOWLP assembly processes. The dummy silicon dies were used instead of functional dies. Therefore, there were not redistribution layers and solder balls. The passivation layers were not lithographed since the lithographing process involved a period of high temperature storage. The final package size was 8.09 mm x 8.09 mm. There were three kinds of FOWLP specimen. Wafer 1 was thick FOWLP. Its dummy silicon die thickness was 370 um and its final package thickness was 490 um. Therefore, Wafer 1 silicon die backside was encapsulated by the epoxy molding compound. Wafer 2 and Wafer 3 were thin FOWLP. They were ground to 200 um from the thickness of 490 um, and their silicon die backside was exposed. Wafer 3 was laminated a backside protection tape before the test, and its final thickness was 225 um.

For the high temperature storage test aspect, we referred to the JEDEC standard – High Temperature Storage Life JESD22-A103. The high temperature storage condition B 150 degree C was used. There were two readout points – 500 hours and 1000 hours.

There are several strength test method such as three-point bending test, four-point bending test [9], ball-on-ring test [10] and ball breaker test [11]. In this work, the three-point bending test method was used. The test machine was Instron universal tester 5569 with Instron 2525-816 static load cell. The three-point bending test fixture was customized.

978-1-5090-6695-7/17 $31.00 © 2017 IEEE

RESULT & DISCUSSION

The three-point bending test was performed at three readout points – zero hours, 500 hours and 1000 hours HTS test. The bending fixture span was fixed at 6 mm to support the specimens. The loading speed was 0.6 mm/min. The whole test was conducted under the room temperature 25 degree C.

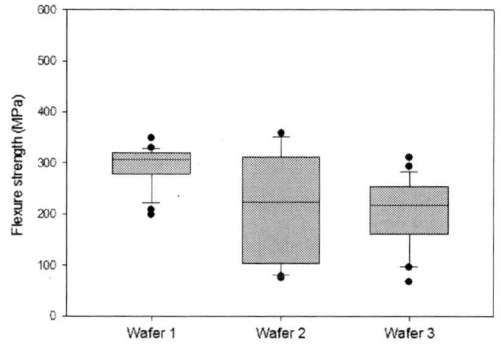

Figure 1: Summary of flexure strength at time zero.

The zero hour three-point bending test result is showed in Figure 1. Wafer 1 average flexure strength was much higher than Wafer 2 and Wafer 3. Wafer 2 and Wafer 3 average flexure strength were almost the same. However, their flexure strength distribution was quite different. Wafer 2 flexure strength dropped obviously after the wafer grinding process, and its flexure strength distribution became scattered. Although Wafer 3 had the same average flexure strength with Wafer 2, it had a tight flexure strength distribution. It was because that Wafer 3 was laminated a backside protection tape after the wafer grinding process. The backside protection tape function was similar to the epoxy molding compound. They were both used to protect the silicon die and enhance the package strength. The backside protection tape was good at its uniformity thickness to compare with the epoxy molding compound.

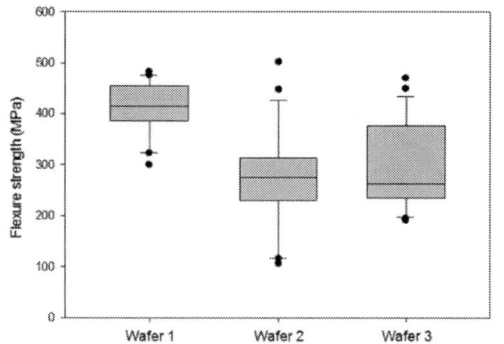

Figure 2: Summary of flexure strength after 500 hours HTS test.

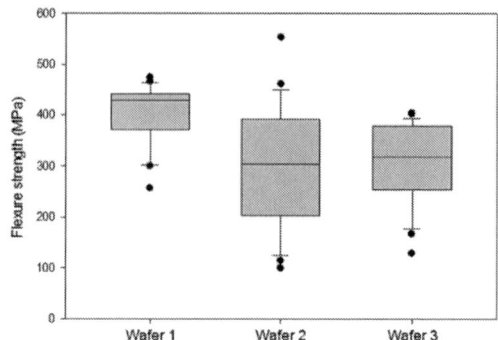

Figure 3: Summary of flexure strength after 1000 hours HTS test.

Figure 2 and Figure 3 shows the three-point bending test results after 500 hours and 1000 hours HTS test. Wafer 1 specimens still had the highest average flexure strength to compare with Wafer 2 and Wafer 3 specimens. Wafer 2 and Wafer 3 average flexure strength were almost the same either after 500 hours HTS test or 1000 hours HTS test. Wafer 3 flexure strength distribution was still tighter than Wafer 2. However, all the specimens average flexure strength increased after the HTS test to compare with time zero test results.

There were two findings from this work. The first finding is the HTS has significant effects on FOWLP strength. Figure 4 shows all the specimen average flexure strength before and after the HTS test. Their average flexure strength increased with the HTS test time extended. The curves trend was increasing, and the specimen flexure strength may still increase once the HTS test is extended to 1500 hours, 2000 hours or more. However, other FOWLP reliability criterion should be considered beyond its strength. One of the common failures after the HTS test is the passivation layer delamination, and it may be observed after 1000 hours HTS test. In this study, all the specimens used dummy silicon dies, and there were not any passivation layers or distribution layers lithographed on the specimen. Therefore, the specimen structure was quite simple, and we cannot found this kind of issues.

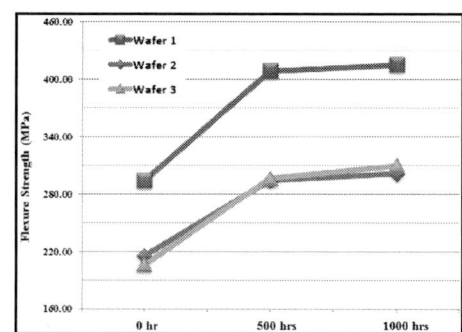

Figure 4: Summary of average flexure strength.

The second finding is the function of backside protection tape. The backside protection tape was considered as it had the ability to enhance the package strength. However, in this work, we find the backside protection tape function represents in reducing the flexure strength distribution. There is no evidence show that the backside protection tape could increase the FOWLP flexure strength. Wafer 1 and Wafer 2 flexure strength distribution had a big difference. It was because of the wafer grinding process. There were many defects created on Wafer 2 surface. The randomly presented defects affected the package strength significantly. However, the silicon die in Wafer 1 was also ground before the molding process, and its flexure strength distribution is not affected. It was because of the epoxy molding compound filled up the defects and created a smooth mold surface above the silicon die. The backside protection tape also had this function. The melted backside protection tape also could fill up the defects during the lamination process, and the backside protection tape surface was smooth enough. However, the backside protection tape thickness was 25 um only, and it cannot have the same performance with the 190 um thickness over-molded epoxy molding compound. Therefore, the backside protection tape cannot enhance the FOWLP strength beyond tightening the strength distribution.

CONCLUSIONS

The three-point bending test was conducted to evaluate the FOWLP strength after the HTS test. There were three kinds of FOWLPs. One 490 um thickness over-molded FOWLP, and two thin FOWLP. One of the thin FOWLP was laminated a backside protection tape. The HTS test condition was 150 degree C, and there were two readout points – 500 hours and 1000 hours. The three-point bending test results showed that the HTS test has significant effects on FOWLP strength. All the specimen flexure strength increased with extending the HTS test time. The thick FOWLP – Wafer 1 always showed the highest flexure strength. The backside protection tape cannot increase the FOWLP flexure strength. However, the backside protection tape has significant effects on flexure strength distribution. The specimens with backside protection tape had tighter flexure strength distribution.

ACKNOWLEDGEMENTS

This work financial support is provided by the Economic Development Board (EDB) Singapore Industrial Post-graduate Programme research grant. The authors are grateful to the support from EDB Singapore and JCET STATS ChipPAC Pte Ltd.

REFERENCES

[1] J. D. Wu, C. Y. Huang, and C. C. Liao. *Microelectronics Reliability,* vol. 43, no. 2, pp. 269-277, 2003.

[2] J.-H. Zhao, J. Tellkamp, V. Gupta, and D. R. Edwards. *IEEE Transactions on Electronics Packaging Manufacturing,* vol. 23, no. 4, pp. 248-255, 2009.

[3] S. Barnat, H. Fremont, A. Gracia, and E. Cadalen. *Microelectronics Reliability,* vol. 52, no. 9-10, pp. 2278-2282, 2012.

[4] S. Schoenfelder, M. Ebert, C. Landesberger, K. Bock, and J. Bagdahn. *Microelectronics Reliability,* vol. 47, no. 2-3, pp. 168-178, 2007.

[5] J. d. Vreugd, K. M. B. Jansen, L. J. Ernst, C. Bohm, and R. Pufall. 2010 11th International Thermal, Mechanical & Multi-Physics Simulation, and Experiments in Microelectronics and Microsystems (EuroSimE), 2010, pp. 1-6.

[6] B. Zhang, M. Johlitz, A. Lion, L. Ernst, K. M. B. Jansen, D. K. Vu, and L. Weiss. 2016 17th International Conference on Thermal, Mechanical and Multi-Physics Simulation and Experiments in Microelectronics and Microsystems (EuroSiME), 2016, pp. 1-6.

[7] C. Xu, Z. W. Zhong, and W.K. Choi. IEEE 23rd International Symposium on the Physical and Failure Analysis of Integrated Circuits, Singapore, 2016, pp. 297-300.

[8] C. Xu, Z. W. Zhong, and W.K. Choi. IEEE 18th Electronics Packaging Technology Conference, Singapore, 2016, pp. 700-703.

[9] B. Yeung, and T.-Y. T. Lee. *IEEE Transactions on Components and Packaging Technologies,* vol. 26, no. 2, pp. 423-428, 2003.

[10] D. K. Shetty, A. R. Rosenfield, P. McGuire, G. K. Bansai, and W. H. Duckworth. *America Ceram Soc Bull,* vol. 59, no. 12, pp. 1193, 1980.

[11] G. Hawkins, H. Berg, M. Mahalingam, G. Lewis, and L. Lofgran. 25th annual reliability physics symposium, 1987, pp. 216-223.

EFFECTS OF COPPER LINE-EDGE ROUGHNESS ON TDDB AT ADVANCED TECHNOLOGY NODES OF 28NM AND BEYOND

Dongyan Tao, Jinling Xu, Yanhui Sun, Wei-Ting Kary Chien, JS Chen, Guan Zhang

Reliability Engineering, Semiconductor Manufacturing International Corporation, No 18, WenChang Rd., Yizhuang Economic and Technological Development Zone, Daxing District, Beijing, China
*Corresponding Author's Email: Dora_Tao@smics.com

ABSTRACT

Ultra low-k films are used in advanced technologies as dielectric interlayers in Cu interconnects. Due to its high porosity, manufacturing reliable low-k films faces many challenges. This paper discusses the reliability of time dependent dielectric breakdown (TDDB). Degradation of the TDDB lifetime can be observed when there is an abnormal I-V breakdown. Our study characterized the interaction of the breakdown leakage to the etch profile. It has shown that the etch profile weak points have impacts on the TDDB lifetime. By characterizing the Cu etching profile and establishing inline correlations to its TDDB lifetime, a new evaluation method was identified to quickly and precisely reflect the TDDB lifetime performance.

KEYWORDS

Trench and via profile; Line-edge roughness; Vbd; TDDB lifetime

INTRODUCTION

In order to reduce the resistance-capacitance (RC) delays, the use of Cu metallization and low-k material is indispensable at the back end of line processes for 28nm and advanced technology node. However, Cu/low-k interconnect systems are vulnerable to breakdown fail because of its materials nature and the complexity of its manufacturing process [1]. Time-dependent dielectric breakdown (TDDB) of damascene structures has to be assessed as a system of a dielectric, diffusion barrier, cap layer (SiC, for example), and Cu interconnect. Low-k materials generally exhibit lower breakdown strength than conventional SiO2 dielectrics, and line-to-line space shrinks with each new generation of technology, resulting in high sensitivity of metal line spacing. Consequently, TDDB fail in Cu damascene structures has become an increasingly serious reliability issue [2].

Some of the factors affect breakdowns are: the lower breakdown strengths of porous low-k materials, the susceptibility of low-k materials to mechanical damage by chemical mechanical polishing (CMP), and the high susceptibility of low-k materials to Cu drift. Due to the porous nature of low-k materials, changes in chemistries can easily alter its surface condition, which can lead to degraded film properties [3]. Meanwhile, Cu etching profile effect was observed while evaluating the TDDB performance. The poor Cu etching profile with weak point

as evidence by the narrowing metal spacing resulted in degraded time dependent breakdown (TDDB) for the dielectrics. Cu trench profile can strongly influence TDDB Weibull distribution characteristics. Furthermore, based on this study, it is found that TDDB and voltage-to-breakdown (Vbd) can reflect each other when optimizing Cu etching profile processes. However, the correlation between Cu etching profile performance and TDDB reliability has not been studied thoroughly.

In this paper, by characterizing the Cu etching profile, a quick inline test was developed to correlates TDDB lifetime for dielectrics. This unique new inline test allows quick evaluation of TDDB performance efficiently and precisely.

EXPERIMENTS

In this work, all the data presented were collected from 300mm wafers fabricated at 28nm and advanced technology node. Dual-damascene Cu processing with TiN hard mask/trench first integration scheme was used to fabricate TDDB test structures with porous (ULK) material (k=2.5-2.6) as the dielectric film. The Comb-Serpentine (C-S) via-via structure is an ideal test structure because it closely resembles the real circuit design. It is composed of comb metal layers connected by serpentine metal, both of which contain vias (Figure 1).

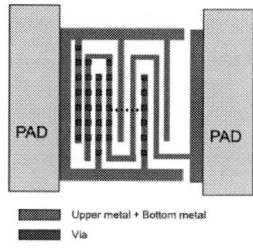

Figure 1: Metal combs with vias: C-S via-via structure

TDDB lifetime was evaluated by applying a constant voltage at 125°C. A 2x increases in measured current was set as the failure criterion (Δ I/I $\geq$2) to detect dielectric breakdown. Weibull statistics, sqrt-E-model voltage acceleration, Arrhenius temperature relationship and Poisson area scaling model were used to analyze the failure time distributions. Weibull parameters were

978-1-5090-6695-7/17 $31.00 © 2017 IEEE

derived by means of a constraint fit that assumes a common shape factor for all measured distributions on the whole sample with a ratio of t63s based on the voltage acceleration model. Typical lifetime target is greater than 10 years at CDF = 1000 ppm. Beta (β) refers to the shape factor or slope of the time to fail plots in Weibull scale. Gamma (γ) refers to the voltage acceleration factor under different stress conditions. Higher beta and gamma values are desirable for longer TDDB lifetimes. Table I shows the values derived from statistical samples.

TABLE I.   BETA AND GAMMA VALUE DERIVED FROM STATISTICAL SAMPLES

Parameter	Beta (β)	Gamma (γ)
Value	0.81	26.51

A voltage-to-breakdown test with 1V/s ramp rate was conducted and breakdown voltage (Vbd) is defined as the voltage when $\Delta I/I \geq 2$ was reached. For physical measurement, the worst spacing verification can be achieved using scanning electron microscopy (SEM) and transmission electron microscope (TEM). To be mentioned, it will be very tough to succeed. so we need take much care of the samples to find the spot with worst spacing.

RESULT AND DISCUSSIONS

A. Vbd Test correlation with TDDB lifetime

Figure 2 shows the line to-line leakage current of the line/space (45nm/72nm), Cu interconnects at 125°C. The leakage current was similar before abrupt breakdown. However, the abrupt Vbd show a wide range. Figure 3 shows the Weibull distribution of the inter-line IMD Vbd stressed at 125℃. Figure 4 shows the Weibull plot of the inter-line TDDB lifetime stressed at 125 ℃ under 3.0MV/cm. Leakage/Vbd and TDDB data are separately collected from half map of the same wafer. We conducted data collection and analysis on more samples, but they are not shown here. From the results we conclude that Vbd and TDDB outliers take place at the same time as being highlighted in red ellipse in Figure 3 and blue ellipse in Figure 4. These two failures can be the same source of weakness in the test structure shown in Figure 5. Figure 5 (a) shows an SEM top view image at breakdown site; Figure 5 (b) shows a TEM cross section at breakdown location. To find out the root cause of earlier failures, we conducted metal spacing measurement on the test structure with and without vias, as shown in Figure 6 (a) and (b). Table II shows the line edge roughness of ~4.5nm variation for metal spacing, leading to higher electric field and earlier failures.

Dielectric leakage mechanisms have been modeled by the Poole-Frankel emission where the conduction of Cu+ ions can be accelerated through traps in dielectrics under an external electric field. When a higher concentration of Cu+ ions is present at the Cu-cap layer interface, the breakdown is prone to happen at lower voltages. Cu + ions with higher energy is more likely to occur [4] at the weak points of metal line spacing

The difficulty to improve TDDB lies in the slow turn-around in getting results from process split experiments due to the long testing time for TDDB lifetime. The large lot-to-lot variation observed also can overwrite the effects of the splits for ULK dielectrics. Thus, a quick inline test to evaluate process splits shall be developed. Instead of doing the TDDB test on test key structures, Vbd on the C-S via-via structures should be used to evaluate splits, because ULK dielectrics TDDB fail is highly correlated to the Vbd of the metal layers.

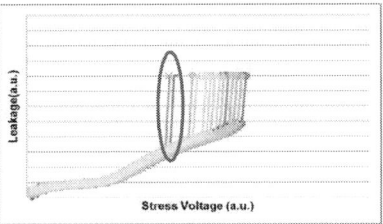

Figure 2: line-to-line leakage current of the line/space=45nm/72nm, Cu interconnects at 125°C

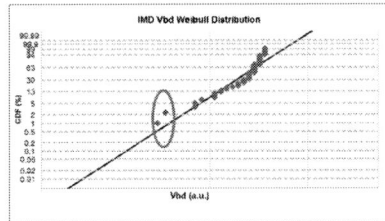

Figure 3: Weibull distribution of the inter-line IMD Vbd of the line/space=45nm/72nm, Cu interconnects at 125°C.

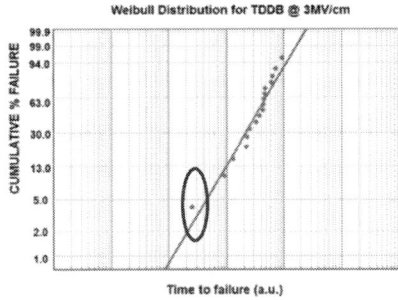

Figure 4: Weibull plot of the inter-line time dependent dielectric breakdown (TDDB) lifetime stressed at 125 ℃ under 3MV/cm

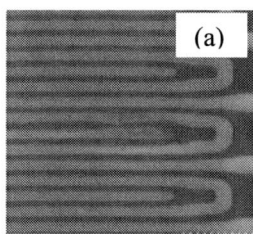 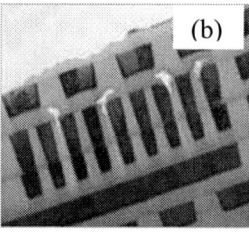

Figure 5: image of the failed test structure after constant voltage stress at 125C. (a) SEM top view image at breakdown location (b) TEM cross section at breakdown location

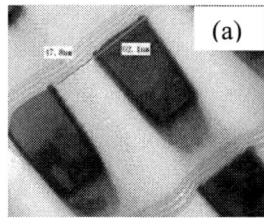 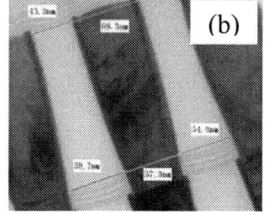

Figure 6: TEM cross section image spacing measurement of the failed test structure (a) without via metal spacing (b) with via metal spacing

TABLE II.   THE LINE EDGE ROUGHNESS FOR METAL SPACING
WITHOUT VIA AND WITH VIA

Pattern	Metal Spacing (nm)	Metal CD (nm)
Without via	47. 8	62. 1
With via	43. 3	69. 3

B. Inline Monitor Optimization for Improving Dielectric TDDB

Understanding the correlation between Vbd and TDDB lifetime was the key to establish a new method to improve TDDB performance for the ULK dielectrics. By analyzing Vbd and its characterization of I-V curves, we found early fail sites caused by line edge roughness always represented normal leakage before the abrupt breakdown. Therefore, the Vbd can be used to monitor dielectric TDDB performance inline. Using Vbd to predict TDDB, or converting Vbd to TDDB, has been done for many years. There is also some work in this aspect for low K TDDB [5-6]. Using the inline actual Weibull parameters which obtained from TDDB data, we setup inline Vbd spec to covering TDDB typical 10 years at CDF = 1000 ppm.

Further, a quick inline SEM measurement of the worst spacing of Cu etching profile can be developed. We define inline parameter of metal line edge roughness to indirectly monitor TDDB performance. Of course, future studies should focus on better understanding of the physical mechanism of TDDB for small spacing with direct tunneling and FN tunneling conduction mechanisms.

CONCLUSIONS

In this paper, we have demonstrated a new method to monitor the ULK dielectric TDDB fail. A strong correlation between Vbd and TDDB was observed, which can be used to evaluate TDDB performance effectively and accurately. Characterization of ULK Cu profile shows that poor TDDB samples often have weakness in metal line spacing. This weakness has shown leading to earlier TDDB fail. In addition, we point out that inline metal line edge roughness is related to wider distribution of the Vbd and early fail of TDDB. To summarize, line edge roughness proves to be a useful inline parameter for earlier prediction and prevention of ULK dielectric TDDB.

ACKNOWLEDGEMENTS

The authors would like to thank the reliability team of SMIC North PRE Department for the support of TDDB/Vbd test and technical discussion with 28nm project team. Valuable discussions with our colleagues from Reliability Engineering are also highly appreciated.

REFERENCES

[1] H.R. Ren et al., "Optimization of 28nm M1 trench etch profile and ILD loss uniformity", Semiconductor Technology International Conference, 2015:1-3.

[2] F. Chen et al., "The Effect of Metal Area and Line Spacing on TDDB Characteristics of 45nm Low- k SiCOH Dielectrics", IEEE International Reliability Physics Symposium, 2007, 31 (3): 382-389.

[3] W.C. Lin, et al., "Effects of Cu surface roughness on TDDB for direct polishing ultra-low k dielectric Cu interconnects at 40nm technology node and beyond", Microelectronic Engineering Vol.92, 27th Annual Advanced Metallization Conference 2010, pp.115-118.

[4] MC Silvestre, et al., "Vertical natural capacitor time dependent dielectric breakdown (TDDB) improvement in 28nm", Advanced Semiconductor Manufacturing Conference, 2015S. Santeria, *Proceedings of Transducers2003*, Boston, June 8-12, 2003, pp. 10-15.

[5] M. Lin, et al., "Correlation between TDDB and VRDB for Low-k Dielectrics with Square Root E Model", Reliability Physics Symposium (IRPS), p.p. 556, 2010.

[6] M. Lin, et al., "New Voltage Ramp Dielectric Breakdown Methodology Based on Square Root E Model for Cu/Low-k Interconnect Reliability", IEEE EDL, Vol. 31, p.p. 494 - 496, 2010.

Study of Safe Operating Area and Improvement for Power Management Integrated Circuit

Sarah Zhou, Yongliang Song, Kary Chien, Canny Chen

Corp. Q&R, Semiconductor Manufacturing International (Shanghai) Corp.
No. 18, Zhangjiang Road, Pudong New Area, Shanghai, China
Email: Sarah_Zhou@smics.com

ABSTRACT

LDMOS (Lateral Double-diffused Metal Oxide Semiconductor) is widely used to smart power management IC, which can be attributed to its high operation voltage and high current driving capability. Furthermore, LDMOS is compatible with conventional CMOS processes. It will be much easier for IC foundries to make it by existing process flows. Operating at both a high drain voltage and a high current, LDMOS is more sensitive to hot carrier degradation than the devices with low operation voltages [1]. Thus, the LDMOS HC-SOA (Hot Carrier Safe Operating Area) is a major reliability concern and requires more attentions. In this paper, the HC-SOA's of conventional core and IO MOS are also illustrated to show different failure mechanisms and we focus on the detailed HC-SOA test method in practice. Additionally, we study the SOA contours for different cores, IO, NPMOS and LDMOS. Finally, we discuss the HC-SOA extension methods for LDMOS.

Key Words: HCI; HC-SOA; LDMOS

INTRODUCTION

The shift of MOSFET parameters due to hot-carrier-induced (HCI) is an important reliability concern in modern ICs. High-energy carriers are generated by large channel electric fields near the drain region, then injected into SiO2 and can be trapped there. The trapping or bond breaking creates oxide charge and interface traps. As a result, it affects the channel carrier mobility and the effective channel potential results in the parameters degradation, such as Idsat, Idlin, and Vtci.

Safe operating area (SOA) is used to define the boundaries that limit the device operating region. It includes electrical and thermal SOAs during a short period and HC-SOA in the long time [2]. This paper focuses on the HC-SOA.

Hot carrier performance is particularly significant for LDMOS. In this paper, we present the detailed stress and the measurement method of both LDNMOS and LDPMOS. The result will help make the SOA analysis more efficient in practice. The comparison between N-type and P-type LDMOS is also adopted here with different failure mechanisms and the corresponding HC-SOA contours. At last, we also discuss the safe operating area extension.

For comparison purposes, we also check the HC-SOA of the conventional core and IO devices at 40nm technology. The failure mechanism is illustrated with different contours.

EXPERIMENT

The device used in this work is a 24V LDNMOS transistor fabricated with 0.18um BCD (Bipolar CMOS DMOS) process with oxide thickness of 15nm. The channel width and length are 40um and 0.4um, respectively. The cross section of the LDMOS is shown in Fig. 1. STI is located in N-drift area near the drain terminal. There is a lateral diffused body region beneath the gate. Power MOS for smart applications typically operates at ON/OFF state and therefore, the Idlin will be selected as the monitor parameter to obtain SOA contours [1].

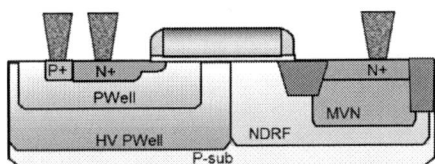

Figure 1: Schematic cross section of a NLDMOS.

In this work, a semi-auto prober and a precision semiconductor parameter analyzer are used for HC-SOA studies. For experimental efficiency and convenience, we select 5 Vgs stress conditions. The details are listed in Table I. The data are only for references, and the maximum Vgs in the plot will be more complicated in practice, which is based on the design requirements, GDI (Gate Dielectric Integrity) and BTI (Bias Temperature Instability) performance. For each Vgs stress condition, three proper Vds' from 110% operation voltage to 80% breakdown voltage are also planned for the degradation plot. Therefore at each Vgs, we can extrapolate the safe operating Vds. The 1/Vd model is adopted here for the lifetime calculation according to Eq. (1), where t_0 and B are fitting parameters. Then we can extrapolate five Vgs -

Vds pairs base on the lifetime model for a specified DC (Direct current) lifetime spec such as 0.2yr. They will be the final data points to draw the HC-SOA contours.

$$t_{TAR} = t_0 \exp\left(B / Vds\right) \qquad (1)$$

TABLE I.   THE 5 STRESS CONDITIONS

Vgs	Vds
Vg1=0V	
Vg2~(Vg1+Vg3)/2	3 Vds 1.1Vop~0.8Vbd produce obvious parameter degradation
Vg3~Isubmax or 1/2*Vop	
Vg4~(Vg3+Vg5)/2	
Vg5~1.1Vop	

Remark: Vg5 can also be defined base on the designer's requirement for special application.

For some cases, the Vds area @ Vg=0V is hard to obtain since the degradation is small or saturation at off-state. We extend the stress time, add more Vds stress conditions, and make estimations based on the test data, break down voltage, and the accumulated GOI (Gate Oxide Integrity) performance.

The HCI stress and analysis model for LDMOS are the same as those for traditional MOSFET's. HC-SOA experiments using conventional core and IO devices of 40nm technology node are also performed for comparison purposes.

RESULTS AND DISCUSSION

Fig. 2 shows SOA plots of conventional core NMOS and PMOS at 40nm technology node. Both contours are similar . For core NMOS, the worst Vgs condition is the same as that of the core PMOS. That is, they have the same performance. The core devices channel length is 0.036um, the worst HCI stress condition is Vgs=Vds at high stress temperature since the e-e scattering mechanism is dominated [3]. Hence, both the core NMOS and PMOS minimum Vds area occurs at the higher Vgs.

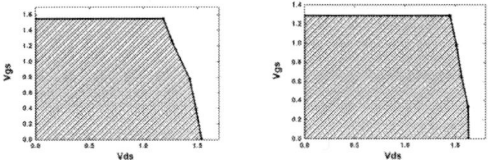

Figure 2: HC-SOA contours of conventional (a) core NMOS and (b) core PMOS

Fig. 3 shows the Isub versus Vgs and HC-SOA contour of a conventional IO NMOS at 40nm technology node with 0.25um channel length. The HC-SOA contour shape is significantly different from the core NMOS due to the longer IO device channel length. The enlarged gate length causes the hot carrier failure mechanism changes. The worst Vds happens at the point close to the Vg@ Isub,max at low temperatures. Lucky electron model can well explain the failure mechanism because the impact ionization is dominated and should be the worst at low temperatures when electrons mean free path is longer.

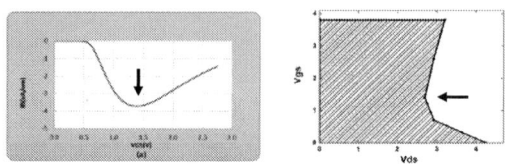

Figure 3: (a) Isub versus Vgs plot and (b) HC-SOA contour of conventional IO NMOS

For this 24V LDNMOS, the Isub versus Vgs plot shows double humps in Fig. 4a. This phenomenon is significantly different from conventional devices. The first peak indicates the maximum impact ionization rate and leads to device degradation; the second peak shows the occurrence of snap-back phenomenon due to kirk effect predicting device destruction [4]. At high Vgs, the Ids becomes higher, and the carrier charges compensate the drift region with N+ dope, the drain edge electric field will increase. A high body current enables the parasitic n-p-n and leads to device snap-back. Therefore, we can see two worse Vds' in the SOA plot (Fig. 4b). For this device, the Vds area at Vg=1.1Vop is smaller than the Vg@Isub max since the second peak is larger than the first one. We studied six LDMOS structures from 10V to 30V, found the maximum Isub occurs at the 1st or the 2nd hump and reached the same conclusion. The worst Vds area attributes to the larger peak in the Isub versus Vgs graph. So the conventional HCI stress condition for Isub max is also applicable for LDMOS.

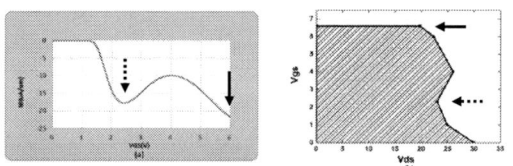

Figure 4: (a) Double humps at Isub versus Vgs curve and (b) HC-SOA contour of 24V NLDMOS

LDMOS safe operating area extension is very critical for smart power applications. The optimization method is to bring down the first peak and prolong the occurrence of the second peak. The increased n-drift concentration can significantly improve the NLDMOS device SOA at a higher gate voltage [5]. For the decreased n-drift concentration case, the depletion width is wider and closer to the drain. The number of carriers will be much easier to exceed the doping concentration at the snapback situation, and hence the snapback voltage is lowered down. The increased n-drift producing results in a higher snapback voltage, thus the safe operation area will be extended. As a result of the change, the maximum Isub will increase. Although the maximum Isub increases according to the increased n-drift dose, the Rds degradation at the 1st peak will not change [5]. Another approach for the HC-SOA improvement is to fine tune the feature size of the LDMOS such as gate to drain, gate to source overlap, and STI size. The STI size will affect the position of the peak electric field. It will be more helpful to improve LDMOS HC-SOA by simulation tools.

CONCLUSION

LDMOS is widely used for smart power applications which can be attributed to its ability of operating at high voltage and high current. Because of this, special attention is needed on the HC-SOA of LDMOS. In this paper, we describe the efficient and convenient method for SOA test and analysis method in detail. We compare the conventional core NMOS, PMOS & IO NMOS. For advanced technologies, such as 40nm node, the core devices HC-SOA show the similar shapes. For IONMOS, the critical point of HC-SOA turns to be the Vg position that induced maximum Isub. For NLDMOS, the Isub versus Vgs curve shows double humps, and the HC-SOA area contours also has two weak points. The methods of extending the long term HC-SOA are also given in detail, including increasing the n-drift concentration, fine tuning the well implant concentration and the gate overlap & STI feature sizes.

REFERENCES

[1] N. Soin, S. S. Shahabuddin and K. K. Goh, "Measurement and Characterization of Hot Carrier Safe Operating Area (HCI-SOA) in 24V n-type Lateral DMOS Transistors" [C]. Semiconductor Electronics (ICSE), 10th IEEE International Conference, 2012, pp. 659 - 663.

[2] P. L. Hower and S. Pendharkar, "Short and long-term safe operating area considerations in LDMOS transistors" [C]. in proc. IRPS, 2005, pp. 545 - 550.

[3] Song Z, Chen ZX, Zhao Y, et al. "The failure mechanism, worst stress condition for hot carrier injection of NMOS"[J].ECS Trans, 2013, 52(1): 947-952.

[4] S. K. Lee; C. J. Kim; J. H. Kim, et al. "Optimization of safe-operating-area using two peaks of body-current in submicron LDMOS transistors" [C]. ISPSD, 2001, pp. 287 - 290.

[5] D. Brisbin, P. Lindorfer and P. Chaparala. "Anomalous Safe Operating Area and Hot Carrier Degradation of NLDMOS Devices" [J]. IEEE Transactions on Device and Materials Reliability, 2006, Vol. 6, no.3, pp. 364 - 370.

Using Static Voltage Propagation Approach to Assist Full Chip LUP and TDDB Physical Verification

Yi-Ting Lee and Frank Feng

Mentor Graphics Corp, 8005 SW Boeckman Road, Wilsonville, OR 97070

Yi-Ting_Lee@mentor.com

ABSTRACT

Latch-up (LUP) [1] and time-dependent dielectric breakdown (TDDB) [2] are well-studied reliability issues. For an individual technology node, LUP is often studied at the transistor/gate level, while interconnect TDDB [3] is often studied from breakdown model and metal-dielectric material composition point of view. However, the complexity of their effects on the function of circuit design during chip assembly is expanding beyond a single device/gate, residing deep inside a chip, and increasing with the advance of technology. Modeling their influence on layouts of larger circuits has proven very challenging; attempts to completely simulate a large design block or full chip with reliable outcomes have not been promising so far, due to the enormous amount of data to be processed and the input of uncertain simulation factors. It is critical for both foundries and independent device manufacturers (IDMs) to provide useful and feasible design guidelines for designers to follow to achieve successful tape-out. In this paper, we present a static voltage propagation approach and design automation methodology to assist and enhance rule-based LUP and TDDB design reliability verification in the chip assembly stage.

INTRODUCTION

Chip design which employs advanced technology takes advantage of shrinkage in both gate size and interconnect width/spacing. The advanced technology enables many more devices to be packed into a die with equal or lesser dimensions. For a 10 nm system-on-chip (SOC) design, it is not surprising to see multi-billions of devices contained in an area under 1 cm^2. During the technology development stage, a gate library is developed with consideration of reliability issues like LUP and TDDB under as-designed operation voltage ranges. For interconnect technology, the co-planar interconnect spacing across dielectrics is defined primarily by optical and etching effects, with consideration given to electrical parasitic impacts. Once the gate library and technology design parameters are developed and packaged into a process design kit (PDK), the place and route (P&R) tool consumes them as input for the P&R process, which is tuned against the customary cost functions to achieve DRC-clean, timing, die size, IR drop, and Electromigration (EM) goals. However, based on the experience of the foundries and IDMs, the resulting layout, still requires further modifications to prevent LUP across gates and TDDB between interconnects of different nets, due to the difference in operation potential on different nets. This design practice is becoming critical for technology at and below 28 nm. Without it, chip reliability failure becomes more likely during burn-in or in silicon.

Using dynamic simulation to analyze these two problems is not feasible for block or whole chip design, due mainly to the large scale of the circuits, which results in too much data to be processed to provide reliable simulation results. From a technology perspective, the foundry and IDM can always provide some sort of design guidelines to schematic and layout designers. However, the requirements of design guidelines in conjunction with practical and accurate means of implementing these design guidelines in a chip assembly automation flow is key to achieving a successful tape-out in a reasonable turnaround time. Employing static design rules is quickly becoming a realistic approach to deal with these two types of reliability issues in the chip/block level design stage.

Whether or not these static design rules can be incorporated into a P&R tool, it is always necessary for a standalone verification tool to carry out such rule checks. Nevertheless, since it is a static approach, designers are expecting to see a human-manageable amount of error markers (within or around the range of hundred) from static rule-checking results that must be further reviewed to identify real physical errors that must be fixed on the layout. Although our proposed static approach (especially for TDDB) yields a small amount of false errors, it is still a valuable design methodology. Due to the dynamic behavior of TDDB, reliability designers accept the need for some reasonable amount of manual work to deal with a significantly reduced number of error markers produced by an automated design verification flow.

EVOLUTION TO STATIC RULE APPROACH FOR RELIABILITY DESIGN VERIFICATION BY FOUNDRY/IDM

In the foundry/IDM technology kit, some critical reliability design rules are defined as vague guidelines, without taking into consideration the feasibility of implementation by an electronic design automation (EDA) tool. For example, to prevent weak interconnect spots burned by high electrostatic discharge (ESD) current surges, the foundry/IDM defines the required wire width of the ESD path to be larger than some criteria to survive an ESD event. Based on this guideline, fabless layout designers usually try to use a traditional design rule checking (DRC) tool to check the wire "width" against the rule criteria. However, a traditional DRC tool can't identify the direction of the electrical current path along polygon shapes to decide "width", and it doesn't know how to sum up the individual widths of a parallel wires network to check against rule criteria. Realistically, this kind of guideline isn't very useful to modern chip assembly. Some designers may choose a dynamic simulation approach, only to find the task is impractical. In an attempt to facilitate automation and meet rule requirement, the foundry/IDM has begun to clearly define a static current density check along ESD paths to find weak interconnect spots.

For preventing LUP, the foundry/IDM provides a DRC approach that performs a spacing check and a check for the presence of a guard ring between aggressor and victim circuits

978-1-5090-6695-7/17 $31.00 © 2017 IEEE

at the chip assembly design stage. However, designers must manually place markers on designated polygons to enable a DRC tool to recognize them, and execute the DRC checks. In an iterative design process, it is difficult for designers to find a reliable way to place markers at the appropriate locations on the layout. For example, the marker methodology requires designers to identify the circuits behind ESD resistors (connected to IO Pads) that are involved in a LUP check. At the same time, the LUP sensitivity of these circuits depends on the effective resistance of the ESD resistors. If the effective resistance is bigger than the rule criteria, then circuits behind the ESD resistors are considered to be isolated from LUP. Some markers also require the identification of delta-operation voltage on polygons of different nets. Depending on the delta-operation voltage range, the spacing check criteria will be different. In an attempt to facilitate automation and meet rule requirement, the foundry/IDM has begun to clearly define non-physical marker methodology that allows logic-driven layout (LDL) [4] checking functionality to find LUP circuits/nets that can be further transformed to physical shapes for property annotated DRC check.

For verifying interconnect TDDB, the foundry/IDM also provides a DRC approach to check spacing of polygons on different nets at the chip assembly design stage, where the spacing criteria is dependent on delta-operation voltage. The old methodology is to use different markers representing different voltage values. Designers struggle to place markers of the appropriate voltage values to polygons in nets of concern across the whole layout. The physical markers don't work well with delta-operation voltage dependent spacing criteria. Assigning voltage from IO/power/ground pads to nets associated with inner circuits is a hassle. Again, the pure dynamic approach is impractical, and yields huge quantities of false voltage values for big blocks or full chip designs. In an attempt to facilitate automation and meet rule requirement, the foundry/IDM has begun to define voltage text/annotation methodology that allows LDL checking functionality to propagate the voltage values into the entire circuit design, and exports nets of TDDB concern, for which the logical nets can be further transformed to physical shapes for property annotated DRC checks.

HOW EDA METHODOLOGY ASSISTS IMPLEMENTATION OF LUP AND TDDB RULE CHECKS

Using the static voltage propagation approach in conjunction with an LDL flow, it is now feasible to implement automated LUP and TDDB rule checking. The LDL flow is described in Fig. 1. The foundry/IDM can now clearly define LUP rules with assurance of 100% coverage based on their rule requirements. For TDDB check, not only is the static voltage propagation approach easier to link with DRC for layout designers (who are mostly not familiar with dynamic simulation), but the quantity of false errors from assigning voltage values to polygons of nets also fall into a human-manageable range if applying appropriate voltage shift circuit constraints. Designers can realistically apply such flows into their production layout design process.

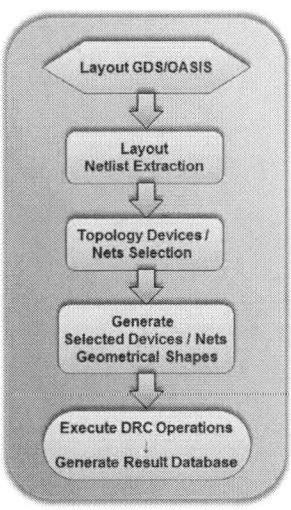

Fig.1. This figure describes a flow chart of distinct steps of a logic-driven layout flow. It has been implemented in the foundry/IDM design kit to replace manual marker flow and facilitate reliability verification automation.

In the previous section, we stated that voltage values are required for LUP and TDDB rule checks. Specifically, for LUP prevention, in addition to inserting guard-rings/guard-straps, spacing among polygons of active diffusion, P+ active, and N-well device layers operated at different potential should be equal or larger than a criteria for which the value increases with increases of potential difference (delta-operation voltage). The circuits/polygons scheme for LUP check is described in Fig. 2. Usually, the criteria and delta-voltages are grouped in a few bins, with one spacing criteria assigned to one delta-voltage range. This delta-voltage dependent spacing requirement for LUP checks is considered to increase the reliability of a design. A user input file is needed to assign the operational voltage value to corresponding IO/power/ground port. The automated flow is as follows:

1. Extract layout netlist (a standard circuit verification step provided by foundry/IDM).

2. Traverse connectivity graph (built from layout netlist) and propagate voltage value from IO/power/ground ports into internal nets based on user-defined constraints. The connectivity traverse engine is also instructed (based on user-defined scripts) to locate and export potential aggressor/victim devices. During the export process, the propagated voltage value on an individual net is annotated as a property to the aggressor/victim device.

3. Generate unique physical layers that correspond to exported aggressor/victim devices. These layers are usually a collection of polygons overlapped with device formation seed shapes. This procedure also annotates the voltage value as a polygon property, and other properties (number or string).

4. Process polygon data, annotated properties, and perform spacing DRC checks to find any edges or polygon pairs that violate spacing criteria according to the delta-voltage range. The voltage propagation performed in step 2 is relatively straightforward. The voltage value penetrates through the device from one pin to another pin (based on user-defined scripts) without alternating the voltage value. This voltage

propagation behavior is reasonable for LUP checks, since the potential LUP occurs on circuits close to IO circuits.

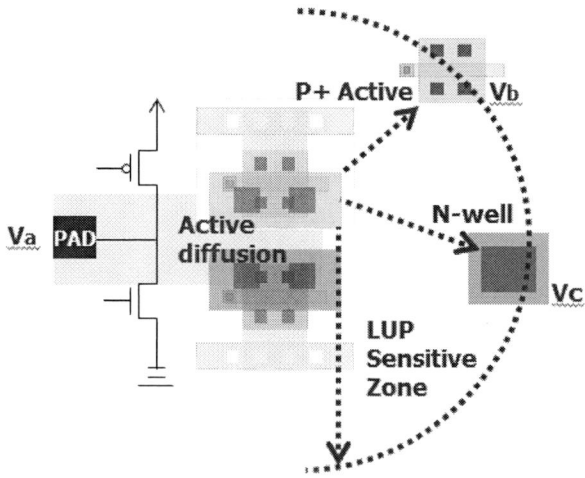

Fig.2. This figure illustrates the schematic and physical LUP sensitive zone. Inside a LUP sensitive zone, a safe spacing among active diffusion, P+ active and N-well polygons is dependent on delta-voltage ("Va - Vb" and "Va - Vc") range.

For interconnect TDDB checks on block or full chip, spacing checks among polygons of the same interconnect layer on different nets are executed against criteria depending on delta-voltage range. There are now multiple DRC engines capable of processing annotated properties and performing spacing checks in consideration of these properties as constraints, and the performance of these engines is continuously improving. The challenge is to propagate voltage value from IO/power/ground ports to internal nets, such that all targeted nets for potential TDDB concerns have the appropriate voltage value to annotate to their corresponding polygons. However, unlike voltage propagation for LUP checks, the voltage values starting propagation from IO/power/ground ports will shift magnitude when going out from some level-shifter related circuits. Without appropriate handling of these voltage shifts, users will eventually get huge quantities of false errors. The static voltage propagation engine doesn't usually recognize this level of circuit function automatically. It requires the user to design a static voltage propagation scheme by introducing additional information like acquiring voltage values on some cell ports related to level-shifter circuits, or defining sub-circuit patterns to enable the voltage propagation engine to manipulate voltage shift. The example of static voltage propagation scheme is described in Fig. 3. There are multiple means by which designers can acquire cell port voltage data. For example, a simplified level of dynamic simulation can be executed to acquire such data. The static voltage propagation engine can be instructed to propagate voltage, break out when encountering the above cell ports or sub-circuit patterns, and then continue propagation based on acquired voltage data on cell port or voltage constraints defined with the sub-circuit patterns.

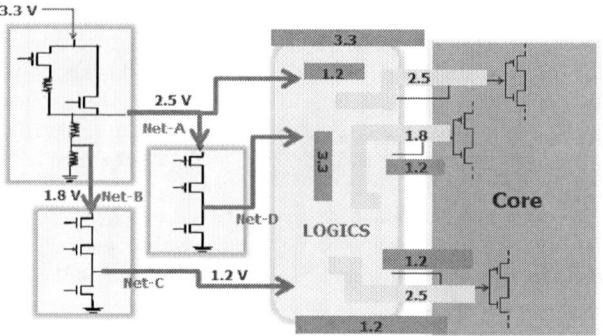

Fig.3. This figure illustrates a 3.3 V on top port, using the static voltage propagation, 2.5, 1.8, or 1.2 V can be assigned to other nets inside the circuit depending on how the voltage propagation scheme is designed. Voltage shift is executed by user-defined sub-circuit patterns, or by acquiring the voltage value on output cell port of a level-shifter circuit by mean of simulation. Upon completing static voltage propagation, the LDL flow is used to annotate the voltage values to polygons of nets of concern for DRC checking.

In a real-world application, if the voltage shift scheme/function is well-defined, designers will see relatively clean run results. There may be some small quantity of false errors, due to the nature of static propagation yielding some pessimistic voltage values on some nets. Experienced designers accept this result as long as the false errors are in a human-manageable range (~ 100), and these false errors will be manually processed toward final tape-out.

SUMMARY

This paper discusses how static voltage propagation is used to assist automated reliability design verification for LUP and TDDB. In conjunction with LDL functionality, the LUP checks defined with delta-voltage dependent spacing criteria are realized in foundry N28 design kit. There is significant opportunity to improve further for interconnect TDDB checks. A complete automated TDDB verification flow is possible when the static voltage propagation engine can recognize the circuits that shift voltage value during the voltage propagation process.

REFERENCES

[1] S.H. Voldman. "CMOS Latchup". ResearchGate. April 20, 2015.

[2] J.W. McPherson and R.B. Khamankar. "Molecular model for intrinsic time-dependent dielectric breakdown in SiO2 dielectrics and the reliability implications for hyper-thin gate oxide". Semiconductor Science and Technology. Volume 15, Number 5. 2000.

[3] Terence K.S. Wong, "Time Dependent Dielectric Breakdown in Copper Low-k Interconnects: Mechanisms and Reliability Models". Materials 2012, 5, 1602-1625. September 12, 2012.

[4] Sridhar Srinivasan, Patrick D. Gibson, Ziyang Lu and Fedor Pikus. "Logic-Driven Layout Verification, United States Patent Application Publication, Pub. No.: 2012/0011480 A1". January 12, 2012.

DEEP LEVEL INVESTIGATION OF INGAAS ON INP LAYER

Chong Wang[1,2,4], Eddy Simoen[1,3]*, Alian AliReza[1], Sonja Sioncke[1], Nadine Collaert[1], Cor Claeys[1,2], and Wei Li[4]

[1]Imec, Kapeldreef 75, B-3001 Leuven, Belgium
[2]KU Leuven, Dep. Electrical Engineering, Kasteelpark 10, B-3001 Leuven, Belgium
[3]Dept. Solid State Sciences, Ghent University, Krijgslaan 281 S1, Gent, Belgium
[4]University of Electronic Science and Technology of China, School of Optoelectronic Information,
610054 Chengdu, Sichuan, China
*Corresponding Author's Email: Eddy.Simoen@imec.be

ABSTRACT

Deep level traps in lattice-matched $In_{0.47}Ga_{0.53}As$ epitaxial layers grown by MBE on InP substrates have been studied by Deep Level Transient Spectroscopy (DLTS) on Al_2O_3/InGaAs Metal-Oxide-Semiconductor (MOS) capacitors. The impact of different surface passivation steps and a post-gate-deposition Forming Gas Annealing (FGA) has been studied. It is shown that spectra are dominated by a near mid gap electron trap in the depletion region, with activation energy in the range 0.37 eV to 0.42 eV. At the same time, a broad background distribution of interface states is found as well, which is significantly reduced by the FGA. Detailed carrier trapping studies have been carried out to identify the origin of the grown-in electron traps, which are shown to be of point defect behavior.

Keywords—high-mobility channels; DLTS; deep level traps; interface states; bulk defects.

INTRODUCTION

For future high-speed low-power logic applications, III–V CMOS is of interest for its superior electron mobility [1, 2]. However, exposure of the III–V surface to air or low vacuum results in the rapid formation of low quality native oxide on the surface, leading to the near midgap Fermi-level pinning and a high D_{it} density of states present at the high-κ/III–V interface [1-7]. The origin of these interface states has been heavily debated on in the past [7] but there are clear indications that a strong relationship exists with native antisite point defects (As_{Ga} or Ga_{As}) [6], as also evidenced by scanning tunneling microscopy [8, 9]. The density D_{it} can be strongly affected by the surface treatment (cleaning, passivation), so that this will determine the degree of Fermi level pinning and dictates whether it will be possible to invert the interface or not.

Effective surface passivation is thus of great importance to obtain a high quality interface. At the same time, as epitaxial deposition is a highly non-equilibrium growth method, the presence of grown-in point and/or extended defects in the InGaAs layer is expected.

Therefore, the purpose of this work is to study the impact of different surface treatments and post-deposition FGA on the deep levels states in Al_2O_3/$In_{0.53}Ga_{0.47}As$/InPMOS capacitors by using Deep Level Transient Spectroscopy [10]

which is a more complete interface-state characterization method than the traditional Capacitance-Voltage C-V and Conductance-Voltage G-V versus frequency and temperature techniques [11-13]. During DLTS Temperature-Scan, different bias pulses will be used to separate electron traps in the depletion region of the n-type InGaAs layer and at the Al_2O_3/InGaAs interface.

EXPERIMENTAL DETAILS

The 300 nm thick n-type $In_{0.47}Ga_{0.53}As$ layers have been grown by Molecular Beam Epitaxy (MBE) lattice matched on n-type InP substrates. Ammonia Sulfide ((NH_4)$_2$S) indicated by AS and HCl surface treatment before 10 nm Al_2O_3 gate oxide by Atomic Layer Deposition (ALD) have been applied, respectively. Post-deposition Forming Gas (10%H_2+90%N_2) annealing at 370°C for 15 min is also investigated. 50nm Pt is used as top metal contact and the back ohmic contact is formed by a Mo metallization stack. More detailed information of the studied samples is summarized in Table I.Top metal pads with area about 0.385mm^2 have been measured by a digital Fast Fourier Transform DLTS system, including a Boonton capacitance bridge operating at a fixed frequency of 1MHz. The gate bias is applied to the gate contact, while the substrate is kept grounded. First, I-V and C-V are measured to define an approximate bias condition for DLTS. Take C-V of AF as an example in Fig. 1. In order to distinguish the different types of defects in the InGaAs layer or at the Al_2O_3/InGaAs interface a reverse bias in depletion and a pulse bias either in depletion (-1 V-->-0.5 V) or in accumulation (-1 V--->0 V) have been applied. The doping density has been derived from the C–V characteristics in depletion. The depletion width in Table I corresponds with V_R=-1 V.

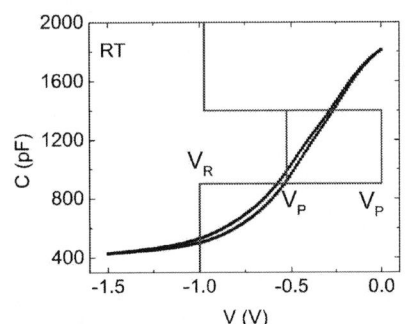

Figure 1: Capacitance-Voltage of the AF capacitor

Temperature-Scan DLTS at a fixed sampling frequency of 1 MHz has been executed to study the deep levels in the MOS system, whereby a voltage pulse from deep depletion to accumulation (V_R-V_P) was applied for a filling time t_p 1ms and with a sampling period t_w of 51.2ms. Pulse Duration Scans are also executed to examine the trap filling kinetics. This reveals information on the point- or extended-defect nature of the observed deep levels. The Arrhenius plots are obtained by using different Fourier coefficients of the numerically filtered transient signal and are used to calculate the activation energy and electron capture cross section.

TABLE I
EXPERIMENTAL DETAILS

Name	AS treated	HCl treated	FGA	Ndop/cm³	W_R/nm
AA	✓			7.41E15	93
AF	✓		✓	1.61E16	116
HA		✓		8.66E15	104
HF		✓	✓	7.46E15	140

*W_R is the depletion width at reverse bias -1V

RESULTS AND DISCUSSION

Figure 2 shows typical DLTS spectra in function of the sample temperature for an n-type InGaAs MOS capacitor, corresponding with different gate voltage pulses from depletion at -1V to accumulation -0.5V and 0V, respectively. It is obvious that the DLTS signal yields a broad peak between 200K-250K, which is associated with grown-in electron traps in the InGaAs epi layer. Changing the pulse bias from -0.5V to 0V, allows to fill more traps at the InGaAs/Al_2O_3 interface with electrons. However, it is observed in Fig. 2 that the DLTS signal intensity only increases and no other new defect signal from interface states appeared.

Comparing Figs 2 and 3 for the HCl-treated samples, one can clearly discern that FGA results in the removal of the broad wing features at higher and lower temperatures, which are thought to be related to the density of interface states (D_{it}), while the central peak at about 200K remains relatively unaffected. As it appears for a bias pulse in depletion, this central peak is associated with defects in the n-type InGaAs depletion region. It is concluded that an effective passivation of the interface states by FGA occurs for both pre-deposition treatments.

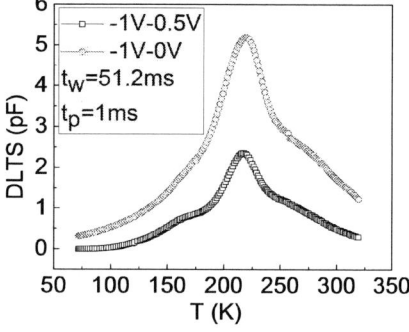

Figure 2: DLTS spectra of a HA; t_w=51.2 ms, t_P=1ms. A bias pulse from -1 V to -0.5 V (depletion) and 0 V (accumulation) has been applied to the gate of the MOS capacitor

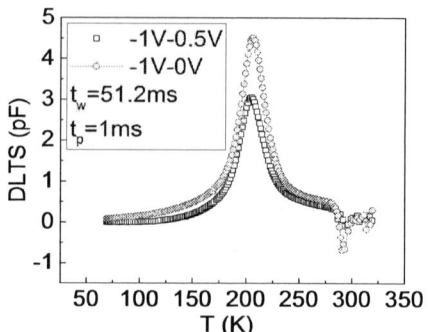

Figure 3: DLTS spectra of a HF capacitor; t_w=51.2 ms, t_P=1 ms. A bias pulse from -1 V to -0.5 V (depletion) and 0 V (accumulation) has been applied to the gate of the MOS capacitor

For the different surface pre-gate treatments Ammonia Sulfide AS and HCl, the DLTS peak is constant, as shown in Fig. 4 (FGA). Interestingly, the FGA appears to affect the bulk traps as well, showing a clear shift of the peak to a lower temperature in Fig. 5. A lower amplitude is also found, although this is mainly related with the removal (passivation) of the interface states.

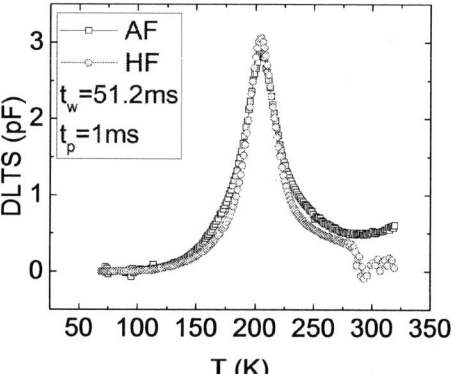

Figure 4: DLTS spectra of an AF and HF capacitor at a pulse from -1 V to -0.5 V, after FGA

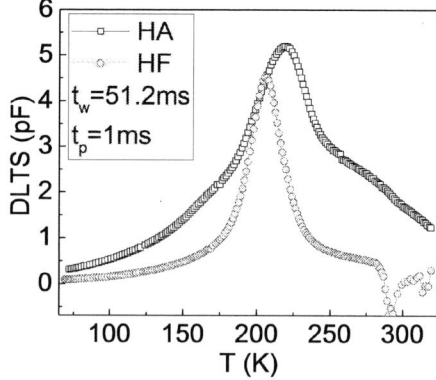

Figure 5: DLTS spectra of a HA and HF capacitor at a pulse from -1 V to 0 V

A different activation energy and electron capture cross

section are derived from the Arrhenius plot in Fig. 6. This indicates a possible reaction of the defects in the InGaAs layer with hydrogen, giving rise to a change in the parameters. A pulse duration scan is measured for AF, as shown in Fig. 7. A mostly point defect behavior is observed for the electron traps in both the InGaAs depletion region and at the Al_2O_3/InGaAs interface. The slow increase of the amplitude at longer t_p filling time could be related to long capture time constant caused by capture in the carrier tails at the end of the depletion region [14].

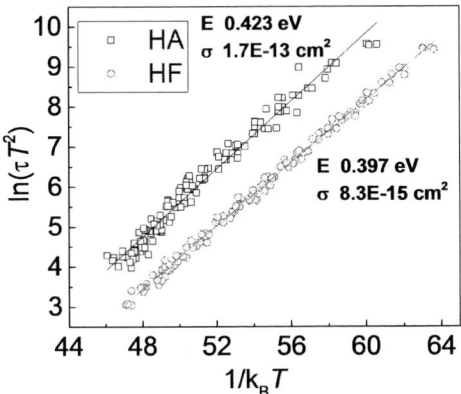

Figure 6: Arrhenius plots of the HA and HF samples

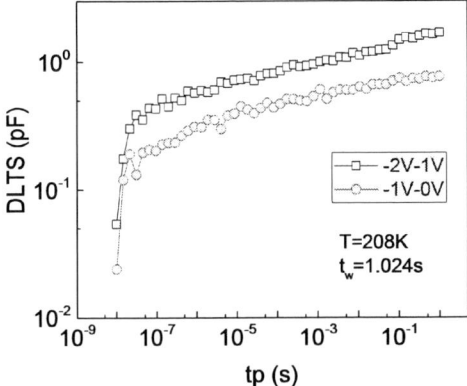

Figure 7: Pulse Duration Scan of AF at 208K;t_w=1.024s

TABLE II EXPERIMENTAL DETAILS

Name	N_{dop}/cm^{-3}	E/eV	σ/cm^2	N_T/cm^{-3}
AA	7.4E16	0.372±0.006	5.7E-13	7.41E15
AF	3.8E17	0.376±0.005	3.3E-15	1.61E16
HA	3.2E17	0.423±0.006	1.7E-13	8.66E15
HF	1.6E17	0.397±0.002	8.3E-15	7.46E15

CONCLUSION

It has been shown that the DLTS spectra of capacitors fabricated on lattice-matched n-type InGaAs epi layers on InP substrates exhibit a broad background distribution of interface states, which is significantly reduced by the FGA. The spectra are dominated by a near mid gap electron trap in the depletion region, with activation energy in the range 0.37 eV to 0.42 eV.

The variation in activation energy can be related to the effect of the electric field on the electron emission, it is also possible that different point defect complexes are present in the InGaAs material. From the trap filling kinetics, it has been shown that this trap behaves as a point defect and should be grown-in point defects possibly related to As antisites [15].

ACKNOWLEDGEMENTS

This work has been carried out in the frame of the Core Partner program on Beyond Silicon CMOS.

REFERENCES:

[1] M. Hong, J. R. Kwo, P. Tsai, Y. Chang, M. Huang, and C. Chen, "III-V Metal-Oxide-Semiconductor Field-Effect Transistors with High κ ," *Japan. J. of Appl. Phys.* vol. 46, p. 3167, 2007.

[2] P. D. Ye, "Main determinants for III-V metal-oxide-semiconductor field-effect transistors," *Birck and NCN Publications*, p. 134, 2008.

[3] A. J. Sambell and J. Wood, "Unpinning the GaAs Fermi level with thin heavily doped silicon overlayers," *IEEE Trans. on Electron Devices*, vol. 37, pp. 88-95, 1990.

[4] M. D. Pashley, K. W. Haberern, R. M. Feenstra, and P. D. Kirchner, "Different Fermi-level pinning behavior on n-and p-type GaAs (001)," *Phys. Rev. B*, vol. 48, p. 4612, 1993.

[5] M. Passlack, R. Droopad, Z. Yu, N. Medendorp, D. Braddock, X. W. Wang, T. P. Ma, and T. Buyuklimanli, "Screening of oxide/GaAs interfaces for MOSFET applications," *IEEE Electron Device Lett.*, vol. 29, pp. 1181-1183, 2008.

[6] J. Robertson, "Model of interface states at III-V oxide interfaces," *Appl.Phys. Lett.*, vol. 94, p. 152104, 2009.

[7] H. Hasegawa and T. Sawada, "On the electrical properties of compound semiconductor interfaces in metal/insulator/semiconductor structures and the possible origin of interface states," *Thin Solid Films*, vol. 103, pp. 119-140, 1983.

[8] B. Grandidier, H. Chen, R. M. Feenstra, D. T. McInturff, P. W. Juodawlkis, and S. E. Ralph, "Scanning tunneling microscopy and spectroscopy of arsenic antisites in low temperature grown InGaAs," *Applied physics letters*, vol. 74, p. 1439, 1999.

[9] B. Grandidier, X. de La Broise, D. Stiévenard, C. Delerue, M. Lannoo, M. Stellmacher, and J. Bourgoin, "Defect-assisted tunneling current: A revised interpretation of scanning tunneling spectroscopy measurements," *Appl. Phys. Lett.*, vol. 76, p. 3142, 2000.

[10] E. R. Simoen, G. Brammertz, J. Penaud, C. Merckling, D. Lin, W. Wang, and M. Meuris, "A DLTS study of Pt/Al$_2$O$_3$/In$_x$Ga$_{1-x}$As capacitors," *ECS Transactions*, vol. 25, pp. 151-161, 2009.

[11] W. Quast, "Small-signal admittance of the insulator-n type-gallium-arsenide interface region," *Electron. Lett.*, vol. 16, pp. 419-421, 1972.

[12] T. Sawada and H. Hasegawa, "Interface state band between GaAs and its anodic native oxide," *Thin Solid Films*, vol. 56, pp. 183-200, 1979.

[13] K. P. Pande, M. L. Chen, M. Yousuf, and B. Lalevic, "Interface characteristics of Ge$_3$N$_4$-(n-type) GaAs MIS devices," *Solid-State Electron.s*, vol. 24, pp. 1107-1109, 1981.

[14] S. H. Segers, J. Lauwaert, P. Clauws, E. Simoen, J.

Vanhellemont, F. Callens, and H. Vrielinck, "Direct estimation of capture cross sections in the presence of slow capture: application to the identification of quenched-in deep-level defects in Ge," *Semicond. Sci. and Technol.,* vol. 29, p. 125007, 2014.

[15] Irvine A C, Palmer D W. First observation of the EL2 lattice defect in indium gallium arsenide grown by molecular-beam epitaxy [J]. *Phys. Rev. Lett.* Vol. 68(14) pp. 2168, 1992

Practical Wafer Level Threshold Voltage Stability Measurement Methodology for the Fast Evaluation of Flash TECHNOLOGY

Gang Niu, Wei-Ting Kary Chien, Jack_Chen, Dennis Zhang, Susie Yu, Daniel Zhao, Silvia Duan, Ming_Li and Alicia_Ding

Semiconductor Manufacturing International（Tianjin）Co., Ltd. Reliability Laboratory

19 Xing hua Avenue, Xi Qing Economic Development Area, P. R. China, 300385

E-mail address:Javen_Niu@smics.com, Phone: 022-23700117

ABSTRACT

Package Level Threshold Voltage Stability (VTS) evaluation on PMOS has emerged as one of the critical reliability concerns in deep sub-micron devices. In this paper, we present a novel method to fast measure VTS at wafer level. Our result shows that changing the Source/Drain IMP species can improve the VTS of a 0.13um Flash.

KEY WORDS

NOR-Flash; Threshold Voltage Stability; Fast Evaluation; Package Level Reliability; Wafer Level Reliability; Ion Implantation

INTRODUCTION

The Threshold Voltage Stability (VTS) (should be "instability" or "stability degradation") occurs when some metal ions and other positive/negative ions around the gate oxide area of an MOSFET are trapped at the Si/SiO2 interface under a high temperature and gate bias stress. This phenomenon is called Ion Drift. When the ions are trapped at the Si/SiO2 interface[1-3], we can observe Vth (Threshold Voltage) shifts.

The Vth is a key parameter for device performance. Most VTS tests are usually performed at package level, with cycle time at 11 days or longer which is too long for supporting process tuning.

To resolve this problem, we proposed a wafer level VTS test. The experiments show that 99% test cycle time could be saved. In this paper, we apply this new method to evaluate the VTS of a 130nm-node Flash.

As we know, device threshold voltage can shift when it is stressed at elevated temperature (e.g., 150°C) and voltage.

It has been reported that the methods below can improve Vth stability:

1) Employ a lower annealing temperature and a more diluted H2/N2 mixture;
2) Purge Nitrogen at SiO2-Ploy interface to reduce Vth stability degradation by preventing Boron penetration;
3) Anneal at BEOL (Back end of line) process to reduce plasma charging;

In this paper, we demonstrate that using BF2 instead of B for source and drain implantation[2-4] is the most effective method to improve VTS. (please fix the format here)

INVESTIGATION AND ANALYSIS

When qualifying low voltage (LV) PMOS device for a Nor-Flash technology, VTS fails with Vt Stability Shift @T168Hrs =16.71% vs. specification of Vt shift (give the specification here), as shown in Fig 1.

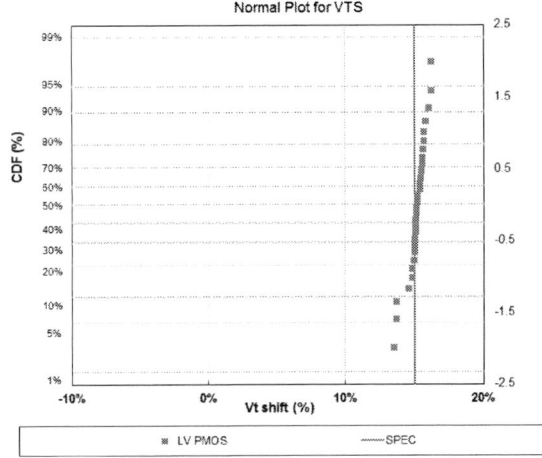

Fig. 1 A normal distribution plot of LV PMOS VTS (2 suggestions: a. adjust the x-axis scale to best present the data; b. I guess the y-axis on the right is the "Z-score". Please either lable it or remove the numbers. Also please fix the format – not center aligned)

Package level VTS test requires assembly(die saw, wire bonding, side braze) and then a stress of 168 hours in the oven. Therefore the total cycle time is at least 11 days. The detailed measurement flow is shown in Fig. 2.

978-1-5090-6695-7/17 $31.00 © 2017 IEEE

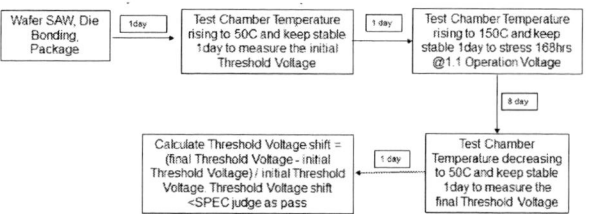

Fig. 2 Package level threshold voltage stability measurement flow (fix the format)

An optimized split condition- of an effective and fast WLR evaluation test was applied to reduce test cycle time to 3.5 hours. The detailed measurement flow is shown in Fig. 3.

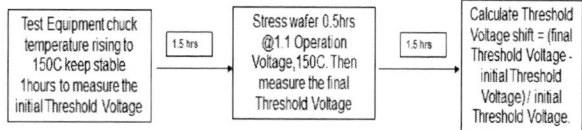

Fig. 3 New method for wafer level threshold voltage stability measurement flow (fix the format)

EXPERIMENTAL RESULTS

We have conducted experiments to optimize the process to improve the device VTS performance. The critical processes that could induce VTS degradation are listed in Fig. 4.

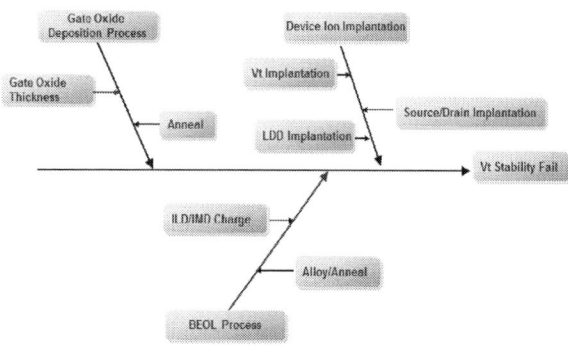

Fig. 4 The Fishbone diagram to analyze root cause

Where are the experimental results?

The experimental results show that increasing the thickness of gate oxide and changing Source/Drain implantation Ion species are effective to improve VTS. A thicker gate oxide have impacts on WAT (wafer acceptance test) results.

Meanwhile, adding fluorine in Source/Drain implantation will integrate interface dangling bonds, leading to a reduced Vt shift.

RESULTS AND DISCUSSIONS

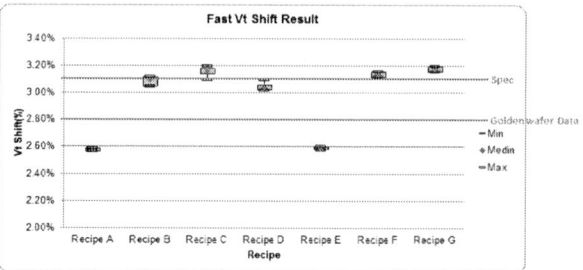

Fig. 5 Fast VTS box plot for process fine tune improvement lot (fix the format, also make sue the text will be within the column edge)

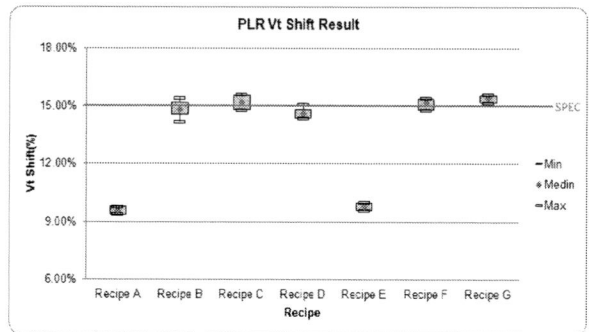

Fig. 6 PLR VTS box plot for process fine tune improvement lot (fix the format)

Recipe A: Gate oxide thickness change (increase thickness) →Ok, but WAT parameter out of SPEC

Recipe B: Gate oxide oxidation →NA

Recipe C: N-Well implantation ion dosage change（decrease the dosage）→NA

Recipe D: PLDD implantation ion dosage change（increase the dosage）→NA

Recipe E: Source/Drain implantation ion species change (add Fluorine) →Ok

Recipe F: ILD/IMD deposition rate change （release the plasma ）→NA

Recipe G: Passivation alloy →NA (it may look better if you make this a table)

According to Fig. 5 & Fig.6, a strong correlation between the wafer level fast VTS and PLR VTS was found. Further, fitting of the mean values of fast VTS shift and PLR VTS shift under 4 different process conditions was investigated. And a good quadratic fitting can be observed as shown in Fig. 7. To summarize, the correlation between fast VTS and PLR VTS can be expressed with a quadratic equations of one vaiable, and the fitting rate (R^2) is 1.

Equation (1): Cubic equations with one variable (a, b, c,d are shape coefficients)

$$y=a*x^3 + b*x^2 + cx +d \qquad (1)$$

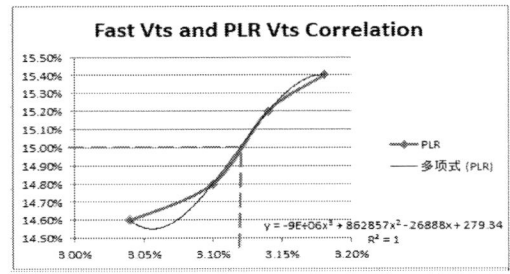

Fig. 7 The correlation between fast VTS and PLR VTS

In addition, the fast VTS shift of golden wafer is 2.8%, so PLR VTS should be able to pass when the corresponding fast VTS shift is lower than 2.8%. Then based on the cubic equations, it is found the fast VTS specification is 3.12%, which corresponds to the specification of PLR VTS shift.

CONCLUSION

We herein conclude from our experiment that changing Source/Drain implantation Ion species is an effective way to improve device Vt stability. It is a proven fast method to measure threshold voltage shift at Wafer level VTS test.

ACKNOWLEDGMENTS

Special thanks to Oliver Guo, Sean Wang, Ming He for informative failure analyses.

REFERENCES:

[1] V. Huard, M. Denais, and C. Parthasarathy, Microelectronics Reliability, 46, PP. 1 (2006).

[2] C.C. Liao, Z.H. Gan, Y.I. Wu , K. Zheng, R. Guo, 1.H. Ju, etc,Factors for Negative Bias Temperature Instability Improvement in Deep Sub-Micron CMOS Technology, 978-1-4244-2186-2/08/$25.00,2008 IEEE

[3] Sang Phill Park, Kunhyuk Kang, etc, Reliability Implications of Bias Temperature Instability in Digital
ICs, Digital Object Indentifier 10.1109/MDT.2009.133, 0740-7475/$26.00 © 2009 IEEE

[4] Udo.Schwalke, D.K. Schroder and lA. Babcock, 1 Appi. Phys.,94,p.l (2003).

GDI FAILURE MECHANISM INVESTIGATION AND IMPROVEMENT IN HK PROCESS

Lingxiao Cheng*, Lijuan Yang, Kai Wang

Semiconductor Manufacturing International (Shanghai) Corporation

No.18 Zhangjiang Road, Pudong New Area Shanghai 201203, China

*Corresponding Author's Email: Flora_Cheng@smics.com

ABSTRACT

Scaling down the complementary metal oxide semiconductor field effect transistors (COMS FET) requires involvement of High K (HK) metal gate technology in sub 45nm nodes. HK enables significant lower leakage at similar effective oxide thickness (EOT) to SiO_2 by effective suppression of direct tunneling. However, stress induced leakage current (SILC) and defects in the ultrathin interlayer in HK stack have been causing severe reliability concerns. In this work, we analyzed the gate dielectric integrity (GDI) performance in 28nm High K metal gate (HKMG) process with Vramp test and discussed the root cause of SILC and Vramp failure. Based on our experiments, we proposed an optimized process, which employs post deposition anneals (PDA) and decoupled plasma nitridation (DPN) process to passivate bulk trap in bulk HK to improve SILC. In-situ steam generation (ISSG) oxide and physical vapor deposition (PVD) TiN work function layer are also main contributors to improve GDI performance.

INTRODUCTION

Ultrathin gate oxide (<15 A) deposition has been a critical process in advanced technologies [1-3]. Scaling down silicon complementary metal-oxide-semiconductor (CMOS) field effect transistors (FET) requires an alternative gate dielectric technology. HKMG allows thicker gate oxides by effectively suppressing direct tunneling [4]. It significantly lowers the leakage comparing with traditional SiO2. Gate dielectrics in HKMG process consist of HK layers and inter-layers (IL) of SiO_2 at atomic level thickness to prevent surface trap charge density [5] caused by lattice mismatch, interface dangling bond, etc. HfO2 has been considered an appropriate candidate for HKMG [6-7]. However, in HK stacks, stress induced leakage current (SILC) has been a reliability concern. SILC degradation often has unacceptable impacts on the long term standby power stability, therefore has greatly limited the wide applications in future products. Also, the ultra-thin SiO_2 interlayers require high performance processes. Defects in non-uniform IL can induce abnormal gate dielectric integrity (GDI).

In this work, we analyzed the GDI performance of 28nm HKMG process by evaluating its Voltage ramp

(Vramp) tests. Several approaches were tested for gate dielectric improvement. Based on our experimental results, we proposed an optimized process, employing post deposition anneals (PDA) and decoupled plasma nitridation (DPN) process to improve SILC. We analyzed the GDI performance of I/O transistors under different experimental conditions of IL formation. We found the work function with different metal deposition methods also contributes to gate dielectric related reliabilities.

(Please note that you only need to show the full name of the short term once at the 1st place where it is used)

EXPERIMENTS

Figure 1 illustrates the 28nm HKMG MOSFETs structure.

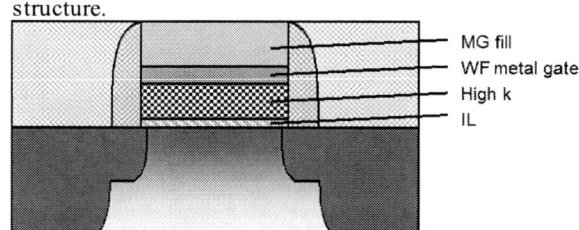

Figure 1 Schematics of the HKMG stack components

Initial un-optimized manufacturing process employs wet furnace IL oxide and atomic layer deposition (ALD) of the work function metal (WFM). Our finally optimized process employs ISSG oxide, DPN process and PDA. PVD WFM was also used in final process. Vramp test was conducted under room temperature in inversion mode to Core and accumulation mode to I/O dielectric until its breakdown. Duing stress, the gate was biased and the source/drain/bulk was grounded, and the stress voltages and operation voltages were recorded. Following the minimum design rules of the channel length and the gate space, a finger structure was used in the Vramp test. Vramp test was performed on Agilent 4072 tester with TEL-P8 prober.

RESULTS AND DISCUSSIONS

SILC phenomenon and the Vbd distribution of an un-optimized HKMG process

To evaluate dielectric reliability performance, a

978-1-5090-6695-7/17 $31.00 © 2017 IEEE

simple, fast and effective Vramp test was employed. Stress voltage was ramped from operation voltage (Vop) to dielectric breakdown at a ramp rate of 1MV/cm·s. HK/IL breakdown was associated with the trap (oxygen vacancies) generated in HK and IL layers. In the first stage of stress test, trap generation in both HK and IL layers were low due to the small Vt shift and minor Gm (??) degradation. In the breakdown stage, a significant amount of traps were generated in the HK/IL layers which triggered the dielectric raptures [8].

SILC has been a critical problem in HKMG process. HK+MG band asymmetry leads to polarity dependent SILC degradation driven by differences in trapping under gate vs substrate injection conditions. In our experiments, NMOS exhibited a more serious SILC increase before the onset breakdown as illustrated in Figure 2. Traps in HK/IL and oxygen vacancies near the interface attributed to the SILC generation [9].

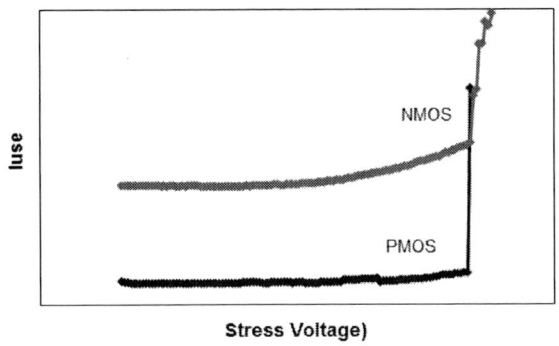

Stress Voltage)

Figure 2 Iuse (why call the leakage current Iuse?)-V curves in inversion mode of Core N/PMOS of un-optimized HK process

A poor GDI performance was observed at I/O gate dielectric. The low Vbd and bimodal Vbd distribution (Figure 3) were also noticed for thick I/O N/PMOS, indicating poor dielectric quality.

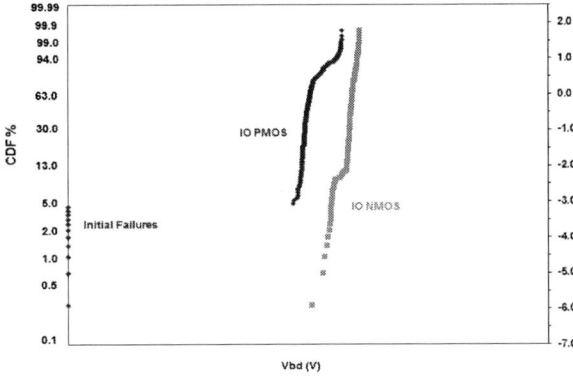

Figure 3 Vbd Weibull distributions of I/O N/PMOS in accumulation mode of un-optimized HK process

Fine-tuned HKMG process and its GDI performance

Based on the SILC and breakdown mechanism, new approachs were taken to solve the problems: improving the uniformity of oxides, increasing the physical thickness of oxides, passivating the defects and traps, and releasing the stresses in HKMG stack. Our optimized process includes the ISSG oxide, DPN process and PDA. PVD WFM was used in the final process.

Our optimized process leds to a much better SILC for NMOS, as illustrated in Figure 4. The slight SILC can be ignored in the fine tune process due to the obviously larger Vbd values.

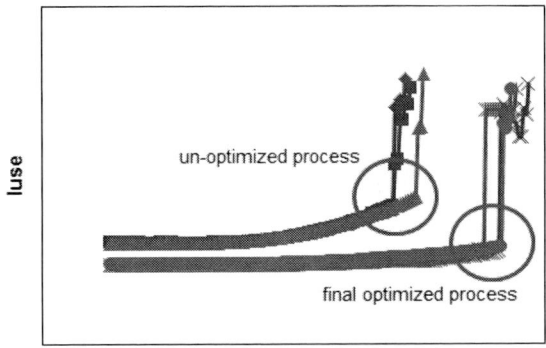

Stress Voltage

Figure 4 Iuse-V curves in inversion mode of Core NMOS of optimized and un-optimized HK process

Thick I/O gate dielectric has showed better Vbd distributions than the un-optimized condition. Bimodal or low Vbd tails were not found as shown in Figure 5.

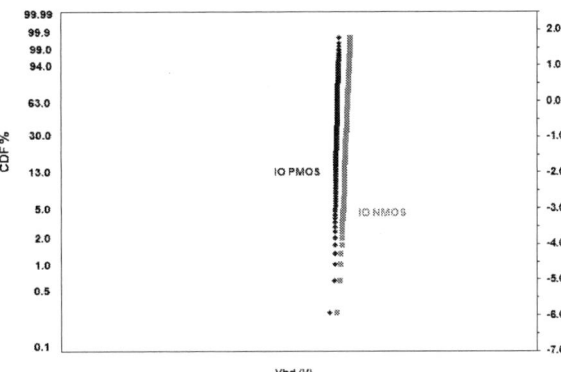

Figure 5 Vbd Weibull distributions of I/O N/PMOS in accumulation mode of final optimized HK process

In our optimized process, PDA process passivates traps and defects.

DPN process provides larger physical oxide thickness with equal EOT. Nitrogen concentration increment enhances the oxide dielectric constant. The physical thickness of oxides increases with increased oxides dielectric constant. This reduces the electric field across dielectric under the same gate stress. Thus, the gate dielectric is strengthened.

The ISSG growth improves the intrinsic quality of gate oxides and reduces structural defects in transition layers, which promotes the GDI performance. PVD formed WFM reveals better tensile stress in metal gate stack.

These optimized processes have provided us better uniformity, higher quality and less stressed gate dielectrics than the conventional methods.

SUMMARY

In conclusion, we reported an optimized process to produce high quality gate dielectrics by measuring and evaluating the Vramp performance of 28nm HfO2-based HKMG gate dielectric. In an un-optimized process, NMOS FETs usually reveals severe SILC problems which were caused by bulk trap in HK layers and weak IL oxides. The thick I/O dielectric also shows Vbd bimodal due to weak IL and stresses in work function metal. By optimizing ISSG oxides, DPN, PDA process and PVD WFM, NMOS SILC is significantly improved, resulting in a uniform Vbd distribution of the thick I/O dielectrics.

ACKNOWLEDGEMENTS

We would like to thank our management, operation modules and reliability group for their supports.

REFERENCES

[1] Y. Taur et al,"CMOS devices below 0.1um: How high will performance go?", IEDM Tech. Dig., 1997, pp. 215-218.

[2] S. H. Lo et al., "Quantum-mechanical modeling of electron tunneling current from the inversion layer of ultrathin-oxide nMOSFETs", IEEE Electron Device Letts., Vol. 18, Apr.1997, pp. 209-211.

[3] J. H. Stathis et al., "Reliability projection for ultrathin oxides at low voltage", IEDM Tech. Dig., 1998, pp. 167-170.

[4] S. Pae et al., "Reliability challenges of high-k and metal gate transistor technology", IRPS 2008, Tutorial.

[5] H. H. Tseng et al., "The impact of interface quality on high-k gate dielectric devices for 32nm technology and beyond". IEDM Tech. Dig., 2008, pp. 1245 – 1248.

[6] G. D. Wilk et al., "High-K gate dielectrics: Current status and materials properties considerations", J. Appl. Phys., Vol. 89 (10), 2001, pp. 5243-5275.

[7] R. Ranjan et al., "A new breakdown failure mechanism in HfO2 gate dielectric", IRPS, 2004, pp. 347-352.

[8] S. C. Chen, et al., "A Reliable TDDB Lifetime Projection Model for Advanced Gate Stack", IIRW FINAL REPORT, 2013, pp. 102-105.

[9] C. Prasad et al, "Dielectric Breakdown in a 45 nm High-K/Metal Gate Process Technology", IRPS, 2008, pp. 667-668.

HIGHLY EFFECTIVE LOW-K DIELECTRIC TEST STRUCTURES AND RELIABILITY ASSESSMENT FOR 28NM TECHNOLOGY NODE AND BEYOND

Zhijuan Wang*, Yueqin Zhu, Kai Wang, Yuzhu Gao, Wei-Ting Kary Chien

Semiconductor Manufacturing International (Shanghai) Corporation

No.18 Zhangjiang Road, Pudong New Area Shanghai 201203, China

*Corresponding Author's Email: emily_wzj@smics.com

ABSTRACT

Low-k/ultra-low-k dielectric is expected to have large-scale implementation in the manufacturing of modern advanced IC technology nodes. The reliability performance of a low-k dielectric must meet the given target lifetime based on semiconductor electrical requirements. Reliability test structures are specifically devised in this paper. We proposed a series of fundamental improvements on test structures in terms of practical applications. Our data shows that intrinsic reliability of the low-k dielectrics can be further optimized through the process tuning without the variation in k value.

INTRODUCTION

Much of the work in advanced CMOS integration is focused on meeting the technological demands of dimensional scaling [1-2]. Low-k dielectrics introduced in copper interconnects associated to critical dimension reduction in sub 28nm technology nodes has a challenge to retain the benefits of interconnect scaling from both process and reliability [3]. The suitable low-k materials as the most efficient choice of signal delay reduction must meet thermo-mechanical properties, reliability performance and strains from the BEOL (Back-End of Line) processing. ULK (ultra-low-k) with k≤2.6 is more likely to have more pores purposely incorporated into it [3]. Its reliability is a typical concern in the applications of high density of interconnects with opposing biases, such as in high speed circuits or capacitors. The reliability performance of the LK/ULK materials used within semiconductor devices must meet the expected operating lifetime and should not suffer from premature electric breakdown [4]. In order to better understand the reliability implications of these dielectrics, suitable test structures are needed to properly assess the reliability. The layouts of inter-level dielectric (ILD) and inter-metal dielectric (IMD) in a real product are rather complicated and would be properly utilized to assess the reliability. The test structures should be designed as the representatives of the structures found in real designs and enable intrinsic studies of the material reliability issues.

In this paper, we investigate a series of test structures to detect and present the worst-case reliability issues based on dielectric breakdown performance. Furthermore, newly designed items are introduced to the test structures in order to avoid leakage paths or breakdown failures from measurement pad due to the continued scaling of dielectrics. With the effectiveness improvements against stress bias, these test structures are more suitable for detecting the intrinsic reliability characteristics. Further efforts have also been made to improve the intrinsic dielectric breakdown since the ULK breakdown has a lower voltage acceleration than the denser low-k [4]. Several approaches have been assessed for improving such process and material dependence of ULK TDDB (Time Dependent Dielectric Breakdown) and an optimal process was identified.

THE EXPERIMENTS

A k=2.5-2.7 low-k SiCOH dielectric was fabricated using a 28nm CMOS process on 300mm wafers. Serpentine-comb with via related structure was used in this study. For interconnect dielectric TDDB studies, the accelerating conditions are given by the applied voltage (or electric field) at high temperature.

TEST STRUCTURES

Novel ILD test structure design

The type of test structre used to assess LK/ULK ILD reliability is the typical comb-to-comb (CC) structure (Figure 1). In dual-damascence integration, a multi-layer dielectric stack (metal barrier, LK/ULK, silicon oxide and silicon nitride) is deposited for covering the transistor gate/contact area [3]. In an actual product, the thickness of the dielectric stack below the first metal layer has variations in terms of polysilicon/metal gate, STI trench, active area of silicon (as in Figure 2). The test structures shown in Figure 1 can only detect the dielectric breakdown between metal and gate, it neglects the dielectric between metal and STI trench as well as the dielectric between metal and active area. Therefor the test structure in Figure 1 would result in a gap between test structures and real circuits. Figure 3, a newly designed ILD test structure, has taken the consideration of all potential situations in chip level and are more sensitive to detect the potential reliability concerns. Part II, Part III, Part IV and Part VI in Figure 3 work as the duplicaitons and representives employed in the real application. In several processes, the thicker metal edge combined with the higher STI step will suffer from the increasd electric field. Thus, the pobability of dielectric breakdown is higher at these weakest areas.

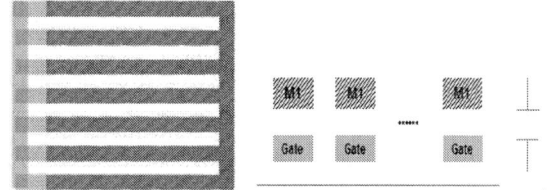

Figure 1: The typical comb-to-comb (CC) structure

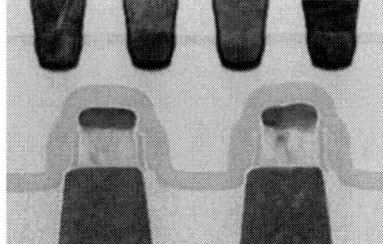

Figure 2: Cross-sectional TEM image of dielectric between metal and gate/active area/STI.

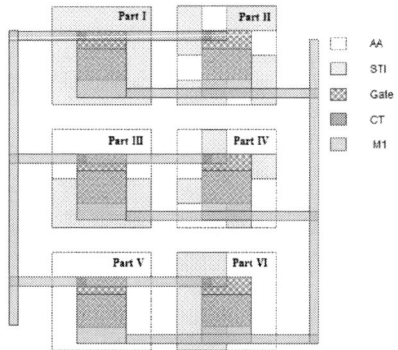

Figure 3: Top view of the designed comb structure

ILD/IMD test structure optimization

From a back-end-of-line dielectric reliability perspective, the ILD/IMD structures are stressed at higher electrical biases through the metal pad connection. The thickness of ILD and the ILD underneath the metal pad become thinnner with the technology demension scaling down. The introduction of ultra low-k dielectric material gives rise to the concern on the leakage path underneath the pad (Figure 4). The implicit capacitor C1 (between M1 and gate) and C2 (between gate and silicon) located at pad bottom showed in Figure 5 need to withstand a continuous electric field bias during the TDDB stress. The breakdown failures under the pad instead of the ILD/IMD test structure main-body can generate confusing data and cause data analysis

difficulties. Our first improvement action was to manually add gate dummy patterns (spicifically designed structure) instead of auto-drawing gate dummy according to auto-layout dummy inserting rule/algorithm, therefore no gate dummy pattern directly underneath M1. The M1 related ILD/IMD test structures with the manual gate pattern prevent the leakage path between M1 and gate (Figure 6). This can also be applied for metal gate related technologies. The second modification is on the active area below pad where N type well is added in order to build an NPN junction below the silicon active area (Figure 7). The last approach is to change the thin gate oxide dummy to the thick oxide dummy pattern. Such dummy pattern change combined with the second modification enables the metal pad area to withdtand a higher voltage stress to avoid the failure pathway under the metal pad.

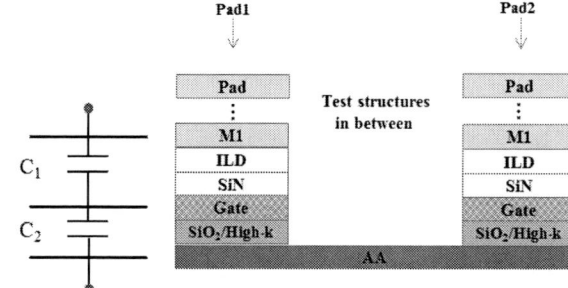

Figure 4: Breakdown failure underneath PAD showed by TEM image

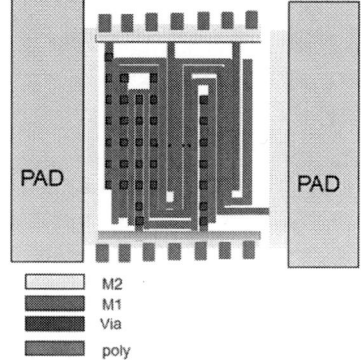

Figure 5: Schematic of implicit capacitors under metal pad bottom

Figure 6: Schematic of M1-Via1-M2 IMD structure with manual gate dummy pattern

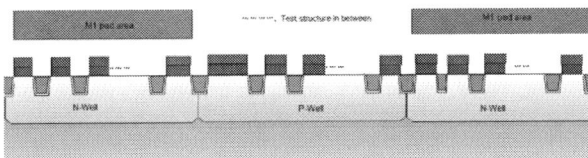

Figure 7: NPN junction Schematic underneath PAD and the test structures

ULK/LK intrinsic reliability improvement

The reliability performance of a given interlevel dielectric, such as a ULK dielectric, must meet not only thermomechanical but also electrical requirements. The TDDB assessment with a fixed accelerating voltage and temperature is a common method to assess dielectric reliability. ULK materials need to be carefully characterized for a successful technology development and qulificaiton. Low-k dielectrics that are investigated have a k value range from 2.0 to 3.0, porpsity of 30%-50% with variable pore sizes [5]. SiCOH (or Si-O-C-H) as one of the pristines makes up the basic atomic ingredients of the ULK/LK dielectrics. The ULK/LK materials with more porosity, less stength and hardness may lead to undesirable TDDB performance. Strengthening the host matrix with more quantity flow of the SiCOH precursor will increase both matetial density and the k value. With the porosity incorporation ingredient adjusment combined with the post-deposition cure, ultraviolet treatment, and CMP process control, we achieved a better intrinc reliability performance (as in Figure 8) with the same k value.

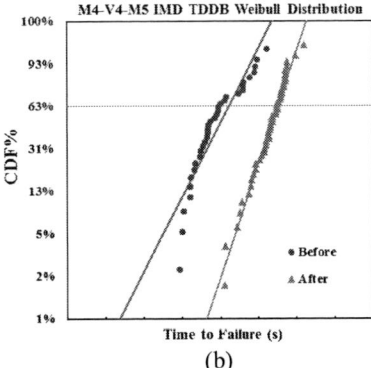

(b)

Figure 8: TDDB distributions comparison of M1-V1-M2 (a) and M4-V4-M4 (b) test structures before and after process tuning.

SUMMARY

In conclusion, several improved ILD/IMD test structures have been presented. These test structures are more suitable in both the product application and reliability assessments. The revised designs take the worst-case into consideration and prevent the external distraction. ULK dielectric related process was verified with improved TDDB performance without changing the k value.

ACKNOWLEDGEMENTS

We would like to thank our management, operation modules and reliability group for their supports in preparing this paper.

REFERENCES

[1] Meindl, J.D., Davis, J.A., Zarkesh-Ha, P., Patel, C.S., Martin, K.P. and Kohl, P.A., Interconnect opportunities for gigascale integration. IBM J. Res. Develop., 2002, pp. 245–263.

[2] List, S., Bamal, M., Stucchi, M. and Maex, K., A global view of interconnects. Microelectron. Engng, 2006, pp. 2200–2207.

[3] Ennis T. Ogawa and Oliver Aubel, Electrical Breakdown in Advanced Interconnect Dielectric, Electronic Materials, Wiley, chapter 11, 2012, pp. 369-434.

[4] F. Chen, M. Shinosky, B. Li, J. Gambino, S. Mongeon, P. Pokrinchak, J. Aitken, D. Badami, Critical Ultra Low-k TDDB Reliability Issues For Advanced CMOS Technologies, IEEE Physics Symposium, 2009, pp. 464-475.

[5] Hualiang Shi, Denis Shamiryan, Jean-Francois de Marneffe, Huai Huang, Paul S. Ho and Mikhail R. Baklanov, Plasma Processing of Low-k Dielectrics, Advanced Interconnects for ULSI Technology, Wiley, 2012, pp. 79-128.

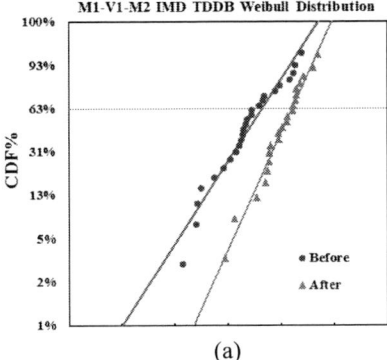

(a)

978-1-5090-6695-7/17 $31.00 © 2017 IEEE

FAIL MECHANISM OF PROGRAM DISTURBANCE FOR ERASE CELLS VT POSITIVE SHIFT IN NAND FLASH TECHNOLOGY

Chunmei Zou[1], Yong Zhao, Wei-Ting Kary Chien, Junyao Tang

Product Reliability Engineering, Semiconductor Manufacturing International Corporation

18 Zhangjiang Road, PuDong New Area, Shanghai 201203, China

Email: Maggie_zou@smics.com

ABSTRACT

Program disturbance is a major intrinsic reliability concern on NAND flash. In this paper, we present that NAND flash E/W (Erase/Write) cycle failures induced by program disturbance for erase cells VT (Threshold Voltage) positive shift. The root cause of program disturbance is the old process of gate re-oxidation issue, which results in ILD (Intra-Layer Dielectric) voids, then Ni fills in the ILD voids and induces the lateral E-field increase between WL's. Interface traps and electrons generated by GIDL (Gate Induce Drain Leakage) are accelerated by the lateral E-field and subsequently injected into the erase cell transistors by HCI effect, therefore erase cells VT positive shift, and program disturbance occurs. The disturbance will get worse than fresh sample as interface traps and couple voltage of WLs increasing after E/W cycles. A new process of gate re-oxidation to depress the program disturbance and enhance NAND Flash E/W cycles performance is provided

INTRODUCTION

NAND flash memories are widely used for data storage because of large capacities, low cost and low power consumption. As technology node shrinks, reliability performance is always a concern for high-density memory components. Program disturbance is impacted by several factors as cells are placed closer to each other with successive generation of physical scaling [1].

Program disturbance is that a certain memory cell is selected to program and the threshold voltage of other unselected cells will increase corespondingly due to weak programming. There are two modes of program disturbance. Mode A happens at the intersection between selected BL (Bit Line) and Pass WL (Word Line). High voltage of pass WL will induce the electron traps, if this condition continues for a long period of time, the program disturbance will occur. Mode B happens at unselected BL's in the selected WL, Vcc will boost the high voltage of the unselected cells channel to couple with programmed voltage VPP (Program Pulse) to inhibit programming. However, the boosted voltage is usually not high enough and there will still be an evident voltage gap between Vch_boost and Vpp. if this condition continues for a long

time, the program disturbance will also occur. See Fig.1 and Table.1.

Table. 1 Program and Program inhibit for NAND Flash

		Mode A	Mode B
	Program	Program Inhibit	Program Inhibit
Sel.WL	VPP	VPASS	VPP
Pass WL	VPASS	VPASS	VPASS1,2
SSL	VCC	VCC	VCC
GSL	0	0	0
BL	0	0	VCC
Bulk	0	0	Vch_boost

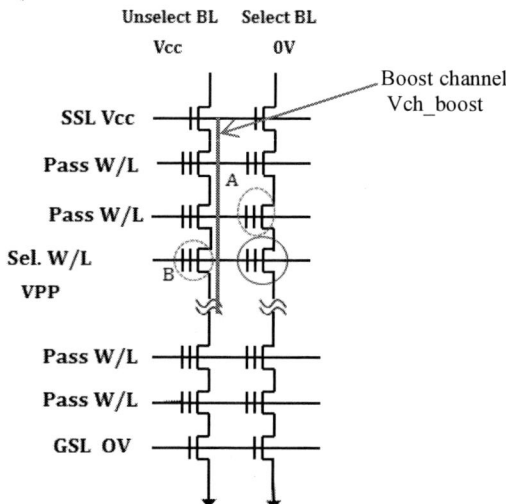

Fig.1 Program disturbance schematics

In this paper, we present Mode B program disturbance for the erase cells VT positive shift, we will discuss the failure mechanism of program disturbance in detail. A new process of gate re-oxidation to depress the program disturb-ance and improve E/W cycles performance is provided.

RESULT AND DISCUSSION

Experiment was conducted on the NAND Flash with double patterning process. There are some inherent challenges in sustaining process control as technology scales. Fig.2 shows CKB (Check Board) and ICKB (Inverse Check Board) pattern schematics during E/W

978-1-5090-6695-7/17 $31.00 © 2017 IEEE

cycles stress in this paper. As shown in fig.2, we collected CKB and All0 pattern VT distributions respectively initial and after E/W cycles stress at FT (function Test). Initial CKB and All0 pattern VT distributions showed normal, after 5K E/W cycles stress with CKB and ICKB pattern, program disturbance occurred at CKB VT distribution, but All0 VT distribution was normal. Erase cells VT positive shift was caused by program disturbance when the neighboring cells were programmed, program disturbance became worse as E/W cycles increasing.

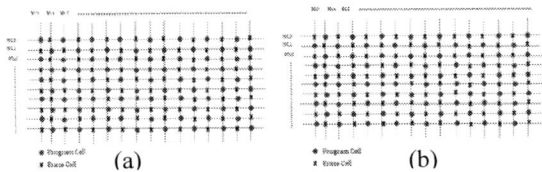

Fig.2 E/W cycles Pattern schematics: (a) CKB, (b) ICKB

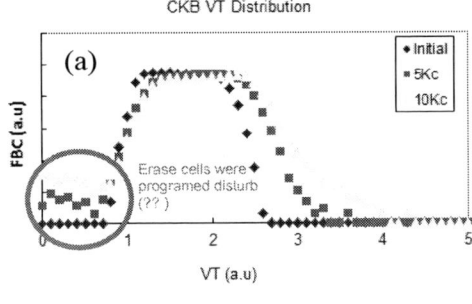

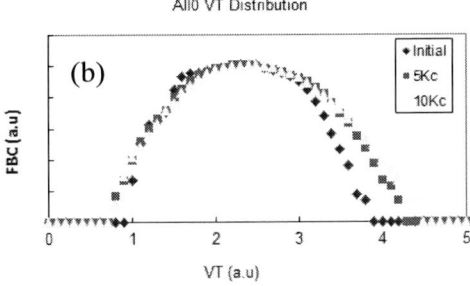

Fig.3 Initial and 10Kc VT distribution:
(a) CKB VT distribution, (b) All0 VT Distribution

Fig.4 shows EDX picture at the failure bits location, Ni filled in the ILD voids. We suspected that the root cause is the old process of the gate re-oxidation, which happens before ILD space oxide. It is used to repair the sidewall etch damage and form the buffer barrier of ILD layer. Gate re-oxidation step uses an old process. The old process temperature is so high, that the gate re-oxidation will be inconsitent and thicker. It makes the ILD layer hard to fill in without ILD void . See Fig.5.

We now discuss the issue caused by ILD voids. Ni will fill in the ILD voids if the top of the voids are open after Ni salicide deposition, as shown in Fig.5. The lateral E-field between WL's will be increased by the Ni filled in the ILD voids. After E/W cycle stress, the lateral E-field increases as coupling voltage gets higher than fresh samples, and interface traps are generated and most of electrons are supplied by GIDL. Interface traps and electrons are accelerated by the lateral E-field and subsequently can be injected into the erase cell transistors by HCI effect[4-5]. Therefore, program disturbance occurs for erase cells VT positive shift . See Fig.3

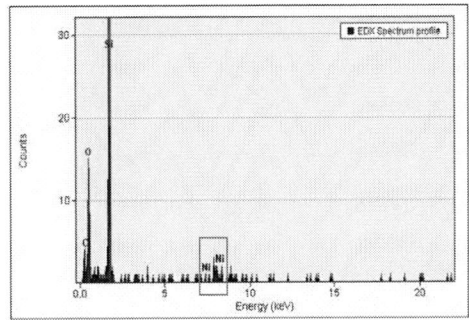

Fig.4 EDX of Counts Vs Energy(Kev) at failure ILD failure location: A few Ni filled in ILD void

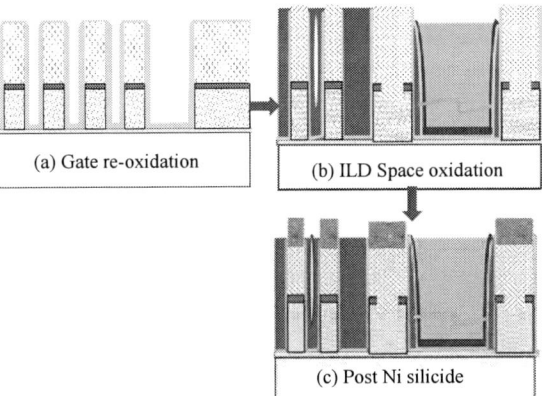

Fig.5 The schematic plots of (a). (b). (c). from gate re-oxidation to Ni silicide process

IMPROVEMENT SOLUTION

As mentioned in old gate re-oxidation process, it has much better repairing capacity to gate etch sidewall damage, but it induces program disturbance due to ILD Voids issue. We then introduced a new process with advantages of low plasma damage, low electron temperature and low interface traps density. Gate re-oxidation uniformity is much better than that of old process. ILD uniformity is much improved. No Ni fills in ILD layer. It also reduces the lateral E-field, so program

978-1-5090-6695-7/17 $31.00 © 2017 IEEE

disturbance can be suppressed effectively in inhibit cells. We performed the cycling stress on new process technology with 10pcs wafers; there are no program disturbance phenomena after 50K E/W cycles. See Fig.6.

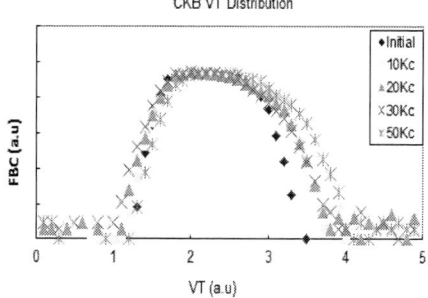

Fig.6 50K E/W cycles VT distribution

CONCLUSION

In this paper, we show that program disturbance issue for erase cell VT positive shift from the old process of gate re-oxidation , the old process results in ILD voids, Ni fills in the ILD voids and increases the lateral E-field between WL's. Interface traps and electrons generated by GIDL are accelerated by the lateral E-field and subsequently can be injected into the erase cell transistors by HCI effect to cause program disturbance. The disturbance will get worse as interface traps and coupling voltage increases after E/W cycles. We then introduced a new improved process; it can improve ILD uniformity, and therefore reduces the lateral E-field of channel. Program disturbance can be suppressed effectively in inhibit cells, and we get better E/W cycles performance.

REFERENCES

[1] Suresh Chandrasekaran, and Srivatsan Venkatesan, "The Effect of Shallow Trench Isolation Improvement on Program Disturbance Response in 20 nm NAND Flash Technology" IEEE Workshop on Microelectronics and Electron Devices (WMED), 2014, pp.1 – 4.

[2] Jaewook Yang, and Wonhyo Cha, "the effect of hydrogen on program disturbance in sub-2ynm Nand flash" Reliability Physics Symposium (IRPS), IEEE International, 2013, pp.1 – 3.

[3] Cristian Zambelli, Fabio Andrian, and Seiichi Aritome; "Compact Modeling of Negative Vt Shift Disturbance in NAND Flash Memories." IEEE Transactions on Electron Devices, 2016, pp.1526~1523.

[4] Youngwoo Park, Jaeduk Lee, and Seong Soon Cho, "Scaling and Reliability of NAND Flash Devices" IEEE International Reliability Physics Symposium,2014, pp.1-4.

[5] Milim Park, Sukkwang Park, and Seokwon Cho; "NAND Flash Reliability Degradation Induced by HCI in Boosted Channel Potential" IEEE Reliability Physics Symposium, 2010, pp.975~976.

THE RESEARCH OF INTELLIGENT FEEDBACK MECHANISM BETWEEN DOCUMENT CONTROL AND PRODUCTION SYSTEM

Zhou Zhenlin (Chad Chou), Guo Yingying (Sherry Guo)

Quality & Reliability, Semiconductor Manufactory International Corporation, Tianjin, China

chad_chou@smics.com

ABSTRACT

In semiconductor industry, strong executive ability reflects not only in one's response time, but also in accurate implementation. Generally speaking, engineers follow the rule defined in document to arrange inline production. No one can ensure zero-missing once the document is updated, and this is mainly due to the "manual talk" between document control center (DCC) and inline production. Currently, there isn't a channel to link DCC and inline production.

This research aims to develop an intelligent feedback mechanism which links DCC and inline production system to replace the traditional "manual talk" mode and achieve accurate implementation.

ABBREVIATIONS AND ACRONYMS

- DAAM: Document Activity Analysis Module
- VCM: Version Conflict Module
- IRAM: Incompatible Risk Analysis Module
- DCC: Document Control Center
- PM: Preventive Maintain
- SPC: Statistics Process Control
- CRM: Customer Relationship Management
- TECN: Temporary engineer change notice

INTRODUCTION

Every semiconductor company has a DCC section to control all engineering and quality related documents, which are published in intranet system. In principle, the setting of inline production system should be based on these official documents. Currently, the inline production system's setting is manually set by engineer according to the official document definition. DCC is mainly responsible for document generation, release, distribution, recall, abolishment, retrieval, and storage. The accurate implementation only relies on engineer's zero-missing operation when a document is newly updated. Otherwise, the production line will still use the old version parameter without any update, and it will probably cause much loss to production. In real internal audit practice, we often observed some of production actual settings conflict with document definition. For example, the control limit setting in SPC system or the PM frequency setting in PM system is in conflict with the newest document definition. Such issue is a kind of serious nonconforming finding and we call it as Say≠Write≠Do issue based on ISO concept [1]. Please refer to Figure 1.

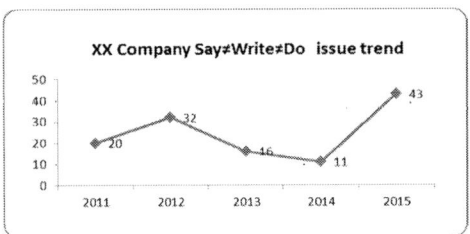

Figure 1: Say ≠Write ≠Do issue chart

No channel to link DCC and inline production is the root cause of this issue. It just relies on engineer's manual control to make sure the setting in production system matching with official document definition. We intend to develop an intelligent feedback mechanism which links DCC and inline production system to prevent the happening of Say≠Write≠Do problem. We also expect this mechanism can affect production by not only giving an alarm message but also stopping the production if the incompatible risk is very high.

THE PROFILE OF INTELLIGENT FEEDBACK MECHANISM

The intelligent feedback mechanism between document control system and production system is designed in Figure 2 intelligent feedback mode.

In this flow, once a document goes into effect after being created, revised, abolished, etc., it will auto trigger DAAM model working. Based on embedded distribution mechanism, DAAM will send action request to corresponding quality management supervisor, system or equipment maintainer, and related e-system owner in production line. Then VCM model will make version conflict judgment between newest document version and current version setting in production e-system. If "conflict", it will go to the next incompatible risk judge; otherwise VCM will be closed. IRAM is an intelligent work model to do incompatible risk judgment, which will trigger corresponding actions, such as inform, alarm, shut down, audit, etc., based on embedded risk analysis rules. In summary, DCC→DAAM→VCM→IRAM forms a close loop to make sure inline production setting match with newest document definition.

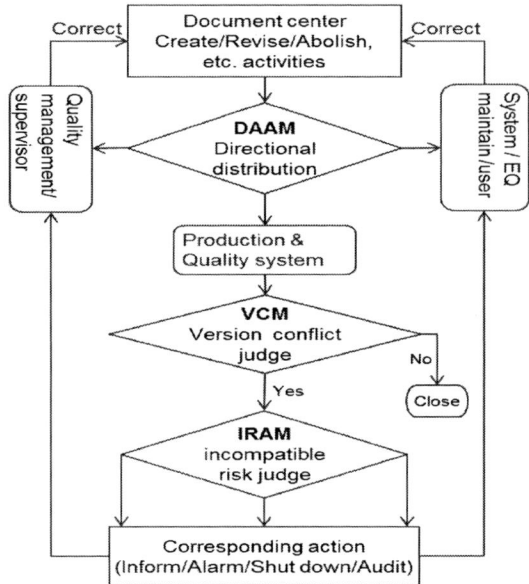

Figure 2: Intelligent feedback mode

To achieve above intelligent feedback mechanism, the document control system's interface should be designed with ability to indicate which area and what kind of change happened, and the index of change information should be indicated as much as possible by document model [2]. For example, once an engineer revises a document, he must select the document change label. The purpose is to lay the foundation for next intelligent module of DAAM, VCM and IRAM. Please refer to Figure 3.

Application>Revise Request	
Process No.	2016-03-15-00045
Purpose.	Purpose. xxx
Purpose. Description	XXXXXXXX

*Document change directional label.

PM system	Equip.1 ⊖	Equip.2 ⊖	Equip.3 ⊖
SPC system	Chat1 ⊖	Chat2 ⊖	Chat3 ⊖
CRM system	CTM1 ⊖	CTM2 ⊖	CTM3 ⊖
Others			

*Document change attribution label.

A	Select	D	Select
Customer related	⊖	Add	⊖
Cross site	⊖	Tighten	⊖
Cross department	⊖	Cancel	⊖
Single Site	⊖	Loosen	⊖
Single department	⊖		
B	Select	C	Select
Inline or Production	⊖	Key monitor items	⊖
Offline or metrology	⊖	Non-key monitor items	⊖
Supporting area	⊖	N/A	⊖

*Document change type label.

Parameter criterion ⊖	Owner change ⊖
Frequency ⊖	Working flow⊖	others ⊖

Apply   Back

Figure 3: Interface design for a document revise application

Intelligent feedback mechanism will be further explained in following chapters with each working mode design in details.

DAAM (DOCUMENT ACTIVITY ANALYSIS MODULE)

DAAM is designed with the ability to automatically identify the document update status, like the document's generation, revision, abolishment, reading ratio, idle status, etc. DAAM can directionally send corresponding document update information to related function owners who need to know. Please refer to Figure 4.

978-1-5090-6695-7/17 $31.00 © 2017 IEEE   382

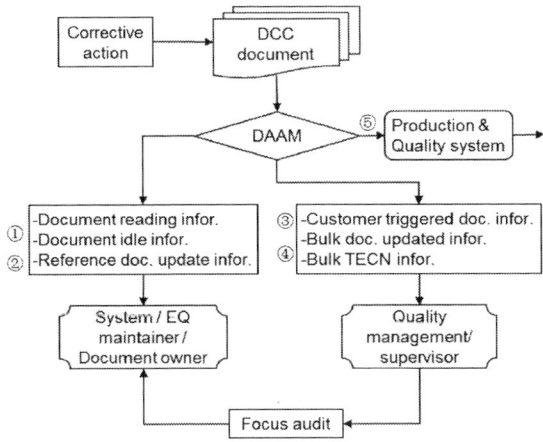

Figure 4: DAAM directionally distribution mode

For example ①, DAAM is designed with the ability to summarize the document reading information and document idle information for the document owner. If a document has few reading ratio or has no change in a long time, the document owner shall trigger document review to evaluate if keep or abolish current document, or revise it to be exactly matched with production in practice.

For example ②, once a document has version update, DAAM ensures its "reference document" should be also reviewed to avoid conflict with its mother or child documents that being updated [3]. DAAM will automatically trigger the version update notice to document owner and remind that "reference document" should be reviewed synchronously. Please refer to Figure 5.

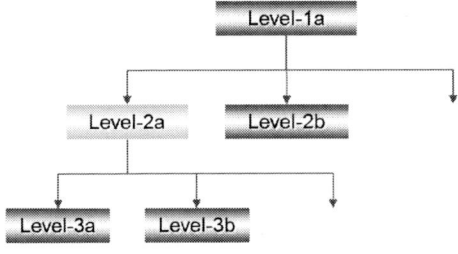

If a document (level-2a) has updated, it will default trigger below related document self review to avoid any conflict:
◆ Adjacent source document (Level-1a)
◆ Adjacent sub-document (Level-3a)
◆ Any reference document if there be (Level-2b)

Figure 5: Tree matrix for document level relationship

For example ③, DAAM is designed with the ability to distinguish whether document update is triggered by

customer transferred document or corrective action from post sales management system. Once such case happens, the quality management/supervisor will receive a reminding message and properly conduct a focus audit to ensure customer transferred document or post sales management system related corrective action being well implemented.

For example ④, DAAM is designed with the ability to know which area encounters bulk document updating or abnormally issues a lot of TECN during a short period. Please refer to Figure 6. Furthermore, DAAM also has ability to identify which type of document has been updated for what kind of abnormal change. Please refer to Figure 7. Once such case happens, the related department manager will be noticed by DAAM and be attracted attention. Meanwhile, the quality management/supervisor will get involved in the investigation of what happened and make sure all the document update activities follow the correct company policy.

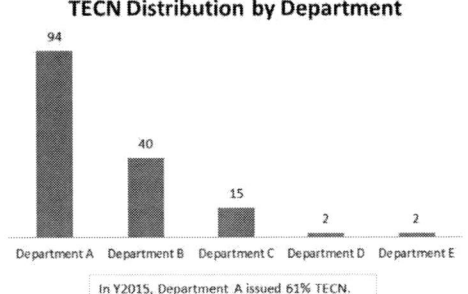

Figure 6: DAAM bulk document updating detection

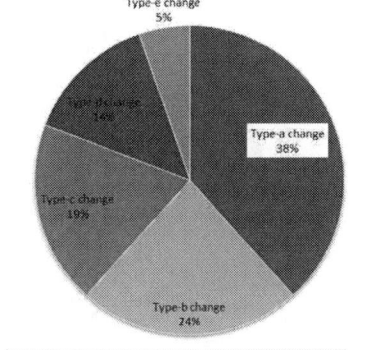

Figure 7: DAAM document updated type detection

The major function of DAAM design is to make document update information to affect inline production system, as ⑤ in Figure 4. The intelligent feedback mechanism designs two modules to realize this function, which are VCM (Version conflict module) and IRAM (incompatible risk analysis module). They can let the document update activities effectively affect production setting to ensure no conflict between document definition and inline production setting. This design will not only give an alarm message but also enable DAAM to stop production if the assessed incompatible risk reaches a high level.

VCM (VERSION CONFLICT MODULE)

VCM is embedded in each inline production and quality system.

For example, for PM system, each type of equipment has an appointed document for PM guideline. In principle, engineer should follow the defined PM frequency and parameter to maintain the equipment. To prevent any human missing, we will initially embed the document number with version for all equipment in PM system. Once a document is updated, VCM will trigger version conflict check between official version in DCC and the version embedded in PM system. If conflicts, it will go to IRAM for further response judgment; if comparable, it means no risk, and the intelligent feedback mechanism can stopped. Please refer to Figure 8.

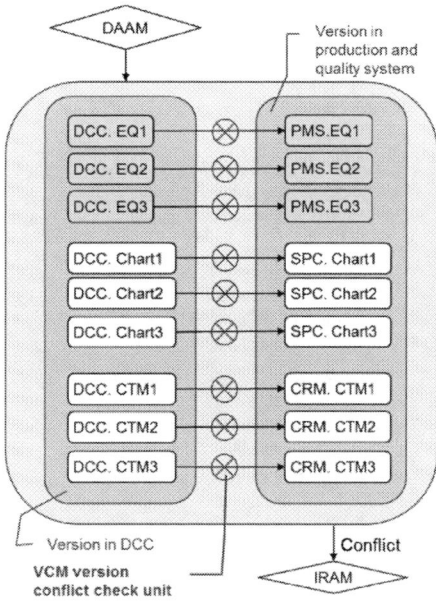

Figure 8: VCM version conflict check mode

IRAM (INCOMPATIBLE RISK ANALYSIS MODULE)

Once VCM sends out a conflict message, the IRAM starts to work. IRAM's major function is to analyze the change risk and judge the change risk level according to a certain assessment rule [4]. Based on the change risk level, IRAM can choose appropriate response to affect production system, like alarm or stop equipment if the risk is very high.

When an engineer applies for the document change, he should firstly identify the change attribution as A & B & C & D. According to the change attribution, IRAM can auto calculate the change risk score E=A*B*C*D based on the risk assessment table. Please refer to Table 1. For example, if a document change involves customer's request to tighten inline key monitor chart, VCM will calculate the change risk score E= A (5)*B (3)*C (3)*D (3) =135

A	Score	D	Score
Customer related	5	Add	4
Cross site	4	Tighten	3
Cross department	3	Cancel	2
Single Site	2	Loosen	1
Single department	1		
B	Score	C	Score
Inline or Production	3	Key monitor items	3
Offline or metrology	2	Non-key monitor items	2
Supporting area	1	N/A	1

Table 1: Change risk assessment table

To effectively affect inline production system, IRAM can auto judge the change risk level and assign corresponding response to related system or equipment. Please refer to Table 2.

Change risk level	Risk score (E)	Response
Low	1~16	e-mail notice engineer,System normal run
Media	16~81	e-mail notice engineer,System alarm
High	81~180	System or equipment stop

Table 2: Change risk level judgment table

Remark:
If a system or equipment suffers several document changes, the risk score should be cumulative multiplication. Such as:
1. Once an equipment suffers 3 changes, the risk score E=e1*e2*e3
2. Once a SPC control chart suffers 2 changes, the risk score E=e1*e2
3. And so on.

CONCLUSION

In this paper, we provide a whole procedure for intelligent feedback mechanism between document control and production system. We use three modules (DAAM, VCM and IRAM) to let document control center have a connection with inline production system or equipment. The traditional manual update mode is replaced by automatically detection and feedback mode.

The procedure defines the intelligent feedback mode, including DAAM directional distribution mode, VCM version conflict check mode, and specific change risk assessment criteria in IRAM incompatible risk analysis mode. With this procedure, company can develop an effective system to establish a channel between document control and inline production, and then prevent Say $\neq$ Write $\neq$ Do issue from happening, thus to improve the accurate implementation of executive ability. We present a spirit of continuous improvement throughout the full text. It is worth mentioning that we also mentioned the customer-oriented concept in risk level adjustment criteria of IRAM module.

ACKNOWLEDGEMENT

The authors would like to extend thanks to SMIC colleagues of the Quality System Management department and Quality Engineering department for their contributions throughout the work.

REFERENCES

[1] ISO, "Quality management system-Requirement ISO9001." 4.2, 2008.

[2] Noronha, M. A., L. G. Golendziner, and C. S. D. Santos. "Extending a Structured Document Model with Version Control.", Database Engineering and Applications Symposium, 1998. Proceedings. IDEAS'98. International, Pages: 234 - 242, 1998.

[3] Martin, S., M. Ahmed-Nacer, and P. Urso. "Controlled conflict resolution for replicated document.", 8th International Conference on Collaborative Computing: Networking, Applications and Worksharing (CollaborateCom), Pages: 471 - 480, 2012.

[4] Zhang, N., et al. "The research of Risk assessment & Sampling methodology combination." , 2015 China Semiconductor Technology International Conference, Pages: 1 - 3, 2015.

A STUDY ON PROBLEM SOLVING STRATEGY USING EXPERIMENT OF DESIGN

Xinyuan Ji[1]*, Sheng Kang[1]

[1]Department of Statistics & Cali. Quality Engineering, Semiconductor Manufacturing International Corporation, Shanghai, P.R. China

*Corresponding Author's Email: Serena_Ji@smics.com

ABSTRACT

This paper presents an approach to solving the problems though the method of design of experiment (DOE). In generally, all problems in semiconductor industry can be summarized as 3 types, called Type X, A and T. The definition is based on whether it's clear for root cause and improvement action of problems. According to the problem types, different solutions are recommended to different problem types for effectively solving the issues. In this paper, we introduced a problem solving strategy to identify the problem types and provided a reasonable solution plan based on DOE method. And one flow is designed in this paper to instruct the problem identification and solution determination. The screening method and optimal design method of DOE are recommended for its wide applications and rigorous statistical theory. Two examples are described to using the flow of real problems solving. The result shows the flow can clearly identify and settle the issues.

KEYWORDS

Design of experiment (DOE); screening method; optimal design; statistical process control (SPC).

INTRODUCTION

In working and real life, we often encounter some problems to solve. Some basic information must be collected, such as the descriptions of this issue, 5-W (what, when, who, where, why) and 2-H (how/ how many). According to this information, we can identify whether the root cause and its improvement actions based on analysis of logical and engineering experiences. In general, we can roughly know whether the root cause and improvement action are clarified. According to this, they can be summarized as 3 types, show in Table I [1]. Type X defines when the root cause of the problem is unclear, and there is sure no any improvement action. Sometime, when the root cause can be clear, but the improvement actions cannot be identified, and this situation calls Type A problem. The clearest problems are Type T as the root causes and improvement actions are clearly known. Many classical methods and theories can be well used here, for example, Seven QC (Quality Control) Tools (including Pareto Chart, Cause and Effect Diagram, Check Sheet, Graph, Histogram, Scatter Diagram, Control Chart) and SPC (Statistical Process Control) [3], [4].

TABLE I.   PROBLEM TYPE SUMMARY

Problem Type	Root Cause	Improvement Action	Solution
Type X	Unknown	Unknown	DOE
Type A	Known	Unknown	DOE
Type T	Known	Known	Seven QC Tools SPC

In this paper, we present a method based on an approach of DOE (Design of Experiment) to solve the problem of Type X & A. The DOE based on statistical & mathematical concept to make your experiment be the most effective and accuracy for model fitting and issue identification.

EXPERIMENTAL DESIGN METHODOLOGY

With development of data analytic method, experimental design presents diversification, as shown in Figure1 [2]. The different experimental methods can be used to solve different problems. Following, it will introduce some commonly used DOE methods respectively.

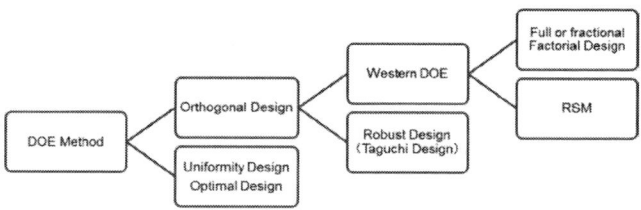

Figure 1 Hierarchy Plot of Experimental Design Method Summary

Full or Fractional Factorial Design

In statistics, a full factorial experiment is an experiment whose design consists of all possible combinations of these levels across all such factors. A full factorial design may also be called a fully crossed design. Such an experiment allows the investigator to study the effect of each factor on the response variable, as well as the effects of interactions between factors on the response variable.

978-1-5090-6695-7/17 $31.00 © 2017 IEEE   386

Fractional factorial designs are experimental designs consisting of a carefully chosen subset (fraction) of the experimental runs of a full factorial design. The subset is chosen so as to reduce resource to exploit the sparsity-of-effects principle to expose information about the most important features of the problem studied. [5]

RSM (Response Surface Method)

The main idea of RSM is to use a sequence of designed experiments to obtain an optimal response. Box and Wilson suggest using a second-degree polynomial model to do this. They acknowledge that this model is only an approximation, but use it because such a model is easy to estimate and apply, even when little is known about the process. [6]

Robust Design (Taguchi Design)

Robust design is statistical methods developed by Genichi Taguchi to improve the quality of manufactured goods, and more recently also applied to engineering, biotechnology, marketing and advertising.

Uniform Design

Uniform design and uniform design experiment method, or space filling design, is a kind of experimental design method. It is only considered sites spread evenly within the scope of the test of an experimental design method.

Optimal Design

In the design of experiments, optimal designs are a class of experimental designs that are optimal with respect to some statistical criterion. In the design of experiments for estimating statistical models, optimal designs allow parameters to be estimated without bias and with minimum-variance. A non-optimal design requires a greater number of experimental runs to estimate the parameters with the same precision as an optimal design. In practical terms, optimal experiments can reduce the costs of experimentation. [7]

The Selection of DOE Methods

Selecting an adequate DOE method can quite reduce time and cost when we solve the problem. As we known, the number of factors is important to show the complexity of a problem. When number of factors and experimental levels are less than 3, the advantage of optimal design and uniformity design are not significant. However, when increasing of number of factors or experimental levels, the merits and recommendations for each experimental design are described as follows.

■ **Screen Method**

2-level full or fractional factorial design is recommended to screen key factors. It calls the screening design. In general, a 2-level factorial design with repeated 2~3 center points is able to analyze main effects and 2-factor interactions; and the lack of fit test is used to assess whether we have selected a suitable model to fit.

The lack-of-fit test for regression model with independent replicate values is a statistical hypothesis test. A significant F_0 of lack-of-fit, indicated in formula (1), implies there may be some systematic factors to be not concerned in the fitting model. It arises when there are exact replicate values of the independent variable in the model that provide an estimation of pure error. The pure error is in essence the amount of error that cannot be explained by any factor.

The statistic F_0 is indicated as following (1) to decide the significant of the lack of fit test.

$$F_0 = \frac{SS_{LoF}/(m-p)}{SS_{PE}/(n-m)} \tag{1}$$

Where,

$SS_{LoF} = \sum_{i=1}^{m} n_i \left(\overline{y_i} - \widehat{y_i}\right)^2$

$SS_{PE} = \sum_{i=1}^{m} \sum_{j=1}^{n_i} (y_{ij} - \overline{y_i})^2$

SS_{PE} : The sum of squares due to pure error

SS_{LoF} : The sum of squares due to lack of fit error

y_{ij} : j^{th} observation of ith experimental condition

$\overline{y_i}$: Average of observations of ith experimental condition

$\widehat{y_i}$: Average of prediction of ith experimental condition

n_i : number of observations of ith experimental condition

n: number of observations

m: number of experimental condition

p: degree-of-freedom of model

■ **Optimization Method**

RSM (Response Surface Methodology) design is recommended to optimize model in surface analysis. RSM is able to analyze model square effects with high precision estimation, but requires more experimental runs.

Optimal design is also a good choice for model optimization. It's enough to estimate high order effects with much less experiment runs, but it is lower precision for model estimation than RSM.

Uniform design is used to complex experiment with many factors at multi levels. It requires too few experiment runs and could analyze non-linear model. Its weakness is lower precision for model estimation.

Robust design (Taguchi design) is to make the stable experimental quality, so that the production process is sensitive to noise. Taguchi's is recommended to apply in industrial experiments maximizing some signal-to-noise ratio (representing the magnitude of the mean of a process compared to its variation).

Base on different experimental purpose and finite resource, we can choose the corresponding design method to solve the problem.

PROBLEM SOLVING STRATEGIES
Transfer the problem from Type X to Type A

In order to solve the problem of Type X, firstly, we should find its root cause. Then it can be transferred Type A. The strategy flow is shown as Figure 2. Fishbone,

5W1H and decision tree are all great tools for root cause investigation. In these methods, firstly we must identify what are the controllable variables and uncontrollable variables. Value of variable could be adjusted, set and controlled that is controllable variable, some like gas flow, pressure, speed and so on. Otherwise is uncontrollable variable, such as atmospheric pressure, external power… For controllable variables, we could use screening design for research the key factors.

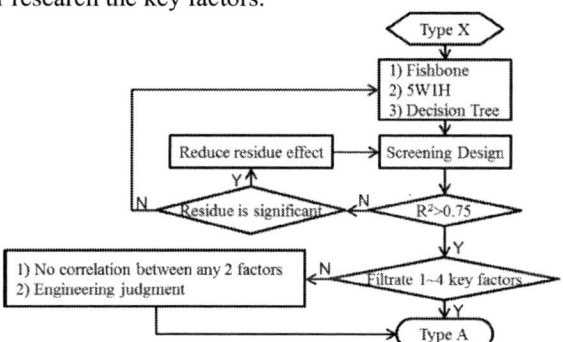

Figure 2 Flow Chart of Type X Problem Solving Strategy

The determination of responses is also very important in experiment. The following guidance can help to get suitable responses.

a) Responses are quantifiable
b) No correlation between any two responses

A guidance of selection of factors is also defined following.

a) No correlation between any two factors
b) Enough and enforceable experimental range

We could choose 2-level factorial design to screen key factors. R^2 (coefficient of determination) is used to evaluate the model linearity.

We indicate observation value as $y'=(y_1, y_2, … ,y_n)$, model error as $\varepsilon_i \sim N(0, \sigma^2)$, hence

$$y_i \sim N\left(\beta_0 + \sum_{j=1}^{k} \beta_j x_{ij}, \sigma^2\right) \qquad (2)$$
$$\beta' = (\beta_1, \beta_2, ..., \beta_k) \qquad (3)$$

is the estimated coefficients of model. Set X as experimental conditions,

$$X = (x_{ij})_{m \times k}$$
$$R^2 = \frac{SS_R}{SS_T} = 1 - \frac{SS_E}{SS_T} \qquad (4)$$
$$SS_T = SS_R + SS_E \qquad (5)$$
$$SS_E = y'y - \widehat{\beta}'X'y \qquad (6)$$
$$SS_T = y'y - \frac{(\sum_{i=1}^{n} y_i)^2}{n} \qquad (7)$$

Where,

SS_T: *Total sum of squares*
SS_R: *Sum of squares of the regression*

The bigger R^2 is, more significant linear model is. When the R^2 is less than 0.75, two issues of model fitting can be suspected. Firstly, the residue is seriously confounding with factors effect. It is recommended to use SOV (Source of Variance), a method of analysis on composition of variations, to improve the process total

variation. Secondly, some key factors excluded in this experiment.

We could conclude that Type X transfer to Type A problem successfully according to below items for judgment:

- $R^2 > 0.75$
- It's better filtrate 1~4 factors for further analysis
- There's no correlation between any 2 factors
- Filtrated factors are made sense in engineering process.

Transfer problem from Type A to Type T

Sometime, even the root causes of problems are clear, we still have no idea to fix the issue, because the effect of factor are trade off with each other. To solve the problem of Type A, we should find out the best set of factor combination. And the reproducibility of the predicting model need also be considered and guaranteed for enough process windows.

When we have no idea about quantifiable relationship of main factors and responses, we are able to use 2-level factional design to model and base on lack of fit result to determine whether there high order effects are needed. Then the RSM or optimal design can be considered to design and optimize the model. On the contrary, we directly use the 2-level factorial design to solve linear model, the RSM or Optimal solve nonlinear model. The complex experiments with too many key factors can use uniformity design. When the object of problem is to solve process stability, the robust design (Taguchi Design) is recommended to be used.

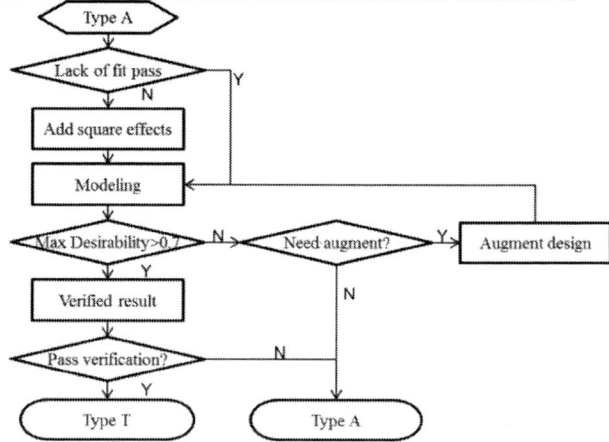

Figure3 Flow Chart of Type A Problem Solving Strategy

In solving the problem of Type A, we need combined with consideration of experimental design factors between the mutual restriction conditions, as well as increase the center design to evaluate whether the model is nonlinear. In model analysis, a suitable analysis method can be selected based on data character of responses, such as yield, which is considered as binominal distribution, we choose Generalized Linear Model to fit it. We can also use polynomial filter method like Stepwise and the desirability weights different to fix the issues.

We use desirability function to optimize model for multi responses. The function transfers predicted values of responses into values between 0 ~ 1, then on the transformation of geometric average to get the final model desirability value. The desirability value is closer to1, the closer to the goal of problem solving, on the other hand, suggests that the model is the more difficult to achieve the goal of current experimental conditions.

Set d_i as desirability value of response, y_i is the predicted value of model, $s_i(max)$ is expectation maximum or maximum predicted value, $s_i(min)$ is expectation minimum or minimum predicted value and T_i is expectation target.

Solution aims at maximum:
$$d_i = \begin{cases} 0, & y_i < s_i(min) \\ \dfrac{[y_i - s_i(min)]}{[s_i(max) - s_i(min)]}, & s_i(min) \le y_i \le s_i(max) \\ 1, & y_i > s_i(max) \end{cases}$$

Solution aims at maximum:
$$d_i = \begin{cases} 1, & y_i < s_i(min) \\ \dfrac{[s_i(max) - y_i]}{[s_i(max) - s_i(min)]}, & s_i(min) \le y_i \le s_i(max) \\ 0, & y_i > s_i(max) \end{cases}$$

Solution aims at meeting target:
$$d_i = \dfrac{\dfrac{1}{\sqrt{2\pi}\sigma} e^{-\frac{(y_i-T)^2}{2\sigma^2}}}{\dfrac{1}{\sqrt{2\pi}\sigma}} = e^{-\frac{(y_i-T)^2}{2\sigma^2}}$$

σ is calculated by (USL-LSL)/6.

Define D as model desirability, therefore
$$D = \sqrt[n]{\prod_{i=}^{n} d_i}$$

If $D < 0.7$ (the reject region can be defined on its own according to the specific project requirement), we need consider whether is necessary to augment experiment, mainly two types of augment design as follows:

a) Experimental scope expansion: Prediction points the optimal combination appeared in boundary of the experimental range, and it is applicable to expand under the engineering considerations so that we could extend related factors range as predicted tend to find the optimal recipe setting.

b) Factor expansion: Confirm factor selection process if any key factor ignore and add the missed factors for expanding experiments.

We have to conclude in regret that we can't find the best combination due to limitation of current production capacity and problem analysis ability. Type A problem-solving fail.

If $D > 0.7$, we have to verify the predicted result by repeating in production environment. Basing on the requirement of engineering precision to select minimum sample size at least to do verification test, we need confirm if the results of the validation are consistent with the prediction. If yes, indicates that Type A problem solving success into Type T. If not, it still needs to continue to explore what reason cause not reproducibility.

DISCUSSIONS

We apply the problem strategy in real productions and find it effectively solve the problems.

Application in Type X to Type A

In PVD (Physical Vapor Deposition) process, often encounter the defect problem. After fishbone diagram analysis, there are 8 controllable factors in Table 2 related to the defect performance, where "H" express high level and "L" express low level. Engineers use DOE to explore what is the real reason influence on defect.

TABLE II.   FACTORS AT EXPERIMENTAL LEVELS

Parameter	Level	Low	High
A	2	L1	H1
B	2	L2	H2
C	2	L3	H3
D	2	L4	H4
E	2	L5	H5
F	2	L6	H6
G	2	L7	H7
H	2	L8	H8

Use resolution IV 2-level fractional factorial design and add 3 repeated center points, total only 22 runs. If we use full factorial design for the experiment with 8 factors, the experimental times reach to 256 at least. For optimal design, the split times also need 46 at least. In other words, it is unnecessary to cost so many to find the major effects. According to sparsity-of-effects and heredity principle, it's enough to use resolution IV fractional factorial design to select main effects. As result shown in Figure 4, we could screen key factors H, C and A.

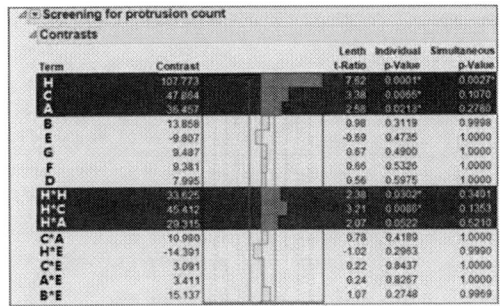

Figure 4 Screening for Defect Count

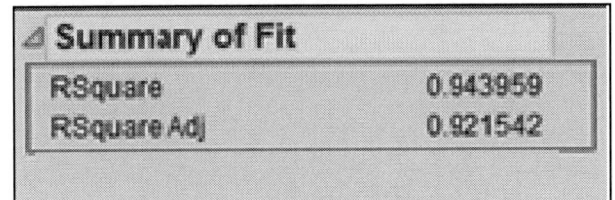

Figure 5 Summary of Model Fit

◢ Lack Of Fit				
Source	DF	Sum of Squares	Mean Square	F Ratio
Lack Of Fit	2	6877.250	3438.63	2.4409
Pure Error	13	18314.000	1408.77	Prob > F
Total Error	15	25191.250		0.1259
				Max RSq
				0.9593

Figure 6 Lack of fit result

As shown in Figure 5 and Figure 6, R^2 is 0.94 > 0.75, it indicates the linear model is significant; Lack of fit test result *Prob> F* is 0.126 > 0.05, it indicates there is no square effects in model. *Prob>F* means the probability is more than a threshold (significance level: 0.05) in *F* distribution, that is lack of fit is not significant to pure error. [Formula (1)].

Through DOE model analysis, we successfully select three significant factors from eight unsure factors and for establish the direction subsequent optimization.

Application in Type A to Type T

The concentration of parameter A is important in CMP (Chemical Mechanical Polishing) process, it is necessary to reduce the usage of concentrate to reduce production cost, meanwhile to guarantee remove rate on target.

As is known to all A, B and C are 3 main factors for CMP. The ratio of the factors mutual restriction with each other and total amount is the same, so use a special design mixture design to do experiment. When experiments in which the response depends on factors that represents proportions of a blend and the proportions sum up to a constant (usually 1 or 100%), we use mixture design. [8]

According to the experience, experimental range of factors lists in Table 3. It defines M, N and K as responses in Table 4. Use mixture design, 3 factors total 9 splits.

TABLE III. FACTORS AT EXPERIMENTAL LEVELS

Parameter	Level	Low	High
A	2	L1	H1
B	2	L2	H2
C	2	L3	H3

TABLE IV. RESPONSES WITH PURPOSE

Parameter	Purpose
M	Maximum
N	Meet Target
K	Minimum

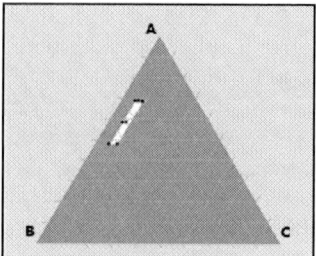

Figure 7 Experiment Splits

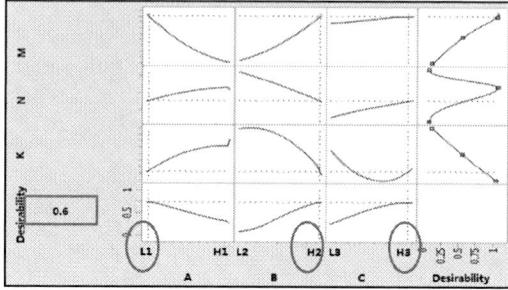

Figure 8 Prediction Plot

Desirability is 0.6 less than 0.7 in Figure 8, the prediction can't meet the expected solutions. On the other hand, we find the forecast values in the boundary of the factors, and combined with the experience, the final concentration of parameter C influent response N most, hence engineers decided to expand the parameter C range to do augment design.

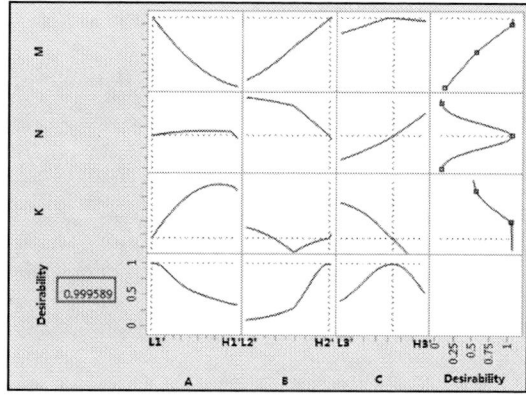

Figure 9 Prediction Plot after Augment Design

After expansion of the experimental we found the predicted desirability could achieve 0.99, which is a very pleasant and surprise result. The optimal combination setting of parameter A, B and C is found. We have to verify the predicted result by repeating in production environment, collected 12 data and found the result is comparable with baseline in Figure10 and Figure11.

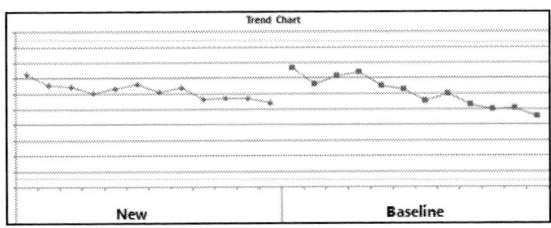

Figure 10 Trend Chart of Response N Verified Result

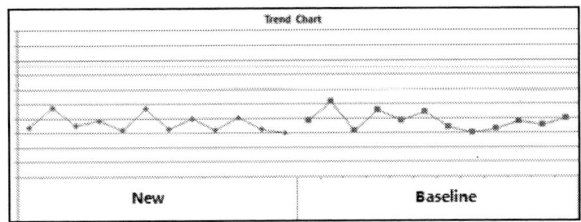

Figure 11 Trend Chart of Response K Verified Result

Through DOE model analysis, we successfully found the optimal setting.

CONCLUSIONS

Through the above two examples, problem solving strategy method seems to be very useful. In the 1st case, we used the DOE method and analysis indicators criterion to identify three significant factors from eight suspected in the case of uncertain root cause; In case 2, we were under the guidance of problem solving strategy to analyze the results of preliminary analysis, according to boundary problem solving proposal managed for sequential expansion experiment to find the best program setting well beyond expectations to achieve the goal of cost reduction.

In summary, it's a good strategy for Type X and Type A problems solving. The project owner can take a clear and detailed train of thought using the DOE tools for the solution of the problem.

Every stage of problem is solved well will help the next phase of problem management, such as Type X is resolved successfully will help quickly and reasonably modeling to solve the Type A problem; and solve A problem correctly will benefits the subsequent management of T issues. According to the different stages of the production process capability and process development needs, we can recycle such a problem solving strategy to achieve the goal of continuous improvement.

REFERENCES

[1] Zhongpu Zhang *Experimental Design Method Learning in An Easy Way*, 5th edition, 2005
[2] Montgomery D C. *Design and Analysis of Experiments,* 6th edition, 2009
[3] Montgomery D C. *Introduction to Statistical Quality Control (I),* 2nd edition, 2009
[4] Andrew Sleeper. *Testing Design for Six Sigma Statistics,* McGraw Hill, 2006
[5] Wiki - Wikipedia, the free encyclopedia.[Online]. Available:
https://en.wikipedia.org/wiki/Fractional_factorial_design
[6] Wiki - Wikipedia, the free encyclopedia. [Online]. Available:
https://en.wikipedia.org/wiki/Response_surface_methodology
[7] Wiki - Wikipedia, the free encyclopedia. [Online]. Available:
https://en.wikipedia.org/wiki/Optimal_design
[8] DOC88 - Online document sharing platform. [Online]. Available:
http://www.doc88.com/p-4823091954525.html

IMPROVEMENT ON THE STRESS MIGRATION IN TUNGSTEN-PLUG VIA

Juan Wen, Wei-Ting Kary Chien, Guan Zhang, Yanhui Sun

Division of Reliability Engineering, Semiconductor Manufacturing International Cooperation

*Corresponding Author's Email: Margaret_Wen@smics.com

ABSTRACT

In this paper, we develop a new via structure (FTV1), to study the reliability performance of SM (Stress Migration). FTV1 structure is different from the conventional Al-based interconnect technology. Here, the metal beneath the tungsten via is Cu (the conventional is Al), while the upper metal above the via is Al. The complex process of the FTV1(single via) combined with the interaction of the metal layer stresses results in that the SM resistance shift after 168hr baking has bi-modal issue. A new failure mechanism in W-plug via was reported in this paper. By changing the anneal step from before FTV1 photo to after passivation etch, the resistance shift becomes smaller than 2% versus the original 8%, which can meet industry specification.

INTRODUCTION

The introduction of CVD tungsten-filled vias for inter-level interconnect brought with it many advantages, great planarity, increased aluminum step coverage, and the ability to reduce the overall via dimension. Currently, highly integrated microprocessor requires dense interconnects which can produces a better yield and more cost effective product. The trend of continuous shrinking in the dimensions of these multilevel interconnects leads to an increase in the current density, which has raised the concern of metal interconnect reliability. One important reliability issue is the metal void at the interface of tungsten via and metal which result from high temperature baking [1-2]. Stress-induced-voiding is one of the predominant failure mechanisms of multi-layer on-chip interconnects. And it is the result of vacancy movement driven by stress-gradients in around vias that connect different metal layers. These gradients are the consequence of a thermal mismatch between the metal and surrounding materials. Clearly, the tungsten plug itself will not migrate under baking, but it is possible for the aluminum to "walk away" from this ideal blocking boundary, which is the cause of failure, and leads to resistance increase and possibly open-circuit in the via [3-5].

In this paper, a new via structure (FTV1) of SM failure mechanism was studied and the methods to improve the SM performance are proposed with high temperature anneal step change. Experiments have shown that after change the anneal step for FTV1, the resistance shift can be significantly decreased after high temperature heating. It attributes to the stress relaxation between the metal layer and FTV1.

EXPERIMENTAL

In this work, all the data presented was from 300mm wafers fabrication technology.

The new via structure (FTV1) to study the reliability performance of SM is shown in Figure 1(a). FTV1 structure is different from the conventional Al-based interconnect technology. Here, the metal beneath the tungsten via is Cu (the conventional is Al), while the upper metal above the via is Al. The interlay dielectrics (ILD) between adjacent metal layers are SiN/TEOS/SiN/TEOS by standard deposition-etch and are planarized by chemical mechanical polishing (CMP). After the via etch, a TiN layer was deposited. The vias were then filled with tungsten. Finally, the Al -0.5Cu covered on the top. Final step is the passivation process.

A single four-terminal kelvin-contact test structure was chosen for investigation of SM performance of vias as shown in Figure 1(b). The line width of upper and bottom metal is 5um. The via size is 0.5um in diameter. SM reliability testing was stressed at 200°C with electrical resistance measurement. Measurements were taken after 7 days of high temperature storage. A sample size of 60~70 was used in this experiment. Resistance shift after high temperature baking is defined as (R (after baking) – R (before baking))/ R (before baking).

978-1-5090-6695-7/17 $31.00 © 2017 IEEE

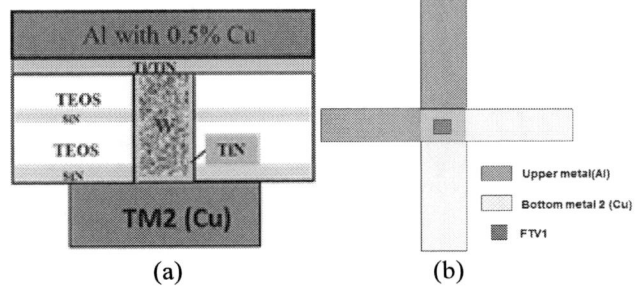

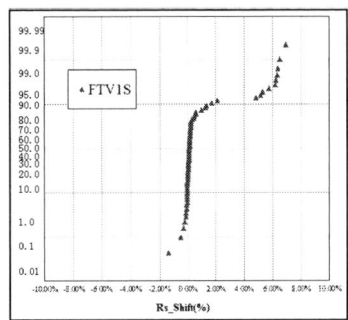

Figure 1: (a) Kelvin test structure; (b) Schematic cross-section of FTV1 Structure

RESULT AND DISCUSSIONS

The complex process of the FTV1(single via) combined with the interaction of the metal layer stresses results in that the SM resistance shift after 168hr baking has bi-modal issue, as shown in Figure 2. The main distribution R shift is smaller than 2%, but several samples have larger shift, even to ~7%. This result is close to the failure criterion.

In order to better understand the SM failure mechanisms, we set about to investigate the process sensitivities around the W-plug process, with emphasis on reducing the mechanical stress in the metal film in the vicinity of the plug. We identify the factors which have important impacts on SM reliability to be the W-plug glue layer (W glue layer Ti is an interfacial adhesion layer between the Cu and FTV1 W and is helpful to relax stress [5], but the original process doesn't have such a layer); FTV1 CD (critical dimension) (The larger the FTV1 CD is , the smaller the film stress is [5]); The width of FTV1 test key structures (The wider metal will bring about more stress for the film stack); and anneal step (Annealing can greatly reduce the film stress) [6-7].

Based on the above factors, experiments were performed. Different wafers were prepared with the split conditions shown in Table I. It is expected that interconnect stress will decrease with decreasing the metal line width and increasing FTV1 CD. However, narrower width and larger FTV1 CD don't improve the SM performance as shown in Figure 3(a). When change the FTV1 CD from 0.5 to 0.55um, and narrow down the metal width from 5 to 1.1um, the SM resistance shift after 168hr baking also presents bi-modal issue, the shift can reach to ~8%.

It is believed that the Ti layer can prevent the stress induced void formation after high temperature baking [5]. However, the high mechanical stress arises because the W-plug produced by CVD deposition produces a high intrinsic stress, which acts on the top and bottom interfaces of the via. This stress, in combination with the high tensile stress in the metal line and the compressive stress in ILD films, acts as a huge destabilizing force. As shown in Figure 3(b), when the process changes the glue layer from TiN to Ti/TiN, the SM performance is still bad as the base line.

From the table, it can be seen that the SM resistance shift after 168hr baking can be significantly reduced by anneal step changing from before FTV1 photo to after passivation etch. The different flow is shown in Figure 4. In "Flow 1", an anneal was carried out before FTV1 photo. And in "Flow 2", the

anneal step was executed after passivation etch.

From the results, the reason for the SM reliability performance improvement can be explained as follows:

In "Flow 1", the anneal step is before FTV1 photo, where there is no tungsten film, while after all films deposition, there is strong tensile stress in the stack films, which can enhance the void nucleation and growth, inducing larger resistance shift. On the other hand, in "Flow 2", the anneal is at the final step after passivation deposition, and the tensile stress in the stack films become smaller after anneal, and the SM failure rate thus decreases [8].

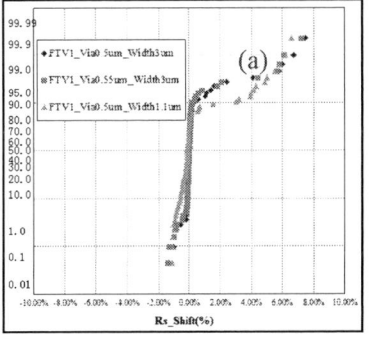

Figure 2: FTV1 (single Via) SM has bi-modal issue

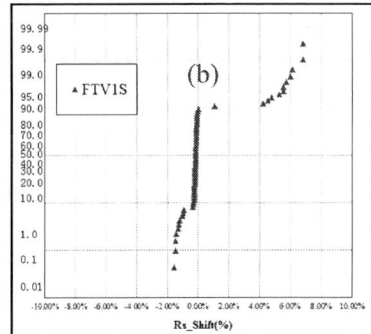

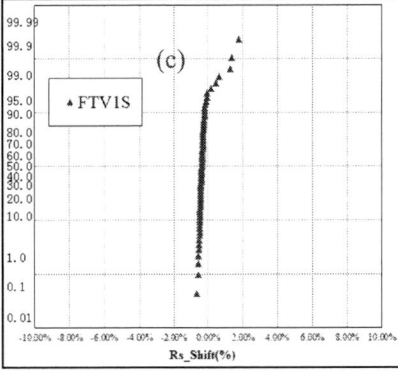

Figure 3: (a)(b) FTV1 SM performance with different FTV1 CDs and metal widths. Larger via size (0.5→0.55um) and narrower down metal width (5→1.1um) can't help for FTV1 SM improvement; (c) FTV1 SM performance with anneal location change from before FTV1 photo to after passivation etch have significantly improve for FTV1 SM. Resistance shift from original 8% to 2%.

978-1-5090-6695-7/17 $31.00 © 2017 IEEE

TABLE I. FTV1 SM test result under different condition

FTV1	Metal width	FTV1 Size	Glue layer	Anneal Step	Test Result
Base Line	5um	0.5um	TiN	Before FTV1 Photo	Bad (Bi-modal Issue)
A	3um	0.5um	TiN	Before FTV1 Photo	Bad (Bi-modal Issue)
B	3um	0.55um	TiN	Before FTV1 Photo	Bad (Bi-modal Issue)
C	1.1um	0.5um	TiN	Before FTV1 Photo	Bad (Bi-modal Issue)
D	5um	0.5um	Ti/ TiN	Before FTV1 Photo	Bad (Bi-modal Issue)
E	5um	0.5um	TiN	After Passivati-on etch	Good (R shift<2%)

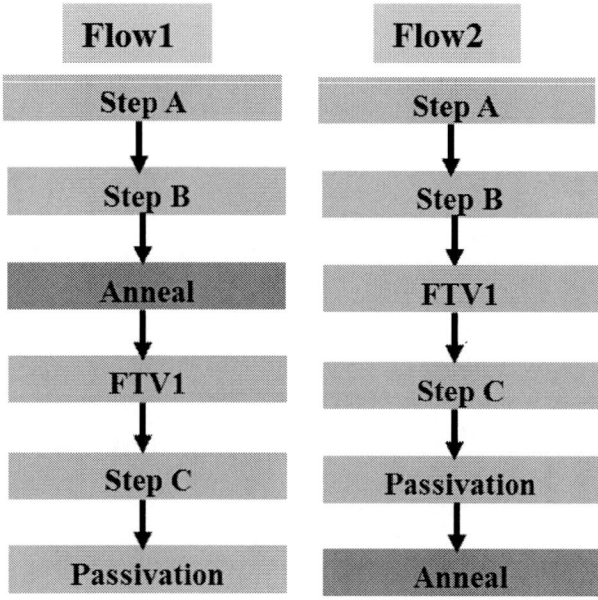

Figure 4: Two process flow described in the work

CONCLUSIONS

In this paper, the new structure FTV1 SM performance was investigated. By changing the anneal step from before FTV1 photo to after passivation etch, resistance shift after high temperature baking, will drastically decrease without degrading device performance and yield. It is attributed to the film stress reduces after anneal.

ACKNOWLEDGEMENTS

The authors would like to thank the reliability team of for the support of SM test and technical discussion with project team. Valuable discussions with our colleagues from Reliability Engineering are also highly appreciated.

REFERENCES

[1] F. Yost, D.Amos, and A.D Roming, F. Yost, D. Amos, and A.D. Romig. IEEE International Reliability Physics Symposium Proceedings, 1989, pp.193.

[2] J. T. Yue, W.P. Funsten, and R.V. Taylor. IEEE International Reliability Physics Symposium Proceedings, 1985, pp. 126.

[3] J. A. Walls. IEEE Transactions on Electron Devices, vol. 44, 1997, pp. 2213.

[4] J. A. Walls. IEEE Electron Device Letters, vol. 16, 1995 pp. 430.

[5] M. Kanazawa, M. Shishino, Y. Hata and T. Umemoto. June 11-12,1991 VMIC conference.

[6] E. T. Ogawa, J. W. McPherson, J. A. Rosal and et al. Reliability Physics Symposium Proceedings, 2002.40th Annual.

[7] B. Wallace, Y.-H. Lee, D.Pantuso, K.Wu, and N. Mielke. IEEE 99CH36296. 37th Annual International Reliability Physics Symposium, 1999.

[8] T. Harada, K. Kobayashi, M. Takahashi, K. Nii, A. Ikeda, T. Ueda and T. Yabu. IEEE International interconnect Technology confenrence, 2003: 92-94.

HIGH-K METAL GATE INLINE MEASUREMENT TECHNIQUE USING XPS

Yang Song, Yi Huang, Yi-Shi Lin*

Technology R&D, Semiconductor Manufacturing International Corp.

Pudong New Area, Shanghai 201203, China

*Corresponding Author's Email: Alien_Lin@smics.com

ABSTRACT

X-ray photoelectron spectroscopy (XPS) can be used for surface layer characterization. Thin film thickness and composition can be determined. This paper presents a study performed to evaluate the benefits of high-k metal gate films measurement using a novel and high throughput in-line XPS technique. Various thin films such as SION and HfO_2 are measured. The strengths and drawbacks of XPS are presented. Repeatability tests were performed in order to check the measurement tool capability.

INTRODUCTION

As integrated circuit scales down to 28nm and beyond, it is of considerable importance to accurately control the film composition and thickness, especially for films thinner than 100A. The electrical properties of the final device can be significantly influenced by ~1A metal gate film thickness change. Traditional inline measurement technique, such as spectroscopic ellipsometry (SE), shows its drawbacks in determining the film properties, as it is a model based technique --- model has to be changed when the process change; in addition, TEM calibration is also needed to setup such ellipsometry models.

X-ray photon-electron spectroscopy (XPS) has proven to be a direct and stable technique to monitor ultra-thin films thickness and composition for any elements with atomic number larger than 2 with very good accuracy and sensitivity. When X-ray strikes a sample material, if the incoming energy is sufficient, it can break an electron free from the orbital and the photon-electrons are ejected. By measuring the kinetic energy and number of electrons that escape from the sample surface, the elements of the sample and the composition and thickness can be determined. The number of electrons ejected is related to the atomic sensitivity factor, which is an element dependent factor. Elements compositions within a film can be quantified by normalizing extracted intensities by appropriate atomic sensitivity factors. Thickness is related to intensity ratio, and is a function of effective attenuation length.

In this paper, we investigated the structural properties, such as composition and thickness of high-K metal gate films, by using a novel, high throughput inline XPS metrology. The strengths and drawbacks of XPS is discussed, as compared to ellipsometry. Long term dynamic precision and short term repeatability of film thickness and composition measurements are also discussed which meet current inline monitoring requirements.

EXPERIMENTS AND DISCUSSIONS
(1) Comparison of XPS and SE

XPS is a direct measurement method, while SE is a model based technique. XPS shows excellent sensitivity, while SE sometimes cannot decouple the composition with thickness. In 2012, IBM has published one paper about SiON DoE test and concluded that ellipsometer fails N dose measurements when thickness changes at the same time. It is because in the SE model, the thickness and n&k value are usually correlated, as shown in the below Figure 1.[1] Thus XPS shows better capability in decoupling elements and thickness.

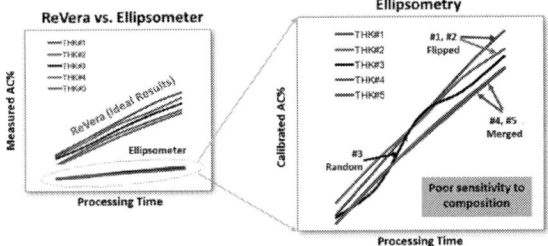

Figure 1: OCD versus TEM trench depth measurement[1]

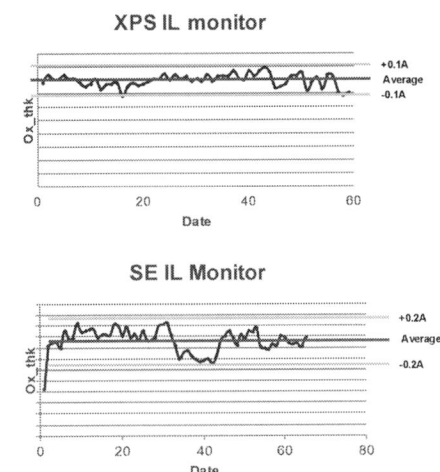

Figure 2: Comparison of stability of XPS and SE

The stability of XPS is also better than SE, as is shown in Figure 2. Two months' data of SiO_2 (IL, interlayer) thickness were collected by XPS and SE, as can be seen from the figure, XPS data is stable in +/-0.1A spec, while SE chart has many unexplained, unpredictable and frequent shift, this maybe due to the SE is sensitive to the environment change (such as temperature and moisture etc.), while XPS is a vacuum based technique. The main challenge of XPS is its throughput is lower than SE.

XPS sensitivity check

Characterization of SION, HfO_2 wafers are important because these are directly correlated to device electrical properties. In order to verify sensitivity of XPS tool, DoE (Design of Experiment) set of wafers are prepared to check the XPS measurement sensitivity, as can be seen from the figures; XPS is sensitive to process changes.

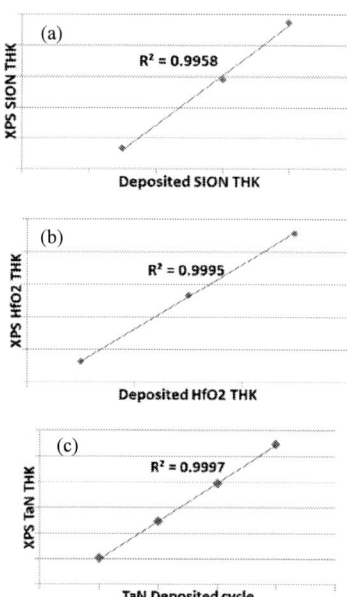

Figure 3: (a) XPS versus deposited SION thickness; (b) XPS versus deposited HfO_2 thickness; (c) XPS versus TaN deposited cycle.

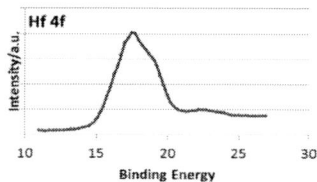

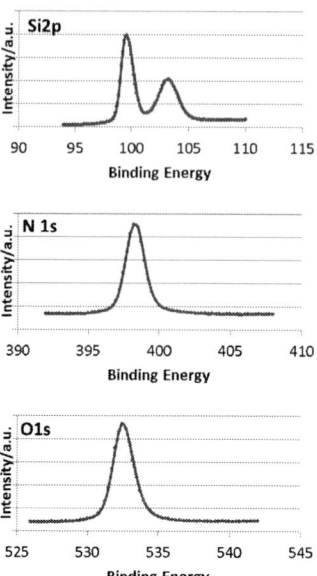

Figure 4: Spectra of SiON, HfO_2 films by inline XPS.

XPS accuracy check

The accuracy was tested by comparing the XPS measured data with TEM results, as shown in Figure 5. The result shows that there is a very good matching between XPS and TEM results (R^2>0.99).

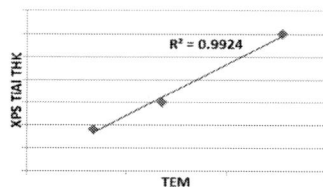

Figure 5: XPS versus TEM measured TiAl thickness.

Gauge Repeatability and Reproducibility (GRR)

GRR is another key quality criterion for metrology tools. GRR test is performed through a 10 times dynamic repetitive measurements on 13 points of the wafer, as is shown in Table I and Table II. Both XPS large beam (LB) on blanket wafer and small beam (SB) on pattern wafer tests are performed and compared with SE. It should be noticed that SE is not able to give Hf% for HfO_2 film, while XPS can do composition analysis with good repeatability. The results show very good repeatability measured by XPS on both thickness and composition (RSD value is within the spec. RSD: relative standard deviation).

TABLE I.   GRR TEST RESULTS OF DPN FILM

RSD	XPS LB	XPS SB	SE
SiON Thk	0.19%	0.38%	0.23%
N Dose	0.27%	0.39%	2.00%

TABLE II.   GRR TEST RESULTS OF IL/HK FILM

RSD	XPS LB	XPS SB	SE
HfO Thk	0.17%	0.27%	0.09%
SiO Thk	0.24%	0.41%	0.41%
O%	0.25%	0.32%	NA
Hf%	0.38%	0.49%	NA

CONCLUSIONS

Inline XPS applications for thickness and composition determination of SION, HfO_2 and metal gate films are discussed in this paper. The sensitivity, accuracy and repeatability of XPS is proved to be good. In comparison to XPS, when measuring HfO_2 films, ellipsometry cannot to give Hf or contamination elements' composition, while XPS is able to do elemental characterization. XPS can be used as the calibration standard of ellipsometry. Overall, it can be seen that the results from XPS can be qualified as accurate and stable. The thickness control of high-k metal gate pattern wafer indicates that the tool provides considerable benefits to process control.

ACKNOWLEDGEMENTS

The authors wish to express their gratitude to Xiaofeng Wang for her support during the whole work. The authors gratefully acknowledge Dr. Rob Trussel and Nono Hou at ReVera for the support and their expertise.

REFERENCES

[1] M. Dai, S. Rangarajan, J. F Shepard etc., "Surface Characterization and Process Control for ALD using Inline XPS Technique", ALD Conference, Dresden, June 18, 2012

APPLICATION STUDY OF QMERIT FUNCTION ON OVERLAY ACCURACY VERIFICATION

Huayong Hu*, Lei Ye, Qiaoqiao Li, Weiming He, Zhicheng Liu
Technology R&D, Semiconductor Manufacturing International Corp.
Pudong New Area, Shanghai 201203, P. R. China
*Corresponding Author's Email: Huayong_Hu@smics.com

ABSTRACT

In advanced semiconductor industries, the overlay (OVL) error budget is getting tighter due to shrinkage in technology. To achieve the tighter overlay requirements, gaining every nanometer of improved overlay is very important, and it is becoming critical to eliminate the smallest imperfections in the metrology targets used for overlay metrology. Conventionally, the performance of overlay metrology is evaluated mainly based on random error contributions such as precision, TIS variability, and total measurement uncertainty (TMU). For image-based overlay (IBO) metrology targets, mark asymmetry (or imperfection) is a common issue which can cause measurement inaccuracy, and it is not captured by traditional methodology. Such measurement error may lead to physical overlay issue, and finally impact the yield run-up.

In this paper, we will present a study of IBO OVL mark measurement accuracy with Qmerit function. Different process integrations are evaluated including normal single layer process with I-line exposure, self-aligned double patterning (SADP) and litho-etch-litho-etch (LELE) double patterning. With the Qmerit function study on SADP process, the OVL mark of previous layer formed with SADP process plays an important role on the OVL accuracy, its effect on current layer OVL mark and itself measurement error due to mark asymmetry during etch process will be discussed in this paper. It is observed that SADP OVL mark design optimization is one important way to improve the OVL residue and reduce the measurement error. It also shows that the measurement light can also be auto-optimized to gain better contrast and less measurement error with Qmerit optimize function.

INTRODUCTION

Optical metrology in general is leaning on symmetric structures. When the metrology target is symmetric, the overlay is well defined. Once an asymmetric effect occurs, the overlay becomes ambiguous. For image based overlay (IBO) targets, the overlay is extracted by center discrepancy between two layers' kernels when the kernels are a summation of intensities of the target's bars in the orthogonal direction to the overlay direction. It is therefore clear that in cases where the IBO target suffers from asymmetry, the overlay isn't well defined and an

inaccuracy error will be embedded within the measurement value. [1-6]

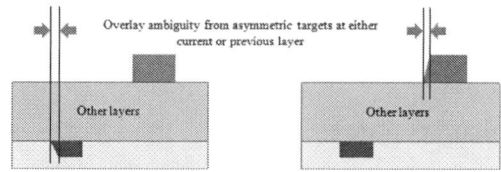

Figure 1: Overlay mark asymmetry leads to a geometrical overlay ambiguity. If the metrology technology has a high sensitivity to mark asymmetry, it may lead to an enhanced asymmetry in the measured signal and generate overlay measurement inaccuracy. [5]

To address this issue, KLA-Tencor developed a quality metric that delivers a quantitative evaluation of the accuracy of an OVL result for a given target under a dedicate OVL measurement condition with several algorithm instead of one only. All the algorithms are guaranteed to give the same overlay estimate if the signal is perfectly symmetric. If the signal is not symmetric each algorithm gives a somewhat different value. The quality metric is related to the width of the distribution of these overlay values. [1-4] Qmerit value is approved to correlated to the OVL measurement accuracy, and it is effected by and can be used to optimize measurement condition/target design in additional to TMU. The detection of defected OVL mark or process induced mark asymmetry with Qmerit is helpful to debug the OVL shift root cause, avoid flyer data input, finally a better OVL control. [5]

In this paper, we will present a study of IBO mark OVL measurement accuracy with Qmerit function on KLA-Tenchor's OVL tool Archer 500. It will be separated into three parts with different process integrations including 1) normal single layer process with I-line exposure with typical mark asymmetry; 2) self-aligned double patterning (SADP) with different mark design; 3)litho-etch-litho-etch (LELE) double patterning on advanced backend-of –line (BEOL).

RESULT AND DISCUSSION

I. I-line exposure with typical mark asymmetry.

It is not common and easy to intentionally form an

asymmetric resist profile during lithography process, as well as for outer layer with ET or CMP process. To verify the Qmerit function on asymmetrical OVL mark, we select an I-line exposure on a thick resist film. Stray light is one common issue for I-line lens if not well maintained; it will introduce obvious resist profile asymmetry and varied with the light transmission circumstance around the mark, especially on thick resist.

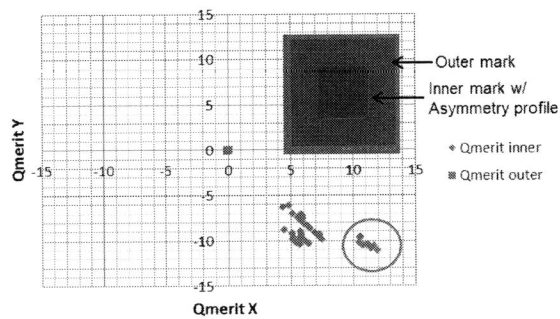

Figure 2: Qmerit test on I-line exposure with current layer resist asymmetry profile.

Figure 2 shows Qmerit value on I-line exposure with current layer resist asymmetry profile. The asymmetrical resist profile is observed on the left and bottom area of the OVL box pad, which is might be induced by the stray light from the left side of the mark since a large open area aside the mark. Large Qmerit inner values in both X and Y is presented corresponding to the resist profile asymmetry, while a very small Qmerit outer value with the outer etched mark. The Qmerit inner value is also varied by group which is correlated with the test mark. The obvious discrete group (circled highlight) in X value is all from one mark with different mark circumstance comparing to other marks, which showed more serious resist profile asymmetry. This mark difference will lead to a larger OVL residue in X than Y direction, even though the OVL accuracy is not trustable in both directions.

II. SADP OVL

SADP approach is one popular method to enable the patterning of pitch less than resolution limitation of i-ArF single exposure like 76nm pitch, which is usually used on memory technology below 3Xnm and FIN formation for logic technology below 2Xnm. [6] As the application of SADP make lithography minimum pitch down to half of design pitch with the remaining spacer aside core, its alignment mark and overlay (OVL) mark have to be well-segmented to ensure enough mark contrast.

In our previous study, comparing to SADP bar-segmental IBO mark, background segmentation was not so suitable for those SADP layers once current layer

was single-layer process (PR only on SADP spacer). The boundary of inner mark, which represented current layer, was defined wrongly for the disturbance of present spacer pattern. [7]

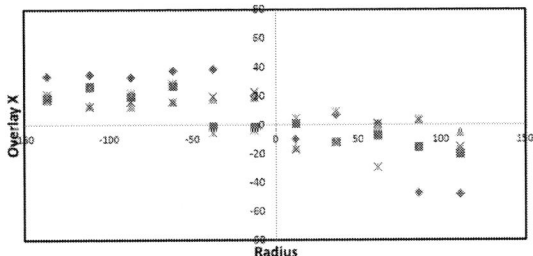

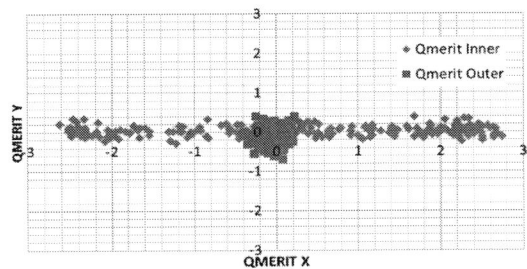

Figure 3: Background-segment IBO OVL with intended X-expansion: a) non-linear OVL X value with radius;[] b)OVL inner mark shows large variation in Qmerit X while much smaller in Y, as well as for out mark.   .

Figure 3b shows Qmerit X shows a large variation for inner OVL mark while relatively stable and small in Y direction, indicating inner mark X direction exist random asymmetrical profile across wafer which leads to the measured OVL value inconsistence to the intended compensation as figure 3a. More test result for normal SADP bar-segment mark provide Qmerit value less than 1nm for inner mark (current layer), indicating no substrate effect.

With this concept, AIM IBO mark with SADP bar-segment was chose for further study. AIM mark provided more position signals than BIB mark that can be averaged to a more trustable OVL value, in additional to its higher pattern density to against process variations. Figure 4 shows an obvious expansion of AEI-ADI OVL correlating to the Qmerit expansion map of outer mark detected at ADI, while the outer mark is formed by spacer post SADP mark segment. Etch process might introduce pattern asymmetry across wafer, mostly decays from center to edge, which will lead to OVL expansion

978-1-5090-6695-7/17 $31.00 © 2017 IEEE

measurement error for the coming layers.[Philippe Leray, IMEC] It should be noted here that all the test are carried with auto Qmerit optimize function to choose the best measurement condition with less Qmerit.

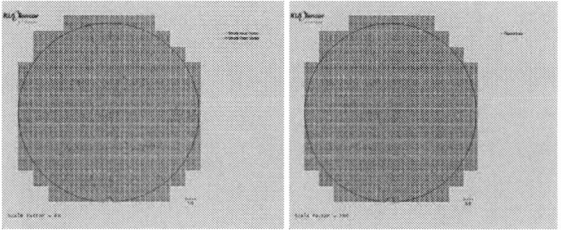

Figure 4: Left is ADI Qmerit value with outer mark expansion (red arrow bar); Right is AEI-ADI OVL value with expansion map.

III. LELE OVL

LELE double patterning method is another approach to realize the extension of resolution limitation, which is mostly used for 2D structure patterning with patterns formed on one thin hard mask then transfer to the substrate. OVL challenge is that the 1st LE formed OVL mark on the thin hard mask usually has weak contrast for 2nd Litho which might result in worse OVL inaccuracy, and thus serious pitch walking post LELE.

Figure 5 shows 2nd OVL measurement condition optimized with Qmerit. Different measurement light was chose to present better thin hard mask visibility, as well as less Qmerit value. Even for both measurement conditions, the estimate OVL residues are small but lead to different real OVL and in consequence different real pitch walking performance post 2nd hard mask etch.

Figure 5: 2nd Litho OVL optimize with Qmerit(right) provide more accurate OVL with more controllable pitch walking on hard mask.

CONCLUSION

In this paper, we have shown the capability of Q merit function on the OVL accuracy detection with different integration process. It is proved to be one effective value to judge the OVL mark profile asymmetry as what we observed on I-line thick photo-resist process. With SADP process, we clarified the OVL error in X is correlated to

Qmerit value disperse in X direction for inner ADI mark due to the mark edge will be mis-located with background segmental spacer. With SADP bar segment OVL mark, we can eliminate this problem. Outer mark Qmerit expansion map seems correlated to AEI-ADI OVL, indicating the spacer formed mark may have asymmetry profile decays from center to edge. Measurement condition can also be optimized with Qmerit auto optimized function to deliver a higher contrast measurement image with less accuracy error. As for accuracy concern, diffraction-based-OVL (DBO) method becomes the mainstream with proved better OVL accuracy but more complicated recipe setup procedure. The Qmerit function is thought to be one effective assistant method with IBO application for OVL control due to its quick setup process, especially for those layers with less mark asymmetry and edge placement requirement.

ACKNOWLEGEMENT

The authors would like to show our appreciation to SMIC TD MMD group and KLA-Tencor for the helpful discussion.

REFERENCES

[1] Daniel Kandel, Vladimir Levinski, Noam Sapiens, Guy Cohen, Eran Amit, Dana Klein, Irina Vakshtein, "Overlay accuracy fundamentals". Proc. of SPIE Vol. 8324 832417. (2012).

[2] Guy Cohen, Eran Amit, Dana Klein, Daniel Kandel, Vladimir B. Levinski, "Overlay Quality Metric". Proc. of SPIE Vol. 8324 832424-1. (2012).

[3] Eran Amita, Dana Kleina, Guy Cohena, Nuriel Amira, Michael Har-Zvia, Cindy Katob, Hiroyuki Kuritab "Overlay accuracy calibration". Proc. of SPIE Vol. 8681 86811G. (2013).

[4] Tetyana Shapoval, Bernd Schulz, Tal Itzkovich, Sean Durrane, Ronny Hauptb, Agostino, etc "Influence of the process-induced asymmetry on the accuracy of overlay measurements". Proc. of SPIE Vol. 9424 94240B. (2015).

[5] Md Zakir Ullah, Mohamed Fazly Mohamed Jazim, Stella Sim, Alan Lim, Biow Hiem, etc "Qmerit-Calibrated Overlay to Improve Overlay Accuracy and Device Performance". Proc. of SPIE Vol. 9424 942425. (2015).

[6] Huayong Hu, Weiming He, Gaorong Li, Nannan_Zhang, Liwan Yue, Lei_Ye, Jinhua_Pei, Qiang Wu, "K=0.266 IMMERSION LITHOGRAPHY PATTERNING AND ITS CHALLENGE FOR NAND FLASH". CSTIC 2015.

[7] Lei Ye, Huayong Hu,Weiming He, "Image-based overlay (IBO) target segment design on self-aligned patterning process". Proc. of SPIE Vol. 9778 9778-77. (2016).

OPTIMIZATION FOR THE MEASUREMENT PARAMETERS OF CD SEM FOR SEVERAL SPECIFIC SITUATIONS

Hanmo Gong[a], Song Bai[a], Qian Ren[a], Qiang Wu[a]

[a]Technology R&D, SMIC Advanced Technology R&D (Shanghai) Corporation
Pudong New Area, Shanghai, P. R. China 201203
+8621 38610000x15177, Hanmo_Gong@smics.com

ABSTRACT

Different measurement parameter settings of CD-measuring scanning electron microscope (CD-SEM) can lead to deviations for identical patterns, especially for photoresist patterns with shrinkage effect. This deviation could not be neglected in the 28 nm and sub-28 nm processes. In this paper, we have studied the following cases: 1) Shrinkage effect of negative tone developing process photoresist; 2) Measurement parameter setting optimization for negative tone developing photoresist 1D&2D patterns; 3) Verification on FEM wafers.

Keywords—CD SEM, measurement parameter, NTD process, signal contrast.

INTRODUCTION

Critical dimension (CD) is regarded as the most important parameter in the design of integrated circuit, which must be accurately measured. CD-measuring scanning electron microscope (CD-SEM) has been the primary measurement tool for this purpose because of its much higher spatial resolution compared to the traditional optical microscope. CD-SEM can be used to image the printed patterns on a wafer surface by collecting the secondary or back-scattered electrons. The detected signals are mainly produced by pattern edges, and the measured CD is defined by the distance of any two edges from a given pattern. However, due to the resolution limit of the electron optics, even for the sharpest edges, the measured signal profile can still have a finite broadening. This will result in some difference in measurement value if different parts of the broadened profile are used. Second, algorithms in slicing the signal profile can also influence the stability of measurement. Proper selection of algorithms can improve the measurement stability which is around 2~3 nm, though small for pre-28 nm generations, can make a big impact to the 28 nm and sub-28-nm processes. In the measurement of photoresist images, several effects can cause CD deviations, such as photoresist shrinkage effect [1-2], parameter settings [3], charging effect [4-5], etc. Due to the fact that the photoresists can shrink in size in response to electron irradiation, the measurement parameter should be carefully chosen, such as, beam acceleration voltage, beam current, and the number of frames, etc. during a measurment scan.

In this paper, we have studied different measurement parameter settings and, through the use of the so-called "Dynamic Repeatability" [3] method, we have also evaluated measurement repeatability of these parameter settings. Our study focuses on the negative tone developing (NTD) process, which has been widely used in 14 nm lithography process due to its notable advantages against positive tone developing (PTD) process, such as, several times higher in energy utilization efficiency and being able to print smaller spaces.

In addition, it has the "good" flatter linewidth though pitch behavior etc. We have compared the shrinkage effect between PTD and NTD processes and find that NTD process shows better stability of CD against electron beam (e-beam) irradiation.

EXPERIMENTAL METHOD

Measurement parameter setting

We have used the standard CD-SEM for the experiment. There are two commonly used methods to define the edges of the detected signals: "threshold" and "linear".

Defined in the threshold method, there is a percentage, which slices the signal peak in the selected area. The percentage parameter ranges from 0% to 100%, where 0% corresponds to the bottom of the signal curve, while 100% to the top of the signal curve, or the peak.

When the linear method is used, we need to define 3 parameters: edge number, base line starting point, and base line area. This algorithm selects the intersections of the slope line and the base line as the edge points of the signal. The slope line is at the point on each curve with the highest slope and the base line starting point is determined by a fixed offset in signal value to the highest slope point. The baseline area is calculated from the base line starting point down to some selected offset in signal value. In this paper, we use X_X_X to express the parameter setting of linear method. For example, 1_2_8 represents the setting of edge number 1, base line start point 2, and baseline area 8.

Evaluation method

In this study, we have used the "Dynamic Repeatability" [3] method, presented by K. Ueda et al. from Hitachi High-Technologies Corporation, to evaluate performance of different measurement parameter settings. According to the method, we have measured the same position of a photoresist pattern for multiple times with fewer frames than normal measurement, and calculated the average CD from each measurement to determine the amount of shrinkage. Through this method, we can detect the variation of measured CD for the identical pattern without the interference of shrinkage effect, which reflects the repeatability of the measurement method.

RESULTS AND DISCUSSION

Shrinkage Effect

As mentioned above, since photoresist shrinkage effect will affect the performance of measurement both on 3-sigma and mean values, the evaluation of photoresist's shrinkage is needed.

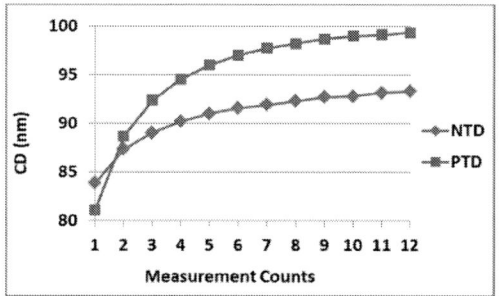

Figure 1: 1D Patterns with similar CD but produced by PTD (red)/NTD (blue) process show different shrinkage. The X-axis represents measurement count with each measuring the same position with 4 frames and multiple times.

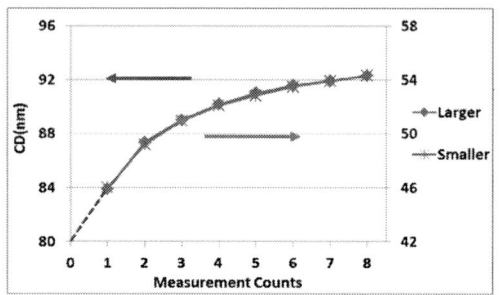

Figure 2: Patterns with larger CD/pitch (blue) and patterns with smaller CD/pitch (red) versus the measurement count. The X-axis represents measurement count with each measuring the same position with 4 frames and multiple times. The Y-axis on the left represents the CD of the larger pattern (blue), and the Y-axis on the right represents the CD of the smaller pattern (red).

We have chosen a few one-dimensional (1D) space patterns and measured the same position of each pattern for multiple times with 4 frames. Figure 1 described CD variation curves as a function of measurement counts for both PTD and NTD photoresists. The figure indicated that the NTD platform is more stable after e-beam scan. By taking the first 4-frame shrinkage into account, which is not shown in Figure 1，we could reason that after 12×4 frames，the PTD photoresist shrinks by more than 20 nm, while the NTD photoresist shrinks by only about 15 nm. If 16-frame is chosen for each measurement, as set in practical applications, the CD has been verified to be close to the average of first four values with 4-frame for each time. As with the shrinkage amount from the original profile with 16-frame，the CD of PTD shrinks by ~ 8 nm, while NTD shrinks only by ~ 4 nm. Figure 2 showed that the NTD photoresist endured the same shrinkage regardless of the pattern CD or pitch, just as the PTD photoresist behaved [1-2]. The dashed line in Figure 2 represents the presumed shrinkage effect of the first 4 frames.

Measurement Parameter Setting

Linear and threshold methods are designed for different circumstances with their characteristic advantages, as discussed below. Five signals in Figure 3 (c) comes from five identical patterns, but the two patterns on both sides have wafer substrates implanted. Therefore, these two groups with different substrates show different signal-to-noise ratio (SNR) in CD-SEM scanning. Both the accuracy and repeatability (3-sigma) of results measured by the threshold method is worse together with lower SNR, while the linear method shows

measurement robustness even when SNR varies.

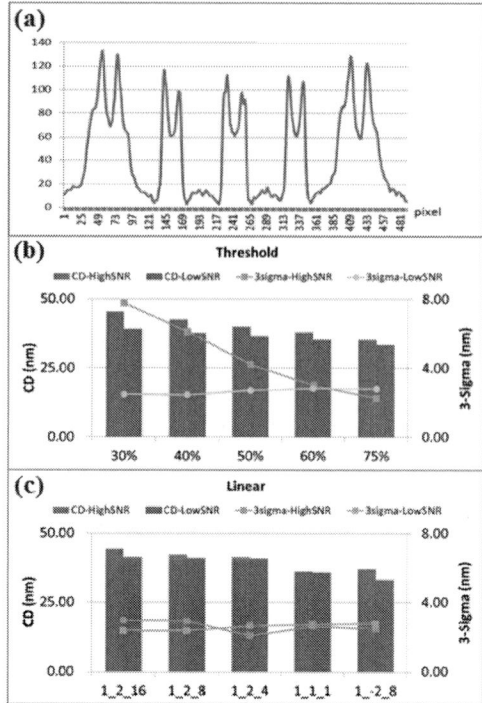

Figure 3: (a) The CD SEM curve, the measured CD trend for two group signals, high signal-noise-ratio (SNR) and low SNR, and their corresponding repeatability using the (b) threshold and (c) linear method, respectively.

When the measured signals become worse for limited sampling, higher smoothing level can be applied to obtain more stable results. However, smoothing level variation can affect the threshold method or linear method unequally.

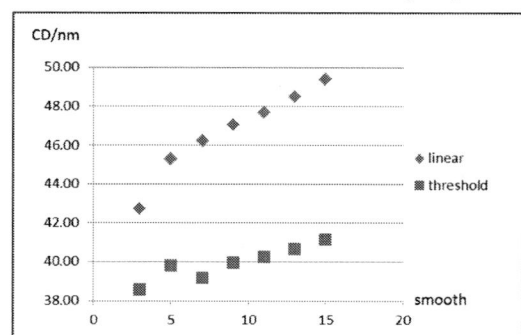

Figure 4: The CD trend with varied smoothing levels. Level 55% is chosen for the threshold, and 1_2_8 is chosen for the linear.

Figure 4 shows that linear and threshold methods respond differently to signal curve smoothing due to their different principles. Because of this, proper selection of measurement parameter setting can be essential to accurate measurement. Since NTD process demonstrated different performance, we have studied it using different settings and evaluated their repeatability in order to obtain appropriate settings.

1D space patterns with different CD/pitch are measured using different parameter settings, and the repeatability of the measured results by these parameter settings are evaluated. Figure 5 indicates that, for 1D pattern, the threshold method is more stable than the linear method, and 55% is the best threshold value in the threshold method. For identical parameter settings, increasing smoothing level is likely to

978-1-5090-6695-7/17 $31.00 © 2017 IEEE

improve the repeatability.

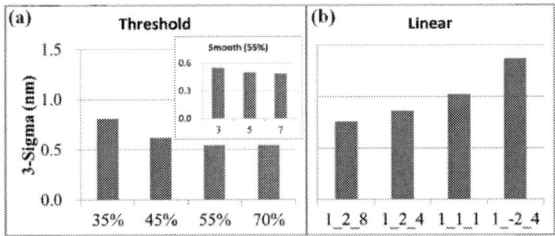

Figure 5: The repeatability (3-sigma) of varied measurement parameter settings for 1D patterns. (a) Various threshold settings: 35%, 45%, 55% and 70% with smoothing level 3, and the inset is threshold 55% with different smoothing levels: 3, 5 and 7. (b) Various linear settings with smoothing level 3.

Measurement Parameter Setting for 2D Pattern

Besides 1D patterns mentioned above, we have also studied typical 2D patterns, such as, the tip-to-tip line pattern.

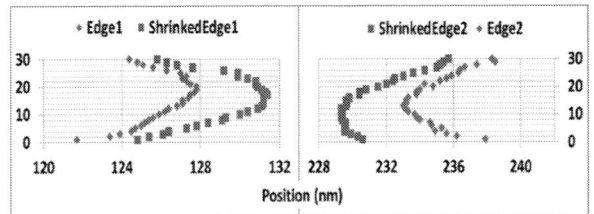

Figure 6: A measured tip-to-tip pattern contour in the X-Y plane. The blue dots represent the contour after the first measurement, while the red dots represent the same pattern after several times of e-beam shot, demonstrating a sizable shrinkage (~8 nm).

In Figure 6, where a tip-to-tip line pattern's measurement is described, with contours plotted in the X-Y plane.

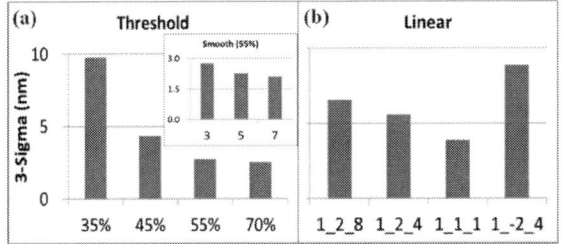

Figure 7: The repeatability (3-sigma) of varied measurement parameter settings for the tip-to-tip pattern shown in Figure 6 with (a) Various threshold settings: 35%, 45%, 55% and 70% with smoothing level 3, and the inset is threshold 55% with different smoothing levels: 3, 5 and 7, (b) Various Linear settings with smoothing level 7.

As shown in Figure 7, for 2D pattern, the threshold method performs much better than the linear one even when both methods use smoothing level 3. As shown in Figure 4, higher smoothing level enhances the performance for the linear method: when we changed the smoothing level from 3 to 7, the 3-sigma value decreased from ~ 50 nm to ~ 6 nm. However, the 3-sigma value of 6 nm is still worse than that of the threshold method.

Based on the results above, we have concluded that for the measured patterns the threshold method is more robust than linear method for the NTD process. For the threshold method, 55% and 70% settings are comparable and the best among all the parameter settings. For the linear method, 1_2_8 and 1_1_1 are better than other settings. In addition, increasing smoothing level is able to improve the repeatability, especially for the linear method.

Finally, we have prepared a focus-exposure matrix (FEM)

wafer and conducted the measurement using different measurement parameter settings, in order to verify the conclusion.

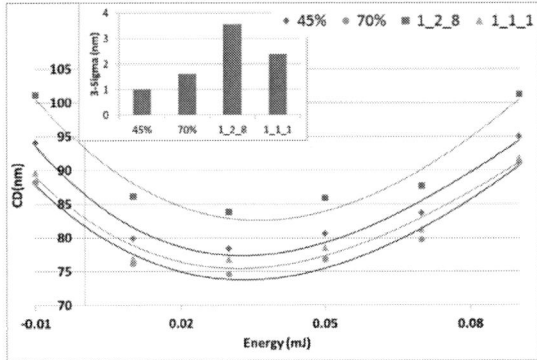

Figure 8: The measured CDs by threshold 45%, 70%, linear 1_2_8, and linear 1_1_1, all with smoothing level 7, where the curves represents the fitted Bossung curves of the measured FEM results. The inset shows the 3-sigma of different measurement parameter settings.

The FEM wafer has been measured once using different parameter settings with the standard 16-frame sampling scheme and the results are displayed in Figure 8. The threshold method and the 1_1_1 fit the Bossung curve well, while the 1_2_8 appears to have rather large deviation. The deviation of each setting from the average of all settings are also calculated, to yield an average 3-sigma value as the metrology capability to reflect the real trend of FEM. This conclusion coincides with the calculated 3-sigma shown in an inset of Figure 8, where the threshold 45% show the best repeatability while linear 1_2_8 the worst.

SUMMARY

In this work, we have studied the shrinkage effect of NTD process photoresist, and evaluated different CD-SEM parameter settings for NTD process measurement. The FEM verification result indicates that the parameter settings do influence the measurement performance, and suitable parameter settings can be found and are able to enhance the repeatability.

Next time, we plan to further study more patterns covering more topologies, and the differences between line and space patterns, etc.

REFERENCES

[1] C. Ke, A. Yen, J. Yee et al., *Proceeding of the SPIE*, 4690, 598-605 (2002).
[2] C. Ke, T. Gau, P. Chen et al., *Proceeding of the SPIE*, 4689, 997-1006 (2002).
[3] K. Ueda , S. Koshihara , T. Mizuno, *Proceeding of the SPIE*, vol.8324, 83242D-10 (2012).
[4] L. Chen, S. Lin, T. Gau et al., *Proceeding of the SPIE*, 5038, 166-176 (2003).
[5] C. Ke, H. Hung, A. Chang et al., *Proceeding of the SPIE*, 5375, 173-182 (2004).

SYSTEMATIC MAINTENANCE AND APPLICATIONS OF FAILURE MODES AND EFFECTS ANALYSIS (FMEA) IN SEMICONDUCTOR MANUFACTURING

Hongtao HT Qian, Ziqian Javaer Liu, Yuhong Betsy Xu

Corporate Quality & Reliability Center, Semiconductor Manufacturing International Corporation, Shanghai, 201203, China

(Javaer_Liu@smics.com)

ABSTRACT

Failure Modes and Effects Analysis (FMEA) is considered as a systematic, learning retention vehicle in semiconductor manufacturing. This paper tries to set up a quantized evaluation approach on FMEA effectiveness for the first time and also presents a systematic maintenance method of FMEA in Foundry (FAB) to instead of traditional paper works which realizes intelligent maintenance to increase the timeliness and accuracy of FMEA. And, cross system applications between e-FMEA system and other FAB manufacture and monitor systems are presented to show the effectiveness improve after systematization.

Keywords—FMEA; effectiveness index; systematic maintenance; cross system application; semiconductor manufacturing

INTRODUCTION

FMEA as an analysis tool was utilized as early as the 1960s by U.S. National Aeronautics and Space Administration (NASA) in programs such as the Apollo, Viking, Voyager, and Skylab explorations. [1-2] The process was also adopted early on by the Society for Automotive Engineers (SAE) in 1967. The use of FMEA spread rapidly to other industries during the 1970s and subsequent years, and is now utilized in a variety of industries including military, semiconductors, healthcare and food service. FMEA is useful in understanding the sources of failure in systems or products, qualifying the effects of failure and aiding in the development of mitigation strategies. FMEA is useful in understanding the sources of failure in systems or products, qualifying the effects of failure and aiding in the development of mitigation strategies. [3-4]

In semiconductor manufacture, a FAB always encounters process changes which mainly come from yield improvement, defect reduction or lesson learns and so on. But each change introduces risk into the factory. FMEA is the right way to do risk management or as the process weakness database. [5-6]However, generally, in semiconductor manufacturing, most FABs are still using traditional paper works to do FMEA maintenance which has low efficiency both on update and sharing. And also, there is no a scientific evaluation method on

FMEA effectiveness index (The users always focus on final RPN values but hard to judge if these values are reasonable or not on wafer manufacturing processes). On this paper the authors try to present a new evaluation methodology on FMEA effectiveness which has been applied on a mature FAB for years. And a system (e-FMEA) also is introduced to realize FMEA intelligent maintenance and systematic application.

This paper is organized as follows: Section II presents the detailed FMEA effectiveness index with actual practice. Section III shows the e-FMEA system with its systematic applications. Then in Section IV achieved benefits from a mature FAB are presented.

FMEA EFFECTIVENESS INDEX (FEI)

Actually, it is hard to generate a reasonable quantized evaluation methodology on FMEA effectiveness, especially on semiconductor manufacturing due to complex applications (one FMEA is not only for one process step but also the whole manufacturing with thousands of steps) and forward-looking (Sometimes a potential failure mode need huge or long time data to verify, which means it is hard to do effectiveness judgment before final result). As we know, semiconductor manufacturing is a very complex process with several kinds of product failures under different severity levels, such as defect high, monitor fail, function test fail and so on. So, these kinds of failure should be the most direct evaluation factors in FMEA effectiveness calculation (Table I). That means if a repeating excursion happen, the effectiveness of related FMEA should be punished basing on detection stage and wafer impact.

TABLE I: S_{basic} value matrix table on different detection stages

Detection Stage	S_{basic}	
	Small Value Scrap	Huge Value Scrap
Defect Scan	3	6
Monitor Alarm	5	8
Electrical Function Fail	7	11
On Board Funtion Fail	9	15

Another factor need be considered during the evaluation is the accuracy of "Occurrence" or "Detection" due to FABs are always requested to do continue improve plan (CIP). Within CIP, the process healthy should be improved which means the detection capability on manufacturing abnormality

978-1-5090-6695-7/17 $31.00 © 2017 IEEE

is improved and abnormal case ratio is keeping decrease. So, the preventive actions on FMEA should be timely updated or the outdated actions will impact the scoring of "Occurrence" or "Detection" then the final FMEA effectiveness scores (Table II).

TABLE II: C$_{FMEA}$ value matrix table on different comparison results

Comparison Result	C$_{FMEA}$
Actual performance matched FMEA	0
Actual occurrence peformance > FMEA	0.3
Actual defection peformance < FMEA	0.5
A new failure mode not in orgrinal FMEA	1.0

By combing the factors (S$_{basic}$ and C$_{FMEA}$) mentioned above, the formula of FMEA Effectiveness Index (FEI) is finally optimized (after several demos) as below:

$$\text{FMEA Effectiveness Index (FEI)} = 100 - \frac{n}{N} * \sum_{i=1}^{n} (S_{basic} * (1 + C_{FMEA}))$$

In this formula, n is the actual production line abnormality case count (monthly base, normally) and N is the average case count on last year at the same time.

With FEI building-up, it becomes easy for FABs to do FMEA weakness finding and improve by setting up a FEI scoring trend chart, especially on advanced technology. As Fig.1 shows, with the aid of FEI, the gap of FMEA (generally, FAB sets up FMEA by technology) between a mature technology and advanced technology is exposed obviously which means some benchmark actions can be internally shared. Additionally, advanced technology also can generate its own improve actions by FEI weakness analysis to improve the risk management after FMEA refinement.

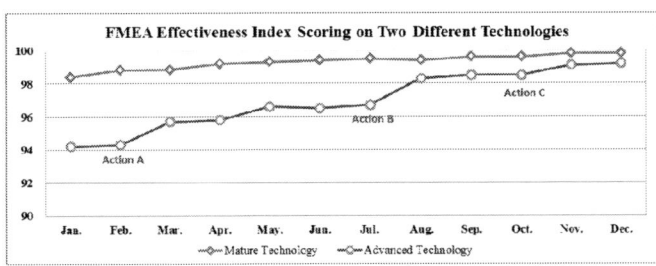

Fig.1. FEI scoring trend chart on two different technologies

E-FMEA SYSTEM AND APPLICATIONS

FMEA should be a dynamically updated document. In semiconductor manufacturing, with the Critical Dimension (CD) narrow down, process window becomes smaller which means lots of CIPs are ongoing and process platform is always under fine tune. So, traditional paper works on FMEA maintenance (Generally, FMEA is maintained as Microsoft Word files and restored in Document Management System. If modification needed, the owner should download this file from

DMS and do paper works to update it then upload to system to publish) cannot meet high efficiency requirements from FAB. That why the authors want to present a systemic concept on FMEA maintenance on this paper. Table III is the comparison summary between traditional FMEA maintenance method and e-FMEA system basing on a mature FAB's real practices. From this table, it is obviously to find the two major benefits of systematically which are high efficiency and intelligence. Due to FMEA standard format has been defined in system (Fig.2), so user can directly do any modification on system through user interface and make it effective real time.

TABLE III: Comparison between traditional method and e-FMEA system

Operation Type	Maintenance Method	
	Traditional	e-FMEA System
Format	Word Files	Directly on System
Modification	Paper Works	Directly on System
Update Cycle Time	Ave. 7 working days	Real Time
Cross Site Sharing	Manually	Automatically
Link to Other Sytems	NA	Yes
Reporting Function	NA	Yes

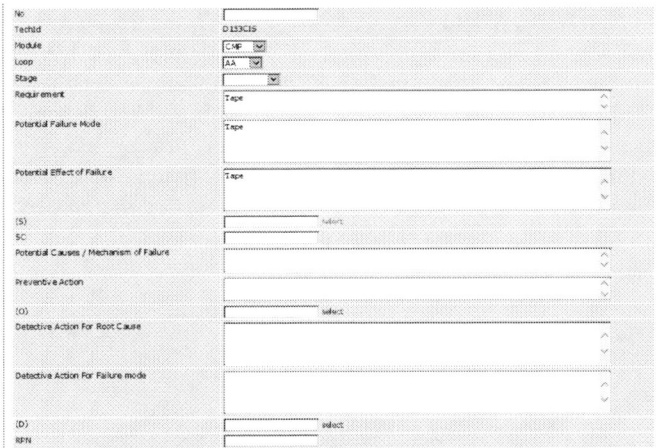

Fig.2. User interface sample from e-FMEA system

The value of FMEA is its implementation on risk management including internal FAB and cross FABs. To make sure exact implementation, a "useful FMEA" must realize "Cross Site Sharing" and "Link to Other Systems" (mentioned on TABLE III). In e-FMEA system, users can define a corresponding matrix table with "FMEA Number", "Technology ID", "Technology Owner" and "Production Line". After that, system can real time automatically trigger a inform message to related owner and FAB when one FMEA has been modified (refer to Fig.3). The receiving site need make a judgment if need follow up or feedback any concern/suggestion to the mother site through e-FMEA system. This modification only can be effective after both sites make alignment to make sure exact sharing.

978-1-5090-6695-7/17 $31.00 © 2017 IEEE   405

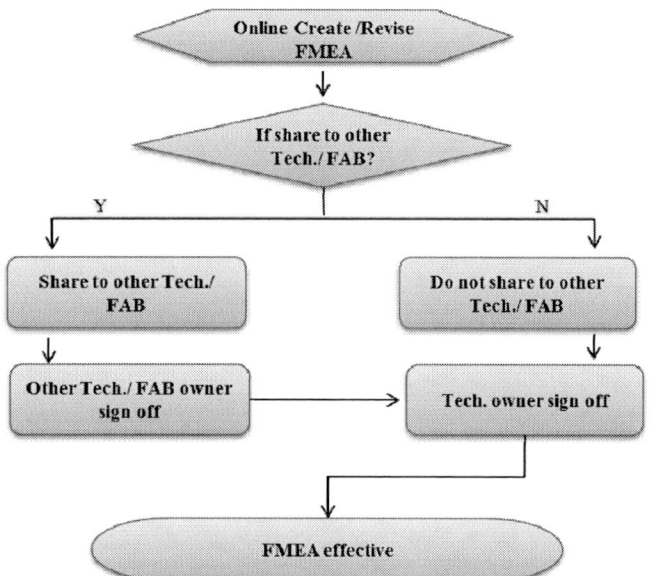

Fig.3. Cross site sharing procedure on e-FMEA system

After getting updated information, the next problem is how to make sure new preventive actions can be implemented on wafer manufacture processes. Only systematization on FMEA and build up a "System Net" can completely solve it. Different from traditional FMEA maintenance method, e-FMEA can realize system talks with other systems (manufacture control, routine monitor or process change control systems). New preventive actions or new detection actions can be automatically transmitted to related systems by system links and the receiving systems can do such as monitor frequency tighten or control limit optimization.

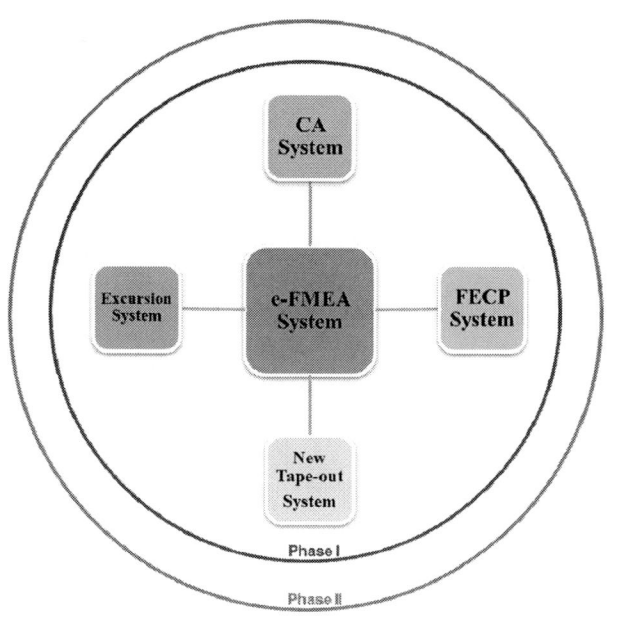

Fig.4. System net by linking to e-FMEA system

As Fig.4 shows, in a mature FAB, e-FMEA has been successfully linked to 4 manufacturing systems:

1. FECP system: FAB Engineering Change Procedure system is used for risk control on all process changes, by linking with FMEA, FECP system can get the latest PRN value from each process stage, the higher RPN value stage means the higher risk on changes which requests more qualification data.

2. NTO System: New Tape-Out System is used for risk gating on new products before being imported to formal manufacture. e-FMEA system will pass process weaknesses by RPN ranking above alarm limit (alarm limit can be adjusted) to NTO system to remind user to collect more verification data on these weaknesses to confirm process robustness before mass production.

3. CA System: Construction Analysis System is used for wafer physical cross-section analysis in order to monitor if baseline performance has shifted or not. By linking to e-FMEA system, CA system can real time queries the risky structure and requests user to get more data to double confirm.

4. Excursion System: All FAB excursion cases with root cause finding, risk assessment on impacted products, containment actions and preventive actions are recorded in this system. e-FMEA system can be treated as a database for lesson learn delivering when user is handling a new case on excursion system.

Specially, FMEA not only used on these four systems but also can be exactly used on every manufacturing system by systematization.

CONCLUSION

With the creation of FMEA effectiveness index (FEI), semiconductor manufacturing can realize the real evaluation on FMEA by FEI scoring monitor and comparison. It makes easy for FAB to continuously do FMEA optimization. And, after systematization instead of traditional paper works, the efficiency of FMEA maintenance has been obviously improved and system net with e-FMEA can make sure FMEA be exactly implemented.

The practice on a world leading Foundry shows all production line excursion case ratios (no matter which technology generation) have been improved due to the enhancement on risk management by FEI usage and e-FMEA system implementation (refer to Fig.5).

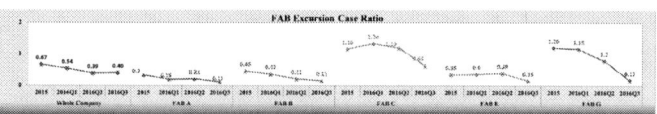

Fig.5. Cross FAB excursion case ratio

REFERENCES

[1] P. Roberts, Application of Hazard and Risk Analysis to Gas Making and Gas Storage Facilities, International Conference on Safety and Loss Prevention in Chemical and Oil Processing Industries, 1989, pp. 23-27.

[2] G. David, Risk Analysis of Single and Dual String Gas-lift Completions, Journal of Petroleum Technology, 1990, vol. 42, PP. 1 1.

[3] C. Carlson, Effective FMEAs: achieving safe, reliable, and economical products and processes using failure mode and effects analysis, Hoboken, 2012.

[4] K. Tay and C. P. Lim, Fuzzy FMEA with A Guided Rules Reduction System For Prioritization Of Failures, International Journal of Quality & Reliability Management, pp. 2006, 1047-1066.

[5] N. Shebl, B. Franklin and N. Barber, Is failure mode and effect analysis reliable, Journal of Patient Safety, 2009, vol. 5, pp. 86-94.

[6] N. Stanton, P. Salmon, L. Rafferty, G. Walker and C. Baber, Human Factors Methods: A Practical Guide for Engineering and Design, Ashgate Publishing Company, 2013.

INVESTIGATION OF MULTIPLE SOFT BREAKDOWN DURING TIME-DEPENDENT DIELECTRIC BREAKDOWN

Qiwei Wu, Binfeng Yin, Ke Zhou, Jiong Wang, Jinde. Gao
Huali Microelectronics Corporation, Shanghai 201203, China
Email: wuqiwei@hlmc.cn

ABSTRACT

In this paper, we studied the multiple soft breakdown phenomena of gate oxide during TDDB（Time Dependent Dielectric Breakdown） . after the soft breakdown, the parameters of the device had been greatly degraded (as a result, the device lost its original functions), although the multiple soft breakdown did not lead the oxide catastrophical failure. Further analysis also verified that the gate oxide had been damaged after the first soft breakdown. We found that there is a very good match between the times of soft breakdown and the failure points fixed using EMMI-OBIRCH system. Moreover, such correspondence was verified using the in -situ failure analysis method. Finally, we proposed an explanation concerning the oxide breakdown based on the phenomenon.

INTRODUCTION

With the development of the technology, It is observed that the soft breakdown phenomenon occurs in thin oxide[1]. And the failure principle of the soft breakdown gate oxide becomes more and more important with the decreasing scale oxide [2].Instead of the single breakdown, we find the multiple soft breakdown phenomena with the TDDB test. This largely affects the failure judgment of the DUT(Device Under Test). To determine the actual TTF (time to failure), it is necessary to study of TDDB's multiple soft breakdown. To do so, one can ensure the data availability and credibility.

This paper deals with the following issues, including understanding the physical mechanism of the multiple soft breakdown, the impact acting on the device, as well as the resulting electrical parameters and TTF distribution. By doing so, we provide a good reference for the future data analysis.

EXPERIMENT

In this work, we used two different structures, i.e., transistor array (with P substrate) and bulk capacitor(with N substrate), respectively. The thickness of the gate oxide are 73Å and 26Å.

It is necessary to notice and identify the stages for the electrical parameters measurement: i) before the stress (fresh); ii) after the first breakdown of multi-BD (multiple breakdown)(i.e., Multi-1st BD); iii) after the last breakdown of multi-BD (namely, the leakage achieve the failure criteria); iv) After the only one-time breakdown (meanwhile the sample become failure) (Single-BD). In particular, one has to position failure point with

EMMI-OBIRCH, and then observe the DUTs using scanning electron microscopy (SEM).

As for the multiple soft breakdown in bulk structure, we did the in-situ EMMI-OBIRCH experiment. In other words, we perform the TDDB test using the EMMI-OBIRCH machine. It is noticeable that we can stop the TDDB test and position the failure points with EMMI-OBIRCH immediately after each soft breakdown.

RESULT AND DISCUSS

In Fig.1 a, the Gate leakages come directly to the failure spec when the sample failed. This is called as single-BD failure. However, one may also observe multiple soft breakdown in some samples. After each breakdown, the leakages are larger than the last time, moreover the leakage keep stable until the next breakdown. As contrast, we define it as multi-BD failure. Two different TTF distribution curves are shown in Fg.1 b. Both the first breakdown and last breakdown are measured separately as the failure time. It is obvious that the gap between these two lines is large. Therefore, one may overestimate the lifetime of the device if the last breakdown is measured. Motivated by this, it is very important to investigate that whether the first few soft breakdown will lead to the DUTs failure.

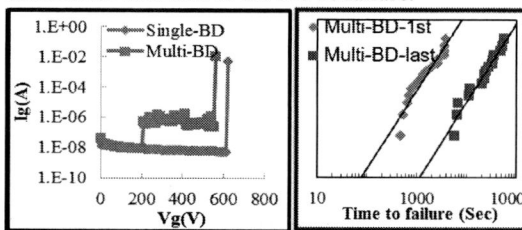

Fig 1: a,I-T curve of single-BD and multi-BD; b, TTF distribution of multi-1st BD and multi-last BD

The electrical characteristics

Since the leakage do not increase too much in the first few breakdown of multi-BD samples, it is hard to figure out the failure criteria. Hence, we measure the electrical characters in four different types of samples.

At Vg = 0V, the device is under off state and the channel is closed. The leakages of the source and drain are very weak. As shown in Fig. 2, This implies that the device is well performed initially. However, the gate leakage of the other three types of the samples increase largely to 1E-3A ~ 1E-2A after any type breakdown. This means the gate lost the control of the channels. In addition, we can see from the plots that the leakages of the source

978-1-5090-6695-7/17 $31.00 © 2017 IEEE

and drain are very close, from which we can argue that the source and drain have already conducted. Further analyzing the gate leakages, we find that the gate leakages of the multi-BD samples (including the Multi-1st BD and the Multi last BD) increase 2 orders after the stress, while the single-BD sample increase 6 orders. The result is in good agreement with the corresponding current failure time of the TDDB (the gate leakage increase dramatically for the single-BD samples, but leakage increasing are limited for the multi-BD oneness).

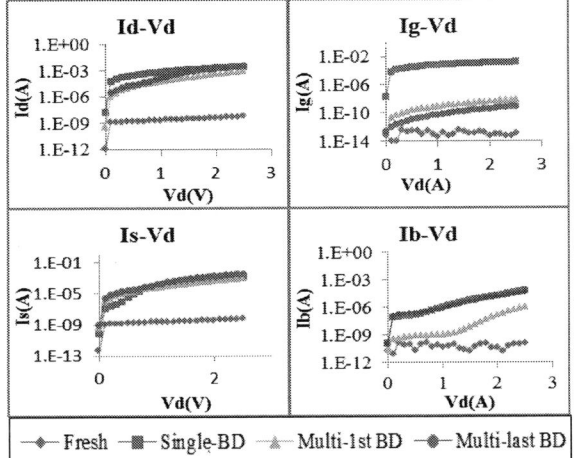

Fig2: Vg=0V, the leakages of all the terminals with Vd sweep.

At Vg=2.5V, the device is turned on, and the channel is inverse. From Fig. 3, the Id-Vd curves of the three kinds of tested samples behave similar. All the leakages are smaller than the fresh DUT. Moreover, the current of the multi-BD is larger than single-BD. The reasons are as follows. For single-BD, as the poly and the bulk are conducted, this induced part of the current between the

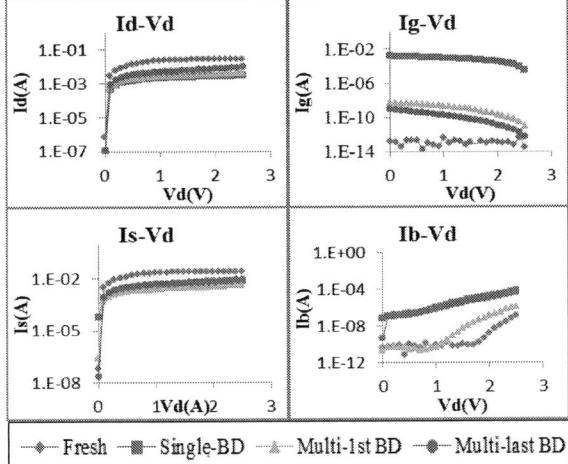

Fig 3: Vg=2.5V, leakage of all terminals with Vd sweep.
source and the drain to flow to the gate/bulk. As for

multi-BD, the drain current becomes smaller than the fresh but stronger than the current at Vg=0V. Moreover the leakages of the gate and the bulk are smaller than single-BD as well. The possible reason is that the gate oxide of multi-BD is a non-catastrophic breakdown.

Based on the above discussions, we conclude that Multi-BD results in the increasing current in off state, although it does not lead to the poly and bulk in short. As a result, the saturation current decrease and thus the device lost its functions finally.

Failure analysis

Fig 4 show the result of failure analysis on Multi-1st BD DUT, It is seen that GOX is damaged largely from the cross section image, and the source/drain salicide diffused to the channel. In principle, the leakage of the gate should be very large. But that is not the case in our experimental result.

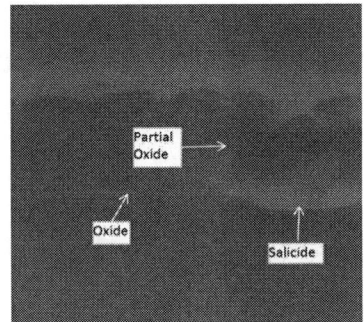

Fig 4: cross section of DUT after multi-1st BD.

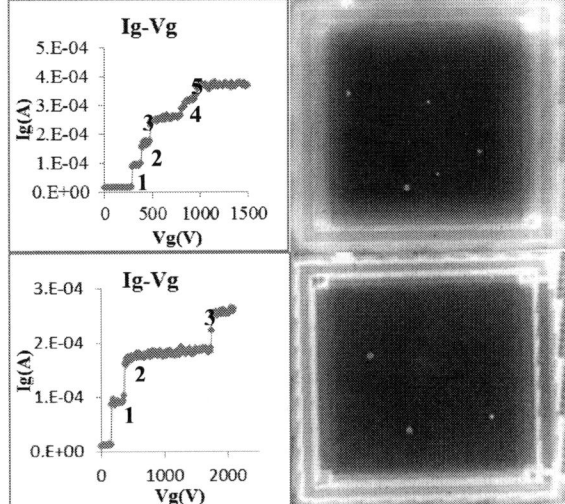

Fig 5: Relationships between soft breakdown and hot spots.

In the analysis of the bulk structure. To our surprise, the number of the failure points is in good agreement with the times of soft breakdown, In more details, we find 5 and 3 points respectively when 5 and 3 times of soft breakdown(corresponding to the left plot in Fig. 5). It is proved that each soft breakdown do correspond to a failure point of oxide.

To further verify the relation between the soft breakdown and the hot spot, as well as the appearing sequence of the hot spot, we design the in-situ EMMI-OBIRCH analysis experiment. In particular, The plots in Fig. 6 indicate that a good correspondence between the soft breakdown and the hotspot exists. The appearance of the subsequent breakdown points implies that the previous breakdown neither lead to catastrophic gate oxide breakdown, nor make the gate and the bulk in short. The increased leakage is so high that it is allowed to be detected by EMMI-ORIRCH.

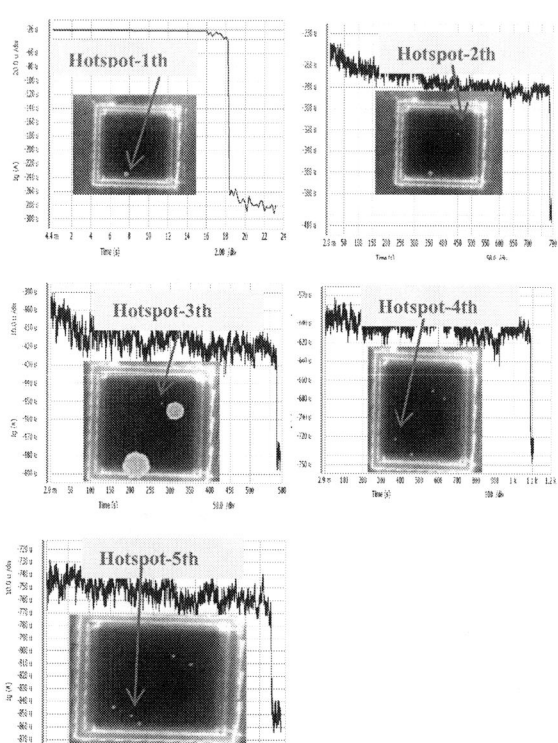

Fig 6: In-situ failure analysis result

Model assumption

Based on the experimental data, we propose a hypothetical model: After the breakdown, the charges continuously accumulated in MOS capacitor release instantaneously, generating huge leakage. The resulting heating effects lead the gate oxide meltdown partially, and then re-oxidized after the charge dissipation. This may repair the conducting path, consequently the gate leakage do not touch the compliance. But, the quality of re-oxide becomes worse comparing with the oxide manufactured using the conventional technology. Besides, it contains mass defects, exhibiting random telegraph noise. For multi-BD, a couple of weak spots of gate oxide appear,

along with the multiple-time breakdown and re-oxidation.

As shown in Fig. 7, percolation theory indicates many defects generate under an applied electric field, forming several potential conducting paths. When the number of the defects achieve to a certain value, the path becomes conductive and current can easily pass through the path (Fig. 7 a b c). Apart from this, the oxide which broken down is re-oxidation again by the self-heat (Fig. 7 d), leading to the fact that the gate oxide do not breakdown fully. When the performance of the re-oxide layer is better than other potential conducting path, the followed breakdowns have a higher probability occurring in the potential breakdown channels (fig 7 d e).

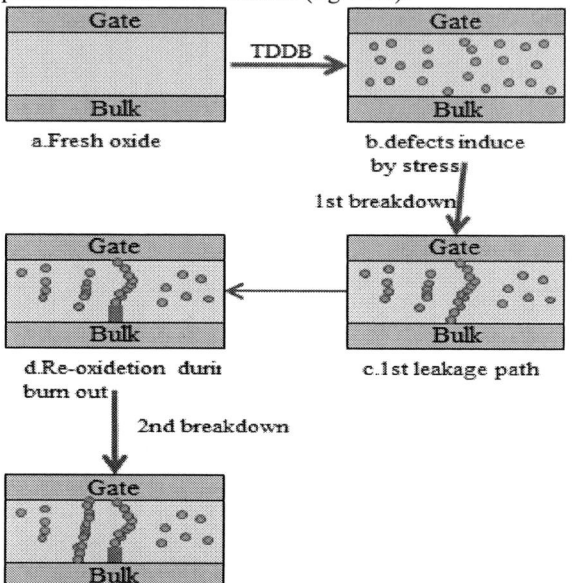

Fig 7: Sketches of model assumption.

CONCLUSION

The device electrical parameters have been modified largely after multi-1st BD. Even if it does not lead the poly and bulk in short (like single-BD) Besides, multi-BD has lead the device lost its former functions. This paper also show the number of soft breakdown and hot spots fixed by EMMI-OBIRCH are in good match. During the measurement process, a couple of hot spots continue to appear after each breakdown. This paper argue that the leakage path is repaired successfully by the resulting self-heating effects of each breakdown.

REFERENCES

[1] S.H.Lee, B.J. Cho, J.C. Kim, and S.H. Choi, "Quasi-breakdown of ultrathin gate oxide under high field stress," in IEDM Tech Dig, p. 605, 1994

[2] J. H. Stathis and D. J. DiMaria, "Reliability projections for ultra-thin oxides at low voltage," in IEDM Tech. Dig., 1998, p. 167.

TECHNOLOGY ADVANCEMENT OF LAMINATE SUBSTRATES FOR MOBILE, IOT, AND AUTOMOTIVE APPLICATIONS

Ken Lee*, Min Sung Kim, Peter Shim, Ica Han, Jack Lee, Jeffery Chun, and Samuel Cha
Simmtech Co., Ltd., Cheongju, Republic of Korea
Corresponding Author's Email: ken.lee@simmtech.co.kr

ABSTRACT

The increasing level of integration in electronic devices requires high density package substrates with good electrical and thermal performance, and high reliability. Organic laminate substrates have been serving these requirements with their continuous improvements in terms of the material characteristics and fabrication process to realize multi-layer fine pattern interconnects and small form factor.

We present the advanced coreless laminate substrates in this paper including 3-layer thin substrate built by ETS (Embedded Trace Substrate) technology, 3-layer SUTC (Simmtech Ultra-Thin substrate with Carrier) for fan-out chip last package, and 3-layer coreless substrate with HSR (High modulus Solder Resist) for reduced warpage. We also present new coreless substrates up to 10 layers and substrate based on EMC. These new laminate substrates are used in many different applications such as application processors, memory, CMOS image sensors, touch screen controllers, MEMS, and RF SIP(System in Package) for over 70GHz applications. One common challenge for all these substrates is to minimize the warpage. The analysis and simulation techniques for the warpage control are presented.

Key words: substrate, package, mobile, IoT, automotive

INTRODUCTION

The fast expansion of Internet of Things (IoT) and increased use of electronics in conventional and electric automobiles to realize autonomous driving are the main factors to push the high integration level of electronic devices together with the mobile devices which has been the main driver for the integration for last couple of decades as shown in Fig.1. In order to support the high level of integration, the package substrates for the semiconductor components has been evolving rapidly in terms of performance and form factor. First, as PoP (Package On Package) gets thinner to support mobile AP

requirements, thin substrate less than 100µm is needed to reduce the overall package thickness. High modulus material is necessary to overcome the warpage risk of thin substrates. In terms of fine patterning technology, ETS is replacing the conventional SAP (Semi Additive Process) with its lower cost[1]. For IoT applications, devices usually have multiple functionalities such as sensing/actuation, computing, and communication. In order to implement these multiple functions in a small form factor, high-layer-count SIP technology has been developed. For upcoming electric and self-driving cars, more electronics will be used for many different sensors including radar and lidar, and high bandwidth GPU with AI capability for maneuver and navigation. High thermal conductivity and reliability is essential for the substrates used for automotive components. And the sensors require good electrical characteristics such as low dielectric constant and low loss. High density substrate for HBM (high bandwidth memory module) and GPU also requires good electrical characteristics for high bandwidth computation.

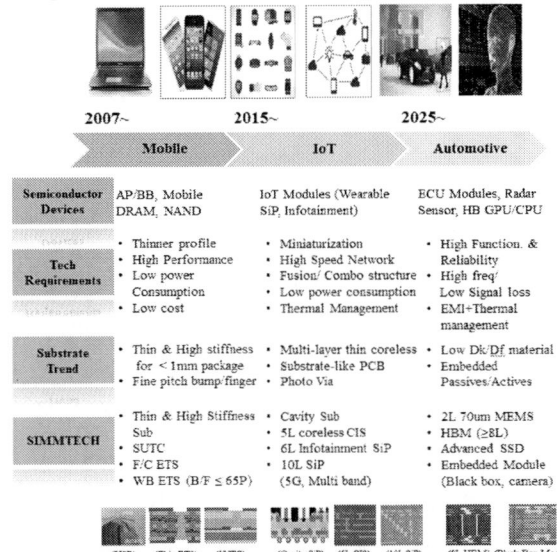

Fig. 1 Technology trend of smart devices

SUBSTRATE TECHNOLOGY ADVANCEMENT

FINE PITCH & THIN SUBSTRATE

We continue to improve the patterning capability of ETS along with the chip technology advancement as shown in Fig. 2. The advantages of ETS over tenting or PSAP are no width reduction of Cu pattern, high reliability by embedding traces into prepreg, and lower process cost over PSAP. Simmtech is in mass production for 13/13μm line/space, mass production ready for 10/10 and can provide sample for 5/5. Currently 4/4 patterning process is under development.

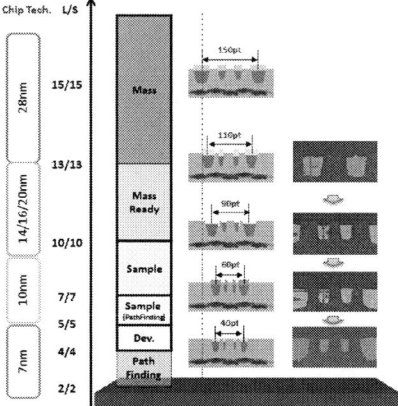

Fig. 2 Fine Patterning status of ETS

The fine pitch ETS is used for the mobile application processors which needs fine pitch patterning and minimal thickness.

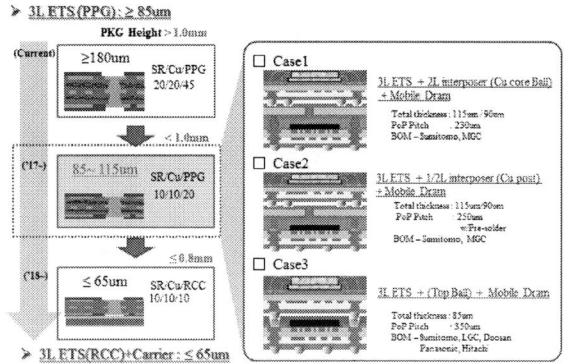

Fig. 3 Thin substrate trends for reducing POP height

Fig.3 shows the substrate thickness trend of 3L ETS used for mobile AP's and the overall package stack in POP

implementations. In order to achieve the total POP height less than 1mm, 3-layer ETS substrate should be 85μm thick[2]. The detailed stack information of this substrate is shown in Table 1.

Table 1. Layer information of 3L 85 μm ETS

Structure		
Total Thickness	85±10μm	
	Dielectric	18±5μm
	Cu	8±3μm
	SR	10±4μm (on Cu)
Application	Application Processor	

This substrate is used as the PoP bottom in three different configurations as shown in Fig. 3. For Case 1 and 2, 2-layer interposer connected to the AP through Cu core ball or Cu post is used to connect the AP and DRAM while for Case 3, top ball directly connects AP and DRAM.

For 3L ETS substrate thinner than 85μm, we employ dielectric material without glass fabric explained in the next section. The target thickness of the substrate is 65μm or less. We will need a carrier for this substrate for the fabrication and assembly. Easy and reliable removal of the carrier during the assembly process is very critical. Several alternative methods are under development as shown in Fig.4.

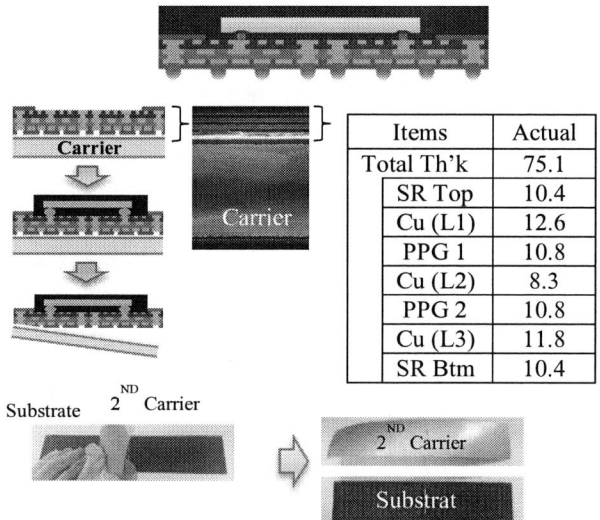

Items	Actual
Total Th'k	75.1
SR Top	10.4
Cu (L1)	12.6
PPG 1	10.8
Cu (L2)	8.3
PPG 2	10.8
Cu (L3)	11.8
SR Btm	10.4

Fig. 4 3L ≤75μm ETS SUTC

978-1-5090-6695-7/17 $31.00 © 2017 IEEE   412

HIGH STIFFNESS THIN SUBSTRATE

HSR (High stiffness Solder Resist) is necessary for thin substrates that need to be fabricated and delivered without carrier. We developed HSR technology with outstanding dimensional stability and low warpage based on its modulus of 15GPa at 25°C and 3GPa at 260°C.

SIMMTECH's proprietary technologies include HSR hot press process and one time HSR etch to open SR for external interconnection. Another advantage of the HSR is strong adhesion with EMC as it is based on the same material as EMC. Due to its mechanical strength, 70μm substrate built with HSR does not need carrier frame in package assembly. Fig.5 shows the application area where HSR provides advantage over the conventional solder resist in terms of ease of assembly as well as mechanical stability of the component. The Table 2 shows the detailed information of 3L 70μm substrate using HSR. The same material for HSR is also used as the prepreg replacement for SUTC as discussed in the previous section.

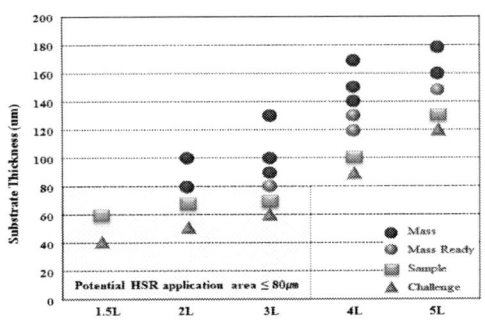

Fig.5 Potential HSR application area

Table 2. Layer information of 3L 70μm HSR

Structure	
Total Thickness	70±7μm
Dielectric	18±5μm
Cu	8±1.6μm
HSR	9.5±5μm (on Cu)
Application	Memory

MULTI LAYER THIN CORELESS

Although coreless substrate is prone to the warpage and damage of panel after panel detachment from the 1st carrier, we developed thin coreless substrates up to 10 layers.

The details of 5L coreless substrate for CMOS image sensor is in Table.3.

Table 3. Layer information of 5L 130μm Coreless

Structure	
Total Thickness	130±15μm
Dielectric	18±5μm
Cu	8±3μm
SR	10±4μm (on Cu)
Application	Camera Module

In order to build 10-layer coreless for SiP(Table 4), we started 4-layer coreless and detach from 1st carrier, then lay-up process repeats three times to get a 10-layer coreless structure. The alignment between adjacent layers is less than +/-25μm. This substrate has signal vias and thermal vias with different sizes. Dielectric thickness tolerance is as tight as +/-6μm.

While this 10L coreless substrate is built based on the conventional materials, we expect that electrical performance of dielectric material needs to improve in terms of Dk and Df. There are materials with lower Dk and Df in the market, but they have poor manufacturability as they have not been optimized for the substrate process[3]. This high-layer-count coreless combined with the new dielectric materials can be the basis for building next generation substrates for automotive applications such as 77GHz radar[4].

Table 4. Layer information of 10L 405μm SiP

Structure	
Total Thickness	405±40μm
Dielectric	30±7μm
Cu	12±6μm
SR	15±7μm (on Cu)
Via/Land	Signal : 60/135μm Thermal : 90/165μm
L to L Alignment	Adj. ±25, All ±75
Application	RFSiP

HIGH THERMAL & RELIABILITY MODLED CORELESS

We developed SIMS (Stud Interconnected Molded Substrate), where EMC material is used as the insulation layer instead of PPG in order to improve thermal conductivity and package warpage. Besides the better thermal conductivity through EMC as shown in Table 5,

SIMS supports Cu filled vias with any shape to enhance the thermal conductivity of the substrate as shown in Fig. 6. Moreover, SIMS has good adhesion with EMC of package because SIMS consists of same insulation material with EMC used in assembly process resulting in high reliability compared with the substrates using conventional materials.

Table 5. Thermal Properties of PPG & EMC

Items	Unit	PPG	EMC	
			A	B
Thermal Conductivity	W/m K	0.35	0.63	3.0
Tg (DMA)	℃	>200	167	200
CTE (α1)	ppm/℃	10	7	15
Dk / Df	@1GHz	3.4 / 0.005	3.2 / 0.006	7.2 / 0.006

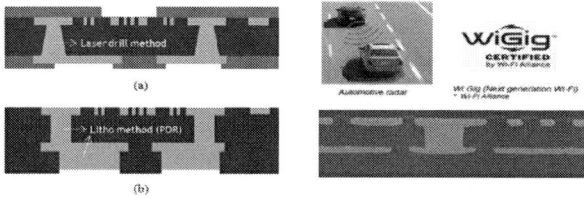

Fig. 6 Section profile of substrate (a) ETS (b) SIMS

WARPAGE REDUCTION

Warpage is critical issue in thin coreless substrates and packages because of the unbalance due to the sequential lay-up. In order to understand and solve the warpage issue, we should measure the warpage behaviors, build mechanism model to simulate the behavior, and verify the model with measured data[5].

Shadow moiré machine is usually used to capture warpage with respect to the temperature. However, this machine cannot detect the deformation, the movement of each point on the specimen.

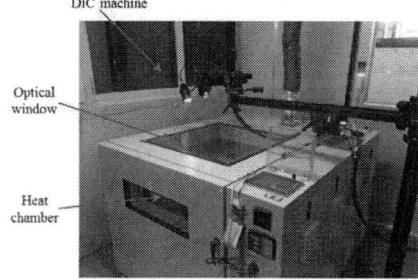

Fig. 7 DIC measuring machine

We use the latest DIC(Digital Image Correlation) machine to measure the deformation of specimen; substrate and/or package as a strip or single unit. DIC is used with a heat chamber with optical window to measure the thermal behavior of substrate and package in horizontal view (Fig.7). After the measurement of warpage, we build a simulation model based on the warpage mechanism with material properties and main factors of manufacturing processes affecting warpage. We then compare simulation results with measured warpage data to verify the warpage mechanism. With the verified model, we can exercise simulation study to minimize the warpage without excessive experimental iterations.

CONCLUSION

The laminate substrate technology continues to advance in terms of the materials and process technology in order to support the higher level of integration in electronic devices driven by the mobile, IoT, and automotive applications. We presented the challenges and new solutions in coreless substrate technologies including ETS, SUTC, HSR, and SIMS. The fundamental understanding of warpage is essential to build these thin substrates.

REFERENCES

[1] Eason Chen, Albert Lan, Jack You and Mark Liao, "Structure reliability and characterization for FC package w/Embedded Trace coreless Substrate", 2014 IEEE 16th Electronics Packaging Technology Conference (EPTC), 10.1109/EPTC.2014.7028314

[2] Ming-Che Hsieh, "Advanced flip chip package on package technology for mobile applications," 2016 17th International Conference on Electronic Packaging Technology (ICEPT), 10.1109/ICEPT.2016.7583181

[3] Takashi Tasaki, Atsushi Shiotani, Takashi, Yamaguchi, Keisuke Sugimoto, "The low Dk/Df polyimide adhesives for low transmission loss substrate", 2016 IEEE 37th International Electronics Manufacturing Technology (IEMT), 10.1109/IEMT.2016.7761910

[4] G. Haubnera, W. Hartnera, S. Pahlkea, M. Niessnerb, "77 GHz automotive RADAR in eWLB package: From consumer to automotive packaging," Microelectronics Reliability, Volume 64, September 2016, Pages 699–704

[5] Cheolgyu Kim, Tae-Ik Lee, Min Sung Kim, and Taek-Soo Kim, "Warpage Analysis of Electroplated Cu Films on Fiber-Reinforced Polymer Packaging Substrates," Polymers 2015, 7, Pages 985-1004

HIGH-BANDWIDTH IC INTERCONNECTS WITH SILICON INTERPOSERS AND BRIDGES FOR 3D MULTI-CHIP INTEGRATION AND PACKAGING

Boping Wu

Intel Research, Hillsboro, OR, USA

Email: bennywu@ieee.org

ABSTRACT

Silicon interposer and bridge is a multi-chip 3D technology that enables high density die-to-die interconnect on a package substrate. It opens a new era for heterogeneous on-package system integration. This paper presents an overview of this packaging architecture and its capabilities from concept to results. The overall components are introduced and discussed including constituent building blocks, embedded elements and structures, die-to-package connections. The high bandwidth signaling performance is analyzed and quantified by using high frequency electromagnetic modeling and full-wave simulation approaches. The inherent cost benefit and advantages, such as scaling and extensibility of this technology, are highlighted among other competing technologies. The assembly process is described in the end at a high level.

Keywords—signal integrity; link; interposer; multi-chip; TSV; silicon; 3D interconnects; IC package; heterogeneous integration; parasitic; IO; on-die; communication channel; embedded

INTRODUCTION

The need for high bandwidth among the inter-chip communications has led to increased focus on multi-chip packaging and on-package IO link [1]. The IO sub-system that delivers high bandwidth interconnect becomes a critical component of the overall computing system [2]. Currently, the highest CPU-Memory bandwidths are pushed by HBM [3], which has a relatively wide bus running at a relatively slow speed compared to GDDR5, which has a narrow bus width running at high speed rate. The other key metric of IO link performance is its power consumption, which also depends on the transceiver circuits and the channel (see Table I for detailed comparison).

Interestingly the memory bandwidth doubles roughly every three years at nearly constant signaling power [4]. The number of data lanes and the data rate speed are two major factors pursuing different directions for bandwidth scaling. The IO link performance peak is the total product of these two factors. Increasing data lanes usually implies scaling the density of all components in the physical layer; such as more routings, interconnects and circuits. Through-Silicon-Via (TSV) and silicon interposers were proposed to continue the scaling in 3D integration [5-11]. The HBM's lower signaling frequency results in less

channel loss and helps improve power efficiency due to the reduction in circuit complexity and voltage scaling [12-14]. The main challenge for enabling this type of links is to keep them away from increased interconnects penalty, such as die area, bump counts, and package form factor and densities. The new generation technology that enables increased on-package interconnects is known as high density Multi-Chip Package (MCP).

TABLE I. COMPARISON OF DIFFERENT HIGH BANDWIDTH MEMORY LINKS

Items	4x GDDR5	4-Die HBM1
Capacities	8 Gb	8 Gb
Data Rate per pin	7 Gbps	1 Gbps
Data Width	128 bits	1,024 bits
Max BW	112 Gb/s	128 Gb/s
Voltage	1.35-1.6 V	1.2-1.3 V
Die Area	1.24 mm^2	1.48 mm^2
Reach	< 130 mm	0.5-10 mm
Power	5.6 W	3.3 W
PCB Area	24 x 28 mm^2	7 x 5 mm^2

One important topic of MCP is to ensure the signal integrity in a cross-talk dominated environment [15]. The fast data rate results in a higher channel loss and signal attenuation. The higher insertion loss in channel is due to the increase of Nyquist frequency. It mandates complex circuits with advanced equalization and sophisticated clocking [16-18]. In general, improved IO power efficiency is inversely proportional to channel loss and data rate. High density interconnects for short reach is an extremely power-efficient approach (<1.5 pJ/b) of transferring massive amounts of data at lower rate on package [19]. Such an interconnect architecture can significantly simplify the transceiver circuits and lead to very low IO power consumption. Although CPU-Memory bandwidth has primarily driven the evolution of MCP, they can also be used for a broader set of applications in heterogeneous chip integration [20-21], where multiple dies coming from different silicon processes and different functionalities are integrated together using high bandwidth and low power links.

IO density or physical breakout capacity is a key metric used to measure the advancement of different packaging technologies. Table II shows major MCP technologies and their capabilities in terms of the IO density and fine pitch.

978-1-5090-6695-7/17 $31.00 © 2017 IEEE

TABLE II. TECHNOLOGY METRICS FOR MCP TECHNOLOGIES

Class	Interconnect density (IO/mm/layer)	Half-line pitch (μm)
FCCSP	20~30	11~15
FCBGA	30~50	5~11
eWLB+ InFO-WLP SWIFT	50~250	2~5
CoWoS NTI / SLIM EMIB	250~500	1~2

ARCHITECTURE AND DESIGN

Intel's Embedded Multi-die Interconnect Bridge (EMIB) technology is unique that uses thin pieces of silicon with multi-layer Back-End-Of-Line (BEOL) wirings, embedded in organic substrates, to enable localized dense interconnect. The elements of the EMIB architecture are described in Fig. 1. A very thin silicon bridge is embedded within the top 2 layers of an organic package and connected to high density chip-pad near the die edge. The remaining portion are still using regular bumps and pitches.

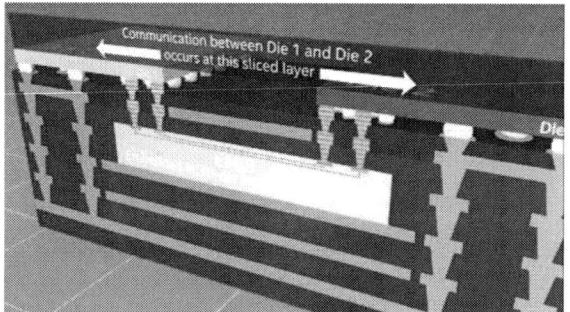

Figure 1: A 3D view of EmIB architecture providing fine pitch for edge-to-edge interconnects and BGA pitch for flip-chip packaging

Dies that have both regular-pitch and fine-pitch bump fields are assembled to a package substrate, such that the fine-pitch bumps bonding to the precision high-density chip pads. During the chip attachment process, the alignment of the dies with respect to each other can be enhanced by a lithographically defined solder mask. From the landside of the package, different kinds of bump sizes and pitches are mixed as shown in Fig.2 and Fig.3.

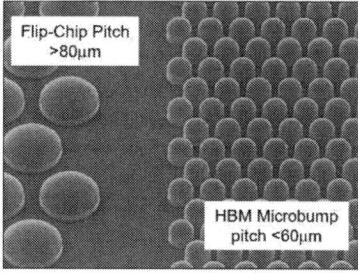

Figure 2: A SEM image of mixed bump sizes and pitches on package

Figure 3: 4-bridge EMIB test vehicle design and layout

The EMIB interconnect channel is connected to a driver and a receiver for testing and validation. The driver is a voltage source with an internal 50 ohm impedance and a pad capacitor. The receiver is represented by a termination resistor of very large value in parallel with a pad capacitor. The channel Inter-Symbol Interference (ISI) and crosstalk can be analyzed using the Peak Distortion Analysis (PDA). Fig.4 shows the eye width opening for a wide range of channel lengths and data rates, assuming source voltage at 1 V, capacitance at 0.4 pf on both ends, and the receiver sensitivity is 200 mV. The results demonstrate an operating region existed, as proper channel length and data rate designed. For instance, if the eye width opening needs to be 50% of the Unit Interval (UI) to accommodate different types of jitter in the system, a 10 mm long EMIB channel can support up to 3 Gb/s signaling. The calculated link power efficiency is as low as 0.7 pJ/b with the help of the dynamic data bus inversion and the low voltage swing terminated logic.

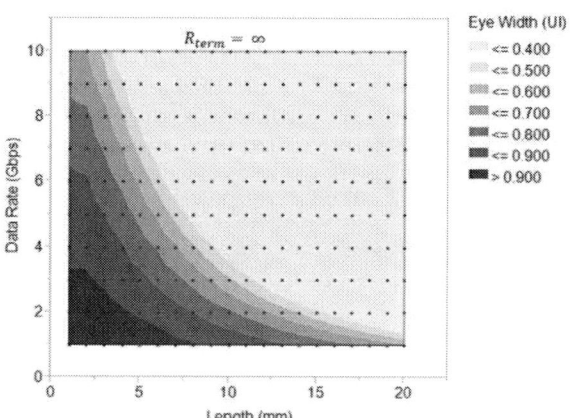

Figure 4: Eye width opening in UI at different channel lengths and data rates

CONCLUSION

A comprehensive picture of the EMIB packaging architecture is presented using localized high density silicon bridges and embedded interposers at a lower cost than competing technologies. EMIB provides a low power and high bandwidth 3D interconnect and integration. Measured signal integrity on test structures fully proves that EMIB could support very high bandwidth using a number of innovative and disruptive technologies.

978-1-5090-6695-7/17 $31.00 © 2017 IEEE

REFERENCES

[1] N. Kurd *et al.*, "Haswell: A family of IA 22 nm processors," *IEEE J. Solid-State Circuits*, vol.50, no.1, pp.49-58, Jan. 2015.

[2] T.O. Dickson *et al.*, "An 8x 10-Gb/s source-synchronous I/O system based on high-density silicon carrier interconnects," *IEEE J. Solid-State Circuits*, vol.47, no.4, pp.884-896, April 2012.

[3] D.U. Lee *et al.*, "A 1.2 V 8 Gb 8-channel 128 GB/s high-bandwidth memory (HBM) stacked DRAM with effective I/O test circuits," *IEEE J. Solid-State Circuits*, vol.50, no.1, pp.191-203, Jan. 2015.

[4] T. Sekiguchi, K. Ono, A. Kotabe and Y. Yanagawa, "1-Tbyte/s 1-Gbit DRAM architecture using 3-D interconnect for high-throughput computing," *IEEE J. Solid-State Circuits*, vol.46, no.4, pp.828-837, April 2011.

[5] I. Ndip *et al.*, "High-frequency modeling of TSVs for 3-D chip integration and silicon interposers considering skin-effect, dielectric quasi-TEM and slow-wave modes," *IEEE Trans. Compon. Packag. Manuf. Tech.*, vol.1, no.10, pp.1627-1641, Oct. 2011.

[6] W.S. Zhao, W.Y. Yin and Y.X. Guo, "Electromagnetic compatibility-oriented study on through silicon single-walled carbon nanotube bundle via (TS-SWCNTBV) arrays," *IEEE Trans. Electromagnetic Compat.*, vol.54, no.1, pp.149-157, Feb. 2012.

[7] H. Kim *et al.*, "Measurement and analysis of a high-speed TSV channel," *IEEE Trans. Compon. Packag. Manuf. Tech.*, vol.2, no.10, pp.1672-1685, Oct. 2012.

[8] K. Kim *et al.*, "Modeling and analysis of a power distribution network in TSV-based 3-D memory IC including P/G TSVs, on-chip decoupling capacitors, and silicon substrate effects," *IEEE Trans. Compon. Packag. Manuf. Tech.*, vol.2, no.12, pp.2057-2070, Dec. 2012.

[9] A.E. Engin and S.R. Narasimhan, "Modeling of crosstalk in through silicon vias," *IEEE Trans. Electromagnetic Compat.*, vol.55, no.1, pp.149-158, Feb. 2013.

[10] A. Todri, S. Kundu, P. Girard, A. Bosio, L. Dilillo and A. Virazel, "A study of tapered 3-D TSVs for power and thermal integrity," *IEEE Trans. VLSI Systems*, vol.21, no.2, pp.306-319, Feb. 2013.

[11] W. Yao, S. Pan, B. Achkir, J. Fan and L. He, "Modeling and application of multi-port TSV networks in 3-D IC," *IEEE Trans. CAD of IC & Systems*, vol.32, no.4, pp.487-496, April 2013.

[12] E. Karl *et al.*, "A 4.6 GHz 162 Mb SRAM design in 22 nm Tri-Gate CMOS technology with integrated read and write assist circuitry," *IEEE J. Solid-State Circuits*, vol.48, no.1, pp.150-158, Jan. 2013.

[13] Y.H. Chen *et al.*, "A 16 nm 128 Mb SRAM in high-k metal-gate FinFET technology with write-assist circuitry for low-Vmin applications," *IEEE J. Solid-State Circuits*, vol.50, no.1, pp.170-177, Jan. 2015.

[14] T. Song *et al.*, "A 14 nm FinFET 128 Mb SRAM with Vmin enhancement techniques for low-power applications," *IEEE J. Solid-State Circuits*, vol.50, no.1, pp.158-169, Jan. 2015.

[15] T.L. Wu, F. Buesink and F. Canavero, "Overview of signal integrity and EMC design technologies on PCB: fundamentals and latest progress," *IEEE Trans. Electromagnetic Compat.*, vol.55, no.4, pp.624-638, Aug. 2013.

[16] T. Toifl *et al.*, "A 2.6 mW/Gbps 12.5 Gbps RX with 8-tap switched-capacitor DFE in 32 nm CMOS," *IEEE J. Solid-State Circuits*, vol.47, no.4, pp.897-910, April 2012.

[17] J.F. Bulzacchelli *et al.*, "A 28-Gb/s 4-tap FFE/15-tap DFE serial link transceiver in 32-nm SOI CMOS technology," *IEEE J. Solid-State Circuits*, vol.47, no.12, pp.3232-3248, Dec. 2012.

[18] H. Kimura *et al.*, "A 28 Gb/s 560 mW multi-standard SerDes with single-stage analog front-end and 14-tap decision feedback equalizer in 28 nm CMOS," *IEEE J. Solid-State Circuits*, vol.49, no.12, pp.3091-3103, Dec. 2014.

[19] X. Wu *et al.*, "Electrical characterization for intertier connections and timing analysis for 3-D ICs," *IEEE Trans. VLSI Systems*, vol.20, no.1, pp.186-191, Jan. 2012.

[20] V. Sukumaran, T. Bandyopadhyay, V. Sundaram and R. Tummala, "Low-cost thin glass interposers as a superior alternative to silicon and organic interposers for packaging of 3-D ICs," *IEEE Trans. Compon. Packag. Manuf. Tech.*, vol.2, no.9, pp.1426-1433, Sept. 2012.

[21] N. Ophir, C. Mineo, D. Mountain and K. Bergman, "Silicon photonic microring links for high-bandwidth-density, low-power chip I/O," *IEEE Micro*, vol.33, no.1, pp.54-67, Jan.-Feb. 2013.

DEVELOPMENT OF WAFER LEVEL HYBRID BONDING PROCESS USING PHOTOSENSITIVE ADHESIVE AND CU PILLAR BUMP

Mingjun Yao[1], Daquan Yu[2]*, Ning Zhao[1]*, Jun Fan[2], Zhiyi Xiao[2], Haitao Ma[1]
[1]Dalian University of Technology, Dalian 116024, China
[2]Huatian Technology (Kunshan) Electronics Co., Ltd., Kunshan 215300, China
*Corresponding Authors' Email: daquan.yu@htkjks.com; zhaoning@dlut.edu.cn

ABSTRACT

Development of low temperature wafer level hybrid bonding process using Cu/SnAg bump and photosensitive adhesive was reported. Two kinds of photosensitive adhesives, i.e., polyimide and dry film, were selected for adhesive bonding. The proposed hybrid bonding method has been successfully applied to 8 inch wafer to wafer bonding. Hybrid bonding using both polyimide and dry film achieved seam-free bonding interface. However, the wafer bonding quality using polyimide is poor and dies were separated during dicing process. As comparison, dry film is more suitable to integrate with Cu/SnAg bump for hybrid bonding and the bonded chip has robust bonding strength.

Key words: Hybrid bonding; 3D packaging; Bump; Polyimide; Dry film

INTRODUCTION

Three-dimensional (3D) integration using through silicon via (TSV) and Cu/Sn micro-joint has been considered to be a promising technology for improving performance and density instead of conventional device scaling [1]. Bonding technology is significant in 3D integration for vertical stack of active layers. Three major 3D integration approaches being pursued are: chip-to-chip, chip-to-wafer and wafer-to-wafer 3D integrations [2]. Although wafer-to-wafer integrations provides high throughput and many performance advantages, but underfill filling becomes extremely challenging especially for ultra-narrow gaps that exist between stacked chips with fine-pitch micro bumps, which may cause the serious reliability issues.

Among various wafer-level bonding scheme, metal and adhesive hybrid bonding has the advantages of simultaneous formation of intrinsic electrical interconnection and adhesive seal around bumps to reinforce reliability. Thus hybrid bonding technology is an enabling technology to address the challenges of underfill filling. Various adhesives such as BCB, SU-8 can be used for hybrid bonding application. Several hybrid schemes, such as Cu/BCB and Cu/oxide, have been developed [2, 3]. Nevertheless, chemical mechanical polishing (CMP) [4] or fly cutting process [5] is usually need to get flat surface, which is complex and costly.

In previous paper, thin WLCSP using via-last scheme with thin wafer handling and low temperature process was successfully demonstrated [6]. In this paper, a low-temperature wafer level hybrid bonding method with simple process using Cu/SnAg bump and photosensitive adhesive was reported. 8 inch wafers to wafer bonding was successfully achieved at optimized bonding conditions. We selected two photosensitive adhesive, i.e., polyimide (PI, HD-4100) and dry film (DF-835P), for adhesive bonding. Based on the proposed hybrid bonding approach, underfill filling and CMP can be skipped. In addition, Cu pillar/solder bump using solid liquid interdiffusion (SLID) gives significant advantages of wider roughness tolerance, low bonding temperature and less bonding duration [7].

PROCESS DEVELOPMENT

Figure 1 presents the process flow of the hybrid bonding approach. For top wafer, Cu/SnAg bumps were formed first by standard sputtering, photolithography, electroplating, seed layer etching and reflow process. The bump diameter and pitch were 50 μm and 150 μm, respectively. Then, adhesive was formed on wafer surface and followed lithography process. PI was spin-coated onto the wafer, and dry film was placed onto the wafer surface and laminated in a vacuum under 95 °C. For the bottom wafer fabrication, Al layer was firstly sputtered on the wafer and then Al RDL was formed using lithographing and etching. Subsequently, the Ni/Au layer with thickness of 2/0.02 μm were fabricated by electroless plating.

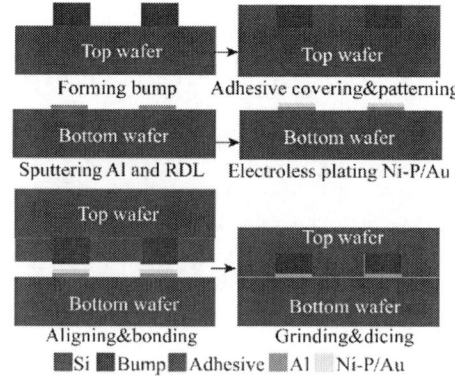

Figure 1: Process flow for the proposed hybrid bonding.

The design specifications for hybrid bonding method are summarized in Table 1.

Table 1: Design specifications for present hybrid bonding

Items	Bonding using PI	Bonding using dry film
Bump height	Cu/SnAg 10/8 μm	Cu/SnAg 10/7 μm
Adhesive	HD-4100 (16 μm)	Dry film (15 μm)

Then the two wafers with Ar plasma surface pre-treatment were bonded face-to-face at 240 °C for 15 min under a force of 10 KN. Then, the adhesive was completely cured at 180 °C for 120 min.

After the top wafer was grinded and polished to a thickness of 150 μm, the bonded wafer was diced into chips 1×1 mm² for bonding interface characterization and reliability evaluation. The individual chips were mounted and cross-sectioned for scanning electron microscopy (SEM) inspection. Energy dispersive x-ray analysis (EDX) was used to examine the compositions at the bonded interface. The shear test was conducted with a shear tester (Dage 4000) using a speed of 150 μm/s.

HYBRID BONDING USING PI ADHESIVE

Figure 2 shows the cross-sectional SEM image of a typical hybrid bonding structure using PI adhesive. As shown in Fig.2, two kinds of IMCs, i.e., $(Cu,Ni)_6Sn_5$ and $(Ni,Cu)_3Sn_4$ were detected by EDX. The formation of this two IMCs during soldering process has been previously reported in solder joints [8, 9].

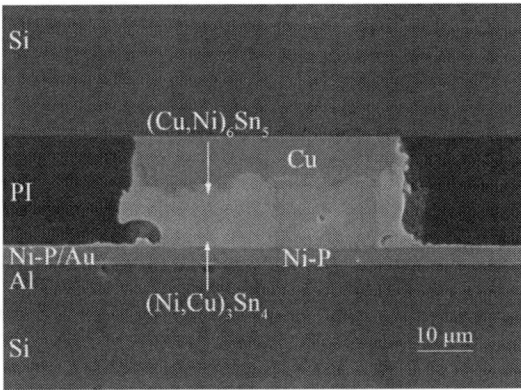

Figure 2: Cross-sectional microstructure of hybrid bonding interface using PI.

The average shear strength of the stacked chip couples was only about 4 MPa, which meant poor bonding strength was achieved for the hybrid bonded chip couples using PI. Figure 3 shows the cross-sectional SEM image of the fractured bump part after shear test. The EDX analysis results show that upon fractured surface of the bump part consists of $(Ni,Cu)_3Sn_4$, indicating that the fracture surface was the interface between the bumps and the electroless Ni-P layer.

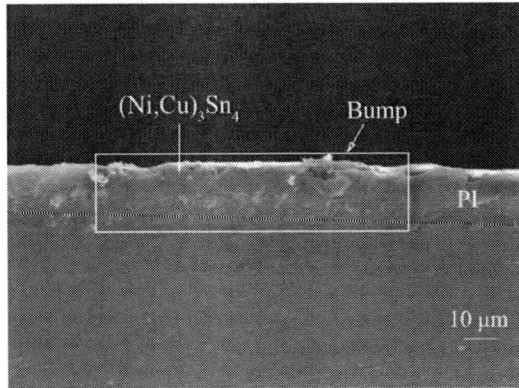

Figure 3: Fracture surface after shear test.

It should be noted that the poor bonding quality resulted in the interface peel in large areas after dicing. In order to removes solvents and volatile substances, the patterned PI has been cured at 350 °C for about 1 hour before bonding process. Because Cu/SnAg-Ni/P bonding condition is less than curing temperature, which is not good for molecule motion of PI during bonding process. We infer that PI is more suitable for Cu-Cu bonding process since its bonding temperature is high enough, but it is not suitable to integrate with Cu/SnAg bump for hybrid bonding application.

HYBRID BONDING USING DRY FILM

As aforementioned, hybrid bonding using PI leads to inadequate bonding quality that cannot stand the dicing process. So we selected photosensitive dry film for adhesive bonding since it can be well bonded at relative low temperature. Figure 4 shows bump morphology after dry film patterning, dry film was successfully laminated on the wafer without any bubbles.

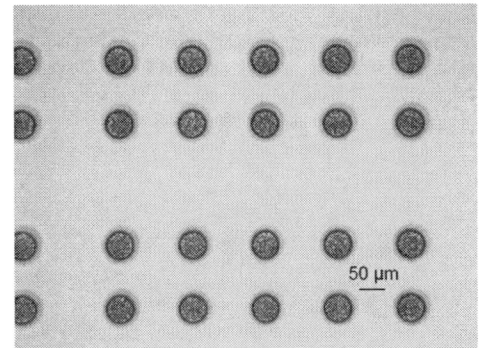

Figure 4: Bump morphology after dry film patterning.

978-1-5090-6695-7/17 $31.00 © 2017 IEEE

Figure 5 shows the cross-sectional SEM image of a typical hybrid bonding structure using dry film adhesive. It shows no voids or seams in the hybrid bonding interface, which indicates a good filling ability and bonding integrity for this fabricated integration results. The vertical interconnection is sealed by the dry film for circumstance isolation and mechanical enhancement. As shown in Fig. 5, $Cu,Ni)_6Sn_5$ on the Cu pillar side and $(Ni,Cu)_3Sn_4$ on the Ni-P layer side were detected by EDX. The results of Cu/SnAg/Ni-P micro interconnect and dry film hybrid bonding condition indicate good bonding integrity with no bonded interface peeling after dicing test except few one around wafer edge.

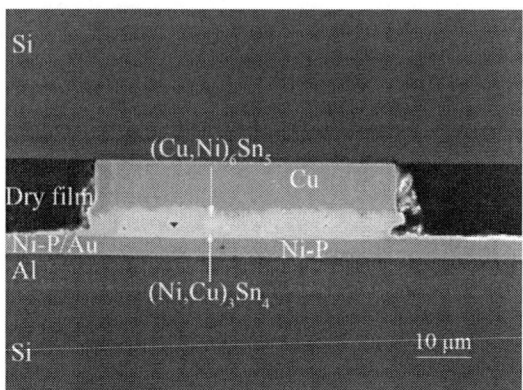

Figure 5: Cross-sectional microstructure of hybrid bonding interface using dry film.

The shear test results show that all the stacked chip couples remain intact even after being shear tested up to 30.0 MPa, which meant robust bonding strength was achieved for these hybrid bonded chip couples. It is clear that dry film has a much greater enhancement on the mechanical property of the hybrid bonded chip couples compared to PI, and it is more suitable to integrate with Cu/SnAg bump for the proposed low-temperature hybrid bonding application. With precise control of bump/dry film thickness, clean metal surface and optimized bonding condition such as appropriate temperature, contact force, duration time, and so on, the hybrid bonding of solder bump and patterned dry film using lithography could be potentially adopted to achieve vertical interconnection for 3D integration.

CONCLUSION

In this study, a simplified low-temperature wafer level hybrid bonding process using Cu/SnAg bump and photosensitive adhesive was developed. Two kinds of photosensitive adhesives, i.e., polyimide and dry film, were evaluated as the hybrid bonding adhesive for combination with Cu/SnAg bumps. The height of the Cu/SnAg bump was 2µm larger than that of the adhesives so as to avoid adhesive trapping in the metal bonding interface. Excess adhesive on the bumps was properly removed using simple and economical lithograph process. Microstructure characterization results show that both PI and dry film can achieve seam free bonding interface. The bonded wafers using PI was peeled in large area during dicing due to poor bonding quality, which did not occur for bonded wafers using dry film. The average shear strength of stacked chips couples using PI and dry film was 4 MPa and 30 MPa, respectively. We conclude that photosensitive dry film is a good candidate for hybrid bonded with Cu/SnAg bumps. The developed fabrication process and bonding structures of hybrid bonding using dry film is promising for future ultra-high density 3D interconnection.

ACKNOWLEDGEMENTS

This research is supported by National Science and Technology Major Project under contract No. 2014ZX02502.

REFERENCES

[1] J.H. Lau. *Microelectron. Int.*, vol. 28, 2011, pp. 8-22.
[2] S.E. Kim and S. Kim. *Microelectron. Eng.*, 2015, vol. 137, pp. 158-163.
[3] Z.C. Hsiao, C.T. Ko, H.H. Chang, H.C. Fu, C.W. Chiang, C.K. Hsu, W.W. Shen and W.C. Lo. *Electronic Packaging and iMAPS All Asia Conference (ICEP-IACC)*, April 14-17, 2015, pp. 834-837.
[4] M. Ohyama, M. Nimura, J. Mizuno, S. Shoji, M. Tamura, T. Enomoto and A. Shigetou. *Electronic Components and Technology Conference (ECTC)*, May 26-29, 2015, pp. 325-330.
[5] R. Agarwal, N. Pham, R. Cotrin, A. Andrei, W. Ruythooren, F. Iker and P. Soussan. *Electronic Components and Technology Conference (ECTC)*, December 9-11, 2009, pp. 267-271.
[6] Z. Xiao, J. Fan, Y. Ren, Y. Li, X. Huang, D. Yu and W. Zhang. *Electronic Components and Technology Conference (ECTC)*, 31 May-3 June, 2016, pp. 302-309.
[7] Y.J. Chang, Y.S. Hsieh, and K. N. Chen. *IEEE. Electr. Device. L.*, vol. 35, 2014, pp. 1118-1120.
[8] C.E. Ho, R.Y. Tsai, Y.L. Lin and C.R. Kao. *J. Electron. Mater.*, vol. 31, 2002, pp. 584-590.
[9] J.Y. Kim, Y.C. Sohn and J. Yu. *J. Mater. Res.*, vol. 22, 2007, pp. 770-776.

978-1-5090-6695-7/17 $31.00 © 2017 IEEE

Novel Leveling Materials for Copper Deposition in Advanced Packaging

Tao Ma[*], Jiang Wang, Zifang Zhu, Peipei Dong

Shinhao Materials LLC, Suzhou, Jiangsu 215200, China

*Corresponding Author's Email: taoma@shinhaomaterials.com

ABSTRACT

Electroplated copper is rapidly becoming the core technology in wafer level packaging. Although copper pillar technology is not new to the semiconductor industry, it is not without significant challenges. One of the most challenges is to obtain desired co-planarity and bump shape under high throughput and various design. Meanwhile, material properties began to gain people's attention. This paper addresses the challenges with a novel class of copper plating leveling composition, L118 system, which has shown remarkable electrochemical modulation abilities on different patterns. XRD (X-ray diffraction spectra) and FIB (focused ion beam) are utilized to investigate the microstructure of the copper pillars, which exhibits that deposited copper with the novel additives is preferentially (111) textured.

INTRODUCTION

The world of semiconductors is at an interesting intersection. 2DIC is transitioning to 3DIC, with or without a 2.5D. Early September 2014 marked the beginning of the commercial 3DIC era when Samsung announced its world first true 3DIC DDR4 modules. Even though 3DIC manufacturing is still in its infancy, research and development in this area has been extremely active for the last decade. Copper pillar, because of its improved electrical and thermal performance, electromigration resistance, as well as potential substrate cost reduction associated with it, has quickly become the core technology that is enabling various possibilities of 3DIC. It has been predicted by industry experts that copper pillar and micro bumping will replace eutectic solder bumping in the near future, and will do away with solder altogether when it gets into ultra fine pitch region.

Copper pillar technology is not new to the semiconductor industry. Intel introduced its first copper pillar flip chip bumps as early as 2005 in its 65nm node CPUs. Since then, the percentage of copper pillar bump to replace eutectic and/or high lead solder bump kept climbing. IBM introduced copper pillar bump in its 32 nm technology node, Samsung in 28 nm node and TSMC in its 22 nm node. While all of this is going on, we saw continued demand for thinner, lighter, fancier, and cheaper mobile devices such as smart phones and tablets. This implies that the chips that are used inside those products have to shrink in size also. For instance, from 2009 to 2015, within five years time span, we saw smart phone thickness shrank by 50%, the processor inside shrank by 30%. The dies and substrates that make the processors would have to become thinner, smaller. When die and substrate become very thin, warpage will overcome the natural forces, and non-contact opens occur during mass reflow. To solve this problem, industry has developed an alternative connection method that utilizes thermal compression bonding. This technology while solves the problem of warpage during mass reflow, it puts more demand on maintaining tight bump height and shape uniformity during the copper pillar deposition step. This demand is challenged further if deposition is taking place at high current densities which are absolutely necessary for HVM. Today, the "industry standard" yields WID (defined as range/2×average) at 5% for copper pillar bumps that are of the dimension of 10 to 200 um in diameter, 50 to 240 um tall. In most of the cases, copper pillar has a domed or dished surface topography, rather than a flat topography with or without PI. Needless to say, the later is much preferred. Not only it is the best bump shape for copper pillar in today's and future's packaging processes, but also because it is easier to make flat bumps consistently and uniformly across the entire wafer, than to make them domed or dished consistently and uniformly. Furthermore, it will also help in terms of on-line quality assurance if an optical method is involved in determining the quality of the plating besides WID. With the rapid advance of material systems, microstructure and mechanical properties are becoming critical for microelectronic reliabilities. In this paper, we shall demonstrate the versatility of the novel leveling system. It could consistently produce flat copper pillar bumps with less than 5% WID uniformity on both recessed and non-recessed patterns with one single bath composition, which commercialized additive systems can NOT. and it also enables conformal growth of copper pillars on recessed patterns. We also characterized the deposited copper by XRD, FIB, nanoindentation, etc. to understand the micro structure and mechanical properties.

MATERIALS AND METHODS

Plating chemicals and method

The plating bath contains 50g/L copper using copper sulfate, 100g/L sulfuric acid and 50mg/L chloride from hydrochloric acid. All the chemicals are reagent grade. Shinhao additives are added to the VMS and plating is conducted in 1L container. Phosphorous copper plate is used as anode and wafer segments with different design are used as cathode. Mass transfer is ensured by paddle agitation and with a 3cm diameter propeller in the center of the cell. The agitation speed depends on the plating requirements. A potentiostat (WaveNow) is used for DC current power supply.

Plating performance characterization

Bump/line shape profile was characterized with 3D laser microscope (KEYENCE, Vk100), co-planarity (WID, WIF) and shape uniformity (Rz) were calculated based on the measuring data. Cross section samples were obtained with grinder-polisher (BUEHLER, Metaserv 250).

Cu film microstructure and mechanical properties

978-1-5090-6695-7/17 $31.00 © 2017 IEEE

The microstructure of electrodeposited copper was observed with focused ion beam (FIB, FEI_Helios600). The growth orientation was identified with X-Ray diffractometer (XRD, BRUKER). The hardness data was obtained with the nanoindentor (Wilson, TUKON1100).

RESULTS AND DISCUSSION
Versatility for a wide range of wafer pattern design

Fig.1 demonstrates 3D laser microscope imaging and cross section view of copper pillars on recessed and non-recessed wafers (mixed with 150 and 120μm openings) with one single bath composition at plating speed >10ASD. Commercialized additive systems were found not been able to produce flat bumps for recessed pattern design.

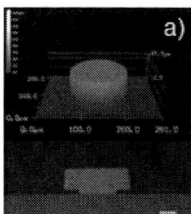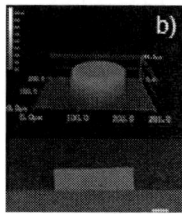

Fig.1 3D laser microscope imaging and cross section view of copper pillars on a) recessed feature; b) non-recessed feature produced by L118 leveling system.

Table 1 and Table 2 summarizes that copper pillars are produced with much preferred co-planarity and flat bump shapes.

For some semiconductor applications, the customers use recessed design to deal with stress problems and they want one single chemistry with high plating speed to meet all the plating requirements on both recessed and non-recessed wafers, as it would significantly reduce their manufacturing and maintaining cost.

In some cases, customer may want conformal growth instead of complete filling on recessed feature design. Fig.2 illustrates a example of conformal growth of electroplated copper on recessed PAD and RDL wafers produced by L118 system.

For some semiconductor applications, the customers use recessed design to deal with stress problems and they want one single chemistry with high plating speed to meet all the plating

requirements on both recessed and non-recessed wafers, as it would significantly reduce their manufacturing and maintaining cost.

In some cases, customer may want conformal growth instead of complete filling on recessed feature design. Fig.2 illustrates a example of conformal growth of electroplated copper on recessed PAD and RDL wafers produced by L118 system.

Customers may have more than one set of specifications for the same wafer pattern design in case of different application purposes. L118 system is versatile as it can meet both specs. with one single bath composition. Figure 3 illustrates 3D laser microscope images and cross section view of copper pillars produced by L118 system on the same wafer

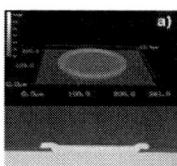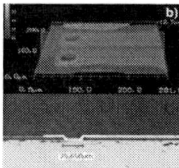

Fig. 2 3D laser microscope imaging and cross section view of conformal growth of electroplated copper on a)PAD wafer; and b) RDL wafer.

patterns for a) copper soldering application (requires 3μm bump height difference between 30 and 17μm openings), and b) copper-to-copper direct bonding application (requires <1μm bump height difference between 30 and 17μm openings).

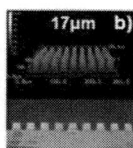

Fig.3 Top view and cross section 3D laser microscope images of copper pillars on the same patterns for a) copper soldering application; b) copper-to-copper direct bond application.

Table 3 and 4 summarizes the bump height differences between 30 and 17μm openings with excellent co-planarity and flat bump shape produced by L118 system for different applications at a plating speed >10ASD.

TABLE 1. Co-planarity and topography uniformity on recessed pattern wafers

Bump size (μm)	Bump height (μm)	WID (%)	Rz (μm)
150	64.93	0.54	-4.29
120	64.00	0.98	-4.62
	64.56	1.97	-4.46

TABLE 2. Co-planarity and topography uniformity on non-recessed pattern wafers

Bump size (μm)	Bump height (μm)	WID (%)	Rz (μm)
150	63.30	0.55	1.31
120	63.80	0.21	1.21
	63.56	1.04	1.26

TABLE 3. Co-planarity and topography uniformity for copper soldering application

Bump size (μm)	Bump height (μm)	WID (%)	Rz (μm)
30	26.07	2.18	-0.81
17	29.19	1.04	0
Δ	-3.12		-0.41

TABLE 4. Co-planarity and topography uniformity for copper-to-copper bonding application

Bump size (μm)	Bump height (μm)	WID (%)	Rz (μm)
30	24.89	1.38	-0.56
17	24.98	0.73	0
Δ	24.82		-0.28

Fig.3a) shows that L118 system is also capable of producing flat bumps with good co-planarity (WID=5.24%) on mega-bumps at a plating speed >15ASD. Fig. 3b) demonstrates 3D laser microscope images that L118 system produces flat bump/line shapes on RDL wafer with mixed patterns.

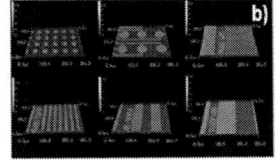

Fig.3 3D laser microscope imaging of a) mega-pillars with good co-planarity at plating speed >15ASD; 3b) flat lines and bricks on RDL with mixed patterns.

Microstructure and mechanical properties

Besides geometry and uniformity control, material property gains growing interests and becomes challenging when the packaging scheme moves to fine pitch. It's not difficult to see that material property is significantly affected by its microstructure. Electrodeposited copper grown with preferred orientation predicts anisotropic chemical and physical properties.

We successfully produced copper with anisotropic structures in nature. FIB results demonstrate the columnar growth of copper in Z-axis obtained by L118 system. XRD profiles in Fig. 4 show that the electrodeposited copper has very strong (111) orientation, and it becomes more pronounced at increasing plating speed.

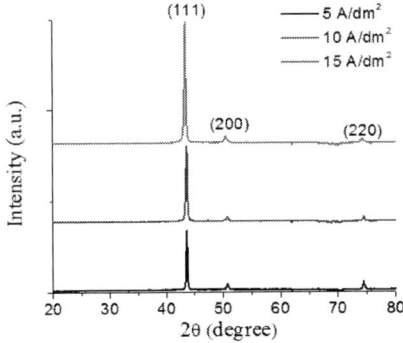

Fig. 4 XRD of copper film produced with L118 chemistry at 5, 10 and 15ASD

Hardness test is performed on electrodeposited copper obtained with L118 chemistry and commercialized additives. Hardness is measured on both the surface and cross section sites. As shown in Fig. 5, it is interesting that L118 system produced copper is about 30% harder perpendicular to its growth direction, while the commercialized chemistry produced copper with little hardness difference between the surface and cross section. It can be most logically explained as the result of the anisotropic nature of its microstructure.

Another example to support advantages of anisotropic electroplated copper is that it can significantly slow down the undercut etching rate. Fig. 6 compares the undercut etching effect between anisotropic copper produced with L118 system and isotropic copper obtained with commercialized additives. Obviously L118 system has less problem with undercut.

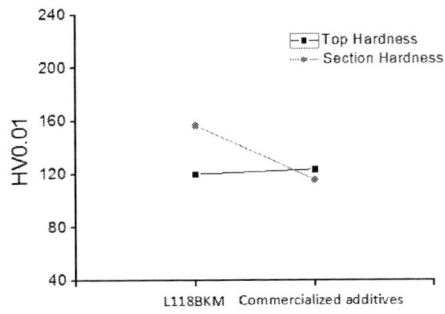

Fig. 5 Surface and cross section hardness of electrodeposited copper obtained with L118 system and commercialized additives.

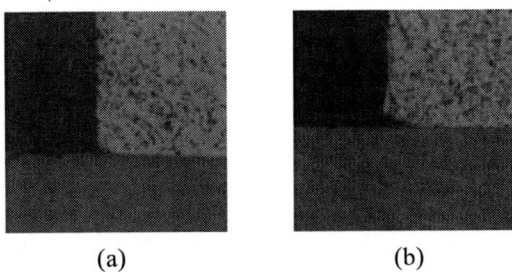

Fig. 6 Undercut characterization of a) anisotropic copper obtained with L118 chemistry; b)isotropic copper obtained with commercialized additives.

CONCLUSION

A novel leveling system was incorporated into electrolytic copper plating chemistries. The additive system is versatile and enables copper pillars/lines with excellent co-planarity and shape uniformity on a wide range of wafer patterns, including recessed and non-recessed pillar patterns, mega-pillar patterns, PADs and RDLs. It can be modulated to meet different specifications for the same wafer design. Furthermore, it produces both flat and conformal grown copper according to different customer needs. The versatility gives the novel leveling system a great potential to reduce manufacturing cost without sacrificing performance. Anisotropic copper is obtained with L118 system. Comprehensive study on material properties and deposited structural characterization should be expected in this paper.

ACKNOWLEDGEMENT

The authors would like to thank FUNSOM Lab, Suzhou University for their assistance on XRD characterization.

REFERENCES

[1] X. Fan, "Wafer Level Packaging (WLP): Fan-in, Fan-out and Three-Dimensional Integration", 11th. Int. Conf. on Thermal, Mechanical and Multiphysics Simulation and Experiments in Micro-Electronics and Micro-Systems, EuroSimE 2010

[2] C. Wang, M. An, P. Yang, J. Zhang, "Prediction of a new leveler (N-butyl-methyl piperidinium bromide) for through-hole electroplating using molecular dynamics simulations", Electrochemistry Communications, vol. 18, 2012, pp. 104-107

978-1-5090-6695-7/17 $31.00 © 2017 IEEE

Production-scale Flux-free Bump Reflow Using Electron Attachment

C. Christine Dong[1]*, Richard E. Patrick[1], Gregory K. Arslanian[1], Tim Bao[1]
Kail Wathne[2], and Phillip Skeen[2]
[1]Air Products and Chemicals, Allentown, PA 18195-1501, USA
[2]Sikama International, Inc., Santa Barbara, CA 93101-2314
*Email: DONGCC@airproducts.com

ABSTRACT

This paper introduces a recent work by a joint effort between Air Products and Sikama International on alpha trials of a production-scale furnace for flux-free wafer bump reflow based on electron attachment (EA).

INTRODUCTION

Packaging technology for electronics devices has advanced rapidly in recent years driven by feature size reduction, new materials developed, and increased demand on device functionality. The most fundamental among the advanced packaging technology is the use of wafer bumping and wafer-level chip scale packaging.

In a wafer bumping process, fine-pitch electroplated solder bumps are formed over an entire silicon wafer on which integrated circuits have been built, the wafer is then reflowed at a temperature above the solder's melting point to complete metallic interconnection of the bumps with underneath metal pads and convert the bumps from a deposited shape into a ball shape. After the wafer bumping, the wafer is cut into individual chips, which then go through subsequent packaging processes. In the packaged devices, the formed bumps serve as electrical, mechanical, and mounting connections. Current study is related to the last step of the wafer bumping process — wafer bump reflow.

One of the keys for successful wafer bump reflow is to remove the native oxide layer and prevent additional oxidation on the bump surface. Any oxide layer on the bump surface will act as a solid skin to constrain molten solder's flow, which in turn causes a non-qualified bump appearance and non-uniform bump shape across a wafer. This oxide elimination is becoming more critical and difficult as the bump size shrinks since the increased surface to volume ratio plus the enlarged surface curvature of the solder bump drives toward a more severe solder oxidation to minimize its surface energy.

Currently, the most common approach is to coat the wafer with a flux and then reflow the wafer in a nitrogen environment. However, such flux-containing reflow process is quite messy since the decomposition of organic fluxes always leaves residues and generates volatiles, which invariably bring contaminants on the wafer and furnace wall. Therefore, a post cleaning of the reflowed wafer is always required. A frequent cleaning of furnace interior surfaces is also needed, causing high maintenance costs and a lot of equipment downtime. In addition, special safety precautions have to be taken for dealing with hazardous disposal of the flux residues and unhealthy exposure of the flux vapor. Besides the cost and inconvenience associated with the cleanings, the flux-containing process directly affects the quality of the reflowed wafer. For example, during reflow the flux can get into the molten solder and create voids inside the bumps, thus degrading mechanical and electrical properties of the solder joints in packed devices. As the pitch and bump sizes are continually decreasing, the need for process cleanliness increases. This has led to increased use of flux-free processing, which is mainly based on using a reactive gas to replace the organic flux for oxide removal.

However, known flux-free technologies all have different problems or limitations. By using formic acid vapor, the process is not completely residue-free and has to be operated in a sealed system. Hydrogen-based flux-free process is clean and non-toxic, but high temperature ($\geq 350°C$) and pure hydrogen (flammable) must be applied to activate and hasten the oxide reduction. Plasma-activated hydrogen can make the oxide reduction efficient at low temperatures, but only vacuum plasma appears to be viable, resulting in a batch process.

Current study is related to a novel flux-free technology based on electron attachment (EA), which can be operated at ambient pressure and normal solder reflow temperatures using non-flammable mixtures of hydrogen (<4 vol%) in nitrogen. The technology is invented by Air Products in recent years, which involves generating a large quantity of low-energy electrons. Some of the electrons can attach to hydrogen molecules, forming active species for oxide removal. The basic concept and the efficiency for oxide removal have been demonstrated in previous studies [1]. The EA-based technology is completely residue-free and has a potential to be widely used in the electronics packaging industry. This paper presents a recent work between Air Products and Sikama International on alpha trials of an EA-enabled prototype furnace for production-scale wafer bump reflow (Fig. 1).

TRIAL RESULTS

As shown in Figure 2, the EA-enabled furnace contains a roller-featured wafer transportation system, which carries wafers through heating and cooling zones with a standard production speed. Before entering a reflow zone, wafers are exposed to EA-activated 4% H_2 in N_2 for removing solder oxides (Fig. 3).

EA-based process can ensure a good bump uniformity across the width of a 12" moving wafer. In addition, the surfaces of the post-reflowed wafers are free of extra solder and foreign materials (Fig. 7).

Figure 1: EA-enabled prototype furnace for production-scale wafer bump reflow

Figure 4: Cross section of the IMC

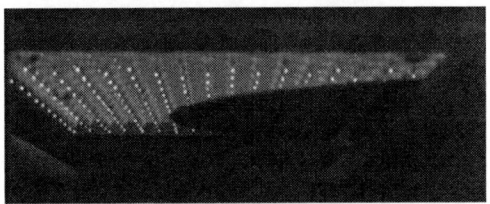

Figure 2: Roller-featured wafer transportation system

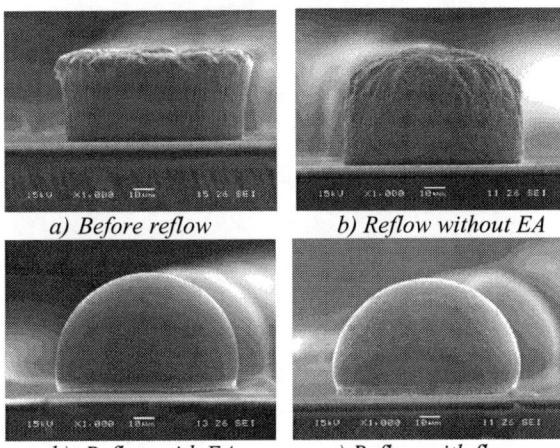

a) Before reflow   b) Reflow without EA

b) Reflow with EA   e) Reflow with flux

Figure 5: Bump shape comparison

Figure 3: Wafer entering an EA zone for oxide removal

Various dummy wafers (8" and 12") with as-plated solder bumps were obtained from different customers and processed in the furnace to evaluate bump reflow quality. Figure 4 shows a cross section of a reflowed tin-based solder bump plated on nickel. The intermetallic compound (IMC) formation controlled by reflow time and temperature is quite acceptable. The effectiveness of EA on oxide removal has been clearly demonstrated in multiple trials. Figure 5 compares bump shapes of a lead-free solder on a wafer undergone different reflow processes. Before reflow, electroplated bumps are in a cylindrical shape (Fig. 5a). Without applying EA in the H_2 and N_2 mixture, reflowed bumps have a rough surface and uncompleted shape conversion (Fig. 5b). With applying EA, reflowed bumps have a smooth surface and spherical shape (Fig. 5c), even better than that of flux-reflowed bumps after cleaning (Fig. 5d). As shown in Figure 6, the

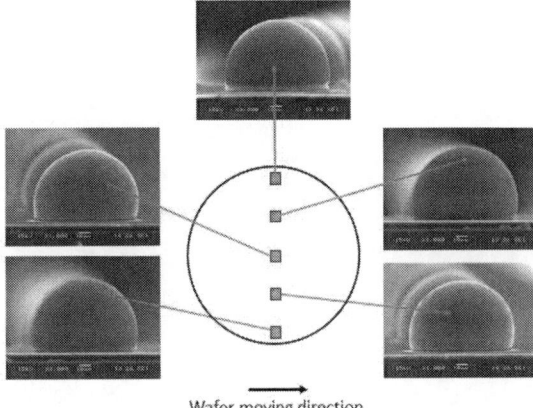

Wafer moving direction

Figure 6: Uniform bump shape by EA-based process

978-1-5090-6695-7/17 $31.00 © 2017 IEEE

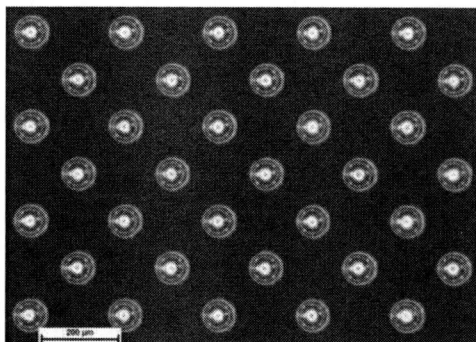

Figure 7: Clean wafer surface after EA-based reflow

Full dummy wafers reflowed in the EA-enabled furnace were also sent back to corresponding customers for standard quality inspections, such as checking bump shape, bump uniformity, shear strength, failure model, and bump voids. Results confirm that the wafers reflowed under the EA-based process indeed meet all specifications. Figures 8 and 9 represent results of automated optical inspection (AOI), which confirm acceptable bump heights (BH) and bump diameters (BD) across an 8" full wafer. Figure 10 shows that all shear failures are within solder bumps and shear strengths well exceed the criterion (> 2 g/mil^2). Figure 11 is an x-ray image of a die on a reflowed wafer, which demonstrates that the number of bump voids (green) is quite low and the size of a typical void is 3% of the bump area, which is much below the specified upper limit (8% of the bump area).

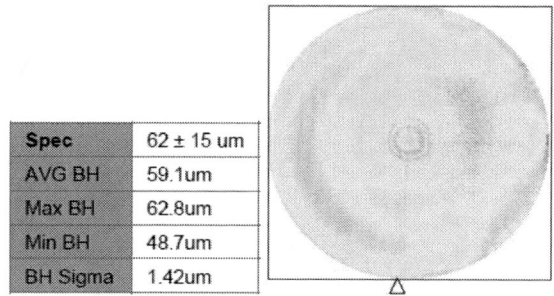

Spec	62 ± 15 um
AVG BH	59.1um
Max BH	62.8um
Min BH	48.7um
BH Sigma	1.42um

Figure 8: BH distribution map and data

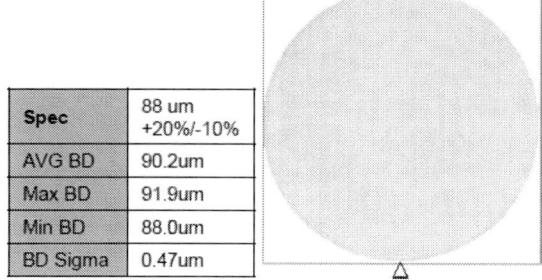

Spec	88 um +20%/-10%
AVG BD	90.2um
Max BD	91.9um
Min BD	88.0um
BD Sigma	0.47um

Figure 9: BD distribution map and data

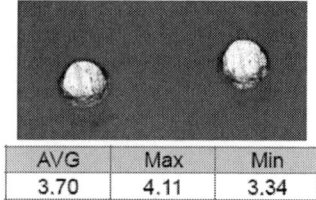

AVG	Max	Min
3.70	4.11	3.34

Spec>2 g/mil2
Figure 10: Bump shear failures and data

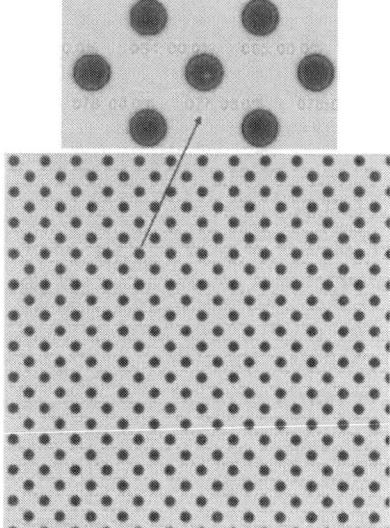

Figure11: X-ray image of a die

CONCLUSIONS

Trial results demonstrate that dummy wafers reflowed in the EA-enabled production-scale furnace meet customer specifications. The EA-based technology offers the following benefits for wafer bump reflow: 1) enhanced bump reflow quality because the flux induced solder voids and wafer contaminations naturally disappear, 2) improved productivity by having in-line process capability, eliminating post wafer cleaning, and avoiding furnace down time cleaning, 3) reduced cost of ownership due to eliminated costs associated with cleaning equipment, solution, labor work, and flux, 4) improved safety by eliminating flux exposure and using a non-toxic and non-flammable gas mixture, and 5) no environmental issues by eliminating organic vapors, hazard residues, and CO_2 emission.

REFERENCES

[1] C. Christine Dong, Richard E. Patrick, Russell A. Siminski, and Tim Bao, "Fluxless soldering in activated hydrogen atmosphere," Proceedings of CSTIC 2013, Shanghai, China, M11010ar. 15-17, 2016, VIIA 1520-1535.

FINE FEATURE SOLDER PASTE PRINTING FOR SIP APPLICATIONS

Sze-Pei Lim*, Kenneth Thum, Dr. Andy Mackie
Indium Corporation, Malaysia/USA
*splim@indium.com

ABSTRACT

Due to the rapid development of IoT "smart" devices, the industry has seen a surge in demand for system-in-package (SiP) devices, which are capable of increased functionality in a smaller package design. This continues to push miniaturization to an even greater level, therefore creating assemblies with smaller components and greater density.

Fine feature solder paste printing for passive component sizes, from 01005 (0.4 x 0.2mm) and now down to 008004 (0.25 x 0.125mm), has become more challenging in SiP assembly. Due to small stencil aperture designs, finer powder size solder pastes, typically type 5 (15–25μm), type 6 (5–15μm), and type 7 (2–11μm), are used for these applications. Since more components are being packed into a SiP, the rheology of the solder paste is an important attribute, as the gap between neighboring pads can be as close as 50μm. Solder paste with minimal slump behavior will be required to minimize bridging in such designs. Consistency in solder paste printing transfer efficiency, stencil life, good wetting, graping resistance, and minimal voiding are other key attributes of solder paste for SiP assembly as well. This paper will focus on the attributes of these finer powder size solder pastes, review the tests conducted for critical-to-function characteristics, and discuss the results.

Key words: SiP, solder paste, 008004, miniaturization

INTRODUCTION

As the trend towards miniaturization in SiP applications continues, from current 01005 components, going down to 008004 ("0201m") or even 0050025 for next generation packages, the printing performance of solder paste becomes critical. The conventional SMT solder paste printing process using type 3 or type 4 powder size, has evolved into a more complicated printing process for SiP, using type 5, 6, or even 7 powder size, with much smaller stencil apertures and thinner stencil thicknesses, as well as more stringent requirements for allowed paste deposit variability.

Other than having to print a smaller and thinner solder paste deposit, the gap between neighboring pads is smaller too. Some customers are already looking at a gap of 50μm between pads. In order to achieve good and consistent printing performance under such challenging conditions, besides having good printer setup and appropriate stencil technology [1] [2], the choice of correct powder size, flux system, rheology, and slump behavior of solder pastes are key.

SOLDER PASTE

Solder Powder

Powder size is classified by type according to IPC J-STD-005A, as shown in Figure 1.

Type	Less than 0.5% larger than	10% Max. between	80% Min. Between	10% Max. Less than
1	160	150-160	75-150	75
2	80	75-80	45-75	45
3	60	45-60	25-45	25
4	50	38-50	20-38	20
5	40	25-40	15-25	15
6	25	15-25	5-15	5
7	15	11-15	2-11	2

Figure 1. Powder type distribution according to IPC J-STD-005A (units in μm)

Table 1 outlines the different solder powder sizes available in the electronics and semiconductor industries. Although 3-5 are increasingly in use for SMT, some OSATs and others are already using powder size type 6 for SiP with 01005 chips, and for next generation packages with 008004 chips, both types 6 and 7 powder sizes are being considered.

Table 1. Powder Sizes

Powder Type	Powder Size (μm)	Minimum Stencil Aperture (μm)	Approximate Surface Area Ratio
3	25–45	270	1.0
4	20–38	230	1.2
5	15–25	150	1.9
6	5–15	90	3.7
7	2–11	66	5.6

It is a general industry guideline that in order to achieve consistent solder paste printing performance, it is important to choose the correct powder size so that a minimum of 5–6 solder particles (the large particle size of the range) can be maintained across the aperture. According to this rule, the recommended minimum stencil aperture for each powder type is shown in Table 1.

978-1-5090-6695-7/17 $31.00 © 2017 IEEE   427

Another concern when choosing a smaller powder size is the increase in the powder surface area. As shown in Table 1, when powder size decreases from Type 3 to Type 7, the powder surface area increases significantly. This increased powder surface area will require more flux activation, or flux with a better oxidation barrier to protect the powder from surface oxidation [1]. Hence it is important to both choose a correctly formulated flux that works with finer powder, as well as a solder powder with a low oxide level of powder. These factors will enhance the stability of the fine-pitch solder paste in terms of shelf life, stencil life, and wetting or solderability performance. Excessive activator levels can also destabilize the paste, so flux formulation is complex.

Flux System

There are different types of flux systems available for solder paste. Typically, the water soluble flux, standard no-clean flux, and more recently, the ultra-low residue (ULR) no-clean flux have seen widespread use, especially in semiconductor assembly applications. Table 2 below compares these different flux systems [3].

Table 2. Comparison of Different Flux Systems

	Standard No-Clean Flux/Paste	Ultra-Low Residue No-Clean Flux/Paste	Water-Soluble Flux/Paste
Typical Flux Residue %	20-60%	0-10%	NA
Cleaning Process	Solvent clean or not needed	Not needed	Water clean
Compatibility with Underfill or Molding Compound	Typically low if residue is not cleaned	Good	Good if completely cleaned

Currently, most applications for SiP use the water-soluble flux system, where flux residue is cleaned away using water after the reflow step, followed by drying and overmolding steps. As the miniaturization trend continues, some SiP applications will soon reach a point where effective cleaning with a very tight gap and standoff becomes challenging, and using an ultra-low residue no-clean paste will be necessary. This new class of solder paste enables all cleaning and chemicals process costs to be eliminated, while reducing both component warpage and cycle time.

Paste Rheology

Decreasing the particle size of the powder component of the solder paste also increases the viscosity, since the flux vehicle itself is thixotropic. The solder paste used in stencil printing is a shear-thinning thixotropic material [4] with a low yield stress, with the rheology being governed by both the flux system and the solder powder. The paste rheology controls a variety of factors: solder paste aperture filling ability, transfer efficiency of solder paste from stencil onto substrate, and the shape of the solder paste deposit after printing. These are keys for good printing performance of a solder paste.

Slump Behavior of Paste

The term "slump" is an x-y direction expansion of the solder paste deposit caused by gravity. A yield stress in the paste can prevent slump. A metal load that is too high can give the paste a very high yield stress and prevent the paste from filling and releasing from stencil apertures. Therefore, optimizing the metal load for the printing process is important and is driven by several factors.

Stencil Life of Paste

A solder paste with long stencil life is important to reduce print-to-print variability over time. Typically, a stencil life of at least 4 to 8 hours will be required for a high-volume manufacturing environment.

PRINTING EXPERIMENT

In this study, several combinations of paste, tooling setup, and stencil aperture sizes were investigated. Three solder paste specimens with different flux chemistries and different powder sizes were studied. Also, a board support system of pallet and vacuum support were compared to investigate whether the use of a pallet alone could allow the printing process to achieve a comparable consistency. Finally, laser cut and electroformed stencils were compared by printing different apertures sizes.

Test Vehicle

A test vehicle was specially designed to mimic a typical substrate size of 237mm in length, 62mm in width, and 0.5mm thick (Figure 2). There are 2 arrays of pads on the test vehicle. Each row has different pad sizes: 150μm x 125μm on row 1; 150μm x 100μm on row 2, and 150μm x 112.5μm on row 3. Each column has a different gap distance between pads, i.e., 50μm, 80μm, 100μm, 130μm, and 150μm (Figure 3). Pads are arranged in horizontal and vertical positions such that different squeegee wiping directions could be simulated. The pad surface finishing is NiAu (ENIG) and are non-solder mask defined (NSMD). The test vehicle also consists of 01005 pads, but it is not the focus of this study.

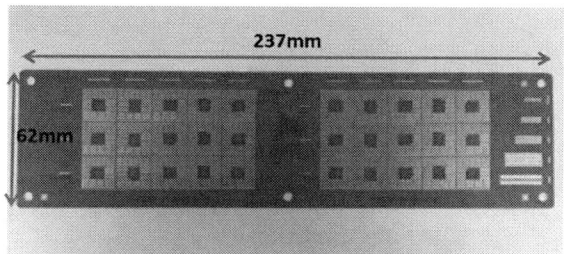

Figure 2. SiP Test Vehicle

978-1-5090-6695-7/17 $31.00 © 2017 IEEE

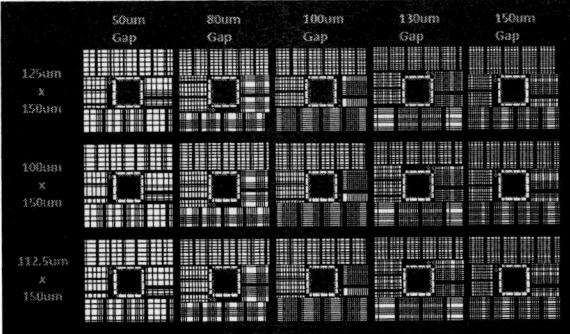

Figure 3. Print Patterns in an Array

Solder Paste Specimens

Three flux vehicles were chosen in this printing test: two water-soluble fluxes and one no-clean flux. These three fluxes were mixed with type 6 and 7 powders. As finer powders yield higher viscosity, a slight tweak in the metal load for type 7 pastes is needed to reduce the viscosity to about the same viscosity as the type 6 solder pastes. Table 3 below shows a summary of the paste specimens utilized.

Table 3. Paste Specimens

Paste Specimen	Flux Vehicle	Powder Size	Metal Load
Paste A	Water Soluble X	T6SG	89.00%
Paste B	Water Soluble Y	T6SG	89.25%
Paste C	No Clean Z	T6SG	90.50%
Paste D	Water Soluble X	T7SG	88.50%
Paste E	Water Soluble Y	T7SG	88.75%
Paste F	No Clean Z	T7SG	90.00%

Stencil Design

Three stencils were prepared for this study: one 50μm-thick laser cut stencil and two electroformed stencils with thicknesses of 35 and 50μm. The laser cut stencil has a 1-to-1 opening with the pad size on the test vehicle, while the electroformed stencil's left array follows a 1-to-1 opening, and the right array has a smaller aperture as shown in Figure 4 and Table 4. The two electroformed stencils have the same aperture designs but only differ in the stencil thicknesses, as discussed above.

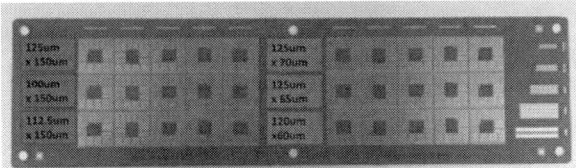

Figure 4. Aperture Size Arrangements

Equipment and Tooling Setup

- DEK Horizon Printer
- Koh Young SPI machine
- 12″ squeegee at 60° wiping angle
- Vacuum support
- Carrier Pallet

Table 4. Aperture Size and Area Ratio

X (μm)	Y (μm)	Stencil Thickness (μm)	Area Ratio
150	125	35	0.97
150	112.5	35	0.92
150	100	35	0.86
125	70	35	0.64
125	65	35	0.61
120	60	35	0.57
150	125	50	0.68
150	112.5	50	0.64
150	100	50	0.60
125	70	50	0.45
125	65	50	0.42
120	60	50	0.40

Printing Parameters

Print speed was fixed in all experiments while the squeegee pressure was adjusted so that minimum pressure was used in each experiment, i.e., enough pressure to just wipe the stencil surface clean (Table 5). A minimum pressure is crucial in achieving consistent prints as excessive pressure could cause movement of the printing surface during the squeegee stroke. Cleaning frequency was set to three pieces, with a cleaning mode of wipe/vacuum/dry (W/V/D).

Table 5. Printing Parameters of Different Solder Pastes

Parameters	Printing Speed	Printing Pressure	Separation Distance
Paste A	20mm/s	5.5Kg	0
Paste B	20mm/s	5Kg	0
Paste C	20mm/s	6Kg	0
Paste D	20mm/s	5.5Kg	0
Paste E	20mm/s	5Kg	0
Paste F	20mm/s	6Kg	0

RESULTS AND DISCUSSIONS

Bridging Between Pads

This comparison demonstrates a clear difference between tooling combinations. By using a carrier pallet as the board support, no bridging was observed down to a 80μm gap distance when using a laser cut stencil.

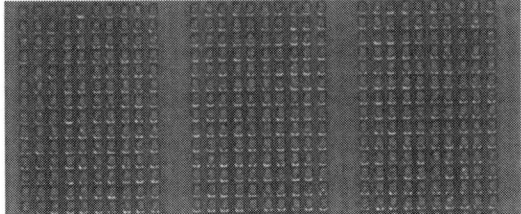

Figure 5. 125μm x 150μm pad and 50μm gap distance

By using a vacuum support, the laser cut stencil was able to achieve non-bridging at the 50μm gap distance column. This suggests that the vacuum support has a much better support with minimal gap between the stencil and PCB during printing. Figure 6 shows the printing on a 125μm x

150μm pad (1-to-1) with a 50μm gap distance between pads.

However, with electroformed stencils, the left array prints encountered bridging. The amount of bridging increased from column 1 (50μm gap) to column 5 (150μm gap). The increased transfer efficiency of the electroformed stencil is most likely the cause for this distinct difference in print performance. The right array with reduced apertures did not give any bridging. Hence, only printing data from the right array was able to be collected and analyzed.

Comparison of Different Solder Pastes

On the right array, the aspect ratio ranges from 0.40 to 0.45 only for a 50μm thick stencil. However, a consistent print is still achievable for these pads. Figures 6, 7, and 8 show the deposit volume comparison of different pastes for different aspect ratios. Paste C ranks the best among the three pastes. Paste A ranks second, and Paste B is third. Paste B has a tighter distribution compared to Paste A and Paste C, however, it has many more insufficients compared to the other two pastes. The rheology plays a big role in printing ultra-fine apertures, and good printability is achievable with both water-soluble and no-clean flux chemistries.

The 35μm thickness stencil gave a higher aspect ratio, hence it was believed that the thinner stencil could yield better printings over the 50μm thick stencil. The aspect ratio improved from range of 0.40–0.45 to 0.57-0.64. The improvement was obvious by using paste sample C, as shown in Figure 9.

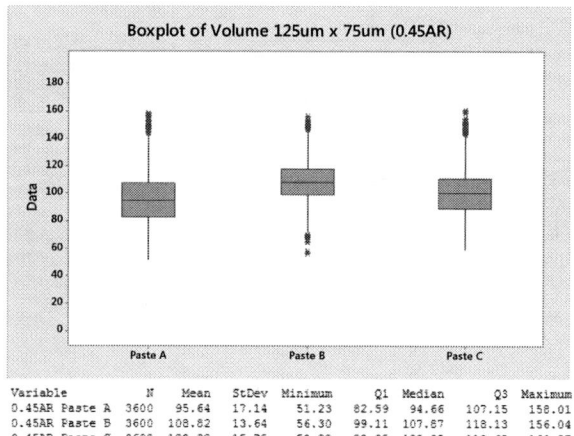

Variable	N	Mean	StDev	Minimum	Q1	Median	Q3	Maximum
0.45AR Paste A	3600	95.64	17.14	51.23	82.59	94.66	107.15	158.01
0.45AR Paste B	3600	108.82	13.64	56.30	99.11	107.87	118.13	156.04
0.45AR Paste C	3600	100.30	15.75	59.20	88.95	100.02	110.69	160.37

Figure 6. Paste Comparison on 125μm x 75μm Pads (0.45AR)

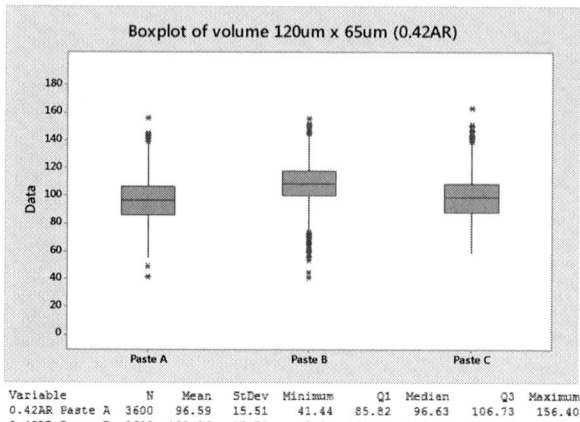

Variable	N	Mean	StDev	Minimum	Q1	Median	Q3	Maximum
0.42AR Paste A	3600	96.59	15.51	41.44	85.82	96.63	106.73	156.40
0.42AR Paste B	3600	108.96	13.78	40.99	100.28	108.90	118.07	155.76
0.42AR Paste C	3600	98.97	14.70	58.83	88.56	98.81	108.67	163.31

Figure 7. Paste Comparison on 120μm x 65μm Pads (0.42AR)

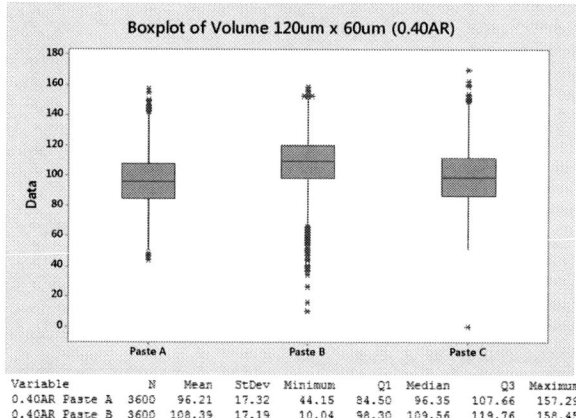

Variable	N	Mean	StDev	Minimum	Q1	Median	Q3	Maximum
0.40AR Paste A	3600	96.21	17.32	44.15	84.50	96.35	107.66	157.29
0.40AR Paste B	3600	108.39	17.19	10.04	98.30	109.56	119.76	158.45
0.40AR Paste C	3600	99.44	17.56	0.00	86.41	99.04	111.36	169.99

Figure 8. Paste Comparison on 120μm x 60μm Pads (0.40AR)

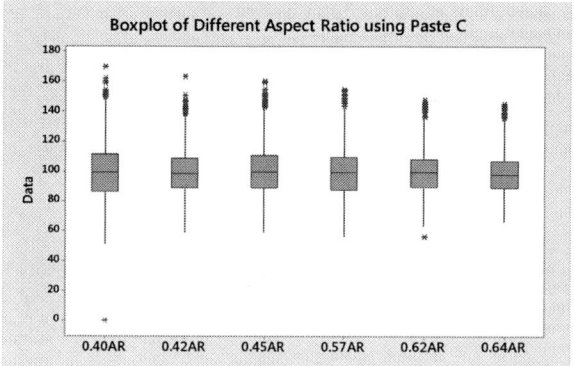

Variable	N	Mean	StDev	Minimum	Q1	Median	Q3	Maximum
0.40AR	3600	99.448	17.563	0.000	86.419	99.049	111.364	169.997
0.42AR	3600	98.974	14.708	58.839	88.567	98.819	108.679	163.317
0.45AR	3600	100.30	15.75	59.20	88.95	100.02	110.69	160.37
0.57AR	3600	99.178	15.550	56.453	87.569	99.344	109.894	154.672
0.62AR	3600	99.134	14.091	56.210	89.831	99.215	108.554	147.521
0.64AR	3600	98.331	12.713	66.484	88.793	97.810	107.167	145.276

Figure 9. Boxplot of Paste C's Solder Paste Volume vs. Different Aspect Ratios (50μm thickness vs. 35μm thickness)

Comparison of T6-SG vs. T7-SG

Figure 10 shows the printing results of T6 and T7 pastes. It was observed that the standard deviation of T7 prints is smaller, so it has a smaller variation. However, it was also observed that more bridges were seen.

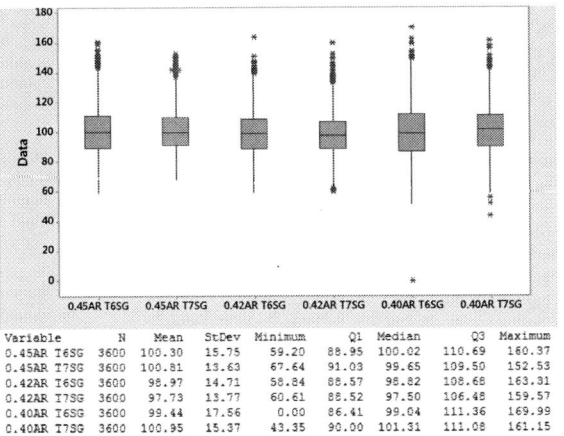

Variable	N	Mean	StDev	Minimum	Q1	Median	Q3	Maximum
0.45AR T6SG	3600	100.30	15.75	59.20	88.95	100.02	110.69	160.37
0.45AR T7SG	3600	100.81	13.63	67.64	91.03	99.65	109.50	152.53
0.42AR T6SG	3600	98.97	14.71	58.84	88.57	98.82	108.68	163.31
0.42AR T7SG	3600	97.73	13.77	60.61	88.52	97.50	106.48	159.57
0.40AR T6SG	3600	99.44	17.56	0.00	86.41	99.04	111.36	169.99
0.40AR T7SG	3600	100.95	15.37	43.35	90.00	101.31	111.08	161.15

Figure 10. Comparison of T6-SG and T7-SG on Various Pad Sizes Using Paste C

Process Capability Study

Cp, Cpk, and Ppk were calculated using Minitab software. The reference specification limit used was 40% to 150% to compare the paste performance. Table 6 indicates that both vacuum support and carrier pallet are capable setups, however, vacuum support has a greater consistency as the Cpk and Ppk values are larger.

Table 6. Capability Comparison between Different Board Support Systems

Pallet	0.68	2.52	1.07	3.20
Pallet	0.64	2.70	1.10	2.97
Pallet	0.60	2.39	1.11	2.61
Vacuum	0.68	2.82	1.29	3.37
Vacuum	0.64	2.91	1.60	3.27
Vacuum	0.60	2.73	1.38	3.19

Table 7 shows the process capability of all pastes using various pad sizes and stencil thickness. The combinations of pad size and stencil thickness were represented as their respective aspect ratio.

Table 7. Cpk, Ppk, and Cp of Different Pastes/Aspect Ratios

	Cpk	Ppk	Cp
Paste A 0.45AR	2.17	1.06	2.20
Paste A 0.42AR	2.06	1.15	2.12
Paste A 0.40AR	1.92	1.04	1.97
Paste B 0.45AR	1.65	1.01	2.21
Paste B 0.42AR	1.49	0.99	1.99
Paste B 0.40AR	1.36	0.81	1.80
Paste C 0.45AR	2.11	1.05	2.33
Paste C 0.42AR	2.19	1.16	2.36
Paste C 0.40AR	2.03	0.96	2.21
Paste C 0.64AR	2.74	1.36	2.91
Paste C 0.62AR	2.82	1.20	3.05
Paste C 0.57AR	2.42	1.09	2.62

CONCLUSIONS

In order to achieve consistently good fine feature printing performance for SiP applications, solder paste attributes, i.e., powder size, flux system, rheology, slump behavior, and stencil life, are important and need to be taken into consideration. Solder paste with suitable rheology, mixed with the correct powder size and flux, should be evaluated and selected accordingly. The appropriate stencil technology, design, and thickness, coupled with a good board support system during printing, are also keys for consistently good solder paste transfer efficiency.

ACKNOWLEDGEMENT

Special thanks to Indium Corporation's Suzhou Simulation Lab team—Ms. Wisdom Qu, Dr. Fiona Chen, and Leon Rao—for their help and support in performing the solder paste printing tests in their lab.

REFERENCES

[1] Ed Briggs, "Meeting Future Stencil Printing Challenges with Ultrafine Powder Solder Pastes" International Conference on Soldering and Reliability, Toronto, Canada, May 2014

[2] Rita Mohanty Ph.D., S. Manian Ramkumar Ph.D., CEMA, Chris Anglin and Toshitake Oda, "Effect of Nano-Coated Stencil on 01005 Printing" APEX 2011

[3] SzePei Lim, Maria Durham, A. Mackie, "No Clean Material For Advance Packaging Assembly", Semicon China Feb 2016

[4] Kravcik and Vehec, "Study of the Rheological Behavior of Solder Pastes" Proc. Scientific Conference of Young Researchers 2010, FEI TU of Kovice

[5] Lim, Thum and Mackie, "Meeting Solder Paste Printing Challenges for SiP in 'Smart' IoT Devices" Chip Scale Review magazine, Jul-Aug 2016

INVESTIGATION OF THERMAL INTERFACE MATERIALS REINFORCED WITH MICRO- AND NANOPARTICLES

Kamil Janeczek[1], Aneta Araźna[1], Yan Zhang[2], Shiwei Ma[2], Janusz Sitek[1], Jingyu Fan[3], Johan Liu[2,4], and Krzysztof Lipiec[1]*

[1]Tele and Radio Research Institute, Ratuszowa 11, 03-450 Warsaw, Poland
[2]SMIT Center, Shanghai University, Shanghai 200072, China
[3]SIAMM, Shanghai University, Shanghai 200072, China
[4]MC2, Chalmers Uni. of Technology, SE-412 96 Gothenburg, Sweden
*Corresponding Author's Email: yzhang@shu.edu.cn

ABSTRACT

Heat management is one of the major challenges in modern electronic devices. The higher performance results in a production of greater amount of heat which needs to be efficiently dissipated so as to ensure the electronic devices operational during the period of lifetime. This paper discusses the application of micro- and nano-materials in thermal interface materials (TIM) used for heat management. Effects of type, size and geometry of different fillers were experimentally investigated. The results showed that it is recommended to utilize silver particles compared to copper ones to achieve higher heat dissipation. And the particles of smaller size may enhance the thermal conductivity of elaborated materials.

INTRODUCTION

In the recent years miniaturization has become one of the main trends in electronic components and systems. At the same time, a strong need has been brought forward to achieve higher performance of these devices and maintain their high reliability in a long period of time [1]. Consequently, efficient methods of thermal management are required to control the temperature of electronic devices during their work conditions.

Currently, the preferred type of package for high performance ASIC and microprocessor devices is a flip chip BGA. Such packages consume and dissipate high amount of power and therefore the generated heat should be removed in an efficient way. To fulfill this task a heat sink with an appropriate thermal interface material (TIM) is the most popular approach used. As TIM materials are responsible for providing a continuous path of heat transfer between heat spreader and BGA component, their properties play a crucial role in achieving suitable degree of heat dissipation in electronic devices [2, 3].

In order to improve TIM properties and thus their thermal performance, carbon nanotubes [1], carbon nanoscroll [4], carbon flakes [5], graphene [6], silver nanoparticles[7], indium matrix polyimide fibers[8] have been utilized as fillers in the TIM composites. This paper fits in this approach of reinforcing TIM with micro- and nanoparticles. Graphite platelet nanofibers (GPN),

graphite powder (GP), silver and copper particles were used for TIM preparation, and the thermal performances were experimental investigated.

EXPERIMENTAL SETUP

Material Preparation

Fig. 1 exhibit images of the fillers used in tested TIM materials, and the filler sizes are summarized in Table I.

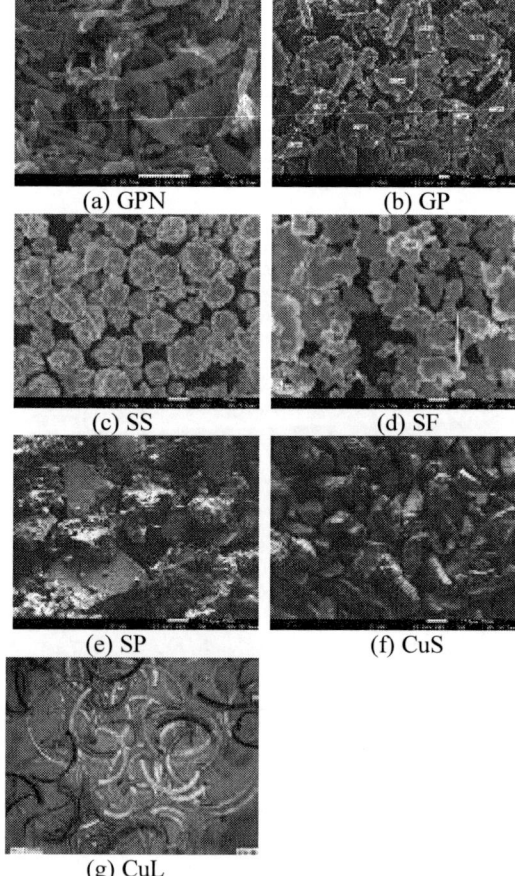

(a) GPN   (b) GP

(c) SS   (d) SF

(e) SP   (f) CuS

(g) CuL

Figure 1: Images of various fillers

978-1-5090-6695-7/17 $31.00 © 2017 IEEE

TABLE I.	PARTICLE SIZES AND MATERIALS

Filler material	Particle size
graphite platelet nanofibers	100 nm x 2.5 µm
graphite powder	44 µm
silver powder (speherical)	1.3-3.2 µm
silver flakes	10 µm
silver particles	160 µm
copper particles (large)	300 x 6000 µm
copper particles (small)	160 µm

Particles used for the TIM reinforcement including the graphite platelet nano-fibers (GPN), graphite powder (GP), spherical silver powder (SS), silver flakes (SF), silver particles (SP), large copper particles (CuL) and small copper particles (CuS). Thermal grease based on silicone (HP) was used as a carrier for preparation of TIM composites. Its thermal conductivity is 1.5 W/mK according to the manufacturer specification [9].

A serious of TIM samples has been prepared and their thermal performances were experimentally measured.

Test Method

The prepared materials were tested by two kinds of measurement setups. One setup consists of two aluminum discs divided by a PMMA plate, as shown in Fig. 2, and the TIM samples were placed between those two discs. The other uses printed circuit board (PCB) with an embedded resistor utilized as a heater, as shown in Fig. 3, and the TIM samples were placed on the PCB.

Temperature distributions were measured by both an infrared camera (FLIR A320) and thermocouples to evaluate the thermal performance of prepared samples.

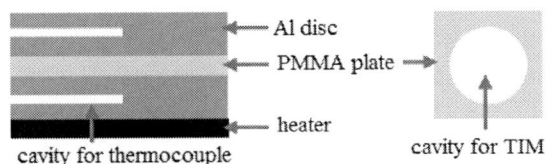

Figure 2: Measurement setup with PMMA

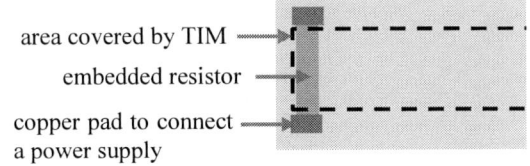

Figure 3: Measurement setup with PCB

RESULTS AND CONLUSIONS

The temperatures for the considered TIM samples have been measured. Fig. 4 shows a temperature distribution recorded by the infrared camera on the top of Al disc. Table II gives the temperature values obtained by the IR measurement on the top surface, and Table III is the values obtained by the thermocouples inside the measurement setup, respectively.

Figure 4: IR image on the top of Al disc for HP/SS.

TABLE II.	TEMPERATURES FOR TIM MEASURED BY IR ON TOP OF AL DISC

TIM	T(°C)		
	100s	*500s*	*1000s*
HP	27.6	44.2	62.7
HP/GPN	27.4	31.7	51.1
HP/GP	29.1	44.6	65.7
HP/SS	29.8	49.1	69.2
HP/SF	30.2	48.0	68.4
HP/SP	31.5	48.5	63.1
HP/CuL	28.5	47.3	63.2
HP/CuS	29.1	47.6	67.4

TABLE III.	TEMPERATURES FOR TIM MEASURED BY THERMOCOUPLES INSIDE

TIM	T(°C)		
	100s	*500s*	*1000s*
HP	27.0	40.1	56.1
HP/GPN	27.6	31.2	49.6
HP/GP	27.7	42.1	58.1
HP/SS	30.1	42.8	59.2
HP/SF	29.6	42.6	59.2
HP/SP	30.3	41.8	55.7
HP/CuL	28.6	42.7	57.4
HP/CuS	26.9	42.1	59.7

Fig.5 is surface temperature field on top of the PCB setup where TIM with spherical silver particles were applied, and Fig.6 is the temperature profile collected along the centerline of the PCB. The temperature values obtained for various TIM sample measurements are presented in Table IV.

The experimental results of all fillers except GPN showed improved thermal properties. The reason why GPN performed differently may be weak dispersion of GPN particles, their geometrical structure or large interface thermal resistance between GPN and thermal grease used in this study [10, 11].

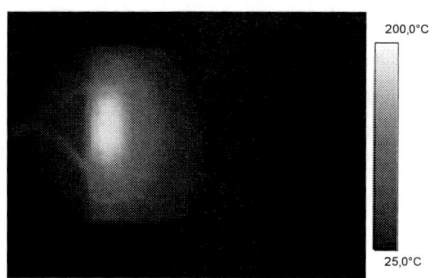

Figure 5: IR image on PCB for HP/SS.

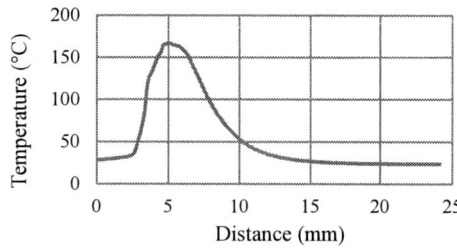

Figure 6: Temperature profile extracted from IR.

TABLE IV.   TEMPERATURE FOR TIM MEASURED BY IR ON PCB WITH EMBEDDED RESISTOR

TIM	T(°C)		
	5 mm*	25 mm	50 mm
HP	205.4	31.6	24.9
HP/GPN	176.0	31.1	23.4
HP/GP	125.3	53.4	33.1
HP/SS	166.4	32.1	23.6
HP/SF	164.8	33.7	23.7
HP/SP	160.8	31.3	24.0
HP/CuL	103.9	57.8	36.7
HP/CuS	173.6	36.8	24.9

* - distance from the left edge of the test PCB

It was also found that the addition of silver particles compared to copper ones exhibited higher heat dissipation capability, and this could be attributed to the higher thermal conductivity of silver [12].

Furthermore, smaller particle size resulted in better thermal performances of the tested TIM samples. And further investigations in this range are planned to be carried out.

ACKNOWLEDGEMENTS

Supports by NSFC (11672171, 11272192) and Poland-China Inter-Governmental S & T Cooperation (36-15) are greatly appreciated.

REFERENCES

[1] J. Idris, T. W. Hon, *36th International Electronics Manufacturing Technology Conf.*, Johor Bahru, Nov 11-13, 2014, pp. 1-5.

[2] L. Li, M. Nagar and J. Xue, *58th Electronic Components and Technology Conf.*, Lake Buena Vista, FL, May 27-30, 2008, pp. 973-978.

[3] L. Larson, Y. Tang, L. Durfee, C. Hale, D. Plante, S. Iruvanti, R. Wagner, T. Davis, H. Longworth, A. Lavoie and R. Langois, *IEEE 64th Electronic Components and Technology Conference*, Orlando, May 27-30, 2014, pp. 236-241.

[4] Y. Wang, Y. Zhang, *IEEE 65th Electronic Components and Technology Conference*, San Diego, May 26-29, 2015, pp. 1234-1239.

[5] I. Sauciuc, R. Yamamoto, J. Culic-Viskota, T. Yoshikawa, S. Jain, M. Yajima, N. Labanok and C. Amoah-Kusi, *IEEE Intersociety Conf. on Thermal and Thermomechanical Phenomena in Electronic Systems*, Orlando, May 27-30, 2014, pp. 426-434.

[6] K. M. F. Shahil, V. Goyal, R. Gulotty and A. A. Balandin, *IEEE Silicon Nanoelectronics Workshop*, Honolulu, un 10-11, 2012, pp. 1-2.

[7] W. Peng, C. Zanden, L. Ye, X. Lu and J. Liu, *IEEE International Conference on Nanotechnology*, Beijing, China, Aug 5-8. 2013, pp. 942-945.

[8] C. Zanden, X. Luo, L. Ye and J. Liu, *19th International Workshop on Thermal Investigations of ICs and Systems*, Berlin, Sept 25-27, 2013, pp. 286-292.

[9] Datasheet of thermal grease HP, AG TermoPasty.

[10] L. Feng, N. Xie and J. Zhong, *Materials*, vol. 7, 2014, pp. 3919-3945.

[11] F. Gardea and D.C. Lagoudas, *Composites Part B*, vol. 56, 2014, pp. 611-620.

[12] http://www.tibtech.com/conductivity.php (accessed: 13.01.2017)

A FAST AND LOW-COST TSV/TGV FILLING METHOD

Jiebin Gu[*], Bingjie Liu, Heng Yang and Xinxin Li
Shanghai Institute of Microsystem and Information Technology, Shanghai, China
*Corresponding Author's Email: j.gu@mail.sim.ac.cn

ABSTRACT

In this paper, we present an alloy via-filling method that does not need pre-metallization of via holes and can be realized on wafer level. Through-Silicon Via(TSV)/Through Glass Via(TGV) with different geometries can be filled simultaneously in few minutes, instead of few hours by electroplating. A specific equipment is made for the alloy via-filling method. The alloy TSV filling method presented in this paper has the potential for industrial applications.

INTRODUCTION

TSV, as a vertical feedthrough interconnection, is the most important interconnection method in advanced packaging, e.g. 3D stacking or silicon interposer. As shown in Fig.1, depends on specific applications, the geometries (diameter and depth of the vias, which are assumed on the same scale) can be in ranges from few microns in stacked memories to hundreds of microns in MEMS and Wireless System-in-Packaging(SiP)[1-2]. In addition, in TGV application, the geometry of vias can also in range of hundreds microns.

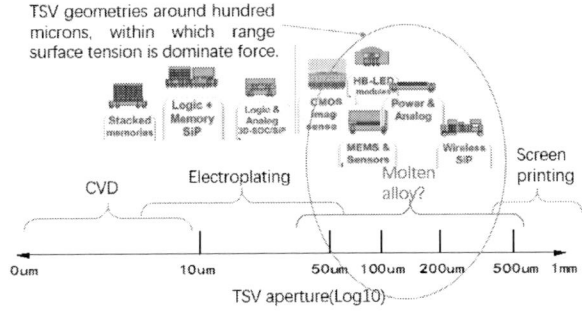

Fig.1. Range of TSV geometries in various applications.

Via-filling is the costliest process in TSV fabrication. Currently, there are three main via-filling methods: chemical-vapor-deposition(CVD), electroplating(EP) and screen printing. However, CVD is limited to vias of few microns because of its extremely low deposition rate. EP is relatively faster, however is also difficult for large vias. In addition, EP uses poisonous electrolyte and is not an environment friendly process. Screen printing is usually for vias in scale of millimeters, and it is also a dirty process. As we can see, for vias in range of tens to hundreds of microns, an efficient and low-cost via-filling

method is still lacking. Via-filling by pressing molten alloy is a promising method to fill that lacking. Alloy via-filling continuously attracts research interest[3-13]. Table.1 compares the pros and cons of alloy and EP via-filling method. However, previous studied alloy via-filling methods still need pre-metallization of via-hole surface and are limited to die-level, therefore they are not fit for industrial applications. In this paper, we present an alloy via-filling method that does not need pre-metallization of via surface and can be scaled-up to wafer-level in any size.

TABLE I.   COMPARASION OF TWO VIA-FILLING METHOD

	Alloy filling	**Electroplating**
Mechanism	Surface tension based	Complex electrochemical reaction
Time cost	In minutes	In hours
Ability to fill different TSV size	Different TSVs can be filled simultaneously	Difficult to fill different TSVs.
filling materials	Inexpensive alloy	Expansive electrolyte, plus additives are needed.
Process complex	No seed layer needed.	Need seed layer.
ECO	non-toxic	Toxic of electrolyte and additives

METHOD, EXPERIMENT AND RESULTS

Method

Compared to previse alloy via-filling methods, the main difference of the method presented in this paper is that, instead of pressing molten metal into via-holes directly, a nozzle wafer is used between molten metal and the TSV wafer. The whole filling process is shown in Fig.2. First, the TSV wafer is sandwiched by a cap wafer and a nozzle wafer, the wafer sandwich is then placed on top of molten alloy surface. The main filling can be separated into two steps:1) Molten alloy is injected into TSV wafer by increasing its pressure. When the pressure of molten alloy is increased to the level that can overcome surface tension and flows into the vias. 2) the pressure of molten alloy is

978-1-5090-6695-7/17 $31.00 © 2017 IEEE   435

then gradually decreased. As pressure dropping, the molten alloy in the TSV wafer is cut off from the alloy surface by surface tension in the nozzle wafer, whose surface has micro-channels. In these micro-channels, molten alloy will break off when its pressure decreasing. The wafer sandwich is then taken down from molten alloy surface. The molten alloy in the TSV wafer is then solidified by cooling. The cap and the nozzle wafer are then released from filled TSV wafer. Finally, the residuals on backside of the TSV wafer is removed by chemical-mechanical polishing(CMP) process.

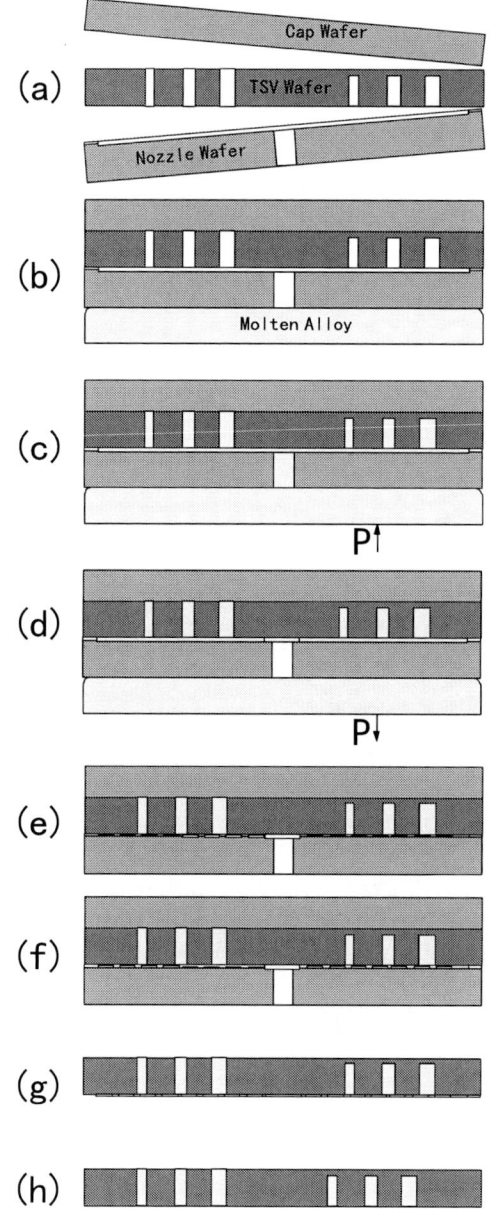

Fig.2. The complete alloy via-filling process. (a) TSV wafer is sandwiched by a cap wafer and a nozzle wafer. (b) the wafer sandwich is place on top of a molten alloy surface. (c) molten alloy is pushed into TSV wafer by increasing alloy pressure. (d) molten alloy in TSV wafer is cut off from outside by decreasing alloy pressure. (e) the wafer sandwich is taken down from molten alloy surface. (f)molten alloy is solidified by cooling. (g) remove cap and nozzle wafer. (h) CMP TSV wafer.

Experiment

A specific equipment is made for the molten alloy via-filling process, as shown in Fig.3. The whole filling process can be controlled automatically and completed in few minutes.

The alloy used is tin alloy, whose melting point is 220C°. The filling process is carried out at 250C°.

Fig.3. The specific-made equipment for the alloy TSV/TGV filling method. The who filling process is automatically controlled by a PLC program.

Results

Fig.4 and Fig.5 shows the TSV and TGV realized by this method.

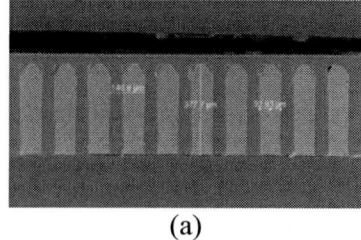

(a)

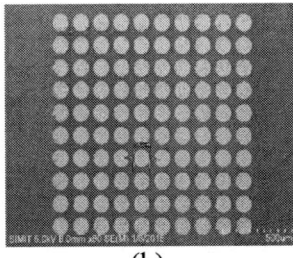

(b)

Fig.4. Blind vias fully filled by the alloy via-filling method.

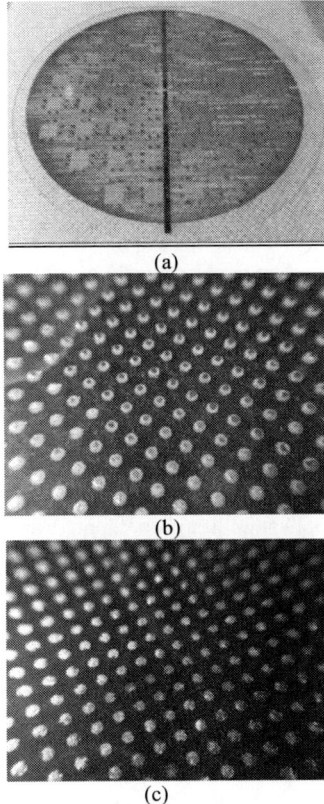

(a)

(b)

(c)

Fig.5. (a) 4" glass wafer after filling without CMP; (b-c) top and bottom view of through-glass vias(TGVs) after CMP. Note (b/c) and (a) are not the same wafer.

CONCLUSION

We successfully demonstrated an alloy via-filling method that does not need pre-metallization and can be done in wafer-level. Large vias can be filled in few minutes without using any poisonous materials. This via-filling method is well fit for relative large TSV/TGV applications, and has the potential for industrial applications.

ACKNOWLEDGEMENTS

We'd like to acknowledge National Natural Science Foundation of China (NSFC 61504158)'s support on this research.

REFERENCES

[1] Bangtao, C., V. N. Sekhar, J. Cheng, L. Ying Ying, J. S. Toh, S. Fernando and J. Sharma (2013). "Low-Loss Broadband Package Platform With Surface Passivation and TSV for Wafer-Level Packaging of RF-MEMS Devices." Components, Packaging and Manufacturing Technology, IEEE Transactions on 3(9): 1443-1452.

[2] Qianwen, C., H. Cui, T. Zhimin and W. Zheyao (2013). "Low Capacitance Through-Silicon-Vias With Uniform Benzocyclobutene Insulation Layers." Components, Packaging and Manufacturing Technology, IEEE Transactions on 3(5): 724-731.

[3] Yamamoto, S., K. Itoi, T. Suemasu and T. Takizawa . "Si through-hole interconnections filled with Au-Sn solder by molten metal suction method. Micro Electro Mechanical Systems". MEMS-03 Kyoto, 2003.

[4] Jiebin Gu, W. T. Pike and W. J. Karl. "A novel capillary-effect-based solder pump structure and its potential application for through-wafer interconnection." Journal of Micromechanics and Microengineering vol.19, p.074005, 2009.

[5] Jiebin, Gu, et al. (2016). "Study of a through-silicon/substrate via filling method based on the combinative effect of capillary action and liquid bridge rupture." Journal of Micromechanics and Microengineering 26(7): 075009.

[6] Young-Ki, K., et al. (2012). Advanced TSV filling method with Sn alloy and its reliability. 3D Systems Integration Conference (3DIC), 2011 IEEE International.

[7] Sato, R., et al. (2012). Study on high performance and productivity of TSV's with new filling method and alloy for advanced 3D-SiP. 3D Systems Integration Conference (3DIC), 2011 IEEE International.

[8] Ko, Y.-K., et al. (2012). "High-speed TSV filling with molten solder." Microelectronic Engineering 89(0): 62-64.

[9] Daquan, Y., et al. (2012). Development of new TSV structure composing of intermetallic compounds. Electronics Packaging Technology Conference (EPTC), 2012 IEEE 14th.

[10] Young-Ki, K., et al. (2011). Advanced solder TSV filling technology developed with vacuum and wave soldering. Electronic Components and Technology Conference (ECTC), 2011 IEEE 61st.

[11] Tsukada, A., et al. (2011). Study on TSV with new filling method and alloy for advanced 3D-SiP. 2011 IEEE 61st Electronic Components and Technology Conference (ECTC).

[12] Ran, H., et al. (2011). Nonlinear thermo-mechanical analysis of TSV interposer filling with solder, Cu and Cu-cored solder. Electronic Packaging Technology and High Density Packaging (ICEPT-HDP), 2011 12th International Conference on.

[13] Jiebin, G., et al. (2011). "Solder Pump Technology for Through-Silicon via Fabrication." Microelectromechanical Systems, Journal of 20(3): 561-563.

978-1-5090-6695-7/17 $31.00 © 2017 IEEE

DEVELOPMENT OF PLATING RESIST FOR FO-WLP

Kenji OKAMOTO

Advanced Electronic Materials Laboratory, JSR Corporation, Yokkaichi 510-8552, Japan

kenji_okamoto@jsr.co.jp

ABSTRACT

We report on the latest ultra-thick photo resist to fabricate high copper pillars over 100um for FO-WLP. The new resist shows excellent coating performance to achieve 100um thickness in a single coat, or over 200um thickness with double coating, on 12 inch wafer. The resist provides good coating uniformity without bubbles, defects, wrinkles or other errors. On top of that, the new material design enables finer resolution for ultra-thick films, with an aspect ratio of 4 and beyond. We also report high copper pillars fabricated with the new THB resist. It is expected that the new ultra-thick resist will be the best candidate for FO-WLP. We will discuss the new material concept in more detail.

Keywords: Negative tone resist, FO-WLP, Ultra high film thickness, Spin coating, Electro-plating

INTRODUCTION

In recent years, novel electronic products, like mobile phones and personal computers, have become dramatically smaller and more highly functionalized. Parallel to these market trends, packaging structures for semiconductors are also required to become smaller, thinner and more complicated. To satisfy various requirements, packaging technologies such as Flip-chip wafer bumping, PoP, 2.5D interposers, and 3D-TSV are under development.[1-5] In recent years, FO-WLP has been the focus of IDM (integrated device manufacture) and OSAT (outsource assembly and test) because of the high process costs of 2.5D and 3D-TSV technologies. Currently the various FO-WLP structures and processes proposed are set to satisfy increases in the number of I/O and complicated and functionalized applications. In this current technology trend, novel materials for finer RDL and higher Cu pillar processes are required to achieve those packaging structures. If the required resolution for the plating pattern is not so tight, normally dry film is applied for lithography. But for the advanced high-density FO-WLP, finer resolution with a higher aspect ratio is required. To achieve good patterning performance for ultra-thick film applications, such as over 100um after development, it is important to control photo-speed, solubility to alkaline developer and

transparency. If transparency is low, irradiation light cannot reach the bottom of the resist due to decrement of light attenuation. From the standpoint of process cost, faster photo-speed and shorter development times are significantly effective to reduce the process cost and also process chemicals cost. By controlling the photo-speed, development speed, and transparency properties of the photoresist, we found the new material design of negative tone resist to support ultra-thick film application. The resist can achieve faster photo-speed (< 2,000mJ/cm2), higher resolution (< 50um), high aspect ratio (>/= 4.0) with straight pattern profiles, and excellent strip-ability. We report the study results of these novel materials with high resolution for ultra-thick application, especially FO-WLP, in this paper.

RESULTS

We have adopted negative tone free-radical type resists because we expect them to have excellent resistance for electroplating. Negative tone acrylic resists have high potential as electro plating resists for various film thickness targets.[6-7] In recent years, high resolution photoresist for ultra-thick application has been required for FO-WLP. We have already developed the ELPAC series as commercial grade for C4 bumping, to achieve process-friendly photo resists with lower exposure doses and shorter development times, novel material design have to be designed.

We prepared the negative tone photo resist which consisted of optimized polymer and cross linker system to achieve thick film thickness application based on keeping high transparency and good solubility for TMAH developer. In order to achieve faster development speed and straight profile, it is important to optimize the solubility unit in the polymer and optimized photo active compound with good efficiency to ghi-line light source. A well-balanced polymer structure should be chosen based on each monomer type, since the impact of wettability and solubility depend on its structure. New negative resist THB-A was developed and evaluated for the performance of coating, lithography, plating, and stripping.

Coating performance evaluation

Figure 1 shows the results of coating performance of new THB-A on a 12 inch bare-Si wafer. A coating film thickness of 179um was achieved with good uniformity. It is notable that the sample was coated with no bubbles and excellent wafer edge profile.

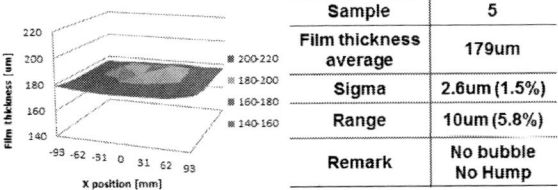

Sample	5
Film thickness average	179um
Sigma	2.6um (1.5%)
Range	10um (5.8%)
Remark	No bubble No Hump

Figure 1. Coating performance of 12 inch bare-Si wafer

Resolution performance evaluation

Resolution performance of new THB-A at 160um film thickness after development is shown in Figure 2. Resolution of 40um squared hole was achieved with high aspect ratio (=4.0) and straight pattern profiles.

	100umSQ	75umSQ	50umSQ	40umSQ	30umSQ
SEM images					

Figure 2. Resolution performance at 160um film thickness

Plating performance evaluation result

Figure 3 shows the x-section profile of THB-A after Cu plating and resist stripping. Cu plating was carried out with a test substrate prepared at 160um film thickness shown in Figure 3. Enthon Microfab SC-50 (w/ R1 and R2), with 4ASD and 10ASD as current density, was applied. After plating, the test substrate was soaked in a typical stripping solution such as DMSO/TMAH for 30 mins at 40°C, and rinsed with DI water. The Cu pillar pattern with a height of 155um was obtained without stripping residue and issue, such as under-plating or rough surfaces. THB-A showed good strip-ability and plating resistance.

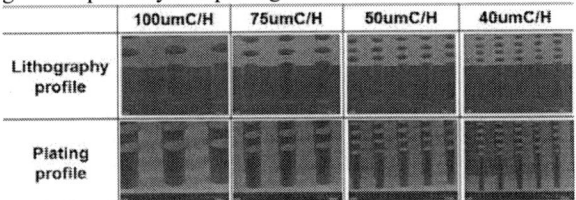

	100umC/H	75umC/H	50umC/H	40umC/H
Lithography profile				
Plating profile				

Figure 3. Pattern profile before plating and after stripping

Evaluation over 200um film thickness condition

Square pattern lithographic and resolution performance of THB-A at 200um film thickness were evaluated and results are shown at Figure 4. At over 200um film thickness, THB-A showed a good, straight pattern profiles. The resolution was achieved with high aspect ratio (=4.0).

	100umSQ	75umSQ	50umSQ	40umSQ
SEM images				

Figure 4. Resolution data at 200um film thickness

SUMMARY

In this study, new negative resist which is consists of solubility controlled polymer and optimized cross-linker/photo- initiator was investigated for ultra-thick film applications. It was found that new negative tone resist, THB-A showed good coating uniformity and lithography performance over 150um film thickness application with straight side wall and clean bottom profile. It was found the key parameters of plating photo resist design for thick application are solubility for THMA developer and controlled transparency which has a major impact on obtaining straight pattern profiles in the ultra-thick film. Based on this result, we developed a novel bump resist which shows excellent photo-speed, developability, strippability and resolution performance at 160um and 200um film thickness conditions in addition to excellent resistance for electroplating which was showed by traditional ELPAC series. Our new resists contribute to the progress of FO-WLP packaging technology.

REFERENCES

[1] G. J. Jung, "Structure and Process Development of Wafer Level Embedded SiP (System in package) for Mobile Applications", 2009 11th EPTC, p.191

[2] Yoichiro Kurita, "A 3D Stacked Memory Integrated on a Logic Device Using SMAFTI Technology", 2007 ECTC, p.821

[3] S. Yoon, et.al., "Mechanical Characterization of Next Generation eWLB (embedded Wafer Level BGA) Packaging", ECTC, pp441-446, 2011

[4] M. Santarini, "Stacked and Loaded: Xilinx SSI, 28-Gbps I/O Yield Amazing FPGAs", Xcell Journal, No.74, pp.8-13, 2011

[5] M. Murugesan, et.al., "Wafer Thinning, Bonding, and Interconnects Induced Local Strain/Stress in 3D-LSIs with Fine-Pitch High-Density Microbumps and Through-Si Vias", IEDM, pp.2.3.1-2.3.4, December 2010

[6] K. Mori, M. Hanamura, T. Kai. 2009, ICEP

[7] H. AKIMARU, H. ISHIKAWA, H. SAKAKIBARA, S. NARUSE, K. OKAMOTO, K. INOMATA. 2014, ICEP

IN-SITU TEM OBSERVATION OF IMC EVOLUTION AT ATOMIC SCALE

Chaolun Wang, Xing Wu*

Key Lab of Multidimensional Info. Processing, Dep. of Elec. Eng., East China Normal Uni.,
Shanghai, China
*Corresponding Author's E-mail: xwu@ee.ecnu.edu.cn

ABSTRACT

The formation mechanism of Cu-Al IMC is critical to control the contact quality of interconnections between Al bond pads and Cu wires. The *in-situ* TEM is a powerful tool to analyze the phase evolution and defects formation of IMC in real time. TEM images reveals that the amount of IMC increases with raising of annealing temperature and duration of time. Energy dispersive spectrometer (EDX) and high resolution TEM (HRTEM) demonstrate that the IMC is composed of $CuAl_2$, $CuAl$, and Cu_9Al_4.

INTRODUCTION

The coming of information and intelligent society demands massive applications of high performance and low cost of integrated circuits. For semiconductor integrated circuits, the high conductivity and low cost of Cu wire is a competing alternative to Au wire. The mechanism of the IMC phase formation during annealing is a key problem to achieve high quality and reliable contact between Cu wire and Al bond pad. Except the mostly reported $CuAl_2$, $CuAl$ and Cu_9Al_4 stable phases, transition phases like Cu_4Al_3, Cu_3Al_2 and Cu_3Al are only presented by a few researchers due to their rapid transformation and small volume[1-2]. The degradation and defects formation around the Cu-Al interface are cracks and Al oxide line, which are related with bonding process parameter, residue stress and intrinsic stress of IMC, or full consumption of Al layer with more Cu-rich phases[3]. However, the reported data are still not enough to fully understand the mechanism of the degradation. Other aspects, for example the influence of the mold material should be considered. Advanced technologies for sample characterization such as *in-situ* TEM and FIB offer the way of analyzing the morphology, structures and composition of the local area of samples at nanoscale in real time. Those technologies are powerful tools to reveal the mechanism of the IMC phase evolution. In this paper, the investigation of Cu-Al IMC evolution and degradation during annealing are presented in detail using *in-situ* TEM technology, and the results are compared with the ex-situ post-annealing counterpart.

EXPERIMENTAL

The Cu wire with the diameter of 22 μm is thermosonically bounded to a metallization pad of AlSiCu with the thickness of 1.5 μm. The bonding force is 25-35 gf. The ultrasonic power is 120-150 with the frequency of 120 kHz, and temperature at 180 ℃, with additional preultrasonic power of 50, then encapsulated with green compound.

The *in-situ* annealing sample is prepared by microsection of the molded sample without mold compound along the ball center using grinding paper and fine polish, followed by FIB cutting to get the TEM samples. The FEI Titan TEM equipped with image aberration corrector and Gantan 628 single tilt heating holder is used for sample analyzation. The samples are heated from 50 ℃ to 200 ℃ for different interval duration, and followed by a temperature 220 ℃ for 240 min.

The *ex-situ* post-annealing samples are prepared by annealing at 150 ℃ for 100, 250, 500 and 1000 h, respectively. Then the samples are sectioned along the ball bond interface, followed by FIB cutting. The morphology and defects of the interface are investigated from FIB images. The IMC phases are identified by the energy dispersive X-ray spectrometer (EDX).

RESULT AND DISCUSSION

Fig.1 shows the Cu-Al IMC growth at different *in-situ* annealing duration. Before annealing, thin layer of IMC with the thickness of 10-30 nm at the Cu-Al interface is presented in Fig.1 (a). The composition of IMC is determined as isolated $CuAl_2$ and Cu_9Al_4 nanocrystals near Al and Cu layer, respectively by indexing Fast Fourier Transform (FFT) of HRTEM images (not shown here).

The growth evolution of IMC behaves differently with temperature as shown in Fig.1 (b)-(f). Below 200 ℃ the growth velocity of IMC is slow, while at 220 ℃ the IMC grows fast. The growth front of IMC maintains its initial profile as-bonded condition. The Al layer is entirely consumed after annealing at 220 ℃ for 240 min, and the IMC continues to grow without any void or crack.

The TEM investigation of Cu-Al IMC annealed at 220 ℃ for 24 h is presented in Fig. 2. The inhomogeneous IMC layer composes of two types of compounds with different contrast. The dark and gray phases in the IMC layer indicated by the larger red squire and two small blue squires shown in Fig.2 (a) are identified as $CuAl_2$ and Cu_9Al_4 phases by HRTEM (b), (e) and corresponding FFT (c), (f), respectively, the same as the IMC before heating. The absence of CuAl phase may be due to the volume is too small to detect. The transition phases such as Cu_4Al_3, Cu_3Al_2 and Cu_3Al are not found. Fig.2 (d) is the HRTEM image of the area indicated by the red ellipse in Fig.2 (a). The amorphous phase indicated by the oval circle is Al

978-1-5090-6695-7/17 $31.00 © 2017 IEEE

oxide confirmed by the EDX. The Al oxide is found embedded in the Cu_9Al_4 phase.

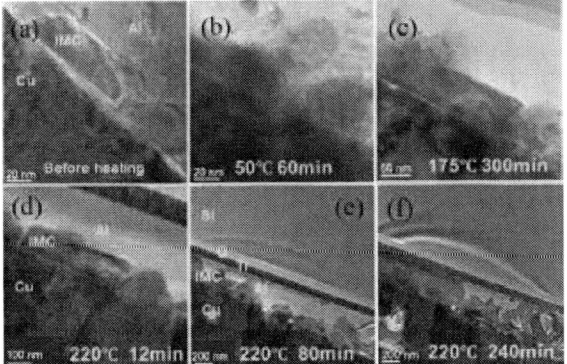

Figure 1: The morphology evolution of IMC during in-situ heating [4].

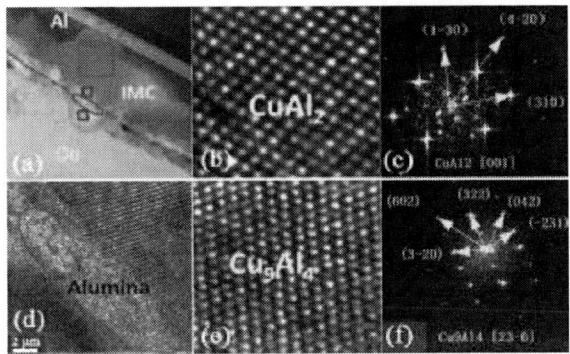

Figure 2: The TEM analysis of morphology and composition of IMC annealed at 220 ℃ for 24 h at the Cu-Al interface [4].

Due to impact between Cu wire and Al pad, the ex-situ annealed samples at 150 ℃ for 100, 250, 500 and 1000 h have concave profiles at the Cr-Al interfaces shown in Fig.3, and parts of Al are squeezed out as demonstrated in Fig.3 (a). At the Cu-Al interface, cracks appears (0.8 μm) after100 h annealing, and they propagate to 16 μm after 1000 h annealing.

Cracks do not exist below the Cu_9Al_4 and Cu interface as shown in Fig.3 (c). This is because that Cu-rich IMC is more fragile than Al-rich IMC. Elements located at the cracks probed by EDX are mainly oxide. The rare vertical crack in Fig.3 (c) may result from the mismatch of the coefficient of thermal expansion. Fig. 3 (c) shows small voids in IMC that do not grow bigger after annealing, therefore they are not Kirkendall voids as reported in the Au-Al IMC system. The voids may be generated by contaminations on the bond pad.

Three different contrast of IMC phases are found in the back scatter electron image shown in Fig.3 (c), and their compositions are analyzed by EDX. Except the

$CuAl_2$ and Cu_9Al_4 phases near Al and Cu layers, respectively, which are similar to the *in-situ* heating study, an extra CuAl phase is discovered between $CuAl_2$ and Cu_9Al_4 phases. Similar to the *in-situ* heating, thin line of Al oxide also exists in IMC between CuAl and Cu_9Al_4 phases as a residue of oxidized Al pad in Fig. 3 (c), which affects the homogeneity growth of IEM and increases the contact resistance. At the initial state, the diffusion of Al atoms into Cu is dominant, which pushes the Al oxide line upward, and stay in the IMC phases[1].

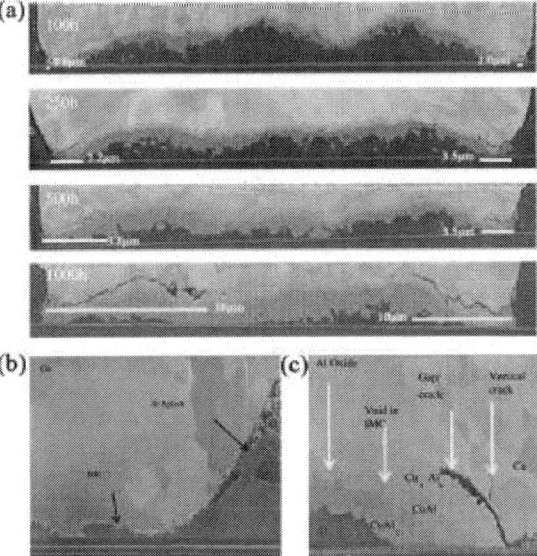

Fig.3. FIB cross section image of ex-situ annealing of Cu-Al interface. (a) Samples annealed at 150 ℃ for different time spans. The yellow lines of (a) show the horizontal lengths of cracks. (b) Sample before annealing. (c) Sample annealed for 1000 h [4].

CONCLUSION

In-situ annealing study of Cu-Al IMC provides the details of IMC evolution presenting the fast growth of IMC at 220 ℃. The composition and defects of in-situ annealed IMC is compared with the ex-situ annealed samples, showing the absence of CuAl phase and cracks. Cu wire forming a more stable IMC without Kirkendall void shows a great potential for wire interconnection.

ACKNOWLEDGEMENTS

We would like to acknowledge supports from NSFC under grant Nos. (11504111, 61574060), Projects of Science and Technology Commission of Shanghai Municipality Grant Nos. (15JC1401800, 14DZ2260800), Program for Professor of Special Appointment (Eastern Scholar), and Shanghai Rising-Star Program

(17QA1401400).

REFERENCES

[1] H. Xu, C. Q. Liu, V. V. Silberschmidt and Z. Chen. *J. Electron. Mater.* vol. 39, 2010, pp. 124-131.

[2] Y. H. Lu, Y. W. Wang, B. K. Appelt, Y. S. Lai and C. R. Kao. *Electronic Components & Technology Conference*, vol. 301, 2011, pp. 1481-1488.

[3] C. F. Yu, C. M. Chan, L. C. Chan and K. C. Hsieh. *Microelectron. Reliab.* vol. 51, 2011, pp. 119-124.

[4] Y. Y. Tan, Q. L. Yang, K. S. Sim, L. T. Sun and X. Wu. *Microelectron. Reliab.* vol. 55, 2015, pp. 2316-2323.

THE STUDY ON THE MOLDABILITY AND RELIABILITY OF EPOXY MOLDING COMPOUND

Wei Tan [1], Hongjie Liu[1], Yangyang Duan[1], Lanxia Li[1], Xingming Cheng[1]*
Dong-en zhang[2], Junyan Gong [2]

[1]Jiangsu HHCK Advanced Materials Co., Ltd., Lianyungang, Jiangsu, 2220004, China
[2]Dept. of Chemical Engineering, Huaihai Institute of Technology, Lianyungang 222005, China
*Corresponding Author's Email: wei.tan@hhck-em.com

ABSTRACT

In this paper, different type of wax, wax content and hot hardness at 175°C on the release force and adhesion of molding compound were studied, The studies reveal that the hot hardness is playing a very important role to balance the conflict of reliability and moldability. With higher hot hardness of epoxy molding compound, the release force can be kept at same level with less wax content which can increase the adhesion significantly. To achieve the high reliability and long moldability, the best wax content should be near 0.2 and the hot hardness should be larger than 85.

Keywords—epoxy molding compound; adhesion; release force; hot hardness; wax; reliability; moldability

INTRODUCTION

With the rapid development of semiconductor packaging, it is essential that the packaging cost should be kept lower and lower. Even for the lower cost, the reliability requirements of the final packaging products must be maintained at the same or even higher level.

There are many methods to reduce the packaging cost. The epoxy molding compound is one of the most important factor. Currently, the long term moldability on SOP8 package has been prolong from 200 shots to 400-500 shots while it can pass JEDEC MSL3 requirement and the cost was decrease about 40%-50%.

To meet such stringent requirements, the epoxy molding compound can't reduce the cost by simply change the raw materials from high-end type (MAR, Biphenyl etc.) to low-end type(ECON，Xylok etc.).

Both effects of waxes and catalysts on the Shore D hardness at 175℃ ,release force and adhesion are studied in this paper.

The waxes selected in this paper are Carnauba wax, ester wax, S wax, stearic acid and Oxidized polyethylene wax[1,2].

The catalysts selected in this paper are TPP, 2MZ, 2MAOK, 2P4MHZ and another organophosphorus type catalyst[3].

EXPERIMENTAL

The waxes and catalysts were formulated into a basic epoxy molding compound system which contains biphenyl resin(CER3000L,Nippon Kayaku Co., Ltd), with MAR resin(MEH7851SS, Meiwa Plastics Ind,Ltd.) as an hardener, the ratio of Epoxy value / hydroxyl value is 1. The filler content is about 88% and the couple agent content is 0.3%.

The compound was prepared through dry blending, followed by melt-mixing, extrusion into sheet form, fine-grinding into powder, and then pelletizing into performs.

Standard transfer molding techniques were used to fabricate the test specimens. In mold cure time was 120sec at175℃ followed by a six-hour post mold cure at the same temperature for all test parts.

A Tab pull sample can be taken as two pieces of leadframes connected by EMC, shown as Figure 1. The enclosed area is encapsulated by Epoxy molding compound and the shadowed area on the leadframe surface is attached by EMC. After transfer-molding, the tap pull samples were post cured in the oven at 175 °C for 6 hours. The tensile test was done with a Universal testing machine (AGS-5kNA,Shimadzu Corporation). The tensile force was reported as the indication of adhesion force between epoxy molding compound and leadframe. The leadframe surface are plated by silver.

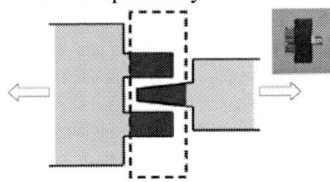

Figure 1: The schematic graph of a tap pull sample for the tensile test. Two pieces of leadframes are connected by EMC.The enclosed area is encapsulated by EMC and the shadowed area on the leadframe surface is attached by EMC. The right graph is a tap pull sample photo taken by camera.

A truncated conic release force test specimen was fabricated by a transfer molding process with a three layer mold chase. When the sample was cured, push the sample out of the mold immediately at the molding temperature. The maximum thrust value was recorded by the pull/push gauge as the release force.

The Shore D hot hardness is measured on an epoxy molding compound specimen at 175℃ which was cured 90 second at the same temperature.

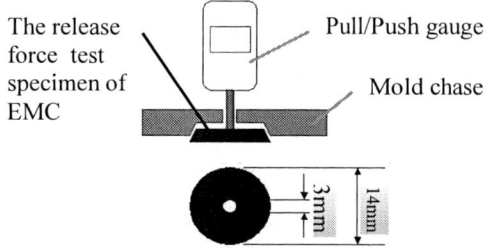

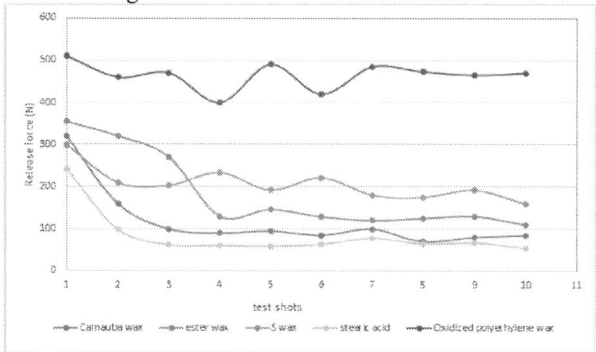

Figure 2: The schematic graph of a release force test.

RESULTS AND DISCUSSION

The effects of different wax on the release force was shown on Figure 3.

Figure 3:the effects of different wax on release force

Figure 3 shows that the Carnauba wax and stearic acid can significantly reduce the release force, the S wax and ester wax are a bit poor while the oxidized polyethylene wax decrease very little.

Compare to the Carnauba wax and Ester Wax, the oxidized polyethylene wax has the feature of high molecular weight and high compatibility with epoxy resin, it is is not easy to come to the mold surface at the transfer molding process. On the contrary, the Carnauba wax can come to the mold surface much easier thus significantly reduce the release force

Figure4 shows that the effect of waxes on the adhesion on the silver leadframe. It reveals the stearic acid can significantly reduce the adhesion while the oxidized polyethylene wax has less influence.This is similar to the effects on release force of mold surface, the amount of oxidized polyethylene wax come to the leadframe surface is much less at transfer process. It is due to the oxidize polyethylene wax has higher molecular weight and better compatibility in the epoxy resin.

Figure 5 shows that the effect of wax on the Shore D hot hardness at 175℃ for 90 seconds. The stearic acid has great influence on the hot hardness due to it can increase the gel time about 30%. The Longer gel time causes the lower crosslinking density at the same curing time and temperature which is shown lower hot hardness.

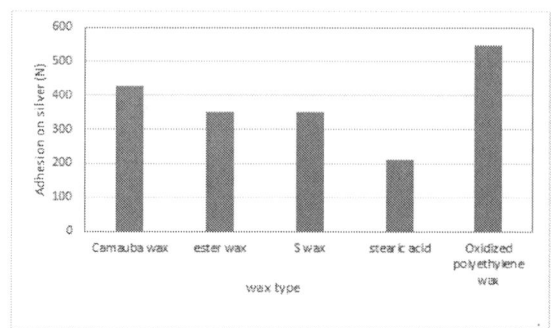

Figure 4: the effect of wax on the adhesion on the silver

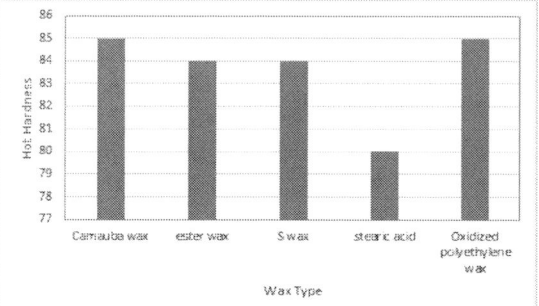

Figure 5: the effect of waxes on hot hardness

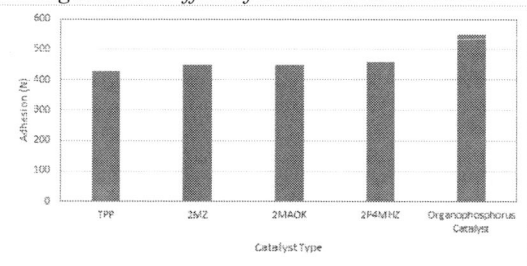

Figure6: the effect of catalyst on the adhesion on the silver

The Figure 6 disclosed that how catalyst effect the adhesion of epoxy molding compound on the silver leadframe. The imidazole type catalyst and TPP are similar. But the organophosphorus type catalyst has great help for the adhesion of epoxy molding compound on the silver leadframe.

At the same time the organophosphorus type catalyst can increase the shore D hot hardness which is shown in the Figure 7. Compare to the organophosphorus type catalyst, The TPP has very low hot hardness although the gel time is same. The lowest hot hardness of 2P4MHZ is caused by the very long gel time compare to 2MZ and 2MAOK.

Figure8 shows that the release force of epoxy molding compound with different catalyst added. The lowest release force of 2P4MHZ and TPP are caused by the lowest hot hardness. It is much softer at high temperature and much easier to come out from the mold due to high shrinkage. On the contrary, the organophosphorus type catalyst has great release force because the highest hot hardness.

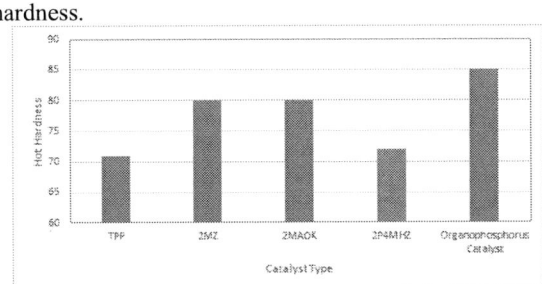

Figure 7: the effect of catalyst on the Hot Hardness

Theoretically, to meet the requirements of long term moldability, the epoxy molding compound must have less release force, the clean and uniformity of waxes on the mold surface after many transfer molding shots. And to meet the requirements of reliability of package, the epoxy molding compound should have less waxes in the formulation.

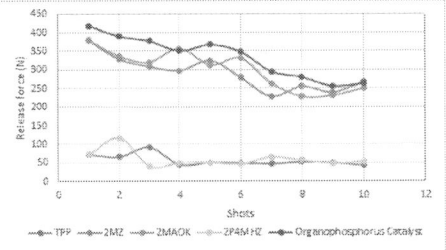

Figure 8: the Release force of different Catalyst

The authors found that higher hot hardness has great help on the moldability. To maintain the same release force, the wax content can be decreased at higher hot hardness. It is disclosed in the table 1. An epoxy molding compound which has both good long term moldability performance and good reliability performance can be made by higher hot hardness and less wax as group 2 in the table 1.

Table 1 the moldability and release force with different hot hardness and wax content

Items	Group1	Group2	Group3	Group4
Hot Hardness at 175℃，90s	80	80	72	72
Wax Content(%)	0.5	0.3	0.5	0.3
Release force after 10 shots (N)	320	400	200	180
Moldability on SOP8 package	mold stain after 200 shots	Less mold stain after 280 shots	mold sticking after 150shots	mold sticking after 100 shots
Reliability	JEDEC MSL3	JEDEC MSL3	JEDEC MSL3	JEDEC MSL3

Figure 9 is the package surface of epoxy molding compound (Group4) after 100shots, the flow mark is obviously found on the package surface after mold sticking happened.

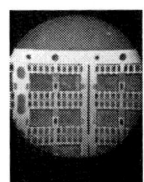

Figure9: SOP8 package surface of group4 in table 1 after180 shots,

Combine the figure 4 to figure7, the author found that with the organophosphorus type catalyst, oxidized polyethylene wax, carnauba wax and ester wax, an epoxy molding compound with highest long term moldability and best reliability performance can be made. Table 2 shows its basic properties.

Table 2 basic properties of epoxy molding compound

Items	Group5
Hot Hardness at 175，90s	85-87
Wax Content(%)	0.15-0.25
Release force after 10shots(N)	300-350
Moldability on SOP8	good mold surface after 400 shots
reliability on SOP8 package	JEDEC MSL3

Figure 10 is the SOP8 package surface of group5 after 400 shots. The package surface is much clean and there is no mold sticking or mold stain found after 400 shots.

 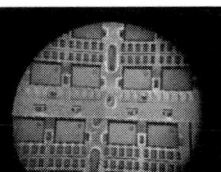

Figure10: SOP8 package surface of group5 in table 2 after 400 shots

CONCLUSION

1. Hot hardness plays a very important role to balance the conflict of reliability and moldability.
2. Different type of catalyst and wax have great effect on the hot hardness, adhesion and release force.
3. To achieve the high reliability and long term moldability , the best wax content should be near 0.2 and the hot hardness should be larger than 85.

REFERENCES

[1] Wei Tan, Hongjie Liu, Lanxia Li,Yanshu Wan. *The Effect of Wax on The Properties of Epoxy Molding Compound, 2011 International Conference on Electronic Packaging Technology & High Density Packaging,* pp.407-409

[2] Clariant GmbH, *Waxes by Clariant (Production, characteristics, and applications),* Division Pigments & Additives, Edition : May 2003/W 320 GB

[3] http://www.shikoku.co.jp/eng/products/curezol.html

978-1-5090-6695-7/17 $31.00 © 2017 IEEE   445

HIGH PRODUCTIVITY PVD SOLUTION FOR AN EVER-EVOLVING ADVANCED PACKAGING INDUSTRY

Frantisek Balon*, Patrick Carazzetti, Juergen Weichart, Mohamed Elghazzali and Mike Hoffmann
Evatec Ltd., Hauptstrasse 1a, CH-9477 Trübbach, Switzerland
*Corresponding Author's Email: Frantisek.Balon@evatecnet.com

ABSTRACT

Advanced Packaging relies heavily on the use of organic/polymer films and mold wafers such as Fan-Out Wafer-Level-Packages (FOWLP). Processing of such wafers poses challenges, potentially resulting in a high contact resistance (Rc), yield loss, increased maintenance costs and low tool productivity. In this paper we focus on novel approaches how to overcome these challenges. It is shown that the combination of Atmospheric Batch Degas and Cooler Modules in connection with the latest generation of Arctic ICP Sputter Etch results in a throughput of more than 45wafers/hour, Etch Shields Kit Lifetime exceeding 30'000 wafers, low and stable Rc values in RDL/UBM processing.

INTRODUCTION

Polymer/organic materials are already well established in RDL/UBM Advanced Packaging providing vital dielectric properties as well as chip passivation [1,2]. However, in the case of FOWLP, the challenge is greater as the wafers are re-constituted out of silicon dies embedded into epoxy mold compound resulting in heavy outgassing during vacuum processes and critical temperature limitations (<120°C) imposed by the polymer material used. Outgassing contaminates the plasma process environment in a conventional PVD tool, which may be detrimental for IC device performance especially in terms of contact resistance (Rc). Rc controls the frequency response of the device, since the charging and discharging of the line resistance is a major cause of power dissipation in high clock rate digital electronics. Rc also causes power dissipation via Joule heating in low frequency and analog circuits and having a low Rc for the contacts and interconnects in the final device package is therefore paramount [3].

One of the main challenges is that the wafer degassing process must be made more efficient and completed within a limited time to avoid a negative impact on the throughput of a PVD tool, which is not acceptable in the very cost-driven packaging industry. Furthermore, the etched polymer films redeposit on the protective chamber shielding and with increasing thickness there is an elevated risk of particle generation, resulting in short maintenance intervals and limiting productivity of PVD tool. All these factors mean that new strategies have to be developed for optimum wafer degassing at a temperature

of 120°C or below as well as for an effective low temperature ICP Sputter Etch to remove the native oxide from the contact pads [4].

EXPERIMENTAL WORK

Most effective degas approach

In general, two different concepts of batch degassers are used, either for vacuum [5] or atmospheric operation [6]. The advantages of the batch degasser in atmosphere are a much faster and more controllable heat-up of the wafers along with a simpler design. Diffusion of volatile components from the bulk material to the surface is driven only by temperature whilst desorption is driven by the concentration gradient which is the same whether desorption takes place in vacuum or in atmosphere. The advantage in atmosphere is that the concentration gradient can always be kept high if a laminar flow of inert gas is applied, while in vacuum the degassing of the volatile components has to take place via less efficient molecular diffusion.

The Vacuum Batch Degasser has much higher risk of contamination of already outgassed wafers coming form the introduction of fresh wafers, to the stack and the associated increase in background pressure from volatiles species. This results in a slowdown of the degasing process and increased risk of Rc instabilities. Eliminating this potential risk would require fresh wafers only to be added once all outgassed wafers have left the degas chamber resulting in very low throughput. In contrast, Atmospheric Batch Degasser with laminar flow concept does not show any such changes in process pressure if fresh (not degased) wafers are loaded into the same environment as already outgassed wafers. Rc therefore remains unaffected and throughput is not limited.

An Atmospheric Batch Degasser (ABD) has been implemented on CLUSTERLINE®300 and HEXAGON for FOWLP processing. The ABD has a maximum 44-wafer capacity and can accommodate Fan-In or Fan-Out wafers with bow up to 8mm. Figure 1 shows the RGA spectra of two plain mold wafers (a) poorly degassed (from a single wafer degassing chamber) and (b) ABD degassed mold wafer during the ICP Sputter Etch plasma process. The badly degased mold wafer shows an increase of partial pressure of H_2O, which can result in contact contamination and therefore Rc deviation from the baseline. In contrast, the ABD degassed wafer clearly

978-1-5090-6695-7/17 $31.00 © 2017 IEEE

shows a decrease of the partial pressure of H2O during the plasma etch process which ensures that resulting Rc will be low and stable within the defined baseline. The decrease of H2O partial pressure is related to proper condensation on the arctic cold chamber shields.

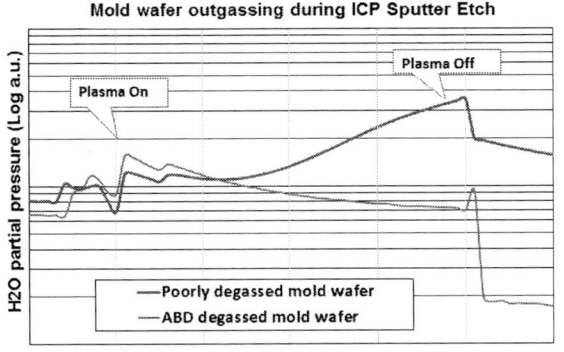

Figure 1: RGA spectra of two wafers (a) poorly degassed and (b) ABD degassed mold wafer during ICP Sputter Etch plasma process

ABD performance was demonstrated by Rc measurements in a production environment. For this purpose a direct comparison between the novel ABD and the conventional single-wafer Vacuum Degas Module (VDM) was made and is shown in Figure 2. It is seen that the mean value of Rc for ABD is 50% lower compared to VDM. In addition, the spread of Rc values for the ABD is narrower, i.e. ±10% in contrast to ±20% achieved by the VDM. Results prove that both the effectiveness of the degassing process and process stability are more efficient in the ABD. The statistical data were collected in production run over a several thousand wafers.

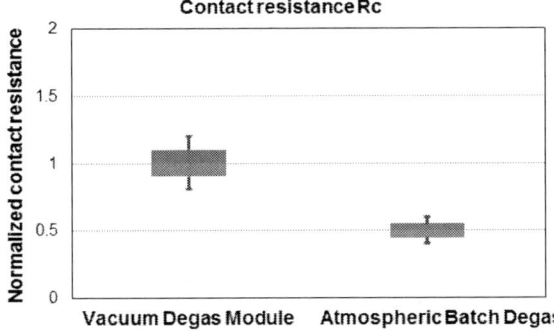

Figure 2: Vacuum Degas Module vs. ABD: Normalized Rc values achieved with different degas strategies

Temperature budget and Arctic Cooler Module
The introduction of the Arctic Cooler Module (ACM) is a very efficient way to meet stringent wafer processing temperature requirements (e.g. <120°C) and still achieve high throughput in FOWLP processing. Figure 3 displays the wafer temperature profile throughout the entire process flow run on a HEXAGON indexer tool. The simulation takes into account a mold wafer thicknesses of 400um and 800um, and is based on the following physical processes: (1) a cooling step in the ACM, (2) ICP-Etch with 300Å removal sequentially split over two ICP chambers to limit peak wafer temperature and enable the requested throughput of 40/50wph, (3) + (4) sputter deposition of Ti-100nm and Cu-200nm, respectively.

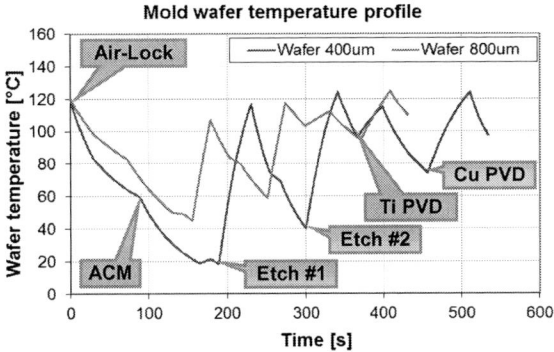

Figure 3: Mold wafer temperature profile across the entire UBM/RDL flow carried out on HEXAGON equipped with ABD and ACM for two wafer thicknesses of 400um and 800um, respectively. The etch removal was 30nm (SiO2 equivalent) for both simulations

For the majority of Fan-In applications the ACM is not mandatory as the allowed thermal budget during RDL/UBM processing is below 200°C (depending on PI/PBO passivation). If ACM is not used the peak wafer temperature will be higher than ~160°C; however, the wafer throughput will be unaffected, thereby resulting in lower cost of ownership (CoO).

Production solution: New Generation of ICP Sputter Etch
Contact pad cleaning prior to the metallization process is a key aspect of UBM/RDL process as it directly influences device contact resistance. In CLUSTERLINE®300 and HEXAGON process modules it is performed by ICP Sputter Etch, where in order to achieve the lowest Rc and the highest tool productivity the process environment is maintained at temperatures considerably lower than room temperature (with setpoint <-20°C also referred as "Arctic" ICP Etch concept). In addition, the latest generation of ICP Sputter Etch uses an optimized set of shields combining highest pumping conductivity to reduce the level of volatile contaminants, with optimum protection of the chamber walls to mitigate the risk of particle generation. The performance of the

optimized shield set was repeatedly verified in production environment where an unprecedented shield lifetime exceeding 30'000 wafers with PI passivation was demonstrated [7]. Furthermore, this particle performance is also valid for other polymer material processing i.e. PBO or alternate PI/PBO production. Figure 4 presents the measured particles performance monitored over a production run of 30'000 wafers and normalized Rc values. The actual levels for particles for bin sizes of both 1-10um and >10um are very low compared to the specified control limits. Rc is stable over entire kit life with values ~50% below the specified level.

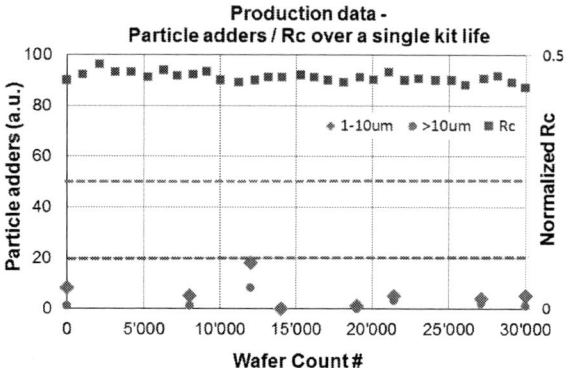

Figure 4: Production data: Normalized Rc and adders count during a single etch kit life of 30'000 wafers

Throughput and Productivity

Evatec's single wafer sputter tools CLUSTERLINE® and HEXAGON optimize throughput in the packaging industry for Fan-Out wafers by combining features of the ABD, ACM and the new generation of ICP Sputter Etch modules.

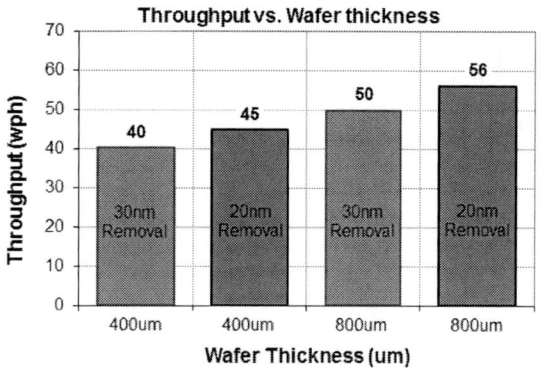

Figure 5: Wafer throughput as a function of wafer thickness and etch removal for Fan-Out process with temperature limitation of 120°C

Figure 5 shows the wafer throughput achieved for

Fan-Out process as a function of wafer thickness (400um and 800um) and etch removal (20nm and 30nm, SiO2 equivalent). It is seen that while the maximum temperature is limited to 120°C the throughput is in the range of 40-45wph for a mold wafer thickness of 400um. For the 800um thick mold wafer the throughput is above 50wph. In the case of Fan-In products the resulting throughput is also well above 50wph because there is no stringent temperature limitation as with Fan-Out products.

With the ever increasing focus on tool productivity and CoO in the advanced packaging market, a long kit life in a process module is a prerequisite. For example, considering a kit life of 30'000 wafers for an ICP Sputter Etch module and assuming a conservative throughput of 40wph, will necessitate maintenance only once per month, i.e. only ~12 shields kit changes with approximately 100 hours of scheduled down time per year.

CONCLUSIONS

Novel and improved solutions for processing of highly outgassing Fan-out wafers have been developed and verified in production. The ABD for Fan-out wafers and an improved ICP Sputter Etch module capable of running high throughput above 50wph with reduced maintenance frequency, result in a low and stable Rc contact resistance. These solutions provide significant improvements in terms of maintenance and CoO, and extend the tool capability for the next generation Fan-In / Fan-Out wafer level packaging requirements.

ACKNOWLEDGEMENTS

The contribution of Oliver Rattunde, Andreas Erhart and Ewald Strolz to this work is highly appreciated.

REFERENCES

[1] P. E. Garrou, W. B. Rogers, D. M. Scheck, A. J. G. Strandjord, Y. Ida and K. Ohba, IEEE Trans. Adv. Packag., vol. 22(3), pp.487-498 (1999).

[2] J. Yota, H. Ly, D. Barone, M. Sun and R. Ramanathan, Proc. of CS Mantec Conference., vol.20, 323 (2008).

[3] J. J. H. Reche and D.-H. Kim, Microelectronics Reliability 43, pp.879–894 (2003).

[4] P. Carazzetti and F. Balon, Chip Semiconductor, Oerlikon, pp.28-31 (2011).

[5] B. Scholte van Mast, Pat. Publ. EP1979930B1, (2006).

[6] M. Elghazzali, A. Erhart, B. Heinz, J. Weichart and A. Koller, Proc. of CSTIC 2016, Semicon China (2016).

[7] F. Balon, P. Carazzetti, J. Weichart, M. Elghazzali, M. Hoffmann and K. Viehweger, Proc. of IWLPC 2016.

ELECTROSTATIC DISCHARGE FAILURE CONTROL OF IC PACKAGE BY EPOXY MOLDING COMPOUND MODIFICATION

Byung-Seon Kong, Sang-Sun Lee, Da Eun Lee, Hyung Ouk Choi, and Hyun Woo Kim*
Central Research Institute, KCC Corporation, Yongin 16891, Republic of Korea
*Corresponding Author's Email: kongku@kccworld.co.kr

ABSTRACT

By changing curing accelerator and modifying the volume resistivity of epoxy molding compound (EMC), electrostatic characteristics of EMC applied package can be improved and electrostatic damage of IC device was reduced. EMC with phosphonium salt accelerator results in much lower ESD failure than EMC with phosphine salt accelerator. Because the volume resistivity of phosphonium salt applied EMC is lower than that of phosphine salt, it could easily dissipate the static electricity that generated inside of package during or after transfer molding process.

INTRODUCTION

As the patterning scale of integrated circuit (IC) is less than 30 nm node, the electrical behavior of packaging materials is important to maintain good performance of packages. Epoxy molding compound (EMC) is a kind of typical insulation encapsulant to protect semiconductor devices from variable environmental factors and it shall not damage on chips at the same time. EMC shows electrostatic behavior as an insulation material based on epoxy resin, so it can lead to an electrostatic charging or discharging during semiconductor packaging process.

This paper introduces a way to prevent electrostatic discharge (ESD) failure of memory (DRAM) packages through just the EMC modification without any change of the process and facilities. Ogata et al. had studied how the physical characteristics of EMC depend on the curing accelerators [1, 2], but there is no report about electrical features of actual application in the semiconductor packages. In this paper, we have discussed ESD characteristics depending on the volume resistivity values of EMCs with different accelerators in the formulation especially, and identified that ESD failure rate can be reduced.

EXPERIMENTAL

Materials

EMC has been designed as the formulation of Table I and manufactured by a twin-screw kneader. Hardener content is controlled to be the same equivalent of epoxy and phenol functionality. Two accelerators (P1, P2) are used independently to compare physical properties of EMC.

TABLE I. FORMULATION OF EPOXY MOLDING COMPOUND

Composition	Raw Materials	Parts by Weight
Epoxy	4,4'-diglycidyloxy-3,3',5,5'-tetramethyl biphenyl	100
Hardener	Phenol-aralkyl resin	90
Filler	Fused silica	1492
Coupling agent	Glycidoxypropyl-trimethoxysilane	7
Releasing agent	Carnauba wax	3
Colorant	Carbon black	3
Accelerator	Phosphine salt (P1) *or* Phosphonium salt (P2)	5

Analysis of Curing Properties

1H-NMR analysis and reaction rate measurement were carried out to compare the epoxy conversion rate of EMC with P1 and P2, respectively. To see epoxy group conversion degrees before and after curing reaction, epoxy ring opening was analyzed by comparing 1H-NMR peak intensities before and after molding for 3 minutes at 180 degrees Celsius. Initial specimen was prepared by mixing epoxy resin, hardener resin, and accelerator for 5 minutes at 100 degrees Celsius. The 1st 1H-NMR was measured with the initial specimen in $CDCl_3$+CH_3OD solvent, and then the 2nd 1H-NMR measurement of cured specimen was done.

The reaction rate of each specimen was measured by differential scanning calorimetry (DSC). Heat flow at different temperatures was checked under three heating rates (5, 10, and 20 degrees Celsius per minute), followed by estimating the epoxy conversion rates versus time at 180 degrees Celsius by using Vyazovkin's Model Free Kinetics [3, 4].

Measurement of Electrical Properties

Volume resistivity has been checked by volume resistivity-meter based on DC bias (1,000 V) and impedance analyzer with AC bias (1.5 V) system at 180 degrees Celsius. ESD failure rates of actual packages have been evaluated by applying BOC (Board on Chip) packages with DRAM devices.

978-1-5090-6695-7/17 $31.00 © 2017 IEEE

RESULTS AND DISCUSSION

Table II shows epoxy ring opening ratios of EMC with P1 and P2, respectively. After measuring 1H-NMR before and after molding, the epoxy ring peak (2.5~3.0 ppm) intensity normalized by aromatic ring peak (6.5~7.2 ppm) intensity has been compared. Based on the normalized peak intensity, epoxy ring opening ratio can be calculated by the equation under Table II. In the results, EMC with P1 (68%) shows higher epoxy conversion rate than EMC with P2 (44%).

Epoxy reaction can be explained by epoxy ring opening mechanism under phenol hardener resin in this study, shown in Figure 1. During molding, epoxy group react with phenol group, and secondary hydroxyl group is generated in the thermosetting network. Thus epoxy NMR peak intensity of a cured specimen can be decreased.

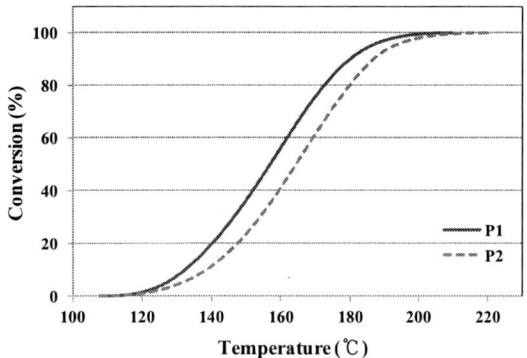

Figure 1: Cure mechanism of epoxy resin by reaction with phenol hardener

To find out more about cure reaction rate, the conversion rate was measured by dynamic mode DSC. Figure 2 represents an integral graph of reaction conversion corresponding to temperature under the 10 degrees Celsius per minute ramp-up rate of EMC with P1 and P2, respectively. EMC with P1 shows higher conversion than EMC with P2 in the same temperature range. Moreover we measured heat flow under ramp-up rate of 5 and 20 degrees Celsius per minute, respectively, and calculated conversion degree versus cure time by Vyazovkin's Model Free Kinetics. As shown in Figure 3, EMC with P1 represents higher conversion degree than EMC with P2 at 180 degrees Celsius.

TABLE II. ANALYSIS OF EPOXY RING OPENING BY 1H-NMR

Molecular Structure		Chemical Shift	Peak Intensity	
			P1	P2
Aromatic Ring		6.5~7.2 ppm	1.0000	1.0000
Epoxy Ring	Before	2.5~3.0 ppm	0.2970	0.2153
	After		0.0948	0.1200
Epoxy Ring Opening Ratio†		-	68%	44%

† Epoxy Ring Opening Ratio (%) = 100 x {1 - (Peak Intensity of Epoxy Ring after Molding / Peak Intensity of Epoxy Ring before Molding)}

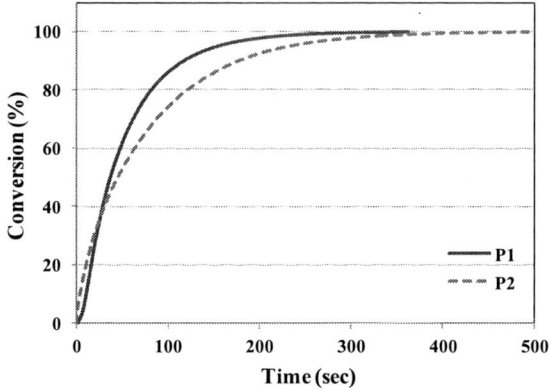

Figure 3: Prediction of reaction conversion versus cure time of EMC with P1 and P2 by DSC

Figure 4 and 5 shows volume resistivity data measured at 180 degrees Celsius under DC 1,000 V and AC 1.5 V applied, respectively. 180 degrees Celsius means the molding temperature of EMC. EMC with P1 shows about five times higher value than EMC with P2 in the case of DC condition and 1.5 ~ 15 times in AC condition versus frequency.

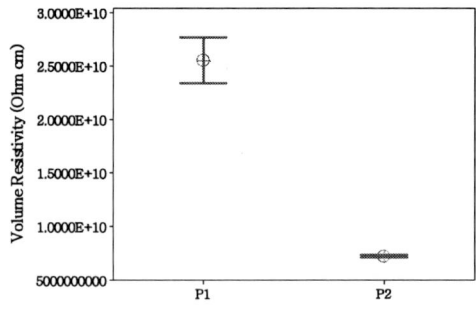

Figure 4: Volume resistivity of EMC with P1 and P2 by DC biased system

Figure 2: Reaction conversion versus temperature of EMC with P1 and P2 by DSC

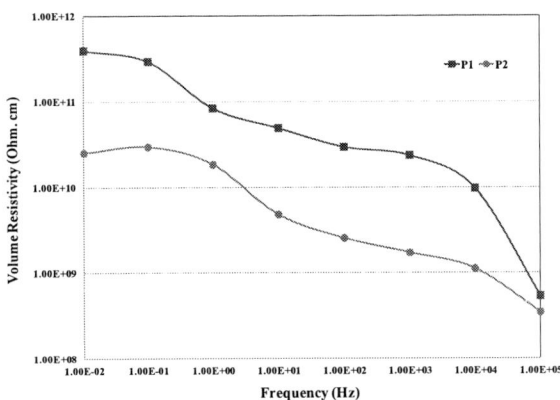

Figure 5: Volume resistivity of EMC with P1 and P2 by AC biased system

This result is basically from a curing behavior difference due to the accelerator of EMC. EMC with higher reaction rate tends to show relatively higher cross-linking density and higher volume resistivity after curing [1].

To assess the final packaging yield of two kinds of EMCs, we have prepared BOC packages (DRAM devices) and evaluated the ESD failure rate with 1,000 units. In the results, the EMC with P1 induces 1,300 ppm failure but EMC with P2 shows 0 ppm.

EMC can cause static electricity by friction with mold surface during transfer molding, especially by the flow of EMC in the mold cavity or by opening mold chase after completion of curing as shown in Figure 6. If the volume resistivity of EMC is low relatively, the static electricity which is generated inside the package can be removed and the EMC will not inflict damage on the device directly. In other words, lower volume resistivity of EMC with P2 leads much less chip damage by ESD energy [5].

Figure 6: Electrostatic charging diagram by EMC; (a) during transfer molding and (b) after opening mold

ACKNOWLEDGEMENTS

ESD evaluation was performed at SK hynix and Hitech Semiconductor.

REFERENCES

[1] M. Ogata, N. Kinjo, S. Eguchi, H. Hozoji, T. Kawata and H. Sashima. *J. Appl. Polym. Sci.*, vol. 44, 1992, pp. 1795-1805.

[2] C. Su, C. Wei and B. Li. *Adv. Mater. Sci. Eng.*, 2013, pp. 1-9.

[3] S. Vyazovkin. *J. Comput. Chem.*, vol. 22, 2001, pp. 178-183.

[4] N. Sbirrazzuoli and S. Vyazovkin. *Thermochim. Acta*, vol. 388, 2002, pp. 289-298.

[5] H. Higuchi, M. Maeda, K. Yamauchi and N. Takahashi. *Kobunshi Ronbunshu*, vol. 48, 1991, pp. 33-39.

NEW ENABLING LASER APPLICATIONS IN ADVANCED CHIP PACKAGING

Dirk Müller[1]*, John Kennedy[1], Dietrich Tönnies[2] and Rainer Pätzel[3]

[1]Coherent Inc., Santa Clara, CA 95054, USA

[2]Coherent-Rofin GmbH, Dieselstr. 15, 85232 Bergkirchen, Germany,

[3]Coherent Laser Systems GmbH & Co. KG, Hans-Böckler-Str. 12, 37079 Göttingen, Germany

*Corresponding Author's Email: dirk.mueller@coherent.com

ABSTRACT

Laser processing is playing an increasing role in advanced packaging. Miniaturization is an obvious driver. As features shrink lasers can play to their core strength of creating intricate features with low damage to surrounding material. But quality is not everything and overshooting the quality goal can mean the process gets more expensive than necessary. The cost of ownership or cost per part has to work out, too. The following laser-based applications have been shown to be offer a cost advantage while simultaneously offering very high quality. They serve as examples of how advances in laser technology are helping to reduce feature size, improve quality and lower cost.

INTRODUCTION

Here we highlight how three applications where fundamental advances in laser technology are impacting advanced packaging applications in the near term. As package sizes shrink, so will μ-via diameters. Being able to drill smaller μ-via holes at a low cost can be achieved with a CO laser. Ultrafast lasers have now been main stream in several other industries and in advanced packaging these lasers are able to deal with any material from ceramics to polymers when cutting packages. Optical debonding has been used in the flexible display industry for years. Demand has significantly driven down the cost of ownership and advanced packaging can ride on the coattails of these developments.

μ-via Drilling

Carbon dioxide (CO_2) lasers have been used in PCB via drilling for over two decades and currently service about 20% of the market. As the industry transitions further toward smaller vias, mechanical drill costs start to increase exponentially, and the use of CO_2 lasers at <100μm hole size is very prevalent.

CO_2 lasers drill vias through a thermal interaction. That is, the material absorbs the infrared light output of the CO_2 laser, which heats it until it vaporizes. Many dielectrics absorb well in the far infrared, while nearly all metals are highly reflective at these wavelengths. As a result, copper layers act as a natural stop when drilling with a CO_2 laser. In order to drill through copper (such as a top clad layer), it often first has to be oxidized to create a dark patina which better absorbs the CO_2 laser light.

While the CO_2 laser can readily produce a smaller via than a mechanical drill, there are limitations on the smallest via diameter it can reach based on the minimum focused spot achievable. Longer (e.g. far infrared) wavelengths, can't be focused as finely as visible or ultraviolet wavelengths. As a result, the practical lower limit on via diameters for CO_2 lasers is about 70 μm. What if the diameter needs to be smaller than 70μm?

CO versus CO2 Laser

This is where carbon monoxide (CO) lasers come in. The reason that CO lasers are of interest is that they output over the 5 μm to 6 μm spectral range, which is about half the CO_2 wavelength and this allows for a smaller focus spot.

For via drilling, this shorter CO laser wavelength provides several important advantages. For example, it lowers the minimum via diameter that can be produced down to about 35 μm (due to diffraction). But, even when producing larger diameter vias, the CO laser has an edge over CO_2.

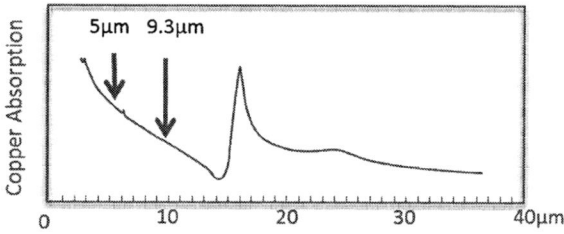

Figure 1: CO laser absorption in copper versus CO_2. The higher absorption of the CO laser wavelength helps to drill non-black copper.

Specifically, the focusing lens used to achieve a 70 μm diameter via with a CO laser has twice the focal length of the lens required to achieve the same via size with a CO_2 laser. This longer focal lens provides greater depth of focus, which increases the field of view. The longer focal length and increased depth of field facilitate an increase in scanning speed, and therefore faster via production, with the shorter wavelength CO laser.

Because the CO laser can be focused to a smaller spot, it's easier to reach higher power densities with it than with a longer wavelength CO_2 laser of the same power. (Since

978-1-5090-6695-7/17 $31.00 © 2017 IEEE

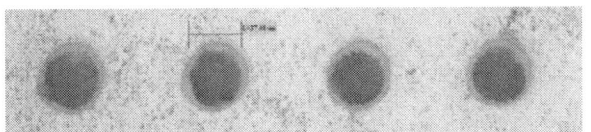

Figure 2: Image of 35μm vias drilled with a CO laser.

the CO laser has roughly half the wavelength of the CO_2, it forms a spot size that is half as big, and which therefore has one quarter the area, or four times the power density.) This higher fluence (intensity) enables it to more readily drill untreated copper. Conversely, achieving a given power density requires only one fourth the total output power with a CO laser as with a CO_2 laser. Depending upon the exact parameters of a particular via drilling task, this makes it possible to use a much lower power CO laser for a specific job. This lowers the cost of the laser and the cost of the electricity.

In addition to the geometric benefits, there are also differences in the light absorption at the shorter wavelengths (Fig.1). The 5μm wavelength of the CO laser is absorbed more strongly by the copper surface making it easier to penetrate through this highly reflective layer.

Ultrafast Laser Processing

Two of the most important parameters to understand when looking at difference in how material is removed are laser wavelength and pulse duration. Most precision industrial microstructuring processes are based on pulsed lasers. Depending on process needs, an industrial laser pulse can go from milliseconds down to picoseconds (10^{-12}s). However, shorter and longer pulsewidth lasers tend to process material by substantially different means.

Many traditional applications rely on lasers, which have pulsewidths in the tens of nanoseconds range, and which remove material via a photothermal interaction. Here, the focused laser beam acts as a spatially confined, intense heat source. Targeted material is heated up rapidly, eventually causing it to be vaporized. Material is essentially boiled away (Fig 3).

The advantage of this approach is that it enables rapid removal of relatively large amounts of target material (particularly considering the multi-kHz repetition rates at which Q-switched lasers typically operate). Furthermore, nanosecond laser technology is well established, and these sources are highly reliable and have attractive cost of ownership characteristics. However, for the most demanding tasks, peripheral HAZ damage (e.g., delamination of surface coatings or charring) and/or the presence of some recast material, can present a limitation. This problem can be somewhat mitigated by employing a laser having output in the ultraviolet.

As laser pulsewidth gets into the picosecond domain, an entirely different mechanism for material removal comes into play, called photoablation. This process

occurs because short laser pulsewidths lead to very high peak powers (megawatts and above). These high peak fluences drive multiphoton absorption which strips electrons from the material that then explodes away because of Coloumb repulsion. Since photoablation involves directly breaking the molecular or atomic bonds which hold the material together, rather than simply heating it, it is intrinsically not a thermal process. Also, when using ultrafast pulses, the laser processed material is removed in such a short timeframe that the ablated material carries away most of the energy before excess energy can spread into the surrounding material. Together, these effects result in significantly reduced HAZ. Plus this is a very clean process, leaving no chunks or droplets of recast material and thereby eliminating the need for elaborate post-cleaning.

Perhaps the most attractive attribute of ultrafast lasers in advanced packaging is the fact that the laser can remove a very broad range of materials, including glass, sapphire or ceramics. Sapphire and glass have low linear optical absorption, and are thus difficult to machine with existing, commercially available lasers. More specifically, the

Photothermal Interaction

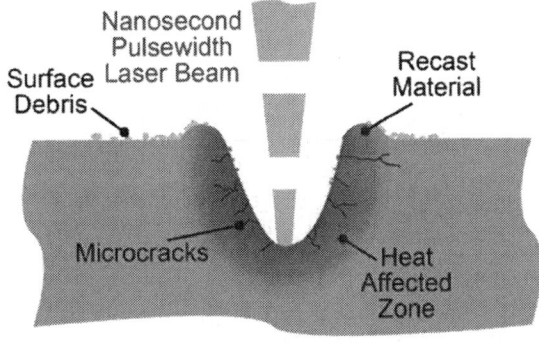

Photoablation

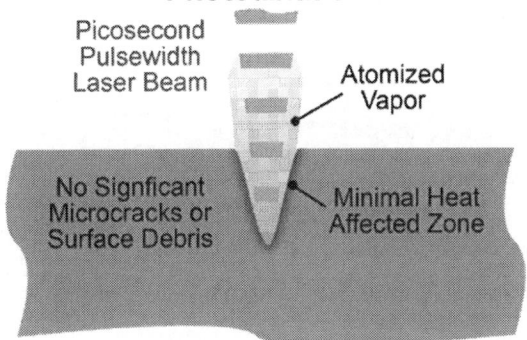

Figure 3: Schematic illustrating the major differences between ultrafast processing and processing with longer pulsewidth lasers.

978-1-5090-6695-7/17 $31.00 © 2017 IEEE

technique is "wavelength neutral," that is, nonlinear absorption can be induced even if the material is nominally transparent at the laser wavelength.

But, if ultrafast pulse laser processing is so wonderful, then why isn't it used for everything? There are two reasons. First, material removal rates can be substantially lower than for longer pulse lasers. Second, generally speaking, laser cost increases as the pulse duration decreases. Thus, ultrafast lasers are best suited for precision applications that require relatively small amounts of volume material removal and present a mixture of materials to be removed.

Package Singulation with Ultrafast Lasers

One important laser application in microelectronics is package cutting. This could be for a SiP package with a complicated perimeter (such as the S1 package in the Apple Watch), a wafer level packaged device with many different materials that has to be separated or a finger print

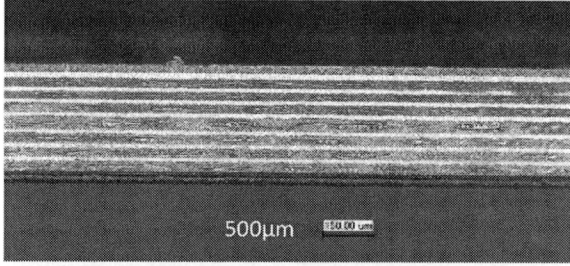

Figure 4: Images of a SiP package cut with a 532nm picosecond laser. The top image shows the full cross section. The middle image shows a close up of the cut circuit board inside the package and the bottom image shows the top view of the mold compound edge.

sensor with a glass or ceramic layer. In all cases multiple devices are typically produced on single sheet/wafer. Mechanical cutting may be adequate for straight cuts in this application, but can't produce the tightly radiused edges or notches. Abrasive water jet cutting is sometimes used for this purpose, but may not be able to cut through all materials and cause delamination with subsequent water damage. The variety of different materials in the package (circuit board, glass, molding compound, sapphire) is a challenge for standard pulsed lasers. This is a clear case where ultrafast lasers have a strong advantage. Figure 4 shows a cut section of a circuit board from a SiP package with molding compound. The roughly 1mm thick package can be cut at an effective cutting speed of more than 10mm/s using a green picoscond laser. The non-thermal ablation process of the picosecond laser prevents the organic material from charring and the copper layers from melting. There are no burrs and the molding compound is cut very cleanly. Depending on the material thickness and perimeter length, the ultrafast laser can singulate devices is about 1 s per device. It is not uncommon to reach effective cutting speeds of 15-20mm/s with a 50W picosecond laser.

Optical Debonding

Wafer level packaging (WLP) often demands processing of very thin dies and redistribution layers on a temporary carrier. Once the package is finished the wafer has to be debonded from the carrier. Thermal, mechanical and chemical debonding mechanisms all have their pros and cons. Optical debonding seems to ameliorate many of those disadvantage, but has been perceived as too costly.

What is known as optical debonding in the advanced packaging industry is known as laser lift off (LLO) in the flexible display industry and in fact a very well established process. Actually, flexible displays would not exist if it was not for the use of LLO. Optical debonding offers very high yield rates and high throughput compared to other debonding techniques. The advantage of using a UV laser versus other wavelength such as IR and green is that the UV wavelength has a very low penetration depth into the organic material, which is bonded to the glass carrier (Fig. 5 and 6). While IR and even green wavelength can penetrate through the polymer and potentially damage delicate circuitry above, the UV wavelength is very strongly absorbed in all common polymers and hence only penetrates a fraction of a micron into the polymer film. The absorbed laser energy creates a vapor film at the glass-polymer interface and this vapor layer detaches the polymer from the glass carrier. The result is a highly reliable and reproducible debonding process that can even work without a sacrificial layer between the polymer and glass offering a throughput of 40-60 wafers/hr for each tool.

Optical debonding can be implemented with different

978-1-5090-6695-7/17 $31.00 © 2017 IEEE   454

UV laser sources. The optimum laser choice depends on throughput requirements and capital investment constraints. Most commonly a diode-pumped-solid-state (DPSS) laser or an Excimer laser is used. The DPSS laser has a fundamental wavelength in the infra-red which is frequency converted to its third harmonic to obtain UV wavelength. An Excimer lasers emit directly in the UV, so there is no need for frequency conversion. Every frequency conversion crystal is subject to degradation over time as the generated UV photons are so energetic that they damage the very crystal that creates them. Hence, DPSS lasers require a crystal replacement every 1-2 years when in 24/7 operation. So the maintenance cost of a DPSS laser can be higher than that of an Excimer laser (Excimer tubes now have lifetimes that amount to 4-5 years of operation for an optical debonding application. 5-10 billion pulses per tube is not uncommon). A DPSS laser solution will likely have a lower initial capital investment and infrastructure cost, but will have a higher cost when calculated over the period of 4-5 years of operation.

Conclusion

New processes in advance packaging can benefit from recent laser developments as well as riding on the coattails of other industries such as the flexible flat panel displays to lower their cost of ownership.

Ultrafast lasers offer a method that can virtually deal with any material from ceramics to polymers and hence is ideally suited for applications where the material stack involves very dissimilar materials and requirements change over time (different production runs).

CO lasers offer a path for drilling μ-via holes at smaller diameters than CO_2 and pretty much the same or lower cost as CO_2.

Laser suppliers are quite used to offer full turn-key optical solutions that can easily be integrated into existing machines with specialized material handling. The laser process can be seen as a self-contained module that can be integrated into existing production flow or machine designs.

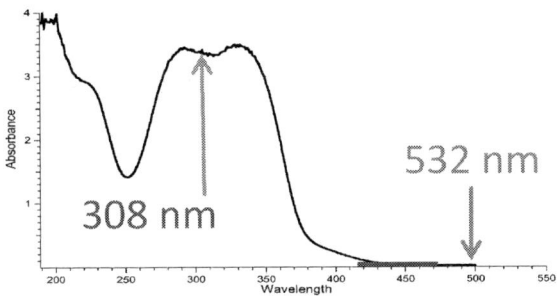

Figure 5: UV wavelength is strongly absorbed in all organic materials used for wafer level packaging.

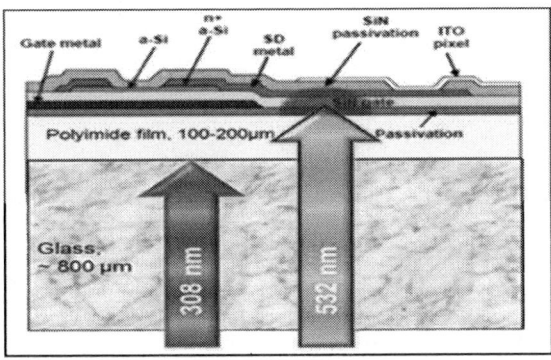

Figure 6: A visible laser wavelength can penetrate through the organic layer allowing for damage to the circuitry inside the package. The strong absorption of the UV laser prevents damage to the circuit and facilitates reliable optical debonding.

MULTI BEAM FULL CUT DICING OF THIN SI IC WAFERS

Jeroen van Borkulo, Richard van der Stam, Won Chul Jung, Paul Verburg

ASMPT Laser Separation International B.V., platinawerf 20, 6641TL Beuningen Netherlands

ABSTRACT

Over the last years singulation of thin semiconductor wafers with (ultra) low-□ top layer has become a challenge in the production process of integrated circuits. The traditional blade dicing process is encountering serious yield issues. These issues can be addressed by applying a laser grooving process prior to the blade dicing, which is the process of reference nowadays. However, as the wafers are becoming thinner this process flow is not providing the yield and productivity required. In this article the unique ASMPT multi beam technology is presented which addresses both concerns and enables a high productivity laser dicing process with a limited heat affected zone and sufficient die strength.

Key words: 3D packaging, dicing, die strength, laser, thin wafer, multiple beam

INTRODUCTION

For a long time the traditional separation technique used in this industry, i.e. blade dicing was able to fulfill the separation requirements of these wafers with a continuously increasing complexity. However, over the last 10 years it became clear that the requirements to process the advanced technology wafers in the semiconductor industry were very hard to follow for the blade dicing technique. We can distinguish two main areas where blade dicing encountered difficulties.

The first one is the transition from 2-D to 3-D IC-packages which requires thin wafers (<50um) to manage the heat dissipation and the geometrical size of the package. The mechanical forces from the blade dicing process pose a serious problem for those thin wafers. The combination of the thinner wafers with ultra low-k top layers does not allow customer to use the nowadays standard hybrid process of laser low-k grooving followed by a blade dicing process. Once Low-K grooved by laser the wafers are so thin that they will likely break prior to applying the blade dicing process. The semiconductor industry is still looking for a good solution which provides a high yield, high productivity and low cost dicing solution.

In the next sections a detailed explanation of the problem will be given. This is followed by the first results achieved for a single step, full cut process for thin Si wafers.

Alternative dicing technologies

To be able to find a solution for the dicing process of thin (ultra) low-K wafers alternative process steps have been developed. One of them is the so called Dice Before Grind (DBG). In the DBG process the wafer is partially diced while still at full thickness. In the subsequent process step the wafer is grinded down to the required thickness which separates the wafer.

Even though this technology is used in production nowadays it does so at the expense of yield loss (cracks and chipouts are observed, which results in unacceptable yield losses) and at the expense of cost. The cost is high due to the various process steps required (retaping, wafer handling, capex).

Another alternative separation technique similar to DBG is Stealth Before Grind (SBG). In this process flow a stealth process is applied prior to grinding the wafer which locally weakens the wafer. Once the grinding process has reached the required thickness the wafer can be separated via expansion of the tape. The constraints for this process flow is the cost of the equipment involved (specifically the stealth system) but as the dies are becoming smaller (<1mm), more and thicker TEG structures are used within the dicing street and wafers are getting thinner (<50um), an SBG process has difficulty to reach the required yield in combination with low cost.

Separation of thin (ultra) low-K wafers

As indicated in the previous section blade dicing is the standard technique to singulate a semiconductor wafer. In modern dicing machines the blade thickness can be reduced to 25 um, which in combination with an average IC-die-size of several square millimeters, has always tempered the need to go to an alternative separation technique. From a yield point of view the saw was meeting the specs and the impact of the street width on the total number of die per wafer was low. This in contrast to, for example, the RFIC or LED industry where due to the smaller dies, street width reduction forms a significant die per wafer yield benefit. The laser dicing full cut process has become the process of reference in these segments with a utilization of >70% for RFIC and close to 50% for LED applications.

However, when the wafer technology approached the 65nm milestone, the use of low-k material instead of SiO_2 became inevitable, and the situation for die singulation changed drastically. Due to the brittleness and the relative poor adhesion between the silicon base material and the low-k layers the low-k layer delaminates and cracks when the mechanical force of the blade is applied. This results in a significant reduction of the dicing quality (*figure 1*) and yield problems.

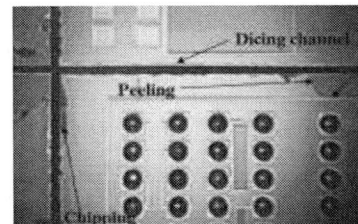

Figure 1: Top view of a blade diced low-k wafer. Severe peeling, chipping is clearly visible.

Laser full cut Si

As various studies have shown the die strength level for a standard laser full cut process tends to be 30-50% below the saw blade reference [1]. Despite that lower die strength level no yield issues were found during assembly process.

In order to improve the die strength for the laser dicing process additional process steps have been added to increase the die strength [2]. Even though these studies showed that it is possible to achieve die strength levels equal to the blade dicing process they did so with the aid of additional process steps (wet etch, dry etch). It is therefore impacting the cost and also the flexibility of the process.

The root cause for the low die strength for a standard laser dicing process is the combination of recast material (molten Si) and the formation of micro cracks along the surface interface (see figure 2 below).

Figure 2: TEM image of the recast interface of a laser diced 100um polished Si wafer

The interface of the crystalline Si and the amorphous recast material creates stress which when placed under pressure can initiate a crack and therefore lower die strength.

By means of etching processes (wet or dry) the recast material is removed and micro cracks generated at the Si recast interface are released of their stress.

Die strength

All results mentioned in this paper have been measured on a 4-point die strength tool both active Structure Up (SU) and Structure Down (SD) (see figure 3).

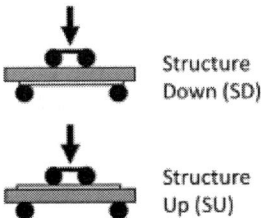

Figure 3: SD and SU measurement method on 4-point bending tool.

In the underneath graph 1 the die strength results are shown for a standard laser dicing process vs. a blade saw process. The wafer used for this test is a 70um polished Si wafer. For reference purposes the same wafer type has been blade diced.

The data shows clearly the reduction in die strength for a standard laser dicing process with respect to a blade dicing process. In the past various parameter changes to the laser dicing process have been applied (pulse duration, power, etc.) however the result on the die strength improvement was marginal (<5%).

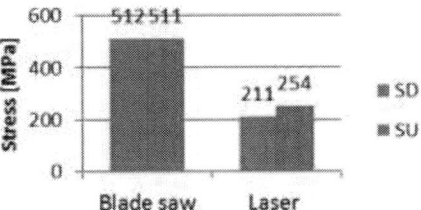

Graph 1: Die strength results Su & SD of a 70um polished Si wafer measured on a 4-point bending tool.

Laser cleaning

It is found that when irradiating the dicing kerf with lower power pulses both the visual appearance and die strength of the die is significantly improved. For a standard laser dicing result on a 100um thick polished Si wafer (see figure 4) irregular and significant amount of recast material is visible which contributes to the lower die strength achieved with this process. When applying the laser cleaning process the recast formation is altered and can be reduced and smoothened at the same time (see figure 5).

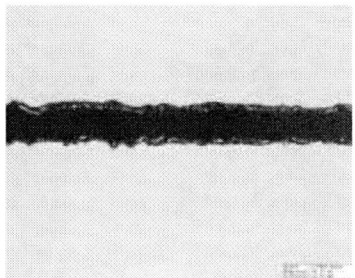

Figure 4: Top view of a full cut laser diced 100um thick polished Si wafer. Abundant and irregular recast is clearly visible.

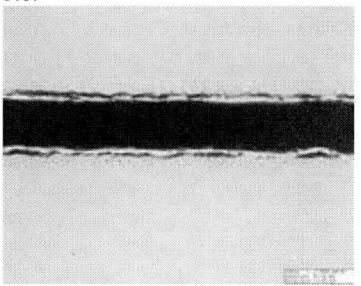

Figure 5: Top view of a full cut laser diced 100um thick polished Si wafer which has been treated with a laser annealing process. Significant reduction of recast formation is visible and the remaining recast is smoother.

Inspection of the side wall structure and roughness also demonstrates that the laser cleaned area has a smoother appearance (see figure 6 & 7 below). The result shown in figure 7 has only been treated for the top section of the side wall.

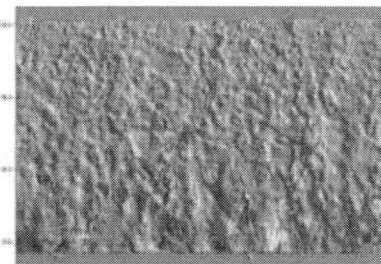

Figure 6: Side wall quality of a standard laser dicing process

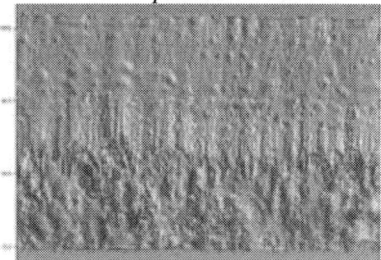

Figure 7: Side wall quality of a top side laser cleaned die. The red arrow indicates the treated area. Clearly visible is the difference in surface structure.

V-Shape DOE

The standard full cut laser dicing process for ASMPT is done with a multi beam laser configuration where all beams are in line with the dicing direction. It was found that lower power beams running over the edge of the dicing kerf will help to clean up the recast and micro cracks and as a result recover the die strength. As the cleaning process is done on the edges of the laser diced area it would require additional process traverse with an offset to both sides of the edges of the dies to apply the cleaning process. This obviously reduces the productivity of the process. ASM Laser Separation International is the inventor of multi beam laser dicing and a multi beam configuration is developed which allows the cleaning process to be done in a single pass.

As shown in figure 8 below the V-Shape DOE (patent pending) consists out of centralized higher intensity beams which dice the wafer and outer lower intensity beams which take care of the cleaning process.

Figure 8: V-Shape DOE design showing the central forward positioned spots which are used to dice through the wafer and the outer lower intensity spots which clean the edges of the die.

The V-shape DOE design enables a single process full cut thin Si laser dicing process with a sufficient die strength.
As no additional equipment is required as well as no additional process steps (e.g. retaping, extra wafer handling, etc.) the cost of ownership of this process will be competitive.

V-DOE Results

Demonstrated below in figure 9 is the result of dicing through a 75um thick polished Si wafer which was taped on a 20um DAF tape. Figure 9 shows the side wall quality of the cut demonstrating that the V-DOE process has diced through the Si and the DAF tape and that the top section of the side wall has been cleaned.

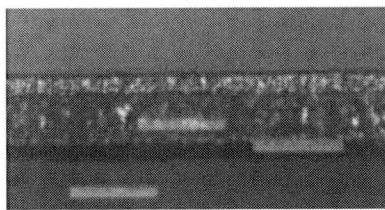

Figure 9: Side wall of a 75um thick polished Si wafer on 20um DAF. The top section of the side wall is clearly cleaned and recast material removed.

In figure 10 the top surface dicing quality is shown using a V-DOE process to dice through a 65um thick NAND Flash memory device mounted on 20um DAF. The die size for this wafer is 8800x10370um and a productivity of 8WPH is achieved.

Figure 10: Top surface quality achieved on NAND Flash memory device 65um thick on 20um DAF. Dicing width 20um.

The die strength results demonstrated in graph 2 and table 1 below for a V-DOE process is showing a 100% die strength improvement with respect to a standard laser dicing process and is therefore similar as the blade dicing process. The results have been achieved on various wafer thickness and DAF tapes but also on various products wafers with and without DAF. Shown in below is the result for only one condition.

The purple points in the graph below show the SD die strength results for a standard laser dicing process. The red points in the graph below show the SD die strength result for the V-DOE process. The die strength for SU situation is already high due to the dicing parameters chosen for the dice through process. Therefore the cleaning is only required on the top side of the die side wall.

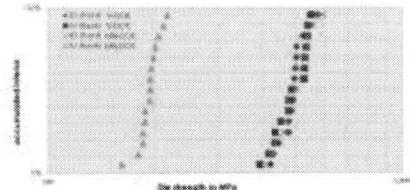

Graph 2: Die strength results SU & SD and for standard Multi Beam (MB) laser dicing process and V-DOE process for a 75um Si wafer on 20um DAF measured on a 4-point bending tool.

	V-DOE Dicing		MB-Dicing	
	Front side	Back side	Front side	Back side
Average (Mpa)	493.2	490.1	193.9	496.3
5% lower bound	412.2	385.9	166.5	391.9
SD (Mpa)	40.5	52.1	13.7	52.2

Table 1: Die strength results SU & SD for standard MB laser dicing process and V-DOE process for a 75um Si wafer on 20um DAF measured on a 4-point bending tool.

CONCLUSIONS

Both the thinning of wafers for 3D stacking and the use of ultra-low-k materials means that the semiconductor IC industry is looking at alternative technologies to separate their wafers.

The V-DOE concept of ASMPT shows that die strength levels can be achieved equal to the blade dicing process which is considered the reference nowadays. The results show that the combined laser dicing and cleaning process allows a high dicing quality in combination with die strengths levels for both SD and SU between 400MPa to 550MPa. Deployment of the technology is currently ongoing with evaluations done together with major players within the semiconductor industry.

REFERENCES

[1] DoHyung Kim, "Evaluation for UV Laser Dicing Process and its Reliability for Various Design of Stack Chip Scale Package", 58th ECTC 2009, pp 1531-1536

[2] Jianhua Li, "Laser Dicing and Subsequent Die Strength Enhancement Technologies For Ultra-thin wafer", 57th ECTC 2007, pp 761-766.

MORPHOLOGY CONTROL OF COPPER NANOMATERIALS FOR IC BONDING

Jiayue Wen, Yanhong Tian*, Zhi Jiang

State Key Lab of Advanced Welding and Joining, Harbin Institute of Tech., Harbin 15000, China

*Corresponding Author's E-mail: tianyh@hit.edu.cn

ABSTRACT

Copper nanoparticles paste was prepared as inconnection materials of power electronics using potassium borohydride (KBH$_4$) as reduction. Various influences on the morpholorgy of copper nanoparticles were discussed, such as reaction time, concentration of precursor ([Cu(NH$_3$)$_4$]SO$_4$) or reducing agent (KBH$_4$) and surfactant (polyvinyl pyrrolidone, PVP). Meanwhile, those influence to preparing pure copper nanomaterials, the growth process and phase evolution of copper nano-materials were discussed. Some new conclusions about controlling the morphology of copper nanowires were drawn about reaction time and reaction temperature. This simple and feasible method of preparing nanomaterials are not limited to devices bonging, which have applications in printing electronics, transparent conducting films and even smart electronics.

INTRODUCTION

In recent years, low-cost, high-efficiency and environmental low-temperature bonding techniques play an important role on Si-based integrated circuit. The low tolerance temperature for packaging of IC, usually below 200°C, requires a new method to prepare nano-solders [1-2]. Because nano-solders are one of the most direct solutions for bonding at a low temperature, higher surface energy of nano-scale materials give it low sintering temperature, and high using temperature. However, SAC alloys or silver nanomaterials solders, universally acknowledged and used in IC bonding, have many shortcomings such as the growth of tin whisker, silver ion migration and high cost [3]. Solders with copper nano-materials using in the bonding of semiconductor IC will be the next wise alternative. Furthermore, nano copper oxides and copper dihydroxide have potential applications in electronics, catalysis, sensors and bioanalysis [4-5].

In this paper, a easy and rapid process of synthesizing copper nanoparticles in solution for solders was come up with. Through controlling morphology and observing by SEM, we analyzed the influence factors of ion concentration, reaction time and reaction temperature and so on to the copper nanoparticles and the reduction route of copper ion. While, copper nanowires, usually prepared by some complex technics, were prepared simply without other surface active agent. Meanwhile, the transformation of copper ion and the process of growth were controlled.

EXPERIMENT AND RESULT

Synthesis of Copper Nanoparticles

Analytical grade copper sulfate (CuSO$_4$·5H$_2$O), Polyvinyl Pyrrolidone (PVP), Sodium hydroxide (NaOH), Ammonium Hydroxide (NH$_3$·H$_2$O), Potassium borohydride (KBH$_4$) were used as received without further purification. Copper sulfate-ammonia complex ([Cu(NH$_3$)$_4$]SO$_4$) was synthesized by adding ammonia to the copper sulfate solution until the solution turning into transparent dark blue. Ammonium is modifier to changing the stability of main salt better in order to obtain copper nanoparticles with uniform grain size.

In a standard synthesis, 10mL of an aqueous solution composed of copper sulfate-ammonia complex (1mmol), Sodium hydroxide (2mmol), and polyvinyl pyrrolidone (0.3g) was stirred in an Erlenmeyer flask at 293K for 3 minutes. Then, potassium borohydride was added into the Erlenmeyer flask stirring at 293K for 1.5 hours. The resulting solution was centrifuged at 4000 rpm for 3min to purify copper nanoparticles, then washed with water and ethanol in turn. Copper nanoparticles were kept in alcohol or evacuated container.

Characterization

XRD pattern was collected with a Rigaku D/max-γ diffractormeter with Cu K$_\alpha$ irradiation. Morphology and structure of the nanoparticles were checked with a field-emission SEM (FEI Quanta 200F). Three sharp peaks at 2θ=43.5°, 50.6°, and 74.3° corresponded to the diffractions of the {111}, {200}, and {220} crystal planes of face-centered cubic (fcc) Cu (PDF#65-9026) (figure 1-b). And such sharp peaks indicated single crystal copper nanoparticles.

The diameter of copper nanoparticles is about 15-30nm and there is only a little of other phases such as Copper(II) oxide and Copper(I) oxide (figure 1), which is appropriate for sintering in size. Those oxide of copper maybe leave due to incomplete reduction or oxidation after prepared, and will be removed by acid pickling before sintering.

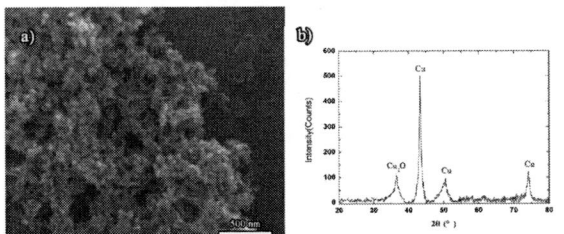

Figure 1. Morphology and phase of Cu nanoparticles in standard synthesis method: a) SEM images; b) XRD pattern.

Reduction reaction asked for strong basicity solution environment, to change copper sulfate-ammonia complex into copper hydroxide. However, without any surfactant, the copper hydroxide presented nanocubes instead of nanoparticles as stable phase. So the copper nanoparticles shoule avoid being synthesized by copper hydroxide nanocubes as the precursor by using fresh precursor.

By simply changing the reaction time, we obtained a series of different products. Nanoparticles grew gradually to nanosheets with the extended response time in figure 2. and Younan Xia et al. [6] had similar research about shape-controlled synthesis of copper nanocrystals. This phenomenon was explained that surfactant can serve as a capping agent binding most strongly to certain facets.

Some bright dots can be counted in figure 2 b) and c), and these dots were usually appeared at endpoints of nanowires. Single crystal copper nanowires were growing along with certain crystal orientation beginning with some high-energy nanoparticles.

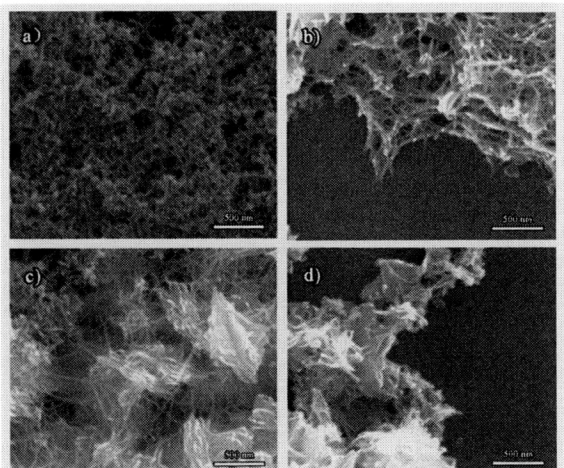

Figure 2. SEM images of a series of Cu nanomaterials at different time point using the standard procedure. a) 10min; b) 30min; c) 1h; d) 2h.

Rising the concentration of reactants, either reductants or precursors, or the temperature of reactions can accelerate the speed of the reaction. For example, by simply increasing the temperature of reaction from 313K to 353K, we obtained nanosheets instead of nanowires in 30 minutes (figure 1(b); firgue 3). All these changing can accelerate the speed of reaction to achieve later period product with thin sheet and purer copper nanomaterials.

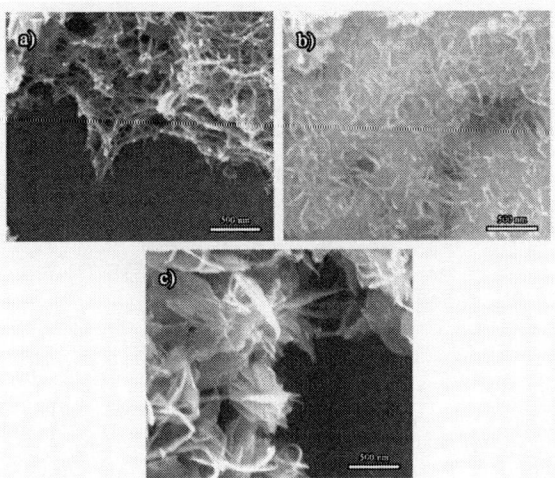

Figure 3. SEM images of copper nanomaterials in differenet react temperature: a) 313K; b) 333K; c) 353K.

The changing of phases also caused our concern. In the period of beginning, cupric sulfate was reduced into copper dihydroxide and Copper(II) oxide (figure 4-a), then into Copper(I) oxide (figure 4-b) and nearly pure copper (figure 1-b). Hardly controlled morphology was just corresponding to the complex reaction process, instead of cupric sulfate reduced to copper directly. Some researches [7] about in situ structural evolution from copper hydroxide nanobelts to copper nanowires also suggested the decomposition followed the sequence of $Cu(OH)_2 \rightarrow CuO \rightarrow Cu_2O \rightarrow Cu$.

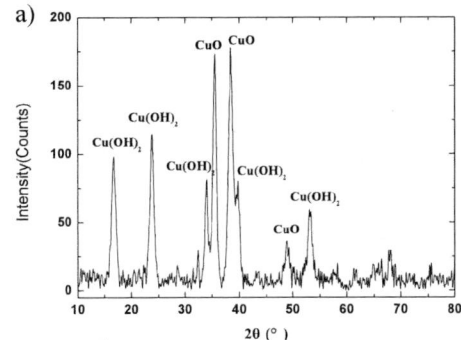

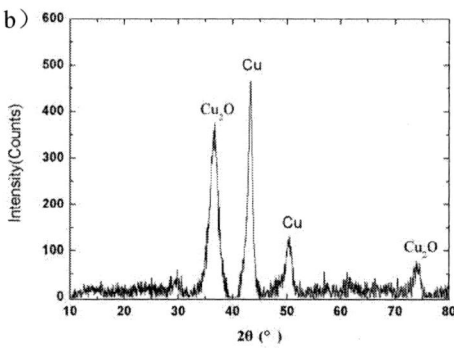

Figure 4. XRD pattern of reduction reaction with different time: a) 2min; b) 5min.

CONCLUSION

15-30nm copper nanoparticles were prepared as inconnection materials of power electronics using potassium borohydride (KBH_4) as reduction. The growth processes and phase evolution of copper nanomaterials were discussed. The changing from particles to sheets were controlled by the temperature of reactions and the concentration of reactants. Other bonding experiments need doing still.

ACKNOWLEDGEMENTS

The authors are grateful for financial support from the National Natural Science Foundation of China (Grant No. 51522503) and support from Program for New Century Excellent Talents in University (NCET-13-0175).

REFERENCES

[1] T. Suga, *IEEE International Workshop on Low Temperature Bonding for 3d Integration*, vol. 35, 2012, pp. 7-10.

[2] V. Dragoi, B. Rebhan, J. Burggraf and N. Razek, *IEEE International Workshop on Low Temperature Bonding for 3d Integration*, 2014, pp. 9-9.

[3] L. Zhang, J. Han, C. He, Y. Guo, *Journal of Materials Science Materials in Electronics*, vol. 23, 2012, pp. 1950-1956.

[4] H. Song, Y. Ni, and S. Kokot, *Colloids & Surfaces A Physicochemical & Engineering Aspects*, vol. 465, 2015, pp. 153-158.

[5] Y. Qiu, G. Tan, P. Xu, Q. Luo, X. Lin, W. Huang, J. Li, *Applied Surface Science*, vol. 347, 2015, pp. 548-552.

[6] M. Tin, G. He, H. Zhang, T. Zeng, Z. Xie, and Y. Xia, *Angewandte Chemie International Edition*, vol. 50, 2011, pp. 10748-10752.

[7] Z. Wang, X. Kong，X. Wen，S. Yang, *Journal of Physical Chemistry B*, vol. 107, 2003, pp. 8275-8280.

Latest material technologies for Fan-Out Wafer Level Package

Itaru Watanabe* ,Masaya Kouda , Koji Makihara , and Hiroki Shinozaki
Electric device materials research laboratory, Sumitomo Bakelite Co., LTD.
NOGATA-CITY, FUKUOKA-PREF, JAPAN
itaru@sumibe.co.jp

ABSTRACT

Currently Wafer Level Package(WLP) is one of famous package structure in mobile consumer electronics industry because of cost, size, density and electrical performance. Recently one of the famous smart phone have on-board new application processor which include Fan-Out Wafer Level Package(FOWLP) as a bottom package in package on package. The selection of process and machines, materials for next-generation FOWLP were settled once. But many players still are looking for a suitable assembly method for FOWLP. So we would like to introduce latest technology and future tasks of the materials which include epoxy molding compound(EMC) and peripheral material.

INTRODUCTION

Now conventional wire bonding structure is changing to flip chip structure and mass production of fan-out wafer level package(FOWLP) with re-distribution technology was started for advanced package with evolution of semi conductor package. Requirement for materials also is changing because of package structure change (e.g., "Figure 1")

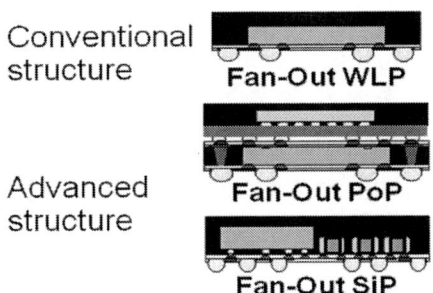

Figure 1: Structure of Fan-Out Wafer Level Package

In addition of warpage and chip sift control, compatibility for re-distribution material, now narrow gap filling and grinding performance, solvent resistance, low pressure molding, optical transparency are needed for EMC with change of process and package structure (e.g., "Figure 2").

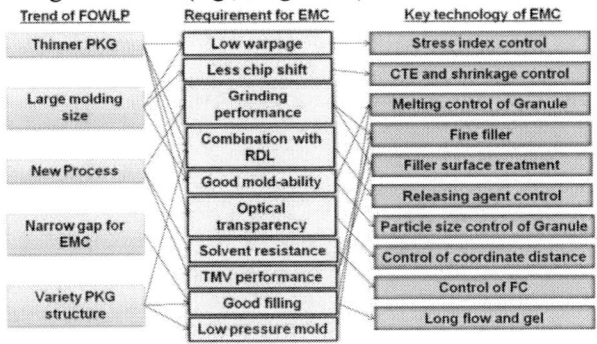

Figure 2: Key technology of EMC

We provide key technology for epoxy molding compound to achieve their requirements.

METHOD

Basically our customer selected compression mold for molding process because of low material yield loss and good control ability of molding pressure. So we study several property of granule EMC by compression mold.

RESULTS

Most important property of EMC in customer process is warpage because if warpage is large over 2mm customer does not treat wafer or panel in their several machine. Wafer warpage was greatly reduced from 30mm to less than 1mm (Figure 3a and 3b) across a 12 inch wafer by tailoring the stress index which is defined as CTE1 times Flexural Modulus at room temperature (Figure 4). Figure 4 shows the correlation between each of the property and warpage. It was found that stress index showed the highest correlation with warpage in comparison with other properties.

 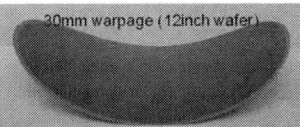

Figure 3: Wafer with low warpage (a) and large warpage (b)

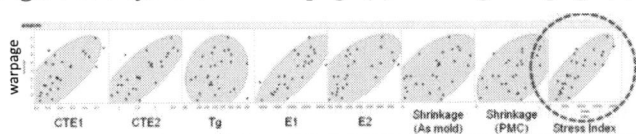

Figure 4: Scatter plot for various properties

In order to reduce the warpage, the CTE and flexural modulus of EMC must be reduced. In general, low CTE is achieved by the method of high filler content and low flexural modulus is obtained by adding some stress additives in the EMC.

For Fan-Out PoP our customers have grinding process to expose the face of Cu pillar on die. We prepared grinding samples of EMC by compression mold and investigated grinding performance of cured EMC. Then we found that size of abrasive grain and contents of low stress additive(LSA) in EMC, filler contents(FC) in EMC are related to grinding performance of EMC (e.g., "Figure 5").

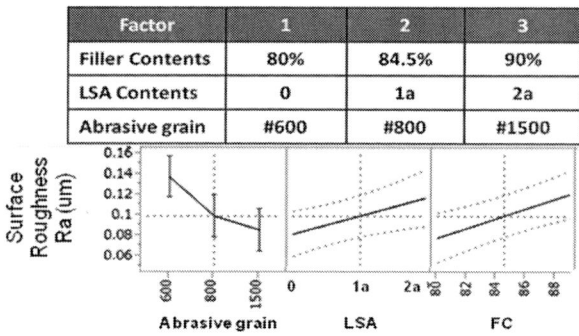

Factor	1	2	3
Filler Contents	80%	84.5%	90%
LSA Contents	0	1a	2a
Abrasive grain	#600	#800	#1500

Figure 5: Grinding test results for EMC

From these results line and space of re-distribution on EMC may be less than 10um. We also investigated chemical grinding for EMC to get more flat surface roughness. It was found that the results of one-sixth can be achieved compared with mechanical grinding (e.g., "Figure 6"). From these results line and space of re-distribution on EMC may be less than 2um.

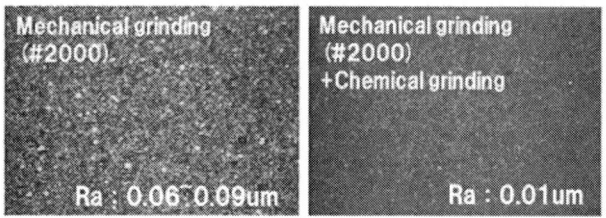

Figure 6: Chemical grinding results for EMC

In FOWLP process cured EMC was exposed to several solvent. So we investigated solvent resistance of EMC (e.g., "Table 1"). We checked EMC weight before and after treatment.

Table 1 Solvent resistance of EMC

	Solvent	Type	Temperature (degC)	Treatment time(h)	Weight change High FC	Weight change Low FC
A	TMAH Tetra Methyl Ammonium Hydroxide (3%)	liquid	RT	1h	+0.01%	+0.04%
B	KOH Potassium Hydroxide (10%)	liquid	RT	1h	+0.02%	+0.02%
C	Propylene Glycol Mono-Methylether (79~81%)	liquid	RT	1h	+0.01%	+0.10%
D	NMP N-Methyl-2-Pyrrolidone (100%)	liquid	60°C	1h	0.00%	+0.71%

In this investigation high filler content EMC shows good solvent resistance and low filler content EMC shows poor solvent resistance especially for NMP.

Now mold pressure is decreasing for advanced System in PKG because they include sensitive component like a saw filter. So mold pressure change from 10MPa to 5MPa or 3MPa and more. Low mold pressure cause void concern. We study influence of EMC formulation to the internal void by low pressure molding (e.g., "Figure 7").

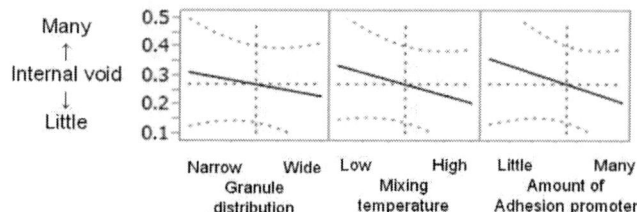

Figure 7: Internal void results by 1MPa molding

From these results we think melting ability of EMC is key property to control internal void because air or out-gassing is easily removed from EMC during decompression. Wide granule distribution is better thermal conductivity than narrow granule distribution because of less contact points of granule each other. High mixing temperature when we make granule EMC provide good dispersion of epoxy resin, so EMC shows good melting ability. Adhesion promoter is liquid material so many amount is good for melting ability. We will continue to investigate correlation between melting ability and internal void.

Thickness of mobile phone is becoming thin year after year and also PKG thickness is same. If die top thickness is thin below 70um, EMC need to control optical transparency. Figure 8 is a molding sample include die after grinding of die top from 100um to 50um. It was seen through the die.

Figure 8: Thin die top molding sample

It's possible to quantify EMC optical transparency by color comparison between on die & around die(e.g., Table 2).

Table 2 Coordinate distance of thin die top molding sample

	L	a	b	Coordinate distance*
1 : On Die	24.7	0.3	-3.3	
2 : Around of die	22.0	-0.1	-4.3	2.8

*Coordinate distance = $[(L1-L2)^2+(a1-a2)^2+(b1-b2)^2]^{0.5}$
L value
Black low---high White
a value
Green low---high Red
b value
Blue low---high Yellow

High coordinate distance value means large difference of color between on die and around die. Table 3 shows coordinate distance of several EMC which include several factor. Increase of coloring agent is effective for shielding of light. Smaller cut size filler means increase of light reflection point. High refraction material is also effective for low coordinate distance.

Factor	EMC-1	EMC-2	EMC-3	EMC-4	EMC-5
Coloring agent	Std	Increase	Increase	Increase	Increase
Filler cut size	Std	Std	Smaller	Smaller	Smaller
High refraction material	No	No	No	Yes	Yes
Coordinate distance	2.8	2.0	1.6	0.1	0.2

Figure 9 shows appearance of EMC-5 after grinding which was not seen through the die.

Figure 9: Appearance of EMC-5 after grinding

We also introduce peripheral materials such as film material and re-distribution material for Fan-Out wafer level package.

Electrostatic Discharge(ESD) issue is increasing with automation of the process in semiconductor field. Especially pick up and place process of the die included in the FOWLP process have high risk of ESD. Figure 10 shows the voltage on the dicing tape after pick up measured by voltage measurement probe.

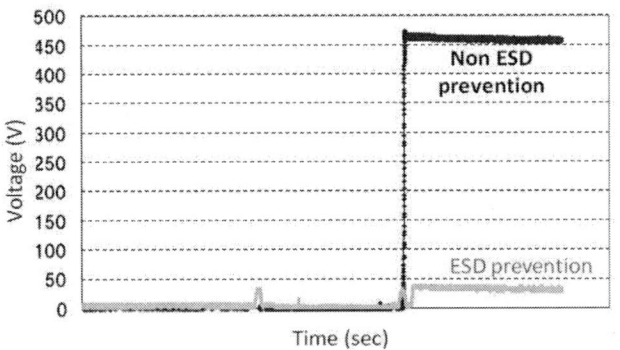

Figure 10: Dicing tape surface voltage after pick up

The voltage on the dicing tape can be lower 50V for ESD prevention dicing tape.

Re-distribution material is also important material to control warpage of package for FOWLP. If number of RDL layer increase and volume of EMC in package decrease such as exposed structure(e.g., "Figure 11"), we need to control warpage by both EMC property and Re-distribution material property. Figure 12 shows relationship of warpage and number of re-distribution layer.

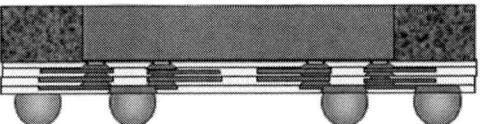

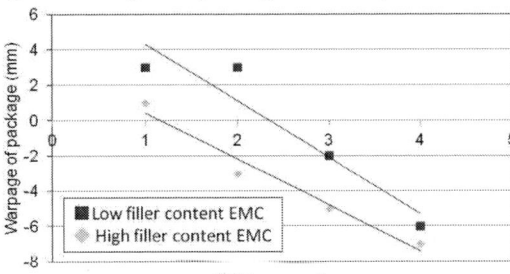

Figure 11: Exposed package with 4 layer re-distribution

Figure 12: Warpage of exposed package

From these results it is impossible to control warpage of multilayer by only EMC property.

Now low temperature curing and low stress re-distribution material is developing to decrease wapage of multilayer FOWLP. We will introduce our new re-distribution material when it's complete.

CONCLUSIONS

We developed EMC for advanced FOWLP especially to reduce warpage. Stress index of EMC was important parameter to control warpage. And we made clear influence of RDL layer to the warpage of exposed structure. We are investigating compatibility between EMC and RDL, and also we need to discuss clean level of granule EMC to the customer.

ACKNOWLEDGEMENTS

I wish to thank the timely help given by M. Kouda, K. Makihara, and H. Shinozaki in experiment and analysis for the large number of samples. This work was supported by TOWA corporation and DISCO corporation.

REFERENCES

[1] B. Rogers, C. Scanlan, and T. Olson, "Implementation of a fully molded fan-out packaging technology", , IWLPC (2013)

[2] Ganesh Hariharan, Raghunandan Chaware, Inderjit Singh, Jeff Lin, Laurene Yip, Kenny Ng, SY Pai, A Comprehensive Reliability Study on a CoWoS 3D IC Package, Electronic Components & Technology Conference, 573 (2015)

USING DOE TO IMPROVE COB BONBABILITY

Wei Xin[*], Sherry Chen, Wei-Ting Kary Chien

Semiconductor Manufacturing International Corporation

18 Zhangjiang Road, PuDong New Area, Shanghai 201203, China

*Corresponding Author's Email:Wei_Xin@smics.com,

ABSTRACT

In recent years, the Chip-on-Board (COB) package technology has become popular in semiconductor industries. The COB technology, in which the dies are directly mounted onto a printed circuit board (PCB) with bonding wires connecting the die and leads. Besides solder bumps, wire bonding is still the most popular interconnect method. With the development of wire bonding technology in the COB package, we can realize advanced processes with good performance by new wire bonding equipment and more powerful software. However, there are still many challenges to be overcome in the bonding process. In this paper, we did experiments to optimize the bonding parameter using 1.0mil Au wire on COB and successfully found the optimal range of bonding parameters through DOE (Design Of Experiment). By the optimal solution, we further improved the bondability; the second bonding quality was also improved by using BSOB (Bond Stitch On Ball）bonding.

INTRODUCTION

Quality of production is important to every company and customer. In assembly processes, the wire bonding is an important stage. In recent years, the Chip-On-Board (COB) has been widely used in semiconductor industries. The main material of the COB PCB (Printed Circuit Board) is FR-4 epoxy glass cloth, which becomes soft at high temperatures. However, we need operate wire bonding at a high temperature. This paper aims to analyze the main factors that affect bondability except for temperature, because the thickness of PCB varies for different products. The three main factors of wire bonding are power, force, and time, which interact with each other. Setting parameters at high levels (high-parameters) on the die can cause pad cratering/ crack, while the parameters at low levels (low-parameters) may lead to NSOP (Non-Stick On Pad). On the other hand, using high-parameters on the lead can cause heel crack or short tail, and low-parameters can result in NSOL (Non-Stick On Lead). In our experiments, we use DOE to find the optimal bonding parameters (1st and 2nd parameters). The output responses are ball size, ball shear, wire pull, stitch pull, crater, and smash ball. Finally, analyzing data by JMP, which is a statistical analysis software, we identify the optimal bonding parameter windows and obtain a very stable wire bonding process for the COB package. It improves the bondability of COB package, and we have successfully used COB to complete ESD and other reliability test for some products.

DOE EXPERIMENT

Different output responses like ball shear, wire pull, ball size, smash ball and crater are collected and measured during the wire bonding DOE experiments. All these data are processed by JMP for the optimal wire bonding parameters. Table 1 lists the wire bonding output responses and requirements.

Table 1 Wire bonding output responses and requirements

	Criteria	Sample
Ball shear	>10g	25 balls
Wire pull	>3g	25 wires
Ball size	2~2.5mil(2~2.5times wire diameter）	25 balls
Crater	0	25 balls
NSOP	0	25 balls
Smash Ball	No over pad/Pad deformation	25 balls

DOE for the parameters of the 1st bonding

The 1st bonding DOE uses a full factorial 3-factor and 2-level design with 11 legs as in Table 2 and the responses are in Table 3. Figure 1 shows the DOE design matrix.

Table 2 The DOE factors and levels

Table 3. The DOE output responses

Responses

Response Name	Goal	Lower Limit	Upper Limit
Ball shear	Match Tarc	15	25
NSOP	Maximize	.	0
Wire pull	Minimize	3	.
Crater	Maximize	.	0
Ball Size	Match Tarc	2	2.5
Smash ball	Maximize	.	0

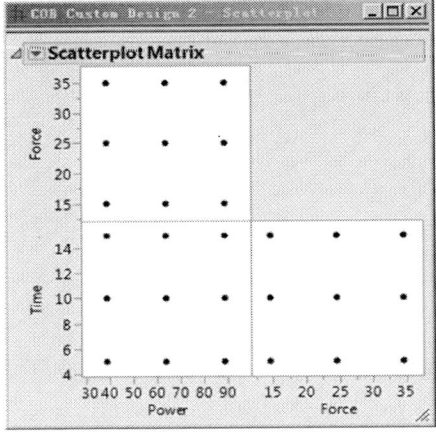

Figure 1. The DOE design matrix

With the same conditions, an experiment with 11 legs was executed and the data were collected. Table 4 shows the experiment data.

Table 4. The DOE experiment data

	Power	Force	Time	Wire Pull(g)	Ball Shear(g)	Ball Size(mil)	NSOP	Crater	Smash ball
Leg1	90	15	15	9.63	28.57	2.56	0	0	3
Leg2	90	35	15	9.59	30.5	2.62	0	0	6
Leg3	90	15	5	9.52	26.44	2.52	0	0	4
Leg4	40	25	5	9.48	13.58	2.11	3	0	0
Leg5	65	35	5	9.69	23.74	2.23	0	0	0
Leg6	90	25	10	9.37	29.09	2.55	0	0	4
Leg7	65	15	10	9.67	21.76	2.42	0	0	0
Leg8	40	35	10	9.43	13.74	2.02	4	0	0
Leg9	40	15	15	9.3	14.3	1.98	6	0	0
Leg10	65	25	10	9.3	24.18	2.32	0	0	0
Leg11	65	25	15	9.42	23.3	2.28	0	0	0

Exploring and analyzing data

Experimental data are analyzed with JMP, which is a statistical analysis software. According to $R^2 > 0.75$ and $P < 0.05$:

A: the ball shear response linearity model shows a good fitting result. The effective parameter is power ($P<0.05$). Figure 2 shows the ball shear response analysis data.

B: the wire pull linearity model shows a good fitting result. The effective parameters are power/force/time ($p<0.05$). Figure 3 shows the wire pull response analysis data.

C: the ball size linearity model shows a good fitting result. The effective parameter is power/($p<0.05$). Figure 4 shows the ball size response analysis data.

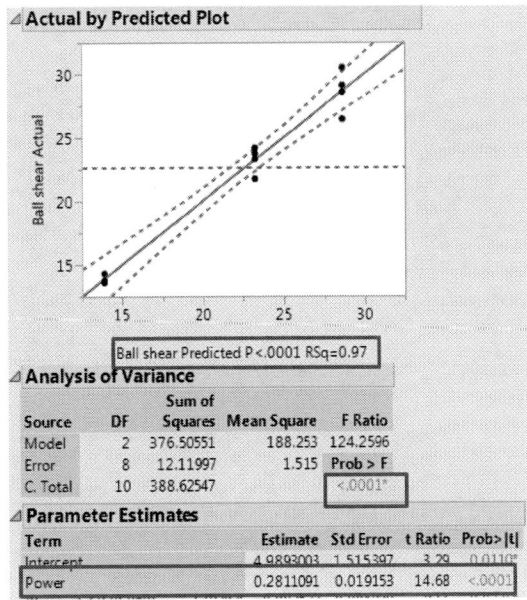

Figure 2. The Ball shear response analysis

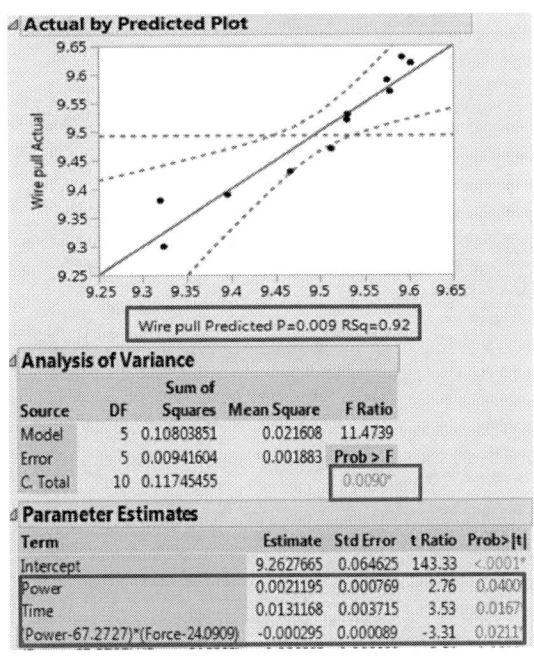

Figure 3. The wire pull responses analysis

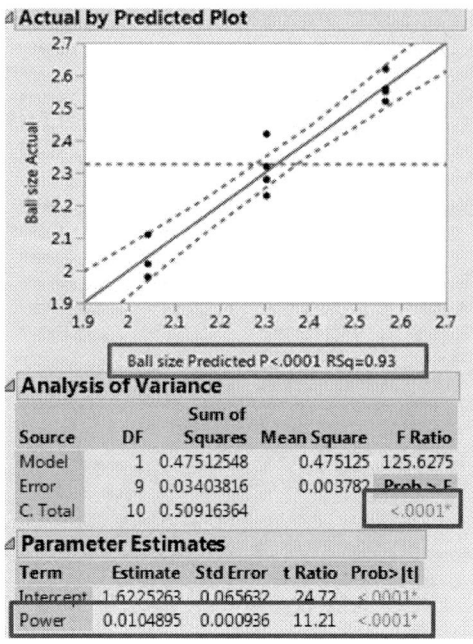

Figure 4. The ball size response analysis

According to the lower value AICc and P<0.005:

D: the effective NSOP model is good, the effective parameter is power(P<0.005) Figure 5 shows the NSOP response analysis data.

E: the effective smash ball model is good, the effective parameter is power(P<0.005) Figure 6 shows the smash ball response analysis data

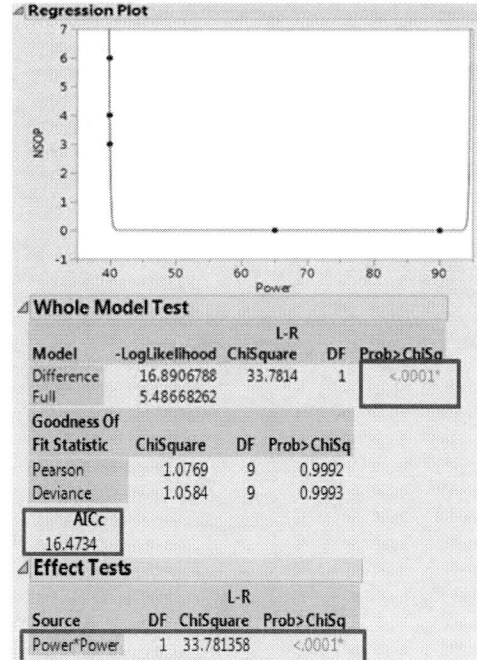

Figure 5. NSOP response analysis

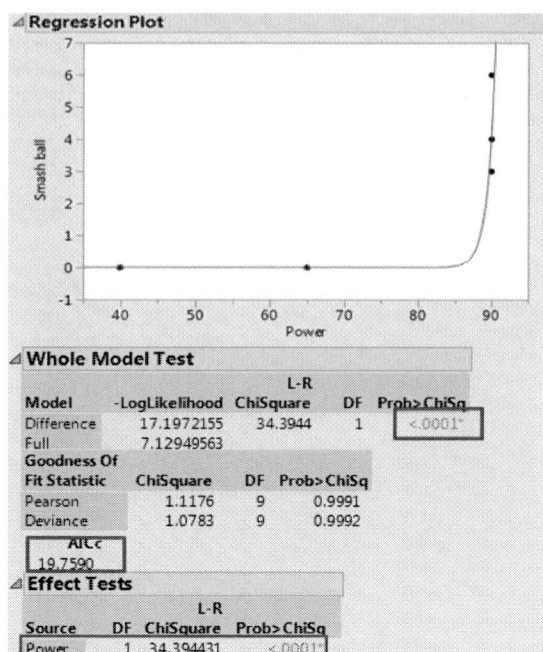

Figure 6. Smash ball response analysis

By selecting main factors and the model fitting analysis, we obtained successful results with power/ force/ time = 66/ 15/ 10 with desirability = 0.633707. All responses meet the requirements with the prediction profiler in Figure 7.

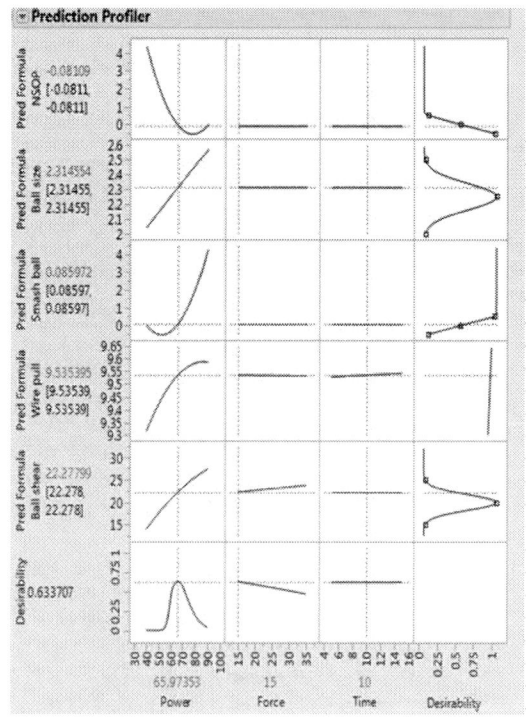

Figure 7. The DOE optimization

Finally process window analysis with JMP

The red area in Figure 8 shows the process window of the 1st bonding parameters, and the data are listed in Table 5:

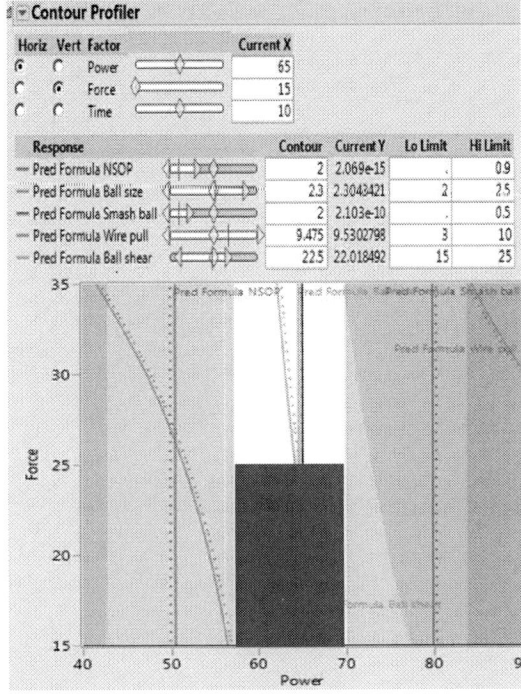

Figure 8. The bonding parameter analysis

Process window confirmation run

To verify parameters and evaluate the bonding performance, main factors were set to high, middle, low; the bonding results (in Table 6) meet the requirements. Figure 9 shows the failure mode using high parameters.

Table 5 The process window

Power	Force	Time
57~70	15~25	10~15

Table 6. The parameter matrix and bonding result

	Power	Time	Force	Wire Pull(g)	Ball Shear(um)	Ball Size(um)	NSOP	Crater	Smash ball	Remark
Low	57	10	15	9.32	21.42	2.26	0	0	0	Pass
Middle	63	12.5	20	9.45	22.79	2.3	0	0	0	Pass
High	70	15	25	9.38	24.37	2.34	0	0	0	Pass

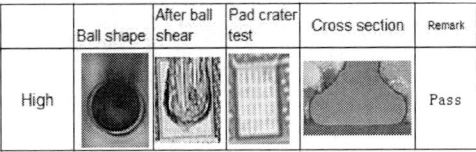

Figure 9. Failure analysis using high parameters

Use DOE to evaluate the 2nd bonding parameters

Similarly, through experiments, it was found that the stitch pull with BSOB was bigger than those from traditional bonding as in Figure 10.

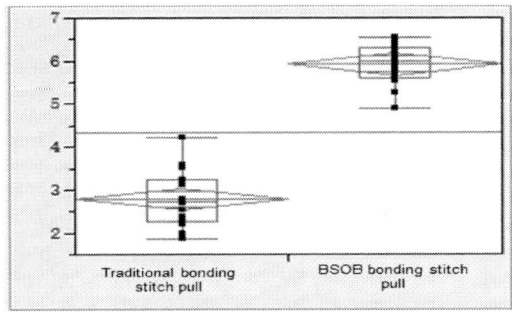

Figure 10. Two types of stitch pull

DOE experiments

Based on the 1st bonding parameters, we evaluated the 2nd bonding parameters (bump parameters) and SSB (Stand-off Stitch Bond) with DOE splits. Table 7 shows the experiment data and results. From the data, we successfully define the process window of the 2nd bonding (bump parameters) and SSB parameters as shown in Table 8.

Table 7. The experiment data and results

Bump	Power	Force	Time	NSOP	Ball Shear(g)	Ball Size(um)
Leg1	65	30	5	9	29.25	53.26
Leg2	65	40	10	7	31.63	52.18
Leg3	85	30	5	0	38.46	64.36
Leg4	85	40	10	0	39.63	66.82
Leg5	105	30	5	0	41.36	74..56
Leg6	105	40	10	0	43.41	75.29

SSB	Power	Force	Time	NSOL(On bump)	Stitch Pull(g)
Leg1	15	15	5	6	4.25
Leg2	20	20	10	0	5.89
Leg3	30	30	15	0	6.12

Table 8. The 2nd bonding parameters process window

	Power	Force	Time
Bump	80~85	35~40	10
SSB	15~20	15~20	10

978-1-5090-6695-7/17 $31.00 © 2017 IEEE   469

CONCLUSION

Using DOE, we obtain the wire bonding process window for the 1st parameter on Die (Table 5) and the 2nd parameter on Lead (Table8). The results of using the parameters within the window for wire bonding have demonstrated good and reliable performance. It improves the bondability of a COB package, and we have successfully used the COB to realize high-pin-count ESD and the associated reliability tests.

REFERENCES

[1] Richard Qian; Yong Liu; Os Jeon; Jerome Teysseyre, "Dynamic Response of A Molded Loaded Package in Wire Bonding Assembly Process", Electronic Packaging Technology (ICEPT), 2015 16th International Conference on 2015, pp. 835 ~ 839.

[2] J. H. L. Pang, C. K. Tan, "Thermal Analysis of A Wire Bond Chip-on-board Package", Thermal and Thermo-mechanical Phenomena in Electronic Systems, ITHERM 1998, The Sixth Intersociety Conference, pp. 481 ~ 487.

[3] G. G. Harman, *Wire Bonding in Microelectronics*, McGraw-Hill, New York, 3rd Edition, 2010.

[4] Xiaoying Wang, Chee-Cheng Chen, "Optimizing Process Conditions Using Design of Experiment – A Wire Bonding Semiconductor Assembly Process Case Study", 2015 12th International Conference on Service Systems and Service Management (ICSSSM), pp. 1 ~ 5.

[5] JMP user guide, Version 12. WWW.jmp.com.

STUDY OF WHITE EPOXY
MOLDING COMPOUND FOR LED BRACKET

Hongjie Liu[1], Wei Tan[1], Lanxia Li[1], Xiaojuan Jiang[1],
Liang Cui[1], Yangyang Duan[1], Xingming Cheng[1], Dongen Zhang[2], and Yuhui Luo[2]

[1]Jiangsu HuaHaiChengKe Advance materials Co. Ltd., Lianyungang 222047, China

[2] D Department of Chemical Engineering, Huaihai Institute of Technology, Lianyungang，China.

*Corresponding Author's Email: Hongjie.liu@hhck-em.com

ABSTRACT

Epoxy molding compound for LED bracket, with the merits of good resistance to heat and light damage as well as good reliability and warpage compared to traditional thermoplastic reflecting materials for LED bracket, such as Polyphthalamide (PPA) and Poly1,4-cyclohexylene dimethylene terephthalate (PCT), has becoming the focus of researchers. Here we studied the reflectance of the white epoxy molding compound from the aspect of epoxy resin, type of titanium dioxide, wax and wetting and dispersing additives. And the results disclosed that epoxy resin A with the structure in Fig.1 exhibited higher reflectance at 450nm after aged at 150 for 1000h with the value of 70% while epoxy resin B with the structure in Fig2. With the value of 50%. At the same time, epoxy resin A displayed higher water absorption than epoxy resin B. Meanwhile, the reflecting material titanium dioxide modified by silica, alumina as well as organics disclosed highest reflectance, and then is the titanium dioxide modified by silica and alumina, and the last one was that only modified by silica or alumina. Wax showed no significant difference on the reflectance due to the tiny addition. When it comes to the wetting and dispersing agent, with the increasing of the content of polyester type wetting and dispersing agent, the reflectance of the white epoxy molding compound decayed.

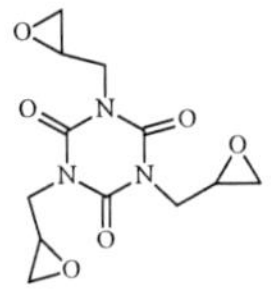

Figure 1: Structure of epoxy resin A

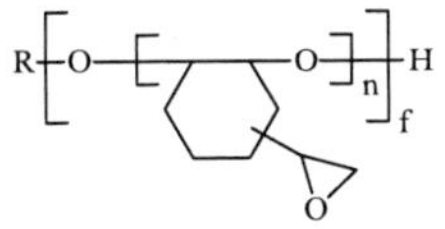

Figure II: Structure of epoxy resin B

Keywords—epoxy molding compound; Led bracket; epoxy resin; wetting and dispersing agent; wax

INTRODUCTION

LED, for its excellent properties like high efficiency, compact size, long lifetime and energy saving, LED has been widely used in many fields as traffic signal lamps, mobile, blue light unit (BLU), display media, and general lighting systems, illumination and automotive[1-2]. As LED is a kind of semiconductor materials that can transform electrical energy into visible light. Much attention has been paid on how efficiency of LED can be improved. The bonding adhesive and silicone encapsulants as one of the major issues affecting the reliability and lifetime of LED packages have been studied by Jemin Kim et al [3]. At the same time, demanding requirements of LED reflector resins on heat and light stability have been drawn forth due to the increase of the brightness and electrical current of LED packaging. For the traditional LED reflector resins as Polyphthalamide (PPA) and Poly1,4-cyclohexylene dimethylene terephthalate (PCT) with the process condition for injection molding, the performance is not robust enough in the aspect of long term reflectance stability under heat and light as well as large warpage after molding. While epoxy molding compound with advantages of good stability, low warpage as well as excellent mechanical performance and low cost compared to silicone, has becoming the main stream of the LED reflector resin market [4-5]. And Fig. III below displayed the SMD packaging process using epoxy molding compound.

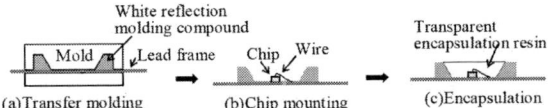

Figure III: SMD packaging process using EMC

Here, we studied the EMC for LED bracket from the aspects of epoxy, reflecting packing type and wetting and dispersing agent

EXPERIMENTAL

Materials

Epoxy resin A and epoxy resin B were provided by Daicel Chemical Ind., Ltd. Hexahydrophthalic anhydride (HHPA) as hardeners was got from New Japan Chemical co., Ltd. Tetra-n -butylphosphonium o,o-diethylphosphorodithioate (HISHICOLIN PX-4ET) as catalyst was from Nippon Chemical Industrial. Silica filler was produced by Tatsumori Ltd. Reflecting packing filler R1, R2, R3 and R4 were got from Kronos international Inc. Coupling agent and wax were received from Momentive and Clariant respectively.

Preparation of EMC

All the ingredients were weighted up according to the formulation and mixed in the high speed mixing machine, after this, the mixture was feeding to double screw extruder under the heating condition for extrusion, and then the discharged material go through sheet formation, cooling, pre-braker, granulation, post blend, and then be stored at -18℃.

Measurement

The specimens for reflectance test with the diameter of 30mm and thickness of 2mms were prepared through transfer molding method with the temperature as 180℃ and cure time as 120s. And then specimens with different aging time of 0h, 300h, 500h, 700h and 1000h at 150℃ were tested by UV-2600 with integration sphere which was got from Shimadzu Corporation. At the same time, the yellow index (YI) value was measured according to ASTMD 1925.

Further more, the gel time (GT), spiral flow (SF), hot hardness (HH) of molding 90 s and the water absorption (WA) at 85℃, 85% for 168h of the EMC samples were tested according to SJ/T 11197-1999.

RESULTS AND DISCUSSION

Influence of epoxy resin

General Layout

Epoxy resin A and epoxy resin B receive from Daicel chemical were all transparent, and in the formulation, the filler content was 83%, reflective packing content was 25%. And the reflectance at 450nm and YI of the resulting EMC were disclosed in Fig.IV. From Fig.IV, we can see that, both EMC with epoxy resin A and epoxy resin B show high reflectance value higher that 90% at 450nm and low YI when the aging time was 0h, however, with the aging time going on, the reflectance at 450nm of EMC with epoxy resin B decreased much rapid than that with epoxy resin A; correspondingly, the YI value of EMC with resin B increased faster that than with resin A. All the information stated that epoxy resin A possess higher performance on heat induced stability.

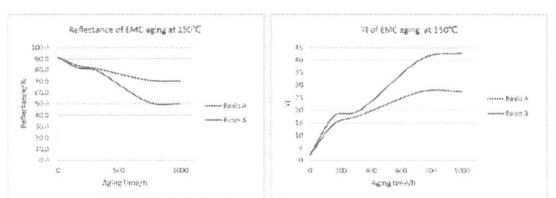

Figure IV: Reflectance and YI of EMC with different epoxy

resin aging at 150℃

Meanwhile, the GT, SF, HH90s, and WA 85℃/85%168h of EMC were also measured, detailed information were in Table I.

TABLE I.   PROPERTIES OF EMC WITH DIFFERENT RESIN

Epoxy resin	Properties of EMC			
	GT/s	SF/cm	HH90s	WA%
Resin A	26	93	60	1.35
Resin B	25	89	61	0.71

From TableI, we can see that there was no significant difference on the GT, SF and HH90s. But EMC with resin A in displayed much higher water absorption value than that with resin B in although it showed higher reflectance and low YI value with the aging time going on

Influence of reflecting packaging type

Reflecting packing R1 only modified by alumina, R2 treated by alumina and silica, R3 modified by alumina and silica and organics together, and R4 treated by alumina and organics were tested in this study. Epoxy resin B was introduced in the testing formulation and the reflecting packing filler content was 22%.

Fig. V indicated that R1and R4 exhibited lowest reflectance at 450nm with the aging time going on, and then was R2, while R3 was the one with the highest value. Meanwhile, YI data gave results in accordance with the reflectance: R1 and R4 with the highest YI value, and then was R2, with the R3 as the lowest.

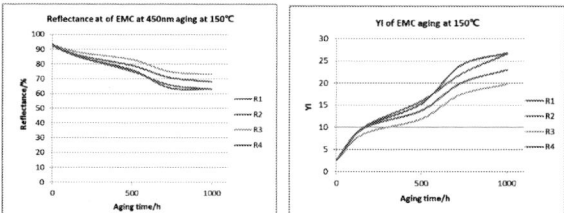

Figure V: Reflectance and YI of EMC with different reflecting packing aging at 150℃

Also, the properties of EMC like GT,SF,HH90s and waterabsortion were carried out as well, and the results were in Table II

TABLE II.   PROPERTIES OF EMC WITH DIFFERENT REFLECTING PACKING

Reflecting packing	Properties of EMC			
	GT/s	SF/cm	HH90s	WA/%
R1	25	89	64	1.33
R2	26	88	65	1.31
R3	25	98	66	1.32
R4	26	96	65	1.33

Table II disclosed that there were no significant difference on the GT, HH90s and WA of the EMC with different reflecting packing, however EMC with R3 and

R4 displayed longer SF than EMC with R1 and R2, perhaps this was caused by the good wetting of R3 and R4 due to their organic modification.

Influencing of wetting and dispersing agent

As the spiral flow shorted significantly with the introducing of reflecting packing filler, here we tried wetting and dispersing agent for lower viscosity and long spiral flow. And in this study, the wetting and dispersing agent with polyester structure with good performance on normal black EMC was chosen

Different dosage of the wetting and dispersing agent were incorporated in the EMC, and the properties of EMC was disclosed in Table III, also the reflectance at 450nm and YI of EMC samples after aging at 150°C for 1000h were also displayed in Fig.VI

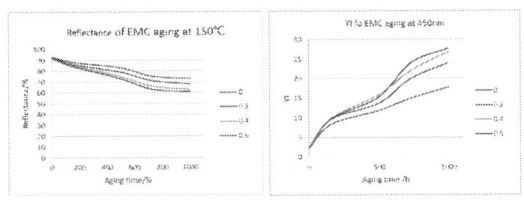

Figure VI: Reflectance and YI of EMC with different wetting and dispersing agent content aging at 150°C
The other properties of EMC samples with wetting and dispersing agent were tested, and the results were showed in table III.

TABLE III. PROPERTIES OF emc WiTH weeting and DISPERSING AGENT

WD content /%	Properties of EMC			
	GT/s	SF/cm	HH90s	WA/%
0	24	80	77	1.31
0.2	25	87	74	1.35
0.4	27	95	69	1.42
0.6	229	110	63	1.53

From the above results, we can see that with the increase of the content of the wetting and dispersing agent, and spiral flow of EMC has been improved signifantly. However, and HH 90s and the water absorption, the reflectance as well as the YI of EMC all getting worse.

CONCLUSIONS

In this study, the influence of epoxy resin type, reflecting packing type and content of wetting and dispersing agent on the properties of white EMC for LED bracket have been discussed. The results indicated epoxy resin A disclosed higher reflectance at 450nm after aged at 150°C for 1000h with the value of 70% while resin B of 50%. However resin A displayed higher water absorption than resin B. At the same time, Reflecting packing R1 only modified by alumina together with R4 which was modified by alumina and organics showed the lowest

reflection after aging test, and then is R2 modified by silica and alumina, while R3 modified by silica, alumina and organics disclosed the highest value. When it comes to the wetting and dispersing agent, also the spiral flow can be improved, the other properties of EMC getting worse, further intensive study must be done later.

Many factors influencing the reflectance and other properties of the white EMC for LED bracket were not included, such as catalyst, wax, antioxidant, stress releasing agent and so on, accordingly, subsequent research will be continued

REFERENCES

[1] Baotan, Zhang, Ru, Li, Xiuning Ch resin Aen, Fan Zhang, Bailing Liu, Gongying Wang and Guofeng He, "Advanced progress of LED encapsulation materials," New Chemicla Material., Vol. 38, No.4(2010), pp23-27

[2] Kwang-Cheol Lee, Sang-Mook Kim, Hwa-Sub Oh and Jong Hyeob Baek, "Whitening Phenomena between Polyphthalamide Sidewall and Silicone Encapsulant in the Light-Emitting Diode Package," 18th IEEE International Symposium on the Physical and Failure Analysis of Integrated Circuits, IPFA 2011

[3] Jemin Kim, Byungjin Ma,* and Kwanhun Lee, "Comparison of Effthe ecfolt of Epoxy and Silicone Adhesive on the Lifetime of Plastic LED Package," Electronic Materials Letters, Vol. 9, No. 4 (2013), pp. 429-432

[4] Sam H.Y. Zou, Mian Tao, Jeffery C. C. Lo, Ricky Lee, "Experimeits ntal Characterization of Adhesion Strength between the alumina anSilicone Encapsulant and the Bottom ofa SMD LED Leadframe Cup," 14th International Conference on Electronic Packaging Technology(ICEPT-HDP), 2013, pp.1212-1216.

[5] Bing Lu, Jianwei Li, KiHoon Lee, Masao Ishijima, "New PCT compound for LED reflector resin," 71st Annual Technical Conference of the Society of Plastics Engineers 2013, v 1, pp. 765-769,

NOVEL ELECTRON DEVICES BASED ON LASER SCRIBED GRAPHENE

Lu-Qi Tao[1], Dan-Yang Wang[1], He Tian[2], Ning-Qin Deng[1], Yi Yang[1,*] and Tian-Ling Ren[1,*]

[1]Institute of Microelectronics & Tsinghua National Laboratory for Information Science and Technology (TNList), Tsinghua University, Beijing 100084, China

[2] Ming Hsieh Department of Electrical Engineering, University of Southern California, Los Angeles, CA, USA

*Corresponding Author's Email: yiyang@tsinghua.edu.cn, RenTL@tsinghua.edu.cn

ABSTRACT

Graphene-based devices have attracted a lot of researchers' interests due to its excellent mechanical and electrical properties. However, graphene fabrication methods, such as CVD and mechanical exfoliation, have obvious drawbacks. Laser scribing technology is a novel method to fabricate graphene. In this paper, some electron devices based on laser scribed graphene, such as acoustic devices, strain sensors, light-emitting devices and memories shows great potentials for commercialization in the future.

INTRODUCTION

Graphene is 2d material which was first found in 2004 by A.K. GEIM [1]. It can be an ideal material for flexible electronics application because of its excellent mechanical and electrical properties. A lot of graphene-based devices, such as transistors [2], memories [3], and sensors [4] have been developed previously. However, the yield and the quality of graphene can't meet the demands of commercialization. It's urgent to develop a large-scale and high efficient method to fabricate graphene.

Laser scribing technology is novel graphene fabrication method [5]. Graphene oxide solution is spin coated on the substrate first. A uniform graphene oxide film will be left after the volatilization of water. Then the graphene oxide film will be converted into graphene by the irradiation of laser.

In this paper, we will introduce some novel devices based on laser scribed graphene including graphene earphone, graphene strain sensor, graphene LED and graphene memory. The graphene earphone has a broad frequency from 100 Hz to 50 kHz, which shows great potentials for communication between human beings and animals. The graphene strain sensors have both large strain range and high gauge factor, so they can be used to monitor both large and subtle motions. The graphene LED has a tunable light spectrum which can vary from blue light to red light. The graphene memory has stable switching performance and flexible characteristics. These graphene-based devices have great potentials for commercialization in the future.

GRAPHENE-BASED DEVICES
Graphene Earphone

The working principle of conventional sound source is based on mechanical vibration. However, graphene can generate sound based on thermal acoustic effect which is completely different from conventional sound sources. When an electrical signal is applied on the graphene, a lot of Joule heating will be generated in the surface of graphene. Graphene has a low heat capacity (≈ 0.03 $J/(m^2 \cdot K)$), so the Joule heating will be emitted into the air near the surface of graphene. The air will expand and contract periodically, causing the generation of sound. The fabrication process is shown in figure 1(a), firstly the graphene oxide solution was spin coated on the DVD disk substrate, then the DVD disk was inserted into a LightScribe DVD driver. Then the graphene oxide film was converted into graphene and finally these graphene elements were packaged into earphone shells.

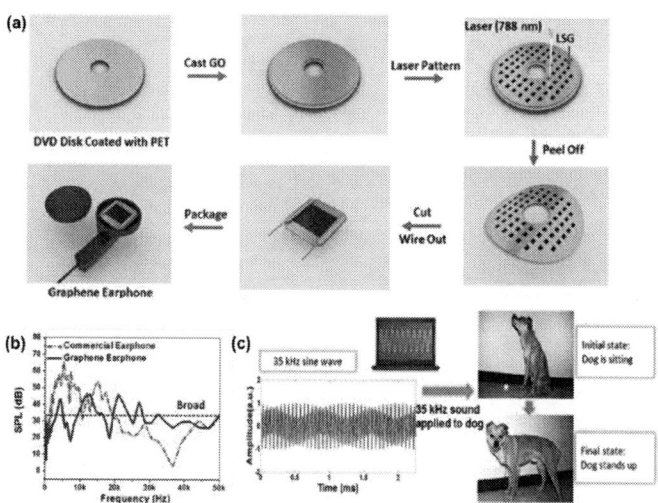

Figure 1: (a) The fabrication process of graphene earphone. (b) The performance comparison between graphene earphone and commercial earphone. (c) The dog can be controlled by graphene earphone.

In order to compare the performance of our graphene earphone and commercial earphone, both are tested under the same audio analysis system. The audio analyzer sweeps the frequency from 100 Hz to 50 kHz. The graphene earphone showed a broad frequency from 100 Hz to 50 kHz with low fluctuation compared with commercial earphone. Conventional sound source based on mechanical vibration will have more obvious resonant peaks, while the graphene earphone would not cause mechanical vibrations, which will result in a low fluctuation. Such a phenomenon shows that the graphene earphone could be widely used in various acoustic applications. Therefore, it can be suitable for communication between human beings and animals. As demonstrated in figure 1(c), a dog worn the graphene earphone can by controlled to stand up when hearing the 35 kHz high-frequency signal played through the graphene earphone.

Graphene Strain Sensor

A lot of strain sensors have been developed to monitor human activities and personal health. However, these strain sensors usually have a large strain range but a low gauge factor (GF) or have a high gauge factor but a small strain range, which means these strain sensors can only detect either large motions or subtle motions. Based on laser scribing technology, a series of strain sensors with large strain range and high gauge factor can be fabricated simultaneously.

As shown in figure 2(a), Ecoflex, an elastic polymer, is chosen to be the substrate because of its excellent stretchability. The mixture of graphene oxide (GO) solution and tetrahydrofuran (THF) was dropped on the substrate. The THF is used to improve the hydrophilicity of the substrate. Then the GO film will be converted into laser patterned graphene flakes (LPGF) with different mesh densities. The copper wires are attached on both side of the LPGF by silver paste. The Ecoflex is then dropped again to encapsulate the whole structures and the strain sensors with different mesh densities are cut down into pieces afterwards.

As shown in figure 2(b), three types of strain sensors with no mesh, low-density mesh and high-density mesh were fabricated. The brown part is GO and the black part is LPGF. The brown GO has been completely broken and cracked under the strain of 40%, but the black LPGF still almost connected. From figure 2(c), we can see that the mesh density will have great effect on the performance of the strain sensors. The strain sensors with different mesh densities present tunable performance. The strain sensor with high-density mesh has a preferable GF of 457 under the strain of 35%, which is higher than that with low-density mesh and no mesh. Figure 2(d) presents that the strain sensor with no mesh will have a preferable strain range of 100% with the GF of 268.

The contact area of the overlapping graphene flakes will decrease when the sensor is stretched, which will cause the increase of the resistance. Besides, the high-density mesh graphene flakes will be easy to form

disconnected state under the same strain because graphene flakes will move to non-overlapping positions.

The performance of the strain sensors can be tuned easily by adjusting the density of the patterned mesh, realizing the fabrication of strain sensors with ultrahigh GF and large strain range at the same time. Therefore, these strain sensors can be used to monitor both large and subtle motions, and they have remarkable and practical potentials in health monitoring, voice recognition, gesture control and many other areas.

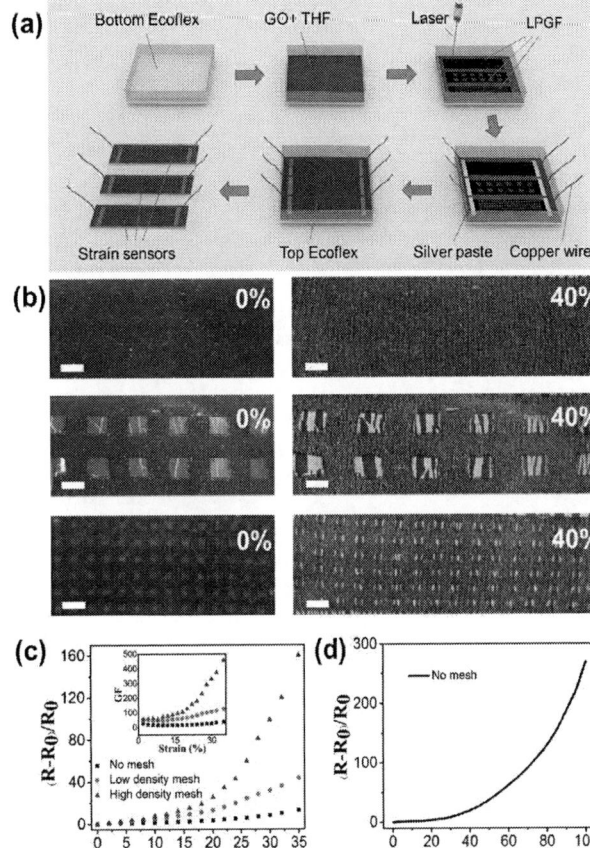

Figure 2: (a) The fabrication process of graphene strain sensors. (b) The optical images of graphene strain sensors. (c) The performance testing of graphene strain sensors.

Graphene LED

The solid-state LED based on the semiconductor industry has been used in the fields of high performance communication, low cost lighting and smart displays. But the emission wavelength of traditional LED can be only adjusted by complex material design and bandgap engineering. We realized the fabrication of laser scribed graphene LED with a desirable combination of a bandgap structure and a bipolar carrier injection in a special type of semi-reduced graphene oxide (GO).

The device formed on a laser-scribed GO surface which is made up of a series of reduced graphene oxide (rGO) nanoclusters with many different sizes, and a single-colour luminescence can be selectively stimulated by controlling the doping level of rGO nanoclusters. The mobility of the semi-reduced GO network is 10 cm^2 V^{-1} s^{-1}. The light-emission spectrum of the graphene LED can be adjusted between the blue (450 nm) and the red (750 nm) by electrical gating or conditioning the environmental doping. In addition, the brightness of the device is up to 6,000 cd m^{-2}, and the efficiency is around 1%.

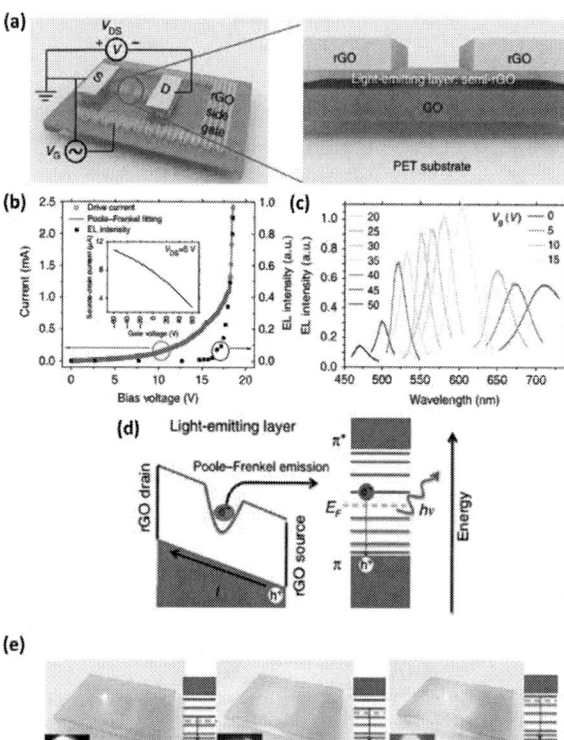

Figure 3: The spectrally tunable all-graphene-based flexible LED device. (a) Schematic of the GFLED. A distinct semi-reduced GO (blue) at the interface between GO (orange) and rGO (gold) is responsible for light emission. (b) Source-drain current (circles) and electroluminescence intensity (solid squares) versus source-drain bias voltage. Inset: source-drain current versus gate voltage of the same device, showing a p-type field effect. (c) Typical electroluminescence spectra of a single GFLED. Gate biases are from 0 to 50V. (d) A schematic of the charge injection process. (e) Schematic of the gate voltage-dependent electroluminescence. Inset: corresponding emission images from a real device.

Figure 3 shows the structure and test result of the graphene LED devices. Figure 3a schematically shows the structure of our all-graphene-based field effect LED (GFLED). The light-emitting layer is the interface between the GO and rGO, and then is uncovered by using current annealing to remove the highly conductive rGO channel. The length of the light-emitting region is 80–120 mm. The light-emitting region is located in the center of the narrow graphene field-effect transistor channel.

Figure 3b shows the source-drain current (circles) and electroluminescence (EL) intensity (solid squares) versus source-drain bias voltage. The inset picture shows the source-drain current versus gate voltage of the same device, which shows a p-type field effect. Figure 3c show the typical electroluminescence spectra of a single GFLED. Gate biases are from 0 to 50V. The peak emission wavelength shifts from 690nm at zero gate voltage to 470nm under 50V gate voltage.

The mechanism of the graphene LED device is shown in the figure 3d and figure 3e, which explains the observed spectral shift. Gating graphene lifts up the chemical potential and thus the energy level of the lowest unoccupied discrete state, and then the excited electron energy is increased. The electroluminescence peak can be adjusted within the whole photo-luminescence range by controlling the gate voltage; and the relative emission intensity is determined by the distribution of density of states of the discrete energy levels. This device opens the possibility of fabricating carbon-based photonic devices due to the availability of graphene-based LEDs.

Graphene Photodetector

Photodetectors that can convert light into electrical signal are crucial for various applications such as laser related sensor networks, imaging techniques, optical communications and other areas.

We fabricated a novel laser scribed graphene photodetector which has a photo responsivity of up to 0.32 A/W and a detectivity of 4.996*10^{10} cmHz$^{1/2}$W^{-1}.

The schematic diagrams and experimental results were illustrated in figure 4. From figure 4a, we can see the fabrication process of laser scribed graphene photodetector. It's a high-efficient and low-cost way to fabricate graphene photodetector. The I-V curves under laser on and off is shown in figure 4b. When the laser was applied, a positive photocurrent was generated. In addition, the experimental setup for testing the photodetector is shown in the inset of the figure 4b. As shown in figure 4c, it can be noticed that the photocurrent will increase with the increase of laser intensity and they have a linear relationship.

These photodetectors can be arranged in the form of 1d array and 2d array. Figure 4d shows the schematic structure of a line array of the photodetectors under a light

978-1-5090-6695-7/17 $31.00 © 2017 IEEE   476

source. The inset picture shows the real photo of the 23 pixels array. The length and width of the device are 250 μm and 3000 μm respectively. The center to center spacing with each devices is 500 μm. And figure 4e shows the intensity of the photo current versus the position of the detector. The inset showing the 1-dimensional pixilation plot, where each pixel is represented by a device in the array. The photocurrent intensity will decrease with the increase of the distance from the light source.

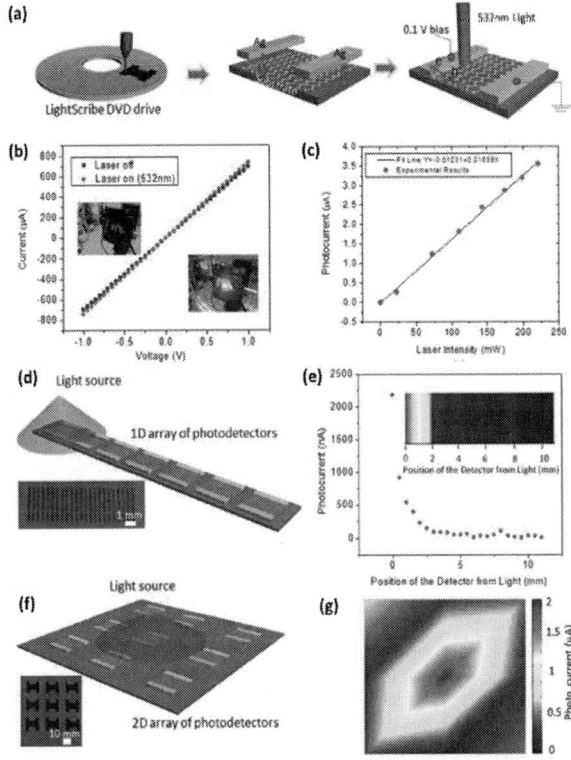

Figure4: Laser scribed graphene photodetector. (a)The fabrication process of a laser scribed graphene photodetector. (b)The I-V cure of the photodetector under laser on and off status. The inset shows the experimental setup. (c) The plot of the photo intensity versus the photo current. The line shows that the photo current increases linearly with the increase of the photo intensity. (d) Schematic structure of a line array of the photodetectors under a light source. Inset shows the real photo of the 23 pixels' array. (e)Photocurrent intensity versus the position of the detector from the light. The inset showing the 1-dimensional pixilation plot. (f)Schematic structure of a 2-dimentional array of the photodetectors under a light source. Inset shows the real photo the 3 ×3 pixels' array. (g)The 2-dimensional pixilation plot of the 3 ×3 array.

Furthermore, figure 4f shows the schematic structure of a 2-dimentional array of the photodetectors under a

light source. The inset picture shows the real photo of the 3×3 pixels' array. Figure 4g shows the 2-dimensional pixilation plot of the 3×3 array. We can see the center part demonstrated the highest photocurrent intensity. The asymmetric pattern may be caused by the dispersion of the light source.

CONCLUSION

In this paper, some novel devices based on laser scribing technology including graphene earphone, graphene strain sensor, graphene LED and graphene photodetector are demonstrated. These devices show excellent performance and flexibility and can be a part of flexible electronics systems. Laser scribing technology is high efficient method for graphene fabrication which has great potentials for commercialization and more practical laser scribed graphene devices will be fabricated in the future.

ACKNOWLEDGEMENTS

This work was supported by the National Natural Science Foundation (61574083, 61434001), National Basic Research Program (2015CB352101), National Key Research and Development Program (2016YFA0200400), National Key Project of Science and Technology (2011ZX02403-002), and Special Fund for Agroscientific Research in the Public Interest (201303107) of China. The authors are also thankful for the support of the Independent Research Program (2014Z01006) of Tsinghua University, and Advanced Sensor and Integrated System Lab of Tsinghua University Graduate School at Shenzhen (ZDSYS20140509172959969).

REFERENCES

[1] K. S. Novoselov, A. K. Geim, S. V Morozov, D. Jiang, Y. Zhang, S. V Dubonos, I. V Grigorieva, and A. A. Firsov, "Electric field effect in atomically thin carbon films," Science (80-.)., vol. 306, no. 5696, pp. 666–669, 2004.
[2] F. Schwierz, "Graphene transistors," Nat. Nanotechnol., vol. 5, no. 7, pp. 487–496, 2010.
[3] X. Wang, W. Xie, and J. Xu, "Graphene Based Non-Volatile Memory Devices," Adv. Mater., vol. 26, no. 31, pp. 5496–5503, 2014.
[4] Y. Shao, J. Wang, H. Wu, J. Liu, I. a. Aksay, and Y. Lin, "Graphene based electrochemical sensors and biosensors: A review," Electroanalysis, vol. 22, no. 10, pp. 1027–1036, 2010.
[5] M. F. El-Kady, V. Strong, S. Dubin, and R. B. Kaner, "Laser scribing of high-performance and flexible graphene-based electrochemical capacitors," Science (80-.)., vol. 335, no. 6074, pp. 1326–1330, 2012.

MATERIALS SCREENING WORKFLOW METHODOLOGIES FOR METAL OXIDES AND CHALCOGENIDES FOR USE IN NOVEL DEVICES

Tony Chiang[1], Karl Littau[2], Stephen L. Weeks, Ashish Pal, Vijay Narasimhan, Greg Nowling, Michael Bowes, Sergey V. Barabash, and Dipankar Pramanik

Intermolecular, Inc.
3011 First St., San Jose, CA 95134 USA
[1](+1) 408-582-5411, tony@intermolecular.com,
[2](+1) 408-582-5673, karl.littau@intermolecular.com

ABSTRACT

Materials are playing an increasingly important role to enable novel device applications beyond dimensional scaling. We describe areas of ferroelectric materials, high E_g dielectrics, chalcogenides, and oxide semiconductors. These materials find potential use in advanced memory, select element and transistor applications. In each area, a material screening workflow methodology is used to garner physical as well as electrical properties. DFT modeling of basic materials properties is used to complement experimental results.

INTRODUCTION

The requirements of scaling, low-power, and advanced applications are combining to challenge the current set of materials used in integrated circuit device manufacturing. New and expanded applications for non-volatile memory are catalyzing consideration of devices such as Ferroelectric FET (FeFET) and Ferroelectric Tunnel Junction (FTJ) memories [1]. Further scaling of DRAM capacitors is forcing evaluation of new capacitor dielectrics and electrodes capable of meeting the challenges of charge storage with low leakage and suitable breakdown strength [2]. 3-D vertical memory integration requires new, non-linear selector elements to reduce sneak currents and power requirements. High mobility oxide semiconductors are of interest for rapidly evolving display applications. All these are leading to development of new thin film materials and material stacks.

ADVANCED MATERIALS

Ferroelectric

The discovery of ferroelectric behavior in doped HfO_2 and mixed HfO_2/ZrO_2 films has stimulated work in FE devices for memory and transistor applications [3]. Amongst variables such as composition, film thickness, substrate, and annealing conditions, we find varying the individual HfO_2/ZrO_2 layer thicknesses in a nanolaminate stack can additionally influence the stabilization of the ferroelectric $Pbc2_1$ phase. By proper choice of processing conditions, a large remanent polarization (P_r) was obtained from these nanolaminates, making the nanolaminate structures relevant for memory applications. Figure 1 shows a TEM cross section, with an inset GIXRD scan and PE curve for an HfO_2/ZrO_2 laminate of 8 nm thickness. The $2P_R$ value of 51 $\mu C/cm^2$ is among the highest reported for this system. A unit layer thickness of ~1nm for HfO_2 and ZrO_2, respectively was used.

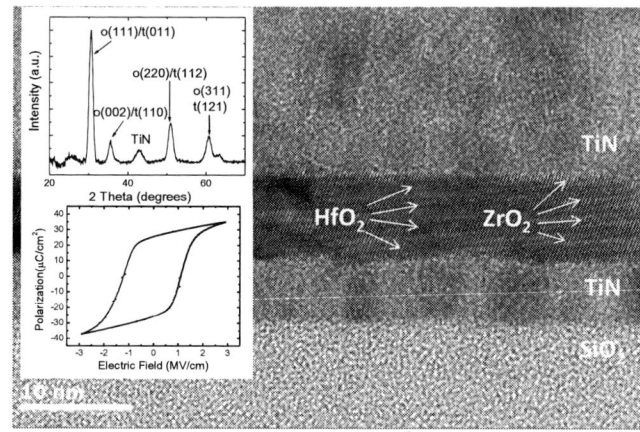

Figure 1: Laminated HfO_2/ZrO_2 ferroelectric film with 1 nm layer thickness showing crystalline GIXRD and a $2P_R$ value of 51 $\mu C/cm^2$ after wakeup.

We have also studied dopant-free HfO_2 ferroelectric M-I-M stacks to understand the fundamental nature of ferroelectricity in HfO_2. Previously it has been shown that ferroelectricity can be promoted in undoped HfO_2 [4]. We explored the impact of ozone dose in the ALD HfO_2 process as well as the bottom electrode material (e.g. titanium nitride vs. iridium) on ferroelectric properties. Using TiN bottom electrodes in HfO_2 M-I-M stacks, we find lowering the ozone dose suppresses the formation of monoclinic HfO_2 (as indicated by x-ray diffraction measurements in Fig 2a) and increases the remnant polarization (Fig 2b), but increases leakage. The increase in leakage is linked to an increase in the oxygen vacancy concentration in the HfO_2 film using trap-assisted tunneling modeling (TAT). Previously ab-initio simulation has shown that oxygen vacancies reduce the formation energy of the tetragonal phase [5]. In the case of ALD Ir as the bottom electrode, Pr increases, while leakage improves with lower ozone dose. It is possible that ALD Ir may provide improved texture for FE formation as compared to TiN.

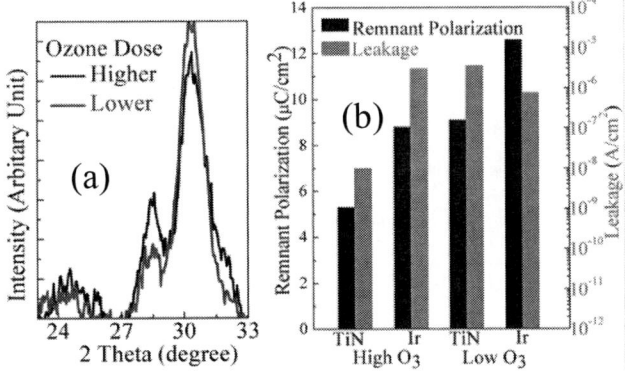

Figure 2: (a) GIXRD spectra of HfO_2 stacks showing suppression of monoclinic phase with lower ozone dose on TiN bottom electrode (b) Remanent polarization and leakage for different cases of bottom electrode and ozone dose. Here higher ozone dose is 40s and lower ozone dose is 5s.

Using DFT simulations, we identify possible metastable $(Hf,Zr)O_2$ phases to serve as a basis to help resolve the phase composition of these thin film stacks despite the structural similarity of the competing phases. DFT-based theoretical modeling also provides a guide to understand relative phase stability vs. the temperature (as illustrated in Fig.3) and thickness. We also find compressive strain can activate ferroelectric switching pathways unavailable in unstrained material [6] indicating potential impact from choice of underlayer/overlayer/thickness/thermal history & order.

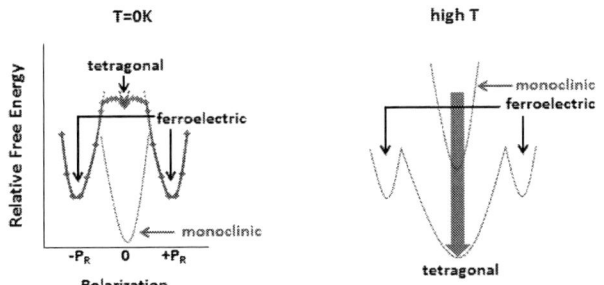

Figure 3: Left: Energy-polarization relationship between the tetragonal and the orthorhombic ferroelectric phases at T=0K, also schematically shown is the curve for the monoclinic phase at T=0K. Right: Schematics of the same relationship at elevated temperatures.

High Band Gap Dielectrics

Scaled capacitors will require high k dielectrics to store sufficient charge in the limited area afforded by advanced node DRAM devices. However, at thicknesses of <6 nm, low band gap dielectrics such as TiO_2 will exhibit excessive leakage currents (>1e-7 A/cm^2) making them unsuitable. Adopting a figure of merit (FOM) that combines k and E_g, we report the results of DFT modeling of several candidate compounds which exhibit comparably higher k while maintaining E_g >4 eV as shown in Table I.

At each composition, the preferred crystalline phase has been predicted by performing DFT simulations for a series of structural candidates using VASP code[7]. Due to the extreme sensitivity of high-k materials to volume (due to the proximity to a ferroelectric instability), k values are compared using both PBE[8] (not shown) and LDA (Table I) functionals. The gap values E_g are determined using hybrid HSE[9] functional. The

figure of merit is defined as $FOM = E_g k$, where E_g can be considered a proxy to the product of the barrier height and the effective tunneling mass[2], thus FOM measuring the effectiveness of the dielectric in suppressing a tunneling leakage at a given material thickness and k value. As a reference point, ZrO_2 dielectric has FOM=220 [2]. For anisotropic materials, we report a range of FOM values corresponding to different orientations of the crystal axes relative to the field direction.

We find the P4/mbm (s.g.127) phase for $SrHfO_3$ to be less energetically favorable vs. Pbmn phase yielding a lower FOM of <140 for this material as indicated in Table I.

Name	Space group name	k (LDA)	E_g (eV) (HSE)	FOM
$SrZrO_3$	Pbnm	30-35	5.5	165-193
$SrHfO_3$	Pbnm	25-28	5.0	125-140
$BaZrO_3$	Pm-3m	54	4.4	238
$BaHfO_3$	Pm-3m	36	4.8	173

Table I. Results of DFT calculations for candidate dielectric materials showing high E_g while maintaining reasonably high k, as combined in the figure-of-merit (FOM) indicating the ability to suppress tunneling leakage.

Results shown in Table I suggests $BaZrO_3$ can possess a sufficiently high k coupled with $E_g > 4$ eV. To realize controlled $BaZrO_3$ deposition via ALD, a self-limiting unit film process for BaO was developed on the Intermolecular A-30 combinatorial ALD reactor, using a Ba metallocene precursor with O_2 as the co-reactant. Fig 4a shows the growth curve of the BaO ALD process, exhibiting a saturated growth rate of ~0.26 Å/cycle. Deposition was performed on 300 mm p-doped ⟨100⟩ Si wafers and on 5 nm ALD ZrO_2, demonstrating compatibility for fabrication of Ba-doped ZrO_2 stacks. Similarly, ZrO_2 growth via ALD was also examined on a 3 nm BaO film.

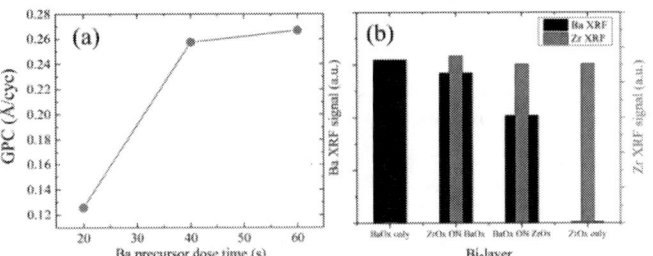

Figure 4: (a) Growth curve of BaO ALD process, showing saturation of GPC at ~0.26 Å/cycle at a process temperature of 260 °C. (b) XRF measurements of Ba and Zr for (from left-to-right) BaO-only, ZrO_2 on BaO_2, BaO_2 on ZrO_2 and ZrO_2-only films.

Fig 4b shows XRF measurements of elemental Ba and Zr for both ZrO_2/BaO and BaO/ZrO_2 bi-layers alongside BaO and ZrO_2 unit films. There is evidence of delayed or slowed growth for BaO deposited on ZrO_2, while ZrO_2 growth on BaO appears equivalent to growth on a Si substrate. Interactions between the

978-1-5090-6695-7/17 $31.00 © 2017 IEEE

two unit film processes during deposition of Ba-doped ZrO_2 must also be considered in order to effectively mitigate undesirable side-reactions such as barium hydroxide/carbonate formation as well as the consumption of Ba precursor by residual co-reactant left over from ZrO_2 ALD cycles.

Screening of Selector Materials

Another area of increasing interest is compounds for non-linear elements such as selectors to reduce sneak currents and manage variability in memory arrays. Use of combinatorial approaches allows rapid screening of candidate material compounds, compositions and stacks. Fig. 5 shows an example of a multi-target sputter chamber capable of controllably forming a variety of compounds in an array across a 300 mm substrate and an example substrate shown at right. The materials can also be deposited on a die to die basis (not shown) over a 300mm wafer test vehicle for direct device testing without the need for patterning. These materials can be chalcogenides, doped oxides, mixed ionic/electronic, semiconducting, and combinations thereof. The effectiveness of the combinatorial screening can be increased by guiding the selection of material compositions using both semi-phenomenological and DFT-based modeling, as well as relating the experimental data to the results obtained from simulated annealing using ab-initio molecular dynamics and further DFT analysis of the simulated quasi-amorphous structures.

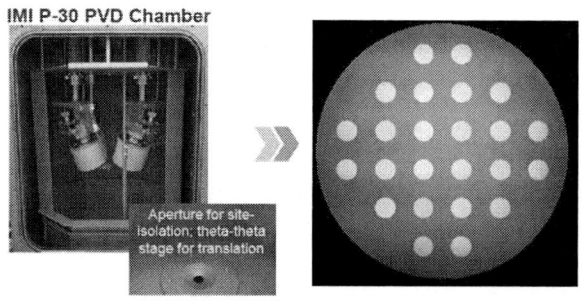

Figure 5: Multi-target PVD chamber with aperture (left), and 300 mm substrate with site isolated compositional variation array (right).

Screening of High Mobility Oxides

In advanced display applications where high mobility, reliability, and optical transparency are of importance, amorphous metal oxides are gaining interest as channel materials for thin-film transistors (TFTs). These materials are often ternary or quaternary compounds. Moreover, the transistor performance of these materials is sensitive to both composition and processing conditions yielding a large phase space for optimization.

In many applications, thin film transistors are implemented in a back-gated configuration, with source and drain contacts patterned above the channel layer. In this configuration, several lithography steps with alignment are required, rendering the rapid screening of a wide compositional space of materials difficult. Here, we present a workflow for the screening of high mobility oxides using a deposit-and-test strategy, as shown in Fig. 6.

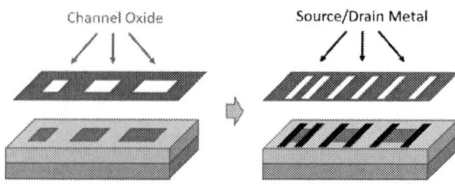

Figure 6: 2-step shadow-mask workflow for rapid screening of oxides for TFTs.

We start with a conductive silicon substrate with a gate oxide material on one side, for example silicon oxide, aluminum oxide, or hafnium oxide. Next, we use a shadow mask to define a set of channels, and deposit the thin channel layer oxide through the mask. We can co-sputter up to 4 targets at once, and each of these can either be elemental or a compound, giving access to a very wide parameter space of materials and processing conditions. After depositing the channel layer, we use a second shadow mask to deposit source and drain electrodes by sputtering. The second shadow mask's electrodes can easily be aligned to the channels using a series of alignment marks. Fig. 7 shows an assembly of the second shadow mask atop a sample with channels already deposited; each device is visible beneath the holes for the source and drain electrodes in the second mask. If required, additional substrate pre-cleaning steps, anneals, and organic and inorganic passivations (sputtered, evaporated, or ALD) can be incorporated into the workflow.

Figure 7: Second shadow mask for source/drain contacts positioned on top of isolated channels generated during first deposition step. Each pair of device electrodes will contact one device, and there are 20 devices in total within a 25 mm site isolated spot.

Fig 8a shows the design of an array of TFT devices formed completely within a 25mm spot similar to one of those shown in Fig. 5. In each array, several sizes of devices can be defined, with multiple repeated devices of a certain size to improve measurement statistics. Using an automated optical microscope, the channel length and width of each device can be extracted. Then, each device is probed from the top on the source and drain contacts, while the entire substrate is contacted through a chuck contact to act as the gate, as shown in Fig 8b. Both transmission-line measurements with a floating substrate contact (for contact resistance) and current-voltage curves with an applied bias on the substrate (for transistor characteristics) can be collected, and characteristics can be automatically extracted.

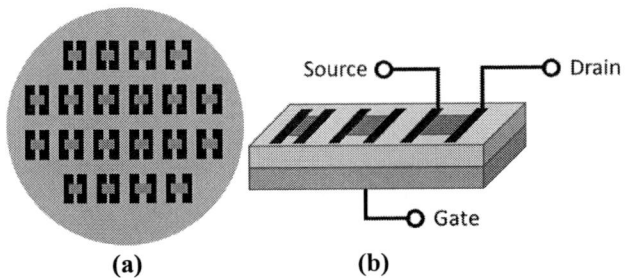

Figure 8: a) TFT array design completely contained in a single 25 mm spot. b) Probe configuration for electrical testing.

REFERENCES

[1] S. Fujii, Y. Kamimuta, T. Ino, Y. Nakasaki, R. Takaishi, and M. Saitoh. "First demonstration and performance improvement of ferroelectric HfO 2-based resistive switch with low operation current and intrinsic diode property." In VLSI Technology, 2016 IEEE Symposium on, pp. 1-2. IEEE, 2016.

[2] K. Yim, Y. Yong, J.Lee, K. Lee, H. Nahm, J. Yoo, C. Lee, C. Hwang, and S. Han. "Novel high-κ dielectrics for next-generation electronic devices screened by automated ab initio calculations." NPG Asia Materials 7, no. 6 (2015): e190.

[3] T.S. Böscke, J. Müller, D. Bräuhaus, U. Schröder, and U. Böttger. "Ferroelectricity in hafnium oxide thin films." Applied Physics Letters 99, no. 10 (2011) 102903.

[4] P. Polakowski and J. Müller, "Ferroelectricity in undoped hafnium oxide", Applied Physics Letters, vol. 106, no. 23, 232905, pp 1-4, 2015.

[5] C-K. Lee, E. Cho, H-S. Lee, C. S. Hwang and S. Han, "First-principles study on doping and phase stability of HfO_2", Physical Review B vol. 78, pp 012102, 2008.

[6] S.V. Barabash, D. Pramanik, Y. Zhai, Blanka Magyari-Kope, and Y. Nishi, "Ferroelectric Switching Pathways and Energetics in $(Hf,Zr)O_2$", in Nonvolatile Memories 5, ECS Transactions (in press).

[7] G.Kresse and J.Hafner, Phys. Rev. B vol.47 (1993), p.558; G. Kresse and J. Furthmüller, Comput. Mater. Sci. vol.6 (1996), p.15; G. Kresse and J. Furthmüller, Phys. Rev. B vol.54 (1996), p.11169.

[8] J. P. Perdew, K. Burke, and M. Ernzerhof, Phys. Rev. Lett. vol.77 (1996), p.3865.

[9] J. Heyd, G. E. Scuseria, and M. Ernzerhof, J. Chem. Phys. vol.118 (2003), p.8207; J. Heyd and G. E. Scuseria, J. Chem. Phys. vol.121 (2004), p.1187; A. V. Krukau , O. A. Vydrov, A. F. Izmaylov, and G. E. Scuseria, J. Chem. Phys. vol.125 (2006), p.224106.

Neutral Beam Technology for Future Nano-device

Seiji Samukawa

Advanced Institute for Materials Research, Tohoku University

Innovative Energy Research Center, Institute of Fluid Science, Tohoku University

2-1-1 Katahira Aoba-ku Sendai, Miyagi 980-8577, Japan

e-mail: samukawa@ifs.tohoku.ac.jp

Abstract — **Advances in plasma process technology have contributed directly to advances in the miniaturization and integration of semiconductor devices. However, in semiconductor devices that encroach on the nanoscale domain, defects or damage can be caused by charged particles and ultraviolet rays emitted from the plasma, severely impairing the characteristics of nano-devices that have a larger surface than bulk areas. It is therefore essential to develop a method for suppressing or controlling charge accumulation and ultraviolet damage in plasma processing. The neutral beam process developed by the authors is a method that suppresses the formation of defects at the atomic layer level in the processed surface, allowing ideal surface chemical reactions to take place at room temperature. This technique is indispensable to develop future innovative nano-devices.**

I. INTRODUCTION

In the fabrication of semiconductor devices, reactive plasmas are widely used in key processes such as microfabrication, surface modification and film deposition, and there are now demands for processing precision at the atomic layer level, and for deposition accuracy that allows the control of structures at the molecular level. However, in ultra-miniature nanoscale devices that will become the mainstream in the future, the use of plasma processes can cause serious problems such as abnormal etching and breakdown of insulation films by the accumulation of ions or electrons emitted from the plasma , also the formation of surface defects (dangling bond) of over a few tens nm in depth by exposure to ultraviolet (UV) emissions from the plasma.[1-4] In particular, since nano-scale devices have a larger surface area compared with the bulk material, plasma processes can have a large influence on the electrical and optical properties of devices due to process-induced defects caused by ultraviolet exposure, since future nano-devices will require size control of three-dimensional structures at the atomic layer level, it will be absolutely essential to control surface chemical reactions with high precision and selectivity at the atomic layer level.

To achieve charge-free and UV photon irradiation damage-free processes, we have developed a new neutral beam generation system based on my discovery that neutral beams can be efficiently generated from the acceleration of negative ions produced in pulsed plasmas. This paper introduce the neutral beam generation technique[5] and discusses its application to atomic layer defect-free etching (ALE), modification (ALM) and deposition (ALD) that have recently been pursued.

II. NEUTRAL BEAM SOURCE

Figure 1 is a conceptual illustration of the proposed neutral beam source, which has evolved from the pulse modulated plasma with an on/off switching time of 50 microseconds.[16] This source uses an inductively coupled plasma (ICP) source, and has carbon ion acceleration electrodes situated at the top and bottom of the quartz plasma chamber. Gas is introduced from the upper electrode in the form of a shower, and ions accelerated from the plasma pass through apertures (1 mm in diameter and 10 mm long) formed in the lower graphite carbon electrode, where they are neutralized by colliding with the aperture sidewalls. In a plasma modulated by pulses of 50 microseconds in duration, the electrons lose energy during the "off" periods, and undergo dissociative attachment with a halogen gas with a large electron affinity (chlorine, bromine or fluorine). As a result, an afterglow plasma during "off" period consisting of mainly both positive and negative ions is formed even in high density and low pressure plasma. We found that a neutral beam formed by the neutralization of mainly negative ions using a pulse-modulated plasma is able to form a neutral beam with higher density and lower energy than when using positive ions. We have already applied this technique to various state-of-the-art sub-25-nm devices, resulting in processes and device characteristics that have not hitherto been possible to achieve.

III NEUTRAL BEAM PROCESSES FOR NANO-DEVICES

Using the neutral beam processing, we successfully demonstrated sub-50nm damage-free gate electrode etching,[5] defect-free Si fin channel etching for 45 nm fin-FETs (as shown in Fig.2)[5,6], ultra-thin gate dielectric film formation for 32 nm fin-FETs, defect-free Ge fin channel etching for sub-10 nm fin-FETs (as shown in Fig.3)[7], transition metal oxidation for ReRAM (as shown in Fig.4)[8], damage-free low dielectric film deposition for 22 nm FETs (as shown in Table 1)[9], damage-free etching of magnetic materials by complexing reactions (as shown in Fig.5)[9] and low-damage surface modification of carbon materials (including nanotubes, graphene and organic molecule)[10] for future nanodevices. More recently we have investigated processing technologies based on the combination of biotechnology with neutral-beam-based nano-processes, i.e., bio-nano processes, for future nanoelectronics devices and successfully achieved the fabrication of sub-10-nm-diameter and high density Si, Ge, GaAs, InGaAs and Graphene nanodisk (nanodot) array structures (Fig.6, Fig.7).[9] The quantum effects of these nano-scaled structures were shown to manifest themselves at room temperature due to the damage-free surfaces made possible by the neutral beam processes. Now, by using these nanodisk structures, we are actively developing "Novel Quantum Effect Devices", such as quantum dot solar cell,[8] quantum dot thermos-electric conversion device and quantum dot laser.[8]

REFERENCES

1) T. Nozawa and T. Kinoshita: Jpn. J. Appl. Phys.34 (1995)pp. 2107.

2) J-P. Carrere, J-C. Oberlin and M. Haond: Proc. Int. Symp. on Plasma Process-Induced Damage (AVS, Monterey, 2000) pp.164.

3) T. Dao and W. Wu: Proc. Int. Symp. on Plasma Process-Induced Damage (AVS, Monterey, 1996) pp.54.

4) Mitsuru Okigawa, Yasushi Ishikawa, Yoshinori Ichihashi and Seiji Samukawa, Journal of Vacuum Science and Technology, B22(6)(2004)pp.2818.

5) S. Samukawa, Japanese Journal of Applied Physics Vol. 45, No. 4A, 2006, pp. 2395.

6) K. Endo, S. Noda, M. Masahara, T. Kubota, T. Ozaki, S. Samukawa, Y. Liu, K. Ishii, Y. Ishikawa, E. Sugimata, T. Matsukawa, H. Takashima, H. Yamauchi, and E. Suzuki, IEEE Transaction on Electron Devices, 53(8), (2006), pp. 1826.

7) E.-T. Lee, S. Noda, W. Mizubayashi, K. Endo, S. Samukawa, Proc. 16th Int. Conf. on Nanotechnology, 2016, p. 816.

8) T. Ohno and S. Samukawa, Appl. Phys. Lett. 106, (2015) pp.173110.

9) S. Samukawa, ECS Journal of Solid State Science and Technology, 4 (6) (2015) pp.N5089.

10) S. Samukawa, Y. Ishikawa, K. Okumura, Y. Sato, K. Tohji and T. Ishida, J. Phys. D: Appl. Phys. 41 (2008) 024006 (6pp)

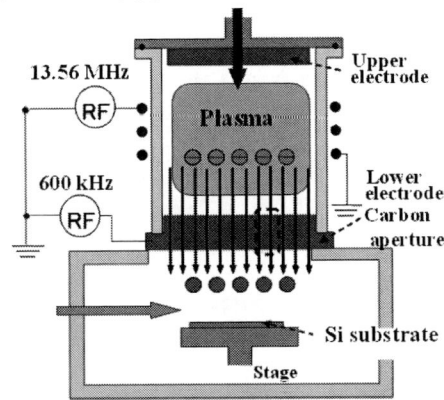

Figure 1. Neutral beam generation device developed based on new concepts. For the first time, we have achieved a practical neutralization rate and energy by using positive ions generated efficiently by pulse-modulated plasma.

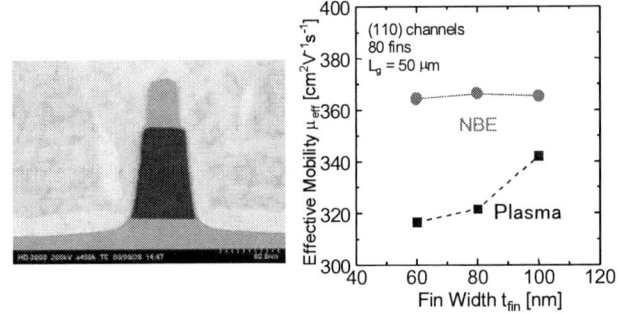

Figure 2. Dependence of electron mobility of Si channel processed by neutral beam etching and plasma etching as function of fin width. The superiority of the neutral beam etching remained unchanged for different values of fin width.

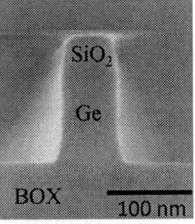

Figure 3. Cross-sectional SEM images of Ge fin after neutral beam etching. Atomically flat surface roughness can be realized by using neutral beam etching.

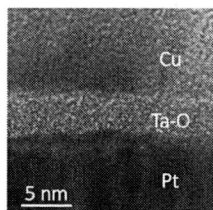

Figure 4. Cu/3.2-nm-thick Ta$_2$O$_5$/Pt structure fabricated by neutral beam oxidation process for ReRAM (at room temperature).

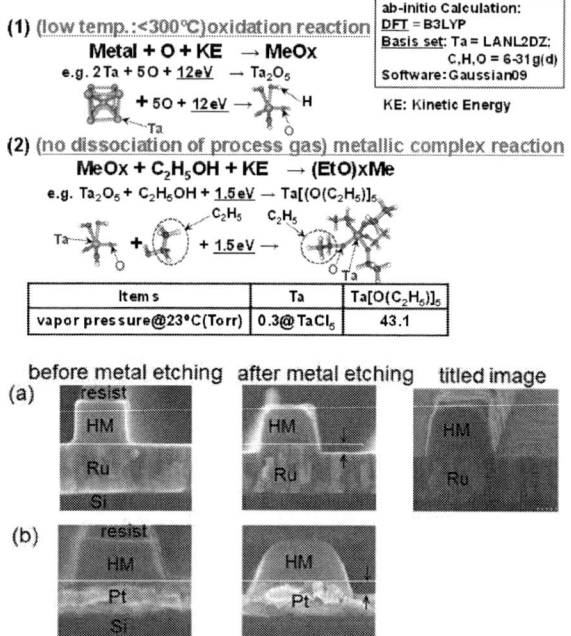

Figure 5. Mechanism of transition metal etching by the metal complexing reaction, and actual etched shapes of Ru and Pt: (a) ruthenium (Ru) (b) platinum (Pt).

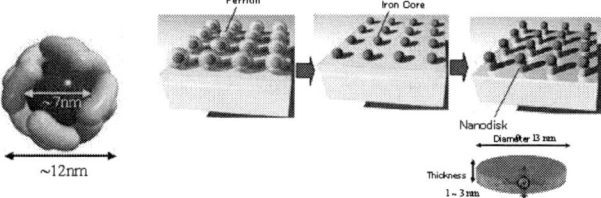

Figure 6. Neutral beam etching process using the biological supermolecule (protein) ferritin, whose self-organizing properties result in two-dimensional crystallization. The etching mask is made from iron cores encapsulated within the ferritin molecules. By using iron cores with a diameter of 7 nm as etching masks, we can form defect-free ultrafine structures with a size of less than 10 nm.

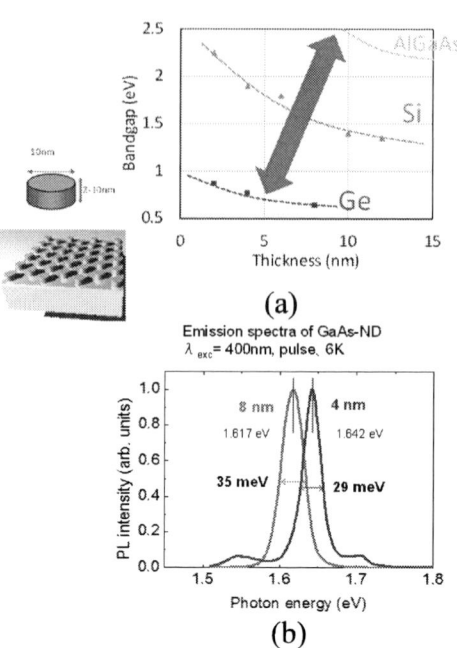

(a)

(b)

Figure 7. (a) Variation of band gap energy with disc thickness in nanodisc structures of silicon, germanium, graphene and aluminum gallium arsenide, and (b) photoluminescence of gallium arsenide nanodisc structures.

Table 1. Comparison of the properties of SiOCH films deposited by plasma vapor deposition and neutral beam excitation

	Metric	Porous SiCO by PECVD	Non-porous SiCO by NBECVD
k-value	Hg-probe	2.6	2.2
Modulus (GPa)	Nano-indenter	6.0	11.7
Density (g/cm3)	XRR	1.27	1.54
Pore size (nm)	SAXS	1.2	No detected

978-1-5090-6695-7/17 $31.00 © 2017 IEEE   484

8 Inches Monolithic CMOS-MEMS Manufacturing platforms for consumer products

Wai Soon, Liew

Shanghai Huahong Grace Semiconductor Manufacturing Corporation
288 Halei Road, Zhangjiang Hi-Tech Park, Shanghai, P.R.China, 201203
(86) 21 38829909

Biography

The author received the BS degree and MS degree in applied physics from University of Malaya, Malaysia, in 1999 and 2001 respectively. He has been working in wafer fabrication and process development for more than 15 years. He is currently working in Technology Department at Shanghai Huahong Grace Semiconductor, in charge on CMOS-MEMS devices and manufacturing platform development.

Abstract

Huahong Grace Semiconductor Manufacturing Corporation (HHGrace) is the largest 8 inch wafer fabrication foundry in China with strong expertise in CMOS fabrication processes. By leveraging on the existing matured CMOS process capabilities, HHGrace has successfully developed MEMS platforms using existing equipments coupled with the combination of MEMS special tool sets. HHGrace has demonstrated its capabilities by providing MEMS customers with advanced monolithic CMOS-MEMS products in stable volume production. 3 majors MEMS platforms: Bulk MEMS, Surface MEMS and magnetic MEMS were developed to serve the major MEMS consumer products in the market.

Keywords—MEMS; CMOS-MEMS; Sensor; wafer fabrication; CMOS foundry;

Introduction

The advancement in mobile gadgets, electronic sensory peripherals and Internet of Thing (IOT) drives a strong demand and growth for the need of advanced MEMS products. Most of the major MEMS suppliers are Integrated Design Manufacturers such as Bosch, STMicroelectronics, Texas Instruments, Freescale and so on, which have been steadily supplying the sensors to automotive and consumer markets since 1980s to 2000s. As for the new players in MEMS arena; namely companies from China, many prefer to start with fabless model due to significant lesser capital investment and faster development timeline. Hence, there is an opportunity for CMOS foundries in China to offer MEMS manufacturing to support these fabless companies demand. As of current, most of the MEMS products are using 0.13um to 0.25um ASIC for signal processing and control; thus, it is an added advantage for 8 inches foundries to leverage on the mature CMOS technologies and processes to develop and manufacture advance monolithic CMOS-MEMS products. Huahong Grace has demonstrated a successful model of synergizing the CMOS technology to MEMS integration, through the setup of several key MEMS platforms and developed monolithic CMOS-MEMS products with China MEMS companies.

CMOS and MEMS fabrication processes synergy

A number of MEMS devices can be fabricated completely within the CMOS process sequence, without requiring any additional complex process steps or special materials. Although some of the MEMS devices might require special tools or new materials in order to form 3-dimensional microstructure and the sensing elements; the major process steps are still using the traditional CMOS process flow. Table I shows several CMOS fabrication processes can be utilized for MEMS fabrication. This offers a great help in reducing the investment needed by CMOS foundries to setup a MEMS fabrication line compared to the investment required in building a brand new standalone MEMS Fab. Besides, Monolithic CMOS-MEMS devices can be fabricated in a CMOS Fab with MEMS device integrated into CMOS circuitry directly on a single chip. This helps in chip miniaturization, performance enhancement and cost reduction on packaging processes compared to conventional CMOS and MEMS chips integration by packaging processes. The monolithic integration of CMOS-MEMS can be realized by fabrication of MEMS on CMOS wafer directly or MEMS and CMOS wafers through bonding processes.

TABLE I
Materials for CMOS and MEMS fabrications [1]

Material	CMOS Applications	MEMS Applications
Silicon (Substrate, Si Epitaxy)	Device fabrication substrate	Substrate, moving structure, supporting layer
Polysilicon, Poly SiGe	Gate	Resistor, Piezoresistor, Moving Structure, structural material
Metallization	Interconnection, Resistor	Interconnection, micro-mirror, Eutectic bonding, sensing material (eg: NiFe, VOx, etc)
Dielectric (SiO2, Si3N4, SiON)	Isolation, Passivation, structural material,	Isolation, passivation/ sealing, sacrificial layer (SiO2 with vapour HF), structural material, optical waveguide, direct bonding Material

		(SiO2-Si)
Polyimide	Passivation	Sacrificial Layer with O2 plasma release
Doping (Diffusion/ Implantation)	N/P Well, source/Drain, Gate, LDD, Halo, contact	Resistor, IR sensing material (eg: Boron doped a-Si)

Monolithic CMOS-MEMS and Challenges

Although MEMS process utilizes most of the CMOS process fabrication steps, some special attentions are required during this CMOS-MEMS integration.

a) Thermal budget control of MEMS processes

There are several classifications in CMOS-MEMS integration such as pre-CMOS, intra-CMOS and post CMOS processes, which depend on the MEMS fabrication steps that incorporated inside the CMOS process sequence. There are pros and cons for these processes depend on the fabrication needs. It is especially important on Intra CMOS and Post CMOS processes, in which these MEMS process steps is incorporated; as not to cause significant effect to the CMOS thermal budget whereby it could result in possible drift of the CMOS devices performance. Any additional thermal process step performed during or after the regular CMOS process sequence have to be carefully studied in order to maintain the CMOS device performance. For post CMOS fabrication, the thermal budget need to be control below 450 degree Celsius to avoid the deformation of aluminum metal interconnects. High temperature process such as polysilicon deposition cannot be used and low temperature poly SiGe shall be employed.

b) Cross contamination Control of new materials

MEMS devices sometimes require different kind of materials for specific function which could result in cross-contamination to CMOS devices. Examples of such materials such as Ni, Fe, Co meant for magnetic field sensing, need a more stringent manufacturing process control compare to conventional CMOS processes. It is not viable to fully dedicate general CMOS tools for MEMS only, in which it will incur higher manufacturing cost and lose flexibility on products mix. There are two types of cross contamination mode on these special material wafer processes in CMOS tools: a) Mechanical handling b) Direct contact of process reactant itself as shown in Table II. For Mechanical handling cross contamination control, wafer backside cleaning and post ICPMS for metallic ions content check for common tools are effective measure to prevent the contamination, while for the other mechanism with direct contact of materials with process reactant, dedicated tool or recovery through maintenance is necessary to avoid cross contamination.

TABLE II
Cross contamination modes and prevention

Cross contamination mode	Contamination mechanism	Control Method
Mechanical handling	Through wafer cassette, robot handler, pedestal, etc	Wafer backside cleaning with cassette change, ICPMS for metallic Ions content check on equipment after tool processing.
Direct contact of process reactant	Solvents, chemical reaction, thermal induced, etc	Dedicated process tools.

c) Film Stress control

For monolithic CMOS-MEMS process sequence, it uses more process steps compare to conventional CMOS-only fabrication. In this sense, the accumulated film stress is higher than the standard CMOS process. Besides, it is also crucial to have low stress thin films to retain the 3-dimensional structures with flat surface in MEMS fabrication. This can be realized by using lower temperature process, PECVD process, stress buffer layers or fine tuning of the process conditions to achieve the desired film properties.

d) Platform standardization

As there are many different types of MEMS devices, hence it is not easy to setup individual processing line just to cater for each MEMS devices' requirements. Furthermore, it is not efficient and economically feasible to setup variety of platforms to enable all MEMS products. HHGrace had chosen 3 major platforms to support major demand on the consumers' products, namely Bulk MEMS, Surface MEMS and Magnetic MEMS. This help to provide standardization with some customization process to cater for customers' requirements and ensure high volume fabrications and platform utilization.

HHGrace's CMOS-MEMS platforms

a) CMOS technologies

HHGrace provides professional and high value-added foundry services covering technology solutions from 1.0μm to 90nm technology nodes, with focus on differentiated technologies which include eNVM (embedded Non-Volatile Memory), power management IC, power discrete, RF, as well as standard logic and mixed-signal. HHGrace's CMOS process offerings are supported by a comprehensive set of one-stop-shop solutions that comprises design service, wafer testing and backend turn-key services, which help customers to reduce overall cost and accelerate time-to-market in order to improve the product competitiveness. For IOT applications, there are four types of major component, namely MCU, sensor, RF communication and power management. All these devices are the core competency of HHGrace's CMOS technologies. In MCU devices, HHGrace provides 0.18um to 90nm Flash, EEPROM, MTP, OTP and ULL eFlash for low power consumption needs. While for RF devices, HHGrace provides variety of RF ICs in wireless communication and wired optical communications, including SiGe Bipolar/BiCMOS, RF LDMOS, logic-compatible RF CMOS and SOI (silicon on insulator) RF CMOS. Besides, HHGrace is also competence in power management with BCD platform for consumer products. Currently, HHGrace is enhancing its MEMS fabrication capabilities to offer the sensor fabrication solutions to the customers. In quality management and compliance, HHGrace is certified under automotive standard ISO/TS16949, AEC Q100, and passed numerous Japanese and US customers and their end customers' quality system audits. HHGrace is able to provide high quality and reliable processes for both CMOS and MEMS fabrication to the customers.

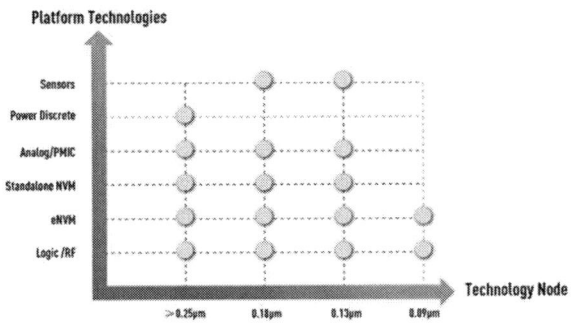

Fig.1. HHGrace's CMOS technologies Platform

b)   MEMS Platforms

i.   Bulk MEMS Platform

HHGrace developed Bulk MEMS platform with SOI-cavity wafer process. This platform provides flexibility on the cavity size and proof-Mass thickness requirements for different types of MEMS product. It is suitable for Motion MEMS products such as accelerator, gyroscope and RF MEMS devices. The bottom wafer is etched by using Si DRIE with desired cavity size and depth, followed by oxidation of the wafers and fusion bonded with another wafer to form the cavity SOI wafer. The proof mass thickness adjustment is through back-grinding and polishing processes to achieved TTV <1um. The moving structure is etched through Si DRIE on the wafer area on top of cavity structure. The CMOS and the MEMS wafers are then electrically connected through subsequent Al-Ge eutectic bonding process. Besides, the Al and Ge frames can be eutectic bonded to form a hermetic structure for MEMS devices as illustrated in Figure 2. The main advantage here is that the MEMS process is completely separated from CMOS until eutectic bonding step; hence there is no major concern on MEMS thermal processes influencing the CMOS fabrication. The Eutectic bond process is about 428-435°C which is also well below the thermal budget impact on CMOS wafer. The bonded wafers are then undergo wafer thinning to desired thickness and follow by contact pad open for subsequent packaging processes. HHGrace has successfully developed with QST for accelerator and gyroscope products and currently in mass production to supply to local consumer market.

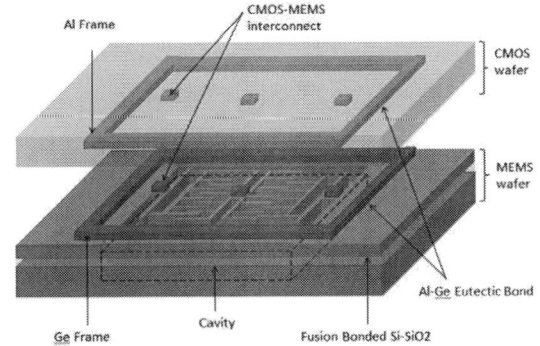

Fig.2. HHGrace's Bulk MEMS platform with CMOS and MEMS wafer bonding process

Figure 3: HHGrace's Motion MEMS Accelerator (Courtesy of QST Corporation)

ii.   Surface MEMS platform

Surface MEMS platform is one of the popular approaches to fabricate monolithic CMOS-MEMS devices. It can be used to fabricate pressure sensor, micro-bolometer, microphone and photonics devices. The key processes of surface MEMS are the sacrificial material and its release method. Besides, for post CMOS' MEMS process, the critical thermal budget need to be control < 450°C to avoid detrimental effect on CMOS performance. HHGrace is equipped with amorphous carbon thin film and polyimide materials as sacrificial layers. The advantage of these materials is easily released by using oxygen plasma process. It is suitable to be used for pressure sensor's diaphragm and micro-bolometer micro bridge structure. Figure 5 is the cross-section of a CMOS-MEMS monolithic pressure sensor, co-developed with LEXVU Opto Microelectronics. The device is suitable for mobile phone and wearable devices.

978-1-5090-6695-7/17 $31.00 © 2017 IEEE   487

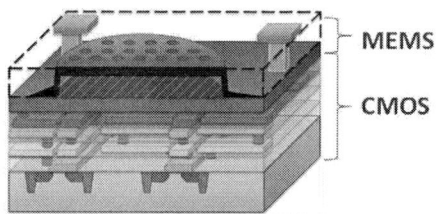

Fig 4: HHGrace's Surface MEMS platform with monolithic

CMOS-MEMS

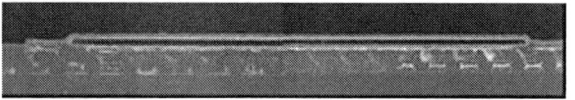

Figure 5: CMOS-MEMS pressure sensor fabricated by HHGrace

(Courtesy of LEXVU)

iii.   Magnetic MEMS platform

HHGrace co-developed with QST Corporation on the AMR magnetic MEMS Platform with special licensing of the AMR technology. Under the licensed company's original device setup; the 3 axis AMR Magnetic sensor is using two AMR chips and package as vertical and horizontal position to form X-, Y- and Z- axis measurement. QST and HHGrace designed 3-axis in a single chip and integrated with CMOS devices as illustrated in Figure 6. This design helps to reduce the footprint to 1.2mmx1.2mm and minimized the complexity of the packaging requirements. The AMR sensors are currently supplied to smart phone and consumer electronics.

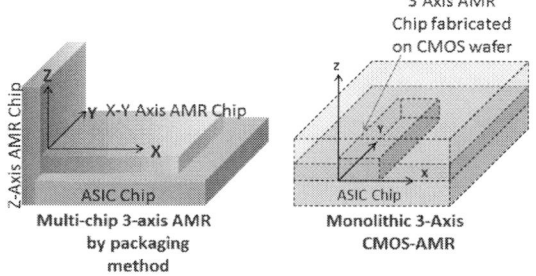

Fig 6: Illustration Multi-chip 3 Axis AMR versus Monolithic 3-Axis

CMOS-AMR

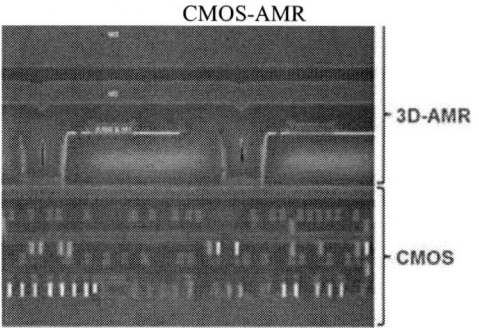

Fig 7: Monolithic CMOS-MEMS 3-axis AMR sensor in single chip

Summary

HHGrace has successfully setup a CMOS compatible MEMS fabrication line to support the local and overseas MEMS companies with both CMOS and MEMS solutions. By having several major MEMS fabrication platforms, HHGrace is able to provide broad solutions to meet the requirements of different standalone MEMS and advanced monolithic CMOS-MEMS fabrications. HHGrace is expanding its technology capabilities to suit for each customer requirements and keep abreast in the technology development needs for the markets. HHGrace is dedicated to help the growing demand of CMOS and MEMS fabrications in China and overseas.

References

[1] H. Baltes, O. Brand, G. K. Fedder, C. Hierold, J. Korvink, O. Tabata, " Advanced Micro and Nanosystems. Vol. 2. CMOS - MEMS," WILEY-VCH Verlag GmbH & Co, pp 12, 2005.

Thin-film Processing of "Exotic" Phase-Change and Ferroelectric Materials for IoT Applications

K. Suu[*], *I. Kimura, H. Kobayashi, Y. Miyaguchi, T. Masuda, Y. Kokaze, and T. Jimbo*

ULVAC, Inc., Hagisono, Chigasaki, Kanagawa, 253-8543, Japan
*Corresponding Author's Email: koukou_suu@ulvac.com

ABSTRACT

We will report our development results of phase-change and ferroelectric thin film processing technologies including sputtering, MOCVD and plasma etching as well as manufacturing processes for PCRAM, FRAM and MEMS/Sensor device applications. Thin-film functional material such as phase-change materials and ferroelectric materials have been utilized to form advanced semiconductor and electronic devices for internet of things (IoT) solutions. We are confident our manufacturing technologies for these materials and devices will contribute to realizing next generation Smart Society.

INTRODUCTION

Smart ICT (Information and Communication Technology) such as "Big Data", "Cloud computing" and "Smart Functionalities" such as Stand-alone self-activating MEMS/Sensors construct smart systems which enable IoT (Internet of Things) and IoE (Internet of Everything) thus "Smart Society". To realize above-mentioned smart technologies, semiconductor devices with high-density, low-power consumption, wide-bandwidth, and fast-operation as well as smart functional devices enabled by integrating functionalities with advanced semiconductor technologies including CMOS technologies are necessary.

Thin-film functional material such as phase-change materials (e.g. $Ge_2Sb_2Te_5$-GST) and ferroelectric materials (e.g. $Pb(Zr,Ti)O_3$-PZT) have been utilized to form advanced semiconductor and electronic devices including Phase-Change Random Access Memory (PCRAM), Ferroelectric Random Access Memory (FRAM), actuators composing gyro meters and advanced MEMS/Sensor devices for internet of things (IoT) solutions.

In this talk, we will report our development results of phase-change and ferroelectric thin film processing technologies including sputtering, Metal organic chemical vapor deposition (MOCVD) and plasma etching as well as manufacturing processes for PCRAM, FRAM and MEMS/Sensor device applications.

Manufacturing technologies for PCRAM

PCRAM uses reversible phase change of chalcogenide material (e.g. GST), which shows two different phases such as crystalline one for set state with low resistance and amorphous one for a reset state with high resistance.

We have been preparing Ge-Sb-Te system thin film by the multi chamber type mass production sputtering system (Entron W300) equipped with an exclusive sputtering module of RF magnetron sputtering. Ge-Sb-Te system thin films have been investigated with its crystallinity and resistivity [1].

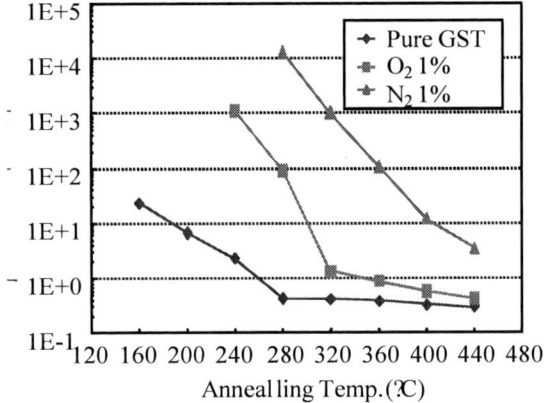

Fig. 1. Resistance change of pure, Nitrogen-doped and Oxygen-doped GST with annealing temperature.

In order to fabricate the mass productive high density PCRAM, it is highly required to reduce down the reset current. To reduce the current, Nitrogen and Oxygen doped GST can be applied instead of conventional GST. We tried to prepare Oxygen-doped and Nitrogen-doped thin films. The test phase change device was fabricated to confirm switching characteristics between crystalline (set) and amorphous (reset) phases. As shown in Fig. 1, the resistance of Nitrogen-doped GST thin film changed gradually and Oxygen-doped one shows a rapid resistance change with annealing temperature, since Nitrogen-doped GST with fcc phase was held to high temperature as compared with the phase transition from fcc to hcp in Oxygen-doped GST.

In order to further integrate PCRAM in the future, a high writing current become a problem caused by heating more than the melting point (e.g. 632°C of GST) at reset state in conventional single GST film. $GeTe/Sb_2Te_3$ super lattice structure has been suggested to resolve this problem by Ge flip-flop switch intercalated between Sb_2Te_3 layers

978-1-5090-6695-7/17 $31.00 © 2017 IEEE

without any melting transition. However, it was difficult to form very thin chalcogenide material film on the large size substrate because chalcogenide material has high vapor pressure. We prepared GeTe/Sb$_2$Te$_3$ super lattice on φ300 mm substrate by a new concept sputtering module which has good controllability substrate temperature and deposition rate. The structure of GeTe/Sb$_2$Te$_3$ super-lattice was observed with cross sectional transmission electron microscope (TEM), the film composition was measured with X-ray fluorescence (XRF) and the film structure measured with X-ray diffraction (XRD). Smooth and close GeTe/Sb$_2$Te$_3$ supper-lattice structure was observed by TEM observation as shown in Fig. 2.

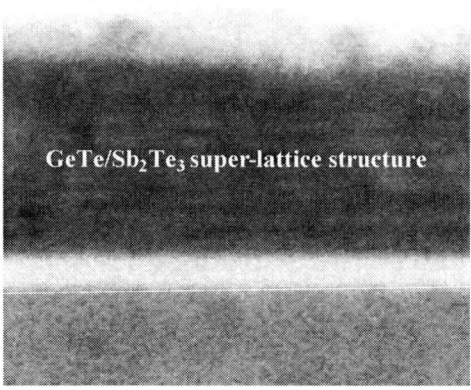

GeTe/Sb$_2$Te$_3$ super-lattice structure

Fig. 2. Sectional view of TEM observation of GeTe/Sb$_2$Te$_3$ super-lattice structure.

We have also been investigating a novel confined PCM cell structure which utilizes a metallic surfactant layer to stabilize the high (and intermediate) resistance state drift in MLC phase change memory technology [2]. The metallic surfactant layer provides an alternative conductive path to the amorphous region during read operation, which makes the cell characteristics immune to amorphous region instabilities such as time- and temperature-dependent resistance drift and noise. A pore with 50nm diameter is formed by the keyhole process on the bottom contact. A pore is first filled by thin ALD metal nitride and then completely filled by ALD GST. We use a multi-chamber cluster (Entron-EX W300) allowing for no vacuum break between metal nitride and GST ALD depositions.

Sputtering technologies for FRAM and MEMS/Sensor devices

We have been developing mass production technology for perovskite oxide thin-film such as PZT for a long period [3-6]. Sputtering method was selected for mass

production technology of perovskite oxide thin-films owing to the following factors: (1) Good compatibility with conventional Si LSI processes. (2) Superb controllability of film quality. (3) Better possibilities of obtaining uniform surfaces. (4) Sputtering as plasma processing was promising for deposition and heat treatment at low temperature. (5) Feasibility of high-speed deposition. (6) Same deposition method as electrodes (Pt, Ir, etc.).

We adapted the RF magnetron sputtering method with ceramic target for perovskite oxide thin-film sputtering. As for the PZT thin-films, sputtering methods included high-temperature deposition, where film deposition was made at substrate temperatures up to 500°C, and low temperature deposition with crystallization by post-annealing process.

It was confirmed that volatile elements within film (e.g. Lead in PZT) were unstable even when the deposition was performed under identical conditions. The conceivable causes for that phenomenon were the instability of the target, fluctuation of temperature in the sputtering chamber, and variation of plasma status over time. Main problem in perovskite oxide thin-film is the change in the volatile element content with the passage of sputtering time due to the change in plasma status. In order to stabilize the status, we installed a stable anode, that is, an a node that avoided charge-up due to the adhesion of insulating PZT film and maintained the role as an anode. Consequently, as can be seen in Fig. 3, stability of Pb content within film in continuous sputtering has been confirmed.

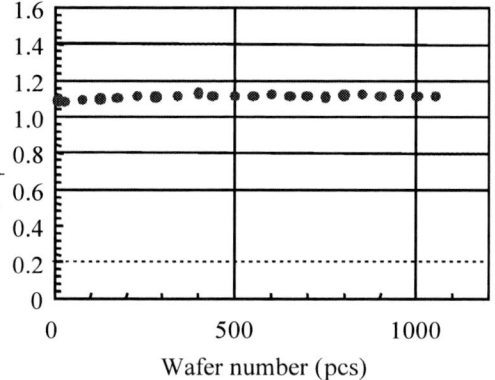

Wafer number (pcs)

Fig. 3. Stable transition of Pb content within film in continuous sputtering.

Sputtered PZT thin-films for non-volatile memory (FRAM)

We used the multi chamber type mass production sputtering system (CERAUS ZX-1000) equipped with an exclusive sputtering module for ferroelectric materials.

This system has features as (1) Deposition of PZT thin-films on 6 and 8 inch substrates are performed using 12 inch single ceramic target. (2) Including the heat chamber, this system has five process chambers, thereby achieving high flexibility. (3) The system uses RF sputtering for ferroelectric deposition and counters RF noises.

Ferroelectric film with high performance is required because scaled thinner capacitor (sub 100 nm) is demanded. Bottom electrode and PZT deposition process were modified to improve the ferroelectric performance and achieve thinner capacitor with good performances [7]. Q_{SW} transition curve of a IrOx/PZT(80 nm)/Pt capacitor was measured at 3 V. Q_{SW} with 1.8 V applied was approximately 24 $\mu C/cm^2$ and $V_{90\%}$ was 1.4 V. Figure 4 shows the fatigue characteristics of voltage application at 2 V. Q_{SW} was not decreased even after switching in 10^9 cycles. From these results, sputter-derived PZT capacitor had been proved to be suitable for 0.18 μm technology node.

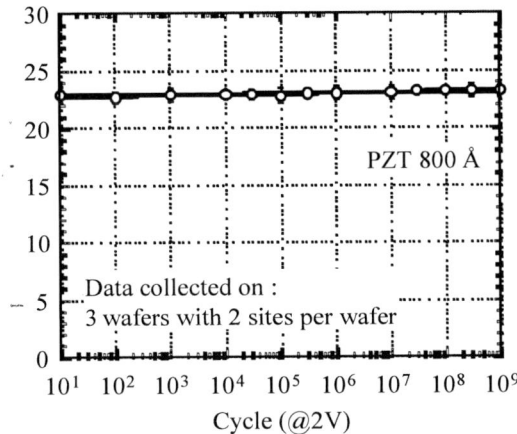

Fig. 4. Fatigue characteristics of IrOx/PZT(80 nm)/Pt capacitor.

Sputtered PZT films for MEMS/Sensor device application

A multi-chamber type mass production sputtering systems for electronic devices SME-200 equipped with an exclusive sputtering module described above. PZT films have been deposited up to 8 inch diameter silicon substrates.

PZT films were deposited by RF sputtering method with ceramic target under Ar/O_2 mixed gas. Substrate temperature was heated up to around 500°C. After the deposition, PZT films were conducted with no thermal treatments. PZT films were deposited with relatively high growth rate about 3.8 $\mu m/h$ and these thicknesses were from 0.5 to 2.0 μm for piezoelectric MEMS/Sensor applications.

For the measurement of the piezoelectric properties of the PZT films, Rectangular beams (cantilevers) with the size of about 30 mm × 3 mm were prepared. Polarization and displacement in these films were simultaneously observed using the laser doppler vibrometer and the laser interferometer. Piezoelectric properties were confirmed for PZT films by checking the cantilever as shown in Fig. 5. While relative large piezoelectric coefficient (e_{31}) with -14.7 C/m^2 was observed even in conventional process, much higher value with -17.3 C/m^2 was successfully obtained by modification of manufacturing process.

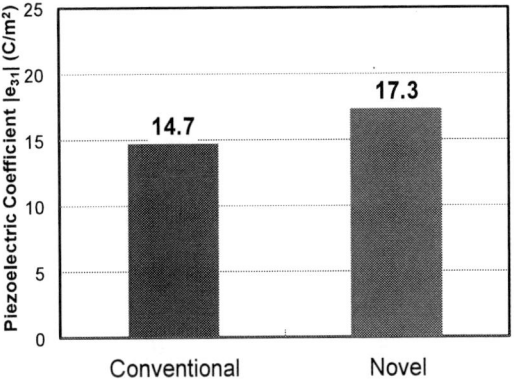

Fig.5.Piezoelectric coefficient of sputtered PZT films with2-μm-thick

MOCVD technology of PZT thin-films for non-volatile memory (FRAM)

For generation of FRAM beyond 0.18 μm, thinner films of ferroelectric associated with shrinking of the thickness of ferroelectric capacitors and the three-dimensional structures associated with shrinking of the capacitor areas were demanded because it was necessary to achieve further larger packing densities and integration with logic devices. In addition, it was necessary to achieve thinner films of ferroelectric in parallel with its higher quality in order to meet the demand of still lower voltage drives for device performance.

MOCVD technology could meet these demands. Denser crystalline films are easily obtainable, and it is possible to achieve thinner films and higher quality. In addition, uniform deposition can be obtained also on three-dimensional structure, and it is considered that good step coverage can be obtained.

For the target of mass-production system development, we adopted the vaporization method. That is, a liquid raw material or solid raw material is melted in an organic solvent. Then, after the solution is vaporized, it is transported with carrier gas. We combined ferroelectric deposition technology for FRAM, module design technology, and CVD equipment technology, which were

978-1-5090-6695-7/17 $31.00 © 2017 IEEE   491

provided to users, and completed full-fledged MOCVD equipment for mass production.

The major features of our MOCVD system are as follows; (1) Reproducibility in the continuous operation of vaporization and the gas transport to a substrate that controls condensation and decomposition were achieved by accurate temperature control for each part of equipment and the optimization of vaporization conditions. Film thicknesses within ±2% and the reproducibility of the PZT composition were kept with no mechanical maintenance, and a running test for 1,000 substrates was successful as shown in Fig. 6 [8,9]. (2) By taking advantage of the progress of the equipment hardware, the optimization of processes was conducted, and PZT thin films can be controlled to ensure preferential orientation in the <111> direction. Consequently, the formation of PZT films within 100 nm that have capacitor properties with a 1.5 V low voltage drive was achieved as shown in Fig. 7 [10,11].

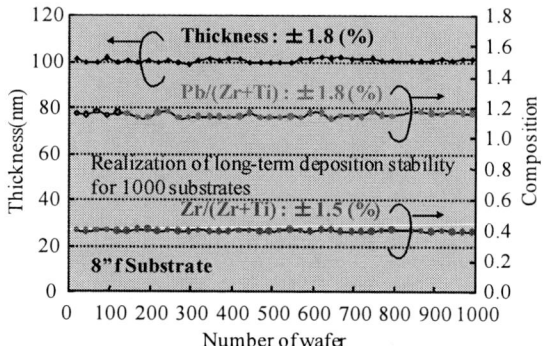

Fig.6 Reproducibility in MOCVD system (film thickness and PZT composition)

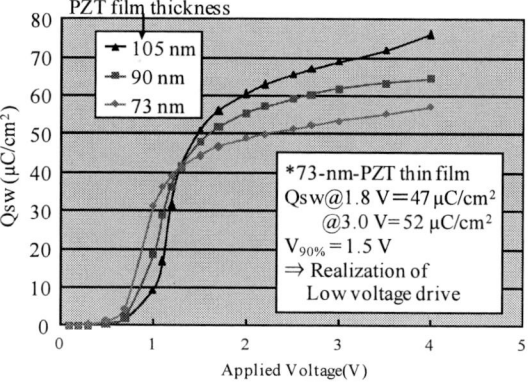

Fig. 7. Characteristics of MOCVD- PZT film (applied voltage dependence of switching charge)

Conclusion

In this talk, we report our development results of phase-change and ferroelectric thin film processing technologies including sputtering, MOCVD and plasma etching as well as manufacturing processes for PCRAM, FRAM and MEMS/Sensor device applications. Thin-film functional material such as phase-change materials and ferroelectric materials have been utilized to form advanced semiconductor and electronic devices for internet of things (IoT) solutions. We are confident our manufacturing technologies for these materials and devices will contribute to realizing next generation Smart Society.

References

[1] S. Kikuchi, D. Y. Oh, I. Kimura, Y. Nishioka, M. Ueda, M. Endo, Y. Kokaze, and K. Suu: Non-Volatile Memory Tech. Symp. (NVMTS) 2006, 81.

[2] S. Kim, N. Sosa, M. BrightSky, D. Mori, W. Kim, Y. Zhu, K. Suu, and C. Lam: Proceeding of IEDM 2013, 13, p765.

[3] K. Suu, A. Osawa, N. Tani, M. Ishikawa, K. Nakamura, T. Ozawa, K. Sameshima, A.Kamisawa and H. Takasu: Jpn. J. Appl. Phys. 35 (1996) 4967.

[4] K. Suu, A. Osawa, N. Tani, M. Ishikawa, K. Nakamura, T. Ozawa, K. Sameshima, A.Kamisawa and H. Takasu: Integr. Ferroelectr. 14 (1997) 59.

[5] K. Suu, A. Osawa, Y. Nishioka and N. Tani: Jpn. J. Appl. Phys. 36 (1997) 5789.

[6] K. Suu, Y. Nishioka, A. Osawa and N. Tani: Oyo Buturi 65 (1996) 1248. (in Japanese)

[7] F. Chu, G. Fox, T. Davenport, Y. Miyaguchi and K. Suu: Integr. Ferroelectr. 48 (2002)161.

[8] T. Yamada, T. Masuda, M. Kajinuma, H. Uchida, M. Uematsu, K. Suu and M. Ishikawa: IFFF2002 abstract, 4 (2002) 37.

[9] T. Masuda, M. Kajinuma, T. Yamada, H. Uchida, M. Uematsu, K. Suu and M. Ishikawa: Integr. Ferroelectr. 46 (2002) 66.

[10] Y. Nishioka, T. Jimbo, T. Yamada, T. Masuda, M. Kajinuma, M. Uematsu, K. Suu and M. Ishikawa: Integr. Ferroelectr., 59 (2003) 1445.

[11] Y. Nishioka, T. Masuda, M. Kajinuma, T. Yamada, M. Uematsu and K. Suu: MRS Fall Meeting Proceedings, 784-C7 (2003) 6.

WAFER SIZE MOS2 WITH FEW MONOLAYER SYNTHESIZED BY H2S SULFURIZATION

Yen-Teng Ho[1], Yung-Ching Chu[1], Lin-Lung Wei[1], Tien-Tung Luong[1], Chih-Chien Lin[2], Chun-Hung Cheng[2], Hung-Ru Hsu[3], Yung-Yi Tu[1] and Edward Yi Chang[1]*

[1]Department of Materials Science and Engineering, National Chiao Tung University, Hsinchu, Taiwan,
[2]Center for Nano Science technology, National Cheng Kung University, Tainan, Taiwan
[3]Industrial Technology Research Institute, Hsinchu, Taiwan
*Corresponding Author's Email: chia500@nctu.edu.tw

ABSTRACT

Wafer sized, high quality continuous films would be a key demand for MoS_2 implemented in circuit application. In this study, the growth of few monolayer Mo_{s2} on 4 inches SiO_2/Si substrate were demonstrated. The MoS_2 thin films were synthesized by sulfurized in a furnace from the ultra-thin MoO_3 starting materials by using H_2S. The obtained MoS_2 thin film examined by Raman analysis and Photoluminescence (PL), shows the semiconductor nature with direct transition peaks of 1.86 eV and 1.99 eV. The 4~5 monolayer of MoS_2 with thickness around 2.6 nm is confirmed by cross-sectional view of transmission electron microscopy (TEM). Additionally, the DC characteristics of MoS_2 MOSFETs exhibit at least 2 order in on/off current ratio, demonstrating the feasibility for circuit application.

INTRODUCTION

Recently, transition metal dichalcogenides (TMDs) with two-dimensional layer have depicted intensive interest due to its great potential in nanoelectronics applications [1]. Among these TMDs, exfoliated monolayer MoS_2 shows a great possibility for n-type FET for next generation due to the ultra-high carrier confinement [2]. For realization in circuit application, MoS_2 obtained from exfoliation is not scalable for large-scale device fabrication. On the other hand, the direct grown MoS_2 by chemical vapor deposition from sulfur and MoO_3, usually exhibit localized discontinuous film [3]. Accordingly, to be implemented in circuit applications, it is meaningful to develop a wafer size MoS_2 synthesis technology with high uniformity.

In this study, the 4 inches wafer size continuous MoS_2 with few monolayer on SiO_2/Si substrate synthesized by using a two steps growth method is demonstrated. The first step is to deposit ultra-thin MoO_3 layer from HV E-gun evaporation as the starting material. Then the high quality MoS_2 can be obtained by sulfurized the MoO_3 in a furnace with H_2S ambient. Raman analysis, PL and cross-sectional view of TEM were used to confirm the quality of MoS_2. Back-gated MOSFET of MoS_2 were fabricated to show the feasibility for circuit application.

EXPERIMENTAL

The synthesis process in this study is performed by a two steps growth method. The ultra-thin (<1nm) starting material of MoO_3 is deposited on a 4 inches size SiO_2/Si wafer from MoO_3 powder with purity of 99.95% (purchased from Gredmann Inc.) by a high vacuum E-gun evaporator. Then the high quality MoS_2 thin films were synthesized by a sulfurization process in a 6 inches furnace with H_2S (10%) as gaseous sulfur source. The formation profile is performed by +18°C /min ramp up to 750°C, keeping steady for 60 min then ramp down by-20°C /min, as shown in the Fig.1. For evaluation the material characteristics of MoS_2, a μ-Raman/PL system equipped with a 532 nm laser source were used. AFM was used to examine the surface morphology. To study the microstructure of MoS_2, the samples were characterized by a JEOL ARM200F transmission electron microscopy system. To study the DC characteristics of MoS_2 MOSFET, back-gated structure FETs were fabricated. Cr/Au metal was used as source/Drain contact, and MoS_2 channel was defined by dry etching. I-V data of MoS_2 MOSFET were collected by using HP4156C system.

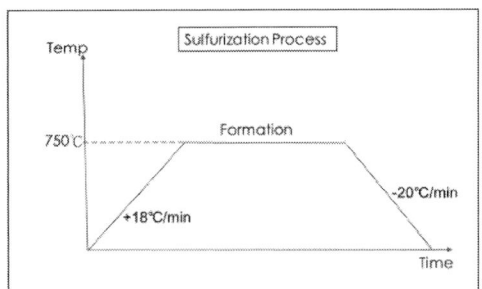

Fig.1. Temperature profile of sulfurization process using 10% H_2S ambient in a furnace.

RESULTS AND DISCUSSION

Fig.2 are the optical image and AFM morphology of MoS_2 which is sulfurized from ultra-thin MoO_3 (<1nm) on SiO_2/Si substrate. As shown in Fig.2, a flat surface with less defect is observed. The root mean square of roughness

(RMS) from AFM is around 1.0 nm. To study the uniformity of MoS_2, we used micro Raman analysis to characterize vibration mode. Fig.3 (a) is the picture of 4 inches MoS_2 wafer. There are 1~5 number indicate the measuring spot position. Fig.3 (b) is Raman spectrum of MoS_2 correlated to the measuring spots. As shown in Fig.3 (b) the significant peaks of E^1_{2g} and A_{1g} are almost the same at 381.0 cm^{-1} and 404.9 cm^{-1}, respectively. Based on the previous studies by C. Lee et.al. [4], around 4~5 monolayers can be estimated by frequency difference of E^1_{2g} and A_{1g} (~24.40 cm^{-1}). Fig. 4 shows the frequency difference of E^1_{2g} and A_{1g} correlated to different measuring position, revealing highly uniform MoS_2 within 4 inches wafer size were achieved.

Fig.2. (a)

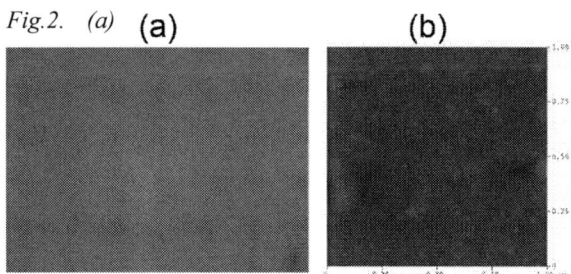

The optical image and (b) AFM morphology of MoS_2 grown on SiO₂/Si substrate.

To examine the microstructure of MoS_2 on SiO_2, we prepared the TEM samples by Focus Ion Beam (FIB). The cross-sectional view TEM images of MoS_2/SiO_2 are shown in Fig.5. As shown in Fig.5 (a) with low magnification, the MoS_2 film exhibit continuous growth without cohesion phenomenon. Fig.5 (b) is high resolution TEM image of MoS_2. The layered structure of MoS_2 with atomic image can be clearly observed, implying highly crystalline MoS_2 obtained. The average thickness of 2.6 nm and 4 monolayer can be estimated, so as to give 0.65 nm per layer.

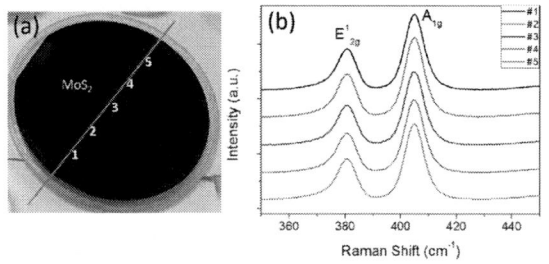

Fig.3 (a) Picture of the 4 inches MoS_2 wafer. The number on wafer indicate the measuring spot position of micro-Raman. (b) Raman spectrum of MoS_2 correlated the measuring spots.

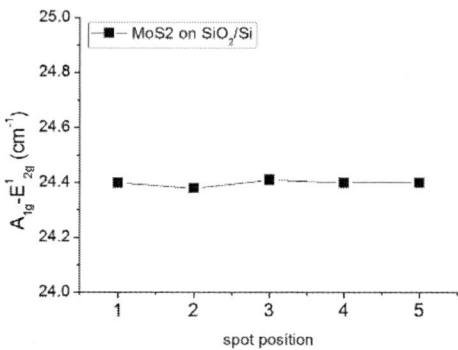

Fig.4. Frequency difference (Δω) of peak A_{1g} and E^1_{2g} in Raman analysis vs. different measuring spot position.

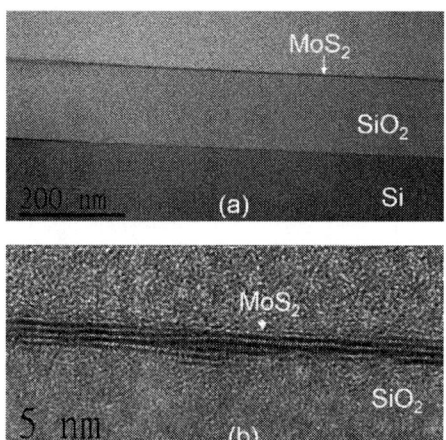

Fig.5 (a) Cross-sectional view of TEM image of MoS_2 grown on SiO₂/Si substrate. (b) High resolution TEM image of MoS_2 with 600K in magnification. The layered structure of MoS_2 are estimated of about 4 monolayers.

In addition, micro photoluminescence (μPL) is a powerful tool for non-destructive analysis on 2D semiconductor materials. The PL spectrum of MoS2 is shown in Fig. 6. As shown in Fig.6, the peaks at 668 nm (1.86 eV) and 623 nm (1.99 eV) are corresponding to the direct transition of A and B transition in MoS_2. The split of transition might come from the Synergistic effect of the coupling between layers and spin-obit coupling [5]. The results suggest that MoS_2 inherent good semiconductor characteristics can be used as channel materials of transistors.

Back-gated MoS_2 MOSFETs were fabricated to study the DC characteristics. The Cr/Au metal contact were deposited as Source/Drain contact and the MoS_2 channel (20um in width/ 3um in length) were defined by dry etch in reactive ion etching method with CF_4. The underneath 150 nm SiO_2 were used as gate dielectrics. The backside highly doped Si substrate were used as gate electrode. The Dc characteristics of MoS_2 MOSFET are shown in Fig.7. As shown in Fig. 7 (a), the Id-Vd curves tend to be saturated at 6V and drain current increased as gate bias increase, suggesting that the transistor function is workable. The transfer curve of MoS_2 MOSFET shown in Fig. 7 (b) reveals n-MOSFET operation with threshold voltage Vth < -10V. That might be due to the small capacitance of thick oxide. The drain current on /off ratio within -10V~10V is near 2 order. The DC characteristics can be improved by fine tuning the device process parameters and reducing the interface states between MoS_2/SiO_2.

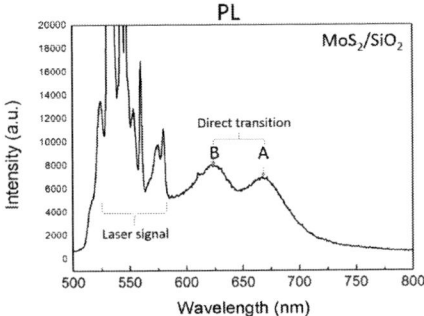

Fig.6. *Room PL spectrum of MoS_2 on SiO_2. The A and B direct transition at 668 nm (1.86 eV) and 623 nm (1.99 eV) are indicated.*

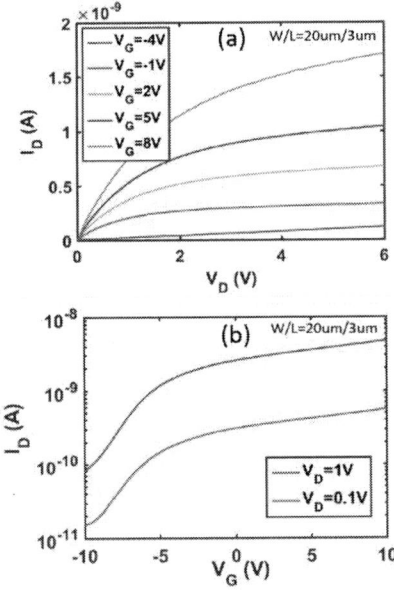

Fig.7. *The DC characteristics of back-gated MoS_2 MOSFET (a) Id-Vd curve vs. various gate bias; (b) transfer curve within gate bias ranging -10V~10V.*

CONCLUSION

In conclusion, 4 inches of 4 monolayer MoS_2 wafer with high uniformity is achieved by a 2-step growth method. The MoS_2 synthesis were performed by a sulfurization process in H_2S ambient from an ultra-thin (<1nm) MoO_3 on SiO_2/Si substrate. The obtained continuous MoS2 thin film with around 4 monolayer were characterized by µRaman/PL, AFM, TEM and fabricated MOSFET. The results demonstrated that the synthesized wafer scale MoS_2 on SiO_2/Si inherent good semiconductor characteristics and can be used as channel materials of transistors.

ACKNOWLEDGEMENTS

This work was sponsored by the NCTU-UCB I-RiCE program, Ministry of Science and Technology, Taiwan, under Grant No. MOST-105-2911-I-009-301-

REFERENCES

[1] Dominik Lembke, Simone Bertolazzi, and Andras Kis, Acc. Chem. Res., vol. 48 (1), 2015, pp. 100–110

[2] B. Radisavljevic, A. Radenovic, J. Brivio, V. Giacometti & A. Kis, Nature Nanotechnology, vol. 6, 2011, pp.147-150

[3] M. Amani, M. L. Chin, A. G. Birdwell, T. P. O'Regan, S. Najmaei, Z. Liu, P. M. Ajayan, J. Lou, and M. Dubey, Appl. Phys. Lett. Vol.102, 2013, pp.193107-1-4.

[4] C. Lee, H. Yan , L. E. Brus, T. F. Heinz, J. Hone, S. Ryu, ACS Nano vol.4, 2010, pp. 2695–2700

[5] K. F. Mak, C. Lee, J. Hone, J. Shan, and T. F. Heinz, Phys. Rev. Lett. Vol.105, 2010, pp.136805-1-4

BUFFER-OPTIMIZED IMPROVEMENT IN RF LOSS OF ALGAN/GAN HEMTS ON 4-INCH SILICON (111)

Tien Tung Luong[1], Franky Lumbantoruan[1], Yen-Yu Chen[1], Yen-Teng Ho[1], Yueh-Chin Lin[1], Shane Chang[1], and Edward-Yi Chang[1,3]*

[1]Department of Materials Science and Engineering, National Chiao Tung University, University Rd. 1001, Hsinchu, 30010 Taiwan

[2] Department of Electronics Engineering, National Chiao Tung University, University Rd. 1001, Hsinchu, 30010 Taiwan

*Corresponding Author's Email: tungluongtien@nctu.edu.tw

ABSTRACT

The effect of different buffer layers on the RF losses of GaN-based high-mobility-transistors (HEMTs) on Si substrate has been studied. It is found that the electron inversion layer induced by the residual tensile stress in AlN buffer is responsible for a dominant loss factor. It is first time such mechanisms of the RF loss of GaN/Si is discussed. It is proven that using a thin high-low-high temperature (HLH) AlN buffer reducing the tensile stress in AlN consequently reduces the RF loss.

INTRODUCTION

Regarding the unique characteristics (high breakdown field, high power density, high efficiency, and broadband) GaN are now broadly recognized as a key technology for many applications; especially, high-frequency power devices. In particular, GaN-based HEMTs are able to operate at high power, high frequencies and high temperatures. GaN-HEMTs on Si technology is expected to drastically reduce the fabrication cost. However, there are still two issues which need to be resolved in order to realize high frequency and high power amplifiers in GaN-HEMT on Si substrate. One of the issues is a parasitic loss.

It was widely believed that the parasitic loss is due to interfacial p-type-doped layer mainly related to Ga and/or Al diffusion into the Si substrate [1-4]. Henson et al. reported that the formation of silicon nitride layer directly on the surface of the Si substrate can function as a diffusion barrier for dopants that can reduce the attenuation of RF signal [3]. However, it is well know that, the formation of SiN_x at the interfacial AlN/Si causes the degradation of AlN and subsequent GaN [5].

Indeed, the attenuation basically relates to interfacial AlN/Si; as well as, the buffers. However, there is a lack of research studying the effect of buffers and AlN/Si interface on the RF loss. In this paper, we have discussed the dependence of the RF losses on the buffers of GaN-HEMT/Si.

EXPERIMENTAL

The epitaxial growth of the AlGaN/GaN HEMT has

been achieved by MOCVD either on high-resistivity (HR) Si(111) (R = 10,000 Ohm.cm) or on p-type low-resistivity (LR) Si(111) (R = 80 Ohm.cm). The GaN-HEMT structures described in this study consist of an AlN buffer layer (with different thickness: ~100 and ~ 200 nm), a three step-graded AlGaN transition layer, a ~1200 nm GaN layer. Also a thick $Al_{0.1}Ga_{0.9}N$, a C-doped GaN, or an unintentionally doped GaN was used as a buffer layer.

Measuring the RF losses on CPW lines in function of frequencies is a straight forward manner to identify and quantify RF loss phenomena. Therefore, we have characterized this effect by measuring transmission line losses in CPWs deposited on the buffer layers and GaN-HEMT/Si (Fig. 1). The total RF losses of CPWs; including radiation loss, conductor loss, dielectric loss (C_{buffer}), substrate losses (R_S, C_S); are determined from scattering (S-) parameter measurements (Agilent N5245A).

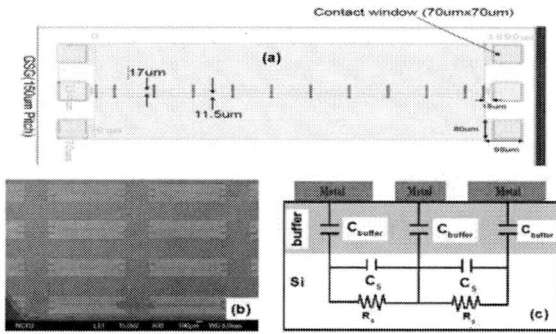

Figure 1: The layout (a); a top-view SEM image (b); and a schematic view of the CPWs on Si substrate (c).

RESULTS AND DISCUSSION

Previously, we have reported that the pre-flow of TMAl eliminated the amorphous SiN_x layer at AlN/Si interface [5]. It is essential to archive a highly crystalline AlN buffer and subsequent GaN to improve the GaN-HEMTs performances. Fig. 2 shows the interface between AlN and Si substrate without (a) and with a TMAl pre-flow (b); as well as, the RF loss of AlN/LR Si (fig. 2c). The attention of the CPW on AlN/Si with pre-Al

978-1-5090-6695-7/17 $31.00 © 2017 IEEE

is below 1 dB/mm across the whole frequency range (250 MHz-40 GHz). On the contrary, the presence of the amorphous SiN_x at the interface AlN/Si leads to the increase of the attenuation. Indeed, the optimization of TMAl pre-flow not only results in the improvement of crystalline quality but also decreases the RF losses. The increase of attenuation relates to reduction of $C_{AlNbuffer}$ (Fig. 2c) due to an increase of AlN buffer leakage, which was caused by the degradation of crystalline quality [5].

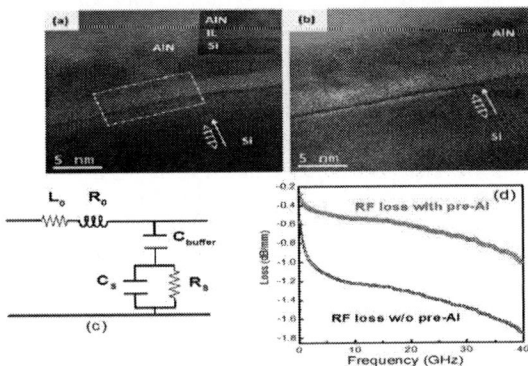

Figure 2: Cross-sectional TEM images of the interface between AlN and Si without (a) and with (b) TMAl pre-flow. An equivalent circuit (c) and RF losses of CPWs on AlN-on-LR Si substrate (d).

Technically, to improve crystallinity and consequent DC performances of GaN-HEMT/Si it is essential to increase the thickness of AlN buffer [7, 8]. However, as increasing the AlN buffer thickness from 100 nm to 200 nm, the RF losses of CPWs lying on AlN/Si drastically increases, even at very low frequency (250 MHz).

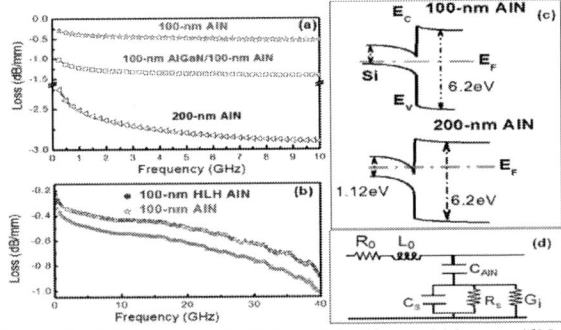

Figure 3: RF losses of different buffers: 100-nm AlN, AlGaN/AlN, and 200-nm AlN (a); and 100nm-HLH AlN on Si substrate (b). Schematic energy band diagram of 100-nm-AlN/Si and 200-nm-AlN/Si (c). An equivalent circuit taking into account a conductive interface channel: G_i (d).

As we know, AlN is a piezoelectric material, the large tensile stress induced by the mismatch between AlN and Si results in a strong polarization field. It can alter the band bending and carrier concentration at the AlN/Si interface

(Fig. 3c). In this way, the free electrons can be confined within a triangular quantum well and exhibit higher mobility than those in bulk. The conductivity of such inversion layer depends on the strength of piezoelectric field and the Fermi energy level (E_F); corresponding to the AlN thickness, and the doping level in Si and AlN. Fig. 3d shows the equivalent circuit taking into account the conductive loss channel at the interface (G_i). In comparison with CPWs on 200-nm-AlN/Si, the reduction of RF losses of CPWs on 100-nm-AlGaN/100-nm-AlN/Si is due to a smaller G_i resulted from a smaller residual tension. It is worth to note that the increase of RF losses with AlN thickness results from the unique property of the single crystal AlN. Such result wasn't observed in CPWs structures fabricated on other di-electric layers; such as PVD-deposited AlN/Si [9], SiO2/Si [10].

Indeed, it is essential to minimize the AlN thickness to suppress G_i. However, it conversely degrades the crystallinity of AlN and subsequent GaN. Therefore, to optimize a GaN-HEMT structure on Si, it is necessary to take into account such trade-off. Our previous works have reported that using a HLH AlN buffer layer resulted in an improvement of the quality of GaN epitaxial layer on Si; as well as enhanced DC performances of GaN-HEMTs [8, 11, 12]. Figure 3b shows the RF loss of CPWs on 100-nm-thick HLH AlN/LR Si, the attenuation remains less than 0.8 dB/mm up to 40 GHz. This attenuation is smaller 0.1 dB/mm than that of CPWs on equally 100-nm-thick HT AlN/Si and is compatible with the attenuation of CPWs directly on HR Si substrate [13].

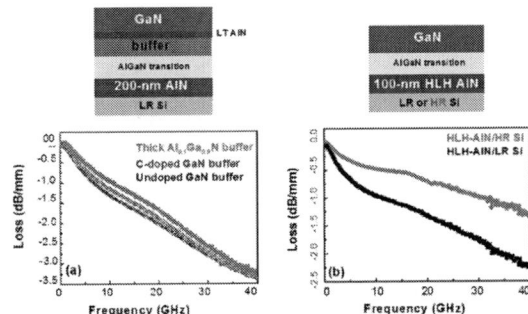

Figure 4: The RF Losses of CPWs on GaN-HEMT structures with 200-nm AlN with different buffer layers on LR Si (a); and with 100-nm HLH AlN on LR and HR Si substrates (b).

Fig. 4a shows the attenuation of the CPWs on GaN with different buffer layers: a thick $Al_{0.1}Ga_{0.9}N$, C-doped GaN, and unintentionally doped GaN on 200-nm AlN/LR Si. The RF attenuations of the structures are similar but drastically increase in comparison with that of the GaN-HEMT/HLH AlN/Si (shown in fig. 4b) even though the DC isolation breakdown voltages are much higher (not shown). It indicates that the electrons inversion channel is a significant loss factor. Obviously using a HLH AlN

978-1-5090-6695-7/17 $31.00 © 2017 IEEE   497

buffer, a stress-engineered buffer, not only improves the crystalline quality but also suppresses the electron inversion layer leading to a reduction of the losses. However, GaN-HEMT/LR Si is still very poor at micro-wave frequencies due to the low resistivity of substrate (Fig. 4b). Using a HR Si substrate is one of approaches to improve RF performances. As shown in fig. 4b, while the RF loss of CPWs on GaN-HEMT/LR Si rapidly increases to 0.95 dB/mm at 10 GHz and to 2.3 dB/mm at 40 GHz, the attenuation of CPWs on GaN-HEMT/HR Si is only 0.4 dB/mm at 10 GHz and remains below 1.2 dB/mm for frequency up to 40 GHz. In comparison with the attenuation of GaN-HEMT/LR Si, the significant improvement in the attenuation of GaN-HEMT/HR Si results from a much higher R_S and the impedance related to C_S, which reduces the substrate loss. However, it should be note that the RF loss of GaN-HEMT on HR Si substrate suffers a strong temperature-dependence. Such effect can be clearly understood when taking into account the effect of 2-dimensional electron gas (2DEG) channel at interface induced by the piezoelectric field caused by the tensile AlN buffer as shown in fig. 3c. Basically, the Fermi energy level of HR Si is higher than that of LR Si (p-type) and it shifts up toward the 2DEG quantum well as temperature slightly increases. Consequently, more electrons fill into 2DEG channel that results in a very conductive layer and consequent a drastic attenuation. It would be conceivable that the very sensitively temperature-dependent loss of GaN-HEMT/HR Si, even at relatively low frequency and low temperature, relates to the sheet mobile electrons at AlN/Si interface.

Indeed, for high frequency applications, it is essential to use ultrahigh resistivity wafers to eliminate the substrate loss. However, in fact that the operating temperature of GaN-HEMT devices is always high enough to limit the advantages of HR Si. Because the loss caused by the sheet mobile electrons become a significant factor as tempera-ture increases. Therefore, to suppress the temperature-dependent attenuation of GaN-HEMT on HR Si, it is required to suppress the sheet mobile electrons underneath the AlN buffer. It turns out that it is necessary to reduce the residual tensile stress in AlN buffer and to deplete the 2DEG channel at Al/Si interface from electrons. In this work, we have introduced a thinner HLH AlN buffer to minimize the residual tensile stress in AlN. We believe that using a thinner p-type AlN buffer would improve in RF loss. The acceptors in AlN will capture the free electrons and deplete the region near the surface of the Si from electrons. More experiments and results will be discussed in other work.

CONCLUSIONS

The interface losses due to induced charges at the interface between AlN buffer and Si, which is caused by the residual tensile stress in AlN buffer, is responsible for the main part of the observed RF loss in GaN-HEMT/Si. Optimization of TMAl pre-flow process eliminated the interfacial amorphous layer leading to improve not only material quality but also RF loss. It is crucial to decrease the buffer thickness to reduce RF loss. We have demonstrated that a thinner HLH AlN buffer, which can reduce the residual tensile stress, results in a loss reduction. We also propose that using a p-type AlN near the interface will help to suppress the inversion layer, it is required to maintain acceptable RF losses of GaN HEMTs on Si for a high frequencies range and temperature.

ACKNOWLEDGEMENTS

This work was sponsored by the NCTU-UCB I-RiCE program, Ministry of Science and Technology, Taiwan, under Grant No. MOST-105-2911-I-009-301-

REFERENCES

[1] H. Shinichi, I. Masanori, M. Toshiharu, O. Hideyuki, M. Yoshiaki, T. Isao, et al., Appl. Phys. Express, vol. 2, 2009, pp. 061001-1-3.

[2] E. Calleja, M. A. Sánchez-García, D. Basak, F. J. Sánchez, F. Calle, P. Youinou, et al., Phys. Rev. B, vol. 58, 1998, pp. 1550-1559.

[3] A. W. Hanson, J. C. Roberts, E. L. Piner, and P. Rajagopal, "III-nitride material structures including silicon substrates," US Patent 20,060,118,819 (2007).

[4] D. Marti, M. Vetter, A. R. Alt, H. Benedickter, and C. R. Bolognesi, Appl. Phys. Express, vol. 3, 2010, pp. 124101-1-3.

[5] F. Lumbantoruan, Y. Y. Wong, Y. H. Wu, W. C. Huang, N. M. Shrestra, T. T. Luong, et al., 2014 IEEE Int. Conf. Semi-cond. Electron., 2014, pp. 20-23.

[6] S. Arulkumaran, T. Egawa, S. Matsui, and H. Ishikawa, Appl. Phys. Lett., vol. 86, 2005, 123503-1-3.

[7] K.-L. Lin, E.-Y. Chang, Y.-L. Hsiao, W.-C. Huang, T. Li, D. Tweet, et al., Appl. Phys. Lett., vol. 91, 2007, 22211-1-3.

[8] S. B. Evseev, L. K. Nanver, and S. Milosavljević, 2013 IEEE Bipolar/BiCMOS Circuits Technol. Meet., 2013, pp. 77-80.

[9] D. Lederer and J.-P. Raskin, Solid-State Electron., vol. 47, 2003, pp. 1927-1936.

[10] H. Yu-Lin, L. Lung-Chi, W. Chia-Hsun, C. Edward Yi, I. K. Chien, M. Jer-Shen, et al., Jpn. J. Appl. Phys., vol. 51, 1012, pp. 025505-1-4.

[11] T. T. Luong, Y. T. Ho, B. T. Tran, Y. Y. Woong, and E. Y. Chang, Chem. Vap. Deposition, vol. 21, 2015, pp. 33-40.

[12] C. Roda Neve, K. Ben Ali, P. Sarafis, E. Hourdakis, A. G. Nassiopoulou, and J. P. Raskin, Microelectron. Eng., vol. 120, 2014, pp.205-209.

SPUTTER DEPOSITION TECHNOLOGY FOR AL$_{(1-X)}$SC$_X$N FILMS WITH HIGH SC CONCENTRATION

Bernd Heinz[1*], Stefan Mertin[2], Oliver Rattunde[1], Marc Alexandre Dubois[3], Sylvain Nicolay[3], Gabriel Christmann[3], Maurus Tschirky[1], Paul Muralt[2]

[1]Evatec AG, Hauptstrasse 1a, CH-9477 Trübbach, Switzerland

[2]Electroceramic Thin Films Group, Ecole Polytechnique Fédérale de Lausanne, CH-1015 Lausanne, Switzerland

[3]CSEM SA, Jaquet-Droz 1, CH-2002 Neuchâtel, Switzerland

*Corresponding Author's Email: bernd.heinz@evatecnet.com

ABSTRACT

Aluminium scandium nitride (Al$_{1-x}$Sc$_x$N) with its strongly enhanced piezoelectric response is the upcoming piezoelectric material of choice in next generation RF filters, sensors, actuators and energy harvesting devices. This paper will concentrate on the deposition technology for Al$_{1-x}$Sc$_x$N films with high Sc content. Films with Sc concentrations close to 43 at% have been grown on 200-mm substrates using a cluster type sputter deposition tool. The piezoelectric response will be discussed and correlated with the deposition parameters and film structural properties. The steps required to deliver a high-volume production solution for high Sc concentration will be described.

INTRODUCTION

Aluminium scandium nitride Al$_{1-x}$Sc$_x$N is a novel thin-film material for various MEMS applications [1,2]. It has the potential for the realization of wide-band RF filters [2] (up to a factor 2 wider than present versions with AlN), and also shows interesting properties for piezoelectric vibration energy harvesting [3,4]. Piezoelectric microphones, speakers and ultrasonic transducers are also expected to be produced in volume soon [5,6]. Despite a well-established theoretical background [7,8], the properties of AlScN thin films for higher Sc concentrations (15 to 30%) expected theoretically have still to be proven experimentally. At first there are technical issues to master, then the issue of reliable methods for characterization comes into play.

We deposited Al$_{1-x}$Sc$_x$N thin films onto 200-mm silicon wafers by reactive pulsed direct current magnetron sputtering using three different hardware/target configurations:

I.) Single target sputtering on stationary substrates using 300-mm compound targets (Sc = 6% and 9.5%)

II.) Single target sputtering on rotating substrates using 100-mm compound targets (Sc = 15% and 28%)

III.) Co-sputtering on rotating substrate using 100-mm Al and Sc targets

Among these, the first method is the most attractive one for industrial use as key requirements for high-yield production, such as tight film stress and thickness uniformity control over target life, can be managed by leveraging existing long-term production experience in FBAR manufacturing for BAW filters with piezoelectric AlN. However, the fabrication of 300-mm compound targets is extremely challenging above x=10%, so the use of smaller compound targets or the co-sputtering of Al and Sc could be seen as an alternative step towards a final solution with high productivity.

EXPERIMENTAL DETAILS & RESULTS

500-nm thick AlScN films were grown on either platinum (Pt) or molybdenum (Mo) electrodes sputtered on silicon wafers coated with a 200-nm thick thermal oxide. The substrate temperature was controlled in the range of 300–350°C during deposition. Ar:N gas flow was kept at a constant ratio of 1:2 for all films sputtered from 100-mm targets.

X-ray diffraction (XRD) pattern verified AlScN growth in the desired piezoelectric wurtzite phase of AlN up to values of x = 39 at% for the given sputter conditions. Pure c-axis orientation with no traces of other orientations or phases were found in the Θ–2Θ scans for co-sputtered as well as for compound target sputtered films.

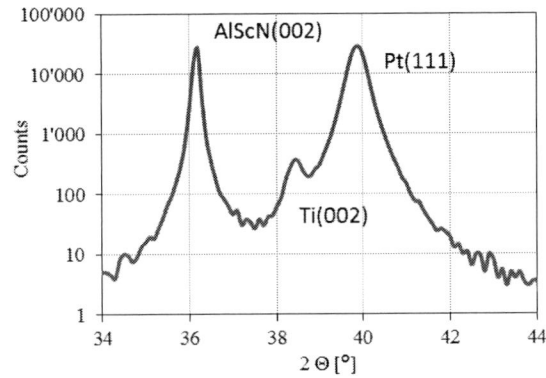

Figure 1: Θ–2Θ scan of co-sputtered AlSc$_{31.5}$N showing the AlN(002) and the Pt(111) diffraction peak. The Ti(002) peak is produced by the adhesion layer below Pt(111).

An example is given in Fig. 1, showing the Θ–2Θ scan of a co-sputtered $AlSc_{31.5}N$ film deposited onto a Pt bottom electrode.

Films with higher Sc concentrations than 39 at% show either very low peak intensities or multiple diffraction peaks indicating the formation of unwanted, non-piezoelectric crystalline phases.

XRD rocking curves of the AlScN(002) diffraction peaks show excellent FWHM values below 2° for the 500-nm films up to 39 at% Sc (see Fig. 2). A trend to slightly higher values can be seen with increasing Sc. Furthermore, the FWHM is typically larger on the Mo electrodes than on the Pt electrodes. This means that films deposited on Pt tend to have a better c-axis orientation than films grown on Mo. However, the main information derived from this chart is the fact that films with the same high crystalline quality can be obtained using either co-sputtering or single target sputtering.

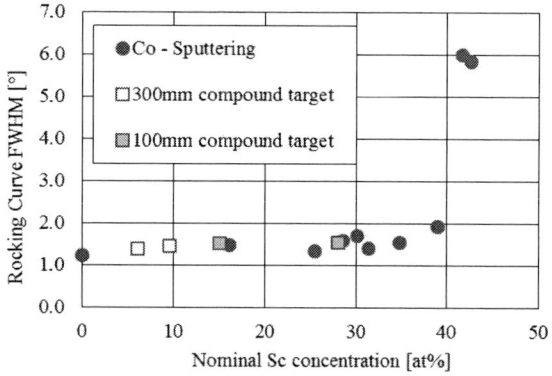

Figure 2: Rocking curve measurements of the (002) diffraction peaks of AlScN films deposited with different target configurations and Sc content. Values were measured in the centre of the substrate.

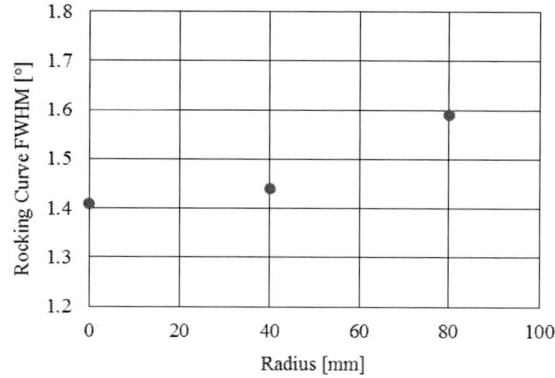

Figure 3: Uniformity of the rocking curve of a co-sputtered $AlSc_{31.5}N$ film across a 200-mm wafer.

One key requirement for high-volume and high-yield production is tight control of film uniformity across the substrate. Fig. 3 and Fig. 4 display the uniformity across the wafer of the c-axis textured growth (rocking curve of the AlScN(002) diffraction peak) and of the scandium concentration for co-sputtered $AlSc_{31.5}N$, respectively.

An excellent c-axis textured crystalline structure is achieved across the whole substrates. The moderate variation of the rocking curve FWHM limited to below 1.6° is not expected to affect the piezoelectric properties. The actual Sc content of the same film was verified by scanning electron microscopy (SEM) based energy dispersive X-ray spectroscopy (EDX). The Sc concentration across the wafer stays within ±1 at% (see Fig. 4).

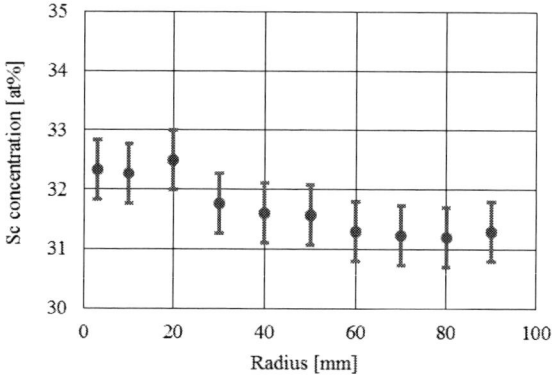

Figure 4: Scandium concentration across the 200-mm substrate measured with EDX on co-sputtered $AlSc_{31.5}N$ film.

Fig. 5 and Fig. 6 depict SEM images of the same $AlSc_{31.5}N$ thin film in cross-sectional and plain view, respectively. Homogeneous regular film growth with a grain size of a few nm can be observed in the plain view. The cross section reveals a dense AlScN film with columnar growth. Below the Pt electrode the thermal silicon oxide layer is visible.

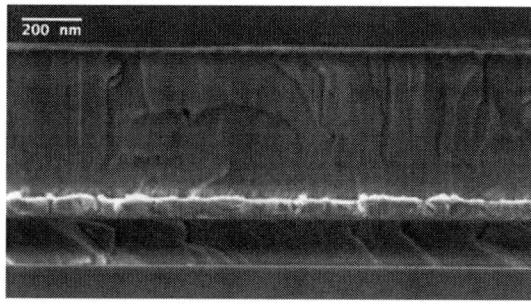

Figure 5: SEM cross section of an $AlSc_{31.5}N$ film grown on a Pt electrode.

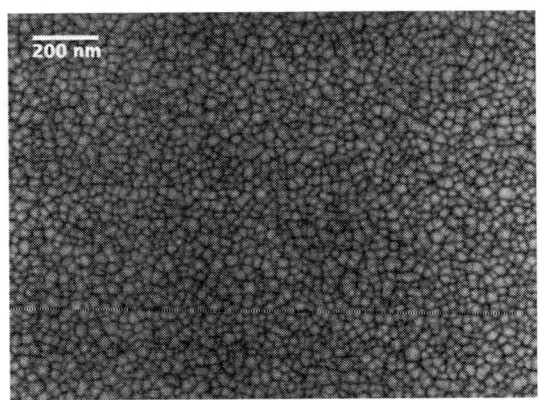

Figure 6: SEM plain view of an AlSc$_{31.5}$N film taken at the wafer centre.

To evaluate basic piezoelectric properties, the transversal piezoelectric coefficient e$_{31,f}$ of thin films was determined with a four-point bending setup of AIXACCT [9], using silicon (100) cantilevers along [110] with a Poisson ratio v_s of 0.064. The bending causes an in-plane strain (S_l) in the piezoelectric thin film, thus producing a charge difference between top and bottom electrode ($\vec{D}$-field). The displacement at the cantilever centre is measured with a laser interferometer and allows the derivation of the curvature. As the out of plane stress is zero, the simple relation

$$D_3 = \frac{e_{31,f} S_1}{1 - v_s}$$

holds, where $e_{31,f} = e_{31} - c_{13}^{E}/c_{33}^{E} \cdot e_{33}$. Fig. 7 plots e$_{31,f}$ for different Sc concentrations. It can be seen that e$_{31,f}$ increases nearly linear with the Sc content. The largest response of e$_{31,f} = -2.37$ C/m^2 is obtained for a film with 31.5 at% Sc concentration.

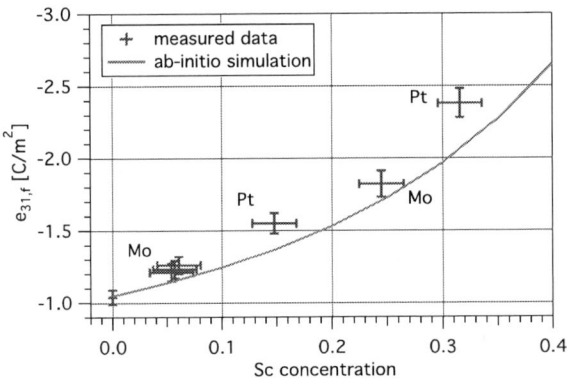

Figure 7: Measured e$_{31,f}$ values for selected films with a Sc concentration between 0 at% and 31.5 at%. For comparison, the ab-initio curve by Caro et al. [8] is plotted. Furthermore, the electrode material is noted for the AlScN films.

The value of pure AlN matches exactly the standard value of -1.05 C/m^2. The value at 31.5 at% Sc is higher than the recently published ab-initio value of -2.05 by Caro et al. [8]. Furthermore, it can been seen in Fig. 7 that the e$_{31,f}$ values of the films deposited on Pt are more above the ab-initio curve than the ones of films deposited on Mo. This observation is consistent with the better crystallographic orientation (c-axis) of the films on Pt, as discussed above.

CONCLUSION

The growth of piezoelectric AlScN films on 200-mm substrates using mass production sputter equipment has been demonstrated successfully. Highly c-axis oriented films are achieved out of compound targets containing up to 28 at% Sc and by co-sputtering from pure Sc and Al targets for compositions up to 39 at%. A maximum value for $|e_{31,f}|$ of 2.37 C/m^2 is achieved in a film with 31.5 at% Sc, corresponding to a factor of 2.3 increase as compared to state-of-the-art AlN thin films.

ACKNOWLEDGEMENTS

The authors would like to thank the Swiss Commission for Technology and Innovation CTI for funding in part this work (contract 18616.1).

REFERENCES

[1] M. Akiyama, K. Kano, and A. Teshegahara, *Appl. Phys. Lett.* 95, 162107, 2009.

[2] R. Matloub, A. Artieda, C. Sandu, E. Milyutin, and P. Muralt, *Appl. Phys. Lett.*, *99*(9), 92903, 2011

[3] R. Matloub, M. Hadad, A. Mazzalai, N. Chidambaram, G. Moulard, C. S. Sandu, T. Metzger and P. Muralt, *Appl. Phys. Lett.* 102 152903, 2013

[4] R. Elfrink, T.M. Kamel, M. Goedbloed, S. Matova, D. Hohlfeld, Y. van Andel, and R. van Schaijk, *J. Micromech. Microeng.* 19(9), 94005 (2009)

[5] M. D. Williams, B. A. Griffin, T. N. Reagan, J. R. Underbrink, and M. Sheplak *J. Microelectromech. Syst.* *21*(2), 270–283, 2012

[6] H. C. Cho, S. C. Ur, M. S. Yoon, and S. H. Yi, *3rd IEEE International Conference on Nano/Micro Engineered and Molecular Systems, NEMS* (pp. 637–640), 2008

[7] A. Alsaad and A. Ahmad. *Eur. Phys. J. B*, *54*(2), 151–156 (2006)

[8] M. A. Caro, S. Zhang, T. Riekkinen, M. Ylilammi, M. A. Moram, O. Lopez-Acevedo, J. Molarius, and T. Laurila, *J. Phys: Condens. Matter* 27 (2015) 245901 (14pp)

[9] K. Prume, P. Muralt, F. Calame, T. Schmitz-Kempen, S. Tiedke, *IEEE transactions on ultrasonics, ferroelectrics, and frequency control* 54.1 (2007), pp. 8–14.

A 10 BIT ANALOG COUNTER IN SPAD PIXEL

Bin Li[1], Yue Xu[1,2]*, RuiMing Luo[1]

[1]College of electronic science & engineering, Nanjing University of Posts and Telecommunications, Nanjing 210003, China

[2]National and Local Joint Engineering Laboratory of RF Integration and Micro-assembly Technology Nanjing 210003, China

*Corresponding Author's Email: yuex@njupt.edu.cn

ABSTRACT

An improved active-quenching circuit and a linear counting circuit used for a silicon single photon avalanche diode (SPAD) are presented in this paper. The proposed quenching circuit is fast and compact. The linear counting circuit can achieve 10 bit large count range with a small capacitor, which can effectively reduce the area of pixel. Due to the significant advantages of low-cost, compactness and high count range, the two pixel circuits are very suitable for the high density SPAD array detector.

INTRODUCTION

In recent years, SPAD detector is widely used in many aspects, such as fluorescence lifetime imaging for medical treatment [1], radar detection technology [2][3]. Nowadays, there are many aspects to be improved in the technology of large scale detector about SPAD, such as pixel cell density, counting scope. Especially, high density is the tendency of SPAD detector in modern standard CMOS technology.

Traditional pixel adopts the digital counting circuit. Although the digital counter features a better suppression of noise and higher sensitivity [4], it requires hundreds of transistors, occupying larger area, which seriously affects the pixel cell fill factor. In order to effectively reduce the counter area, and improve the fill factor, an analog counter is researched in this paper.

The main problem of an analog counter is the big occupation area of counter capacitance. While the count range is proportional to the capacitor area, so by using this huge-range counter, even with a smaller count capacitor the circuit can also provide a satisfying performance. Aiming at this feature, this paper proposed a huge-range analog counter, which can be applied to high-density pixel.

CIRCUIT DESIGN

The entire pixel includes an active quenching circuit and an analog counting circuit. Figure 1 shows a schematic diagram of a transistor level circuit for an improved active quenching solution [5].

The operating principle of the circuit is described as follows. Initially circuit waiting for photon signal, the cathode (node k) is set at high potential and the node q is set at low potential. When an incident photon is detected,

the SPAD rapidly starts to avalanche and large current flows through the M1 and M2 branches. Since the node k in the initial state is set to the supply voltage (VDD), the large avalanche current causes the voltage of node k dropping immediately, whereby the voltage of node q (the output of the inverter which is made of M3 and M4) increases rapidly, namely the voltage of gate M1 becomes higher, along with the channel resistance of the M1 becomes larger, which leads to the rapid drop of the node k potential, the way accelerating the voltage reduction of node k forms a positive feedback of the circuit, which speeds up the quenching process. At this state the voltage across the SPAD drops below the avalanche breakdown voltage, so the avalanche current is quenched.

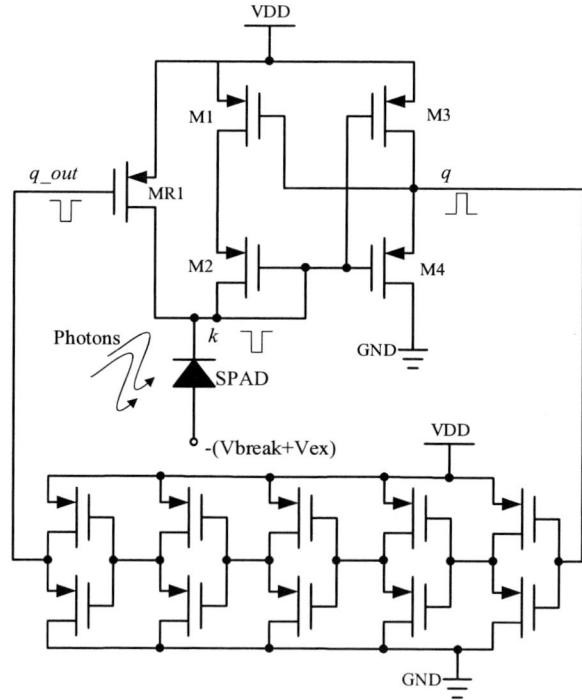

Figure 1: Diagram of active-quenching circuit

After the quenching process, the circuit needs to restore to the Geiger mode waiting for the next avalanche

978-1-5090-6695-7/17 $31.00 © 2017 IEEE

process triggered by photons, so a MR1 transistor is connected in parallel with a signal q_out for reset control. To complete the active quenching scheme, a simple hold-off circuit which is made up of 5 inverters is given in the lower part of Figure 1. The hold-off circuit connects node q and q_out, which delays the signal of node q. By means of the hold-off circuit, the high voltage in node q transfers to a low voltage in node q_out, so the reset MOSFET (MR1) is turned on, and the voltage of k node is reset again to supply voltage (VDD). In this state, the voltages of node q and q_out keep low and high respectively before the next quenching action. Finally, it can be seen that, during every triggered avalanche process, the node q_out is a negative pulse, which is chosen as the output of the quenching circuit.

The proposed improved linear analog counting circuit is shown in Figure 2. The whole circuit uses only five MOSFETs, which is simple and compact. The input pulse of the counter circuit is the node q_out (the output of the quenching circuit), and the node reset is the reset signal for count capacitor C which will be reset to zero potential before each count period. The charges on the capacitor increase the same amount after each pulse. As a result, the voltage on capacitor is proportional to the number of count pulses, so it is called linear counter.

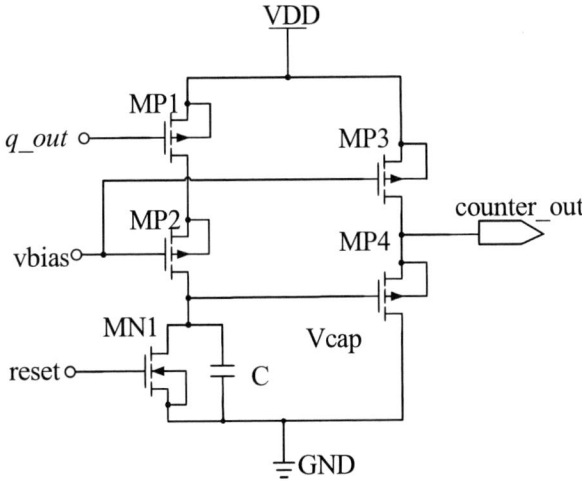

Figure 2: Analog linear counter for SPAD

The working process of the counter circuit is shown in Figure 3, which can be divided into 3 steps, namely reset step, count step and readout step.

During the reset step, MP1 is in the off state. The node reset goes to high potential, so the charges on the upper electrode plate of the count capacitor C flow to GND, as a result, the voltage of the count capacitor is restored to the initial state.

During the count step, MP1 is in the on state and the node reset goes to zero potential, so the charges on the count capacitor is determined by the number of quenching pulses. Every time an avalanche pulse comes, the MP1 switch is turned on during the negative pulse and little amount of current flows from the supply voltage to the capacitor. Along this current path, there is a MOSFET MP2 whose bias voltage is set at a suitable high voltage in order to limit the current size. So every avalanche pulse adds just few charges on the capacitor.

During the readout step, MP1 is in the off state and the node reset is still in zero potential. The voltage follower reads the voltage on the count capacitor, and by simple calculation, the number of photons detected by SPAD pixel during the count period can be obtained easily.

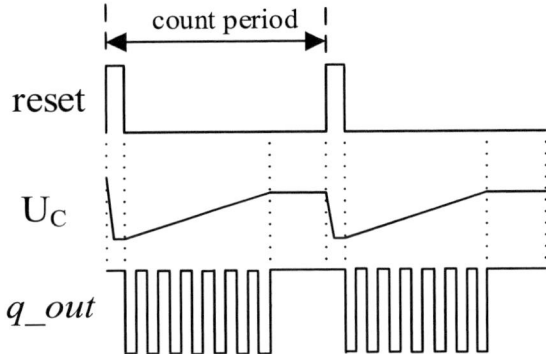

Figure 3: the working process of the counter circuit

Some advantages of the proposed linear analog counter circuit are shown as following:

1) The MP2 transistor operates on the sub-threshold region (near the linear region), and the appropriate bias voltage can be adopted to make the count range of the linear counter increase significantly.

2) Linear counting circuit eliminates the substrate bias effect by using PMOS source follower. In SMIC 0.18μm process, PMOS has N-well, while NMOS does not have P-well. Hence, every substrate node of NMOSFET in the NMOS source follower must be attached to GND, as a result, the threshold voltage becomes larger and the output voltage is much lower than the capacitor voltage. Fortunately, there is no extra voltage loss by using PMOS follower since the source node and substrate node of PMOSFET can be connected together.

3) The excellent linearity of the output voltage following the capacitor voltage is another advantage. As can be seen from Figure 2, MP4 is the source follower whose load MP3 is set the same bias voltage as MP2. The advantage is that the branch viewed from the count capacitor has approximately the same voltage drop as the branch seen upwards from the MP4 transistor.

SIMULATION RESULTS

Based on SMIC 0.18μm CMOS technology, the quenching circuit and linear counter circuit are designed and simulated on SPECTURE in Cadence environment.

The simulation results of quenching circuit are shown in Figure 4 where a voltage pulse is used to simulate the arrival of a photon signal with its amplitude of 5 V and its pulse width of 100 ps. When a photon arrives, the cathode voltage of SPAD drops rapidly, shown as Figure 4 (b), in which the quenching time is about 8 ns, then the voltage drop on node k is converted to the voltage up on node q. After the signal goes through the hold-off circuit, a reset signal (node q_out) is generated, and the avalanche diode is again reset to Geiger mode. After that, the node k is reset to high voltage, at the same time, the node q turns to low voltage and the node q_out is turned to low voltage, waiting for the next avalanche action.

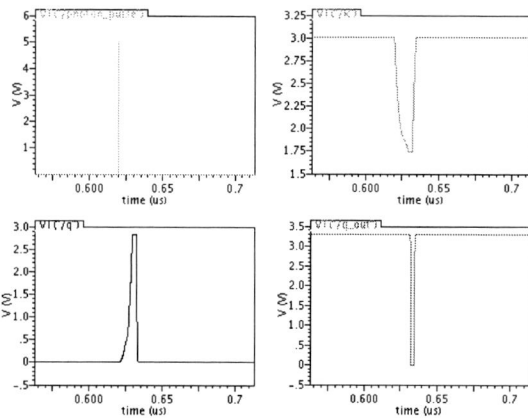

Figure 4: Simulation results of quenching circuit

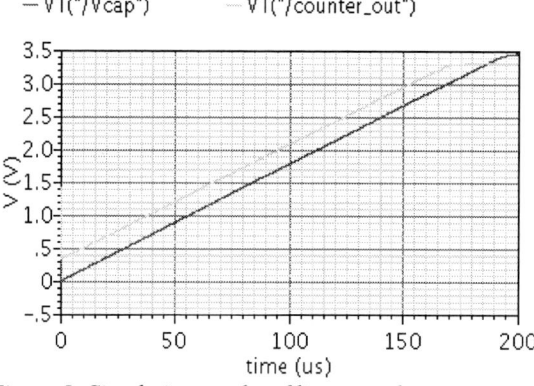

Figure 5: Simulation results of linear analog counter circuit

Figure 5 shows the results of a transient simulation, in which the q_out signal has a pulse period of 100 ns, a pulse width of 10 ns. The voltage on the capacitor (shown as the blue line) starts to increase from 0 V, and does not rise again when it reaches the supply voltage. While the output voltage (shown as the green line) increases with the capacitance voltage, the voltage difference between them is only 0.3 V. The whole simulation time is 200 μs, and the output voltage increases to supply voltage before stopped, and at this time, the simulation time is 167μs, so the circuit can count 1670 pulses (10 bit) with a small count capacitor of 0.5pF.

CONCLUSIONS

A compact quenching circuit and a linear analog counter circuit suitable for driving a SPAD image sensor have been presented. The quenching circuit provides a quenching time of 8 ns, and the linear analog counter circuit achieves an excellent performance of 10 bit count range with a small count capacitor.

ACKNOWLEDGEMENTS

This work is supported by National Science Foundation of China (No. 61571235) Project and QingLan Project and Graduate Research and Innovation Projects of China Jiangsu Province (No. SJLX16_0325).

REFERENCES

[1] M. W. Seo *et al.*, "11.2 A 10.8ps-time-resolution 256×512 image sensor with 2-Tap true-CDS lock-in pixels for fluorescence lifetime imaging," *2015 IEEE International Solid-State Circuits Conference - (ISSCC) Digest of Technical Papers*, San Francisco, CA, 2015, pp. 1-3.

[2] G. Boso, M. Buttafava, F. Villa and A. Tosi, "Low Cost and Compact Single-Photon Counter Based on a CMOS SPAD Smart Pixel," in IEEE Photonics Technology Letters, vol. 27, no. 23, pp. 2504-2507, Dec.1, 1 2015.

[3] S. Jahromi, J. Jansson, I. Nissinen, J. Nissinen and J. Kostamovaara, "A single chip laser radar receiver with a 9×9 SPAD detector array and a 10-channel TDC," *ESSCIRC Conference 2015 - 41st European Solid-State Circuits Conference (ESSCIRC)*, Graz, 2015, pp. 364-367.

[4] E. Panina, L. Pancheri, G. F. Dalla Betta, N. Massari and D. Stoppa, "Compact CMOS Analog Counter for SPAD Pixel Arrays," in IEEE Transactions on Circuits and Systems II: Express Briefs, vol. 61, no. 4, pp. 214-218, April 2014.

[5] R. Mita and G. Palumbo, "High-Speed and Compact Quenching Circuit for Single-Photon Avalanche Diodes," in IEEE Transactions on Instrumentation and Measurement, vol. 57, no. 3, pp. 543-547, March 2008.

CONTROLLABLE SHRINKING OF SILICON OXIDE NANOPORES BY HIGH TEMPERATURE ANNEALING

Jian Chen[1], Tao Deng[2], Zewen Liu[3], and Haizhi Song[1]*

[1] Department of Laser Photoelectric Technology, Southwest Institute of Technical Physics, Chengdu 610041, China

[2]School of Electronic and Information Engineering, Beijing Jiaotong University, Beijing 100044, China

[3]Institute of Microelectronics, Tsinghua University, Beijing 100084, China

*Corresponding Author's Email: gzuchenjian@126.com

ABSTRACT

This paper presents a novel method for fabricating silicon oxide nanopores. First, pores of 80-400 nm were fabricated in a free-standing silicon membrane by anisotropic wet etching process. After thermal oxidation of 90 nm silicon oxide, the pores can be reduced to 35-300 nm. Finally, high temperature annealing promotes the viscous flow of the silicon dioxide membrane and results in shrinking the pores to sub-15 nm, with an estimated precision of 1 nm. Our results are in agreement with the surface-tension-driven model.

INTRODUCTION

It has recently been shown that solid-state nanometer-sized pores (nanopores) have potential applications in genetics, medical diagnostics and biomolecule detection due to their flexibility in size, robustness and durability [1]. Control over the nanopore size is critical as the pore must be comparable in size to the analyte molecule in question. Fabrication and modification of nanopores are the most widely methods for desired size [2]. Among them, electron beams in transmission electron microscope (TEM) [2,3] and scanning electron microscope (SEM) [4] or focused ion beam (FIB) [5] are utilized to fabricate and shrink nanopores in silicon dioxide and silicon nitride membranes down to several nanometers. Furthermore, many effective methods have also been successfully used to shrink small solid-state pores, for example, atomic layer deposition (ALD) of alumina [6], electron beam induced local deposition of silicon dioxide [7]. Even though very small pores have been realized using these approaches, several issues remain. These include the requirement of realize massive production of nanopore arrays, and the fabrication cost.

We have previously reported the combination of anisotropic wet etching and dry etching to fabricate nanopore arrays in a silicon membrane [8]; however, it's difficult to realize the nanopores below 30 nm. Inspired by Waseem Asghar et.al, who shrinking nanopores drilled in pure silicon dioxide membrane by direct thermal heating [9], we extend the investigation to solid-state nanopores. The existing nanopores are etched in a silicon membrane

using anisotropic wet etching and can be thermal oxidation and then shrunk down to sub-15 nm with a precision of 1 nm by high temperature annealing.

METHOD

The fabrication procedure of silicon oxide nanopore arrays is represented in Fig. 1. Initially, (50 × 50) square and rectangle dots array was patterned on Cr etching mask layer prepared on Si (100) substrate and inverted hollow pyramid array was successively fabricated by anisotropic Si etching as illustrated in Fig. 1(a). Next, an array of 1200 μm × 1200 μm square windows on the back of Si3N4 etching mask layer and etching until the residual thickness reached about 15 μm as illustrated in Fig. 1(b). An unconventional, low-temperature wet etching at 30 °C was performed to precisely remove the silicon. When KOH thinning reached the tip of the inverted pyramid, pores were created as illustrated in Fig. 1(c). Then, the mask layers were removed from both sides, giving the silicon nanopore arrays illustrated in Fig. 1(d). The detailed experimental procedure for the fabrication of nanopore arrays can be found in our previous work [10]. In this

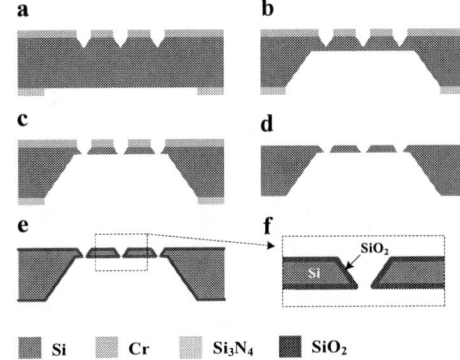

Figure 1: Fabrication process of silicon dioxide nanopores. (a) Fabrication of an inverted pyramid by anisotropic Si wet etching using KOH. (b) Etching from back side. (c) Nanopores formation using low-temperature wet etching. (d) Remove the mask layers. (e) Thermal oxidation. (f) Enlarged, cross-section view of the silicon oxide pore shown in panel (e).

978-1-5090-6695-7/17 $31.00 © 2017 IEEE

experiment, an additional process of thermal oxidation at 900 °C was carried out, the surface of the pore and the membrane will be covered with silicon oxide layer, as described in Figs. 1(e) and 1(f).

RESULTS AND DISCUSSIONS

We discuss a number of experiments performed to test the fluidized silicon oxide drives the morphological changes induced by the high temperature annealing. For a rectangle pore with initial size of 129 nm × 63 nm was thermal oxidized at 900 °C, and a thickness of 90 nm silicon dioxide was grown to shrink the pore to 96 nm × 34 nm [Fig. 2(a)]. This silicon dioxide is visible as a dark gray region around the black pore in the SEM image. These silicon dioxide nanopore with initial size of 96 nm × 34 nm, processed at 1050 °C, treated for 3, 8 and 14 min, respectively, in Figs. 2(b)-2(d). The SEM images of the nanopore show the shrinking process, the horizontal direction and vertical direction are all shrink. At last, the pore shrink to 67 nm × 14 nm. Extended high temperature annealing can reduce the size of the nanopore to less than 10 nm, or close the pore completely.

Fig. 3 shows a rectangle silicon dioxide nanopore expansion by high temperature annealing. An initial size of 189 nm × 239 nm nanopore in a 90 nm thick membrane was processed at 1050 °C for 14 min. The pore expanded to 212 nm × 277 nm [Fig. 3(b)]. We also note that the initial size of 85 nm × 110 nm nanopore shows very little changes after 14 min of thermal shrinking at 1050 °C in the SEM microscope, the initial size is about equal to the

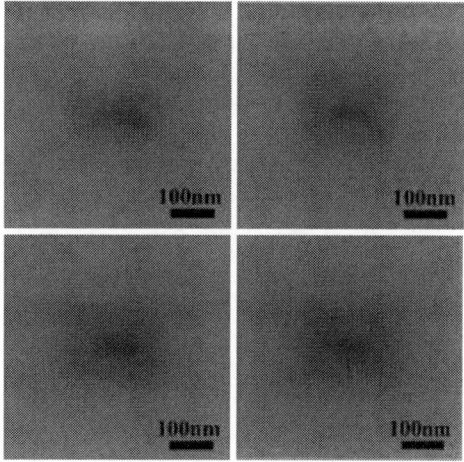

Figure 2: SEM images of silicon dioxide rectangular nanopores shrinking by high temperature annealing. (a) SEM image of silicon dioxide nanopore after the final thermal oxidation with a size of 96 nm × 34 nm. (b) SEM image of nanopore after 3 min of thermal shrinking at 1050 °C with a size of 82 nm × 30 nm. (c) Nanopore after 8 min with a size of 78 nm × 22 nm. (d) Nanopore after 14 min with a size of 67 nm × 14 nm.

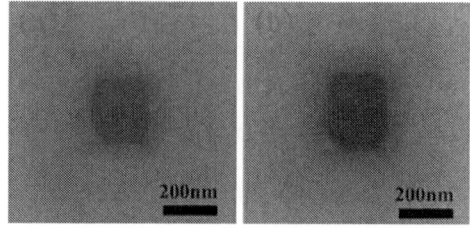

Figure 3: SEM images of silicon dioxide nanopore expansion by high temperature annealing. (a) SEM image of silicon dioxide nanopore after the final thermal oxidation with a size of 189 nm × 239 nm. (b) Nanopore after 14 min of thermal shrinking at 1050 °C with a size of 212 nm × 277 nm.

thickness of the silicon oxide.

Fig. 4 (inset) shows the size of a closing silicon dioxide nanopore as a function of annealing time. In this experiment, the average rate of closing was roughly constant at a rate of 1.4 nm/min. This experiment clearly demonstrates that the rate of closing is slow enough to stop at any desired size, with single-nanometer precision. We found that the closing rate for nanopores is reproducible from pore to pore, which allowed to shrink multiple nanopore arrays in parallel.

An obvious concern in the pore shrinking process is the possibility of hydrocarbon contamination being involved in the pore shrinkage process, a common phenomenon in electron microscopy shrinking process. All the chips were rinsed with acetone and alcohol respectively, and then cleaned with oxygen plasma for 5 min before and after each shrinking step. The local energy dispersive X-ray spectroscopy (EDS) analysis after processing step showed no traces of hydrocarbons around the pore, as shown in Fig. 4. Moreover, cleaning the chips prior to insertion into the heating furnace with oxygen plasma did not alter the observed pore shrinkage behavior

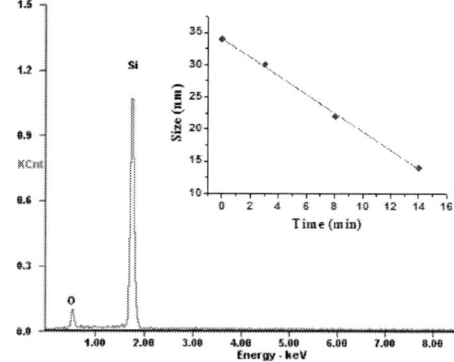

Figure 4: EDS spectrum around the silicon dioxide nanopore after thermal shrinking, presence only Si and oxygen. The inset the size versus time for a shrinking pore, the line is obtained through linear fitting of the experimental data.

978-1-5090-6695-7/17 $31.00 © 2017 IEEE

[6]. In total, the pore shrinkage process is not associated with hydrocarbon contamination.

The magnitude of forces generated by surface tension depends strongly on the geometry of a surface. We explained the pore shrinkage and expansion by surface-tension-driven model [2]. The silicon oxide is softened by high temperature, and thus will deform as a viscous fluid to find a configuration with a lower surface free energy E. The model assume that the pore as a cylindrical hole with radius r in a film with uniform thickness h. The free energy change ΔE is proportional to the change in surface area ΔA:

$$\Delta E = \gamma \Delta A = 2\pi\gamma(rh - r^2) \qquad (1)$$

Where γ is the surface tension of the liquid. So, pores with $2r < h$ can lower their free energy by shrinking while larger pores can do so by expanding. Further, the Vogel-Fulcher-Tammann (VFT) equation was used to discover the γ at 1050 °C [10]. The VFT equation is accurate for temperatures of a few thousand centigrade. It is given as follows:

$$Log_{10}(\eta) = A + [B/(T - T_0)] \qquad (2)$$

where T is temperature, and $A = -7.925$, $B = 31282.9$ °C, and $T_0 = -415$ °C are physical parameters inherent to the glass, and $\gamma = 10\eta$.

Equation (2) is used to calculate γ during experiments, at 1050°C, the value of γ is 2.682×10^{14} Pa-s, with these dates and for h=60 nm, 90 nm, 120 nm, 150 nm , Eq. (2) can be used to describe ΔE as function of r, which is shown in Fig. 5. The trend shows that the overall gradient of the curve increases as membrane thickness increases, means the rate of shrinkage is slower. So, increasing the membrane thickness can provide much more control to the

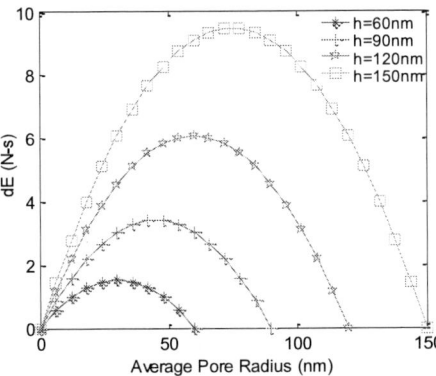

Figure 5: (Color online) Change in ΔE with respect to nanopore radius at 1050 °C based on differing values of pore membrane thickness h. The curve to the left of the maximum means the pore shrinking, while right expanding.

shrinking process.

Within this model the ratio of nanopore radius and membrane thickness is the important factor on whether the pore will shrink or expand. This experiment clearly demonstrates the existence of a "critical size": Pores below a certain size shrink, while larger ones expand. We estimate that for our experiment, the critical size is 90 ±20 nm, in good agreement with the model, considering the approximations made.

CONCLUSION

In summary, this article presents a new technique to fabricate and shrink nanopores in silicon dioxide membrane to sub-15 nm with a precision of 1 nm. The shrinking process is well-controlled and reproducible. Silicon dioxide membrane is softened at high temperature condition and the configuration is changed to find a lower surface free energy, such an effect is explained by surface-tension-driven model. Whether the nanopore shrinks or expands only based on ratio of pore size and membrane thickness. The technique here can be used to controllable shrink nanopore arrays in parallel and avoid hydrocarbon contamination surrounding the pore compared to TEM or FESEM shrinking method.

ACKNOWLEDGEMENTS

This work was partially supported by the Recruitment Program of Global Experts, China, and the 1000 Talents Plan of Sichuan Province, China.

REFERENCES

[1] B. Venkatesan and R. Bashir. *Nat. Nanotech.*, vol. 6, 2011, pp. 1-10.

[2] A. Storm, J. Chen, X. Ling, H. Zandbergen and C. Dekker. *J. Appl. Phys.*, vol. 98, 2005, pp. 014307.

[3] S. Liu, Q. Zhao, Q. Li, H. Zhang, L. You, J, Zhang and D. Yu. *Nanotechnology*, vol. 22, 2011, pp. 115302.

[4] A. Prabhu, K. Freedman, J. Robertson, Z. Nikolov, J. Kasianowicz and M. Kim. *Nanotechnology*, vol. 22, 2011, pp. 425302.

[5] P. Spinney, D. Howitt, R. Smith and S. Collins. *Nanotechnology*, vol. 21, 2010, pp. 375301.

[6] A. Andreozzi, L. Lamagna, G. seguini, M. Fanciulli, S. Schamm, C. Castro and M. Perego. *Nanotechnology*, vol. 22, 2011, pp. 335303.

[7] C. Danelon, C. Santschi, J. Brugger and H. Vogel. *Langmuir*, vol. 22, 2006, pp. 10711-10715.

[8] T. Deng, J. Chen, M. Li, Y. Wang, C. Zhao, Z. Zhang and Z. Liu. *Nanotechnology*, vol. 24, 2013, pp. 505303.

[9] J. Billo, J. Jones, W. Asghar, R, Carter and S. Iqbal. *Appl. Phys. Lett.*, vol. 100, 2012, pp. 233107.

[10] J. Chen, T. Deng, C. Wu and Z. Liu. *ECS Trans.*, Vol. 52, 2013, pp. 371-376.

A NOVEL DUAL-FREQUENCY TERAHERTZ ANTENNA IN STANDARD CMOS TECHNOLOGY

Jingyu Peng, Xiaoli Ji*, Xingxing Zhang, Yiming Liao and Feng Yan

Institute of the electronic Science and Engineering, Nanjing University, Nanjing, China

*Corresponding Author's Email: xji@nju.edu.cn

ABSTRACT

A dual-frequency antenna resonated at 0.32 THz and 0.65 THz has been analyzed and simulated for CMOS THz imaging and sensing systems. It includes the two rectified bowtie structures designed at metal and poly-Si layers in standard CMOS technology. Simulation results show that the antenna has the high gain and radiation efficiency and a great impedance matching comparing to the conventional metal dual-band antenna. The demonstrated design opens a brand new way for ease realization of multi-band on-chip THz antenna in CMOS technologies.

INTRODUCTION

Terahertz (THz) waves lie between the far infrared band and the micro-wave band. It has the characteristics of the small photon energies, the fine resolution, the large information carrying capacity and penetrating the non-polar compounds easily. In the application of CMOS THZ imaging, traditional one-band antenna had been widely designed [1,2], and dual-and multi-frequency antennas also have been designed [3,4]. However, all these designs are restricted to the low frequency regime and suffer from the poor radiation efficiency. Fig. 1 shows the typical geometrical configuration of the rectangular micro-strip patch for CMOS THz detector [5]. Through the change of the feed position, it can realize the dual-band resonation. However, the antenna exhibits the low peak gain and the span size between the designed two respond frequency could not adjust easily restricted to the area of the patch size. The implementation of the dual- or multi- band THz antenna is still the matter of research for CMOS THz detectors.

In this paper, we demonstrate a novel dual-band THz antenna structure responded at 0.65 THz and 0.32 THz stimulately. This novel antenna consists of radio wave and optical structures. Compared with the traditional dual- band antenna, the area size of antenna and input impedance matching difficulty were significantly reduced.

DESIGN OF DUAL - BAND ANTENNA

Fig. 2 shows the top (a) and cross (b) views of the designed antenna sturture. It consists of a radio wave part for 0.65 THz in yellow and an optical part for 0.32 THz in bule in Fig. 2 (a). The radio wave structure is designed by bowtie-shaped aluminum which is made of sixth metal layer in CMOS technology. The separation of the bowtie and the ground plane, as well as the thickness of the patch metallization are determined by the process given values. To match the impedance of the radio wave structure to the source and drain ports of MOSFET, the input resistance and reactance of MOSFET detectors are calculated as 162Ω and -448Ω respectively according to TCAD simulations. The structure size has been calculated and listed in Fig. 2 (b). The length of the structure is nearly equal to free space half wavelength at 0. 65 THz.

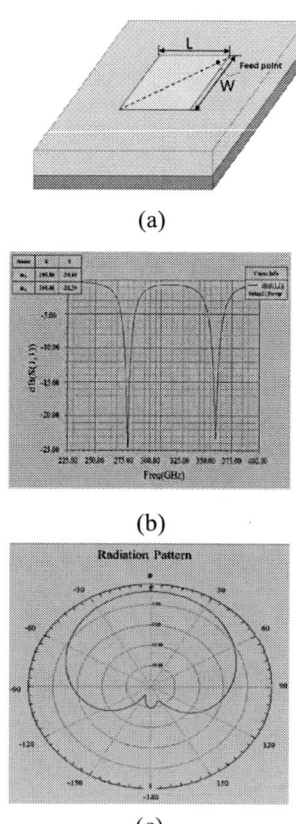

(a)

(b)

(c)

Fig. 1 (a) The typical the patch strucutre of two band antenna. The feed position moved to the angle of the patch planar enable two kinds syntony corresponding to

978-1-5090-6695-7/17 $31.00 © 2017 IEEE   508

the length and width of the patch planar. (b) Return loss and (c) radiation pattern of antennas near 0.3 THz.

The optical structure of the antenna for 0.32THz is designed at poly-Si layer using bowtie and rot mixed structure with a tunned dimentions, shown in Fig. 2 (a). When THz signals incident upon a Poly-Si materials, surface plasmon polaritons (SPPs) can be excited in accordance with the momentum conservation law. For poly-Si materials doping with 1×10^{20} / cm^3, the wavelength of estimated SPP for $f = 0.32$ THz is about 324μm, which is shorter than the half wavelength in the air.

The geometrical parameters of antenna have an influence on the optical structure performance. For obtaining the local field enhancement of the optimized optical and radio wave structures, we performed the theoretical simulation.

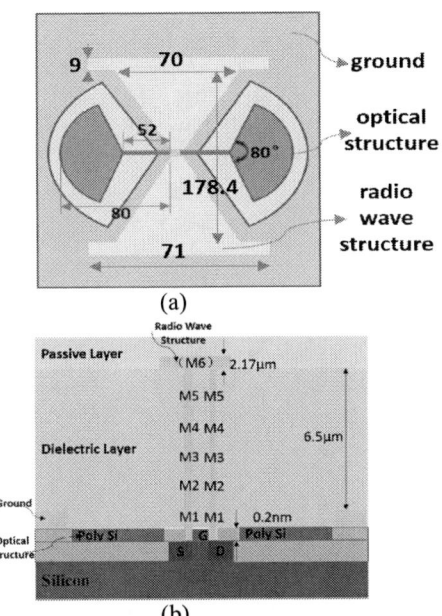

(a)

(b)

Fig. 2 (a) A top view of dual-band antennas response at 0.32 THZ and 0.65 THZ; (b) A cross view of the antenna based on 0.18 μm CMOS technology. The MOSFETs located at the center of the designed antenna, its source port and drain port are connected to the two side of the radio wave antenna by the metal vias, respectively.

SIMULATION RESULTS

We perform the High Frequency Structure Simulator (HFSS) calculation for the optimization of the antenna parameters. For radio wave part, lumped port is established in the middle of structure as the incentive source. Fig. 3 shows the return loss (a) and radiation patterns (b) including the peak gain at 0 degree and 90 degree with the parameter listed in Fig. 2 (a). It is clear that the resonance occurs at 0.65 THz. At this frequency,

the return loss is better than -30 dB. As shown in Fig. 3 (b), the antenna improves the peak gain at 0 degree to 6.5 dB for the designed structure.

The second resonance at 0.32 THz has been provided by optical structure. Its geometrical parameters have influence on the SPP excitation. By monitoring the SPP modes observed as maximum in electrical field, we can find the most proper structural parameters shown in Fig. 2 (a). Fig. 4 (a) shows the calculated electric field distribution (normalized to the incidence) in the gap of antenna with the parameter sizes. SPP resonance appears along the center edges of rods and its surroundings. Fig. 4 (b) shows the response frequency and the magnification of the center electric field strengthn near 0.32 THz. It is seen that the center electric field of the optical antenna is enlarged 257 times than the incident one, which indicates that the optical stucture can resonate at designed frequency.

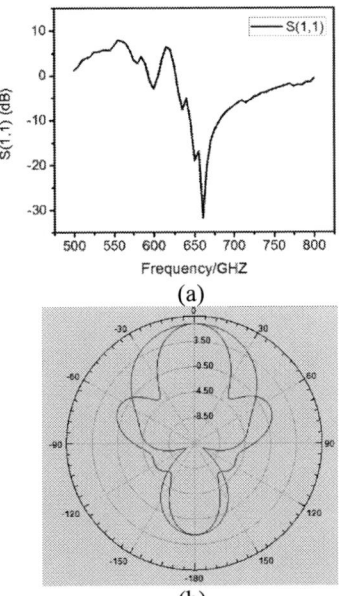

(a)

(b)

Fig. 3 (a) Return loss and (b) radiation pattern which include of peak gain at 0 degree and 90 degree for radio wave structure at 0.65 THz.

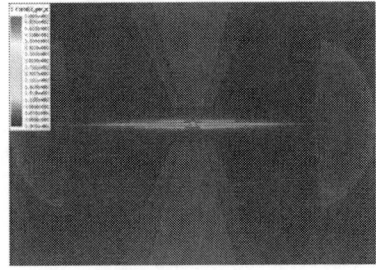

978-1-5090-6695-7/17 $31.00 © 2017 IEEE

(a)

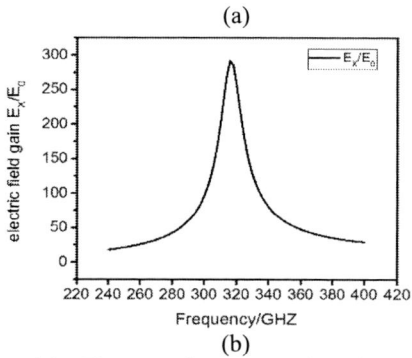

(b)

Fig.4 (a) The simulation result of the response magnification near the structure center at 0.32THz, (b) The frequency dependence of the response magnification at center location of MOSFETs.

CONCLUSION

In this paper, a dual-frequency antenna has been analyzed and simulated at the terahertz frequency regime for CMOS THz imaging and sensing systems. It includes the two bowtie structures designed at metal and poly-Si layers in standard CMOS technology. Simulation results shows that the antenna has the high gain and radiation efficiency and a great impedance matching comparing to the conventional metal dual-band antenna. The potential application of this antenna would find in terahertz surveillance system.

References

[1] E. Öjefors. "A 0.65 THz Focal-Plane Array in a Quarter-Micron CMOS Process Technology". *Journal of Solid-State Circuits*, vol. 44, No. 7, July, 2009, pp. 1968 – 1976.

[2] R. Hadi, H. Sherry, J. Grzyb and U. Pfriffer. "A 1 k-Pixel Video Camera for 0.7-1.1 Terahertz Imaging Applications in 65-nm CMOS". *IEEE Journal of Solid-State Circuits*, vol. 47, No. 12, pp. 2999-3012 (2012).

[3] Kumud Ranjan Jha and Ghanshyam Singh. "Dual-Frequency Terahertz Rectangular Microstrip Patch Antenna on Photonic Crystal Substrate". *Applied Electromagnetics Conference (AEMC)*, 2009, pp. 1-3.

[4] Kumud Ranjan Jha · G. Singh. "Dual-band rectangular microstrip patch antenna at terahertz frequency for surveillance system". *Journal of Computational Electronics*, 2010, pp. 31-41.

[5] Xiaoli Ji, Ying Zhou, Fuwei Wu, Feng Yan and Yiming Liao. "A dual-band terahertz antenna based on standard CMOS technology". Patent number: 2015102440960, 2015.

MULTI-SENSORY COMBINED INTEGRATED LOW POWER TUNABLE-GAIN INTERFACE CIRCUIT

Chun-Te Tung and Kuei-Ann Wen

National Chiao Tung University, Hsinchu 30010, Taiwan, R.O.C

Department of Electronic Engineering

Email: c.t.tung@ieee.org; twtstella@gmail.com

Abstract—This paper presents MEMS multi-sensors with low power tunable-gain interface circuit that can be monolithically integrated in the ASIC compatible standard CMOS process. A high gain ultra-low power sustaining TIA amplifier circuit with PLL compactly has been integrated with the resonator-based core sensing structure. The proposed low-power readout circuit adopts Correlated Double Sampling (CDS) to suppress low frequency noise and compensate DC offset. The gyroscope sensitivity is designed to be 1.8 aF/°/sec within ±100 °/sec. The tunable sensitivity can be adjusted from 28 mV/fF to 224 mV/fF by fully-differential programmable-gain amplifier (PGA). The interface circuit has 61.12dB SNR under 500 KHz sampling rate.

Index Terms—silicon resonator; CMOS MEMS sensor; readout; Gyroscope; Correlated Double Sampling; Tunable-Gain; Interface Circuit

I. INTRODUCTION

RESONATOR based MEMS sensors have been widely studied for various applications. Most of MEMS resonator-based sensors adopt MEMS processes incompatible with amplifier circuitry [1][2] and sensor readout. To reduce overall chip size, monolithic integration of mechanical structure with CMOS amplifier or readout circuitry has been reported [3-5]. In this work, a readout circuitry for single resonator-based sensing core is proposed for multiple sensing functions including temperature, pressure, gyro sensor and accelerometer under 0.18µm 1-poly-6-metal (1P6M) standard ASIC-MEMS process. The readout circuits widely used in MEMS sensor is switched-capacitor (SC) charge integration method [1]. The main drawback of the SC circuit is the related high kT/C noise with small feedback capacitor. While, the Correlated Double Sampling (CDS) technique has the advantage to significantly reduce the DC offset and low frequency noise. These noise and offset reduce techniques could be integrated for high performance capacitive sensing readout circuits [6] [7].

II. STANDARD ASIC/MEMS FABRICATION

The ASIC-MEMS process starts with the standard ASIC 1P6M CMOS process. After the readout circuitry and MEMS structures fabricated under the process, two etching steps are followed to obtain the final MEMS sensors. As shown in Fig. 1, MEMS structure is etched by anisotropic oxide etching (DRIE) and followed by silicon substrate isotropic etching to release the structure.

The proposed resonator-based sensing structure compatible to the process consists of two comb-finger structures for driving

and sensing electrodes with movable shutter structure for mechanical resonance and gyroscope, as shown in Fig. 2. The four symmetrical springs are connected to individual anchors through two Clamp-Clamp beams for compensation of the curvature of long shutter due to residual stress.

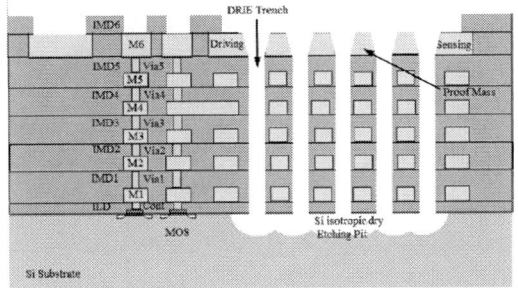

Fig.1: Cross-section view of ASIC/MEMS process

Fig. 2. The Comb finger structured gyroscope and resonator and implemented in the standard 0.18µm 1P6M ASIC/MEMS process

III. RESONATOR-BASED SENSING CORE UNDER STANDARD ASIC/MEMS FABRICATION

The structure based resonators shown in Fig.3 reveals that the stiffness constant of the resonator correlates the environment temperature. To measure the stiffness constant change, a PLL is used to provide a sustaining loop to drive the resonator and to track the resonant frequency shifts due to variation of the stiffness constant. The temperature can therefore be estimated from the control voltages of the PLL VCO. Figure 4 shows the measured linearity between the ambient temperatures based on the structure illustrated in Fig.4. Linearity between the resonant frequency and temperature is measured to be - 6.58Hz/K.

Based on the similar core structure, the resonator based pressure sensor has been studied. Fig. 5 shows the measurement results of the quality factor, displacement and the air pressure. With the proposed structure, the displacement of the sensing fingers are proportional to the Q factor and thus are inversely

978-1-5090-6695-7/17 $31.00 © 2017 IEEE

proportional to the air pressure [8][9]. The air pressure changes from 78 to 1600 Pa, Q factor will change from 1042 to 256, and the displacement will be from 11μm to 2μm.With the transducer designed, the displacement of the sensing figures are proportional to the Q factor and thus are proportional to the air pressure.

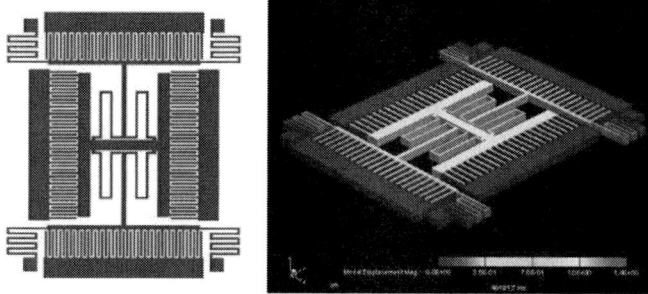

Fig. 3. The dual function resonator for acceleroeter and gyro sensor

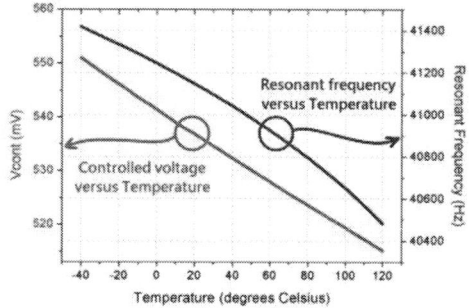

Fig. 4.The linearity between temperature and resonant frequency and VCO control voltage

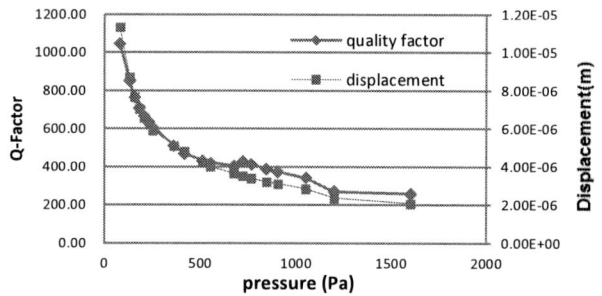

Fig. 5. The relation between ambient pressure and resonator characteristics

Fig.6 shows the dual function resonator and schematic of the PLL. The CMOS PLL clock synthesizer targets for sensor applications that require low power consumption. Based on the integer-N architecture, the PLL clock synthesizer produces 40 kHz output signal to drive the MEMS resonator. The loop has 101.06 Hz bandwidth with phase margin 53.7°. The charge pump output current is 4.8nA, and the gain of the voltage controlled oscillator is 597.456 kHz/V.

A sustaining TIA amplifier is required to convert the current induced by the resonator finger for input voltage swing of the PLL. The output current induced by the MEMS resonator can be calculated [4]

$$i_o \approx V_{dc}\frac{\partial c_s}{\partial t} + C_p\frac{\partial v_{ac}}{\partial t} \approx V_{dc}\frac{\partial C}{\partial t} = V_{dc}\frac{\partial C}{\partial x}f\Delta x \qquad (1)$$

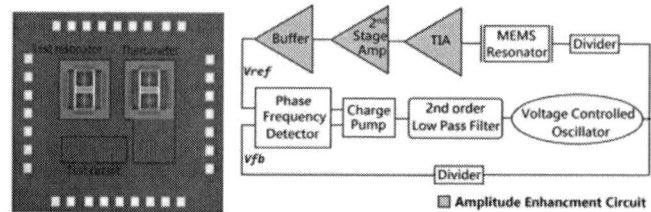

Fig. 6. The dual function resonator and PLL schematic

The sustaining amplifier circuit for the resonator adopts capacitive-feedback amplifier structure [10]. The trans-impedance DC gain of the designed TIA is 108.4MΩ with 221-kHz bandwidth in order to compensate high insertion loss of the CMOS MEMS resonator. The trans-impedance gain provides 50mV output swing for PLL input.

Fig. 8 shows the simulation result of the multi-sensory readout circuit. The PLL circuitry provides a stable clock to drive the resonator. The output resonant frequency F_{ref} of the resonator shifts when the environment temperature changes. The PLL will track the F_{ref} and the VCO control voltage changes accordingly. The micro-controller may obtain the temperature information in real-time from the control voltages of the VCO.

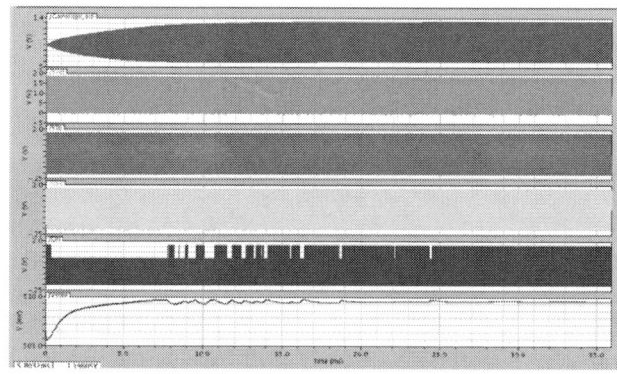

Fig. 8. The co-simulation result of PLL and the sensor model.

IV. CMOS MEMS GYROSCOPE

A. Architecture of Sensor

The 3-axis CMOS MEMS gyroscope is illustrated in Fig.7. It includes mass, spring and comb finger design. The size of 3-asix CMOS-MEMS gyroscope is 1.5 × 1.5 mm^2

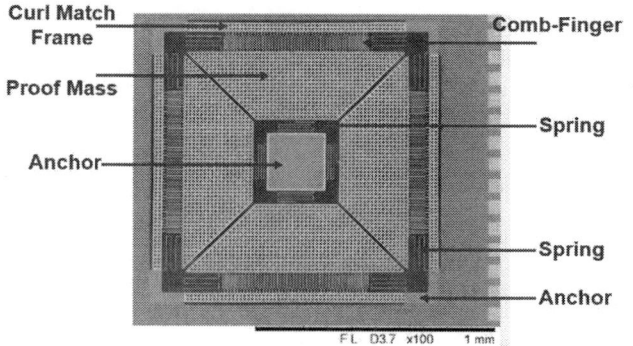

Fig. 7. Sensor Architecture

978-1-5090-6695-7/17 $31.00 © 2017 IEEE

B. Sensor Operation due to Coriolis Force

The operation of 3-axis gyroscope is shows in Fig.8. When gyroscope suffered the angular velocity (blue-line) on driving direction (red-line), then the proof mass would generate displacement (green-line) due to Coriolis Effect

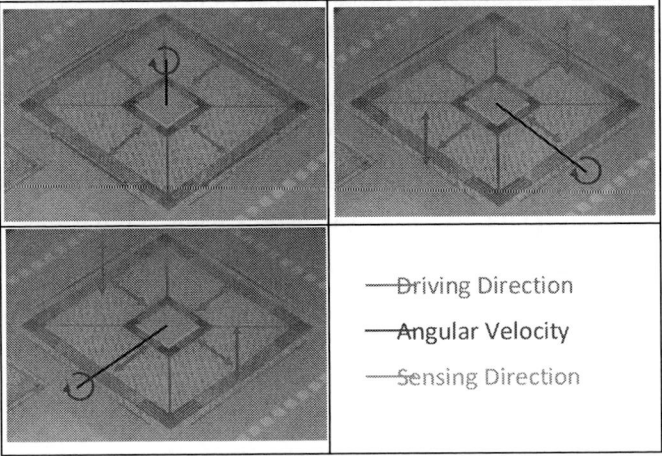

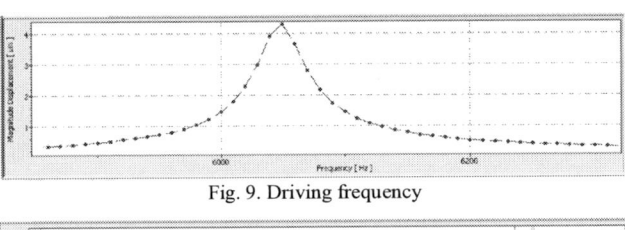

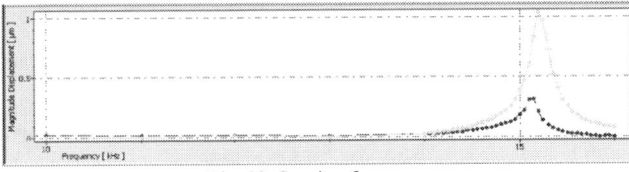

Fig. 8. Gyroscope operations

C. Driving and Sensing Frequency

The gyroscope driving frequency is designed to be 6.1 KHz and sensing frequencies are designed in 15.3 KHz and 15.4 KHz. The measurement results are shown in Fig.9 and Fig.10.

Fig. 9. Driving frequency

Fig. 10. Sensing frequency

D. Sensitivity and Linear Range

The linear range of gyroscope sensing is ±100 °/sec, the total sensing capacitance is 400fF and the sensitivity caused by Coriolis force is 1.8aF/°/sec, the simulation result for 100°/sec and the simulation results is show in Fig.11

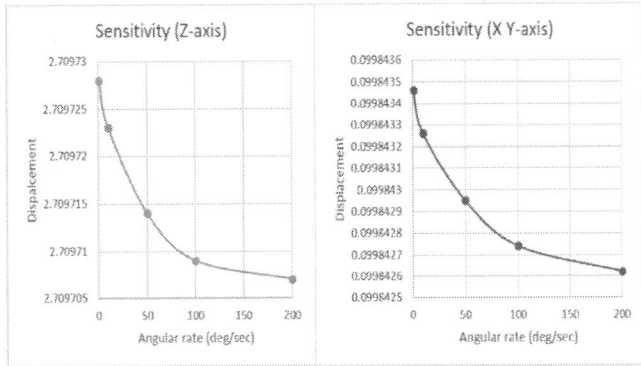

Fig.11 Sensitivity of gyroscope

V. INTERFACE READOUT CIRCUIT

A. Interface Circuit Design

The readout circuit is designed in the fully-differential configuration. The schematic of the circuit is shown in Fig. 12. The C/V converter acts as a differential charge integrator [11], converting the capacitance changes in a capacitive full-bridge caused by proof mass deflections to a voltage output. The pre-amplifier is an operational trans-conductance amplifier with capacitive feedback. It offers process-insensitive additional gain to amplify the small signal variation from the C/V output. To cope with the severe flicker noise of OTAs in the C/V converter and the pre-amplifier, the holding capacitor C_{CDS} and some extra switches are configured for passive CDS. This eliminates the flicker noise and DC offset voltage of OTAs. In order to maintain flicker noise cancellation, the CDS clock rate must be higher than the corner frequency of flicker noise. The CDS clock rate is set to 500 KHz. A higher CDS clock rate helps further reducing flicker noise but also it will increase the OTA power consumption. The post-sim spectrum of interface circuit is listed in Fig. 13.

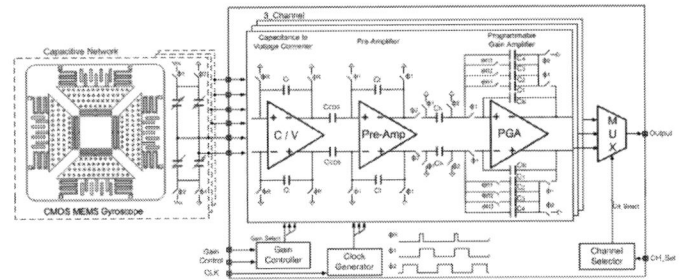

Fig. 12. Architecture of interface circuit

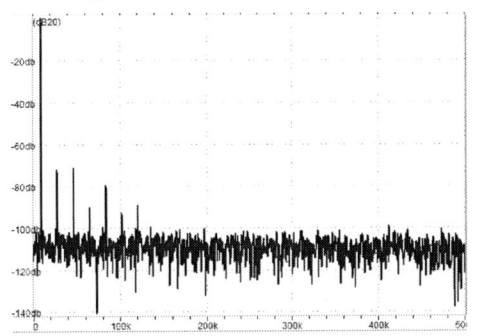

Fig. 13. Spectrum of interface circuit

B. Correlated Double Sampling (CDS)

In order to provide large sensitivity, the readout circuits need to provide a larger capacitance-to-voltage (C/V) conversion gain compared to that of non-CMOS-MEMS accelerometers. Large gain amplifies noise and degrades the output signal to-noise ratio (SNR). Another degradation comes from the flicker noise due to the random trapping and de-trapping processes of charges in the oxide traps near the silicon/silicon dioxide interface. Flicker noise will be severe in CMOS devices when operated in low frequency [12]. To cope with low-frequency flicker noise, correlated double sampling (CDS) techniques is adopted for noise and offset reduction [13]-[15].

978-1-5090-6695-7/17 $31.00 © 2017 IEEE

C. Programmable Gain Amplifier(PGA)

To provide high dynamic range support, a PGA is cascaded after the pre-amplifier to provide 1× to 8× voltage gain. The schematic of the circuit is shown in Fig. 14 and the PGA also acts as a low-pass filter [16]. The −3dB frequency of this low-pass filter is approximated as:

$$f_{-3bB} = f_s \cdot C_{fb} / 2\pi \cdot C_{fil} \qquad (2)$$

where fs is the sampling frequency of the readout circuit, C_{fil} is the feedback capacitance of the PGA as shown in Fig. 14, and C_{fb} denotes the total equivalent capacitance in the feedback path, which is a combination of C_1, C_2, C_3, and C_4 and depends on the PGA gain.

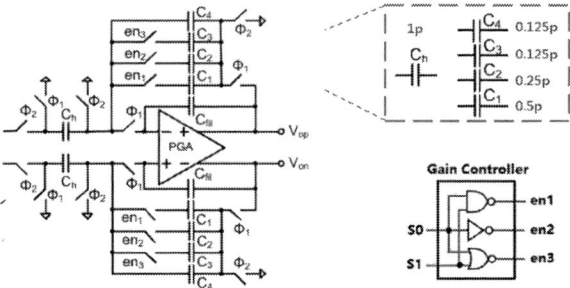

Fig. 14. Schematic of the PGA

D. Experiment Result

The chips were wire-bounded in ceramic dual-in-line packages (DIPs) and mounted on a PCB for chip tests. Fig. 115 shows the timing waveform of the PGA differential output.

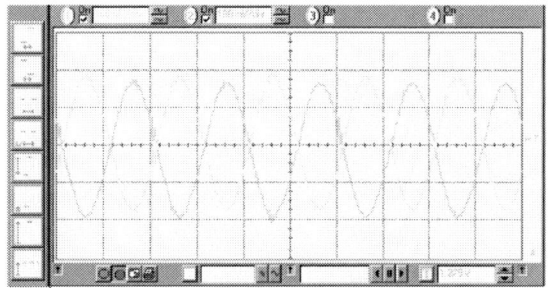

Fig. 15. Measured timing waveform of circuit output

Fig. 16 shows the output spectrum of the PGA differential output. When the sampling frequency fs is 500 kHz. Higher sampling frequency can decrease output thermal noise floor but increase power consumption of readout circuits.

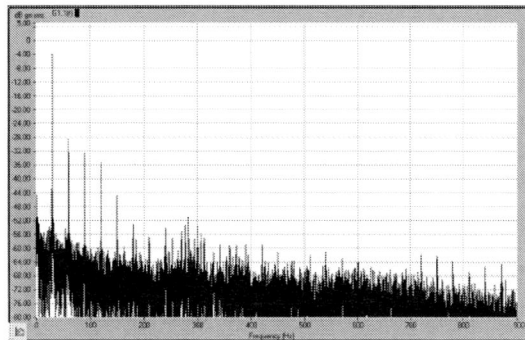

Fig. 16. Measured output spectrum

VI. CONCLUSION

This paper provides an effective solution for single chip sensor SoC design and wafer-level 0.18-μm CMOS MEMS process is suitable for integrated inertial sensors and presents the low power interface readout circuitry of the single MEMS sensing core that can be manufactured and monolithically integrated in the ASIC compatible standard CMOS process with low power dissipation, low noise and low cost.

ACKNOWLEDGMENT

This design was sponsored in part by the National Science Council of Taiwan under a grant of MOST 104-3115-E-009-022. The authors appreciate the National Chip Implementation Center (CIC), Taiwan, for supporting the chip manufacturing.

REFERENCES

[1] D. E. Serrano, R. Tabrizian, and F. Ayazi, "Tunable piezoelectric MEMS resonators for real-time clock," Proc. IEEE FCS 2011, pp. 1-4.

[2] B. Razavi, "A 622 Mb/s 4.5 pA//spl radic/Hz CMOS transimpedance amplifier," ISSCC. 2000, pp.162-163.

[3] C.T.-C. Nguyen and R.T. Howe, "An integrated CMOS micromechanical resonator high-Q oscillator," IEEE J. Solid-State Circuits, vol. 34, 1999, pp. 440-455.

[4] W.-C. Chen and W. Fang, S.-S. Li, "A generalized CMOS-MEMS platform for micromechanical resonators monolithically integrated with circuits," J. Micromech. Microeng., vol. 21, no. 6, 2011, pp. 065012.

[5] H. G. Barrow, R.A. Schneider, T.O. Rocheleau, V. Yeh, Ren Zeying and C.T.-C. Nguyen, "A real-time 32.768-kHz clock oscillator using a 0.0154-mm² micromechanical resonator frequency-setting element," Proc. IEEE FCS 2012, pp. 1-6.

[6] M. Paavola, et al., "A micropower interface ASIC for a capacitive 3- axis micro-accelerometer," Ieee Journal of Solid-State Circuits, vol. 42, pp. 2651-2665, Dec 2007.

[7] M. Paavola, et al., "A Micropower Delta Sigma-Based Interface ASIC for a Capacitive 3-Axis Micro-Accelerometer," Ieee Journal of SolidState Circuits, vol. 44, pp. 3193-3210, Nov 2009.

[8] Li, Q.; Goosen, J.F.L.; van Keulen, F.; van Beek, J.T.M., "Gas ambient dependence of quality factor in MEMS resonators," in Sensors, 2009 IEEE , vol., no., pp.1040-1043, 25-28 Oct. 2009

[9] F. R. Blom, S. Bouwstra, M. Elwenspoek, J. H. J. Fluitman,"Dependence of the quality factor of micromachined silicon beam resonators on pressure and geometry", J. Vac. Sci. Technol. B, vol. 10, Issue 1, pp. 19-26, Jan 1992.

[10] B. Razavi, "A 622 Mb/s 4.5 pA//spl radic/Hz CMOS transimpedance amplifier," ISSCC. 2000, pp.162-163.

[11] M. Lemkin, e al.,"A three-axis micromachined accelerometer with a CMOS position-sense interface and digital offset-trim electronics," IEEE J. Solid-State Circuits, vol. 34, no. 4, pp. 456–468, Apr. 1999.

[12] Y. Nemirovsky, "1/f noise in CMOS transistors for analog applications," IEEE Trans. Electron Devices, vol. 48, no. 5, pp. 921–927, May 2001.

[13] V. P. Petkov and B. E. Boser, "A fourth-order ΔΣ interface for micromachined inertial sensors," IEEE J. Solid-State Circuits, vol. 40, no. 8, pp. 1602–1609, Aug. 2005.

[14] L. He and M. Palaniapan, "A CMOS readout circuit for SOI resonant accelerometer with 4-μg bias stability and 20-μg/√Hz resolution," IEEE J. Solid-State Circuits, vol. 43, no. 6, pp. 1480–1490, Jun. 2008.

[15] M. Paavola,"A micropower interface ASIC for a capacitive 3-axis micro-accelerometer," IEEE J. Solid-State Circuits, vol. 42, no. 12, pp. 2651–2665, Dec. 2007.

[16] M. Schipani, P. Bruschi, G. C. Tripoli, and T. Ungaretti, "A low power CMOS interface circuit for three-axis integrated accelerometers," in Proc. PRIME, Bordeaux, France, 2007, pp. 117–120.

TSV INDUCTOR OPTIMIZATION AND ITS DESIGN IMPLICATION

Cheng Zhuo[1]* and Baixin Chen[2]

[1]College of Information Science & Electronic Engineering, Zhejiang University, Hangzhou 310027, China
[2]School of Microelectronics, Xidian University, Xian 710126, China
*Corresponding Author's Email: czhuo@zju.edu.cn

ABSTRACT

Due to the relatively slow scaling of packaging technology, on chip inductor has shown promising potential to enable more compact design and smaller parasitics for inductor-central designs, such as voltage regulator, resonant clocking, filter, etc. However, conventional on-chip spiral inductor has to be placed on the top few metal layers, thereby consuming significant routing area for global interconnects. Moreover, it typically needs more dedicated shielding to prevent unnecessary coupling, and further increases its occupied area. With the popularity of 2.5D and 3D design architecture, TSVs have been widely used, a significant portion of which are dummy and placed for thermal/DFM/reliability purposes. Thus, TSV inductor has been proposed to utilize those dummy TSVs to form the on-chip inductor for 2.5D/3D designs, with different characteristics from the conventional spiral inductor. This paper first reviews the concept of TSV inductor, investigates its physics and then discusses the compact models to enable more efficient design and trade off. Based on that, we study its design impacts on signal integrity and application for filter designs.

INTRODUCTION

With the technology advancement, three-dimensional (3D) integrated circuits (ICs) have received more and more attentions. On one hand, 3D IC enables higher integration density to mitigate the challenges in technology scaling and continue the Moore's Law at Post-Moore era. On the other hand, the use of vertical Through-Silicon-Vias (TSVs) reduces the global interconnect and hence saves the energy for data transfer. [1] However, due to the demands from manufacturability and thermal concerns, particular amount of TSVs need to be placed within certain area, leaving a lot of dummy and thermal TSVs in 3D ICs. Thus, it remains one of the most important research topic to effectively place and allocate dummy and thermal TSVs in 3D IC design.

Unlike investigating the TSV allocation from physical design perspective, there are also a few works exploring the possibility of constructing important circuit elements with those TSVs. One essential on-die circuit element is inductor. With the popularity of integrated voltage regulator and filter, the use of inductor in circuits is more common than ever. However, package or PCB inductor do not scale well and hence can hardly fit increasingly smaller die area at sub-22nm regime. Thus, it is more crucial than ever to develop on-chip inductor technique with good scalability. In tradition, the implementation of on-chip inductors uses multi-turn planar spiral structure. This structure demands a significant area and global routing resources, causing challenges for power and signal routes on the top metal layers. Moreover, it requires additional measures to reduce the coupling effect, such as patterned ground shield, and further increases the area cost. The development of the 3D on-chip TSV based inductor may help mitigate the aforementioned problems and hence enable designs with more compact area and smaller parasitics. In prior works, the use of dummy TSVs can create very compact on-chip inductor to achieve 3X area saving compared with the conventional spiral inductor with similar performance.

In this paper, we will first review the existing TSV inductor architectures, and then discuss the compact model for several key measures of on-chip inductor design. After that, we will further investigate its impact on the signal integrity and its application in filter design.

BACKGROUND AND MOTIVATION

In general, unlike PCB/package inductor, on-chip inductors are with the advantages of lower area cost, smaller power consumption, higher reliability, accuracy and design flexibility. On the other hand, TSV based three-dimensional integrated circuits (3D ICs) are considered to be a promising alternative by making use of the vertical dimension for higher integration density, shorter wire length, smaller footprint, and lower power consumption [1, 7].

However, what is TSV inductor? One example of a toroidal TSV inductor in a two-tier 3-D IC is shown in Fig. 1. It utilizes the TSVs and thick routing metals (on top metal layer or RDL layer) to form the inductor loop. With multiple TSVs, multiple vertical loops can be constructed and connected. The advantages of such an inductor are the minimal footprint on routing layers and accordingly higher inductance density. In this paper, on the basis of [2], we will investigate tradeoffs among several key measures of on-chip inductor design to enable more efficient circuit design, and a few guidelines from signal perspective for a more robust design.

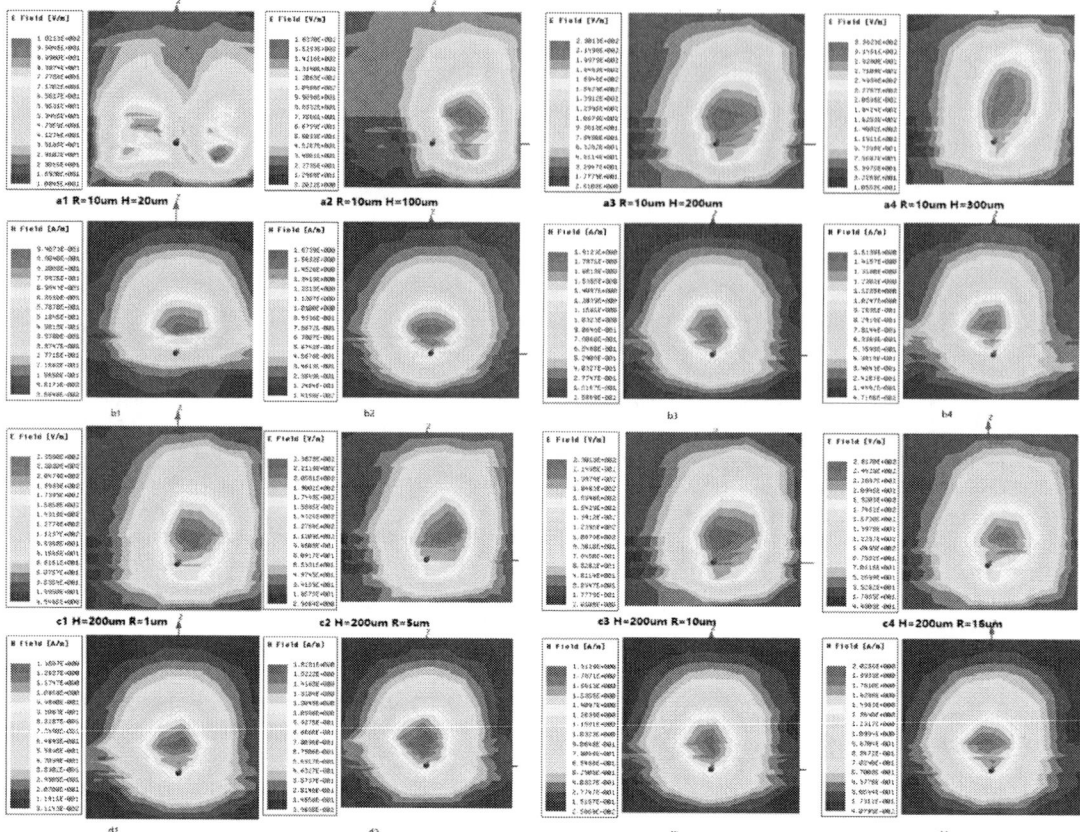

Fig. 3: E field and H field v.s. TSV height and TSV radius

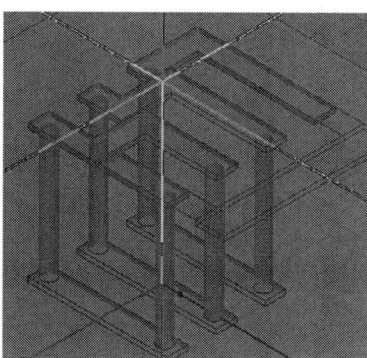

Fig. 1: 3D TSV inductor

PHYSICS BEHIND TSV INDUCTOR

The properties of an inductor are highly associated with the magnetic field it creates. The magnetic field intensity is represented by the magnetic flux density (B) as in Eq (1), where μ is absolute magnetic permeability:

$$B = \mu \cdot H \qquad (1)$$

Similar as capacitors storing electric charge, an inductor stores magnetic energy within the core of its windings where the flux density is greatest. Its inductance is in general a function of its geometric shape, and can be determined by the ratio of flux linkages to the current that creates the magnetic flux as in Eq(2) for the N-Loop solenoid in Fig. 2 [3].

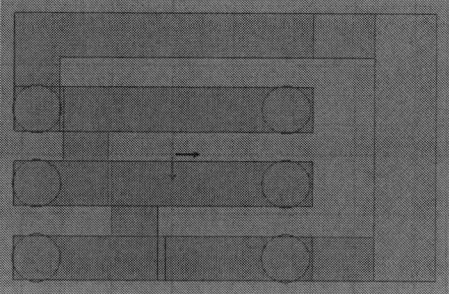

Fig. 2: Three-loop TSV inductor top view

$$L = \frac{\lambda}{I} = \frac{\Phi_T \cdot N}{I} = \frac{\mu \cdot N^2 \cdot \pi \cdot a^2}{h} \propto N^2 \qquad (2)$$

It is important for us to further investigate and understand the various factors that affect TSV inductor performance. In order to have a perceptual understanding of various parameters, we, here, will use TSV inductor height (h) and radius (r) as an example to demonstrate

978-1-5090-6695-7/17 $31.00 © 2017 IEEE

their impacts on the TSV inductor cross section' electric field and magnetic field, as simulated by a full-wave high-frequency structural simulator [4]. As shown in Fig. 3, we can directly observe the electric and magnetic fields change *w.r.t.* h and r. For example, with an increased h, the electric field and the magnetic field happen to be stronger. The field contours also changed accordingly, from two peaks to one peak, with growing central region. However, it is difficult to directly apply 3D solver for circuit design and optimize for different circuit metrics. In the next section, we will present a compact model for the inductor key physical parameters, pitch (p), radius (r), and number of loops (N), to enable more efficient circuit design.

OPTIMIZATION FOR TSV-INDUCTOR

In this section, we will study how various design parameters impact the inductance (L), quality factor (Q), and AC resistance (R) of the TSV inductor. All the simulations in this paper are conducted with a full-wave high-frequency structural simulator [5] with mixed order basis function. For clarity, we outline the key parameters of the study in Table I, as most other parameters are determined by process and difficult to be changed at circuit design stage. The ranges for the parameters of interest are also included.

There are a few key design constraints that we apply in our design: First, p is the pitch distance between the two TSVs. Second, in order to achieve maximum quality factor, the cross-sectional area should be square. Third, the operating frequency of interest is 1GHz. Thus, in our following discussion, the variables for optimization are only p, r, and N.

Table I: List of parameters and their ranges of interest

Type	Notation	Definition	Range
Process	h(um)	Substrate height	20-300
	r(um)	TSV radius	1-16
Design	N	Number of turns	3-6
	p(um)	Loop pitch	35-105

The quality factor and inductance for different TSV radius, pitch, and N are shown in Fig. 4. We first explore their relationships with each individual parameter by assuming the rest of parameters are constant. While the metrics are shown to have exponential or multi-order dependence with r and p, their relationships with N is very consistent and similar to spiral inductor. By sweeping p, r and N with the 3D-solver, we can fit the following compact model to the simulated data:

$$L(r,p,N)= p_{L3}\cdot r^{p_{L1}}\cdot p^{p_{L2}}\cdot N+ p_{L4}\cdot r^{p_{L1}}\cdot N+ \\ p_{L5}\cdot p^{p_{L2}}\cdot N+ p_{L6}\cdot N+ p_{L7} \quad (3)$$

where p_{L1}=-0.1508, p_{L2}=-1.475, p_{L3}=3.304, p_{L4}=0.6647, p_{L5}=22.04, p_{L6}=0.08547, p_{L7}=-0.3703. The fitting goodness for N=3 is shown in Fig. 5, and Fig. 6 with

average relative error around 0.97%.

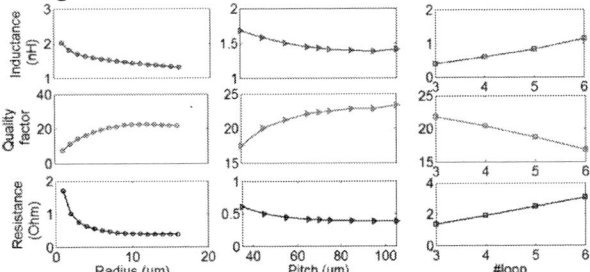

Fig. 4: Simulated inductor/quality factor/resistance w.r.t. r, p and N

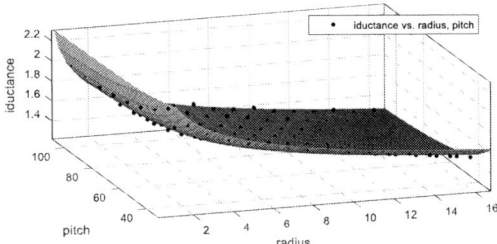

Fig. 5: Inductor v.s. radius and pitch

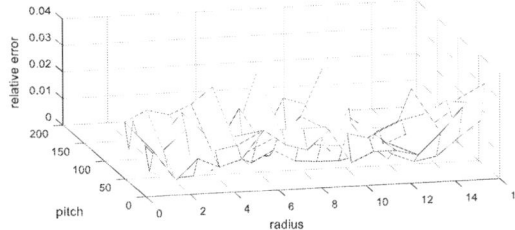

Fig. 6: Inductor relative error v.s. radius and pitch

Similarly, we can achieve the following compact model for Q:

$$Q(r,p,N)= p_{Q3}\cdot r^{p_{Q1}}\cdot p^{p_{Q2}}\cdot N+ p_{Q4}\cdot r^{p_{Q1}}\cdot N+ \\ p_{Q5}\cdot p^{p_{Q2}}\cdot N+ p_{Q6}\cdot N+ p_{Q7} \quad (4)$$

where p_{Q1}=-0.3470, p_{Q2}=-2.325, p_{Q3}=2.432e+04, p_{Q4}=-10.20, p_{Q5}=-1.904e+04, p_{Q6}=3.17, p_{Q7}=26.89, with an average error of 5.26%;
and R:

$$R(r,p,N)= p_{R3}\cdot r^{p_{R1}}\cdot p^{p_{R2}}\cdot N+ p_{R4}\cdot r^{p_{R1}}\cdot N+ \\ p_{R5}\cdot p^{p_{R2}}\cdot N+ p_{R6}\cdot N+ p_{R7} \quad (5)$$

where p_{R1}=-1.012, p_{R2}=-2.462, p_{R3}=-390, p_{R4}=-0.4913, p_{R5}=590, p_{R6}=0.2051, p_{R7}=-0.3911, with an average error of 5.61%.

With the aforementioned compact models and design specs, we can easily formulate optimization problem for inductor design and make reasonable tradeoffs among area, quality factor and inductor performance.

TSV-INDUCTOR'S IMPACT ON SIGNAL INTEGRITY AND ITS IMPLICATION TO DESIGNS

A. Signal Integrity Analysis

Fig. 7 and Fig.8 show the S parameters for the near-end and far-end couplings with the victim interconnect at different location on the same plane as the spiral inductor (or the thick metal wires of TSV-inductor). The two inductors have similar inductance and quality factor. However, due to the vertical structure of TSV-inductor, it showed slightly better immunity to coupling.

Fig. 9 shows the coupling of the TSV-inductor with vertical TSVs, which has stronger coupling compared with metal wires and need to be taken care of during the design stage to avoid unnecessary coupling.

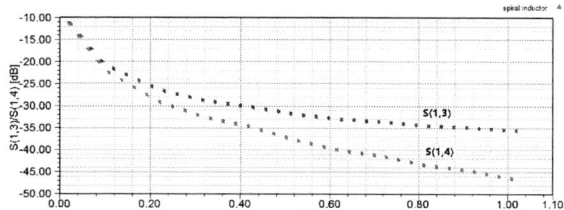

Fig. 7: Spiral inductor coupling with metal wire

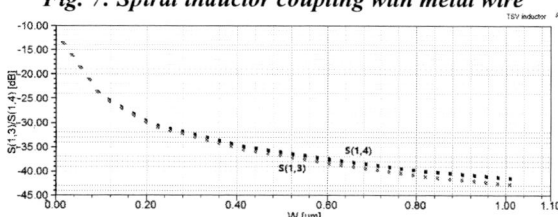

Fig. 8: TSV inductor coupling with metal wire

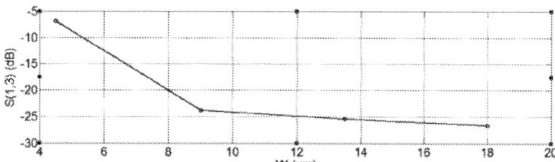

Fig. 9: TSV inductor coupling with TSV

B. Design of Butterworth Filter and Design Implication

This section we present the use of TSV-inductor for a Butterworth filter as a design example. An ideal reference design can be created through the ADS Design-Guide with $A_p = 1$db, $A_s = 25$db, $F_p = 3$GHz, $F_s = 8$GHz [5]. For the inductors used in the ideal filter, we can apply the compact models in the previous section to design the TSV-inductor, and finally replace the ideal inductors in the inductor with TSV-inductors. Fig. 10 demonstrates the differences between the ideal filter and TSV-inductor based filter.

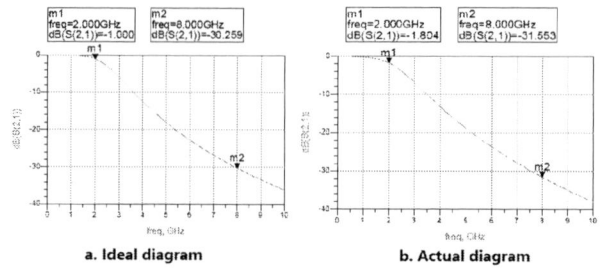

a. Ideal diagram   b. Actual diagram

Fig. 10: Frequency response for ideal and TSV-inductor based filter designs

As can be seen from the figure, the performance of the filter almost remained the same for the two designs, with about 3% degradation at 8GHz. Thus, the TSV inductor is applicable for small inductor (1-2nH) based designs with smaller area, better immunity to coupling, higher reliability and larger design flexibility. However, due to substrate loss, its quality factor may drop faster than the spiral inductor and need to be limited to specific applications [6, 7].

CONCLUSION

The paper introduces the concept of TSV-inductor and briefly discusses its physical nature. The compact models are built based on the 3D solver characterization data, which demonstrates the changes of physical parameters on the inductor metrics. The coupling analysis and comparison with spiral inductors show that TSV-inductor may have better immunity to coupling. Finally, a filter design with the TSV-inductor is presented showing comparable performance as the reference design.

REFERENCES

[1] Tida, U. R., Yang, R., Zhuo, C., & Shi, Y. On the efficacy of through-silicon-via inductors. *IEEE Transactions on Very Large Scale Integration Systems, 23*(7), 2014.

[2] Tida, U. R., Zhuo, C., & Shi, Y. Through-silicon-via inductor: Is it real or just a fantasy? *Asia and South Pacific Design Automation Conference*, pp. 837-842, 2014.

[3] Vanackern, G., Design Guide for CMOS Process On-Chip 3D Inductor using Thru-Wafer Vias. *Boise State University Theses and Dissertations*, 2011.

[4] HFSS. [Online]. Available: http://www.ansys.com/.

[5] ADS. [Online]. Available: http://adsabs.harvard.edu/.

[6] Franzon, P., Davis, W., & Thorolfsson, T. Creating 3D Specific Systems: Architecture, Design and CAD. *Design, Automation & Test in Europe Conference & Exhibition*, pp. 1684-1688, 2010.

[7] Tida, U. R., Zhuo, C., & Shi, Y., Novel Through-Silicon-Via Inductor-Based On-Chip DC-DC Converter Designs in 3D ICs, *ACM Journal on Emerging Technologies in Computing Systems*, 11(2), 2014.

ENERGY EFFICIENT SOC POWER DELIVERY USING FULLY-INTEGRATED VOLTAGE REGULATORS WITH HIGH-FREQUENCY SWITCH CONTROL

Boping Wu

Intel Research, Hillsboro, OR, USA

Email: bennywu@ieee.org

ABSTRACT

Intel® introduced an energy efficient SoC power delivery scheme utilizing fully-integrated high-frequency voltage regulators along the roadmap of Moore's law scaling. From 22nm process to 14nm or even 10nm, circuit blocks shrink and the embedded passives are scaled sequentially in the similar manner. A major challenge in the on-die VR design is to achieve sufficient integration and minimization of the required components, while still maintaining high power efficiency and multi-phase switching capability. This allows SoC to continue delivering a compelling power performance benefit to support the scaling process. In this paper, the optimized performance metrics of the silicon integrations are presented with measured implications and correlated simulations. The new generation microprocessor is demonstrated to be powered by a highly configurable VR solution of wide voltage and frequency range that facilitates potentially 50% more energy saving and peak available power increase.

Keywords—integrated voltage regulator; embedded passives; system-on-chip; silicon integration; PDN; energy efficiency; power supply; DC-DC conversion; buck converter; system-in-package

INTRODUCTION

Modern microprocessors [1-3] use integrated Voltage Regulators (VR) which are high-frequency multi-phase synchronous switching DC-DC converters with phase shedding and LC filters. On-die VR, as with other buck converters, includes power transistors, compensation and control circuitry, all of which are integrated on the System-on-Chip (SoC) [4-6]. This enables high integration thereby allowing for small form factors while increasing battery life through Dynamic Voltage and Frequency Scaling (DVFS) techniques [7]. Digital controllers [8] provide scalability, synthesizability, programmability and enable seamless mode hopping and help create fine grain voltage domains to lower size and power profiles of SoC. A very high bandwidth controller with light load frequency modulation preserves the best efficiency and also prevents large droops from high di/dt load events [9]. The controller monitors the feedback and compares it with the internally generated voltage reference. Compensated error amplifier is used with Pulse Width Modulation (PWM) signals to control the duty cycle of the regulators and operates the power stage in the frequency range of 20MHz-200MHz. The converter also features current limit protection to avoid over-current in each phase and zero current detection in the event of low load Discontinuous Conduction Mode (DCM) operation [10]. The passives at the output of the buck help mitigate the voltage droop and noise ripple [11-12]. The buck converter provides high efficiency while the Low Drop-Out regulator (LDO) provides fine-grained spatiotemporal power management [13] as well as good Power Supply Rejection (PSR) for the IO drivers [14]. The ultimate purpose of using on-die VR are increased burst power performance, improved transient response and reduced power consumption [15-19].

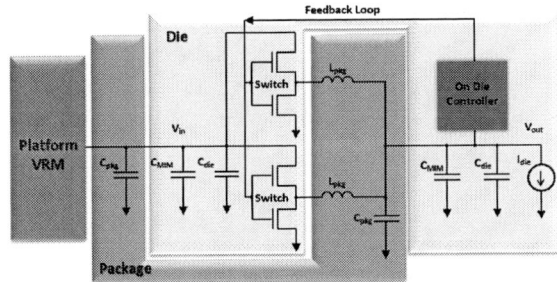

Figure 1: Block diagram of an on-die VR implementation

ARCHITECTURE AND DESIGN

A first stage VR is on the motherboard which converts from Voltage Regulator Module (VRM) or battery voltage. The second conversion stage is comprised of multiple VR units, which are synchronous multiphase buck converters. The simplified block diagram of Power Delivery Network (PDN) along with the partitioning of components between the die, package and motherboard, is shown in Fig. 1. By using a buck conversion topology, the on-die VR allows the usage of one single input domain to feed a large number and wide range of derived high-current power domain, while maintaining the efficiency. The phase number is generally optimized for a particular output domain with the consideration the efficiency, load current, allowed ripple and droop at the output. Each power domain of the die is supplied using a dedicated VR unit. Each unit is independently programmed to achieve optimal operation given the

978-1-5090-6695-7/17 $31.00 © 2017 IEEE

requirements of the domain it is powering. The settings are optimized by the Power Control Unit (PCU), which specifies the input voltage, output voltage, number of operating phases, and a variety of other settings to minimize the total power consumption of the die. This allows for per-core voltage control, and fine grained power management of different IP blocks such as for the ring, PLLs and IO. The feedback loop is designed using a type-III op-amp compensator on the die. The inductor of the LC filter of the DC-DC converter is implemented using the Air Core Inductors (ACI) [20] in package. Fig.2 gives one 3D illustration of the L structures in package. Other potential alternatives are Magnetic Core Inductor (MCI) [21-25] and possibly with on-die 3D-stacked Through-Silicon Via (TSV) inductor [26-28] as shown in Fig.3. These innovations may further improve the granularity, power efficiency and scaling at cost of process complexity. The capacitance of the LC stabilizer could be achieved by the Metal-Insulator-Metal (MIM) capacitor [29] on die. The MIM capacitors are high frequency capacitors inside silicon. For few domains, mid-frequency capacitors are also added on the package for improved decoupling. The performance of integrated VR is largely influenced by the die-package co-design effort.

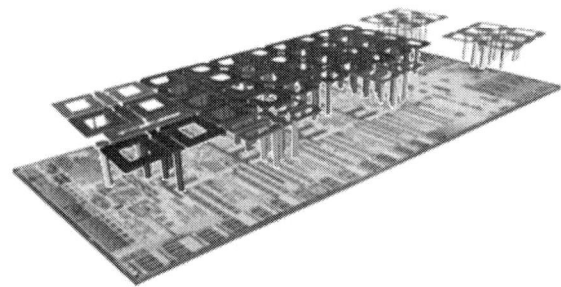

Figure 2: Flip-chip package inductor coils laying on the back side of the corresponding microprocessor die

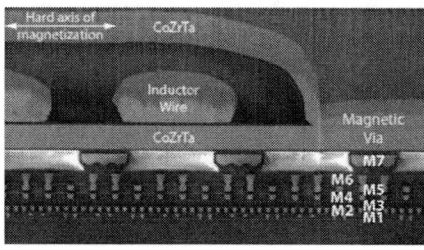 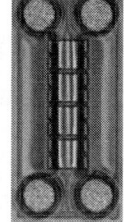

Figure 3: Views of magnetic core inductors under research

A systematic *in-situ* approach to measure the efficiency was developed using an external current load. The clock trees in the domain are operated at varying frequencies to generate an adjustable load current. The change in the input current can be used to precisely determine the value of this load current by calibration to the external load. The output voltage was measured at package top side test points, and at each loading point,

which essentially provides the output power. Fig. 4 shows the efficiency as measured using this procedure for varying numbers of phases, in each case showing a peak efficiency of approximately 89% at 0.9A/phase. By employing a phase shedding scheme, it is possible to keep the efficiency of the domain within a few percent of the peak efficiency of the domain from 0.5A to 20A. This is managed by the PCU which can phase shed when the efficiency needs improvement. It also has the capability to avoid phase shedding when a large load transient happens problematically.

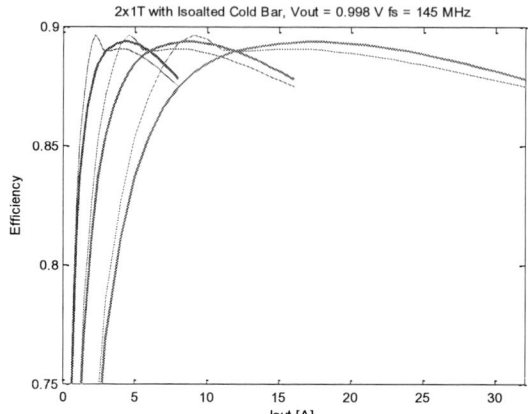

Figure 4: Measured efficiency of a voltage domain for different number of active phases (Red: 16 phases, Green: 8 phases, Blue: 4 phases. The bold line shows the hard switch; the thin line shows the soft switch)

CONCLUSION

Fully integrated on-die VR promise efficient and wide-range local power management with fast transient response while reducing losses in the off-die high voltage power delivery network. Intel new generation Core microprocessors are powered by these high-frequency integrated switching voltage regulators. This technology creates profound energy-saving and economic value. New research of other materials and structures are still undergoing to continue the Moore's law.

REFERENCES

[1] P. Hammarlund *et al.*, "Haswell: the fourth-generation Intel core processor," *IEEE Micro*, vol.34, no.2, pp.6-20, Mar.-Apr. 2014.

[2] D. Jacquet *et al.*, "A 3 GHz dual core processor ARM Cortex™-A9 in 28 nm UTBB FD-SOI CMOS with ultra-wide voltage range and energy efficiency optimization," *IEEE J. Solid-State Circuits*, vol.49, no.4, pp.812-826, April 2014.

[3] E.J. Fluhr *et al.*, "The 12-core POWER8™ processor with 7.6 Tb/s IO bandwidth, integrated voltage regulation, and resonant clocking," *IEEE J. Solid-State Circuits*, vol.50, no.1, pp.10-23, Jan. 2015.

[4] L. Chang *et al.*, "Practical strategies for

power-efficient computing technologies," *Proc. IEEE*, vol.98, no.2, pp.215-236, Feb. 2010.

[5] C.J. Shih, K.Y. Chu, Y.H. Lee, W.C. Chen, H.Y. Luo and K.H. Chen, "A power cloud system (PCS) for high efficiency and enhanced transient response in SoC," *IEEE Trans. Power Electronics*, vol.28, no.3, pp.1320-1330, March 2013.

[6] B. Zimmer *et al.*, "A RISC-V vector processor with simultaneous-switching switched-capacitor DC-DC converters in 28 nm FDSOI," *IEEE J. Solid-State Circuits*, vol.51, no.4, pp.930-942, April 2016.

[7] W. Kim, D. Brooks and G.Y. Wei, "A fully-integrated 3-level DC-DC converter for nanosecond-scale DVFS," *IEEE J. Solid-State Circuits*, vol.47, no.1, pp.206-219, Jan. 2012.

[8] Y.K. Ramadass, A.A. Fayed and A.P. Chandrakasan, "A fully-integrated switched-capacitor step-down DC-DC converter with digital capacitance modulation in 45 nm CMOS," *IEEE J. Solid-State Circuits*, vol.45, no.12, pp.2557-2565, Dec. 2010.

[9] C. Huang and P.K.T. Mok, "An 84.7% efficiency 100-MHz package bondwire-based fully integrated buck converter with precise DCM operation and enhanced light-load efficiency," *IEEE J. Solid-State Circuits*, vol.48, no.11, pp.2595-2607, Nov. 2013.

[10] C.H. Chia, R.C.H. Chang, P.S. Lei and H.M. Chen, "A two-phase fully-integrated DC-DC converter with self-adaptive DCM control and GIPD passive components," *IEEE Trans. Power Electronics*, vol.30, no.6, pp.3252-3261, June 2015.

[11] M. Bathily, B. Allard and F. Hasbani, "A 200-MHz integrated buck converter with resonant gate drivers for an RF power amplifier," *IEEE Trans. Power Electronics*, vol.27, no.2, pp.610-613, Feb. 2012.

[12] Y. Ahn, H. Nam and J. Roh, "A 50-MHz fully integrated low-swing buck converter using packaging inductors," *IEEE Trans. Power Electronics*, vol.27, no.10, pp.4347-4356, Oct. 2012.

[13] Y. Lu, W.H. Ki and C.Patrick Yue, "An NMOS-LDO regulated switched-capacitor DC-DC converter with fast-response adaptive-phase digital control," *IEEE Trans. Power Electronics*, vol.31, no.2, pp.1294-1303, Feb. 2016.

[14] J. Zarate-Roldan, M. Wang, J. Torres and E. Sánchez-Sinencio, "A capacitor-less LDO with high-frequency PSR suitable for a wide range of on-chip capacitive loads," *IEEE Trans. VLSI Systems*, vol.24, no.9, pp.2970-2982, Sept. 2016.

[15] M. Wens and M.S.J. Steyaert, "A fully integrated CMOS 800-mW four-phase semiconstant ON/OFF-time step-down converter," *IEEE Trans. Power Electronics*, vol.26, no.2, pp.326-333, Feb. 2011.

[16] S.S. Kudva and R. Harjani, "Fully-integrated on-chip DC-DC converter with a 450X output range," *IEEE J. Solid-State Circuits*, vol.46, no.8, pp.1940-1951, Aug. 2011.

[17] G. Villar-Piqué, H.J. Bergveld and E. Alarcón, "Survey and benchmark of fully integrated switching power converters: switched-capacitor versus inductive approach," *IEEE Trans. Power Electronics*, vol.28, no.9, pp.4156-4167, Sept. 2013.

[18] F.C. Lee and Q. Li, "High-frequency integrated point-of-load converters: overview," *IEEE Trans. Power Electronics*, vol.28, no.9, pp.4127-4136, Sept. 2013.

[19] S.R. Sanders, E. Alon, H.P. Le, M.D. Seeman, M. John and V.W. Ng, "The road to fully integrated DC-DC conversion via the switched-capacitor approach," *IEEE Trans. Power Electronics*, vol.28, no.9, pp.4146-4155, Sept. 2013.

[20] W. Liang, L. Raymond and J. Rivas, "3D-printed air-core inductors for high-frequency power converters," *IEEE Trans. Power Electronics*, vol.31, no.1, pp.52-64, Jan. 2016.

[21] D.S. Gardner, G. Schrom, F. Paillet, B. Jamieson, T. Karnik and S. Borkar, "Review of on-chip inductor structures with magnetic films," *IEEE Trans. Magnetics*, vol.45, no.10, pp.4760-4766, Oct. 2009.

[22] C. Ó Mathúna, N. Wang, S. Kulkarni and S. Roy, "Review of integrated magnetics for power supply on chip (PwrSoC)," *IEEE Trans. Power Electronics*, vol.27, no.11, pp.4799-4816, Nov. 2012.

[23] N. Sturcken *et al.*, "A 2.5D integrated voltage regulator using coupled-magnetic-core inductors on silicon interposer," *IEEE J. Solid-State Circuits*, vol.48, no.1, pp.244-254, Jan. 2013.

[24] I. Vaisband and E.G. Friedman, "Heterogeneous methodology for energy efficient distribution of on-chip power supplies," *IEEE Trans. Power Electronics*, vol.28, no.9, pp.4267-4280, Sept. 2013.

[25] C.R. Sullivan, D.V. Harburg, J. Qiu, C.G. Levey and D. Yao, "Integrating magnetics for on-chip power: a perspective," *IEEE Trans. Power Electronics*, vol.28, no.9, pp.4342-4353, Sept. 2013.

[26] U.R. Tida, Z. Cheng, and Y. Shi, "Novel through-silicon-via inductor based on-chip DC-DC converter designs in 3D ICs," *ACM J. Emerging Tech. in Comp. Systems*, no.11 vol.2, pp.16:1-16:14, 2014.

[27] U.R. Tida, C. Zhuo and Y. Shi, "Through-silicon-via inductor: is it real or just a fantasy?" *in Proc. Asia & South Pacific Design Automation Conference (APDAC)*, pp.837-842, 2014.

[28] U.R. Tida, R. Yang, C. Zhuo and Y. Shi, "On the efficacy of through-silicon-via inductors," *IEEE Trans. VLSI Systems*, no.23, vol.7, pp.1322-1334, 2015.

[29] R. Jain *et al.*, "A 0.45-1V fully-integrated distributed switched capacitor DC-DC converter with high density MIM capacitor in 22 nm Tri-Gate CMOS," *IEEE J. Solid-State Circuits*, vol.49, no.4, pp.917-927, April 2014.

978-1-5090-6695-7/17 $31.00 © 2017 IEEE

Geometry Effect with Respect to ESD and Radiative Charged Particles in SoC

C.-Z. Chen[a*] and David Y. Hu [b]

[a]Qualchip Technologies, Inc., Wuxi, Jiangsu, China 214072
[*]Mobile: (+86) 13910199301, Email: czchen126@126.com
[b]MetroSilicon Microsystems, Kunshan, Jiangsu, China 215300
Mobile : (+86) 13764219094, Email: david@metroSilicon.com

ABSTRACT

Thermal grown silicon dioxide (SiO_2) used as gate oxide are found commonly in a SoC (system-on-chip) design for the CMOS technology of 40nm and above, which is vulnerable under ESD (electrostatic discharge) stress typically known as CDM (charged device model) events. Reliability in SoC designs towards ESD protection and RadHard (radiation hardening) against single event effects (SEE) displays a key measure and desired feature in high-end applications such as automotive and aeronautical electronic systems. Using the calculated values of linear energy transfer *LET* and *Range* of radiative Alpha particles in SiO_2, in relation to the geometrical sizes in an SoC design, we continue analyze the potential ionizing radiation damage to transistor gate of CMOS in analogous to ESD damage described by CDM and TLP (transmission-line pulse) method. In this paper we present TLP testing structures with various rise times up to 10 ns on the thermal grown oxide from 70nm to 400nm with focus on the PMOS device, which is more likely damaged in the event of CDM stress due to its hot carrier penetrations to the gate oxide from the source area. Comparative results of Alpha particles are also presented and discussed as in a previous work using radiative particles of protons.

Keywords: silicon dioxide (SiO_2), electrostatic discharge (ESD), charged device model (CDM), ionizing radiation, radiation hardening (RadHard), single event effects (SEE), transmission-line pulse (TLP).

INTRODUCTION

There are increased concerns on safety and reliability in modern system-on-chip (SoC) designs adopted in advanced applications such as automotive and aeronautical electronic systems. It has been qualitatively described that electrostatic discharge (ESD) and single event effects (SEE), induced by ionizing radiation particles coming from solar or galactic cosmic rays, display similar damage phenomenon by causing voltage pulses or glitches that propagate the circuit. Some recent studies focus on single event upset (SEU) of SEE in SoC in relation to damage described by the charged device model (CDM) of ESD due to similar behavior in physics [1-4].

In the ESD prevention, analysis and testing, CDM has been considered playing a critical role, especially due to its peak current of up to >10A within a short rise time of <0.5 ns, even a few pF electric charges and up to kV voltage CDM can be observed [5], see *Table 1*. This phenomenon has lead to study of finding the corresponding types of SEE, see the study summary in *Table 2* [6]. While SEL can be prevented with epitaxial substrates, such as silicon on insulator (SOI)

or silicon in sapphire (SOS) technology, both SEB and SEGR types can cause permanent damages by large dose of ionizing radiation, thus we limit the discussion below on SEU and SET.

Table 1: Measured ESD Parameters

Models	Measured ESD Parameters		
	$Time_{rise}$ (ns)	$Time_{decay}$ (ns)	V_{peak} (V)
HBM	<10	150±20	± 2000~15000
MM	6~7.5	66~90	± 100~400
CDM	<0.2~0.4	0.4~2.0	± 250~2000
IEC	0.7~1.0	~80	± 2000~15000

Table 2: Summary of SEE types and their behavior

SEE Type	Description	Degree of Severity	Notes
SEU	single event upset	non-destructive	may cause wrong logic turnover
SEFI	SE function interrupt	non-destructive	derivative of SEU
SET	single event transient	transient	may cause a transient logic
SEL	single event latchup	destructive	preventable with epitaxial
SEB	single event burnout	permanent damage	at huge radiation dose
SEGR	single event gate rupture	permanent damage	at huge radiation dose

The primary ionizing radiation particles are protons and electrons in the Van-Allen belts; and protons, Alphas and other heavy (HZE) ions in the Solar and Galactic cosmic rays (*Table 3*). Thus in the literatures, the SEE together with the total ionizing dose (TID, its SI unit is Gy, 1 Gy=1 J/kg) effect by protons are widely studied for IC or SoC designs. For example, it has shown that protons of 60 MeV energy can lose or transfer a few keV energy in a micrometer (μm) distance in a transistor gate of SiO_2 media [6].

Table 3: Ionizing radiation particles in/from the Space

Environment	Composition	Energy	Flux, 1/(cm²·s)
Inner Van Allen belt Alt. (1,000-6,000) km	protons (99%) electrons (1%)	10-50 MeV ≥100 keV	(1-2)×10⁶ (≥50MeV) 3×10⁶ (≥1 MeV)
Outer Van Allen belt Alt. (13,000-60,000) km	protons (1%) electrons (99%)	1 MeV (0.1-10) MeV	2×10⁶
Solar/Galactic cosmics At Earth's surface	protons (90%) alphas (9%) HZE ions (1%) photons (x-, g-)	1×10⁹ (1 GeV) 1×10¹² (1 TeV) 1×10¹⁵ (10 PeV) 1×10²⁰ (100 EeV)	1×10⁴ 1×10⁰ 1×10⁻⁷ [a] 1×10⁻⁹ [b]

[a] a few times a year; [b] once a century

To continue investigate and compare the short pulse of CDM in ESD study (*Table 1*) and using SET in SEE, a transmission-line pulse (TLP) test for ESD measurement is proposed. Based on previous work on linear energy transfer (LET) and Range (R_{CSDA}) for protons and electrons, we continue to do the calculation for Alpha particles, such that a comparison can be performed.

METHODOLOGY
TLP Experiment and Layout Scheme

We firstly propose ESD design structures for future TLP testing. *Figs. 1 & 2* (Cases A-D) illustrate the four layout examples of a PMOSCAP (in pF) of a gate oxide of 70Å (Angstrom) with area of 1600µm2 (0.25~10.0pF) in the 0.35 µm HV process technology. To analyze the CDM damage to the gate oxide and its potential dependence of the layout geometry, we have chosen a PMOSCAP with area to perimeter A/P as the variable.

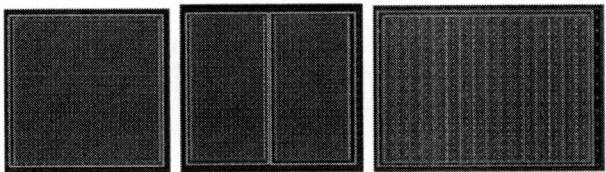

Figure 1: (Case A, Left) W=40, L=40, the total area A is 40x40 µm²; the perimeter P=160, PMOSCAP becomes A/P=10. (Case B, Middle) Using multiple fingers M=2 with W=40, L=20, the total P=240, PMOSCAP becomes A/P=2 x 40 x 20/240=6.67. (Case C, Right) Using multiple fingers M=16 with W=40, L=2.5, the total P=1360, PMOSCAP becomes A/P=1.17

Figure 2: (Case D) Using multiple fingers M=80 with W=40, L=0.5, the total P=, PMOSCAP becomes A/P=0.25

Calculation of LET and Range

As discussed previously [6], the ionizing radiation of charged particles (protons, Alphas, heavy ions), when passing through matter, can be described with the *linear* stopping power $S(E)$; when solid media are studied, the *mass* stopping power S_m or simply S ($MeV\text{-}cm^2/g$) can be obtained from $S=S(E)/\rho$, where ρ is the density of the stopping material. The classic Bethe formula with corrections is used to calculate the S,

$$S = k_1 \frac{Zz^2}{M_a\beta^2}[\ln \frac{2\mu\beta^2\Delta}{I^2(1-\beta^2)} - \frac{(1-\beta^2)\Delta}{2\mu} - \beta^2 - 2\frac{C}{Z} - \delta] \quad (1)$$

Alternatively, the linear energy transfer (LET or L) is used, we can define the unrestricted LET, conventionally expressed as $S(E)=-dE/dx=L$, where the minus sign means energy loss. The unit L (keV/µm) is the energy loss per unit length, it is the product of S (in general) or S_{tot} (electronic and nuclear fractions) and the density ρ,

$$L = S_{tot} \bullet \rho \quad (2)$$

Charged particles having initial energy E_0 lose energy through many ionizing interactions in passing through the matter, due to the straggling, in a continuous slowing down approximation (CSDA), until their energy is about zero, the total distance travelled which may not be in a straight line path, the CSDA Range, R_{CSDA} (g/cm²), or simply Δx is

$$\Delta x = \int_0^{E_0}[1/S(E)]dE \quad (3)$$

RESULTS AND DISCUSSION

From the computing programs available at National Institute of Standards and Technology (NIST) of USA [7], we have performed several calculation works, the results are presented below [*Figs. 3 & 4, Table 4*].

Results

The total stopping power (electronic and nuclear) in SiO₂ using Eq. (1) is shown in *Fig. 3*, for Alphas energies E(a) between 1 keV and 1 GeV. Similarly, the CSDA Range (maximum Range versus projected Range) in SiO₂ using Eq. (3) are calculated and shown in *Fig. 4*.

SILICON DIOXIDE

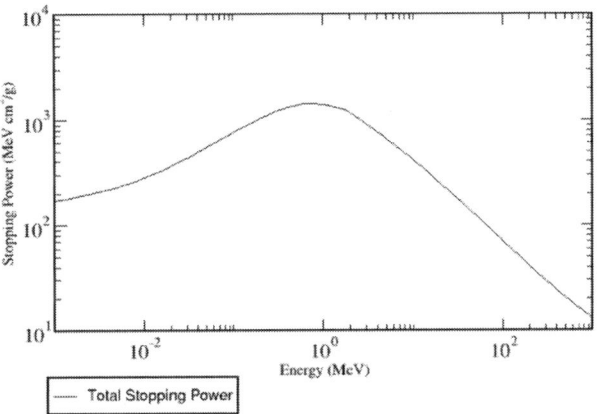

Figure 3: Total Stopping Power of Radiation Particles in Silicon Dioxide, for Alphas at 1keV-1GeV

SILICON DIOXIDE

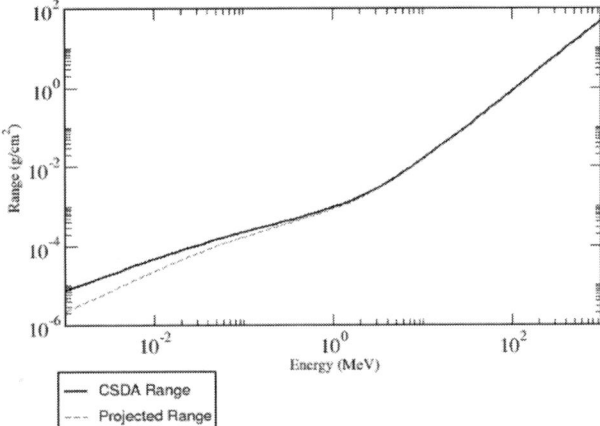

Figure 4: CSDA Range (Δx) in Silicon Dioxide, for Alphas at 1keV-1GeV

The stopping power S using Eq. (1), linear energy transfer L using Eq. (2), and Ranges (R_{CSDA}, R_{CSDA}/ρ) from Eq. (3) are derived for Alphas at various energy E(a); as a comparison, protons of E(p) is also listed (Table 4). Only selected values are listed for the discussion and analysis purpose.

Table 4: S, L, and R of Alphas and Protons

E(a)	S	L	R_{CSDA}	R_{CSDA}/ρ
MeV	MeV-cm^2/g	keV/um	g/cm^2	um
0.001	170	39.03	7.31E-06	0.0318
0.01	282	64.79	4.66E-05	0.2025
0.1	759	174.6	2.20E-04	0.9565
1	1390	319.5	9.26E-04	4.027
6	583.3	134.2	6.91E-03	30.05
10	408.9	94.05	1.53E-02	66.35
60	105.3	24.22	0.3226	1403
100	70.3	16.17	0.7985	3472
600	18.02	4.145	19.2	8.35E+04
1000	13.01	2.992	45.91	2.00E+05

E(p)	S	L	R_{CSDA}	R_{CSDA}/ρ
MeV	MeV-cm^2/g	keV/um	g/cm^2	um
0.001	101.9	23.44	1.39E-05	0.06
0.01	276.2	63.53	6.14E-05	0.267
0.1	533.9	122.8	2.56E-04	1.114
1	189.8	43.65	3.49E-03	15.16
6	53.85	12.38	0.06489	282.1
10	36.38	8.367	0.1573	683.9
60	8.874	2.041	3.791	1.65E+04
100	6.036	1.388	9.391	4.08E+04
600	2.147	0.4938	185.2	8.05E+05
1000	1.859	0.4276	388.1	1.69E+06

Discussion

To prepare CDM for ESD test, the proposed transmission line pulse (TLP) layouts (*Figs. 1 & 2*, with common W=40, L=40, 20, 2.5 and 0.5μm and M=1, 2, 16 and 80 respectively) can be linked to energy depositions and traveling ranges of radiation particles in SiO$_2$. With the varying A/P values (10, 6.67, 1.17 and 0.25 pF), an RCL circuit can be applied to measure CDM behavior, such as rise times, pulse duration at high voltage tolerance, in TLP experiments in future. Primarily TLP test showed the damage to the oxide by very fast voltage/current purge is inversely proportional to the perimeter to area ratio. That also correlates well with the experimental findings that thick gate oxide is more susceptible to radiation damage than ultra thin gate oxide.

For the SEU or SET types of SEE study, the results of the stopping power S, linear energy transfer L, the Ranges (R_{CSDA} and R_{CSDA}/ρ) for Alpha particles in CMOS gate media SiO$_2$ are obtained, these data can be used as references for RadHard study such that energy deposition to be evaluated for either SEE or TID, as a comparison to visualize the TLP tests (*Figs. 1 & 2*) or physical tracks of Alphas in a CMOS device (see *Fig. 4*) at known energies.

For example, for Alphas of 60 MeV, the total stopping power S_{tot} is 105.3 MeV-cm^2/g, L is 0.32 keV/μm, the R_{CSDA} is only 1.4 mm. Compared with that of proton of 60 MeV, its S_{tot} is 8.874 MeV-cm2/g, L is 2.04 keV/μm, and R_{CSDA} is 16.5 mm. This shows us a 60 MeV proton can travel more than 10 times distance, or lose 1/3 energy per unit length, of that of Alphas in SiO$_2$. In comparative experiments for SEU or SET

in accumulated TID studies, from Table 4, L and R_{CSDA} can provide geometrical guidance; as the Range is monotonic as a function of the particle energy, therefore it is more meaningful to use them to relate to physical application of CMOS gate.

For an SET which may propagate a pulse signal as in a coupling effect in a pair of parallel routing paths, to result in a signal integrity issue, if to result in an incorrect value being latched in a sequential logic unit, it is then considered an SEU. Thus understanding the differences between SET and SEU impose significant meaning in preventing them. Both SET and CDM may exhibit sudden short pulse rise in electric charges or current to cause a CMOS device failure. It is anticipated that the SET has a similar behavior as CDM. These glitches can cause a failure in a SoC design which can lead to life threatening accent if it is applied in an automotive electronic system, or aviation disaster such as in a civil air flight or advanced satellite [2, 3].

Conclusion

Primary study on geometries of TLP layouts and Ranges of Alpha particles present a linkage for the study between CDM of ESD and SEU and/or SET of SEE. Study of CMOS physical dimensions versus the energy depositions and Ranges of the protons and electrons [6] has been extended to Alpha particles in current work. The method of CMOS gate geometries relating to monotonic values of Range of charged particles is established for future work.

FFUTURE WORK

The calculation of Alpha particles may be extended to other heavy (HZE) ions and non-charged particles such as neutrons, results in SoC design for automobile applications for high reliability, including spacecraft and aviation. TLP measurement vs. SEU or SET to be determined in future tape-out chips.

REFERENCES

[1] N. Wakai *et. al*, "Consideration for CDM breakdown and reliability designing in the latest semiconductor technology" Reliability and Maintainability Symposium, 2009. RAMS 2009. Annual Conference 26-29 Jan. 2009, pp.509-514.

[2] D. Kobayashi, E. Simoen and S. Put et al., Proton-Induced Mobility Degradation in FinFETs with Stressor Layers and Strained SOI Substrates, *IEEE Trans. Nucl. Sci.*, **58**(3), 800-807 (2011).

[3] D.Y. Hu, and C. -Z. Chen, (Invited) Reliability Aspects of Advanced IC Technologies with ESD and Anti-radiation Capabilities, *ECS Trans.*, **60**(1), 1185-1190 (2014).

[4] M. Bertoldo, A. V. de Oliveira, P. G. Der Agopian, E. Simoen, C. Claeys and J. A. Martino, Proton Radiation Effects on the Analog Performance of Bulk n- and p-FinFETs, *ECS Trans.*, **66**(5), 295-301 (2015).

[5] D. Y. Hu, ESD Challenges due to CMOS Technology Scaling, Invited paper that presented on SMIC Tech. Forum, 23-25 Jun., Shanghai, China (2012).

[6] C. -Z. Chen and D.Y. Hu, Analysis of ESD Effect and Radiation Damage in SoC Design, Semiconductor Technology International Conference (CSTIC, March 13-14, Shanghai), 2016 China. Page(s): 1-4. Accepted and unpublished (full paper available).

[7] NIST, National Institute of Standards and Technology, http://www.physics.nist.gov/PhysRefData/Star/Text/programs.html

A LOW NOISE SPAD PIXEL ARRAY WITH ANALOG READOUT METHOD

Ruiming Luo[1], Yue Xu[1,2], Bin Li[1]*

[1]College of electronic science & engineering, Nanjing University of Posts and Telecommunications, Nanjing 210003, China

[2]National and Local Joint Engineering Laboratory of RF Integration and Micro-assembly Technology, Nanjing 210003, China

*Corresponding Author's Email: yuex@njupt.edu.cn

ABSTRACT

This paper presents a parallel readout circuit for high density single photon avalanche diode (SPAD) pixel array. Each pixel consists of analog quenching circuit and counting circuit. Column parallel readout method is adopted and every eight columns of array pixel shares one multiplexer where the analog output signals of these pixels are selected to pass it subsequently. After that, the signals will be sent to correlated double sampling (CDS) circuit for elimination of fixed pattern noise (FPN). Then the processed signals will be inputted to off-chip high speed ADC for analog to digital conversion. The shared CDS can reduce chip area without lowering performance.

INTRODUCTION

Single photon detection technology provides a high sensitivity for detecting extremely weak optical signals in a wide range of spectral range [1].Single photon detector has been widely used in the fields of biomedical, military and optical communication. Now, SPAD can be manufactured in CMOS technology perfectly. Benefited from development of advanced application specific integrated circuit (ASIC), an efficient array peripheral readout circuit especially more reliable readout circuit is highly demanded [2].

However, analog readout suitable for high-density arrays has not been proposed. This paper presents a high density and low noise array readout architecture for single photon detector applications based on CMOS technology. The core chip of detector includes pixel array and readout circuit. In pixel control circuit of SPAD usually adopts digital counting circuit. Recently, analog counting has been proposed. Because of its small size, low cost and semiconductor materials and technology maturity, the integration of SPAD array detector can be greatly improved. But many proposed analog array readout circuits are not very satisfactory because noise is the biggest problem in analog circuit. This paper focuses on the research of high density SPAD array and presents a parallel readout structure with low noise.

CIRCUIT DESCRIPTIONS

The proposed readout circuit is apt to high density pixel array with analog counting. Peripheral readout circuit adopts the method of parallel readout and can

efficiently eliminate noise and realize individual reset. The circuit structure presented in this paper is shown in Figure 1, including pixel array, multiplexer, correlated double sampling (CDS), row address decoder, decoder 3-8, row reset control (RRC), Gray encoder, clock circuit and analog-digital converter (ADC). The key module is correlated double sampling for noise reduction and other modules work together with CDS. ADC can be an off-chip module to simplify the circuit design. In consideration of competition and adventure in assembled logic circuit, Gray encoder is required.

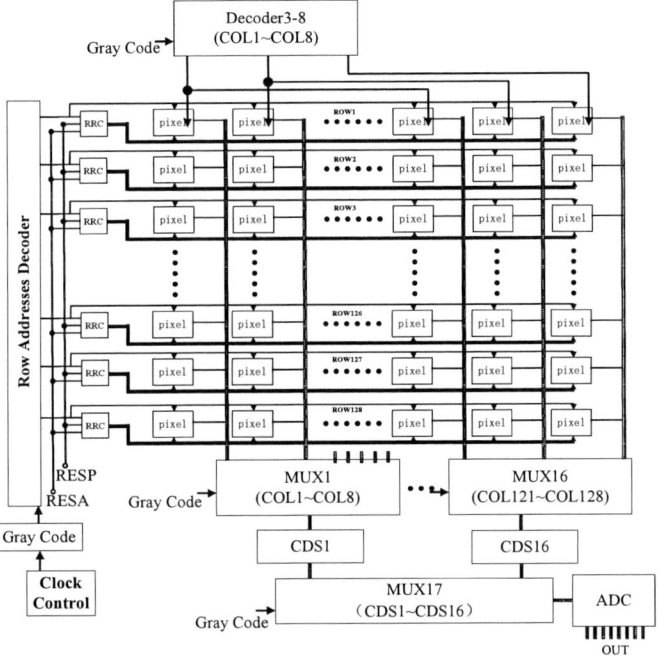

Figure 1: Readout circuit structure of SPAD array

The connections of all modules in the diagram are as follows. There is a pixel array (128 × 128) where each pixel in the array has 5 pins: output of row address decoder, exposure enable signal, output of decoder3-8, output of row reset control, and data output. The pixel circuit will be detailed in next paragraph. The whole array is divided into 128 columns by the output of row address decoder. There are three inputs of row reset control: row address decoder, global reset signal (RESA) and local reset signal (RESP).

978-1-5090-6695-7/17 $31.00 © 2017 IEEE   525

If RESA is high, the module outputs high, and if RESA is low, it will output the result of RESP multiplies row address decoder. Decoder 3-8 wires to the C1 (column 1)~C8, C9~C16... and the output data of C1~C8 connects to the MUX1, C9~C16 connects to the MUX2... , then all MUX output to CDS.

In each pixel (Figure2), it contains SPAD, quenching circuit, counting circuit, reset control circuit and gate circuit (output selection). In order to achieve individual reset, every pixel requires a reset control circuit composed of logic gates. The input of reset control circuit connects to the output end of decoder 3-8 and row reset control. When the output signal of decoder 3-8 and row reset control is high, the reset control is enabled and this particular pixel is in a reset state. Every pixel contains the quenching circuit for switching off or initializing the SPAD and analog counting module for registering the photon single during photon exposure. When reset control outputs high voltage, the voltage of counting becomes zero and reset is completed. The gate circuit is connected to the pixel and outer circuit is controlled by row address decoder. First, at exposure stage (after global reset), the sensor (SPAD) will generate avalanche pulse when incident photons arrive and the density of pulse is proportional to the number of photons. Therefore, the number of avalanche pulses represents the light intensity. Meanwhile analog counting circuit converts the avalanche pulses to a voltage signal and stores it in a capacitor.

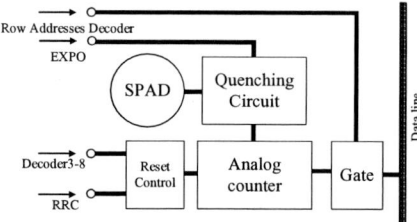

Figure 2: Modules of pixel

After exposure, the readout circuit starts work. First, row address decoding convert the Gray code, and every single row of the whole array will be chosen in sequence (Figure 3), then the counting signal in the pixel of this selected row will be transmitted to the multiplexer. It is assumed that the first row is selected. Since every row includes 128 pixels and just 16 CDSs in the chip, the 128 columns are divided into 16 groups, and every adjacent 8 columns share one CDS. Therefore, all the pixels in one row will be selected at the same time and be read after 8 clock cycle of multiplexer. When multiplexer selects the data in sequence the decoder 3-8 also selects the column synchronously to coordinate CDS achieving local reset (Figure 4). Then after analog signal of counting circuit is processed by CDS, the output signals will pass through another multiplexer to a high speed ADC. The structure will save area by several columns sharing one CDS.

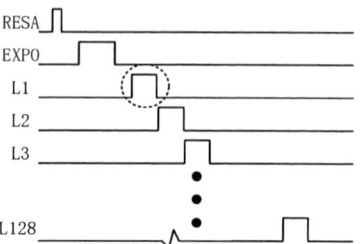

Figure 3: Sequence diagram of row parallel readout

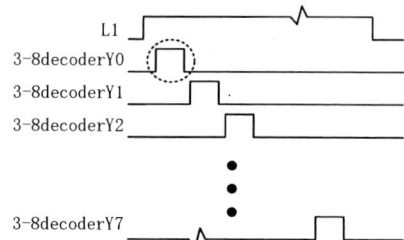

Figure 4: Sequence diagram of one row

Correlated double sampling is used to reduce noise. Mismatch due to deviation in process and temperature dependence causes variation of threshold voltage of MOSFET and fixed pattern noise (FPN) occurs in output signals. In addition, as a result of dark count inherent in SPAD sensors coupled with the FPN imaging quality becomes unsatisfactory. CDS techniques are often used in analog circuits, which are designed to reduce mismatch and eliminate 1/f noise in MOS circuit. Therefore, CDS is required in the array readout circuit as most significant module to reduce noise. This paper presents a CDS design based on the design of the switched capacitor circuit [3].

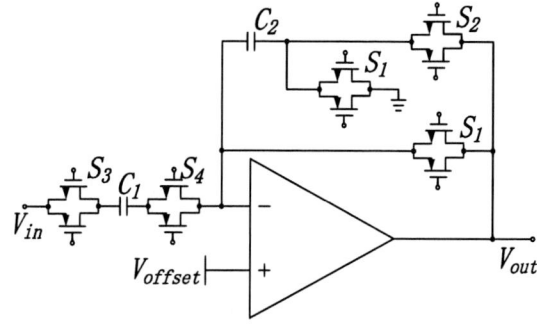

Figure 5: Diagram of optimized CDS

As shown in Figure 5, the circuit includes one operational amplifier, two capacitances and several CMOS transmission gate switches [4]. It will sample the input signal Vin1 and Vin2 in sequence by switches (Figure 6), and the output signal equals the difference of two sample voltage ($V_{out} = V_{in1} - V_{in2}$) through charge changing cross two capacitance. In fact, these switches in the circuit usually have resistance. For example, a NMOS transmission gate (0.18μm technology, minimum size) has about 8 k Ω when its gate connects to the highest potential.

Because of the resistance of switches, the output will decrease and will not equal to Vin1-Vin2.Two methods can resolve this problem. First we can increase the value of C1 to make the gain of CDS greater than 1. However, it increases the area consume because of precise capacitor in chip costs much area. Also adding resistance in the input side can be a solution. The CDS circuit in Figure 2 uses two transmission gate S3, S4 in input side to compensate the voltage loss.

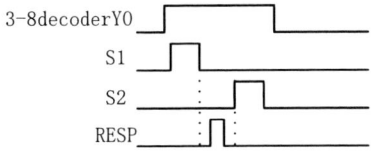

Figure 6: Sequence diagram of CDS

SIMULATION RESULTS

The proposed array readout circuit is designed in standard SMIC 0.18 μm technology and simulated on Cadence virtuoso simulator. Figure 7 (a) and (c) are two opposite phase switch signals S1 and S2 of CDS. In first stage when S1 is turn-on (high voltage), the CDS samples the input voltage. Then, next stage S2 turns on and S1 turns off, and CDS samples another voltage moreover outputs the difference. As Figure 7 d shows, the first square wave level Vi is roughly equal to $V_a - V_b$ (Figure 3 b).

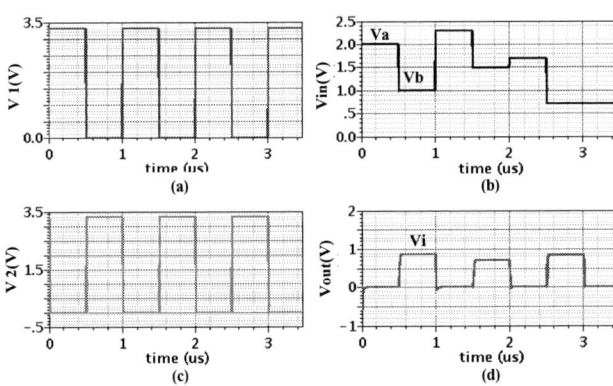

Figure 7: Simulation of CDS

Figure 4 is one pixel of the simulation, where the photon pulse period was set to 0.5μs, 1μs and 1.5μs respectively. The counting result of each pixel unit is transmitted to the correlated double sampling circuit through the multiplexer, and finally converted to the digital signal by ADC. ADC collects output analog signal of pixel from counting capacitance which is set to 100fF and using logarithmic count method. Because of the

unreadable output binary number, the final simulation results are converted to decimal numbers shown in Figure 8. It can be seen from the simulation results that the shorter the photon pulse period, the faster the voltage drop of the counting capacitor, thus the purpose of detecting the photon is reached.

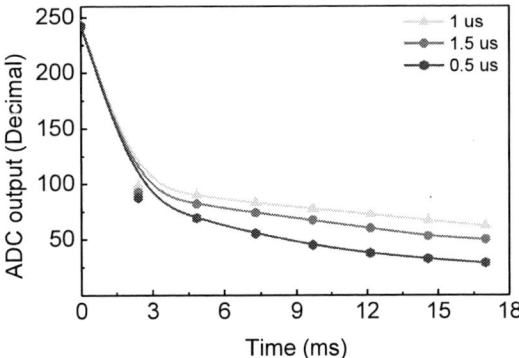

Figure 8: Simulation of readout circuit

CONCLUSION

Readout architecture of SPAD pixel array with low noise analog mode has been proposed and analyzed in this paper. Simulation results show that the CDS can effectively reduce the noise, so that the output is a pure photon signal.

ACKNOWLEDGEMENTS

This work was supported by National Science Foundation of China (No. 61571235) and Natural Science Foundation of China Jiangsu Province (No. BK20131379) and QingLan Project.

REFERENCES

[1] Palubiak D P, Deen M J. CMOS SPADs: Design Issues and Research Challenges for Detectors, Circuits, and Arrays[J]. IEEE Journal of Selected Topics in Quantum Electronics, 2014, 20(6):409-426.

[2] Sudharsanan R, Yuan P, Boisvert J, et al. Single photon counting Geiger mode InGaAs (P)/InP avalanche photodiode arrays for 3D imaging[C].SPIE Defense and Security Symposium. International Society for Optics and Photonics, 2008: 69500N-69500N-9.

[3] H. Yue, Y. Xu, Y. Huang and X. Xie, "Fully integrated high density SPAD array detector," Solid-State and Integrated Circuit Technology (ICSICT), 2014 12th IEEE International Conference on, Guilin, 2014, pp. 1-3.

[4] Demierre M. Improvements of CMOS Hall microsystems and application for absolute angular position measurements[D]. Ecole Polytechnique Federale De Lausanne, 2003.

A NOVEL FMEA TOOL APPLICATION IN SEMICONDUCTOR MANUFACTURE

Lijuan Sun[1]*, Liping Peng[1], Guihong Deng[1], and Kary Chien[1]

[1]Semiconductor Manufacturing International Corporation, Shanghai 201210, China
*Corresponding Author's Email: Sunny_Sun@smics.com

ABSTRACT

In this Paper, we introduce a novel FMEA (Failure Mode and effect Analysis) system, It can achieve FMEA more practicable and valuable compared with current FMEA application status as record archives. The novel FMEA basic unit is module and related modules are combined to form a whole FMEA, Six new link/experience function modules are introduced into the standard FMEA format to integrate the database, and the module unit exists independently so that different FMEA file cross share the similar failure modules. Three new link function modules can connect FMEA system with other related production systems to embed FMEA useful resource into production as guidance, this link function can prevent potential and old failure modes timely occurring timely, further reduce defect and improve production efficiency and quality. This novel FMEA tool can make FMEA more value and important functions in semiconductor process.

KEYWORDS: FMEA, cross sharing, experience module, integrated database, link module, production.

INTRODUCTION

FMEA was originally developed by the aerospace industry in the 1960s as a tool for improving hardware design, then its application has been expanded to other industries for the analysis of process/design/system, such as automotive, semiconductor, and computer. etc.[1]

Especially in the high-technology intensive semiconductor industry, FMEA play an important role in each tech in each technology life cycle, to forecast the all potential failed modes, and take actions to solve develop and production problem in advance, reduce cost and improve product yield performance. Unfortunately, in current semiconductor industry, mostly FMEA exists as the written word files, and FMEA reference manual of fourth edition [2] defines the basic FMEA template, and each failure contains total 20 items in Figure 1, so they are huge FMEA files in word version files, even up to more than 20000 items, it is difficult for content search and new item update, and inconvenient for document management.

Besides, once one FMEA file is completed, it doesn't be updated until develop or process suffers some abnormal events, actually, it become one record archives after the event. Furthermore, current FMEA as archives files are absolutely isolated from production systems in Figure 2,

there is no any auto link to make FMEA embedded in manufacture as guidance using its valuable resources.[3]

Process/ Function	Require ment	Potential Failure Mode	Potential Effect of Failure	(S)	SC	Potential Causes/Mechanism of Failure		Preventive Action	
Detective Actions					Recommended Action	Responsibility and Target Due Date	Action Results		
For root cause	For Failure Mode	(D)	RPN		Recommended Action	Responsibility and Target Due Date	Actions Taken	S	O	D	RPN

Figure 1: FMEA reference manual template of fourth edition.

The most important characteristic of FMEA is to forecast timely all potential failure modes and take correct actions timely before occurring in production, so it means FMEA is the tool before the event, not current after the event.

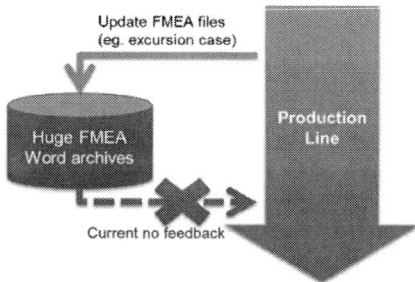

Figure 2: FMEA current connection production line

In this paper, considering current FMEA application weakness, we set up a novel FEMA system, it has integrated database, well self-maintenance and auto-link with other related production systems, it can introduce FMEA valuable database into process to guide production, at the same time, production abnormal feedback can enrich and optimize FMEA database.

A NOVEL FMEA DATABASE SYSTEM

This paper introduce a novel FMEA database system, compared with current FMEA control method, this novel FMEA system displays more advantaged aspects.

Firstly, its contents exist in the form of "Module", and related modules combine into a FMEA file. As for users, it is more advantageous and convenient to search and update content.

Secondly, it introduces 3 new experience modules to enrich FMEA database, such as "process weakness", "weakness product", "lesson learn report", attached report function is a novel function to provide one position to put the detailed lesson learn history use for future prevention.

Thirdly, also add 3 new link modules to realize

collection link function with other production systems, such as "Tech. ID", "Depart." , and "Stage", instead of current FMEA isolated condition with other systems.

On the basis of original items, new link modules and new function modules with blue mark in Figure 3 are added to set up integrated FMEA database, and divide the basic item "Process/function" into two items "department" and "Stage", these modules' aim with "Tech. ID" is to realize well maintenance optimization and correction link with other related production systems. [4]

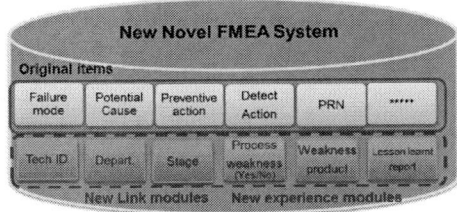

Figure 3: The novel FMEA database system

Semiconductor manufacture mostly has electronic systems to control production, such as MES system (Manufacturing Execution System), NTO system (New Tape Out System), FECP system (Engineering Change System), Tool Recover System. etc., new added three link modules in Figure 4, Technology (Tech. ID), Department (Depart.), Stage, can set up connection link with these related production systems to introduce FMEA valuable content into the real mass production as guidance function.

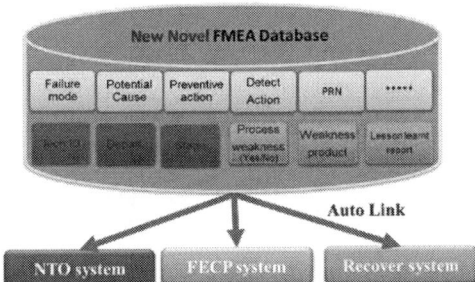

Figure 4: FMEA system connections with production systems

FMEA New Link Function Modules

Current, FMEA files generally is treated as the record archive after the event and isolated with production systems. The new link function modules (Technology, Department, Stage) can build the linkage with production, and embedded FMEA resource in the manufacture, especially the valuable experience modules as guideline, which are the summary of previous lesson learn cases with process and product weakness.

The link modules are introduced to terminate current isolated FMEA status, make it exist as the real living FMEA database and provide more information for production.

FMEA New Experience Function Modules

One FMEA file is the full-scale risk evaluation and actions for happened or potential failure modes in a whole technology flow, especially for the major excursion cases (e.g. MRB), more attention should be paid to avoid reoccurrence, so these experience modes (If process weakness, Weakness product, Report) is necessary to record these lesson learn cases in FMEA database as reminder in the future, especially weakness process and products, it is just current FMEA shortcoming in the application.

So the experience modules are newly introduced in FMEA system to play the lesson learn role, they are the summary of major excursion cases with report attachment function, and can fully enrich FMEA database.

Synchronously, experience modules can be embedded into related production systems on the connection bridge by the new link modules, so that the past experience resource can provide clear directions in the current and future production to escape the same excursion happening, and reduce production impact and cost lost.

FMEA Cross Sharing Function

Semiconductor manufacture has a kind of technologies for various applications, such as 0.35um Logic, 0.18um Flash, 0.13um MEMS, etc. Every technology flow need set up a completed FMEA file, however, different technologies exists the same process stages, e.g. Metal line in the back-end process, the same processes have the same failure modes and actions, so it is necessary for different technologies to share the FMEA contents update to obtain new information timely.

However, current all process FMEA files solely exist and need be updated by technology respectively, so induce lesson learn can't be shared to the related technologies' FMEA in time   to avoid case reoccurring.

In the novel FMEA, all content exist in the form of independence modules, not like current entire files, and can achieve the same failure modes sharing between different technologies. As shown in Figure 5, three technologies FMEA (A/B/C) have the same layer Metal Etch, so them can share the same failure modes and actions modules at metal etch stage, at the same time, if update failure modes of Tech. A, other 2 technologies can share the same update contents, so this new function can timely realize resource sharing between the different technologies.[5]

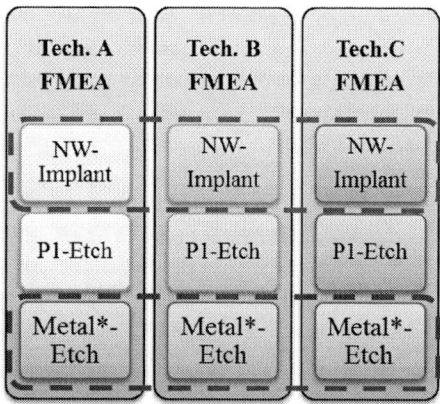

Figure 5: FMEA cross sharing function

FMEA SYSTEM CONNECTIONS WITH PRODUCTION SYSTEMS

The new FMEA database can connect with related production systems by link function modules, such as tool recover system, NTO system, FECP system, etc. These link production systems need also maintain the same link modules with FMEA system, so that the linkage bridge can be really built to achieve the communication between FMEA and production.[6]

FMEA Connection with Tool Recover System

In semiconductor, production tool must have highest performance with the best precise, accurate and stability. The higher recover successful ratio is also necessary after routine maintenance or trouble shooting,

So tool recover need select the representative products to pi-run small quantity and then mass production for risk control, and weakness products are more sensitive for the tool condition difference, tool few bias can induce obvious abnormal result on these weakness products. so these product can be treated as representative products to judge if the tool is recovered to the normal condition

Here, weakness product from the past experience resource is added in FMEA system as experience modules, and the important information can be embedded into tool recover systems on the bridge of link modules to escape the same excursion happening. Additionally, report experience module is attachment function to get detailed lesson learn information.[7]

In Figure 6, the three mutual link modules, Technology (Tech.), Department (Depart.), Stage, set up the communication bridge between FMEA system and tool recover system. For example, abnormal tool A01 need recover to production and select recover condition (Tech. B, Depart Etch, Stage Poly), but mostly engineers are confused which representative product can be selected in the large product pool to pi-lot this tool successfully.

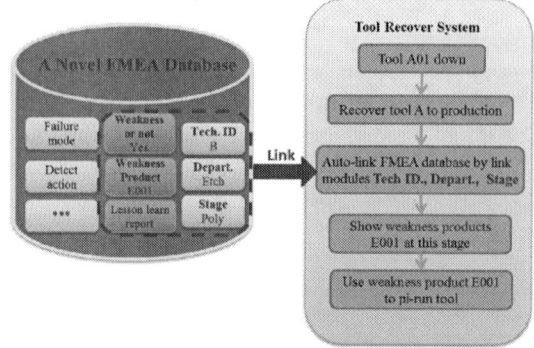

Figure 6: FMEA system connections with tool recover system

At this time, system can auto connect the novel FMEA system by the link modules, then detect Tech. B has weakness product E001 at Poly Etch Stage, and show weakness result in recover system. so that tool A01 can select weakness product E001 to pi-run small quantity during recover. Sometimes, it is possibility that there is no product E001 nearly poly etch stage, so product E001 can be constrained on tool A01 until its arrival, then do pi-run test to control sightless pi-run risk.

FMEA Connection with NTO System

One technology process flow generally contains twenty or thirty stages, from the front-end device process to the back-end metal line and via process. Some special stages are the weakest in the whole flow, even significant impact event, such as MRB. So these layers need be enhanced quality control when a new product start of this technology, risk evaluation is more important before start.

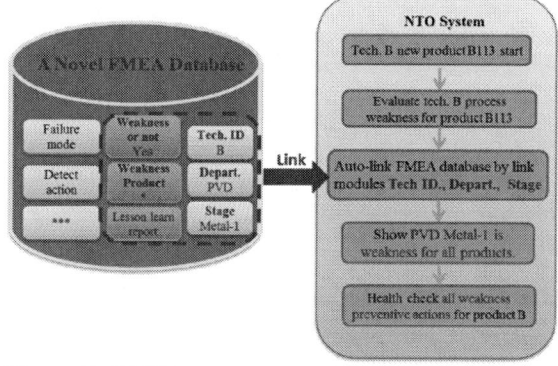

Figure 7: FMEA system connections with NTO system

In Figure 7, Tech. B new product B113 starts need evaluate process weakness and preventive actions readiness for product B113, NTO system auto-link FMEA database by link module "Tech.ID" to obtain all weakness layers of this tech flow, and result shows PVD Metal-1 is weakness stage for all products of tech. B process at NTO system interface, then product owner and related responsible teams need risk health check if preventive actions of this stage weakness are taken for this new product B113, so order to avoid abnormal event reoccurring in mass production and reduce impact and cost lost.

FMEA Connection with FECP system

Currently, semiconductor process develops with rapid change, every process flow change possible exist some unexpected risks, such as electronic value, yield and reliability impact. Process FMEA resource is the summary of the past excursions occurring and all potential failures in one whole flow, it play a very important reference role for current and future process change, so FMEA is introduced into process change assessment procedure as guidance, it can make sure the change assessment more comprehensive and accurate.[8]

For instance, In Figure 8, Tech. C need do CVD process E change qualification at ILD layers, and risk evaluation is necessary to decide if this change is feasibility to be implemented without the same problem appearing during or after process change. Three link modules can interlink FMEA system to obtain this layer is weak process with weak product F125, Once process change is determined, weakness product F125 can be selected as the pilot product to evaluate the new process performance.

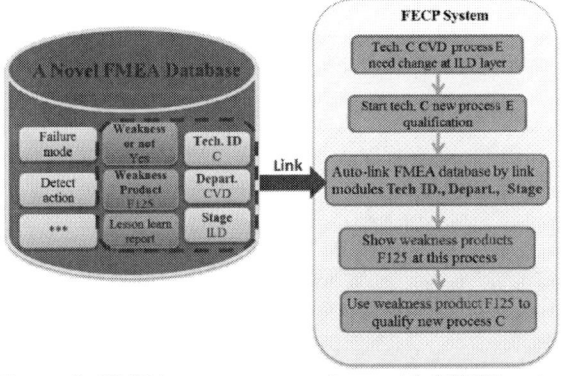

Figure 8: FMEA system connections with FECP system

The above three examples simply display FMEA connection function with production system by new link modules. Of course, here, take these three production systems as samples to explain the connection, and don't mean only these three systems.

All related production systems can set up the linkage to use FMEA resource, such as SPC system (Statistics process control), it can provide statistical techniques and graphical control charts to enable the quality and process stability be monitored timely. But sometimes, process adjusting decision or adjusting priority may be hard to determine according to the knowledge of engineers when the quality problem has been detected by SPC, so set up the linkage between SPC system and the novel FMEA system can overcome this difficulty, based on the novel FMEA database, engineers' decision of process adjusting can be supported by valuable and intelligent FMEA database, so that the quality control process can also be facilitated and manufacturing process can be continually improved.[9]

ACKNOWLEDGEMENTS

We would like to thank Mrs. Catherine Liu of SMIC IT center who completed the novel FMEA system development.

CONCLUSIONS

In a summary, compared with current huge FMEA word files with inconvenient update and isolated with production systems, this novel FMEA system provide integrated database, it can achieve well maintenance and cross sharing function, secondly, it set up the communication bridge with related production systems by the link modules to introduce valuable resource (experiences modes) into production order to improve equipment, new product and process quality control, such as recover, NTO, FECP system.

At the same time, the novel FMEA system can be continually enriched by excursion events during production, process optimization, and customer complaint and so on, this is a mutual optimization process between FMEA database and mass production.

REFERENCES

[1] R. Whitcomb and M. Rioux, "Failure modes and effects analysis (FMEA) system deployment in a semiconductor manufacturing environment," in Proc. 1994 IEEE/SEMI Advanced Semiconductor Manufacturing Conference and Workshop, 1994, pp. 136-139.

[2] FMEA reference manual, fourth edition, Ford Co., General Motor Co., 2008.

[3] C. S. Cai, "FMW design and failure analysis database setup construction," Quanta Quality, vol. 19, pp. 23-25, Nov. 2009.

[4] K.Onodera, "Effective techniques of FMEA at each life-cycle stage," in Proc. Reliability and Maintainability Symposium, 1997, pp. 50-56.

[5] T. Hu, Y Jian, and S. Z. Wang, "Research on complex system FMEA method based on functional modeling," in Proc. 2009 ICRMS, 2009, pp. 63-66.

[6] Z. G. Liu and F. Q. Li, "The research of application of craftwork FMEA," Quality and Reliability, vol. 20, pp. 42-46, 2005.

[7] M. H. Wang, "A cost-based FMEA decision tool for product quality design and management," in Proc. ISI 2011 IEEE International Conference, 2011, pp. 297-302.

[8] H. Zhang and Z. B. Zhou, "The process of FMEA," Technology of Avigate Manufacture, no. 8, pp. 66, 2007.

[9] X. X. Zhao, Y. B. Ma, R. Cai; X. L. Bai, and L. Y. Ning, "Research and application of intelligent quality control system based on FMEA repository," in Proc. ITCS2009 International Conference, 2009, pp. 514-517.

KA-BAND LOW NOISE AMPLIFIER USING 70NM MHEMT PROCESS FOR WIDEBAND COMMUNICATION

Xu Cheng[1], Liang Zhang[1], and Xianjin Deng[1]*

[1]Microelectronic and Terahertz Research Center, CAEP, Chengdu 610000, China

*Corresponding Author's Email: chengxu@mtrc.ac.cn

ABSTRACT

The paper proposes a Ka band low noise amplifier (LNA) using 70nm GaAs metamorphic high electron mobility transistor (mHEMT). An ultra low noise three stage amplifier demonstrated an average noise figure of 1.1dB between 24GHz and 30GHz, an average small signal power gain of 27dB, a compact chip area of 1.5mm^2 (with dicing streets) and a power consumption of around 80mW@1.5V power supply. The LNA features broad bandwidth, relatively high gain, low noise figure and compact size, and will be further used in broadband receivers for communications.

INTRODUCTION

As a critical component in microwave transceiver front-end, the LNA needs to provide high gain, low noise figure and reasonable linearity performance simultaneously while these merits generally can only be pursued in III-V MMIC semiconductor technologies. Although high performance III-V devices and technologies are developing rapidly, there are still many insufficiencies which hamper the fulfillment of circuit design target. In fact, novel efficient design methodologies should be developed and implemented to match III-V technological improvements.

In this contribution, the complete design of a Ka band LNA using 70nm mHEMT process with design approach, design target and post simulation results will be presented in the following sections.

TECHNOLOGY

The D007IH process is based on a high Indium content epitaxial active layer, grown on a metamorphic buffer layer creating a smooth transition with the GaAs substrate. This process shows very high cut off frequency and extremely low noise. Its profile is given in figure 1. For a typical double mushroom gate mHEMT, the measured f_T and f_{max} are higher than 300GHz and a minimum noise figure of 0.55dB with an associated gain of 12.5dB is obtained for a 4*15um transistor.

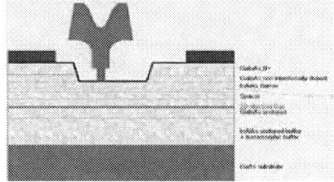

Figure 1. D007IH Profile

LNA DESIGN METHODOLOGY

This LNA is targeted to cover 20 GHz to 30 GHz with a small signal power gain higher than 25 dB, a noise figure less than 1.5dB and a saturated power higher than 10dBm. Therefore, transistor size, amplifier architecture and stability analysis should be carefully verified in this design.

In the first place, the output stage transistor determines the general linearity such as P_{-1dB} and output saturated power. According to the requirement of a P_{-1dB} higher than 10dBm, we conduct a loadpull analysis of a single transistor as seen in figure 2 and set the final stage transistor size to be 4 finger*45um.

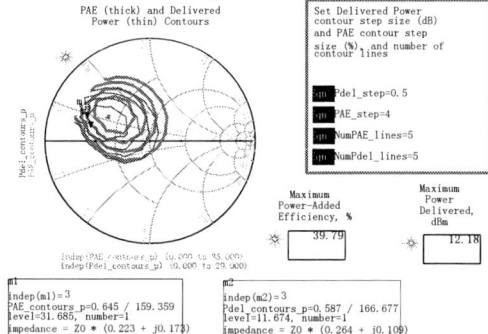

Figure 2. Final stage transistor P-1dB simulation

In the second place, for a typical transistor in D007IH, the available power gain is lower than 11dB and the overall power gain requirement of higher than 25dB requires an amplifier architecture of at least three cascaded stages. In a similar associated power gain analysis, the first and second stage transistor size is set to be 4 finger*25um for an acceptable balance between power gain and linearity performance.

In the third place, we allocate the power gain requirement between three stages and their respective passive networks as demonstrated in figure 3. The general architecture of the LNA is given in figure 4 and the resistor-capacitor stability network is given in figure 5. In figure 4, inter-stage matching networks select typical T-shape transmission lines for relatively wide band matching and the gated resistor-capacitor network brings in better low frequency stability. According to [], the resonance frequency is set to be in the higher end of the operating frequency and in-band stability with low frequency stability is simultaneously guaranteed. The quality factor of the resistor-capacitor network is given as follows.

$$Real(Z_{ins}) = R_{in} + \frac{R_g}{1 + (R_g C_g \omega)^2} \qquad (1)$$

$$Imag(Z_{ins}) = \left(-\frac{1}{C_{in}\omega}\right) \cdot [1 + \frac{R_g^2 C_g C_{in} \omega^2}{1 + (R_g C_g \omega)^2}] \qquad (2)$$

$$Q(Z_{ins}) = (\frac{1}{R_{in}C_{in}\omega}) \cdot (\frac{1 + R_g^2 C_g (C_{in} + C_g)\omega^2}{1 + (R_g C_g \omega)^2 + \frac{R_g}{R_{in}}}) \qquad (3)$$

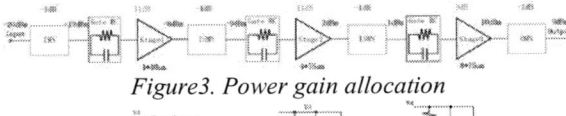

Figure3. Power gain allocation

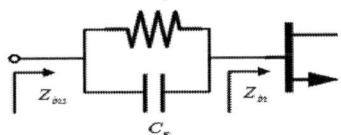

Figure4. Amplifier achitecture

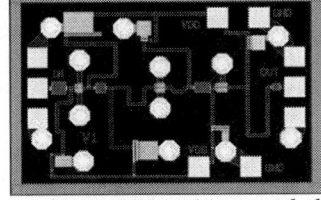

Figure 5. resistor-capacitor stability network

As seen in equations (1)~(3), the quality factor of Z_{ins} is higher than that of Z_{in} when $R_g C_g R_{in} C_{in} > 1$ and thus, bandwidth will be weakened.

In addition, the on-chip filter needs careful design in order to resist the relatively large inductance of off-chip bondwires and packaging influence should be pre-considered.

LAYOUT AND POST-SIM RESULTS

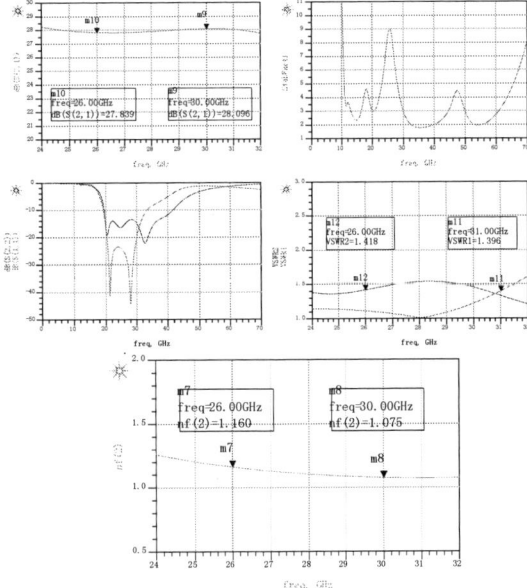

This low noise amplifier adopts GaAs 0.13um mHEMT process to conduct front-end and layout post simulations. The layout is given in figure 6. The small signal s-parameter analysis is conducted and post simulation result is given in figure 7.

Figure 7. LNA post simulation results (power gain higher than 27dB, stability guaranteed, S11 and S22 better than -10dB from 20GHz to 30GHz, noise figure lower than 1.16dB@26GHz)

In conclusion, the LNA features an operating frequency range of 20GHz~30GHz, a small signal gain higher than 27dB, a noise figure less than 1.26dB from 26GHz to 30GHz and power consumption lower than 83mW@1.5V power supply.

ACKNOWLEDGEMENTS

Many thanks to colleagues in MTRC for their technical support.

REFERENCES

[1] Weng S H, Chang H Y, Chiong C C, et al. Cryogenic evaluation of a 30–50 GHz 0.15-μm MHEMT low noise amplifier for radio astronomy applications[C]//Microwave Conference (EuMC), 2011 41st European. IEEE, 2011: 934-937.

[2] Ciccognani W, Limiti E, Longhi P E, et al. MMIC LNAs for radioastronomy applications using advanced industrial 70 nm metamorphic technology[J]. IEEE Journal of Solid-State Circuits, 2010, 45(10): 2008-2015.

[3] Cuadrado-Calle D, George D, Fuller G. A GaAs Ka-band (26–36 GHz) LNA for radio astronomy[C]//2014 IEEE International Microwave and RF Conference (IMaRC). IEEE, 2014: 301-303.

[4] Hacker J B, Bergman J, Nagy G, et al. An ultra-low power InAs/AlSb HEMT W-band low-noise amplifier[C]//IEEE MTT-S International Microwave Symposium Digest, 2005. IEEE, 2005: 4 pp.

Figure 6. LNA layout (1.5mm*1mm with dicing street)

978-1-5090-6695-7/17 $31.00 © 2017 IEEE

DESIGN OF K/KA-BAND PASSIVE HEMT SPDT SWITCHES WITH HIGH ISOLATION

Liang Zhang[1, 2], Xu Cheng[1, 2], Xianjin Deng[1, 2] and Xinxin Li[3]*

[1]Institude of Electronic Engineering, China Academy of Engineering Physics, Mianyang Sichuan 621999, China

[2]Microsystem and Terahertz Research Center, China Academy of Engineering Physics, Chengdu Sichuan 610200, China

[3] Patent Examination Cooperation Center of The Patent Office, SIPO, Sichuan, China

*Corresponding Author's Email: zhangliang@mtrc.ac.cn

ABSTRACT

In this paper, high isolation K/Ka band monolithic microwave integrated circuit (MMIC) single pole double throw (SPDT) switches with different topologies are proposed and discussed. By using a 0.07 μm GaAs high electron mobility transistor (HEMT) process, the shunt type and series-shunt type of passive HEMT SPDT switches are designed and compared. In 24~27GHz, both switches demonstrate an isolation better than 39dB. The shunt type of switch demonstrates an insertion loss less than 1.5dB and the series-shunt type of switch less than 2dB. With a compact chip size smaller than 1.5*2 mm², the designed switches can be used to form the MMIC T/R systems in the future.

INTRODUCTION

Microwave switches are extensively used in the front end of the communication systems as well as radar based subsystems. They play a vital role in a variety of applications, such as channel filter selection, and choosing the transmitting or receiving RF paths in time division duplex (TDD) systems with common antenna. Desirable features of microwave switches include smaller size, higher power-handling capability, lower insertion loss, higher isolation and so on.

Many approaches and technologies can be used to design microwave switches [1]. Monolithic PIN diodes switches provide low insertion loss with high switching speed, but at the expense of a complex bias circuitry. MEMS switches can work up to Ka band with low insertion loss, but their actuation voltage requirement is high. Waveguide switches can offer extremely low insertion loss and high isolation, while the volume of waveguide switches is unacceptable in lower frequencies. Nowadays, GaAs-based passive HEMT switches are widely studied for their higher working frequencies, higher power budget and nearly zero quiescent DC current with comparable performance. What's more, GaAs-based passive HEMT switches can be easily integrated with other building blocks to fulfill the T/R MMICs [2].

For different design specifications, different circuit topologies can be applied to build the passive HEMT switches [3~5]. As the most common type, series or shunt HEMT switches show good performance with simple structure. Broadband characteristics can be obtained by the travelling wave switches. High isolation performance can be achieved by the resonant type HEMT switches.

In this paper, high isolation K/Ka band shunt type and series-shunt type of SPDT switches are designed. The characteristics and simulated results of these designed switches are compared. With low insertion loss, high isolation and small chip size, the passive HEMT switches lay the foundation for future T/R MMICs.

CIRCUIT DESIGN

The switches are fabricated in OMMIC D007IH process, which is based on a high Indium content epitaxial active layer, grown on a metamorphic buffer layer creating a smooth transition with the GaAs substrate. This process shows very high cut off frequency and extremely low noise.

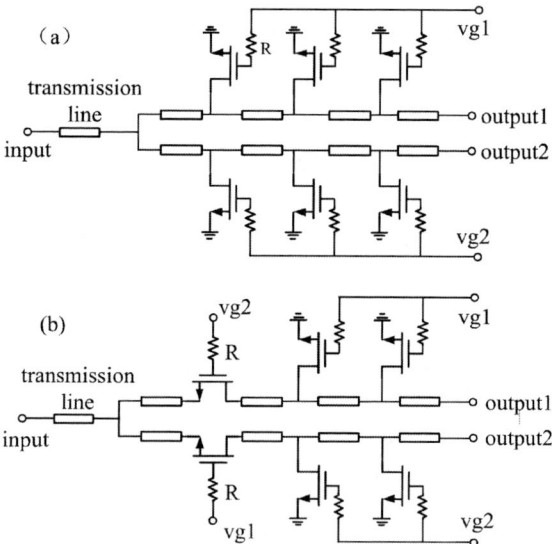

Figure 1: schematics of (a) shunt type and (b) series-shunt type of passive HEMT SPDT switches

The circuit schematics of shunt type and series-shunt type of passive HEMT SPDT switches are showed in

978-1-5090-6695-7/17 $31.00 © 2017 IEEE

figure 1. Both topologies are composed of two symmetric signal paths, which can also be seen as single pole single throw (SPST) switches. For the shunt type HEMT switch, all the transistors' drain terminals are connected to the transmission lines while the source terminals connected to ground. The gate terminals are biased by two different voltages vg1 and vg2 using a large resistor R. The circuit structure of the series-shunt type of HEMT switch is similar to the former, except that two transistors are in series with the transmission lines.

Different states of HEMTs can result in different impedance matching networks in the SPDT switches. Thus, the signal flow can be changed between the two output terms in figure 1. The simplified models for passive HEMT are showed in figure 2. The state of HEMT changes with two different controlling voltages (0/-2V) added to the gate of HEMT. When 0V is applied, the on state HEMT can be approximated by a small on-state resistor. When -2V is applied, the off state HEMT can be approximated by a small off-state capacitor shunted with an off-state resistor.

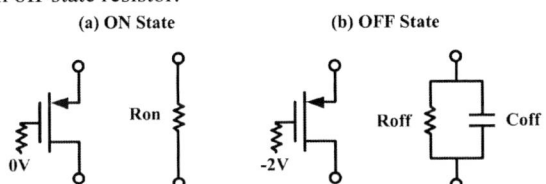

Figure 2: models for (a) on- and (b) off-state of passive HEMTs

At vg1=0V and vg2=-2V, the input impedance of the upper signal path in the shunt type of HEMT switch is very low and the input signal will be reflected, while the lower signal path can be equivalent to a 50Ω transmission lien. Thus, the input signal flows mainly through the lower signal path. Similarly, the upper signal path in the series-shunt type of switch turns off for impedance mismatch, while the lower signal path transmits the input RF signal to the output term.

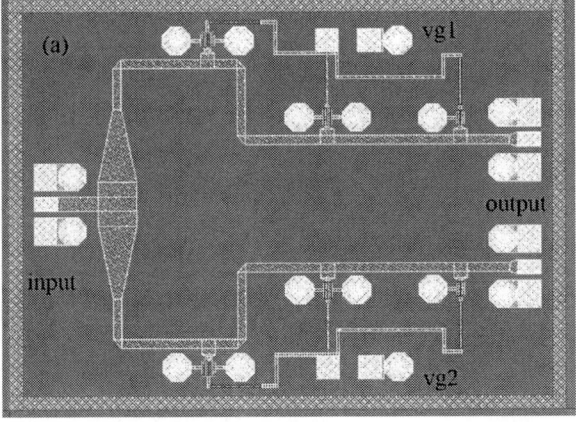

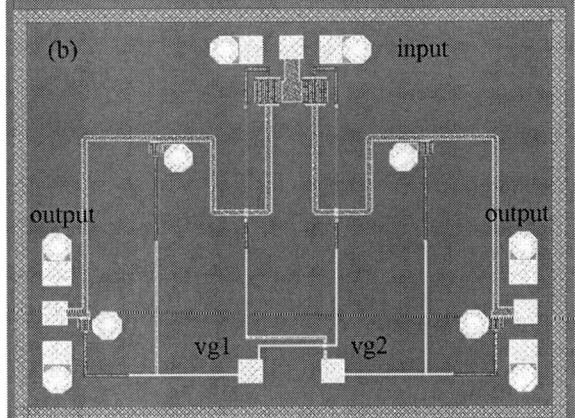

Figure 3: layout photos of (a) shunt type and (b) series-shunt type of HEMT SPDT switches

For desired isolation and insertion loss, the device size should be carefully selected. Larger transistor has smaller on-state resistor and larger off-state capacitor, resulting in better isolation but poor insertion loss. The switch stage number also needs to be properly considered. Switches with more stages have better isolation but larger chip size. So, trade-offs should be done in the design of a passive HEMT SPDT switch.

In this paper, three stages shunt type and series-shunt type of passive HEMT SPDT switches are simulated and compared. Figure 3 shows the layout of the switches, both have a compact chip size smaller than 1.5*2 mm².

RESULTS AND DISCUSSION

The designed switches are simulated using the software of advanced design systems. Figure 4 shows the isolation of the switches. It can be seen that both the shunt type and series-shunt type of HEMT SPDT switches demonstrate an isolation better than 39dB in 24~27GHz. We can also draw a conclusion that the number of shunted transistors used to form the low input impedance affects the isolation performance significantly.

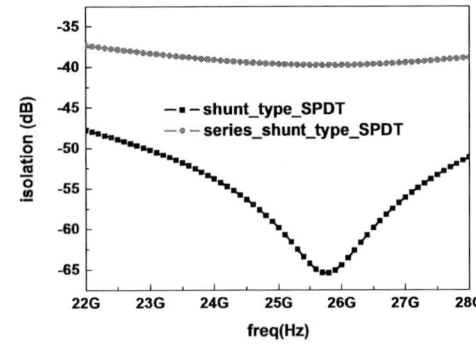

Figure 4: simulated isolation of the HEMT SPDT switches

Figure 5 compares the insertion loss of the two types of switches. The shunt type of switch demonstrates an insertion loss less than 1.5dB and the series-shunt type of switch less than 2dB in 24~27GHz. The on-state resistor of HEMT in series with the transmission line consumes some energy from the input signal, so the series-shunt type of switch shows poor insertion loss compared with the shunt type of switch.

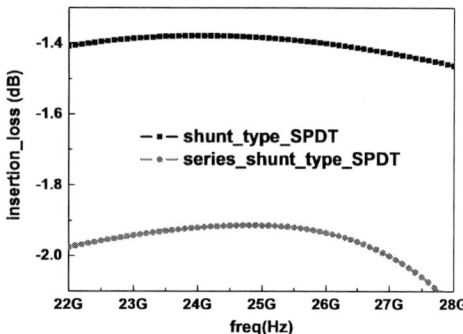

Figure 5: simulated insertion loss of the HEMT SPDT switches

Figure 6 depicts the return loss of the designed switches. With return loss better than 15dB, the energy of input RF signal can be mostly conducted to the output term.

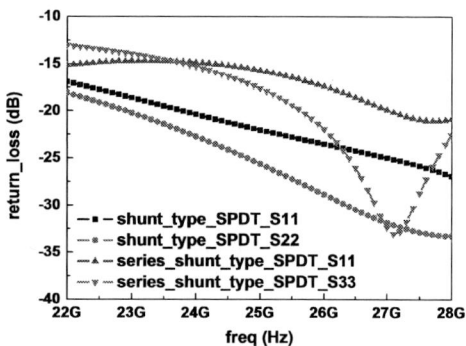

Figure 6: simulated return loss of the HEMT SPDT switches

The characteristics of both the shunt type and series-shunt type of passive HEMT SPDT switches are compared and discussed above. We can learn that high isolation and low insertion loss performance can be obtained by these designed switches. What's more, with more stages of transistors shunted to form the impedance matching network, the shunt type of switch overwhelms the series-shunt type of switch in the circuit performance with comparable circuit size.

CONCLUSION

In this paper, high isolation K/Ka band shunt type and series-shunt type of passive HEMT SPDT switches are designed and discussed. The simulated results show isolation better than 29dB and insertion loss better than 2dB for both the designed circuits with 3GHz bandwidth. With proper circuit design, isolation better than 50dB and insertion loss better than 1.5dB can be achieved by the shunt type of switch with a compact circuit size smaller than $1.5*2$ mm^2. This work laid the foundation for future T/R MMICs, in which the designed switches can be used to choose the transmitting or receiving RF paths with common antenna.

REFERENCES

[1] Verma Prolay, Kumar Ajay, Bhattacharya A.N. and Arora Rajkumar . IEEE Int. Conf. Electron., Comput. Commun. Technol., CONECCT 2015.

[2] Ingram D L, Sjogren L, Kraus J, et al. IEEE MTT-S International Microwave Symposium, 1998, 1:227-230.

[3] Ruei Bin Lai, Shih Fong Chao, Zuo Min Tsai, J. Lee, Huei Wang. IEEE Transactions on Microwave Theory and Techniques 2008, Vol 56, Iss 7, pp: 1545-1554.

[4] Lin K Y, Tu W H, Chen P Y, et al. IEEE Transactions on Microwave Theory Techniques, 2004, 52(8):1798 - 1808.

[5] Tsai, Yi Chien, J. L. Kuo, and H. Wang. Microwave Conference, 2008. APMC 2008. Asia-Pacific IEEE, 2008:1-4.

The multi-segment adaptive control of the high temperature heat source in a MOCVD vacuum reactor

Jung- Ching Chiu[1], Chih-Kai Hu[2], Pi-Cheng Tung[2], Tomi T. Li[2]*

[1] Graduate institute of Opto- Mechatronics Engineering, National Central University, Taoyuan 320, Taiwan (R.O.C)

[2] Department of Mechanical Engineering, National Central University, Taoyuan 320, Taiwan (R.O.C)

*Corresponding Author's Email: a17365182@gmail.com

ABSTRACT

This research focuses on the adaptive control for a high temperature heat source. The significant characteristic of adaptive control is self-regulation. The adaptive control is very important in a MOCVD process since different parameter values to control temperature in each temperature step are required. Simulations and experiments are performed to verify the feasibility of the proposed control scheme for the MOCVD process. The simulation results are compared with those of the experiment. The results show that the steady-state error in proportional-integral (PI) control is about 3°C, and in adaptive control is 0.3°C. It can be found that the performance of the adaptive control is much better than that of the traditional PI control which is commonly used in industry.

Keywords—MOCVD reactor; heater; susceptor; temperature control; adaptive control

INTRODUCTION

Gallium nitride (GaN) is a complex semiconductor material to manufacture optoelectronic devices, such as light-emitting and laser diodes. [1-3]. For growing high quality GaN based on blue epitaxial layers, metal-organic chemical vapor deposition (MOCVD) is the major technique.

The uniformity of thin film is closely related to temperature distribution on the susceptor. Since the stable heating temperature is significant for GaN film quality, a perfect heating system needs to provide uniform heat flux and the stable heat source [4-5].

In MOCVD process, generally when heating at different operating temperature, PID (Proportion - Integral - Differential) control is the convenient tool for temperature adjustment. However, if we use this traditional method [6-7], the controller will cause the heating fluctuation for the system to produce maximum overshoot.

Therefore, intelligent controller [8-9] is the best way to solve this issue. It can not only regulate the system parameters at different operating temperature but also suppress the interference which is caused by the external environment.

As the result, this study demonstrates the adaptive control method for application of high temperature control. In order to implement the adaptive control for tracking the desired temperature during MOCVD process, at the very start, the full program parameter is developed by the commercial software MATLAB. The best heating curve is obtained after we calculate the data based on the optimized initial parameter value. And then, we apply it to the man-machine interface to perform adaptive temperature control. In other words, when the user inputs temperature command, the man-machine interface will response to the best parameters for the control immediately, and adaptive control is executed by the system controller.

SYSTEM DESCRIPTION AND MATHMATICAL MODE SYSTEM

Adaptive control combined with Smith predictor

The multi-segment heating system is a system with the delay and coupling. It is difficult to obtain system parameters, but the process model is simple to derive. Therefore, the controller uses the adaptive controller to solve the coupling and the uncertain parameters. The adaptive controller will then combine with Smith predictor because the response of temperature will not react immediately at the temperature rising. The structure of adaptive system is shown in Figure 1 where G is the system plant including heater, $\hat{G}_m$ is the estimation system, $\hat{T}_m$ is the estimation time y_{ref} is target temperature namely operating temperature, y is the output, u_{ff} is the feedforward controller, u_{fb} is the feedback controller and u is control input.

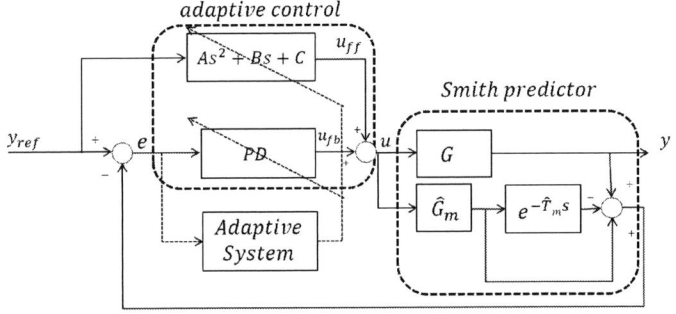

Fig 1: System structure. The left side shows adaptive structure and right side shows Smith predictor.

The Model Reference Adaptive Controller (MRAC) derivation is shown as follows:

The original system dynamic equation can be expressed as

$$G(s) = \frac{Y(s)}{U(s)} = \frac{b}{s+a} \times \frac{1}{\tau s+1} = \frac{b}{\tau s^2 + (1+\tau a)s + a}$$

$$u(t) = A^* \ddot{y} + B^* \dot{y} + C^* y , \qquad (1)$$

where $\frac{b}{s+a}$ is transfer function of the controlled plant, $\frac{1}{\tau s+1}$ is the transfer function of the heater, τ is time constant,

$$A^* = \frac{\tau}{b}, B^* = \frac{1+\tau a}{b}, C^* = \frac{a}{b}$$

$$e = y_{ref} - y.$$

The controller equation is given by

$$u_{fb}(t) = k_p e(t) + k_d \dot{e}(t) + F \qquad (2)$$

$$u_{ff}(t) = A\ddot{y}_{ref} + B\dot{y}_{ref} + Cy_{ref} \qquad (3)$$

$$u(t) = u_{fb}(t) + u_{ff}(t) \qquad (4)$$

$$A^*\ddot{y} + B^*\dot{y} + C^*y = k_p e + k_d \dot{e} + F + (A\ddot{y}_{ref} + B\dot{y}_{ref} + Cy_{ref})$$

$$\ddot{y}_{ref} - \ddot{y} = \frac{1}{A^*}[-(B^* + k_d)(\dot{y}_{ref} - \dot{y}) - (C^* + k_p)(y_{ref} - y) - F +$$

$$(C^* - C)y_{ref} + (B^* - B)\dot{y}_{ref} + (A^* - A)\ddot{y}_{ref}]. \qquad (5)$$

Defining the 2×1 matrix

$$\mathbf{z} = \begin{bmatrix} e \\ \dot{e} \end{bmatrix}$$

and let $z_1 = e$, $z_2 = \dot{z}_1 = \dot{e}$, $\dot{z}_2 = \ddot{e}$, will get

$$\dot{z} = \begin{bmatrix} 0 & 1 \\ \frac{-1}{A^*}(C^* + k_p) & \frac{-1}{A^*}(B^* + k_d) \end{bmatrix} z + \begin{bmatrix} 0 \\ \frac{-F}{A^*} \end{bmatrix} + \begin{bmatrix} 0 \\ \frac{C^* - C}{A^*} \end{bmatrix} y_{ref}$$

$$+ \begin{bmatrix} 0 \\ \frac{B^* - B}{A^*} \end{bmatrix} \dot{y}_{ref} + \begin{bmatrix} 0 \\ \frac{A^* - A}{A^*} \end{bmatrix} \ddot{y}_{ref} \qquad (6)$$

$$= \begin{bmatrix} 0 & 1 \\ -\Delta_1 & -\Delta_2 \end{bmatrix} z + \begin{bmatrix} 0 \\ -\Delta_3 \end{bmatrix} + \begin{bmatrix} 0 \\ \Delta_4 \end{bmatrix} y_{ref} + \begin{bmatrix} 0 \\ \Delta_5 \end{bmatrix} \dot{y}_{ref} + \begin{bmatrix} 0 \\ \Delta_6 \end{bmatrix} \ddot{y}_{ref},$$

where z is system state-space matrix and

$$\Delta_1 = \frac{C^* + k_p}{A^*}, \Delta_2 = \frac{B^* + k_d}{A^*}, \Delta_3 = \frac{F}{A^*}, \Delta_4 = \frac{C^* - C}{A^*}, \Delta_5 = \frac{B^* - B}{A^*},$$

$$\Delta_6 = \frac{A^* - A}{A^*}.$$

Define a reference model (It is stable and existent in symmetric positive definite matrix.)

$$\dot{z}_\mathbf{m} = -\mathbf{D_m z_m}, \qquad (7)$$

where $\mathbf{D_m}$ is the 2×2 matrix $\begin{bmatrix} 0 & 1 \\ -D_1 & -D_2 \end{bmatrix}$ D_1 and D_2 are constants, and the subscript 'm' denotes the reference model. Since the reference model is stable, there exists a 2×2 symmetric positive definite matrix

$$\mathbf{P} = \begin{bmatrix} P_1 & P_2 \\ P_2 & P_3 \end{bmatrix},$$

which satisfies the Lyapunov equation

$$\mathbf{PD + D'P = -Q}, \qquad (8)$$

where $\mathbf{Q}$ is a symmetric positive definite 2×2 constant matrix. The state error is shown as follow

$$\mathbf{E = z_m - z} \qquad (9)$$

$$\dot{\mathbf{E}} = \dot{z}_\mathbf{m} - \dot{z}$$

$$= \begin{bmatrix} 0 & 1 \\ -D_1 & -D_2 \end{bmatrix} E + \begin{bmatrix} 0 & 1 \\ \Delta_1 - D_1 & \Delta_2 - D_2 \end{bmatrix} z + \begin{bmatrix} 0 \\ -\Delta_3 \end{bmatrix} + \begin{bmatrix} 0 \\ -\Delta_4 \end{bmatrix} y_{ref}$$

$$+ \begin{bmatrix} 0 \\ -\Delta_5 \end{bmatrix} \dot{y}_{ref} + \begin{bmatrix} 0 \\ -\Delta_6 \end{bmatrix} \ddot{y}_{ref}. \qquad (10)$$

Defining parameter error ϕ

$$\phi = \begin{bmatrix} 0 & 1 \\ \Delta_1 - D_1 & \Delta_2 - D_2 \end{bmatrix} z + \begin{bmatrix} 0 \\ -\Delta_3 \end{bmatrix} + \begin{bmatrix} 0 \\ -\Delta_4 \end{bmatrix} y_{ref} + \begin{bmatrix} 0 \\ -\Delta_5 \end{bmatrix} \dot{y}_{ref}$$

$$+ \begin{bmatrix} 0 \\ -\Delta_6 \end{bmatrix} \ddot{y}_{ref}. \qquad (11)$$

To make sure state error stable, define Lyapunov equation

$$\mathbf{V = PE}^2 + \mathbf{Q\phi}^2 \qquad (12)$$

$$\dot{\mathbf{V}} = 2\mathbf{PE\dot{E}} + 2\mathbf{Q\phi\dot{\phi}}$$

$$= 2\mathbf{PE}(-\mathbf{AE} + \varphi) + 2\mathbf{Q\phi\dot{\phi}}$$

$$= -2\mathbf{PAE}^2 + 2R\varphi + 2\mathbf{Q\phi\dot{\phi}}$$

$$= -2\mathbf{PAE}^2 + 2\varphi(R + 2\mathbf{Q\dot{\phi}}). \qquad (13)$$

If equation (13) is negative definite function, the state of the adjustable system tends to that of the reference model asymptotically.

Finally we obtain adaptive law:

$$F(t) = F(0) + \delta_2 R + \delta_1 \int_0^t R(t)dt \qquad (14)$$

$$K_P(t) = K_P(0) + \alpha_2 R(t)E(t) + \alpha_1 \int_0^t R(t)E(t)dt \qquad (15)$$

$$K_D(t) = K_D(0) + \beta_2 R(t)\dot{E}(t) + \beta_1 \int_0^t R(t)\dot{E}(t)dt \qquad (16)$$

$$C(t) = C(0) + \nu_2 R(t)\theta(t) + \nu_1 \int_0^t R(t)\theta(t)dt \qquad (17)$$

$$B(t) = B(0) + \gamma_2 R(t)\dot{\theta}(t) + \gamma_1 \int_0^t R(t)\dot{\theta}(t)dt \qquad (18)$$

$$A(t) = A(0) + \lambda_2 R(t)\ddot{\theta}(t) + \lambda_1 \int_0^t R(t)\ddot{\theta}(t)dt, \qquad (19)$$

where $\{ \delta_1, \alpha_1, \beta_1, \nu_1, \gamma_1, \lambda_1 \}$ and $\{ \delta_2, \alpha_2, \beta_2, \nu_2, \gamma_2, \lambda_2 \}$ are positive and zero/ positive scalar adaptation gains, and $R(t)$ is the "weighted" error defined as

$$R(t) = p_2 E(t) + p_3 \dot{E}(t). \qquad (20)$$

Smith predictor

The system can be divided into $\hat{G}_m$ and $e^{-\hat{T}_m s}$ as shown in Figure 2, where o $\hat{G}_m$ is the transfer function without delay time, $e^{-\hat{T}_m s}$ is the delay time of G. After Smith predictor, Y_m is the real output without delay time.

$$Y_m = Y - Y_{s1} + Y_{s2}. \qquad (21)$$

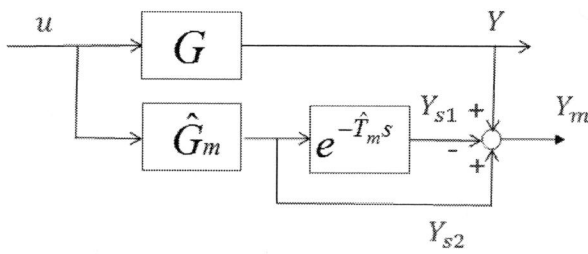

Fig 2: Smith predictor structure.

Visual Basic human–machine interaction

The derive formula is imported to Visual Basic code. When temperature is rising, the real temperature will feedback to the calculation. After inputting target temperature, the computer executes the adaptive control and calculates the control input in real time with Visual Basic human–machine interaction which is shown in Figure 3.

978-1-5090-6695-7/17 $31.00 © 2017 IEEE

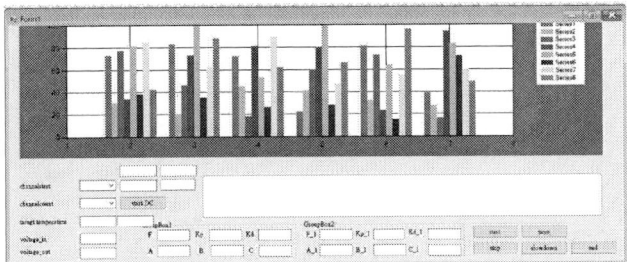

Fig.3: Visual Basic human–machine interaction

RESULTS AND DISCUSSION

First, using PI (proportional-integral) controller and the target temperature of heater is set to 950°C. The result is shown in Figure 4. There is steady-state error about ± 3°C.

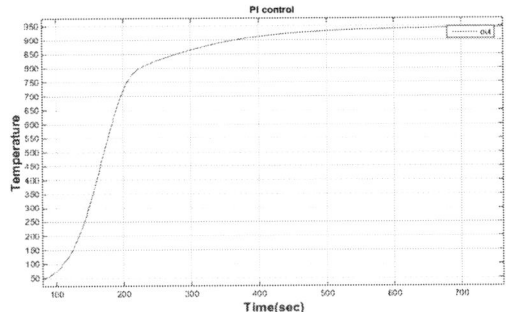

Fig 4: 950°C results of PI control.

Then, the same system is controlled by the adaptive controller, and the target temperature is set to 950°C. The result is shown in Figure 5. Obviously, the system is more stable, and reduces steady-state error to ±0.3°C

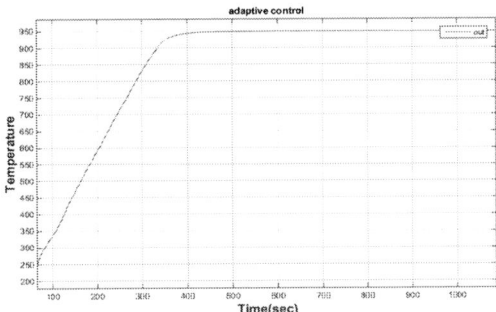

Fig.5: 950°C results of adaptive control.

At last, the adaptive controller is imported to two-segment heating system, which consists of inner and outer heater as shown in Figure 6. The target temperature of inner heater is set 950°C-850°C-1000°C, outer heater is set 900°C-800°C-950°C. The simulation result and experiment results are shown in Figures 7-9.

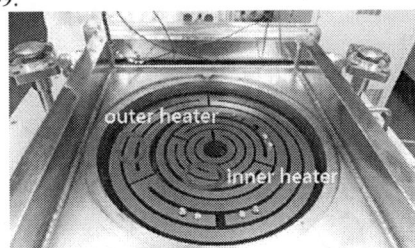

Fig.6: The position of inner and outer heater.

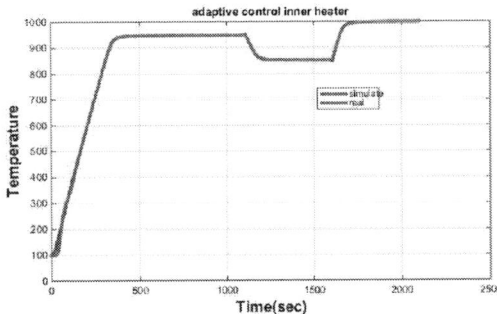

Fig 7: Temperature results of adaptive control from the MATLAB simulation and the real experiment of inner heater. The pressure of the thermal experiment in the reactor is about 1 tor.

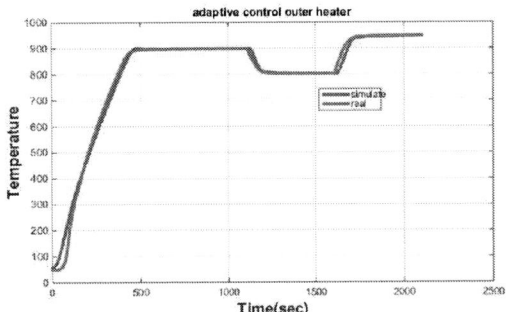

Fig 8: Temperature results of adaptive control from the MATLAB simulation and the real experiment of outer heater. The pressure of the thermal experiment in the reactor is about 1 tor.

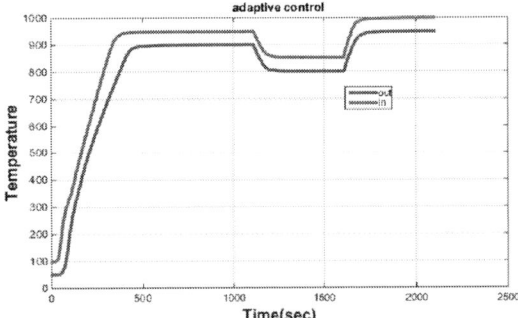

Fig 9: Temperature results from the actual heating situation.
According to the results, which can be found that the response of temperature is smooth to achieve target temperature. Also the temperatures of inner and outer heater are close to simulation result. Additionally, it can be achieved that the curve of rising temperature shows no overshoot and the steady-state error is ±0.3°C.

CONCLUSION

In MOCVD process, the control of heating system is necessary. This study shows that the adaptive controller is much better than PI control. The adaptive controller can be self-tuning to fit each temperature layer, inhibit the noise of environment, and solve the coupling problem between the two heaters.

978-1-5090-6695-7/17 $31.00 © 2017 IEEE

References

[1]F. A. Ponce and D.P. Bour, Nature, vol. 386, 1997, pp.351-359.

[2] S.J. Chang, W.C. Lai, Y.K. Su, J.F. Chen, C.H. Liu and U.H. Liaw, IEEE J. Sel. Top. Quant. Electron., vol.8, 2002, pp. 278-283.

[3]G.S. Padma, S. Singh, M. Mathew, K. Singh, B.C. Joshi, S.Das and C. Dhanavantri, J. Opt., vol.41, 2012, pp.198-200.

[4] M. Dauelsberg, C. Martin, H. Protzmann, A.R. Boyd, E.J. Thrush J. Käppeler, M. Heuken, R.A. Talalaev, E.V. Yakovlev and A.V. Kondratyev, J. Cryst. Growth, vol.298, 2007, pp. 418-424.

[5] D. G. Zhao, J. J. Zhu, D. S. Jiang, Hui Yanga, J.W. Liang, X.Y. Li and H.M. Gong, J. Cryst. Growth, vol.289, 2006, pp. 72-75.

[6] J. C. Basilio and S. R. Matos, IEEE Trans. Educ., vol. 45, 2002, pp.364-370.

[7] E. Aridhi, M. Abbes and A. Mami, 15th International Conference on Sciences and Techniques of Automatic Control and Computer Engineering (STA), Dec. 2014, pp.123-128.

[8]T. Fukuda, and T. Shibata, IEEE Transactions on Industrial Electronics, vol.39, 1992, pp.472-489.

[9]H. Seraji, J. Dyn. Sys., Meas., Control, vol. 109, 1987, pp.193-202.

HYBRID THERMAL AWARE RECONFIGURABLE 3D IC WITH DYNAMIC POWER GATING ARCHITECTURE

Chun-chen Liu[1], Yilei Li[2], Yuan Du[2], Li Du[2] and Tianchen Wang[3]*

[1]University of California, Los Angeles, USA
[2]Kneron, Inc., San Diego, California, USA
[3]University of Notre Dame, Notre Dame, Indiana, USA
*Corresponding Author's Email: twang9@nd.edu

ABSTRACT

In this paper we propose an innovative 3D IC architecture that combines reconfigurable 2D structure with monolithic 3D. This new architecture not only resolves Power Distributive Network (PDN) design and thermal management issues of traditional 3D-IC, but also provides additional power control and programmable routing capability. It provides a cost effective way to integrate different modules together using stacked interposer structure. With power rails and signal paths that can be routed dynamically using reconfigurable peripheral switches, the new system is adjustable. Moreover, area saving is achieved by using monolithic 3D to realize the modules. With the corresponding new thermal aware hierarchical simulated annealing floorplan algorithm designed for our hybrid reconfigurable architecture, the thermal problem can be further alleviated. Our testing results on 15 benchmarks show that we obtain an average 1.69X lower temperature and average 2.82X smaller power compared with traditional 2D SoC structure, 1.3X lower temperature compared to traditional 3D structure.

INTRODUCTION

While 3D IC provides high vertical interconnection density between device ties using Through-silicon-via (TSV), which makes global interconnect shorter and faster, another problem rises as the adoption of 3D architecture leads to increased power densities that can result from placing one high power density block over another in the multi-layered 3D stack. Besides thermal problem, power consumption has always been a major concern in IC design, and the low power challenge remains in 3D design architecture. With the scaling down of device size, a big portion of power consumption is now due to leakage power. TSVs can potentially be used to implement inductors in 3-D integrated systems for minimal footprint and large inductance [1, 2], which also provide heat removal [3] and substrate loss reduction [4].

Floorplan aiming at solving 3D thermal problem have been investigated by many researchers. Cong [5] also proposed a thermal-driven floorplanning algorithm for 3D ICs. A novel 3D floorplan representation called Combined Bucket and 2D Array (CBA) was introduced, and thermal consideration was realized by calculating the temperature distribution of the circuit and putting thermal constraint on the simulated annealing cost function. Hsien-Te Chen [6] designed a new architecture for power network in 3D IC, which included a thermal distribution network model. Based on the thermal network, thermal TSVs were placed at locations with maximum temperature gradient vertically among tiers to provide thermal dissipation paths.

In this paper, we propose an innovative 2D and monolithic 3D hybrid reconfigurable integrated circuit architecture that takes the advantage of 2D reconfigurable structure and monolithic 3D interconnection. On architecture level, our structure allows different modules to integrate together using the stacked interposer. Moreover, modules are dynamically connected together using reconfigurable peripheral switches. And Power rails and signal paths can be routed to different modules similar to field programmable gate array. On block level and device level, our structure saves area by employing 3D architecture concept, and the power network problem 3D-IC suffers are solved by routing power rails through peripheral switches.

3D-IC HYBRID RECONFIGURABLE ARCHITECTURE

3D-IC Hybrid Reconfigurable Architecture is shown in Figure 1. The dies are stacked together using interposer and connected using redistribution layer in the horizontal direction. Instead of routing the power and signal through the dies, additional reconfigurable peripheral switches are used so that all the power rails and signal paths are routed through the switches from the lower level to upper one in vertical direction. Based on different requirements, the power rails and signal paths can be routed to various dies in the same package, therefore, the single package can be used for multiple purposes through the routing modification only. This approach can also bypass the false die and reroute all the connection.

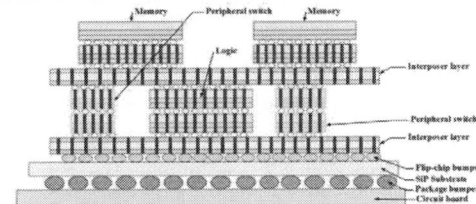

Figure 1: Our proposed 3D ICs Hybrid Reconfigurable Architecture

978-1-5090-6695-7/17 $31.00 © 2017 IEEE

Reconfigurable peripheral switches with 3D structure not only provide horizontal connections on each interposer layer, but also offers vertical interconnections that connect the upper and lower interposer layers. The vertical connections are provided by the TSVs through dies, which are controlled by the device layer. Meanwhile, the device layer also provides the horizontal connections within the interposer layer.

Benefit from the reconfigurable structure, the identical die can be used for any packages without modification. It significantly reduces the design effort and shorten the development cycle. With this approach, it is easy to stack more dies in the same package without limitation. The expandable capability is important which can create different modules suitable for different requirements. Take in-house server as an example, the original design placed 2 processor and 2 memory chips in the package. With 3D reconfigurable architecture, 2 more memory chips can be easily added into the package without significant modification.

3D-IC THERMAL AWARE FLOORPLAN ALGORITHM

Our proposed 3D reconfigurable architecture has provided a reliable implementation approach. Moreover, the problem of thermal dissipation can be further alleviated by our proposed thermal aware floorplan. The thermal aware floorplan is a simulated annealing algorithm based on a novel 3D IC floorplan tree and block power correlation.

Novel 3D IC floorplan tree

Existing 3D floorplan trees are usually designed based on 2D floorplan B*-trees. A widely used 3D floorplan tree is called 3D B*-tree, in which a 2D B*-tree is used for each tier of the 3D IC. And the modules within one tier are placed by the 2D B*-tree placement.

In this paper we use a novel 3D tree structure called ternary 3D tree, which consists of 4 ternary trees. Ternary 3D tree structure has the advantage of exchanging blocks in the same layer and different layer in the same step, instead of using a special inter-tier swap step to realize interlayer blocks changing in 3D B*-tree.

The ternary tree is a tree data structure in which each node has at most three child nodes. Although two ternary trees are enough to determine a unique 3D floorplan, it is not enough to easily identify invalid cases. Thus we use four ternary trees to represent one unique floorplan structure. As shown in Figure 2, block A and F are diagonally adjacent, and four ternary trees were grown started from four non-adjacent vertexes to describe this floorplan. The red lines indicate the paths that connect four pairs of the cubic diagonal vertexes.

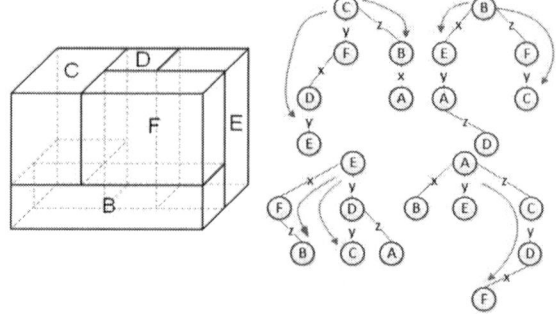

Figure 2: Our proposed 3D ICs Hybrid Reconfigurable Architecture

Power Correlation Calculation

3D IC design reduces the global interconnect power, but the stacking of multiple active layers leads to higher power densities, which becomes a major obstacle for the adoption of this new structure. With the fast development of 3D IC technology, the thermal problem has drawn significant attention.

Instead of directly putting temperature into SA cost function, our thermal aware 3D floorplan algorithm explores the power correlation between different blocks, based on which we add more specific objectives to be optimized in the SA cost function. By providing power correlation coefficient, we avoid calculating the chip peak temperature over again to evaluate the value of cost function. The utilization correlation between blocks can be defined by Pearson correlation coefficient as Eqn. 1:

$$\rho(i,j) = cov(i,j)/(\sigma_i \cdot \sigma_j) \quad (1)$$

After fetching data from simulation, a utilization per block table can be created as Table I. $p_{i\tau}$ is the power block i consumes during time t_τ.

TABLE I.   POWER CORRELATION TABLE

	t_1	t_2	t_3	t_4	...	t_N
$block_1$	p_{11}	p_{12}	p_{13}	p_{14}	...	p_{1N}
$block_2$	p_{21}	p_{22}	p_{23}	p_{24}	...	p_{2N}
...						
$block_M$	p_{M1}	p_{M2}	p_{M3}	p_{M4}	...	p_{MN}

So $cov(i,j)$ can be calculated as Eqn. 2:

$$cov(i,j) = \sum_{\tau=1}^{N} p_{\tau i} \cdot q_{\tau j} - \frac{1}{N}\sum_{\tau=1}^{N} p_{\tau i} \sum_{\tau=1}^{N} q_{\tau j} \quad (2)$$

Eqn. 2 defines the co-variance matrix between blocks. And the standard deviation of $block_i$ and $block_j$ are represented by σ_i and σ_j, which can be calculated as:

978-1-5090-6695-7/17 $31.00 © 2017 IEEE   542

$$\sigma_i = \sqrt{\sum_{\tau=1}^{N} \frac{p_{\tau i}^2}{N} - \left(\sum_{\tau=1}^{N} \frac{p_{\tau i}}{N}\right)^2} \qquad (3)$$

After calculation, we now have the power correlation coefficient between any two blocks in the design. If the correlation coefficient is positive and near to 1, it implies that $block_i$ and $block_j$ tend to have similar utilization.

EXPERIMENT RESULT

To test the performance of our new 3D hybrid reconfigurable structure, a set of SPEC 2000 benchmarks was used for evaluation. We run 20 SPEC 2000 applications and get the average value for different blocks on traditional 2D SoC model and our 3D hybrid reconfigurable model separately under the same operating frequency.

The lef/def files we used were provided by TSMC 28HPC. The peak temperature result is shown in Figure 3, where the new architecture decreases the peak temperature of the chip on average by 1.69X. The temperature improvement we obtained even compared with 2D SoC indicates the superior of our thermal-aware floorplan algorithm.

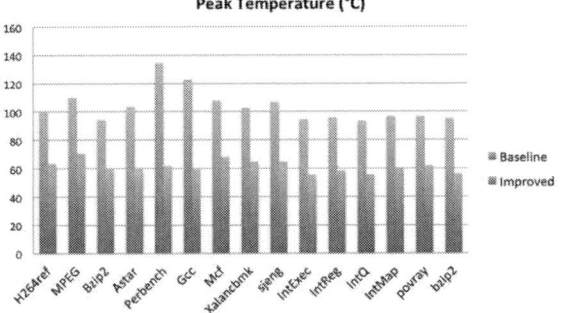

Figure 3: Peak temperature comparison between baseline (2D) and improved (3D hybrid reconfigurable)

The effect of thermal-aware algorithm can be illustrated in detailed temperature distribution map as shown in Figure 4, which is the average temperature map of H264ref block. Without the thermal-aware floorplan process, regular floorplan has the possibility of putting high positive power correlated blocks together, thus causing three hot spots with peak temperature over 90°C on the chip. While in our design the high positive power correlated blocks are separated as far as possible, to avoid the hot spot and achieve a more balanced temperature distribution where no place with peak temperature higher than 65°C is shown on the map. Beside temperature improvement, saving on power consumption is also achieved using our new architecture and corresponding thermal-aware floorplan algorithm.

For comparison with traditional 3D architecture, we compare our result with [7]. They generated 10 random DFG based applications, and the tasks in these applications

are randomly selected from the SPEC 2000 benchmarks. The benchmarks are implemented in a 3D multi-core architecture and the average peak temperature for all cores exceeds 80°C, while test result on our 3D hybrid reconfigurable result shows that the average peak temperature of all the blocks is 61.3°C, which means compared to 3D architecture, our new structure reduces peak temperature more than 1.3X.

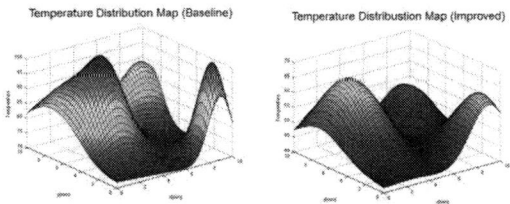

Figure 4: Peak Temperature distribution map for baseline (2D) and 3D hybrid reconfigure

CONCLUSION

In this paper, we proposed a 3D IC hybrid reconfigurable architecture with peripheral switch support that will revolute current 3D IC development. Multiple dies with different functionalities can be integrated into single package and reprogrammable for different applications, which can significantly reduce the overall development cost. We also introduced a thermal aware floorplan algorithm corresponding to our reconfigurable 3D architecture. Our results show that the new architecture has a better performance compared to both 2D and traditional 3D architecture.

REFERENCES

[1] U. R. Tida, C. Zhuo, and Y. Shi. *19th Asia and South Pacific Design Automation Conference (ASP-DAC) IEEE.* January, 2014, pp. 837-842.

[2] H. Wang, J. Kim, Y. Shi, and J. Fan. *Proc. of Asia-Pacific EMC Symposium. Jeju Island, Korea.* 2011.

[3] U. R. Tida, C. Zhuo, and Y. Shi. *ACM Journal on Emerging Technologies in Computing Systems (JETC)*, vol. 11(2), 2014, article 16.

[4] U. R. Tida, R. Yang, C. Zhuo, and Y. Shi. *IEEE Transactions on Very Large Scale Integration (VLSI) Systems*, vol. 23(7), 2015, pp. 1322-1334.

[5] J. Cong, J. Wei, and Y. Zhang, *IEEE/ACM International Conference on Computer Aided Design.* ICCAD-2004., 2004, pp. 306-313.

[6] H.-T. Chen, H.-L. Lin, Z.-C. Wang, and T. Hwang. *Design, Automation & Test in Europe, IEEE.* 2011, pp. 1-6.

[7] J. Li, M. Qiu, J. Hu, and E. H.-M. Sha. *Signal Processing Systems (SIPS), 2010 IEEE Workshop.* 2010, pp. 323-326.

THE APPLICATION OF A HEATING BAFFLE IN A HIGH TEMPERATURE VACUUM REACTOR

Kuei-Fang Chen[1], Jun-Ching Chiu[1], Chih-Kai Hu[2], Tomi T. Li[2*], and Pi-Chen Tung[2]

[1]Department of Opto-Mechanical Engineering, National Central University, Taoyuan 320, Taiwan (R.O.C)
[2]Department of Mechanical Engineering, National Central University, Taoyuan 320, Taiwan (R.O.C)
* Corresponding Author's Email: fh100sh@gmail.com

ABSTRACT

The temperature uniformity of thin film process is an important issue for susceptor in high temperature vacuum system. When the temperature of process is extremely high, the uniformity is not only difficult to control, but also hard to maintain. In order to raise the temperature utilization at high temperature which is higher than 1300°C, the heating baffle is introduced for this purpose. As the result, the thermal radiation is economized about 1% in the chamber which contains six pieces of two-inch wafers. On the other hand, the heating baffle also improves the temperature uniformity of the wafer area on susceptor. The results show the temperature difference decreases to 6°C after using the heating baffle, and it shows that an optimum temperature uniformity can be achieved in a self-assembly reactor.

INTRODUCTION

Metalorganic chemical vapor deposition (MOCVD) is a key process in manufacturing of compound semiconductor devices such as light-emitting diodes and laser diodes. Among these, the uniformity of temperature on susceptor plays an important role for growing the group III-nitride thin film [1]. Moreover, critical growth temperature 1050°C is needed in various illumination industry and application [2-3].

On the other hand, the interest in the ultraviolet (UV) light emitting diodes has grown rapidly in recent years [4], and the growth temperature of ultraviolet light-emitting layer needs to be higher than the blue one. As the result, the critical temperature for each layer is listed as GaN: 1050°C [5], AlInGaN: 1100°C [6-7], and AlGaN: 1400°C [8]. Design on simulation, which is based on commercial numerical software is now widely used in the semiconductor industry in the optimization and troubleshooting of epitaxial growth processes that occur on the substrate. Furthermore, it is commonly employed in reactor design [9-10]. According to those listed references, the boundary effect is raised as the process temperature gradually increases, however, the uniformity of temperature descends. It is unfavorable to grow the thin film from above situation. By integrating the above mentioned works, in this study, a numerical analysis procedure for thermal utilization by using the heating baffle is presented in order to raise the thermal radiation efficiency. In addition, the result shows that the non-uniformity of temperature on susceptor is also reduced by introducing the heating baffle.

EXPERIMENTAL METHOD

The high temperature experiment model and mesh

The self-assembly high temperature reactor (Figure 1) is a horizontal-type MOCVD, and the heating system which included a two-zone heater and heating baffle. In this study, the commercial software COMSOL Multiphysics is used for this purpose. First, the model is built and the initial and boundary conditions are set for this study (Table 1). The pressure of the thermal experiment in the reactor is about 1 torr, the initial temperature in the reactor, before the experiment is room temperature, and the wall temperature maintains 40°C by a cooling system control. Second, the heat transfer module is imported and the parameter of this study is set by the user. Third, the numerical mesh of the three-dimension model is established (Figure 2). Final, the numerical result is obtained by the heat transfer model.

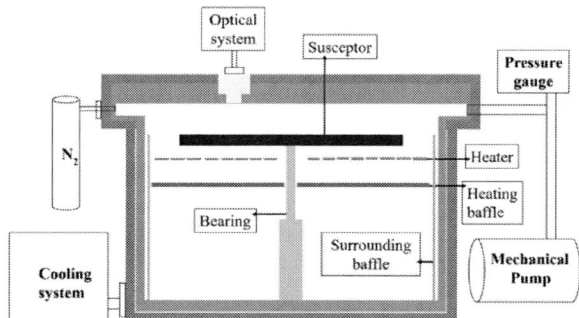

Figure 1: Schematic diagram of self-assembly high temperature reactor

TABLE I.   THE HEATING EXPERIMENT ENVIRONMENT

Model	Parameter		
	Original temperature	*Boundary temperature*	*Pressure*
3D	25°C	40°C	1 torr

The simulation of heating baffle model

The heating baffle in the system has two portions, which are divided into the below heating baffle and the surrounding baffle, the below one is beneath the heater and the surrounding baffle is close to the wall but does not cling it (Figure 3). The study of subject one is directing to the below baffle affect the distribution of the thermal radiation and the temperature uniformity. The study of subject two is to modify the boundary and see its effect.

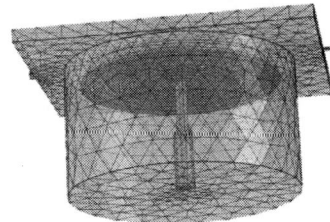

Figure 2: The mesh profile for the three-dimension horizontal reactor

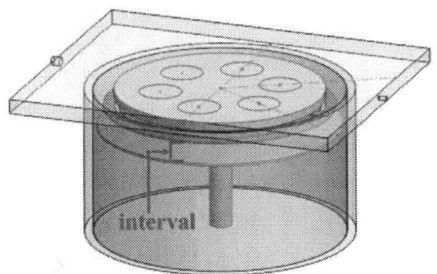

Figure 3: The three-dimension heating baffle model

The subject 1 involves five cases and studies the different intervals of heater to the below heating baffle from 10mm to 30mm increasing 5mm per interval. The subject 2 involves three cases and studies the boundary effect on the temperature distribution of skirts on susceptor. For the purpose of observing the temperature distribution on the cross-section of reactor and susceptor, the temperature uniformity on wafer for both needs to be simulated and further analyzed.

RESULTS AND DISCUSSION

The heating baffle below the heater

The thermal radiation is varied by with or without the heating baffle setting, thus the temperature on susceptor is also varied. The space-span of the thermal radiation becomes decreasing and narrowing when the thermal energy is close to upper level of the heating baffle, (Figures 4 and 5). The temperature of susceptor has conspicuously different distribution with respect to different intervals of heater to the below heating baffle (Figure 6). The results of wafer temperature on susceptor are recorded in Table II. Inspection of the temperature difference is deviated from the minimum to maximum temperature on the wafer, and the standard deviation of temperature for case III, IV, and V, is better than the contrast case (Figure 7). The case V which has interval for

30mm is the best case for all, and the temperature difference is smaller than 6°C, in other words, the temperature deviation is less than 1%.

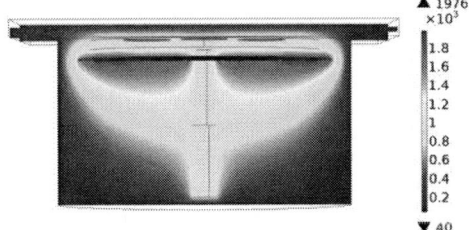

Figure 4: The thermal radiation of the high temperature reactor without the below heating baffle

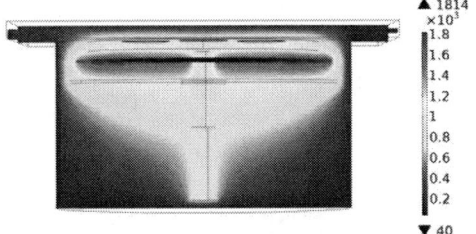

Figure 5: The thermal radiation of the high temperature reactor with the below heating baffle

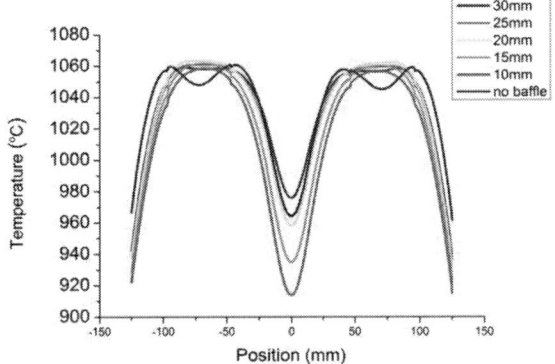

Figure 6: The temperature curve on the susceptor for the subject one

TABLE II.   THE ANALYZING TEMPERATURE OF THE WAFER

Case	Subject one for the below heating baffle		
	Interval	*Temperature difference*	*Standard deviation*
I	10mm	17.559	4.2689
II	15mm	14.747	3.4629
III	20mm	9.4253	2.0167
IV	25mm	8.637	1.8207
V	30mm	5.6302	0.9555
contrast	No baffle	12.7824	4.2322

978-1-5090-6695-7/17 $31.00 © 2017 IEEE

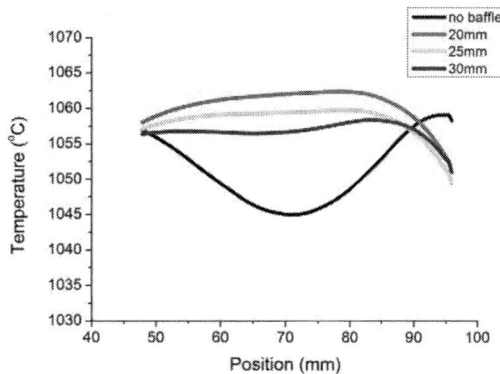

Figure 7: The temperature distribution of the wafer for case III to V and the contrast.

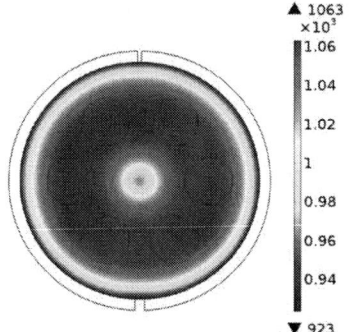

Figure 8: The temperature distribution on susceptor

In Figure 7, the temperature of wafer decreases starting from the 85mm in position. It is found that the periphery of susceptor's temperature distribution is lower than other areas (Figure 8). This indicates that there is an effect by the boundary; therefore, the surrounding baffle in heating experiment is used to reduce the deviation of temperature.

The correcting boundary effect with surrounding heating baffle

The boundary temperature is decreasing, because the cooling system is used in the wall and the thermal radiation is declining. Inspection of the case VI to case VIII to compare the case III to case V, adding the surrounding baffle in the self- assembly high temperature reactor can improve the uniformity of temperature on wafer (See Table III). Comparing with Table II, the temperature standard deviation reduces slightly, but the temperature difference is decreased about 2°C in the same of interval with the below heating baffle. In addition, the thermal radiation is economized around 19.6%.

TABLE III. THE EFFECT OF SURROUNDING BAFFLE

Case	the application of surrounding baffle		
	Interval	*Temperature difference*	*Standard deviation*
VI	20mm	6.5265	1.8107
VII	25mm	7.0638	1.6581
VIII	30mm	5.2311	1.1216

CONCLUSION

The thermal radiation distribution and the temperature uniformity are progressive that the heating baffle is applied to the self-assembly high temperature reactor. It is beneficial that the below heating baffle improves the temperature uniformity of the wafer, on the other hand, the quality of thin film is also preferable than before. Although the surrounding heating baffle improves the boundary effect less, it decreases the lost thermal radiation in a high temperature reactor. As a whole, the heating baffle which is used in a high temperature reactor like MOCVD raises the benefits for both the production and the film quality.

REFERENCES

[1] S. Nakamura, Jpn. J. Appl. Phys., vol. 30, 1991, pp. L1705-L1707.

[2] S. Nakamura, M. Senoh, and T. Mukai, Jpn. J. Appl. Phys., vol. 30, 1991, pp. L1708-L1711.

[3] S. Nakamura, T. Mukai, M. Senoh, and N. Iwasa, Jpn. J. Appl. Phys., vol. 31, 1992, pp.L139-L142.

[4] J. Yan, J. Wang, Y. Zhang, P. Cong, L. Sun, Y. Tian, C. Zhao, and J. Li, J. Cryst. Growth, vol. 414, 2015, pp. 254-257.

[5] S. Nakamura, T. Mukai, and M. Senoh, Jpn. J. Appl. Phys., vol. 30, 1991, pp. L1998-L2001.

[6] Y. Liu, T. Egawa, H. Ishikawa, B. Zhang and M. Hao, Jpn. J. Appl. Phys., vol. 43, 2004, pp. 2414-2418.

[7] R. Loganathan, M. Balaji, K. Prabakaran, R. Ramesh, M. Jayasakthi, P. Arivazhagan, S. Singh, and K. Baskar, J. Mater. Sci.: Mater. Electron., vol. 26, 2015, pp.5373-5380.

[8] N. Okada, N. Fujimoto, T. Kitano, G. Narita, M. Imura, K. Balakrishnan, M. Iwaya, S. Kamiyama, H. Amano, I. Akasaki, K. Shimono, T. Noro, T. Takagi, and A. Bandoh, Jpn. J. Appl. Phys., vol. 45, 2006, pp. 2502-2504.

[9] H. C. Chiu, C. K. Hu, H. I Chien, Tomi T. Li, P. C. Tung, *Semiconductor Technology International Conference*, Shanghai, 2015.

[10] C. P. Chang, C. K. Hu, H. C. Chiu, Tomi T. Li Semiconductor Technology International Conference, Shanghai, 2016

EVALUATION OF ULTRA-LOW POWER TUNNELING FIELD EFFECT TRANSISTOR POWER MANAGEMENT UNIT

Haifang Lu[1], Xin'an Wang[1]*, Jipan Huang[1], Zhiqiang Yang[1], Yuqian Huang[1] and Jijia Guo[1]

[1]The Key Laboratory of Integrated Micro-system Science and Engineering Applications, Peking University Shenzhen Graduate School, Shenzhen, 518000, China

* Email: anxinwang@pku.edu.cn

ABSTRACT

In this paper, we propose a TFET (Tunneling Field Effect Transistor) PMU (power management unit) of R80515 for ultra-low power. Both the dynamic power and leakage power are evaluated by HSPICE circuit simulation with Verilog-A models. From the simulation, we find the dynamic power of TFET circuits can be reduced by 80% and leakage power reduction can be nearly 30% compared with 130nm CMOS (Complementary Metal Oxide Semiconductor) implementation. The results indicate that TFET can achieve much higher power efficiency and the replacement can be vital to the whole design.

INTRODUCTION

As Moore's law goes on, the problem of power consumption in traditional integrated circuit design is more challenging because of quite large leakage current caused by short channel effect. Two usual ways for improvement, one is reducing the source voltage and the other is cutting down the leakage current [1]. TFET can cut down the power consumption in both ways by the BTBT (band-to-band tunneling) conduction mechanism. R80515 is a widely used single-chip 8-bit microcontroller, which executes all ASM51 instructions and now, is a microcode-free design. As an important module in R80515, PMU is the highest energy-consumption part. Considering the advantages TFET has, we use TFETs to build up the PMU instead of CMOS to investigate and make a comparison of the energy reduction.

CHARACTERISTICS OF DEVICE

Though the structure of TFET looks like the traditional MOSFET, it is a new type of devices as its conductive mechanism is band-to-band tunneling. Figure 1 presents the energy band of nTFET at off-state and on-state [2] and shows how BTBT works. During the off-state the width of energy barrier is too large to tunneling through for the carriers. However, when proper Vg turns on the device, the width of the energy barrier became much smaller. Then the carries can tunnel into the channel, which forms a current from source to drain.

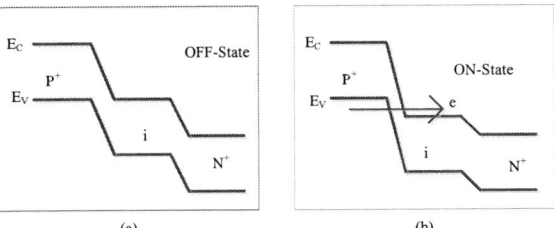

Figure 1: The energy band of nTFET. (a)off-state: VS=Vg=0V,Vd>0.(b)on-state: Vs=0V, Vg=Vd>0

Figure 2 shows the Vg-Id curves of TFET and MOSFET [3-4]. First the vertical line in figure 2 shows that at a low gate voltage TFET has a bigger Id than MOSFET which means TFET can work in a low voltage. Besides, from the picture below the SS (sub-threshold slop) of TFET is much steeper than that of MOSFET (Metal Oxide Semiconductor Field Effect Transistor) which is constrained by carrier thermionic emission [5].

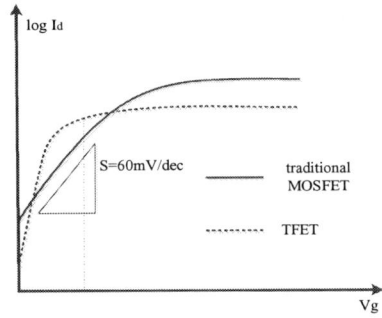

Figure 2: The comparison of sub-threshold charicterics of TFET and CMOS

The smaller leakage current and the lower operating current both make TFET have significant power reduction and better performance, which make TFET be considered to be one of the most promising devices to replace CMOS [6].

There are many studies about finding materials to achieve a better TFET performance and the InAs double-gate TFET model used in our design is provided by

University of Notre Dame [7]. TABLE I shows the technology parameters of the TFET model used in PMU. And the CMOS model used in the simulation is 130nm standard technology.

TABLE I. TECHNOLOGY PARAMETERS OF THE TFET MODEL USED IN PMU

Technology Parameters	Value
Gate Length	130nm
Channel Thickness	5nm
Equivalent Oxide Thickness	0.2nm
Semiconductor Band Gap	0.354eV
Threshold Voltage	0.145V
OFF Voltage	0V

STRUCTURES OF CIRCUITS

In R80515, power management unit is designed to reduce power consumption by totally cut down the input signals including clock signal when some parts of R80515 are not working. So it uses logical circuit to generate enable signals to control clocks to the core and the peripheral part of R80515, which can turn off and turn on the modules differently at each time. The detailed function is described as Figure 3.

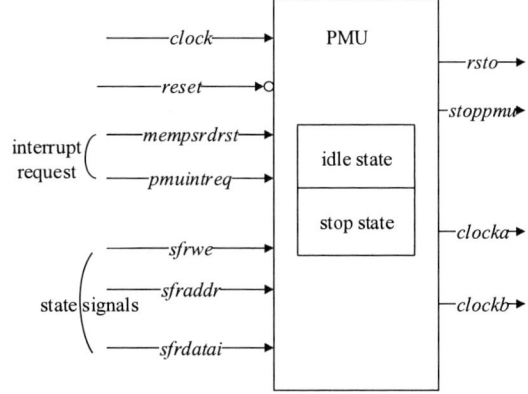

Figure 3:the block of PMU

The memory read reset signal *mempsrdrst* decides the output *rsto* at the positive edge of clock signal;

When the reset signal is logic 1, all the output will be reset and the two output clocks are just the same as the input clock;

If the interrupt request signals *pmuintreq* is logic 1, the two output clocks are the same with the input clock while the output *stoppmu* is decided by other inputs;

Besides the above conditions, the outputs are influenced by input signals *sfraddr*, *sfrwe*, *sfrdatai* and the

formal state of the outputs.

TABLE II shows part data of the report gotten from Design Compiler with the slow library of SMIC13. In the table we can see the power consumption of PMU takes 4.4% of the total chip, which is far more than any module. The power management unit aims to improve energy efficiency of the whole chip, but itself is most energy-intensive since the ever-converted clock signal, especially the dynamic power.

TABLE II. CMOS CIRCUIT POWER REPORT FROM DC

Hierarchy	Dynamic power	Leakage Power	percentage
R80515_CHIP	7.6330	0.0895	100.0
U_CHIP_R80515	1.9440	0.0849	26.3
U_R80515	1.2250	0.0118	16.0
U_TIMER2	0.1167	0.0013	1.5
U_PMU	0.3408	0.0004	4.4
U_CPU	0.0443	0.0004	0.6
U_ALU	0.1317	0.0025	1.7

At the same time, we can get the gate-level netlist of the PMU from the Design Compiler, which can be used in the later simulation.

SIMULATION AND COMPARISON

Firstly, all gate-level circuits in the netlist gotten from Design Compiler are built up and are checked the functions are right. Then put them together to realize the PMU function and the circuits are taken simulation in HSPICE.

The power consumption in a design contains two parts including dynamic power and leakage power. In HSPICE dynamic power is measured at every edge of the input signals under many conditions. For so many states and conditions should be tested, we choose several energy data under special conditions to be contained in TableIII. Two kinds of data are shown. The former eight lines are data tested at transiting edge of different input signals the moment they are not really changing the output. The later six lines are tested at the transiting edge of clock signal which means real work as the FF (flip-flop) in PMU.

The state input (HL) means the input transits from logic 1 to logic 0, and input (LH) means the input transits from logic 0 to logic 1. !signal has the definition that the signal is logic 0 and signal is just logic 1 but pay attention to !*sfrdatai* and !*sfraddr*. As the signal *sfrdatai* and *sfraddr* are multi bit binary data,!*sfrdatai* means the signal is not equal to logic 6'b00_0011 and the !*sfraddr* means the input is not equal to 7'b000_0111. These definitions are decided by the function described before.

978-1-5090-6695-7/17 $31.00 © 2017 IEEE   548

TABLE III. Dynamic power of TFET PMU under different conditions

INPUT	CONDITIONS	ENERGY (J)
mempsrdrst (LH)	none	2.66E-16
mempsrdrst (HL)	none	4.13E-17
rst(LH)	none	4.72E-17
rst(HL)	none	1.66E-16
pmuintreq (LH)	none	4.00E-16
pmuintreq (HL)	none	4.24E-17
sfrwe(LH)	none	4.25E-17
sfrwe(HL)	none	2.45E-16
clk(HL)	*mempsrdrst&rst &pmuintreq&!sfrwe &!sfraddr&!sfrdatai*	2.672E-15
clk(HL)	*mempsrdrst&!rst &pmuintreq&sfrwe &!sfraddr&sfrdatai*	2.428E-15
clk(HL)	*!mempsrdrst&!rst &!pmuintreq&sfrwe &sfraddr&sfrdatai*	1.166E-15
clk(LH)	*mempsrdrst&rst &pmuintreq&!sfrwe &!sfraddr&!sfrdatai*	2.428E-15
clk(LH)	*mempsrdrst&!rst &pmuintreq&sfrwe &!sfraddr&sfrdatai*	1.546E-15
clk(LH)	*!mempsrdrst&!rst &!pmuintreq&sfrwe &sfraddr&sfrdatai*	9.480E-16

Take the average of the above data and change them to the same unit. Table IV shows the comparison of dynamic power of TFET and CMOS. We can see the dynamic power of TFET circuit is nearly ten times as much as that of CMOS PMU.

TABLE IV. Comparison of Power Consumption

Power Consumption	TFET	CMOS
Dynamic Power(mW)	0.037	0.339
Leakage Power(mW)	2.81e-05	4.00e-05

The leakage power is measured under every state of input signals and output signals. Just like the comparison of dynamic power, we take the average of the leakage power above in the table and the final comparison is in below TABLE IV.

CONCLUSION

In this paper, a TFET power management unit of R80515 is proposed and we explore its function and the power reduction. The use of TFET in PMU can reduce 80% dynamic power and 30% leakage power at VDD=0.8V. The paper presents that the TFET circuit has a better power efficiency compared with the same size MOSFET implementations. Large-scale circuits like TFET R80515 will be evaluated in the future.

ACKNOWLEDGEMENTS

This research was supported by the fundamental research key project of Shenzhen Science & Technology Program (Grant No: JCYJ20140717102743108) and the fundamental research project of Shenzhen Science & Technology Program (Grant No: JCYJ20140417144423206).

REFERENCES

[1] Y. Khatami and K. Banerjee, "Steep Sub-threshold Slope n- and p-Type Tunnel-FET Devices for Low-Power and Energy-Efficient Digital Circuits," in IEEE Transactions on Electron Devices, vol. 56, no. 11, pp. 2752-2761, Nov. 2009.

[2] Q. Huang et al., "A novel Si tunnel FET with 36mV/dec sub-threshold slope based on junction depleted-modulation through striped gate configuration," Electron Devices Meeting (IEDM), 2012 IEEE International, San Francisco, CA, 2012, pp. 8.5.1-8.5.4.

[3] T. Nirschl et al., "The tunneling field effect transistor (TFET) as an add-on for ultra-low-voltage analog and digital processes," Electron Devices Meeting, 2004. IEDM Technical Digest. IEEE International, 2004, pp. 195-198.

[4] U. E. Avci et al., "Energy efficiency comparison of nanowire heterojunction TFET and Si MOSFET at Lg=13nm, including P-TFET and variation considerations," Electron Devices Meeting (IEDM), 2013 IEEE International, Washington, DC, 2013, pp. 33.4.1-33.4.4.

[5] L. Barboni, M. Siniscalchi and B. Sensale-Rodriguez, "TFET-Based Circuit Design Using the Transconductance Generation Efficiency Method," in IEEE Journal of the Electron Devices Society, vol. 3, no. 3, pp. 208-216, May 2015.

[6] H. Lu, J. W. Kim, D. Esseni and A. Seabaugh, "Continuous semiempirical model for the current-voltage characteristics of tunnel fets," Ultimate Integration on Silicon (ULIS), 2014 15th International Conference on, Stockholm, 2014, pp. 25-28.

[7] B. Senale-Rodríguez et al., "Perspectives of TFETs for low power analog ICs," Subthreshold Microelectronics Conference (SubVT), 2012 IEEE, Waltham, MA, 2012, pp. 1-3.

Design and Implementation of a digital HBC coordinator for Body Area Network

Ying Zhang[1], Hao chen[1], Zhongmin Lin[1], Xin-an Wang[1]*, Xing Zhang[2]

[1]The Key Lab of Integrated Microsystems Peking University Shenzhen Graduate School, Shenzhen 518055

[2]Institute of Microelectronics, Peking University, Bejing 100871, CHINA

* Email:wangxa@pkusz.edu.cn

ABSTRACT

Considering the bottleneck of energy efficiency and reliability in the pervasive and personalized wearable healthcare management based on Human Body Communication (HBC), an all-digital extraordinary HBC coordinator is proposed, as a configurable hub in the baseband system for Body Area Network (BAN). The unique HBC coordinator in a BAN is capable of enhancing the control, and programming of the network system, as the core of the BAN and gateway between body sensor nodes and the devices outside of body. With emphasis on power consumption and high data rate, The implementation of network management unit in the HBC coordinator, equipped with a dynamic lookup table, not only keeps memory for the nodes paired before, but also makes improvements in retransmission mechanism in multipoint communication, and can be initialized whenever you want. Thereby the HBC coordinator chip has been typed out and achieves miniaturized physical size of 0.8mm^2 in 0.13-um standard CMOS technology. It periodically monitors the data transmission with 15 nodes with less than 126.75uW and the operation is successfully demonstrated on our test system board.

INTRODUCTION

Driven by the ubiquitous mobile, wearable electronic devices, prevalence of application-specific integrated circuits and skyrocketing medical cost, the healthcare system is undergoing a paradigm which shifts from the conventional hospital-centered to an individual-centered system nowadays. In order to address the problems of real-time and energy efficiency in the wearable monitoring system. IEEE802.15.6 provides a solution of Wireless Body Area Network (WBAN), in a promising attempt to collect bio-data, such as electrocardiogram (ECG), electroencephalograph (EEG), oxygen saturation (SpO2) and so on, has been recognized as an accessible and affordable technology for monitoring health condition continuously. As a kind of communication medium, HBC is efficient in terms of data rate, bit error rate (BER) [1], low attenuation [2] and interference-free [1] in contrast to RF techniques. So on the basis of simplified MAC protocol designed by Li N[3], An energy-efficient individual BAN is designed, shown in Figure 1, keeping the HBC coordinator as its center and communication bridge between various sensor nodes and external PC. Various nodes are placed in the specific parts of body to detect bio-data, than all the bio-data is processed in the HBC coordinator and transmitted to devices outside of body. Utilizing the human body as the communication channel acquires high data rate of 2M-b/s in half-duplex mode with a system clock of 16MHZ.

Previous to our work, several endeavors related to HBC coordinator for BAN have been made. Song, S. J provided a 2M-b/s, 0.2mW digital transceiver based on wideband signal [4]. Later, Song, S. J presented a 0.9V Body-Coupled scalable PHY transceiver for body sensor applications considering the limitation for scalability [5], but it dissipates 2.6mW power. A 2mW star-topology body area network controller was proposed by Choi, S. [6], consisting of 16bit RISC processor, all these designs appear a little poor in power consumption.

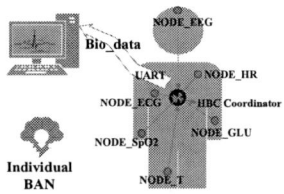

Figure 1.   Individual Body Area Network

So, in order to solve the problem, we design an all-digital HBC coordinator, adopting a master-slave architecture, where the nodes send out pair request to the coordinator and ask for joining the BAN zealously, which revolutionized the master-slave framework compared with Lee, H [7], where the hub (coordinator) issues a beacon at every start of super-frame, which would dissipate much power for the HBC coordinator. The HBC coordinator, containing four all-digital blocks, consumes only 126.75uW power in the communication with various sensor nodes, which is far less than power of the body area network controller [6]. Additionally, without any other processor, all network management operations are implemented with hardware in the HBC coordinator.

DESCRIPTION OF HBC SYSTEM

Human Body Communication mode

In our BAN prototype, when the power is on, the HBC coordinator keeps in working state, and waits for the pair request of the nodes. Some frame structures are defined to assist the process of the BAN, including pair request frame (PRF), pair assignment frame (PAF), acknowledgment frames (ACK), synchronization time request frame (STRF), synchronization time assignment frame (STAF) and data frame (DF). The communication mode is a kind of handshake protocol, as shown in Figure 2. As we can see, the HBC coordinator is defined to transmit PAF, STAF and ACK, and to receive PRF, ACK, STRF and DF.

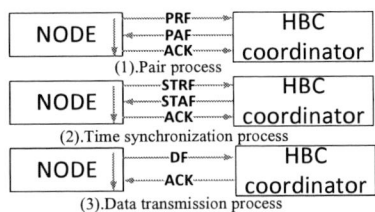

Figure 2. Handshake communication mode

System design

A simplified HBC system, including a HBC coordinator, a node, communication analog front module (AFE) and a collection AFE. The collection AFE collects bio-data firstly and received by the receiver of the node. Then the transmitter of the node transmits the processed data frames to the HBC coordinator through the body channel. The receiver of the HBC coordinator is utilized to receive frames from the node and its transmitter is utilized to reply to the node. Finally, the UART of the HBC coordinator transmits the bio-data to the PC. All these communication mechanisms are controlled in the network management unit (NMU).

As is shown in Figure 3, the HBC coordinator consists of an all-digital receiver and transmitter, a clock data recover module (CDR) and a NMU. The transmitter contains seven modules. The primary frame structure is assembled in the Frame Assemble module, controlled by the NMU. The temporary frame information is stored in the Tx_reg module. Then the parallel data is converted into serial data in the p2s module before generating CRC16 check codes and being encoded respectively in the Crc16 Generator and Fm0 Encode module. The preamble is added in the Frame length module. The CDR module is utilized to eliminate the jitter and skew, and aligns the received data with the local clock. Then the binary data is feeding into the receiver. The Pream Detect module detects the frame information after the CDR

module, and judges whether the received frame is among these frames described in Figure 2. After that, it is decoded and checked respectively in the Fm0 Decode and CRC Check module. After storing in the SRAM, the Frame Identify module is used to confirm the frame eventually and export critical information related to the received frame to the NMU. Under the control of NMU for the receiver and transceiver, the HBC coordinator will be in operation orderly.

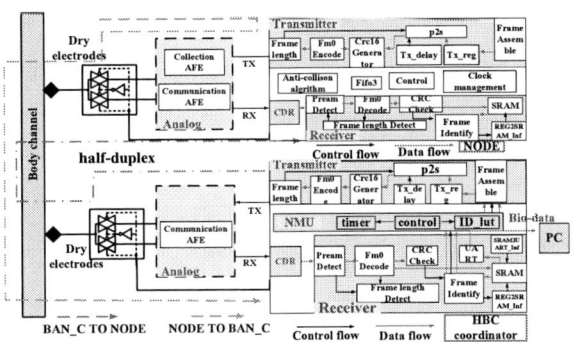

Figure 3. Proposed HBC System architecture

Network management mechanism

In this design, the HBC coordinator can be on line with 15 nodes simultaneously, with option to add more nodes as required.

As is shown in Table1, EUI_NODE can present physical address of nodes. NODE ID (NID) is a unique address of nodes assigned by the HBC coordinator. Sequence number (SN) can prevent the HBC coordinator from receiving the same frame twice. ADDR is directly bounded up with the received EUI_NODE, NID, and SN. When a node asks for pairing with the HBC coordinator for the first time, the NMU assigns NID (ADDR+1) to the node and stores the EUI_NODE and NID in the corresponding ADDR column. If the node sends frames to the HBC coordinator again after a brief retreat from communication conflicts, the NMU will exploit polling program to search the dynamic lookup table for the EUI_NODE and NID and detect whether the EUI_NODE and NID match well with information of the lookup table or not. If they match, the SN will be updated in the lookup table. Then, the NMU will manage the transmitter to reply to the node.

The SN functions in the retransmission mechanism, Every time the node sends out a frame to the HBC coordinator, the value of SN would increase 1. After the node sends a frame to the HBC coordinator, the coordinator would reply corresponding frame to the node. If the node does not receive the replied frame, then the node would retransmit the identical frame. At that time,

978-1-5090-6695-7/17 $31.00 © 2017 IEEE   551

when the HBC coordinator receives the same frame once again, NMU will check the SN of the received frame, If the SN received now is no more than the SN of the lookup table, the HBC coordinator will reject to receive and reply the frame, which avoids frame loss and assures fewer conflicts and conserves power.

Table 1. Dynamic Lookup Table

ADDR	EUI_NODE	NID	SN
0	$0xM_1N_1$	0x01	X_1
1	$0xM_2N_2$	0x02	X_2
2	$0xM_3N_3$	0x03	X_3
3	$0xM_4N_4$	0x04	X_4
…(m)	…	…(m+1)	…
13	$0xM_{13}N_{13}$	0x0E	X_{13}
14	$0xM_{14}N_{14}$	0x0F	X_{14}

System implementation

Figure 4(a) shows two nodes communicate with the HBC coordinator in the individual BAN simultaneously, the sensors we adopt are ADS1191(ECG), DS18B20 (Temperature) and they are attached with nodes, named NODE1 and NODE2. MSG-201 is utilized to create ECG signal. DMM4050 is exploited to analyze power consumption. The verification results shown in Figure 5 are in accordance with the communication mode in Figure 2. As we can see, NODE2 sends out a DF, when the receiver of the HBC coordinator receives the frame successfully, it will create a flag of RX_OK with high pulse and reply a ACK frame to NODE2 immediately. The STRF is sent by nodes every 2ms for synchronizing the time of the node and the HBC coordinator. Similarly, when the coordinator received the STRF from NODE1 successfully, it replies the STAF with precise time to NODE1. After receiving the frame of STAF and calibrating time, NODE1 will send out an acknowledgment frame, named ACK.

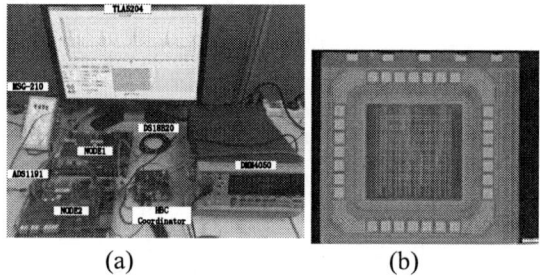

(a)   (b)

Figure 4(a). test system board

Figure 4(b). HBC coordinator chip microphotography

The chip is fabricated in 0.13-um standard CMOS technology and occupies area of 0.8mm² including core

area of 0.6mm², as shown in Figure 4(b). The work voltage and current of the chip are 1.2451V and 101.8uA respectively, resulting in power consumption of 126.75-uW in all.

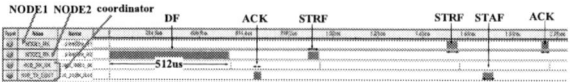

Figure 5. Waveform of verification

SUMMARY

A HBC coordinator for low power BAN is presented and implemented in the HBC system. The test verification results prove that the HBC coordinator can communicate with multiple nodes smoothly and continuously for a long time. The power consumption of the HBC coordinator is totally 126.75uW, appearing a great advantages over the body area network controller in reference[6], whose power dissipation is 2mW.

ACKNOWLEDGEMENTS

This work is supported by R&D project of Shenzhen Government (Project NO.JSGG20130918140947999) and Natural Science Foundation of China (Project NO.61471011).

REFERENCES

[1] Ansari, A. R. and Cho, S., Human body: The future communication channel for WBAN. The IEEE International Symposium on Consumer Electronics, 2014.p.1-3.

[2] Callejon, M. A. and Naranjo-Hernandez, A comprehensive study into intra body communication measurements. IEEE Transactions on Instrumentation & Measurement, 2013, p.62.

[3]. Li N, Lin K, Yong S, et al. Design and implementation of a MAC protocol for a wearable monitoring system on human body[C]// IEEE, International Conference on Asic. IEEE, 2015.

[4] Song, S. J. and Cho, N., A 0.2-mw 2-mb/s digital transceiver based on wideband signaling for human body communications. IEEE Journal of Solid-State Circuits, 2007, p.49.

[5] Song, S. J. and Cho, A 0.9v 2.6mw body-coupled scalable phy transceiver for body sensor applications. 2007, p366-609.

[6] Choi, S., Song, S. J., A low-power star-topology body area network controller for periodic data monitoring around and inside the human body. 2006, p139-140.

[7] Lee, H., Cho, H., & Yoo, H. J. A 33μW/node Duty Cycle Controlled HBC Transceiver system for medical BAN with 64 sensor nodes. IEEE Custom Integrated Circuits Conference – Cicc, 2014, pp.1-8.

978-1-5090-6695-7/17 $31.00 © 2017 IEEE

The Design and Implementation of A Reconfigurable Convolution Operator based on APU

Yuqian Huang, Jipan Huang, Xin'an Wang* , Zhiqiang Yang,Haifang Lu,Miren Tian

The Key Laboratory of Integrated Micro-system Science and Engineering Applications, Peking University
Shenzhen Graduate School, Shenzhen, China
（+86）13410938507 Email: anxinwang@pku.edu.cn

Biography

Yuqian Huang, a postgraduate from Peking University, works in the laboratory of integrated micro-system science and engineering applications.

Abstract

In order to achieve fast IC design, reduce development cycle and cost, Key Lab of Integrated Microsystems in Peking University proposed Array Processing for Unification Architecture(APU) which consists of four kinds of operators: computation, data path, control and MEM operators to replace the configurable logic block in current FPGA fabric. In this paper, a reconfigurable convolution operator based on APU, is presented which is used to convolutional neural network computing. The reconfigurable convolution operator is with more coarse-grain and function changeable than APU. And the process of operator synthesis is verified by simulation-based verification and formal verification separately. With the verification method proposed in this paper, certainty and completeness have been achieved. The results show that compared with the hardware design at the cost of resources based APU, this methodology can obtain hardware of suitable performance with regular structure with the cost of 34.08% less area and 25.26% lower power.

Keywords—Reconfigurable Convolution Operator, Operator Design Methodology

Introduction

In the domain of the image recognition and speech analysis, convolutional neural network has become hotspot areas[1]. Its advantages as we known, are its weights shared network structure making it more classes similar to biological neural networks, and reducing the complexity of the network model. Also it reduce the number of weights. Convolutional neural network structure is shown in Figure 1. Convolutional network is to identify two-dimensional shape and a special design a multi-layer perceptron, the network structure of the translation, scaling, tilting oblique or other forms of deformation is highly invariant.

However, convolution accounting cellular neural networks use and training is more than 90% of the amount of computation[2]. The accuracy required to achieve the most advanced cellular neural networks

depends not only on more layers, and the right to one million filter weights, such as size of the filter, the number of the filters and channels. In fact, as a result of the neural network in the application process is relatively lack of guidance, in order to get a good effect CNN, its parameters need continuous testing, evaluation, modification and improvement, thereby it is of great importance of reducing the training time which is usually to be able to achieve better results.

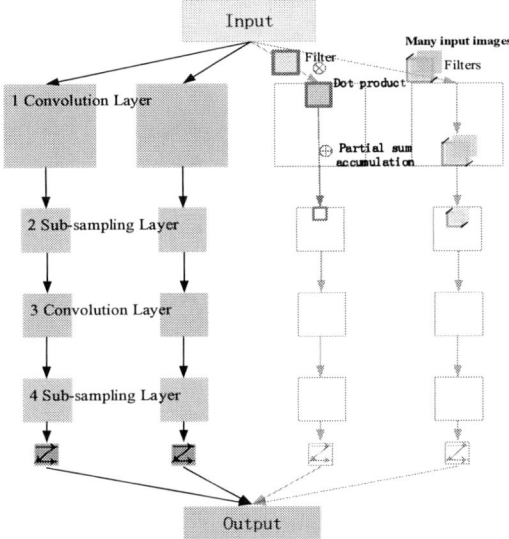

Figure 1. Convolutional neural network

Then it becomes very necessary reducing the computing time through specialized hardware to accelerate the calculation of convolution. Based on the study on the operator design method and continuous research of structure in coarse-grain operators array architecture（APU）[3]，we comes up with larger coarse-grain reconfigurable operators(ReOps) for convolutional neural network, which we call RCO. Expand the type of the basic reconfigurable operators[4] and optimize the design of operators. We introduce the IC design methodology for reconfigurable convolutional operator as follow. Then we describe the architecture for convolutional operators array based on APU. What's more, we have a comparison between RCO and the basic operators. A conclusion is given in our paper at last.

978-1-5090-6695-7/17 $31.00 © 2017 IEEE

The architecture of reconfigurable convolutional operator

Operator design methodology is a fast hardware design method based operator, it draws HLS thought based on this operator compared the standard unit larger design elements for rapid hardware design[5].

In CNN, each map is composed of multiple neural units. And all the same map neural units share a convolution kernels (weight) which is shown as the filters in Figure 1(b). Convolution kernels tend to represent a feature. For an example, if a convolution sum represents an circle, after the convolution kernels rolling on the whole image, the larger convolution value area is more likely a circle.

Now we try to use the reconfigurable operators in order to implement the application of convolution and the function of data storage and reading[6] as Figure 2 shown. GSM（General Switch Matrix）is used to implement long distance data transfer. C operators complete multiplication and A operators complete addition.

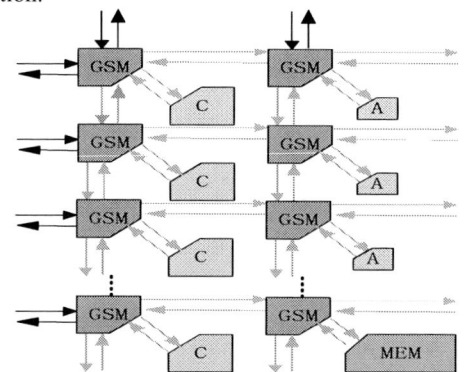

Figure 2 convolution array based on the reconfigurable

operators

The reconfigurable Convolutional operators are defined as follows:

Based on the operators array architecture，GSM、IM and IOGSM operators set up the interconnection structure. IM (Input Mux) is used to perform input selection from GSM and adjacent other operators. IOGSM(Input and Output GSM) is used to connect internal logic and IO. According to the functions of Convolutional layer, we adapt its data bits.

The calculation section is realized by the reconfigurable convolutional operators and other basic arithmetic operators. The reconfigurable Convolutional operator RCO(Figure 3) is in charge of computation in reconfigurable circuit consist of configure unit(configure register) and Convolution unit. Convolution unit carries out the basic arithmetic which is controlled by the configure information in the configure unit.

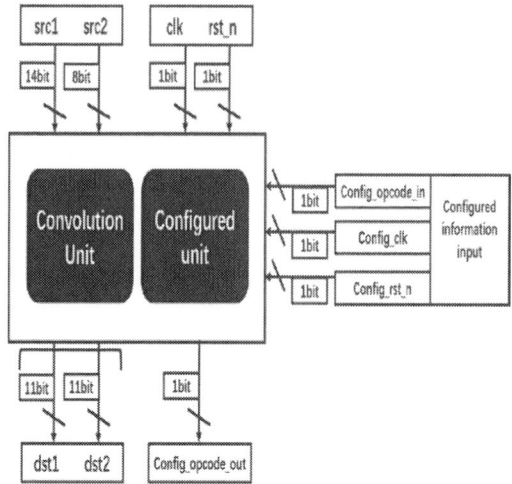

Figure 3 RCO architecture

The configure information in RCO contains the operation code for RCO. RCO parses the operation code and carries out the indicated function. The logic meanings of the configure information can be summarized as a format below:

$$dst=（dst2,dst1）=src1*src2 \qquad (1)$$

RCO calculates the convolution of the image pixel which gray value is in the range 0-255 that is 8bit src2 input shown in the reference table and kernels. Single numerical convolution kernel accurates to five decimal places. The RCO can be designed to perform not only the multiplications but also complex functions according to the convolution by changing the input and output bits width. Configuration unit controlled by the configuration of clock updates the internal configuration information, then decode the information to control operator performing the corresponding processing. Operation code is 1bit wide.0 indicates that the output result is not registered, else 1 indicates that the output result is registered.

Table 1. RCO Interface specification

name	IO	width	description
src1	I	14bit	Convolution kernel
src2	I	8bit	Input pixel value
config_opcode_in	I	1bit	Configured information input
config_clk	I	1bit	Configured clock input
config_rst_n	I	1bit	Configured reset input
clk	I	1bit	Working clock input
rst_n	I	1bit	Working reset input
dst1	O	11bit	Low 11-bit output calculation result
dst2	O	11bit	High 11-bit output calculation result
config_opcode_out	O	1bit	Configured information output

In the convolution unit, the calculation combinational logic circuit is described as follow: we use eight alternative multiplexers to get the part of produce. Then use four shift adders which move 4 bit data left to add achieving primary results. We use two shift adders which move 2 bit data left to add achieving secondary results afterwards. Last, use a shift adder which move 1 bit data left to add achieving the final results.

Simulation and Synthesization

Using verilog vhdl to complete the digital circuit design of RCO operator, the design is verified through extensive functional simulation.

And based on 130nm CMOS standard cell library, we use EDA tools to complete the digital front-end logic circuit synthesis and backend physical layout design shown in Figure 5. The synthetical results show that the operator's area is about $14560.15\mu m^2$ which is 34.08% less than the same circuit fuction that APU operators complete and the circuit power is 9.752 µw reduced about 25.26% at the clock frequency of 100 MHz.

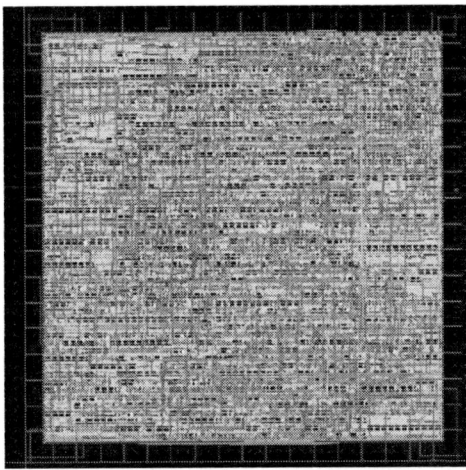

Figure 5 RCO backend physical layout

Compared to the APU operators architecture completing the same function of convolution arithmetic, RCO is more suitable as a operator of convolution layer because it has smaller area of consumption with more flexible data configuration and less power. According to the area report of the ASIC backend physical layout design, we can get the horizontal and vertical length scales of operators. With the wiring channel design of connecting operators, we determine the distribution of pins which will be equally distributed on each side.

Conclusion

Based on APU, a reconfigurable convolution operator RCO which has less area and lower power consumption is proposed in this paper. The design could improve operators utilization, also reduce the wiring and operators area in sum. As the fewer operators we use, the lower power consumption we could get. Optimization of the operator suit for convolutional neural network computing is our major concern while designing a circuit. By verifying and synthesizing RCO, we confirm that the design for convolution has a huge advantage which is 34.08% less area and 25.26% lower power than the APU build. With more simple configuration information and more coarse-grain convolution calculative ability, it is efficiently for us to improve the design speed based on RCO convolution array.

Acknowledgments

This work is supported by the Science and Technology Project of Shenzhen Government (Grant No. JCYJ 20160229094148396).

References

[1] Ting Rui et al,"Head detection based on convolutional neural network with multi-stage weighted feature", IEEE ChinaSIP, p.147-150 (2015).

[2] Yu-Hsin Chen et al, "Eyeriss: An Energy-Efficient Reconfigurable Accelerator for Deep Convolutional Neural Networks", ISSCC, p.262-263 (2016).

[3] Shanshan Yong et al, "An integrated development environment for reconfigurable operators array", IEEE ASIC,p.1-4 (2013).

[4] Peng Dai, Xin'an Wang, Xing Zhang, "A novel reconfigurable operator based IC design methodology for multimedia processing", IEEE TENCON, p. 1-5 (2009).

[5] Hu Zi Yi, Zhao Yong, Wang Xin An, "Theory and Verification of Operator Design Methodology", Science China, Information Sciences,55(2), p.480 (2012).

[6] Jaehyeong Sim et al, "14.6 A 1.42TOPS/W deep convolutional neural network recognition processor for intelligent IoE systems" ISSCC, p.264-265 (2016).

A NOVEL OLED-ON-SILICON MICRODISPLAY DRIVE CIRCUIT WITH THE DIGITAL ANALOG HYBRID SCAN STRATEGY

Yongnan Chu[1], Tingzhou Mu[1], Yuan Ji[1*], Yunsen Yu[1], Feng Ran[1], Jiao Li[1]

[1]Microelectronic Research and Development Center, Shanghai University, Shanghai, 200072, China

*Corresponding author: jiyuan@shu.edu.cn

ABSTRACT

A novel AMOLED-on-silicon microdisplay driver circuit with the digital-analog-hybrid scan strategy is designed for high resolution and high frame refresh rate display. The strategy of the digital-analog-hybrid scan is analyzed. A particularly designed multiplex column driver circuit is proposed. The column driver circuit is simulated. The experimental results show that the digital-analog-hybrid scan method can effectively reduce data flow.

INTRODUCTION

OLED microdisplay has two scanning strategies: the analog amplitude modulation and the digital pulse width modulation [1] [2]. The analog modulation strategy requires digital-to-analog converter (DAC) with high conversion speed and high accuracy [3] [4]. In case of the high refresh rates and high resolution, the digital pulse width modulation strategy requires high data flow [5].

A digital-analog-hybrid scan strategy is proposed to overcome the limitations on the resolution and the refresh rate caused by the data flow under the digital scan strategy. Based on this strategy, a novel OLED-on-silicon microdisplay driver architecture has been designed. The simulation results show that the OLED microdisplay driver architecture meets the timing requirements of the proposed scan strategy, and effectively reduce the data flow.

SCAN STRATEGY

As shown in the Figure 1, The gray scale data is divided into high and low bits. In digital-analog-hybrid scan strategy, the gray-scale modulation of high bits is realized by digital pulse width modulation strategy, that is, the sub-field scan strategy is adopted in this work. Analog amplitude modulation strategy is utilized to complete the gray-scale modulation of low bits, which means, digital-to-analog converters are used to accomplish this process. A full image is divided into several multiple subfields which includes analog frame and digital frames. In the digital frames, the highest or lowest level is driven to pixel column directly corresponding to the high-bits; In the analog frame, DACs convert the low-bits to analog voltage output, which is then driven to pixel column. The analog frame can be extended to reduce the data flow in this frame.

DRIVER ARCHITECTURE

System architecture for microdisplay

The diagram of the OLED-on-silicon microdisplay driver architecture based on the digital-analog-hybrid scan strategy is shown in Figure 2. It contains row driver, column driver and pixel arrays. Row driver is mainly used for address decoding and row selection. Column driver is the focus of this work, which has four tasks: (1) digital to analog conversion; (2) Digital and analog frame switching; (3) column selection; (4) Driving output voltage to pixel array.

Architecture for column driver

The circuit area consuming and the response time of the column driver must be taken into consider. There are two measures to meet the area restriction of the column driver. The pixels are arrayed horizontally to extend the width of each column, as shown in Figure 3. A scheme of the multiplexed column driver is proposed, in which multiple columns of the pixels are driven by a single column driver channel sequentially when one of the rows is being set to the voltage. As shown in Figure 4, compared with the traditional analog driver architecture [1], the pixel columns are divided into several groups, each column share a single DAC-buffer channel through multiplexers. In addition, a structure to implement switching operations between digital and analog frames is integrated in the column driver, which is controlled by the signal "D / A SELECT". Signal "COL SELECT" selects a column of pixels that needs to be written to the voltage.

We using rail-to-rail buffer consisting of input differential stage, output stage and inverter, shown in the Figure 5, to ensure precision and drive capability.

RESULTS AND DISCUSSION

In the case of the 256 gray scale levels, 100Hz refresh rate, proposed circuit is simulated and analyzed. The parasitic resistances and capacitances on the pixel columns were measured to be approximately 1 kΩ and 4 pF. The time budget of driving a single column is simulated, as shown in Table 1. The total area of DAC and buffer is $60 \times 65\mu m^2$. Table 2 shows the comparison of several different schemes. The resolution, the number of analog modulation bits and the number of sub-fields are considered. The grouping of the columns is determined depends on the sub-field scan budget, the response time and the circuit area of the column driver. The data flow of the digital frame is also shown in Table 2.

With the traditional digital scan method, the data amount is about 7.6Gbps in the 19 subfields (luminous efficiency is 83.88%) [5]. With the proposed scheme, the data amount is only about 3.6Gbps in the case of 9 subfields (luminous efficiency is 86.11%), which is lower than that of the former 50% or so. Overall accuracy error of DAC and buffer is about 0.6% in proposed scheme.

ACKNOWLEDGMENT

This work is supported by the China National Natural Science Fund under Grant Number 61376028 and 61674100.

REFERENCE

[1] Wacyk I, Prache O, Ghosh A. Ultra-high resolution

AMOLED. Proc Spie, 2011, 8042(1)

[2]Ji Y, Ran F R, Xu H, et al. A Digitally Driven Pixel Circuit with OLED Degradation Compensation for AMOLED Microdisplays. Journal of the Society Information Display, 2015, vol.22(9), pp.465–472.

[3] Park I, Kim T W, Lee J Y, et al. Data driver architecture and driving scheme of AMOLED microdisplay for mobile projectors. IEEE Transactions on Consumer Electronics, 2009, 55(4).

[4] Onoyama Y, Yamashita J, Kitagawa H, et al., 0.5-inch XGA micro-OLED display on a silicon backplane with high-definition technologies, Proceedings SID Technical Digest, 2012, pp. 950–95

[5] Ji Y, Ran F, Xu H, et al. Design on AM-OLED display control ASIC with high gray scale levels. Journal of Shanghai University (English Edition), 2011, 15: 310-315.

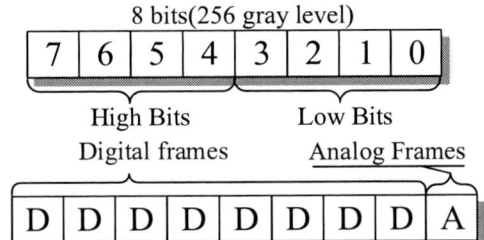

Figure 1: Scanning strategy diagram

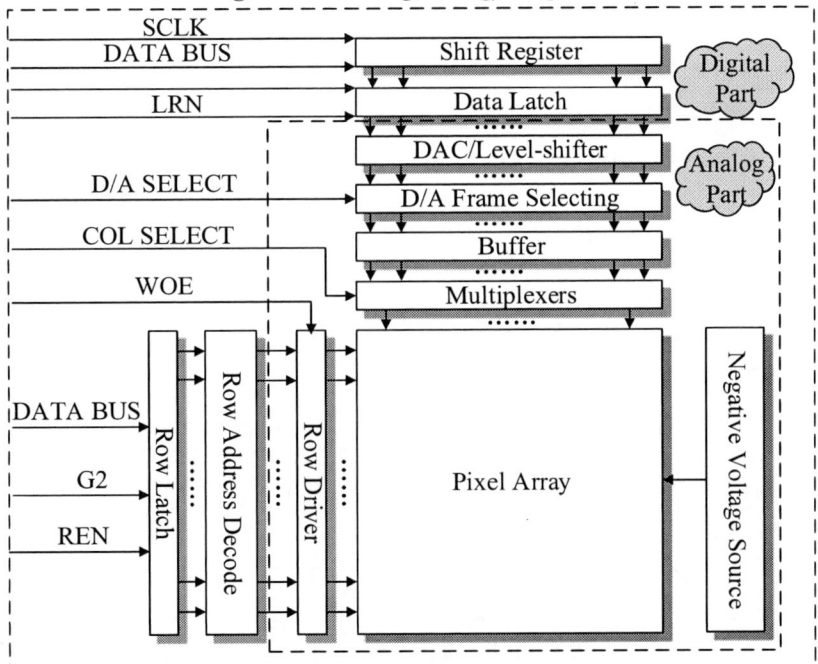

Figure 2: System diagram of the OLED-on-silicon microdis-play driver architecture

978-1-5090-6695-7/17 $31.00 © 2017 IEEE

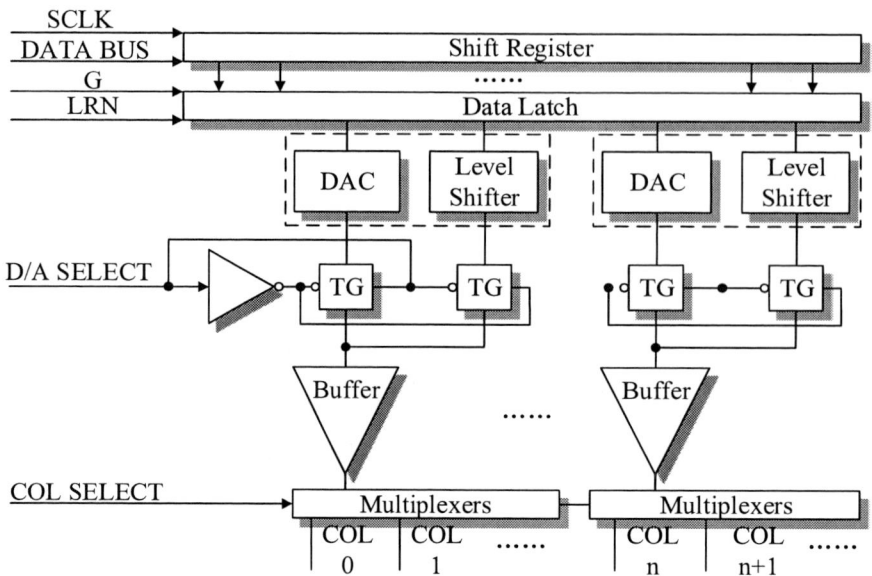

Figure 4: Column driver structure

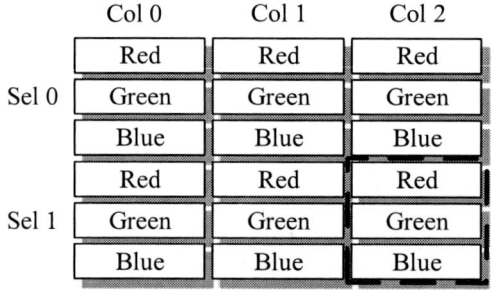

Figure 3: Pixel array diagram

TABLE I.   TIME BUDGET

Function	Times(ns)
Digital-to-analog Conversion	20
Output Buffer Voltage Setting	50
Multiplexer Switching	10
Total Response Time	60

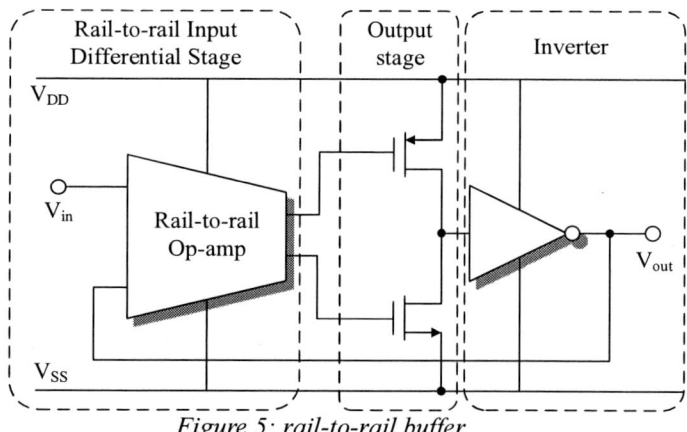

Figure 5: rail-to-rail buffer

TABLE II.   COMPARISON OF SEVERAL SCHEMES

Resolution	2k*2k				1.6k*1.6k			
Number of Analog Modulation Bits	3		4		3		4	
Number of Sub-fields	7	10	6	9	7	10	6	9
Luminous Efficiency(%)	56.25	78.75	64.59	86.11	56.25	78.75	64.59	86.11
Single Row&sub-field Scan Budget(μs)	0.714	0.500	0.834	0.556	0.893	0.625	1.042	0.695
Number of Columns that can Group	8~11	8	8~13	8~9	8~14	8~10	8~17	8~11
Data flow (Gbps)	2.8	4.0	2.4	3.6	2.5	3.2	2.0	2.9

978-1-5090-6695-7/17 $31.00 © 2017 IEEE

A CONCISE AND PRECISE MODEL OF THE GATE DELAY FOR EDA SIMULATION

Zhipeng Yue, Zhuoquan Huang, Dihu Chen, and Tao Su[*]
School of Electronics and Information Technology, Sun Yat-sen University,
Guangzhou 510006, China
*Corresponding Author's Email: sutao@mail.sysu.edu.cn

ABSTRACT

This paper considers the relationship between the static gate delay and the supply voltage of the digital integrated circuit. An empirical equation with physical implication is proposed for calculating the static gate delay. The expression of the gate delay is very simple. It contains only three constants. The calculation includes only one subtraction step, one division step and one addition step. Transistor level simulations are performed to verify the equation. The model matches the experiment results precisely. It is valid for various technologies, gate types, and operation temperatures. The equation can be applied in EDA tools to simulate the timing of the circuit under the PVT variation and the electromagnetic interference.

INTRODUCTION

Modern integrated circuits suffer supply variation due to the imperfect package, on-chip power supply network, transistor structure, and the operation of active devices. The increment of the integration density, the signal frequency, and the signal power cause strong electromagnetic interference on integrated digital circuits. The timing behavior of circuits under those non-ideal situations should be considered in the design phase of the circuit. The gate delay is fundamental for timing analysis of the digital integrated circuit. The relationship between the gate delay and the supply voltage is an important data of the gate for EDA-based design and is thereby worth studying. Current EDA tool, like developed by Synopsys [1], supply a linear model for the gate delay and supply voltage: $t_D = t_{D0} + kV * (V_{DD} - V_{DD0})$. It is valid only for very small variation on the supply. Other models [2-5] in literatures are too complex, which requires large computation efforts.

The motivation of the paper is to provide a simple calculation formula for delay calculation and show how accurate the calculation results are. Section 2 gives the form of the formula. Section 3 and 4 verifies the model through transistor-level simulations and practical measurements. Section 5 checks the validation of the model for various types of gate.

DELAY MODEL

The delay proposed in this paper is following:

$$t_D = a + b/(v_{DD} - c) \qquad (1)$$

The equation is empire. It is extracted by fitting the observed experimental results. However, the three parameters in the equation have physics implications. Parameter c correspond to a threshold voltage. If the supply voltage is below c, the gate delay is negative, meaning beyond infinite. Parameter a is the down limit of the gate delay. When the supply voltage approach infinite, the gate switches with its fast speed. Damage due to large supply voltage is not considered here. Parameter b is a coefficient for converting the voltage into the delay.

EXPERIMENT

Transistor level simulation using HSPICE are performed to verify the model.

Simulation Setup

To perform the simulation, inverters are cascaded into a ring oscillator, as shown in Fig. 1. The ring oscillator is operated at various supply voltage. The gate delay at each supply voltage is obtained by diving the cycle time of the ring oscillator by twice of the stage account.

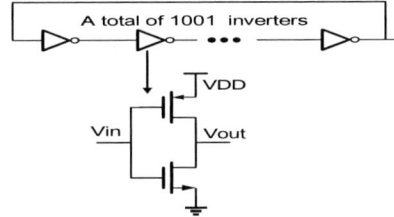

Figure 1: Device under test for gate delay extraction

Transistor model of various process technologies are used for the simulation. Those technologies are:
- SMIC 180 nm,   - SMIC 130 nm
- SMIC 90 nm,   - SMIC 65 nm
- SMIC 40 nm,   - SMIC 28 nm
- TSMC180 nm,   - TSMC130 nm

Simulation Results

The simulated delay are fitted with (1). The fitted

978-1-5090-6695-7/17 $31.00 © 2017 IEEE

results are compared with the simulated results in Fig. 2. Table I summarizes the fitting parameters.

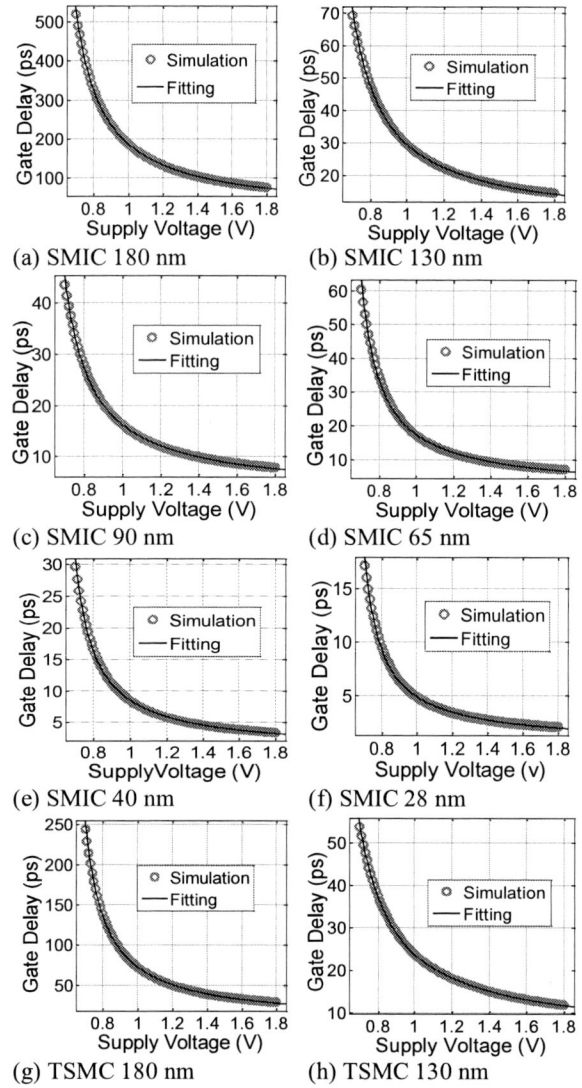

Figure 2: Simulated gate delay for various technologies.

TABLE I.   FITTING RESULTS FOR VARIOUS TECHNOLOGIES

Technology	Fitting Parameter		
	a (ps)	b (ps · V)	c (V)
SMIC 180 nm	9.415	81.27	0.5441
SMIC 130 nm	4.126	12.53	0.5103
SMIC 90 nm	2.972	5.836	0.5618
SMIC 65 nm	1.030	6.690	0.5943

Technology	Fitting Parameter		
	a (ps)	b (ps · V)	c (V)
SMIC 40 nm	0.5982	3.229	0.5943
SMIC 28 nm	0.4911	1.762	0.5999
TSMC180 nm	4.303	28.27	0.5876
TSMC130 nm	4.282	9.579	0.5097

The R-square, representing the matching quality, is above 99.99% for all fitting results. It indicates that the relationship between the supply voltage and the gate can be precisely described with (1).

DISCUSSION

A standard cell library contains gates other than the simple inverter. To check the validation of the model on those gates, additional simulation are performed. Fig. 4 shows the selected DUTs. Gate NAND and NOR are representatives of single transistor level gates. XOR and NXOR are representatives of multiple transistor level gates. The gates are cascaded into ring oscillators. HSPICE simulation are performed on those circuits with transistor model from SMIC 130 nm.

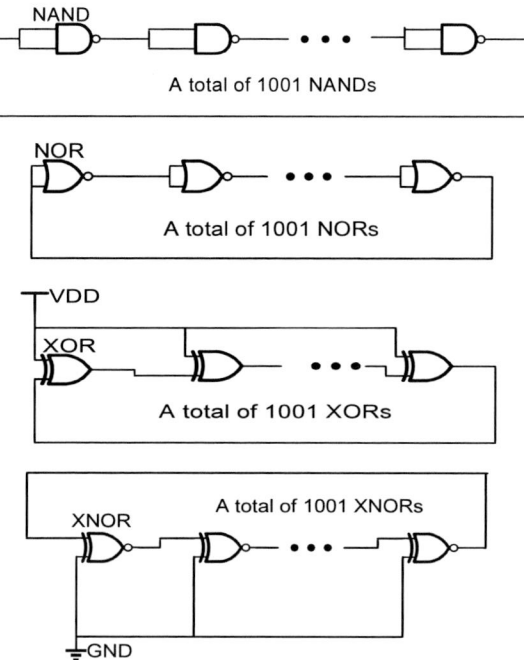

Figure 4: Ring oscillators of various gate types.

The simulated delay are fitted with (1). The fitted results are compared with the simulated results in Fig. 5. Table III summarizes the fitting parameters.

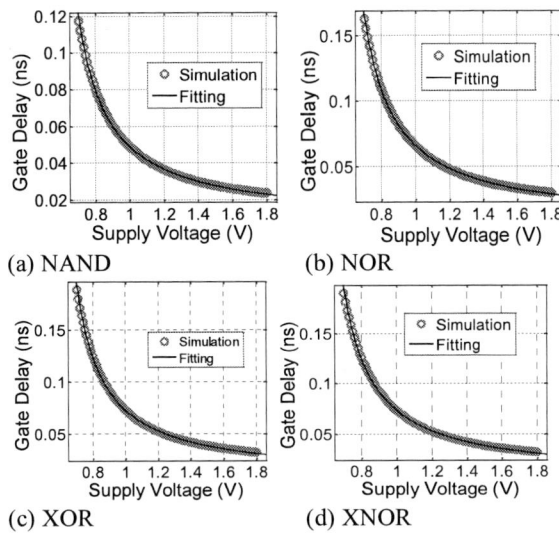

Figure 5: Simulated gate delay for various gate types.

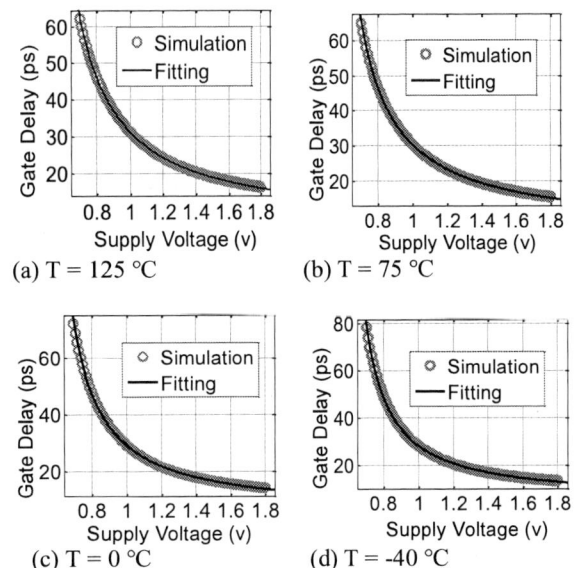

Figure 6: Simulated gate delay for various temperatures.

TABLE II.   FITTING RESULTS FOR VARIOUS GATE TYPES

Technology	Fitting Parameter		
	a (ps)	*b* (ps • V)	*c* (V)
NAND	7.417	19.86	0.5236
NOR	6.604	28.23	0.5238
XOR	6.473	31.12	0.5361
NXOR	6.573	30.50	0.5382

The R-square is above 99.99% for all fitting results. Model described with (1) is valid not only for simple inverters but also for other gate types.

The environment may influence the behavior of the gate. Fig. 6 presents the static gate delay of the inverter of SMIC 130 nm technology at temperature from -40 °C to 125 °C. All the simulated results are well matched to the fitting results, with R-square above 99.99%. Refer to Table III, as temperature decreases, the gate delay, according to parameters *a* and *b*, is reduced. It is a reasonable behavior.

TABLE III.   FITTING RESULTS FOR VARIOUS TEMPERATURE

T（°C）	*a* (ps)	*b* (ps • V)	*c* (V)
125	5.602	13.9	0.4575
75	5.357	12.82	0.4889
0	4.967	11.21	0.538
-40	4.745	10.28	0.5651

CONCLUSION

The three-parameter model offer a way to calculate the static gate delay at supply voltages above the threshold. Simulation with transistor model from various process technologies fit the model precisely. The measurement results match the model too. The model is accurate and simple. It is suitable for fast delay calculation in the EDA-based circuit design procedure.

ACKNOWLEDGEMENTS

Supported by the Fundamental Research Funds for the Central Universities of China (15lgpy32) and by the Project Science and Technology of Guangdong Province of China (2016B010123005, 2015B090912001, and 2015B090909002).

REFERENCES

[1] Synopsys, *Library Compiler User Guide*, Synopsys, Inc, 2016.

[2] M. Masanori, R. Nair, *Power Integrity for Nanoscale Integrated Systems*, McGraw-Hill, 2014.

[3] X. Gao, C. Sui, etc. *IEEE Trans. Electromagn. Compat.*, vol. 57, 2015, pp. 1179-1187.

[4] J. Zhu, L. Pan, etc. *IEEE Trans. Very Large Scale Integr. (VLSI) Syst.*, vol. 22, 2014, pp. 2629-2634.

[5] F. Frustaci, P. Corsonello, etc. *IEEE Trans. Circuits Syst. II, Exp. Briefs*, vol. 59, 2012, pp. 168-172.

DESIGN OF A NOVEL AC LED DRIVER WITH NO CURRENT GLITCH BASED ON SOFT SWITCHING OPERATION

Yang Boxin[1]*, Liang Zhiming[1], Wu Zhaohui[1], and Li Guoyuan[1]

[1]School of Electronic and Information Engineering, South China University of Technology

Guangzhou, People's Republic of China

*Tel: 86-18126744767, Email: youngboxin@163.com

ABSTRACT

In this paper, a novel design on AC LED driver with no current glitch for multiple-string LEDs is presented. The LED driver can run LEDs directly from a pulsating voltage obtained by rectifying an AC voltage without any other AC-DC converter. In order to avoid current glitch due to hard switching operation, negative feedback control is used and the driver operates on a soft switching mode. Without AC-DC converter and complicated control circuit, the driving system is suitable for integration and the driver can be designed on a simple chip. Simulation results show that the lifespan is prolonged, and the quality of lighting and the efficiency of the system are improved because of no more current glitch. The proposed AC LED driver was fabricated in 1 um SPDM 5V40V700V BCD process. The measured results show that the efficiency is about 91% and the power factor is 98.5% under 3.3W 220V~50Hz condition.

Keywords—AC LED driver; current glitch; soft switching

INTRODUCTION

As the next generation lighting product, the light-emitting diode (LED) has been developed rapidly in recent years by both industry and academia. When compared to the traditional lighting module, the LED reduces energy dissipation and has high luminous efficiency. Moreover, there are many other advantages that make it more attractive to the market, such as high density of integration, fast response, ultralong life (above 5,000 h), small size, no mercury, etc. [1-3]. To improve the performance of LED lighting module, a lot of researches have been carried out in driver circuit design and light sources. These LED drivers are designed to improve the current state of LED operation generally. However, AC-DC drivers need bulky transformer and electrolytic capacitor inherently whose lifetime is much shorter than LED. Therefore, these passive components increase the cost, size, and energy dissipation of the LED module while decreasing the reliability [4]. Thus, running LEDs directly from AC voltage is meaningful and attractive. The simplest AC LED module is the LED strings operating directly from a pulsating voltage obtained by rectifying an AC voltage, and the peak current is limited by a series resistance. Unfortunately, such a system doesn't work well owing to high current distortion [5, 6]. In order to realize good performance, AC LED driver is needed to control the current state for upgrading power factor while degrading total harmonic distortion. Nowadays, Some AC LED drivers operate on hard-switching mode, which switching operation is under the control of switching signals. And there is a time deviation between two signals, leading to current glitch which causes lower power factor and greater total harmonic distortion in turn. And worse still, the current glitch has an impact on LED, so it would shorten the lifetime of LED and decrease the quality of lighting [7]. Hence, it is meaningful to design a driver with no current glitch by using soft-switching mode, which switching operation works smoothly and automatically. In this work, an AC LED driver with no current glitch was designed based on soft switching operation mode, and the electrical properties were simulated.

PROPOSED AC LED DRIVER

The proposed electric circuit of the AC LED driver with no current glitch is depicted in Figure.1. In the electric circuit, it can be seen that there are four surface mounted LED strings, four resistances with the same value, and a full-wave rectified bridge outside chip. The rectified bridge works on the AC supply source and the output rectified voltage directly drives LEDs whose current is controlled by the in-chip circuit. The LEDs are designed to operate in constant current status because negative feedback loops consisted of the operational amplifiers, the analog adders, the high-voltage transistors, and the resistances were designed in circuit. The high-voltage transistors are driven by the operational amplifiers. The drains of the transistors are connected to LEDs, while the sources are connected to resistances. By using resistances outside chip, the LEDs' current can be adjusted to the required value. The positive terminals of the operational amplifiers are connected to a constant voltage reference denoted as V_{REF}, while the negative terminals are connected to the outputs of the analog adders. With the negative feedback system, the voltages of the negative terminals are forced to be V_{REF}.

It is worth to note that, with the help of negative feedback loops, the AC LED driver works automatically. Thus, controller is no longer needed for high voltage transistors to turn them on/off. This can be explained as below. Assuming the voltages of the drains and sources of the high voltage transistors are denoted as $V_{D1} \sim V_{D4}$, $V_{S1} \sim V_{S4}$ respectively, the turn on voltage and forward voltage in constant current state of the LED strings are indicated by V_{on} and V_{LED}. If $V_{IN} < V_{on}$, the LEDs are in cutoff status, so there are no currents through high voltage transistors. Thus, high-voltage transistors are all turned on, because operational amplifiers output high voltage.

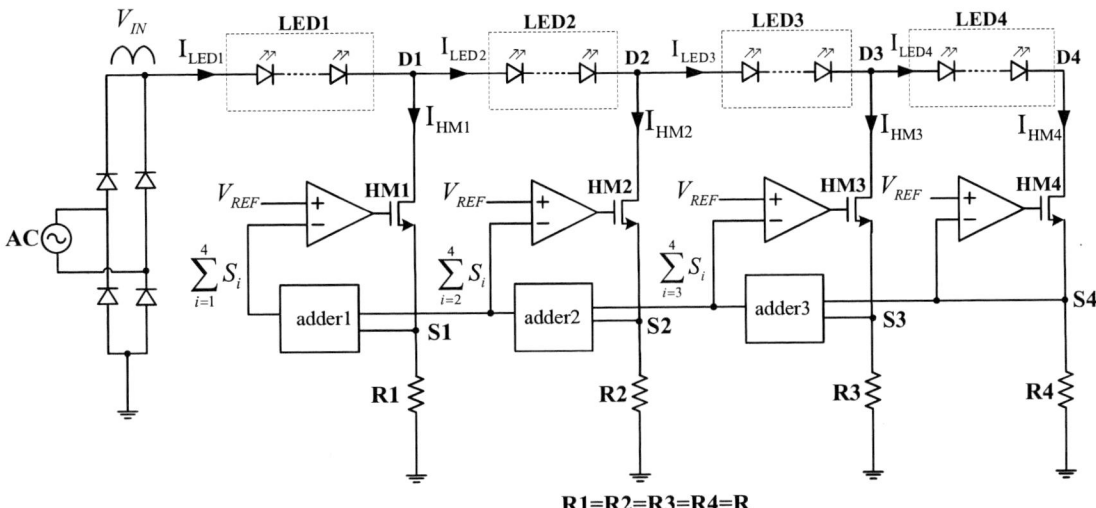

Figure 1: Schematic of the proposed AC LED driver

As V_{IN} increases and gets at the turn on voltage of LED1, the current I_{LED1} flowed through LED1, high-voltage transistor HM1, and resistance R1, increases exponentially with the increase in V_{IN}, and remains other LEDs in off. During this period, the voltages on R2~R4 are zero, and thus the output voltage of the adder1 module is $I_{LED1} * R$. Under the control of constant current module (CCM) that consists of an operational amplifier, a high-voltage transistor, a resistance, and an adder, I_{LED1} exponentially increases from zero to V_{REF}/R and then holds steady. As I_{LED1} is constant, the voltage across the LED1 V_{LED1} is also steady, where $V_{LED1} = V_{IN} - V_{D1}$. Therefore, V_{D1} starts rising and flows by the increase in V_{IN}. As V_{IN} continues increasing, LED2 will turn on if V_{D1} is larger than V_{on}, but LED3 and LED4 are still turn-off. And likewise, the current I_{LED2} through LED2 is going to rise exponentially to V_{REF}/R under the control of CCM2. During this transfer, as the I_{LED2} flows through LED2, HM2 and resistance R2, the voltage on R2 is bound to increase. As a result, the current I_{HM1} flowing through HM1 will decrease because the output voltage of adder1 is forced to V_{REF}, i.e. $I_{HM1}*R + I_{LED2}*R = V_{REF}$. As I_{LED2} increases from zero to V_{REF}/R, I_{HM1} decreases from V_{REF}/R to zero and HM1 consumes nothing any more. When V_{IN} goes on increasing further, other CCMs work in a similar way. When V_{IN} starts falling and is not larger enough to hold the current through all LEDs steady, the current of LED4 I_{LED4} will decrease from V_{REF}/R to zero, forcing HM3 to turn on and the current through this transistor will rise from zero to V_{REF}/R. Because CCM3 makes sure that the sum of these two currents is V_{REF}/R, i.e. $I_{HM3} + I_{LED4} = V_{REF}/R$, which is exactly the current flowing through LED3, I_{LED3}. As a result, the first three LEDs are still working in constant current status. Like this way, the lighted LEDs are going to be cut off one by one with the falling V_{IN}.

Figure.2 shows the working principle of the driver described above. During the whole period, there is no hard switching working on HMs, which causes current glitch because time deviation of the switch control signals [8]. Hence, the proposed driver based on soft switching operation can avoid current glitch during current switching. Besides, as not all HMs work at the same time, the proposed driver is good for reducing dissipation.

SIMULATED RESULTS AND LAYOUT

The proposed AC LED driver with four LED strings has been simulated under 3.3W 220V/50Hz condition. Figure.3 shows that LEDs operate in the constant current status gradually and then in cut-off status one after the other with the change of V_{IN}. As can be seen from the simulated results, there is no current glitch during current switching because of no hard switching. Besides, the

978-1-5090-6695-7/17 $31.00 © 2017 IEEE

simulated results show that the efficiency of the proposed driver is 91% and the PF is 98.5%.

The proposed AC LED driver was fabricated in 1 um SPDM 5V40V700V BCD process. Figure.4 shows the core layout of the driver circuit without reference sources and PADs.

TABLE I
COMPARISON BETWEEN SIMILAR RESEARCH AND THIS WORK

	Ref.[3]	Ref.[5]	This Work
PF	0.974	0.97	0.985%
Efficiency	90.64%	82%	91%
Output Power	9W	1.5W	3.3W

CONCLUSION

A novel AC LED driver with no current glitch based on soft switching operation is proposed in this paper. Under the control of constant current module by using negative feedback loops, the LEDs operate in the constant current status asynchronously and automatically with the changing V_{IN}. The designed driver does not need the control circuit for switching signals. Hence, there is no current glitch due to the time deviation of the switch control signals. The simulated results show that the efficiency is about 91% with four LED strings under 3.3W 220V/50Hz condition, and the PF is 98.5%. In addition, the driver is suitable for working on the 110V/60Hz condition as well.

ACKNOWLEDGMENT

This research is supported by the Planned Science and Technology Project of Guangdong Province, China (No. 2014A010103016).

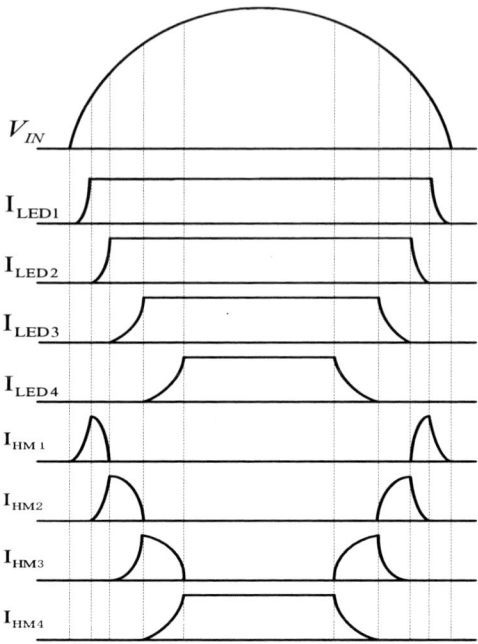

Figure.2: V_{IN} and currents flowing through LEDs and HMs

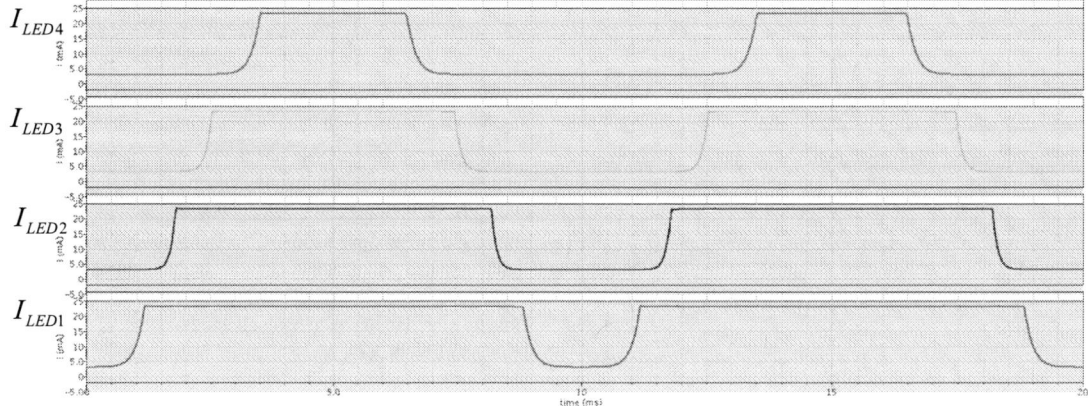

Figure.3.1: The measured currents of LEDs during two periods

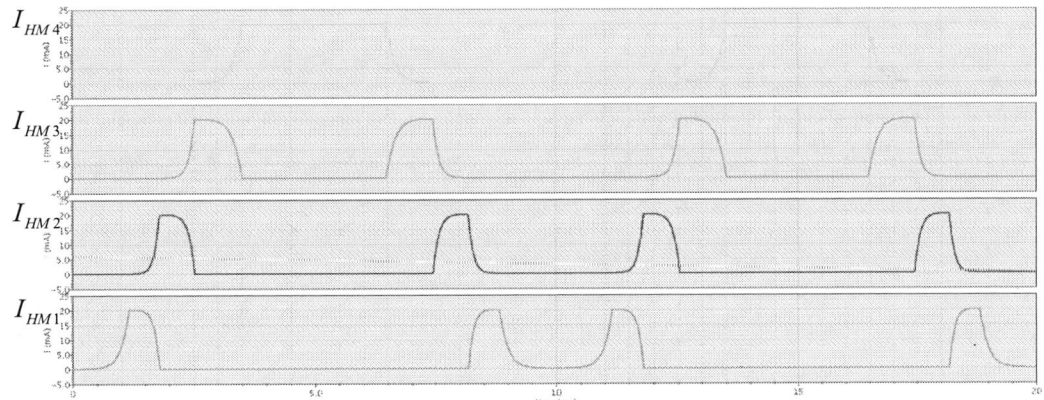

Figure.3.2: The measured currents of HMs during two periods

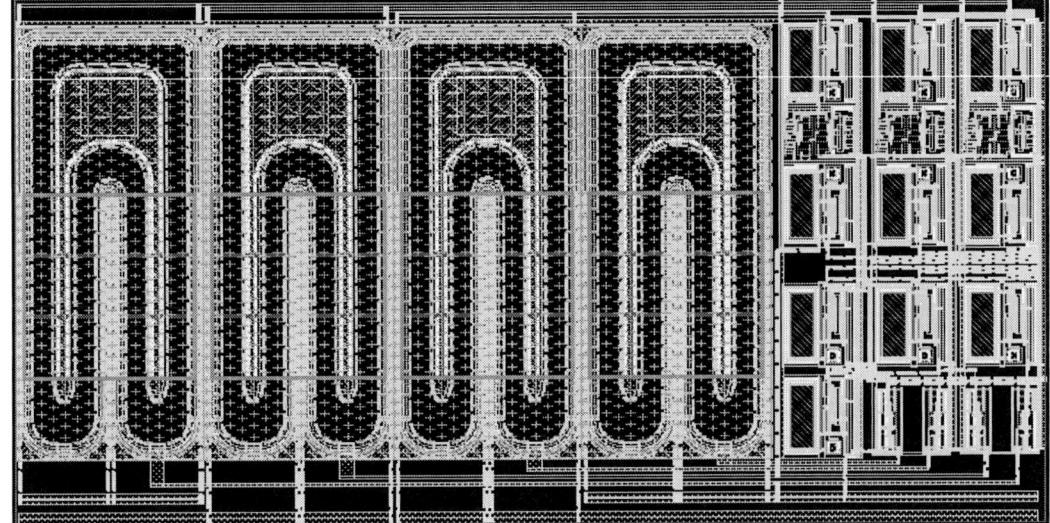

Figure.4: The core layout of the driver circuit

REFERENCE

[1] Kim, J., Lee, S., and Park., "A soft self-commutating method using minimum control circuitry for multiple-string LED drivers," IEEE international solid-state circuits conference (ISSCC), pp. 376–377, 2013.

[2] K. I. Hwu and Jenn-Jong Shieh, "Dimmable AC LED Driver Based on Series Drive," JOURNAL OF DISPLAY TECHNOLOGY, VOL. 12, NO. 10, page. 1097-1105, OCTOBER 2016.

[3] Hongbo Gao, Kexu Sun, Jingyuan Chen, Xufeng Wu, Yahui Leng, Jianxiong Xi, and Lenian He, "An Electrolytic-capacitorless and Inductorless AC Direct LED Driver with Power Compensation," IEEE 2nd International Future Energy Electronics Conference (IFEEC), pp. 1-5, 2015.

[4] John C. W. Lam and Praveen K. Jain, "A High Power Factor, Electrolytic Capacitor-Less AC-Input LED Driver Topology With High Frequency Pulsating Output Current," IEEE TRANSACTIONS ON POWER ELECTRONICS, VOL. 30, NO. 2, page. 943-945, FEBRUARY 2015.

[5] Dayal, R., Modepalli, K., and Parsa, L.: "A direct AC LED driver with high power factor without the use of passive components,". Proc. IEEE Energy Conversion Congress Expo., Raleigh, NC, USA, September 2012, pp. 4230–4234.

[6] Ye, Z., Greenfeld, F., and Liang, Z., " A topology study of single-phase offline ac/dc converters for high brightness white LED lighting with power factor pre-regulation and brightness dimmable," In Proceedings of 34th IEEE Annual Conference of Industrial Electronics (pp. 1961–1967).

[7] Hwu, K.I., and Tu, W.C.: "A high brightness

978-1-5090-6695-7/17 $31.00 © 2017 IEEE

light-emitting diode driver with power factor and total harmonic distortion improved," Proc. IEEE Applications Power Electronics Conf., Fort Worth, TX, USA, March 2011, pp. 713–717.

[8] Ning Ning, Wen Bin Chen, De Jun Yu, Chun Yi Feng and Cheng Bi Wang, "Self-adaptive load technology for multiple-string LED drivers," ELECTRONICS LETTERS, Vol. 49 No. 18, page 1170-1171, August 2013.

HIGH-PERFORMANCE SINGLE-PHASE FULL-BRIDGE INVERTER USING GALLIUM NITRIDE FIELD EFFECT TRANSISTORS

Chih-Chiang Wu [1], and Shyr-Long Jeng [2]*

[1] Department of Mechanical Engineering, National Chiao Tung University, Hsinchu, Taiwan, R.O.C.

[2] Department of Electrical and Electronic Engineering, Ta Hua University of Science and Technology Hsinchu, Taiwan, R.O.C.

*Corresponding Author's Email: aetsl@tust.edu.tw

ABSTRACT

This paper presents the performance of a single-phase full-bridge inverter based on wide-bandgap devices. The control strategy for the full-bridge inverter applies unipolar sinusoidal pulse width modulation. The experimental results demonstrated that a smaller figure of merit is preferred for a more efficient design; specifically, the full-bridge inverter using gallium nitride field effect transistors inside could easily reach 96% efficiency or more within a 100- to 1000-W range.

INTRODUCTION

The single-phase full-bridge inverter topology has been widely adopted in various applications, such as motor drivers or solar inverters, because of its low output current distortion and high efficiency performance [1-3]. Using wide-bandgap-based devices (e.g., gallium nitride [GaN] and silicon carbide [SiC]) in power conversion can further improve the efficiency of the technology. However, distinctive characteristics and a lack of appropriate gate drives for GaN field effect transistors (FETs) prevent them from entering the switching power area. Moreover, the current collapse phenomenon affects the voltage rating under the conditions of fast switching and high input voltage in circuit applications [4-5]. Thus, determining the optimal performance for circuit applications is a critical concern [6].

In this study, a single-phase full-bridge inverter with LCL filter topology was simulated in software for Power Electronics Simulation (PSIM); subsequently, the H-Bridge inverter topologies based on SiC [7] and GaN FETs [8] with unipolar switching method, were used to test transistor performance.

DEVICES CHARACTERISTICS

SiC and GaN FETs were used in the present experiments (Fig. 1). A comparison of these two devices in terms of their electrical information is presented in Table I. Notably, GaN FETs have an overall superior figure of merit (FOM) compared with SiC FETs (GaN: 312; SiC: 2400). The total loss of transistors are comprising of switching losses and conduction losses, which are influenced by the charge Q_{GD} and the on-resistance ($R_{DS(on)}$); therefore, the power transistors with lower figure of merit [9-10], a higher quality circuit

application is expected.

TEST CONDITION

The single-phase full-bridge inverter with LCL filter topology shown in Fig. 1 was simulated in PSIM software, as outlined in Fig. 2, and the control strategy for the inverter applies unipolar SPWM, as depicted in Fig. 3. Specifically, the SPWM uses two sinusoidal modulating waves, namely V_{ctrl1} and V_{ctrl2}, with the same magnitude and frequency but 180° out of phase. The two modulating waves are compared with a triangular carrier wave, V_{tri}, to generate two gating signals. In contrast to the bipolar SPWM, where all four devices are switched at the same time, the unipolar SPWM prevents the upper two devices from switching simultaneously. Instead, the inverter output voltage switches between either zero and $+V_{dc}$ during the positive half cycle or zero and $-V_{dc}$ during negative half cycle of the fundamental frequency. Logic control modified the conventional unipolar gate signals [3], which switches Q1 and Q2 at high frequency and switches Q3 and Q4 at low frequency in the positive half cycle; and switches Q1 and Q2 at low frequency and switches Q3 and Q4 at high frequency in the negative half cycle, which reduces switching losses and enhances efficiency.

RESULTS

The DC–AC inverter test setup is shown in Fig. 4. The input power from the 165-V_{dc} source and output power to a resistive load were measured with a power analyzer. Fig. 5 illustrates the experimental waveforms of the inverter with an LCL filter, and Fig. 6 presents the output power versus efficiency and output power versus loss results. Notably, a high efficiency of 96% or more was obtained within a 100- to 1000-W range.

TABLE I.   COMPARISON OF SiC AND GaN DEVICES UNDER TEST CONDITIONS

Parameter \ Devices	SiC [7]	GaN [8]
V_{DS} (V)	1200	600
I_D Continuous (25°C)	41 A	36 A
Package	TO-247-3L	TO-247
$R_{DS(on)}$ (T_J = 25°C)	60 mΩ	52 mΩ

Input Capacitance (C_{ISS})	1800 pF	2200 pF
Output Capacitance (C_{OSS})	104 pF	115 pF
Gate Charge (Q_{GD})	40 nC	6 nC
Total Gate Charge (Q_G)	98 nC	28 nC
Figure of Merit (FOM) ($Q_{GD} \times R_{DS(on)}\ 10^{-12}\Omega\cdot C$)	2400	312

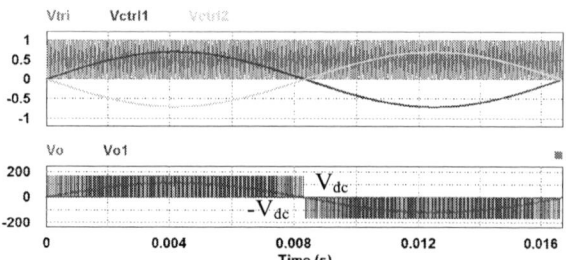

Figure 3: Control strategy for the full-bridge inverter using unipolar SPWM. (Vtri: carrier waveform; Vctrl1, Vctrl2: modulation waveforms). The output (Vo1) switching frequency was 50 kHz, and the AC output (Vo) was 60 Hz (110 V).

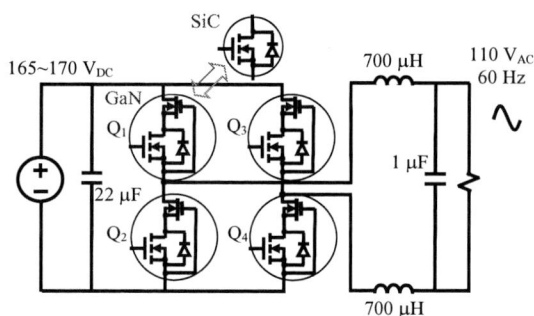

Figure 1: Single-phase full-bridge inverter with SiC-based and GaN-based topology.

Figure 4: DC–AC full-bridge inverter test setup with GaN FETs.

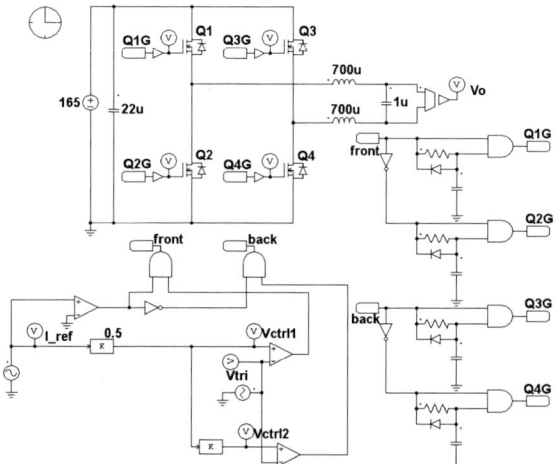

Figure 2: Single-phase full-bridge inverter with an LCL filter topology is simulated in PSIM software. The two filter inductors used in this study both had an inductance of 700 µH, and the capacitor had a capacitance of 1 µF. The input DC source voltage was set to 165 V, and the output voltage was regulated to 60 Hz and 110 V_{ac}.

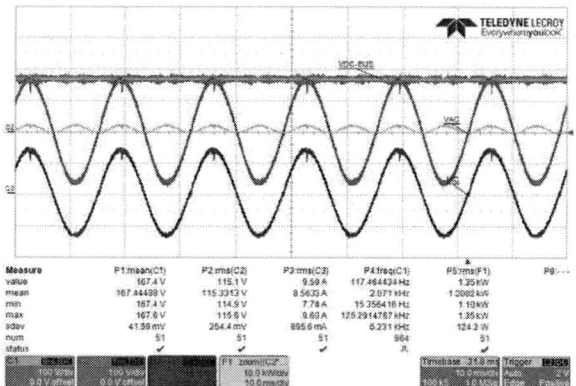

Figure 5: Experimental waveforms of the inverter with an LCL filter (red: output voltage 115.1 V (rms), blue: output current 9.59 A (rms)).

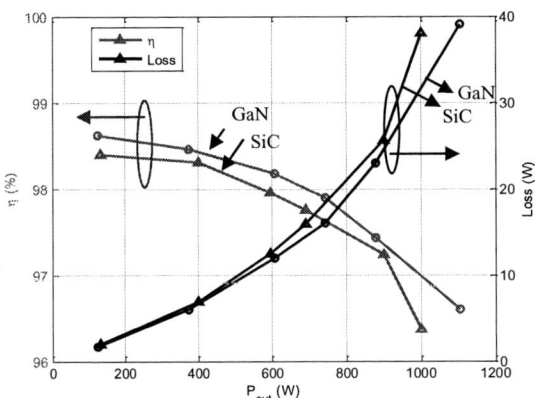

Figure 6: Performance of the SiC-based and GaN-based inverter. Notably, a high efficiency of 96% or more was obtained within the 100- to 1000-W range.

ACKNOWLEDGEMENTS

This work was supported by the National Chung-Shan Institute of Science and Technology (Project NCSIST-102-V211(105)) in Taiwan, R.O.C. The authors thank Hestia Power Inc. for providing the SiC devices; Fu-Jen Hsu for assisting in designing the SiC-based circuit; Nan-Hsiung Tseng at the Industrial Technology Research Institute in Hsinchu, Taiwan, for providing support with the electrical measurements; and Professor Wei-Hua Chieng of National Chiao Tung University for their helpful suggestions and technical support in completing this manuscript.

REFERENCES

[1] Z. Wang, Y. Wu, J. Honea, and L. Zhou, "Paralleling GaN HEMTs for diode-free bridge power converters," *IEEE Proc. of Applied Power Electronics Conference and Exposition (APEC)*, pp. 752–758, Mar. 2015.

[2] A. Morsy, and P. Enjeti, "Comparison of Active Power Decoupling Methods for High Power Density Single Phase Inverters Using Wide band Gap FETS for Google Little Box Challenge," *IEEE Journal of Emerging and Selected Topics in Power Electronics, vol. 4, no. 3,* pp. 790–798, May, 2016.

[3] A. Namboodiri, and H. S. Wani, "Unipolar and bipolar PWM inverter. *International Journal for Innovative Research in Science and Technology, vol. 1, no. 7,* pp. 237–243, 2015.

[4] E. A. Jones, F. F. Wang, & D. Costinett, "Review of Commercial GaN Power Devices and GaN-Based Converter Design Challenges," *IEEE Journal of Emerging and Selected Topics in Power Electronics, vol. 4, no. 3,* pp. 707–719, 2016.

[5] T. Ishibashi, M. Okamoto, E. Hiraki, T. Tanaka, T. Hashizume, D. Kikuta, and T. Kachi, "Experimental Validation of Normally-On GaN HEMT and Its Gate Drive Circuit," *Industry Applications, IEEE Transactions on, vol.51, no. 3,* pp. 2415–2422, 2015.

[6] S. L. Jeng, C. C. Wu, and W. H. Chieng, "Gallium Nitride Electrical Characteristics Extraction and Uniformity Sorting," *Journal of Nanomaterials*, 501, 478375, 2015.

[7] HestiaPower, H1M120F060 Silicon Carbide Power MOSFET, [Online]. Available: www.hestia-power.com.

[8] Transphorm, TPH3205WS Cascode GaN FET, [Online]. Available: www.transphormusa.com.

[9] B. J. Baliga, "Power Semiconductor Device Figure-of-Merit for High-frequency Applications," *IEEE Electron Device Letters*, vol. 10, pp. 455-457, 1989.

[10] D. Reusch & J. Glaser, *DC-DC Converter Handbook: A Supplement to GaN Transistors for Efficient Power Conversion.* Power Conversion Publications, pp.16-20, 2015.

THE AIR QUALITY EVALUATION BASED ON GAS SENSOR ARRAY

Chang-yong Chiu[*], Zhun Zhang

College Of Optoelectronic Engineering, Shenzhen University, Shenzhen518060, China
*Corresponding Author's Email: qiuchangyong@sina.com

ABSTRACT

Nowadays, air quality is more and more getting attention in our daily-life, not only outdoors, but also indoors like workplace, smart home system. Air quality is to a great extent closely related to people's health and comfort, so its evaluation is meaningful and significant to avoid danger or even harmful influence to life safety. The quality of air people inhale should be under the full monitoring scope, however, such generalized air quality evaluation is not defined. If it could be defined, that would bring people some helpful tips and cautions. In this paper, several evaluation methods of air quality are proposed, with sensing signals of all sensor dimensions acquired, air quality at different levels can be evaluated by calculated results. Pollutant type can be identified based on different sensors response, provided that more and more pollutant types are trained.

INTRODUCTION

The air quality is becoming more and more concerned about with the deterioration of the industrial environment and emission from automobile exhaust. With the continuing growth of economy in the next 10-15 years, developing country like China will face long-term air pollution issues and challenges in future [1]. However, to evaluate air quality we mainly use a specific gas sensor such as VOCs, combustible gas, pollutant gases or PM2.5 detection dedicated sensor. In general, the evaluation should take into account all the cases that may influence air quality. So only one type sensor is not adequate, and while more than one sensor are used, does each sensor works in standalone mode or do all the sensors work in combination mode? The answer is "No" now for air quality evaluation.

Gas sensors are widely used for air pollutants monitoring; more sensors can work together to smell different kinds of air pollutants like human's nose, electronic nose is such a kind of system. A narrow definition of electronic nose is a system of bionics can give unique response to different gaseous substances; its representatives are FOX series of ALPHA MOS with a gas sensor array [2]. Modeling after that, we mock up a simple electronic nose system consists of a gas sensor array, signal conditioning circuit and signal acquisition module. Sensing data is acquired by ADC collection module with signal conditioning circuit enhanced. In this paper, the air quality is evaluated by a gas sensor array which is composed of eight MOX (metal oxide) gas sensors, as

listed in Table I. Each gas sensor is connected in series with a load resistance of 10 kΩ to make up a voltage divider. Voltage applied on two ends is 3.3V, measured voltage on each gas sensor is lead to 2x7 pinheads and can be easily connected out to signal conditioning circuit and signal acquisition module provided that pins match with them. A picture of gas sensor array's matrix board is shown in Fig. 1.

The voltage on sensor changes due to sensing element's resistance variation. Resistance variation gives characterization pattern which can be used for air quality evaluation. To implement a signal processing system for air quality evaluation, we developed an app named GasSensorsViewer which can run in Android cell phone.

One of the air quality assessment methods, which have been widely used, is the air quality index, abbreviated as AQI [3], the main pollutants in the air quality evaluation are fine particulate matter (PM10), inhaled particulate matter (PM2.5), sulfur dioxide, nitrogen dioxide, ozone, carbon monoxide. It was released in 2012. Earlier air quality evaluation standard was API (Air Pollution Index) released in 1996, which only took into account PM10, sulfur dioxide, and nitrogen dioxide's concentrations. The AQI calculation is based on Individual Air Quality Index (abbreviated as IAQI) through formula as following:

$$IAQI_p = \frac{IAQI_{Hi} - IAQI_{Lo}}{BP_{Hi} - BP_{Lo}} (C_p - BP_{Lo}) + IAQI_{Lo} \quad (1)$$

Where $IAQI_p$ evaluates only one air pollutant. $IAQI_{Hi}$ and $IAQI_{Lo}$ are high value and low value of divided levels, corresponding to concentration limits BP_{Hi} and BP_{Lo} respectively. A table of divided levels gives such a relationship. C_p is mass concentration with a range between BP_{Hi} and BP_{Lo}. Thus $IAQI_p$ value changes proportionally to its mass concentration in each divided level. AQI is the maximum IAQI of each main monitoring air pollutant:

$$AQI = \max\{IAQI_1, IAQI_2, IAQI_3, \dots, IAQI_n\} \quad (2)$$

The formula definition of AQI means always only the maximum air pollutant is considered, while other air pollutants' effects on air quality are ignored. So this evaluation is not comprehensive but intelligible. Generalized air quality evaluation should cover actually more than the range AQI deals with. It should include other cases like cigarette smoke or hateful smell or nauseating odors or anything else. It should include combustible and explosive gases which are hazardous to living environment [4]. So an electronic nose like

978-1-5090-6695-7/17 $31.00 © 2017 IEEE

intelligent system is useful and welcome if it could detect more kinds of air pollutants.

It is necessary to take account into all kinds of air pollutants which may affect air quality, so we try to give several new evaluation methods in next section. J. C. Chang and S.R.Hanna mentioned that there could be three components to the evaluation of air quality models: scientific, statistical, and operational [5]. The following evaluation methods given are mainly on statistical and operational aspects.

As for air pollutant type identification, it is meaningful because different air pollutants have uneven effects to air quality. For example, some air pollutants with low concentrations could give severe effects on air quality, while some unpleasant smell is no harm to human, thus its effect on air quality should be evaluated to minor one from a health perspective.

TABLE I. COMPOSITION OF A GAS SENSOR ARRAY USED FOR AIR QUALITY EVALUATION

Sensor no.	Sensor type	Sensor applications
1	MQ-2	For smoke detection
2	MQ-3	For alcohol detection
3	MQ-7	For carbon monoxide detection
4	MQ-138	For formaldehyde detection
5	MQ-137	For ammonia detection
6	MQ-136	For hydrogen sulfide detection
7	MQ-135	For air pollutants detection
8	MQ-4	For methane detection

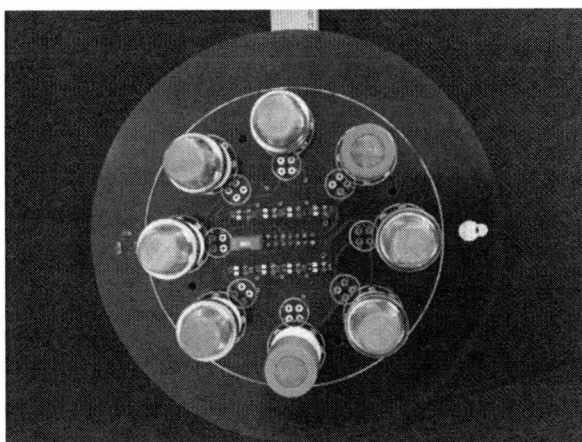

Figure 1: The picture of gas sensor array's matrix board, with eight MOX gas sensors distributed evenly around a circle. These sensors are all MQ-series produced by Winsensor Technologies, Zhengzhou, China.

SEVERAL METHODS OF AIR QUALITY EVALUATION

A. Generalized evaluation method

The basic calculation formula for air quality evaluation is defined below:

$$Q_A(N, \Delta s, S) = \sqrt[2]{\sum_{n=1}^{N} \left(\frac{2\Delta s_n}{S_n}\right)^2 / N} \qquad (3)$$

Where Δs_n is the nth sensor's voltage deviation from baseline state, S_n is the full range of variable voltage change, N is total gas sensors number. So the value of Q_A is actually confined to a range [0, 1), the smaller value of Q_A denotes the better air quality. In such evaluation systems, S_n is a constant value, equivalent to total applied voltage, provided that baseline state lies in the middle position of sensor response by selecting appropriate load resistance, or S_n is twice the voltage measured on the nth sensor in baseline state. As shown in Fig. 2, voltage deviation in sensor response curves. This picture is generated by screenshot of GasSensorsViewer.

This method doesn't need to train with air pollutant types before use, so the evaluation result is relative and rough.

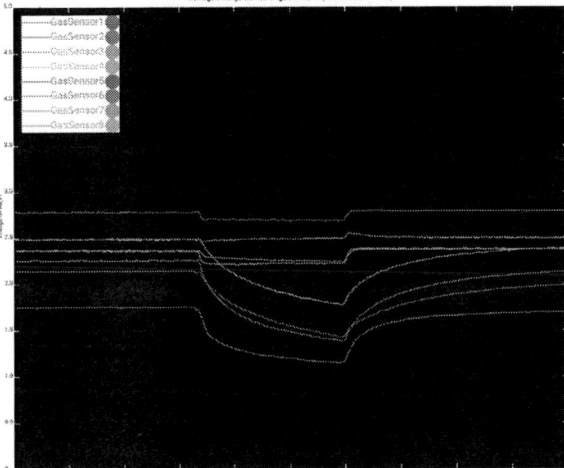

Figure 2: Voltage deviation in sensor response curves. Compared to baseline state, the sensor response curves change due to sensors' response to air pollutants.

B. Evaluation method with weighted coefficients

Considering each sensor dimension gives different contribution to air quality. An improved method is proposed. Its calculation formula is changed to:

$$Q_B(W, N, \Delta s, S) = \sqrt[2]{\left[\sum_{n=1}^{N} W_n \left(\frac{2\Delta s_n}{S_n}\right)^2\right] / \sum_{n=1}^{N} W_n} \qquad (4)$$

Where W_n is the nth sensor's weighted coefficient, Δs_n and S_n are the same as in formula (3). This has the advantage of distinguishing harmful levels of different pollutant types effectively, not only depending on how much Δs_n is. Weighted coefficients W_n (n=1, ..., N) can be assessed by air quality specialists. They are fixed values after sensing elements of a gas sensor array are determined. The larger value of W_n denotes the more remarkable contribution to the value of Q_B.

This method doesn't need to train before use too, but appropriate W_n coefficients can make the evaluation result more reliable and credible compared to method A.

C. Evaluation method with type classification

This method is an improvement to method B. There are some main pollutants which are indicative for bad air quality. Sensor array could give a unique fingerprint such as vector $Vs = [s_1, s_2, ..., s_n]$ with n dimensions. Let it be normalized by dividing its modulus (2 norm of vector) $\|Vs\|_2$, it becomes form of vector $Vm = [m_1, m_2, ..., m_n]$. Its modulus (also called vector's length or magnitude) is related to level of air quality evaluation. The higher modulus denotes the higher concentration, and thus the more severe air quality. Vm is an important characteristic for identifying the air pollutant type.

If an air pollutant type is identified, and its concentration (in ppm or mg/m^3) is calculated by vector's modulus, then air quality evaluation can be calculated from a well-trained model.

This method need to train different pollutant types with gas sensor array before use, so the evaluation result is quantified and accurate. Training details are presented in the next section.

THE TRAINING AND IDENTIFICATION PROCESS FOR NEWLY POLLUTANT TYPE

As shown in Fig. 3. The training UI of GasSensorsViewer app. Training is a process of data sampling , machine learning and classification. It has several steps:

1. One newly air pollutant must be sampled into a gas chamber to react with gas sensor array adequately.
2. Select gas type of air pollutant to train, and specialists can input evaluation criterion such as level index or mass concentration or volume concentration, etc.
3. Observe voltage change curves of gas sensors, if sensors response becomes stable, all characteristic parameters are extracted as a training set used for training model.

Air quality evaluation system can learn fingerprints of different air pollutant types and give evaluation results based on training sets. It's a process of classification and pattern recognition. Data treatment methods such as PCA (principal component analysis), DFA (discriminant factor analysis), Euclidian distance, KNN (k-nearest neighbor), and ANN (artificial neural network) [10-12] can be used here.

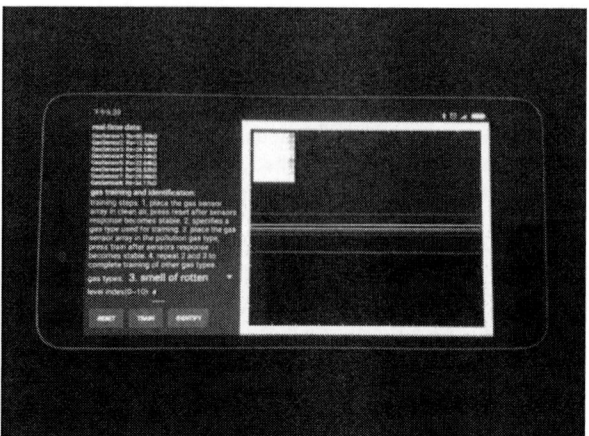

Figure 3: This is a UI picture of training a newly air pollutant. It currently shows gas sensor array is in baseline state (in clean air). If trained with gas type "3. Smell of rotten", specialist could give a level index value which could be a value between 0 and 10, say 4, then this system could learn the characteristic state.

EXPERIMENTAL RESULTS AND ANALYSIS

To demonstrate the above designed gas sensor array and proposed evaluation methods for practical use, two common application scenarios are chosen for gathering sensor data:

a. Air quality inside car

The unpleasant odor in a newly produced car is usually toxic substances mainly composed of VOCs emitted by interior materials. Sensor data of a response process is exported from GasSensorsViewer to redraw sensor response curves, as shown in Fig. 4a. However, some perfume odor or body odor in an aged car should not be deemed as air pollutants although it can also cause sensor response as shown in Fig. 4b.

b. Air quality affected by refuse burning

Refuse burning as one of the MSW (municipal solid waste) disposal technologies may result in some bad effects especially to nearby residential area. If not properly controlled and disposed, generated secondary pollutants (including dioxin, heavy metal in fine particulates, fly ash, and acid gases) could destroy environment and ecosystem [13, 14]. Sensor response curves are redrawn based on exported sensor data from GasSensorsViewer, as shown in Fig. 4c.

The sensor response data of steady state is exported from GasSensorsViewer and shown in Table II.

For scenario a, evaluation method C need to be used. To distinguish if the odor comes from VOCs or other odors like perfume odor or body odor, feature abstraction must be done by constructing a vector from voltage or resistance in each sensor dimension. Here we employ voltage as the basic feature element due to its limited dynamic range. We have a feature vector $Vs = [2.66, 0.75,$

2.15, 1.52, 1.14, 1.18, 2.09, 2.48] for VOCs, modulus $\|Vs\|_2 = 5.27$, Vm = [0.50, 0.14, 0.41, 0.29, 0.22, 0.22, 0.40, 0.47]. Similarly, for perfume odor, we can get a different modulus and Vm from its feature vector. After training with the pollutant types, normalized feature vectors, moduli and specified air quality levels, a correct identification result is obtained from a trained model mentioned in the above section by a pattern recognition process.

For scenario b, evaluation method A or B should be used. Secondary pollutants exist in the form of smoke which is actually a complex mixture composed of multiple components and each component has variable proportion, so it is difficult to identify the content of each component by a gas sensor array. Here we give a comprehensive evaluation. By calculation with method A, we can get a rounded result $Q_A = 0.53$. If given W_1 to W_8 of {5, 1, 3, 3, 1, 4, 5, 2}, by calculation with method B, we can get a rounded result $Q_B = 0.48$. The experimental data is gathered near the smoke source, while the calculation result should be relatively small if data is gathered farther away.

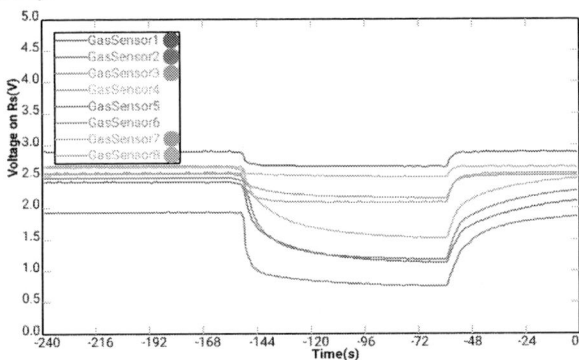

Figure 4a: Sensor response caused by VOCs in a newly produced car

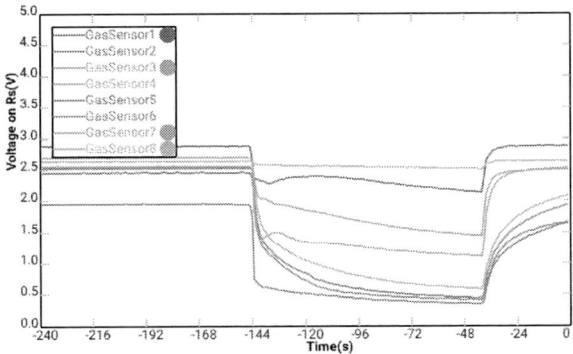

Figure 4b: Sensor response caused by perfume odor in an aged car

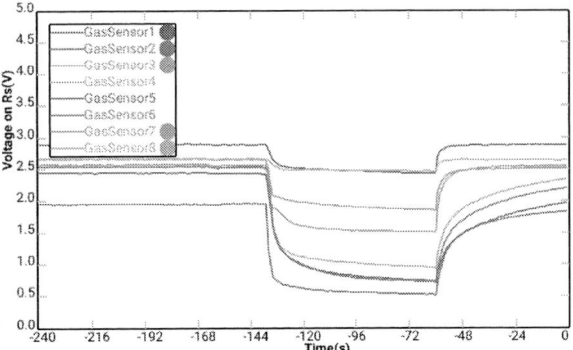

Figure 4c: Sensor response caused by smoke from refuse burning

TABLE II.   SENSOR RESPONSE DATA OF STEADY STATE

Sensor dimension	Voltage (V) and resistance (kΩ) in each sensor dimension			
	Clean air	*VOCs*	*Perfume*	*Smoke*
1	2.89, 70.45	2.66, 41.19	2.15, 18.63	2.44, 28.18
2	2.02, 15.69	0.75, 2.94	0.35, 1.18	0.53, 1.91
3	2.55, 33.88	2.15, 18.76	1.45, 7.82	1.86, 12.91
4	2.69, 44.50	1.52, 8.55	0.59, 2.19	0.95, 4.07
5	2.50, 31.20	1.14, 5.25	0.45, 1.57	0.74, 2.91
6	2.55, 34.17	1.18, 5.56	0.42, 1.47	0.73, 2.84
7	2.56, 34.46	2.09, 7.25	1.12, 5.14	1.51, 8.40
8	2.67, 41.98	2.48, 30.46	2.51, 31.97	2.47, 29.75

SUMMARY

Compared to common air quality monitoring methods where sensors work in standalone mode, however, as a pattern recognition method, this evaluation method is able to learn newly odors which could affect air quality or not. It can avoid evaluation mistakes, e.g., some harmless odors could more or less cause sensors response due to sensor's cross-sensitivity, but this case should not be attributed to air pollution case. If such cases are excluded, thus evaluation accuracy could be improved. It could be used for low cost detector devices and also portable devices. Air quality data given by department of environmental protection is collected at fixed observation sites in a certain time interval, so it may not be able to represent real-time status of air quality and distributed status of changing locations. Through full-time

978-1-5090-6695-7/17 $31.00 © 2017 IEEE

monitoring, we can obtain air quality of a certain area and view the trend curve of the air pollutants. Deploying more distributed air quality monitoring devices can constitute big data and full coverage network of air quality; combining it with geographic information would make daily-life related sharing possible.

Low oxygen conditions could make people feel dizzy and tired. Excessive carbon monoxide or carbon dioxide content of indoor air can threat human life safety. In a newly refurnished house, certain amounts of formaldehyde and other toxic gases can affect people health. In all the above cases, air quality should be monitored. The typical applications include cars, rooms, cellars, workshops, underground mines, surroundings of plants and so on. These cases are common, thus the demand for such lightweight products should be urgent and huge.

In this paper, gas sensors used are all MOX semiconductors. The merit of these sensors is low cost and easy for integration. However, there are also evident disadvantages of these sensors and corresponding solutions are listed:

1. Sensitivity is not high enough to sense trivial change of air quality. Specific functionality related sensors need to be developed and used due to common sensor's selectivity, resulting in incomplete coverage of all cases.

2. Sensor size should be reduced to fit portable applications. MEMS technology can lead to the development of miniaturized and micro-scale sensors, being an active research area [15, 16].

3. Long-time heating is needed to make sensors reach working temperature and stable state from cold state, and to sustain the temperature, heating is still need. This is a process of high power consumption. Therefore, in a long time of mobile monitoring, the power supply by battery may be a problem due to insufficient power supply time. The development of low power consumption sensors, "microhotplate" structure is a practical approach. With characteristics of rapid heating/cooling [17], it can be a good solution to power consumption issue.

REFERENCES

[1] Wang, S., & Hao, J. *Air quality management in china: issues, challenges, and options.* Journal of Environmental Sciences, 24.1(2012):2-13.

[2] Röck, F., Nicolae Barsan, A., & Weimar, U. *Electronic nose: current status and future trends.* Chemical Reviews, 108.2(2008):705-25.

[3] *Definition of air quality index.* http://baike.baidu.com/view/3251379.htm

[4] DS Lee, DD Lee, SW Ban, M Lee, YT Kim. *SnO₂ Gas Sensing Array for Combustible and Explosive Gas Leakage Recognition.* IEEE Sensors

Journal, 2.3(2002):140-149.

[5] Chang, J. C., & Hanna, S. R. *Air quality model performance evaluation.* Meteorology & Atmospheric Physics, 87.1(2004):167-196.

[6] Finlayson-Pitts, B. J., &Jr, P. J. *Tropospheric air pollution: ozone, airborne toxics, polycyclic aromatic hydrocarbons, and particles.* Science, 276.5315(1997):1045-52.

[7] Chang, J. C., & Hanna, S. R. *Air quality model performance evaluation.* Meteorology & Atmospheric Physics, 87.1(2004):167-196.

[8] Kalberer, M., Paulsen, D., Sax, M., Steinbacher, M., Dommen, J., &Prevot, A. S., et al. *Identification of polymers as major components of atmospheric organic aerosols.* Science, 303.5664(2004):1659-62.

[9] Fraser, M. P., Cass, G. R., Simoneit, B. R. T., & Rasmussen, R. A. *Air quality model evaluation data for organics: 5 c6-c22 nonpolar and semipolar aromatic compounds.* Environmental Science & Technology, 32.12(1998):1760-1770.

[10] Zou, H. Q., Lu, G., Liu, Y., Bauer, R., Tao, O., & Gong, J. T., et al. *Is it possible to rapidly and noninvasively identify different plants from asteraceae, using electronic nose with multiple mathematical algorithms?.* Journal of Food & Drug Analysis, 23.4(2015):788-794.

[11] Ampuero, S., &Bosset, J. O. *The electronic nose applied to dairy products: a review.* Sensors & Actuators B Chemical, 94.1(2003):1-12.

[12] Rodriguez-Lujan, Irene, et al. *"Analysis of pattern recognition and dimensionality reduction techniques for odor biometrics."* Knowledge-Based Systems 52.6(2013):279-289.

[13] Zhong J, & Zhu G F. *"Control and Disposal Technology of Secondary Pollutant from Power Generation by Incineration of Rubbish."* Pollution Control Technology, (2007).

[14] J Zhang, Q Yao, & Z Lu. *Generation and control technology of secondary pollutants for incinerator.* Environmental Protection, 5 (2001):17-18.

[15] Afridi, M., Hefner, A., Berning, D., Ellenwood, C., Varma, A., & Jacob, B., et al. *Mems-based embedded sensor virtual components for system-on-a-chip (soc).* Solid-State Electronics, 48.10(2004):1777-1781.

[16] Benkstein, K. D., Martinez, C. J., Li, G., Meier, D. C., Montgomery, C. B., &Semancik, S. *Integration of nanostructured materials with memsmicrohotplate platforms to enhance chemical sensor performance.* Journal of Nanoparticle Research, 8.6(2006):809-822.

[17] Semancik, S., Cavicchi, R. E., Wheeler, M. C., Tiffany, J. E., Poirier, G. E., & Walton, R. M., et al. *Microhotplate platforms for chemical sensor research.* Sensors & Actuators B Chemical, 77.1(2001):579-591.

DESIGN AND IMPLEMENTATION OF A HIGH QUALITY R-PEAK DETECTION ALGORITHM

Zhongmin Lin, Bo Wang*, Hao Chen, Ying Zhang, Xin-An Wang

The Key Lab of Integrated Microsystems, Peking University Shenzhen Graduate School, Shenzhen
518055, China

* Email:wangbo@pkusz.edu.cn

ABSTRACT

In modern medicine, electrocardiogram (ECG) is an important way to diagnose cardiovascular disease and monitor health information. The detection of R-peak is very important in ECG signal processing. To improve the accuracy and sensitivity of detection, a compound algorithm with high quality is presented in this paper. The algorithm removes high frequency noise and power frequency noise through an IIR low-pass filter, then do wavelet transform to the filtered signal. Adaptive threshold was used to extract modulus maxima. Rechecking is applied when there are mistakes. Additionally, template matching method is exploited in the rechecking to false detection. The algorithm is evaluated by using MIT-BIH arrhythmia database [1]. Finally, we obtained sensitivity of 99.79% and accuracy of 99.81%.

INTRODUCTION

With the aging problem becoming increasingly serious, people pay more attention to cardiovascular disease, and wearable real-time ECG monitoring equipment is gradually becoming a hot research topic.

ECG signal processing depends heavily on R-peak detection. Characteristic parameters such as heart rate (HR), heart rate variation rate (HRV) all depend on R-peak detection. The detection of QRS complex, P wave, T wave also depend on it. Due to the less number of leads and the patient's movement, wearable real-time monitoring equipment obtains ECG signal in lower quality comparing to professional equipment in hospital. Consequently, better R-peak detection algorithm is required.

A lot of algorithms were proposed, but each algorithm has its limitations. To improve the quality, the compound algorithm, a combination of several algorithms is becoming the most appropriate choice. [2] puts forward a QRS detection method using combined adaptive threshold. [3] presents a method using multiresolution wavelet analysis based on selective coefficient method. [4] puts forward a method based on empirical mode decomposition. [5] presents a method based on Hilbert transform. [6] presents a novel method based on many different methods. Even though these algorithms obtain high accuracy and sensitivity, we need a more adaptive and less complicated algorithm and it is the feature of the algorithm presented in this paper.

Besides, our algorithm also has high accuracy and sensitivity.

The flow of the algorithm is shown in Figure 1.

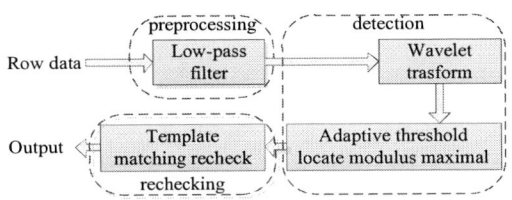

Figure 1: ECG signal processing flow

R-PEAK DETECTION FLOW

Low-pass filtering

Noises of ECG signal mainly include baseline drift, myoelectricity, 50Hz power frequency noise and so on, so filtering processing should be conducted before further processing. 90% frequency spectrum energy of ECG concentrate on 0.25Hz to 35 Hz, so a low-pass filter with the cut-off frequency 35Hz is required. 50Hz power frequency noise and high-frequency noise are filtered.

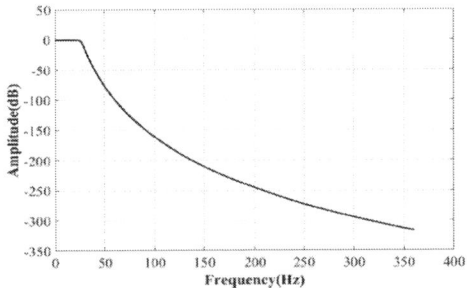

Figure 2: Amplitude frequency response

In order to reduce costs and delay, we chose the IIR filter here. The squared amplitude function of Butterworth low-pass filter is shown as formula (1), and the system function is shown as formula (2).

$$|H_a(j\Omega)|^2 = \frac{1}{1+(\frac{\Omega}{\Omega_c})^{2N}} \qquad (1)$$

$$H(z) = \frac{\sum_{j=0}^{M} b_j z^{-r}}{1+\sum_{k=1}^{N} a_k z^{-k}} \qquad (2)$$

Coefficients b_j and a_k of system function are calculated by Pulse Response Method, and amplitude frequency response is shown in Figure 2.

978-1-5090-6695-7/17 $31.00 © 2017 IEEE

Wavelet transform

After low-pass filtering, the wavelet transform method is adopted in this paper to detect singular values and deal with the baseline drift.

Wavelet transform is a kind of common used signal processing method which can do time-frequency analysis, filtering and data compression. The singular value detection function of wavelet transform is used here, and the wavelet transform formula is shown as formula (3). $\varphi(t)$ is the mother wavelet function. Discrete form of wavelet transform is shown as formula (4).

$$WT_x(a,\tau) = \frac{1}{\sqrt{a}} \int x(t)\, \varphi^*\left(\frac{t-\tau}{a}\right) dt \qquad (3)$$

$$WT_x(a,t) = \frac{1}{\sqrt{a}} \sum_{n=-\infty}^{\infty} x(n)\, \varphi^*\left(\frac{n-\tau}{a}\right) \qquad (4)$$

Chose the second derivative of Gaussian function as mother wavelet function.

$$\varphi(t) = \frac{1}{\sqrt{2\pi}}(1 - t^2)e^{\frac{-t^2}{2}} \qquad (5)$$

Comparing the effects of decompositions of several different scales, optimal scale 2^3 is selected. Because of the band-pass characteristic of second derivative of Gaussian function, the baseline drift and high frequency noise are inhibited. The result of the wavelet transform is shown in figure 3, and the location of wavelet transform modulus maximum is the location of R-peak.

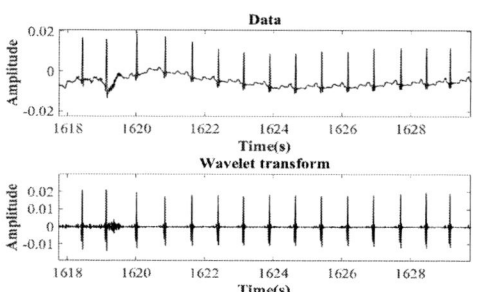

Figure 3: Wavelet transform result

Adaptive threshold

After the wavelet transform, we detected wavelet transform modulus maxima by using adaptive threshold method. The initial thresholds are obtained from the first 5s data. We get maximum value from each second time of this 5s data, then averaging the 5 value except the maximum and the minimum one, and choose 0.5 times of the average value as the positive threshold. Negative threshold is obtained from minimum values in the same way. After every modulus maximum is detected, update the positive threshold as the value of the last detected modulus maximum, and decrease the positive threshold at slope 0.6 after 0.2s [7]. The negative threshold is updated in the same way. The principle is shown as figure 4. If the time between two detected points less than 0.2s, we take the bigger one as correct. Every modulus maximum is corresponding to one R-peak in the signal, and modulus minimum is used to verify if the modulus maximum is detected correctly. When a location detected is shifted from the exact location of R-peak, find the first peak at the side with a rise trend of the location.

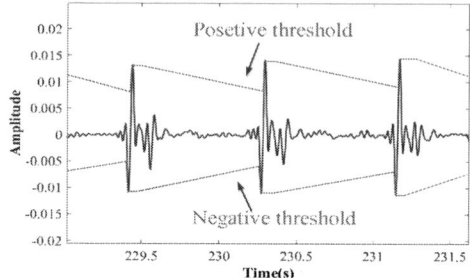

Figure 4: Adaptive threshold

For higher accuracy and sensitivity, a recheck mechanism is needed for signal. Supposing R_m is the number m R-peak, and $t_{m,m+1}$ means RR interval between R_m and R_{m+1}. If $t_{m,m+1}>1.5s$ or no R-peaks are detected after 1.6s, we detected the data from R_m to R_{m+1} or the point $t_m+1.6s$ again with thresholds which are 0.5 times of normal, and if still no R-peak is detected, reduce thresholds to 0.5 times again and recheck again. If $t_{m,m+1}<0.3s$, it means that one R-peak of R_m and R_{m+1} is false. To determine which one is detected mistakenly, template matching method is introduced, and it will be explained in the next section.

Template matching rechecking

In template matching method, specific waveform data should be saved in advance as template, and compare it with ECG signal. We determine whether there is a waveform which is the same kind as the template according to the similarity of the data and template. In the algorithm presented in this paper, template matching method is used to recheck the data that may be detected falsely, then locates the error and corrects it.

The template is not limited to the right R-peak, it should also include some of the abnormal waveform which may lead to a false detection. Then we can detect the right R-peak and mark the false one to facilitate the subsequent analysis. We take the number of template into an odd number, supposing that we want to verify the location x_0, and dif is the difference between data and template as shown in formula(6). A low dif means a high similarity.

$$dif = \frac{\sum_{K=1}^{N}\left[\left|x_{k-\frac{N+1}{2}}-t_k\right|-\sum_{K=1}^{N}\left|x_{k-\frac{N+1}{2}}-t_k\right|/N\right]^2}{N} \qquad (6)$$

978-1-5090-6695-7/17 $31.00 © 2017 IEEE   576

As shown in figure 5, peak A and B are detected as R-peaks. Because $t_{A,B}<0.3s$, one of these two is detected falsely. We use template to recheck A and B, and determine which one is false according to dif_A and dif_B. A is eliminated finally.

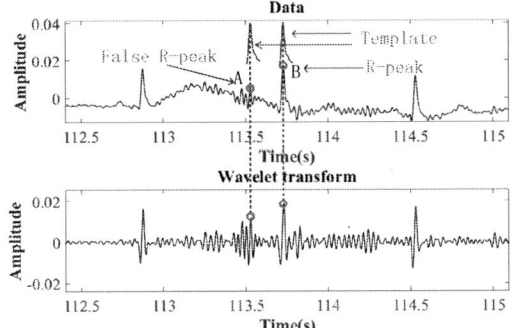

Figure 5: Template matching rechecking

RESULT AND DISCUSSION

As shown in figure 6, the algorithm is evaluated by using several sets of data in MIT-BIH arrhythmia database, and the location with a point being circled means there is a R-peak detected. The result is shown in Table 1. Num represents the actual number of R-peaks in the data. TP represents the number of R-peaks that detected correctly. FP represents the number of R-peaks detected falsely. FN represents the number of R-peaks missed. Sen represents sensitivity and calculated by equation (7). Acc represents accuracy and calculated by equation (8).

$$Sen = \frac{TP}{TP+FN} \times 100\% \qquad (7)$$

$$Acc = \frac{TP}{TP+FP} \times 100\% \qquad (8)$$

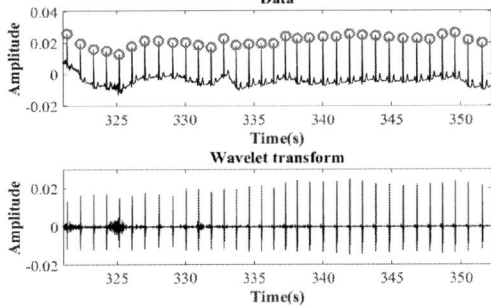

Figure 6: Peak detection result

From the evaluation result listed in TABLE I, an average sensitivity of 99.79% and an average accuracy of 99.81% can be obtained. Compared with paper [1][2][3][4][5], the sensitivity and accuracy of the algorithm presented in this paper are comparable, and the applicability is good. The computational complexity of this algorithm is relatively low, so the limited resources of wearable real-time monitoring devices can be saved.

TABLE I. EVALUATION To ALGORITHM

Data	Num	TP	FN	FP	Sen(%)	Acc(%)
100	2273	2273	0	0	100	100
101	1865	1862	3	4	99.84	99.79
102	2187	2187	0	0	100	100
103	2084	2084	0	0	100	100
106	2027	2018	9	7	99.56	99.65
107	2137	2117	20	19	99.06	99.11
109	2532	2528	4	0	100	99.84
111	2124	2123	1	2	99.95	99.91
Total	17229	17192	37	32	99.79	99.81

SUMMARY

The algorithm for ECG signal processing presented in this paper is designed for wearable real-time monitoring devices. High sensitivity and accuracy are important, but the strong adaptability to ECG signal and the low computational complexity is the primary target. This algorithm shows a good performance according to the evaluation result.

ACKNOWLEDGMENTS

This work is supported by R&D project of Shenzhen Government (Project NO.JSGG 2013091840947999 and NO.JCYJ20150331102721193).

REFERENCES

[1] Mark R, Moody G: MIT-BIH Arrhythmia data base directory. Cambridge: Massachusetts Institute of Technology 1988.
[2] Christov I I. Real time electrocardiogram QRS detection using combined adaptive threshold[J]. Biomedical Engineering Online, 2004, 3: 28.
[3] Pal, Saurabh, and M. Mitra. Detection of ECG characteristic points using Multiresolution Wavelet Analysis based Selective Coefficient Method. Measurement., vol. 43, 2010, pp.255-261.
[4] Slimane, Zine Eddine Hadj, and A. Naït-Ali. QRS complex detection using Empirical Mode Decomposition. Digital Signal Processing., vol. 20, 2010, pp.1221-1228.
[5] Valluraiah, P., and B. Biswal. ECG signal analysis using Hilbert transform. IEEE Power, Communication and Information Technology Conference IEEE, 2015.
[6] MLAManikandan, M. Sabarimalai, and K. P. Soman. A novel method for detecting R-peaks in electrocardiogram (ECG) signal. Biomedical Signal Processing & Control., vol. 7, 2012, pp.118-128.
[7] Shin, H. S., Lee, C. and Lee, M., Adaptive threshold method for the peak detection of photoplethysmographic waveform. Computers in Biology and Medicine, 39(12), vol. 3, 2009. pp.1145-1152.

A

Abe Tamotsu	II-18
AliReza Alian	VI-24
Ameen T	I-31
Ang D S	I-34
Araźna Aneta	VII-2
Arslanlan Gregory K	VII-15
Auger Robert	V-22

B

Babu S V	V-9
Bai Fan	V-16
Bai Song	II-11, VI-43
Baklanov Michael R	IV-39
Balon Frantisek	VII-5
Bao Tim	VII-15
Bao Yu	IV-28, IV-29
Barabash Sergey V	VIII-23
Bosund Markus	IV-36
Bowes Michael	VIII-23
Boxin Yang	IX-15

C

Cai Zhenhua	IV-28
Cao L	I-31
Cao Zigui	I-1, V-5
Capodieci Luigi	II-45
Carazetti Patrick	VII-5
Cha Samuel	VII-32
Chan Sifei	III-29, III-34
Chaney A	I-31
Chang Edward Yi	VIII-25, VIII-26
Chang Rong-Chao	III-11
Chang Shane	VIII-26
Chao Tai Fong	V-12
Chao Taifong	V-7
Che Dongchen	IV-39
Chen Baixin	IX-27
Chen Canny	VI-15
Chen C Z	IX-16
Chen Dihu	IX-14
Chen F	I-31
Chen Fang	I-2
Chen Fuhong	III-29, III-34
Chen Hao	IX-2, IX-19
Chen Hong	I-1

Chen Hua	VI-9
Chen Huiping	VI-7
Chen Hunglin	VI-39
Chen Jack	VI-4
Chen Jian	VIII-19
Chen Jing	I-21
Chen Kuei-Fang	IX-7
Chen Kun	V-19
Chen Ping	VI-32
Chen Qiaoli	II-29
Chen Quan	I-38
Chen Sherry	VII-3
Chen Xinhua	IV-25
Chen Yen-Yu	VIII-26
Chen Yongyue	IV-26
Chen Zhufan	III-25
Chen Zhuo-Fan	III-10
Cheng Chun-Hung	VIII-25
Cheng Lingxiao	VI-10
Cheng Shiqiu	VI-7
Cheng Xingming	VII-4, VII-6
Cheng Xinhua	IV-27
Cheng Xu	IX-11, IX-12
Chi Mingwa	IV-42
Chi Min-hwa	I-4
Chiang Tony	VIII-23
Chien Kary	IV-4, VI-15, VI-16, VI-17, IX-9
Chien Wei-Ting Kary	I-24, I-26, VI-3, VI-4, VI-11, VI-14, VII-3
Chiu Chang-Yong	IX-18
Chiu Jung-Ching	IX-5, IX-7
Cho Hyun Yong	II-24
Cho Y	I-31
Choi Hyung Ouk	VII-8
Choi W K	VI-31
Choi Woo Young	I-7
Christmann Gabriel	VIII-2
Chu Yongnan	IX-13
Chu Yun-Ching	VIII-25
Chun Jeffery	VII-32
Claeys Cor	VI-9, VI-24
Collaert Nadine	VI-24
Cui Liang	VII-6

D

Den Tao	VIII-19
Deng Guihong	IX-9

Deng Hai   II-8, II-9
Deng Nin-Qin   VIII-16
Deng Xianjin   IX-11, IX-12
Digiovanni D   I-31
Ding Alicia   VI-4
Dong C Christine   VII-15
Dong Lisong   II-1
Dong Peipei   VII-10
Du Feifan   VI-38
Du Li   IX-6
Du Ling   I-21
Du Yuan   IX-6
Du Yaojun   II-4
Duan Silvia   VI-4
Duan Wenting   I-33, I-35
Duan Yangyang   VII-4, VII-6
Dubois Marc Alexandre   VIII-2

E
Elghazzali Mohamed   VII-5

F
Fan Jingyu   VII-2
Fan Yub   VII-23
Fan Pengwei   VI-18
Fan Rongwei   VI-39
Fan Wenbin   V-19
Fan Xuedong   II-13
Fang Jingxun   IV-25, IV-26, IV-27, IV-28, IV-29, V-19
Fang Qiang   V-7
Fang Xiaogong   IV-27
Fang Zhou   II-19
Fay P   I-31
Feng Frank   VI-23
Feng Tong   IV-42, V-26
Feng Yulin   V-20
Fogle Jeffrey D   II-24
Fu Zubiao   VI-2
Fujii Koichi   II-13
Fujiwara Koichi   II-35

G
Gao Jinde   VI-21
Gao Lin   IV-28, IV-29
Gao Ying   IV-28
Gao Yuzhu   VI-11

Ge Jun	V-26
Ge Qiang	III-36
Gong Junyan	VII-4
Gong Hanmo	II-11
Gonga Hanmo	VI-43
Grunow Stephan	IV-14
Gu Jiebin	VII-16
Gu Man	IV-14
Guo Jijia	IX-1
Guo Wei William	I-23
Guo Yi	V-22
Guo Yufeng	I-21
Guoyuan Li	IX-15

H

Hailan Yi	IV-26
Han Guoqing	I-1
Han Hongpeng	VI-41
Han Ica	VII-32
Han Ja-Hyung	V-7
Han Qiuhua	III-21, III-30
Han Qiu-Hua	III-14
Han Tao	IV-14
Han Wei	I-26
Han Yemei	V-20
Hao Jing'an	II-23
He Liang	VI-9
He River	I-29
He Qiyang	III-12
He Weiming	II-25, VI-36
He Yanghua	V-25
He Yonggen	IV-43
Heinz Bernd	VIII-2
Ho Yen-Teng	VIII-25, VIII-26
Hoffmann Mike	VII-5
Hong Ruijin	IV-44
Hori Tsukasa	II-18
Hsieh Yu-Lin	IV-15, IV-19
Hsu Hung-Run	VIII-25
Hu Chih-Kai	IX-5, IX-7
Hu David Y	IX-16
Hu Huayong	II-25, VI-36
Hu Jun	I-33
Hu Minda	III-12
Hu Wei	IV-26, IV-28
Hu Yin	VI-9

Huang Haigou	V-7
Huang Hao	V-5
Huang Jipan	IX-1, IX-3
Huang Qiming	IV-23
Huang Ruixuan	III-21
Huang Song	III-36
Huang Yi	VI-19, VI-2
Huang Ying	III-36
Huang Yong	III-43
Huang Yuqian	IX-1, IX-3
Huang Zhuoquan	IX-14
Huo Ruifang	V-20

I

Ikeda Junji	II-13
Ilatikhameneh H	I-31
Islam S M	I-31

J

Janeczek Kamil	VII-2
Jena D	I-31
Jeng Shyr-Long	IX-17
Ji Shiliang	III-21
Ji Xiaoli	VIII-24
Ji Xinchun	I-21
Ji Xinyuan	VI-13
Ji Yuan	IX-13
Yuan Xiaofent	II-23
Ji Z	I-20
Jia Yutao	VI-7
Jiang Bin-Jie	II-2
Jiang Binjie	II-10
Jiang Haitao	VI-25
Jiang Xiaojuan	VII-6
Jiang Zhi	VII-17
Jimbo T	VIII-17
Jin Feng	I-33
Jing Xuezhen	IV-10
Jing Wang	III-15
Jou Sheng- Kai	IV-19
Ju Jianhua	I-42
Jun He	IV-10
Jung Won Chul	VII-20

K

Kan Ho-Jung	I-14

Kang Jin	I-34, V-26
Kang Jun	V-5
Kang Junlong	IV-27
Kang Sheng	VI-13
Kawashima T	I-34
Kawasuji Yasufumi	II-18
Keller S	I-31
Kennedy John	VII-11
Kim Hyun Woo	VII-8
Kim Min Sung	VII-32
Kimura I	VIII-17
Klimeck G	I-31
Kobayashi H	VIII-17
Kodama Takeshi	II-18
Kokaze Y	VIII-17
Koli Dinesh	V-7, V-12
Kong Byung-Seon	VII-8
Kouda Masaya	VII-14

L

Lau W S	IV-22
Lee Byunghak	I-2, I-29
Lee Cheng-Kuei	I-41
Lee Chien-Chieh	IV-15, IV-19
Lee Da-Eun	VII-8
Lee Jack	VII-32
Lee Jong-Ho	I-14
Lee Ken	VII-32
Lee Sang-Sun	VII-8
Lee Yi-Ting	VI-23
Leighton Jamie	V-25
Leng Jianghua	IV-28
Li Bin	VIII-3, IX-10
Li Cheng	IV-17, IV-18
Li Dan	II-30
Li Haihua	VI-39
Li Jiao	IX-13
Li Jie	II-8, II-9
Li Jing	I-15
Li Junfeng	VI-9
Li Lanxia	VII-4, VII-6
Li Li	II-14
Li Ming	VI-4
Li Qiang	V-26
Li Qiaoqiao	II-25, VI-36
Li Quanbo	III-29, III-33, III-34

Li Ruoyuan	I-2
Li Ruzhang	VI-12
Li SenSheng	I-42
Li Tiehu	VI-12
Li Tomi T	IV-15, IV-19, IX-5, IX-7
Li W	I-31
Li Wei	VI-24
Li Xinxin	VII-16, IX-12
Li Xuemiao	II-8, II-9
Li Yulong	II-26, II-30
Li Yilei	IX-6
Liao Yiming	VIII-24
Lim SzePei	VII-38
Lin Alien	VI-2
Lin Chih-Chien	VIII-25
Lin Jie	V-4
Lin Man	I-21
Lin Shawn	I-42
Lin Yi-Shi	VI-18, VI-19
Lin Yueh-Chin	VIII-26
Lin Zhongmin	IX-2, IX-19
Lipiec Krzystof	VII-2
Littau Karl	VIII-23
Liu Bingjie	VII-16
Liu Biqiu	II-30
Liu Chang	II-11, II-27, II-19
Liu Chun-chen	IX-6
Liu Donghua	I-33, I-35
Liu Hia	I-29
Liu Hongjie	VII-4, VII-6
Liu Huang	IV-14
Liu Huanxin	III-43
Liu Jia	VI-12
Liu JiaLei	III-43
Liu Jinping	IV-14
Liu Johan	VII-2
Liu Jun	VI-12
Liu Longhai	VI-25
Liu Pan-pan	III-31
Liu Xuan	II-14
Liu Yansong	II-1
Liu Yifei	II-13
Liu Yingming	IV-23, IV-28, IV-29
Liu Zewen	VIII-19
Liu Zhicheng	VI-36
Liu Zhiqian	I-28
Liu Ziqian	VI-5

Long Yin	VI-39
Lu Haifang	IX-1, IX-3
Lu Lian	III-29, III-34
Lu Wen Yin	V-12
Lu Yaqi	VI-25
Lu Ying Emily	I-23
Lube Michael	V-25
Lumbantoruan Franky	VIII-26
Lund C	I-31
Luo Huaming	IV-27
Luo Jun	VI-9
Luo RuiMing	VIII-3, IX-10
Luo Yuhui	VII-6
Luong Tien-Tung	VIII-25, VIII-26
Lv Zhengyon	I-2

M

Ma Guiying	IV-1
Ma Haitao	VII-23
Ma J	I-20
Ma Shiwei	VII-2
Ma Tao	VII-10
Ma Yuanzhao	II-13
Mackie Andy	VII-38
Majlis Burhanuddin Yeop	I-6
Makihara Koij	VII-14
Mao Gang	IV-17, IV-18
Mao Zhibiao	II-26, II-29, II-30, III-29
Mao Zhi-Biao	II-2, II-10
Masuda T	VIII-17
Mavliev Rashid	V-23
Meng Xiangguo	III-29
Mertin Stefan	VIII-2
Miao Nie	III-15
Mishima Kazuhiko	II-32
Miyaguchi Y	VIII-17
Mizoguchi Hakaru	II-18
Mongilnikov Konstantin P	IV-39
Mosley David	V-22
Mu Tingzhou	IX-13
Müller Dirk	VII-11
Muralt Paul	VIII-2

N

Nakajima Makoto	II-17
Nakarai Hiroaki	II-18

Narasimhan Vijay | VIII-23
Nicolay Sylvain | VIII-2
Niu Gang | VI-4
Noor Mimiwaty Mohd | I-6
Nowak Krzysztof | II-18
Nowling Greg | VIII-23

O

Okamoto Kenji | VII-12
Okamoto Takeshi | II-18

P

Pa Shyam | II-14
Pal Ashish | VIII-23
Pan Zhoujun | II-19
Pang Albert | III-29, III-34
Patrick Richard E | VII-15
Pätzel Rainer | VII-11
Peng Jingyu | VIII-24
Peng Lipin | IX-9
Penigalapati Dinesh | V-12
Piloux Yannick | III-40
Pramanik Dipankar | VIII-23

Q

Qian Hongtao | I-28, VI-5
Qian Jun | I-30
Qian Wensheng | I-33, I-35
Qin Long | II-26, II-29
Qin Weijun | VI-6
Qin Xiaoting | VI-9
Qin Yusen | VI-12

R

Rader W Scott | V-4
Ran Feng | IX-13
Ranhman R | I-31
Rastegar Vahid | V-9
Rattunde Olivier | VIII-2
Reddy Arun | V-22
Ren Jia | III-32
Ren Qian | VI-43
Ren Tian-Ling | VIII-16
Rhoades Robert L | V-13

S

Saitou Takashi	II-18
Sakamoto Rikimaru	II-17
Samakawa Seiji	VIII-1
Shanghai Liew	VIII-6
Sankaran Sujatha	IV-14
Shao Chun	II-13
Shao Lingling	I-26
Sharma Ram	II-24
Shen Haibo	III-33
Shen Li	I-42
Shen Yiijiang	II-22
Shi Gang	IV- 28, IV-29
Shi Yaoming	VI-7, VI-2
Shibayama Wataru	II-17
Shigaki Shuhei	II-17
Shim Peter	VII-32
Shinozaki Hiroki	VII-14
Shiraishi Yutaka	II-18
Shou Qinghua	VI-25
Sioncke Sonja	VI-24
Simoen Eddy	VI-9, VI-24
Sitek Janusz	VII-2
Skeen Philip	VII-15
Sneck Sami	IV-36
Söderlund Mikko	IV-36
Soininen Pekka	IV-36
Song Haizhi	VIII-19
Song Yang	VI-18, VI-19
Song Yibin	III-13
Song Yongliang	VI-15
Soon Wai	VIII-6
Su Tao	IX-14
Su Xiaojing	II-1
Sugandi Gandi	I-6
Sun Chang	I-30
Sun Lei	III-33
Sun Lijuan	IX-9
Sun Ling	I-2
Sun Lu	I-22
Sun Xiaoyan	II-27
Sun Yanhui	VI-14, VI-16
Sun Yibin	VI-32
Suu K	VIII-17

T

Takahashi Kazuhiro	II-32
Takeuchi Hajime	II-32
Tan Wei	VII-4, VII-6
Tan Yiqun	I-38
Tanaka Hiroshi	II-18
Tang Junyao	VI-17
Tang Lian	I-2
Tang Long-Juan	III-14
Tang Poren	IV-42
Tao Chunxian	IV-44
Tao Dongyan	VI-14
Tao Lu-Qi	VIII-16
Tao Yuan	II-13
Thum Kenneth	VII-38
Tian He	VIII-16
Tian Miren	IX-3
Tian Peng	IV-28
Tian Yanhong	VII-17
Tong Lei	IV-26
Tönnies Dietrich	VII-11
Tschirky Maurus	VIII-2
Tseng Chien-Lung	I-40
Tseng Ching-Lin	IV-15
Tu Yung-Yi	VIII-25
Tung Chun-Te	VIII-11
Tung Pi-Cheng	IX-5, IX-7

V

van Borkulo Jeroen	VII-20
van der Stam Richard	VII-20
Verburg Paul	VII-20

W

Wang Anni	IV-10
Wang Bo	IX-19
Wang Chaolun	VII-37
Wang Chong	VI-24
Wang Dan-Yang	VIII-16
Wang Fang	V-20
Wang Guilei	VI-9
Wang Hongdi	V-16
Wang Hui	I-1, V-5
Wang Huihui	I-33
Wang Jia	V-16
Wang Jiang	VII-10
Wang Jinxia	IV-44

Wang Jiong	VI-21
Wang Judy	III-36
Wang Kai	VI-10, VI-11
Wang Qingpeng	IV-17
Wang Tianchen	IX-6
Wang Wuping	II-26
Wang Xin'an	IX-1, IX-2, IX-3, IX-19
Wang Yan	I-41, III-30
Wang Yansheng	I-30
Wang Yanyun	II-29
Wang Ying	VI-25
Wanf Yuan Sheng	I-40
Wang Yuxin	VI-12
Wang Zhijuan	VI-11
Wang Zidong	III-12
Watanabe Itaru	VII-14
Watanabe Yukio	II-18
Wathne Kail	VII-15
Weeks Stephen L	VIII-23
Wei Lin-Lung	VIII-25
Wei Yayi	II-1
Wei Zhengying	I-30
Weichart Jürgen	VII-5
Wen Jiayue	VII-17
Wen Juan	VI-16
Wen Kuei-Ann	VIII-11
Wen Zhenping	IV-25
Wu Boping	VII-7, IX-8
Wu Chih-Chianf	IX-17
Wu Chingluan Jenny	I-23
Wu Hanming	I-4, IV-42, V-26
Wu Lei	II-27
Wu Qiang	II-11, II-20, VI-43
Wu Qiwei	VI-21
Wu Weiwei	I-38
Wu Xing	VII-37
Wei Zhiqiang	I-15

X

Xi Fuchun	IV-28
Xiao Fangyuan	IV-17
Xiao Zhiyi	VII-23
Xie Jianhua	VI-25
Xin Wei	VII-3
Xu Cheng	VI-31
Xu Jian	II-29

Xu Jianhua	IV-10
Xu Jinling	VI-14
Xu Kaidong	IV-39
Xu Kangning	IV-39
Xu Tao	I-1
Xu Weizhong	I-2
Xu Yiping	VI-7, VI-2
Xu Yue	VIII-3, IX-10
Xu Yuhong	VI-5
Xu Zhaozhao	I-33

Y

Yahui Huang	III-15
Yamada Masanori	II-32
Yamagadi Tatsuya	II-18
Yamazaki Taku	II-18
Yan Junhua	V-19
Yan Feng	VIII-24
Yan weiwei	I-28
Yang Heng	VII-16
Yang Ji Chul	V-12
Yang Jiye	I-33
Yang Kelly	VI-3
Yang Lijuan	VI-10
Yang Rex	IV-17
Yang Weidong	VI-12
Yang Wenqing	I-33
Yang Yongsheng	I-2
Yang Zhengkai	II-26, II-29, II-30
Yang Zhenyu	II-13
Yang Zhigang	IV-26, IV-28
Yang Zhiqiang	IX-1, IX-3
Yang Zhiyong	III-43
Yao Jiafei	I-21
Yao Mingjun	VII-23
Ye Kang	IV-26, IV-29
Ye Lei	VI-36
Yew K S	I-34
Yin Ziqing	V-26
Binfeng Yin	VI-21
Yingying Guo	VI-1
Yoshimura Keiji	II-32
Yu Daquan	VII-23
Yu Haibin	II-10
Yu Hsiang-Chih	IV-15

Yu Hui	VI-38
Yu Min-Lun	IV-19
Yu Shaofeng	IV-17
Yu Shi-Rui	II-2
Yu Shirui	I-38, II-10
Yu Susie	VI-4
Yu TzuChian	I-2
Yu Tzuchiang	IV-1
Yu Yunsen	IX-13
Yuan Kefang	III-12
Yuan Yujie	V-20
Yue Zhipeng	IX-14
Yuling Liu	V-26

Z

Zha Yue	I-15
Zhang Beichao	IV-10
Zhang Changchun	I-21
Zhang Cheng-Long	III-31
Zhang David Wei	I-33, IV-44
Zhang Dennis	VI-4
Zhang Dongen	VII-4, VII-6
Zhang Guan	VI-14, VI-16
Zhang Haiyang	III-12, III-13, III-24, III-25, III-30, III-32
Zhang Hai-Yang	III-10, III-1I, III-14, III-31
Zhang J F	I-20
Zhang Jin-Yu	I-41
Zhang Kailiang	V-20
Zhang Lei	V-19
Zhang Liang	IX-11, IX-12
Zhang Libin	II-1
Zhang Qiang	II-23
Zhang W	I-20
Zhang Wuzhi	I-30
Zhang Xiaolin	VI-2
Zhang Xing	IV-42, V-26, IX-2
Zhang Xingxing	VIII-24
Zhang Yan	VII-2
Zhang Yanyan	IV-28, IV-29
Zhang Ying	IX-2, IX-19
Zhang Yiying	III-13, III-24, III-25, III-32
Zhang Yi-Ying	III-10, III-11
Zhang Yu	II-26, II-30, III-29, III-34, IV-27
Zhang Yueyu	II-10
Zhang Zhijie	V-16
Zhang Ziying	IV-10

Zhang Zhun	IX-18
Zhao Baojun	IV-28
Zhao Chao	VI-9
Zhao Daniel	VI-4
Zhao Guangyan	I-24
Zhao Hai	IV-17, IV-18
Zhao Jing	III-43
Zhao Jiuzhou	I-28
Zhao Lijun	II-1
Zhao Ning	VII-23
Zhao Xiang Fu	VI-3
Zhao Yong	I-24, VI-17
Zhao Yong Atman	I-26
Zhao Yuhang	I-30
Zhaohui Wu	IX-15
Zheng Erhu	III-24
Zhenlin Zhou	VI-1
Zhi Hui	II-29
Zhiming Liang	IX-15
Zhong Bin	IV-29
Zhong Fengjiao	VI-7
Zong Z W	VI-31
Zhongwei Jiang	III-15
Zhou Haifeng	IV-23, IV-26, IV-28, IV-29
Zhou Junqing	III-12
Zhou Ke	VI-21
Zhou Sarah	VI-15
Zhou Wei	I-30
Zhou Y	I-34
Zhu Yefang	V-19
Zhu YiZheng	III-34
Zhu Yueqin	VI-11
Zhu Zifang	VII-10
Zuo Cheng	IX-27
Zou Chunmei	VI-17

IEEE
445 Hoes Lane
Piscataway, NJ 08854-4141

ISBN 978-1-5090-6695-7

2025 IEEE International Conference on Distributed Computing, VLSI, Electrical Circuits and Robotics (DISCOVER 2025)

Mangalore, India
17-18 October 2025

IEEE Catalog Number: CFP25F62-POD
ISBN: 979-8-3315-3899-6